219

THE McGRAW-HILL HANDBOOK OF ESSENTIAL ENGINEERING INFORMATION AND DATA

Other McGraw-Hill Reference Books of Interest

Avallone and Baumeister • MARKS' STANDARD HANDBOOK FOR MECHANICAL ENGINEERS

Brady and Clauser • MATERIALS HANDBOOK

Corbitt • STANDARD HANDBOOK OF ENVIRONMENTAL ENGINEERING

Freeman • STANDARD HANDBOOK OF HAZARDOUS WASTE TREATMENT AND DISPOSAL

Gaylord and Gaylord • STRUCTURAL ENGINEERING HANDBOOK

Gieck • ENGINEERING FORMULAS

Grimm and Rosaler • HANDBOOK OF HVAC DESIGN

Harris • HANDBOOK OF NOISE CONTROL

Harris • SHOCK AND VIBRATION HANDBOOK

Hicks • STANDARD HANDBOOK OF ENGINEERING CALCULATIONS

Juran and Gryna • JURAN'S QUALITY CONTROL HANDBOOK

Karassik et al. • PUMP HANDBOOK

Merritt • STANDARD HANDBOOK FOR CIVIL ENGINEERS

Perry and Green • PERRY'S CHEMICAL ENGINEERS' HANDBOOK

Rohsenow, Hartnett, and Ganic • HANDBOOK OF HEAT TRANSFER FUNDAMENTALS

Rohsenow, Hartnett, and Ganic • HANDBOOK OF HEAT TRANSFER APPLICATIONS

Rosaler and Rice • STANDARD HANDBOOK OF PLANT ENGINEERING

Shigley and Mischke • STANDARD HANDBOOK OF MACHINE DESIGN

Tuma • ENGINEERING MATHEMATICS HANDBOOK

Tuma • HANDBOOK OF NUMERICAL CALCULATIONS IN ENGINEERING

Wadsworth • HANDBOOK OF STATISTICAL METHODS FOR ENGINEERS AND SCIENTISTS

Young • ROARK'S FORMULAS FOR STRESS AND STRAIN

For more information about other McGraw-Hill materials, call 1-800-2-MCGRAW in the United States. In other countries, call your nearest McGraw-Hill office.

THE McGRAW-HILL HANDBOOK OF ESSENTIAL ENGINEERING INFORMATION AND DATA

Ejup N. Ganić, Sc.D. Editor in Chief

Professor of Mechanical Engineering
University of Sarajevo, Yugoslavia
Director, UNIS Research Institute, Sarajevo
Member: American Society of Mechanical Engineers

Tyler G. Hicks, P.E. Editor

International Engineering Associates
Member: American Society of Mechanical Engineers
and Institute of Electrical and Electronics Engineers

McGRAW-HILL, INC.

New York St. Louis San Francisco Auckland Bogotá
Caracas Hamburg Lisbon London Madrid
Mexico Milan Montreal New Delhi Paris
San Juan São Paulo Singapore
Sydney Tokyo Toronto

Library of Congress Cataloging-in-Publication Data

The McGraw-Hill handbook of essential engineering information and data / edited by Ejup N. Ganić, Tyler G. Hicks.

p. cm.
ISBN 0-07-022764-0
1. Engineering—Handbooks, manuals, etc. I. Ganić, Ejup N. II. Hicks, Tyler Gregory.
TA151.M34 1991
620—dc20 90-35667
CIP

Copyright © 1991 by McGraw-Hill, Inc. All rights reserved. Printed in the United States of America. Except as permitted under the United States Copyright Act of 1976, no part of this publication may be reproduced or distributed in any form or by any means, or stored in a data base or retrieval system, without the prior written permission of the publisher.

1 2 3 4 5 6 7 8 9 0 DOC/DOC 9 5 4 3 2 1 0

ISBN 0-07-022764-0

The sponsoring editor for this book was Harold B. Crawford, the editing supervisor was David E. Fogarty, the designer was Naomi Auerbach, and the production supervisor was Thomas G. Kowalczyk. It was set in Times Roman by McGraw-Hill's Professional Publishing composition unit.

Printed and bound by R. R. Donnelly & Sons Company.

Information contained in this work has been obtained by McGraw-Hill, Inc. from sources believed to be reliable. However, neither McGraw-Hill nor its authors guarantees the accuracy or completeness of any information published herein and neither McGraw-Hill nor its authors shall be responsible for any errors, omissions, or damages arising out of use of this information. This work is published with the understanding that McGraw-Hill and its authors are supplying information but are not attempting to render engineering or other professional services. If such services are required, the assistance of an appropriate professional should be sought.

For more information about other McGraw-Hill materials, call 1-800-2-MCGRAW in the United States. In other countries, call your nearest McGraw-Hill office.

CONTENTS

Chapter 18. Agricultural, Food, and Nutrition Engineering 18.1

Chapter 19. Reliability Engineering, Systems Engineering, and Safety Engineering 19.1

PREFACE

This engineering work integrates a spectrum of knowledge extending from the pure physical sciences to the most advanced and specialized technologies. Today, engineering, both as art and science, touches almost every facet of human life and can be said to have created the physical structure of civilization.

The plan for this Handbook was conceived in the belief that a great amount of essential engineering information and data could be assembled from key reference works in the major engineering fields and presented in a compact one-volume book which would be valuable as ready reference for practicing engineers and for students.

Having this Handbook on the desk or bench, the practicing engineer will be able to solve most of the applied problems met in daily activities of design, operation, analysis, or economic evaluation anywhere in the world.

As in any handbook, emphasis has been placed on ready-to-use material. The Handbook contains many formulas and tables which give immediate answers to questions arising from practical work. However engineering is an involved and complicated subject, and more than facts and formulas are needed to apply it properly. Therefore, each chapter of this book contains a brief description of the essential information and lines of thought of its subject, thus enabling the reader to understand the logical background of the ready results and to think beyond them. For that purpose the first part of the book is devoted to basics of engineering and the second part covers fields of engineering. It has been assumed that the users of the book are familiar with general engineering theory and have a fundamental knowledge of the use of ordinary equations, diagrams, tables, charts, and other common methods of presenting facts and coordinating functions. Further, the editors assume that a member of one branch of the profession (i.e., a mechanical engineer, a civil engineer, or an electrical engineer) will require information which normally belongs to one of the other branches. For this reason, as far as possible, specific facts and definite methods have been presented in such a way as to require the exercise of a minimum of judgment on the part of the user.

The engineering sections give much applied information. With such data in hand, an engineer can quickly solve a variety of everyday problems. Thus, the data in the engineering sections cover design, operation, analysis, and trouble-shooting topics. The equations and tables presented in these sections develop and apply earlier analyses given in the Handbook.

A variety of worked-out examples are presented throughout the Handbook. Such an approach is popular with working engineers because it helps both in understanding concepts and in applying them to practical situations. Each chapter is uniformly outlined, ending—in many instances—with a nomenclature section which includes a list of all symbols and their SI units.

Meticulous care has been exercised to minimize errors, but it is impossible in a work of this magnitude to achieve an error-free publication. Accordingly we would appreciate being informed of any errors so that they may be eliminated from subsequent printings. We would also appreciate suggestions from readers on possible improvements in the usefulness of the Handbook so that they may be included in future editions.

Ejup N. Ganić
Tyler G. Hicks

ACKNOWLEDGMENTS

Parts of this Handbook were drawn from the McGraw-Hill books listed below, with updating and metric units added by the editors.

J. A. Dean, *Lange's Handbook of Chemistry*, 13th ed., 1985. (Part of Chap. 4)

W. Flugge, *Handbook of Engineering Mechanics*, 1962. (Part of Chap. 5)

K. Gieck, *Engineering Formulas*, 5th ed., 1986. (Part of Chap. 3)

T. G. Hicks, *Standard Handbook of Engineering Calculations*, 2d ed., 1985. (Part of Chap. 12)

E. Kasner, *Essentials of Engineering Economics*, 1979. (Part of Chap. 20)

R. Lewis, *An Introduction to Reliability Engineering*, 1970. (Part of Chap. 19)

McGraw-Hill Encyclopedia of Physics, 1983. (Part of Chap. 11)

McGraw-Hill Encyclopedia of Science and Technology, 6th ed., 1987. (Parts of Chaps. 13, 15, 18, and 19)

Mark's Standard Handbook for Mechanical Engineers (T. Baumeister, E. A. Avallone, and T. Baumeister, III, eds.), 8th and 9th eds., 1978 and 1987. (Parts of Chaps. 3, 6, 12, 13, and 16)

J. F. Mulligan, *Practical Physics*, 1980. (Part of Chap. 11)

C. E. O'Rourke, *General Engineering Handbook*, 2d ed., 1940. (Parts of Chaps. 5, 6, 13, 14, and 16)

R. H. Perry, *Engineering Manual*, 3d ed., 1976. (Parts of Chaps. 3, 14, 15, and 17)

R. H. Perry, D. W. Green, and J. O. Maloney, *Perry's Chemical Engineers' Handbook*, 6th ed., 1984. (Parts of Chaps. 3 and 12)

W. M. Rohsenhow, J. P. Hartnett, and E. N. Ganic, *Handbook of Heat Transfer Fundamentals*, 2d ed., 1985. (Parts of Chaps. 9 and 10)

R. C. Rosaler and J. O. Rice, *Plant Equipment Reference Guide*, 1987. (Part of Chap. 19)

H. A. Rothbart, *Mechanical Design and Systems Handbook*, 2d ed., 1985. (Part of Chap. 3)

W. Staniar, *Plant Engineering Handbook*, 1950. (Part of Chap. 20)

D. A. Wells and H. S. Slusher, *Physics for Engineering and Science*, 1983. (Part of Chap. 11)

CHAPTER 1
ENGINEERING UNITS

DIMENSIONS AND UNITS

There are as many *dimensions* as there are kinds of physical quantities. Each new physical quantity gives rise to a new dimension. There can be only one dimension for each physical quantity.

A *unit* is a particular amount of the physical quantity. There are infinite possibilities for choosing a unit of a single physical quantity. All the possible units of the same physical quantity must be related by purely numerical factors.

Derived units are algebraic combinations of base units with some of the combinations being assigned special names and symbols.

SYSTEMS OF UNITS

There are still several systems of common units in use throughout the world. Transition from the others to Système International d'Unités (International System of Units), or SI, will proceed at a rational pace to accommodate the needs of the professions or industries involved and the public. The transition period will be long and complex, and duality of units probably will be demanded for at least a decade after the change is introduced.

1. SI Units. In October 1960, the Eleventh General (International) Conference on Weights and Measures redefined some of the original metric units and expanded the SI system to include other physical and engineering units.

The Metric Conversion Act of 1975 codifies the voluntary conversion of the United States to the SI system. It is expected that in time all units used in the United States will be in SI. For that reason, this chapter includes tables showing SI units, prefixes, and equivalents.

SI consists of seven base units, two supplementary units, a series of derived units consistent with the base and supplementary units, and a series of approved prefixes for the formation of multiples and submultiples of the various units. Multiple and submultiple prefixes in steps of 1000 are recommended.

Table 1.1 shows SI base and supplementary quantities (dimensions) and units. Tables 1.2 and 1.3 and Fig. 1.1 include additional derived units of SI. Table 1.4 shows SI prefixes.

TABLE 1.1 SI Base and Supplementary Quantities and Units

Quantity or "dimension"	SI unit	SI unit symbol ("abbreviation");
Base quantity or "dimension"		
length	meter	m
mass	kilogram	kg
time	second	s
electric current	ampere	A
thermodynamic temperature	kelvin	K
amount of substance	mole	mol
luminous intensity	candela	cd
Supplementary quantity or "dimension"		
plane angle	radian	rad
solid angle	steradian	sr

Source: From Perry, Green, and Maloney.[1]

TABLE 1.2 Derived Units of SI which Have Special Names

Quantity	Unit	Symbol	Formula
frequency (of a periodic phenomenon)	hertz	Hz	1/s
force	newton	N	$(kg \cdot m)/s^2$
pressure, stress	pascal	Pa	N/m^2
energy, work, quantity of heat	joule	J	$N \cdot m$
power, radiant flux	watt	W	J/s
quantity of electricity, electric charge	coulomb	C	$A \cdot s$
electric potential, potential difference, electromotive force	volt	V	W/A
capacitance	farad	F	C/V
electric resistance	ohm	Ω	V/A
conductance	siemens	S	A/V
magnetic flux	weber	Wb	$V \cdot s$
magnetic-flux density	tesla	T	Wb/m^2
inductance	henry	H	Wb/A
luminous flux	lumen	lm	$cd \cdot sr$
illuminance	lux	lx	lm/m^2
activity (of radionuclides)	becquerel	Bq	1/s
absorbed dose	gray	Gy	J/kg

Source: From Perry, Green, and Maloney.[1]

2. U.S. Customary System Units. The U.S. Customary System (USCS) is the system of units most commonly used for measures of weight and length in the United States. They are identical for practical purposes with the corresponding English units (see Sec. 4), but the capacity measures differ from those now in use in the British Commonwealth, the U.S. gallon being defined as 231 in^3 and the bushel as 2150.42 in^3, whereas the corresponding British Imperial units are, respectively, 277.42 in^3, and 2219.36 in^3 (1 Imp gal = 1.2 U.S. gal, approx.; 1 Imp bu = 1.03 U.S. bu, approx.). Table 1.5*a* shows USCS units, the corresponding SI units, and the numerical factors used to convert USCS values into SI. Table 1.5*b* defines the abbreviations used.

3. Metric System of Units. In the United States, the *metric* is commonly taken to refer to a system of length and mass units developed in France about 1800. The unit of length was equal to 1/10,000,000 of a quarter meridian (north pole to equator) and named the *meter*. A cube 0.1 meter on a side was the *liter*,

TABLE 1.3 Additional Common Derived Units of SI

Quantity	Unit	Symbol
acceleration	meter per second squared	m/s^2
angular acceleration	radian per second squared	rad/s^2
angular velocity	radian per second	rad/s
area	square meter	m^2
concentration (of amount of substance)	mole per cubic meter	mol/m^3
current density	ampere per square meter	A/m^2
density, mass	kilogram per cubic meter	kg/m^3
electric-charge density	coulomb per cubic meter	C/m^3
electric-field strength	volt per meter	V/m
electric-flux density	coulomb per square meter	C/m^2
energy density	joule per cubic meter	J/m^3
entropy	joule per kelvin	J/K
heat capacity	joule per kelvin	J/K
heat-flux density irradiance	watt per square meter	W/m^2
luminance	candela per square meter	cd/m^2
magnetic-field strength	ampere per meter	A/m
molar energy	joule per mole	J/mol
molar entropy	joule per mole-kelvin	$J/(mol \cdot K)$
molar-heat capacity	joule per mole-kelvin	$J/(mol \cdot K)$
moment of force	newton-meter	$N \cdot m$
permeability	henry per meter	H/m
permittivity	farad per meter	F/m
radiance	watt per square-meter–steradian	$W/(m^2 \cdot sr)$
radiant intensity	watt per steradian	W/sr
specific-heat capacity	joule per kilogram-kelvin	$J/(kg \cdot K)$
specific energy	joule per kilogram	J/kg
specific entropy	joule per kilogram-kelvin	$J/(kg \cdot K)$
specific volume	cubic meter per kilogram	m^3/kg
surface tension	newton per meter	N/m
thermal conductivity	watt per meter-kelvin	$W/(m \cdot K)$
velocity	meter per second	m/s
viscosity, dynamic	pascal-second	$Pa \cdot s$
viscosity, kinematic	square meter per second	m^2/s
volume	cubic meter	m^3
wave number	1 per meter	1/m

Source: From Perry, Green, and Maloney.[1]

the unit of volume. The mass of water filling this cube was the *kilogram*, or standard of mass; i.e., 1 liter of water = 1 kilogram of mass. Metal bars and weights were constructed conforming to these prescriptions for the meter and kilogram. One bar and one weight were selected to be the primary representations. The kilogram and the meter are now defined independently, and the liter, although for many years defined as the volume of a kilogram of water at the temperature of its maximum density, 4°C, and under a pressure of 76 cm of mercury, is now equal to 1 cubic decimeter (1 dm^3).

In 1866, the U.S. Congress formally recognized metric units as a legal system, thereby making their use permissible in the United States. In 1893, the Office of Weights and Measures (now the National Bureau of Standards), by executive order, fixed the values of the U.S. yard and pound in terms of the meter and kilogram, respectively, as 1 yd = 3600/3937 m and 1 lb = 0.453 592 4277 kg. By agreement in 1959 among the national standards laboratories of the English-speaking nations, the relations in use now are: 1 yd = 0.9144 m, whence 1 in = 25.4 mm exactly; and 1 lb = 0.453 592 37 kg, or 1 lb = 453.59 g (nearly).

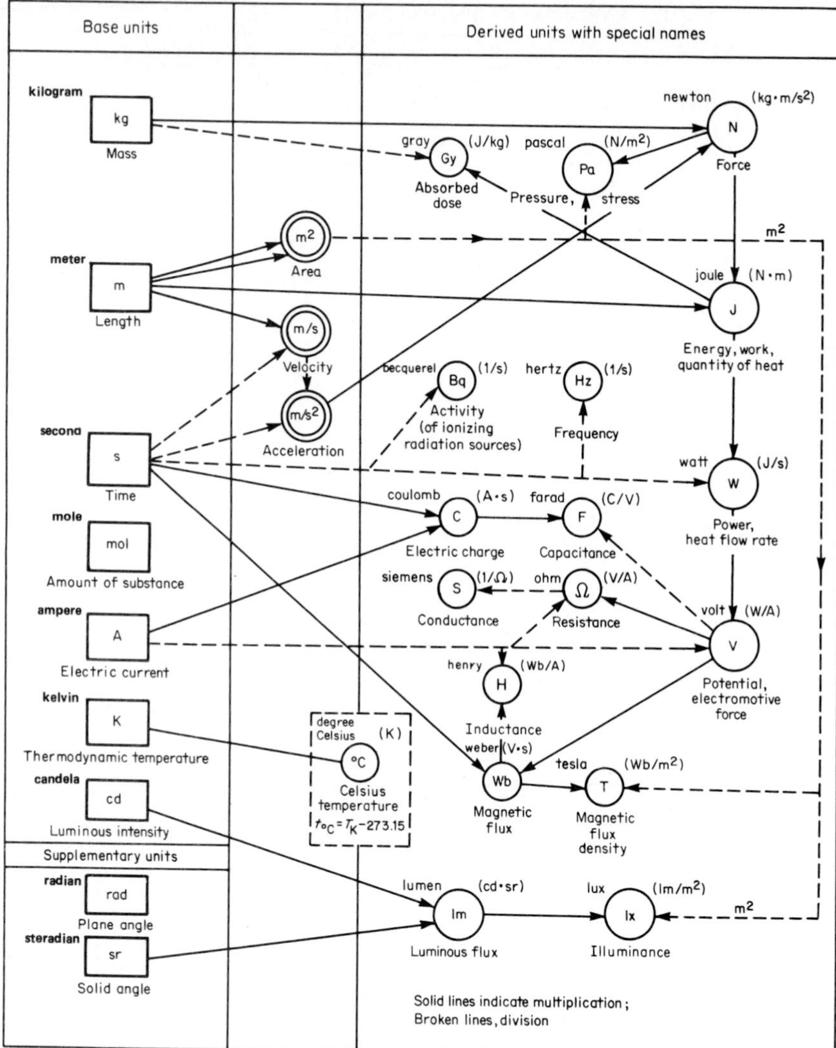

FIGURE 1.1 Graphic relationships of SI units with names. *(From Perry, Green, and Maloney.[1])*

4. English Units. The base units for the English engineering units are given in Table 1.6 (third column). The unit of force in English units is the pound force (lb_f). However, the use of the pound mass (lb) and pound force in engineering work causes considerable confusion in the proper use of these two fundamentally different units. A clear understanding of the units of mass and force can be gained by examining Newton's second law of motion. With any system of units, a conversion factor g_c must be introduced into the newtonian dynamics equation so that both sides of the equation will have the same units. Thus,

TABLE 1.4 SI Prefixes

Factor	Prefix	Symbol
10^{18}	exa	E
10^{15}	peta	P
10^{12}	tera	T
10^{9}	giga	G
10^{6}	mega	M
10^{3}	kilo	k
10^{2}	hecto *	h
10	deka *	da
10^{-1}	deci *	d
10^{-2}	centi	c
10^{-3}	milli	m
10^{-6}	micro	μ
10^{-9}	nano	n
10^{-12}	pico	p
10^{-15}	femto	f
10^{-18}	atto	a

*Generally to be avoided.

Source: From Rohsenow, Hartnett, and Ganić.[2]

$$F = \frac{ma}{g_c}$$

in which the numerical value and units of g_c depend on the units chosen for mass, force, length, and time.

The units of pound mass and pound force are related by the standard gravitational acceleration, which has a value of 32.174 ft/s^2. When a 1-lb mass is held at a location on the earth's surface where the gravitational acceleration is 32.174 ft/s^2, the mass weighs 1 lb$_f$. With this system of units, the value of g_c is determined as follows:

$$F = \frac{ma}{g_c}$$

$$1 \text{ lb}_f = \frac{1 \text{ lb} \times 32.174 \text{ ft/s}^2}{g_c}$$

whence

$$g_c = 32.174 \text{ lb} \cdot \text{ft/(lb}_f \cdot \text{s}^2)$$

Thus, g_c is merely a conversion factor and it should not be confused with the gravitational acceleration g. The numerical value of g_c is a constant depending only on the system of units involved, and not on the value of the gravitational acceleration at a particular location. Values of g_c corresponding to different systems of units found in engineering literature are given in Table 1.6.

Note that the SI units comprise a rigorously coherent form of the metric system; i.e., all remaining units may be derived from the base units using formulas which do not involve any numerical factors. For example, the unit of force is the

TABLE 1.5a Conversion Factors: U.S. Customary and Commonly Used Units to SI Units

Quantity	Customary or commonly used unit	SI unit	Alternative SI unit	Conversion factor*; multiply customary unit by factor to obtain SI unit†
		Space,‡time		
Length	naut mi	km		1.852* E + 00
	mi	km		1.609 344* E + 00
	yd	m		9.144* E − 01
	ft	m		3.048* E − 01
		cm		3.048* E + 01
	in	mm		2.54* E + 01
	in	cm		2.54 E + 00
	mil	μm		2.54* E + 01
Area	mi^2	km^2		2.589 988 E + 00
	acre	ha		4.046 856 E − 01
	ha	m^2		1.000 000* E + 04
	yd^2	m^2		8.361 274 E − 01
	ft^2	m^2		9.290 304* E − 02
	in^2	mm^2		6.451 6° E + 02
		cm^2		6.451 6* E + 00
Volume	cubem	km^3		4.168 182 E + 00
	acre · ft	m^3		1.233 482 E + 03
		ha · m		1.233 482 E − 01
	yd^3	m^3		7.645 549 E − 01
	bbl (42 U.S. gal)	m^3		1.589 873 E − 02
	ft^3	m^3		2.831 685 E − 02
		dm^3	L	2.831 685 E + 01
	U.K. gal	m^3		4.546 092 E − 03
		dm^3	L	4.546 092 E + 00
	U.S. gal	m^3		3.785 412 E − 03
		dm^3	L	3.785 412 E + 00
	U.K. qt	dm^3	L	1.136 523 E + 00
	U.S. qt	dm^3	L	9.463 529 E − 01
	U.S. pt	dm^3	L	4.731 765 E − 01
	U.K. fl oz	cm^3		2.841 307 E + 01
	U.S. fl oz	cm^3		2.957 353 E + 01
	in^3	cm^3		1.638 706 E + 01
Plane angle	rad	rad		1
	deg (°)	rad		1.745 329 E − 02
	min (′)	rad		2.908 882 E − 04
	sec (″)	rad		4.848 137 E − 06
Solid angle	sr	sr		1
Time	year	a		1
	week	d		7.0* E + 00
	h	s		3.6* E + 03
		min		6.0* E + 01
	min	s		6.0* E + 01
		h		1.666 667 E − 02
	mμs	ns		1

TABLE 1.5a Conversion Factors: U.S. Customary and Commonly Used Units to SI Units (*Continued*)

Quantity	Customary or commonly used unit	SI unit	Alternative SI unit	Conversion factor*; multiply customary unit by factor to obtain SI unit†
		Mass, amount of substance		
Mass	U.K. ton	Mg	t	1.016 047 E + 00
	U.S. ton	Mg	t	9.071 847 E − 01
	U.K. cwt	kg		5.080 234 E + 01
	U.S. cwt	kg		4.535 924 E + 01
	lb	kg		4.535 924 E − 01
	oz (troy)	g		3.110 348 E + 01
	oz (av)	g		2.834 952 E + 01
	gr	mg		6.479 891 E + 01
Amount of substance	lb · mol	kmol		4.535 924 E − 01
	std m^3(0°C, 1 atm)	kmol		4.461 58 E − 02
	std ft^3(60°F, 1 atm)	kmol		1.195 30 E − 03
		Enthalpy, calorific value, heat, entropy, heat capacity		
Calorific value, enthalpy (mass basis)	Btu/lb	MJ/kg		2.326 000 E − 03
		kJ/kg	J/g	2.326 000 E + 00
		kWh/kg		6.461 112 E − 04
	cal/g	kJ/kg	J/g	4.184* E + 00
	cal/lb	J/kg		9.224 141 E + 00
Caloric value, enthalpy (mole basis)	kcal/(g · mol)	kJ/kmol		4.184* E + 03
	Btu/(lb · mol)	kJ/kmol		2.326 000 E + 00
Calorific value (volume basis—solids and liquids)	Btu/U.S. gal	MJ/m^3	kJ/dm^3	2.787 163 E − 01
		kJ/m^3		2.787 163 E + 02
		kWh/m^3		7.742 119 E − 02
	Btu/U.K. gal	MJ/m^3	kJ/dm^3	2.320 800 E − 01
		kJ/m^3		2.320 800 E + 02
		kWh/m^3		6.446 667 E − 02
	Btu/ft^3	MJ/m^3	kJ/dm^3	3.725 895 E − 02
		kJ/m^3		3.725 895 E + 01
		kWh/m^3		1.034 971 E − 02
	cal/mL	MJ/m^3		4.184* E + 00
	(ft · lb_f)/U.S. gal	kJ/m^3		3.581 692 E − 01
Calorific value (volume basis—gases)	cal/mL	kJ/m^3	J/dm^3	4.184* E + 03
	$kcal/m^3$	kJ/m^3	J/dm^3	4.184* E + 00
	Btu/ft^3	kJ/m^3	J/dm^3	3.725 895 E + 01
		kWh/m^3		1.034 971 E − 02
Specific entropy	Btu/(lb · °R)	kJ/(kg · K)	J/(g · K)	4.186 8* E + 00
	cal/(g · K)	kJ/(kg · K)	J/(g · K)	4.184* E + 00
	kcal/(kg · °C)	kJ/(kg · K)	J/(g · K)	4.184* E + 00
Specific-heat capacity (mass basis)	Btu/(lb · °F)	kJ/(kg · K)	J/(g · K)	4.186 8* E + 00
	kcal/(kg · °C)	kJ/(kg · K)	J/(g · K)	4.184* E + 00
Specific heat capacity (mole basis)	Btu/(lb · mol · °F)	kJ/(kmol · K)		4.186 8* E + 00
	cal/(g · mol · °C)	kJ/(kmol · K)		4.184* E + 00

TABLE 1.5a Conversion Factors: U.S. Customary and Commonly Used Units to SI Units (*Continued*)

Quantity	Customary or commonly used unit	SI unit	Alternative SI unit	Conversion factor*, multiply customary unit by factor to obtain SI unit†	
	Temperature, pressure, vacuum				
Temperature (absolute)	°R	K		5/9	
	K	K		1	
Temperature (traditional)	°F	°C		5/9(°F − 32)	
Temperature (difference)	°F	K, °C		5/9	
Pressure	atm (760 mmHg at 0°C or 14,696 lb_f/in^2)	MPa		1.013 250*	E − 01
		kPa		1.013 250*	E + 02
		bar		1.013 250*	E + 00
	bar	MPa		1.0*	E − 01
		kPa		1.0*	E + 02
	mmHg (0°C) = torr	MPa		6.894 757	E − 03
		kPa		6.894 757	E + 00
		bar		6.894 757	E − 02
	μmHg (0°C)	kPa		3.376 85	E + 00
	μ bar	kPa		2.488 4	E − 01
	mmHg = torr (0°C)	kPa		1.333 224	E − 01
	cmH_2O (4°C)	kPa		9.806 38	E − 02
	lb_f/ft^2 (psf)	kPa		4.788 026	E − 02
	mHg (0°C)	Pa		1.333 224	E − 01
	bar	Pa		1.0*	E − 01
	dyn/cm^2	Pa		1.0*	E − 01
Vacuum, draft	inHg (60°F)	kPa		3.376 85	E + 00
	inH_2O (39.2°F)	kPa		2.490 82	E − 01
	inH_2O (60°F)	kPa		2.488 4	E − 01
	mmHg (0°C) = torr	kPa		1.333 224	E − 01
	cmH_2O (4°C)	kPa		9.806 38	E − 02
Liquid head	ft	m		3.048*	E − 01
	in	mm		2.54*	E + 01
		cm		2.54*	E + 00
Pressure drop/length	$lb_f/in^2/ft$	kPa/m		2.262 059	E + 01
	Density, specific volume, concentration, dosage				
Density	lb/ft^3	kg/m^3		1.601 846	E + 01
		g/m^3		1.601 846	E + 04
	lb/U.S. gal	kg/m^3		1.198 264	E + 02
		g/cm^3		1.198 264	E − 01
	lb/U.K. gal	kg/m^3		9.977 633	E + 01

TABLE 1.5a Conversion Factors: U.S. Customary and Commonly Used Units to SI Units (*Continued*)

Quantity	Customary or commonly used unit	SI unit	Alternative SI unit	Conversion factor*, multiply customary unit by factor to obtain SI unit†	
		Density, specific volume, concentration, dosage			
	lb/ft³	kg/m³		1.601 846	E + 01
		g/cm³		1.601 846	E − 02
	g/cm³	kg/m³		1.0*	E + 03
	lb/ft³	kg/m³		1.601 846	E + 01
Specific volume	ft³/lb	m³/kg		6.242 796	E − 02
		m³/g		6.242 796	E − 05
	ft³/lb	dm³/kg		6.242 796	E + 01
	U.K. gal/lb	dm³/kg	cm³/g	1.002 242	E + 01
	U.S. gal/lb	dm³/kg	cm³/g	8.345 404	E + 00
Specific volume (mole basis)	L/(g · mol)	m³/kmol		1	
	ft³/(lb · mol)	m³/kmol		6.242 796	E − 02
Concentration (mass/volume)	lb/bbl	kg/m³	g/dm³	2.853 010	E + 00
	g/U.S. gal	kg/m³		2.641 720	E − 01
	g/U.K. gal	kg/m³	g/L	2.199 692	E − 01
	lb/1000 U.S. gal	g/m³	mg/dm³	1.198 264	E + 02
	lb/1000 U.K. gal	g/m³	mg/dm³	9.977 633	E + 01
	gr/U.S. gal	g/m³	mg/dm³	1.711 806	E + 01
	gr/ft³	mg/m³		2.288 351	E + 03
	lb/1000 bbl	g/m³	mg/dm³	2.853 010	E + 00
	mg/U.S. gal	g/m³	mg/dm³	2.641 720	E − 01
	gr/100 ft³	mg/m³		2.288 351	E + 01
Concentration (mole/volume)	(lb · mol)/U.S. gal	kmol/m³		1.198 264	E + 02
	(lb · mol)/U.K. gal	kmol/m³		9.977 644	E + 01
	(lb · mol)/ft³	kmol/m³		1.601 846	E + 01
		Flow rate			
Flow rate (mass basis)	lb/s	kg/s		4.535 924	E − 01
	lb/min	kg/s		7.559 873	E − 03
	lb/h	kg/s		1.259 979	E − 04
Flow rate (volume basis)	bbl/day	m³/d		1.589 873	E − 01
		L/s		1.840 131	E − 03
	ft³/day	m³/d		2.831 685	E − 02
	bbl/h	L/s		3.277 413	E − 04
		m³/s		4.416 314	E − 05
	ft³/h	L/s		4.416 314	E − 02
		m³/s		7.865 791	E − 06
	ft³/min	dm³/s	L/s	6.309 020	E − 02
	ft³/s	dm³/s	L/s	4.719 474	E − 01
		dm³/s	L/s	2.831 685	E + 01
Flow rate (mole basis)	(lb · mol)/s	kmol/s		4.535 924	E − 01
	(lb · mol)/h	kmol/s		1.259 979	E − 04

TABLE 1.5a Conversion Factors: U.S. Customary and Commonly Used Units to SI Units (*Continued*)

Quantity	Customary of commonly used unit	SI unit	Alternative SI unit	Conversion factor,* multiply customary unit by factor to obtain SI unit†	
		Flow rate (*continued*)			
Flow rate/area (mass basis)	lb/(s · ft^2)	kg/(s · m^2)		4.882 428	E + 00
	lb/(h · ft^2)	kg/(s · m^2)		1.356 230	E − 03
Flow rate/area (volume basis)	ft^3/(s · ft^2)	m/s	m^3/(s · m^2)	3.0 048*	E − 01
	ft^3/(min · ft^2)	m/s	m^3/(s · m^2)	5.08*	E − 03
	U.K. gal/(h · ft^2)	m/s	m^3/(s · m^2)	1.359 270	E − 05
	U.S. gal/(h · ft^2)	m/s	m^3/(s · m^2)	1.131 829	E − 05
		Energy, work, power			
Energy, work	therm	MJ		1.055 056	E + 02
		kJ		1.055 056	E + 05
		kWh		2.930 711	E + 01
	hp · h	MJ		2.684 520	E + 00
		kJ		2.684 520	E + 03
		kWh		7.456 999	E − 01
	ch · h or CV · h	MJ		2.647 780	E + 00
		kJ		2.647 780	E + 03
		kWh		7.354 999	E − 01
	kWh	MJ		3.6*	E + 00
		kJ		3.6*	E + 03
	Chu	kJ		1.899 101	E + 00
		kWh		5.275 280	E − 04
	Btu	kJ		1.055 056	E + 00
		kWh		2.930 711	E − 04
	kcal	kJ		4.184*	E + 00
	cal	kJ		4.184*	E − 03
	ft · lb_f	kJ		1.355 818	E − 03
	lb_f · ft	kJ		1.355 818	E − 03
	J	kJ		1.0*	E − 03
	(lb_f · ft^2)/s^2	kJ		4.214 011	E − 05
	erg	J		1.0*	E − 07
Impact energy	kg_f · m	J		9.806 650*	E + 00
	lb_f · ft	J		1.355 818	E + 00
Surface energy	erg/cm^2	mJ/m^2		1.0*	E + 00
Specific-impact energy	(kg_f · m)/cm^2	J/cm^2		9.806 650*	E − 02
	(lb_f · ft)/in^2	J/cm^2		2.101 522	E − 03
Power	million Btu/h	MW		2.930 711	E − 01
	ton of refrigeration	kW		3.516 853	E + 00
	Btu/s	kW		1.055 056	E + 00
	hydraulic horsepower—hhp	kW		7.460 43	E − 01
	hp (electric)	kW		7.46*	E − 01
	hp [(550 ft · lb_f)/s]	kW		7.456 999	E − 01

TABLE 1.5a Conversion Factors: U.S. Customary and Commonly Used Units to SI Units (*Continued*)

Quantity	Customary of commonly used unit	SI unit	Alternative SI unit	Conversion factor,* multiply customary unit by factor to obtain SI unit†	
	Energy, work, power (*continued*)				
	Btu/min	kW		1.758 427	E − 02
	(ft · lb_f)/s	kW		1.355 818	E − 03
	kcal/h	W		1.162 222	E + 00
	Btu/h	W		2.930 711	E − 01
	(ft · lb_f)/min	W		2.259 697	E − 02
Power/area	Btu/(s · ft^2)	kW/m^2		1.135 653	E + 01
	cal/(h · cm^2)	kW/m^2		1.162 222	E − 02
	Btu/(h · ft^2)	kW/m^2		3.154 591	E − 03
Heat-release rate, mixing power	hp/ft^3	kW/m^3		2.633 414	E + 01
	cal/(h · cm^3)	kW/m^3		1.162 222	E + 00
	Btu/(s · ft^3)	kW/m^3		3.725 895	E + 01
	Btu/(h · ft^3)	kW/m^3		1.034 971	E − 02
Cooling duty (machinery)	Btu/(bhp · h)	W/kW		3.930 148	E − 01
Specific fuel consumption (mass basis)	lb/(hp · h)	mg/J	kg/MJ	1.689 659	E − 01
		kg/kWh		6.082 774	E − 01
Specific fuel consumption (volume basis)	m^3/kWh	dm^3/MJ	mm^3/J	2.777 778	E + 02
	U.S. gal/(hp · h)	dm^3/MJ	mm^3/J	1.410 089	E + 00
Fuel consumption	U.K. gal/mi	dm^3/100 km	L/100 km	2.824 807	E + 02
	U.S. gal/mi	dm^3/100 km	L/100 km	2.352 146	E + 02
	mi/U.S. gal	km/dm^3	km/L	4.251 437	E − 01
	mi/U.K. gal	km/dm^3	km/L	3.540 064	E − 01
Velocity (linear), speed	knot	km/h		1.852*	E + 00
	mi/h	km/h		1.609 344*	E + 00
	ft/s	m/s		3.048*	E − 01
		cm/s		3.048*	E + 01
	ft/min	m/s		5.08*	E − 03
	ft/h	mm/s		8.466 667	E − 02
	ft/day	mm/s		3.527 778	E − 03
		m/d		3.048*	E − 01
	in/s	mm/s		2.54*	E + 01
	in/min	mm/s		4.233 333	E − 01
Corrosion rate	in/year (ipy)	mm/a		2.54*	E + 01
	mil/year	mm/a		2.54*	E − 02
Rotational frequency	r/min	r/s		1.666 667	E − 02
		rad/s		1.047 198	E − 01
Acceleration (linear)	ft/s^2	m/s^2		3.048*	E − 01
		cm/s^2		3.048*	E + 01

TABLE 1.5a Conversion Factors: U.S. Customary and Commonly Used Units to SI Units (*Continued*)

Quantity	Customary or commonly used unit	SI unit	Alternative SI unit	Conversion factor;* multiply customary unit by factor to obtain SI unit†	
	Energy, work, power (*continued*)				
Acceleration (rotational)	rpm/s	rad/s^2		1.047 198	E − 01
Momentum	(lb · ft)/s	(kg · m)/s		1.382 550	E − 01
Force	U.K. ton_f	kN		9.964 016	E + 00
	U.S. ton_f	kN		8.896 443	E + 00
	kg_f (kp)	N		9.806 650*	E + 00
	lb_f	N		4.448 222	E + 00
	dyn	mN		1.0	E − 02
Bending moment, torque	U.S. ton_f · ft	kN · m		2.711 636	E + 00
	kg_f · m	N · m		9.806 650*	E + 00
	lb_f · ft	N · m		1.355 818	E + 00
	lb_f · in	N · m		1.129 848	E − 01
Bending moment/length	(lb_f · ft)/in	(N · m)/m		5.337 866	E + 01
	(lb_f · in)/in	(N · m)/m		4.448 222	E + 00
Moment of inertia	lb · ft^2	kg · m^2		4.214 011	E − 02
Stress	U.S. ton_f /in^2	MPa	N/mm^2	1.378 951	E + 01
	kg_f /mm^2	MPa	N/mm^2	9.806 650*	E + 00
	U.S. ton_f /ft^2	MPa	N/mm^2	9.576 052	E − 02
	lb_f /in^2 (psi)	MPa	N/mm^2	6.894 757	E − 03
	lb_f /ft^2 (psf)	kPa		4.788 026	E − 02
	dyn/cm^2	Pa		1.0*	E − 01
Mass/length	lb/ft	kg/m		1.488 164	E + 00
Mass area structural loading, bearing capacity (mass basis)	U.S. ton/ft^2	Mg/m^2		9.764 855	E + 00
	lb/ft^2	kg/m^2		4.882 428	E + 00
	Miscellaneous transport properties				
Diffusivity	ft^2/s	m^2/s		9.290 304*	E − 02
	m^2/s	mm^2/s		1.0*	E + 06
	ft^2/h	m^2/s		2.580 64*	E − 05
Thermal resistance	(°C · m^2 · h)/kcal	(K · m^2)/kW		8.604 208	E + 02
	(°F · ft^2 · h)/Btu	(K · m^2)/kW		1.761 102	E + 02
Heat flux	Btu/(h · ft^2)	kW/m^2		3.154 591	E − 03
Thermal conductivity	(cal · cm)/ (s · cm^2 · °C)	W/(m · K)		4.184*	E + 02
	(Btu · ft)/ (h · ft^2 · °F)	W/(m · K)		1.730 735	E + 00
		(kJ · m)/ (h · m^2 · K)		6.230 646	E + 00
	(kcal · m)/ (h · m^2 · °C)	W/(m · K)		1.162 222	E + 00

TABLE 1.5a Conversion Factors: U.S. Customary and Commonly Used Units to SI Units (*Continued*)

Quantity	Customary or commonly used unit	SI unit	Alternative SI unit	Conversion factor;* multiply customary unit by factor to obtain SI unit†
	Miscellaneous transport properties (*continued*)			
	(Btu · in)/ (h · ft^2 · °F)	W/(m · K)		1.442 279 E − 01
	(cal · cm)/ (h · cm^2 · °C)	W/(m · K)		1.162 222 E − 01
Heat-transfer coefficient	cal/(h · cm^2 · °C)	kW/(m^2 · K)		1.162 222 E − 02
	Btu/(h · ft^2 · °F)	kW/(m^2 · K)		5.678 263 E − 03
		kJ/(h · m^2 · K)		2.044 175 E + 01
	Btu/(h · ft^2 · °R)	kW/(m^2 · K)		5.678 263 E − 03
	kcal/(h · m^2 · °C)	kW/(m^2 · K)		1.162 222 E − 03
Volumetric heat-transfer coefficient	Btu/(s · ft^3 · °F)	kW/(m^3 · K)		6.706 611 E + 01
	Btu/(h · ft^3 · °F)	kW/(m^3 · K)		1.862 947 E − 02
Surface tension	dyn/cm	N/m		1. E + 03
Viscosity (dynamic)	(lb$_f$ · s)/in^2	Pa · s	(N · s)/m^2	6.894 757 E + 03
	(lb$_f$ · s)/ft^2	Pa · s	(N · s)/m^2	4.788 026 E + 01
	(kg$_f$ · s)/m^2	Pa · s	(N · s)/m^2	9.806 650* E + 00
	lb/(ft · s)	Pa · s	(N · s)/m^2	1.488 164 E + 00
	(dyn · s)/cm^2	Pa · s	(N · s)/m^2	1.0* E − 01
	cP	Pa · s	(N · s)/m^2	1.0* E − 03
	lb/(ft · h)	Pa · s	(N · s)/m^2	4.133 789 E − 04
Viscosity (kinematic)	ft^2/s	m^2/s		9.290 304* E 02
	in^2/s	mm^2/s		6.451 6* E + 02
	m^2/h	mm^2/s		2.777 778 E + 02
	ft^2/h	m^2/s		2.580 64* E − 05
	cSt	mm^2/s		1
Permeability	darcy	μm^2		9.869 233 E − 01
	millidarcy	μm^2		9.869 233 E − 04
Thermal flux	Btu/(h · ft^2)	W/m^2		3.152 E + 00
	Btu/(s · ft^2)	W/m^2		1.135 E + 04
	cal/(s · cm^2)	W/m^2		4.184 E + 04
Mass-transfer coefficient	(lb · mol)/[h · ft^2 (lb · mol/ft^3)]	m/s		8.467 E − 05
	(g · mol)/[s · m^2 (g · mol/L)]	m/s		1.0 E + 01
	Electricity, magnetism			
Admittance	S	S		1
Capacitance	μF	μF		1
Charge density	C/mm^3	C/mm^3		1
Conductance	S	S		1
	Ω(mho)	S		1

TABLE 1.5a Conversion Factors: U.S. Customary and Commonly Used Units to SI Units (*Continued*)

Quantity	Customary or commonly used unit	SI unit	Alternative SI unit	Conversion factor;* multiply customary unit by factor to obtain SI unit†
	Electricity, magnetism (*continued*)			
Conductivity	S/m	S/m		1
	Ω/m	S/m		1
	mΩ/m	mS/m		1
Current density	A/mm^2	A/mm^2		1
Displacement	C/cm^2	C/cm^2		1
Electric charge	C	C		1
Electric current	A	A		1
Electric-dipole moment	C · m	C · m		1
Electric-field strength	V/m	V/m		1
Electric flux	C	C		1
Electric polarization	C/cm^2	C/cm^2		1
Electric potential	V	V		1
	mV	mV		1
Electromagnetic moment	$A \cdot m^2$	$A \cdot m^2$		1
Electromotive force	V	V		1
Flux of displacement	C	C		1
Frequency	cycles/s	Hz		1
Impedance	Ω	Ω		1
Linear-current density	A/mm	A/mm		1
Magnetic-dipole moment	Wb · m	Wb · m		1
Magnetic-field strength	A/mm	A/mm		1
	Oe	A/m		7.957 747 E + 01
	gamma	A/m		7.957 747 E − 04
Magnetic flux	mWb	mWb		1
Magnetic-flux density	mT	mT		1
	G	T		1.0* E − 04
	gamma	nT		1
Magnetic induction	mT	mT		1
Magnetic moment	$A \cdot m^2$	$A \cdot m^2$		1
Magnetic polarization	mT	mT		1

TABLE 1.5a Conversion Factors: U.S. Customary and Commonly Used Units to SI Units (*Continued*)

Quantity	Customary or commonly used unit	SI unit	Alternative SI unit	Conversion factor;* multiply customary unit by factor to obtain SI unit†	
		Electricity, magnetism (*continued*			
Magnetic potential difference	A	A		1	
Magnetic-vector potential	Wb/mm	Wb/mm		1	
Magnetization	A/mm	A/mm		1	
Modulus of admittance	S	S		1	
Modulus of impedance	Ω	Ω		1	
Mutual inductance	H	H		1	
Permeability	μH/m	μH/m		1	
Permeance	H	H		1	
Permittivity	μF/m	μF/m		1	
Potential difference	V	V		1	
Quantity of electricity	C	C		1	
Reactance	Ω	Ω		1	
Reluctance	H^{-1}	H^{-1}		1	
Resistance	Ω	Ω		1	
Resistivity	Ω · cm Ω · m	Ω · cm Ω · m		1 1	
Self-inductance	mH	mH		1	
Surface density of change	mC/m^2	mC/m^2		1	
Susceptance	S	S		1	
Volume density of charge	C/mm^3	C/mm^3		1	
		Acoustics, light, radiation			
Absorbed dose	rad	Gy		1.0*	E − 02
Acoustical energy	J	J		1	
Acoustical intensity	W/cm^2	W/m^2		1.0*	E + 04
Acoustical power	W	W		1	
Sound pressure	N/m^2	N/m^2		1.0*	
Illuminance	fc	lx		1.076 391	E + 01

TABLE 1.5a Conversion Factors: U.S. Customary and Commonly Used Units to SI Units (*Continued*)

Quantity	Customary or commonly used unit	SI unit	Alternative SI unit	Conversion factor;* multiply customary unit by factor to obtain SI unit†	
		Acoustics, light, radiation (*continued*)			
Illumination	fc	lx		1.076 391	E + 01
Irradiance	W/m^2	W/m^2		1	
Light exposure	fc · s	lx · s		1.076 391	E + 01
Luminance	cd/m^2	cd/m^2		1	
Luminous efficacy	lm/W	lm/W		1	
Luminous exitance	lm/m^2	lm/m^2		1	
Luminous flux	lm	lm		1	
Luminous intensity	cd	cd		1	
Radiance	$W/m^2 \cdot sr$	$W/m^2 \cdot sr$		1	
Radiant energy	J	J		1	
Radiant flux	W	W		1	
Radiant intensity	W/sr	W/sr		1	
Radiant power	W	W		1	
Wavelength	Å	nm		1.0*	E − 01
Capture unit	$10^{-3}cm^{-1}$	m^{-1}		1.0*	E + 01
			$10^{-3}cm^{-1}$	1	
	m^{-1}	m^{-1}		1	
Radioactivity	Ci	Bq		3.7*	E + 10

*An asterisk indicates that the conversion factor is exact.
†Or multiply SI unit by reciprocal value of factor to obtain USCS unit.
‡Conversion factors for length, area, and volume are based on the international foot. The international foot is longer by 2 parts in 1 million than the U.S. Survey foot (land-measurement use).
Source: From Perry, Green, and Maloney.[1]

newton (N); a 1-N force will accelerate a 1-kg mass at 1 m/s^2. Hence 1 N = 1kg · m/s^2. The unit of pressure is the N/m^2 and is often referred to as the *pascal* (Pa). In the SI system, there is one unit of energy, whether the energy is thermal, mechanical, or electrical: the joule (J). (1 J = 1 N · m.) The unit for energy rate, or power, is the J/s, where one joule per second is equivalent to one watt (W) (1 J/s = 1 W).

In the English system of units, it is necessary to relate thermal and mechanical energy via the mechanical equivalent of heat, J_c. Thus

$$J_c \times \text{thermal energy} = \text{mechanical energy}$$

TABLE 1.5b Unit Symbols Used in Table 1.5a

Unit symbol	Name	Unit symbol	Name
A	ampere	lm	lumen
a	annum (year)	lx	lux
Bq	becquerel	m	meter
C	coulomb	℧	mho
cd	candela	min	minute
Ci	curie	′	minute
d	day	N	newton
°C	degree Celsius	naut mi	U.S. nautical mile
°	degree	Oe	oersted
dyn	dyne	Ω	ohm
F	farad	Pa	pascal
fc	footcandle	rad	radian
G	gauss	r	revolution
g	gram	S	siemens
gr	grain	s	second
Gy	gray	″	second
H	henry	sr	steradian
h	hour	St	stokes
ha	hectare	T	tesla
Hz	hertz	ton_f	ton force
J	Joule	t	tonne
K	kelvin	V	volt
L, *l*, l	liter	W	watt
lb	pound mass	Wb	weber
lb_f	pound force		

The unit of heat in the English system is the British thermal unit (Btu). When the unit of mechanical energy is the pound-force-foot ($lb_f \cdot ft$), then

$$J_c = 778.16\ lb_f \cdot ft/Btu$$

as 1 Btu = 778.16 $lb_f \cdot ft$. Happily, in the SI system the units of heat and work are identical and J_c is unity.

5. SI Learning and Usage. The technical and scientific community throughout the world accepts SI units for use in both applied and theoretical calculation. With such widespread acceptance, every engineer must become proficient in the use of this system if he or she is to remain up to date. For this reason, most calculation procedures in this handbook are given in both SI and USCS. This will help all engineers become proficient in using both systems. However, in some cases results and tables are given in one system, mostly to save space, and conversion factors are printed at the end of such results (or tables) for the reader's convenience.

Engineers accustomed to working in USCS are often timid about using SI. There are really no sound reasons for these fears. SI is a logical, easily understood, and readily manipulated group of units. Most engineers grow to prefer SI, once they become familiar with it and overcome their fears.

TABLE 1.6 Conversion Factor g_c for the Common Unit Systems

Quantity	SI	English engineering*	cgs†	Metric engineering
Mass	kilogram, kg	pound mass, lb	gram, g	kilogram mass, kg
Length	meter, m	foot, ft	centimeter, cm	meter, m
Time	second, s	second, s, or hour, h	second, s	second, s
Force	newton, N	pound force, lb_f	dyne, dyn	kilogram force, kg_f
g_c	1 $kg \cdot m/(N \cdot s^2)$‡	32.174 $lb \cdot ft/(lb_f \cdot s^2)$ or 4.1698×10^2 $lb \cdot ft/(lb_f \cdot h^2)$	1 $g \cdot cm/(dyn \cdot s^2)$	9.80665 $kg \cdot m/(kg_f \cdot s^2)$

* In this system of units the temperature is given in degrees Fahrenheit (°F).
† Centimeter-gram-second: this system of units has been used mostly in scientific work.
‡ Since 1 $kg \cdot m/s^2$ = 1 N, then g_c = 1 in the SI system of units.
Source: From Rohsenow, Hartnett, and Ganic.[2]

Overseas engineers who must work in USCS because they have a job requiring its usage will find the dual-unit presentation of calculation procedures most helpful. Knowing SI, they can easily convert to USCS.

An efficient way for the USCS-conversant engineer to learn SI follows these steps:

1. List units of measurement commonly used in one's daily work.
2. Insert, opposite each USCS unit, the usual SI unit used; Table 1.5 shows a variety of commonly used quantities and the corresponding SI units.
3. Find, from a table of conversion factors, such as Table 1.5, the value to use to convert the USCS unit to SI, and insert it in the list. (Most engineers prefer a conversion factor that can be used as a multiplier of the USCS unit to give the SI unit.)
4. Apply the conversion factor whenever the opportunity arises. Think in terms of SI when an USCS unit is encountered.
5. Recognize—here and now—that the most difficult aspect of SI is becoming comfortable with the names and magnitudes of the units. Numerical conversion is simple once a conversion table has been set up. So think *pascal* whenever pounds per square inch pressure are encountered, *newton* whenever a force in pounds is being dealt with, etc.

CONVERSION FACTORS

Conversion factors between SI and USCS units are given in Table 1.5. Note that E indicates an exponent, as in scientific notation, followed by a positive or negative number representing the power of 10 by which the given conversion factor is to be multiplied before use. Thus, for the square feet conversion factor, 9.290 304 × 1/100 = 0.092 903 04, the factor to be used to convert square feet to square meters. For a positive exponent, as in converting British thermal units per cubic foot to kilojoules per cubic meter, 3.725 895 × 10 = 37.258 95.

Where a conversion factor cannot be found, simply use the dimensional substitution. Thus, to convert pounds per cubic inch to kilograms per cubic meter, find 1 lb = 0.453 592 4 kg and 1 m^3 = 0.000 016 387 06 m^3. Then, 1 lb/in^3 = 0.453 592 4 kg/0.000 016 387 06 m^3 = 2.767 990 E + 04.

SELECTED PHYSICAL CONSTANTS

A list of selected physical constants is given in Table 1.7.

DIMENSIONAL ANALYSIS

Dimensional analysis is the mathematics of dimensions and quantities and provides procedural techniques whereby the variables that are assumed to be significant in a problem can be formed into dimensionless parameters, the number of parameters being less than the number of variables. This is a great advantage, because fewer experimental runs are then required to establish a relationship between the parameter than between the variables. While the user is not presumed to have any knowledge of the fundamental physical equations, the more knowledgeable the user, the better the results. If any significant variable or variables are omitted, the relationship obtained from dimensional analysis will not apply to the physical problem. On the other hand, inclusion of all possible variables will result in losing the principal advantage of dimensional analysis, i.e., the reduction of the amount of experimental data required to establish a relationship. Formal methods of dimensional analysis are given in Chap. 10.

REFERENCES*

1. R. H. Perry, D. W. Green, and J. O. Maloney (eds.), *Perry's Chemical Engineers Handbook*, 6th ed., McGraw-Hill, New York, 1984.
2. W. M. Rohsenow, J. P. Hartnett, and E. N. Ganić (eds.), *Handbook of Heat Transfer Fundamentals*, 2d ed., McGraw-Hill, New York, 1985.
3. E. A. Avallone and T. Baumeister III (eds.), *Mark's Standard Handbook for Mechanical Engineers*, 9th ed., McGraw-Hill, New York, 1987.
4. O. W. Eshbach and M. Souders, *Handbook of Engineering Fundamentals*, John Wiley & Sons, New York, 1975.
5. T. G. Hicks, *Standard Handbook of Engineering Calculation*, 2d ed., McGraw-Hill, New York, 1985.
6. R. H. Perry, *Engineering Manual*, 3d ed., McGraw-Hill, New York, 1976.

*Those references listed above but not cited in the text were used for comparison between different data sources, clarification, clarity of presentation, and, most important, reader's convenience when further interest in the subject exists.

TABLE 1.7 Fundamental Physical Constants

Constant	Remarks
1 sec. = 1.00273791 sidereal seconds	sec. = mean solar second
g_0 = 9.80665 m./sec.^{2}	Definition: g_0 = standard gravity
1 liter = 0.001 cu. m.	
1 atm. = 101,325 newtons/sq. m.	Definition: atm. = standard atmosphere
1 mm. Hg (pressure) = (1/760) atm.	mm. Hg (pressure) = standard millimeter mercury
= 133.3224 newtons/sq. m.	
1 int. ohm = 1.000495 ± 0.000015 abs. ohm	int. = international; abs. = absolute
1 int. amp. = 0.999835 ± 0.000025 abs. amp.	amp. = ampere
1 int. coul. = 0.999835 ± 0.000025 abs. coul.	coul. = coulomb
1 int. volt = 1.000330 ± 0.000029 abs. volt	
1 int. watt = 1.000165 ± 0.000052 abs. watt	
1 int. joule = 1.000165 ± 0.000052 abs. joule	
$T_{0°C.}$ = 273.150 ± 0.010°K.	Absolute temperature of the ice point, 0°C.
R = 8.31439 ± 0.00034 abs. joule/deg. mole	R = gas constant per mole
= 1.98719 ± 0.00013 cal./deg. mole	
= 82.0567 ± 0.0034 cu. cm. atm./deg. mole	
= 0.0820567 ± 0.0000034 liter atm./deg. mole	
ln 10 = 2.302585	ln = natural logarithm (base e)
R ln 10 = 19.14460 ± 0.00078 abs. joule/deg. mole	
= 4.57567 ± 0.00030 cal./deg. mole	
$N = (6.02283 \pm 0.0022) \times 10^{23}$/mole	N = *Avogadro number*
$h = (6.6242 \pm 0.0044) \times 10^{-34}$ joule sec.	h = Planck constant
$c = (2.99776 \pm 0.00008) \times 10^{8}$ m./sec.	c = velocity of light
$(h^2/8\pi^2k) = (4.0258 \pm 0.0037) \times 10{-}39$ g. sq. cm. deg.	Constant in rotational partition function of gases
$(h/8\pi^2c) = (2.7986 \pm 0.0018) \times 10^{-39}$ g. cm.	Constant relating wave number and moment of inertia
$Z = Nhc$ = 11.9600 ± 0.0036 abs. joule cm./mole	Z = constant relating wave number and energy per mole
= 2.85851 ± 0.0009 cal. cm./mole	
$(Z/R) = (hc/k) = c_2$ = 1.43847 ± 0.00045 cm. deg.	c_2 = second radiation constant
$\mathfrak{F}$ = 96,501.2 ± 10.0 int. coul./g.-equiv. or int. joule/int. volt g.-equiv.	$\mathfrak{F}$ = Faraday constant
= 96,485.3 ± 10.0 abs. coul./g.-equiv. or abs. joule/abs. volt g.-equiv.	
= 23,068.1 ± 2.4 cal./int. volt g.-equiv.	
= 23,060.5 ± 2.4 cal./abs. volt g.-equiv.	
$e = (1.60199 \pm 0.00060) \times 10^{-19}$ abs. coul.	e = electronic charge
$= (1.60199 \pm 0.00060) \times 10^{-20}$ abs. e.m.u.	
$= (4.80239 \pm 0.00180) \times 10^{-10}$ abs. e.s.u.	

1 int. electron-volt/molecule = 96,501.2 ± 10 int. joule/mole	
= 23,068.1 ± 2.4 cal./mole	
1 abs. electron-volt/molecule = 96,485.3 ± 10. abs. joule/mole	
= 23,060.5 ± 2.4 cal./mole	
1 int. electron-volt = (1.60252 ± 0.00060) × 10^{-12} erg	
1 abs. electron-volt = (1.60199 ± 0.00060) × 10^{-12} erg	
hc = (1.23916 ± 0.00032) × 10^{-4} int. electron-volt cm.	Constant relating wave number and energy per molecule
= (1.23957 ± 0.00032) × 10^{-4} abs. electron-volt cm.	
k = (8.61442 ± 0.00100) × 10^{-5} int. electron-volt/deg.	k = Boltzmann constant
= (8.61727 ± 0.00100) × 10^{-5} abs. electron-volt/deg.	
= (R/N) = (1.38048 ± 0.00050) × 10^{-23} joule/deg.	
1 I.T. cal. = ($^1/_{860}$) = 0.00116279 int. watt-hr.	Definition of I.T. cal.: I.T. = International steam tables
= 4.18605 int. joule	
= 4.18674 abs. joule	
= 1.000654 cal.	cal. = thermochemical calorie
1 cal. = 4.1840 abs. joule	Definition: cal. = thermochemical calorie
= 4.1833 int. joule	
= 41.2929 ± 0.0020 cu. cm. atm.	
= 0.0412929 ± 0.0000020 liter atm.	
1 I.T. cal./g. = 1.8 B.t.u./lb.	Definition of B.t.u.: B.t.u. = I.T. British Thermal Unit
1 B.t.u. = 251.996 I.T. cal.	
= 0.293018 int. watt-hr.	
= 1054.866 int. joule	
= 1055.040 abs. joule	
= 252.161 cal.	cal. = thermochemical calorie
1 horsepower = 550 ft.-lb. (wt.)/sec.	Definition of horsepower (mechanical): lb. (wt.) = weight of 1 lb. at standard gravity
= 745.578 int. watt	
= 745.70 abs. watt	
1 in. = (1/0.3937) = 2.54 cm.	Definition of in.: in. = U.S. inch
1 ft. = 0.304800610 m.	ft. = U.S. foot (1 ft. = 12 in.)
1 lb. = 453.5924277 g.	Definition; lb. = avoirdupois pound
1 gal. = 231 cu. in.	Definition; gal. = U.S. gallon
= 0.133680555 cu. ft.	
= 3.785412 × 10^{-3} cu. m.	
= 3.785412 liter	

Source: From Perry, Green, and Maloney.[1]

CHAPTER 2

GENERAL PROPERTIES OF MATERIALS

All materials have properties which must be known in order promote their proper use. Knowing these properties is also essential to selecting the best material for a given application. This chapter includes general properties widely used in the field of chemical, mechanical, civil, and electrical engineering.

Note that results are given in SI units. Use Table 1.5 of Chap. 1 to obtain results in USCS units.

CHEMICAL PROPERTIES

Every elementary substance is made up of atoms which are all alike and which cannot be further subdivided or broken up by chemical processes. There are as many different classes or families of atoms as there are chemical elements (Table 2.1).

Two or more atoms, either of the same kind or of different kinds, are, in the case of most elements, capable of uniting with one another to form a higher order of distinct particles called *molecules*. If the molecules or atoms of which any given material is composed are all exactly alike, the material is a pure substance. If they are not all alike, the material is a mixture.

If the atoms which compose the molecules of any pure substances are all of the same kind, the substance is, as already stated, an elementary substance. If the atoms which compose the molecules of a pure chemical substance are not all of the same kind, the substance is a compound substance.

It appears that some substances which cannot by any available means be decomposed into simpler substances and which must, therefore, be defined as elements, are continually undergoing *spontaneous* changes or radioactive transformation into other substances which can be recognized as physically different from the original substance. The view generally accepted at present is that the atoms of all the chemical elements, including those not known to be radioactive, consist of several kinds of still smaller particles, three of which are known as protons, neutrons, and electrons. The protons are bound together in the atomic nucleus with other particles, including neutrons, and are positively charged. The neutrons are particles having approximately the mass of a proton but no charge. The electrons are negatively charged particles, all alike, external to the nucleus; and sufficient in number to neutralize the nuclear charge in an atom. The differ-

TABLE 2.1 Chemical Elements[a]

Element	Symbol	Atomic No.	Atomic weight[b]
Actinium	Ac	89	
Aluminum	Al	13	26.9815
Americium	Am	95	
Antimony	Sb	51	121.75
Argon[c]	Ar	18	39.948
Arsenic[d]	As	33	74.9216
Astatine	At	85	
Barium	Ba	56	137.34
Berkelium	Bk	97	
Beryllium	Be	4	9.0122
Bismuth	Bi	83	208.980
Boron[d]	B	5	10.811[l]
Bromine[e]	Br	35	79.904[m]
Cadmium	Cd	48	112.40
Calcium	Ca	20	40.08
Californium	Cf	98	
Carbon[d]	C	6	12.01115[l]
Cerium	Ce	58	140.12
Cesium[k]	Cs	55	132.905
Chlorine[f]	Cl	17	35.453[m]
Chromium	Cr	24	51.996[m]
Cobalt	Co	27	58.9332
Columbium (see Niobium)			
Copper	Cu	29	63.546[m]
Curium	Cm	96	
Dysprosium	Dy	66	162.50
Einsteinium	Es	99	
Erbium	Er	68	167.26
Europium	Eu	63	151.96
Fermium	Fm	100	
Fluorine[g]	F	9	18.9984
Francium	Fr	87	
Gadolinium	Gd	64	157.25
Gallium[k]	Ga	31	69.72
Germanium	Ge	32	72.59
Gold	Au	79	196.967
Hafnium	Hf	72	178.49
Helium[c]	He	2	4.0026
Holmium	Ho	67	164.930
Hydrogen[h]	H	1	1.00797[l]
Indium	In	49	114.82
Iodine[d]	I	53	126.9044
Iridium	Ir	77	192.2
Iron	Fe	26	55.847[m]
Krypton[c]	Kr	36	83.80
Lanthanum	La	57	138.91
Lead	Pb	82	207.19
Lithium[i]	Li	3	6.939
Lutetium	Lu	71	174.97
Magnesium	Mg	12	24.312
Manganese	Mn	25	54.9380
Mendelevium	Md	101	
Mercury[e]	Hg	80	200.59
Molybdenum	Mo	42	95.94
Neodymium	Nd	60	144.24
Neon[c]	Ne	10	20.183
Neptunium	Np	93	
Nickel	Ni	28	58.71
Niobium	Nb	41	92.906
Nitrogen[f]	N	7	14.0067
Nobelium	No	102	
Osmium	Os	76	190.2
Oxygen[f]	O	8	15.9994[l]
Palladium	Pd	46	106.4
Phosphorus[d]	P	15	30.9738

TABLE 2.1 Chemical Elements (*Continued*)

Element	Symbol	Atomic No.	Atomic weight[b]
Platinum	Pt	78	195.09
Plutonium	Pu	94	
Polonium	Po	84	
Potassium	K	19	39.102
Praseodymium	Pr	59	140.907
Promethium	Pm	61	
Protactinium	Pa	91	
Radium	Ra	88	
Radon[j]	Rn	86	
Rhenium	Re	75	186.2
Rhodium	Rh	45	102.905
Rubidium	Rb	37	85.47
Ruthenium	Ru	44	101.07
Samarium	Sm	62	150.35
Scandium	Sc	21	44.956
Selenium[d]	Se	34	78.96
Silicon[d]	Si	14	28.086[l]
Silver	Ag	47	107.868[m]
Sodium	Na	11	22.9898
Strontium	Sr	38	87.62
Sulphur[d]	S	16	32.064[l]
Tantalum	Ta	73	180.948
Technetium	Tc	43	
Tellurium[d]	Te	52	127.60
Terbium	Tb	65	158.924
Thallium	Tl	81	204.37
Thorium	Th	90	232.038
Thulium	Tm	69	168.934
Tin	Sn	50	118.69
Titanium	Ti	22	47.90
Tungsten	W	74	183.85
Uranium	U	92	238.03
Vanadium	V	23	50.942
Xenon[c]	Xe	54	131.30
Ytterbium	Yb	70	173.04
Yttrium	Y	39	88.905
Zinc	Zn	30	65.37
Zirconium	Zr	40	91.22

[a]All the elements for which atomic weights are listed are metals, except as otherwise indicated. No atomic weights are listed for most radioactive elements, as these elements have no fixed value.

[b]The atomic weights are based upon nuclidic mass of C^{12} = 12.

[c]Inert gas. [d]Metalloid. [e]Liquid. [f]Gas. [g]Most active gas. [h]Lightest gas. [i]Lightest metal. [j]Not placed. [k]Liquid at 25°C.

[l]The atomic weight varies because of natural variations in the isotopic composition of the element. The observed ranges are boron, ±0.003; carbon, ±0.00005; hydrogen, ±0.00001; oxygen, ±0.0001; silicon, ±0.001; sulfur, ±0.003.

[m]The atomic weight is believed to have an experimental uncertainty of the following magnitude: bromine, ±0.001; chlorine, ±0.001; chromium, ±0.001; copper, ±0.001; iron, ±0.003; silver, ±0.001. For other elements, the last digit given is believed to be reliable to ±0.5.

Source: From Avallone and Baumeister.[1]

ences between the atoms of different chemical elements are due to the different numbers of these smaller particles composing them.

In a hydrogen atom, there is one proton and one electron; in a radium atom, there are 88 electrons surrounding a nucleus 226 times as massive as the hydrogen nucleus. Only a few, in general the outermost or *valence* electrons of such an atom, are subject to rearrangement within, or ejection from, the atom, thereby enabling it, because of its increased energy, to combine with other atoms to form molecules of either elementary substances or compounds. The atomic number of an element is the number of excess positive charges on the nucleus of the atom.

The essential feature that distinguishes one element from another is this charge of the nucleus. It also determines the position of the element in the periodic table (Table 2.2). Modern research has shown the existence of isotopes, that is, two or more species of atoms having the same atomic number and thus occupying the same place in the periodic system, but differing somewhat in atomic weight. These isotopes are chemically identical and are merely different species of the same chemical element.

Data for solubility of inorganic substances and gases in water are given in Tables 2.3 and 2.4, respectively. See Refs. 1 and 3 for information on other chemical properties of materials.

THERMOPHYSICAL PROPERTIES

Most frequently used thermophysical properties in engineering practice are

Density (ρ)
Specific heat (c)
Specific heat at constant pressure (c_p)
Thermal conductivity (k)
Thermal diffusivity (α)
Dynamic viscosity (μ)
Kinematic viscosity (ν)
Surface tension (σ)
Coefficient of thermal expansion (β)

The kinematic viscosity of a fluid is its dynamic viscosity divided by its density, or $\nu = \mu/\rho$. Its units are m^2/s. The surface tension of a fluid is the work done in extending the surface of a liquid one unit of area or work per unit area. Its units are N/m. Also, note that $\alpha = k/\rho c$ and

$$\beta = -\frac{1}{\rho}\left(\frac{\partial \rho}{\partial t}\right)_p = \text{const}$$

In general, all thermophysical properties are strong functions of temperature.

Table 2.5 shows properties of metallic solids. Table 2.6 shows properties of nonmetallic solids. Table 2.7 shows properties of saturated liquids. (Note that the Prandtl number $\Pr = \nu/\alpha$.) Table 2.8 shows properties of gases at atmospheric pressure. Table 2.9 shows data of surface tension of various liquids.

MECHANICAL PROPERTIES

Mechanical properties commonly used by engineers are

Ultimate tensile strength
Tensile yield strength

Atomic number — Element
Symbol
Valence — Atomic weight based on $C^{12} = 12.00$
( ) denotes mass number of most stable known Isotope

Light metals		Brittle metals					Ductile metals				Low melting	Nonmetalic elements					Inert gases
																1 Hydrogen H 1.00797 1	2 Helium He 4.0026 0
3 Lithium Li 6.939 1	4 Beryllium Be 9.0122 2											5 Boron B 10.811 3	6 Carbon C 12.01115 2,4	7 Nitrogen N 14.0067 3,5	8 Oxygen O 15.9994 2	9 Fluorine F 18.9984 1	10 Neon Ne 20.183 0
11 Sodium Na 22.9898 1	12 Magnesium Mg 24.312 2											13 Aluminum Al 26.9815 3	14 Silicon Si 28.086 4	15 Phosphorus P 30.9738 3,5	16 Sulphur S 32.064 2,4,6	17 Chlorine Cl 35.453 1,3,5,7	18 Argon Ar 39.948 0
19 Potassium K 39.102 1	20 Calcium Ca 40.08 2	21 Scandium Sc 44.956 3	22 Titanium Ti 47.90 3,4	23 Vanadium V 50.942 1,2,3,4,5	24 Chromium Cr 51.996 2,3,6	25 Manganese Mn 54.938 2,3,4,6,7	26 Iron Fe 55.847 2,3	27 Cobalt Co 58.9332 2,3	28 Nickel Ni 58.71 2,3,4	29 Copper Cu 63.546 1,2	30 Zinc Zn 65.37 2	31 Gallium Ga 69.72 2,3	32 Germanium Ge 72.59 2,4	33 Arsenic As 74.9216 3,5	34 Selenium Se 78.96 2,4,6	35 Bromine Br 79.904 1,3,5	36 Krypton Kr 83.80 0
37 Rubidium Rd 85.47 1	38 Strontium Sr 87.62 2	39 Yttrium Y 88.905 3	40 Zirconium Zr 91.22 4	41 Niobium Nb 92.906 2,3,4,5	42 Molybdenum Mo 95.94 3,4,5,6	43 Technetium Tc (99)	44 Ruthenium Ru 101.07 3,4,6,8	45 Rhodium Rh 103.905 3,4	46 Palladium Pd 106.4 2,4	47 Silver Ag 107.868 1	48 Cadmium Cd 112.40 2	49 Indium In 114.82 1,2,3	50 Tin Sn 118.69 2,4	51 Antimony Sb 121.75 3,5	52 Tellurium Te 127.60 2,4,6	53 Iodine I 126.9044 1,3,5,7	54 Xenon Xe 131.30 0
					LANTHANIDE SERIES		Rare earth elements										
55 Cesium Cs 132.905 1	56 Barium Ba 137.34 2	57 Lanthanum La 137.91 3	58 Cerium Ce 140.12 3,4	59 Praseodymium Pr 140.907 3	60 Neodymium Nd 144.24 3	61 Promethium Pm (147) 5	62 Samarium Sm 150.35 3	63 Europium Eu 151.96 2,3	64 Godalinium Gd 157.25 3	65 Terbium Tb 158.924 3	66 Dysprosium Dy 162.50 3	67 Holmium Ho 164.93 3	68 Erdium Er 167.26 3	69 Thulium Tm 168.934 3	70 Ytterbium Yb 173.04 2,3	71 Lutetium Lu 174.97 3	
			72 Hafnium Hf 178.49 4	73 Tantalum Ta 180.948 4,5	74 Tungsten W 183.85 3,4,5,6	75 Rhenium Re 186.2 1,4,7	76 Osmium Os 190.2 2,3,4,6,8	77 Iridium Ir 192.2 2,3,4,6	78 Platinum Pt 195.09 2,4	79 Gold Au 196.967 1,3	80 Mercury Hg 200.59 1,2	81 Thallium Tl 204.37 1,3	82 Lead Pb 207.19 2,4	83 Bismuth Bi 208.98 3,5	84 Polonium Po (210) 2,4	85 Astatine At (210)	86 Radon Rn (212) 0
					ACTINIDE SERIES		Transuranium elements										
87 Francium Fr (223)	88 Radium Ra (227) 2	89 Actinium Ac (227)	90 Thorium Th 232.038	91 Protactinium Pa (231)	92 Uranium U 238.03	93 Neptunium Np (237)	94 Plutonium Pu (242)	95 Americium Am (243)	96 Curium Cm (245)	97 Berkelium Bk (249)	98 Californium Cf (249)	99 Einsteinium Es (254)	100 Fermium Fm (252)	101 Mendelevium Md (256)	102 Nobelium No (254)	103 Lawreneium Lw (257)	

Source: From Avallone and Baumeister.[1]

TABLE 2.3 Solubility of Inorganic Substances in Water

(Number of grams of the anhydrous substance soluble in 1000 g of water. The common name of the substance is given in parentheses.)

	Composition	Temperature, °F (°C) 32 (0)	122 (50)	212 (100)
Aluminum sulfate	$Al_2(SO_4)_3$	313	521	891
Aluminum potassium sulfate (potassium alum)	$Al_2K_2(SO_4)_4 \cdot 24H_2O$	30	170	1540
Ammonium bicarbonate	NH_4HCO_3	119		
Ammonium chloride (sal ammoniac)	NH_4Cl	297	504	760
Ammonium nitrate	NH_4NO_3	1183	3440	8710
Ammonium sulfate	$(NH_4)_2SO_4$	706	847	1033
Barium chloride	$BaCl_2 \cdot 2H_2O$	317	436	587
Barium nitrate	$Ba(NO_3)_2$	50	172	345
Calcium carbonate (calcite)	$CaCO_3$	0.018*		0.88
Calcium chloride	$CaCl_2$	594		1576
Calcium hydroxide (hydrated lime)	$Ca(OH)_2$	1.77		0.67
Calcium nitrate	$Ca(NO_3)_2 \cdot 4H_2O$	931	3561	3626
Calcium sulfate (gypsum)	$CaSO_4 \cdot 2H_2O$	1.76	2.06	1.69
Copper sulfate (blue vitriol)	$CuSO_4 \cdot 5H_2O$	140	334	753
Ferrous chloride	$FeCl_2 \cdot 4H_2O$	644§	820	1060
Ferrous hydroxide	$Fe(OH)_2$	0.0067‡		
Ferrous sulfate (green vitriol or copperas)	$FeSO_4 \cdot 7H_2O$	156	482	
Ferric chloride	$FeCl_3$	730	3160	5369
Lead chloride	$PbCl_2$	6.73	16.7	33.3
Lead nitrate	$Pb(NO_3)_2$	403		1255
Lead sulfate	$PbSO_4$	0.042†		
Magnesium carbonate	$MgCO_3$	0.13‡		
Magnesium chloride	$MgCl_2 \cdot 6H_2O$	524		723
Magnesium hydroxide (milk of magnesia)	$Mg(OH)_2$	0.009‡		
Magnesium nitrate	$Mg(NO_3)_2 \cdot 6H_2O$	665	903	
Magnesium sulfate (Epsom salts)	$MgSO_4 \cdot 7H_2O$	269	500	710
Potassium carbonate (potash)	K_2CO_3	893	1216	1562
Potassium chloride	KCl	284	435	566
Potassium hydroxide (caustic potash)	KOH	971	1414	1773
Potassium nitrate (saltpeter or niter)	KNO_3	131	851	2477
Potassium sulfate	K_2SO_4	74	165	241
Sodium bicarbonate (baking soda)	$NaHCO_3$	69	145	
Sodium carbonate (sal soda or soda ash)	$NaCO_3 \cdot 10H_2O$	204	475	452
Sodium chloride (common salt)	$NaCl$	357	366	392
Sodium hydroxide (caustic soda)	$NaOH$	420	1448	3388
Sodium nitrate (Chile saltpeter)	$NaNO_3$	733	1148	1755
Sodium sulfate (Glauber salts)	$Na_2SO_4 \cdot 10H_2O$	49	466	422
Zinc chloride	$ZnCl_2$	2044	4702	6147
Zinc nitrate	$Zn(NO_3)_2 \cdot 6H_2O$	947		
Zinc sulfate	$ZnSO_4 \cdot 7H_2O$	419	768	807

*59°F (15°C).
§50°F (10°C).
‡In cold water.
†68°F (20°C).
Source: From Avallone and Baumeister.[1]

Elongation
Modulus of elasticity
Compressive strength
Shear strength
Endurance limit

Ultimate tensile strength is defined as the maximum load per unit of original cross-sectional area sustained by a material during a tension test. It is also called *ultimate strength*.

TABLE 2.4 Solubility of Gases in Water

(By volume, at atmospheric pressure)

t, °F (°C)	32 (0)	68 (20)	212 (100)
Air	0.032	0.020	0.012
Acetylene	1.89	1.12	
Ammonia	1250	700	
Carbon dioxide	1.87	0.96	0.26
Carbon monoxide	0.039	0.025	
Chlorine	5.0	2.5	0.00
Hydrogen	0.023	0.020	0.018
Hydrogen sulfide	5.0	2.8	0.87
Hydrochloric acid	560	480	
Nitrogen	0.026	0.017	0.0105
Oxygen	0.053	0.034	0.0185
Sulfuric acid	87	43	

Source: From Avallone and Baumeister.[1]

Tensile yield strength is defined as the stress corresponding to some permanent deformation from the modulus slope, e.g., 0.2 percent offset in the case of heat-treated alloy steels.

Elongation is defined as the amount of permanent extension in a ruptured tensile test specimen; it is usually expressed as a percentage of the original gage length. Elongation is usually taken as a measure of ductility.

Modulus of elasticity is the property of a material which indicates its rigidity. This property is the ratio of stress to strain within the elastic range.

On a stress-strain diagram, the modulus of elasticity is usually represented by the straight portion of the curve when the stress is directly proportional to the strain. The steeper the curve, the higher the modulus of elasticity and the stiffer the material.

Compressive strength is defined as the maximum compressive stress that a material is capable of developing based on the original cross-sectional area. The general design practice is to assume the compressive strength of a steel is equal to its tensile strength, although it is actually somewhat greater.

Shear strength is defined as the stress required to produce fracture in the plane of cross section, the conditions of loading being such that the directions of force and of resistance are parallel and opposite although their paths are offset a specified minimum amount. The ultimate shear strength is generally assumed to be three-fourths the material's ultimate tensile strength.

Endurance limit is defined as the maximum stress to which the material can be subjected for an indefinite service life. Although the standards vary for various types of members and different industries, it is common practice to assume that carrying a certain load for several million cycles of stress reversals indicates that the load can be carried for an indefinite time.

Hardness measures the resistance of the material to indentation. Hardness tests measure the plastic deformation (the size or depth) of an indentation. Brinell hardness tests use spheres as indenters; the Vickers test uses pyramids. Rockwell tests use cones or spheres. Microhardness tests for specimens are also available, using the Knoop method with miniature pyramid indenters. Another hardness scale is Mohs' scale, which lists materials in order of their hardness, beginning with talc and ending with diamond.

TABLE 2.5 Properties of Metallic Solids

Metal	Properties at 20°C				Thermal conductivity, k(W/m · K)									
	ρ (kg/m^3)	c_p (J/kg · K)	k (W/m · K)	α (10^{-6}m^2/s)	−170°C	−100°C	0°C	100°C	200°C	300°C	400°C	600°C	800°C	1000°C
Aluminum														
Pure	2,707	905	237	9.61	302	242	236	240	238	234	228	215	~ 95 (liq)	
99% pure			211		220	206	209							
Duralumin (~4% Cu)	2,787	883	164	6.66		126	164	182	194					
Chromium	7,190	453	90	2.77	158	120	95	88	85	82	77	69	64	62
Copper and Cu alloys														
Pure	8,954	384	398	11.57	483	420	401	391	389	384	378	366	352	336
Brass (30% Zn)	8,522	385	109	3.32	73	89	106	133	143	146	147			
Bronze (25% Sn)	8,666	343	26	0.86	(Data on this and other bronzes vary by a factor of about 2)									
Constantan (40% Ni)	8,922	410	22	0.61	17	19	22	26	35					
German silver (15% Ni, 22% Zn)	8,618	394	25	0.73	18	19	24	31	40	45	48			
Gold	19,320	129	315	12.64			318		309					
Ferrous metals														
Pure iron	7,897	447	80	2.26	132	98	84	72	63	56	50	39	30	29.5
Cast iron (0.4% C)	7,272	420	52	1.70										
Steels (C < 1.5%)														
0.5% carbon (mild)	7,833	465	54	1.47			55	52	48	45	42	35	31	29
1.0% carbon	7,801	473	43	1.17			43	43	42	40	36	33	29	28
1.5% carbon	7,753	486	36	0.97			36	36	36	35	33	31	28	28

Stainless steel, type:														
304	8,000	400	13.8	0.4				15	17	19	21	25		
316	8,000	460	13.5	0.37		12		15	16	17	19	21	24	26
347	8,000	420	15	0.44		13		16	18	19	20	23	26	28
410	7,700	460	25	0.7				25	26	27	27	28		
414	7,700	460	25	0.7							29	29		
Lead	11,373	130	35	2.34	40	37	36	34	33	31	17 (liq.)	20 (liq.)		
Magnesium	1,746	1023	156	8.76	17	16	157	154	152	150	148		90 (liq.)	
Mercury (polycrystalline)					32	30	7.8 (liq.)							
Nickel														
Pure	8,906	445	91	2.30	156	114	94	83	74	67	64	69	73	78
Nichrome (24% Fe, 16% Cr)	8,250	448		0.34				13						
Nichrome V (20% Cr)	8,410	466	13	0.33			12	14	15	17	19			
Platinum	21,450	133	71	2.50	78	73	72	72	72	73	74	77	80	84
Silver														
99.99% pure	10,524	236	427	17.19	449	431	428	422	417	409	401	386	370	176 (liq.)
99.9% pure	10,524	236	411	16.55		422	405		373	367	364			
Tin (polycrystalline)	7,304	\~220	67	4.17	85	76	68	63						
Titanium (polycrystalline)	4,540	523	22	0.93	31	26	22	21	20	20	19	21	21	22
Tungsten	19,350	133	178	6.92	235	223	182	166	153	141	134	125	122	114
Uranium	18,700	116	28	1.29	22	24	27	29	31	33	36	41	46	
Zinc	7,144	388	121	4.37	124	122	122	117	110	106	100	60 (liq.)		

Source: From Lienhard.[2] Portions of the original table have been omitted where not relevant to this chapter. The data can also be found in Refs. 1, 2, and 4 through 9.

TABLE 2.6 Properties of Nonmetallic Solids

Material	Temperature range, °C	Density ρ, kg/m³	Specific heat c, J/kg · °C	Thermal conductivity k, W/m · °C	Thermal diffusivity α, m²/s
Asbestos					
Cement board	20			0.6	
Fiber (properties vary	20	1930		0.8	
with packing)	20	980		0.14	
Asphalt	20–25			0.75	
Beef	25				1.35×10^{-7}
Brick					
B&W, K-28 insulating	300			0.3	
B&W, K-28 insulating	1000			0.4	
Cement	10	720		0.34	
Common	0–1000			0.7	
Chrome	100			1.9	
Firebrick	300	2000	960	0.1	5.4×10^{-8}
Firebrick	1000			0.2	
Carbon					
Diamond (type II b)	20	~3250	510	1350.	8.1×10^{-4}
Graphite	20	~2100	~2090	Highly variable structure	
Cardboard	0–20	790		0.14	
Clay					
Fireclay	500–750			1.	
Sandy clay	20	1780		0.9	
Coal					
Anthracite	900	~1500		~0.2	
Brown coal	900			~0.1	
Bituminous in situ		~1300		0.5–0.7	3 to 4×10^{-7}
Concrete					
Limestone gravel	20	1850		0.6	
Portland cement	90	2300		1.7	
Sand:cement (3:1)	230			0.1	
Slag cement	20			0.14	
Corkboard (medium ρ)	30	170		0.04	
Egg white	20				1.37×10^{-7}
Glass					
Lead	36			1.2	
Plate	20			1.3	
Pyrex	60–100	2210	753	1.3	7.8×10^{-7}
Soda	20			0.7	
Window	46			1.3	
Glass wool	20	64–160		0.04	
Ice	0	917	2100	2.215	1.15×10^{-6}
Ivory	80			0.5	
Kapok	30			0.035	
Magnesia (85%)	38			0.067	
	93			0.071	
	150			0.074	
	204			0.08	

TABLE 2.6 Properties of Nonmetallic Solids (*Continued*)

Material	Temperature range, °C	Density ρ, kg/m³	Specific heat c, J/kg · °C	Thermal conductivity k, W/m · °C	Thermal diffusivity α, m²/s
Lunar surface dust (high vacuum)	250	1500 ± 300	~600	~0.0006	~7 × 10⁻¹⁰
Rock wool	−5	~130		0.03	
	93			0.05	
Rubber (hard)	0	1200	2010	0.15	6.2 × 10⁻⁸
Silica aerogel	0	140		0.024	
	120	136		0.022	
Silo-cel (diatomaceous earth)	0	320		0.061	
Soil (mineral)					
Dry	15	1500	1840	1.	4 × 10⁻⁷
Wet	15	1930		2.	
Stone					
Granite (NTS)	20	~2640	~820	1.6	~7.4 × 10⁻⁷
Limestone (Indiana)	100	2300	~900	1.1	~5.3 × 10⁻⁷
Sandstone (Berea)	25			~3	
Slate	100			1.5	
Wood (perpendicular to grain)					
Ash	15	740		0.15–0.3	
Balsa	15	100		0.05	
Cedar	15	480		0.11	
Fir	15	600	2720	0.12	7.4 × 10⁻⁸
Mahogany	20	700		0.16	
Oak	20	600	2390	0.1–0.4	(0.7–2.8) × 10⁻⁷
Pitch pine	20	450		0.14	
Sawdust (dry)	17	128		0.05	
Spruce	20	410		0.11	
Wool (sheep)	20	145		0.05	

Source: From Lienhard.[2] Portions of the original table have been omitted where not relevant to this chapter. The data can also be found in Refs. 1, 2, and 4 through 9.

Table 2.10 shows typical mechanical properties of some metals and alloys. See Ref. 11 for more data.

ELECTRICAL PROPERTIES

The most frequently used properties of materials in the field of electrical and electronics engineering are specific resistance (ρ_s) and the temperature coefficient of resistance (α_t).

Specific resistance (or *resistivity*) is the resistance of a sample of material having both a length and cross section of unity. The two most common resistivity

TABLE 2.7 Thermophysical Properties of Saturated Liquids

Temp., K	Temp., °C	ρ, kg/m^3	c_p, J/kg · K	k, W/m · K	α, m^2/s	ν, m^2/s	Pr	β, K^{-1}
Ammonia (there is considerable disagreement among sources)								
220	−53	706	4426	0.66	2.11×10^{-7}			
240	−33	682	4484	0.61	2.00	4.17×10^{-7}	2.09	
260	−13	656	4547	0.57	1.91	3.27	1.71	
280	7	629	4625	0.52	1.79	2.68	1.50	0.00025
300	27	600	4736	0.470	1.65	2.32	1.41	
320	47	568	4962	0.424	1.50	2.06	1.37	
340	67	533	5214	0.379	1.36	1.79	1.32	
360	87	490	5635	0.335	1.21	1.55	1.28	
380	107	436		0.289		1.34		
400	127	345		0.245		1.19		
CO_2								
250	−23	1046	1990	0.135	6.49×10^{-8}			
260	−13	998	2110	0.123	5.84	1.15×10^{-7}	1.97	
270	−3	944	2390	0.113	5.09	1.08	2.12	
280	7	883	2760	0.102	4.19	1.04	2.48	
290	17	805	3630	0.090	3.08	0.99	3.20	0.014
300	27	676	7690	0.076	1.46	0.88	6.04	
303	30	604						
D_2O (heavy water)								
589	316	740	2034	0.0509	0.978×10^{-7}	1.23×10^{-7}	1.257	
Freon 11 (trichlorofluoromethane)								
220	−53		829	0.110				
240	−33	1607	841	0.105	7.8×10^{-8}	4.78×10^{-7}	6.1	
260	−13	1564	855	0.099	7.4	4.10	5.5	
280	7	1518	871	0.093	7.0	3.81	5.4	0.00154
300	27	1472	888	0.088	6.7	2.82	4.2	0.00163
320	47	1421	906	0.082	6.4			
340	67	1369	927	0.076	6.0			
Freon 12 (dichlorodifluoromethane)								
160	−113			0.133				
180	−93	1664	834	0.124	8.935×10^{-8}			
200	−73	1610	856	0.1148	8.33			
220	−53	1555	873	0.1057	7.79	3.2×10^{-7}	4.11	0.00263
240	−33	1498	892	0.0965	7.22	2.60	3.60	
260	−13	1438	914	0.0874	6.65	2.26	3.40	
280	7	1374	942	0.0782	6.04	2.06	3.41	
300	27	1305	980	0.0690	5.39	1.95	3.62	
320	47	1229	1031	0.0599	4.72	1.9	4.03	
340	67		1097	0.0507				
Glycerin (or glycerol)								
273	0	1276	2200	0.282	1.00×10^{-7}	0.0083	83,000	
293	20	1261	2350	0.285	0.962	0.001120	11,630	0.00048
303	30	1255	2400	0.285	0.946	0.000488	5,161	0.00049
313	40	1249	2460	0.285	0.928	0.000227	2,451	0.00049
323	50	1243	2520	0.285	0.910	0.000114	1,254	0.00050

TABLE 2.7 Thermophysical Properties of Saturated Liquids (*Continued*)

Temp., K	Temp., °C	ρ, kg/m³	c_p, J/kg · K	k, W/m · K	α, m²/s	ν, m²/s	Pr	β, K⁻¹
				Lead				
644	371	10,540	159	16.1	1.084×10^{-5}	2.276×10^{-7}	0.024	
755	482	10,442	155	15.6	1.223	1.85	0.017	
811	538	10,348	145	15.3	1.02	1.68	0.017	
				Mercury				
234	−39		141.5	6.97	3.62×10^{-6}			
250	−23		140.5	7.32	3.83			
300	27	13,611	139.1	8.34	4.41	1.2×10^{-7}	0.027	
350	77	13,489	137.7	5.29	4.91	1.0	0.020	
400	127	13,367	136.7	5.69	5.83×10^{-6}	0.95×10^{-7}	0.016	
500	277	13,128	135.6	6.36	6.00	0.80	0.013	
600	327		135.4	6.93	6.55	0.68	0.010	
700	427		136.1	7.34				
800	527			7.40				
				Methyl alcohol (methanol)				
260	−13	823	2336	0.2164	1.126×10^{-7}	$\sim 1.3 \times 10^{-6}$	~11.5	
280	7	804	2423	0.2078	1.021	~0.9	~8.8	0.00114
300	27	785	2534	0.2022	1.016	~0.7	~6.9	
320	47	767	2672	0.1965	0.959	~0.6	~6.3	
340	67	748	2856	0.1908	0.893	~0.44	~4.9	
360	87	729	3036	0.1851	0.836	~0.36	~4.3	
380	107	710	3265	0.1794	0.774	~0.30	~4.1	
				Oxygen				
54		1276	1648	0.191	9.08×10^{-8}	6.5×10^{-7}	7.15	
60	−213		1649	0.185				
80	−193		1653	0.1623				
90	−183	1114	1655	0.1501	8.14×10^{-8}	1.75×10^{-7}	2.15	
120	−153			0.1096				
150	−123			0.061				
				Oils (some approximate viscosities)				
273	0	MS-20				0.0076	100,000	
339	66	California crude (heavy)				0.00008		
289	16	California crude (light)				0.00005		
339	66	California crude (light)				0.000010		
289	16	Light machine oil				0.0007		
339	66	Light machine oil				0.00004		
289	16	Light machine oil (ρ = 907)				0.00016		
339	66	Light machine oil (ρ = 907)				0.000013		
289	16	SAE 30				0.00044	~ 5,000	
339	66	SAE 30				0.00003		
289	16	SAE 30 (Eastern)				0.00011		
339	66	SAE 30 (Eastern)				0.00001		
289	16	Spindle oil (ρ = 885)				0.00005		
339	66	Spindle oil (ρ = 885)				0.000007		

TABLE 2.7 Thermophysical Properties of Saturated Liquids (*Continued*)

Temp., K	Temp., °C	ρ, kg/m³	c_p, J/kg · K	k, W/m · K	α, m²/s	ν, m²/s	Pr	β, K⁻¹
				Water				
273	0	999.8	4205	0.5750	1.368×10^{-7}	1.753×10^{-6}	12.81	
280	7	999.9	4196	0.5818	1.386	1.422	10.26	
300	27	996.6	4177	0.6084	1.462	0.826	5.65	0.000275
320	47	989.3	4177	0.6367	1.541×10^{-7}	0.566×10^{-6}	3.67	0.000435
340	67	979.5	4187	0.6587	1.606	0.420	2.61	
360	87	967.4	4206	0.6743	1.657	0.330	1.99	
373	100	957.2	4219	0.6811	1.683	0.290	1.72	
400	127	937.5	4241	0.6864	1.726	0.229	1.33	
420	147	919.9	4306	0.6836	1.726×10^{-7}	2.000×10^{-7}	1.16	
440	167	900.5	4391	0.6774	1.713	1.786	1.04	
460	187	879.5	4456	0.6672	1.703	1.626	0.955	
480	207	856.6	4534	0.6530	1.681	1.504	0.894	
500	227	831.5	4647	0.6348	1.463	1.412	0.859	
520	247	803.9	4831	0.6123	1.577×10^{-7}	1.345×10^{-7}	0.853	
540	267	773.0	5099	0.5857	1.486	1.298	0.873	
560	287	738.2	5487	0.555	1.370	1.269	0.926	
580	307	697.6	6010	0.520	1.240	1.240	1.000	
600	327	648.8	6691	0.481	1.108	1.215	1.097	
620	347	586.3				1.213×10^{-7}		
640	367	482.1				1.218		
647.3	374.2	306.8				1.356		

Source: From Lienhard.[2] Portions of the original table have been omitted where not relevant to this chapter. The data can also be found in Refs. 1, 2, and 4 through 9.

samples are the centimeter cube and the cir mil · ft. If l is the length of a conductor of uniform cross section A, then the resistance is

$$R = \rho_s \cdot \frac{l}{A}$$

where ρ_s is the resistivity. A *circular mil* is a unit of area equal to that of a circle whose diameter is 1 mil (0.001 in).

The resistance of the pure metals increases with temperature as

$$R = R_0(1 + \alpha_t \cdot t)$$

where R_0 is the resistance at 0°C and α_t is the *temperature coefficient of resistance*. Over a narrow range of temperature, the electrical resistivity changes approximately linearly with temperature:

$$\rho_s(t_2) = \rho_s(t_1)[1 + \alpha_{t_1}(t_2 - t_1)]$$

where $\rho_s(t_1)$ is the resistivity at t_1, etc.

When the temperature of reference t_1 is changed to some other value of t, the coefficient α_{t_1} will change to a new value α_t:

$$\alpha_t = \frac{\alpha_{t_1}}{1 + \alpha_{t_1}(t - t_1)}$$

Values of ρ_s are given in Tables 2.11 to 2.13 and Fig. 2.1. Values of α_t are given in Table 2.11. Table 2.11 also includes values of melting point and density.

TABLE 2.8 Thermophysical Properties of Gases at Atmospheric Pressure

T, K	ρ, kg/m^3	c_p, J/kg · K	μ, kg/m · s	ν, m^2/s	k, W/m · K	α, m^2/s	Pr
				Air			
100		1009	0.706×10^{-5}	$\sim 0.2 \times 10^{-5}$	0.00922		
150		1005	1.038	~ 0.4	0.01375		
200	1.79	1003	1.336	0.746	0.01810	1.01×10^{-5}	0.74
250	1.43	1003	1.606	1.123	0.02226	1.55	0.724
300	1.183	1003	1.853	1.566	0.02614	2.203	0.711
350	1.009	1008	2.081	2.062	0.02970	2.920	0.706
400	0.8826	1013	2.294	2.599	0.03305	3.697	0.703
450	0.7846	1020	2.493	3.177	0.03633	4.540	0.700
500	0.7061	1029	2.682×10^{-5}	3.798×10^{-5}	0.03951	5.438×10^{-5}	0.698
550	0.6419	1039	2.860	4.456	0.0426	6.387	0.698
600	0.5884	1051	3.030	5.150	0.0456	7.374	0.698
650	0.5432	1063	3.193	5.878	0.0484	8.382	0.701
700	0.5044	1075	3.349	6.64	0.0513	9.461	0.702
750	0.4707	1087	3.498	7.43	0.0541	10.57	0.703
800	0.4413	1099	3.643	8.26	0.0569	11.73	0.704
850	0.4154	1110	3.783	9.11	0.0597	12.95	0.704
900	0.3923	1121	3.918	9.99	0.0625	14.22	0.703
950	0.3716	1131	4.049	10.90	0.0649	15.44	0.706
1000	0.3531	1142	4.177×10^{-5}	11.83×10^{-5}	0.0672	16.67×10^{-5}	0.710
1100	0.3210	1159	4.42	13.8	0.0717	19.27	0.716
1200	0.2942	1175	4.65	15.8	0.0759	29.96	0.720
1300	0.2716	1189	4.88	18.0	0.0797	24.7	0.729
1400	0.2522		5.09	20.2	0.0835	27.5	0.734
1500	0.2354		5.30	22.5	0.0870	30.3	0.74
				Ammonia (NH_3)			
220	0.3828	2.198×10^3	7.255×10^{-6}	1.90×10^{-5}	0.0171	0.2054×10^{-4}	0.93
273	0.7929	2.177	9.353	1.18	0.0220	0.1308	0.90
323	0.6487	2.177	11.035	1.70	0.0270	0.1920	0.88
373	0.5590	2.236	12.886	2.30	0.0327	0.2619	0.87
423	0.4934	2.315	14.672	2.97	0.0391	0.3432	0.87
473	0.4405	2.395	16.49	3.74	0.0467	0.4421	0.84

TABLE 2.8 Thermophysical Properties of Gases at Atmospheric Pressure (*Continued*)

T, K	ρ, kg/m^3	c_p, J/kg · K	μ, kg/m · s	ν, m^2/s	k, W/m · K	α, m^2/s	Pr
			Carbon dioxide				
220	2.4733	0.783×10^3	11.105×10^{-6}	4.490×10^{-6}	0.010805	0.05920×10^{-4}	0.818
250	2.1657	0.804	12.590	5.813	0.012884	0.07401	0.793
300	1.7973	0.871	14.958	8.321	0.016572	0.10588	0.770
350	1.5362	0.900	17.205	11.19	0.02047	0.14808	0.755
400	1.3424	0.942	19.32	14.39	0.02461	0.19463	0.738
450	1.1918	0.980	21.34	17.90	0.02897	0.24813	0.721
500	1.0732	1.013	23.26	21.67	0.03352	0.3084	0.702
550	0.9739	1.047	25.08	25.74	0.03821	0.3750	0.685
600	0.8938	1.076	26.83	30.02	0.04311	0.4483	0.668
			Helium				
3		5.200×10^3	8.42×10^{-7}		0.0106		
33	1.4657	5.200	50.2	3.42×10^{-6}	0.0353	0.04625×10^{-4}	0.74
144	3.3799	5.200	125.5	37.11	0.0928	0.5275	0.70
200	0.2435	5.200	156.6	64.38	0.1177	0.9288	0.694
255	0.1906	5.200	181.7	95.50	0.1357	1.3675	0.70
366	0.13280	5.200	230.5	173.6	0.1691	2.449	0.71
477	0.10204	5.200	275.0	269.3	0.197	3.716	0.72
589	0.08282	5.200	311.3	375.8	0.225	5.215	0.72
700	0.07032	5.200	347.5	494.2	0.251	6.661	0.72
800	0.06023	5.200	381.7	634.1	0.275	8.774	0.72
900	0.05286	5.200	413.6	781.3	0.298	10.834	0.72
			Hydrogen				
30	0.84722	10.840×10^3	1.606×10^{-6}	1.805×10^{-6}	0.0228	0.0249×10^{-4}	0.759
50	0.50955	10.501	2.516	4.880	0.0362	0.0676	0.721
100	0.24572	11.229	4.212	17.14	0.0665	0.2408	0.7122
150	0.16371	12.602	5.595	34.18	0.0981	0.475	0.718
200	0.12270	13.540	6.813	55.53	0.1282	0.772	0.719
250	0.09819	14.059	7.919	80.64	0.1561	1.130	0.713
300	0.08185	14.314	8.963	109.5	0.182	1.554	0.706

350	0.07016	14.436	9.954	141.9	0.206	2.031	0.697
400	0.06135	14.491	10.864	177.1	0.228	2.568	0.690
450	0.05462	14.499	11.779	215.6	0.251	3.164	0.682
500	0.04918	14.507	12.636	257.0	0.272	3.817	0.675
550	0.04469.	14.532	13.475	301.6	0.292	4.516	0.668
600	0.04085	14.537×10^{-3}	14.285×10^{-6}	349.7×10^{-6}	0.315	5.306×10^{-4}	0.664
700	0.03492	14.574	15.89	455.1	0.351	6.903	0.659
800	0.03060	14.675	17.40	569	0.384	8.563	0.664
900	0.02723	14.821	18.78	690	0.412	10.21	0.675
1000	0.02451	14.968	20.16	822	0.445	12.13	0.678
1100	0.02227	15.165	21.46	965	0.488	14.45	0.668
1200	0.02050	15.366	22.75	1107	0.528	16.76	0.661
1300	0.01890	15.575	24.08	1273	0.568	19.3	0.660
1333	0.01842	15.638	24.44	1328	0.58	20.1	0.661
				Nitrogen			
100	3.439	1.0722×10^{3}	6.862×10^{-6}	1.995×10^{-6}	0.00958	0.026×10^{-4}	0.767
200	1.688	1.0429	12.947	7.67	0.0183	0.104	0.738
300	1.1233	1.0408	17.84	15.88	0.0259	0.222	0.715
400	0.8425	1.0459	21.98	26.1	0.0327	0.371	0.704
500	0.6739	1.0555	25.70	38.1	0.0389	0.547	0.696
600	0.5615	1.0756	29.11	51.8	0.0446	0.738	0.702
700	0.4812	1.0969	32.13	66.8	0.0499	0.945	0.707
800	0.4211	1.1225	34.84	82.7	0.0548	1.16	0.713
900	0.3743	1.1464	37.49	100.2	0.0597	1.39	0.721
1000	0.3368	1.1677	40.00	119.	0.0647	1.65	0.723
1100	0.3062	1.1857	42.28	138.	0.0700	1.93	0.716
1200	0.2807	1.2037	44.50	158.	0.0758	2.24	0.704
				Oxygen			
100	3.9918	0.9479×10^{2}	7.768×10^{-6}	1.946×10^{-6}	0.00903	0.0239×10^{-4}	0.815
150	2.6190	0.9178	11.490	4.387	0.01367	0.0569	0.773
200	1.9559	0.9131	14.850	7.593	0.01824	0.1021	0.745
250	1.5618	0.9157	17.87	11.45	0.02259	0.1579	0.725
300	1.3007	0.9203	20.63	15.86	0.02676	0.2235	0.709

TABLE 2.8 Thermophysical Properties of Gases at Atmospheric Pressure (*Continued*)

T, K	ρ, kg/m^3	c_p, J/kg · K	μ, kg/m · s	ν, m^2/s	k, W/m · K	α, m^2/s	Pr
			Oxygen (*continued*)				
350	1.1133	0.9291	23.16	20.80	0.03070	0.2968	0.702
400	0.9755	0.9420	25.54	26.18	0.03161	0.3768	0.695
450	0.8682	0.9567	27.77	31.99	0.03828	0.4609	0.694
500	0.7801	0.9722	29.91	38.34	0.04173	0.5502	0.697
550	0.7096	0.9881	31.97	45.05	0.04517	0.6441	0.700
600	0.6504	1.0044	33.92	52.15	0.04832	0.7399	0.704
			Steam (H_2O vapor)				
373.15	0.597	2030	12.28×10^{-6}	21.28×10^{-6}	0.0237	2.023×10^{-5}	1.052
393.15	0.547	1997	13.04	23.85	0.0251	2.298	1.038
413.15	0.520	1980	13.81	26.56	0.0265	2.574	1.032
433.15	0.494	1972	14.59	29.53	0.0280	2.874	1.027
453.15	0.473	1963	15.38	32.52	0.0294	3.166	1.027
473.15	0.452	1963	16.18	35.80	0.0309	3.483	1.029
493.15	0.433	1968	17.00	39.25	0.0323	3.790	1.036
513.15	0.416	1972	17.81	42.82	0.0338	4.120	1.039
533.15	0.400	1976	18.63	46.58	0.0354	4.479	1.040
553.15	0.386	1985	19.46×10^{-6}	50.42×10^{-6}	0.0369	4.816×10^{-5}	1.047
573.15	0.372	1997	20.29	54.54	0.0385	5.183	1.052
593.15	0.359	2010	21.12	58.84	0.0401	5.557	1.059
613.15	0.348	2022	21.95	63.09	0.0416	5.912	1.067

Source: From Lienhard.[2] Portions of the original table have been omitted where not relevant to this chapter. The data can also be found in Refs. 1, 2, and 4 through 9.

TABLE 2.9 Surface Tension of Liquids

	Temperature, K										
Substance	250	260	270	280	290	300	320	340	360	380	400
Acetone	0.0292	0.0280	0.0267	0.0254	0.0241	0.0229	0.0203	0.0178	0.016	0.014	0.012
Benzene			0.0321	0.0307	0.0293	0.0279	0.0253	0.0228	0.0204	0.0180	0.0156
Bromine	0.047	0.046	0.045	0.044	0.0425	0.041	0.038	0.035	0.032	0.030	0.027
Butane	0.0176	0.0164	0.0152	0.0140	0.0128	0.0116	0.0092	0.0069	0.0049	0.0031	0.0016
Chlorine	0.0243	0.0227	0.0212	0.0197	0.0182	0.0167	0.0137	0.0107	0.0079	0.0051	0.0037
Decane	0.0278	0.0269	0.0260	0.0251	0.0241	0.0233	0.0215	0.0196	0.0178	0.0161	0.0145
Diphenyl						0.0416	0.0388	0.0362	0.0338	0.0316	0.0295
Ethane	0.0061	0.0049	0.0037	0.0026	0.0015	0.0007	—	—	—	—	—
Ethanol			0.0247	0.0239	0.0231	0.0222	0.0204	0.0186	0.0167	0.0148	0.0126
Ethylene	0.0033	0.0020	0.0009	0.0002	—	—	—	—	—	—	—
Heptane	0.0242	0.0233	0.0224	0.0214	0.0204	0.0194	0.0175	0.0156	0.0137	0.0118	0.0100
Hexane	0.0230	0.0219	0.0207	0.0198	0.0187	0.0176	0.0154	0.0134	0.0116	0.0096	0.0077
Methanol	0.0266	0.0257	0.0248	0.0238	0.0229	0.0221	0.0204	0.0187	0.0169	0.0150	0.0129
Nonane	0.0270	0.0261	0.0251	0.0242	0.0232	0.0223	0.0204	0.0186	0.0167	0.0148	0.0129
Octane	0.0256	0.0247	0.0237	0.0228	0.0219	0.0210	0.0191	0.0173	0.0155	0.0138	0.0123
Pentane	0.0210	0.0198	0.0186	0.0175	0.0164	0.0153	0.0131	0.0108	0.0088	0.0069	0.0053
Propane	0.0128	0.0114	0.0101	0.0088	0.0076	0.0064	0.0043	0.0025	0.0007	—	—
Propanol	0.0274	0.0266	0.0258	0.0249	0.0241	0.0232	0.0214	0.0198	0.0182	0.0168	0.0155
Propylene	0.0132	0.0119	0.0105	0.0090	0.0077	0.0064	0.0041	0.0022	0.0005	—	—
R 12	0.0147	0.0134	0.0121	0.0108	0.0095	0.0082	0.0057	0.0034			
Toluene	0.0345	0.0330	0.0315	0.0301	0.0288	0.0275	0.0251	0.0227	0.0205	0.0185	0.0165
Water	—	—	—	0.0747	0.0733	0.0717	0.0685	0.0651	0.0615	0.0576	0.0536

Tabular values in N/m.

Source: From Rohsenow, Hartnett, and Ganić.[3]

TABLE 2.10 Mechanical Properties of Some Metals and Alloys at Room Temperature

Material	Composition	Condition	Yield strength (or 0.2% proof stress), MPa	Ultimate tensile stress, MPa	Elongation on 2 in, %	Hardness*
Ag	99.9	Annealed 600–650°C	7.6†	137	50	26 VHN
		Hard	...	380	4	90 VHN
Al	99.95	Rolled rod 0	...	55	61	17 B
Au	99.99	Soft, cast	0	121	30	33 B
		Hard, 60% red.	21.2	228	4	58 B
Co	99.9	Soft	190	240	4–8	124 B
		Hard	...	675	2–8	165 B
Cu	99.997	Annealed	340	351	60	
		Rod, cold-drawn	34	213	14	37 RB
Ni	>99.0	Annealed	138	482	40	100 B
		Cold-drawn	482	654	25	170 B
Pt	99.99	Annealed	...	120–130	25–40	38–40 VHN
		50% cold-rolled	180	200	3	92 VHN
Pd	99.9	Annealed	35	190	40	37 VHN
		50% cold-drawn	...	320	1.5	106 VHN
Ta	99.98	Annealed	180	200	36	90 VHN
	99.95	Cold-rolled	330	410	5	160 VHN
W		Swaged, recrys.	195	405	16	(200 VHN)
	99.9	Swaged	...	1,750	1–4	450–490 VHN
Aluminum alloys:						
1100	1% Si	O	27.5	69	45	19 B
1100		H18	124	131	15	35 B
3003	1–1.5% Mn	O	41	110	40	28 B
3003		H18	186	200	10	55 B
5056	4.5–5.6% Mg	O	152	290	35	65 B
5056		H38	345	415	15	100 B
7075	1.2–2% Cu, 2.1–2.9% Mg 5.1–5.6% Z	O	104	227	...	60 B
7075		T6	505	570	...	150 B
Copper alloys:						
OHFC, copper		Wire, soft	...	241	35	
		Wire, hard	...	380	1.5	
Gilding metal	5% Zn	Annealed, strip	69	234	45	46 RF
		Extra hard	390	435	4	73 RB
Red brass	15% Zn	Annealed	70–130	280–320	48	
		Half hard	350	405	12	65 RB
		Extra hard	435	555	4	83 RB
Yellow brass	35% Zn	Annealed	100–157	326–376	54–65	58–78 RF
		Half hard	426	526	8	80 RB
		Extra hard	44	605	5	87 RB
Phosphor bronze	5% Sn	Annealed	150–220	340–390	57–48	33–46 RB
		Hard	560	575	8	89 RB
		Extra spring	710	725	3	98 RB
Beryllium copper	1.9% Be	Annealed	...	430–550	35	45–78 RB
		HT (cold worked and precipitation-hardened)	1,070	1,420	2	42 RC
Steels:						
C1010	0.08–0.13 C	Hot-rolled	188	340	28	95 B
		Cold-drawn	318	383	20	105 B
C1080	0.77–0.88%	Hot-rolled	440	810	10	229 B
12% Mn steel	12% Mn	Tempered 600°F	1,480	1,520	10	
Stainless steel	9% Ni, 19% Cr	Annealed	340	590	60	160 B
Type 304		Cold-rolled	1,110	1,280	...	400 B

*VHN = Vickers hardness number; B = Brinell; RB = Rockwell B; RC = Rockwell C; RF = Rockwell F.
†0.01% proof stress.

Source: From Fink and Christiansen.[4]

TABLE 2.11 Electrical Resistivity and Temperature Coefficients of Resistivity of Pure Metals

Metal	mp, °C	Density at 20°C, $kg/m^3 \times 10^{-3}$	Resistivity at 20°C, $\Omega m \times 10^8$	Temp. coeff. of resistivity, 0–100°C,[a] $K^{-1} \times 10^3$
Aluminum	660.1	2.70	2.67	4.5
Antimony	630.5	6.68	40.1	5.1
Barium	729	3.5	60[b]	...
Beryllium	1287	1.848	3.3	9.0
Bismuth	271	9.80	117	4.6
Cadmium	320.9	8.64	7.3	4.3
Calcium	839	1.54	3.7	4.57
Cerium	798	6.75	85.4	8.7
Cesium	28.5	1.87	20	4.8
Chromium	1860	7.1	13.2	2.14
Cobalt	1492	8.9	6.34	6.6
Copper	1083.4	8.96	1.694	4.3
Gallium	29.7	5.91	c	...
Germanium	937	5.32	...	...
Gold	1063	19.3	2.20	4.0
Hafnium	2227	13.1	32.2	4.4
Indium	156.4	7.3	8.8	5.2
Iridium	2454	22.4	5.1	4.5
Iron	1536	7.87	10.1	6.5
Lead	327.4	11.68	20.6	4.2
Lithium	181	0.534	9.29	4.35
Magnesium	649	1.74	4.2	4.25
Manganese	1244	7.4	160(α)	...
Mercury	−38.87	13.546	95.9	1.0
Molybdenum	2615	10.2	5.7	4.35
Nickel	1455	8.9	6.9	6.8
Niobium	2467	8.6	16.0	2.6
Osmium	3030	22.5	8.8	4.1
Palladium	1552	12.0	10.8	4.2
Platinum	1769	21.45	10.58	3.92
Potassium	63.2	0.86	6.8	5.7
Radium	700	5		
Rhenium	3180	21.0	18.7	4.5
Rhodium	1966	12.4	4.7	4.4
Rubidium	38.8	1.53	12.1	4.8
Ruthenium	2310	12.2	7.7	4.1
Silicon	1412	2.34	...	...
Silver	960.8	10.5	1.63	4.1
Sodium	97.8	0.97	4.7	5.5
Strontium	770	2.6	23[b]	...

TABLE 2.11 Electrical Resistivity and Temperature Coefficients of Resistivity of Pure Metals (*Continued*)

Metal	mp, °C	Density at 20°C, $kg/m^3 \times 10^{-3}$	Resistivity at 20°C, $\Omega m \times 10^8$	Temp. coeff. of resistivity, 0–100°C,[a] $K^{-1} \times 10^3$
Tantalum	2980	16.6	13.5	3.5
Tellurium	450	6.24	1.6×10^{5b}	...
Thallium	304	11.85	16.6	5.2
Thorium	1755	11.5	14	4.0
Tin	231.9	7.3	12.6	4.6
Titanium	1667	4.5	54	3.8
Tungsten	3400	19.3	5.4	4.8
Uranium	1132	19.05(α),18.89(β)	27	3.4
Vanadium	1902	6.1	19.6	3.9
Zinc	419.5	7.14	5.96	4.2
Zirconium	1852	6.49	44	4.4

[a]By convention, the signs on orders of magnitude are reversed when they are moved from column to stub (leftmost column) or from column to column heading in a table; i.e., the first value in the temperature-coefficient-of-resistivity column is read $4.5 \times 10^{-3} K^{-1}$. The convention is not always observed by different authors, but it has been followed throughout this table.

[b]0°C.

[c]17.4 | *a* axis, 8.1 | *b* axis, 54.3 | *c* axis.

Source: From Fink and Christiansen.[4]

TABLE 2.12 Electrical Resistivity of Some Alloys and Compounds

Material	Resistivity, $\Omega m \times 10^8$	Material	Resistivity, $\Omega m \times 10^8$
92.5 Ag–7.5 Cu	2		
60 Ag–40 Pd	23	TaN	135
97 Ag–3 Pt	3.5	ZrN	13.6
90 Pt–10 Ir	25	TiN	21.7
96 Pt–4 W	36	VN	85.9
70 Pd–30 Ag	40	TaC	30
90 Au–10 Cu	10.8	ZrC	63.4
75 Au–25 Ag	10.8	SiC	100–200
78.5 Ni–20 Cr–1.5 Si	108.05	WC	12
71 Ni–29 Fe	19.95	$MoSi_2$	37
80 Ni–20 Cr	112.2	$CbSi_2$	6.5
Carbon steel 0.65% C	18	ZrB_2	9
Electrical Si sheet steels	18–52	LaB_6	15
Stainless steel, type 302	72	TiB_2	20
Type 316	74		

Source: From Fink and Christiansen.[4]

TABLE 2.13 Electrical Resistivity of Insulating Materials

Material	Volume resistivity, M Ω · cm
Asbestos board (ebonized)	10^7
Bakelite	$5\text{–}30 \times 10^{11}$
Epoxy	10^{14}
Fluorocarbons:	
Fluorinated ethylene propylene	10^{18}
Polytetrafluoroethylene	10^{18}
Glass	17×10^9
Magnesium oxide	
Mica	$10^{14} - 10^{17}$
Nylon	$10^{14} - 10^{17}$
Neoprene	
Oils:	
Mineral	21×10^6
Paraffin	10^{15}
Paper	
Paper, treated	
Phenolic (glass filled)	$10^{12} - 10^{13}$
Polyethylene	$10^{15} - 10^{18}$
Polyimide	$10^{16} - 10^{17}$
Polyvinyl chloride (flexible)	$10^{11} - 10^{15}$
Porcelain	3×10^8
Rubber	$10^{14} - 10^{16}$
Rubber (butyl)	10^{18}

Source: From Avallone and Baumeister.[1]

REFERENCES*

1. E. A. Avallone and T. Baumeister III (eds.), *Mark's Standard Handbook for Mechanical Engineers*, 9th ed., McGraw-Hill, New York, 1987.
2. J. H. Lienhard, *A Heat Transfer Textbook*, Prentice-Hall, Englewood Cliffs, N.J., 1981.
3. W. M. Rohsenow, J. P. Hartnett, and E. N. Ganić (eds.), *Handbook of Heat Transfer Fundamentals*, McGraw-Hill, New York, 1985, chap. 3.
4. D. G. Fink and D. Christiansen (eds.), *Electronics Engineers' Handbook*, 3d ed., McGraw-Hill, New York, 1989.
5. E. R. G. Eckert and R. M. Drake, Jr., *Analysis of Heat and Mass Transfer*, McGraw-Hill, New York, 1972.
6. H. Gartmann (ed.), *De Laval Engineering Handbook*, 3d ed., McGraw-Hill, New York, 1970.
7. T. F. Irvine, Jr., and J. P. Hartnett (eds.), *Steam and Air Tables in S.I. Units*, Hemisphere Publishing Corp./McGraw-Hill, Washington, D.C., 1976.
8. R. H. Perry, *Engineering Manual*, 3d ed., McGraw-Hill, New York, 1976.
9. R. H. Perry, D. W. Green, and J. O. Maloney (eds.), *Perry's Chemical Engineers' Handbook*, 6th ed., McGraw-Hill, New York, 1984.

*Those references listed above but not cited in the text were used for comparison between different data sources, clarification, clarity of presentation, and, most importantly, reader's convenience when further interest in subject exists.

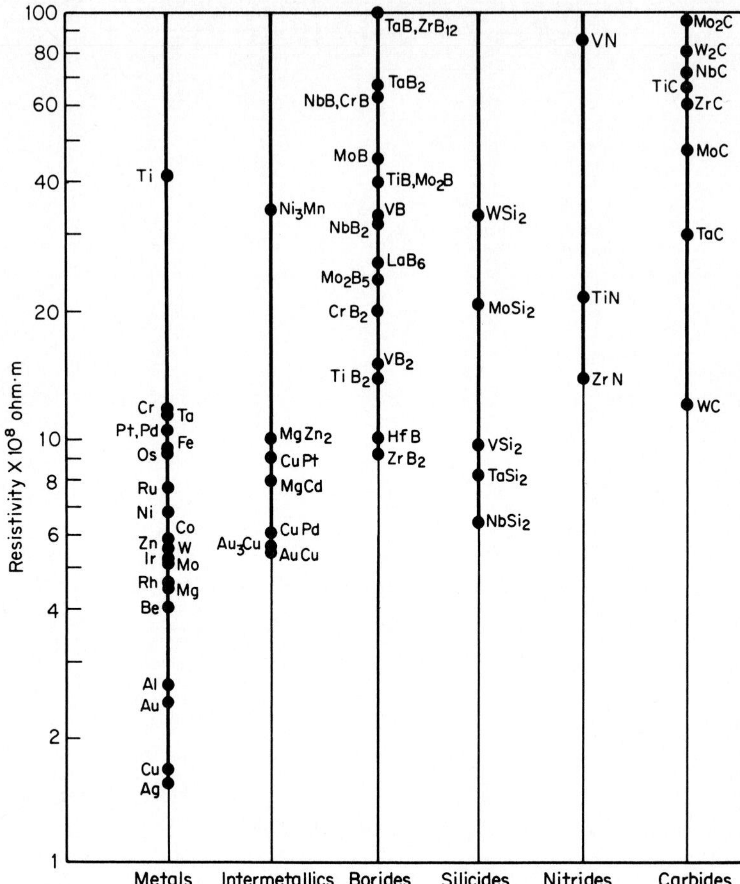

FIGURE 2.1 Electrical Resistivity of Various Materials. (*From Fink and Christiansen.*[4])

10. K. Raznjevic, *Handbook of Thermodynamic Tables and Charts*, McGraw-Hill, New York, 1976.
11. R. C. Reid, J. M. Prausnitz, and T. K. Sherwood, *The Properties of Gases and Liquids*, 3d ed., McGraw-Hill, New York, 1977.
12. Y. S. Touloukian, *Thermophysical Properties of Matter*, vols. 1–6, 10, and 11, Purdue University, West Lafayette, Ind., 1970–1975.

CHAPTER 3
MATHEMATICS

ALGEBRA

1. Basic Laws

Commutative law:

$$a + b = b + a \qquad ab = ba$$

Associative law:

$$a + (b + c) = (a + b) + c \qquad a(bc) = (ab)c$$

Distributive law:

$$a(b + c) = ab + ac$$

2. Sums of Numbers

The sum of the first n numbers:

$$\sum_{1}^{n} (n) = \frac{n(n + 1)}{2}$$

The sum of the squares of the first n numbers:

$$\sum_{1}^{n} (n^2) = \frac{n(n + 1)(2n + 1)}{6}$$

The sum of the cubes of the first n numbers:

$$\sum_{1}^{n} (n^3) = \frac{n^2(n + 1)^2}{4}$$

3. Progressions

Arithmetic Progression

$$a, a + d, a + 2d, a + 3d, \ldots$$

where a = first term
d = common difference
n = number of terms

S = sum of n terms
l = last term
$l = a + (n - 1)d$
$S = (n/2)(a + l)$
$(a + b)/2$ = arithmetic mean of a and b

Geometric Progression

$$a, ar, ar^2, ar^3, \ldots$$

where a = first term
r = common ratio
n = number of terms
S = sum of n terms
l = last term
$l = ar^{n-1}$

$$S = a\frac{r^n - 1}{r - 1} = \frac{rl - a}{r - 1}$$

$$S = \frac{a}{1 - r} \qquad \text{for } r^2 < 1 \text{ and } n = \infty$$

$\sqrt{ab}$ = geometric mean of a and b

4. Powers and Roots

$$a^x a^y = a^{x+y}$$

$$\frac{a^x}{a^y} = a^{x-y}$$

$$(ab)^x = a^x b^x$$

$$(a^x)^y = a^{xy}$$

$$a^0 = 1 \text{ if } a \neq 0$$

$$a^{-x} = 1/a^x$$

$$a^{x/y} = \sqrt[y]{a^x}$$

$$a^{1/y} = \sqrt[y]{a}$$

$$\sqrt[x]{ab} = \sqrt[x]{a}\sqrt[x]{b}$$

$$\sqrt[x]{a/b} = \sqrt[x]{a}/\sqrt[x]{b}$$

5. Binomial Theorem

$$(a \pm b)^n = a^n \pm na^{n-1}b + \frac{n(n-1)}{2!}a^{n-2}b^2 \pm \frac{n(n-1)(n-2)}{3!}a^{n-3}b^3 + \cdots + (\pm 1)^m \frac{n(n-1)\cdots(n-m+1)}{m!}a^{n-m}b^m + \cdots$$

where $m! = 1 \cdot 2 \cdot 3 \cdots (m - 1)m$

The series is finite if n is a positive integer. If n is negative or fractional, the series is infinite and will converge for $|b| < |a|$ only.

6. Absolute Values. The numerical or absolute value of a number n is denoted by $|n|$ and represents the magnitude of the number without regard to algebraic sign. For example, $|-3| = |+3| = 3$.

7. Logarithms. Definition of a logarithm: If $N = b^x$, the exponent x is the logarithm of N to the base b and is written $x = \log_b N$. The number b must be positive, finite, and different from unity. The base of common, or briggsian, logarithms is 10. The base of natural, napierian, or hyperbolic logarithms is 2.71 828 18..., denoted by e.

Laws of Logarithms

$$\log_b MN = \log_b M + \log_b N \qquad \log_b 1 = 0$$

$$\log_b \frac{M}{N} = \log_b M - \log_b N \qquad \log_b b = 1$$

$$\log_b N^m = m \log_b N \qquad \log_b 0 = +\infty,\ 0 < b < 1$$

$$\log_b \sqrt[r]{N^m} = m/r \log_b N \qquad \log_b 0 = -\infty,\ 1 < b < \infty$$

Important Constants

$$\log_{10} e = 0.434\ 294\ 481\ 9$$

$$\log_{10} x = 0.434\ 3 \log_e x = 0.434\ 3 \ln x$$

$$\ln 10 = \log_e 10 = 2.302\ 585\ 093\ 0$$

$$\ln x = \log_e x = 2.302\ 6 \log_{10} x$$

8. Permutations. The number of possible permutations or arrangments of n different elements is $1 \times 2 \times 3 \times \cdots \times n = n!$ (read: "n factorial").

If among the n elements there are p equal ones of one sort, q equal ones of another sort, r equal ones of a third sort, etc., then the number of possible permutations is $(n!)/(p! \times q! \times r! \times \cdots)$, where $p + q + r + \cdots = n$.

9. Combinations. The number of possible combinations or groups of n elements taken r at a time (without repetition of any element within any group) is $[n(n-1)(n-2)(n-3)\cdots(n-r+1)]/(r!) = (n)_r$. If repetitions are allowed, so that a group, for example, may contain as many as r equal elements, then the number of combinations of n elements taken r at a time is $(m)_r$, where $m_r = n + r - 1$.

Note that $(n)_1 + (n)_2 + \cdots + (n)_n = 2^n - 1$.

10. Equations in One Unknown*

Roots of an Equation. An equation containing a single variable x will in general be true for some values of x and false for other values. Any value of x for which the equation is true is called a *root of the equation*. To *solve* an equation means

*This section is in part taken from Ref. 1, *Marks' Standard Handbook for Mechanical Engineers*, 9th ed., by E. A. Avallone and T. Baumeister III (eds.). Copyright © 1987. Used by permission of McGraw-Hill, Inc. All rights reserved.

to find all its roots. Any root of an equation, when substituted therein for x, will *satisfy* the equation. An equation which is true for all values of x, like $(x + 1)^2 = x^2 + 2x + 1$, is called an *identity* [often written $(x + 1)^2 \equiv x^2 + 2x + 1$].

Types of Equations

1. Algebraic equations

   Of the first degree (linear), e.g., $2x + 6 = 0$ (root: $x = -3$)

   Of the second degree (quadratic), e.g., $x^2 - 2x - 3 = 0$ (roots: $-1, 3$)

   Of the third degree (cubic), e.g., $x^3 - 6x^2 + 5x + 12 = 0$ (roots: $-1, 3, 4$)

2. Transcendental equations

   Exponential equations, e.g., $2^x = 32$ (root: $x = 5$); $2^x = -32$ (no real root)

   Trigonometric equations, e.g., $10 \sin x - \sin 3x = 3$ (roots: 30°, 150°).

Equations of First Degree. These are linear equations. *Solution*: Collect all the terms involving x on one side of the equation, thus: $ax = b$, where a and b are known numbers. Then divide through by the coefficient of x, obtaining $x = b/a$ as the root.

Equations of Second Degree. These are quadratic equations. *Solution*: Throw the equation into the standard form $ax^2 + bx + c = 0$. Then the two roots are

$$x_1 = \frac{-b + \sqrt{b^2 - 4ac}}{2a} \qquad x_2 = \frac{-b - \sqrt{b^2 - 4ac}}{2a}$$

The roots are real and distinct, coincident, or imaginary, according as $b^2 - 4ac$ is positive, zero, or negative. The sum of the roots is $x_1 + x_2 = -b/a$; the product of the roots is $x_1 x_2 = c/a$.

Graphical solution: Write the equation in the form $x^2 = px + q$, and plot the parabola $y_1 = x^2$, and the straight line $y_2 = px + q$. The abscissas of the points of intersection will be the roots of the equation. If the line does not cut the parabola, the roots are imaginary.

Equations of Third Degree. The general cubic equation $x^3 + bx^2 + cx + d = 0$ is reducible (substitute $x = y - b/3$) to the form $y^3 + vy + w = 0$, where $v = (3c - b^2)/3$ and $w = (2b^3 - 9bc + 27d)/27$. The roots of the reduced equation are

$$x_1 = A + B \quad \text{and} \quad x_{2,3} = -\frac{1}{2}(A + B) \pm \frac{\sqrt{3}}{2}(A - B)i$$

where

$$A = \left[-\frac{w}{2} + \left(\frac{w^2}{4} + \frac{v^3}{27}\right)^{1/2}\right]^{1/3}$$

$$B = \left[-\frac{w}{2} - \left(\frac{w^2}{4} + \frac{v^3}{27}\right)^{1/2}\right]^{1/3}$$

If $(w^2/4 + v^3/27) > 0$, there are one real root and two conjugate complex roots.

11. Determinants*

Evaluation of Determinants. Of the second order:

$$\begin{vmatrix} a_1 b_1 \\ a_2 b_2 \end{vmatrix} = a_1 b_2 - a_2 b_1$$

Of the third order:

$$\begin{vmatrix} a_1 b_1 c_1 \\ a_2 b_2 c_2 \\ a_3 b_3 c_3 \end{vmatrix} = a_1 \begin{vmatrix} b_2 c_2 \\ b_3 c_3 \end{vmatrix} - a_2 \begin{vmatrix} b_1 c_1 \\ b_3 c_3 \end{vmatrix} + a_3 \begin{vmatrix} b_1 c_1 \\ b_2 c_2 \end{vmatrix}$$

$$= a_1(b_2 c_3 - b_3 c_2) - a_2(b_1 c_3 - b_3 c_1) + a_3(b_1 c_2 - b_2 c_1)$$

Of the fourth order:

$$\begin{vmatrix} a_1 b_1 c_1 d_1 \\ a_2 b_2 c_2 d_2 \\ a_3 b_3 c_3 d_3 \\ a_4 b_4 c_4 d_4 \end{vmatrix}$$

$$= a_1 \begin{vmatrix} b_2 c_2 d_2 \\ b_3 c_3 d_3 \\ b_4 c_4 d_4 \end{vmatrix} - a_2 \begin{vmatrix} b_1 c_1 d_1 \\ b_3 c_3 d_3 \\ b_4 c_4 d_4 \end{vmatrix} + a_3 \begin{vmatrix} b_1 c_1 d_1 \\ b_2 c_2 d_2 \\ b_4 c_4 d_4 \end{vmatrix} - a_4 \begin{vmatrix} b_1 c_1 d_1 \\ b_2 c_2 d_2 \\ b_3 c_3 d_3 \end{vmatrix}$$

etc. In general, to evaluate a determinant of the nth order, take the elements of the first column with signs alternately plus and minus, and form the sum of the products obtained by multiplying each of these elements by its corresponding minor. The minor corresponding to any element a_1 is the determinant (of next lower order) obtained by striking out from the given determinant the row and column containing a_1.

Properties of Determinants

1. The columns may be changed to rows and the rows to columns:

$$\begin{vmatrix} a_1 b_1 c_1 \\ a_2 b_2 c_2 \\ a_3 b_3 c_3 \end{vmatrix} = \begin{vmatrix} a_1 a_2 a_3 \\ b_1 b_2 b_3 \\ c_1 c_2 c_3 \end{vmatrix}$$

2. Interchanging two adjacent columns changes the sign of the result.
3. If two columns are equal, the determinant is zero.
4. If the elements of one column are m times the elements of another column, the determinant is zero.
5. To multiply a determinant by any number m, multiply all the elements of any one column by m.

---

*This section taken in part from Ref. 1, *Marks' Standard Handbook for Mechanical Engineers*, 9th ed., by E. A. Avallone and T. Baumeister (eds.). Copyright © 1987. Used by permission of McGraw-Hill, Inc. All rights reserved.

6

$$\begin{vmatrix} a_1 + p_1 + q_1, b_1c_1 \\ a_2 + p_2 + q_2, b_2c_2 \\ a_3 + p_3 + q_3, b_3c_3 \end{vmatrix} = \begin{vmatrix} a_1b_1c_1 \\ a_2b_2c_2 \\ a_3b_3c_3 \end{vmatrix} + \begin{vmatrix} p_1b_1c_1 \\ p_2b_2c_2 \\ p_3b_3c_3 \end{vmatrix} + \begin{vmatrix} q_1b_1c_1 \\ q_2b_2c_2 \\ q_3b_3c_3 \end{vmatrix}$$

7

$$\begin{vmatrix} a_1b_1c_1 \\ a_2b_2c_2 \\ a_3b_3c_3 \end{vmatrix} = \begin{vmatrix} a_1 + mb_1, b_1c_1 \\ a_2 + mb_2, b_2c_2 \\ a_3 + mb_3, b_3c_3 \end{vmatrix}$$

Solution of Simultaneous Equations by Determinants. If

$$a_1x + b_1y + c_1z = p_1$$

$$a_2x + b_2y + c_2z = p_2$$

$$a_3x + b_3y + c_3z = p_3$$

where

$$D = \begin{vmatrix} a_1b_1c_1 \\ a_2b_2c_2 \\ a_3b_3c_3 \end{vmatrix} \neq 0$$

then

$$x = D_1/D$$
$$y = D_2/D$$
$$z = D_3/D$$

where

$$D_1 = \begin{vmatrix} p_1b_1c_1 \\ p_2b_2c_2 \\ p_3b_3c_3 \end{vmatrix}, \quad D_2 = \begin{vmatrix} a_1p_1c_1 \\ a_2p_2c_2 \\ a_3p_3c_3 \end{vmatrix}, \quad D_3 = \begin{vmatrix} a_1b_1p_1 \\ a_2b_2p_2 \\ a_3b_3p_3 \end{vmatrix}$$

Similarly for a larger (or smaller) number of equations.

*GEOMETRY**

Areas of Plane Figures

square

$A = a^2$

$a = \sqrt{A}$

$d = a\sqrt{2}$

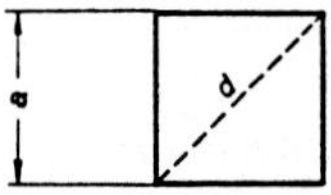

*Material in this section taken from Ref. 2, *Engineering Formulas*, 4th ed., by K. Gieck. Copyright © 1983. Used by permission of McGraw-Hill, Inc. All rights reserved.

rectangle

$$A = a\,b$$

$$d = \sqrt{a^2 + b^2}$$

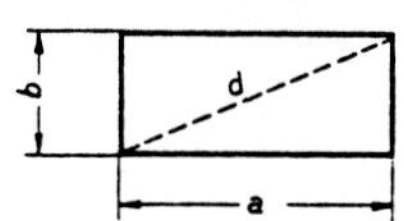

parallelogram

$$A = a\,h = a\,b\,\sin\alpha$$

$$d_1 = \sqrt{(a + h\cot\alpha)^2 + h^2}$$

$$d_2 = \sqrt{(a - h\cot\alpha)^2 + h^2}$$

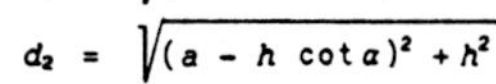

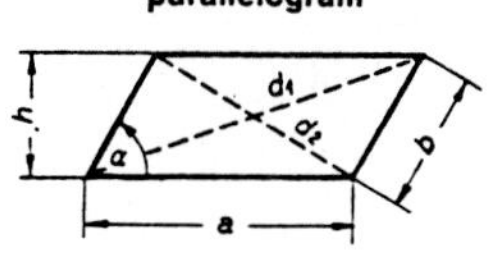

trapezoid

$$A = \frac{a+b}{2}\,h = m\,h$$

$$m = \frac{a+b}{2}$$

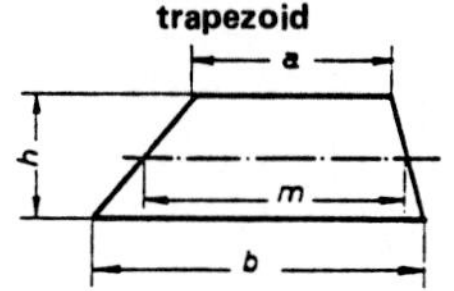

triangle

$$A = \frac{a \cdot h}{2} = \varrho\,s$$

$$= \sqrt{s(s-a)(s-b)(s-c)}$$

$$s = \frac{a+b+c}{2}$$

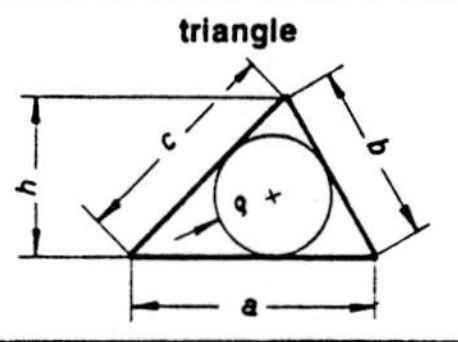

equilateral triangle

$$A = \frac{a^2}{4}\sqrt{3}$$

$$h = \frac{a}{2}\sqrt{3}$$

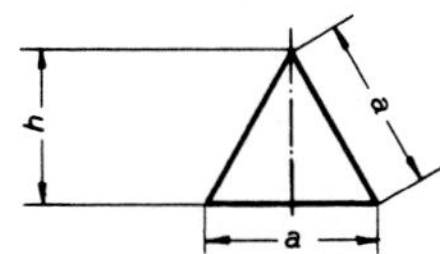

pentagon

$$A = \frac{5}{8}\,r^2\sqrt{10 + 2\sqrt{5}}$$

$$a = \frac{1}{2}\,r\sqrt{10 - 2\sqrt{5}}$$

$$\varrho = \frac{1}{4}\,r\sqrt{6 + 2\sqrt{5}}$$

construction:
$\overline{AB} = 0{\cdot}5\,r$, $\overline{BC} = \overline{BD}$, $\overline{CD} = \overline{CE}$

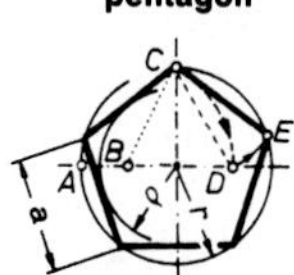

hexagon

$$A = \frac{3}{2}\,a^2\sqrt{3}$$

$$d = 2\,a$$

$$= \frac{2}{\sqrt{3}}\,s \approx 1{\cdot}155\,s$$

$$s = \frac{\sqrt{3}}{2}\,d \approx 0{\cdot}866\,d$$

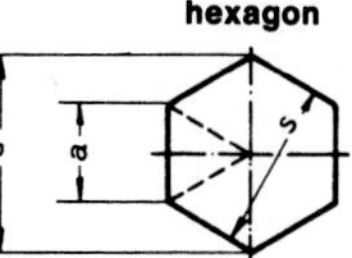

octagon

$$A = 2\,a\,s \approx 0{\cdot}83\,s^2$$

$$= 2\,s\sqrt{d^2 - s^2}$$

$$a = s\tan 22{\cdot}5^0 \approx 0{\cdot}415\,s$$

$$s = d\cos 22{\cdot}5^0 \approx 0{\cdot}924\,d$$

$$d = \frac{s}{\cos 22{\cdot}5^0} \approx 1{\cdot}083\,s$$

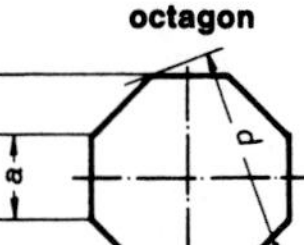

polygon

$$A = A_1 + A_2 + A_3$$

$$= \frac{a\,h_1 + b\,h_2 + b\,h_3}{2}$$

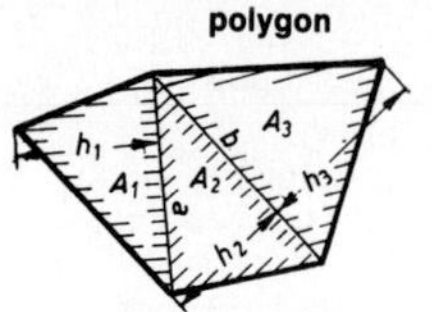

---

circle

$$A = \frac{\pi}{4}d^2 = \pi r^2$$

$$\approx 0{\cdot}785\,d^2$$

$$U = 2\pi r = \pi d$$

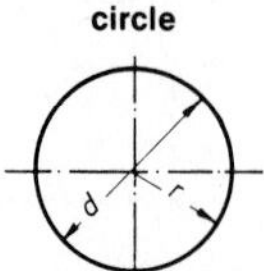

---

annulus

$$A = \frac{\pi}{4}(D^2 - d^2)$$

$$= \pi\,(d + b)b$$

$$b = \frac{D - d}{2}$$

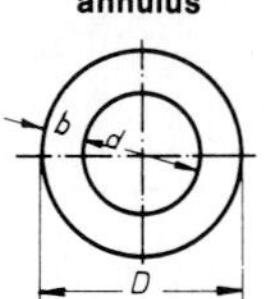

---

sector of a circle

$$A = \frac{\pi}{360^0} r^2 \alpha = \frac{\hat{\alpha}}{2} r^2$$

$$= \frac{b\,r}{2}$$

$$b = \frac{\pi}{180^0}\,r\,\alpha$$

$$\hat{\alpha} = \frac{\pi}{180^0}\,\alpha \quad (\hat{\alpha} = \alpha \text{ in circular measure})$$

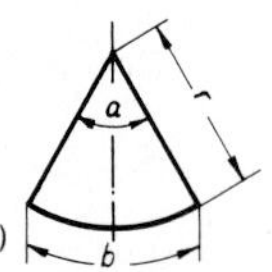

---

segment of a circle

$$s = 2\,r\,\sin\frac{\alpha}{2}$$

$$A = \frac{h}{6\,s}(3\,h^2 + 4\,s^2) = \frac{r^2}{2}(\hat{\alpha} - \sin\alpha)$$

$$r = \frac{h}{2} + \frac{s^2}{8\,h}$$

$$h = r(1 - \cos\frac{\alpha}{2}) = \frac{s}{2}\tan\frac{\alpha}{4}$$

$\hat{\alpha}$ see formula b 39

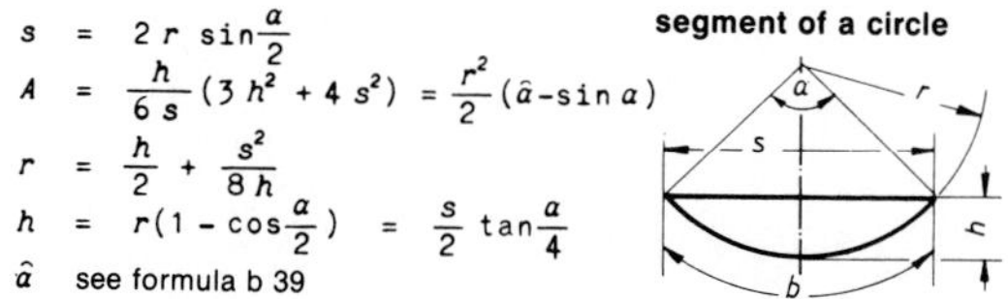

---

ellipse

$$A = \frac{\pi}{4} D\,d = \pi\,a\,b$$

$$U \approx \pi\,\frac{D + d}{2}$$

$$= \pi(a+b)\left[1 + \frac{1}{4}\lambda^2 + \frac{1}{64}\lambda^4 + \frac{1}{256}\lambda^6 + \frac{25}{16384}\lambda^8 + \ldots\right], \text{ where } \lambda = \frac{a - b}{a + b}$$

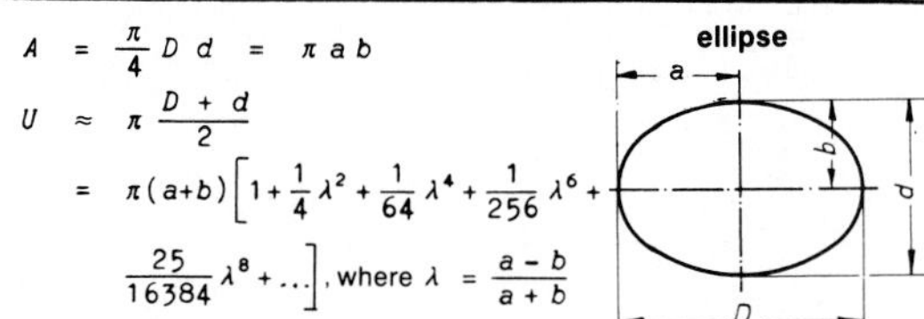

Volumes and areas of solid bodies. For these, the following symbols are used: V = volume, A_o = surface area, A_m = generated surface.

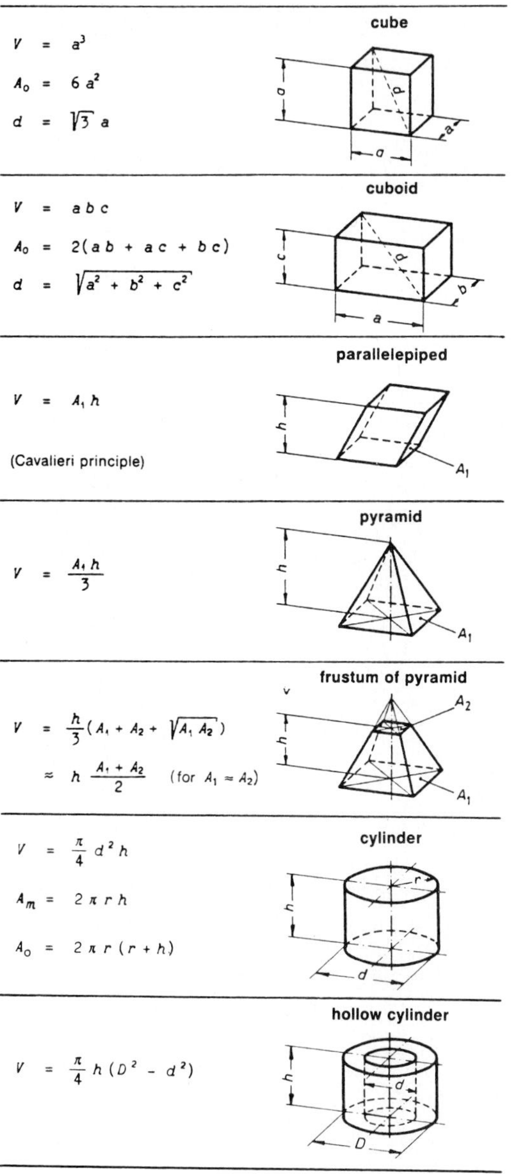

$V = a^3$ $A_o = 6a^2$ $d = \sqrt{3}\,a$	cube
$V = abc$ $A_o = 2(ab + ac + bc)$ $d = \sqrt{a^2 + b^2 + c^2}$	cuboid
$V = A_1 h$ (Cavalieri principle)	parallelepiped
$V = \dfrac{A_1 h}{3}$	pyramid
$V = \dfrac{h}{3}(A_1 + A_2 + \sqrt{A_1 A_2})$ $\approx h\,\dfrac{A_1 + A_2}{2}$ (for $A_1 \approx A_2$)	frustum of pyramid
$V = \dfrac{\pi}{4} d^2 h$ $A_m = 2\pi r h$ $A_o = 2\pi r(r + h)$	cylinder
$V = \dfrac{\pi}{4} h(D^2 - d^2)$	hollow cylinder

cone

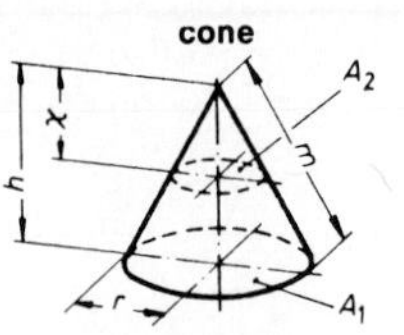

$$V = \frac{\pi}{3} r^2 h$$

$$A_m = \pi r m$$

$$A_o = \pi r (r + m)$$

$$m = \sqrt{h^2 + r^2}$$

$$A_2 : A_1 = x^2 : h^2$$

frustum of cone

$$V = \frac{\pi}{12} h (D^2 + Dd + d^2)$$

$$A_m = \frac{\pi}{2} m (D + d) = 2 \pi p m$$

$$m = \sqrt{\left(\frac{D - d}{2}\right)^2 + h^2}$$

sphere

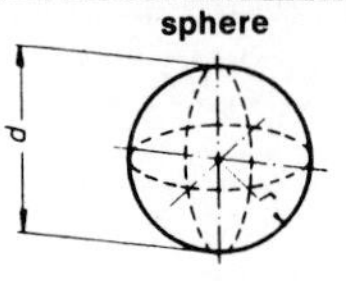

$$V = \frac{4}{3} \pi r^3 = \frac{1}{6} \pi d^3$$

$$\approx 4 \cdot 189\, r^3$$

$$A_o = 4 \pi r^2 = \pi d^2$$

zone of a sphere

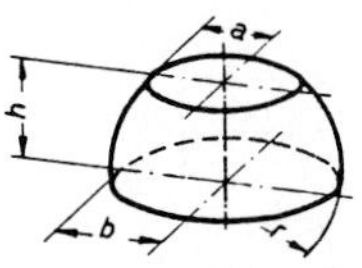

$$V = \frac{\pi}{6} h (3 a^2 + 3 b^2 + h^2)$$

$$A_m = 2 \pi r h$$

$$A_o = \pi (2 r h + a^2 + b^2)$$

segment of a sphere

$$V = \frac{\pi}{6} h \left(\frac{3}{4} s^2 + h^2\right)$$

$$= \pi h^2 \left(r - \frac{h}{3}\right)$$

$$A_m = 2 \pi r h$$

$$= \frac{\pi}{4} (s^2 + 4 h^2)$$

sector of a sphere

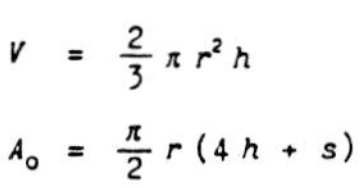

$$V = \frac{2}{3} \pi r^2 h$$

$$A_o = \frac{\pi}{2} r (4 h + s)$$

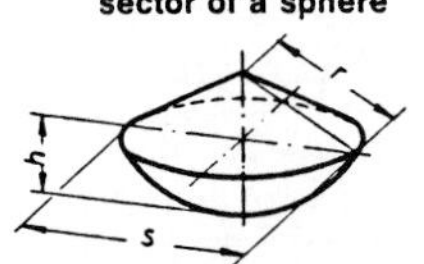

sphere with cylindrical boring

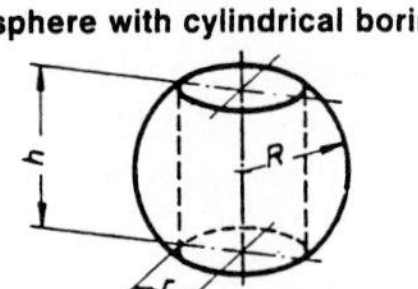

$$V = \frac{\pi}{6} h^3$$

$$A_o = 2 \pi h (R + r)$$

sphere with conical boring

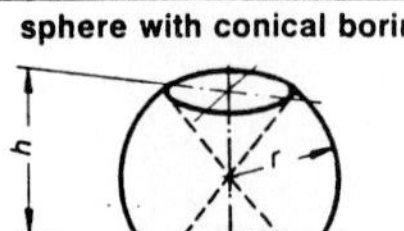

$$V = \frac{2}{3} \pi r^2 h$$

$$A_o = 2 \pi r \left(h + \sqrt{r^2 - \frac{h^2}{4}}\right)$$

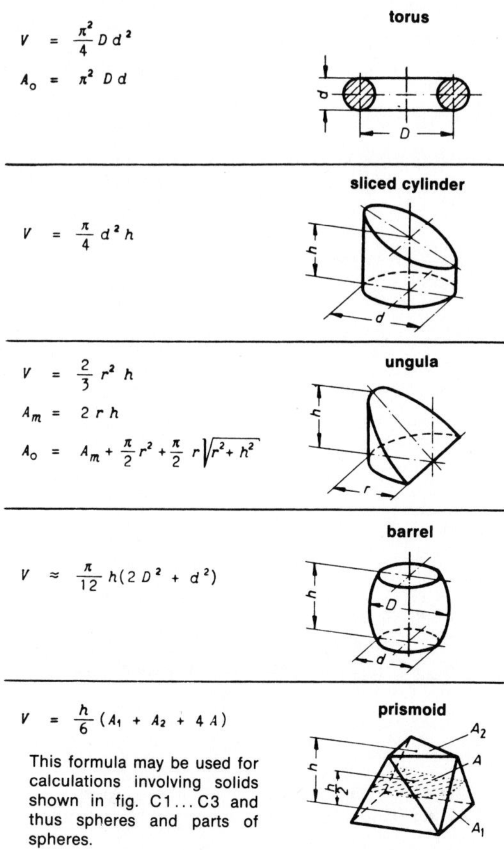

torus
$V = \frac{\pi^2}{4} D d^2$
$A_o = \pi^2 D d$
d
D
sliced cylinder
$V = \frac{\pi}{4} d^2 h$
h
d
ungula
$V = \frac{2}{3} r^2 h$
$A_m = 2 r h$
$A_o = A_m + \frac{\pi}{2} r^2 + \frac{\pi}{2} r \sqrt{r^2 + h^2}$
h
r
barrel
$V \approx \frac{\pi}{12} h (2 D^2 + d^2)$
h
D
d
prismoid
$V = \frac{h}{6} (A_1 + A_2 + 4 A)$
This formula may be used for calculations involving solids shown in fig. C1...C3 and thus spheres and parts of spheres.
A_2
A
A_1
h
$\frac{h}{2}$

ANALYTIC GEOMETRY

1. Rectangular Coordinate Systems. The rectangular coordinate system in space is defined by three mutually perpendicular coordinate axes which intersect at the origin O as shown in Fig. 3.1. The position of a point $P(x, y, z)$ is given by the distances x, y, z from the coordinate planes ZOY, XOZ, and XOY, respectively.

2. Cylindrical Coordinate Systems. The position of any point $P\ (r, \theta, z)$ is given by the polar coordinates r and θ, the projection of P on the XY plane, and by z, the distance from the XY plane to the point (Fig. 3.2).

3. Spherical Coordinate Systems. The position of any point $P\ (r, \theta, \phi)$ (Fig. 3.3) is given by the distance $r\ (= \overline{OP})$, the angle θ which is formed by the intersection of the X coordinate and the projection of OP on the XY plane, and the angle ϕ which is formed by $\overline{OP}$ and the coordinate z.

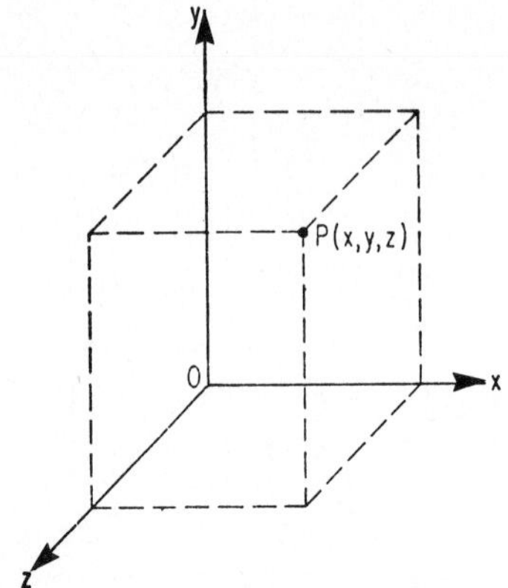

FIGURE 3.1 Rectangular coordinate system. (*From Rothbart.*[3])

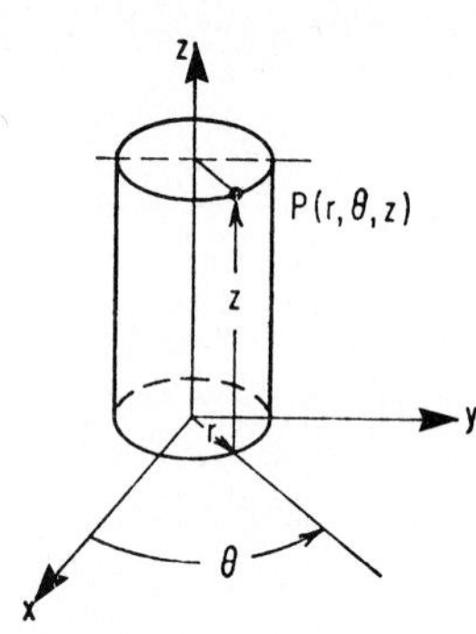

FIGURE 3.2 Cylindrical coordinate system. (*From Rothbart.*[3])

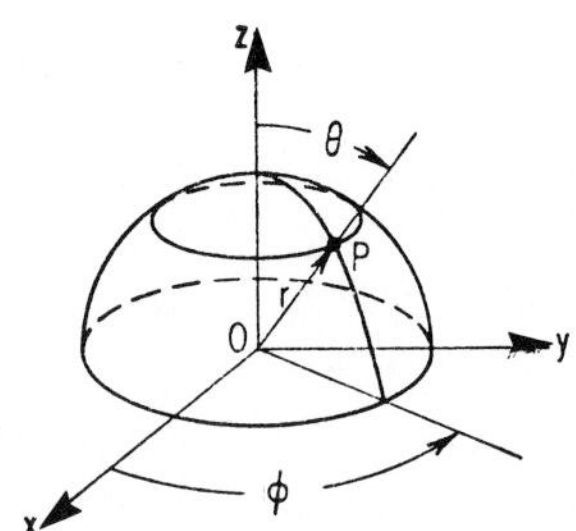

FIGURE 3.3 Spherical coordinate system. (*From Rothbart.*[3])

Relations between Coordinate Systems. Rectangular and cylindrical:

$$x = r\cos\theta \qquad y = r\sin\theta \qquad z = z \qquad \overline{OP} = r^2 + z^2$$

Rectangular and spherical:

$$x = r\sin\theta\cos\phi \qquad y = r\sin\theta\sin\phi \qquad z = r\cos\theta$$

4. Equations of a Straight Line*. Straight line distance (Fig. 3.4): Let $P_1 = (x_1, y_1)$, $P_2 = (x_2, y_2)$. Then, distance $P_1P_2 = (x_2 - x_1)^2 + (y_2 - y_1)^2$; slope of $P_1P_2 = m = \tan\alpha = (y_2 - y_1)/(x_2 - x_1)$; coordinates of the midpoint are $x = 1/2(x_1 + x_2)$, $y = 1/2(y_1 + y_2)$.

Let m_1, m_2 be the slopes of two lines; then, if the lines are parallel, $m_1 = m_2$; if the lines are perpendicular to each other, $m_1 = -1/m_2$.

*This section is taken in part from Ref. 1, *Marks' Standard Handbook for Mechanical Engineers*, 9th ed., by E. A. Avallone and T. Baumeister III (eds.). Copyright © 1987. Used by permission of McGraw-Hill, Inc. All rights reserved.

Various forms of a straight line equation:

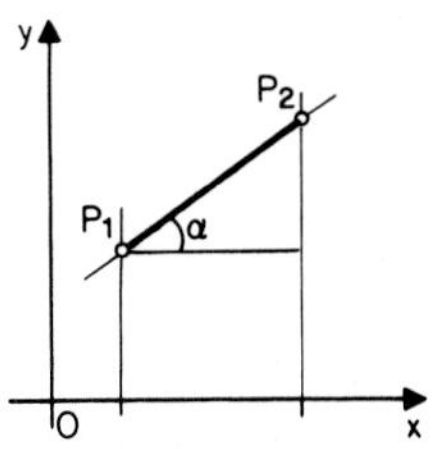

FIGURE 3.4

Intercept form (Fig. 3.5)

$$\frac{x}{a} + \frac{y}{b} = 1$$

a, b = intercepts of the line on the axes.

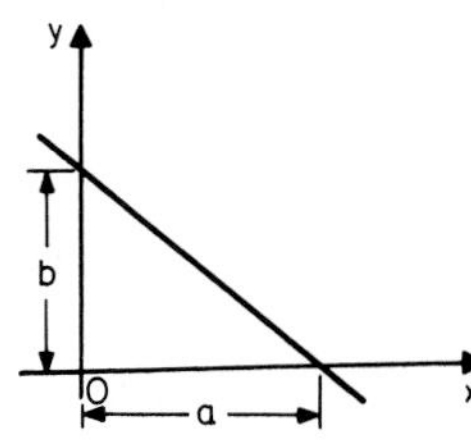

FIGURE 3.5

Slope form (Fig. 3.6)

$$y = mx + b$$

$m = \tan \alpha =$ slope; $b =$ intercept on the y axis.

Normal form (Fig. 3.7)

$$x \cos \beta + y \sin \beta = p$$

p = perpendicular from origin to line; β = angle from the x axis to p.

Parallel-intercept form (Fig. 3.8)

$$\frac{y - b}{c - b} = \frac{x}{k}$$

b, c = intercepts on two parallels at distance k apart.

General form

$$Ax + By + C = 0$$

Here $a = -C/A$, $b = -C/B$, $m = -A/B$, $\cos \beta = A/R$, $\sin \beta = B/R$, $p = -C/R$, where $R = \pm A^2 + B^2$ (sign to be so chosen that p is positive).

Line through (x_1, y_1) with slope m

$$y - y_1 = m(x - x_1)$$

Line through (x_1, y_1) and (x_2, y_2)

$$y - y_1 = \frac{y_2 - y_1}{x_2 - x_1}(x - x_1)$$

Line parallel to x axis and to y axis, respectively

$$y = a \qquad x = b$$

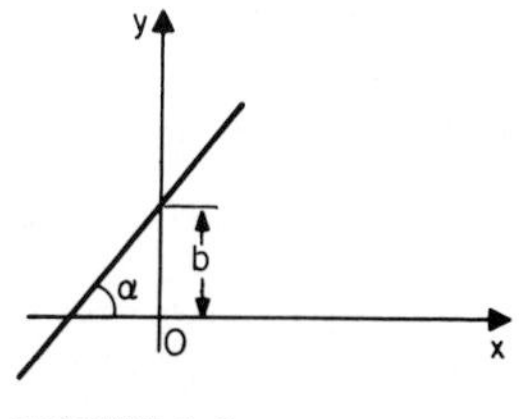

FIGURE 3.6

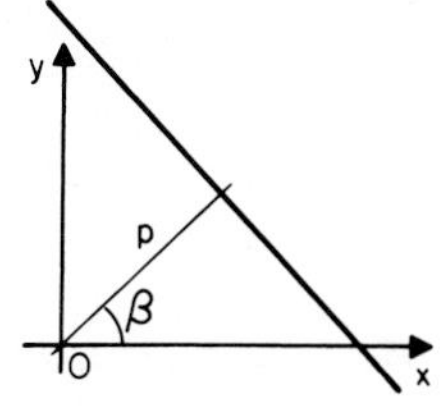

FIGURE 3.7

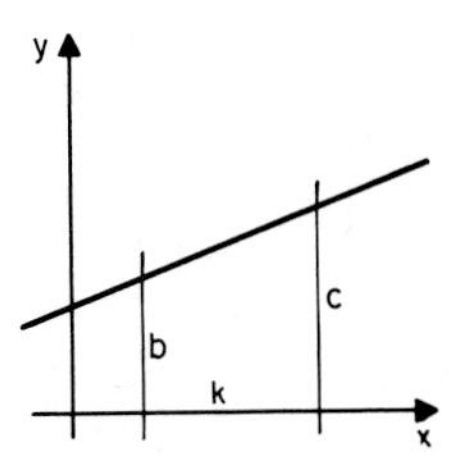

FIGURE 3.8

Angles: If α = angle formed by the line with slope m_1 and the line with slope m_2, then

$$\tan \alpha = \frac{m_2 - m_1}{1 + m_2 m_1}$$

If parallel, $m_1 = m_2$. If perpendicular, $m_1 m_2 = -1$.

5. Circle. See Fig. 3.9. Equations of a circle:

Center at the origin:

$$x^2 + y^2 = r^2$$

Center at (x_o, y_o):

$$(x - x_o)^2 + (y - y_o) = r^2$$

General equation of a circle:

$$x^2 + y^2 + ax + by + c = 0$$

Radius of a circle:

$$r = (x_o^2 + y_o^2 - c)^{1/2}$$

Coordinates of the center M:

$$x_o = -\frac{a}{2} \qquad y_o = -\frac{b}{2}$$

Tangent T at point P_1 (x_1, y_1):

$$y = r^2 - \frac{(x - x_o)(x_1 - x_o)}{(y_1 - y_2) + y_o}$$

6. Parabola. See Fig. 3.10. Equations of a parabola:

Vertex at the origin:

$$x^2 = 2py$$

Vertex at (x_o, y_o):

$$(x - x_o)^2 = 2p(y - y_o)$$

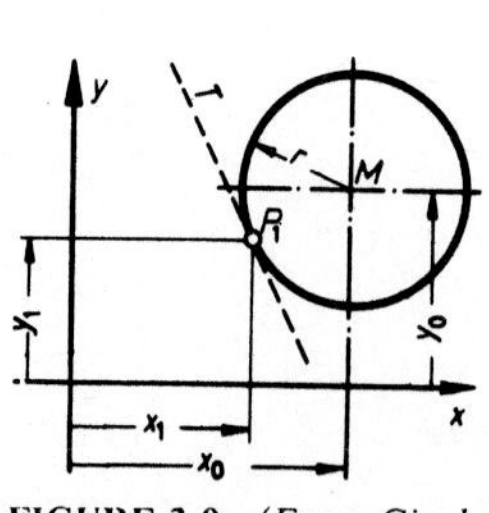

FIGURE 3.9 (*From Gieck.*[2])

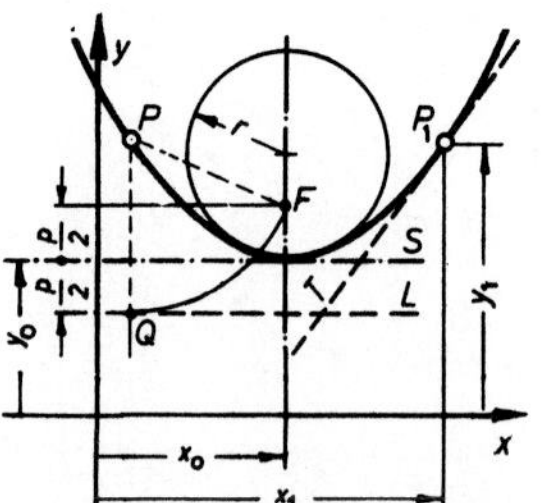

FIGURE 3.10 (*From Gieck.*[2])

General equation of a parabola:

$$y = ax^2 + bx + c$$

Vertex radius:

$$r = p$$

Basic property:

$$\overline{PF} = \overline{PQ}$$

Tangent at point P_1 (x_1, y_1):

$$y = \frac{2(y_1 - y_o)(x - x_1)}{(x_1 - y_o) + y_1}$$

(F = focus, L = directrix, S = tangent at the vertex.)

7. Hyperbola. See Fig. 3.11. Equations of a hyperbola:

Point of intersection of asymptotes at the origin:

$$\frac{x^2}{a^2} - \frac{y^2}{b^2} - 1 = 0$$

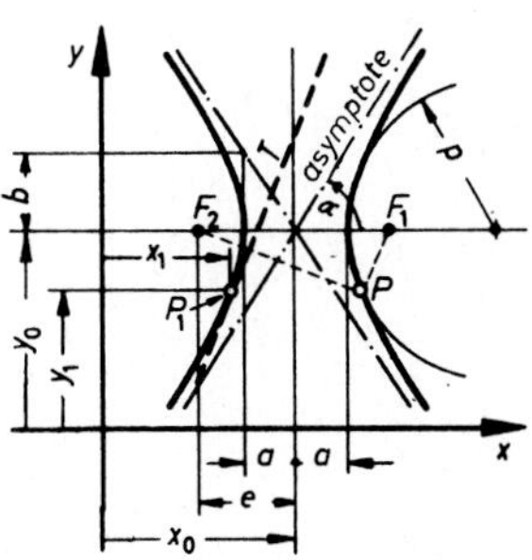

FIGURE 3.11 (*From Gieck.*[2])

Point of intersection of asymptotes at (x_o, y_o):

$$\frac{(x - x_o)^2}{a^2} - \frac{(y - y_o)^2}{b^2} - 1 = 0$$

General equation of a hyperbola:

$$ax^2 + by^2 + cx = dy + e = 0$$

Basic property:

$$\overline{F_2P} = \overline{F_1P} = 2a$$

Eccentricity:

$$e = (a^2 + b^2)^{1/2}$$

Gradient of asymptote:

$$\tan\alpha = m = \pm\frac{b}{a}$$

Vertex radius:

$$p = \frac{b^2}{a}$$

Tangent T at point P_1 (x_1, y_1):

$$y = \frac{b^2}{a^2}\frac{(x_1 - x_o)(x - x_1)}{(y_1 - y_o)} + y_1$$

8. Ellipse. See Fig. 3.12. Equation of an ellipse:

Point of interception of axes at the origin:

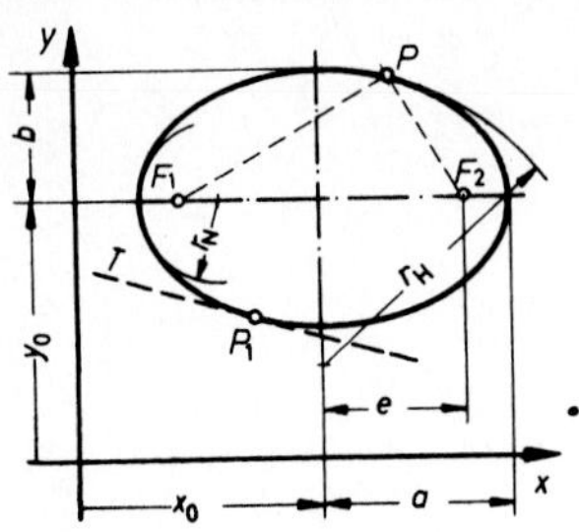

FIGURE 3.12 (*From Gieck.*[2])

$$\frac{x^2}{a^2} + \frac{y^2}{b^2} - 1 = 0$$

Point of interception of axes at (x_o, y_o):

$$\frac{(x - x_o)^2}{a^2} + \frac{(y - y_o)^2}{b^2} - 1 = 0$$

Vertex radii:

$$r_N = \frac{b^2}{a} \qquad r_H = \frac{a^2}{b}$$

Eccentricity:
$e = (a^2 - b^2)^{1/2}$

Basic property:

$$\overline{F_1P} + \overline{F_2P} = 2a$$

Tangent T at $P_1(x_1, y_1)$:

$$y = -\frac{b^2}{a^2}\frac{(x_1 - x_o)(x - x_1)}{(y_1 - y_o)} + y_1$$

(F_1 and F_2 are focal points.)

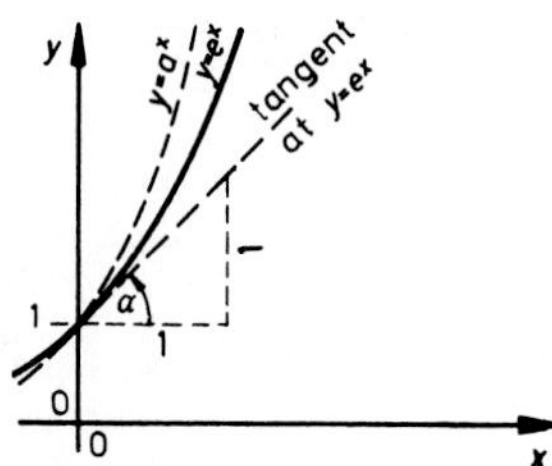

FIGURE 3.13 (*From Gieck.*[2])

9. Exponential curve. See Fig. 3.13.

Basic equation of an exponential curve: $y = a^x$. Here a is a positive constant, and x is a number.

All exponential curves pass through the point $(x = 0, y = 1)$.

The derivative of the curve passing through this point with a gradient of 45° ($\tan\alpha = 1$) is equal to the curve itself. The constant a now becomes e (e = base of the natural logarithm).

TRIGONOMETRY

1. Circular and Angular Measure of a Plane Angle. Circular measure is the ratio of the distance d measured along the arc to the radius r (Fig. 3.14). It is given in *radians* (rad), which have no dimensions:

$$\alpha = \frac{d}{r} \quad \text{rad}$$

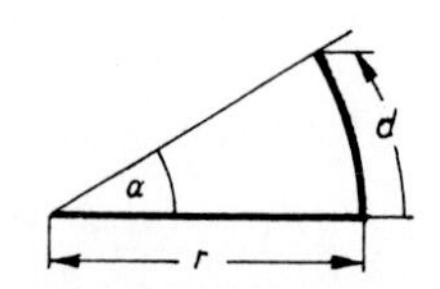

FIGURE 3.14 (*From Gieck.*[2])

Angular measure is obtained by dividing the angle α subtended at the center of a circle into 360 equal divisions known as *degrees* (°). A degree is divided into 60 minutes (unit: ′); a minute is divided into 60 seconds (unit: ″). Relationships between the more important circular and angular measures are shown in Table 3.1.

TABLE 3.1 Relationships between Circular and Angular Measures

$360° = 2\pi$ radians

or 1 rad $= 57.2958°$

degrees	0°	30°	45°	60°	75°	90°	180°	270°	360°
radians	0	$\frac{\pi}{6}$	$\frac{\pi}{4}$	$\frac{\pi}{3}$	$\frac{5}{12}\pi$	$\frac{\pi}{2}$	π	$\frac{3}{2}\pi$	2π
	0	0·52	0·79	1·05	1·31	1·57	3·14	4·71	6·28

Source: From Gieck.[2]

2. Basic Relations between Trigonometric Functions. The principal trigonometric functions of the angle α (see right-angle triangle in Fig. 3.15) are:

$$\sin\alpha = \frac{\text{opposite}}{\text{hypotenuse}} = \frac{a}{c} \qquad \tan\alpha = \frac{\text{opposite}}{\text{adjacent}} = \frac{a}{b}$$

$$\cos\alpha = \frac{\text{adjacent}}{\text{hypotenuse}} = \frac{b}{c} \qquad \cot\alpha = \frac{\text{adjacent}}{\text{opposite}} = \frac{b}{a}$$

Values of the trigonometric functions for the more important angles are shown in Table 3.2. Trigonometric conversions are shown in Table 3.3. The relationships between sine and cosine functions are shown in Fig. 3.16, and those between tangent, sine, cotangent, and cosine functions are shown in Fig. 3.17.

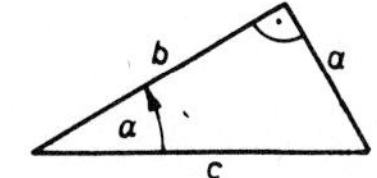

FIGURE 3.15 (*From Gieck.*[2])

3. Hyperbolic Functions. The hyperbolic functions are certain combinations of exponentials e^x *and* e^{-x}:

$$\cosh x = \frac{e^x + e^{-x}}{2}$$

$$\sinh x = \frac{e^x - e^{-x}}{2}$$

$$\tanh x = \frac{\sinh x}{\cosh x} = \frac{e^x - e^{-x}}{e^x + e^{-x}}$$

$$\coth x = \frac{e^x + e^{-x}}{e^x - e^{-x}} = \frac{1}{\tanh x} = \frac{\cosh x}{\sinh x}$$

$$\operatorname{sech} x = \frac{1}{\cosh x} = \frac{2}{e^x + e^{-x}}$$

$$\operatorname{csch} x = \frac{1}{\sinh x} = \frac{2}{e^x - e^{-x}}$$

TABLE 3.2 Values of Functions for the More Important Angles

angle α	0°	30°	45°	60°	75°	90°	180°	270°	360°
sin α	0	0,500	0,707	0,866	0,966	1	0	−1	0
cos α	1	0,866	0,707	0,500	0,259	0	−1	0	1
tan α	0	0,577	1,000	1,732	3,732	∞	0	∞	0
cot α	∞	1,732	1,000	0,577	0,268	0	∞	0	∞

Source: From Gieck.[2]

TABLE 3.3 Trigonometric Conversions

$$\sin^2\alpha + \cos^2\alpha = 1$$

$$1 + \tan^2\alpha = \frac{1}{\cos^2\alpha}$$

$$\tan\alpha \quad \cot\alpha = 1$$

$$1 + \cot^2\alpha = \frac{1}{\sin^2\alpha}$$

$$\sin(\alpha \pm \beta) = \sin\alpha \quad \cos\beta \pm \cos\alpha \quad \sin\beta$$

$$\cos(\alpha \pm \beta) = \cos\alpha \quad \cos\beta \mp \sin\alpha \quad \sin\beta$$

$$\tan(\alpha \pm \beta) = \frac{\tan\alpha \pm \tan\beta}{1 \mp \tan\alpha \quad \tan\beta};$$

$$\cot(\alpha \pm \beta) = \frac{\cot\alpha \quad \cot\beta \mp 1}{\pm \cot\alpha + \cot\beta}$$

$$\sin\alpha + \sin\beta = 2\sin\frac{\alpha+\beta}{2} \quad \cos\frac{\alpha-\beta}{2}$$

$$\sin\alpha - \sin\beta = 2\cos\frac{\alpha+\beta}{2} \quad \sin\frac{\alpha-\beta}{2}$$

$$\cos\alpha + \cos\beta = 2\cos\frac{\alpha+\beta}{2} \quad \cos\frac{\alpha-\beta}{2}$$

$$\cos\alpha - \cos\beta = -2\sin\frac{\alpha+\beta}{2} \quad \sin\frac{\alpha-\beta}{2}$$

$$\tan\alpha \pm \tan\beta = \frac{\sin(\alpha \pm \beta)}{\cos\alpha \quad \cos\beta}$$

$$\cot\alpha \pm \cot\beta = \frac{\sin(\beta \pm \alpha)}{\sin\alpha \quad \sin\beta}$$

$$\sin\alpha \quad \cos\beta = \frac{1}{2}\sin(\alpha + \beta) + \frac{1}{2}\sin(\alpha - \beta)$$

$$\cos\alpha \quad \cos\beta = \frac{1}{2}\cos(\alpha + \beta) + \frac{1}{2}\cos(\alpha - \beta)$$

$$\sin\alpha \quad \sin\beta = \frac{1}{2}\cos(\alpha - \beta) - \frac{1}{2}\cos(\alpha + \beta)$$

$$\tan\alpha \quad \tan\beta = \frac{\tan\alpha + \tan\beta}{\cot\alpha + \cot\beta} = -\frac{\tan\alpha - \tan\beta}{\cot\alpha - \cot\beta}$$

$$\cot\alpha \quad \cot\beta = \frac{\cot\alpha + \cot\beta}{\tan\alpha + \tan\beta} = -\frac{\cot\alpha - \cot\beta}{\tan\alpha - \tan\beta}$$

$$\cot\alpha \quad \tan\beta = \frac{\cot\alpha + \tan\beta}{\tan\alpha + \cot\beta} = -\frac{\cot\alpha - \tan\beta}{\tan\alpha - \cot\beta}$$

Source: From Gieck.[2]

Basic equations

Sine function	$y = A \sin(k\alpha - \varphi)$
Cosine function	$y = A \cos(k\alpha - \varphi)$

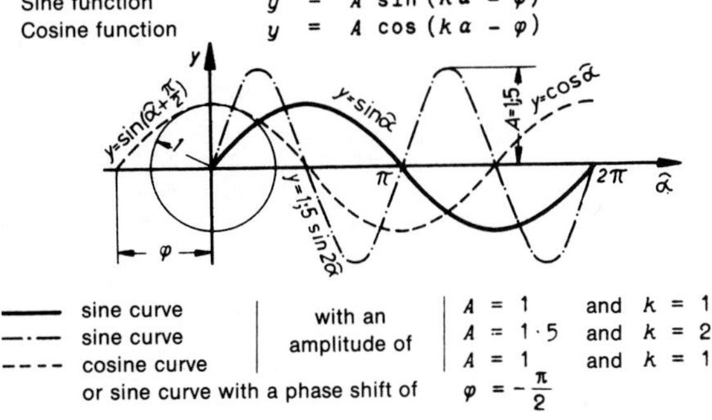

FIGURE 3.16 Relations between sine and cosine functions. (*From Gieck.*[2])

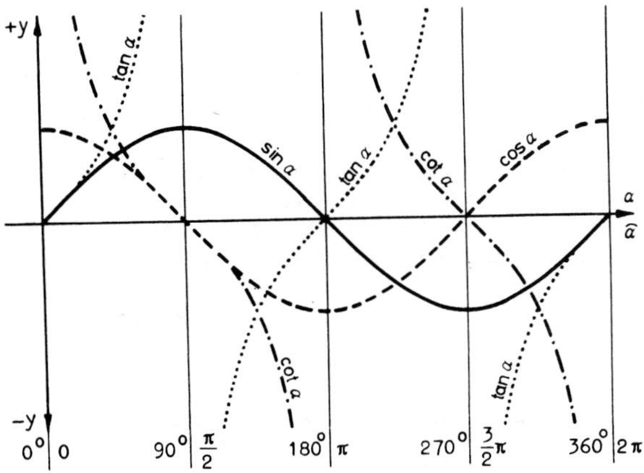

FIGURE 3.17 Relations between trigonometric functions. (*From Gieck.*[2])

In regard to inverse hyperbolic functions: If $x = \sinh y$, then y is the inverse hyperbolic sine of x written

$$y = \sinh^{-1} x \qquad \text{or} \qquad \text{arcsinh}\, x \qquad \text{etc.}$$

DIFFERENTIAL AND INTEGRAL CALCULUS

1. Differential Calculus*

Derivative. The derivative of a function $y = f(x)$ of a single variable x is

$$\frac{dy}{dx} = \lim_{\Delta x \to 0} \frac{\Delta y}{\Delta x} = \lim_{\Delta x \to 0} \frac{f(x + \Delta x) - f(x)}{\Delta x} = f'(x)$$

*This section taken partly from Ref. 3, *Mechanical Design and Systems Handbook*, 2d ed., by H. A. Rothbart. Copyright © 1985. Used by permission of McGraw-Hill, Inc. All rights reserved.

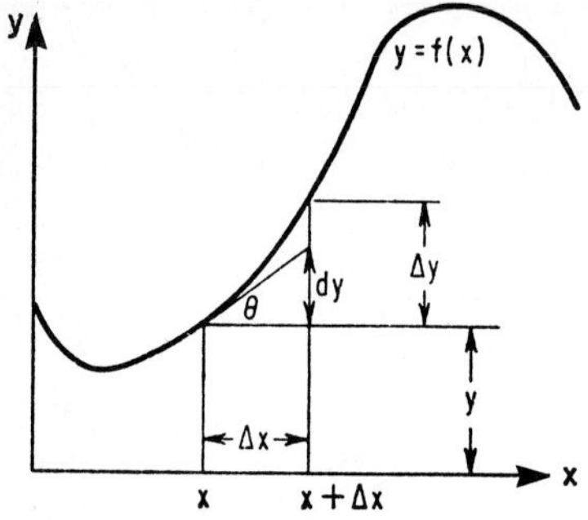

FIGURE 3.18 (*From Rothbart.*[3])

i.e., the derivative of the function y is the limit of the ratio of the increment of the function y to the increment of the independent variable x as the increment of x varies and approaches zero as a limit. The derivative at a point can also be shown to equal the slope of the tangent line to the curve at the same point; i.e., $dy/dx = \tan\alpha$, as shown in Fig. 3.18. The derivative of $f(x)$ is a function of x and may also be differentiated with respect to x. The first differentiation of the first derivative yields the second derivative of the function d^2y/dx^2 or $f''(x)$. Similarly, the third derivative d^3y/dx^3 or $f'''(x)$ of the function is the first derivative of d^2y/dx^2, and so on.

Conditions of Maxima and Minima. From Fig. 3.18 it is seen that the function $f(x)$ possesses a maximum value where the derivative is zero and the concavity is downward, and the function possesses a minimum value where the slope is zero and the curve has an upward concavity. If the function $f(x)$ is concave upward, the second derivative will have a positive value; if negative, the curve will be concave downward. If the second derivative equals zero at a point, that point is a point of inflection. More particularly, where the nature of the curve is not well known, Table 3.4 may be used to determine the significance of the derivatives.

TABLE 3.4 Significance of the Derivative of $f(x)$

$f'(x)\Big]_{x_0}$	$f''(x)\Big]_{x_0}$	$f'''(x)\Big]_{x_0}$	$f''''(x)\Big]_{x_0}$	Comment
0	<0			x_0 a maximum point
0	>0			x_0 a minimum point
0	0	$\neq 0$		x_0 a point of inflection
0	0	0	<0	x_0 a maximum point
0	0	0	>0	x_0 a minimum point

Source: From Rothbart.[3]

The *partial derivative* of a function $u = u(x,y)$ of two variables, taken with respect to the variable x, is defined by

$$\frac{\partial u}{\partial x} = \lim_{\Delta x \to 0} \frac{u(x + \Delta x, y) - u(x,y)}{\Delta x} = u_x$$

The partial derivatives are taken by differentiating with respect to one of the variables only, regarding the remaining variables as momentarily constant. Thus the partial derivative of u with respect to x of the function $u = 2xy^2$ is equal to $2y^2$; similarly $\partial u/\partial y = 4xy$. Higher derivatives are similarly formed. Thus

$$\frac{\partial^2 u}{\partial y^2} = u_{yy} = 4x \qquad \frac{\partial^2 u}{\partial x^2} = u_{xx} = 0$$

$$\frac{\partial^2 u}{\partial x \partial y} = u_{xy} = 4y \qquad \frac{\partial^2 u}{\partial y \partial x} = 4y$$

The order of differentiation in obtaining the mixed derivatives is immaterial if the derivatives are continuous.

Table 3.5 gives the conditions required to determine maxima, minima, and saddle points for $u = u(x,y)$.

Implicit Functions. If y is an implicit function of x, as, for example, in

$$xy - 5x^3y^2 = 3$$

and if it is difficult to solve the equation for y (or x), differentiate the terms as given, treating y as a function of x and solving for dy/dx. Thus, taking the derivatives of the above expression,

$$\frac{d}{dx}xy - \frac{d}{dx}5x^3y^2 = \frac{d}{dx}3$$

$$x\frac{dy}{dx} + y - 5x^3y^2\frac{dy}{dx} - y^2 15x^2 = 0$$

$$\frac{dy}{dx} = \frac{15^2y^2 - y}{x - 10x^3y}$$

Another approach utilizes the relationship $dy/dx = -f_x/f_y$. Thus, in the foregoing example,

$$f_x = y - y^2 15x^2 \qquad f_y = x - 10x^3y$$

Curvature. The curvature K at a point P of a curve $y = f(x)$ (Fig. 3.19) is

$$K = \left|\lim_{\Delta s \to 0} \frac{\Delta \gamma}{\Delta s}\right| = \frac{d\gamma}{ds}$$

A working expression for the curvature is

$$K = \frac{y''}{[1 + (y'^2)]^{3/2}}$$

where the derivatives are evaluated at the point P. For a curve described in polar coordinates, the corresponding expression for the curvature is

$$K = \frac{\rho^2 + 2\rho'^2 - \rho\rho''}{(\rho^2 + \rho'^2)^{3/2}}$$

where ρ' and ρ'' represent the first and second derivatives of ρ with respect to θ.

The circle of curvature tangent, on its concave side, is curve $y = f(x)$ at P. The

TABLE 3.5 Significance of the Derivative of $u(x, y)$

$\left.\frac{\partial u}{\partial x}\right]_{x_0,y_0}$	$\left.\frac{\partial u}{\partial y}\right]_{x_0,y_0}$	$\left.\frac{\partial u^2}{\partial x^2}\right]_{x_0,y_0}$	$\left.\frac{\partial^2 u}{\partial y^2}\right]_{x_0,y_0}$	Comments
0	0	<0	<0	**$u(x,y)$ maximum at x_0, y_0 if $(u_{xx})(u_{yy}) - (u_{xy})^2 > 0$**
0	0	>0	>0	**$u(x,y)$ minimum at x_0, y_0 if $(u_{xx})(u_{yy}) - (u_{xy})^2 > 0$**
0	0			**Saddle point if u_{xx} and u_{yy} are of different sign**

Source: From Rothbart.[3]

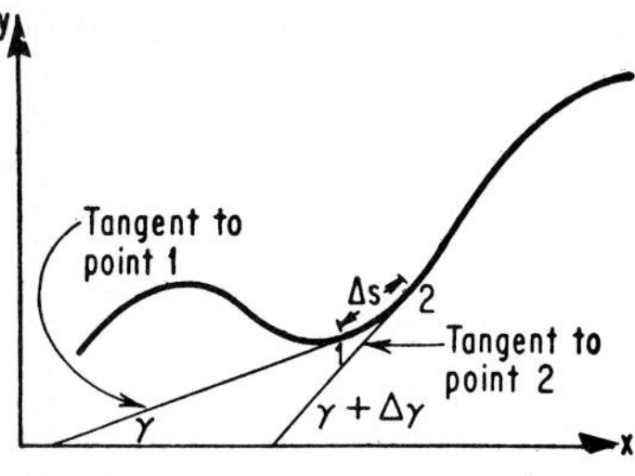

FIGURE 3.19 *(From Rothbart.[3])*

circle of curvature and the given curve have equal curvature at P. The circle of curvature is described by a radius of curvature R located at the center of curvature:

$$R = \frac{1}{K} = \frac{(1 + y'^2)^{3/2}}{y''}$$

The center of curvature is located at $(\bar{\alpha}, \bar{\beta})$ given by the following expressions:

$$\bar{\alpha} = x - \frac{y'(1 + y'^2)}{y''} \qquad \bar{\beta} = y + \frac{(1 + y'^2)}{y''}$$

where the expressions are evaluated at the point P. Table 3.6 shows derivatives of functions.

Differentials. The differential of a function is equal to the derivative of the function multiplied by the differential of the independent variable. Thus,

$$dy = \frac{dy}{dx}\,dx = f'(x)dx$$

The total differential dz of a function of two variables $z = z(x,y)$ is

$$dz = \frac{\partial z}{\partial x}dx + \frac{\partial z}{\partial y}dy$$

$$\frac{dz}{dt} = \frac{\partial z}{\partial x}\frac{dx}{dt} + \frac{\partial z}{\partial y}\frac{dy}{dt}$$

For a function of three variables $u = u(x,y,z)$,

$$du = \frac{\partial u}{\partial x}dx + \frac{\partial u}{\partial y}dy + \frac{\partial u}{\partial z}dz$$

$$\frac{du}{dt} = \frac{\partial u}{\partial x}\frac{dx}{dt} + \frac{\partial u}{\partial y}\frac{dy}{dt} + \frac{\partial u}{\partial z}\frac{dz}{dt}$$

x, y, z being functions of the independent variable t.

TABLE 3.6 List of Derivatives

(Note: u, v, and w are functions of x)

$$\frac{d}{dx}(a) = 0 \qquad a = \text{constant}$$

$$\frac{d}{dx}(x) = 1$$

$$\frac{dy}{dx} = \frac{dy}{dv}\frac{dv}{dx} \qquad y = y(v)$$

$$\frac{d}{dx}(au) = a\frac{du}{dx}$$

$$\frac{dy}{dx} = \frac{1}{dx/dy} \qquad \text{if } \frac{dx}{dy} \neq 0$$

$$\frac{d}{dx}(\pm u \pm v \pm \cdots) = \pm\frac{du}{dx} \pm \frac{dv}{dx} \pm \cdots$$

$$\frac{d}{dx}(u^n) = nu^{n-1}\frac{du}{dx}$$

$$\frac{d}{dx}(uv) = u\frac{dv}{dx} + v\frac{du}{dx}$$

$$\frac{d}{dx}\frac{u}{v} = \frac{v\,du/dx - u\,dv/dx}{v^2}$$

$$\frac{d}{dx}(u^v) = vu^{v-1}\frac{du}{dx} + u^v \ln u\frac{dv}{dx}$$

$$\frac{d}{dx}(a^u) = a^u \ln a\frac{du}{dx}$$

$$\frac{d}{dx}(e^u) = e^u\frac{du}{dx}$$

$$\frac{d}{dx}(\ln u) = \frac{1}{u}\frac{du}{dx}$$

$$\frac{d}{dx}(\log_a u) = \frac{\log_a e}{u}\frac{du}{dx}$$

$$\frac{d}{dx}(\sin u) = \cos u\frac{du}{dx}$$

$$\frac{d}{dx}(\cos u) = -\sin u\frac{du}{dx}$$

$$\frac{d}{dx}(\tan u) = \sec^2 u\frac{du}{dx}$$

$$\frac{d}{dx}(\csc u) = -\csc u \cot u\frac{du}{dx}$$

$$\frac{d}{dx}(\sec u) = \sec u \tan u\frac{du}{dx}$$

$$\frac{d}{dx}(\cot u) = -\csc^2 u\frac{du}{dx}$$

$$\frac{d}{dx}(\text{vers } u) = \sin u\frac{du}{dx}$$

$$\frac{d}{dx}\sin^{-1} u = \frac{1}{\sqrt{1-u^2}}\frac{du}{dx} \qquad \frac{-\pi}{2} \leqq \sin^{-1} u \leqq \frac{\pi}{2}$$

$$\frac{d}{dx}\cos^{-1} u = -\frac{1}{\sqrt{1-u^2}}\frac{du}{dx} \qquad 0 \leqq \cos^{-1} u \leqq \pi$$

$$\frac{d}{dx}\tan^{-1} u = \frac{1}{1+u^2}\frac{du}{dx}$$

Source: From Rothbart.[3]

TABLE 3.6 List of Derivatives (*Continued*

$\frac{d}{dx}\sinh^{-1} u = \frac{1}{\sqrt{u^2+1}}\frac{du}{dx}$	
$\frac{d}{dx}\cosh^{-1} u = \frac{1}{\sqrt{u^2-1}}\frac{du}{dx}$	$u > 1$
$\frac{d}{dx}\tanh^{-1} u = \frac{1}{1-u^2}\frac{du}{dx}$	
$\frac{d}{dx}\operatorname{csch}^{-1} u = -\frac{1}{u\sqrt{u^2+1}}\frac{du}{dx}$	
$\frac{d}{dx}\operatorname{sech}^{-1} u = -\frac{1}{u\sqrt{1-u^2}}\frac{du}{dx}$	$u > 0$
$\frac{d}{dx}\coth^{-1} u = \frac{1}{1-u^2}\frac{du}{dx}$	
$\frac{d}{dx}\csc^{-1} u = -\frac{1}{u\sqrt{u^2-1}}\frac{du}{dx}$	$-\pi < \csc^{-1} u \leqq -\frac{\pi}{2}, 0 < \csc^{-1} u \leqq \frac{\pi}{2}$
$\frac{d}{dx}\sec^{-1} u = \frac{1}{u\sqrt{u^2-1}}\frac{du}{dx}$	$-\pi \leqq \sec^{-1} u < -\frac{\pi}{2}, 0 \leqq \sec^{-1} u < \frac{\pi}{2}$
$\frac{d}{dx}\cot^{-1} u = \frac{-1}{1+u^2}\frac{du}{dx}$	
$\frac{d}{dx}\operatorname{vers}^{-1} u = \frac{1}{\sqrt{2u-u^2}}\frac{du}{dx}$	$0 \leqq \operatorname{vers}^{-1} u \leqq \pi$
$\frac{d}{dx}\sinh u = \cosh u \frac{du}{dx}$	
$\frac{d}{dx}\cosh u = \sinh u \frac{du}{dx}$	
$\frac{d}{dx}\tanh u = \operatorname{sech}^2 u \frac{du}{dx}$	
$\frac{d}{dx}\operatorname{csch} u = -\operatorname{csch} u \coth u \frac{du}{dx}$	
$\frac{d}{dx}\operatorname{sech} u = -\operatorname{sech} u \tanh u \frac{du}{dx}$	
$\frac{d}{dx}\coth u = -\operatorname{csch}^2 u \frac{du}{dx}$	

In the following relationships for differential of arc in rectangular coordinates, *ds* represents the differential of arc and β is the angle of the tangent drawn at the point in question; i.e., tanβ = slope:

$$ds^2 = dx^2 + dy^2$$

$$ds = \left[1 + \left(\frac{dy}{dx}\right)^2\right]^{1/2} dx = \left[1 + \left(\frac{dx}{dy}\right)^2\right]^{1/2} dy$$

$$\frac{dx}{ds} = \cos\beta = \frac{1}{(1 + y'^2)^{1/2}} \qquad \frac{dy}{ds} = \sin\beta = \frac{y'}{(1 + y'^2)^{1/2}}$$

In the following relationships for differential of arc in polar coordinates for the function $\rho = \rho(\theta)$, *ds* represents the differential of arc:

$$ds = \sqrt{d\rho^2 + \rho^2 d\theta^2} \qquad ds = \left[\rho^2 + \left(\frac{d\rho}{d\theta}\right)^2\right]^{1/2} d\theta$$

Indeterminate Forms. The function $f(x) = u(x)/v(x)$ has an indeterminate form 0/0 at $x = a$ if $u(x)$ and $v(x)$ each approach zero as x approaches a through values greater than $a(x \to a+)$. The function $f(x)$ is not defined at $x = a$, and therefore it is often useful to assign a value to $f(a)$. L'Hôpital's rule is readily applied to indeterminacies of the form 0/0:

$$\lim_{x \to a^+} \frac{u(x)}{v(x)} = \lim_{x \to a^+} \frac{u'(x)}{v'(x)}$$

L'Hôpital's rule may be reapplied as often as necessary, but it is important to remember to differentiate numerator and denominator separately. The above discussion is equally valid if $x \to a-$.

Other indeterminate forms such as ∞/∞, $0 \cdot \infty$, $\infty - \infty$, 0^0, ∞^0, and 1^∞ may also be evaluated by L'Hôpital's rule by changing their forms. For example, in order to evaluate the indeterminate form $0 \cdot \infty$, the function $u(x)v(x)$ may be written $u(x)/[1/v(x)]$ and the same technique employed as before.

2. Expansion in Series*. The range of values of x for which each of the series is convergent is stated at the right of the series. Arithmetic and geometrical series and the binomial theorem were given earlier, near the beginning of the chapter.

Exponential and Logarithmic Series

$$e^x = 1 + \frac{x}{1!} + \frac{x^2}{2!} + \frac{x^3}{3!} + \frac{x^4}{4!} + \cdots \qquad [-\infty < x < +\infty]$$

$$a^x = e^{mx} = 1 + \frac{m}{1!}x + \frac{m^2}{2!}x^2 + \frac{m^3}{3!}x^3 + \cdots \qquad [a > 0, -\infty < x < +\infty]$$

where $m = \ln a = (2.302\,6 \times \log_{10} a)$

$$\ln(1 + x) = x - \frac{x^2}{2} + \frac{x^3}{3} - \frac{x^4}{4} + \frac{x^5}{5} + \cdots \qquad [-1 < x < +1]$$

$$\ln(1 - x) = -x - \frac{x^2}{2} - \frac{x^3}{3} - \frac{x^4}{4} - \frac{x^5}{5} - \cdots \qquad [-1 < x < +1]$$

$$\ln\left(\frac{1 + x}{1 - x}\right) = 2\left(x + \frac{x^3}{3} + \frac{x^5}{5} + \frac{x^7}{7} + \cdots\right) \qquad [-1 < x < +1]$$

$$\ln\left(\frac{x + 1}{x - 1}\right) = 2\left(\frac{1}{x} + \frac{1}{3x^3} + \frac{1}{5x^5} + \frac{1}{7x^7} + \cdots\right) \qquad [x < -1 \text{ or } +1 < x]$$

$$\ln x = 2\left[\frac{x - 1}{x + 1} + \frac{1}{3}\left(\frac{x - 1}{x + 1}\right)^3 + \frac{1}{5}\left(\frac{x - 1}{x + 1}\right)^5 + \cdots\right] \qquad [0 < x < \infty]$$

*This section partly is taken from Ref. 1, *Marks' Standard Handbook for Mechanical Engineers*, 9th ed., by E. A. Avallone and T. Baumeister III (eds.). Copyright © 1987. Used by permission of McGraw-Hill, Inc. All rights reserved.

$$\ln(a+x) = \ln a + 2\left[\frac{x}{2a+x} + \frac{1}{3}\left(\frac{x}{2a+x}\right)^3 + \frac{1}{5}\left(\frac{x}{2a+x}\right)^5 + \cdots\right] \qquad [0 < a < +\infty, -a < x < +\infty]$$

Series for the Trigonometric Functions

$$\sin x = x - \frac{x^3}{3!} + \frac{x^5}{5!} - \frac{x^7}{7!} + \cdots \qquad [-\infty < x < +\infty]$$

$$\cos x = 1 - \frac{x^2}{2!} + \frac{x^4}{4!} - \frac{x^6}{6!} + \frac{x^8}{8!} - \cdots \qquad [-\infty < x < +\infty]$$

$$\tan x = x + \frac{x^3}{3} + \frac{2x^3}{15} + \frac{17x^7}{315} + \frac{62x^9}{2835} + \cdots \qquad [-\pi/2 < x < +\pi/2]$$

$$\cot x = \frac{1}{x} - \frac{x}{3} - \frac{x^3}{45} - \frac{2x^5}{945} - \frac{x^7}{4725} - \cdots \qquad [-\pi < x < +\pi]$$

$$\sin^{-1} y = y + \frac{y^3}{6} + \frac{3y^5}{40} + \frac{5y^7}{112} + \cdots \qquad [-1 \le y \le +1]$$

$$\tan^{-1} y = y - \frac{y^3}{3} + \frac{y^5}{5} - \frac{y^7}{7} + \cdots \qquad [-1 \le y \le +1]$$

$$\cos^{-1} y = \tfrac{1}{2}\pi - \sin^{-1} y \qquad \cot^{-1} y = \tfrac{1}{2}\pi - \tan^{-1} y$$

In these formulas, *all angles must be expressed in radians*. If D = the number of degrees in the angle, and x = its radian measure, then $x = 0.017\,453D$.

Series for the Hyperbolic Functions

$$\sinh x = x + \frac{x^3}{3!} + \frac{x^5}{5!} + \frac{x^7}{7!} + \cdots \qquad [-\infty < x < \infty]$$

$$\cosh x = 1 + \frac{x^2}{2!} + \frac{x^4}{4!} + \frac{x^6}{6!} + \cdots \qquad [-\infty < x < \infty]$$

$$\sinh^{-1} y = y - \frac{y^3}{6} + \frac{3y^5}{40} - \frac{5y^7}{112} + \cdots \qquad [-1 < y < +1]$$

$$\tanh^{-1} y = y + \frac{y^3}{3} + \frac{y^5}{5} + \frac{y^7}{7} + \cdots \qquad [-1 < y < +1]$$

General Formulas of Maclaurin and Taylor. If $f(x)$ and all its derivatives are continuous in the neighborhood of the point $x = 0$ (or $x = a$), then, for any value of x in this neighborhood, the function $f(x)$ may be expressed as a power series arranged according to ascending powers of x (or of $x - a$), as follows:

$$f(x) = f(0) + \frac{f'(0)}{1!}x + \frac{f''(0)}{2!}x^2 + \frac{f'''(0)}{3!}x^3 + \cdots + \frac{f^{(n-1)}(0)}{(n-1)!}x^{n-1} + (P_n)x^n \quad \text{(Maclaurin)}$$

$$f(x) = f(a) + \frac{f'(a)}{1!}(x - a) + \frac{f''(a)}{2!}(x - a)^2 + \frac{f'''(a)}{3!}(x - a)^3 + \cdots$$

$$+ \frac{f^{(n-1)}(a)}{(n - 1)!}(x - a)^{n-1} + (Q_n)(x - a)^n \text{ (Taylor)}$$

Here $(P_n)x^n$, or $(Q_n)(x - a)^n$, is called the *remainder term*; the values of the coefficients P_n and Q_n may be expressed as follows:

$$P_n = \frac{[f^{(n)}(sx)]}{n!} = \frac{(1 - t)^{n-1} f^{(n)}(tx)}{(n - 1)!}$$

$$Q_n = \frac{f^{(n)}[a + s(x - a)]}{n!} = \frac{(1 - t)^{n-1} f^{(n)}[a + t(x - a)]}{(n - 1)!}$$

where s and t are certain unknown numbers between 0 and 1; the s form is due to Lagrange, the t form to Cauchy.

The error due to neglecting the remainder term is less than $(\overline{P}_n)x^n$, or $(\overline{Q}_n)(x - a)^n$, where $\overline{P}_n$, or $\overline{Q}_n$, is the largest value taken on by P_n, or Q_n, when s or t ranges from 0 to 1. If this error, which depends on both n and x, approaches 0 as n increases (for any given value of x), then the general expression with remainder becomes (for that value of x) a convergent infinite series.

The sum of the first few terms of Maclaurin's series gives a good approximation to $f(x)$ for values of x near $x = 0$; Taylor's series gives a similar approximation for values near $x = a$.

Fourier's Series. Let $f(x)$ be a function which is finite in the interval from $x = -c$ to $x = +c$ and whose graph has finite arc length in that interval (see note below). Then, for any value of x between $-c$ and c,

$$f(x) = \tfrac{1}{2}a_0 + a_1 \cos\frac{\pi x}{c} + a_2 \cos\frac{2\pi x}{c} + a_3 \cos\frac{3\pi x}{c} + \cdots + b_1 \sin\frac{\pi x}{c}$$

$$+ b_2 \sin\frac{2\pi x}{c} + b_3 \sin\frac{3\pi x}{c} + \cdots$$

where the constant coefficients are determined as follows:

$$a_n = \frac{1}{c}\int_{-c}^{c} f(t) \cos\frac{n\pi t}{c}\,dt \qquad b_n = \frac{1}{c}\int_{-c}^{c} f(t) \sin\frac{n\pi t}{c}\,dt$$

In case the curve $y = f(x)$ is symmetrical with respect to the origin, the a's are all zero, and the series is a sine series. In case the curve is symmetrical with respect to the y axis, the b's are all zero, and a cosine series results. (In this case, the series will be valid not only for values of x between $-c$ and c, but also for $x = -c$ and $x = c$.) A Fourier series can always be integrated term by term; but the result of differentiating term by term may not be a convergent series.

Note. If $x = x_0$ is a point of discontinuity, $f(x_0)$ is to be defined as $\tfrac{1}{2}[f_1(x_0) + f_2(x_0)]$, where $f_1(x_0)$ is the limit of $f(x)$ when x approaches x_0 from below, and $f_2(x_0)$ is the limit of $f(x)$ when x approaches x_0 from above.

3. Integral Calculus

Indefinite Integrals. An integral of $f(x)\,dx$ is any function whose differential is $f(x)\,dx$, and is denoted by $\int f(x)\,dx$. All the integrals of $f(x)\,dx$ are included in the expression $\int f(x)\,dx + C$, where $\int f(x)\,dx$ is any particular integral and C is an arbitrary constant. The process of finding (when possible) an integral of a given function consists in recognizing by inspection a function which, when differentiated, will produce the given function, or in transforming the given function into a form in which such recognition is easy. The most common integrable forms are collected in Table 3.7.

Definite Integrals. The definite integral of $f(x)\,dx$ from $x = a$ to $x = b$ is denoted by

$$\int_a^b f(x)\,dx$$

The fundamental theorem for the evaluation of a definite integral is the following:

$$\int_a^b f(x)\,dx = \int f(x)\,dx\Big|_{x=b} - \int f(x)\,dx\Big|_{x=a}$$

i.e., the definite integral is equal to the difference between two values of any one of the indefinite integrals of the function in question. In other words, the limit of a sum can be found whenever the function can be integrated.

Properties of definite integrals:

$$\int_a^b = -\int_b^a$$

$$\int_a^c + \int_c^b = \int_a^b$$

The mean value of $f(x)$ between a and b:

$$\bar{f} = \frac{1}{b-a}\int_a^b f(x)\,dx$$

If b is variable, then

$$\frac{d}{db}\int_a^b f(x)\,dx = f(b)$$

If c is a parameter, then

$$\frac{\partial}{\partial c}\int_a^b f(x, c)\,dx = \int_a^b \frac{\partial f(x, c)}{\partial c}\,dx$$

The following definite integrals have received special names, and their values are tabulated in standard references:

1. Elliptic integral of the first kind:

$$F(u, k) = \int_0^u \frac{dx}{\sqrt{1 - k^2 \sin^2 x}} \qquad \text{when } k^2 < 1$$

TABLE 3.7 List of Most Common Integrals

$$\int a\,du = a\int du = au + C.$$

$$\int (u + v)\,dx = \int u\,dx + \int v\,dx.$$

$$\int u\,dv = uv - \int v\,du. \qquad \text{(integration by parts)}$$

$$\int dy \int f(x, y)\,dx = \int dx \int f(x, y)\,dy.$$

$$\int x^n\,dx = \frac{x^{n+1}}{n+1} + C, \text{ when } n \neq -1.$$

$$\int \frac{dx}{x} = \ln x + C = \ln cx.$$

$$\int e^x\,dx = e^x + C.$$

$$\int \sin x\,dx = -\cos x + C.$$

$$\int \cos x\,dx = \sin x + C.$$

$$\int \frac{dx}{\sin^2 x} = -\cot x + C.$$

$$\int \frac{dx}{\cos^2 x} = \tan x + C.$$

$$\int \frac{dx}{\sqrt{1 - x^2}} = \sin^{-1} x + C = -\cos^{-1} x + c.$$

$$\int \frac{dx}{1 + x^2} = \tan^{-1} x + C = -\cot^{-1} x + c.$$

$$\int (a + bx)^n\,dx = \frac{(a + bx)^{n+1}}{(n+1)b} + C.$$

$$\int \frac{dx}{a + bx} = \frac{1}{b}\ln(a + bx) + C = \frac{1}{b}\ln c(a + bx).$$

$$\int \frac{ax}{(a + bx)^2} = -\frac{1}{b(a + bx)} + C.$$

$$\int \frac{dx}{1 - x^2} = \tfrac{1}{2}\ln\frac{1 + x}{1 - x} + C$$
$$= \tanh^{-1} x + C, \text{ when } x < 1.$$

$$\int \frac{dx}{x^2 - 1} = \tfrac{1}{2}\ln\frac{x - 1}{x + 1} + C$$
$$= -\coth^{-1} x + C, \text{ when } x > 1.$$

23. $$\int \frac{dx}{a + 2bx + cx^2} = \frac{1}{\sqrt{ac - b^2}}\tan^{-1}\frac{b + cx}{\sqrt{ac - b^2}} + C \qquad [ac - b^2 > 0].$$

$$= \frac{1}{2\sqrt{b^2 - ac}}\ln\frac{\sqrt{b^2 - ac} - b - cx}{\sqrt{b^2 - ac} + b + cx} + C$$
$$= -\frac{1}{\sqrt{b^2 - ac}}\tanh^{-1}\frac{b + cx}{\sqrt{b^2 - ac}} + C \qquad [b^2 - ac > 0].$$

TABLE 3.7 List of Most Common Integrals (*Continued*)

$$\int \sqrt{a+bx}\,dx = \frac{2}{3b}(\sqrt{a+bx})^3 + C.$$

$$\int \frac{dx}{\sqrt{a+bx}} = \frac{2}{b}\sqrt{a+bx} + C.$$

$$\int \frac{(m+nx)\,dx}{\sqrt{a+bx}} = \frac{2}{3b^2}(3mb - 2an + nbx)\sqrt{a+bx} + C.$$

$$\int \sqrt{a+2bx+cx^2}\,dx = \frac{b+cx}{2c}\sqrt{a+2bx+cx^2} + \frac{ac-b^2}{2c}\int \frac{dx}{\sqrt{a+2bx+cx^2}} + C.$$

$$\int a^x\,dx = \frac{a^x}{\ln a} + C.$$

$$\int x^n e^{ax}\,dx = \frac{x^n e^{ax}}{a}\left[1 - \frac{n}{ax} + \frac{n(n-1)}{a^2x^2} - \cdots \pm \frac{n!}{a^n x^n}\right] + C.$$

$$\int \ln x\,dx = x\ln x - x + C.$$

$$\int \frac{\ln x}{x^2}\,dx = -\frac{\ln x}{x} - \frac{1}{x} + C.$$

$$\int \frac{(\ln x)^n}{x}\,dx = \frac{1}{n+1}(\ln x)^{n+1} + C.$$

$$\int \sin^2 x\,dx = -\tfrac{1}{4}\sin 2x + \tfrac{1}{2}x + C = -\tfrac{1}{2}\sin x\cos x + \tfrac{1}{2}x + C.$$

$$\int \cos^2 x\,dx = \tfrac{1}{4}\sin 2x + \tfrac{1}{2}x + C = \tfrac{1}{2}\sin x\cos x + \tfrac{1}{2}x + C.$$

$$\int \sin mx\,dx = -\frac{\cos mx}{m} + C.$$

$$\int \cos mx\,dx = \frac{\sin mx}{m} + C.$$

$$\int \sin mx\cos nx\,dx = -\frac{\cos(m+n)x}{2(m+n)} - \frac{\cos(m-n)x}{2(m-n)} + C.$$

$$\int \sin mx\sin nx\,dx = \frac{\sin(m-n)x}{2(m-n)} - \frac{\sin(m+n)x}{2(m+n)} + C.$$

$$\int \cos mx\cos nx\,dx = \frac{\sin(m-n)x}{2(m-n)} + \frac{\sin(m+n)x}{2(m+n)} + C.$$

$$\int \tan x\,dx = -\ln\cos x + C.$$

$$\int \cot x\,dx = \ln\sin x + C.$$

TABLE 3.7 List of Most Common Integrals (*Continued*)

$$\int \frac{dx}{\sin x} = \ln \tan \frac{x}{2} + C.$$

$$\int \frac{dx}{\cos x} = \ln \tan \left(\frac{\pi}{4} + \frac{x}{2}\right) + C.$$

$$\int \frac{dx}{1 + \cos x} = \tan \frac{x}{2} + C.$$

$$\int \frac{dx}{1 - \cos x} = -\cot \frac{x}{2} + C.$$

$$\int \sin x \cos x \, dx = ½ \sin^2 x + C.$$

$$\int \frac{dx}{\sin x \cos x} = \ln \tan x + C.$$

$$\int \frac{\cos x \, dx}{a + b \cos x} = \frac{x}{b} - \frac{a}{b} \int \frac{dx}{a + b \cos x} + C.$$

$$\int \frac{\sin x \, dx}{a + b \cos x} = -\frac{1}{b} \ln (a + b \cos x) + C.$$

$$\int \frac{A + B \cos x + C \sin x}{a + b \cos x + c \sin x} dx = A \int \frac{dy}{a + p \cos y}$$
$$+ (B \cos u + C \sin u) \int \frac{\cos y \, dy}{a + p \cos y} - (B \sin u$$
$$- C \cos u) \int \frac{\sin y \, dy}{a + p \cos y}, \text{ where } b = p \cos u, c$$
$$= p \sin u \text{ and } x - u = y.$$

$$\int e^{ax} \sin bx \, dx = \frac{a \sin bx - b \cos bx}{a^2 + b^2} e^{ax} + C.$$

$$\int e^{ax} \cos bx \, dx = \frac{a \cos bx + b \sin bx}{a^2 + b^2} e^{ax} + C.$$

$$\int \sinh x \, dx = \cosh x + C.$$

$$\int \tanh x \, dx = \ln \cosh x + C.$$

$$\int \cosh x \, dx = \sinh x + C.$$

$$\int \coth x \, dx = \ln \sinh x + C.$$

2. Elliptic integral of the second kind:

$$E(u, k) = \int_0^u \sqrt{1 - k^2 \sin^2 x} \, dx \qquad \text{when } k^2 < 1$$

3, 4. Complete elliptic integrals of the first and second kinds: put $u = \pi/2$ in the two equations just given.

5. The probability integral

$$\frac{2}{\sqrt{\pi}} \int_0^x e^{-x^2} dx$$

6. The gamma function

$$\Gamma(n) = \int_0^\infty x^{n-1} e^{-x}\, dx$$

Multiple Integrals. Multiple integrals are of the form

$$\iint f(x, y) \qquad \iiint f(x, y, z) \qquad \text{etc.}$$

Two successive integrations, for example, an integration with respect to y holding x constant, and an integration with respect to x between constant limits, will yield the value for the double integral

$$\int_a^b \int_{y_1(x)}^{y_2(x)} f(x, y)dy\, dx$$

Similarly a triple integral is evaluated by three successive single integrations. The order of integration can be reversed if the function $f(x, y, \ldots)$ is continuous.

DIFFERENTIAL EQUATIONS

1. General. An ordinary differential equation is one which contains a single independent variable, or argument, and a single dependent variable, or function, with its derivatives of various orders. A partial differential equation is one which contains a function of several independent variables, and its partial derivatives of various orders. The order of a differential equation is the order of the highest derivative which occurs in it. A solution of a differential equation is any relation between the variables, which, when substituted in the given equation, will satisfy it. The general solution of an ordinary differential equation of the nth order will contain n arbitrary constants. A differential equation is usually said to be solved when the problem is reduced to simple quadratures, i.e., intergrations of the form $y = \int f(x)\, dx$.

2. Methods of Solving Ordinary Differential Equations

*Differential Equations of the First Order**

1. If possible, separate the variables; i.e., collect all the x's and dx on one side, and all the y's and dy on the other side; then integrate both sides, and add the constant of integration.

2. If the equation is homogeneous in x and y, the value of dy/dx in terms of x and y will be of the form $dy/dx = f(y/x)$. Substituting $y = xt$ will enable the variables to be separated.

Solution:

$$\ln x = \int \frac{dt}{f(t) - t} + C$$

3. The expression $f(x, y)\, dx + F(x, y)\, dy$ is an *exact differential* if

*This section taken in part from Ref. 1, *Marks' Standard Handbook for Mechanical Engineers*, 9th ed., by E. A. Avallone and T. Baumeister III (eds.). Copyright © 1987. Used by permission of McGraw-Hill, Inc. All rights reserved.

$$\frac{\partial f(x, y)}{\partial y} = \frac{\partial F(x, y)}{\partial x} \; (= P, \text{ say}).$$

In this case the solution of $f(x, y)\,dx + F(x, y)\,dy = 0$ is

$$\int f(x, y)\,dx + \int [F(x, y) - \int P\,dx]\,dy = C$$

or

$$\int F(x, y)\,dy + \int [f(x, y) - \int P\,dy]\,dx = C$$

4. Linear differential equation of the first order:

$$\frac{dy}{dx} + f(x) \cdot y = F(x)$$

Solution: $y = e^{-P}[\int e^{P}F(x)\,dx + C]$, where $P = \int f(x)\,dx$.

5. Bernoulli's equation:

$$\frac{dy}{dx} + f(x) \cdot y = F(x) \cdot y^{n}$$

Substituting $y^{1-n} = v$ gives $(dv/dx) + (1 - n)f(x) \cdot v = (1 - n)F(x)$, which is linear in v and x.

6. Clairaut's equation: $y = xp + f(p)$, where $p = dy/dx$. The solution consists of the family of lines given by $y = Cx + f(C)$, where C is any constant, together with the curve obtained by eliminating p between the equations $y = xp + f(p)$ and $x + f'(p) = 0$, where $f'(p)$ is the derivative of $f(p)$.

Differential Equations of the Second Order

7.

$$\frac{d^2y}{dx^2} = -n^2y$$

Solution: $y = C_1 \sin (nx + C_2)$ or $y = C_3 \sin nx + C_4 \cos nx$.

8.

$$\frac{d^2y}{dx^2} = +n^2y$$

Solution: $y = C_1 \sinh (nx + C_2)$ or $y = C_3e^{nx} + C_4e^{-nx}$

9.

$$\frac{d^2y}{dx^2} = f(y)$$

Solution:

$$x = \int \frac{dy}{\sqrt{C_1 + 2P}} + C_2$$

where $P = \int f(y)\,dy$

10.

$$\frac{d^2y}{dx^2} = f(x)$$

Solution:

$$y = \int P\,dx + C_1x + C_2,$$

where $P = \int f(x)dx$, *or*

$$y = xP - \int xf(x)\,dx + C_1x + C_2.$$

11.

$$\frac{d^2y}{dx^2} = f\left(\frac{dy}{dx}\right)$$

Putting

$$\frac{dy}{dx} = z \qquad \frac{d^2y}{dx^2} = \frac{dz}{dx} \qquad x = \int \frac{dz}{f(z)} + C_1 \qquad y = \int \frac{zdz}{f(z)} + C_2$$

Then eliminate z from these two equations.

12. The equation for damped vibration:

$$\frac{d^2y}{dx^2} + 2b\frac{dy}{dx} + a^2y = 0$$

CASE 1: If $a^2 - b^2 > 0$, let $m = \sqrt{a^2 - b^2}$. *Solution:* $y = C_1e^{-bx} \sin (mx + C_2)$ or $y = e^{-bx}[C_3 \sin (mx) + C_4 \cos (mx)]$.

CASE 2: If $a^2 - b^2 = 0$, solution is $y = e^{-bx}(C_1 + C_2x)$.

CASE 3: If $a^2 - b^2 < 0$, let $n = \sqrt{b^2 - a^2}$. *Solution:* $y = C_1e^{-bx} \sinh (nx + C_3)$ or $y = C_3e^{-(b+n)x} + C_4e^{-(b-n)x}$.

13.

$$\frac{d^2y}{dx^2} + 2b\frac{dy}{dx} + a^2y = c$$

Solution: $y = c/a^2 + y_1$, where y_1 = the solution of the corresponding equation with second member zero (see item 12 above).

14.

$$\frac{d^2y}{dx^2} + 2b\frac{dy}{dx} + a^2y = c \sin (kx)$$

Solution:

$$y = R \sin (kx - S) + y_1$$

where $R = c/\sqrt{(a^2 - k^2)^2 + 4b^2k^2}$, $\tan S = 2bk/(a^2 - k^2)$, and y_1 = the solution of the corresponding equation with second member zero (see item 12 above).

15.

$$\frac{d^2y}{dx^2} + 2b\frac{dy}{dx} + a^2y = f(x)$$

Solution: $y = y_0 + y_1$, where y_0 = any particular solution of the given equation, and y_1 = the general solution of the corresponding equation with second member zero (see item 12 above).

If $b^2 < a^2$,

$$y_0 = \frac{1}{2\sqrt{b^2 - a^2}}[e^{m_1x}\int e^{-m_1x}f(x)\,dx - e^{m_2x}\int e^{-m_2x}f(x)\,dx]$$

where $m_1 = -b + \sqrt{b^2 - a^2}$ and $m_2 = -b - \sqrt{b^2 - a^2}$.

If $b^2 < a^2$, let $m = \sqrt{a^2 - b^2}$; then

$$y_0 = \frac{1}{m}e^{-bx}[\sin(mx)\int e^{bx}\cos(mx)\cdot f(x)\,dx$$

$$- \cos(mx)\int e^{bx}\sin(mx)\cdot f(x)\,dx]$$

If $b^2 = a^2$, $y_0 = e^{-bx}[x\ e^{bx}f(x)\,dx - \int x\cdot e^{bx}f(x)\,dx]$.

Types 12 to 15 are examples of linear differential equations with constant coefficients. The solutions of such equations are often found most simply by the use of Laplace transforms.

Linear Equations (Constant Coefficients). For the linear equation of the nth order

$$\frac{A_n(x)d^ny}{dx^n} + \frac{A_{n-1}(x)d^{n-1}y}{dx^{n-1}} + \cdots + \frac{A_1(x)dy}{dx} + A_0(x)y = E(x)$$

the general solution is $y = u + c_1u_1 + c_2u_2 + \cdots + c_nu_n$. Here u, the particular integral, is any solution of the given equation, and $u_1, u_2, \ldots, u_n$ form a fundamental system of solutions of the homogeneous equation obtained by replacing $E(x)$ by zero.

To solve the homogeneous equation of the nth order $A_nd^ny/dx^n + A_{n-1}d^{n-1}y/dx^{n-1} + \cdots + A_1dy/dx + A_0y = 0$, $A_n \neq 0$, where $A_n, A_{n-1}, \ldots, A_0$ are constants, find the roots of the auxiliary equation

$$A_np^n + A_{n-1}p^{n-1} + \cdots + A_1p + A_0 = 0$$

For each simple real root r, there is a term ce^{rx} in the solution. The terms of the solution are to be added together. When r occurs twice among the n roots of the auxiliary equation, the corresponding term is $e^{rx}(c_1 + c_2x)$. When r occurs three times, the corresponding term is $e^{rx}(c_1 + c_2x + c_3x^2)$, and so forth. When there is a pair of conjugate complex roots $a + bi$ and $a - bi$, the real form of the terms in the solution is $e^{ax}(c_1 \cos bx + d_1 \sin bx)$. When the same pair occurs twice, the corresponding term is $e^{ax}[(c_1 + c_2x)\cos bx + (d_1 + d_2x)\sin bx]$, and so forth.

The general nonhomogeneous linear differential equation of order n, with constant coefficients, or

$$\frac{A_n d^n y}{dx^n} + \frac{A_{n-1} d^{n-1} y}{dx^{n-1}} + \cdots + \frac{A_1 dy}{dx} + A_0 y = E(x)$$

may be solved by adding any particular integral to the complementary function, or general solution, of the homogeneous equation obtained by replacing $E(x)$ by zero. The complementary function may be found from the rules just given. And the particular integral may be found assuming that $E(x)$ is a single term or a sum of terms each of which is of the type k, $k \cos bx$, $k \sin bx$, ke^{ax}, where a and b are any real numbers. (See Ref. 1 for more details.)

Solutions of Partial Differential Equations. The only means of solution that will be discussed is the separation-of-variables method. For the following equation,

$$f_1 \frac{\partial^2 z}{\partial x^2} + f_2 \frac{\partial z}{\partial x} + f_3 z + g_1 \frac{\partial^2 z}{\partial y^2} + g_2 \frac{\partial x}{\partial y} + g_3 z = 0$$

the solution of $z = z(x, y)$ will be assumed to have the form of a product of two functions $X(x)$ and $Y(y)$, which are functions of x and y only, respectively:

$$z = z(x, y) = X(x)Y(y)$$

Substituting $z = XY$ into the original differential equation, we obtain

$$-\frac{1}{X}\left[f_1 \frac{\partial^2 X}{\partial x^2} + f_2 \frac{\partial X}{\partial x} + f_3 X\right] = \frac{1}{Y}\left[g_1 \frac{\partial^2 Y}{\partial y^2} + g_2 \frac{\partial Y}{\partial y} + g_3 Y\right]$$

Note that the left-hand side contains the function of x only and that the right-hand side contains functions of y only. Since the right- and left-hand sides are independent of x and y, respectively, they must be equal to a common constant, called a separation constant α; thus,

$$f_1 \frac{d^2 X}{dx^2} + f_2 \frac{dX}{dx} + [f_3 + \alpha]X = 0$$

and

$$g_1 \frac{d^2 Y}{dy^2} + g_2 \frac{dY}{dy} + [g_3 - \alpha]Y = 0$$

Once the solutions to the above have been obtained, the product solution XY is obtained. The method outlined may be extended to additional variables and it is applicable also when f_1, f_2, f_3 are functions of x and g_1, g_2, and g_3 are functions of y.

LAPLACE TRANSFORMATION

The Laplace transformation of a function $f(t)$ is

$$F(s) = \mathcal{L}[f(t)] = \int_0^\infty e^{-st} f(t)\, dt$$

where $f(t)$ = a function of a real variable (usually t = time)
s = a complex variable of the form $(\sigma + j\omega)$
$F(s)$ = an equation expressed in the transform variable s, resulting from operating on a function of time with the Laplace integral
$\mathcal{L}$ = an operational symbol indicating that the quantity which it prefixes is to be transformed into the frequency domain

Example

$$f(t) = A \qquad f(t) = \int_0^\infty Ae^{-st}\,dt = \frac{A}{s}$$

Table 3.8 lists the transforms of common functions.

An inverse transformation is represented symbolically as

$$\mathcal{L}^{-1}F(s) = f(t)$$

For any $f(t)$ there is only one direct transform, $F(s)$. For any given $F(s)$ there is only one inverse transform $f(t)$. Therefore, tables are generally used for determining inverse transforms.

COMPLEX VARIABLES

1. Complex Numbers. A complex number z consists of a real part x and an imaginary part y and is represented as

$$z = x + iy$$

where $i = \sqrt{-1}(i^2 = -1)$

The conjugate $\bar{z}$ of a complex number is defined as

$$\bar{z} = x - iy$$

Two complex numbers are equal only if their real parts are equal and their imaginary parts are equal; i.e.,

$$x_1 + iy_1 = x_2 + iy_2$$

only if

$$x_1 = x_2 \qquad \text{and} \qquad y_1 = y_2$$

Also

$$x + iy = 0$$

only if

$$x = 0 \qquad \text{and} \qquad y = 0$$

Complex numbers satisfy the distributive, associative, and commutative laws of algebra.

TABLE 3.8 Laplace Transforms

$f(t)$	$F(s) = [f(t)]$
A	A/s
$f_1(t) + f_2(t)$	$F_1(s) + F_2(s)$
$e^{-\alpha t}$	$\dfrac{1}{s + \alpha}$
$\dfrac{1}{\tau} e^{-t/\tau}$	$\dfrac{1}{rs + 1}$
$Ae^{-\alpha t}$	$\dfrac{A}{s + \alpha}$
$\sin \beta t$	$\dfrac{\beta}{s^2 + \beta^2}$
$\cos \beta t$	$\dfrac{s}{s^2 + \beta^2}$
$\dfrac{1}{\beta} e^{-\alpha t} \sin \beta t$	$\dfrac{1}{s^2 + 2\alpha s + \alpha^2 + \beta^2}$
$\dfrac{e^{-\alpha t}}{\beta - \alpha} - \dfrac{e^{-\beta t}}{\beta - \alpha}$	$\dfrac{1}{(s + \alpha)(s + \beta)}$
$\dfrac{Ae^{-\alpha t} - Be^{-\beta t}}{C}$ where $A = a - \alpha$, $B = a - \beta$, $C = \beta - \alpha$	$\dfrac{s + a}{(s + \alpha)(s + \beta)}$
$\dfrac{e^{-\alpha t}}{A} + \dfrac{e^{-\beta t}}{B} + \dfrac{e^{-\delta t}}{C}$ where $A = (\beta - \alpha)(\delta - \alpha)$ $B = (\alpha - \beta)(\delta - \beta)$ $C = (\alpha - \delta)(\beta - \delta)$	$\dfrac{1}{(s + \alpha)(s + \beta)(s + \delta)}$
t	$\dfrac{1}{s^2}$
t^2	$\dfrac{2}{s^3}$
t^n	$\dfrac{n!}{s^{n+1}}$
$d/dt[f(t)]$	$sF(s) - f(0^+)$*
$d^2/dt^2[f(t)]$	$s^2F(s) - sf(0^+) - \dfrac{df}{dt}(0^+)$
$d^3/dt^3[f(t)]$	$s^2F(s) - s^2f(0^+) - s\,\dfrac{df(0^+)}{dt} - \dfrac{d^2f(0^+)}{dt^2}$
$\int f(t)dt$	$\dfrac{1}{s}[F(s) + \int f(t)dt\|0^+\|$
$\dfrac{1}{\alpha} \sinh \alpha t$	$\dfrac{1}{s^2 - \alpha^2}$
$\cosh \alpha t$	$s/(s^2 - \alpha^2)$

*$f(0^+)$ = initial value of $f(t)$, evaluated as t approaches zero from positive values.
Source: From Avallone and Baumeister.[1]

Complex numbers may be graphically represented on the $z(x - y)$ plane or in polar (r,θ) coordinates. The polar coordinates of a complex number are

$$r = \sqrt{x^2 + y^2} = |z| = \text{mod } z, r \geq 0$$

where mod = modulus, and

$$\theta = \tan^{-1}\frac{y}{x} = \arg z = \text{amp } z$$

where arg = argument and amp = amplitude. arg z is multiple-valued, but for an angular interval of range 2π there is only one value of θ for a given z.

2. Elementary Complex Functions

Polynomials. A polynomial in z, $a_n z^n + a_{n-1} z^{n-1} + \cdots + a_0$, where n is a positive integer, is simply a sum of complex numbers times integral powers of z which have already been defined. Every polynomial of degree n has precisely n complex roots provided each multiple root of multiplicity m is counted m times.

Exponential Functions. The exponential function e^z is defined by the equation $e^z = e^{x+iy} = e^x \cdot e^{iy} = e^x(\cos y + i \sin y)$. Properties: $e^0 = 1$; $e^{z1} \cdot e^{z2} = e^{z1+z2}$; $e^{z1}/e^{z2} = e^{z1-z2}$; $e^{z+2k\pi i} = e^z$.

Trigonometric Functions. $\sin z = (e^{iz} - e^{-iz})/2i$; $\cos z = (e^{iz} + e^{-iz})/2$; $\tan z = \sin z/\cos z$; $\cot z = \cos z/\sin z$; $\sec z = 1/\cos z$; $\cos z = 1/\sin z$. Fundamental identities for these functions are the same as their real counterparts. Thus $\cos^2 z + \sin^2 z = 1$, $\cos(z_1 \pm z_2) = \cos z_1 \cos z_2 \pm \sin z_1 \sin z_2$, $\sin(z_1 \pm z_2) = \sin z_1 \cos z_2 \pm \cos z_1 \sin z_2$. The sine and cosine of z are periodic functions of period 2π; thus $\sin(z + 2\pi) = \sin z$. For computation purposes, $\sin z = \sin (x + iy) = \sin x \cosh y + i \cos x \sinh y$, where $\sin x$, $\cosh y$, etc., are the real trigonometric and hyperbolic functions. Similarly, $\cos z = \cos x \cosh y - i \sin x \sinh y$. If $x = 0$ in the results given, $\cos iy = \cosh y$, $\sin iy = i \sinh y$.

VECTORS

1. Representation.* A vector quantity has magnitude and direction; a scalar quantity has magnitude only. Common vector quantities are acceleration, alternating currents, voltages, force, and velocity. A vector can be represented graphically by a straight line with an arrowhead, as in Fig. 3.20. (Length represents magnitude; direction is determined from the position of the line and the arrowhead.)

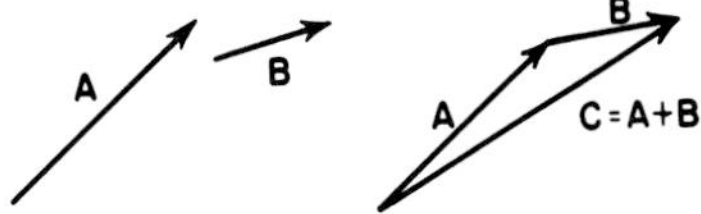

FIGURE 3.20 *(From Perry.*[4]*)*

Vectors are usually indicated by boldface type (**A**), or by an arrow over the symbol ($\vec{A}$), or by a bar ($\bar{\mathbf{A}}$).

A vector **V** in three dimensions can be represented by its projections along three mutually perpendicular lines, the x, y, and z axes. Vectors of unit magnitude, directed in the positive sense along these three axes, are denoted by **i**, **j**, and **k**, respectively.

*This section and the next one on algebra are partly taken from Ref. 4, *Engineering Manual*, 3d ed., by R. H. Perry (ed.). Copyright © 1976. Used by permission of McGraw-Hill, Inc. All rights reserved.

If a, b, and c represent the lengths of the projections of **V** along these axes, we may represent **V** as $\mathbf{V} = a\mathbf{j} + b\mathbf{j} + c\mathbf{k}$ (Fig. 3.20). The length (magnitude) of **V** is

$$\mathbf{V} = (a^2 + b^2 + c^2)^{1/2}$$

2. Algebra

Equality. $\mathbf{A} = \mathbf{B}$ if and only if both have the same magnitude and the same direction.

Addition and Subtraction. $\mathbf{A} + \mathbf{B} = \mathbf{B} + \mathbf{A}$ (commutative law); $\mathbf{A} + \mathbf{B} + \mathbf{C} = (\mathbf{A} + \mathbf{B}) + \mathbf{C} = \mathbf{A} + (\mathbf{B} + \mathbf{C})$ (associative law). If $\mathbf{A} = a_1\mathbf{i} + a_2\mathbf{j} + a_3\mathbf{k}$, $\mathbf{B} = b_1\mathbf{i} + b_2\mathbf{j} + b_3\mathbf{k}$, $\mathbf{A} \pm \mathbf{B} = (a_1 \pm b_1)\mathbf{i} + (a_2 \pm b_2)\mathbf{j} + (a_3 \pm b_3)\mathbf{k}$.

Product of Vector V and Scalar s. $s\mathbf{V} = \mathbf{V}s = (sa)\mathbf{i} + (sb)\mathbf{j} + (sc)\mathbf{k}$.

Scalar Product of Two Vectors $\mathbf{V}_1$, $\mathbf{V}_2$. The scalar (dot or inner) product, indicated by $\mathbf{V}_1 \cdot \mathbf{V}_2$, is a scalar defined by $\mathbf{V}_1 \cdot \mathbf{V}_2 = |\mathbf{V}_1||\mathbf{V}_2| \cos\theta$, where θ = angle between the vectors. $\mathbf{V}_1 \cdot \mathbf{V}_2 = a_1a_2 + b_1b_2 + c_1c_2$; $(\mathbf{V}_1 + \mathbf{V}_2) \cdot \mathbf{V}_3 = \mathbf{V}_1 \cdot \mathbf{V}_3 + \mathbf{V}_2 \cdot \mathbf{V}_3$; $\mathbf{V}_1 \cdot (\mathbf{V}_2 + \mathbf{V}_3) = \mathbf{V}_1 \cdot \mathbf{V}_2 + \mathbf{V}_1 \cdot \mathbf{V}_3 = (\mathbf{V}_2 + \mathbf{V}_3) \cdot \mathbf{V}_1$ (commutative); $\mathbf{i} \cdot \mathbf{i} = \mathbf{j} \cdot \mathbf{j} = \mathbf{k} \cdot \mathbf{k} = 1$; $\mathbf{i} \cdot \mathbf{j} = \mathbf{i} \cdot \mathbf{k} = \mathbf{j} \cdot \mathbf{k} = 0$.

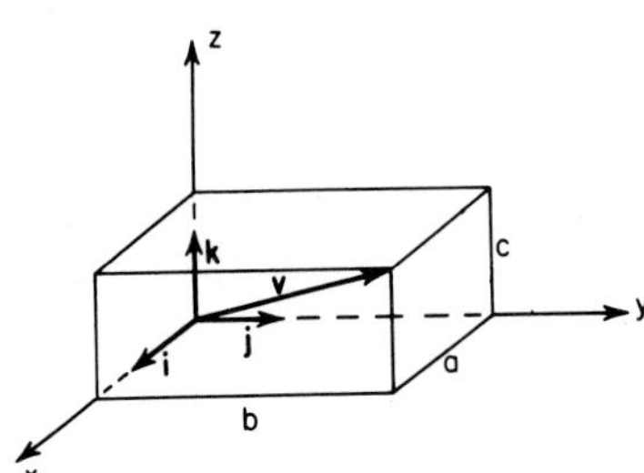

FIGURE 3.21 (*From Perry.*[4])

Vector Product. With reference to Fig. 3.21, the vector (outer) product of $\mathbf{V}_1$ and $\mathbf{V}_2$ is defined as the vector $\mathbf{V} = \mathbf{V}_1 \times \mathbf{V}_2$, $\mathbf{V}_1 \times \mathbf{V}_2 = (b_1c_2 - b_2c_1)\mathbf{i} + (c_1a_2 - c_2a_1)\mathbf{j} + (a_1b_2 - a_2b_1)\mathbf{k}$. $|\mathbf{V}| = |\mathbf{V}_1||\mathbf{V}_2| \sin\theta$; $\mathbf{V}_1 \times \mathbf{V}_2 = -\mathbf{V}_2 \times \mathbf{V}_1$; $\mathbf{V}_1 \times (\mathbf{V}_2 + \mathbf{V}_3) = \mathbf{V}_1 \times \mathbf{V}_2 + \mathbf{V}_1 \times \mathbf{V}_3$; $(\mathbf{V}_1 + \mathbf{V}_2) \times \mathbf{V}_3 = \mathbf{V}_1 \times \mathbf{V}_3 + \mathbf{V}_2 \times \mathbf{V}_3$; $\mathbf{i} \times \mathbf{i} = \mathbf{j} \times \mathbf{j} = \mathbf{k} \times \mathbf{k} = 0$; $\mathbf{i} = \mathbf{j} \times \mathbf{k} = -\mathbf{k} \times \mathbf{j}$; $\mathbf{j} = \mathbf{k} \times \mathbf{i} = -\mathbf{i} \times \mathbf{k}$; $\mathbf{k} = \mathbf{i} \times \mathbf{j} = -\mathbf{j} \times \mathbf{i}$; $\mathbf{V} \times \mathbf{V} = 0$.

Multiple Products. (1) $\mathbf{A}(\mathbf{B} \cdot \mathbf{C})$; here $\mathbf{B} \cdot \mathbf{C}$ is a scalar, so that $\mathbf{A}(\mathbf{B} \cdot \mathbf{C})$ is a vector parallel to **A**. Clearly, $\mathbf{A}(\mathbf{B} \cdot \mathbf{C}) \neq (\mathbf{A} \cdot \mathbf{B})\mathbf{C}$. (2) $\mathbf{A} \cdot (\mathbf{B} \times \mathbf{C}) = \mathbf{B} \cdot (\mathbf{C} \times \mathbf{A}) = \mathbf{C} \cdot (\mathbf{A} \times \mathbf{B})$. (3) $\mathbf{A} \times (\mathbf{B} \times \mathbf{C}) = \mathbf{B}(\mathbf{A} \cdot \mathbf{C}) - \mathbf{C}(\mathbf{A} \cdot \mathbf{B})$.

Vector Function. $\mathbf{r} = \mathbf{F}(x,y,z)$ gives a vector **r** as a function of scalars x, y, and z.

3. Differential Operations. By definition,

$$\nabla = \text{del} = \mathbf{i}\frac{\partial}{\partial x} + \mathbf{j}\frac{\partial}{\partial y} + \mathbf{k}\frac{\partial}{\partial z}$$

and

$$\nabla^2 = \text{Laplacian } \nabla^2 = \text{Laplacian} = \nabla \cdot \nabla = \frac{\partial^2}{\partial x^2} + \frac{\partial^2}{\partial y^2} + \frac{\partial^2}{\partial z^2}$$

$$\nabla s = \text{grad } s = \frac{\partial s}{\partial x}\mathbf{i} + \frac{\partial s}{\partial y}\mathbf{j} + \frac{\partial s}{\partial z}\mathbf{k},$$

the gradient of a scalar function $S(x,y,z)$. For a vector function $\mathbf{V}(x,y,z) = P\mathbf{i} + Q\mathbf{j} + R\mathbf{k}$, the divergence of **V** is

$$\nabla \cdot \mathbf{V} = \frac{\partial P}{\partial x} + \frac{\partial Q}{\partial y} + \frac{\partial R}{\partial z}$$

And the curl of **V** is

$$\nabla \times \mathbf{V} = \text{curl } \mathbf{V} = \text{rot } \mathbf{V} = \begin{vmatrix} \mathbf{i} & \mathbf{j} & \mathbf{k} \\ \frac{\partial}{\partial x} & \frac{\partial}{\partial y} & \frac{\partial}{\partial z} \\ P & Q & R \end{vmatrix}$$

The divergence theorem states that if **F** is a vector function and V is a volume bounded by a surface S, then

$$\iiint_V \text{div } \mathbf{F}\, dv = \iiint_V \nabla \cdot \mathbf{F}\, dv = \iint_S \mathbf{F} \cdot d\mathbf{S}$$

The integrations are to be carried out over the volume V and the surface S. And if **n** is the unit outward normal and $dS = |d\mathbf{S}|$ is the scalar element of surface, $d\mathbf{S} = \mathbf{n}\, dS$, $\mathbf{F} \cdot d\mathbf{S} = \mathbf{F} \cdot \mathbf{n}\, dS$.

Stokes' theorem states that if **F** is a vector function and S is a surface bounded by a simple closed curve **C**, then

$$\int_C \mathbf{F} \cdot d\mathbf{r} = \iint_S \text{curl } \mathbf{F} \cdot d\mathbf{S} = \iint_S (\nabla \times \mathbf{F}) \cdot d\mathbf{S}$$

Here, $d\mathbf{r} = dx\,\mathbf{i} + dy\,\mathbf{j} + dz\,\mathbf{k}$. The theorem implies that the surface integral of $(\nabla \times \mathbf{F})$ over any surface S which is bounded by C is equal to the line integral of **F** over the contour C, taken in the direction related to that of **n** (in $d\mathbf{S} = \mathbf{n}\, dS$) by the right-hand rule.

STATISTICS AND PROBABILITY

This section includes topics related to measurements and errors.

1. Error of Observation. The error of an observation is $e_i = m_i - m$, $i = 1, 2, \ldots, n$, where m_i are the observed values, e_i the errors, and m the mean value, that is, the arithmetic mean of a very large number (theoretically infinite) of observations.

In a large number of measurements, random errors are as often negative as positive and have little effect on the arithmetic mean. All other errors are classed as systematic. If due to the same cause, they affect the mean in the same sense and give it a definite bias.

Best Estimate and Measured Value. If all systematic errors have been eliminated, it is possible to consider the sample of individual repeated measurements of a quantity with a view to securing the "best" estimate of the mean value m and assessing the degree of reproducibility which has been obtained. The final result will then be expressed in the form $E \pm L$, where E is the best estimate of m, and L the characteristic limit of variation associated with a certain risk. The value measured is not merely E, but the entire result $E \pm L$.

The Arithmetic Mean. If a large number of measurements have been made to determine directly the mean m of a cerain quantity, all measurements having been made with equal skill and care, the best estimate of m from a sample of n is the arithmetic mean $\overline{m}$ of the measurements in the sample,

$$\overline{m} = \frac{1}{n} \sum_{i=1}^{n} m_i$$

2. Standard Deviation. Standard deviation is the root-mean-square of the deviations e_i of a set of observations from the mean,

$$\sigma = \left(\frac{1}{n} \sum_{i=1}^{n} e_i^2 \right)^{1/2}$$

Since neither the mean m nor the errors of observation e_i are ordinarily known, the deviations from the arithmetic mean, or the residuals, $x_i = m_i - \overline{m}$, $i = 1, 2, \ldots, n$, will be referred to as errors. Likewise, for the unbiased value

$$\sigma = (n - 1)^{-1/2} \left[\sum_{i=1}^{n} (m_i - \overline{m})^2 \right]^{1/2} = (n - 1)^{-1/2} \left(\sum_{i=1}^{n} e_i^2 \right)^{1/2}$$

will be used, in which n is replaced by $n - 1$ since one degree of freedom is lost by using $\overline{m}$ instead of m, $\overline{m}$ being related to the m_i.

3. Normal Distribution

Relative Frequency of Errors. The Gauss-Laplace, or normal, distribution of frequency of errors is (Fig. 3.22)

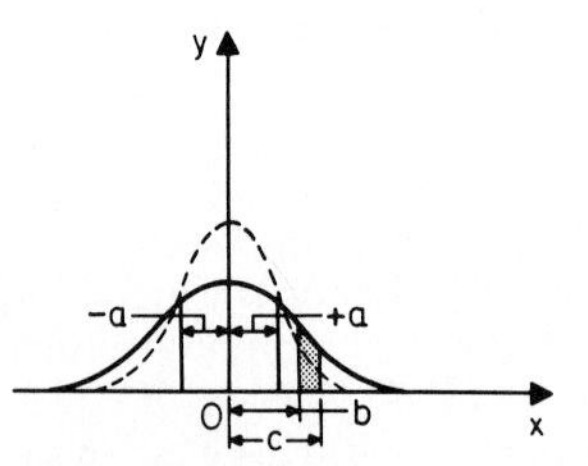

FIGURE 3.22

$$y = \frac{1}{\sigma\sqrt{2\pi}} e^{-x^2/2\sigma^2}$$

or

$$y = \frac{1}{\sqrt{\pi}} h e^{-h^2x^2}$$

where $2h^2\sigma^2 = 1$, or $h = 1/(\sqrt{2}\sigma)$, and y represents the proportionate number of errors of value x. The area under the curve in Fig. 3.22 is unity. The dotted curve is also an error distribution curve with a greater value of the precision index h which measures the concentration of observations about their mean.

Probability. The fraction of the total number of errors whose values lie between $x = -a$ and $x = a$ is

$$P = \frac{h}{\sqrt{\pi}} \int_{-a}^{+a} e^{-h^2x^2}\, dx = \frac{2}{\sqrt{\pi}} \int_0^{ha} e^{-h^2x^2}\, d(hx) \tag{3.1}$$

that is, P is the probability of an error x having a value between $-a$ and a. Similarly, the shaded area in Fig. 3.22 represents the probability of errors between b and c.

Probable Error. Results of measurements are sometimes expressed in the form $E \pm r$, where r is the probable error of a single observation and is defined as the number which the actual error may with equal probability be greater or less than. From Eq. (3.1) and

$$\frac{2}{\sqrt{\pi}} \int_0^{hr} e^{-h^2x^2}\, d(hx) = 0.50$$

$$hr = 0.47694$$

or

$$r = 0.4769x\sqrt{2}\sigma = 0.6745\sigma$$

NUMERICAL METHODS

Numerical techniques do not always yield exact results. The goal of approximate and numerical methods is to provide convenient techniques for obtaining useful information from mathematical formulations of physical problems. Often these mathematical statements are not solvable by analytical means. Or perhaps analytic solutions are available but in a form that is inconvenient for direct interpretation numerically.

1. Approximation Identities. For the following relationships the sign $\cong$ means approximately equal to, when X is small:

$$\frac{1}{1 \pm X} \cong \mp X \qquad \frac{1+Y}{1 \mp X} \cong 1 + Y \pm X$$

$$(1 \pm X)^n \cong 1 \pm nX \qquad (a \pm X)^2 \cong a^2 \pm 2aX$$

$$\sin X \cong X (X \text{ rad})$$

$$\sqrt{Y(Y+X)} \cong \frac{2Y+X}{2} \qquad \sqrt{1 \pm X} \cong 1 \pm \frac{X}{2}$$

$$(1 \pm X)^{-n} \cong 1 \mp nX \qquad (1 \pm X)^{-1/2} \cong 1 \mp \frac{X}{2}$$

$$e^2 \cong 1 + X \qquad \tan X \cong X$$

$$\sqrt{Y^2 + X^2} \cong Y + \frac{X^2}{2Y} \left(\frac{X}{Y} \text{ small}\right)$$

$$\sum_{1}^{m} \sqrt{n} \cong \tfrac{2}{3} m^{3/2} + \frac{\sqrt{m}}{2} - 0.245$$

$$n! \cong e^{-n} n \sqrt{2\pi n} \qquad n! \cong \sqrt{2\pi} \left\{ \frac{\sqrt{n^2 + n + \tfrac{1}{4}}}{e} \right\}^{n+1/2}$$

2. Interpolation and Finite Differences. Let $y_0, y_1, y_2, \ldots, y_n$ denote a set of values of a function $y = f(x)$. Then $y_1 - y_0, y_2 - y_1$, etc., are called the *first differences* of y and are written with the notation

$$\Delta y_n = y_{n+1} - y_n$$

The differences of these first differences are called *second differences*, so that

$$\Delta^2 y_n = \Delta y_{n+1} - \Delta y_n - y_{n+2} - 2y_{n+1} + y_n$$

Likewise,

$$\Delta^3 y_n = \Delta^2 y_{n+1} - \Delta^2 y_n = y_{n+3} - 3y_{n+2} + 3y_{n+1} - y_n$$

etc.

The relation between differences and derivatives is

$$\Delta^n f(x) = (\Delta x)^n f^{(n)}(x + \theta n \,\Delta x) \qquad 0 < \theta < 1 \tag{3.2}$$

Consequently,

$$\lim_{\Delta x \to 0} \frac{\Delta^n f(x)}{(\Delta x)^n} = f^{(n)}(x)$$

The process of interpolation is used not only for finding in-between values of a function from a numerical table, but also for replacing complicated functions by simpler ones which can be made to coincide with them as closely as desired over a given interval.

Since polynomials are the simplest functions, interpolating functions are usually polynomials, the coefficients of whose terms are expressed in differences of ascending order.

The most important and useful of the polynomial interpolation formulas are Newton's, Stirling's, and Bessel's (see Refs. 5 and 6). Newton's formula for forward interpolation reads:

$$y = y_0 + u\Delta y_0 + \frac{u(u-1)}{2!}\Delta^2 y_0 + \frac{u(u-1)(u-2)}{3!}\Delta^3 y_0 + \frac{u(u-1)(u-2)(u-3)}{4!}\Delta^4 y_0 + \cdots + \frac{u(u-1)(u-2)\cdots(u-n+1)}{n!}\Delta^n y_0 \tag{3.3}$$

where $u = (x - x_0)/h$, and it is called by that name because it employs y's from y_0 forward to the right.

y_0 may, of course, be any tabular value, but the formula will contain only values of y which come after the value chosen as the starting point.

3. Numerical Differentiation. The derivative of a tabulated function can be found at any point by representing the function by an appropriate interpolation formula and then differentiating the formula with respect to u. Then the derivative is given by the relation

$$\frac{dy}{dx} = \frac{1}{h}\frac{dy}{du}$$

where dy/du can be obtained, for example, from Eq. (3.3) by differentiation.

4. Numerical Integration. The numerical value of a definite integral can be found, to any desired degree of accuracy, by means of any of several quadrature formulas which express the integral as a linear function of given values of the integrand. In any problem, the definite integral should be set up before a quadrature formula is applied.

When the values of the integrand are given at equidistant intervals of width h of the independent variable, various quadrature formulas can be derived by integrating any of the standard interpolation formulas between given limits. By integrating Eq. (3.3), for example, and retaining differences of successively higher orders, the following formulas can be derived:

Trapezoidal Rule

$$\int_a^b y\,dx = \int_a^b f(x)\,dx = h(\tfrac{1}{2}y_0 + y_1 + y_2 + \cdots + y_{n-2} + y_{n-1} + \tfrac{1}{2}y_n) \qquad (3.4)$$

where $h = (b - a)/n$, and n is either even or odd:

$$\text{Correction term} = \text{error} = -\frac{nh^3}{12}f''(\xi) \qquad a \le \xi \le b$$

Equation (3.4) is the simplest of the quadrature formulas, but is also the least accurate. The accuracy can be increased by decreasing the interval h.

Simpson's Rule

$$\int_a^b y\,dx = \int_a^b f(x)\,dx = \frac{h}{3}(y_0 + 4y_1 + 2y_2 + 4y_3 + 2y_4 + \cdots + 2y_{n-2} + 4y_{n-1} + y_n) \qquad (3.5)$$

$$\text{Error} = -\frac{nh^5}{180}f^{iv}(\xi) \qquad a \le \xi \le b$$

$$= -\frac{(b-a)}{180}h^4 f^{iv}(\xi) \approx -\frac{h}{90}\Sigma\,\Delta^4 y$$

In (3.5), n must be *even*.

Because of its simplicity, flexibility, and relatively high accuracy, Simpson's rule is the most useful of all quadrature formulas.

5. Numerical Solution of Ordinary Differential Equations.* A variety of methods have been devised to solve ordinary differential equations numerically.

Equations of the First Order. Let the first-order differential equation be

$$\frac{dy}{dx} = f(x,y)$$

with the initial condition (x_0, y_0), that is, $y = y_0$ when $x = x_0$. (Note that any differential equation of the first order in the variables x and y can be written in this form.)

The procedure, called the *modified Euler method*, is as follows:

Step 1. From the given initial conditions (x_0, y_0) compute $y_0' = f(x_0, y_0)$ and

$$y_0'' = \frac{\partial f(x_0, y_0)}{\partial x} + \frac{\partial f(x_0, y_0)}{\partial y} y_0'$$

Then determine $y_0 = y_0 + hy_0' + (h^2/2)y_0''$, where h = subdivision of the independent variable.

Step 2. Determine $y_1' = f(x_1, y_1)$. $(x_1 = x_0 + h)$. These prepare us for:

Predictor Steps

Step 3. For $n \geq 1$ calculate $(y_{n+1})_1 = y_{n-1} + 2hy_n'$.

Step 4. Calculate $(y'_{n+1})_1 = f[x_{n+1}, (y_{n+1})_1]$.

Corrector Steps

Step 5. Calculate $(y_{n+1})_2 = y_n + (h/2)\,[(y'_{n+1})_1 + y_n']$, where y_n, y_n' without the subscripts are the previous values obtained by this process (or by steps 1 and 2).

Step 6. $(y'_{n+1})_2 = f[x_{n+1}, (y_{n+1})_2]$.

Step 7. Repeat the corrector steps 5 and 6 if necessary until the desired accuracy is produced in y_{n+1}, y'_{n+1}.

Example. Consider the equation $y' = 2y^2 + x$ with the initial conditions $y_0 = 1$ when $x_0 = 0$. Let $h = 0.1$. A few steps of the computation are illustrated.

Step 1

$$y_0' = 2y_0^2 + x_0 = 2$$

$$y_0'' = 1 + 4y_0'' = 1 + 8 = 9$$

*This section partly taken from Ref. 7, *Perry's Chemical Engineer's Handbook*, 6th ed., by R. H. Perry, D. W. Green, and J. O. Maloney. Copyright © 1984. Used by permission of McGraw-Hill, Inc. All rights reserved.

$$y_1 = 1 + (0.1)(2) + \left[\frac{(0.1)^2}{2}\right]9 = 1.245$$

Step 2

$$y_1' = 2y_1^2 + x_1 = 3.100 + 0.1 = 3.210$$

Step 3

$$(y_2)_1 = y_0 + 2hy_1' = 1 + 2(0.1)3.210 = 1.642$$

Step 4

$$(y_2')_1 = 2(y_2)_1^2 + x_2 = 5.592$$

Step 5

$$(y_2)_2 = y_1 + \frac{0.1}{2}[(y_2')_1 + y_1'] = 1.685$$

Step 6

$$(y_2')_2 = 2(y_2)_2^2 + x_2 = 5.878$$

Step 5 (repeat)

$$(y_2)_3 = y_1 + (0.05)[(y_2')_2 + y_1'] = 1.699$$

Step 6 (repeat)

$$(y_2')_3 = 2(y_2)_3^2 + x_3 = 5.974$$

and so forth. This procedure may be programmed for a computer. A discussion of the truncation error of this process may be found in Refs. 6 and 7.

Equations of the Second and Higher Orders. The substitution of $dy/dx = y'$ reduces a second-order equation of the form

$$\frac{d^2y}{dx^2} + P\frac{dy}{dx} + Qy = f(x)$$

to the two first-order equations

$$\frac{dy}{dx} = y' \qquad y'' = f(x) - Py' - Qy$$

A similar procedure will reduce an equation of any order to a system of first-order equations, and these can be solved numerically by the method already given or by those given in Refs. 5 and 6.

6. Numerical Solution of Partial Differential Equations. The techniques will be introduced by an example. Consider the typical linear-diffusion problem with given initial and boundary conditions:

$$\frac{\partial z}{\partial t} = \frac{\partial^2 z}{\partial x^2} \qquad 0 < x < 1,\ 0 < t \le T$$

$$z(x,0) = f(x) \qquad 0 < x < 1$$

$$z(0,t) = g(t) \qquad 0 < t \leq T$$

$$z(1,t) = h(t) \qquad 0 < t \leq T$$

A finite-difference analog for this problem is developed by introducing a net whose mesh points are denoted by $x_i = ih$, $t_j = jk$, where $i = 0, 1, 2, \ldots, M$; $j = 0, 1, \ldots, N$; $h = \Delta x = 1/M$; and $k = \Delta t = T/N$. The boundaries are specified by $i = 0$ and $i = M$ and any "false" boundaries by $i = -1, -2, \ldots, i = M + 1, M + 2, \ldots$, and so forth. The initial is denoted by $j = 0$, and the discrete approximation at $x_i = \pm h$, $t_j = jk$ is designated $Z_{i,j}$.

If an approximate solution $Z_{i,j}$ is assumed to be known at all the mesh points up to the t_j, a method must be specified to advance the solution to time t_{j+1}.

The value of $Z_{i,j+1}$ at $x = 0$ and $x = 1$ should be selected as those boundary conditions specified above, that is,

$$Z_{0,j+1} = g(t_{j+1}) \qquad Z_{M,j+1} = h(t_{j+1})$$

At other points $0 < i < M$, the partial differential equation will be replaced by some difference equation. The simplest replacement consists in approximating the space derivative by a centered second difference and the time derivative by a forward difference at (x_i, t_j).

Based on formulas given in Sec. 2, Interpolation and Finite Differences, earlier, the partial derivatives may be approximated by

$$\frac{\partial z}{\partial t} = \frac{1}{k}(Z_{i,j+1} - Z_{i,j})$$

$$\frac{\partial^2 z}{\partial x^2} = \frac{1}{h^2}(Z_{i+1,j} + Z_{i-1,j} - 2Z_{i,j})$$

The resulting difference equation is now

$$\frac{1}{k}(Z_{i,j+1} - Z_{i,j}) = \frac{1}{h^2}(Z_{i+1,j} - 2Z_{i,j} + Z_{i-1,j}) \qquad i = 1, \ldots, M - 1$$

Upon solving for $Z_{i,j+1}$, we obtain the explicit equation for "marching" ahead in time:

$$Z_{i,j+1} = rZ_{i-1,j} + (1 - 2r)\,Z_{i,j} + rZ_{i+1,j} \qquad i = 1, \ldots, M - 1$$

where $r = k/h^2 = \Delta t/(\Delta x)^2$.

For a discussion of error and convergence and consistency of this computational procedure see Refs. 7 and 8.

REFERENCES*

1. E. A. Avallone and T. Baumeister III (eds.), *Mark's Standard Handbook for Mechanical Engineers*, 9th ed., McGraw-Hill, New York, 1987, sec. 2.

*Those references listed above but not cited in the text were used for comparison between different data sources, clarification, clarity of presentation, and, most important, reader's convenience when further interest in the subject exists.

2. K. Gieck, *Engineering Formulas*, 4th ed., McGraw-Hill, New York, 1983.
3. H. A. Rothbart, *Mechanical Design and Systems Handbook*, 2d ed., McGraw-Hill, New York, 1985, part 1.
4. R. H. Perry (ed.), *Engineering Manual*, 3d ed., McGraw-Hill, New York, 1976.
5. W. Flugge (ed.), *Handbook of Engineering Mechanics*, McGraw-Hill, New York, 1962, part 1.
6. G. Dahlquist and A. Bjorck, *Numerical Methods*, Prentice-Hall, Englewood Cliffs, N.J., 1974.
7. R. H. Perry, D. W. Green, and J. O. Maloney (eds.), *Perry's Chemical Engineer's Handbook*, 6th ed., McGraw-Hill, New York, 1984, sec. 2.
8. S. V. Patankar, *Numerical Heat Transfer and Fluid Flow*, Hemisphere/McGraw-Hill, New York, 1980.
9. O. W. Eshbach and M. Souders, *Handbook of Engineering Fundamentals*, John Wiley & Sons, New York, 1975.
10. W. M. Rohsenow, J. P. Hartnett, and E. N. Ganić (eds.), *Handbook of Heat Transfer Fundamentals*, 2d ed., McGraw-Hill, New York, 1985, chap. 2.
11. O. C. Zienkiewicz, *The Finite Element Method in Structural and Continuum Mechanics*, McGraw-Hill, New York, 1967.

CHAPTER 4
CHEMISTRY

COMMON DEFINITIONS

Atoms. Atoms contain a dense nucleus of positively charged protons and uncharged neutrons, surrounded by negatively charged orbital electrons that occupy discrete energy levels and orbital configurations. In a stable atom, the positively charged protons and the negatively charged electrons are equal. Each type of atom, or *element*, is assigned a *mass number* based on the number of protons and neutrons contained in its nucleus. Mathematically these relationships are expressed as:

$$A = \text{mass number}$$
$$Z = \text{number of positively charged protons}$$
$$A - Z = \text{number of uncharged neutrons}$$

Each element is also assigned an *atomic number* equal to Z. Gain or loss of electrons results in ionized atoms with a net charge. Chemical reactions are concerned only with changes in electrical structure, while nuclear reactions involve changes in the constitution of the nucleus. Isotopes are atoms having the same atomic number, the same number of electrons, and the same chemical reactions, but a different mass number because their nuclei contain different numbers of neutrons.

Molecules. A molecule consists of a group of two or more atoms held together through chemical bonds (represented diagrammatically as geometric arrangements). Chemical bonds in turn are based on the electrical configurations of the individual atoms involved, and occur when electrons in one atom occupy vacant orbital sites in another atom. Chemical bonds between atoms result in a joint electrical configuration of greater stability, or lower energy, than that possessed by the individual atoms. The total negative charge of a molecule's combined electrons equals the total positive charge of its combined protons.

Avogadro's Number. Avogadro's number is the number of molecules (6.023×10^{23}) in 1 g · mol of a substance.

Gram Equivalent Weight. This involves two kinds of reactions:

- *Nonredox reaction:* the mass in grams of a substance equivalent to 1 g · atom of hydrogen, 0.5 g · atom of oxygen, or 1 g · ion of the hydroxyl ion. It can be determined by dividing the molecular weight of the substance by the number of hydrogen or oxygen atoms or hydroxyl ions (or their equivalent) supplied or required by the molecule in a given reaction.

- *Redox reaction:* the molecular weight in grams divided by the change in oxidation state.

Molality. Expressed mathematically by the symbol m, molality is defined as:

$$m = \frac{\text{number of gram moles of solute}}{\text{number of kilograms of solvent}}$$

Molarity. Expressed mathematically by the symbol M, molarity is defined as:

$$M = \frac{\text{number of gram moles of solute}}{\text{number of liters of solution}}$$

Normality. Expressed mathematically as N, normality is defined as:

$$N = \frac{\text{number of gram equivalents of solute}}{\text{number of liters of solution}}$$

Oxidation. Oxidation is the loss of electrons by an atom or group of atoms.

Reduction. Reduction is the gain of electrons by an atom or group of atoms.

Ion Product of Water. Ion product of water (K_w) is the product of the hydrogen ion (H^+) and hydroxyl ion (OH^-) concentrations in gram-ions per liter:

$$K_w = [H^+]\,[OH^-]$$

pH Values. pH value (see Fig. 4.1) is expressed as the negative logarithm (base 10) of the hydrogen ion concentration in gram-ions per liter:

$$\text{pH} = -\log\,[H^+]$$

See Table 4.1 for establishing pH values by using suitable indicators.

Note. A list of common chemicals is given in Table 4.2. A list of chemical reactions for the preparation of some chemicals is given in Table 4.3. A procedure for the preparation of freezing mixtures in a certain temperature range is given in Table 4.4. A list of common laboratory solutions and reagents is given in Table 4.5.

STOICHIOMETRY

Stoichiometry is the theory of the proportions in which chemical species combine with one another. The stoichiometric equation of a chemical reaction is a statement of the relative number of molecules or moles of reactants and products that participate in the reaction. The ratios obtained from the numerical coefficients in

[H+]	1	10^{-1}	10^{-2}	...	10^{-7}	...	10^{-12}	10^{-13}	10^{-14}
pH value	0	1	2		7		12	13	14
	Acid				Neutral		Alkaline		

FIGURE 4.1 pH values.

TABLE 4.1 List of Acid-Base Indicators

Indicator	*pH-* Range	Colour change from	to
thymol blue [benz.	1·2...2·8	red	yellow
p-dimethylamino-azo-	2·9...4·0	red	orange-yellow
bromophenolblue	3·0...4·6	yellow	red-violet
congo red	3·0...4·2	blue-violet	red-orange
methyl orange	3·1...4·4	red	yellow-(orange)
brom cresol green	3·8...5·4	yellow	blue
methyl red	4·4...6·2	red	(orange)-yellow
litmus	5·0...8·0	red	blue
bromocresol purple	5·2...6·8	yellow	purple
brom phenol red	5·2...6·8	orange yell.	purple
bromothymol blue	6·0...7·6	yellow	blue
phenol red	6·4...8·2	yellow	red
neutral red	6·4...8·0	(blue)-red	orange-yellow
cresol red	7·0...8·8	yellow	purple
meta cresol purple	7·4...9·0	yellow	purple
thymol blue	8·0...9·6	yellow	blue
phenolphtalein	8·2...9·8	colourless	red-violet
alizarin yellow 66	10·0...12·1	light-yellow	light brown-yellow

Source: From Gieck.[1] Data shown in this table can also be found in Refs. 2 through 5.

the chemical equation are the stoichiometric ratios that permit one to calculate the moles of one substance as related to the moles of another substance in the chemical equation. For example, the stoichiometric equation for the combustion of heptane

$$C_7H_{16} + 11O_2 \rightarrow 7CO_2 + 8H_2O$$

indicates that 1 mol (not kilogram) of heptane will react with 11 mol of oxygen to give 7 mol of carbon dioxide plus 8 mol of water. These may be pound-moles, kilogram-moles, gram-moles or any other type of mole, as shown in Fig. 4.2. One mol of CO_2 is formed from each $\frac{1}{7}$ mol of C_7H_{16}. Also, 1 mol of H_2O is formed with each $\frac{7}{8}$ mol of CO_2. Thus the equations tells us in terms of moles (not kilograms) the ratios among reactants and products.

TABLE 4.2 List of Common Chemicals

trade name	chemical name	chemical formula
acetone	acetone	$(CH_3)_2 \cdot CO$
acetylene	acetylene	C_2H_2
ammonia	ammonia	NH_3
ammonium (hydroxide of)	ammonium hydroxide	NH_4OH
aniline	aniline	$C_6H_5 \cdot NH_2$
bauxite	hydrated aluminium oxides	$Al_2O_3 \cdot 2\ H_2O$
bleaching powder	calcium hypochlorite	$CaCl\ (OCl)$
blue vitriol	copper sulfate	$CuSO_4 \cdot 5\ H_2O$
borax	sodium tetraborate	$Na_2B_4O_7 \cdot 10\ H_2O$
butter of zinc	zinc chloride	$ZnCl_2 \cdot 3\ H_2O$
cadmium sulfate	cadmium sulfate	$CdSO_4$
calcium chloride	calcium chloride	$CaCl_2$
carbide	calcium carbide	CaC_2
carbolic acid	phenol	C_6H_5OH
carbon dioxide	carbon dioxide	CO_2
carborundum	silicon carbide	SiC
caustic potash	potassium hydroxide	KOH
caustic soda	sodium hydroxide	$NaOH$
chalk	calcium carbonate	$CaCO_3$
cinnabar	mercuric sulfide	HgS
ether	di-ethyl ether	$(C_2H_5)_2O$
fixing salt or hypo	sodium thiosulfate	$Na_2S_2O_3 \cdot 5\ H_2O$
glauber's salt	sodium sulfate	$Na_2SO_4 \cdot 10\ H_2O$
glycerine or glycerol	glycerine	$C_3H_5\ (OH)_3$
graphite	crystalline carbon	C
green vitriol	ferrous sulfate	$FeSO_4 \cdot 7\ H_2O$
gypsum	calcium sulfate	$CaSO_4 \cdot 2\ H_2O$
heating gas	propane	C_3H_8
hydrochloric acid	hydrochloric acid	HCl
hydrofluoric acid	hydrofluoric acid	HF
hydrogen sulfide	hydrogen sulfide	H_2S
iron chloride	ferrous chloride	$FeCl_2 \cdot 4\ H_2O$
iron sulfide	ferrous sulfide	FeS
laughing gas	nitrous oxide	N_2O
lead sulfide	lead sulfide	PbS
limestone	calcium carbonate	$CaCO_3$
magnesia	magnesium oxide	MgO
marsh gas	methane	CH_4
minimum or red lead	plumbate	$2\ PbO \cdot PbO_2$
nitric acid	nitric acid	HNO_3
phosphoric acid	ortho phosphoric acid	H_3PO_4
potash	potassium carbonate	K_2CO_3
potassium bromide	potassium bromide	KBr
potassium chlorate	potassium chlorate	$KClO_3$
potassium chloride	potassium chloride	KCl
potassium chromate	potassium chromate	K_2CrO_4
potassium cyanide	potassium cyanide	KCN
potassium dichromate	potassium dichromate	$K_2Cr_2O_7$
potassium iodide	potassium iodide	KI
prussic acid	hydrogen cyanide	HCN
pyrolusite	manganese dioxide	MnO_2
quicklime	calcium monoxide	CaO
red prussiate of potassium	potassium ferrocyan.	$K_3Fe(CN)_6$
salammoniac	ammonium chloride	NH_4Cl
silver bromide	silver bromide	$AgBr$

Source: From Gieck.[1] Data shown in this table can also be found in Refs. 2 through 5.

TABLE 4.2 List of Common Chemicals (*Continued*)

trade name	chemical name	chemical formula
silver nitrate	silver nitrate	$AgNO_3$
slaked lime	calcium hydroxide	$Ca(OH)_2$
soda ash	hydrated sodium carb.	$Na_2CO_3 \cdot 10\ H_2O$
sodium monoxide	sodium oxide	Na_2O
soot	amorphous carbon	C
stannous chloride	stannous chloride	$SnCl_2 \cdot 2\ H_2O$
sulfuric acid	sulfuric acid	H_2SO_4
table salt	sodium chloride	NaCl
tinstone, tin putty	stannic oxide	SnO_2
trilene	trichlorethylene	C_2HCl_3
urea	urea	$CO(NH_2)_2$
white lead	basic lead carbonate	$2\ PbCO_3 \cdot Pb\ (OH)_2$
white vitriol	zinc sulphate	$ZnSO_4 \cdot 7\ H_2O$
yellow prussiate of potass.	potass. ferrocyanide	$K_4Fe(CN)_6 \cdot 3\ H_2O$
zinc blende	zinc sulfide	ZnS
zinc or chinese white	zinc oxide	ZnO

Source: From Gieck.[1] Data shown in this table can also be found in Refs. 2 through 5.

TABLE 4.3 List of Chemical Reactions for Preparation of Some Chemicals

to prepare	use reaction
ammonia	$CO(NH_2)_2 + H_2O \rightarrow 2\ NH_3 + CO_2$
ammonium chloride	$NH_4OH + HCl \rightarrow NH_4Cl + H_2O$
ammonium hydroxide	$NH_3 + H_2O \rightarrow NH_4OH$
cadmium sulfide	$CdSO_4 + H_2S \rightarrow Cds + H_2SO_4$
carbon dioxide	$CaCO_3 + 2\ HCl \rightarrow CO_2 + CaCl_2 + H_2O$
chlorine	$CaOCl_2 + 2\ HCl \rightarrow Cl_2 + CaCl_2 + H_2O$
hydrogen	$H_2SO_4 + Zn \rightarrow H_2 + ZnSO_4$
hydrogen sulfide	$FeS + 2\ HCl \rightarrow H_2S + FeCl_2$
lead sulfide	$Pb(NO_3)_2 + H_2S \rightarrow PbS + 2\ HNO_3$
oxygen	$2\ KClO_3 \rightarrow 3\ O_2 + 2\ KCl$
sodium hydroxide	$Na_2O + H_2O \rightarrow 2\ NaOH$
zinc sulfide	$ZnSO_4 + H_2S \rightarrow ZnS + H_2SO_4$

Source: From Gieck.[1] Data shown in this table can also be found in Refs. 2 through 5.

TABLE 4.4 Preparation of Freezing Mixtures

Drop in temperature from °C	to °C	Mixture (The figures stand for proportions by mass)
+ 10	− 12	$4\ H_2O + 1\ KCl$
+ 10	− 15	$1\ H_2O + 1\ NH_4NO_3$
+ 8	− 24	$1\ H_2O + 1\ NaNO_3 + 1\ NH_4Cl$
0	− 21	3·0 ice (crushed) + 1 NACl
0	− 39	1·2 ice (crushed) + $2\ CaCl_2 \cdot 6\ H_2O$
0	− 55	1·4 ice (crushed) + $2\ CaCl_2 \cdot 6\ H_2O$
+ 15	− 78	1 methyl alcohol + 1 CO_2 solid

Source: From Gieck.[1] Data shown in this table can also be found in Refs. 2 through 5.

TABLE 4.5 List of Laboratory Solutions, Reagents, and Indicators

Unless otherwise stated, the term g per liter *signifies grams of the formula indicated dissolved in water and made up to a liter of solution.*

Acetic acid, $HC_2H_3O_2$—6*N*: 350 mL glacial acetic acid per liter.

Alcohol, amyl, $C_5H_{11}OH$: use as purchased.

Alcohol, ethyl, C_2H_5OH; 95% alcohol, as purchased.

Alizarin, dihydroxyanthraquinone (indicator): dissolve 0.1 g in 100 mL alcohol; pH range yellow 5.5—6.8 red.

Alizarin yellow R, sodium *p*-nitrobenzeneazosalicylate (indicator): dissolve 0.1 g in 100 mL water; pH range yellow 10.1—violet 12.1.

Alizarin yellow GG, salicyl yellow, sodium *m*-nitrobenzeneazosalicylate (indicator); dissolve 0.1 g in 100 mL 50% alcohol; pH range yellow 10.0—12.0 lilac.

Alizarin S, alizarin carmine, sodium alizarin sulfonate (indicator): dissolve 0.1 g in 100 mL water; pH range yellow 3.7—5.2 violet.

Aluminon (qualitative test for aluminum). The reagent consists of 0.1% solution of the ammonium salt of aurin tricarboxylic acid. A bright red precipitate, persisting in alkaline solution, indicates aluminum.

Aluminum chloride, $AlCl_3$—0.5*N*: 22 g per liter.

Aluminum nitrate, $Al(NO_3)_3 \cdot 7.5H_2O$—0.5*N*: 58 g per liter.

Aluminum sulfate, $Al_2(SO_4)_3 \cdot 18H_2O$—0.5*N*: 55 g per liter.

Ammonium acetate, $NH_4C_2H_3O_2$—3*N*: 231 g per liter.

Ammonium carbonate, $(NH_4)_2CO_3 \cdot H_2O$—3*N*: 171 g per liter; for the anhydrous salt: 144 g per liter.

Ammonium chloride, NH_4Cl—3*N*: 161 g per liter.

Ammonium hydroxide, NH_4OH—15*N*: the concentrated solution which contains 28% NH_3; for 6*N*: 400 mL per liter.

Ammonium molybdate, $(NH_4)_2MoO_4$—*N*: dissolve 88.3 g of solid $(NH_4)_6Mo_7O_{24} \cdot 4H_2O$ in 100 mL 6*N* NH_4OH. Add 240 g of solid NH_4NO_3 and dilute to one liter. Another method is to take 72 g of MoO_3, add 130 mL of water and 75 mL of 15*N* NH_4OH; stir mechanically until nearly all has dissolved, then add it to a solution of 240 mL concentrated HNO_3 and 500 mL of water; stir continuously while solutions are being mixed; allow to stand 3 days, filter, and use the clear filtrate.

Ammonium nitrate, NH_4NO_3—*N*: 80 g per liter.

Ammonium oxalate, $(NH_4)_2C_2O_4 \cdot H_2O$—0.5*N*: 40 g per liter.

Ammonium polysulfide (yellow ammonium sulfide), $(NH_4)_2S_x$: allow the colorless $(NH_4)_2S$ to stand, or add sulfur.

Ammonium sulfate, $(NH_4)_2SO_4$—0.5 *N*: 33 g per liter; saturated: dissolve 780 g of $(NH_4)_2SO_4$ in water and make up to a liter.

Source: From Dean.[2]

Ammonium sulfide (colorless), $(NH_4)_2S$—saturated: pass H_2S through 200 mL of concentrated NH_4OH in the cold until no more gas is dissolved, add 200 mL NH_4OH and dilute with water to a liter; the addition of 15 g of sulfur is sufficient to make the polysulfide.

Antimony pentachloride, $SbCl_5$—0.5*N*: 30 g per liter.

Antimony trichloride, $SbCl_3$—0.5*N*: 38 g per liter.

Aqua regia: mix 3 parts of concentrated HCl and 1 part of concentrated HNO_3 just before ready to use.

Arsenic acid, $H_3AsO_4 \cdot 0.5H_2O$—0.5*N* ($= \frac{1}{2}H_3AsO_4 \div 5$): 15 g per liter.

Arsenous oxide, As_2O_3—0.25*N*: 8 g per liter for saturation.

Aurichloric acid, $HAuCl_4 \cdot 3H_2O$: dissolve in ten parts of water.

Aurin, see *rosolic acid.*

Azolitmin solution (indicator): make up a 1% solution of azolitmin by boiling in water for 5 minutes; it may be necessary to add a small amount of NaOH to make the solution neutral; pH range red 4.5—8.3 blue.

Bang's reagent (for glucose estimation): dissolve 100 g of K_2CO_3, 66 g of KCl, and 160 of $KHCO_3$ in the order given in about 700 mL of water at 30°C. Add 4.4 g of copper sulfate and dilute to 1 liter after the CO_2 is evolved. This solution should be shaken only in such a manner as not to allow the entry of air. After 24 hours 300 mL are diluted to a liter with saturated KCl solution, shaken gently and used after 24 hours; 50 mL $\Leftrightarrow$ 10 mg glucose.

Barfoed's reagent (test for glucose): dissolve 66 g of cupric acetate and 10 mL of glacial acetic acid in water and dilute to one liter.

Barium chloride, $BaCl_2 \cdot 2H_2O$—0.5*N*: 61 g per liter.

Barium hydroxide, $Ba(OH)_2 \cdot 8H_2O$—0.2*N*: 32 g per liter for saturation.

Barium nitrate, $Ba(NO_3)_2$—0.5*N*: 65 g per liter.

Baudisch's reagent: see *cupferron.*

Benedict's qualitative reagent (for glucose): dissolve 173 g of sodium citrate and 100 g of anhydrous sodium carbonate in about 600 mL of water, and dilute to 850 mL; dissolve 17.3 g of $CuSO_4 \cdot 5H_2O$ in 100 mL of water and dilute to 150 mL; this solution is added to the citrate-carbonate solution with constant stirring. See also the *quantitative reagent* below.

Benedict's quantitative reagent (sugar in urine): This solution contains 18 g copper sulfate, 100 g of anhydrous sodium carbonate, 200 g of potassium citrate, 125 g of potassium thiocyanate, and 0.25 g of potassium ferrocyanide per liter; 1 mL of this solution $\Leftrightarrow$ 0.002 sugar.

Benzidine hydrochloride solution (for sulfate determination): mix 6.7 g of benzidine $[C_{12}H_8(NH_2)_2]$ or 8.0 g of the hydrochloride $[C_{12}H_8(NH_2)_2 \cdot 2HCl]$ into a paste with 20 mL of water; add 20 mL of HCl (sp. gr. 1.12) and dilute the mixture to one liter with water; each mL of this solution is equivalent to 0.00357 g H_2SO_4.

Benzopurpurine 4B (indicator): dissolve 0.1 g in 100 mL water; pH range blue-violet 1.3—4.0 red.

Benzoyl auramine (indicator): dissolve 0.25 g in 100 mL methyl alcohol; pH range violet 5.0—5.6 pale yellow. Since this compound is not stable in aqueous solution, hydrolyzing slowly in neutral medium, more rapidly in alkaline and still more rapidly in acid solution, the indicator should not be added until one is ready to titrate. The acid quinoid form of the compound is dichroic, showing a red-violet in thick layers and blue in thin. At a pH of 5.4 the indicator appears a neutral gray color by daylight or a pale red under tungsten light. The change to yellow is easily recognized in either case. Cf. Scanlan and Reid, *Ind. Eng. Chem., Anal. Ed.* 7, 125 (1935).

Bertrand's reagents (glucose estimation): (*a*) 40 g of copper sulfate diluted to one liter; (*b*) rochelle salt 200 g, NaOH 150 g, and sufficient water to make one liter; (*c*) ferric sulfate 50 g, H_2SO_4 200 g, and sufficient water to make one liter; (*d*) $KMnO_4$ 5 g and sufficient water to make one liter.

Bial's reagent (for pentoses): dissolve 1 g of orcinol in 500 mL of 30% HCl to which 30 drops of a 10% ferric chloride solution have been added.

Bismuth chloride, $BiCl_3$—0.5*N*: 52 g per liter, using 1:5 HCl in place of water.

Bismuth nitrate, $Bi(NO_3)_3{\cdot}5H_2O$—0.25*N*: 40 g per liter, using 1:5 HNO_3 in place of water.

Bismuth standard solution (quantitative color test for Bi): dissolve 1 g of bismuth in a mixture of 3 mL of concentrated HNO_3 and 2.8 mL of H_2O and make up to 100 mL with glycerol. Also dissolve 5 g of KI in 5 mL of water and make up to 100 mL with glycerol. The two solutions are used together in the colorimetric estimation of Bi.

Boutron-Boudet solution: see *soap solution.*

Bromchlorophenol blue, dibromodichlorophenol-sulfonphthalein (indicator): dissolve 0.1 g in 8.6 mL 0.02 *N* NaOH and dilute with water to 250 mL; pH range yellow 3.2—4.8 blue.

Bromcresol green, tetrabromo-*m*-cresol-sulfonphthalein (indicator): dissolve 0.1 g in 7.15 mL 0.02 *N* NaOH and dilute with water to 250 mL; or, 0.1 g in 100 mL 20% alcohol; pH range yellow 4.0—5.6 blue.

Bromcresol purple, dibromo-*o*-cresol-sulfonphthalein (indicator): dissolve 0.1 g in 9.5 mL 0.02 *N* NaOH and dilute with water to 250 mL; or, 0.1 g in 100 mL 20% alcohol; pH range yellow 5.2—6.8 purple.

Bromine water, saturated solution: to 400 mL water add 20 mL of bromine; use a glass stopper coated with petrolatum.

Bromphenol blue, tetrabromophenol-sulfonphthalein (indicator): dissolve 0.1 g in 7.45 mL 0.02 *N* NaOH and dilute with water to 250 mL; or, 0.1 g in 100 mL 20% alcohol; pH range yellow 3.6—4.6 violet-blue.

Bromphenol red, dibromophenol-sulfonphthalein (indicator): dissolve 0.1 g in 9.75 mL 0.02 *N* NaOH and dilute with water to 250 mL; pH range yellow 5.2—7.0 red.

Bromthymol blue, dibromothymol-sulfonphthalein (indicator): dissolve 0.1 g in 8.0 mL 0.02 *N* NaOH and dilute with water to 250 mL; or, 0.1 g in 100 mL of 20% alcohol; pH range yellow 6.0—7.6 blue.

Brucke's reagent (protein precipitant): dissolve 50 g of KI in 500 mL of water, saturate with HgI_2 (about 120 g), and dilute to one liter.

Cadmium chloride, $CdCl_2$—0.5*N*: 46 g per liter.

Cadmium nitrate, $Cd(NO_3)_2 \cdot 4H_2O$—0.5*N*: 77 g per liter.

Cadmium sulfate, $CdSO_4 \cdot 4H_2O$—0.5*N*: 70 g per liter.

Calcium chloride, $CaCl_2 \cdot 6H_2O$—0.5*N*: 55 g per liter.

Calcium hydroxide, $Ca(OH)_2$—0.04*N*: 10 g per liter for saturation.

Calcium nitrate, $Ca(NO_3)_2 \cdot 4H_2O$—0.5*N*: 59 g per liter.

Calcium sulfate, $CaSO_4 \cdot 2H_2O$—0.03*N*: mechanically stir 10 g in a liter of water for 3 hours; decant and use the clear liquid.

Carbon disulfide, CS_2: commercial grade which is colorless.

Chloride reagent: dissolve 1.7 g of $AgNO_3$ and 25 g KNO_3 in water, add 17 mL of concentrated NH_4OH and make up to one liter with water.

Chlorine water, saturated solution: pass chlorine gas into small amounts of water as needed; solutions deteriorate on standing.

Chloroform, $CHCl_3$: commercial grade.

Chloroplatinic acid, $H_2PtCl_6 \cdot 6H_2O$—10% solution: dissolve 1 g in 9 mL of water; keep in a dropping bottle.

Chlorphenol red, dichlorophenol-sulfonphthalein (indicator): dissolve 0.1 g in 11.8 mL 0.02 *N* NaOH and dilute with water to 250 mL; or, 0.1 g in 100 mL 20% alcohol; pH range yellow 5.2—6.6 red.

Chromic chloride, $CrCl_3$—0.5*N*: 26 g per liter.

Chromic nitrate, $Cr(NO_3)_3$—0.5*N*: 40 g per liter.

Chromic sulfate, $Cr_2(SO_4)_3 \cdot 18H_2O$—0.5*N*: 60 g per liter.

Cobaltous nitrate, $Co(NO_3)_2 \cdot 6H_2O$—0.5*N*: 73 g per liter.

Cobaltous sulfate, $CoSO_4 \cdot 7H_2O$—0.5*N*: 70 g per liter.

Cochineal (indicator): triturate 1 g with 75 mL alcohol and 75 mL water, let stand for two days and filter; pH range red 4.8—6.2 violet.

Congo red, sodium tetrazodiphenyl-naphthionate (indicator): dissolve 0.1 g in 100 mL water; pH range blue 3.0—5.2 red.

Corallin (indicator): see *rosolic acid.*

Cresol red, *o*-cresol-sulfonphthalein (indicator): dissolve 0.1 g in 13.1 mL 0.02 *N* NaOH and dilute with water to 250 mL; or, 0.1 g in 100 mL 20% alcohol; pH range yellow 7.2—8.8 red.

***o*-Cresolphthalein** (indicator): dissolve 0.1 g in 250 mL alcohol; pH range colorless 8.2—10.4 red.

Cupferron (iron analysis): dissolve 6 g of ammonium nitrosophenyl-hydroxylamine (cupferron) in water and dilute to 100 mL. This solution is stable for about one week if protected from light.

Cupric chloride, $CuCl_2 \cdot 2H_2O$—0.5*N*: 43 g per liter.

Cupric nitrate, $Cu(NO_3)_2 \cdot 6H_2O$—0.5*N*: 74 g per liter.

Cupric sulfate, $CuSO_4 \cdot 5H_2O$—0.5*N*: 62 g per liter.

Cuprous chloride, CuCl—0.5*N*: 50 g per liter, using 1:5 HCl in place of water.

Cuprous chloride, acid (for gas analysis, absorption of CO): cover the bottom of a two-liter bottle with a layer of copper oxide 3/8 inch deep, and place a bundle of copper wire an inch thick in the bottle so that it extends from the top to the bottom. Fill the bottle with HCl (sp. gr. 1.10). The bottle is shaken occasionally, and when the solution is colorless or nearly so, it is poured into half-liter bottles containing copper wire. The large bottle may be filled with hydrochloric acid, and by adding the oxide or wire when either is exhausted, a constant supply of the reagent is available.

Cuprous chloride, ammoniacal: this solution is used for the same purpose and is made in the same manner as the acid cuprous chloride above, except that the acid solution is treated with ammonia until a faint odor of ammonia is perceptible. Copper wire should be kept with the solution as in the acid reagent.

Curcumin (indicator): prepare a saturated aqueous solution; pH range yellow 6.0—8.0 brownish red.

Dibromophenol-tetrabromophenol-sulfonphthalein (indicator): dissolve 0.1 g in 1.21 mL 0.1*N* NaOH and dilute with water to 250 mL; pH range yellow 5.6—7.2 purple.

Dimethyl glyoxime, $(CH_3CNOH)_2$—0.01*N*: 6 g in 500 mL of 95% alcohol.

2,4-Dinitrophenol (indicator): dissolve 0.1 g in a few mL alcohol, then dilute with water to 100 mL; pH range colorless 2.6—4.0 yellow.

2,5-Dinitrophenol (indicator): dissolve 0.1 g in 20 mL alcohol, then dilute with water to 100 mL; pH range colorless 4—5.8 yellow.

2,6-Dinitrophenol (indicator): dissolve 0.1 g in a few mL alcohol, then dilute with water to 100 mL; pH range colorless 2.4—4.0 yellow.

Esbach's reagent (estimation of proteins): dissolve 10 g of picric acid and 20 g of citric acid in water and dilute to one liter.

Eschka's mixture (sulfur in coal): mix 2 parts of porous calcined MgO with 1 part of anhydrous Na_2CO_3; not a solution but a dry mixture.

Ether, $(C_2H_5)_2O$—use commercial grade.

***p*-Ethoxychrysoidine,** *p*-ethoxybenzeneazo-*m*-phenylenediamine (indicator): dissolve 0.1 g of the base in 100 mL 90% alcohol; or, 0.1 g of the hydrochloride salt in 100 mL water; pH range red 3.5—5.5 yellow.

Ethyl bis-(2,4-dinitrophenyl) acetate (indicator): the stock solution is prepared by saturating a solution containing equal volumes of alcohol and acetone with the indicator; pH range colorless 7.4—9.1 deep blue. This compound is available commercially. The preparation of this compound is described by Fehnel and Amstutz, *Ind. Eng. Chem., Anal. Ed.* 16, 53 (1944), and by von Richter, *Ber.* 21, 2470 (1888), who recommended it for the titration of orange- and red-colored solutions or dark oils in which the end-point of phenolphthalein is not easily visible. The indicator is an orange solid which after crystallization from benzene gives pale yellow crystals melting at 150–153.5°C, uncorrected.

Fehling's solution (sugar detection and estimation): (*a*) Copper sulfate solution: dissolve 34.639 g of $CuSO_4 \cdot 5H_2O$ in water and dilute to 500 mL. (*b*) Alkaline tartrate solution: dissolve 173 g of rochelle salts ($KNaC_4H_4O_6 \cdot 4H_2O$) and 125 g of KOH in water and dilute to 500 mL. Equal volumes of the two solutions are mixed just prior to use. The Methods of the Assoc. of Official Agricultural Chemists give 50 g of NaOH in place of the 125 g KOH.

Ferric chloride, $FeCl_3$—0.5*N*: 27 g per liter.

Ferric nitrate, $Fe(NO_3)_3 \cdot 9H_2O$—0.5*N*: 67 g per liter.

Ferrous ammonium sulfate, Mohr's salt, $FeSO_4 \cdot (NH_4)_2SO_4 \cdot 6H_2O$—0.5*N*: 196 g per liter.

Ferrous sulfate, $FeSO_4 \cdot 7H_2O$—0.5*N*: 80 g per liter; add a few drops of H_2SO_4.

Folin's mixture (for uric acid): dissolve 500 g of ammonium sulfate, 5 g of uranium acetate, and 6 mL of glacial acetic acid, in 650 mL of water. The volume is about a liter.

Formal or Formalin: use the commercial 40% solution of formaldehyde.

Froehde's reagent (gives characteristic colorations with certain alkaloids and glycosides): dissolve 0.01 g of sodium molybdate in 1 mL of concentrated H_2SO_4; use only a freshly prepared solution.

Gallein (indicator): dissolve 0.1 g in 100 mL alcohol; pH range light brown-yellow 3.8—6.6 rose.

Glyoxylic acid solution (protein detection): cover 10 g of magnesium powder with water and add slowly 250 mL of a saturated oxalic acid solution, keeping the mixture cool; filter off the magnesium oxalate, acidify the filtrate with acetic acid and make up to a liter with water.

Guaiacum tincture: dissolve 1 g of guaiacum in 100 mL of alcohol.

Gunzberg's reagent (detection of HCl in gastric juice): dissolve 4 g of phloroglucinol and 2 g of vanillin in 100 mL of absolute alcohol; use only a freshly prepared solution.

Hager's reagent (for alkaloids): this reagent is a saturated solution of picric acid in water.

Hanus solution (for determination of iodine number): dissolve 13.2 g of iodine in a liter of glacial acetic acid that will not reduce chromic acid; add sufficient bromine to double the halogen content determined by titration (3 mL is about the right amount). The iodine may be dissolved with the aid of heat, but the solution must be cold when the bromine is added.

Hematoxylin (indicator): dissolve 0.5 g in 100 mL alcohol; pH range yellow 5.0—6.0.

Heptamethoxy red, 2,4,6,2′,4′,2″,4″-heptamethoxytriphenyl carbinol (indicator): dissolve 0.1 g in 100 mL alcohol; pH range red 5.0—7.0 colorless.

Hydriodic acid, HI—0.5*N*: 64 g per liter.

Hydrobromic acid, HBr—0.5*N*: 40 g per liter.

Hydrochloric acid, HCl—5*N*: 182 g per liter; sp. gr. 1.084.

Hydrofluoric acid, H_2F_2—48% solution: use as purchased, and keep in the special container.

Hydrogen peroxide, H_2O_2—3% solution: use as purchased.

Hydrogen sulfide, H_2S: prepare a saturated aqueous solution.

Indicator solutions: a number of indicator solutions are listed in this section under the names of the indicators; e.g., alizarin, aurin, azolitmin, et al., which follow alphabetically. See also various index entries.

Indigo carmine, sodium indigodisulfonate (indicator): dissolve 0.25 g in 100 mL 50% alcohol; pH range blue 11.6—14.0 yellow.

Indo-oxine, 5,8-quinolinequinone-8-hydroxy-5-quinoyl-5-imide (indicator): dissolve 0.05 g in 100 mL alcohol; pH range red 6.0—8.0 blue. Cf. Berg and Becker, *Z. Anal. Chem.* 119, 81 (1940).

Iodeosin, tetraiodofluorescein (indicator): dissolve 0.1 g in 100 mL ether saturated with water; pH range yellow 0—about 4 rose-red; see also under *methyl orange.*

Iodic acid, HIO_3—0.5*N* ($HIO_3/12$): 15 g per liter.

Iodine: see *tincture of iodine.*

Lacmoid (indicator): dissolve 0.5 g in 100 mL alcohol; pH range red 4.4—6.2 blue.

Litmus (indicator): powder the litmus and make up a 2% solution in water by boiling for 5 minutes; pH range red 4.5—8.3 blue.

Lead acetate, $Pb(C_2H_3O_2)_2 \cdot 3H_2O$—0.5*N*: 95 g per liter.

Lead chloride, $PbCl_2$—saturated solution is 1/7*N*.

Lead nitrate, $Pb(NO_3)_2$—0.5*N*: 83 g per liter.

Lime water: see *calcium hydroxide.*

Magnesia mixture: 100 g of $MgSO_4$, 200 g of NH_4Cl, 400 mL of NH_4OH, 800 mL of water; each cc ≎ 0.01 g phosphorus P).

Magnesium chloride, $MgCl_2 \cdot 6H_2O$—0.5*N*: 50 g per liter.

Magnesium nitrate, $Mg(NO_3)_2 \cdot 6H_2O$—0.5*N*: 64 g per liter.

Magnesium sulfate, epsom salts, $MgSO_4 \cdot 7H_2O$—0.5*N*: 62 g per liter; saturated solution dissolve 600 g of the salt in water and dilute to one liter.

Manganous chloride, $MnCl_2 \cdot 4H_2O$—0.5*N*: 50 g per liter.

Manganous nitrate, $Mn(NO_3)_2 \cdot 6H_2O$—0.5*N*: 72 g per liter.

Manganous sulfate, $MnSO_4 \cdot 7H_2O$—0.5*N*: 69 g per liter.

Marme's reagent (gives yellowish-white precipitate with salts of alkaloids): saturate a boiling solution of 4 parts of KI in 12 parts of water with CdI_2; then add an equal volume of cold saturated KI solution.

Marquis reagent (gives a purple-red coloration, then violet, then blue with morphine, codeine, dionine, and heroine): mix 3 mL of concentrated H_2SO_4 with 3 drops of a 35% formaldehyde solution.

Mayer's reagent (gives white precipitate with most alkaloids in a slightly acid solution): dissolve 13.55 g of $HgCl_2$ and 50 g of KI in a liter of water.

Mercuric chloride, $HgCl_2$—0.5*N*: 68 g per liter.

Mercuric nitrate, $Hg(NO_3)_2$—0.5*N*: 81 g per liter.

Mercuric sulfate, $HgSO_4$—0.5*N*: 74 g per liter.

Mercurous nitrate, $HgNO_3$: mix 1 part of $HgNO_3$, 20 parts of H_2O, and 1 part of HNO_3.

Metacresol purple, *m*-cresol-sulfonphthalein (indicator): dissolve 0.1 g in 13.6 mL 0.02*N* NaOH and dilute with water to 250 mL; acid pH range red 0.5—2.5 yellow, alkaline pH range yellow 7.4—9.0 purple.

Metanil yellow, diphenylaminoazo-*m*-benzene sulfonic acid (indicator): dissolve 0.25 g in 100 mL alcohol; pH range red 1.2—2.3 yellow.

Methyl green, hexamethylpararosaniline hydroxymethylate (component of mixed indicator): dissolve 0.1 g in 100 mL alcohol; when used with equal part of hexamethoxytriphenyl carbinol gives color change from violet to green at a titration exponent (pI) of 4.0.

Methyl orange, orange III, tropeolin D, sodium *p*-dimethylaminoazobenzene-sulfonate (indicator): dissolve 0.1 g in 100 mL water; pH range red 3.0—4.4 orange-yellow. If during a titration where methyl yellow is being used a precipitate forms which tends to remove the indicator from the aqueous phase, methyl orange will be found to be a more suitable indicator. This occurs, for example, in titrations of soaps with acids. The fatty acids, liberated by the titration, extract the methyl yellow so that the end-point cannot be perceived. Likewise methyl orange is more suitable for titrations in the presence of immiscible organic solvents such as carbon tetrachloride or ether used in the extraction of alkaloids for analysis. Iodeosin (*q.v.*) has also been proposed as an indicator for such cases. Cf. Mylius and Foerster, *Ber.* 24, 1482 (1891); *Z. Anal. Chem.* 31, 240 (1892).

Methyl red, *p*-dimethylaminoazobenzene-*o*′-carboxylic acid (indicator): dissolve 0.1 g in 18.6 mL 0.02 *N* NaOH and dilute with water to 250 mL; or, 0.1 g in 60% alcohol; pH range red 4.4—6.2 yellow.

Methyl violet (indicator): dissolve 0.25 g in 100 mL water; pH range blue 1.5—3.2 violet.

Methyl yellow, *p*-dimethylaminoazobenzene, benzeneazodimethylaniline (indicator): dissolve 0.1 g in 200 mL alcohol; pH range red 2.9—4.0 yellow. The color change from yellow to orange can be perceived somewhat more sharply than the change of methyl orange from orange to rose, so that methyl yellow seems to deserve preference in many cases. See also under *methyl orange.*

Methylene blue, *N,N,N′,N′*-tetramethylthionine (component of mixed indicator): dissolve 0.1 g in 100 mL alcohol; when used with equal part of methyl yellow gives color change from blue-violet to green at a titration exponent (pI) of 3.25; when used with equal part of 0.2% methyl red in alcohol gives color change from red-violet to green at a titration exponent (pI) of 5.4; when used with an equal part of neutral red gives color change from violet-blue to green at a titration exponent (pI) of 7.0.

Millon's reagent (gives a red precipitate with certain proteins and with various phenols): dissolve 1 part of mercury in 1 part of HNO_3 (sp. gr. 1.40) with gentle heating, then add 2 parts of water; a few crystals of KNO_3 help to maintain the strength of the reagent.

Mohr's salt: see *ferrous ammonium sulfate.*

α-Naphthol solution: dissolve 144 g of α-naphthol in enough alcohol to make a liter of solution.

α-Naphtholbenzein (indicator): dissolve 0.1 g in 100 mL 70% alcohol; pH range colorless 9.0—11.0 blue.

α-Naphtholphthalein (indicator): dissolve 0.1 g in 50 mL alcohol and dilute with water to 100 mL; pH range pale yellow-red 7.3—8.7 green.

Nessler's reagent (for free ammonia): dissolve 50 g of KI in the least possible amount of cold water; add a saturated solution of $HgCl_2$ until a very slight excess is indicated; add 400 mL of a 50% solution of KOH; allow to settle, make up to a liter with water, and decant.

Neutral red, toluylene red, dimethyldiaminophenazine chloride, aminodimethylaminotoluphenazine hydrochloride (indicator): dissolve 0.1 g in 60 mL alcohol and dilute with water to 100 mL; pH range red 6.8—8.0 yellow-orange.

Nickel chloride, $NiCl_2 \cdot 6H_2O$—0.5*N*: 59 g per liter.

Nickel nitrate, $Ni(NO_3)_2 \cdot 6H_2O$—0.5*N*: 73 g per iter.

Nickel sulfate, $NiSO_4 \cdot 6H_2O$—0.5*N*: 66 g per liter.

Nitramine, picrylmethylnitramine, 2,4,6-trinitrophenylmethyl nitramine (indicator): dissolve 0.1 g in 60 mL alcohol and dilute with water to 100 mL; pH range colorless 10.8—13.0 red-brown; the solution should be kept in the dark as nitramine is unstable; on boiling with alkali it decomposes quickly. Fresh solutions should be prepared every few months.

Nitric acid, HNO_3—5*N*: 315 g per liter; sp. gr. 1.165.

Nitrohydrochloric acid: see *aqua regia.*

***p*-Nitrophenol** (indicator): dissolve 0.2 g in 100 mL water; pH range colorless at about 5—7 yellow.

Nitroso-β-naphthol, $HOC_{10}H_6NO$—saturated solution: saturate 100 mL of 50% acetic acid with the solid.

Nylander's solution (detection of glucose): dissolve 40 g of rochelle salt and 20 g of bismuth subnitrate in 1000 mL of an 8% NaOH solution.

Obermayer's reagent (detection of indoxyl in urine): dissolve 4 g of $FeCl_3$ in a liter of concentrated HCl.

Orange III (indicator): see under *methyl orange.*

Oxalic acid, $H_2C_2O_4 \cdot 2H_2O$: dissolve in ten parts of water.

Pavy's solution (estimation of glucose): mix 120 mL of Fehling's solution and 300 mL of ammonium hydroxide (sp. gr. 0.88), and dilute to a liter with water.

Perchloric acid, $HClO_4$—60%: use as purchased.

Phenol solution: dissolve 20 g of phenol (carbolic acid) in a liter of water.

Phenol red, phenol-sulfonphthalein (indicator): dissolve 0.1 g in 14.20 mL 0.02*N* NaOH and dilute with water to 250 mL; or, 0.1 g in 100 mL 20% alcohol; pH range yellow 6.8—8.0 red.

Phenolphthalein (indicator): dissolve 1 g in 60 mL of alcohol and dilute with water to 100 mL; pH range colorless 8.2—10.0 red.

Phenol sulfonic acid (determination of nitrogen as nitrate; water analysis for nitrate): dissolve 25 g pure, white phenol in 150 mL of pure concentrated H_2SO_4, add 75 mL of fuming H_2SO_4 (15% SO_3), stir well and heat for two hours at 100°C.

Phosphoric acid, *ortho,* H_3PO_4—0.5*N*: 16 g per liter.

Poirrer blue C4B (indicator): dissolve 0.2 g in 100 mL water; pH range blue 11.0—13.0 red.

Potassium acid antimonate, KH_2SbO_4—0.1*N*: boil 23 g of the salt with 950 mL of water for 5 minutes, cool rapidly and add 35 mL of 6*N* KOH; allow to stand for one day, filter dilute filtrate to a liter.

Potassium arsenate, K_3AsO_4—0.5*N* (K_3AsO_4/10): 26 g per liter.

Potassium arsenite, $KAsO_2$—0.5*N* ($KAsO_2$/6): 24 g per liter.

Potassium bromate, $KBrO_3$—0.5*N* ($KBrO_3$/12): 14 g per liter.

Potassium bromide, KBr—0.5*N*: 60 g per liter.

Potassium carbonate, K_2CO_3—3*N*: 207 g per liter.

Potassium chloride, KCl—0.5*N*: 37 g per liter.

Potassium chromate, K_2CrO_4—0.5*N*: 49 g per liter.

Potassium cyanide, KCN—0.5*N*: 33 g per liter.

Potassium dichromate, $K_2Cr_2O_7$—0.5*N* ($K_2Cr_2O_7$/8): 38 g per liter.

Potassium ferricyanide, $K_3Fe(CN)_6$—0.5*N*: 55 g per liter.

Potassium ferrocyanide, $K_4Fe(CN)_6 \cdot 3H_2O$—0.5*N*: 53 g per liter.

Potassium hydroxide, KOH—5*N*: 312 g per liter.

Potassium iodate, KIO_3—0.5*N* (KIO_3/12): 18 g per liter.

Potassium iodide, KI—0.5*N*: 83 g per liter.

Potassium nitrate, KNO_3—0.5*N*: 50 g per liter.

Potassium nitrite, KNO_2—6*N*: 510 g per liter.

Potassium permanganate, $KMnO_4$—0.5*N* ($KMnO_4$/10): 16 g per liter.

Potassium pyrogallate (oxygen in gas analysis): weigh out 5 g of pyrogallol (pyrogallic acid), and pour upon it 100 mL of a KOH solution. If the gas contains less than 28% of oxygen, the KOH solution should be 500 g KOH in a liter of water; if there is more than 28% of oxygen in the gas, the KOH solution should be 120 g of KOH in 100 mL of water.

Potassium sulfate, K_2SO_4—0.5*N*: 44 g per liter.

Potassium thiocyanate, KCNS—0.5*N*: 49 g per liter.

Precipitating reagent (for group II, anions): dissolve 61 g of $BaCl_2 \cdot 2H_2O$ and 52 g of $CaCl_2 \cdot 6H_2O$ in water and dilute to one liter. If the solution becomes turbid, filter and use filtrate.

Quinaldine red (indicator): dissolve 0.1 g in 100 mL alcohol; pH range colorless 1.4—3.2 red.

Quinoline blue, cyanin (indicator): dissolve 1 g in 100 mL alcohol; pH range colorless 6.6—8.6 blue.

Rosolic acid, aurin, corallin, corallinphthalein, 4,4′-dihydroxy-fuchsone, 4,4′-dihydroxy-3-methyl-fuchsone (indicator): dissolve 0.5 g in 50 mL alcohol and dilute with water to 100 mL.

Salicyl yellow (indicator): see *alizarin yellow GG.*

Scheibler's reagent (precipitates alkaloids, albumoses and peptones): dissolve sodium tungstate in boiling water containing half its weight of phosphoric acid (sp. gr. 1.13); on evaporation of this solution, crystals of phosphotungstic acid are obtained. A 10% solution of phosphotungstic acid in water constitutes the reagent.

Schweitzer's reagent (dissolves cotton, linen, and silk, but not wool); add NH_4Cl and NaOH to a solution of copper sulfate. The blue precipitate is filtered off, washed, pressed, and dissolved in ammonia (sp. gr. 0.92).

Silver nitrate, $AgNO_3$—0.25*N*: 43 g per liter.

Silver sulfate, Ag_2SO_4—*N*/13 (saturated solution): stir mechanically 10 g of the salt in a liter of water for 3 hours; decant and use the clear liquid.

Soap solution (for hardness in water): (*a*) *Clark's or A.P.H.A. Stand. Methods*—prepare stock solution of 100 g of pure powdered castile soap in a liter of 80% ethyl alcohol; allow to stand over night and decant. Titrate against $CaCl_2$ solution (0.5 g $CaCO_3$ dissolved in a concentrated HCl, neutralized with NH_4OH to slight alkalinity using litmus as the indicator, make up to 500 mL; 1 mL of this solution is equivalent to 1 mg $CaCO_3$) and dilute with 80% alcohol until 1 mL of the resulting solution is equivalent to 1 mL of the standard $CaCl_2$ making due allowance for the lather factor (the lather factor is that amount of standard soap solution required to produce a permanent lather in a 50-mL portion of distilled water). One mL of this solution after subtracting the lather factor is equivalent to 1 mg of $CaCO_3$. (*b*) *Boutron-Boudet*—dissolve 100 g of pure castile soap in about 2500 mL of 56% ethyl alcohol and adjust so that 2.4 mL will give a permanent lather with 40 mL of a solution containing 0.59 g $Ba(NO_3)_2$ per liter of water; 2.4 mL of this solution is equivalent to 22 French degrees or 220 parts per million of hardness (as $CaCO_3$) on a 40-mL sample of water.

Sodium acetate, $NaC_2H_3O_2 \cdot 3H_2O$: dissolve 1 part of the salt in 10 parts of water.

Sodium acetate, acid: dissolve 100 g of sodium acetate and 30 mL of glacial acetic acid in water and dilute to one liter.

Sodium bismuthate (oxidation of manganese): heat 20 parts of NaOH nearly to redness in an iron or nickel crucible, and add slowly 10 parts of basic bismuth nitrate which has been previously dried. Add 2 parts of sodium peroxide, and pour the brownish-yellow fused mass on an iron plate to cool. When cold break up in a mortar, extract with water, and collect on an asbestos filter.

Sodium carbonate, Na_2CO_3—3*N*: 159 g per liter; one part Na_2CO_3, or 2.7 parts of the crystalline $Na_2CO_3 \cdot 10H_2O$ in 5 parts of water.

Sodium chloride, NaCl—0.5*N*: 29 g per liter.

Sodium chloroplatinite, Na_2PtCl_4: dissolve 1 part of the salt in 12 parts of water.

Sodium cobaltinitrite, $Na_3Co(NO_2)_6$—0.3*N*: dissolve 230 g of $NaNO_2$ in 500 mL of water, add 160 mL of 6*N* acetic acid and 35 g of $Co(NO_3)_2 \cdot 6H_2O$. Allow to stand one day, filter, and dilute the filtrate to a liter.

Sodium hydrogen phosphate, $Na_2HPO_4 \cdot 12H_2O$—0.5*N*: 60 g per liter.

Sodium hydroxide, NaOH—5*N*: 220 g per liter.

Sodium hydroxide, alcoholic: dissolve 20 g of NaOH in alcohol and dilute to one liter with alcohol.

Sodium hypobromite: dissolve 100 g of NaOH in 250 mL of water and add 25 mL of bromine.

Sodium nitrate, $NaNO_3$—0.5*N*: 43 g per liter.

Sodium nitroprusside (for sulfur detection): dissolve about one gram of sodium nitroprusside in 10 mL of water; as the solution deteriorates on standing, only freshly prepared solutions should be used. This compound is also called sodium nitroferricyanide and has the formula $Na_2Fe(NO)(CN)_5 \cdot 2H_2O$.

Sodium polysulfide, Na_2S_x: dissolve 480 g of $Na_2S \cdot 9H_2O$ in 500 mL of water, add 40 g of NaOH and 18 g of sulfur, stir mechanically and dilute to one liter with water.

Sodium sulfate, Na_2SO_4—0.5*N*: 35 g per liter.

Sodium sulfide, Na_2S; saturate NaOH solution with H_2S, then add as much NaOH as was used in the original solution.

Sodium sulfite, $Na_2SO_3 \cdot 7H_2O$—0.5*N*: 63 g per liter.

Sodium sulfite, acid (saturated): dissolve 600 g of $NaHSO_3$ in water and dilute to one liter; for the preparation of addition compounds with aldehydes and ketones: prepare a saturated solution of sodium carbonate in water and saturate with sulfur dioxide.

Sodium tartrate, acid, $NaHC_4H_4O_6$: dissolve 1 part of the salt in 10 parts of water.

Sodium thiosulfate, $Na_2S_2O_3 \cdot 5H_2O$: one part of the salt in 40 parts of water.

Sonnenschein's reagent (alkaloid detection): a nitric acid solution of ammonium molybdate is treated with phosphoric acid. The precipitate so produced is washed and boiled with aqua regia until the ammonium salt is decomposed. The solution is evaporated to dryness and the residue is dissolved in 10% HNO_3.

Stannic chloride, $SnCl_4$—0.5*N*: 33 g per liter.

Stannous chloride, $SnCl_2 \cdot 2H_2O$—0.5*N*: 56 g per liter. The water should be acid with HCl and some metallic tin should be kept in the bottle.

Starch solution (iodine indicator): dissolve 5 g of soluble starch in cold water, pour the solution into 2 liters of water and boil for a few minutes. Keep in a glass-stoppered bottle.

Starch solution (other than soluble): make a thin paste of the starch with cold water, then stir in 200 times its weight of boiling water and boil for a few minutes. A few drops of chloroform added to the solution acts as a preservative.

Stoke's reagent: dissolve 30 g of ferrous sulfate and 20 g of tartaric acid in water and dilute to one liter. When required for use, add strong ammonia until the precipitate first formed is dissolved.

Strontium chloride, $SrCl_2 \cdot 6H_2O$—0.5*N*: 67 g per liter.

Strontium nitrate, $Sr(NO_3)_2$—0.5*N*: 53 g per liter.

Strontium sulfate, $SrSO_4$: prepare a saturated solution.

Sulfanilic acid (for detection of nitrites): dissolve 8 g of sulfanilic acid in one liter of acetic acid (sp. gr. 1.04).

Sulfuric acid, H_2SO_4—5*N*: 245 g per liter, sp. gr. 1.153.

Sulfurous acid, H_2SO_3: saturate water with sulfur dioxide.

Tartaric acid, $H_2C_4H_4O_6$: dissolve one part of the acid in 3 parts of water; for a saturated solution dissolve 750 g of tartaric acid in water and dilute to one liter.

Tannic acid: dissolve 1 g tannic acid in 1 mL alcohol and make up to 10 mL with water.

Tetrabromophenol blue, tetrabromophenol-tetrabromosulfonphthalein (indicator): dissolve 0.1 g in 5 mL 0.02*N* NaOH and dilute with water to 250 mL; pH range yellow 3.0—4.6 blue.

Thymol blue, thymol-sulfonphthalein (indicator): dissolve 0.1 g in 10.75 mL 0.02*N* NaOH and dilute with water to 250 mL; or, dissolve 0.1 g in 20 mL warm alcohol and dilute with water to 100 mL; pH range (acid) red 1.2—2.8 yellow, and (alkaline) yellow 8.0—9.6 blue.

Thymolphthalein (indicator): dissolve 0.1 g in 100 mL alcohol; pH range colorless 9.3—10.5 blue.

Tincture of iodine (antiseptic): add 70 g of iodine and 50 g of KI to 50 mL of water; make up to one liter with alcohol.

***o*-Tolidine solution** (for residual chlorine in water analysis): dissolve 1 g of pulverized *o*-tolidine, m.p. 129°C., in one liter of dilute hydrochloric acid (100 mL conc. HCl diluted to one liter).

Toluylene red (indicator): see *neutral red.*

Trichloroacetic acid: dissolve 100 g of the acid in water and dilute to one liter.

Trinitrobenzene, 1,3,5-trinitrobenzene (indicator): dissolve 0.1 g in 100 mL alcohol; pH range colorless 11.5—14.0 orange.

Trinitrobenzoic acid, 2,4,6-trinitrobenzoic acid (indicator): dissolve 0.1 g in 100 mL water; pH range colorless 12.0—13.4 orange-red.

Tropeolin D (indicator): see *methyl orange.*

Tropeolin O, sodium 2,4-dihydroxyazobenzene-4-sulfonate (indicator): dissolve 0.1 g in 100 mL water; pH range yellow 11.0—13.0 orange-brown.

Tropeolin OO, orange IV, sodium *p*-diphenylamino-azobenzene sulfonate, sodium 4′-anilino-azobenzene-4-sulfonate (indicator): dissolve 0.1 g in 100 mL water; pH range red 1.3—3.2 yellow.

Tropeolin OOO, sodium α-naphtholazobenzene sulfonate (indicator): dissolve 0.1 g in 100 mL water; pH range yellow 7.6—8.9 red.

Turmeric paper (gives a rose-brown coloration with boric acid); wash the ground root of turmeric with water and discard the washings. Digest with alcohol and filter, using the clear filtrate to impregnate white, unsized paper, which is then dried.

Uffelmann's reagent (gives a yellow coloration in the presence of lactic acid): add a ferric chloride solution to a 2% phenol solution until the solution becomes violet in color.

Wagner's solution (phosphate rock analysis): dissolve 25 g citric acid and 1 g

TABLE 4.5 List of Laboratory Solutions, Reagents, and Indicators *(Continued)*

salicylic acid in water, and make up to one liter. Twenty-five to 50 mL of this reagent prevents precipitation of iron and aluminum.

Wijs solution (for iodine number): dissolve 13 g resublimed iodine in one liter of glacial acetic acid (99.5%), and pass in washed and dried (over or through H_2SO_4) chlorine gas until the original thio titration of the solution is not quite doubled. There should be only a slight excess of iodine and no excess of chlorine. Preserve the solution in amber colored bottles sealed with paraffin. Do not use the solution after it has been prepared for more than 30 days.

Xylene cyanole-methyl orange indicator, Schoepfle modification (for partially color blind operators): dissolve 0.75 g xylene cyanole FF (Eastman No. T 1579) and 1.50 g methyl orange in 1 liter of water.

***p*-Xylenol blue,** 1,4-dimethyl-5-hydroxybenzene-sulfonphthalein (indicator): dissolve 0.1 g in 250 mL alcohol; pH range (acid) red 1.2—2.8 yellow, and (alkaline) yellow 8.0—9.6 blue.

Zinc chloride, $ZnCl_2$—0.5*N*: 34 g per iter.

Zinc nitrate, $Zn(NO_3)_2 \cdot 6H_2O$—0.5*N*: 74 g per liter.

Zinc sulfate, $ZnSO_4 \cdot 7H_2O$—0.5*N*: 72 g per liter.

Example 4.1. Determine the mass of oxygen that can be produced by the complete decomposition of 490 g of potassium chlorate.

Solution. The solution is based on the following information: There is 490 g $KClO_3$. Also,

$$2KClO_3 \rightarrow 2KCl + 3O_2$$

$$\text{Mol wt of } KClO_3 = (39 + 35.5 + 48) = 122.5 \text{ g/g} \cdot \text{mol}$$

$$2KClO_3 = 2 \times 122.5 = 245$$

$$\text{Mol wt of } O_2 = (16)(2) = 32 \text{ g/g} \cdot \text{mol}$$

$$3O_2 = 3(32) = 96$$

Let X be the mass of O_2 liberated.

The proportion dictated by the law of combining weights is expressed by this scheme:

$$2KClO_3 \rightarrow 2KCl + 3O_2$$

$$\frac{2(122.5)\text{ g}}{490\text{ g}} = \frac{3(32)\text{ g}}{X\text{ g}}$$

Set a proportion

$$\frac{2(122.5)}{490} = \frac{3(32)}{X}$$

C_7H_{16}	+	$11O_2$	→	$7CO_2$	+	$8H_2O$
			Qualitative description			
Heptane	reacts with	oxygen	to give	carbon dioxide	and	water
			Quantitative description			
1 molecule of heptane	reacts with	11 molecules of oxygen	to give	7 molecules of carbon dioxide	and	8 molecules of water
6.023×10^{23} molecules of C_7H_{16}	+	$11(6.023 \times 10^{23})$ molecules of O_2	→	$7(6.023 \times 10^{23})$ molecules of CO_2	+	$8(6.023 \times 10^{23})$ molecules of H_2O
1 g · mol of C_7H_{16}	+	11 g · mol of O_2	→	7 g · mol of CO_2	+	8 g · mol of H_2O
1 lb · mol of C_7H_{16}	+	11 lb · mol of O_2	→	7 lb · mol of CO_2	+	8 lb · mol of H_2O
1 kg · mol of C_7H_{16}	+	11 kg · mol of O_2	→	7 kg · mol of CO_2	+	8 kg · mol of H_2O
1(100) g of C_7H_{16}	+	11(32) g of O_2	→	7(44) g of CO_2	+	8(18) g of H_2O
100 g		352 g		308 g		144 g
	452 g		=		452 g	
	452 kg		=		452 kg	
	452 lb		=		452 lb	

Note:

Component	Mol wt
C_7H_{16}	100
O_2	32
CO_2	44
H_2O	18

FIGURE 4.2 The chemical equation.

or

$$X = \frac{192 \text{ g } O_2}{490 \text{ g } KClO_3}$$

or expressed in another form: The chemical equation states that every 245 g of $KClO_3$ produces 96 g of O_2. How much O_2 can be produced with 490 g of $KClO_3$?

$$\frac{245 \text{ g } KClO_3}{490 \text{ g } KClO_3} = \frac{96 \text{ g } O_2}{X}$$

and

$$X = \frac{96 \text{ g } O_2 \times 490}{245}$$
$$= 192 \text{ g } O_2$$

CHEMICAL THERMODYNAMIC RELATIONS

Chemical Potential, $\hat{\mu}$. Chemical potential at constant temperature and pressure is Gibbs free energy per mole, expressed in J/mol. For a multicomponent system, $\hat{\mu}$ is the summation of potentials (partial molal free energies) of all the components present:

$$\hat{\mu} = \Sigma \frac{G_i}{n_i} = \Sigma \hat{g}_i \tag{4.1}$$

where G_i is Gibbs free energy of component i, and n_i is moles of i present.

Chemical Energy. Like other forms of energy, all products of intensive and an extensive properties, chemical energy added to a system is $\hat{\mu}_i \cdot dn_i$ for one component, and $\Sigma \hat{\mu}_i \cdot dn_i$ for all components of the system.

Chemical Thermodynamic Equations. Basic differential equations of engineering thermodynamics are expanded by adding the chemical energy and other appropriate energy terms.

Relation of chemical potential to internal energy:

$$dU = T\,dS - P\,dV + \Sigma\,\hat{\mu}_i \cdot dn_i \tag{4.2}$$

Relation to enthalpy:

$$dH = T\,dS + V\,dP + \Sigma\,\hat{\mu}_i \cdot dn_i \tag{4.3}$$

Relation to Gibbs free energy:

$$dG = -S\,dT + V\,dP + \Sigma\,\hat{\mu}_i \cdot dn_i \tag{4.4}$$

Relation to Helmholtz function:

$$dF = -S\,dT - P\,dV + \Sigma\,\hat{\mu}_i \cdot dn_i \tag{4.5}$$

Activity, a. Activity is a thermodynamically effective concentration used in lieu of actual concentrations to compensate for deviations of gases from ideality, incomplete ionization of strong electrolytes in solution, and other discrepancies between calculated and experimental behavior.

Fugacity, f. Fugacity (escaping tendency) of a gas equals its activity. Fugacity is an effective partial pressure, expressed in atmospheres. For ideal gases fugacity exactly equals partial pressure. Since real gases in the standard state (in their usual phase at 1 atm pressure and 25°C) deviate slightly from ideality, activity of real gases is defined as the ratio of fugacity f to fugacity f^0, in a standard state where $f^0 = 1$ atm. For all practical purposes, f^0 is 1 atmosphere pressure. Nonidealities of real gases at other temperatures and pressures are handled by an activity coefficient, γ.

Activity Coefficient, γ. The activity coefficient is the ratio of fugacity to partial pressure, or the ratio of activity to molal concentration. For ideal gases: $\gamma = 1$.

Calculation of ΔG. At constant temperature and constant number of moles $dT = 0$ and $dn_i = 0$, Eq. (4.4) reduces to $dG = +V\,dP$. The calculation of ΔG for ideal gases, real gases, solids, and liquids at any pressure is dependent only on an expression for V as $f(P)$.

For ideal gases, integrating between limits gives:

$$\Delta G = RT \ln \frac{P_2}{P_1} \tag{4.6}$$

$$\Delta G = RT \ln \frac{f_2}{f_1} \tag{4.7}$$

For incompressible liquids and solids, V is constant, so

$$\Delta G = V(P_2 - P_1) \tag{4.8}$$

using appropriate units. The magnitude of this change for all condensed phases is very small.

THERMOCHEMISTRY*

1. Heat of Chemical Reaction. The heat of reaction or enthalpy of reaction $\Delta\hat{H}$, (T,P) is the difference $H_{\text{products}} - H_{\text{reactants}}$ for a reaction that takes place under the following circumstances:

1. Stoichiometric quantities of the reactants are fed, and the reaction proceeds to completion.
2. The reactants are fed at temperature T and pressure P, and the products emerge at the same temperature and pressure.

For example, for the reaction

$$CaC_2(s) + 2H_2O(l) \rightarrow Ca(OH)_2(s) + C_2H_2(g)$$

the heat of reaction at 25°C and a total pressure of 1 atm is

$$\Delta\hat{H}(25°\text{C, 1 atm}) = -125.4 \text{ kJ/mol}$$

If the reaction is run under conditions such that the energy balance reduces to $Q = \Delta H$, then 125.4 kJ is emitted by the reaction system in the course of the reaction. (A negative Q implies flow of heat out of the system.)

If ν_A is the stoichiometric coefficient of a reactant or reaction product A, and n_A moles of A are consumed or produced at $T = T_0$, the total enthalpy change is

*Parts of this section based on material taken from *Tehnička hemija* (*Chemistry for Engineers—Lecture Notes*), by E. N. Ganić. Copyright © 1984 by the author. Parts are also taken from *Elementary Principles of Chemical Processes*, by R. M. Felder and R. W. Rousseau. Copyright © 1978. Used by permission of Wiley. All rights reserved.

$$\Delta H = \frac{\Delta \hat{H}_r(T_0)}{\nu_A} n_A \tag{4.9}$$

Several definitions and properties of heats of reaction are summarized below:

1. The standard heat of reaction $\Delta \hat{H}_r^0$ is the heat of reaction when both the reactions and products are at a specified temperature and pressure, usually 25°C and 1 atm.
2. If $\Delta \hat{H}_r(T)$ is negative, the reaction is said to be *exothermic* at temperature T; if $\Delta \hat{H}_r(T)$ is positive, the reaction is *endothermic* at T.
3. At low and moderate pressure, $\Delta \hat{H}_r$ is nearly independent of pressure.
4. The value of a heat of reaction depends on how the stoichiometric equation is written: for example, $\Delta \hat{H}_r^0$ for $A \rightarrow B$ is half of $\Delta \hat{H}_r^0$ for $2A \rightarrow 2B$. (This should seem reasonable taking into account the definition of $\Delta \hat{H}_r$.)
5. The value of a heat of reaction depends on the states of aggregation (gas, liquid, or solid) of the reactants and products.

Various types of heats of reaction are found in thermochemical systems. They include heats of formation, combustion, neutralization, solution, dilution, dissociation, and polymerization, among others. Since all are particular types of heats of reaction, they are all subject to similar treatments and thermodynamic requirements.

Hess's Law of Heat Summation. Hess showed in 1840 that the overall heat evolved or absorbed in a chemical reaction proceeding in several steps is equal to the algebraic sum of the enthalpies of the various stages. These chemical equations may be treated like ordinary algebric equations.

2. Formation Reactions and Heat of Formation. A formation reaction of a compound is the reaction in which the compound is formed from its atomic constituents as they normally occur in nature (e.g., O_2 from O). The standard heat of such a reaction is the standard heat of formation, $\Delta \hat{H}_f^0$, of the compund.

Standard heats of formation for many compunds are listed in Tables 9.1 and 9.2 of Ref. 1. For example, $\Delta \hat{H}_f^0$ for crystalline ammonium nitrate is given in Table 9.1 of Ref. 2 as −365.14 kJ/mol, signifying

$$N_2(g) + 2H_2(g) + \tfrac{3}{2}O_2(g) \rightarrow NH_4NO_3(c){:}\ \Delta \hat{H}_r^0 = -365.14 \text{ kJ/mol}$$

Similarly, for liquid benzene $\Delta \hat{H}_f^0 = 48.66$ kJ/mol, or

$$6C(s) + 3H_2(g) \rightarrow C_6H_6(l){:}\ \Delta \hat{H}_r^0 = 48.66 \text{ kJ/mol}$$

The standard heat of formation (pressure 1 atm and a temperature of 25°C) of an element is conveniently chosen to be zero.

It may be shown using Hess's law that if ν_i is the stoichiometric coefficient of the ith species participating in a reaction and $(\Delta \hat{H}_f^0)$ is the standard heat of formation of this species, the standard heat of the reaction is

$$\Delta \hat{H}_r^0 = \sum_{\text{products}} \nu_i (\Delta \hat{H}_f^0)_i - \sum_{\text{reactants}} \nu_i (\Delta \hat{H}_f^0)_i \tag{4.10}$$

3. Energy Balance on Reactive Processes. Several approaches to setting up energy balances on reactive systems are often adopted, which differ in the reference conditions used for enthalpy calculations. Suppose $\Delta \hat{H}_r^0$ for each reaction occurring in a system is known or calculable at a temperature T_0 (which is usually but not always 25°C). The following are two common choices of reference conditions, along with the procedures for calculating ΔH that correspond to each choice.

Choice 1. Reference conditions: the elements that constitute the reactants and products at 25°C and the nonreactive molecular species at any convenient temperature. A table of n_i's (n = number of moles) and $\hat{H}_i$'s for all stream components is constructed, only now $\hat{H}_i$ for a reactant or product is the sum of the heat of formation of the species at 25°C and any sensible and latent heats required to bring the species from 25°C to its inlet or outlet state. The overall enthalpy change for the process is then

$$\Delta H = \sum_{\substack{\text{outlet} \\ \text{(products)}}} n_i \hat{H}_i - \sum_{\substack{\text{inlet} \\ \text{(reactants)}}} n_i \hat{H}_i \tag{4.11}$$

Example 4.2 Energy balance on a methane oxidation reactor. Methane is oxidized with air to produce formaldehyde in a continuous reactor. A competing reaction is the combustion of methane to form CO_2.

1. $CH_4(g) + O_2 \rightarrow HCHO(g) + H_2O(g)$
2. $CH_4 + 2O_2 \rightarrow CO_2 + 2H_2O(g)$

A flow chart of the process for an assumed basis of 100 mol of methane fed to the reactor is shown in Fig. 4.3.

The pressure is low enough for ideal gas behavior to be assumed. If the methane enters the reactor at 25°C and the air enters at 100°C, how much heat must be withdrawn for the product stream to emerge at 150°C?

Solution. The problem is solved as follows:

CH_4(150°C):

$$(\hat{H}_{CH_4})_{out} = (\Delta \hat{H}_f^0)_{CH_4} + \int_{25}^{150} (C_p)_{CH_4}\, dT$$

$$= -74.85 + 4.90 = -69.95 \text{ kJ/mol}$$

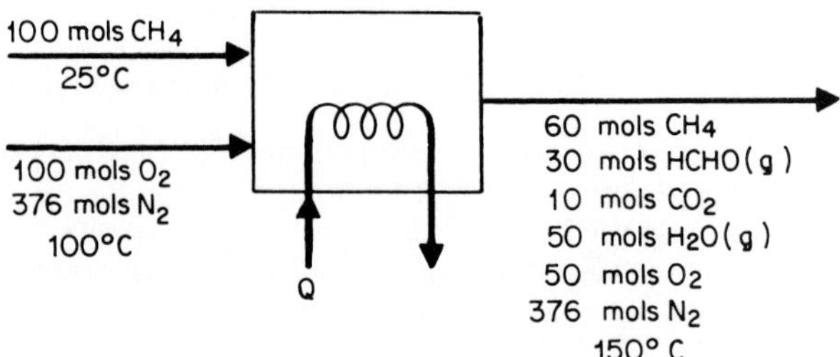

FIGURE 4.3 Flow chart of methane oxidation process.

HCHO(g, 150°C):

$$\hat{H} = (\Delta \hat{H}_f^0)_{HCHO(g)} + \int_{25}^{150} (C_p)_{HCHO}\, dT$$
$$= -115.90 + 4.75 = -111.15 \text{ kJ/mol}$$

In the last two integrals C_p is substituted from Eq. (4.17) (see below) (see Table 4.6 for values of coefficients of a, b, and c).

CO_2(150°C):

$$\hat{H} = (\Delta \hat{H}_f^0)_{CO_2(g)} + [125\overline{C}_p(150°C)]$$

Taking $\Delta \hat{H}_f^0$ from Table 9.2 of Ref. 1 and $\overline{C}_p$ from Table 4.7,

$$(\hat{H}_{CO_2})_{out} = -393.5 + 4.94 = -388.6 \text{ kJ/mol}$$

H_2O(g, 150°C):

$$\hat{H} = (\Delta \hat{H}_f^0)_{H_2O(g)} + 125\overline{C}_p(150°C)$$

Taking $\Delta \hat{H}_f^0$ from Table 9.1 of Ref. 1 and $\overline{C}_p$ from Table 4.7,

$$\hat{H} = -241.83 + 4.27 = -237.56 \text{ kJ/mol}$$

Evaluate ΔH. From Eq. (4.11)

$$\Delta H = \sum_{out} n_i \hat{H}_i - \sum_{in} n_i \hat{H}_i = -15{,}300 \text{ kJ}$$

For the energy balance, neglecting kinetic energy changes

$$Q = \Delta H = -15{,}300 \text{ kJ}$$

Choice 2. Reference conditions: reactant and product species at T_0 in the state of aggregation for which $\Delta \hat{H}_r^0$ is known, and nonreactive species at any convenient temperature (such as the reactor inlet temperature, or the reference temperature of a mean heat capacity table).

TABLE 4.6 Molar Heat Capacities of Some Gases

C_p(J/mol · °C) $= a + bT + cT^2 + dT^3$; T in °C

Compound	Formula	a	$b \cdot 10^2$	$C \cdot 10^5$	$d \cdot 10^9$
Oxygen	O_2	29.10	1.158	−0.6076	1.311
Hydrogen	H_2	28.84	0.00765	0.3288	−0.8698
Nitrogen	N_2	29.00	0.2199	0.5723	−2.871
Carbon monoxide	CO	28.95	0.4110	0.3548	−2.220
Carbon dioxide	CO_2	36.11	4.233	−2.887	7.464
Water	$H_2O(g)$	33.46	0.6880	0.7604	−3.593
Methane	CH_4	34.28	4.268	0.0	−8.694
Formaldehyde	CH_2O	34.31	5.469	0.366	−11.0

Note: Data shown in this table can also be found in Refs. 2 through 5.

TABLE 4.7 Mean Heat Capacities of Combustion Gases

$\overline{C}_p$(J/mol · °C); reference state: P_{ref} = 1 atm, T_{ref} = 25°C

T(°C)	Air	O_2	N_2	H_2	CO	CO_2	H_2O
0	28.94	29.24	29.03	28.84	29.00	36.63	33.55
25	29.05	29.39	29.06	28.84	29.06	37.15	33.63
100	29.21	29.80	29.16	28.86	29.23	38.63	33.92
200	29.45	30.32	29.32	28.90	29.47	40.45	34.34
300	29.71	30.80	29.52	28.95	29.72	42.10	34.80
400	29.97	31.24	29.74	29.03	29.99	43.59	35.29
500	30.25	31.65	29.98	29.12	30.27	44.93	35.81
600	30.53	32.02	30.24	29.23	30.56	46.14	36.36
700	30.81	32.39	30.51	29.35	30.85	47.23	36.92
800	31.10	32.71	30.79	29.48	31.14	48.20	37.49
900	31.38	33.02	31.07	29.63	31.42	49.07	38.08
1000	31.65	33.30	31.34	29.78	31.70	49.85	38.66
1100	31.92	33.55	31.62	29.94	31.97	50.54	39.24
1200	32.18	33.79	31.88	30.12	32.23	51.18	39.81
1300	32.42	34.02	32.13	30.29	32.47	51.75	40.37
1400	32.65	34.23	32.37	30.47	32.69	52.28	40.91
1500	32.85	34.42	32.58	30.66	32.89	52.77	41.42

Source: From Felder and Rousseau,[7] which in turn was taken from data presented in Himmelblau.[8]

If these reference conditions are chosen, the enthalpy change for the process may be determined by setting up and filling in a table of inlet and outlet stream component flow rates n_i and specific enthalpies $\hat{H}_i$, and calculating

$$\Delta H = \frac{n_{AR}\Delta \hat{H}_r^0}{\nu_A} + \sum_{\text{outlet}} n_i\hat{H}_i - \sum_{\text{inlet}} n_i\hat{H}_i \tag{4.12}$$

where A = any reactant or product
n_{AR} = moles of A produced or consumed in the process (not necessarily moles fed or moles present in the product)
ν_A = stoichiometric coefficient of A

Both n_{AR} and ν_A are positive numbers.

If multiple reactions occur, a term of the form $n_A\Delta \hat{H}_r^0/\nu_A$ must be included in Eq. (4.12) for each reaction. (Method 1 is generally easier to use in such a case.)

Example 4.3 Simultaneous material and energy balances. The ethanol dehydrogenation reaction is carried out with the feed entering at 300°C. The feed contains 90 mol % ethanol and the balance acetaldehyde. To keep the temperature from dropping too rapidly and hence quenching the reaction at a low conversion, heat is added to the reactor. Observe that when the heat addition rate is 5300 kJ per 100 mol of the feed gas, the outlet temperature is 265°C. Calculate the fractional conversion of ethanol achieved in the reactor using the heat capacity data given in Tables 4.6 and 4.7. A flow chart of the process is shown in Fig. 4.4.

Solution. Basis: 100 mol feed gas; also,

$$C_2H_5OH(g) \rightarrow CH_3CHO(g) + H_2(g){:}\ \Delta \hat{H}_r^0 = 68.95 \text{ kJ/mol}$$

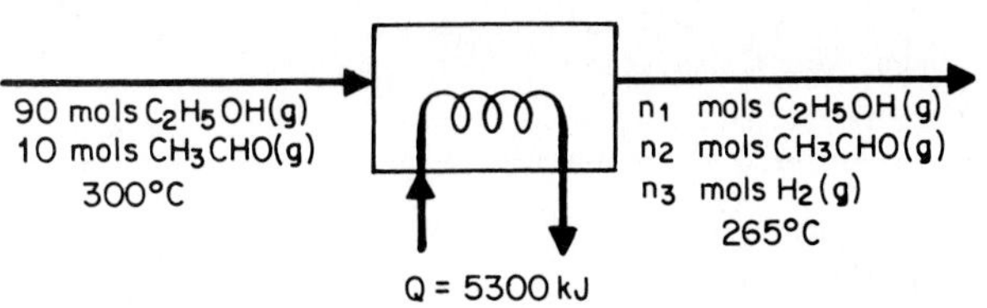

FIGURE 4.4 Flow chart of ethanol dehydrogenation process.

There are three unknowns, and no three material balance equations will enable one to solve for n_i, n_2, and n_3. The problem must be solved by writing and simultaneously solving two material balance equations and an energy balance equation.

Balance on C:

$$(90)(2) + (10)(2) = 2n_1 + 2n_2$$

$$n_1 + n_2 = 100 \tag{1}$$

Balance on H:

$$(90)(6) + (10)(4) = 6n_1 + 4n_2 + 2n_3$$

$$6n_1 + 4n_2 + 2n_3 = 580 \tag{2}$$

Energy balance (at 25°C):

Substance	n_{in}	$\hat{H}_{in}$	n_{out}	$\hat{H}_{out}$	
$C_2H_5OH(g)$	90	30.3	n_1	26.4	n in mol
$CH_3CHO(g)$	10	22.0	n_2	19.2	$\hat{H}$ in kJ/mol
$H_2(g)$			n_3	7.0	

$\hat{H} = \overline{C}_p(T - 25)$: heat capacities from Table 4.7. The table provides the following information:

$(C_2H_5OH)_{inlet}$:

$$\hat{H} = (0.110 \text{ kJ/mol} \cdot {}^\circ\text{C})(300^\circ\text{C} - 25^\circ\text{C}) = 30.3 \text{ kJ/mol}$$

$(CH_3CHO)_{inlet}$:

$$\hat{H} = (0.080)(275) = 22.0 \text{ kJ/mol}$$

$(C_2H_5OH)_{outlet}$:

$$\hat{H} = (0.110)(265 - 25) = 26.4 \text{ kJ/mol}$$

$(CH_3CHO)_{outlet}$:

$$\hat{H} = (0.080)(240) = 19.2 \text{ kJ/mol}$$

$(H_2)_{outlet}$:

$$\hat{H} = (0.029)(240) = 7.0 \text{ kJ/mol}$$

$Q = \Delta H$. Thus from Eq. (4.12):

$$Q = \frac{(n_{H_2})_{out}\,\Delta \hat{H}_r^0}{\nu_{H_2}} + \sum_{out} n_i\hat{H}_i - \sum_{in} n_i\hat{H}_i$$

$$5300 = \frac{n_3(68.95)}{1} + 26.4n_1 + 19.2n_2 + 7.0n_3 - (90)(30.3) - (10)(22.0)$$

$$26.4n_1 + 19.2n_2 + 7.6n_3 = 8247 \tag{3}$$

Solving Eqs. (1) through (3) simultaneously yields

$$n_1 = 7.5 \text{ mol } C_2H_5OH(g)$$

$$n_2 = 92.5 \text{ mol } CH_3CHO(g)$$

$$n_3 = 82.5 \text{ mol } H_2(g)$$

The fractional conversion of ethanol is

$$x = \frac{(n_{C_2H_5OH})_{in} - (n_{C_2H_5OH})_{out}}{(n_{C_2H_5OH})_{in}} = \frac{90 - 7.5}{90} = 0.917$$

Note on Heat Capacity. If the change in specific enthalpy of a substance that goes from T_1 to T_2 is $\hat{H}_2 - \hat{H}_1$, a mean heat capacity $\overline{C}_p$ may be defined as

$$\overline{C}_p = \frac{\hat{H}_2 - \hat{H}_1}{T_2 - T_1} \tag{4.13}$$

Substituting for $\hat{H}_2 - \hat{H}_1$ from the relation

$$\Delta\hat{H} = \int_{T_1}^{T_2} C_p(T)\,dT$$

one gets

$$\overline{C}_p(T_1 \rightarrow T_2) = \frac{\int_{T_1}^{T_2} C_p(T)\,dT}{(T_2 - T_1)} \tag{4.14}$$

Once $\overline{C}_p$ is known for the specified change in temperature (e.g., from a table of mean heat capacities), one may calculate the enthalpy change as

$$\Delta\hat{H} = \overline{C}_p\Delta T \tag{4.15}$$

A simple multiplication thus replaces integration of $C_p\,dT$ in the calculation of $\Delta\hat{H}$. [Note, however, that if $\overline{C}_p$ is not known and the calculation of $\Delta\hat{H}$ is to be done only once, there is little point in first determining $\overline{C}_p$ from Eq. (4.13) since the numerator of this equation is the desired quantity.]

Table 4.7 give mean heat capacities for the transitions of several common gases from a single base temperature T_{ref} to higher temperatures. These tables may be used to calculate the specific enthalpy of any of the tabulated species relative to the reference state as $\hat{H} = \overline{C}_p(T - T_{ref})$; the table may also be used to calculate $\Delta\hat{H}$ for a change between any two temperatures T_1 and T_2:

$$\Delta\hat{H}(T_1 \rightarrow T_2) = \hat{H}(T_2) - \hat{H}(T_1) = (\overline{C}_p)_{T_2}(T_2 - T_{\text{ref}}) - (\overline{C}_p)_{T_1}(T_1 - T_{\text{ref}}) \tag{4.16}$$

Introducing

$$C_p = a + bT + cT^2 + dT^3 \tag{4.17}$$

in Eq. (4.14) allows one to calculate $\overline{C}_p$ when coefficients a, b, c, and d are given.

CHEMICAL EQUILIBRIUM

1. Equilibrium Constant. The equilibrium constant for the reaction

$$aA + bB = cC + dD$$

where the reaction is in solution is

$$K_c = \frac{[C]^c[D]^d}{[A]^a[B]^b} \tag{4.18}$$

([ ] refers to molarity). If reaction is in the gas phase, the equilibrium constant is

$$K_p = \frac{p_C^c \cdot p_D^d}{p_A^a \cdot p_B^b} \tag{4.19}$$

where p = partial pressure.

Note that equilibrium constants are dimensional; their units are useful in reconstructing the equilibrium constant expression and the reaction equation on which it is based.

2. Gibbs Free Energy Change and Reactivity. For the reaction

$$a_1A + cB \leftrightarrows cC + dD$$

standard free energy is

$$(\Delta G)^0 = -RT \ln \frac{(a_C^c)\,(a_D^d)}{(a_A)^{a_1}(a_B)^b} = -RT \ln K_a \tag{4.20}$$

where a = activity, ( ) refers to molality, and the superscript ^{0} designates standard state at 25°C (198 K) and unit activity or 1-atm fugacity.

- If $(\Delta G)^0$ is negative, the reaction can occur as written (a forward driving force).
- If $(\Delta G)^0$ is zero, no driving force exists (the system is at equilibrium).
- If $(\Delta G)^0$ is positive, a reverse reaction can occur (a reverse driving force).

Values of $(\Delta G)^0$ are tabulated in many handbooks under the "standard free energy change" for the reaction (see Refs. 2 and 3).

PHASE EQUILIBRIA*

See Chap. 7 for information on temperature-pressure phase relations, equation of state, gas mixture, and Gibb's phase rule.

Ideal Solutions. The activity of each constituent of ideal liquid solutions is equal to its mole fraction under all conditions of temperature, pressure, and concentration. The total volume of the solution exactly equals the sum of the volumes of its components. The enthalpy when the components are mixed is zero. The total vapor pressure is the sum of the contribution of the individual components following Raoult's law: the vapor pressure contribution of each individual component is the product of its mole fraction and the vapor pressure of the pure component. This also applies to the vapor pressure of solutions containing nonvolatile components. The freezing point of the solvent in ideal solutions occurs at the temperature where the vapor pressure of the solution equals the vapor pressure of the solid solvent.

Real Solutions. Actual liquid solutions are seldom ideal, often showing deviations from the conditions of ideality described above. Most significant are positive or negative deviations in the direct summation of vapor pressure component contributions; these affect distillation behavior in the separation of components. Deviations from ideality increase with solute concentration; i.e., dilute solutions behave reasonably ideally.

Henry's Law. At a constant temperature, the concentration of a gas dissolved in a liquid is directly proportional to the partial pressure of the gas above the liquid.

Raoult's Law. This states that

$$p_a = x_a P_a \tag{4.21}$$

where p_a = partial pressure of component A in vapor
x_a = mole fraction of A in liquid solution
P_a = vapor pressure of pure liquid A

Binary Solution Vapor-Liquid Equilibria. In the vapor phase:

$$P_{\text{total}} = p_i + p_j \tag{4.22}$$

$$n_{\text{total}} = n_i + n_j$$

$$y_i = \frac{p_i}{P_{\text{total}}}$$

$$y_i + y_j = 1$$

*Parts of this section are based on material taken from *Tehnička hemija* (*Chemistry for Engineers—Lecture Notes*), by E. N. Ganić. Copyright © 1984 by the author. Parts are also taken from *Elementary Principles of Chemical Processes*, by R. M. Felder and R. W. Rousseau. Copyright © 1978. Used by permission of Wiley. All rights reserved.

where p_i is the partial pressure, y_i is the mole fraction, and n_i is the number of moles of i.

In the liquid phase:

$$x_i = \frac{n_i}{n_{\text{total}}} \tag{4.23}$$

$$x_i + x_j = 1$$

$$n_{\text{total}} = n_i + n_j$$

where x_i is the mole fraction of i.

Ideal Solutions. Each component of an ideal solution obeys Raoult's law [Eq. (4.21)] relating concentrations in vapor and liquid phases.

Real Solutions. Numerical distillation calculations often use a vapor-liquid equilibrium ratio k for each component:

$$k_i = \frac{y_i}{x_i} = \frac{f \text{ of pure } i \text{ at its vapor pressure at } T \text{ of the system}}{f \text{ of pure } i \text{ at } T, P \text{ of the system}} \tag{4.24}$$

The volatility ratio between components of binary solutions is:

$$\alpha = \frac{k_i}{k_j} = \frac{y_i x_j}{x_i y_j} \tag{4.25}$$

where α is the volatility ratio. For ideal binary solutions, where α is constant, manipulation of Eqs. (4.24) and (4.25) relates mole fraction in the vapor phase y_i to volatility ratio α and mole fraction in the liquid phase x_i:

$$y_i = \frac{\alpha x_i}{1 + x_i(\alpha - 1)} \tag{4.26}$$

Binary Solution Vapor Pressure Composition Diagrams

Ideal Solutions. These follow Raoult's law [Eq. (4.21)]. As shown in Fig. 4.5*a*, the vapor pressure of each component is linear and proportional to the mole fraction, and the vapor pressure of the mixture is the simple sum of the component vapor pressures. The diagrams shown represent one fixed temperature.

Real Solutions. These show significant deviations from linearity of individual vapor pressures with the mole fraction. At low mole fractions in the liquid phase, Henry's law is followed, while at high mole fractions, Raoult's law tends to be followed. These deviations from linearity are shown in Fig. 4.5*b* and *c*.

Boiling Point Composition Diagrams. Figure 4.6 shows three types of boiling point composition diagrams. These are drawn for a single pressure. Figure 4.6*a* is the usual case that exists for ideal solutions; it corresponds with the vapor pressure diagram shown in Fig. 4.5*a*. At any given temperature, vapor composition y is in equilibrium with liquid composition x. This ideal type of boiling point diagram exists for many combinations of chemically similar materials.

Figure 4.6*b* and *c* corresponds to the vapor pressure diagrams shown in Fig. 4.5*b* and *c*.

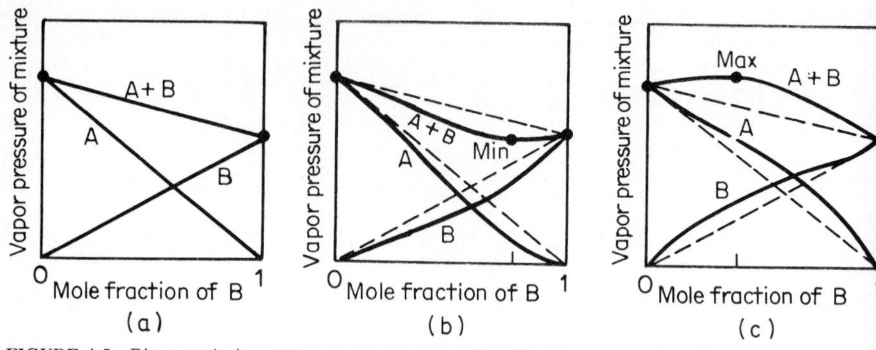

FIGURE 4.5 Binary solution vapor pressure composition diagrams.

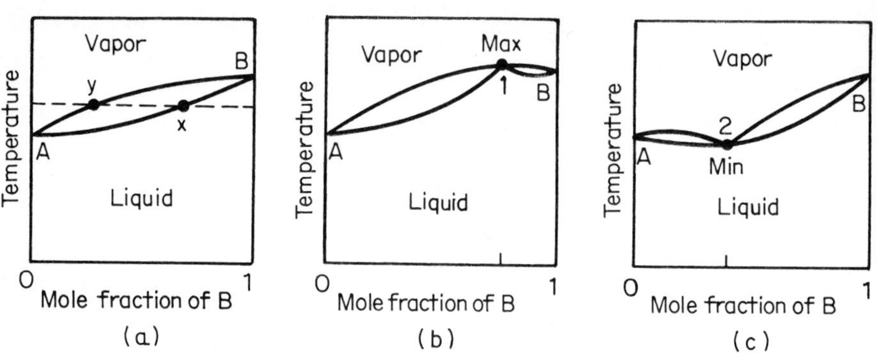

FIGURE 4.6 Binary solution boiling point composition diagrams.

Azeotropes (constant boiling mixtures) exist at 1 and 2. Their composition can be altered by changing pressure. These arise from nonideality of the solutions. For more data on azeotropes, see *Handbook of Chemistry and Physics* and Refs. 2, 9, and 10.

CHEMICAL REACTION RATES*

1. Kinetics

Rate of Reaction. The rate of reaction of any component A based on unit volume of fluid is

$$r_A = \frac{1}{V}\frac{dN_A}{dt} \tag{4.27}$$

*Parts of this section based on material taken from *Tehnička hemija* (*Chemistry for Engineers—Lecture Notes*), by E. N. Ganić. Copyright © 1984 by the author. Parts are also taken from *Elementary Principles of Chemical Processes*, by R. M. Felder and R. W. Rousseau. Copyright © 1978. Used by permission of Wiley. All rights reserved.

and where density remains unchanged

$$r_A = \frac{dC_A}{dt} \tag{4.28}$$

Frequently

$$r_A = \begin{pmatrix}\text{temperature-}\\ \text{dependent term}\end{pmatrix}\begin{pmatrix}\text{concentration-}\\ \text{dependent term}\end{pmatrix} = kf(C_A, C_B, \ldots) \tag{4.29}$$

N_A = number of moles of component A

Order, Molecularity, Elementary Reactions. Where the rate can be expressed as

$$-r_A = kC_A^a C_B^b \cdots \tag{4.30}$$

the reaction is ath order with respect to A and nth order overall where $n = a + b + \cdots$.

Note. $a, b, \ldots$ are empirically observed and are not necessarily equal to the stoichiometric coefficients. In the special case where $a, b, \ldots$ are the stoichiometric coefficients, the reaction is elementary: unimolecular ($n = 1$), bimolecular ($n = 2$), trimolecular ($n = 3$).

Rate Constant k and Temperature Dependency of a Reaction

$$k = (\text{conc})^{1-n}(\text{time})^{-1}$$

From Arrhenius's law the variation with temperature is

$$k = k_0 e^{-E/RT} \qquad \text{or} \qquad \ln\frac{k_2}{k_1} = \frac{E}{R}\left[\frac{1}{T_1} - \frac{1}{T_2}\right] \tag{4.31}$$

where E is the activation energy of the reaction.

2. Homogeneous, Constant Fluid Density, Bath Kinetics

Irreversible First-Order Reaction. For the reaction $A \rightarrow$ products, with rate

$$-\frac{dC_A}{dt} = kC_A \qquad \text{or} \qquad \frac{dX_A}{dt} = k(1 - X_A) \tag{4.32}$$

the integrated form is

$$-\ln\left(\frac{C_A}{C_{A_0}}\right) = -\ln(1 - X_A) = kt \tag{4.33}$$

Irreversible Second-Order Reaction. For the reaction $A + B \rightarrow$ products with rate

$$-\frac{dC_A}{dt} = kC_A C_B \tag{4.34}$$

when $M = C_{B_0}/C_{A_0} \neq 1$, the integrated form is

$$\ln\left(\frac{C_B\,C_{A_0}}{C_{B_0}\,C_A}\right) = \ln\left(\frac{M - X_A}{M(1 - X_A)}\right) = (C_{B_0} - C_{A_0})kt$$

When $C_{A_0} = C_{B_0}$, the integrated form is

$$\frac{1}{C_A} - \frac{1}{C_{A_0}} = \frac{1}{C_{A_0}}\left(\frac{X_A}{1 - X_A}\right) = kt \tag{4.35}$$

Irreversible nth-Order Reaction. For the reaction with rate

$$-r_A = -\frac{dC_A}{dt} = kC_A^n \tag{4.36}$$

the integrated form for $n \neq 1$ is

$$C_A^{1-n} - C_{A_0}^{1-n} = (n - 1)kt \tag{4.37}$$

Reversible First-Order Reaction. For the reaction

$$A \underset{k_2}{\overset{k_1}{\rightleftharpoons}} R \qquad K = \frac{k_1}{k_2} \tag{4.38}$$

with rate

$$-\frac{dC_A}{dt} = \frac{dC_R}{dt} = k_1C_A - k_2C_R$$

the integrated form is

$$-\ln\left(\frac{X_{Ae} - X_A}{X_{Ae}}\right) = -\ln\left(\frac{C_A - C_{Ae}}{C_{A_0} - C_{Ae}}\right) = (k_1 + k_2)t \tag{4.39}$$

Integration of Rate in General. For the reaction with rate

$$-r_A = -\frac{dC_A}{dt} = k \cdot f(C_A, C_B, \ldots) \tag{4.40}$$

the integration is

$$t = C_{A_0}\int_0^{X_A}\frac{dX_A}{(-r_A)} = \int_{C_{A_0}}^{C_A}\frac{dC_A}{kf(C_A, C_B, \ldots)} \tag{4.41}$$

which is solved analytically or graphically.

3. Bath Reaction with Changing Fluid Density. Where density change is proportional to the fractional conversion of any reactant A (isothermal system),

$$\frac{C_A}{C_{A_0}} = \frac{1 - X_A}{1 + \epsilon_A X_A} \tag{4.42}$$

where

$$\epsilon_A = \frac{(\text{volume where } X_A = 1) - V_{x_A = 0}}{V_{x_A = 0}} \tag{4.43}$$

The rate for any reactant A is then

$$-r_A = -\frac{1}{V}\frac{dN_A}{dt} = \frac{C_{A_0}}{(1 + \epsilon_A X_A)}\frac{dX_A}{dt} = kf(C_A, C_B, \ldots) \tag{4.44}$$

Integrating in the general case:

$$t = C_{A_0}\int_0^{X_A} \frac{dX_A}{(1 + \epsilon_A X_A)(-r_A)} \tag{4.45}$$

ELECTROCHEMISTRY*

1. General. The chemical energy of redox reactions can be equated at constant T, P with Gibbs free energy using Eq. (4.4).

Since dT and dP are both zero at constant T, P then $dG = \Sigma \hat{\mu}_i \, dn_i$. Gibbs free energy can then be manifested as electric energy by an electron flow through an external conductor provided there is electrical isolation of the oxidation reaction (electron source) and reduction reaction (electron sink). The external conductor transfers electrons from source to sink, and a liquid junction permits internal migration of ions for preservation of charge neutrality of the system as a whole.

Chemical energy at constant T, P can be considered convertible to electric and/or thermal energy, for it can be shown by conservation of energy that $\Sigma \hat{\mu}_i \, dn_i = -E\,dq - T\,dS$. In the absence of $T\,dS$ thermal energy:

$$dG = -E\,dq \tag{4.46}$$

where E is volts, dq is coulombs/gram-mole, and dG is energy in joules/gram-mole.

2. Redox Cells

Definitions of Anode and Cathode

Anode: Oxidation occurs at this electrode; it is the site where the reducing agent is oxidized; it is the source of electrons to the external circuit whether in a battery or an electrolytic cell.

Cathode: Reduction occurs at this electrode; its characteristics are the reverse of those of the anode.

*Parts of this section based on material taken from *Tehnička hemija* (*Chemistry for Engineers—Lecture Notes*), by E. N. Ganić. Copyright © 1984 by the author. Parts are also taken from *Elementary Principles of Chemical Processes*, by R. M. Felder and R. W. Rousseau. Copyright © 1978. Used by permission of Wiley. All rights reserved.

Batteries. Batteries are redox cells that are physically arranged for external flow and internal ion mobility. When different electrolytes are used, gel structures or a porous membrane prevent mixing. The electrolyte(s) must permit ionization of the reactant species and be conductive. Mobility of ions in solution provides the mechanism for maintaining a charge balance and electrical neutrality within the electrolyte. By convention, batteries are labeled negative at the anode of the spontaneous discharge reaction.

Battery reactions on discharge are listed below. Only a few are reversible.

1. *Lead storage battery (H_2SO_4 electrolyte) (reversible):* Oxidation reaction at lead plates, labeled electrically negative:

$$Pb + HSO_4^- \rightarrow PbSO_4 + 2e^- + H^+ \qquad E_o = +0.36\ V$$

Reduction reaction at lead dioxide plates, labeled electrically positive:

$$PbO_2 + 3H^+ + HSO_4^- + 2e^- \rightarrow PbSO_4 + 2H_2O \qquad E^0 = +1.68V$$

2. *Mercury cell (KOH electrolyte saturated with ZnO):* Oxidation (negative terminal):

$$Zn + 4OH^- \rightarrow ZnO_2^{2-} + 2H_2O + 2e^- \qquad E_B^0 = +1.22\ V$$

Reduction (positive terminal):

$$HgO + H_2O + 2e^- \rightarrow Hg + 2OH^- \qquad E_B^0 = +0.10\ V$$

3. *Zinc-silver peroxide cell (KOH electrolyte saturated with ZnO):* Oxidation (negative terminal):

$$Zn + 4OH^- \rightarrow ZnO_2^{2-} + 2H_2O + 2e^- \qquad E_B^0 = +1.22\ V$$

Reduction (positive terminal):

$$Ag_2O + H_2O + 3e^- \rightarrow 2Ag + 2OH^- \qquad E_B^0 = +0.34\ V$$

4. *LeClanche cell (flashlight battery, NH_4Cl electrolyte):* Oxidation (negative terminal):

$$Zn \rightarrow Zn^{2+} + 2e^- \qquad E^0 = +0.76\ V$$

Reduction (positive terminal):

$$2MnO_2 + 2NH_4^+ + 2e^- \rightarrow Mn_2O_3 + H_2O$$

$$+ 2NH_3 \text{ (for complexing } Zn^{2+}) \qquad E^0 = +0.74\ V$$

5. *Alkaline-zinc-manganese dioxide "alkaline flashlight battery" (KOH electrolyte):* Oxidation (negative terminal):

$$Zn + 4OH^- \rightarrow ZnO_2^{2-} + 2H_2O + 2e^- \qquad E_B^0 = +1.22\ V$$

Reduction (positive terminal):

$$MnO_2 + 2H_2O + e^- \rightarrow Mn(OH)_3 + OH^- \qquad E_B^0 = +0.35\ V$$

6. *Nickel-cadmium storage battery (KOH electrolyte) (reversible):* Oxidation (negative terminal):

$$Cd + 2OH^- \rightarrow Cd(OH)_2 + 2e^- \qquad E_B^0 = +0.81\ V$$

Reduction (positive terminal):

$$NiO_4 + 2H_2O + 2e^- \rightarrow Ni(OH)_2 \qquad E_B^0 = +0.49\ V$$

In all these equations

$$E^0 = \text{standard potential (acid solution)}$$

$$E_B^0 = \text{standard potential (basic solution)}$$

ORGANIC CHEMISTRY

Reactions. Organic reactions relate to changes in the bonding to carbon atoms; their course is greatly influenced by reaction conditions that are experimentally established on the basis of reaction mechanism studies. Complications arise because several different reactions may occur simultaneously, accompanied by cyclization, molecular rearrangements, and oxidation-reduction reactions. All occur in the direction of minimum energy and increased stability. There is an encyclopedic literature on the subject, replete with tens of thousands of reactions, several hundred of which are generally useful in synthesis and have been given names in honor of early investigators in the field.

General Classes of Compounds. These include

1. The straight and branched chain types of compounds (see Table 4.8)
2. Cyclic compounds (see Table 4.9)

Note. For conciseness the following symbols are used in diagrammatic representations of compounds:

R = H atom or saturated hydrocarbon group
R' = hydrocarbon group only
X = halogen
n = an integer

NUCLEAR REACTIONS

The conventional notation for a nuclear reaction is

$$({}_zB^A)_1 + ({}_zB^A)_2 \rightarrow ({}_zB^A)_3 + ({}_zB^A)_4 + Q$$

in which z is the number of protons, A is the mass number, B is the chemical symbol for the atom, electron, or nucleon, and Q is the energy released.

A specific example is the fission reaction:

$${}_{92}\text{U}^{235} + {}_0\text{n}^1 \rightarrow {}_{92}\text{U}^{236} \rightarrow {}_{38}\text{Sr}^{94} + {}_{54}\text{Xe}^{140} + 2{}_0\text{n}^1 + \text{Q}$$

The strontium and xenon products are highly radioactive and decay further into other products. The final result is a spectrum of products.

The conservation equations are

$$\Sigma z_i = 0 \qquad \text{and} \qquad \Sigma A_i = 0$$

$$\text{Initial mass} - \text{final mass} = Q$$

TABLE 4.8 Straight and Branched Chain Types of Compounds

Type of name	General formula
1. Alkane or paraffin (also saturated hydrocarbons)	$R_2C R_2$ (C bonded to four R groups)
2. Alkene or olefin (unsaturated hydrocarbons)	$R_2C{=}CR_2$
3. Alkyne	$R{-}C{\equiv}C{-}R$
4. Alcohol	$R_3C{-}OH$
5. Ether	$R{-}O{-}R'$
6. Aldehyde	$R{-}C(H){=}O$
7. Ketone	$R'{-}C({=}O){-}R'$
8. Carboxylic acid	$R{-}C({=}O){-}OH$
9. Grignard reagent	$R_3C{-}Mg{-}X$
10. Acyl halide	$R'{-}C({=}O){-}X$
11. Anhydride	$R{-}C({=}O){-}O{-}C({=}O){-}R$
12. Ester	$R{-}C({=}O){-}O{-}CR_3$

TABLE 4.8 Straight and Branched Chain Types of Compounds (*Continued*)

Type of name	General formula
13. Amide	$R{-}C({=}O){-}NH_2$
14. Amine (base)	$R_3C{-}NH_2$
15. Nitrile	$R_3C{-}C{\equiv}N$

BIOCHEMISTRY

The elements that make up the major organic constituents of the body are C, H, O, N, P, and S. These organic constituents are of two classes: very large molecules, which may be classed as elements of storage, structure, or function, and small molecules which are (1) the constituents of the macromolecules, (2) derivatives of these constituents with special functions (e.g., coenzymes, vitamins), and (3) sources of energy, through fermentation or oxidation processes.

The inorganic elements are also of two classes: (1) monovalent and divalent ions, which are major constituents of the body fluids: Na^+, K^+, Cl^-, HCO_3^-, Ca^{2+}, Mg^{2+}, HPO_4^{2-}, $H_2PO_4^-$; and (2) trace elements: Fe, Cu, Mn, Zn, I, Co, Mo, Si, F.

The most abundant single compound in the body is water, which makes up 65 to 70 percent of the lean body mass, or 55 percent of the whole body weight.

Cell. The unit of organization and of metabolic operations in the body is the cell. Every cell is bounded by a plasma membrane, which actively guards the internal environment of the cell by excluding certain materials from, allowing some to pass into, and expending energy to facilitate the transport of others through the cell. The fabric of the membrane is a bimolecular layer of lipid, mainly phosphoglyceride molecules presenting their polar ends toward the surface of the membrane and their hydrocarbon tails toward the interior and one another.

Proteins. The principal nitrogen-containing compounds of plant and animal tissues are the proteins. Their essential nature is illustrated in their role as biological catalysts, or enzymes, but they act also as constituents of membranes; as prominent components of connective tissues, bone matrix, cartilage, hair, horn, claws, nails, as parts of hormones, antigens, and antibodies; as osmotic regulators; and as both specific and general carriers, not only of the respiratory gases but also of a host of simpler organic compounds and many inorganic ions. Their elementary composition is simple: C = 50 to 55 percent; H = 6.0 to 7.3 percent; O = 19 to

TABLE 4.9 Cyclic Compounds

Type or name	General formula
1. Cycloparaffin (Naphthene)	see structure 1 below
2. Cycloalkene	see structure 2 below
3. Aromatic	see structure 3 below
4. Naphthalenic	see structure 4 below

1. Cycloparaffin (Naphthene)

```
     R                    R
     |                    |
R — C ———————————————————— C — R
     |         R          |
     |      [  |  ]       |
     └——————[— C —]———————┘
            [  |  ]
               R   n
```

2. Cycloalkene

```
R— C ==================== C—R
     |         R          |
     |      [  |  ]       |
     └——————[— C —]———————┘
            [  |  ]
               R   n
```

3. Aromatic

```
          R   R
          |   |
         C = C
R — C  ⟋       ⟍ C — R
    ⟍ ⟍       ⟋ ⟋
         C — C
         |   |
         R   R
```

4. Naphthalenic

```
        R       R
        |       |
        C       C
     ⟋⟋   ⟍   ⟋   ⟍⟍
R — C       C       C — R
    |       ‖       |
R — C       C       C — R
     ⟍⟍   ⟋   ⟍   ⟋⟋
        C       C
        |       |
        R       R
```

24 percent; N = 13 to 19 percent; S = 0 to 4 percent. Many other elements are associated with proteins in more or less firm combination, but only iodine is found as a constituent of an α-amino acid in a derived protein, thyroglobulin.

Proteins are polymers of α-amino acids. Their sizes vary enormously, from tiny proteins such as insulin (mol wt approximately 6000) to very large aggregates with molecular weights of many millions. The structure common to them all is the peptide bond, formed by the loss of water between the carboxyl group of one α-amino acid and the α-amino group of another. This yields a repeating backbone structure as follows:

```
    H    H    0    H    H    0
    |    |    ‖    |    |    ‖
... N —— C —— C —— N —— C —— C ...
         |              |
         R              R
```

Compounds formed in this way are called *peptides*. Proteins, containing many such bonds, are called *polypeptides*.

In addition to the proteins, the other class of nitrogenous high-molecular-weight constituents of cells is the *nucleic acids*.

Enzymes. All of the chemical reactions in the body take place at a constant, relatively low temperature, and most of them would be infinitely slow unless they were catalyzed. A major function of proteins is to serve as the biological catalysts synthesized in living cells known as enzymes. The characteristics of a differentiated cell are determined not only by its structural proteins but also by the nature, variety, and amounts of its constituent enzymes and by their ordering into enzyme systems. Over 1500 enzymes are known, and many of these have been isolated in crystalline form and studied in detail. All enzymes have proven to be proteins.

Energy Exchange and Production. Living organisms are enclaves of order open to an environment continually tending to disorder. Organisms are maintained by the constant expenditure of energy derived from foodstuffs by processes that are, in the long run, oxidative. All of the life processes of growth, reproduction, repair, maintenance, and chemical, electrical, and mechanical work are supported by chemical reactions which, taking the entire system that can be described as animal + input + output, proceed with a net loss of free energy, that is, of the component of the total energy that is available for useful work. Thus, although an organism may maintain a constancy of composition, internal order, and total internal energy for a large part of its lifetime, it does so at the continuous expense of its environment.

The quantitative aspects of the energy exchanges of the whole organism and of the energy of individual chemical reactions are important to an understanding of bioenergetics in human beings.

NOMENCLATURE

Symbol = *Definition, SI unit (U.S. Customary unit)*

C_A = concentration of A, mol A/volume

C_{A_0} = initial concentration of A, mol A/volume

C_p = specific heat, J/mol · °C (Btu/mol · °F)

e^- = electron (sign $^-$ indicates negative charge)

F = Helmholtz function, J/mol (Btu/mol)

f = fugacity, N/m^2 (lb$_f$/ft^2)

G = Gibbs free energy, J/mol (Btu/mol)

H = enthalpy, J/mol (Btu/mol)

H = enthalpy (in "Thermochemistry" section), J (Btu)

ΔH = enthalpy difference, J (Btu)

$\hat{H}$ = enthalpy (in "Thermochemistry" section), J/mol (Btu/mol)

N_A = number of moles (in "Thermochemistry" section)

n = number of moles

p_a = partial pressure, N/m^2 (lb$_f$/ft^2)
P = pressure, N/m^2 (lb$_f$/ft^2)
Q = heat, J (Btu)
R = gas constant, J/mol · K (Btu/mol · °R)
S = entropy, J/mol · K (Btu/mol · °F)
T = temperature, K (°R)
t = time, s
V = volume, m^3 (ft^3)
U = internal energy, J/mol (Btu/mol)
X_A = fraction of reactant A converted
X_a = mole fraction

Greek

$\hat{\mu}$ = chemical potential, J/mol (Btu/mol)
γ = activity coefficient

Subscripts

a = component A
A = component A
f = reaction
0 = standard condition
0 = initial conditions (1 atm, 25°C)
s = solid
g = gas
l = liquid

Others

[ ] = concentration (C_A = [A])
$\bar{\ }$ = average

Note. Other symbols used in this chapter are defined in the text.

REFERENCES*

1. K. Gieck, *Engineering Formulas*, 5th ed., McGraw-Hill, New York, 1986.
2. J. A. Dean, *Lange's Handbook of Chemistry*, 13th ed., McGraw-Hill, New York, 1985.

*Those references listed above but not cited in the text were used for comparison between different data sources, clarification, clarity of presentation, and, most important, reader's convenience when further interest in the subject exists.

3. R. H. Perry and D. W. Green, *Perry's Chemical Engineers' Handbook*, 6th ed., McGraw-Hill, New York, 1984.
4. N. Radosevic, *Priručnik za hemičare i tehnologe* (*Handbook for Chemists*), Tehnička knjiga, Belgrade, 1968.
5. R. H. Perry, *Engineering Manual*, 3d ed., McGraw-Hill, New York, 1976.
6. E. N. Ganić, *Tehnička hemija* (*Chemistry for Engineers—Class Notes*), Dept. of Mechanical Engineering, University of Sarajevo, 1984.
7. R. M. Felder and R. W. Rousseau, *Elementary Principles of Chemical Processes*, Wiley, New York, 1978.
8. D. M. Himmelblau, *Basic Principles and Calculations in Chemical Engineering*, 3d ed., Prentice-Hall, Englewood Cliffs, N.J., 1974.
9. E. U. Condon and H. Odishaw, *Handbook of Physics*, 2d ed., McGraw-Hill, New York, 1967.
10. S. P. Parker (ed.), *McGraw-Hill Encyclopedia of Physics*, McGraw-Hill, New York, 1983.
11. O. Levenshpiel, *Chemical Reaction Engineering*, 2d ed., Wiley, New York, 1969.
12. J. F. Mulligan, *Practical Physics*, McGraw-Hill, New York, 1980.
13. J. M. Smith and H. C. Van Ness, *Introduction to Chemical Engineering Thermodynamics*, 2d ed., McGraw-Hill, New York, 1959.

CHAPTER 5
MECHANICS OF RIGID BODIES

STATICS

1. Definitions

Statics. That branch of mechanics which deals with the equilibrium of forces on bodies at rest (or moving at a uniform velocity in a straight line).

Force. The action of one body on another. It is a vector quantity defined by its magnitude, direction, and point of application (P) (Fig. 5.1). The forces by which the individual particles of a body act on each other are *internal*. All other forces are *external*. The forces exerted by the body on its supports are called *reactions*. They are equal in magnitude and opposite in direction to the forces, called *supporting forces*, with which the supports act on the body.

Gravitation. The force of the earth's attraction directed toward the earth's center.

Moment. The moment of a force $\mathbf{F}$ about a point O (Fig. 5.7) is

$$\mathbf{M} = \mathbf{F} \cdot l \qquad (5.1)$$

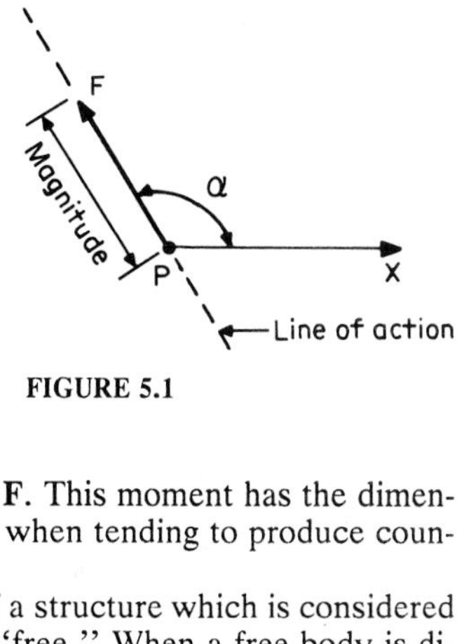

FIGURE 5.1

where l = perpendicular distance (called level arm l) from point O to the line of action of force $\mathbf{F}$. This moment has the dimension of force × length and is considered positive when tending to produce counterclockwise rotation around point O.

Free-Body Diagram. A structure or a part of a structure which is considered to be separated from everything else and mode "free." When a free body is diagrammed, vector arrows are placed representing the forces which the other parts exert upon the free body. The forces which the free body exerts upon the other parts are not represented. See Fig. 5.2. (The roller at B in Fig. 5.2*a* is replaced in Fig. 5.2*b* by a vertical force F_B, since the roller offers constraint only in the vertical direction. Similarly, the pin at A is replaced by horizontal and vertical reactions F_{XA} and F_{YA}, since the pin offers two degrees of constraint. The drawing shown in Fig. 5.2*b*, in which the

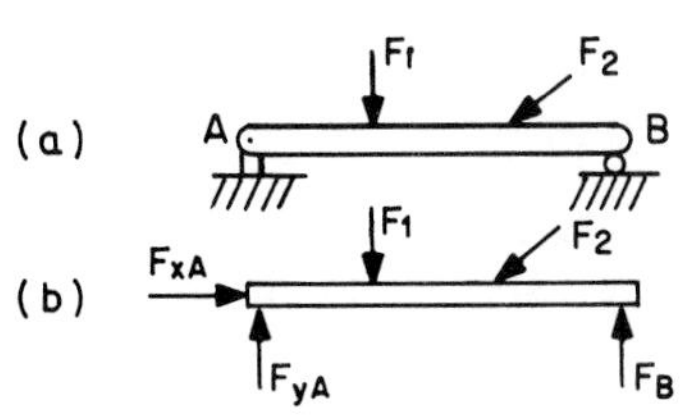

FIGURE 5.2

beam is shown completely isolated from its supports and where all forces are shown by vectors, is the actual *free-body diagram*.)

Basic Laws

Newton's Laws

1. If a body is at rest, it will remain at rest, or if in motion, it will move uniformly in a straight line, until acted on by some force.
2. If a body is acted on by several forces, it will obey each as though the others did not exist, and this will happen whether the body is at rest or in motion. Change of the motion of a body is proportional both to the force exerted against it and to the time during which the force is exerted, and is in the same direction as the force.
3. If a force acts to change the state of a body with respect to rest or motion, the body will offer a resistance equal to and directly opposed to the force. Or, to every action there is opposed an equal and opposite reaction.

Linear Momentum. The product of mass and the linear velocity of a particle; it is a vector. The moment of the linear momentum vector about a fixed axis is the *angular momentum* of the particle about that fixed axis. For a rigid body rotating about a fixed axis, angular momentum is defined as the product of the moment of inertia and angular velocity, each measured about the fixed axis.

The relation between mass (m), acceleration (a), and force ($\mathbf{F}$) is contained in Newton's second law of motion:

$$\mathbf{F} = m \cdot \mathbf{a} \tag{5.2}$$

The direction of $\mathbf{F}$ is the same as the direction of $\mathbf{a}$. An alternative form of Newton's second law states that the resultant force is equal to the time rate of change of momentum:

$$\mathbf{F} = \frac{d(m \cdot \mathbf{v})}{dt} \tag{5.3}$$

The linear momentum of a system of bodies is unchanged if there is no resultant external force on the system. The angular momentum of a system of bodies about a fixed axis is unchanged if there is no resultant external moment about this axis.

2. Composition of Forces

Classification of System of Forces. The classification of a system of forces is made according to the arrangement of the forces' action lines. If the action lines lie in the same plane, the system is *coplanar*; otherwise it is *noncoplanar*. If they pass through the same point, the system is *concurrent*; otherwise it is *nonconcurrent*. If two or more forces have the same action line, they are *collinear*. A system of two equal forces, parallel, opposite in sense, and having different action lines is a *couple*.

Graphical Composition of Forces. A force may be represented by a straight line in a determined position, and its magnitude by the length of the straight line. The direction in which it acts may be indicated by an arrow.

A parallelogram of two forces intersecting each other (Fig. 5.3*a*) leads directly to the graphical composition by means of a triangle of forces. In Fig. 5.3*b*, F_R is

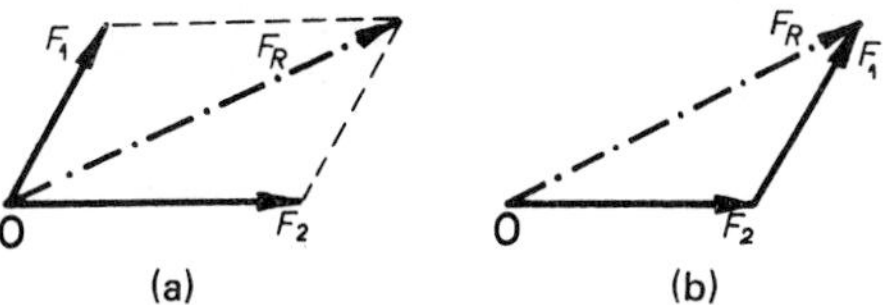

FIGURE 5.3 (*a*) Diagram of forces; (*b*) force polygon. (*From Gieck.*[1])

called the *closing side*, and represents the resultant of the forces of F_1 and F_2 in magnitude and direction. Its position is given by the point of application O. By means of the repeated use of the triangle of forces and by omitting the closing sides of the individual triangles, the magnitude and direction of the resultant F_R of any number of forces in the same plane and intersecting at a single point can be found (see Fig. 5.4). If the forces are in equilibrium, F_R must be equal zero, i.e., the force polygon must close.

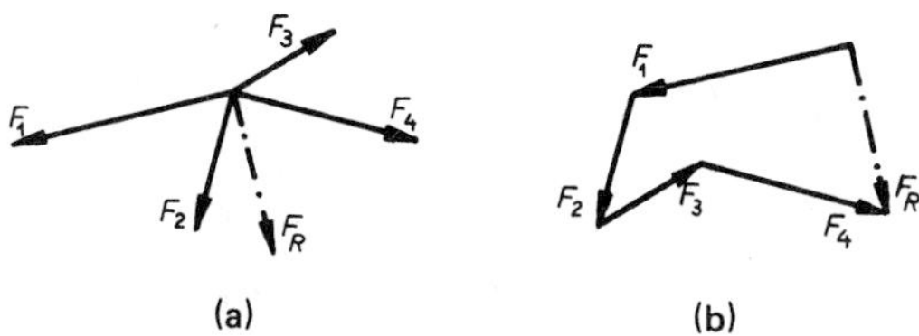

FIGURE 5.4 (*a*) Diagram of forces; (*b*) force polygon. (*From Gieck.*[1])

If in a closed polygon one of the forces is reversed in direction, this force becomes the resultant of all the others.

If the forces do not all lie in the same plane, the diagram becomes a polygon in space. The resultant F_R of this system may be obtained by adding the forces in space. The resultant is the vector which closes the space polygon. The space polygon may be projected onto three coordinate planes, giving three related plane polygons. Any two of these projections will involve all the static equilibrium conditions and will be sufficient for a full description of the force system (see Fig. 5.5).

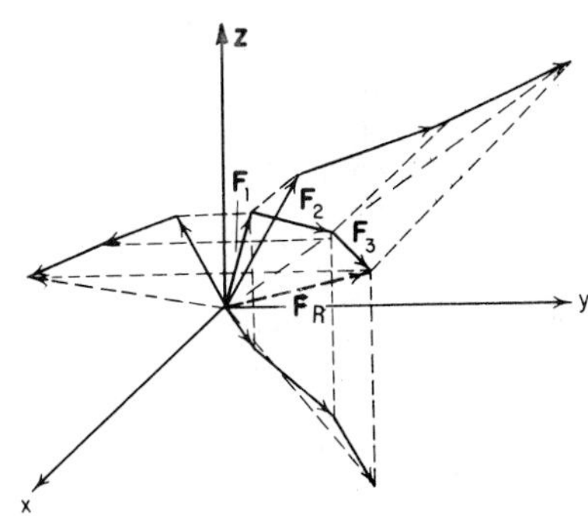

FIGURE 5.5 (*From Avallone and Baumeister.*[2])

Mathematical Composition of Forces. The resultant F_R of two forces F_1 and F_2 applied at the same point is defined analytically as follows (Fig. 5.3*a*):

$$F_R = (F_1^2 + F_2^2 + 2F_1F_2 \cos \alpha)^{1/2} \tag{5.4}$$

where
$$\tan \alpha = \frac{F_2 \sin \alpha}{F_1 + F_2 \cos \alpha} \tag{5.5}$$

Force **F** may be resolved into two component forces (Fig. 5.6) as:

$$F = (F_x^2 + F_y^2)^{1/2} \tag{5.6}$$

where
$$F_x = F \cdot \cos \alpha$$

$$F_y = F \cdot \sin \alpha$$

$$\tan \alpha = \frac{F_y}{F_x}$$

Moment $\mathbf{M}_0$ of a force **F** about a point O (Fig. 5.7) is

$$M_0 = F \cdot l_0 = F_y \cdot \bar{x} - F_x \cdot \bar{y} \tag{5.7}$$

The resultant force F_R of any given forces is (Fig. 5.8):

$$F_R = (F_{Rx}^2 + F_{Ry}^2)^{1/2} \tag{5.8}$$

$$F_{Rx} = \Sigma F_x \qquad F_{Ry} = \Sigma F_y$$

$$\tan \alpha_R = \frac{F_{Ry}}{F_{Rx}}$$

$$\Sigma M_0 = F_R \cdot l_R \tag{5.9}$$

where l_R is distance of F_R from reference point.

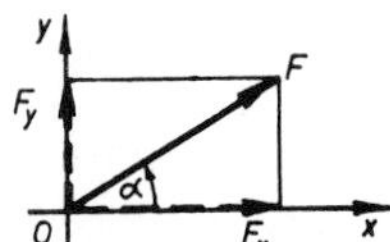

FIGURE 5.6 (*From Gieck.*[1])

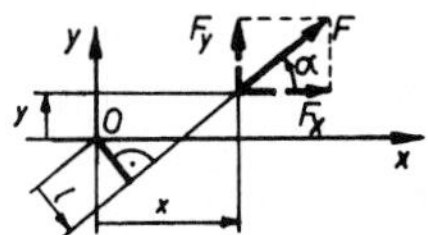

FIGURE 5.7 (*From Gieck.*[1])

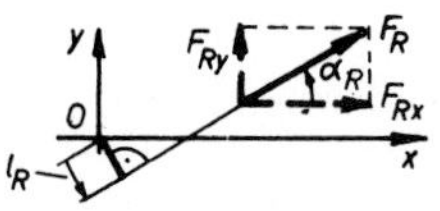

FIGURE 5.8 (*From Gieck.*[1])

3. Conditions of Equilibrium*. A body is in equilibrium when both the resultant force and the sum of the moments of all external forces about any random point are equal to zero.

Concurrent Forces in a Plane. If a system of concurrent forces in a plane is in equilibrium, the sum of the components of the forces along any axis in the plane of the forces is equal to zero. Each force may be resolved into two rectangular components along the X and Y axes. Let F_x and F_y represent the X and Y components of force **F**, and let ΣF_x and ΣF_y represent the summation of all of the X and Y components, respectively. Then $\Sigma F_x = 0$ and $\Sigma F_y = 0$ are the two independent equations of equilibrium by means of which two unknown quantities may be determined.

*Examples 5.1 through 5.5 in this section taken partly from *General Engineering Handbook*, 2d ed., by C. E. O'Rourke. Copyright © 1940. Used by permission of McGraw-Hill, Inc. All rights reserved.

Example 5.1. In Fig. 5.9, $\overline{BC}$ is a boom which is hinged at C and supported by the tie bar $\overline{AB}$. The stresses in the boom $\overline{BC}$ and in the tie bar $\overline{AB}$ due to the load of $F_N = 1000\ \text{lb}_f$ (4448 N) are required. Point B is the free body. Let the X axis be horizontal and the Y axis vertical:

$$\Sigma F_x = 0.966F_{AB} - 0.866F_{BC} = 0$$

$$\Sigma F_y = 0.500F_{BC} - 0.2588F_{AB} - F_N = 0$$

where $\alpha_1 = 15°$, $\alpha_2 = 30°$, and $l = 40$ ft (12.2 m); cos 15° = 0.966, and cos 30° = 0.866, sin 15° = 0.2588, and sin 30° = 0.500.

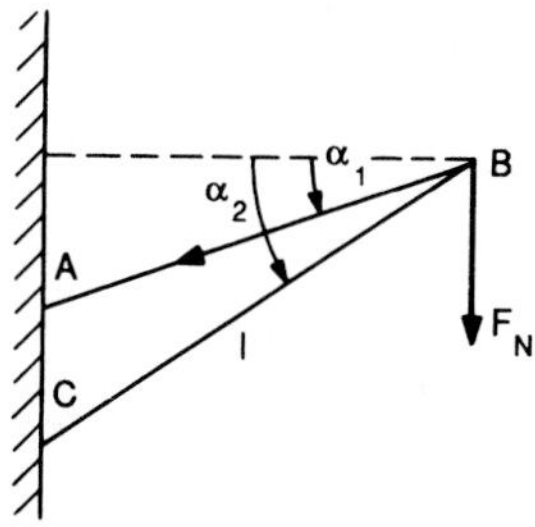

FIGURE 5.9

Solution. Solving for these two equations gives

$$F_{AB} = 3350\ \text{lb}_f\ (14{,}900\ \text{N}) - \text{tension}$$

$$F_{BC} = 3730\ \text{lb}_f\ (16{,}591\ \text{N}) - \text{compression}$$

Example 5.1 could also have been solved by the method of moments. Since the free body is in equilibrium, the summation of the moments with respect to any axis is equal to zero. The equations $\Sigma M_A = 0$ and $\Sigma M_C = 0$ give the two unknown forces:

$$\Sigma M_A = F_N \times 34.64 - F_{BC} \times 9.3 = 0$$

$$\Sigma M_C = F_N \times 34.64 - F_{AB} \times 10.33 = 0$$

Thus

$$F_{BC} = 3730\ \text{lb}_f\ (16{,}591\ \text{N}) \qquad F_{AB} = 3350\ \text{lb}_f\ (14{,}900\ \text{N})$$

Parallel Forces in a Plane. If a system of parallel forces in a plane is in equilibrium, the equation $\Sigma F = 0$ gives only one equation. The equation $\Sigma M = 0$ gives another independent equation, however, so two unknown quantities can be determined. The moment equation should be written first, with a point on the line of action of one of the unknown forces as the center of moments. The resulting equation has only one unknown quantity which is thus determined. The equation $\Sigma F = 0$ completes the solution.

Example 5.2. Figure 5.10 represents a beam 32 ft (9.754 m) long with three vertical loads and two unknown vertical reactions, F_{R_1} and F_{R_2}. If

$F_1 = 2000\ \text{lb}_f$ (8896 N)

$F_2 = 5000\ \text{lb}_f$ (22,240 N)

$F_3 = 6000\ \text{lb}_f$ (26,688 N)

$l_1 = 16$ ft (4.87 m)

$l_2 = 12$ ft (3.65 m)

$l_3 = 4$ ft (1.22 m)

$l_4 = 7$ ft (2.13 m)

$l_5 = 25$ ft (7.62 m)

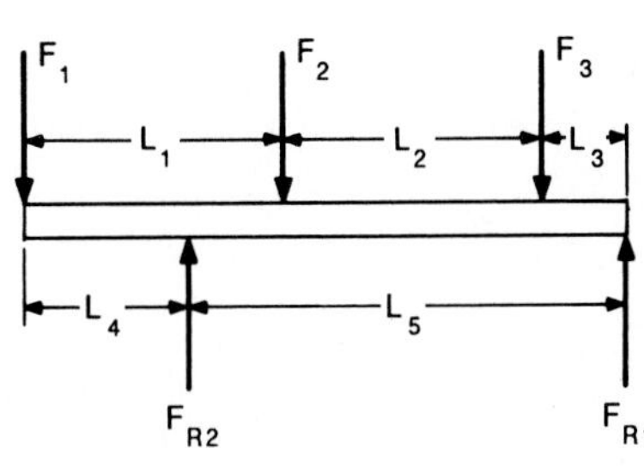

FIGURE 5.10

then equation $\Sigma M_{R_2} = 0$ gives the following:

$$25F_{R_1} - 2000 \times 32 - 5000 \times 16 - 6000 \times 4 = 0$$

or

$$F_{R_1} = 6720 \text{ lb}_f \text{ (29,890 N)}$$

Equation $\Sigma F_y = 0$ gives the following:

$$F_{R_2} + 6720 - 13{,}000 = 0 \qquad \text{or} \qquad F_{R_2} = 6280 \text{ lb}_f \text{ (27,933 N)}$$

This result may be checked by using the equation $\Sigma M_{R_1} = 0$:

$$25F_{R_2} + 2000 \times 7 - 5000 \times 9 - 6000 \times 21 = 0$$

or

$$F_{R_2} = 6280 \text{ lb}_f \text{ (27,933 N)}$$

Any Force System in a Plane. If a coplanar system of forces in equilibrium is nonconcurrent and nonparallel, three independent equations of equilibrium may be written, and therefore three unknown quantities may be determined. The three equations may consist of one moment equation and two force equations, of two moment equations and one force equation, or of three moment equations. Obtaining the solution is simplified if each equation is written in such a way as to contain only one unknown quantity.

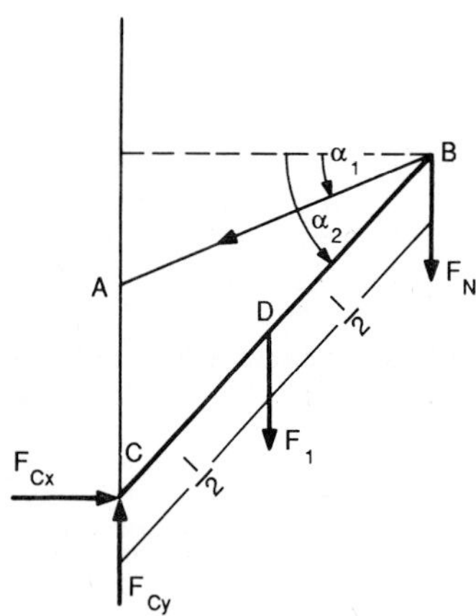

FIGURE 5.11

Example 5.3. In Fig. 5.11, the boom $\overline{CB}$ is hinged at C and supported by the tie bar $\overline{AB}$. The boom is the free body, and the three unknown quantities consist of the stress in AB and the amount and direction of reaction C. If reaction C is replaced by its vertical and horizontal components, the unknown quantities are three forces, whose directions are all known.

The equation $\Sigma M_C = 0$ gives the following [note that $l = 40$ ft (12.2 m), $F_1 = 2000$ lb$_f$ (8896 N), $F_N = 3000$ lb$_f$ (13,344 N), and $\alpha_1 = 15°$, $\alpha_2 = 30°$):

$$F_{AB} \times 40 \times 0.2588 - F_1 \times 20 \times 0.866 - F_N \times 40 \times 0.866 = 0$$

or $F_{AB} = 13{,}400$ lb$_f$ (59,603 N). The equation $\Sigma F_x = 0$ gives $F_{Cx} - 13{,}400 \times 0.966 = 0$, or $F_{Cx} = 12{,}900$ lb$_f$ (57,379 N).

The equation $\Sigma F_y = 0$ gives $F_{Cy} = 5000 - 13{,}400 \times 0.2588 = 0$, or $F_{Cy} = 8460$ lb$_f$ (37,630 N). Reaction $F_C = (F_{Cx}^2 + F_{Cy}^2)^{1/2} = 15{,}400$ lb$_f$ (68,499 N).

Values of F_{Cx} and F_{Cy} could have been obtained by writing the equations $\Sigma M_A = 0$ and $\Sigma M_B = 0$.

Concurrent Forces in Space. If a system of concurrent forces in space is in equilibrium, three independent equations may be written, and therefore three unknown quantities may be determined. The three equations may consist of three moment equations, of two moment equations and one force equation, of one moment equation and two force equations, or of three force equations. The solution is simplified if equations can be written containing only one unknown quantity.

Example 5.4. If the weights of the members of the shear-legs crane (Fig. 5.12) are neglected, the force system is concurrent. The stresses in members $\overline{AB}$, $\overline{BC}$, and $\overline{BE}$ by the 1000-lb$_f$ (4448-N) load are required. Point B is the free body.

Given

$$l_1 = 40 \text{ ft } (12.2 \text{ m})$$

$$l_2 = 20 \text{ ft } (6.1 \text{ m})$$

$$l_3 = 6 \text{ ft } (1.83 \text{ m})$$

$$l_4 = 8 \text{ ft } (2.44 \text{ m})$$

$$\text{Length } \overline{BD} = (20^2 - 6^2)^{1/2} = 19.08 \text{ ft } (5.81 \text{ m})$$

$$\overline{BF} = (19.08^2 - 8^2)^{1/2} = 17.32 \text{ ft } (5.28 \text{ m})$$

$$\overline{FA} = (40^2 - 17.32^2)^{1/2} = 36.06 \text{ ft } (10.99 \text{ m})$$

$$\overline{DA} = 28.06 \text{ ft } (8.55 \text{ m})$$

and

$$F_N = 1000 \text{ lb}_f \text{ } (4448 \text{ N})$$

Solution. With line CE as the axis, the moment equation is

$$F_{AB} \times 28.06 \times \frac{17.32}{40} - F_N \times 8 = 0$$

or $F_{AB} = 660$ lb$_f$ (2935 N), $F_{BC} = F_{BE}$ by symmetry.

The equation $\Sigma F_x = 0$ gives:

$$2F_{BC} \times 0.4 - 660 \times \frac{36.06}{40} = 0 \qquad \text{or} \qquad F_{BC} = 745 \text{ lb}_f \text{ } (3313 \text{ N})$$

Instead of the equation $\Sigma F_x = 0$, either equation $\Sigma F_y = 0$ or a moment equation could have been used. An axis through E parallel to $\overline{AD}$ would have been a satisfactory axis of moments.

Parallel Forces in Space. If a system of parallel forces in space is in equilibrium, three independent equations may be written; hence three unknown quantities

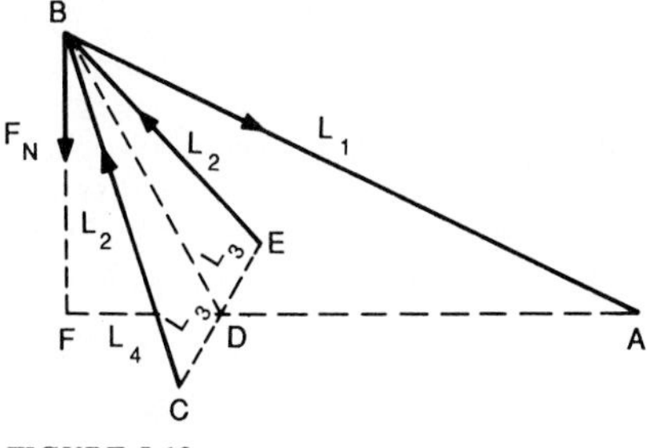

FIGURE 5.12

may be determined. These three equations may consist of one force equation and two moment equations, or of three moment equations. The moment equation with respect to an axis intersecting two of the unknown forces will give the value of the third unknown force.

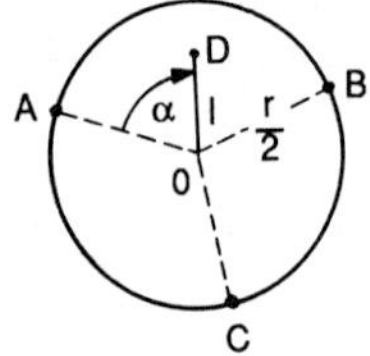

FIGURE 5.13

Example 5.5. The table of diameter r shown in Fig. 5.13 is supported by legs A, B, and C, 120° apart. A load F_N is applied at point D. Find reactions due to the F_N load if:

$$r = 5 \text{ ft } (1.52 \text{ m})$$

$$l = 8 \text{ in } (0.2032 \text{ m})$$

$$\alpha = 45°$$

$$F_N = 200 \text{ lb}_f \ (889.6 \text{ N})$$

Solution. Use BC as the axis of moments:

$$3.75F_A - 200 \times 1.721 = 0 \qquad \text{or} \qquad F_A = 91.8 \text{ lb}_f \ (408.3 \text{ N})$$

Next, use an axis through C parallel to OA as the axis of moments. Then

$$4.33F_B + 91.8 \times 2.165 - 200 \times 2.636 = 0 \qquad \text{or} \qquad F_B = 75.9 \text{ lb}_f \ (337.6 \text{ N})$$

$$F_y = 91.8 + 75.9 + F_C - 200 = 0 \qquad \text{or} \qquad F_C = 32.3 \text{ lb}_f \ (145.6 \ N)$$

Any Force System in Space. If a force system in equilibrium consists of forces which are noncoplanar, nonconcurrent, and nonparallel, six independent equations of equilibrium may be written, three force equations, and three moment equations:

$$\Sigma F_x = 0 \qquad \Sigma F_y = 0 \qquad \Sigma F_z = 0 \tag{5.10}$$

$$\Sigma M_x = 0 \qquad \Sigma M_y = 0 \qquad \Sigma M_z = 0$$

This makes it possible to solve for six unknown quantities in such a system.

It is also convenient to apply the algebraical condition ΣM about every axis = 0 instead of using so many resolution equations. For graphical presentation, the projection of the system on any plane is in equilibrium, and algebraical and graphical (force polygon) conditions can be used to solve such projected systems.

Note. Combining the solutions discussed in the above five subsections, one can solve other more elaborate problems.

Flexible Cord: Load Uniformly Distributed Horizontally. Flexible cords, ropes, cables, or chains which are suspended horizontally between two points and are loaded uniformly horizontally take the shape of a parabola. If the cord carries equally concentrated loads evenly spaced horizontally, the smooth curve through the points of application of the loads will take the shape of a parabola, and the cord itself will not deviate very much from the curve. The suspension bridge is a good example of this type of loading, the chief weight carried being that of the roadway.

If F_N is the weight carried by a cord, l the span between supports, F the tension at the lowest point, and s the length of the cord, the length s is given by the rapidly converging series

$$s = l + \frac{F_N^2 \cdot l}{24F^2} - \frac{F_N^4 \cdot l}{640F^4} + \cdots \tag{5.11}$$

In terms of the span l and the sag d,

$$s = l + \frac{8d^2}{3l} - \frac{32d^4}{5l^3} + \cdots \tag{5.12}$$

Figure 5.14 shows the right half of a loaded cord as the free body. The three forces must act through the common point B.
The equation $\Sigma F_x = 0$ gives:

$$F_R \cdot \cos \alpha = F$$

The equation $\Sigma F_y = 0$ gives:

$$F_R \cdot \sin \alpha = \frac{F_N}{2}$$

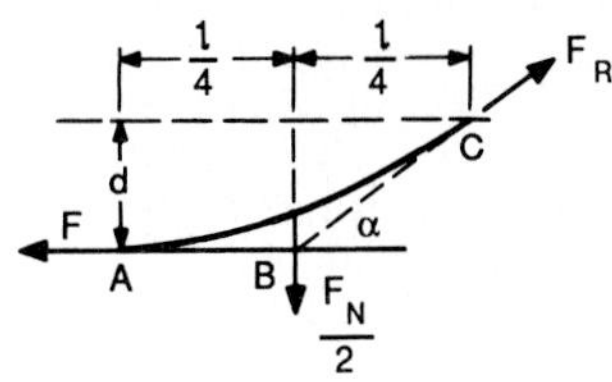

FIGURE 5.14

If these two equations are squared and added, the resulting equation is

$$F_R^2 = F^2 + \left(\frac{F_N}{2}\right)^2$$

The equation $\Sigma M_c = 0$ gives:

$$F \cdot d - \frac{F_N \cdot l}{8} = 0$$

4. Center of Gravity

Definitions. The attraction of the earth for each particle of a body constitutes a system of forces with fixed application points. No matter how a body is turned, the resultant pull of the earth upon it always acts through a certain fixed point, called its *center of gravity*. The centroid of a geometric solid coincides with the center of gravity of a homogeneous body which would occupy the same volume. The centroid of a surface is the limiting position of the center of gravity of a homogeneous thin plate, one face of which coincides with the surface as the thickness of the plate approaches zero. The centroid of a line is the limiting position of the center of gravity of a homogeneous thin rod, the axis of which coincides with the line as the thickness of the rod approaches zero.

Planes of Symmetry and Axes of Symmetry. If a solid, surface, or line has a plane of symmetry, this plane contains the centroid. The line of intersection of two planes of symmetry contains the centroid. The point of intersection of three planes of symmetry locates the centroid.

Centroids by Integration. For a line of length l,

$$x_c = \frac{1}{l} \int x \, dl \tag{5.13}$$

For a surface of area A,

$$x_c = \frac{1}{A}\int x\, dA \tag{5.14}$$

For a solid of volume V,

$$x_c = \frac{1}{V}\int x\, dV \tag{5.15}$$

Similar expressions hold true for y_c and z_c. Centroids of some simple figures are given in Tables 5.1 and 5.2. Note that

- If a line, surface, or solid is composed of several simple parts whose centroids are known, the moment of the entire line, surface, or solid is equal to the sum of the moments of its several simple parts.
- If a mass is made up of several component parts of different material, and if the centers of gravity of these component masses are known, the moment equation for the component weights will locate the center of gravity of the entire mass.

5. Moment of Inertia*

Moment of Inertia and Radius of Gyration of an Area. The moment of inertia of an area with respect to any axis is the sum of the products of the differential areas and the squares of their distances from the given axis. Moment of inertia is denoted by I, usually with a subscript to indicate the axis. Thus

$$I_x = \int y^2\, dA \qquad \text{and} \qquad I_y = \int x^2\, dA \tag{5.16}$$

are the rectangular moments of inertia with respect to the X and Y axes in the plane of the area. Also,

$$I_p = \int \rho^2\, dA \tag{5.17}$$

is the polar moment of inertia with respect to the Z axis normal to the area.

In terms of the total area, $I = A\mathbf{r}^2$, $\mathbf{r}^2$ being the mean value of x^2 or y^2. Since $I = A\mathbf{r}^2$,

$$\mathbf{r} = \sqrt{\frac{I}{A}} \tag{5.18}$$

The quantity $\mathbf{r}$ is called the *radius of gyration*.

If X is any axis in the plane of an area A, X_0 the parallel centroidal axis, and d the distance between the axes, the relationship between the two moments of inertia is given by the transfer formula

$$I_x = I_{x_0} + Ad^2 \tag{5.19}$$

---

*This section partly taken from *General Engineering Handbook*, 2d ed., by C. E. O'Rourke. Copyright © 1940. Used by permission of McGraw-Hill, Inc. All rights reserved.

TABLE 5.1 Properties of Plane Sections

Figure	Area and centroid	Moment of inertia	r^2
Triangle	$A = \frac{1}{2}bh$ $x_c = \frac{1}{3}(a + b)$ $y_c = \frac{1}{3}h$	$I_{x_c} = \frac{bh^3}{36}$ $I_{y_c} = \frac{bh}{36}(b^2 - ab + a^2)$ $I_x = \frac{bh^3}{12}$ $I_y = \frac{bh}{12}(b^2 + ab + a^2)$	$r^2_{x_c} = \frac{1}{18}h^2$ $r^2_{y_c} = \frac{1}{18}(b^2 - ab + a^2)$ $r^2_x = \frac{1}{6}h^2$ $r^2_y = \frac{1}{6}(b^2 + ab + a^2)$
Rectangle	$A = bh$ $x_c = \frac{1}{2}b$ $y_c = \frac{1}{2}h$	$I_{x_c} = \frac{bh^3}{12}$ $I_{y_c} = \frac{b^3h}{12}$ $I_x = \frac{bh^3}{3}$ $I_y = \frac{b^3h}{3}$ $I_p = \frac{bh}{12}(b^2 + h^2)$	$r^2_{x_c} = \frac{1}{12}h^2$ $r^2_{y_c} = \frac{1}{12}b^2$ $r^2_x = \frac{1}{3}h^2$ $r^2_y = \frac{1}{3}b^2$ $r^2_p = \frac{1}{12}(b^2 + h^2)$
Parallelogram	$A = ab \sin\theta$ $x_c = \frac{1}{2}(b + a\cos\theta)$ $y_c = \frac{1}{2}(a\sin\theta)$	$I_{x_c} = \frac{a^3b}{12}\sin^3\theta$ $I_{y_c} = \frac{ab}{12}\sin\theta(b^2 + a^2\cos^2\theta)$ $I_x = \frac{a^3b}{3}\sin^3\theta$ $I_y = \frac{ab}{3}\sin\theta\,(b + a\cos\theta)^2$ $- \frac{a^2b^2}{6}\sin\theta\cos\theta$	$r^2_{x_c} = \frac{1}{12}(a\sin\theta)^2$ $r^2_{y_c} = \frac{1}{12}(b^2 + a^2\cos^2\theta)$ $r^2_x = \frac{1}{3}(a\sin\theta)^2$ $r^2_y = \frac{1}{3}(b + a\cos\theta)^2$ $- \frac{1}{6}(ab\cos\theta)$

TABLE 5.1 Properties of Plane Sections (*Continued*)

Figure	Area and centroid	Moment of inertia	r^2
y, a, h, G, b, x Trapezoid	$A = \frac{1}{2}h\,(a + b)$ $y_c = \frac{1}{3}h\,\frac{2a + b}{a + b}$	$I_{x_c} = \frac{h^3(a^2 + 4ab + b^2)}{36(a + b)}$ $I_x = \frac{h^3(3a + b)}{12}$	$r_{x_c}^2 = \frac{h^2(a^2 + 4ab + b^2)}{18(a + b)^2}$ $r_x^2 = \frac{h^2(3a + b)}{6(a + b)}$
y, a, G, b, x Annulus	$A = \pi(a^2 - b^2)$ $x_c = a$ $y_c = a$	$I_{x_c} = I_{y_c} = \frac{\pi}{4}(a^4 - b^4)$ $I_x = I_y = \frac{5}{4}\pi a^4 - \pi a^2 b^2 - \frac{\pi}{4}b^4$ $I_P = \frac{\pi}{2}(a^4 - b^4)$	$r_{x_c}^2 = r_{y_c}^2 = \frac{1}{4}(a^2 + b^2)$ $r_x^2 = r_y^2 = \frac{1}{4}(5a^2 + b^2)$ $r_P^2 = \frac{1}{2}(a^2 + b^2)$
y, G, 2a, x Semicircle	$A = \frac{1}{2}\pi a^2$ $x_c = a$ $y_c = \frac{4a}{3\pi}$	$I_{x_c} = \frac{a^4(9\pi^2 - 64)}{72\pi}$ $I_{y_c} = \frac{1}{8}\pi a^4$ $I_x = \frac{1}{8}\pi a^4$ $I_y = \frac{5}{8}\pi a^4$	$r_{x_c}^2 = \frac{a^2(9\pi^2 - 64)}{36\pi^2}$ $r_{y_c}^2 = \frac{1}{4}a^2$ $r_x^2 = \frac{1}{4}a^2$ $r_y^2 = \frac{5}{4}a^2$

Circular sector	$A = a^2\theta$ $x_c = \frac{2a}{3}\frac{\sin\theta}{\theta}$ $y_c = 0$	$I_x = \frac{1}{4}a^4(\theta - \sin\theta\cos\theta)$ $I_y = \frac{1}{4}a^4(\theta + \sin\theta\cos\theta)$	$r_x^2 = \frac{1}{4}a^2\frac{\theta - \sin\theta\cos\theta}{\theta}$ $r_y^2 = \frac{1}{4}a^2\frac{\theta + \sin\theta\cos\theta}{\theta}$
Circular segment	$A = a^2(\theta - \frac{1}{2}\sin 2\theta)$ $x_c = \frac{2a}{3}\frac{\sin^3\theta}{\theta - \sin\theta\cos\theta}$ $y_c = 0$	$I_x = \frac{Aa^2}{4}\left[1 - \frac{2\sin^3\theta\cos\theta}{3(\theta - \sin\theta\cos\theta)}\right]$ $I_y = \frac{Aa^2}{4}\left(1 + \frac{2\sin^3\theta\cos\theta}{\theta - \sin\theta\cos\theta}\right)$	$r_x^2 = \frac{a^2}{4}\left[1 - \frac{2\sin^3\theta\cos\theta}{3(\theta - \sin\theta\cos\theta)}\right]$ $r_y^2 = \frac{a^2}{4}\left(1 + \frac{2\sin^3\theta\cos\theta}{\theta - \sin\theta\cos\theta}\right)$
Ellipse	$A = \pi ab$ $x_c = a$ $y_c = b$	$I_{x_c} = \frac{\pi}{4}ab^3$ $I_{y_c} = \frac{\pi}{4}a^3b$ $I_x = \frac{5}{4}\pi ab^3$ $I_y = \frac{5}{4}\pi a^3b$ $I_P = \frac{\pi ab}{4}(a^2 + b^2)$	$r_{x_c}^2 = \frac{1}{4}b^2$ $r_{y_c}^2 = \frac{1}{4}a^2$ $r_x^2 = \frac{5}{4}b^2$ $r_y^2 = \frac{5}{4}a^2$ $r_P^2 = \frac{1}{4}(a^2 + b^2)$

TABLE 5.1 Properties of Plane Sections (*Continued*)

Figure	Area and centroid	Moment of inertia	r^2
Semiellipse (y, x, G, b, 2a)	$A = \frac{1}{2}\pi ab$ $x_c = a$ $y_c = \dfrac{4b}{3\pi}$	$I_{x_c} = \dfrac{ab^3}{72\pi}(9\pi^2 - 64)$ $I_{y_c} = \dfrac{\pi}{8}a^3b$ $I_x = \dfrac{\pi}{8}ab^3$ $I_y = \frac{5}{8}\pi a^3b$	$r_{x_c}^2 = \dfrac{b^2}{36\pi^2}(9\pi^2 - 64)$ $r_{y_c}^2 = \frac{1}{4}a^2$ $r_x^2 = \frac{1}{4}b^2$ $r_y^2 = \frac{5}{4}a^2$
Parabola (y, x, G, a, b, b)	$A = \frac{4}{3}ab$ $x_c = \frac{3}{5}a$ $y_c = 0$	$I_{x_c} = I_x = \frac{4}{15}ab^3$ $I_{y_c} = \frac{16}{175}a^3b$ $I_y = \frac{4}{7}a^3b$	$r_{x_c} = r_x^2 = \frac{1}{5}b^2$ $r_{y_c}^2 = \frac{12}{175}a^2$ $r_y^2 = \frac{3}{7}a^2$
Semiparabola (y, x, G, a, b)	$A = \frac{2}{3}ab$ $x_c = \frac{3}{5}a$ $y_c = \frac{3}{8}b$	$I_x = \frac{2}{15}ab^3$ $I_y = \frac{2}{7}a^3b$	$r_x^2 = \frac{1}{5}b^2$ $r_y^2 = \frac{3}{7}a^2$

Note: A = area
x_c, y_c = coordinates of centroid of section in XY coordinate system
r_x, r_y = radius of gyration of the section with respect to the centroidal axes parallel to the X and Y axes
r_{x_c}, r_{y_c} = radius of gyration of the section with respect to the centroidal axes parallel to the X and Y axes
r_p = radius of gyration of the section about the polar axis passing through the centroid
I_x, I_y = moment of inertia with respect to the X and Y axes shown
I_p = polar moment of inertia about an axis passing through the centroid
I_{x_c}, I_{y_c} = moment of inertia about an axis through the centroid parallel to the X and Y axes
G = marks the centroid

Source: From Rothbart.[4]

TABLE 5.2 Properties of Homogeneous Bodies

Body	Mass and centroid	Moment of inertia	r^2
Thin rod	$m = \rho l$ $x_c = \frac{1}{2}l$ $y_c = 0$ $z_c = 0$	$I_x = I_{x_c} = 0$ $I_{y_c} = I_{z_c} = \frac{m}{12} l^2$ $I_y = I_z = \frac{m}{3} l^2$	$r_x^2 = r_{x_c}^2 = 0$ $r_{y_c}^2 = r_{x_c}^2 = \frac{1}{12} l^2$ $r_y^2 = r_z^2 = \frac{1}{3} l^2$
Thin circular rod	$m = 2\rho R\theta$ $x_c = \frac{R \sin\theta}{\theta}$ $y_c = 0$ $z_c = 0$	$I_x = I_{x_c}$ $= \frac{mR^2(\theta - \sin\theta\cos\theta)}{2\theta}$ $I_y = \frac{mR^2(\theta + \sin\theta\cos\theta)}{2\theta}$ $I_z = mR^2$	$r_x^2 = r_{x_c}^2 = \frac{R^2(\theta - \sin\theta\cos\theta)}{2\theta}$ $r_y^2 = \frac{R^2(\theta + \sin\theta\cos\theta)}{2\theta}$ $r_z^2 = R^2$
Thin hoop	$m = 2\pi\rho R$ $x_c = R$ $y_c = R$ $z_c = 0$	$I_{x_c} = I_{y_c} = \frac{m}{2} R^2$ $I_{z_c} = mR^2$ $I_x = I_y = \frac{3}{2} mR^2$ $I_z = 3mR^2$	$r_{x_c}^2 = r_{y_c}^2 = \frac{1}{2} R^2$ $r_{z_c}^2 = R^2$ $r_x^2 = r_y^2 = \frac{3}{2} R^2$ $r_z = 3R^2$

TABLE 5.2 Properties of Homogeneous Bodies (*Continued*)

Body	Mass and centroid	Moment of inertia	r^2
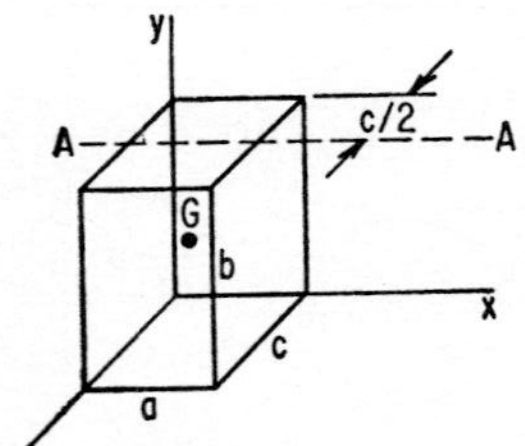 Rectangular prism	$m = \rho abc$ $x_c = \frac{1}{2}a$ $y_c = \frac{1}{2}b$ $z_c = \frac{1}{2}c$	$I_{x_c} = \frac{1}{12}m(b^2 + c^2)$ $I_x = \frac{1}{3}m(b^2 + c^2)$ $I_{AA} = \frac{1}{12}m(4b^2 + c^2)$	$r^2_{x_c} = \frac{1}{12}(b^2 + c^2)$ $r^2_x = \frac{1}{3}(b^2 + c^2)$ $r^2_{AA} = \frac{1}{12}(4b^2 + c^2)$
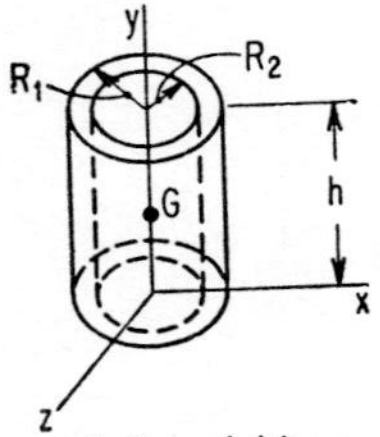 Hollow right circular cylinder	$m = \pi\rho h(R_1^2 - R_2^2)$ $x_c = 0$ $y_c = \frac{1}{2}h$ $z_c = 0$	$I_{x_c} = I_{z_c} = \frac{1}{12}m(3R_1^2 + 3R_2^2 + h^2)$ $I_{y_c} = I_y = \frac{1}{2}m(R_1^2 + R_2^2)$ $I_x = I_z = \frac{1}{12}m(3R_1^2 + 3R_2^2 + 4h^2)$	$r^2_{x_c} = r^2_{z_c} = \frac{1}{12}(3R_1^2 + 3R_2^2 + h^2)$ $r^2_{y_c} = r^2_y = \frac{1}{2}(R_1^2 + R_2^2)$ $r^2_x = r^2_z = \frac{1}{12}(3R_1^2 + 3R_2^2 + 4h^2)$

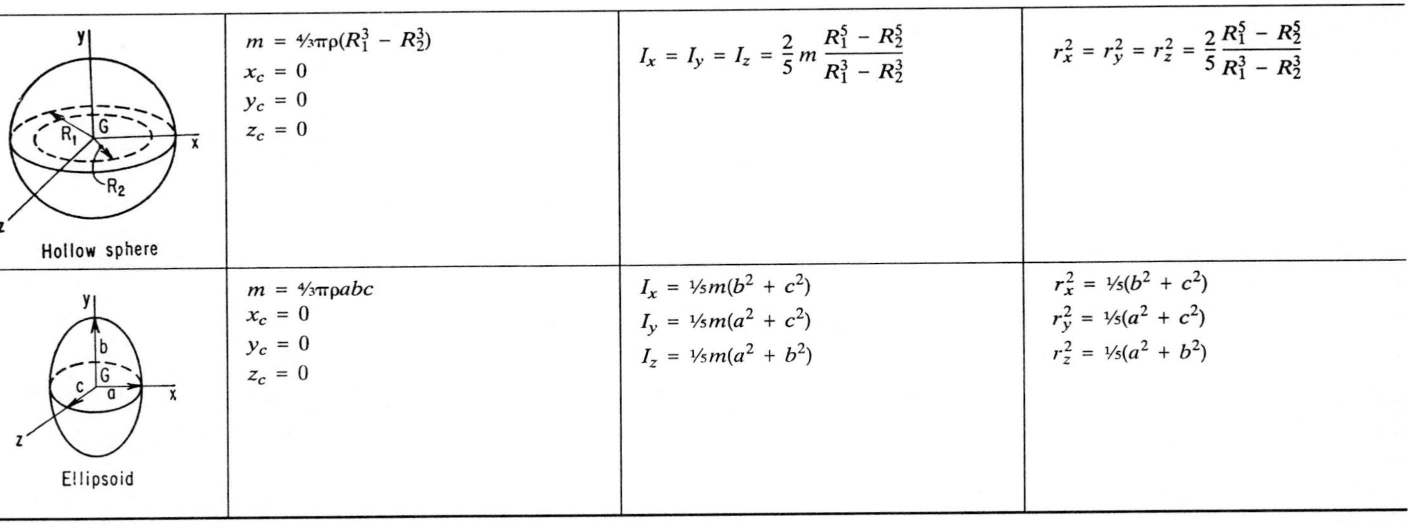

Hollow sphere	$m = \frac{4}{3}\pi\rho(R_1^3 - R_2^3)$ $x_c = 0$ $y_c = 0$ $z_c = 0$	$I_x = I_y = I_z = \frac{2}{5} m \frac{R_1^5 - R_2^5}{R_1^3 - R_2^3}$	$r_x^2 = r_y^2 = r_z^2 = \frac{2}{5} \frac{R_1^5 - R_2^5}{R_1^3 - R_2^3}$
Ellipsoid	$m = \frac{4}{3}\pi\rho abc$ $x_c = 0$ $y_c = 0$ $z_c = 0$	$I_x = \frac{1}{5}m(b^2 + c^2)$ $I_y = \frac{1}{5}m(a^2 + c^2)$ $I_z = \frac{1}{5}m(a^2 + b^2)$	$r_x^2 = \frac{1}{5}(b^2 + c^2)$ $r_y^2 = \frac{1}{5}(a^2 + c^2)$ $r_z^2 = \frac{1}{5}(a^2 + b^2)$

Note: ρ = mass density
m = mass
x_c, y_c, z_c = coordinates of centroid in *XYZ* coordinate system
$r_{x_c}, r_{y_c}, r_{z_c}$ = radius of gyration of the body with respect to the centroidal axes parallel to the *X, Y*, and *Z* axes shown
r_x, r_y, r_z = radius of gyration of the body with respect to the *X, Y*, and *Z* axes shown
I_x, I_y, I_z = moment of inertia with respect of *X, Y*, and *Z* axes shown
$I_{x_c}, I_{y_c}, I_{z_c}$ = moment of inertia about an axis through the centroid parallel to the *X, Y*, and *Z* axes shown
I_{AA} = moment of inertia with respect to special axes shown
G = marks the centroid

Source: From Rothbart.[4]

If Z is any polar axis of an area A, Z_0 the centroidal axis, and d the distance between the two axes,

$$I_p = I_{p_0} + Ad^2 \tag{5.20}$$

If X and Y are two rectangular axes in the plane of the area and Z is the polar axis through their point of intersection,

$$I_p = I_x + I_y \tag{5.21}$$

If an area is composed of several simple parts, the moment of inertia of each part with respect to any given axis may be computed separately and added to obtain the moment of inertia of the entire area.

Moment of Inertia of Mass. In problems involving the rotation of a mass about an axis, it is necessary to compute the quantity $\int \rho^2\, dm$. In this expression, dm is the mass of any differential particle and ρ is its distance from the axis of rotation. The quantity $\int \rho^2\, dm$ is called the *moment of inertia of the mass* and is represented by I, usually with a subscript to indicate the axis. The radius of gyration **r** is the distance from the axis at which all of the mass could be concentrated and has the same moment of inertia; or, in other words, r^2 is the mean value of the variable ρ^2. Thus

$$I = \int \rho^2\, dm = mr^2 \tag{5.22}$$

The relation between the moment of inertia with respect to the centroidal axis O and that with respect to any parallel axis C is given by the transfer formula

$$I_c = I_0 + md^2 \tag{5.23}$$

d being the perpendicular distance between the axes.

Values of the moment of inertia for different plane sections and homogeneous bodies are given in Tables 5.1 and 5.2.

FRICTION

1. Definitions of Sliding and Static Friction. Friction is the resistance that is encountered when two solid surfaces slide or tend to slide over each other.

In stationary systems, friction manifests itself as a force equal and opposite to the shear force applied to the interface. Thus, as shown in Fig. 5.15, if a force F_0 is applied, a friction force F will be generated, equal and opposite to F_0, so that the surfaces remain at rest. F can take any magnitude up to a limiting value F', and can therefore prevent sliding whenever F_0 is less than F'. If F_0 exceeds F', slipping of the body occurs.

The friction force is proportional to the normal force F_N (note that $F_N = -W$, where W is equal to the weight of the body), and the coefficient of proportionality is defined as the *static* friction coefficient f_0. This is expressed by the equation

$$f_0 = \frac{F'}{F_N} \tag{5.24}$$

The angle of static friction α (the largest angle relative to the horizontal at which a surface may be tilted, so that the body placed on the surface does not slide down) is defined by

$$\tan \alpha = \frac{F'}{F_N} = f_0 \tag{5.25}$$

If one surface slides over the other, a frictional force resulting from the motion must be overcome. This force is usually smaller, by about 20 percent, than F' and is expressed as

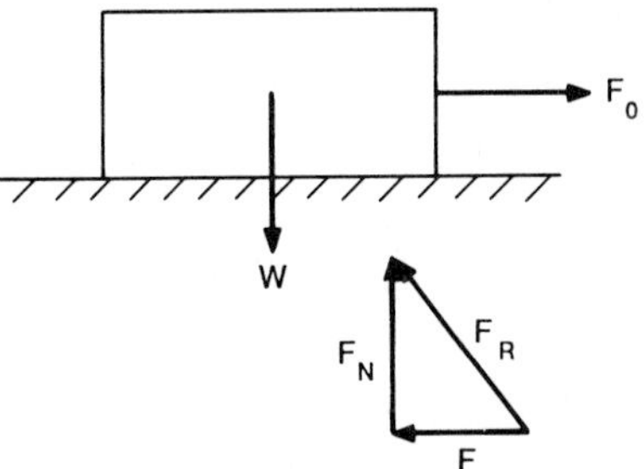

FIGURE 5.15

$$F = f \cdot F_N \tag{5.26}$$

where f is the *sliding* friction coefficient or kinetic friction. Angle of kinetic friction is defined as

$$\tan \alpha = \frac{F}{F_N} \tag{5.27}$$

In the range of practical velocities of sliding, the coefficients of sliding friction are smaller than the coefficients of static friction. With small velocities of sliding and very clean surfaces, the two coefficients do not differ appreciably.

When the surfaces are free from contaminating fluids, or films, the resistance is called *dry friction*. When the surfaces are separated from each other by a very thin film of lubricant, the friction is that of *boundary* (or *greasy*) *lubrication*.

Values of sliding and static coefficients are given in Table 5.3. Additional data for sliding coefficients are given in Table 5.4.

2. Rolling Friction. If a sphere touches a plane or a cylindrical surface, the contact is only a point, while if a cylinder touches a plane or another cylindrical surface, the contact is a line. Assuming that all materials are elastic, the sphere or cylinder under pressure and its supporting surface are distorted so that there is an area of contact. As the sphere or cylinder is rolled along under pressure, it is assumed that the material of the supporting surface is rolled ahead as a miniature wave and the center of contact is a small distance a in front of the normal radius, as shown in Fig. 5.16. With the center of contact as the center of moments,

$$F \frac{r}{2} = Wa \tag{5.28}$$

where F is the horizontal force and W is the normal force (in this case the weight of the body). With hard rollers and hard bearings, the value of a is very small and the resistance to rolling is very much smaller than the resistance to sliding. Values of a as recommended in the literature are given in Table 5.5.

3. Friction of Flexible Belts and Bands. See Fig. 5.17. Let T_A be the tension in the taut side of a belt over a pulley, T_2 the tension in the slack side, f the coeffi-

TABLE 5.3 Coefficients of Static and Sliding Friction
(Reference letters indicate the lubricant used; see footnote)

Materials	Static		Sliding	
	Dry	Greasy	Dry	Greasy
Hard steel on hard steel	0.78	0.11 (*a*)	0.42	0.029 (*h*)
	....	0.23 (*b*)	....	0.081 (*c*)
	....	0.15 (*c*)	....	0.080 (*i*)
	....	0.11 (*d*)	....	0.058 (*j*)
	....	0.0075 (*p*)	....	0.084 (*d*)
	....	0.0052 (*h*)	....	0.105 (*k*)
	....	....	....	0.096 (*l*)
	....	....	....	0.108 (*m*)
	....	....	....	0.12 (*a*)
Mild steel on mild steel	0.74	....	0.57	0.09 (*a*)
	....	....	....	0.19 (*u*)
Hard steel on graphite	0.21	0.09 (*a*)		
Hard steel on babbitt (ASTM No. 1)	0.70	0.23 (*b*)	0.33	0.16 (*b*)
	....	0.15 (*c*)	....	0.06 (*c*)
	....	0.08 (*d*)	....	0.11 (*d*)
	....	0.085 (*e*)		
Hard steel on babbitt (ASTM No. 8)	0.42	0.17 (*b*)	0.35	0.14 (*b*)
	....	0.11 (*c*)	....	0.065 (*c*)
	....	0.09 (*d*)	....	0.07 (*d*)
	....	0.08 (*e*)	....	0.08 (*h*)
Hard steel on babbitt (ASTM No. 10)	....	0.25 (*b*)	....	0.13 (*b*)
	....	0.12 (*c*)	....	0.06 (*c*)
	....	0.10 (*d*)	....	0.055 (*d*)
	....	0.11 (*e*)		
Mild steel on cadmium silver	....	....	....	0.097 (*f*)
Mild steel on phosphor bronze	....	....	0.34	0.173 (*f*)
Mild steel on copper lead	....	....	....	0.145 (*f*)
Mild steel on cast iron	....	0.183 (*c*)	0.23	0.133 (*f*)
Mild steel on lead	0.95	0.5 (*f*)	0.95	0.3 (*f*)
Nickel on mild steel	....	....	0.64	0.178 (*x*)
Aluminum on mild steel	0.61	....	0.47	
Magnesium on mild steel	....	....	0.42	
Magnesium on magnesium	0.6	0.08 (*y*)		
Teflon on Teflon	0.04	....	....	0.04 (*f*)
Teflon on steel	0.04	....	....	0.04 (*f*)
Tungsten carbide on tungsten carbide	0.2	0.12 (*a*)		
Tungsten carbide on steel	0.5	0.08 (*a*)		
Tungsten carbide on copper	0.35			
Tungsten carbide on iron	0.8			
Bonded carbide on copper	0.35			
Bonded carbide on iron	0.8			
Cadmium on mild steel	....	....	0.46	
Copper on mild steel	0.53	....	0.36	0.18 (*a*)
Nickel on nickel	1.10	....	0.53	0.12 (*w*)
Brass on mild steel	0.51	....	0.44	
Brass on cast iron	....	....	0.30	

TABLE 5.3 Coefficients of Static and Sliding Friction (*Continued*)

Materials	Static		Sliding	
	Dry	Greasy	Dry	Greasy
Zinc on cast iron	0.85	....	0.21	
Magnesium on cast iron	....	....	0.25	
Copper on cast iron	1.05	....	0.29	
Tin on cast iron	....	....	0.32	
Lead on cast iron	....	....	0.43	
Aluminum on aluminum	1.05	....	1.4	
Glass on glass	0.94	0.01 (*p*)	0.40	0.09 (*a*)
	....	0.005 (*q*)	....	0.116 (*v*)
Carbon on glass	....	....	0.18	
Garnet on mild steel	....	....	0.39	
Glass on nickel	0.78	....	0.56	
Copper on glass	0.68	....	0.53	
Cast iron on cast iron	1.10	....	0.15	0.070 (*d*)
	....	....	....	0.064 (*n*)
Bronze on cast iron	....	....	0.22	0.077 (*n*)
Oak on oak (parallel to grain)	0.62	....	0.48	0.164 (*r*)
	....	....	....	0.067 (*s*)
Oak on oak (perpendicular)	0.54	....	0.32	0.072 (*s*)
Leather on oak (parallel)	0.61	....	0.52	
Cast iron on oak	....	....	0.49	0.075 (*n*)
Leather on cast iron	....	....	0.56	0.36 (*t*)
	....	....	....	0.13 (*n*)
Laminated plastic on steel	....	....	0.35	0.05 (*t*)
Fluted rubber bearing on steel	....	....	....	0.05 (*t*)

(*a*) Oleic acid; (*b*) Atlantic spindle oil (light mineral); (*c*) castor oil; (*d*) lard oil; (*e*) Atlantic spindle oil plus 2 percent oleic acid; (*f*) medium mineral oil; (*g*) medium mineral oil plus ½ percent oleic acid; (*h*) stearic acid; (*i*) grease (zinc oxide base); (*j*) graphite; (*k*) turbine oil plus 1 percent graphite; (*l*) turbine oil plus 1 percent stearic acid; (*m*) turbine oil (medium mineral); (*n*) olive oil; (*p*) palmitic acid; (*q*) ricinoleic acid; (*r*) dry soap; (*s*) lard; (*t*) water; (*u*) rape oil; (*v*) 3-in-1 oil; (*w*) octyl alcohol; (*x*) triolein; (*y*) 1 percent lauric acid in paraffin oil.

Source: From Avallone and Baumeister.[2]

TABLE 5.4 Values of f for Mild Steel on Medium Steel—Effect of Sliding Velocity

in/s	0.0001	0.001	0.01	0.1	1	10	100
cm/s	0.000254	0.00254	0.0254	0.254	2.54	25.4	254
f	0.53	0.48	0.39	0.31	0.23	0.19	0.18

Source: Avallone and Baumeister.[2]

cient of friction, and α the angle of contact in radians. Then the expression giving the limiting relation between T_1 and T_2 when slipping impends is

$$T_1 = T_2 e^{\alpha f} \tag{5.29}$$

The quantity e is the base of the natural system of logarithms and its value is 2.71828. The same relation holds true for a flexible band brake and for a rope

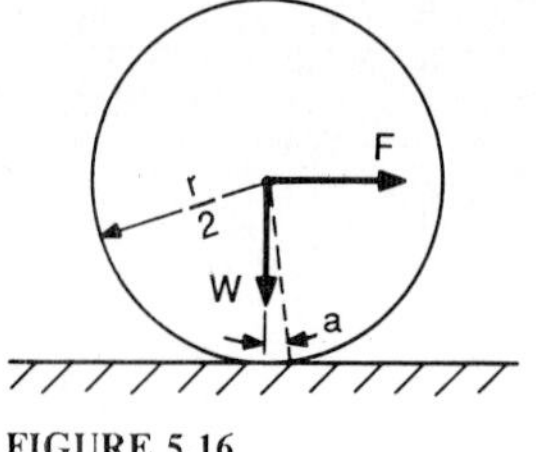

FIGURE 5.16

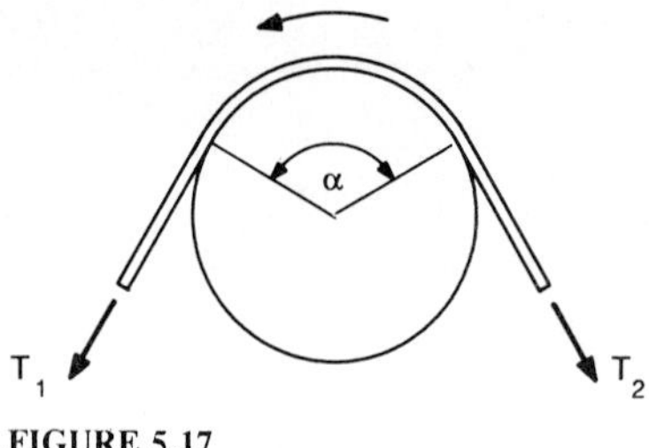

FIGURE 5.17

TABLE 5.5 Rolling Resistance

Materials	a, inches (cm)
Hardwood on hardwood	0.02 (0.05)
Iron on iron (steel on steel)	0.002 (0.005)
Hard polish steel on hard polish steel	0.0002 to 0.0004 (0.0005 to 0.001)
Pneumatic tires on good road of asphalt	0.02 to 0.022 (0.0508 to 0.0558)
Pneumatic tires on heavy mud of asphalt	0.04 to 0.06 (0.1016 to 0.1524)
Iron or steel wheels on wood track	0.06 to 0.10 (0.1524 to 0.254)

Source: Updated from O'Rourke.[3]

over a spar or around a snubbing post. When slipping occurs, the quantity f is the kinetic coefficient of friction.

Note that $T_1 - T_2$ is equal to the circumferential force transfered by friction. Values of T_1/T_2 for various values of f and α are shown in Table 5.6.

Note. Values of friction for different machine elements are given in Ref. 2.

KINEMATICS

1. Definitions

Kinematics. The study of the motion of bodies without reference to the forces causing that motion or the mass of the bodies.

Motion. Motion of a particle with respect to other particles or objects is its state of continual changing of position with respect to them.

Rectilinear Motion. Motion along a straight path.

Curvilinear Motion. Motion along a curved path which may be either planar or skewed.

Displacement. Displacement of a particle is its change of position and is a vector quantity. If point A is the position of a particle at a time t, and B its position at a later time t_2, its displacement in the time interval $t_2 - t_1 = \Delta t$ is the vector **AB**, no matter whether the path is straight or curved.

Velocity. Velocity of a particle is its time rate of displacement and is a vector quantity.

Speed. The magnitude of velocity without reference to direction.

Acceleration. Acceleration of a particle is its time rate of change of velocity and is a vector quantity.

TABLE 5.6 Values of T_1/T_2

$\frac{a^\circ}{360^\circ}$	*f*								
	0.1	0.15	0.2	0.25	0.3	0.35	0.4	0.45	0.5
0.1	1.06	1.1	1.13	1.17	1.21	1.25	1.29	1.33	1.37
0.2	1.13	1.21	1.29	1.37	1.46	1.55	1.65	1.76	1.87
0.3	1.21	1.32	1.45	1.60	1.76	1.93	2.13	2.34	2.57
0.4	1.29	1.46	1.65	1.87	2.12	2.41	2.73	3.10	3.51
0.425	1.31	1.49	1.70	1.95	2.23	2.55	2.91	3.33	3.80
0.45	1.33	1.53	1.76	2.03	2.34	2.69	3.10	3.57	4.11
0.475	1.35	1.56	1.82	2.11	2.45	2.84	3.30	3.83	4.45
0.5	1.37	1.60	1.87	2.19	2.57	3.00	3.51	4.11	4.81
0.525	1.39	1.64	1.93	2.28	2.69	3.17	3.74	4.41	5.20
0.55	1.41	1.68	2.00	2.37	2.82	3.35	3.98	4.74	5.63
0.6	1.46	1.76	2.13	2.57	3.10	3.74	4.52	5.45	6.59
0.7	1.55	1.93	2.41	3.00	3.74	4.66	5.81	7.24	9.02
0.8	1.65	2.13	2.73	3.51	4.52	5.81	7.47	9.60	12.35
0.9	1.76	2.34	3.10	4.11	5.45	7.24	9.60	12.74	16.90
1.0	1.87	2.57	3.51	4.81	6.59	9.02	12.35	16.90	23.14
1.5	2.57	4.11	6.59	10.55	16.90	27.08	43.38	69.49	111.32
2.0	3.51	6.59	12.35	23.14	43.38	81.31	152.40	285.68	535.49
2.5	4.81	10.55	23.14	50.75	111.32	244.15	535.49	1,174.5	2,575.9
3.0	6.59	16.90	43.38	111.32	285.68	733.14	1,881.5	4,828.5	12,391
3.5	9.02	27.08	81.31	244.15	733.14	2,199.90	6,610.7	19,851	59,608
4.0	12.35	43.38	152.40	535.49	1,881.5	6,610.7	23,227	81,610	286,744

Source: From Avallone and Baumeister.[2]

Relative Motion. The displacement, velocity, and acceleration of a body with respect to a fixed point on the earth are called its *absolute* displacement, velocity, and acceleration, respectively. The displacement, velocity, and acceleration of one body with respect to another which may also have displacement, velocity, and acceleration with respect to the earth are called its *relative* displacement, velocity, and acceleration, respectively.

2. Rectilinear Motion

Velocity

$$\text{Instantaneous velocity} = v = \frac{ds}{dt} = \lim_{\Delta t \to 0} \frac{\Delta s}{\Delta t} \tag{5.30}$$

$$\text{Average velocity} = \bar{v} = \frac{\Delta s}{\Delta t} \tag{5.31}$$

where s = distance measured along the path of a particle.

Acceleration

$$\text{Instantaneous acceleration} = a = \frac{dv}{dt} = \frac{d^2s}{dt^2} = \lim_{\Delta t \to 0} \frac{\Delta v}{\Delta t} \tag{5.32}$$

$$\text{Average acceleration} = \bar{a} = \frac{\Delta v}{\Delta t} \tag{5.33}$$

Relations among a, v, s, t

$$v = \frac{ds}{dt} \tag{5.34}$$

$$a = \frac{dv}{dt} = \frac{d^2s}{dt^2} \tag{5.35}$$

$$\frac{a}{v} = \frac{dv}{ds} \tag{5.36}$$

$$s_2 - s_1 = \int_{t_1}^{t_2} v\,dt \tag{5.37}$$

$$v_2 - v_1 = \int_{t_1}^{t_2} a\,dt \tag{5.38}$$

$$t_2 - t_1 = \int_{s_1}^{s_2} \frac{ds}{v} = \int_{v_1}^{v_2} \frac{dv}{a} \tag{5.39}$$

$$v_2^2 - v_1^2 = 2\int_{s_1}^{s_2} a\,ds \tag{5.40}$$

where s_1 = distance from origin of time t_1
s_2 = distance at a later time t_2
v_1 = velocity of particle at time t_1
v_2 = velocity at later time t_2

Motion Graphs. These include

- A *distance-time curve*, which offers a convenient means for the study of motion of a point. The slope of the curve at any point will represent the velocity at that time.
- A *velocity-time curve*, which offers a convenient means for the study of acceleration. The slope of the curve at any point will represent the acceleration at that time.
- An *acceleration-time curve*, which may be constructed by plotting accelerations as ordinates, and times as abscissas. The area under this curve between any two ordinates will represent the total increase in velocity during the time interval.

Examples of Rectilinear Motion

Uniform Acceleration (a = constant). Let v_0 be initial velocity and s_0 initial distance. Then from Eqs. (5.38) through (5.40):

$$v = at + v_0 \tag{5.41}$$

$$s = \tfrac{1}{2}at^2 + v_0t \tag{5.42}$$

$$v^2 = 2a(s - s_0) + v_0^2 \tag{5.43}$$

Free Fall. If a body falls from rest in a vacuum, $v_0 = 0$, $s_0 = 0$, and $a = g$ (= acceleration due to gravity). Then:

$$v = gt = (2gs)^{1/2} \tag{5.44}$$

$$s = \tfrac{1}{2}gt^2 \tag{5.45}$$

If a body is projected upward at an initial velocity v_0, then

$$a = -g$$

$$v = -gt + v_0 = (-2gs + v_0^2)^{1/2} \tag{5.46}$$

$$s = -\tfrac{1}{2}gt^2 + v_0 t \tag{5.47}$$

$$H \text{ (total ascent to highest position)} = \frac{v_0^2}{2g} \quad \text{and} \quad t_H \text{ (time required)} = \frac{v_0}{g}$$

Crank and Connecting-Rod Mechanism. The problem is to find expressions for the velocity and acceleration of any point in the crosshead A, as shown in Fig. 5.18. Such a point describes rectilinear motion. Let $\lambda = (r/l) = \tfrac{1}{4} \cdots \tfrac{1}{6}$, n = revolutions-per-second assumed constant, ω = radians of angle described by crank per second, and s = distance of A from its extreme position O. Then

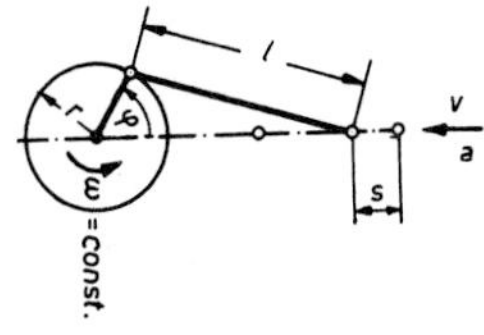

FIGURE 5.18 (*From Gieck.*[1])

$$s = r(1 - \cos\varphi) + \frac{\lambda}{2} r \sin^2\varphi \tag{5.48}$$

$$v = \omega r \sin\varphi(1 + \lambda\cos\varphi) \tag{5.49}$$

$$a = \omega^2 r(\cos\varphi + \lambda\cos 2\varphi) \tag{5.50}$$

$$\varphi = \omega t = 2\pi n t$$

Simple Harmonic Motion. This type of motion has wide application in physics and engineering (for example, a body supported by a spring performs a linear harmonic oscillation). For this kind of motion, quantities s, v, and a are defined as:

$$s = A\sin(\omega t + \varphi_0) \tag{5.51}$$

$$v = \frac{ds}{dt} = A\omega\cos(\omega t + \varphi_0) \tag{5.52}$$

$$a = \frac{dv}{dt} = A\omega\cos(\omega t + \varphi_0) \tag{5.53}$$

where A = amplitude (maximum displacement)
$\varphi = (\omega t + \varphi_0)$ = angular position at time t
φ_0 = angular position at $t = 0$
s = displacement
ω = angular frequency, radians per unit time

Sliding Motion on an Inclined Plane. See Fig. 5.19. Here the quantities are defined as:

$$a = g(\sin\alpha - f\cos\alpha) \tag{5.54}$$

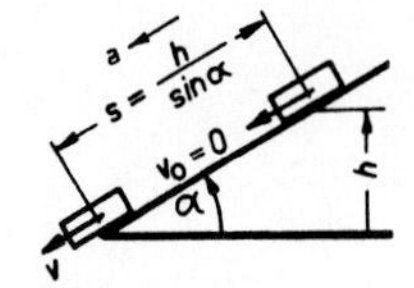

FIGURE 5.19 (*From Gieck.*[1])

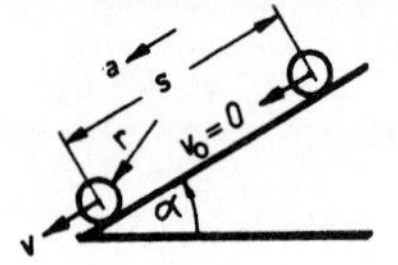

FIGURE 5.20 (*From Gieck.*[1])

$$v = (2as)^{1/2} \tag{5.55}$$

$$s = \frac{v^2}{2a} \tag{5.56}$$

where f = coefficient of sliding friction.

Rolling Motion on an Inclined Plane. See Fig. 5.20. Here the quantities are defined as:

$$a = gr^2 \frac{\sin \alpha - (r_0/r) \cos \alpha}{r^2 + K^2} \tag{5.57}$$

v = value defined by Eq. (5.55)

s = value defined by Eq. (5.56)

r_0 = values given in Table 5.5, i.e., use $r_0 = a$

$$K^2 = \tfrac{2}{5}r^2 \qquad = \frac{r^2}{2} \qquad = r^2$$

for ball, solid cylinder, and pipe with low wall thickness, respectively.

3. Curvilinear motion

Velocity and Acceleration. The magnitude of velocity at any instant is

$$v = \frac{ds}{dt} \tag{5.58}$$

where s is measured along the curved path of a particle (Fig. 5.21*a*). The linear direction of the velocity is the tangent to the path of the point. In Fig. 5.21*a*, let *ABC* be the path of a moving point and $\mathbf{v}_1$, $\mathbf{v}_2$, $\mathbf{v}_3$ velocity vectors at *A*, *B*, and *C*. If *O* is taken as a pole (Fig. 5.21*b*) and vectors $\mathbf{v}_1$, $\mathbf{v}_2$, $\mathbf{v}_3$ representing the velocities at corresponding points are drawn, the curve connecting the terminal points of these vectors is known as the *hodograph* of the motion.

The acceleration is given in terms of its tangential and normal components a_t and a_n, respectively. The tangential component a_t may be positive, zero, or negative, assuming the direction of the velocity is positive. Its amount is

$$a_t = \frac{dv}{dt} \tag{5.59}$$

the *normal* component is always directed toward the center of curvature along the radius ρ at the point. Its amount is

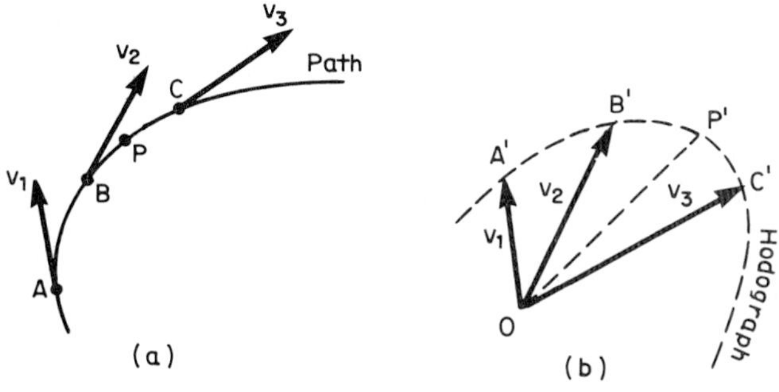

FIGURE 5.21

$$a_n = \frac{v^2}{\rho} \tag{5.60}$$

where v is the velocity at the point. If the radius of curvature ρ increases and becomes infinity, the path of the point becomes a straight line. If the radius of curvature ρ ceases to vary and becomes the constant radius r, the path of the point becomes a circle and the motion is rotation. The normal acceleration is

$$a_n = \frac{v^2}{r} \tag{5.61}$$

Resultant acceleration is

$$a = (a_t^2 + a_n^2)^{1/2} \tag{5.62}$$

Components of Velocity and Acceleration in Different Coordinate Systems. See Figs. 3.1 to 3.3 in Chap. 3.

Rectangular, $P(x,y,z)$

$$v = (v_x^2 + v_y^2 + v_z^2)^{1/2} \tag{5.63}$$

$$a = (a_x^2 + a_y^2 + a_z^2)^{1/2} \tag{5.64}$$

$$v_x = \dot{x} \qquad v_y = \dot{y} \qquad v_z = \dot{z} \tag{5.65}$$

$$a_x = \ddot{x} \qquad a_y = \ddot{y} \qquad a_z = \ddot{z} \tag{5.66}$$

Cylindrical, $P(r,\theta,z)$

$$v = (v_r^2 + v_\theta^2 + v_z^2)^{1/2} \tag{5.67}$$

$$a = (a_r^2 + a_\theta^2 + a_z^2)^{1/2} \tag{5.68}$$

$$v_r = \dot{r} \qquad v_\theta = r\dot{\theta} \qquad v_z = \dot{z} \tag{5.69}$$

$$a_r = \ddot{r} - r\dot{\theta}^2 \qquad a_\theta = r\ddot{\theta} + 2\dot{r}\dot{\theta} \qquad a_z = \ddot{z} \tag{5.70}$$

Spherical, P(r, θ, φ)

$$v = (v_r^2 + v_\theta^2 + v_\phi^2)^{1/2} \tag{5.71}$$

$$a = (a_r^2 + a_\theta^2 + a_\phi^2)^{1/2} \tag{5.72}$$

$$v_r = \dot{r} \qquad v_\phi = r\dot{\phi} \qquad v_\theta = r\dot{\theta} \sin \phi \tag{5.73}$$

$$a_r = \ddot{r} - r\dot{\phi}^2 - r\dot{\theta}^2 \sin^2 \phi \tag{5.74}$$

$$a_\phi = 2\dot{r}\dot{\phi} + r\dot{\phi} - r\dot{\theta}^2 \sin \phi \cos \phi \tag{5.75}$$

$$a_\theta = 2\dot{r}\dot{\theta} \sin \phi + r\ddot{\theta} \sin \phi + 2r\dot{\theta}\dot{\phi} \cos \phi \tag{5.76}$$

Note. The above relations show that velocities and acceleration (like forces) may be composed or resolved according to the parallelogram and parallelepipedon laws discussed above in "Graphical Composition of Forces."

Motion of a Projectile (Example). Assuming that the resistance of air is neglected, the vertical component of the motion of a projectile is the same as that of a falling body, and the horizontal component of the motion is that of a body with constant velocity. Let v_0 be the initial velocity, at an angle α with the horizontal (Fig. 5.22). In the vertical direction,

$$a_y = -g$$

$$v_y = v_0 \sin \alpha - g \cdot t$$

$$y = v_0 t \sin \alpha - \frac{gt^2}{2} \tag{5.77}$$

In the horizontal direction,

$$a_x = 0$$

$$v_x = v_0 \cos \alpha$$

$$x = v_0 t \cos \alpha \tag{5.78}$$

Eliminating t from Eqs. (5.77) and (5.78),

$$y = x \tan \alpha - \frac{gx^2}{2v_0^2 \cos^2 \alpha} \tag{5.79}$$

Also,

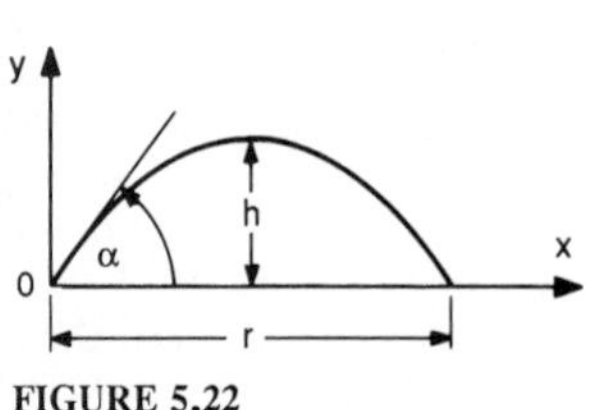

FIGURE 5.22

$$h = \frac{(\sin^2 \alpha)v_0^2}{2g}$$

$$r = \frac{(\sin 2\alpha)v_0^2}{g}$$

$$T = \frac{2v_0 \sin \alpha}{g}$$

where $\alpha = \alpha_1 = 45° =$ value of α for maximum r
$h =$ greatest height attained
$T =$ time of flight

Note. Motion graphs can be constructed for curvilinear motion of a particle, as they can for rectilinear motion.

4. Rotation

***Angular Displacement* θ.** If a body rotates about a fixed axis, each particle of the body describes a circle (Fig. 5.23). The angle described by any radius (or by any straight line in a plane normal to the axis) is called the *angular displacement*. The unit of angular displacement is the radian, the angle at the center subtended by an arc s equal in length to the radius r.

If θ is the angle of displacement, the arc $s = r\theta$. One circumference or $360° = 2\pi$ rad, or 1 rad $= 180°/\pi$.

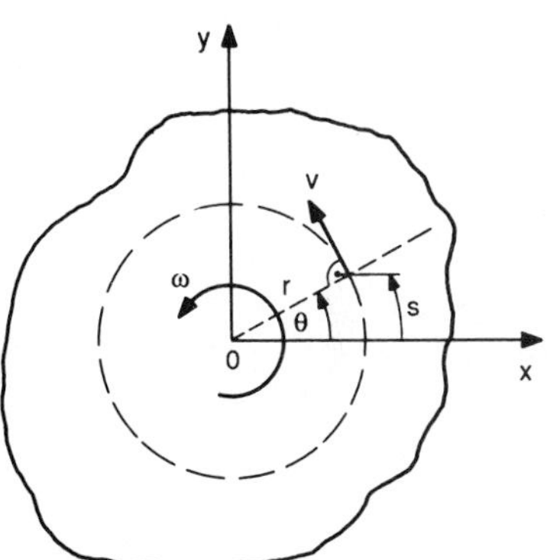

FIGURE 5.23

Angular Velocity. The angular velocity ω is the time rate of angular displacement:

$$\omega = \frac{d\theta}{dt} = \dot{\theta} \tag{5.80}$$

As $s = r\theta$, by differentiation with respect to t the following relation is obtained:

$$v = r\omega \tag{5.81}$$

Angular Acceleration. The angular acceleration α is the time rate of change of angular velocity:

$$\alpha = \frac{d\omega}{dt} = \dot{\omega} = \ddot{\theta} \tag{5.82}$$

As $v = r\omega$, by differentiation with respect to t the following relation is obtained:

$$a_t = r \cdot \alpha \tag{5.83}$$

As $a_n = v^2/r$ [Eq. (5.60)] and $v = r\omega$, then

$$a_n = r\omega^2 \tag{5.84}$$

If $\alpha = 0$, ω and a_n are constant. The angle moved through in time t is

$$\theta = \omega t \tag{5.85}$$

and the period of one complete rotation is

$$T = \frac{2\pi}{\omega} \tag{5.86}$$

The frequency f is the reciprocal of the period T:

$$f = \frac{1}{T}$$

In general, as with rectilinear motion, the following apply:

$$\frac{\alpha}{\omega} = \frac{d\omega}{d\theta} \tag{5.87}$$

$$\theta_2 - \theta_1 = \int_{t_1}^{t_2} \omega \, dt \tag{5.88}$$

$$\omega_2 - \omega_1 = \int_{t_1}^{t_2} \alpha \, dt \tag{5.89}$$

$$t_2 - t_1 = \int_{\theta_1}^{\theta_2} \frac{d\theta}{\omega} = \int_{\omega_1}^{\omega_2} \frac{d\omega}{\alpha} \tag{5.90}$$

$$\omega_2{}^2 - \omega_1{}^2 = 2 \int_{\theta_1}^{\theta_2} \alpha \, d\theta \tag{5.91}$$

If α = const., the following relations are obtained ($\omega_0 = \omega_1$, $\omega_2 = \omega$, $t_1 = 0$, $t_2 = t$):

$$\omega = \omega_0 + \alpha t \tag{5.92}$$

$$\theta = \omega_0 t + \frac{\alpha t^2}{2} \tag{5.93}$$

$$\omega^2 = \omega_0^2 + 2\alpha\theta \tag{5.94}$$

Note also that for rotation of a body about a fixed axis through the center of gravity, if a constant unbalanced moment M is applied,

$$M = I\alpha \tag{5.95}$$

where I = moment of inertia (mass).

DYNAMICS

1. Definitions

Dynamics. That branch of mechanics which deals with the effects of unbalanced external forces in modifying the motion of bodies. Dynamics also deals with the terms "work," "energy," and "power."

Force (Gravitational). The force F is the product of mass m and acceleration a:

$$F = ma \tag{5.96}$$

The gravitational force W is the force acting on a mass m due to the earth's acceleration g:

$$W = mg \tag{5.97}$$

From Eqs. (5.96) and (5.97)

$$F = \frac{W}{g} a \tag{5.98}$$

Being a gravitational force, the weight W is measured by means of a spring balance.

Conservation of Mass. The mass of a body remains unchanged by ordinary physical changes to which it may be subjected.

Conservation of Energy. The principle of the conservation of energy requires that the total mechanical energy of a system remain unchanged if it is subjected only to forces which depend on position or configuration.

Conservation of Momentum. The linear momentum of a system of bodies remains unchanged if there is no resultant external force on the system. The angular momentum of a system of bodies about a fixed axis is unchanged if there is no resultant external momentum about this axis.

Mutual Attraction (Gravitation). Two particles attract each other with a force F proportional to their masses m_1 and m_2 and inversely proportional to the square of the distance r between them, or

$$F = \frac{Km_1m_2}{r^2} \tag{5.99}$$

where K (= 6.673×10^{-11} m^3/kg · s^2 or 3.44×10^{-8} ft^3/lb · s^2) is the gravitational constant.

2. Particle Dynamics—Selected Problems*. In many engineering problems involving the motion of a particle or the motion of a mass center, the magnitude and direction of the force are specified, and the solution of the problem is given by the integration of the component differential equations

$$F_x = m\ddot{x} \qquad F_y = m\ddot{y} \qquad F_z = m\ddot{z} \qquad \text{etc.}$$

The differential equations and the solutions of some commonly discussed problems in particle dynamics are given below.

1. Body of mass m under a constant gravitational acceleration g:

$$m\ddot{y} = mg \qquad\qquad m\ddot{x} = 0$$

$$\dot{y} = gt + \dot{y}_0 \qquad\qquad \dot{x} = \dot{x}_0$$

$$y = \frac{1}{2gt^2} + \dot{y}_0 t + y_0 \qquad x = \dot{x}_0 t + x_0$$

$$y - y_0 = ½g\left(\frac{x - x_0}{\dot{x}_0}\right)^2 + \dot{y}_0 \frac{x - x_0}{\dot{x}_0}$$

The details of the motion depend on the initial values ($t = 0$) of the displacements and velocities. For the particular case $y_0 = \dot{y}_0 = x_0 = \dot{x}_0 = 0$:

*This section taken in part from *Handbook of Engineering Mechanics*, by W. Flügge. Copyright © 1962. Used by permission of McGraw-Hill, Inc. All rights reserved.

$$\dot{y} = gt \qquad y = \frac{1}{2gt^2} \qquad \dot{y} = \sqrt{2gy}$$

2. Body falling under constant gravitational force and a drag force proportional to the velocity ($F_d = -kv$):

$$m\ddot{y} = mg - k\dot{y} \qquad \frac{\dot{y}k}{mg} = 1 - \left(1 - \frac{\dot{y}_0 k}{mg}\right) e^{-kt/m}$$

$$\frac{yk^2}{m^2 g} = \frac{y_0 k^2}{m^2 g} + \frac{kt}{m} - \left(1 - \frac{\dot{y}_0 k}{mg}\right)(1 - e^{-kt/m}) \qquad \dot{y}_{\max} = \frac{mg}{k} = \text{terminal velocity}$$

For a sphere falling in a fluid medium, Stokes' law states that $F_d = -3\pi d\mu v$, where μ = dynamic viscosity, d = diameter, and $\rho v d/\mu < 1.0$, ρ being the density of the medium.

3. Body falling with drag force proportional to velocity squared ($F_d = -kv^2$), and $y_0 = \dot{y}_0 = 0$:

$$m\ddot{y} = mg - k\dot{y}^2 \qquad \dot{y}\left(\frac{k}{mg}\right)^{1/2} = \tanh\left(\frac{kgt^2}{m}\right)^{1/2} \qquad \frac{yk}{m} = \ln\cosh\left(\frac{kgt^2}{m}\right)^{1/2}$$

4. Projectile retarded by resisting force $F_d = -a - bv^2$, and $x_0 = \dot{y}_0 = 0$ (Poncelet's penetration problem):

$$m\ddot{x} = -a - b\dot{x}^2 \qquad m\dot{y} = 0$$

$$\tan^{-1}\dot{x}\left(\frac{b}{a}\right)^{1/2} = (ab)^{1/2}\frac{t}{m} + \tan^{-1}\dot{x}_0\left(\frac{b}{a}\right)^{1/2}$$

$$\frac{bx}{m} = \tfrac{1}{2}\ln\frac{(b/a)\dot{x}_0^2 + 1}{(b/a)\dot{x}^2 + 1}$$

5. Projectile with drag force $F_d = -kv^2$, and $x_0 = y_0 = 0$; approximate solution for flat trajectory:

$$m\ddot{x} = -kv^2\cos\phi = -k\dot{x}^2\left[1 + \left(\frac{\dot{y}}{\dot{x}}\right)^2\right]^{1/2} \simeq -k\dot{x}^2$$

$$m\ddot{y} = -mg - kv^2\sin\phi = -mg - k\dot{x}\dot{y}\left[1 + \left(\frac{\dot{y}}{\dot{x}}\right)^2\right]^{1/2} \simeq -mg - k\dot{x}\dot{y}$$

$$\frac{\dot{x}}{\dot{x}_0} = \left(1 + \frac{k\dot{x}_0}{m}t\right)^{-1} \qquad \frac{2k\dot{x}_0\dot{y}}{mg} = \left(1 + \frac{2k\dot{x}_0\dot{y}_0}{mg}\right)\left(1 + \frac{k\dot{x}_0 t}{m}\right)^{-1} - \left(1 + \frac{k\dot{x}_0 t}{m}\right)$$

6. Projectile with drag force $F_d = -kv$, and $x_0 = y_0 = 0$:

$$m\ddot{x} = -k\dot{x} \qquad m\ddot{y} = -k\dot{y} - mg$$

$$\frac{xk}{\dot{x}_0 m} = 1 - e^{-kt/m} \qquad \frac{yk^2}{gm^2} = \left(1 + \frac{\dot{y}_0 k}{gm}\right)(1 - e^{-kt/m}) - \frac{k}{m}t$$

7. Mass m performing forced vibrations under action of force F:

$$m\ddot{x} + c\dot{x} + kx = F(t)$$

$$x = \frac{T}{2\pi m}\int_0^t F(r)e^{-c(t-r)/2m}\sin\frac{2\pi}{T}(t-r)\,dr + \frac{\dot{x}_0 T}{2\pi}\sin\frac{2\pi}{T}t + x_0\cos\frac{2\pi}{T}t$$

$$T = \frac{2\pi}{[k/m - (c/2m)^2]^{1/2}}$$

8. Planetary motion of a particle of mass m about a fixed point with inverse-square attraction. For plane motion (polar coordinates):

$$m(2\dot{r}\dot{\theta} + r\ddot{\theta}) = 0 \qquad m(\ddot{r} - r\dot{\theta}^2) = -\frac{k}{r^2}$$

$$mr^2\dot{\theta} = h = \text{constant angular momentum}$$

$$\frac{1}{2}m(\dot{r}^2 + r^2\dot{\theta}^2) - \frac{k}{r} = E = \text{constant energy}$$

$$r = \frac{(h/m)^2}{E(1 + e\cos\theta)}\ \text{(orbit)} \qquad e^2 = 1 + \frac{2Eh^2}{mk^2}$$

The orbital equation is that of a conic having a focus at the origin. If $e < 1$, the orbit is an ellipse; if $e = 1$, it is a parabola; and if $e > 1$, it is a hyperbola; e depends on the magnitudes of the initial E and h but not on the direction of initial velocity.

3. Centrifugal and Centipetal Forces. Let a particle of mass m move in a circle of radius r about a fixed axis (Fig. 5.24). The resultant of all forces acting on the particle has a normal component

$$F_c = mr\omega^2 = \frac{mv^2}{r} \tag{5.100}$$

and a tangential component $= mr\alpha$. If ω is constant and $\alpha = 0$, the resultant force acting on the particle to make it rotate in its circular path is $mr\omega^2$ toward the axis, and is called *centripetal* force. *Centrifugal* force for the particle is equal and opposite to centripetal force, and is exerted by the particle upon its neighboring particles, or upon the axis of rotation.

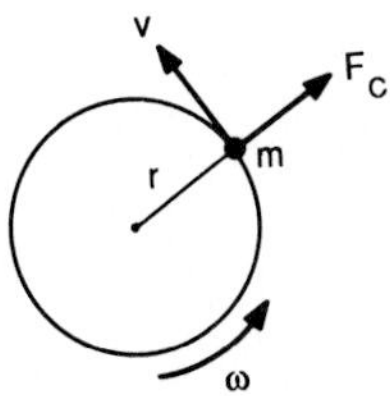

FIGURE 5.24

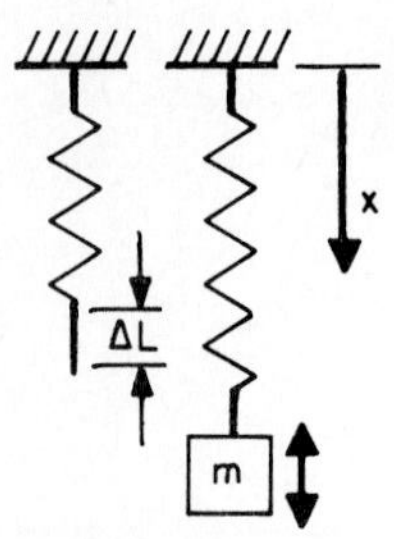

FIGURE 5.25

4. Harmonic Oscillation

Mechanical Oscillation. For the spring-mass system (Fig. 5.25), motion is described by the following equation:

$$m\ddot{x} = -kx \tag{5.101}$$

and the solution for period of oscilation and frequency are

$$T = 2\pi\sqrt{\frac{m}{k}} \tag{5.102}$$

$$f = \frac{1}{T} = \frac{1}{2\pi}\sqrt{\frac{k}{m}} \tag{5.103}$$

where k = (spring stiffness) = $G/\Delta l = mg/\Delta l$

Also, the angular frequency (angular velocity) $\omega = 2\pi f$. Values of k [Eq. (5.103)] for various types of springs are given in Table 5.7.

Pendulum. Basic relations are given for simple, conical, and compound pendulums only.

Simple Pendulum. See Fig. 5.26. The basic equation for this is

$$T = 2\pi\sqrt{\frac{l}{g}} \tag{5.104}$$

Conical Pendulum. See Fig. 5.27. The basic equations are

$$T = 2\pi\sqrt{\frac{h}{g}} = 2\pi\sqrt{\frac{l\cos\beta}{g}} \tag{5.105}$$

$$\tan\beta = \frac{r}{h} = \frac{r\omega^2}{g} \tag{5.106}$$

Compound Pendulum. See Fig. 5.28. The basic relations are

$$T = 2\pi\sqrt{\frac{I_0}{mg\bar{l}}} \tag{5.107}$$

$$I_0 = I_G + m\bar{l}^2 \tag{5.108}$$

$$I_G \approx mg\bar{l}\left(\frac{T^2}{4\pi^2} - \frac{\bar{l}}{g}\right)$$

Point G in Fig. 5.28 is the center of gravity of mass m.

TABLE 5.7 Stiffness of Various Types of Springs

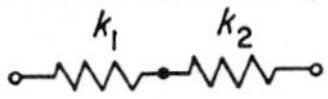

$$k = \frac{1}{1/k_1 + 1/k_2}$$

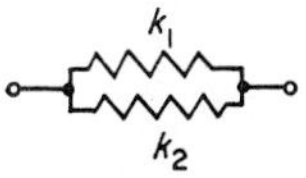

$$k = k_1 + k_2$$

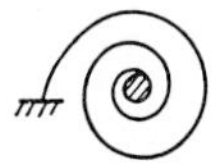

$$k = \frac{EI}{l}$$

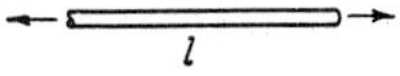

$$k = \frac{EA}{l}$$

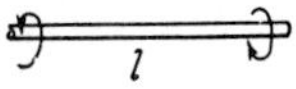

$$k = \frac{GJ}{l}$$

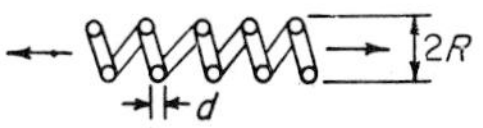

$$k = \frac{Gd^4}{64nR^3}$$

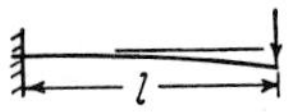

$$k = \frac{3EI}{l^3}$$

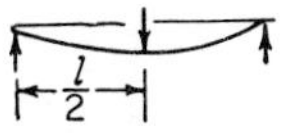

$$k = \frac{48EI}{l^3}$$

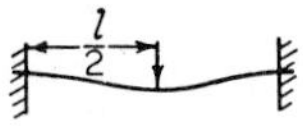

$$k = \frac{192EI}{l^3}$$

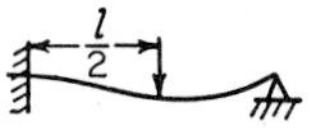

$$k = \frac{768EI}{7l^3}$$

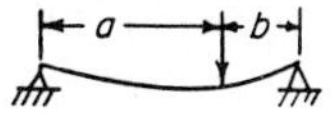

$$k = \frac{3EIl}{a^2b^2}$$

Note:
I = moment of inertia of cross-sectional area
l = total length
A = cross-sectional area
J = torsion constant of cross section (= $\frac{1}{2}\pi r^4$ for circular cross section)
n = number of turns
E = modulus of elasticity
G = shear modulus ≈ $0.385E$
Source: From Flügge.[5]

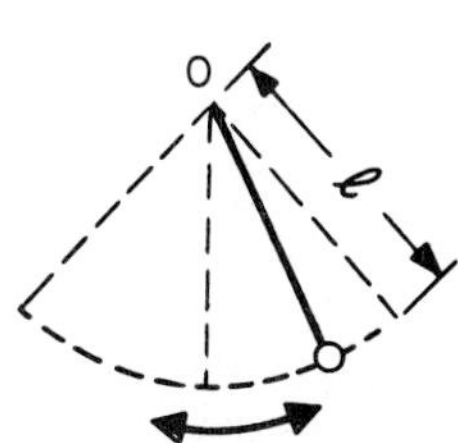

FIGURE 5.26

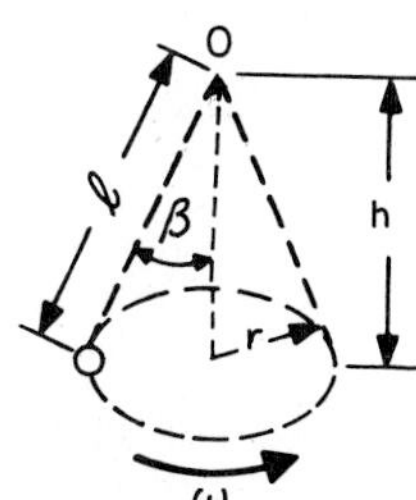

FIGURE 5.27

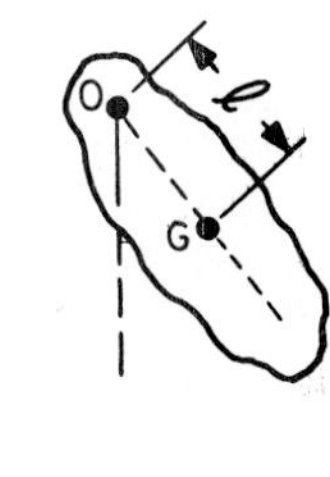

FIGURE 5.28

5. Work

Work of a Force. The work of a variable force in moving a body through the distance $\Delta s = s_2 - s_1$ is given as

$$W = \int_{s_1}^{s_2} F \cos \alpha \, ds \tag{5.109}$$

where F is the variable force, ds is the elementary length of the path, and α is angle between the force and the element ds.

Work of Gravity. The work of gravity on a body in any motion is equal to the product of the weight and the change in height of the mass center.

Work of a Torque. The work of a torque (twisting moment) T on a rotating body for an angular displacement $d\theta$ (in radians) is

$$W = \int_{\theta_1}^{\theta_2} T \, d\theta \tag{5.110}$$

If T is constant, $W = T(\theta_2 - \theta_1)$.

Mechanical Efficiency. Mechanical efficiency η of a machine is the ratio of useful output to total input of work:

$$\eta = \frac{W_u}{W_a} \tag{5.111}$$

and $W_a = W_u + W_f$ where W_a = work applied to the machine, W_u = useful work performed, and W_f = work required to overcome friction or other types of resistance.

6. Energy. Energy (potential and kinetic) of a body is the amount of work it can do by virtue of its position or its motion against forces applied to it. The unit of energy is same as the unit of work.

Potential Energy. The potential energy of a body is that possessed by virtue of its configuration. For example, a body of weight W, located at a height above the earth's surface such that its mass center can descend h, has a potential energy

$$E_p = Wh \tag{5.112}$$

Kinetic Energy of Translation. If a body has a motion of translation, each particle of the body is moving with the same velocity; so at each instant of velocity v, the kinetic energy of the body is

$$E_k = \tfrac{1}{2}mv^2 \tag{5.113}$$

Kinetic Energy of Rotation. In a motion of rotation, each particle of the body is moving with a different velocity, but if ρ represents the radius of any particle and ω the angular velocity of the body, the kinetic energy of the body is

$$E_k = \tfrac{1}{2}\int \rho^2\omega^2\, dm = \tfrac{1}{2}I\omega^2 \tag{5.114}$$

where I = moment of inertia (mass) about the axis of rotation.

Kinetic Energy of Translation and Rotation. A body which has a motion of comined translation and rotation has both a kinetic energy of translation and a kinetic energy of rotation. If v is the velocity of the center of gravity at any instant and ω is the angular velocity, the kinetic energy of the body is

$$E_k = \tfrac{1}{2}mv^2 + \tfrac{1}{2}I\omega^2 \tag{5.115}$$

7. Power. Power is the rate at which work is being done. The unit of power is one unit of work performed in one unit of time. If a force F is acting upon a body and moving it in the direction of the force with a velocity v, the power is

$$P = Fv \tag{5.116}$$

The power of a torque at any instant is

$$P = T\omega \tag{5.117}$$

where ω is instantaneous angular velocity of the body.

8. Impulse and Impact

Linear Impulse and Momentum. The *impulse* of a force is the product of the force and the time during which it acts. The total impulse during time t is

$$\int_0^t F\, dt$$

If the force is constant, this becomes $F \cdot t$. Impulse is a vector quantity, its direction being the same as the direction of the force F.

Momentum is the product of a mass m and velocity v:

$$\text{momentum} = mv$$

Momentum is also a vector quantity, its direction being the same as the direction of the velocity v.

An alternative statement of Newton's second law of motion is that the resultant of an unbalanced force system must be equal to the time rate of change of linear momentum:

$$\Sigma F = \frac{d(mv)}{dt} \tag{5.118}$$

Also,

$$\int F\, dt = m(v_1 - v_2) = \text{the change of the momentum of the body}$$

Direct Central Impact. If two inelastic bodies collide in a direct central impact, they have a common velocity v after their impact. Since the impulses are equal and opposite, there is no change in momentum, or

$$m_1v_1 + m_2v_2 = (m_1 + m_2)v$$

If the two bodies are elastic, they separate, but their relative velocities after impact are less than before impact. The ratio of the relative velocity of each after impact to that before is e, the coefficient of restitution. There is no change in linear momentum, so

$$m_1v_1 + m_2v_2 = m_1v_1' + m_2v_2' \tag{5.119}$$

Also,

$$e = \frac{(v_2' - v_1')}{(v_1 - v_2)} \tag{5.120}$$

Angular Impulse and Angular Momentum. The moment of an impulse is the product of the impulse and the distance from the force to the center of moments. It is also called *angular impulse*. Angular impulse and linear impulse cannot be added.

The moment of a momentum is the product of the momentum and the distance from the center of mass to the center of moments. It is also called *angular momentum*. Its unit is $I\omega$.

In any motion of rotation, the initial angular momentum plus the positive angular impulse minus the negative angular impulse is equal to the final angular momentum. In any mutual action between two rotating bodies or two parts of the same rotating body, the angular impulses are equal in amount and opposite in direction; hence for the entire system the change in angular momentum must be zero.

NOMENCLATURE

Symbol = *Definition, SI units (U.S. Customary unit)*

A = area, m^2 (ft^2)

a = acceleration, m/s^2 (ft/s^2)

E_k = kinetic energy, J ($lb_f \cdot ft$)

E_p = potential energy, J ($lb_f \cdot ft$)

F = force, N (lb_f)

f = sliding friction coefficient

f_0 = static friction coefficient

f = frequency of oscillation, s^{-1} or $1/s$

g = acceleration of gravity, m/s^2 (ft/s^2)

I = moment of inertia of an area, m^4 (ft^4)

I = moment of inertia of mass, $kg \cdot m^2$ ($lb \cdot ft^2$)

l = distance, m (ft)

m = mass, kg (lb)

M = momentum, N · m (lb_f · ft)
P = power, J/s (lb_f · ft/s)
r = radius (or diameter), m (ft)
r = radius of gyration, m (ft)
s = distance (displacement), m (ft)
T = time period of oscillation, s
T = torque (twisting moment), N · m (lb_f · ft)
t = time, s
v = velocity, m/s (ft/s)
V = volume, m^3 (ft^3)
W = work, N · m (lb_f · ft)
W = weight of a body, N (lb_f)
x = rectangular coordinate, m (ft)
y = rectangular coordinate, m (ft)
z = rectangular coordinate, m (ft)

Subscripts

0 = initial condition
t = tangential
n = normal
x = x direction
y = y direction
z = z direction
N = normal direction
R = reaction

Greek

α = angular acceleration, s^{-2}, rad/s^2 (in "Rotation" subsection)
α = angle, rad, deg
ρ = mass density, kg/m^3 (lb/ft^3)
ρ = distance, m (ft) (in "Moment of Inertia" subsection)
μ = dynamic viscosity, N · s/m^2 [lb/(ft · s)]
ω = angular velocity, s^{-1}, rad/s

Superscripts

′ = ft
″ = inch
$\bar{}$ = time average ($\bar{x}$ is time average of x)

Mathematical Operation Symbols

d/dt = derivative with respect to t, s^{-1}

$\dot{}$ = first derivative ($\dot{x} = dx/dt$)

$\ddot{}$ = second derivative ($\ddot{x} = d^2x/dt^2$)

Note. Other symbols are defined in the text. Boldfaced symbols in the text denote vector quantities.

REFERENCES*

1. K. Gieck, *Engineering Formulas*, 5th ed., McGraw-Hill, New York, 1986.
2. E. A. Avallone and T. Baumeister III (eds.), *Mark's Standard Handbook for Mechanical Engineers*, 9th ed., McGraw-Hill, New York, 1987.
3. C. E. O'Rourke, *General Engineering Handbook*, 2d ed., McGraw-Hill, New York, 1940.
4. H. A. Rothbart, *Mechanical Design and Systems Handbook*, 2d ed., McGraw-Hill, New York, 1985.
5. W. Flügge, *Handbook of Engineering Mechanics*, McGraw-Hill, New York, 1962.
6. F. P. Beer and E. R. Johnson, *Vector Mechanics for Engineers*, McGraw-Hill, New York, 1964.
7. N. H. Cook, "Mechanics of Solids," Course 2.01, MIT class notes, 1973–1976.
8. S. H. Crandall, "Dynamics," Course 2.032, MIT class notes, 1973–1976.
9. S. H. Crandall, et al., *Dynamics of Mechanical and Electromechanical Systems*, McGraw-Hill, New York, 1968.
10. J. P. Den Hartog, *Mechanics*, Dover, New York, 1961.
11. O. W. Eshbach and M. Souders, *Handbook of Engineering Fundamentals*, 3d ed., John Wiley & Sons, New York, 1975.
12. C. W. Ham, E. J. Crane, and W. L. Rogers, *Mechanics of Machinery*, 4th ed., McGraw-Hill, New York, 1958.
13. J. L. Synge and B. A. Griffith, *Principles of Mechanics*, 3d ed., McGraw-Hill, New York, 1959.
14. S. Timoshenko and D. H. Young, *Engineering Mechanics*, 4th ed., McGraw-Hill, New York, 1956.
15. S. Timoshenko and D. H. Young, *Theory of Structures*, McGraw-Hill, New York, 1945.
16. D. N. Wormley, "Dynamic Systems," Course 2.023, MIT class notes, 1973–1976.

*Those references listed above but not cited in the text were used for comparison between different data sources, clarification, clarity of presentation, and, most important, reader's convenience when further interest in the subject exists.

CHAPTER 6
MECHANICS OF DEFORMABLE BODIES

STATIC STRESSES*

1. Compression and Tension. If a bar with a cross-sectional area of A is acted upon by two equal and oppositely directed axial forces P, P, the load per unit area, called the *unit stress* S, is given by the expression

$$S = \frac{P}{A} \tag{6.1}$$

If the forces are acting toward each other, the stress is *compression* and is denoted by S_c. If the forces are acting away from each other, the stress is *tension* and is denoted by S_t.

2. Shear. If two equal and oppositely directed forces P, P, are applied normal to the axis of a bar and in different planes of action, the part of the bar between the planes of action of the forces is subjected to a shearing action in which any cross section tends to slide over the one next to it. Unit *shearing stress* is denoted by S_s. The shearing stress is not uniform across the cross section, but the average unit stress is given by the expression $S_s = P/A$. (See Fig. 6.1.)

3. Modulus of Elasticity and Shearing Modulus of Elasticity. Elasticity is the ability of a material to return to its original dimension after the removal of stresses. Nearly all of the materials used in engineering work are elastic, and within certain limits obey fairly well *Hooke's law* of proportionality of stress to deformation or strain. If l is the length of a bar and e its total change in length, the *unit deformation or strain* is

$$\delta = \frac{e}{l} \tag{6.2}$$

The ratio of the unit stress S to the corresponding unit strain δ is called the *modulus of elasticity* and is denoted by E:

$$E = \frac{S}{\delta} \tag{6.3}$$

*This section is taken in part from *General Engineering Handbook*, 2d ed., by C. E. O'Rourke. Copyright © 1940. Used by permission of McGraw-Hill, Inc. All rights reserved.

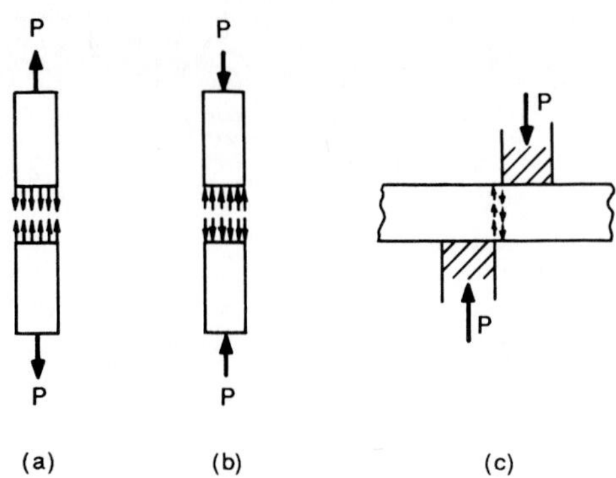

FIGURE 6.1 (*a*) Tension; (*b*) compression; (*c*) shear.

For nearly all materials, the values of E in tension and compression are practically the same. The modulus of elasticity in tension is also known as *Young's modulus*.

In a similar way, the ratio of the unit shearing stress S_s to the corresponding unit shearing strain δ_s is called the *shearing modulus of elasticity* E_s:

$$E_s = \frac{S_s}{\delta_s} \tag{6.4}$$

For homogeneous materials, E_s is about two-fifths of the value of E. Symbol G is often used instead of E_s and is often referred to as the *modulus of rigidity*.

4. Elastic Limit, Yield Point, and Ultimate Strength. The maximum unit stress for which Hooke's law is valid is called the *proportional elastic limit*. The elastic limit is also defined as the maximum stress to which the body may be subjected without permanent deformation or set. As the axial dimension of a bar is changed by stress, any lateral dimension is changed oppositely. The ratio of the unit lateral strain to the unit axial strain is called *Poisson's ratio* μ:

$$\mu = \frac{\text{lateral strain}}{\text{longitudinal strain}} \tag{6.5}$$

If the load on a bar of ductile material is increased above the elastic limit, it soon reaches a value at which the deformation continues to increase with little or no increase of the load. The unit stress at which this occurs is called the *yield point*. Materials which are not ductile have no yield point.

The greatest load a bar will hold is called the *maximum load*, and the corresponding unit stress is the *ultimate strength*. Ductile materials tested in tension form a neck at about the time the ultimate strength is reached, and the total load carried then decreases. The stretching continues, and the bar breaks at a total load less than the maximum. This load is called the *rupture* or *breaking load*, and the unit stress at rupture is called the *rupture strength* (see Fig. 6.2). Unless specified as actual stresses, all of these four stresses—the elastic limit, the yield point, the ultimate strength, and the rupture strength—are computed by dividing the loads by the original cross-sectional area.

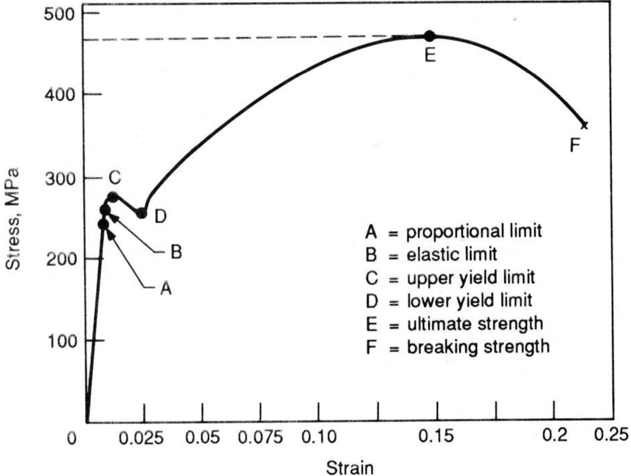

FIGURE 6.2 Stress-strain diagram for mild steel (1 Pa = 1 N/m^2 = 1.45 x 10^{-4} lb_f/in^2).

5. Bulk Modulus of Elasticity, Reliance, Thermal Stress, and Design Stress. The *bulk modulus of elasticity* K is the ratio of normal stress, applied to all six faces of a cube, to the change of volume,

$$K = \frac{S}{\Delta V/V_0} \tag{6.6}$$

where V_0 is the original volume and ΔV is the change in the volume equal to $V_0 - V$.

The following relations are useful for calculations (Ref. 2):

$$G = \frac{E}{2(1 + \mu)} \tag{6.7}$$

and

$$\frac{1}{E} = \frac{1}{9K} + \frac{1}{3G} \tag{6.8}$$

Reliance U is the potential energy stored in a deformed body.

Thermal stress is developed if expansion or contraction is prevented; it is equal to

$$S = E \cdot \alpha \cdot \Delta t \tag{6.9}$$

where α is the linear coefficient of thermal expansion and $\Delta t = t_2 - t_1$ is the temperature rise from t_1 to t_2.

The design stress is determined by dividing the applicable material property yield stress, ultimate stress) by a factor of safety.

DYNAMIC STRESSES

Dynamic stresses occur where the dimension of time is necessary in defining the loads. They include creep, fatigue, and impact stresses.

Creep stresses occur when either the load or the deformation progressively varies with time. They are usually associated with noncyclic phenomena.

Fatigue stresses occur when the cyclic variation of either load or strain is coincident with respect to time.

Impact stresses occur in regard to loads that are transient in respect to time. The duration of the load application is of the same order of magnitude as the natural period of vibration of the specimen.

For *steady vibration stresses*, the deflection of the bar, or beam, is increased by the dynamic magnification factor K_d:

$$S_{\text{dynamic}} = S_{\text{static}} \cdot K_d \tag{6.10}$$

and

$$K_d = \frac{1}{1 - (\omega/\omega_n)^2}$$

where ω the frequency of oscillation of the load and ω_n is the natural frequency of the bar determined by

$$\omega_n = \frac{3\,EIg}{L^3W} \tag{6.11}$$

where L = length of bar
I = moment of inertia
W = weight of the oscillating load

BEAMS

1. Types of Beams. A beam is a bar or structural member subjected to transverse loads and reactions that tend to bend it. Usually beams are horizontal bars designed to carry vertical loads, but any structural member acts as a beam if bending is induced by external transverse forces.

Beams freely supported at both ends are called *simple beams* (Fig. 6.3*a*). A *fixed beam* is rigidly fixed at both ends or rigidly fixed at one end and simply supported at the other (Fig. 6.3*b*). A *continuous beam* is one that is resting on more than two supports (Fig. 6.3*c*). Beams fixed at one end and unsupported at the other are called *cantilever beams* (Fig. 6.3*d*). Loads are usually concentrated, uniformly distributed over part or all of the length of the beam, or uniformly varying.

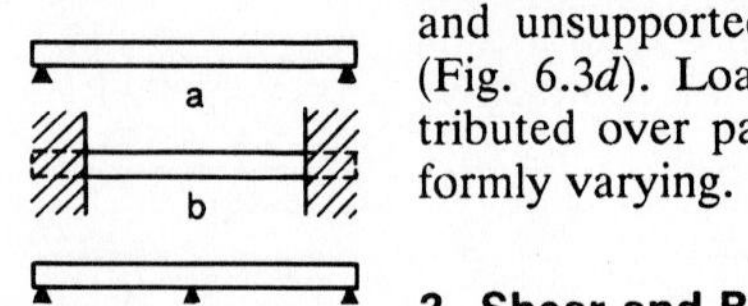

FIGURE 6.3

2. Shear and Bending Moments in Beams

Vertical Shear. At any cross section of a beam the resultant of the external vertical forces acting on one side of the section is equal and opposite to the resultant of the external vertical forces on the other side of the section. These forces tend to cause the beam to shear vertically along the section. The value of either resultant, which is a measure of the shearing tendency, is known as the *vertical shear* (V) at the section considered. It is computed by finding the algebraic sum of the vertical forces to the left of the section; i.e., it is equal to the left reaction minus the sum of the vertical downward forces acting between the left support and the section. The vertical shear V is also given as the sum of the transverse shear stresses (S) acting on the section:

$$V = \int S\,dA \tag{6.12}$$

A *shear diagram* is a graphical representation of the vertical shear at all cross sections of the beam.

Bending Moment. The bending moment, or moment, at any cross section of a beam is the algebraic sum of the moments of the external forces acting on either side of the section. It is considered positive when it causes the beam to bend convex downward, hence causing compression in upper fibers and tension in lower fibers of beam. The shear V is the first derivative of moment with respect to distance x along the beam:

$$V = \frac{dM}{dx} \tag{6.13}$$

Also,

$$M = \int V\,dx \tag{6.14}$$

A *moment diagram* is a line drawn to show the magnitude and character of the bending moment.

Figure 6.4 illustrates a simple beam subjected to a uniform load w per unit length. Then

$$M = R_1 x - wx \cdot \frac{x}{2} = \frac{wlx}{2} - \frac{wx^2}{2}$$

$$V = R_1 - wx = \frac{wl}{2} - wx$$

Also,

$$V = \frac{d}{dx}\left(\frac{wlx}{2} - \frac{wx^2}{2}\right) = \frac{wl}{2} - wx$$

Shear and moment diagrams are also shown in Fig. 6.4. The moment curve is always parabolic under uniformly distributed loads. R_1 = reaction at A.

Table 6.1 gives the reactions, bending moment equations, vertical shear equations, and deflection of some of the more common types of beams.

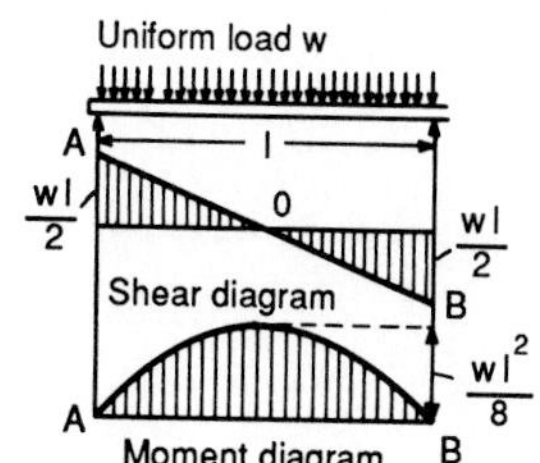

FIGURE 6.4

TABLE 6.1 Bending Moment, Vertical Shear, and Deflection of Beams of Uniform Cross Section and Various Conditions of Loading

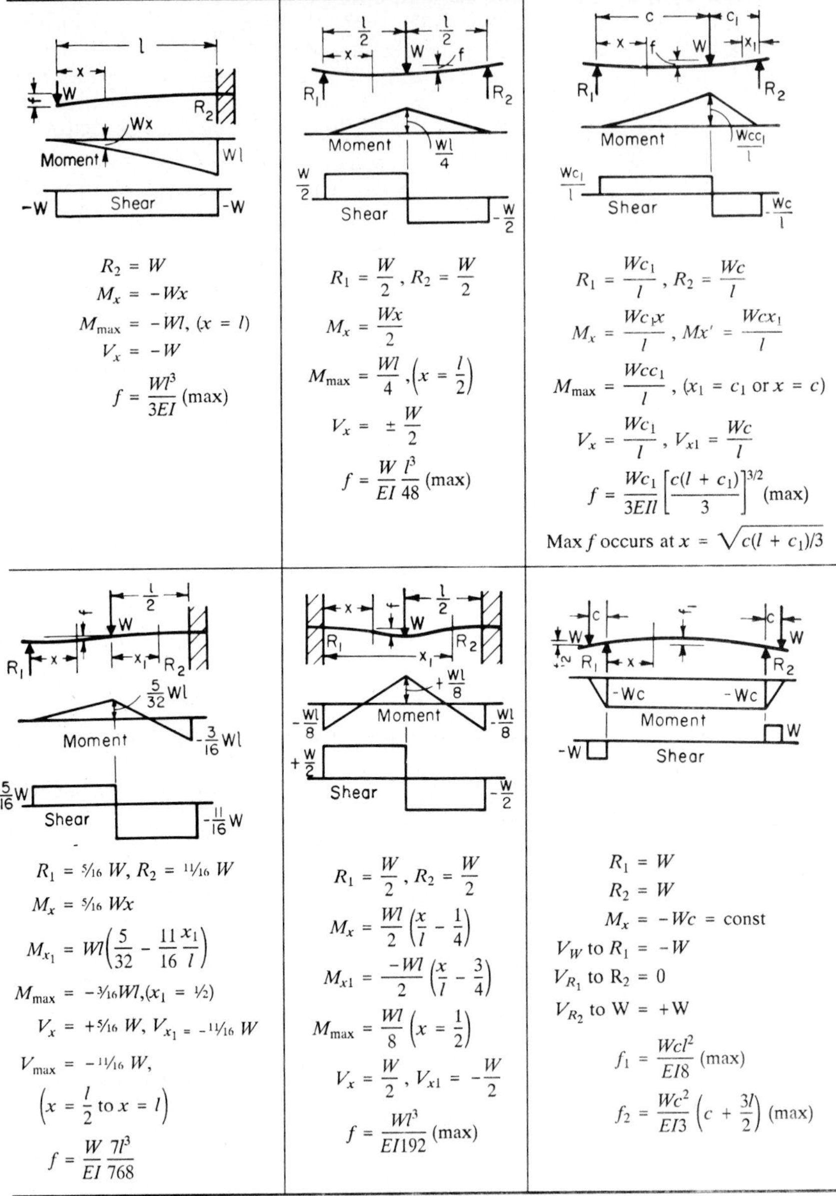

$R_2 = W$ $M_x = -Wx$ $M_{max} = -Wl,\ (x = l)$ $V_x = -W$ $f = \dfrac{Wl^3}{3EI}$ (max)	$R_1 = \dfrac{W}{2},\ R_2 = \dfrac{W}{2}$ $M_x = \dfrac{Wx}{2}$ $M_{max} = \dfrac{Wl}{4},\left(x = \dfrac{l}{2}\right)$ $V_x = \pm\dfrac{W}{2}$ $f = \dfrac{W}{EI}\dfrac{l^3}{48}$ (max)	$R_1 = \dfrac{Wc_1}{l},\ R_2 = \dfrac{Wc}{l}$ $M_x = \dfrac{Wc_1x}{l},\ Mx' = \dfrac{Wcx_1}{l}$ $M_{max} = \dfrac{Wcc_1}{l},\ (x_1 = c_1 \text{ or } x = c)$ $V_x = \dfrac{Wc_1}{l},\ V_{x1} = \dfrac{Wc}{l}$ $f = \dfrac{Wc_1}{3EIl}\left[\dfrac{c(l + c_1)}{3}\right]^{3/2}$ (max) Max f occurs at $x = \sqrt{c(l + c_1)/3}$
$R_1 = 5/16\ W,\ R_2 = 11/16\ W$ $M_x = 5/16\ Wx$ $M_{x_1} = Wl\left(\dfrac{5}{32} - \dfrac{11}{16}\dfrac{x_1}{l}\right)$ $M_{max} = -3/16 Wl, (x_1 = 1/2)$ $V_x = +5/16\ W,\ V_{x_1} = -11/16\ W$ $V_{max} = -11/16\ W,$ $\left(x = \dfrac{l}{2} \text{ to } x = l\right)$ $f = \dfrac{W}{EI}\dfrac{7l^3}{768}$	$R_1 = \dfrac{W}{2},\ R_2 = \dfrac{W}{2}$ $M_x = \dfrac{Wl}{2}\left(\dfrac{x}{l} - \dfrac{1}{4}\right)$ $M_{x1} = \dfrac{-Wl}{2}\left(\dfrac{x}{l} - \dfrac{3}{4}\right)$ $M_{max} = \dfrac{Wl}{8}\left(x = \dfrac{1}{2}\right)$ $V_x = \dfrac{W}{2},\ V_{x1} = -\dfrac{W}{2}$ $f = \dfrac{Wl^3}{EI192}$ (max)	$R_1 = W$ $R_2 = W$ $M_x = -Wc = \text{const}$ V_W to $R_1 = -W$ V_{R_1} to $R_2 = 0$ V_{R_2} to $W = +W$ $f_1 = \dfrac{Wcl^2}{EI8}$ (max) $f_2 = \dfrac{Wc^2}{EI3}\left(c + \dfrac{3l}{2}\right)$ (max)

TABLE 6.1 Bending Moment, Vertical Shear, and Deflection of Beams of Uniform Cross Section and Various Conditions of Loading (*Continued*)

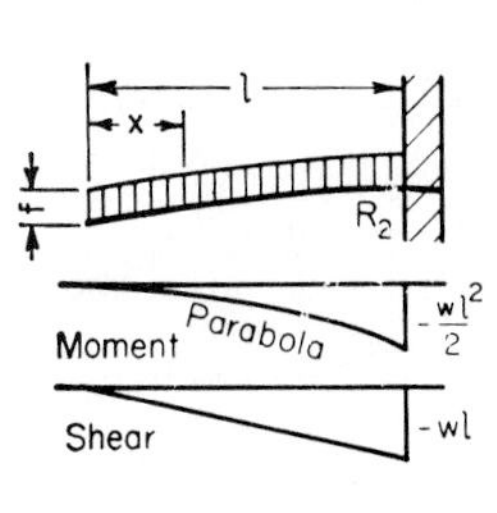

$R_2 = W = wl$

$M_x = -\dfrac{wx^2}{2}$

$M_{max} = -\dfrac{wl^2}{2}\ (x = l)$

$V_x = -wx$

$V_{max} = -wl\ (x = l)$

$f = \dfrac{Wl^2}{EI8}$ (max)

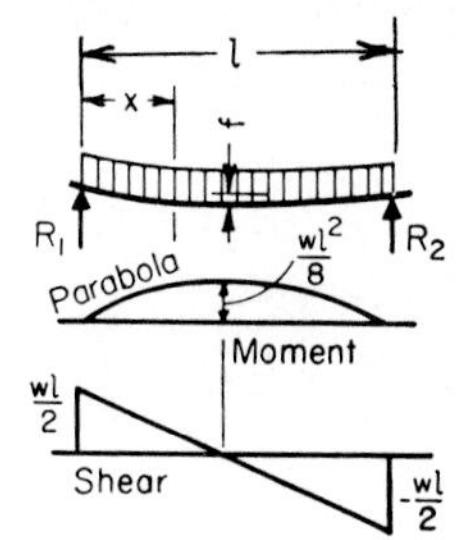

$R_1 = \dfrac{W}{2} = \dfrac{wl}{2}$

$R_2 = \dfrac{W}{2} = \dfrac{wl}{2}$

$M_x = \dfrac{wx}{2}(l - x)$

$M_{max} = \dfrac{wl^2}{8}\ (x = \tfrac{1}{2}l)$

$V_x = \dfrac{wl}{2} - wx$

$V_{max} = \dfrac{wl}{2}\ (x = 0)$

$f = \dfrac{W5l^3}{EI384}$ (max)

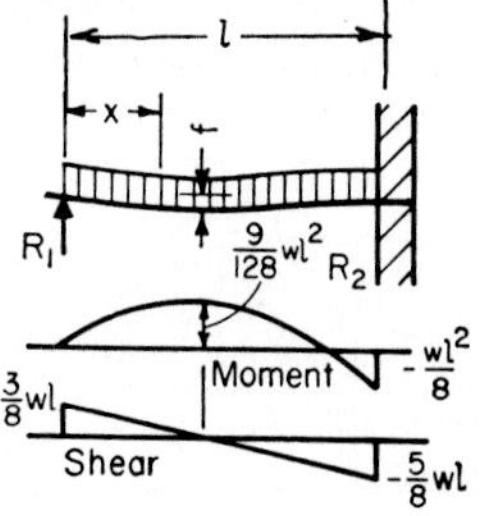

$R_1 = \tfrac{3}{8}W = \tfrac{3}{8}wl$

$R_2 = \tfrac{5}{8}W = \tfrac{5}{8}wl$

$M_x = \dfrac{wx}{2}\left(\dfrac{3}{4}l - x\right)$

$M_{max} = \tfrac{9}{128}wl^2\ (x = \tfrac{3}{8}l)$

$M_{max} = \dfrac{wl^2}{8}\ (x = l)$

$V_x = \tfrac{3}{8}wl - wx$

$V_{max} = -\tfrac{5}{8}wl$

$f = \dfrac{W\,l^3}{EI185}$ (max)

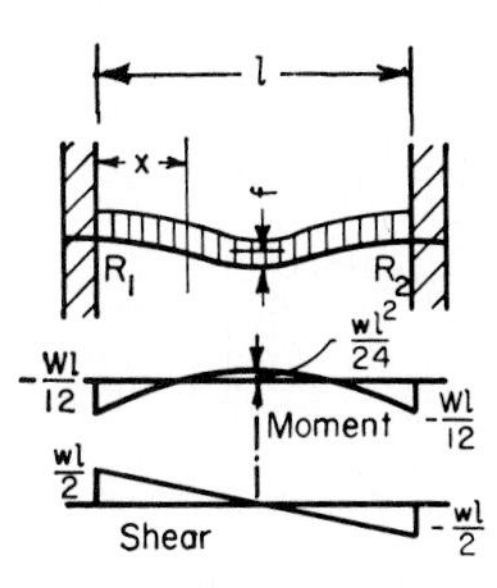

$R_1 = \dfrac{W}{2} = \dfrac{wl}{2},\ R_2 = \dfrac{W}{2} = \dfrac{wl}{2}$

$M_x = -\dfrac{wl^2}{2}\left(\dfrac{1}{16} - \dfrac{x}{l} + \dfrac{x^2}{l^2}\right)$

$M_{max} = -\tfrac{1}{12}wl^2$, $(x = 0$, or $x = l)$

$V_x = \dfrac{wl}{2} - wx$

$V_{max} = \pm\dfrac{wl}{2}$

$f = \dfrac{W\,l^2}{EI384}$ (max)

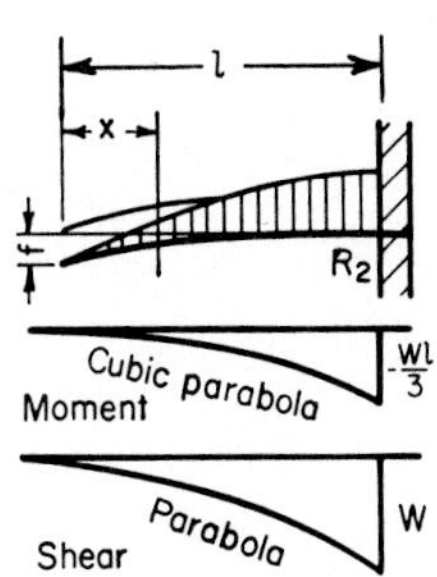

$R_2 = W =$ total load

$M_x = -\dfrac{Wx^3}{3\,l^2}$

$M_{max} = -\dfrac{Wl}{3}$

$V_x = -\dfrac{Wx^2}{l^2}$

$V_{max} = -W$

$f = \dfrac{W}{EI}\dfrac{l^2}{15}$ (max)

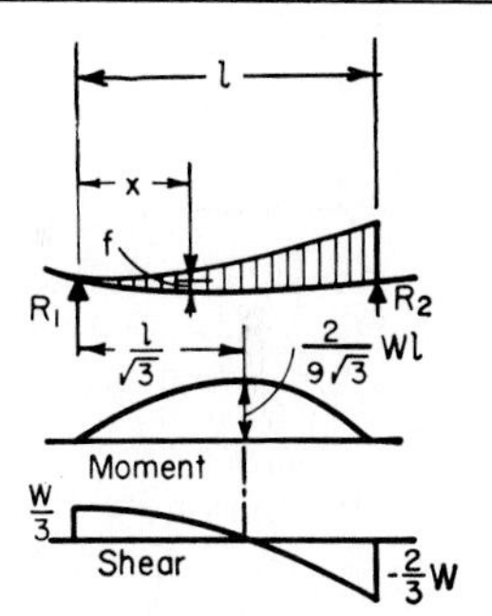

$R_1 = \tfrac{1}{3}W,\ R_2 = \tfrac{2}{3}W$

$M_x = \dfrac{Wx}{3}\left(1 - \dfrac{x^2}{l^2}\right)$

$M_{max} = \dfrac{2}{9\sqrt{3}}Wl\left(x = \dfrac{1}{\sqrt{3}}\right)$

$V_x = W\left(\dfrac{1}{3} - \dfrac{x^2}{l^2}\right)$

$V_{max} = -\tfrac{2}{3}W\ (x = l)$

$f = 0.01304\,\dfrac{Wl^2}{EI}$ (max)

TABLE 6.1 Bending Moment, Vertical Shear, and Deflection of Beams of Uniform Cross Section and Various Conditions of Loading (*Continued*)

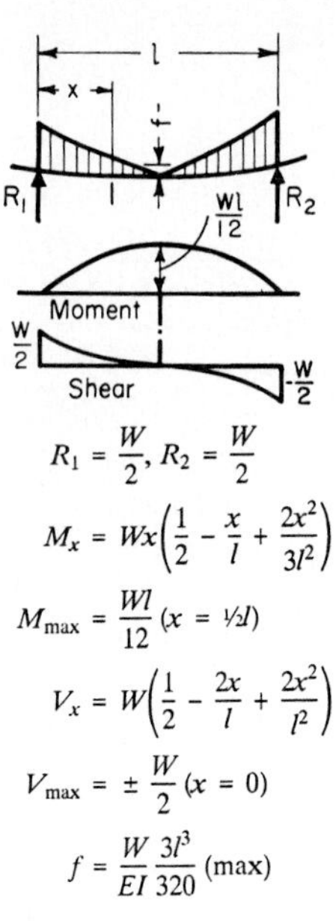

$$R_1 = \frac{W}{2},\ R_2 = \frac{W}{2}$$

$$M_x = Wx\left(\frac{1}{2} - \frac{x}{l} + \frac{2x^2}{3l^2}\right)$$

$$M_{max} = \frac{Wl}{12}\ (x = \tfrac{1}{2}l)$$

$$V_x = W\left(\frac{1}{2} - \frac{2x}{l} + \frac{2x^2}{l^2}\right)$$

$$V_{max} = \pm\frac{W}{2}\ (x = 0)$$

$$f = \frac{W}{EI}\frac{3l^3}{320}\ \text{(max)}$$

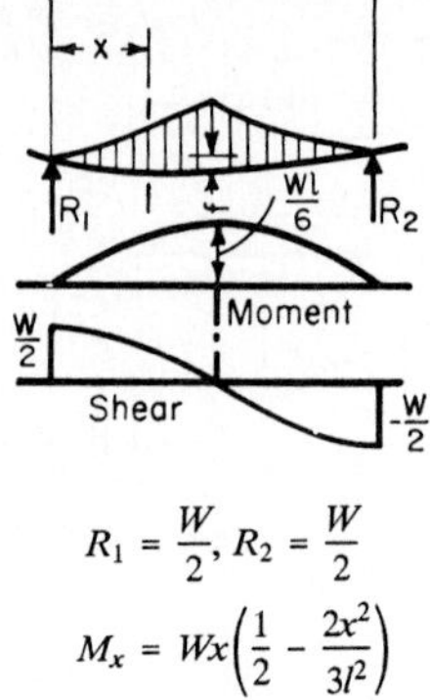

$$R_1 = \frac{W}{2},\ R_2 = \frac{W}{2}$$

$$M_x = Wx\left(\frac{1}{2} - \frac{2x^2}{3l^2}\right)$$

$$M_{max} = \frac{Wl}{6}\ (x = \tfrac{1}{2}l)$$

$$V_x = W\left(\tfrac{1}{2} - \frac{2x^2}{l^2}\right)$$

$$V_{max} = \pm\frac{W}{2}\ (x = 0)$$

$$f = \frac{Wl^3}{EI60}\ \text{(max)}$$

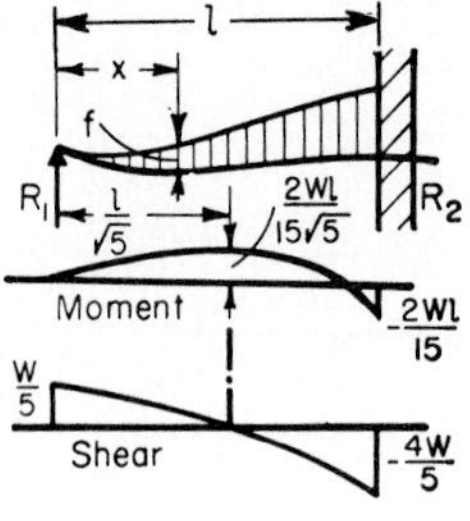

$$R_1 = \frac{W}{5},\ R_2 = \frac{4W}{5}$$

$$M_x = Wx\left(\frac{1}{5} - \frac{x^2}{3l^2}\right)$$

$M_{max} = -\tfrac{2}{15}Wl$ at support 2

$$V_x = W\left(\frac{1}{5} - \frac{x^2}{l^2}\right)$$

$$V_{max} = -\frac{4W}{5}$$

$$f = \frac{16Wl^2}{1500\sqrt{5EI}} = \frac{0.00477Wl^3}{EI}\ \text{(max)}$$

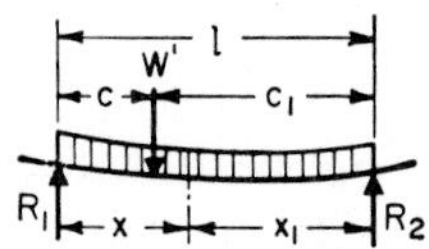

Concentrated load W'

Uniformly dist. load $W = wl$; $c < c_1$

$$R_1 = W'\frac{c_1}{l} + \frac{W}{2}$$

$$R_2 = W'\frac{c}{l} + \frac{W}{2}$$

(*a*) $\dfrac{W'}{W} < \dfrac{c_1 - c}{2c}$

$$M_{max} = R_2\frac{x_1}{2} = \frac{R_2^2 l}{2W}\left(x_1 = \frac{R_2 l}{W}\right)$$

(*b*) $\dfrac{W'}{W} > \dfrac{c_1 - c}{2c}$

$$M_{max} = \left(W' + \frac{W}{2}\right)\frac{cc_1}{l}\ (x_1 = c_1)$$

Deflection of beam under W':

$$f = \left(W' + \frac{l^2 + cc_1}{8cc_1}W\right)\frac{c^2c_1^2}{3EIl}$$

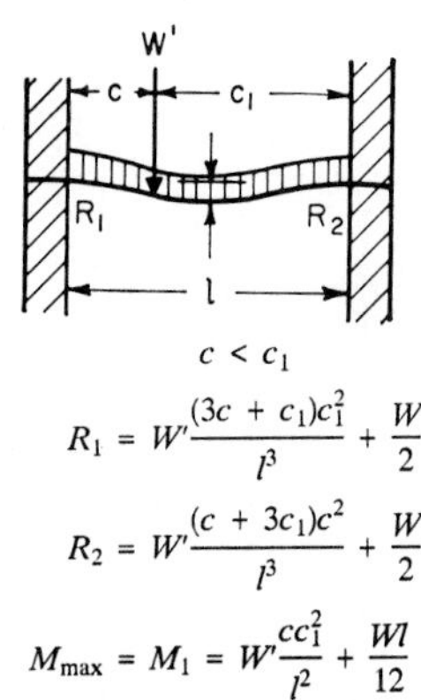

$c < c_1$

$$R_1 = W'\frac{(3c + c_1)c_1^2}{l^3} + \frac{W}{2}$$

$$R_2 = W'\frac{(c + 3c_1)c^2}{l^3} + \frac{W}{2}$$

$$M_{max} = M_1 = W'\frac{cc_1^2}{l^2} + \frac{Wl}{12}$$

Deflection under W'

$$f = \frac{1}{EI}\left(W'\frac{c^3c_1^3}{3l^3} + W\frac{c^2c_1^2}{24l}\right)$$

TABLE 6.1 Bending Moment, Vertical Shear, and Deflection of Beams of Uniform Cross Section and Various Conditions of Loading (*Continued*)

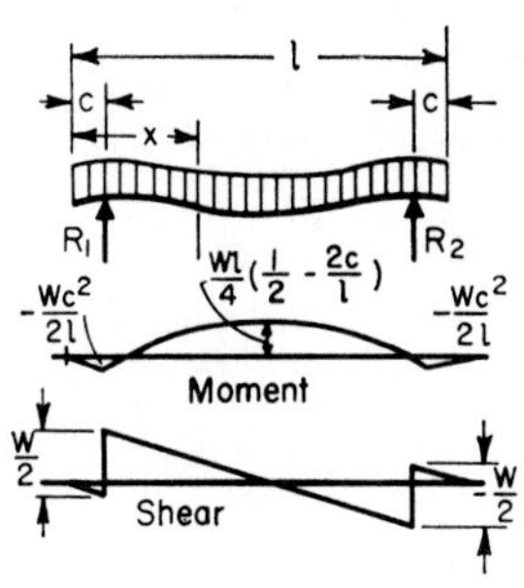

$$R_1 = \frac{W}{2} = \frac{wl}{2},\quad R_2 = \frac{W}{2} = \frac{wl}{2}$$

$$M_x = \frac{Wx}{2}\left(1 - \frac{c}{x} - \frac{x}{l}\right) \quad (x > c)$$

$$M_x = -\frac{Wx^2}{2l} \quad (x \le c)$$

$$M_{\max} = \frac{Wl}{4}\left(\frac{1}{2} - \frac{2c}{l}\right),\quad c \le \left(\frac{\sqrt{2} - 1}{2}\right)l$$

$$V_x = \frac{W}{2} - wx \quad (x > c)$$

$$V_x = -wx \quad (x \le c)$$

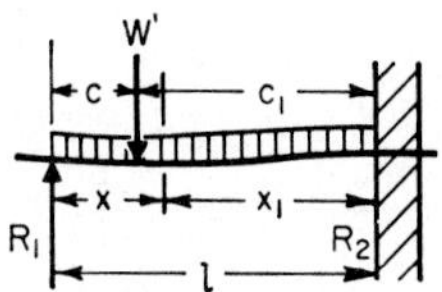

Concentrated load W'
Uniformly dist. load $W = wl$

$$R_1 = W' \frac{c_1^2(3c + 2c_1)}{2l^3} + \tfrac{3}{8}W$$

$$R_2 = W' \frac{(2c^2 + 6cc_1 + 3c_1^2)c}{2l^3} + \tfrac{5}{8}W$$

$$M_2 = W' \frac{cc_1(2c + c_1)}{2l^3} + W\frac{l}{8}$$

$$Mw' = W' \frac{cc_1^2(3c + 2c_1)}{2l^3} + W\frac{(3c_1 - c)c}{8l}$$

$$(a)\ \frac{W}{W'} < \frac{l^2}{4c_1^2}\,\frac{5c - 3c_1}{3c + 2c_1}$$

$$M_{c\,\max} = \frac{R_1^2}{2W}l \left(x = \frac{R_1 l}{W}\right)$$

$$(b)\ \frac{W'}{W} < \frac{l^2(3c_1 - 5c)}{4c(2c^2 + 6cc_1 + 3c_1^2)}$$

$$M_{c1\,\max} = W'c + \frac{(R_1 - W')^2}{2W} l \left(x = \frac{R_1 - W'}{W} l\right)$$

Deflection under W'

$$f = \frac{W'}{EI}\,\frac{c^2c_1^3(4c + 3c_1)}{12l^3} + \frac{W}{EI}\,\frac{cc_1^2(3c + c_1)}{48l}$$

Note:
R_1, R_2 = reactions
w = distributed load per longitudinal unit
W = total distributed load
M = bending moment
M_x = local value of bending moment
$M_{\max}$ = maximum value of M
V = vertical shear
V_x = local vertical shear (shear at any section)
$V_{\max}$ = maximum value of V
f = deflection
E = modulus of elasticity
I = moment of inertia
l = distance between supports
W' = concentrated load

Source: Avallone and Baumeister.[3]

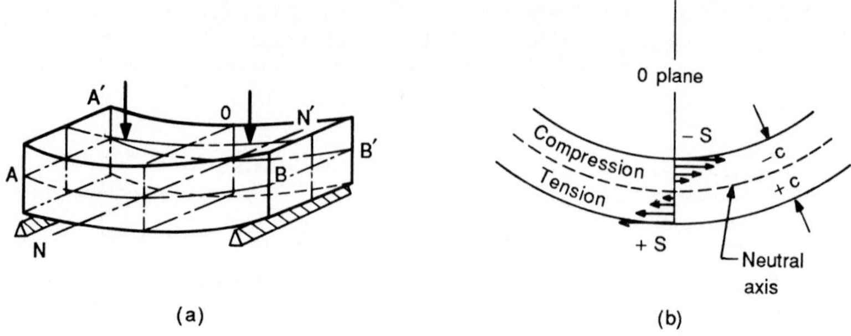

FIGURE 6.5

Flexule Formula. The concave side of a bent beam (Fig. 6.5*a* and *b*) is in compression and the convex side in tension. They are divided by the neutral plane of zero stress $A'B'BA$. The intersection of the neutral plane with the face of the beam is in the neutral line or elastic curve AB. The neutral axis NN' is the intersection of the neutral plane with the cross section.

The neutral axis contains the center of gravity of the cross section.

The flexule formula

$$S = \frac{Mc}{I} \tag{6.15}$$

is basic to the design and investigation of beams. It holds only when the maximum horizontal fiber stress S does not exceed the proportional limit of the material. c = distance of that fiber from the neutral axis. The I/c factor, the section modulus, is a measure of the capacity of a section to resist any bending moment M to which it may be subjected. I = moment of inertia of the cross section with respect to its neutral axis.

Values of I and I/c for simple shapes used as beams are given in Table 6.2.

Rolling Loads.* These may change their position on the beam. Figure 6.6 shows a beam with two equal concentrated moving loads (example: two wheels on a crane girder, or the wheels of a truck on a bridge). As the maximum moment occurs where the shear is zero, it is evident from the shear diagram that the maximum moment will occur under the wheel. As $x < a/2$ (Fig. 6.6), then

$$R_1 = P\left(1 - \frac{2x}{l} + \frac{a}{l}\right)$$

$$M_2 = \frac{Pl}{2}\left(1 - \frac{a}{l} + \frac{2x}{l}\frac{a}{l} - \frac{4x^2}{l^2}\right)$$

$$R_2 = P\left(1 + \frac{2x}{l} - \frac{a}{l}\right)$$

*From *Marks' Standard Handbook for Mechanical Engineers*, 9th ed., by E. A. Avallone and T. Baumeister III (eds.). Copyright © 1987. Used by permission of McGraw-Hill. All rights reserved.

TABLE 6.2 Properties of Sections of Beams

Section of beam	Moment of inertia I	Section modulus IC	Radius of gyration r
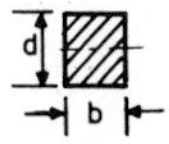	$\frac{bd^3}{12}$	$\frac{bd^3}{6}$	$\frac{d}{\sqrt{12}} = 0.289d$
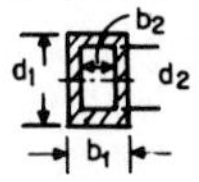	$\frac{b_1d_1^3 - b_2d_2^3}{12}$	$\frac{b_1d_1^3 - b_2d_2^3}{6d_1}$	$\sqrt{\frac{b_1d_1^3 - b_2d_2^3}{12(b_1d_1 - b_2d_2)}}$
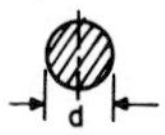	$\frac{\pi d^4}{64}$	$\frac{\pi d^3}{32}$	$\frac{d}{4}$
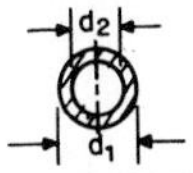	$\frac{\pi(d_1^4 - d_2^4)}{64}$	$\frac{\pi(d_1^4 - d_2^4)}{32d_1}$	$\frac{\sqrt{d_1^2 + d_2^2}}{4}$
	$\frac{bd^3}{36}$	$\frac{bd^2}{24}$ (min.)	$\frac{d}{\sqrt{18}} = 0.236d$

Source: From Avallone and Baumeister.[3]

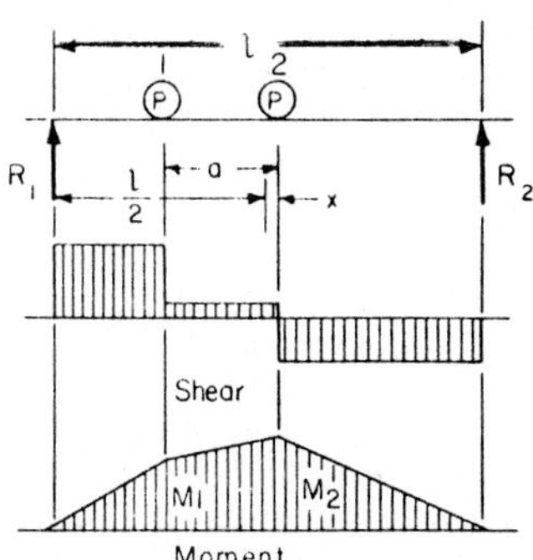

FIGURE 6.6 *(From Avallone and Baumeister.[3])*

$$M_1 = \frac{Pl}{2}\left(1 - \frac{a}{l} - \frac{2a^2}{l^2} + \frac{2x}{l}\frac{3a}{l} - \frac{4x^2}{l^2}\right)$$

$$M_{2_{\max}} = M_2\left(x = \frac{a}{4}\right)$$

$$M_{1_{\max}} = M_1\left(x = \frac{3a}{4}\right)$$

$$M_{max} = \frac{Pl}{2}\left(1 - \frac{a}{2l}\right)^2 = \frac{P}{2l}\left(l - \frac{a}{2}\right)^2$$

COLUMNS*

A column is a bar which is loaded axially in compression. A column is shortened by the compression, and it also tends to deflect laterally, owing partly to the fact that the load cannot be applied symmetrically with respect to the longitudinal axis of the column, and partly to the fact that the material of which the column is made is not perfectly homogeneous. The lateral deflection takes place usually in the direction of the least resisting moment of the section, and ultimate failure is caused by a combination of compression, shearing, and bending stresses. The load that will produce the ultimate failure of a given column is dependent upon the ratio between the length and the lateral resistance of the column. A long column of a given cross section will not support as much load as a shorter column of the same cross section.

If a column has round ends, so that the bending is not restrained, the equation of its elastic curve is

$$EI\frac{d^2y}{dx^2} = -Py \tag{6.16}$$

when the origin of the coordinate axis is at the top of the column, the positive direction of x being taken downward and the positive direction of y being taken in the direction of the deflection. P = axial load, I = least moment of inertia ($I = Ar^2$) in m^4 (ft^4), and E = modulus of elasticity in kg_f/m^2 (lb_f/in^2). Integrating the above expression twice and determining the constants of integration results in

$$P = \frac{n\pi^2 EI}{l^2} \tag{6.17}$$

which is Euler's formula for long columns. l = the length of the column.

The coefficient n in Eq. (6.17) accounts for end conditions. When the column is pivoted at both ends, $n = 1$; when one end is fixed and the other rounded, $n = 2$; when both are fixed, $n = 4$; and when one end is fixed with the other free, $n = \frac{1}{4}$. If, under load P, a slight deflection is produced, the column will not return to its original position; if P is decreased, the column will approach its original position; but if P is increased, the deflection will increase until the column fails by bending. For columns with a value of l/r (r = least radius of gyration) less than about 150, Eq. (6.17) gives results distinctly higher than those observed in tests. A theoretical equation for a short column has not been derived. Some empirical formulas for short columns are given in Ref. 3, p. 5-43.

*This section is taken in part from *General Engineering Handbook*, 2d ed., by C. E. O'Rourke. Copyright © 1940. Used by permission of McGraw-Hill, Inc. All rights reserved.

TORSION

A cylindrical bar or shaft which is being twisted about its own axis is said to be *in torsion* (see Fig. 6.7). The twisting moment or torque T is

$$T = \frac{S_s J}{r} \tag{6.18}$$

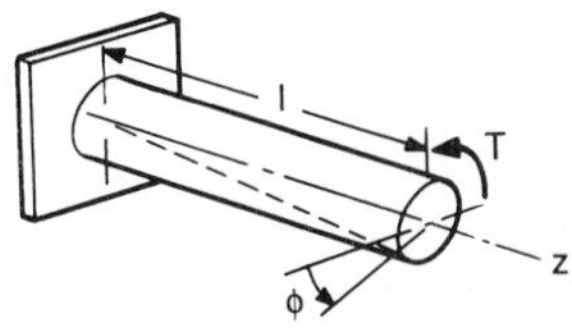

FIGURE 6.7

where S_s = unit shearing stress, r = radius of the shaft, and J = moment of inertia of the cross section with respect to its center. (For a round shaft $J = \pi r^4/2$.)

If ϕ is the angle of twist in radians,

$$T = \frac{GJ\phi}{l} \tag{6.19}$$

where G = shearing modulus of elasticity and l = the length of a bar.

The relationship between the torque in a shaft and the power (P) transmitted by it is given by the equation

$$T = \frac{P}{\omega} = \frac{P}{2\pi n} \tag{6.20}$$

where n = number of revolutions per minute on the shaft.

Table 6.3 gives approximate formulas for the maximum shearing stress and angle of twist in members subjected to torsion.

TABLE 6.3 Approximate Formulas for Maximum Shearing Stress and Angle of Twist in Members Subjected to Torsion

Shape	Maximum unit stress S_s	Angle of twist ϕ
d	$\frac{16T}{\pi d^2}$	$\frac{Tl}{GJ}$
d_1, d	$\frac{16Td}{\pi(d^4 - d_1^4)}$	$\frac{32Tl}{\pi(d^4 - d_1^4)G}$
2d, 2a	$\frac{2T}{\pi ab^2}$	$\frac{T(a^2 + b^2)l}{\pi a^3 b^3 G}$
b	$\frac{20T}{b^3}$	$\frac{46.2Tl}{b^4 G}$
t_1, t, a_1, b_1, b, a	$\frac{T}{2t(a - t)(b - t_1)}$	$\frac{Tl(at + bt_1 - t^2 - t_1^2)}{2tt_1(a - t)^2(b - t_1)^2 G}$

Source: From Avallone and Baumeister.[3]

COMBINED STRESSES*

1. Combined Direct and Flexural Stresses. If a bar is subjected to a direct axial loading and also to transverse loading, the stress at any point is given by the algebraic sum of the direct stress P/A and the bending stress My/I. The maximum and minimum stresses at the outer fibers are given by the equation

$$S = \frac{P}{A} \pm \frac{Mc}{I} \tag{6.21}$$

in which c is the distance from the neutral axis to the extreme fiber in question.

If a load is applied to a bar in a direction parallel to the axis but at a distance e from it, the single load produces the double effect of direct stress and flexural stress. The flexural effect is caused by the moment Pe, and the maximum and minimum stresses are given by the equation

$$S = \frac{P}{A} \pm \frac{Pec}{I} \tag{6.22}$$

In the upper chord of a bridge, the compressive load may be applied with enough eccentricity so as just to balance the flexural effect of the weight of the member at the middle point.

2. Combined Shearing and Flexural Stresses. In a beam the direct shearing stress, the induced shearing stress, and the direct flexural stress at any point combine to cause a resultant shearing stress on some diagonal plane which is larger than the direct shearing stress, and a resultant tensile or compressive stress along some diagonal plane which is larger than the direct flexural stress at that point. Let S_s' represent this maximum diagonal unit shearing stress, and let S_t' and S_c' represent the maximum diagonal tensile and compressive unit stresses, respectively. For any point in the beam where the direct flexural stress is tension,

$$S_s' = \sqrt{\left(\frac{S_t}{2}\right)^2 + S_s^2} \quad \text{and} \quad S_t' = \frac{S_t}{2} + S_s' \tag{6.23}$$

For any point in the beam where the direct flexural stress is compression, S_c and S_c' replace S_t and S_t', respectively. In a beam the maximum values of S_s and S_t do not occur at the same place.

3. Combined Torsional and Flexural Stresses. In a shaft which is subjected to both torsion and bending, the maximum values of S_s and S_t (or S_c) occur at the same point; so the values of S_s' and S_t' are greater than the maximum values of S_s and S_t, respectively.

CYLINDERS AND PLATES

1. Thin Cylinder under Internal Pressure. A cylinder is regarded as thin if $t/d \simeq 0$, i.e., when the thickness of the wall t is small compared with the diameter

*Adapted from *General Engineering Handbook*, 2d ed., by C. E. O'Rourke. Copyright © 1940. Used by permission of McGraw-Hill, Inc. All rights reserved.

d. Assuming that the tensile stress across a longitudinal section is uniformly distributed over the thickness of the wall,

$$pdl = 2Stl$$

or

$$S = \frac{pd}{2t} \tag{6.24}$$

For tensile stress across a transverse section

$$p\,\frac{\pi d^2}{4} = St\pi d$$

or

$$S = \frac{pd}{4t} \tag{6.25}$$

The last equation applies also to the stresses in the walls of a thin *hollow sphere, hemisphere* or *dome*. Here, p = internal pressure, l = length of cylinder, t = thickness of wall, d = diameter of cylinder, and S = tensile stress.

2. Circular and Elliptical Flat Plates. A relation for the maximum stress at the center for a circular flat plate of radius r, uniformly loaded, edge simply supported, is

$$S = \frac{3}{8}\,\frac{wr^2}{t^2}\,(3 + \mu) \tag{6.26}$$

and the maximum deflection at the center f is

$$f = \frac{2wr^4}{3Et^3} \tag{6.27}$$

An approximate formula for the maximum stress in elliptical plates, simply supported at the edge (major axis $2a$, minor axis $2b$), is

$$S = \frac{(3a - 2b)}{a}\,\frac{wb^2}{t^2} \tag{6.28}$$

3. Rectangular and Square Plates. For a distributed load w, with supports along the four sides, the unit stress is

$$S = \frac{a^2b^2w}{2t(a^2 + b^2)} \tag{6.29}$$

where a = long side and b = short side. If $a = b$ (square plate),

$$S = \frac{wa^2}{4t^2} \tag{6.30}$$

In the above equations, w is the uniformly distributed load per unit area and t the thickness of the wall (plate).

NOMENCLATURE

Symbol = *definition, SI units (U.S. Customary units)*

A = cross-sectional area, m^2 (ft^2)

a = distance, m (ft)

b = distance, m (ft)

c = distance, m (ft)

d = diameter, m (ft)

E = modulus of elasticity, N/m^2, kg_f/cm^2 (lb_f/in^2)

E_s = shearing modulus of elasticity, N/m^2, kg_f/cm^2 (lb_f/in^2)

e = distance, m (ft)

f = deflection, m (ft)

G = shearing modulus of elasticity, N/m^2, kg_f/cm^2 (lb_f/in^2)

g = acceleration of gravity, m/s^2 (ft/s^2)

I = moment of inertia, m^4 (ft^4)

J = moment of inertia of the cross section with respect to its center, m^4 (ft^4)

K = bulk modulus of elasticity, N/m^2, kg_f/cm^2 (lb_f/in^2)

l = distance, m (ft)

M = bending moment, $N \cdot m$ ($lb_f \cdot in$)

M_x = local value of bending moment, $N \cdot m$ ($lb_f \cdot in$)

P = load, N (lb_f)

P = axial force, N (lb_f)

P = axial load [Eq. (6.17)], N (lb_f)

P = power [Eq. (6.20)], W, $N \cdot m/s$ ($lb_f \cdot ft/s$)

p = pressure, N/m^2 (lb_f/in^2)

R = reaction, N (lb_f)

r = radius, m (ft)

r = radius of gyration, m (ft)

S = unit stress, N/m^2, kg_f/cm^2 (lb_f/in^2)

S_c = unit stress—compression, N/m^2, kg_f/cm^2 (lb_f/in^2)

S_s = unit stress—shear, N/m^2, kg_f/cm^2 (lb_f/in^2)

S_t = unit stress—tension, N/m^2, kg_f/cm^2 (lb_f/in^2)

T = torque (twisting moment), $N \cdot m$ ($lb_f \cdot in$)

t = wall thickness, m (ft)

U = reliance, $N \cdot m$, $kg_f \cdot cm$ ($lb_f \cdot in$)

V = vertical shear [Eq. (6.12)], N, kg_f (lb_f)

W = weight, N, kg_f (lb_f)
W' = concentrated load, N, kg_f (lb_f)
w = load per unit length, N/m, kg_f/cm (lb_f/in)
w = uniformly distributed load per unit area, N/m^2 (lb_f/in^2)
x = distance, m (ft)
y = distance, m (ft)

Greek

α = linear coefficient of thermal expantion, 1/°C (1/°F)
ϕ = angle of twist, rad (deg)
μ = Poisson's ratio (lateral strain/longitudinal strain)
ω = frequency of oscillation, 1/s
ω_n = frequency of natural oscillation, 1/s
δ = strain (unit deformation)
δ_s = shearing strain

Subscripts

max = maximum
1 = position at 1
2 = position at 2

REFERENCES*

1. C. E. O'Rourke, *General Engineering Handbook*, 2d ed., McGraw-Hill, New York, 1940.
2. Z. D. Jastrzebski, *The Nature and Properties of Engineering Materials*, John Wiley & Sons, New York, 1977.
3. E. A. Avallone and T. Baumeister III (eds.), *Marks' Standard Handbook for Mechanical Engineers*, 9th ed. McGraw-Hill, New York, 1987.
4. N. H. Cook, "Mechanics of Solids," Course 2.01, MIT class notes, 1973–1976.
5. J. P. Den Hartog, *Mechanics*, Dover, New York, 1961.
6. O. W. Eshbach and M. Souders, *Handbook of Engineering Fundamentals*, 3d ed., John Wiley & Sons, New York, 1975.
7. W. Flügge, *Handbook of Engineering Mechanics*, McGraw-Hill, New York, 1962.
8. D. M. Parks, "Solid Mechanics," Course 2.073, MIT class notes, 1973–1976.
9. S. Timoshenko and D. H. Young, *Engineering Mechanics*, 4th ed., McGraw-Hill, New York, 1956.
10. J. H. Williams, Jr., "Applied Elasticity," Course 2.083, MIT class notes, 1973–1976.

*Those references listed above but not cited in the text were used for comparison between different data sources, clarification, clarity of presentation, and, most importantly, reader's convenience when further interest in the subject exists.

CHAPTER 7
THERMODYNAMICS

INTRODUCTION

1. Definitions

Thermodynamics. The branch of science that embodies the principles and restrictions of energy transformation in macroscopic systems.

System and Surroundings. A system is taken to be any quantity of matter under consideration (or any object or region associated with the quantity of matter under consideration) selected for study and set apart (imaginary) from everything else, the latter then being called the *surroundings* (Fig. 7.1).

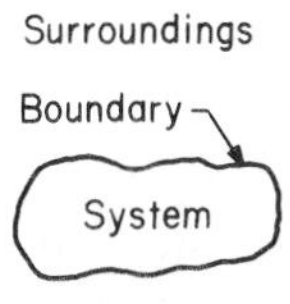

Surroundings
Boundary
System

FIGURE 7.1

Boundary. The imaginary envelope which encloses the system and separates it from its surroundings is called the *boundary of the system*. With a *closed system* there is no interchange of matter through the boundary between the system and its surroundings. With an *open system* there is such an interchange (Fig. 7.2). An *isolated system* can exchange neither matter nor energy with its surroundings.

Processes. Any change that the system may undergo is known as a *process*. Engineering thermodynamics considers chiefly those processes in which energy transformation occurs by means of changes in the physical state of fluids. The processes are classified into either *reversible* or *irreversible*. Another classification is *nonflow* and *steady flow* processes.

A reversible process is one in which both the system and the surroundings may be returned to their original states. With an irreversible process, this is not possible. All actual processes are irreversible. Among the conditions which contribute to the irreversibility of a process are the following: heat flow from a higher to a lower temperature, mixing of fluids at different temperatures, fluid turbulence, fluid or solid friction, inelastic deformation, etc.

Nonflow processes are those occurring in a container or a space in such a way that the fluid does not flow in or out of the container or space during the process (closed system). An example is the expansion of steam in a cylinder during the period when the valves are closed.

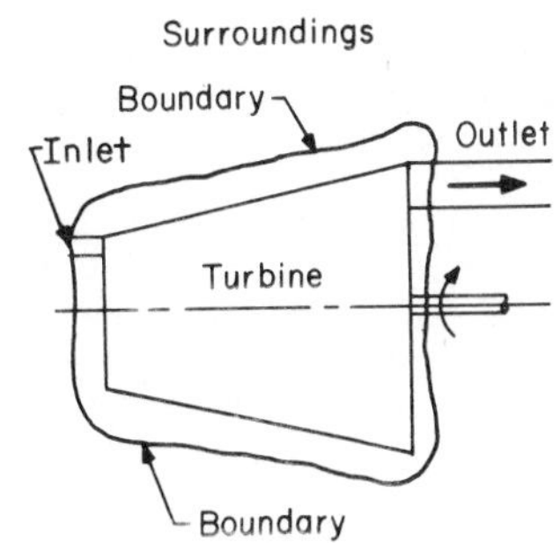

FIGURE 7.2

Steady-flow processes are those in which the fluid passes continuously through a region in a steady flow (open system). The steady-flow process or a process which closely approximates steady flow exists in most of the devices and machines employed in engineering practice. Examples are steam engines, turbines, condensers, pumps, boilers, nozzles, valves, and most heat-exchange appliances.

Other common types of processes are defined as follows:

Constant-pressure process, in which the pressure of the fluid is constant throughout the process.
Constant-volume process, in which the volume of the fluid is constant throughout the process.
Isothermal process, in which the temperature of the fluid is constant throughout the process.
Adiabitic process, in which no heat is added to or removed from the fluid during the process.

Heat. The energy in transit through the system boundary under the influence of a temperature difference or gradient. Its symbol is Q. The quantity of heat is not a property of the system.

Work. Also energy in transit between a system and its surroundings, but resulting from the displacement of an external force acting on the system. Its symbol is W. Like heat, it is not a property of the system. The usual convention with respect to signs for Q and W are shown in Fig. 7.3.

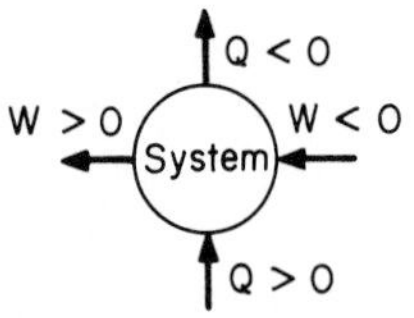

Q < 0
W > 0
W < 0
System
Q > 0

FIGURE 7.3

Properties of a System. A system has an identifiable, reproducible state when all its properties are fixed. Properties include the system's internal energy (symbol U), its entropy (symbol S), and its volume (symbol V), all of which are *extensive*, or dependent on the system's size. Other properties, such as temperature (symbol T), pressure (symbol p), entropy per unit mass (symbol s), and internal energy per unit mass (symbol u), are *intensive*, or independent of the system's size.

Properties of a Process. When a system is displaced from an equilibrium state, it undergoes a process during which its properties change until a new equilibrium state is reached. During such a process, the system may interact with its surroundings so as to interchange energy in the forms of heat and work. The amount of energy interchange that occurs during the process is dependent on the specific path along which a process proceeds. Therefore, heat and work are properties of the process, i.e., Q and W are path functions.

FIRST LAW OF THERMODYNAMICS*

1. Law of Conservation of Energy. The first law of thermodynamics is the law of the conservation of energy: During any process, the total energy of any system and its surroundings is conserved.

For a closed (nonflow) system, the first law is expressed as:

$$Q - W = U_2 - U_1 \tag{7.1}$$

*This section is based in part on *Engineering Thermodynamics*, 2d ed., by W. C. Reynolds and H. C. Perkins, copyright © 1977; in part on *Engineering Manual*, 3d ed., by R. H. Perry, copyright © 1976;

where Q = heat supplied to the system, W = work output, and $U_2 - U_1$ = gain in internal energy.

For an open system (*steady-flow process*), where the fluid flow rate through a machine or piece of apparatus is constant, the first law is expressed as

$$\dot{Q} - \dot{W} = \dot{m}\left[(h_2 - h_1) + \frac{V_2^2 - V_1^2}{2} + g(z_2 - z_1)\right] \tag{7.2}$$

where $\dot{m}$ = mass flow rate.

For many processes, the last two terms in Eq. (7.2) are often negligible.

Work done on overcoming a fluid pressure is expressed as:

$$W = \int p\, dv \tag{7.3}$$

where p is the pressure effectively applied to the surroundings for doing work and dv represents the change in volume of the system.

For steady-flow processes involving only mechanical effects, the total work done by or on a unit amount of fluid is made up of that done on the two diaphragms (flow cross sections):

$$p_2v_2 - p_1v_1 \tag{7.4}$$

and that done on the rest of the surroundings:

$$\int p\, dv - (p_2v_2 - p_1v_1) \tag{7.5}$$

By differentiation,

$$p\, dv - d(pv) = -v\, dp \tag{7.6}$$

The net useful work done on the surroundings (often called the *shaft work*) is expressed as:

$$-\int v\, dp \tag{7.7}$$

The net useful or shaft work differs from the total work by $p_2v_2 - p_1v_1$.

2. The Joule-Thomson Coefficient. If a fluid is passed adiabatically through a conduit, without doing any net or useful work,

$$h_2 = h_1 \tag{7.8}$$

---

in part on *Perry's Chemical Engineers' Handbook*, 6th ed., by R. H. Perry and D. W. Green, copyright © 1984; and in part on *Marks' Standard Handbook for Mechanical Engineers*, 9th ed., by E. A. Avallone and T. Baumeister III (eds.), copyright © 1987. All material used by permission of McGraw-Hill, Inc. All rights reserved. The section is also based in part on "Heat Transfer Characteristics of Working Fluids for OTEC," by E. N. Ganić, copyright © 1980 by ASME. Used by permission of ASME. All rights reserved.

as velocity and potential effects are negligible. A process of this type is called *Joule-Thomson flow* and the ratio

$$\frac{\partial T}{\partial p} \tag{7.9}$$

for such a flow is the Joule-Thomson coefficient. If a fluid is passed through a nonadiabatic conduit without doing any net or useful work,

$$q = h_2 - h_1 \tag{7.10}$$

assuming $V_2 = V_1$, $z_2 = z_1$. This equation is used to calculate heat balances on different types of flow apparatus (condensers, coolers, and other types of heat exchangers).

THE SECOND LAW OF THERMODYNAMICS

The second law of thermodynamics is related to the limitation of energy conversion by the temperature at which the conversion occurs. It may be shown that the efficiency of all reversible cycles absorbing heat (Q_1) only at a single constant higher temperature T_1 and rejecting heat (Q_2) only at a single constant lower temperature T_2 must be the same. For such cycles

$$\eta = \frac{W}{Q_1} = \frac{T_1 - T_2}{T_1} \tag{7.11}$$

This is called the *Carnot cycle efficiency*.

Also

$$\frac{Q_1}{T_1} + \frac{Q_2}{T_2} = 0 \tag{7.12}$$

1. Entropy. The second law of thermodynamics (for open or closed systems) is also expressed as

$$Q_{1-2} = \int_1^2 T\,dS \tag{7.13}$$

where the property S ($S = ms$) is the entropy. (The entropy is a point function, meaning that the change in its value for processes depends only on the end points of the process and not at all upon the particular path taken by the process between the end points.)

Also for any reversible process,

$$\Delta S = S_2 - S_1 = \int_1^2 \frac{dQ_{rev}}{T} \tag{7.14}$$

For any reversible process, the change in entropy of the system and surroundings is zero, whereas for any irreversible process, the net entropy change is positive.

2. Other Relations of General Principles of Thermodynamics

Specific Heat. Specific heat at constant volume:

$$c_v = \left.\frac{\partial u}{\partial T}\right|_v \qquad (7.15)$$

Specific heat at constant pressure:

$$c_p = \left.\frac{\partial h}{\partial T}\right|_p \qquad (7.16)$$

Mean specific heats for constant pressure and constant volume, respectively, in the temperature range between t_1 and t_2:

$$\bar{c}_p = \bar{c}_p\Big|_{t_1}^{t_2} = \frac{\bar{c}_p\Big|_0^{t_2} \cdot t_2 - \bar{c}_p\Big|_0^{t_1} \cdot t_1}{t_2 - t_1} \qquad (7.17)$$

$$\bar{c}_v = \bar{c}_v\Big|_{t_1}^{t_2} = \bar{c}_p\Big|_{t_1}^{t_2} - R \qquad (7.18)$$

Enthalpy

$$h = u + pv \qquad (7.19)$$

Free Energy (Helmholtz Function)

$$f = u - Ts \qquad (7.20)$$

Free Enthalpy (Gibbs Function)

$$g = h - Ts \qquad (7.21)$$

Availability of System

$$g_0 = h - T_0 s \qquad (7.22)$$

If velocity and potential are not negligible,

$$g_0 = h - T_0 s + \frac{V^2}{2} + gz \qquad (7.23)$$

where T_0 = lowest temperature available for heat discard.

Note. The availability function g_0 is useful for determining thermodynamic efficiencies of a turbine or similar devices, i.e., the ratio of actual work performed during a process to that which theoretically should have been performed.

Maxwell Relations. See Table 7.1.

Other Relations. There are also equations of state:

$$\left.\frac{\partial u}{\partial v}\right|_T = -\left[p - T\left(\frac{\partial p}{\partial T}\right)_v\right] \qquad (7.24)$$

$$\left.\frac{\partial h}{\partial p}\right|_T = \left[v - T\left(\frac{\partial v}{\partial T}\right)_p\right] \qquad (7.25)$$

TABLE 7.1 Maxwell Relations

Function	Differential	Maxwell relation
$\Delta u = q + W$	$du = T\,ds - p\,dv$	$\left(\frac{\partial T}{\partial v}\right)_s = -\left(\frac{\partial p}{\partial s}\right)_v$
$h = u + pv$	$dh = T\,ds + v\,dp$	$\left(\frac{\partial T}{\partial p}\right)_s = \left(\frac{\partial v}{\partial s}\right)_p$
$f = u - Ts$	$df = -s\,dT - p\,dv$	$\left(\frac{\partial s}{\partial v}\right)_T = \left(\frac{\partial p}{\partial T}\right)_v$
$g = h - Ts$	$dg = -s\,dT + v\,dp$	$\left(\frac{\partial s}{\partial p}\right)_T = -\left(\frac{\partial v}{\partial T}\right)_p$

By holding certain variables constant, a second set of relations is obtained:

Differential	Independent variable held constant	Relation
$du = T\,ds - p\,dv$	s	$\left(\frac{\partial u}{\partial v}\right)_s = -p$
	v	$\left(\frac{\partial u}{\partial s}\right)_v = T$
$dh = T\,ds + v\,dp$	s	$\left(\frac{\partial h}{\partial p}\right)_s = v$
	p	$\left(\frac{\partial h}{\partial s}\right)_p = T$
$df = -s dT - p\,dv$	T	$\left(\frac{\partial f}{\partial v}\right)_T = -p$
	v	$\left(\frac{\partial f}{\partial T}\right)_v = -s$
$dg = -s\,dT + v\,dp$	T	$\left(\frac{\partial g}{\partial p}\right)_T = v$
	p	$\left(\frac{\partial g}{\partial T}\right)_p = -s$

Source: From Avallone and Baumeister.[4]

$$T\,ds = du + pv \tag{7.26}$$

Note. The last equation expresses the joint statement of first and second laws.

*IDEAL GASES**

1. Equation of state. The defining equation of state of an ideal gas is

$$pv = RT \tag{7.27}$$

*This section is based in part on *Engineering Thermodynamics*, 2d ed., by W. C. Reynolds, copyright © 1977; in part on *Engineering Manual*, 3d ed., by R. H. Perry, copyright © 1976; in part on *Perry's Chemical Engineers' Handbook*, 6th ed., by R. H. Perry and D. W. Green, copyright © 1984;

or

$$pV = mRT \tag{7.28}$$

or

$$p = \rho RT \tag{7.29}$$

If V_m is the volume of one molecular weight of gas,

$$pV_m = R_m T \tag{7.30}$$

where

$$R_m = M \cdot R$$

and R_m = 8314.3 J/(kmol · K) is the universal gas constant and is the same for all ideal gases in any chosen system of units.

Each change of state of an ideal gas may be represented by the equation

$$p\,v^n = \text{const} \tag{7.31}$$

See Table 7.2 for various values of n.

2. Changes of State of Ideal Gases. Using the symbols for initial state (p_1, v_1, T_1) and final state (p_2, v_2, T_2), the equation for *internal energy* reads as:

$$u_2 - u_1 = \bar{c}_v(T_2 - T_1) = \frac{p_2v_2 - p_1v_1}{k - 1} \tag{7.32}$$

The equation for *enthalpy* reads as:

$$h_2 - h_1 = \bar{c}_p(T_2 - T_1) = \frac{k(p_2v_2 - p_1v_1)}{k - 1} \tag{7.33}$$

And the equation for *entropy* reads as:

$$s_2 - s_1 = \bar{c}_v \ln\frac{T_2}{T_1} + R\ln\frac{v_2}{v_1} = \bar{c}_p \ln\frac{T_2}{T_1} - R\ln\frac{p_2}{p_1} = \bar{c}_p \ln\frac{v_2}{v_1} + \bar{c}_v \ln\frac{p_2}{p_1} \tag{7.34}$$

Note that

$$k = \frac{c_p}{c_v} \tag{7.35}$$

3. Processes. Table 7.2 includes the equations applicable for determining the relationships between states, work, and heat for the simplest processes where an ideal gas is the medium and in which conditions throughout the process are idealized.

Graphical representation of the change of state of a substance may be done by taking any two of the six variables p, v, T, S, U, and H as independent coordi-

---

and in part on *Marks' Standard Handbook for Mechanical Engineers*, 9th ed., by E. A. Avallone and T. Baumeister III (eds.), copyright © 1987. All material used by permission of McGraw-Hill, Inc. All rights reserved. The section is also based in part on "Heat Transfer Characteristics of Working Fluids for OTEC," by E. N. Ganić, copyright © 1980 by ASME. Used by permission of ASME. All rights reserved.

TABLE 7.2 Relations for Ideal-Gas Processes

Process	p, v, T relations between states 1 and 2	$w_{1\text{-}2}$, work per unit mass (closed system)—Eq. (7.3); reversible	$w_{1\text{-}2}$, work per unit mass (open system)—Eq. (7.7); reversible	$q_{1\text{-}2}$, heat transfered per unit mass	p-v Diagram	T-S diagram
Isothermal T = const n = 1	$\frac{p_2}{p_1} = \frac{v_1}{v_2}$	$RT \ln \frac{v_2}{v_1}$ $= RT \ln \frac{p_1}{p_2}$ $= p_1 v_1 \ln \frac{v_2}{v_1}$	$w_{1\text{-}2}$	$w_{1\text{-}2}$		
Isobaric p = const n = 0	$\frac{v_2}{v_1} = \frac{T_2}{T_1}$	$p(v_2 - v_1)$ $= R(T_2 - T_1)$	0	$\bar{c}_p(T_2 - T_1)$		
Isochoric v = const n = ∞	$\frac{p_2}{p_1} = \frac{T_2}{T_1}$	0	$v(p_1 - p_2) = R(T_1 - T_2)$	$\bar{c}_v(T_2 - T_1)$		

Isentropic s = const n = k	$\frac{p_2}{p_1} = \left(\frac{v_1}{v_2}\right)^k$ $\frac{v_2}{v_1} = \left(\frac{T_1}{T_2}\right)^{1/(k-1)}$	$u_1 - u_2 = \bar{c}_v(T_1 - T_2)$ $= -\bar{c}_v T_1 \times \left[\left(\frac{p_2}{p_1}\right)^{(k-1)/k} - 1\right]$ $= -\frac{1}{k-1} RT_1 \times \left[\left(\frac{p_2}{p_1}\right)^{(k-1)/k} - 1\right]$	$h_1 - h_2 = \bar{c}_p(T_2 - T_1)$ $= -\bar{c}_p T_1 \times \left[\left(\frac{p_2}{p_1}\right)^{(k-1)/k} - 1\right]$ $= -\frac{k}{k-1} RT_1 \times \left[\left(\frac{p_2}{p_1}\right)^{(k-1)/k} - 1\right]$	0		
Politropic n = const	$\frac{p_2}{p_1} = \left(\frac{v_1}{v_2}\right)^n$ $\frac{p_2}{p_1} = \left(\frac{T_2}{T_1}\right)^{n/(n-1)}$ $\frac{v_2}{v_1} = \left(\frac{T_1}{T_2}\right)^{1/(n-1)}$	$\frac{1}{n-1} R(T_1 - T_2)$ $= \frac{1}{n-1}(p_1v_1 - p_2v_2)$ $= -\frac{1}{n-1} RT \times \left[\left(\frac{p_2}{p_1}\right)^{(n-1)/n} - 1\right]$	$\frac{n}{n-1} R(T_1 - T_2)$ $= -\frac{n}{n-1} RT_1 \times \left[\left(\frac{p_2}{p_1}\right)^{(n-1)/n} - 1\right]$	$\bar{c}_v \frac{n-k}{n-1}(T_2 - T_1)$		

Source: From Perry,[2] Avallone and Baumeister,[4] and Gieck.[5]

nates. While any pair may be chosen, there are three systems of graphical representation that are widely used:

- The pressure-volume (p-v) diagram
- The T-S diagram
- The H-S diagram (Mollier diagram)

Figures 7.14 through 7.16 include the diagrams for water.

4. Mixtures of Gases. There are a number of equations for mixtures, depending on the property of the mixture involved:

Mass m of a mixture of components $m_1, m_2, \ldots$:

$$m = m_1 + m_2 + \cdots + m_n = \sum_{i=1}^{n} m_i \tag{7.36}$$

Mass fraction x_i of a mixture:

$$x_i = \frac{m_i}{m} \quad \text{and} \quad \sum_{i=1}^{n} x_i = 1 \tag{7.37}$$

Mole fraction y_i of a mixture:

$$y_i = \frac{n_i}{n} \quad \text{and} \quad \sum_{i=1}^{n} y_i = 1 \tag{7.38}$$

$$n = n_1 + n_2 + \cdots + n_n = \sum_{i=1}^{n} n_i \tag{7.39}$$

Equivalent molecular mass M of a mixture:

$$M_i = \frac{m_i}{n} \quad \text{and} \quad M = \frac{m}{n} \tag{7.40}$$

$$M = \sum_{i=1}^{n} M_i \cdot y_i \quad \text{and} \quad \frac{1}{M} = \sum_{i=1}^{n} \frac{x_i}{M_i} \tag{7.41}$$

and

$$x_i = \frac{M_i}{M} y_i \tag{7.42}$$

Pressure p of the mixture and the partial pressure p_i of its components:

$$p = \sum_{i=1}^{n} p_i \qquad p_i = y_i \cdot p \tag{7.43}$$

Volume fraction ξ_i of a mixture:

$$\xi_i = \frac{\mathbf{V}_i}{\mathbf{V}} = y_i \quad \text{and} \quad \sum_{i=1}^{n} \xi_i = 1 \tag{7.44}$$

Partial volume $\mathbf{V}_i$:

$$\mathbf{V}_i = \frac{m_i R_i T}{p} = \frac{n_i R_m T}{p} \tag{7.45}$$

$$\sum_{i=1}^{n} \mathbf{V}_i = \mathbf{V} \tag{7.46}$$

Internal energy and enthalpy of a mixture:

$$u = \sum_{i=1}^{n} x_i u_i \tag{7.47}$$

$$h = \sum_{i=1}^{n} x_i h_i \tag{7.48}$$

In this relation the temperature of the mixture is given as: For the calculation of u:

$$t = \frac{\bar{c}_{v_1} m_1 t_1 + \bar{c}_{v_2} m_2 t_2 + \cdots + \bar{c}_{v_n} m_n t_n}{\bar{c}_v \cdot m} \tag{7.49}$$

For calculation of h:

$$t = \frac{\bar{c}_{p_1} m_1 t_1 + \bar{c}_{p_2} m_2 t_2 + \cdots + \bar{c}_{p_n} m_n t_n}{\bar{c}_p \cdot m} \tag{7.50}$$

For the mixture:

$$\bar{c}_v = \bar{c}_p - R \tag{7.51}$$

where

$$\bar{c}_p = \sum_{i=1}^{n} x_i \bar{c}_{p_i} \tag{7.52}$$

The *relative humidity* (R.H.) of an ideal-gas mixture is defined as

$$\text{R.H.} = \frac{p_v}{p_{\text{sat}}} \tag{7.53}$$

where p_v is the actual pressure of the vapor phase and p_{sat} is the pressure exerted by the same vapor when saturated at the mixture temperature.

The *absolute humidity* [or *specific humidity* (S.H.)] is the ratio of the mass of vapor in the mixture (m_v) to the mass to the dry gas (m_g):

$$\text{S.H.} = \frac{m_v}{m_g} \tag{7.54}$$

Observations of the concentrations of these two components are usually made in terms of three temperatures:

1. The dew-point temperature (the temperature at which condensation of the vapor takes place if the mixture is cooled at constant pressure)
2. The wet-bulb temperature (the temperature achieved by the mixture if it is saturated by evaporating liquid into it so that the latent heat of vaporization comes from the mixture, thereby depressing its temperature)
3. The dry-bulb temperature (the normal temperature of the mixture)

REAL GASES*

In many cases, especially for gases at high pressures, the ideal-gas law does not adequately represent the relationship between the properties. For that reason it is modified to apply to real gases by the introduction of a correction factor, the compressibility factor C:

$$p\mathbf{V} = CnR_mT \tag{7.55}$$

where $\mathbf{V}$ = volume of n moles.

Values of C are given in Fig. 7.4 in terms of reduced pressure (p_r) and reduced temperature T_r, where

$$T_r = \frac{T}{T_c} \quad \text{and} \quad p_r = \frac{p}{p_c}$$

Values of critical pressure p_c and critical temperature T_c are shown in Table 7.3.

1. Systems Containing More than One Phase. When several phases exist in equilibrium together, the thermodynamic condition of equilibrium is described in terms of a system property g, or Gibbs function, already given as Eq. (7.21).

In the case where a single component exists in several phases, such as ice and steam, the equilibrium conditions are given as

$$g_{\text{I}} = g_{\text{II}} = g_{\text{III}} \tag{7.56}$$

*This section is based in part on *Engineering Thermodynamics*, 2d ed., by W. C. Reynolds and H. C. Perkins, copyright © 1977; in part on *Engineering Manual*, 3d ed., by R. H. Perry, copyright © 1976; in part on *Perry's Chemical Engineers' Handbook*, 6th ed., by R. H. Perry and D. W. Green, copyright © 1984; and in part on *Marks' Standard Handbook for Mechanical Engineers*, 9th ed., by E. A. Avallone and T. Baumeister III (eds.), copyright © 1987. All material used by permission of McGraw-Hill, Inc. All rights reserved. The section is also based in part on "Heat Transfer Characteristics of Working Fluids for OTEC," by E. N. Ganić, copyright © 1980 by ASME. Used by permission of ASME. All rights reserved.

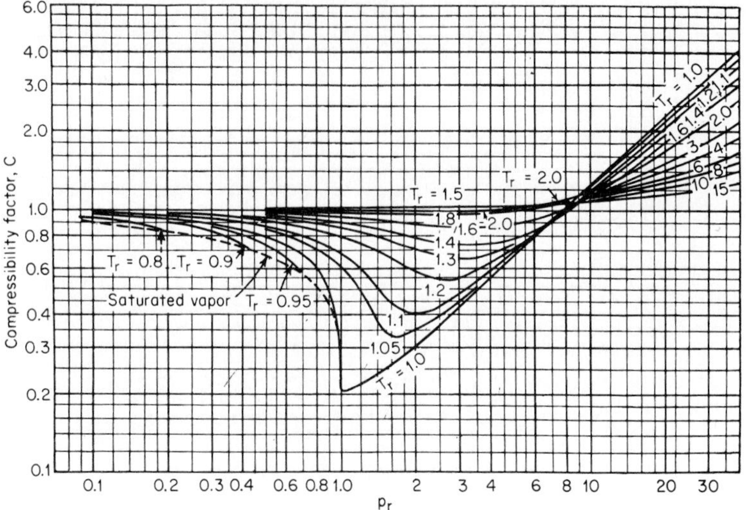

FIGURE 7.4 Compressibility factors of gases. (*From Perry.*[2])

where g_I represents the specific Gibbs function for phase I. The different conditions of existence are best represented graphically by a phase diagram and a pressure-volume (p-v) diagram, as illustrated in Fig. 7.5.

2. Vapor Pressure. The vapor pressure or saturation pressure at any given temperature can often be obtained from property tables (see Tables 7.4 and 7.5), or may be approximated by the following relation:

$$\ln p = \frac{A}{T} + B \tag{7.57}$$

where A and B are constants whose values depend on the substance.

3. Clapeyron Equation. The Clapeyron equation is an important relationship useful in calculations relating to the constant-pressure evaporation of pure substances. In such cases the equation may be written as

$$v_{fg} = \frac{h_{fg}}{T} \frac{1}{dp/dT} \tag{7.58}$$

4. Properties of Mixtures of Liquid and Vapor. The properties of a unit mass of a mixture of liquid and vapor of quality x (mass fraction) are given by the following expressions:

$$v = v_f + x \cdot v_{fg} \tag{7.59}$$

$$h = h_f + x \cdot h_{fg} \tag{7.60}$$

$$u = u_f + x \cdot u_{fg} \tag{7.61}$$

$$s = s_f + x \cdot s_{fg} \tag{7.62}$$

TABLE 7.3 Critical Properties of Gases

Gas	Symbol	Approx. mol wt	Critical pressure, psia	Critical pressure, (MPa)	Critical temperature, °F	Critical temperature, (°C)
Acetylene	C_2H_2	26	911	(6.28)	96.3	(35.7)
Air	...	29	546	(3.76)	−220.3	(−140.2)
Ammonia	NH_3	17	1640	(11.30)	270.3	(132.4)
Argon	Ar	40	706	(4.86)	−187.7	(−122)
Benzene	C_4H_6	78	702	(4.84)	551.4	(288.5)
Butane	C_4H_{10}	58	530	(3.65)	307.4	(153)
Carbon dioxide	CO_2	44	1073	(7.39)	88.0	(31.1)
Carbon monoxide	CO	28	515	(3.55)	−220.3	(−140.2)
Dichlorodifluoromethane (R12)	CCl_2F_2	121	597	(4.11)	233.6	(112)
Ethane	C_2H_6	30	718	(4.95)	90.0	(32.2)
Ethylene	C_2H_4	28	748	(5.15)	49.3	(9.6)
Helium	He	4	33	(0.22)	−450.2	(−267.8)
Heptane	C_7H_{16}	100	394	(2.71)	517.1	(269.5)
Hydrogen	H_2	2	188	(1.29)	−399.8	(−239.8)
Methane	CH_4	16	674	(4.64)	−116.5	(−82.5)
Nitrogen	N_2	28	493	(3.39)	−232.8	(−147.1)
Oxygen	O_2	32	731	(5.04)	−181.8	(−118.7)
Propane	C_3H_8	44	632	(4.35)	206.3	(96.8)
Water	H_2O	18	3106	(21.41)	705.5	(374.2)

Source: From Perry.[2]

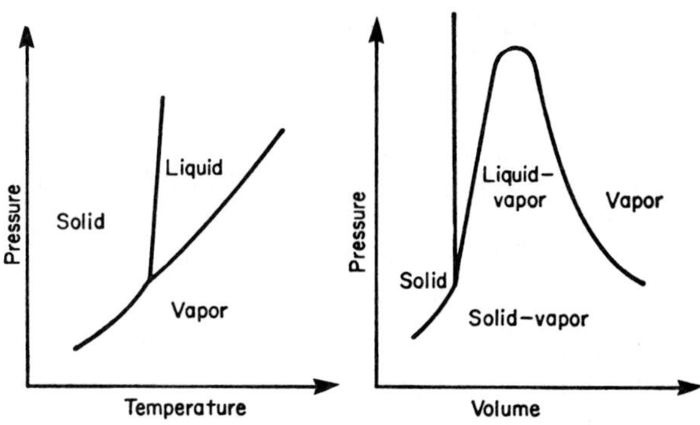

FIGURE 7.5 Phase and pressure-volume diagrams. (*From Perry.*[2])

where $v_{fg} = v_g - v_f$ = increase of volume during vaporization
$h_{fg} = h_g - h_f$ = heat of vaporization (heat required to vaporize unit mass of liquid at constant pressure and temperature)
$u_{fg} = u_g - u_f$ = increase of internal energy during vaporization
$s_{fg} = s_g - s_f = \dfrac{h_{fg}}{T}$ = increase of entropy during vaporization
$p\,v_{fg}$ = work performed during vaporization

Note that the subscripts g and f are related to the gas (vapor) and liquid states, respectively. Tables 7.4 through 7.6 and Figs. 7.14 through 7.16 are useful for calculating these properties.

5. The Phase Rule. For equilibrium states, not all variables which describe the state of the system are independent, and fixing a limited number of them automatically establishes the others. This number of independent variables (pressure, temperature, mole fractions, etc.) is given by the phase rule and is called the *number of degrees of freedom of the system*. The following relationship (the actual phase rule) applies:

$$F = 2 - \pi + m - r \tag{7.63}$$

where π = the number of phases
m = the number of chemical species
r = the number of independent chemical reactions
F = the number of degrees of freedom

For example, for the liquid-vapor system $F = 1$ when $\pi = 2$, $m = 1$, and $r = 0$, i.e., at a given pressure two phases coexist only at one temperature.

6. Systems with Flow of Compressible Fluids

Flow through Orifices and Nozzles. As a compressible fluid passes through a nozzle, its pressure drops and simultaneously its velocity increases. By assuming that flow is adiabatic, it is possible to calculate from the properties of the fluid the required area for the cross section of the nozzle at any point. The smallest cross

TABLE 7.4 Properties of Saturated Water and Water Vapor-Temperature Table

		Volume, m^3/kg		Internal energy, kJ/kg		Enthalpy, kJ/kg			Entropy, kJ/(kg · K)		
P, MPa	T, °C	v_f	v_g	u_f	u_g	h_f	h_{fg}	h_g	s_f	s_{fg}	s_g
0.000611	0.01	0.001000	206.1	0.0	2375.3	0.0	2501.3	2501.3	0.0000	9.1571	9.1571
0.0008	3.8	0.001000	159.7	15.8	2380.5	15.8	2492.5	2508.3	0.0575	9.0007	9.0582
0.001	7.0	0.001000	129.2	29.3	2385.0	29.3	2484.9	2514.2	0.1059	8.8706	8.9765
0.0012	9.7	0.001000	108.7	40.6	2388.7	40.6	2478.5	2519.1	0.1460	8.7639	8.9099
0.0014	12.0	0.001001	93.92	50.3	2391.9	50.3	2473.1	2523.4	0.1802	8.6736	8.8538
0.0016	14.0	0.001001	82.76	58.9	2394.7	58.9	2468.2	2527.1	0.2101	8.5952	8.8053
0.0018	15.8	0.001001	74.03	66.5	2397.2	66.5	2464.0	2530.5	0.2367	8.5259	8.7626
0.002	17.5	0.001001	67.00	73.5	2399.5	73.5	2460.0	2533.5	0.2606	8.4639	8.7245
0.003	24.1	0.001003	45.67	101.0	2408.5	101.0	2444.5	2545.5	0.3544	8.2240	8.5784
0.004	29.0	0.001004	34.80	121.4	2415.2	121.4	2433.0	2554.4	0.4225	8.0529	8.4754
0.006	36.2	0.001006	23.74	151.5	2424.9	151.5	2415.9	2567.4	0.5208	7.8104	8.3312
0.008	41.5	0.001008	18.10	173.9	2432.1	173.9	2403.1	2577.0	0.5924	7.6371	8.2295
0.01	45.8	0.001010	14.67	191.8	2437.9	191.8	2392.8	2584.6	0.6491	7.5019	8.1510
0.012	49.4	0.001012	12.36	206.9	2442.7	206.9	2384.1	2591.0	0.6961	7.3910	8.0871
0.014	52.6	0.001013	10.69	220.0	2446.9	220.0	2376.6	2596.6	0.7365	7.2968	8.0333
0.016	55.3	0.001015	9.433	231.5	2450.5	231.5	2369.9	2601.4	0.7719	7.2149	7.9868
0.018	57.8	0.001016	8.445	241.9	2453.8	241.9	2363.9	2605.8	0.8034	7.1425	7.9459
0.02	60.1	0.001017	7.649	251.4	2456.7	251.4	2358.3	2609.7	0.8319	7.0774	7.9093
0.03	69.1	0.001022	5.229	289.2	2468.4	289.2	2336.1	2625.3	0.9439	6.8256	7.7695
0.04	75.9	0.001026	3.993	317.5	2477.0	317.6	2319.1	2636.7	1.0260	6.6449	7.6709
0.06	85.9	0.001033	2.732	359.8	2489.6	359.8	2293.7	2653.5	1.1455	6.3873	7.5328
0.08	93.5	0.001039	2.087	391.6	2498.8	391.6	2274.1	2665.7	1.2331	6.2023	7.4354
0.1	99.6	0.001043	1.694	417.3	2506.1	417.4	2258.1	2675.5	1.3029	6.0573	7.3602
0.12	104.8	0.001047	1.428	439.2	2512.1	439.3	2244.2	2683.5	1.3611	5.9378	7.2989
0.14	109.3	0.001051	1.237	458.2	2517.3	458.4	2232.0	2690.4	1.4112	5.8360	7.2472

140	0.3613	0.001080	0.5089	588.7	2550.0	589.1	2144.8	2733.9	1.7395	5.1912	6.9307
150	0.4758	0.001090	0.3928	631.7	2559.5	632.2	2114.2	2746.4	1.8422	4.9965	6.8387
160	0.6178	0.001102	0.3071	674.9	2568.4	675.5	2082.6	2758.1	1.9431	4.8079	6.7510
170	0.7916	0.001114	0.2428	718.3	2576.5	719.2	2049.5	2768.7	2.0423	4.6249	6.6672
180	1.002	0.001127	0.1941	762.1	2583.7	763.2	2015.0	2778.2	2.1400	4.4466	6.5866
190	1.254	0.001141	0.1565	806.2	2590.0	807.5	1978.8	2786.4	2.2363	4.2724	6.5087
200	1.554	0.001156	0.1274	850.6	2595.3	852.4	1940.8	2793.2	2.3313	4.1018	6.4331
210	1.906	0.001173	0.1044	895.5	2599.4	897.7	1900.8	2798.5	2.4253	3.9340	6.3593
220	2.318	0.001190	0.08620	940.9	2602.4	943.6	1858.5	2802.1	2.5183	3.7686	6.2869
230	2.795	0.001209	0.07159	986.7	2603.9	990.1	1813.9	2804.0	2.6105	3.6050	6.2155
240	3.344	0.001229	0.05977	1033.2	2604.0	1037.3	1766.5	2803.8	2.7021	3.4425	6.1446
250	3.973	0.001251	0.05013	1080.4	2602.4	1085.3	1716.2	2801.5	2.7933	3.2805	6.0738
260	4.688	0.001276	0.04221	1128.4	2599.0	1134.4	1662.5	2796.9	2.8844	3.1184	6.0028
270	5.498	0.001302	0.03565	1177.3	2593.7	1184.5	1605.2	2789.7	2.9757	2.9553	5.9310
280	6.411	0.001332	0.03017	1227.4	2586.1	1236.0	1543.6	2779.6	3.0674	2.7905	5.8579
290	7.436	0.001366	0.02557	1278.9	2576.0	1289.0	1477.2	2766.2	3.1600	2.6230	5.7830
300	8.580	0.001404	0.02168	1332.0	2563.0	1344.0	1405.0	2749.0	3.2540	2.4513	5.7053
310	9.856	0.001447	0.01835	1387.0	2546.4	1401.3	1326.0	2727.3	3.3500	2.2739	5.6239
320	11.27	0.001499	0.01549	1444.6	2525.5	1461.4	1238.7	2700.1	3.4487	2.0883	5.5370
330	12.84	0.001561	0.01300	1505.2	2499.0	1525.3	1140.6	2665.9	3.5514	1.8911	5.4425
340	14.59	0.001638	0.01080	1570.3	2464.6	1594.2	1027.9	2622.1	3.6601	1.6765	5.3366
350	16.51	0.001740	0.008815	1641.8	2418.5	1670.6	893.4	2564.0	3.7784	1.4338	5.2122
360	18.65	0.001892	0.006947	1725.2	2351.6	1760.5	720.7	2481.2	3.9154	1.1382	5.0536
370	21.03	0.002213	0.004931	1844.0	2229.0	1890.5	442.2	2332.7	4.1114	0.6876	4.7990
374.136	22.088	0.003155	0.003155	2029.6	2029.6	2099.3	0.0	2099.3	4.4305	0.0000	4.4305

Source: From Reynolds and Perkins.[1]

TABLE 7.5 Properties of Saturated Water and Water Vapor-Pressure Table

		Volume, m^3/kg		Internal energy, kJ/kg		Enthalpy, kJ/kg			Entropy, kJ/(kg · K)		
T, °C	P, MPa	v_f	v_g	u_f	u_g	h_f	h_{fg}	h_g	s_f	s_{fg}	s_g
0.010	0.0006113	0.001000	206.1	0.0	2375.3	0.0	2501.3	2501.3	0.0000	9.1571	9.1571
2	0.0007056	0.001000	179.9	8.4	2378.1	8.4	2496.6	2505.0	0.0305	9.0738	9.1043
5	0.0008721	0.001000	147.1	21.0	2382.2	21.0	2489.5	2510.5	0.0761	8.9505	9.0266
10	0.001228	0.001000	106.4	42.0	2389.2	42.0	2477.7	2519.7	0.1510	8.7506	8.9016
15	0.001705	0.001001	77.93	63.0	2396.0	63.0	2465.9	2528.9	0.2244	8.5578	8.7822
20	0.002338	0.001002	57.79	83.9	2402.9	83.9	2454.2	2538.1	0.2965	8.3715	8.6680
25	0.003169	0.001003	43.36	104.9	2409.8	104.9	2442.3	2547.2	0.3672	8.1916	8.5588
30	0.004246	0.001004	32.90	125.8	2416.6	125.8	2430.4	2556.2	0.4367	8.0174	8.4541
35	0.005628	0.001006	25.22	146.7	2423.4	146.7	2418.6	2565.3	0.5051	7.8488	8.3539
40	0.007383	0.001008	19.52	167.5	2430.1	167.5	2406.8	2574.3	0.5723	7.6855	8.2578
45	0.009593	0.001010	15.26	188.4	2436.8	188.4	2394.8	2583.2	0.6385	7.5271	8.1656
50	0.01235	0.001012	12.03	209.3	2443.5	209.3	2382.8	2592.1	0.7036	7.3735	8.0771
55	0.01576	0.001015	9.569	230.2	2450.1	230.2	2370.7	2600.9	0.7678	7.2243	7.9921
60	0.01994	0.001017	7.671	251.1	2456.6	251.1	2358.5	2609.6	0.8310	7.0794	7.9104
65	0.02503	0.001020	6.197	272.0	2463.1	272.0	2346.2	2618.2	0.8934	6.9384	7.8318
70	0.03119	0.001023	5.042	292.9	2469.5	293.0	2333.8	2626.8	0.9549	6.8012	7.7561
75	0.03858	0.001026	4.131	313.9	2475.9	313.9	2321.4	2635.3	1.0155	6.6678	7.6833
80	0.04739	0.001029	3.407	334.8	2482.2	334.9	2308.8	2643.7	1.0754	6.5376	7.6130
85	0.05783	0.001032	2.828	355.8	2488.4	355.9	2296.0	2651.9	1.1344	6.4109	7.5453
90	0.07013	0.001036	2.361	376.8	2494.5	376.9	2283.2	2660.1	1.1927	6.2872	7.4799
95	0.08455	0.001040	1.982	397.9	2500.6	397.9	2270.2	2668.1	1.2503	6.1664	7.4167
100	0.1013	0.001044	1.673	418.9	2506.5	419.0	2257.0	2676.0	1.3071	6.0486	7.3557
110	0.1433	0.001052	1.210	461.1	2518.1	461.3	2230.2	2691.5	1.4188	5.8207	7.2395
120	0.1985	0.001060	0.8919	503.5	2529.2	503.7	2202.6	2706.3	1.5280	5.6024	7.1304
130	0.2701	0.001070	0.6685	546.0	2539.9	546.3	2174.2	2720.5	1.6348	5.3929	7.0277

0.16	113.3	0.001054	1.091	475.2	2521.8	475.3	2221.2	2696.5	1.4553	5.7472	7.2025
0.18	116.9	0.001058	0.9775	490.5	2525.9	490.7	2211.1	2701.8	1.4948	5.6683	7.1631
0.2	120.2	0.001061	0.8857	504.5	2529.5	504.7	2201.9	2706.6	1.5305	5.5975	7.1280
0.3	133.5	0.001073	0.6058	561.1	2543.6	561.5	2163.8	2725.3	1.6722	5.3205	6.9927
0.4	143.6	0.001084	0.4625	604.3	2553.6	604.7	2133.8	2738.5	1.7770	5.1197	6.8967
0.6	158.9	0.001101	0.3157	669.9	2567.4	670.6	2086.2	2756.8	1.9316	4.8293	6.7609
0.8	170.4	0.001115	0.2404	720.2	2576.8	721.1	2048.0	2769.1	2.0466	4.6170	6.6636
1	179.9	0.001127	0.1944	761.7	2583.6	762.8	2015.3	2778.1	2.1391	4.4482	6.5873
1.2	188.0	0.001139	0.1633	797.3	2588.8	798.6	1986.2	2784.8	2.2170	4.3072	6.5242
1.4	195.1	0.001149	0.1408	828.7	2592.8	830.3	1959.7	2790.0	2.2847	4.1854	6.4701
1.6	201.4	0.001159	0.1238	856.9	2596.0	858.8	1935.2	2794.0	2.3446	4.0780	6.4226
1.8	207.2	0.001168	0.1104	882.7	2598.4	884.8	1912.3	2797.1	2.3986	3.9816	6.3802
2	212.4	0.001177	0.09963	906.4	2600.3	908.8	1890.7	2799.5	2.4478	3.8939	6.3417
3	233.9	0.001216	0.06668	1004.8	2604.1	1008.4	1795.7	2804.1	2.6462	3.5416	6.1878
4	250.4	0.001252	0.04978	1082.3	2602.3	1087.3	1714.1	2801.4	2.7970	3.2739	6.0709
6	275.6	0.001319	0.03244	1205.4	2589.7	1213.3	1571.0	2784.3	3.0273	2.8627	5.8900
8	295.1	0.001384	0.02352	1305.6	2569.8	1316.6	1441.4	2758.0	3.2075	2.5365	5.7440
9	303.4	0.001418	0.02048	1350.5	2557.8	1363.3	1378.8	2742.1	3.2865	2.3916	5.6781
10	311.1	0.001452	0.01803	1393.0	2544.4	1407.6	1317.1	2724.7	3.3603	2.2546	5.6149
12	324.8	0.001527	0.01426	1472.9	2513.7	1491.3	1193.6	2684.9	3.4970	1.9963	5.4933
14	336.8	0.001611	0.01149	1548.6	2476.8	1571.1	1066.5	2637.6	3.6240	1.7486	5.3726
16	347.4	0.001711	0.009307	1622.7	2431.8	1650.0	930.7	2580.7	3.7468	1.4996	5.2464
18	357.1	0.001840	0.007491	1698.9	2374.4	1732.0	777.2	2509.2	3.8722	1.2332	5.1054
20	365.8	0.002036	0.005836	1785.6	2293.2	1826.3	583.7	2410.0	4.0146	0.9135	4.9281
22.088	374.136	0.003155	0.003155	2029.6	2029.6	2099.3	0.0	2099.3	4.4305	0.0000	4.4305

Source: From Reynolds and Perkins.[1]

TABLE 7.6 Properties of Superheated Water Vapor

P, MPa (T_{sat}, °C)		Temperature, °C												
		50	100	150	200	250	300	350	400	500	600	700	800	900
0.002	v, m³/kg	74.52	86.08	97.63	109.2	120.7	132.3	143.8	155.3	178.4	201.5	224.6	247.6	270.7
(17.5)	u, kJ/kg	2445.2	2516.3	2588.3	2661.6	2736.2	2812.2	2889.8	2969.0	3132.3	3302.5	3479.7	3663.9	3855.1
	h, kJ/kg	2594.3	2688.4	2783.6	2879.9	2977.6	3076.7	3177.4	3279.6	3489.1	3705.5	3928.8	4159.1	4396.5
	s, kJ/(kg · K)	8.9227	9.1936	9.4328	9.6479	9.8442	10.0251	10.1935	10.3513	10.6414	10.9044	11.1465	11.3718	11.5832
0.005	v, m³/kg	29.78	34.42	39.04	43.66	48.28	52.90	57.51	62.13	71.36	80.59	89.82	99.05	108.3
(32.9)	u, kJ/kg	2444.7	2516.0	2588.1	2661.4	2736.1	2812.2	2889.8	2968.9	3132.3	3302.5	3479.6	3663.9	3855.0
	h, kJ/kg	2593.6	2688.1	2783.3	2879.8	2977.5	3076.6	3177.3	3279.6	3489.1	3705.4	3928.8	4159.1	4396.5
	s, kJ/(kg · K)	8.4982	8.7699	9.0095	9.2248	9.4212	9.6022	9.7706	9.9284	10.2185	10.4815	10.7236	10.9489	11.1603
0.01	v, m³/kg	14.87	17.20	19.51	21.83	24.14	26.45	28.75	31.06	35.68	40.29	44.91	49.53	54.14
(45.8)	u, kJ/kg	2443.9	2515.5	2587.9	2661.3	2736.0	2812.1	2889.7	2968.9	3132.3	3302.5	3479.6	3663.8	3855.0
	h, kJ/kg	2592.6	2687.5	2783.0	2879.5	2977.3	3076.5	3177.2	3279.5	3489.0	3705.4	3928.7	4159.1	4396.4
	s, kJ/(kg · K)	8.1757	8.4487	8.6890	8.9046	9.1010	9.2821	9.4506	9.6084	9.8985	10.1616	10.4037	10.6290	10.8404
0.02	v, m³/kg		8.585	9.748	10.91	12.06	13.22	14.37	15.53	17.84	20.15	22.45	24.76	27.07
(60.1)	u, kJ/kg		2514.5	2587.3	2660.9	2735.7	2811.9	2889.5	2968.8	3132.2	3302.4	3479.6	3663.8	3855.0
	h, kJ/kg		2686.2	2782.3	2879.1	2977.0	3076.3	3177.0	3279.4	3488.9	3705.3	3928.7	4159.1	4396.4
	s, kJ/(kg · K)		8.1263	8.3678	8.5839	8.7807	8.9619	9.1304	9.2884	9.5785	9.8417	10.0838	10.3091	10.5205
0.05	v, m³/kg		3.418	3.889	4.356	4.820	5.284	5.747	6.209	7.134	8.057	8.981	9.904	10.83
(81.3)	u, kJ/kg		2511.6	2585.6	2659.8	2735.0	2811.3	2889.1	2968.4	3131.9	3302.2	3479.5	3663.7	3854.9
	h, kJ/kg		2682.5	2780.1	2877.6	2976.0	3075.5	3176.4	3278.9	3488.6	3705.1	3928.5	4158.9	4396.3
	s, kJ/(kg · K)		7.6955	7.9409	8.1588	8.3564	8.5380	8.7069	8.8650	9.1554	9.4186	9.6608	9.8861	10.0975
0.07	v, m³/kg		2.434	2.773	3.108	3.441	3.772	4.103	4.434	5.095	5.755	6.415	7.074	7.734
(89.9)	u, kJ/kg		2509.6	2584.5	2659.1	2734.5	2811.0	2888.8	2968.2	3131.8	3302.1	3479.4	3663.6	3854.9
	h, kJ/kg		2680.0	2778.6	2876.7	2975.3	3075.0	3176.1	3278.6	3488.4	3704.9	3928.4	4158.8	4396.2
	s, kJ/(kg · K)		7.5349	7.7829	8.0020	8.2001	8.3821	8.5511	8.7094	8.9999	9.2632	9.5054	9.7307	9.9422
0.1	v, m³/kg		1.696	1.936	2.172	2.406	2.639	2.871	3.103	3.565	4.028	4.490	4.952	5.414
(99.6)	u, kJ/kg		2506.6	2582.7	2658.0	2733.7	2810.4	2888.4	2967.8	3131.5	3301.9	3479.2	3663.5	3854.8
	h, kJ/kg		2676.2	2776.4	2875.3	2974.3	3074.3	3175.5	3278.1	3488.1	3704.7	3928.2	4158.7	4396.1
	s, kJ/(kg · K)		7.3622	7.6142	7.8351	8.0341	8.2165	8.3858	8.5442	8.8350	9.0984	9.3406	9.5660	9.7775

P, MPa (T_{sat}, °C)		Temperature, °C 150	200	250	300	350	400	450	500	550	600	700	800	900
0.15	v, m³/kg	1.285	1.444	1.601	1.757	1.912	2.067	2.222	2.376	2.530	2.685	2.993	3.301	3.609
(111.4)	u, kJ/kg	2579.8	2656.2	2732.5	2809.5	2887.7	2967.3	3048.4	3131.1	3215.6	3301.6	3479.0	3663.4	3854.6
	h, kJ/kg	2772.6	2872.9	2972.7	3073.0	3174.5	3277.3	3381.7	3487.6	3595.1	3704.3	3927.9	4158.5	4395.9
	s, kJ/(kg · K)	7.4201	7.6441	7.8446	8.0278	8.1975	8.3562	8.5057	8.6473	8.7821	8.9109	9.1533	9.3787	9.5903
0.2	v, m³/kg	0.9596	1.080	1.199	1.316	1.433	1.549	1.665	1.781	1.897	2.013	2.244	2.475	2.706
(120.2)	u, kJ/kg	2576.9	2654.4	2731.2	2808.6	2886.9	2966.7	3047.9	3130.7	3215.2	3301.4	3478.8	3663.2	3854.5
	h, kJ/kg	2768.8	2870.5	2971.0	3071.8	3173.5	3276.5	3381.0	3487.0	3594.7	3704.0	3927.7	4158.3	4395.8
	s, kJ/(kg · K)	7.2803	7.5074	7.7094	7.8934	8.0636	8.2226	8.3723	8.5140	8.6489	8.7778	9.0203	9.2458	9.4574
0.4	v, m³/kg	0.4708	0.5342	0.5951	0.6548	0.7139	0.7726	0.8311	0.8893	0.9475	1.006	1.121	1.237	1.353
(143.6)	u, kJ/kg	2564.5	2646.8	2726.1	2804.8	2884.0	2964.4	3046.0	3129.2	3213.9	3300.2	3477.9	3662.5	3853.9
	h, kJ/kg	2752.8	2860.5	2964.2	3066.7	3169.6	3273.4	3378.4	3484.9	3592.9	3702.4	3926.5	4157.4	4395.1
	s, kJ/(kg · K)	6.9307	7.1714	7.3797	7.5670	7.7390	7.8992	8.0497	8.1921	8.3274	8.4566	8.6995	8.9253	9.1370
0.6	v, m³/kg		0.3520	0.3938	0.4344	0.4742	0.5137	0.5529	0.5920	0.6309	0.6697	0.7472	0.8245	0.9017
(158.9)	u, kJ/kg		2638.9	2720.9	2801.0	2881.1	2962.0	3044.1	3127.6	3212.5	3299.1	3477.1	3661.8	3853.3
	h, kJ/kg		2850.1	2957.2	3061.6	3165.7	3270.2	3375.9	3482.7	3591.1	3700.9	3925.4	4156.5	4394.4
	s, kJ/(kg · K)		6.9673	7.1824	7.3732	7.5472	7.7086	7.8600	8.0029	8.1386	8.2682	8.5115	8.7375	8.9494
0.8	v, m³/kg		0.2608	0.2931	0.3241	0.3544	0.3843	0.4139	0.4433	0.4726	0.5018	0.5601	0.6181	0.6761
(170.4)	u, kJ/kg		2630.6	2715.5	2797.1	2878.2	2959.7	3042.2	3125.9	3211.2	3297.9	3476.2	3661.1	3852.8
	h, kJ/kg		2839.2	2950.0	3056.4	3161.7	3267.1	3373.3	3480.6	3589.3	3699.4	3924.3	4155.7	4393.6
	s, kJ/(kg · K)		6.8167	7.0392	7.2336	7.4097	7.5723	7.7245	7.8680	8.0042	8.1341	8.3779	8.6041	8.8161
1	v, m³/kg		0.2060	0.2327	0.2579	0.2825	0.3066	0.3304	0.3541	0.3776	0.4011	0.4478	0.4943	0.5407
(179.9)	u, kJ/kg		2621.9	2709.9	2793.2	2875.2	2957.3	3040.2	3124.3	3209.8	3296.8	3475.4	3660.5	3852.2
	h, kJ/kg		2827.9	2942.6	3051.2	3157.7	3263.9	3370.7	3478.4	3587.5	3697.9	3923.1	4154.8	4392.9
	s, kJ/(kg · K)		6.6948	6.9255	7.1237	7.3019	7.4658	7.6188	7.7630	7.8996	8.0298	8.2740	8.5005	8.7127
1.5	v, m³/kg		0.1325	0.1520	0.1697	0.1866	0.2030	0.2192	0.2352	0.2510	0.2668	0.2981	0.3292	0.3603
(198.3)	u, kJ/kg		2598.1	2695.3	2783.1	2867.6	2951.3	3035.3	3120.3	3206.4	3293.9	3473.2	3658.7	3850.8
	h, kJ/kg		2796.8	2923.2	3037.6	3147.4	3255.8	3364.1	3473.0	3582.9	3694.0	3920.3	4152.6	4391.2
	s, kJ/(kg · K)		6.4554	6.7098	6.9187	7.1025	7.2697	7.4249	7.5706	7.7083	7.8393	8.0846	8.3118	8.5243

TABLE 7.6 Properties of Superheated Water Vapor (*Continued*)

P, MPa (T_{sat}, °C)		Temperature, °C 250	300	350	400	450	500	550	600	650	700	750	800	900
2	v, m³/kg	0.1114	0.1255	0.1386	0.1512	0.1635	0.1757	0.1877	0.1996	0.2114	0.2232	0.2350	0.2467	0.2700
(212.4)	u, kJ/kg	2679.6	2772.6	2859.8	2945.2	3030.4	3116.2	3203.0	3290.9	3380.2	3471.0	3563.2	3657.0	3849.3
	h, kJ/kg	2902.5	3023.5	3137.0	3247.6	3357.5	3467.6	3578.3	3690.1	3803.1	3917.5	4033.2	4150.4	4389.4
	s, kJ/(kg · K)	6.5461	6.7672	6.9571	7.1279	7.2853	7.4325	7.5713	7.7032	7.8290	7.9496	8.0656	8.1774	8.3903
3	v, m³/kg	0.07058	0.08114	0.09053	0.09936	0.1079	0.1162	0.1244	0.1324	0.1404	0.1484	0.1563	0.1641	0.1798
(233.9)	u, kJ/kg	2644.0	2750.0	2843.7	2932.7	3020.4	3107.9	3196.0	3285.0	3375.2	3466.6	3559.4	3653.6	3846.5
	h, kJ/kg	2855.8	2993.5	3115.3	3230.8	3344.0	3456.5	3569.1	3682.3	3796.5	3911.7	4028.2	4146.0	4385.9
	s, kJ/(kg · K)	6.2880	6.5398	6.7436	6.9220	7.0842	7.2346	7.3757	7.5093	7.6364	7.7580	7.8747	7.9871	8.2008
4	v, m³/kg		0.05884	0.06645	0.07341	0.08003	0.08643	0.09269	0.09885	0.1049	0.1109	0.1169	0.1229	0.1347
(250.4)	u, kJ/kg		2725.3	2826.6	2919.9	3010.1	3099.5	3189.0	3279.1	3370.1	3462.1	3555.5	3650.1	3843.6
	h, kJ/kg		2960.7	3092.4	3213.5	3330.2	3445.2	3559.7	3674.4	3789.8	3905.9	4023.2	4141.6	4382.3
	s, kJ/(kg · K)		6.3622	6.5828	6.7698	6.9371	7.0908	7.2343	7.3696	7.4981	7.6206	7.7381	7.8511	8.0655
6	v, m³/kg		0.03616	0.04223	0.04739	0.05214	0.05665	0.06101	0.06525	0.06942	0.07352	0.07758	0.08160	0.08958
(275.6)	u, kJ/kg		2667.2	2789.6	2892.8	2988.9	3082.2	3174.6	3266.9	3359.6	3453.2	3547.6	3643.1	3837.8
	h, kJ/kg		2884.2	3043.0	3177.2	3301.8	3422.1	3540.6	3658.4	3776.2	3894.3	4013.1	4132.7	4375.3
	s, kJ/(kg · K)		6.0682	6.3342	6.5415	6.7201	6.8811	7.0296	7.1685	7.2996	7.4242	7.5433	7.6575	7.8735
8	v, m³/kg		0.02426	0.02995	0.03432	0.03817	0.04175	0.04516	0.04845	0.05166	0.05481	0.05791	0.06097	0.06702
(295.1)	u, kJ/kg		2590.9	2747.7	2863.8	2966.7	3064.3	3159.8	3254.4	3349.0	3444.0	3539.6	3636.1	3832.1
	h, kJ/kg		2785.0	2987.3	3138.3	3272.0	3398.3	3521.0	3642.0	3762.3	3882.5	4002.9	4123.8	4368.3
	s, kJ/(kg · K)		5.7914	6.1309	6.3642	6.5559	6.7248	6.8786	7.0214	7.1553	7.2821	7.4027	7.5182	7.7359
10	v, m³/kg			0.02242	0.02641	0.02975	0.03279	0.03564	0.03837	0.04101	0.04358	0.04611	0.04859	0.05349
(311.1)	u, kJ/kg			2699.2	2832.4	2943.3	3045.8	3144.5	3241.7	3338.2	3434.7	3531.5	3629.0	3826.3
	h, kJ/kg			2923.4	3096.5	3240.8	3373.6	3500.9	3625.3	3748.3	3870.5	3992.6	4114.9	4361.2
	s, kJ/(kg · K)			5.9451	6.2127	6.4197	6.5974	6.7569	6.9037	7.0406	7.1696	7.2919	7.4086	7.6280
12	v, m³/kg			0.01721	0.02108	0.02412	0.02680	0.02929	0.03164	0.03390	0.03610	0.03824	0.04034	0.04447
(324.8)	u, kJ/kg			2641.1	2798.3	2918.8	3026.6	3128.9	3228.7	3327.2	3425.3	3523.4	3621.8	3820.6
	h, kJ/kg			2847.6	3051.2	3208.2	3348.2	3480.3	3608.3	3734.0	3858.4	3982.3	4105.9	4354.2
	s, kJ/(kg · K)			5.7604	6.0754	6.3006	6.4879	6.6535	6.8045	6.9445	7.0757	7.1998	7.3178	7.5390

P, MPa (T_{sat}, °C)		Temperature, °C												
		400	450	500	550	600	650	700	750	800	850	900	950	1000
15	v, m³/kg	0.01565	0.01845	0.02080	0.02293	0.02491	0.02680	0.02861	0.03037	0.03210	0.03379	0.03546	0.03711	0.03875
(342.2)	u, kJ/kg	2740.7	2879.5	2996.5	3104.7	3208.6	3310.4	3410.9	3511.0	3611.0	3711.2	3811.9	3913.2	4015.4
	h, kJ/kg	2975.4	3156.2	3308.5	3448.6	3582.3	3712.3	3840.1	3966.6	4092.4	4218.0	4343.8	4469.9	4596.6
	s, kJ/(kg · K)	5.8819	6.1412	6.3451	6.5207	6.6784	6.8232	6.9580	7.0848	7.2048	7.3192	7.4288	7.5340	7.6356
20	v, m³/kg	0.00994	0.01270	0.01477	0.01656	0.01818	0.1969	0.02113	0.02251	0.02385	0.02516	0.02645	0.02771	0.02897
(365.8)	u, kJ/kg	2619.2	2806.2	2942.8	3062.3	3174.0	3281.5	3386.5	3490.0	3592.7	3695.1	3797.4	3900.0	4003.1
	h, kJ/kg	2818.1	3060.1	3238.2	3393.4	3537.6	3675.3	3809.1	3940.3	4069.8	4198.3	4326.4	4454.3	4582.5
	s, kJ/(kg · K)	5.5548	5.9025	6.1409	6.3356	6.5056	6.6591	6.8002	6.9317	7.0553	7.1723	7.2839	7.3907	7.4933
22.088	v, m³/kg	0.00818	0.01104	0.01305	0.01475	0.01627	0.01768	0.01901	0.02029	0.02152	0.02272	0.02389	0.02505	0.02619
(374.136)	u, kJ/kg	2552.9	2772.1	2919.0	3043.9	3159.1	3269.1	3376.1	3481.1	3585.0	3688.3	3791.4	3894.5	3998.0
	h, kJ/kg	2733.7	3015.9	3207.2	3369.6	3518.4	3659.6	3796.0	3929.2	4060.3	4190.1	4319.1	4447.9	4576.6
	s, kJ/(kg · K)	5.4013	5.8072	6.0634	6.2670	6.4426	6.5998	6.7437	6.8772	7.0024	7.1206	7.2330	7.3404	7.4436
30	v, m³/kg	0.00279	0.00674	0.00868	0.01017	0.01145	0.01260	0.01366	0.01466	0.01562	0.01655	0.01745	0.01833	0.01920
	u, kJ/kg	2067.3	2619.3	2820.7	2970.3	3100.5	3221.0	3335.8	3447.0	3555.6	3662.6	3768.5	3873.8	3978.8
	h, kJ/kg	2151.0	2821.4	3081.0	3275.4	3443.9	3598.9	3745.7	3886.9	4024.3	4159.0	4291.9	4423.6	4554.7
	s, kJ/(kg · K)	4.4736	5.4432	5.7912	6.0350	6.2339	6.4066	6.5614	6.7030	6.8341	6.9568	7.0726	7.1825	7.2875
40	v, m³/kg	0.00191	0.00369	0.00562	0.00698	0.00809	0.00906	0.00994	0.01076	0.01152	0.01226	0.01296	0.01365	0.01432
	u, kJ/kg	1854.5	2365.1	2678.4	2869.7	3022.6	3158.0	3283.6	3402.9	3517.9	3629.8	3739.4	3847.5	3954.6
	h, kJ/kg	1930.8	2512.8	2903.3	3149.1	3346.4	3520.6	3681.3	3833.1	3978.8	4120.0	4257.9	4393.6	4527.6
	s, kJ/(kg · K)	4.1143	4.9467	5.4707	5.7793	6.0122	6.2063	6.3759	6.5281	6.6671	6.7957	6.9158	7.0291	7.1365
60	v, m³/kg	0.00163	0.00208	0.00296	0.00396	0.00483	0.00560	0.00627	0.00689	0.00746	0.00800	0.00851	0.00900	0.00948
	u, kJ/kg	1745.3	2053.9	2390.5	2658.8	2861.1	3028.8	3177.2	3313.6	3441.6	3563.6	3681.0	3795.0	3906.4
	h, kJ/kg	1843.4	2179.0	2567.9	2896.2	3151.2	3364.5	3553.6	3726.8	3889.1	4043.3	4191.5	4335.0	4475.2
	s, kJ/(kg · K)	3.9325	4.4128	4.9329	5.3449	5.6460	5.8838	6.0832	6.2569	6.4118	6.5523	6.6814	6.8012	6.9135
80	v, m³/kg	0.00152	0.00177	0.00219	0.00276	0.00339	0.00398	0.00452	0.00502	0.00548	0.00591	0.00632	0.00671	0.00709
	u, kJ/kg	1687.0	1944.9	2218.9	2483.9	2711.8	2904.7	3073.2	3225.3	3365.7	3497.3	3622.3	3742.1	3857.8
	h, kJ/kg	1808.3	2086.9	2393.9	2704.9	2982.7	3222.8	3434.7	3626.6	3803.8	3970.1	4127.9	4279.1	4425.2
	s, kJ/(kg · K)	3.8338	4.2328	4.6432	5.0331	5.3609	5.6284	5.8521	6.0445	6.2137	6.3652	6.5026	6.6289	6.7459

Source: From Reynolds and Perkins.[1]

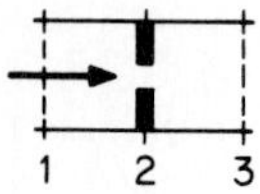

FIGURE 7.6

section of the nozzle is called its *throat*, and the pressure at the throat is the critical flow pressure p_m. If the nozzle is cut off at the throat with no diverging section and the pressure at the discharge end is progressively decreased, with fixed inlet pressure, the amount of fluid passing increases until the discharge pressure equals the critical pressure, but a further decrease in discharge pressure does not result in increased flow. For gases $p_m/p_1 \simeq 0.53$, for saturated steam $p_m/p_1 \simeq 0.575$, and for moderately superheated steam $p_m/p_1 \simeq 0.55$.

For orifice computation (see Fig. 7.6), the general relation from the first law gives

$$\frac{V_2^2 - V_1^2}{2} = h_1 - h_2 \tag{7.64}$$

where section 2 is at the orifice and section 3 is beyond the orifice. Then

$$V_2 = \frac{C\sqrt{2(h_1 - h_2)}}{\sqrt{1 - (A_2/A_1)^2 (v_1/v_2)^2}} \tag{7.65}$$

where coefficient of discharge $C \simeq 0.50$ to 0.60 (for steam nozzles this may be as high as 0.95). The volume flow is V_2A_2 and mass flow is $V_2A_2\rho$.

For an ideal gas, assuming reversible adiabatic expansion through the orifice,

$$V_2 = \frac{C\sqrt{2p_1v_1[k/(k-1)][1 - (p_2/p_1)^{k-1/k}]}}{\sqrt{1 - (A_2/A_1)^2 (p_2/p_1)^{2/k}}} \tag{7.66}$$

Throttling. When a fluid flows from a region of higher pressure into a region of lower pressure through a valve or similar constricted passage, it is said to be *throttled* or *wiredrawn*. Equation (7.64) is applicable to the throttling process and since V_2V_1 are practically equal,

$$h_1 = h_2 \tag{7.67}$$

i.e., in a throttling process there is no change in enthalpy.

For a mixture of liquid and vapor, the equation of throttling is

$$h_{f_1} + x_1 \cdot h_{fg_1} = h_{f_2} + x_2 \cdot h_{fg_2} \tag{7.68}$$

POWER CYCLES*

1. Heat Engine. According to the first law of thermodynamics, when a system undergoes a complete cycle, the net heat supplied is equal to the net work done. This is based on the conservation of energy principle, which follows from the observation of natural events. The second law of thermodynamics, which is also a

*This section is based in part on *Engineering Thermodynamics*, 2d ed., by W. C. Reynolds and H. C. Perkins, copyright © 1977; in part on *Engineering Manual*, 3d ed., by R. H. Perry, copyright © 1976; in part on *Perry's Chemical Engineers' Handbook*, 6th ed., by R. H. Perry and D. W. Green, copyright

natural law, indicates that, although the net heat supplied in a cycle is equal to the net work done, the gross heat supplied must be greater than the net work done [see Eq. (7.11)]; some heat must always be rejected by the system. To explain and analyze the second law and analyze power cycles, this chapter next discusses heat engine.

A heat engine is a system operating in a complete cycle and developing net work from a supply of heat. The second law implies that a source of heat supply and the rejection of heat are both necessary, since some heat must always be rejected by the system. Therefore, it is impossible for a heat engine to produce net work in a complete cycle if it exchanges heat only with bodies (reservoirs) at a single fixed temperature. A diagrammatic representation of a heat engine is shown in Fig. 7.7. The first and second laws apply equally well to cycles working in the reverse direction of those applicable to a heat engine. In the case of a reversed cycle, net work is done on the system which is equal to the net heat rejected by the system. Such cycles occur in heat pumps and refrigerators. A diagrammatic representation of a heat pump (or refrigerator) is shown in Fig. 7.8.

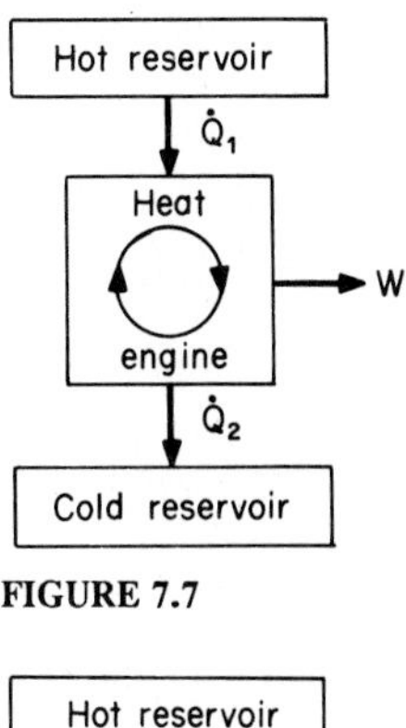

FIGURE 7.7

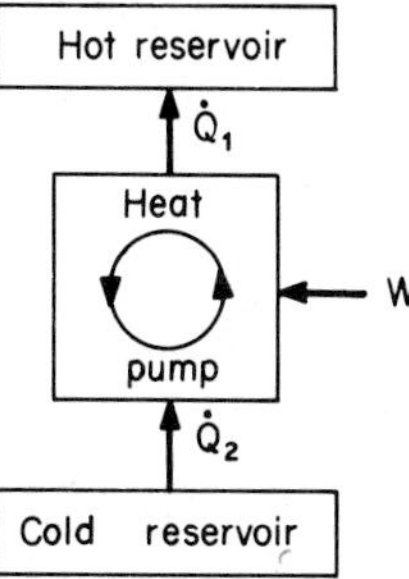

FIGURE 7.8

2. Ideal Gas Cycles. The analysis of real power cycles can often be approximated by idealized cycles, using ideal gases as working fluids. Several such approximations are of interest.

Carnot Cycle. This cycle consists of four processes, two isothermal and two isentropic, as shown in Fig. 7.9. A detailed analysis of the cycle shows that

$$q_{\text{net}} = w_{\text{net}} = (T_3 - T_4)R \ln \frac{p_2}{p_3}$$

$$q_{1\text{-}2} = q_{3\text{-}4} = 0$$

$$q_{2\text{-}3} = w_{2\text{-}3} = p_2 v_2 \ln \frac{p_2}{p_3}$$

$$w_{1\text{-}2} = \bar{c}_v(T_2 - T_1)$$

$$w_{3\text{-}4} = \bar{c}_v(T_4 - T_3)$$

$$u_3 - u_2 = 0 \qquad u_1 - u_4 = 0$$

© 1984; and in part on *Marks' Standard Handbook for Mechanical Engineers*, 9th ed., by E. A. Avallone and T. Baumeister III (eds.), copyright © 1987. All material used by permission of McGraw-Hill, Inc. All rights reserved. The section is also based in part on "Heat Transfer Characteristics of Working Fluids for OTEC," by E. N. Ganić, copyright © 1980 by ASME. Used by permission of ASME. All rights reserved.

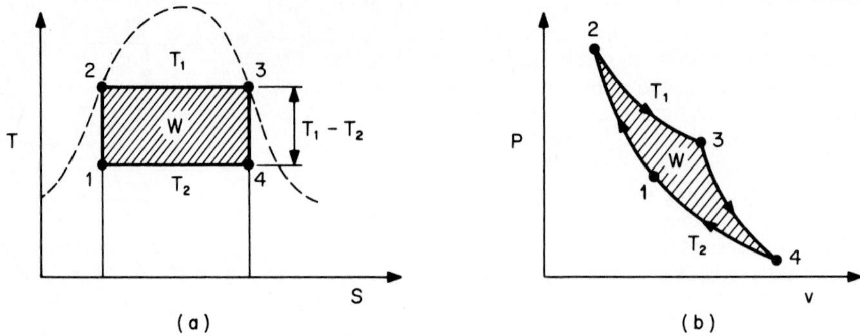

FIGURE 7.9 The Carnot cycle. (*a*) T-S diagram, (*b*) p-v diagram.

$$q_{1\text{-}4} = w_{4\text{-}1} = p_4 v_4 \ln \frac{p_4}{p_1}$$

$$\eta = \frac{T_1 - T_2}{T_1} = \frac{T_3 - T_4}{T_3}$$

Otto Cycle. This cycle consists of two isentropic and two constant-volume processes, as shown in Fig. 7.10, and is often used as a representation of a spark-ignition engine.

Diesel Cycle. This cycle consists of two isentropic, one constant-volume, and one constant-pressure process, as shown in Fig. 7.11. It is representative of a diesel engine.

Brayton Cycle. This cycle consists of two isentropic and two constant-pressure processes. It is representative of the gas turbine, and is shown in Fig. 7.12.

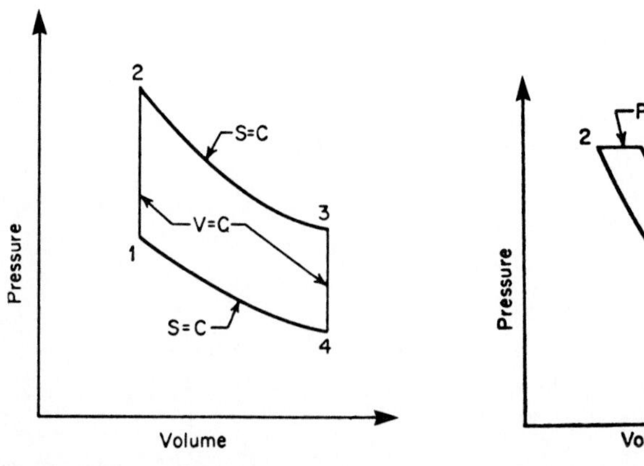

FIGURE 7.10 Otto cycle. (*From Perry.*[2])

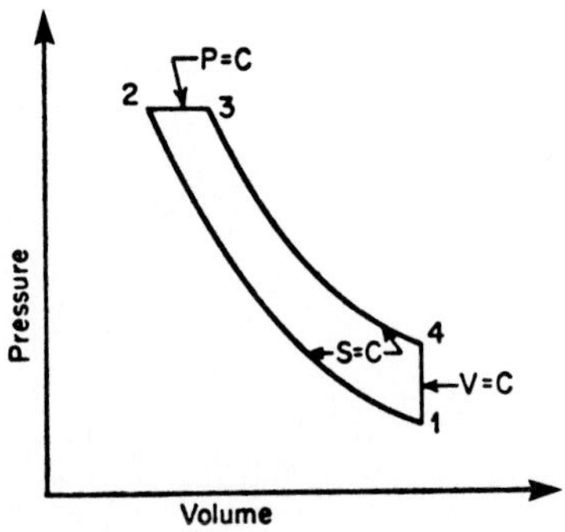

FIGURE 7.11 Diesel cycle. (*From Perry.*[2])

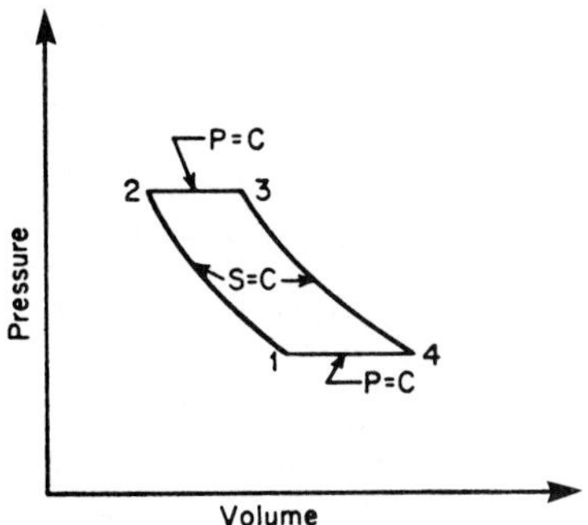

FIGURE 7.12 Brayton cycle. (*From Perry.*[2])

Stirling Cycle. The stirling cycle consists of two isothermal and two constant-volume processes. Practical examples of this cycle have been developed recently, and depend on regeneration to achieve practical efficiencies. Figure 7.13 shows this cycle. Practical cycles use the heat rejected in process 4-1 to partly regenerate the gas during process 2-3. Cycle analysis must depend on the degree of regeneration.

3. Vapor Cycles. Analyses of vapor cycles are done using Figs. 7.14 through 7.16 and Tables 7.4 through 7.6.

Rankine Cycle. The Rankine cycle is the ideal representation for the vapor power cycle. It consists of five processes—two isothermal, two isentropic, and one constant-pressure. The cycle is shown in Fig. 7.17 and the corresponding apparatus diagram in Fig. 7.18.

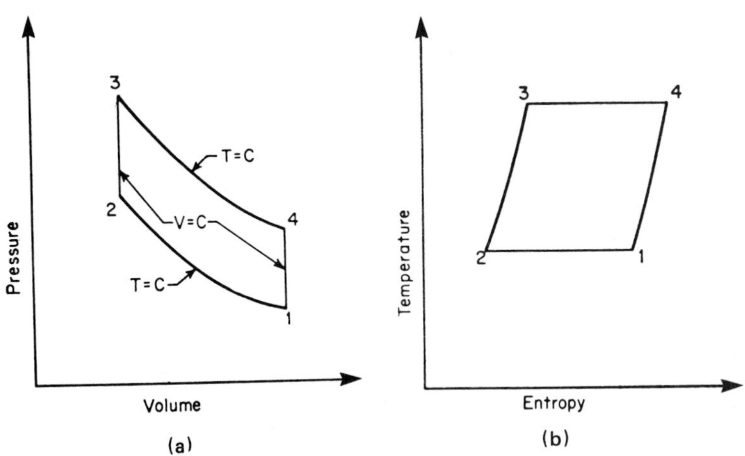

FIGURE 7.13 Stirling cycle. (*a*) p-v diagram; (*b*) T-S diagram. (*From Perry.*[2])

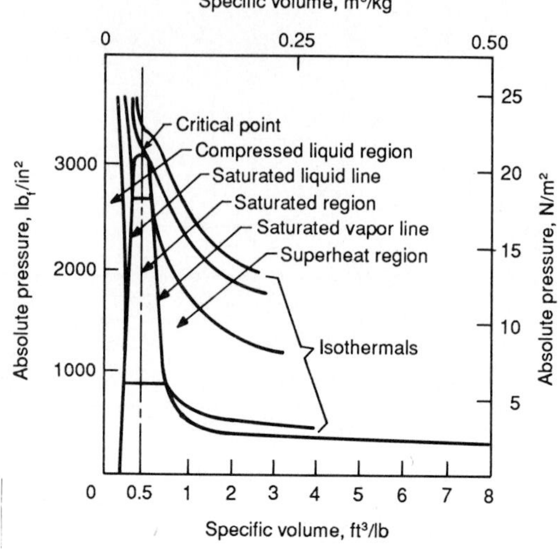

FIGURE 7.14 p-v diagram, water and water vapor.

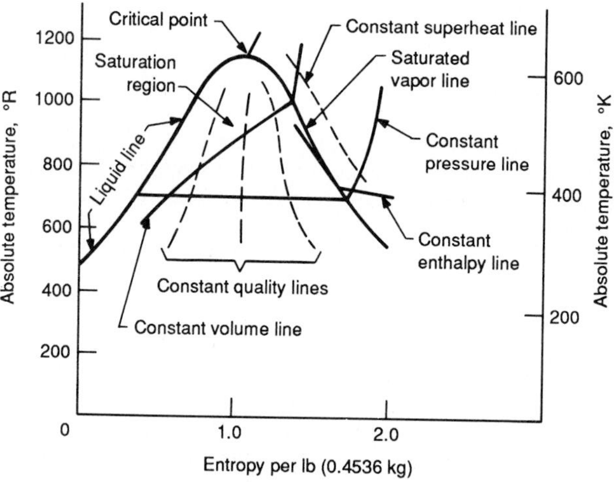

FIGURE 7.15 T-S diagram, water and water vapor.

The results of a cycle analysis show

$$q_{2\text{-}3} = h_4 - h_2$$

$$q_{5\text{-}1} = h_1 - h_5$$

$$w_{4\text{-}5} = h_5 - h_4$$

$$w_{1\text{-}2} = h_1 - h_2$$

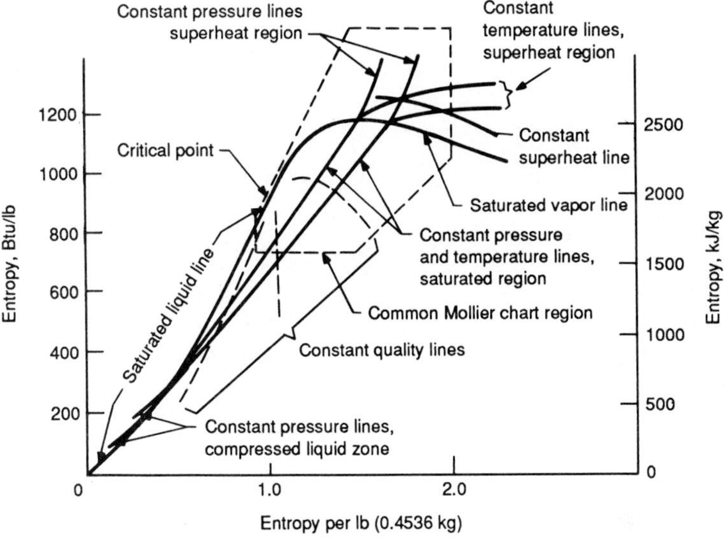

FIGURE 7.16 H-S diagram, water and water vapor.

$$\eta = \frac{(h_5 - h_4) + w_{1\text{-}2}}{h_4 - h_2}$$

Compression Refrigeration Cycle. The idealized compression refrigeration cycle consists of two constant-pressure processes, an isentropic process, and an irreversible throttling process. The cycle is shown in Fig. 7.19 and the corresponding apparatus in Fig. 7.20.

Note that C in Figs. 7.10 through 7.13 and 7.17 through 7.20 represent the constant property line.

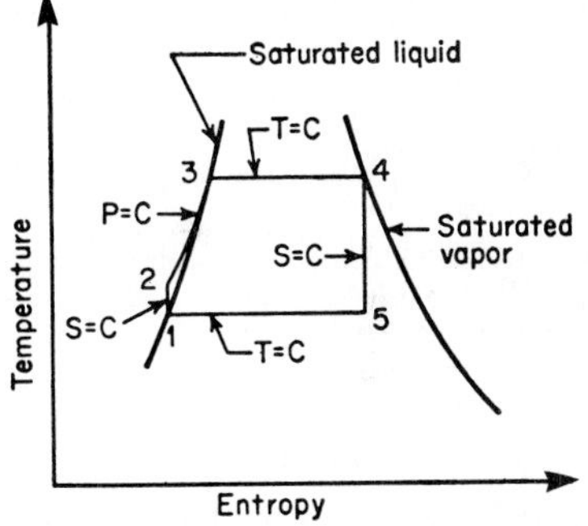

FIGURE 7.17 Rankine cycle. (*From Perry.*[2])

4. Working Fluid Selection. Different fluids are used as the "working fluid" in various power cycles and other energy conversion cycles. The most desirable properties of a working fluid are:

1. Large enthalpy of evaporation—to minimize the mass-flow rate for a given power output
2. High critical temperature—to permit evaporation at a high temperature
3. Low saturation pressures at the maximum temperatures—to minimize pressure vessel and piping costs
4. Pressure just above 1 atm at condensing temperature—to eliminate air leakage problems
5. Rapidly diverging pressure lines on the *h-s* diagram—to minimize reheating

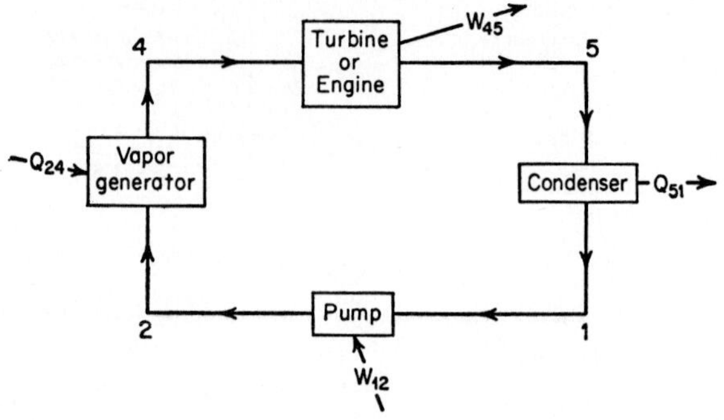

FIGURE 7.18 Vapor power system. (*From Perry.*[2])

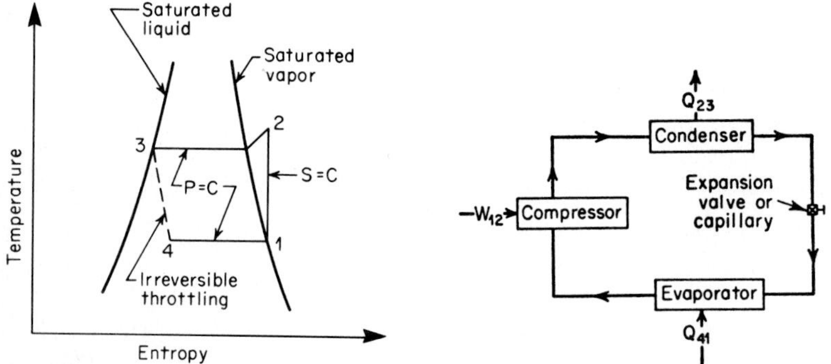

FIGURE 7.19 Compression-refrigeration cycle. (*From Perry.*[2])

FIGURE 7.20 Compression-refrigeration system. (*From Perry.*[2])

6. No degrading aspects—to prevent corrosion and clogging
7. No hazardous features—to prevent toxicity and flammability
8. Low cost and ready availability

Note that water is rather good only in regard to items 1, 5, 7, and 8. Other fluids are better for other situations. However, water remains a predominant choice for ground-based power plants.

NOMENCLATURE

Symbol = *definition, SI units (U.S. Customary units)*

c_v = specific heat at constant volume, J/kg · K (Btu/lb · °F)

c_p = specific heat at constant pressure, J/kg · K (Btu/lb · °F)

g = acceleration due to gravity, m/s^2 (ft/s^2)

h = enthalpy per unit mass, J/kg (Btu/lb)

H = enthalpy, J (Btu)
k = c_p/c_v
m = mass, kg (lb)
$\dot{m}$ = mass flow rate, kg/s (lb/s)
p = pressure, N/m^2, Pa (lb$_f$/ft^2)
Q = heat, J (Btu)
$\dot{Q}$ = heat rate, J/s (Btu/s)
q = heat per unit mass, J/kg (Btu/lb)
R = gas constant, J/kg · K (Btu/lb · °R)
R_m = universal gas constant, J/mol · K (Btu/mol · °R)
s = entropy per unit mass J/kg · K (Btu/lb · °R)
S = entropy, J/K (Btu/°R)
T = temperature, K (°R)
t = temperature, °C (°F)
u = internal energy per unit mass, J/kg (Btu/lb)
U = internal energy, J (Btu)
v = specific volume, m^3/kg (ft^3/lb)
$\mathbf{V}$ = volume, m^3 (ft^3)
$\mathbf{V}_m$ = molar volume, m^3/mol (ft^3/mol)
V = velocity, m/s (ft/s)
W = work, J (Btu)
w = work per unit mass, J/kg (Btu/lb)
W = work per unit time, J/s (Btu/s)
z = distance, m (ft)

Greek

ρ = density, kg/m^3 (lb/ft^3)
Δ = finite change
η = thermal efficiency

Subscripts

g = gas (vapor)
g = dry gas (in "Clapeyron Equation" subsection)
f = liquid
fg = $f - g$, or change from liquid to gas
p = at constant p
v = at constant v
s = at constant s
v = vapor (in "Clapeyron Equation" subsection)

T = at constant T
rev = reversible
sat = saturation
1 = state 1
2 = state 2
1-2 = change from 1 to 2

Superscript

$\bar{}$ = mean value

REFERENCES*

1. W. C. Reynolds and H. C. Perkins, *Engineering Thermodynamics*, 2d ed., McGraw-Hill, New York, 1977.
2. R. H. Perry, *Engineering Manual*, 3d ed., McGraw-Hill, New York, 1976.
3. R. H. Perry and D. W. Green, *Perry's Chemical Engineers' Handbook*, 6th ed., McGraw-Hill, New York, 1984.
4. E. A. Avallone and T. Baumeister III (eds.), *Mark's Standard Handbook for Mechanical Engineers*, 9th ed., McGraw-Hill, New York, 1987.
5. E. N. Ganić, "Heat Transfer Characteristics of Working Fluids for OTEC," *ASME Publication HTD*, vol. 12, pp. 55–66, 1980.
6. K. Gieck, *Engineering Formulas*, 5th ed., McGraw-Hill, New York, 1986.
7. E. G. Cravalo and J. L. Smith, Jr., *Engineering Thermodynamics*, Pitman, Boston, 1981.
8. T. D. Eastop and A. McConkey, *Applied Thermodynamics for Engineering Technologists*, 3d ed., Longman, London, 1978.
9. O. W. Eshbach and M. Sounder, *Handbook of Engineering Fundamentals*, 3d ed., John Wiley & Sons, New York, 1975.
10. V. M. Faires and C. M. Simmang, *Thermodynamics*, 6th ed., Macmillan, New York, 1978.
11. H. Gartmann, *De Laval Engineering Handbook*, 3d ed., McGraw-Hill, New York, 1970.
12. J. H. Keenan, *Thermodynamics*, MIT Press, Cambridge, Mass., 1970.
13. C. E. O'Rourke, *General Engineering Handbook*, 2d ed., McGraw-Hill, New York, 1940.
14. A. Shapiro, *Dynamics and Thermodynamics of Compressible Fluid Flow*, Ronald Press, New York, 1953.
15. M. W. Zemansky, *Heat and Thermodynamics*, 5th ed., McGraw-Hill, New York, 1968.
16. M. W. Zemansky, M. M. Abbott, and H. C. Van Ness, *Basic Engineering Thermodynamics*, 2d ed., McGraw-Hill, New York, 1975.

*Those references listed above but not cited in the text were used for comparison between different data sources, clarification, clarity of presentation, and, most important, reader's convenience when further interest in the subject exists.

CHAPTER 8
MECHANICS OF FLUIDS

NATURE OF FLUIDS

Fluid is a substance that deforms continuously when subjected to a shear stress, no matter how small that shear stress may be. A *shear force* is the force component tangent to a surface, and this force divided by the area of the surface is the average shear stress over the area. Shear stress at a point is a limiting value of shear force to area as the area is reduced to the point.

The relation between the shear stress τ and the rate of angular deformation for one-dimensional flow of a fluid du/dy is given as

$$\tau = \mu \frac{du}{dy} \tag{8.1}$$

where the proportionality factor μ is the viscosity of fluid.

The viscosity of a gas increases with temperature, but the viscosity of a liquid decreases with temperature. For ordinary pressures, viscosity is independent of pressure and depends on temperature only. For very great pressures, gases and most liquids have shown erratic variations of viscosity with pressure. The viscosity μ is frequently referred to as the *absolute viscosity* or the *dynamic viscosity* to avoid confusing it with the *kinematic viscosity* ν, which is the ratio of viscosity to mass density:

$$\nu = \frac{\mu}{\rho} \tag{8.2}$$

Fluids may be classified as newtonian or nonnewtonian. In a *newtonian fluid*, there is a linear relation between the magnitude of applied shear stress and the resulting rate of deformation [μ constant in Eq. (8.1)], as shown in Fig. 8.1. In a *nonnewtonian fluid*, there is a nonlinear relation between the magnitude of applied shear stress and the rate of angular deformation.

An *ideal plastic* has a definite yield stress and a constant linear relation of τ to du/dy. A *thixotropic* substance, such as printer's ink, has a viscosity that is dependent on the immediately prior angular deformation of the substance and has a tendency to take a set when at rest. Gases and liquids tend to be newtonian fluids, whereas thick, long-chained hydrocarbons may be nonnewtonian.

For purposes of analysis, the assumption is frequently made that a fluid is nonviscous (inviscid). With zero viscosity, the shear stress is always zero, re-

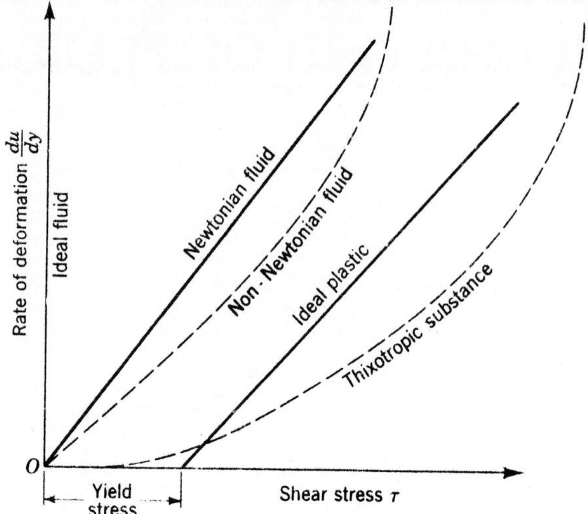

FIGURE 8.1 Deformation characteristics of substances. (*From Streeter and Wylie,[1] p. 5.*)

gardless of the motion of the liquid. If the fluid is also considered to be incompressible, it is then called an *ideal fluid* and plots as the ordinate in Fig. 8.1.

The properties of density and viscosity play principal roles in open- and closed-channel flow and in flow around immersed objects. Surface-tension effects are important in the formation of droplets, in the flow of small jets, and in situations where liquid-gas-solid or liquid-liquid-solid interfaces occur, as well as in the formation of capillary waves. Values of density and surface tension for different fluids are given in Chap. 2.

Capillary attraction is caused by surface tension and by the relative value of adhesion between liquid and solid to cohesion of the liquid. A liquid that wets the solid has greater adhesion than cohesion. The action of surface tension in this case is to cause the liquid to rise within a small vertical tube that is partially immersed in it. For liquids that do not wet the solid, surface tension tends to depress the meniscus in a small vertical tube. When the contact angle between liquid and solids is known, the capillary rise can be computed for an assumed shape of meniscus. Figure 8.2 shows the capillary rise for water and mercury in circular glass tubes in air.

If the interface is curved, a mechanical balance shows that there is a pressure difference across the interface, the pressure being higher on the concave side. This is illustrated in Fig. 8.3. The pressure increase in the interior of a spherical droplet balances a ring of surface-tension force

$$\pi r^2 \Delta p = 2\pi r \sigma \tag{8.3}$$

or

$$\Delta p = \frac{2\sigma}{r} \tag{8.4}$$

An important surface effect is also the contact angle θ that appears when a liquid interface intersects with a solid surface, as in Fig. 8.4. The force balance

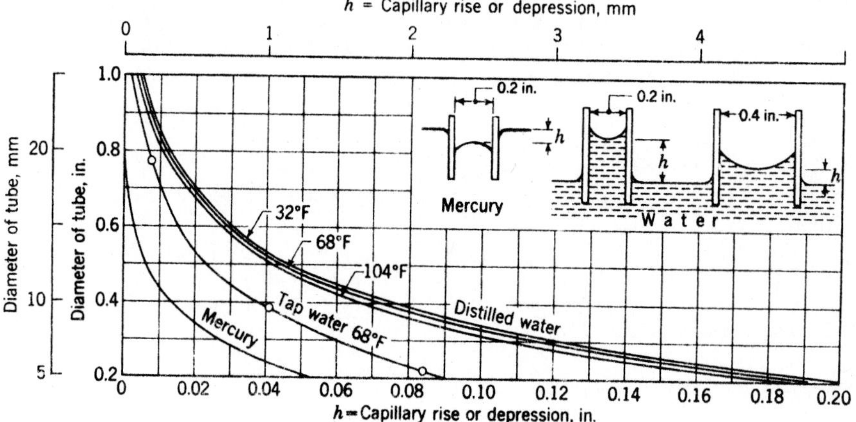

FIGURE 8.2 Capillarity in circular glass tubes. (*From Streeter and Wylie,*[1] *p. 18.*)

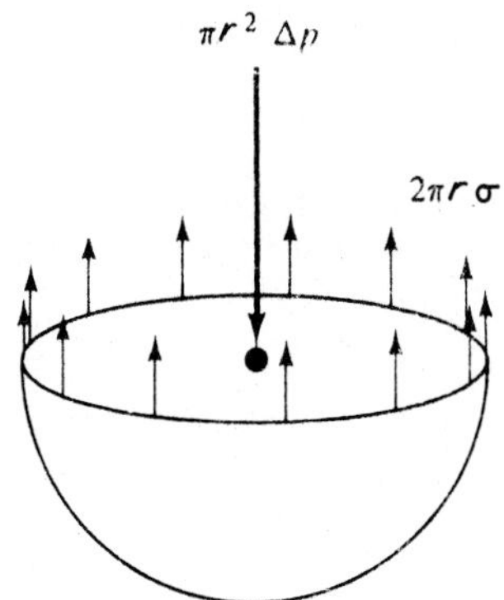

FIGURE 8.3 Pressure change across a curved interface resulting from surface tension. (*From White,*[2] *p. 32.*)

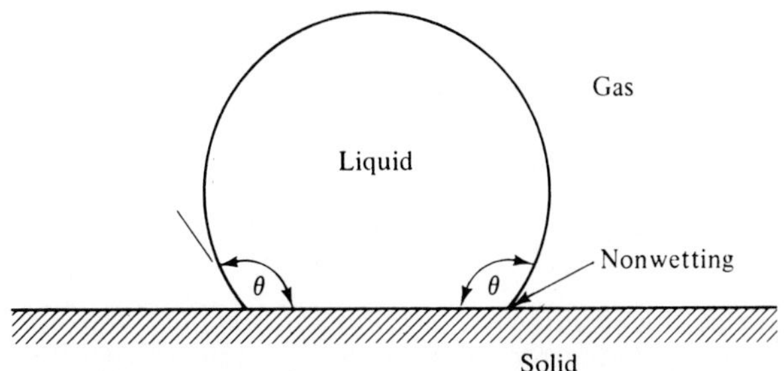

FIGURE 8.4 Contact-angle effects at a liquid-gas-solid interface. (*From White,*[2] *p. 33.*)

would then involve both σ and θ. If the contact angle is less than 90°, the liquid is said to *wet* the solid; if $\theta > 90°$, the liquid is termed *nonwetting*.

The *specific weight* γ of a substance is its weight per unit volume. It changes with location, that is,

$$\gamma = \rho g \tag{8.5}$$

depending on gravity. It is a convenient property when dealing with fluid statics or with liquids with a free surface.

Specific gravity (sp gr) of a substance is the ratio of its weight to the weight of an equal volume of water at standard conditions. It also may be expressed as a ratio of a substance's density or specific weight to that of water.

FLUID STATICS

1. Introduction. Fluid statics basically treats pressure and its variations throughout a fluid and pressure forces on finite surfaces. The average pressure is calculated by dividing the normal force pushing against a plane area by the area. The pressure at a point is the limit of the ratio of normal force to area as the area approaches zero size at the point. At a point, a fluid at rest has the same pressure in all directions.

The fluid static law of variation of pressure is given by the relation

$$\nabla p = \rho g \tag{8.6}$$

In component form (assuming that coordinate z is "up"), this equation becomes

$$\frac{\partial p}{\partial x} = 0 \qquad \frac{\partial p}{\partial y} = 0 \qquad \frac{\partial p}{\partial z} = -\rho g$$

Since pressure is a function of z, only

$$\frac{dp}{dz} = -\rho g \tag{8.7}$$

or

$$p_2 - p_1 = -\int_1^2 \rho g \, dz \tag{8.8}$$

Equation (8.8) is the solution to the hydrostatic problem.

2. Hydrostatic Pressure in Liquids. Assuming constant density in liquid hydrostatic calculation, Eq. (8.8) becomes

$$p_2 - p_1 = -\rho g(z_2 - z_1) \tag{8.9}$$

or

$$z_1 - z_2 = \frac{p_2}{\rho g} - \frac{p_1}{\rho g} \tag{8.10}$$

The quantity $p/\rho g$ is a length called the *pressure head* of the liquid.

For oceans and atmospheres, the coordinate system is usually chosen as in Fig. 8.5 with $z = 0$ at the free surface, where p equals the surface atmospheric pressure p_a.

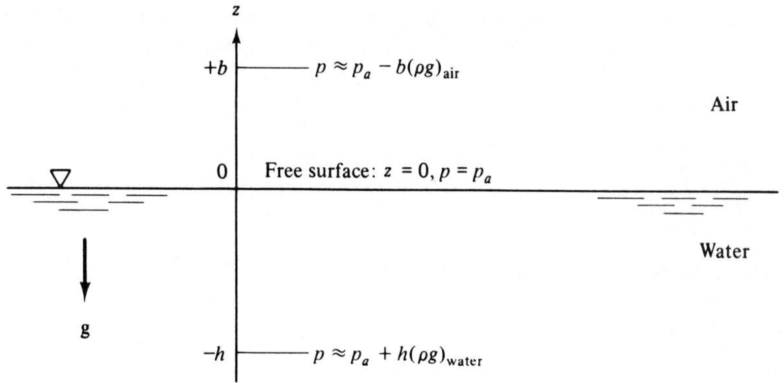

FIGURE 8.5 Hydrostatic-pressure distribution in oceans and atmospheres. (*From White,*[2] *p. 63.*)

3. Hydrostatic Pressure in Gases. Gases are compressible, with density proportional to pressure. It is sufficiently accurate to introduce the perfect-gas law

$$p = \rho RT$$

in Eq. (8.8). Then,

$$\int_1^2 \frac{dp}{p} = \ln \frac{p_2}{p_1} = -\frac{g}{R}\int_1^2 \frac{dz}{T} \tag{8.11}$$

One common approximation is the isothermal atmosphere, where $T = T_0$,

$$p_2 = p_1 \exp\left[-\frac{g(z_2 - z_1)}{RT_0}\right] \tag{8.12}$$

4. Scales of Pressure. Pressure may be expressed with reference to any arbitrary datum. The usual data are absolute zero and local atmospheric pressure. When a pressure is expressed as a difference between its value and a complete vacuum, it is called an *absolute pressure*. When it is expressed as a difference between its value and the local atmospheric pressure, it is called a *gage pressure*.

Figure 8.6 illustrates the data and the relations of the common units of pressure measurements. A pressure expressed in terms of the length of a column of liquid ($p = \gamma h$) is equivalent to the force per unit area at the base of the column.

5. Pressure-Sensing Devices. The two principal devices using liquids are the *barometer* and the *manometer*. The barometer senses absolute pressures, and the manometer senses pressure differential.

Manometers are a direct application of the basic equation of fluid statics. For a *U-tube manometer* (Fig. 8.7),

$$p_1 - p_2 = (\gamma_m - \gamma_f)h \tag{8.13}$$

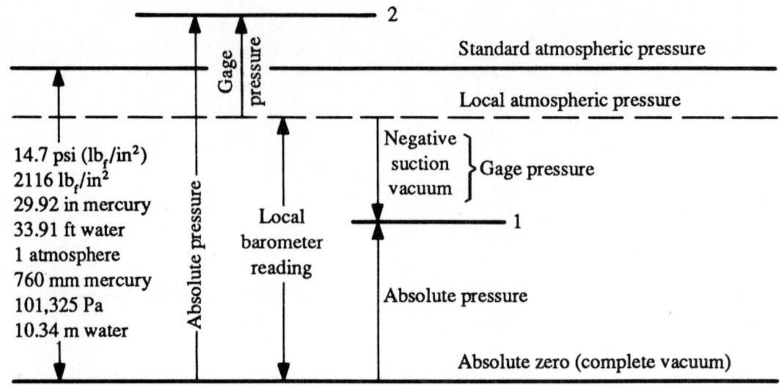

FIGURE 8.6 Units and scales for pressure measurements. (*From Streeter and Wylie,*[1] *p. 30.*)

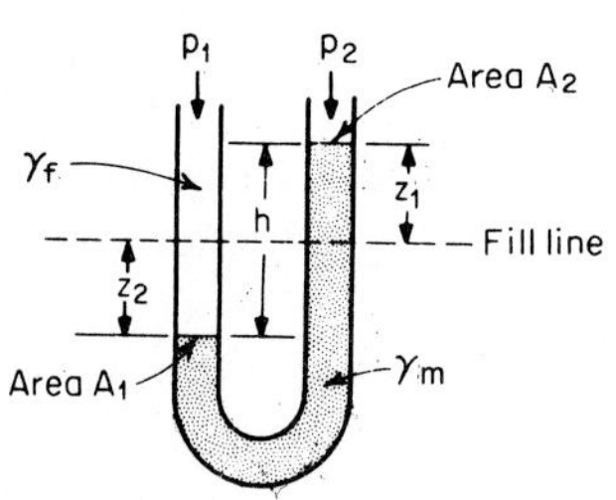

FIGURE 8.7 U-tube manometer. (*From Avallone and Baumeister,*[3] *p. 3-39.*)

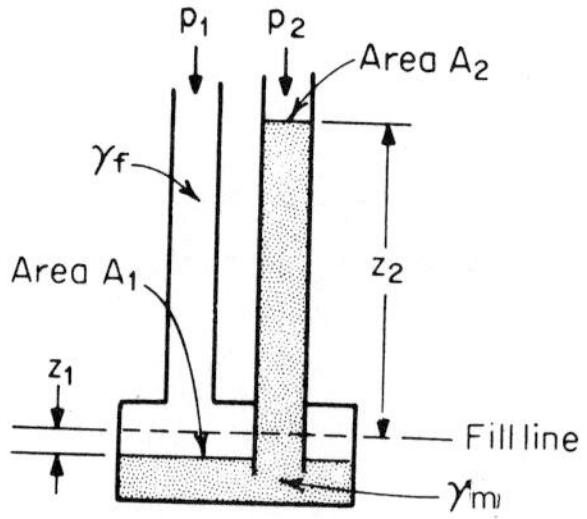

FIGURE 8.8 Well or cistern-type manometer. (*From Avallone and Baumeister,*[3] *p. 3-39.*)

For a *well* or *cistern-type manometer* (Fig. 8.8),

$$p_1 - p_2 = (\gamma_m - \gamma_f)(z_2)\left(1 + \frac{A_2}{A_1}\right) \tag{8.14}$$

For an *inclined manometer* (Fig. 8.9),

$$p_1 - p_2 = (\gamma_m - \gamma_f)\left(\frac{A_2}{A_1} + \sin\theta\right)R \tag{8.15}$$

6. Hydrostatic Forces. The force exerted by a liquid (i.e., caused by the weight of the fluid) on a plane submerged surface (Fig. 8.10) is given by

$$F = \int_A p\,dA = \gamma\int_A \sin\theta y\,dA = \gamma\sin\theta\bar{y}A = \gamma\bar{h}A = p_G A \tag{8.16}$$

where $\bar{y}\sin\theta = \bar{h}$ and $p_G = \gamma\bar{h}$

i.e., the force F is the product of the area and the pressure at its centroid G.

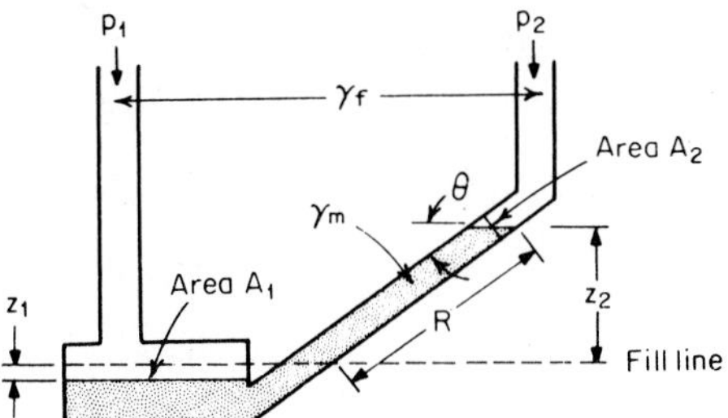

FIGURE 8.9 Inclined manometer. (*From Avallone and Baumeister,*[3] *p. 3-39.*)

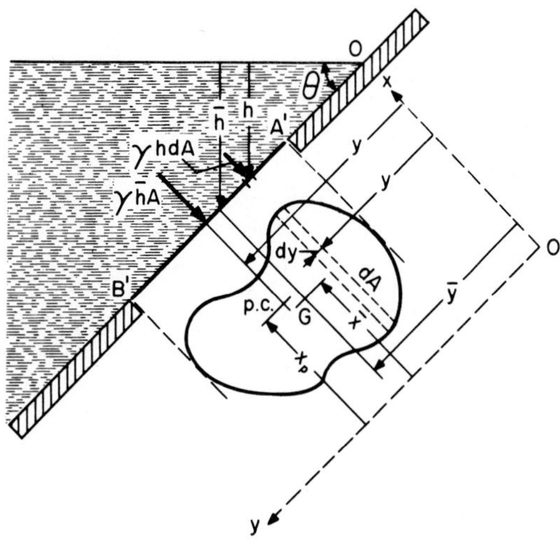

FIGURE 8.10 Notation for force of liquid on one side of a plane inclined area. (*From Streeter and Wylie,*[1] *p. 40.*)

The line of action of the resultant force F has its piercing point in the surface at the point called the *pressure center* (p.c.) with coordinates (x_p, y_p) given as

$$x_p = \frac{1}{\bar{y}A} \int_A xy \, dA \tag{8.17}$$

$$y_p = \frac{1}{\bar{y}A} \int_A y^2 \, dA \tag{8.18}$$

Note that the pressure center (p.c.) is always below the centroid of the surface.

The hydrostatic force acting on the curved surface AB is resolved into the horizontal component F_H and the vertical component F_V (Fig. 8.11). Note that F_V is equal to the weight of the fluid above the area AB. F_H is equal to the hydrostatic pressure force acting on the projection of the considered surface AB on the plane perpendicular to F_H. Calculation of F_V and F_H is accomplished also by Eq. (8.16).

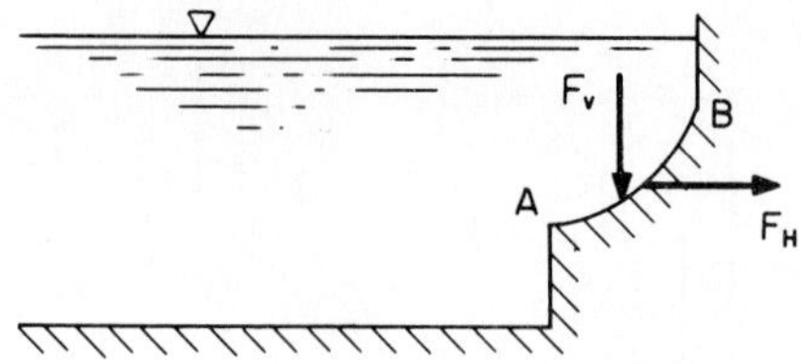

FIGURE 8.11 Forces on a curved surface.

7. Buoyant Force. The resultant force exerted on a body by a static fluid in which it is submerged or floating is called the *buoyant force*. The buoyant force always acts vertically upward. There can be no horizontal force of the resultant because the projection of the submerged body or submerged portion of the floating body on a vertical plane is always zero.

For the system in Fig. 8.12, the buoyancy force F is equal to the weight of the displaced fluids of densities ρ_l and ρ_g, that is,

$$F = g\rho_l V + g\rho_g V' \tag{8.19}$$

If the fluid of density ρ_g is a gas, then

$$F \simeq g\rho_l V \tag{8.20}$$

If $\rho_l > \rho_m$ (where ρ_m is the density of the body), the body will float.

If $\rho_l = \rho_m$, the body will remain suspended.

If $\rho_l < \rho_m$, the body will sink.

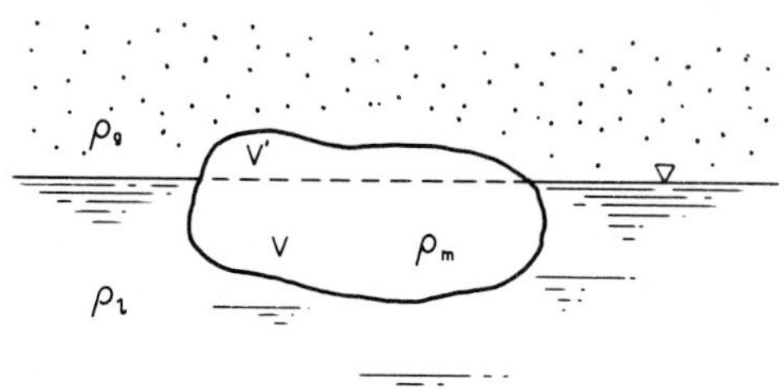

FIGURE 8.12 Buoyant force on the body.

FLUID-FLOW CHARACTERISTICS

Flow may be classified in many ways, such as turbulent, laminar, real, ideal, uniform, nonuniform, steady, unsteady, reversible, or irreversible.

1. Turbulent Flow. Turbulent flow situations are most prevalent in engineering practice. In *turbulent flow*, the fluid particles move in very irregular paths, causing an exchange of momentum from one portion of the fluid to another. The fluid particles can range in size from very small (e.g., a few thousand molecules) to very large (e.g., thousands of cubic feet in a large swirl in a river or an atmospheric gust). Also, in turbulent flow, the momentum losses vary about 1.7 to 2 powers of the velocity; in laminar flow, they vary about the first power of the velocity.

2. Laminar Flow. In *laminar flow*, fluid particles move along smooth paths in *laminas*, or layers, with one layer gliding smoothly over an adjacent layer. Laminar flow is governed by Newton's law of viscosity [Eq. (8.1)] or extensions of it to three-dimensional flow, which relates shear stress to rate of angular deformation. Laminar flow is not stable in situations involving combinations of low viscosity, high velocity, or large flow passages and breaks down into turbulent flow. Figure 8.13 illustrates how an unstable laminar flow may turn into a turbulent flow. Figure 8.13*a* shows that a local disturbance causes an increase in the velocity of particle *A*. This increase in velocity will cause the pressure at *A* to fall below that at *B*. The pressure difference will further increase the velocity difference across the surface of discontinuity, as shown in Fig. 8.13*b*. The result is the formation of eddies at the surface of discontinuity and the initiation of turbulent flow.

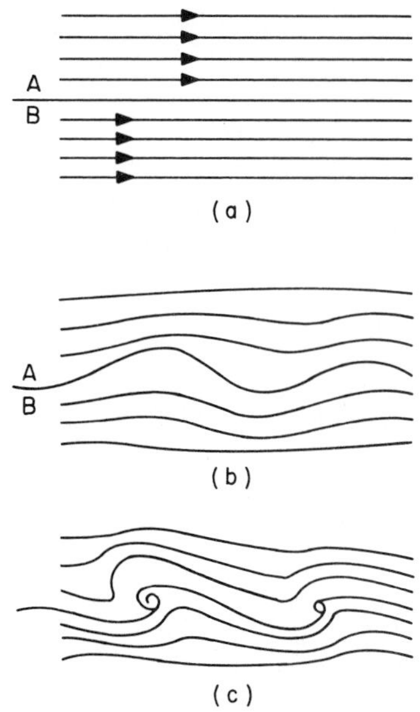

FIGURE 8.13 Eddies formed in an unstable laminar flow. (*From Probstein.*[4])

The turbulent mixing also results in a more rapid transfer of momentum between different layers of fluid. Thus the velocity distribution in a turbulent flow is more uniform than that in a laminar flow, as shown in Fig. 8.14.

An equation similar in form to Newton's law of viscosity may be written for turbulent flow:

$$\tau = \eta \frac{du}{dy} \tag{8.21}$$

The eddy viscosity η, however, is not a fluid property alone; it depends on the fluid motion and the density. In many practical flow situations, both viscosity and turbulence contribute to the shear stress:

$$\tau = (\mu + \eta) \frac{du}{dy} \tag{8.22}$$

and experiments are required to determine this type of flow.

3. Boundary Layer. An *ideal fluid* is frictionless (inviscid) and incompressible and should not be confused with a perfect (ideal) gas. The assumption of an ideal fluid is helpful in analyzing flow situations involving large expanses of fluids, as in the motion of an airplane or a submarine. A frictionless fluid is nonviscous, and its flow processes are reversible.

The layer of fluid in the immediate neighborhood of an actual flow boundary

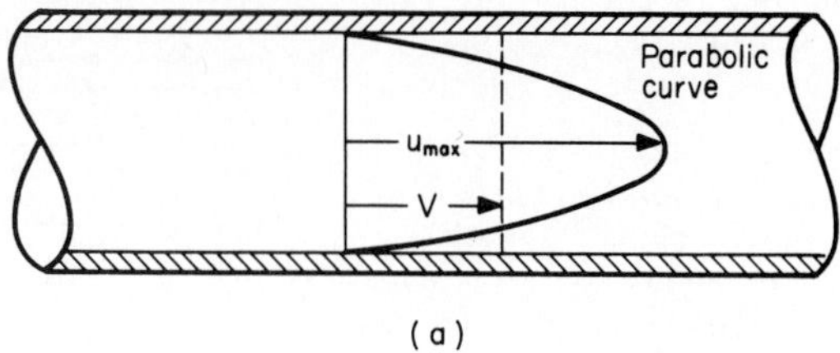

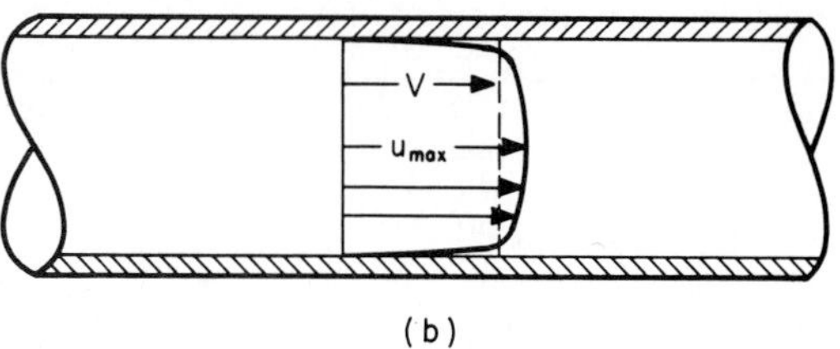

FIGURE 8.14 Comparison of velocity distribution for (*a*) laminar and (*b*) turbulent pipe flow. (*From Probstein.*[4])

that has had its velocity relative to the boundary affected by viscous shear is called the *boundary layer*. Boundary layers may be laminar or turbulent (Fig. 8.15) depending generally on their length, the viscosity, the velocity of the flow near them, and the boundary roughness.

The *boundary-layer thickness* δ may be defined as the region in which the fluid velocity changes from its free-stream, or inviscid-flow, value to zero at the body surface (Fig. 8.16). Developing boundary layers in the entrance of a duct flow are shown in Fig. 8.17.

A particularly interesting phenomenon connected with transition in the boundary layer occurs with blunt bodies, e.g., a sphere or a circular cylinder. In the region of adverse pressure gradient (Fig. 8.18), the boundary layer separates from the surface. At this location, the shear stress goes to zero, and beyond this point

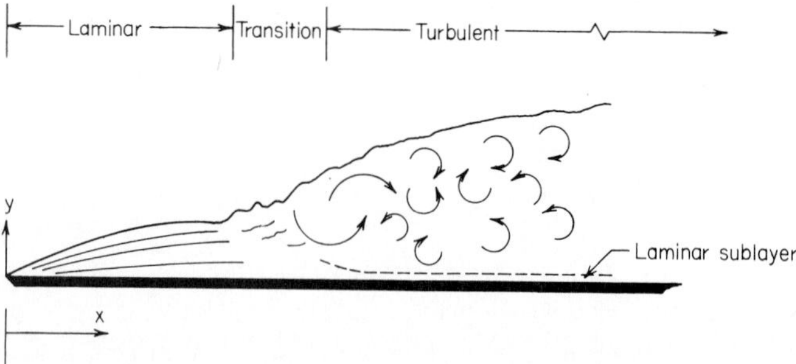

FIGURE 8.15 Laminar, transition, and turbulent boundary-layer-flow regimes in flow over a flat plate. (*From Rohsenow, Hartnett, and Ganić,*[5] *p. 1-9.*)

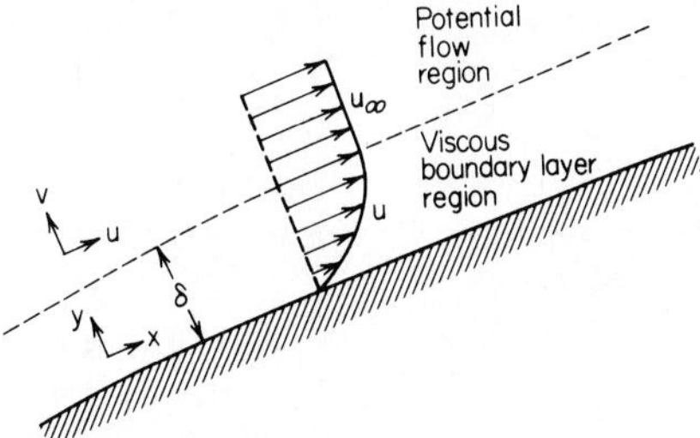

FIGURE 8.16 Boundary-layer flow past an external surface. (*From Rohsenow, Hartnett, and Ganić,*[5] *p. 1-8.*)

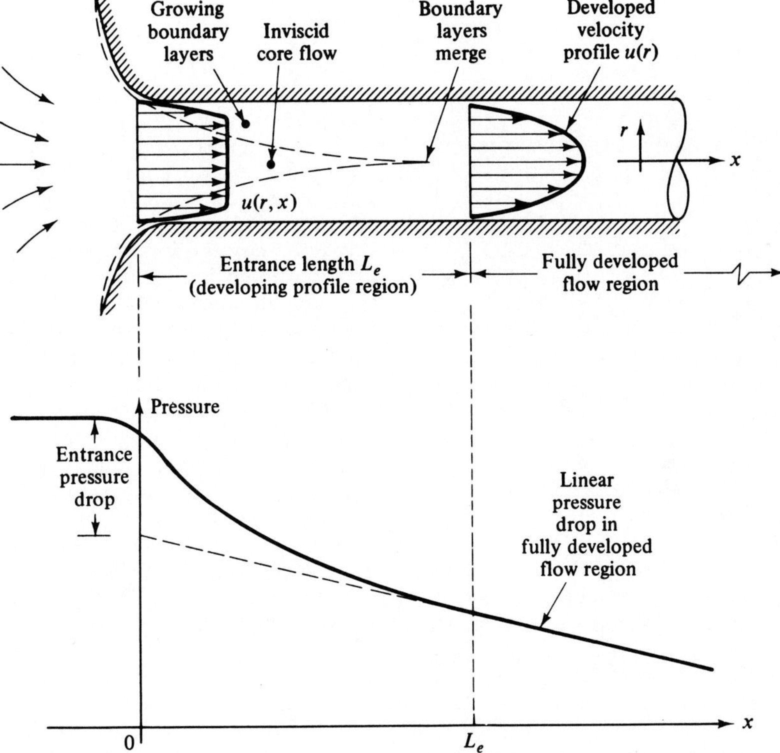

FIGURE 8.17 Developing velocity profiles and pressure changes in the entrance of a duct flow. (*From White,*[2] *p. 312.*)

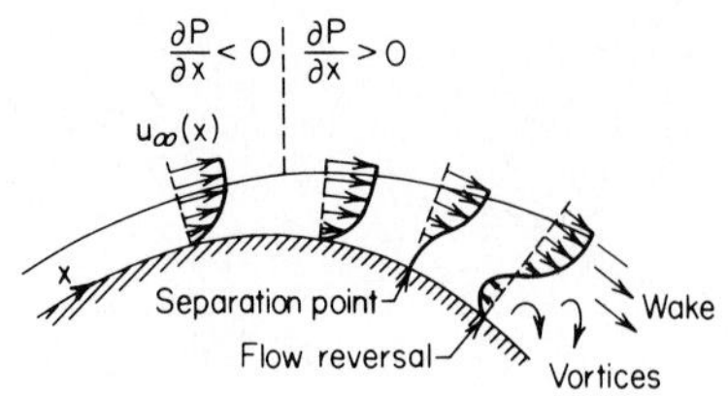

FIGURE 8.18 Velocity profile associated with separation on a circular cylinder in cross flow. (*From Rohsenow, Hartnett, and Ganić,*[5] *p. 1-12.*)

there is a reversal of flow in the vicinity of the wall, as shown in Fig. 8.18. In this separation region, analysis of the viscous flow is very difficult, and emphasis is placed on the use of experimental methods.

4. Steady and Unsteady Flow. *Steady flow* occurs when conditions at any point in the fluid do not change with time. This can be expressed as $\partial V/\partial t = 0$, in which space ($x$, y, z coordinates of the point) is held constant. Likewise, in steady flow there is no change in density ρ, pressure p, or temperature T with time at any point. Thus

$$\frac{\partial \rho}{\partial t} = 0 \qquad \frac{\partial p}{\partial t} = 0 \qquad \frac{\partial T}{\partial t} = 0 \tag{8.23}$$

In turbulent flow, a very efficient mixing takes place; i.e., macroscopic chunks of fluid move across streamlines and transport energy and mass as well as momentum vigorously. The most essential feature of a turbulent flow is the fact that at a given point in it, the flow property X (e.g., velocity component, pressure, temperature, etc.) is not constant with time but exhibits very irregular, high-frequency fluctuations (Fig. 8.19). At any instant, X may be represented as the sum of a time-mean value $\overline{X}$ and a fluctuating component X'. The average is taken over a time that is large compared with the period of typical fluctuation, and if $\overline{X}$ is independent of time, the time-mean flow is said to be *steady*.

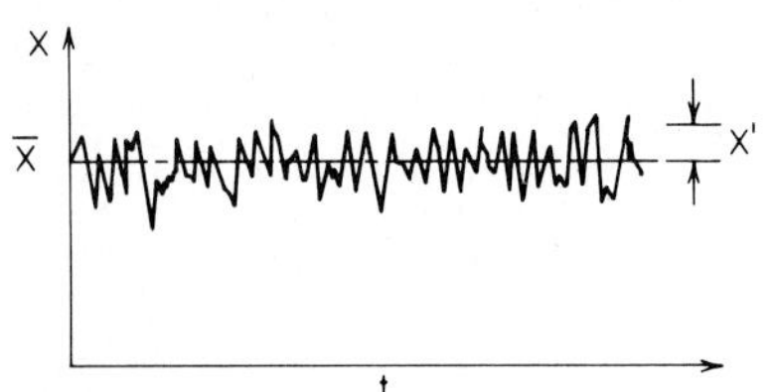

FIGURE 8.19 Property variation with time at some point in steady turbulent flow. (*From Rohsenow, Hartnett, and Ganić,*[5] *p. 1-10.*)

The flow is *unsteady* when conditions at any point change with time, that is, $\partial V/\partial t \neq 0$. Water being pumped through a fixed system at a constant rate is an example of steady flow. Water being pumped through a fixed system at an increasing rate is an example of unsteady flow.

Uniform flow occurs when, at every point, the velocity vector is identically the same (in magnitude and direction) for any given instant.

One-dimensional flow neglects variations or changes in velocity, pressure, etc. transverse to the main flow direction. Conditions at a cross section are expressed in terms of average values of velocity, density, and other properties. Flow through a pipe, for example, usually may be characterized as one dimensional.

5. Streamlines and Stream Tubes. Velocity is a vector, and hence it has both magnitude and direction. A *streamline* is a line that gives the direction of the velocity of a fluid particle at each point in the flow stream. When streamlines are

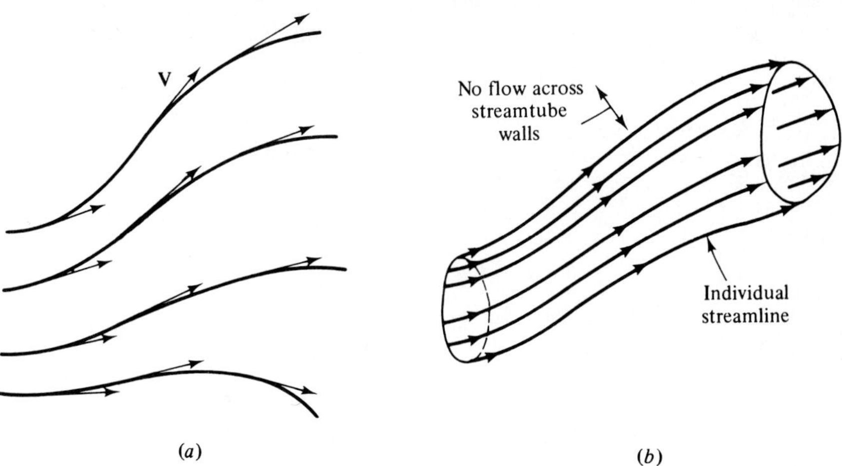

FIGURE 8.20 The most common method of flow-pattern presentation: (*a*) streamlines are everywhere tangent to the local velocity vector; (*b*) a stream tube is formed by a closed collection of streamlines. (*From White,*[2] *p. 40.*)

connected by a closed curve in steady flow, they will form a boundary through which the fluid particles cannot pass. The space between the streamlines becomes a *stream tube* (Fig. 8.20).

The stream-tube concept broadens the application of fluid-flow principles; for example, it allows treating the flow inside a pipe and the flow around an object with the same laws.

Velocity V along a streamline is a function of distance s and time t, or $V = f(s, t)$. Also,

$$dV = \frac{\partial V}{\partial s} ds + \frac{\partial V}{\partial t} dt$$

Acceleration dV/dt may be obtained for the last equation:

$$\frac{dV}{dt} = \frac{\partial V}{\partial s}\frac{ds}{dt} + \frac{\partial V}{\partial t} \tag{8.24}$$

For steady flow, $\partial V/\partial t = 0$.

FLUID DYNAMICS

This section includes basic relations for steady one-dimensional flow.

1. Continuity Equation. (See Fig. 8.21.)

$$\rho_1 A_1 V_1 = \rho A V = \rho_2 A_2 V_2 = \dot{m} \tag{8.25}$$

where $\dot{m}$ is mass flow rate, which is constant across every section of the tube. Equation (8.25) is a law of conservation of mass.

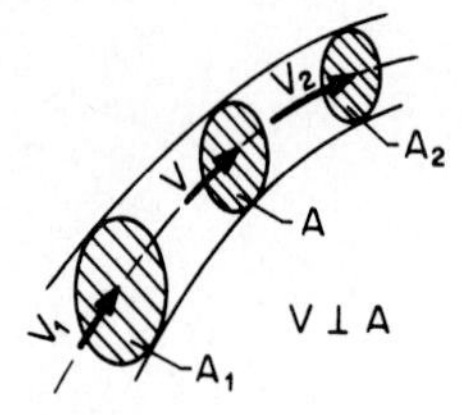

FIGURE 8.21 Steady-flow tube.

The continuity equation also may be written as

$$\frac{dA}{A} = -\frac{dV}{V} - \frac{d\rho}{\rho} \tag{8.26}$$

For incompressible fluids, $d\rho = 0$, so

$$\frac{dA}{A} = -\frac{dV}{V} \tag{8.27}$$

From this equation,

If the area increases, the velocity decreases.
If the area is constant, the velocity is constant.
There are no critical values.

For the frictionless flow of compressible fluids, it can be shown that

$$\frac{dA}{A} = -\frac{dV}{V}\left[1 - \left(\frac{V}{c}\right)^2\right] \tag{8.28}$$

where c is the velocity of sound. From Eq. (8.28),

In *subsonic* velocity, $V < c$. If the area increases, the velocity decreases, same as for incompressible flow.

In *sonic* velocity, $V = c$. Sonic velocity can exist only where the change in area is zero, i.e., at the end of a convergent passage or at the exit of a constant-area duct.

In *supersonic* velocity, $V > c$. If area increases, the velocity increases, the reverse of incompressible flow. Also, supersonic velocity can exist only in the expanding portion of a passage after a contraction where sonic velocity existed.

2. Bernoulli's Equation (Law of Conservation of Energy)

1. For *nonviscous fluid* (no friction),

$$\frac{p_1}{\rho} + gz_1 + \frac{V_1^2}{2} = \frac{p_2}{\rho} + gz_2 + \frac{V_2^2}{2} \tag{8.29}$$

where p/ρ = pressure energy per unit mass
gz = potential energy per unit mass
$V^2/2$ = kinetic energy per unit mass

Figure 8.22 gives energy relations for Eq. (8.29). The energy grade line at any point is $\Sigma(p/\rho g + V^2/2g + z)$ and the hydraulic grade line is $\Sigma(p/\rho g + z)$. In hydraulic practice, each type of energy is referred to as a *head*. The *static pressure head* is $p/\rho g$. The *velocity head* is $V^2/2g$, and the *potential head* is z.

2. For *real fluid* (including losses),

$$\frac{p_1}{\rho} + gz_1 + \frac{V_1^2}{2} = \frac{p_2}{\rho} + gz_2 + \frac{V_2^2}{2} + \bar{h}_{1-2} \tag{8.30}$$

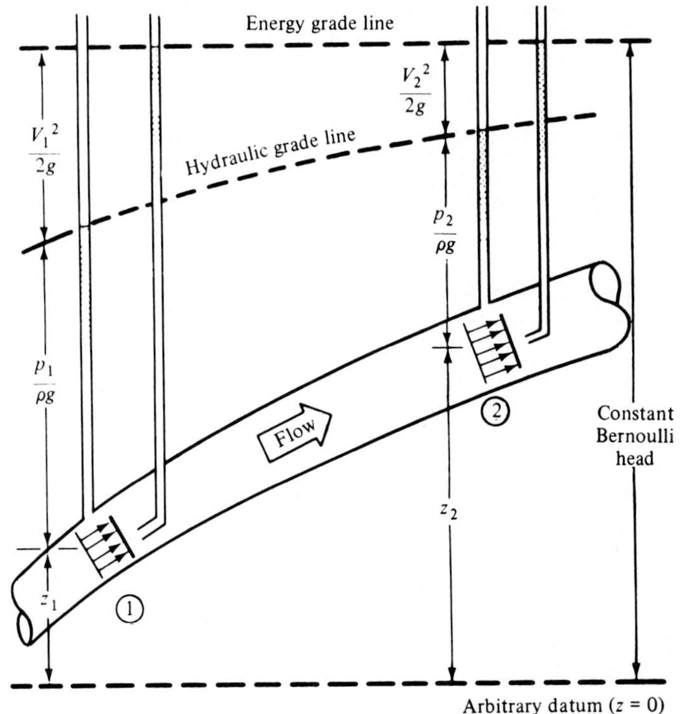

FIGURE 8.22 Hydraulic and energy grade lines for frictionless flow in a duct. *(From White,[2] p. 164.)*

where $\bar{h}_{1-2}$ = resistance losses along path from 1 to 2 per unit mass. The energy loss between sections $h_{1-2} = \bar{h}_{1-2}/g$ is called the *friction head*.

Note: The power P of a hydraulic machine is

$$P = \dot{m}w_{1-2} \tag{8.31}$$

and work per unit mass is

$$w_{1-2} = \frac{1}{\rho}(p_2 - p_1) + g(z_2 - z_1) + \frac{1}{\rho}(V_2^2 - V_1^2) - \bar{h}_{1-2} \tag{8.32}$$

For pumps, $w_{1-2} < 0$. For hydraulic machines, $w_{1-2} > 0$.

Example 8.1. Liquid flows from a large tank, open to the atmosphere, through a small, well-rounded aperture into the atmosphere, as shown in Fig. 8.23*a*. Neglecting losses (e.g., assume frictionless fluid), find an expression for the velocity of the efflux from the tank. Assume that the pressure at 2, the aperture exit, is atmospheric.

From Eq. (8.29),

$$\frac{p_1 - p_2}{\rho} + g(z_1 - z_2) = gh = \frac{V_2^2 - V_1^2}{2}$$

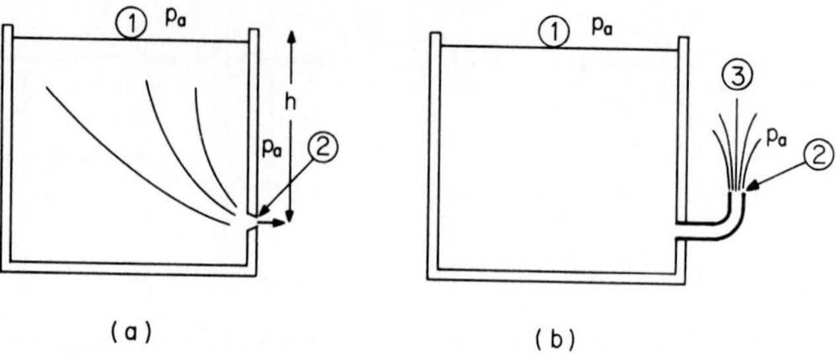

FIGURE 8.23 *(Adapted from Probstein[4] and Sonin.[6])*

For this case, $p_1 = p_2 = p_a$ and V_1^2 is approximately zero, since the cross-sectional area of the tank A_1 is much greater than the aperture A_2. Therefore,

$$V_2 = \sqrt{2gh}$$

Next, place a 90° elbow at the tank exit (Fig. 8.23*b*). Determine how high the water jet will reach. Again, assume frictionless flow, and take aperture 2 to be distance h below the water surface.

The velocity at the exit remains unchanged in magnitude and is given by (for a frictionless elbow) the last equation. Next, use Bernoulli's equation between points 2 and 3. At point 3, the position of maximum elevation of the jet, the velocity of the jet V_3 is zero. Thus

$$\frac{p_3 - p_2}{\rho} + g(z_3 - z_2) = gz = \frac{V_2^2 - V_3^2}{2}$$

Hence $p_2 = p_3 = p_a$; therefore, $gz = V_2^2/2$. With $V_2^2 = 2gh$, it follows that $gz = gh$ and $z = h$. Thus, if there are no losses, the water jet reaches the initial level of the water in the tank. With losses, the maximum elevation of the water jet will be below the water level in the tank.

Example 8.2. The siphon in Fig. 8.24 is filled with water and is disharging at 150 L/s ($1 \text{ m}^3 = 1000$ L). Find the losses from point 1 to point 3 in terms of velocity head. Find the pressure at point 2 if two-thirds of the losses occur between points 1 and 2. From Eq. (8.30) for points 1 and 3,

$$\frac{V_1^2}{2g} + \frac{p_1}{\rho g} + z_1 = \frac{V_3^2}{2g} + \frac{p_3}{\rho g} + z_3 + h_{1\text{-}2}$$

or for $p_1 = p_3 = p_a$, $V_1 \simeq 0$, and $z_1 - z_3 = 1.5$ m,

$$1.5 = \frac{V_3^2}{2g} + h_{1\text{-}2}$$

So

$$V_3 = \frac{Q}{A} = \frac{150}{(3.14)(0.1^2)(1000)} = 4.77 \text{ m/s}$$

Substituting for V_3 in the last equation,

$$h_{1\text{-}2} = 0.34 \text{ m} \cdot \text{N/N}$$

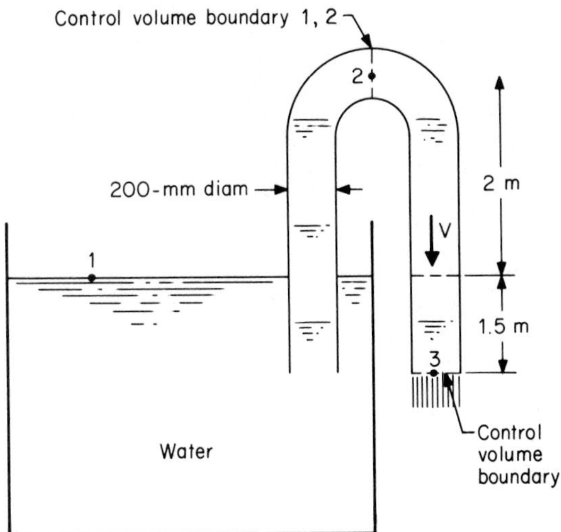

FIGURE 8.24 Siphon. *(From Streeter and Wylie,[1] p. 114.)*

Pressure at point 2 is

$$0 = 1.16 + \frac{p_2}{\rho g} + 2 + \frac{2}{3} h_{1-2}$$

or

$$p_2 = -33.2 \text{ kPa}$$

3. Linear Momentum Equation. For a fluid flowing through a stationary reference volume, the following linear momentum equation is valid:

$$\Sigma F = \dot{m}(V_2 - V_1) \tag{8.33}$$

This equation is derived from Newton's second law and is used to calculate the resultant force exerted on a solid boundary by a fluid stream. Actually, ΣF is the sum of the forces acting on the fluid contained in the reference volume. These can be volume forces (e.g., weight), pressure forces, and friction forces.

V_2 is the exit velocity of the fluid leaving the reference control volume, V_1 is the entrance velocity of the fluid entering the reference volume, and $\dot{m}$ is the mass flow per unit time. In selecting the arbitrary control volume, it is generally advantageous to take the surface normal to velocity whenever it cuts across the flow.

As a scalar equation for any direction, such as the x direction, Eq. (8.33) becomes

$$\Sigma F_x = \dot{m}(V_{x_2} - V_{x_1}) \tag{8.34}$$

These equations are restricted to steady flow in the simple forms given above.

Example 8.3. We wish to evaluate the force coming onto the reducing elbow shown in Fig. 8.25 as a result of an internal steady flow of fluid. The flow may be assumed to be one-dimensional. The average values of the flow characteristics at the inlet and

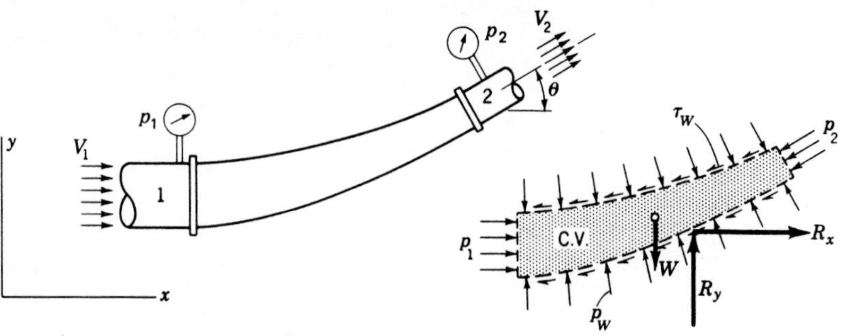

FIGURE 8.25 (*From Shames,*[7] *p. 98.*)

outlet are known, as is the geometry of the reducer. A control volume chosen at the interior of the reducer will enable us to relate known quantities at the inlet and outlet with the force **R** on the fluid from the reducer wall.

The reaction to this latter force is the quantity to be computed. This is shown in the diagram, where the control volume has been separately illustrated. All the forces acting on the fluid in the control volume at any time t have been designed. The surface forces include the effects of pressures p_1 and p_2 at the entrance and exit of the reducer as well as distributions of normal and shear stresses p_W and τ_W, whose resultant force is **R**. The body force is simply the weight of the fluid inside the control volume at time t and is indicated in the diagram as W. The x and y components of the resultant force on the fluid may be expressed as

$$\Sigma F_x = p_1 A_1 - p_2 A_2 \cos\theta + R_x$$

$$\Sigma F_y = -p_2 A_2 \sin\theta - W + R_y$$

where R_x and R_y are the net force components of the reducer wall on the fluid. R_x and R_y, being unknown, have been selected positive.

The continuity equation for this control volume, meanwhile, is

$$\rho_1 V_1 A_1 = \rho_2 V_2 A_2 = \dot{m}$$

Now, substituting the preceding results into the momentum equations in x and y directions, we get

$$p_1 A_1 - p_2 A_2 \cos\theta + R_x = \dot{m}(V_2 \cos\theta - V_1) \quad - p_2 A_2 \sin\theta - W + R_y = \dot{m}(V_2 \sin\theta)$$

One may now solve for R_x and R_y. Changing the sign of these results will then give the force components on the elbow from the fluid.

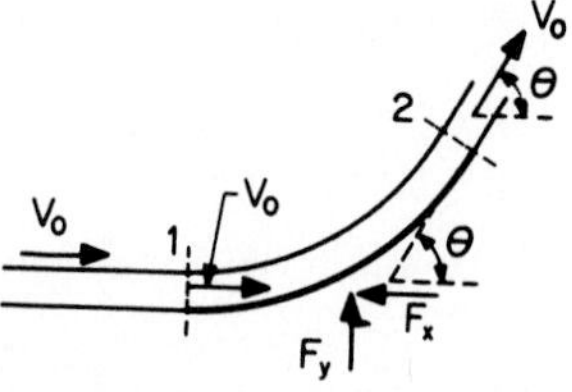

FIGURE 8.26 Flow over a fixed vane. (*Adapted from Probstein*[4] *and Sonin.*[6])

Example 8.4. In Fig. 8.26, the fluid is deflected through the angle θ by the fixed vane. To find the force components F_x and F_y exerted on the fluid by the vane, the momentum equation is applied to the free body of fluid between sections 1 and 2,

$$-F_x = \dot{m}(V_0 \cos\theta - V_0)$$

or

$$F_x = \dot{m}V_0(1 - \cos\theta)$$

In the y direction (neglecting W),

$$F_y = \dot{m}V_0 \sin\theta$$

where $\dot{m} = Q\rho$ (where Q is volume flow rate).

For a vane moving in the x direction with velocity u (Fig. 8.27),

$$F_x = \dot{m}(V_0 - u)(1 - \cos\theta)$$

$$F_y = \dot{m}(V_0 - u)\sin\theta$$

The force acting on the vanes in the direction of motion is equal and opposite to F_x and yields the power given up to the wheel when multiplied by u (a series of vanes is mounted on a wheel such that a vane always intercepts the jet):

$$P = \dot{m}(V_0 - u)u(1 - \cos\theta)$$

Note that P has maximum value for $\theta = 180°$ and $u = V_0^2/2$. The impulse turbines are designed based on these relations.

4. Moment-of-Momentum Equation. The *moment of momentum*, or *angular momentum*, is useful in analyzing the the flow in steady rotating channels, such as in a centrifugal pump or a reaction turbine. Since momentum is a vector quantity, its moment about an axis may be determined analogous to the moment of a force about the axis. The component of the momentum in the radial direction contributes nothing to the moment of momentum. The equation is

$$T = \dot{m}(V_{t_2}r_2 - V_{t_1}r_1) \tag{8.35}$$

where T is the resultant torque on the fluid, $\dot{m}$ is mass flow having its angular momentum changed, and V_t is the tangential component of velocity, with subscript 2 referring to the final condition and subscript 1 referring to the initial condition. Expressing the tangential velocities in terms of the absolute velocities (see Fig. 8.28) at entrance and exit, we have

$$T = \dot{m}(V_2r_2\cos\alpha_2 - V_1r_1\cos\alpha_1) \tag{8.36}$$

The power obtained from the fluid is then

$$P = T \cdot \omega \tag{8.37}$$

where ω is angular velocity.

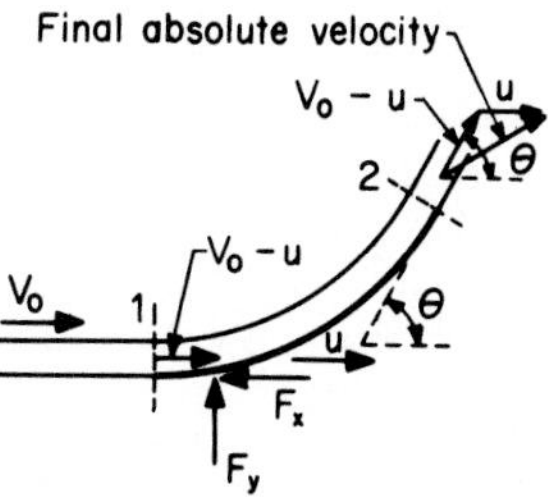

FIGURE 8.27 Deflection of jet by a moving vane. (*Adapted from Probstein*[4] *and Sonin.*[6])

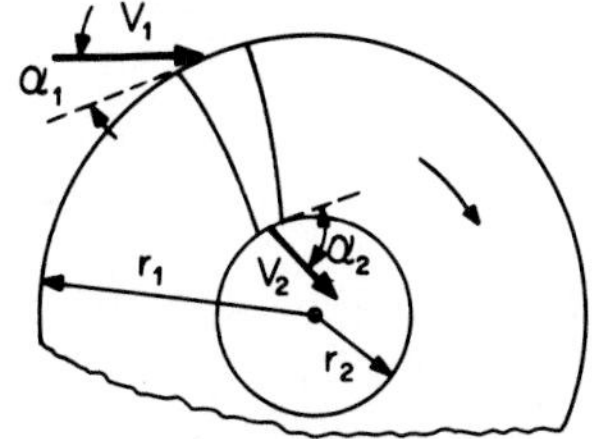

FIGURE 8.28 Turbine runner (rotating channel). (*Adapted from Probstein*[4] *and Sonin.*[6])

BOUNDARY-LAYER FLOWS

1. Laminar Flow. The *boundary-layer thickness* is defined as the locus of points where the velocity u parallel to the plate reaches 99 percent of the external velocity U. The boundary-layer thickness for incompressible flow over smooth, flat plates (Fig. 8.29) may be calculated from the following relation:

$$\frac{\delta}{x} = \frac{5.0}{\mathrm{Re}_x} \tag{8.38}$$

where $\mathrm{Re}_x = \rho U x/\mu$ is called the *local Reynolds number* of the flow along the flat plate.

The *fluid shear stress* at the surface of the plate is

$$\tau_w = \mu \left(\frac{\partial u}{\partial y}\right)_{y=0} \tag{8.39}$$

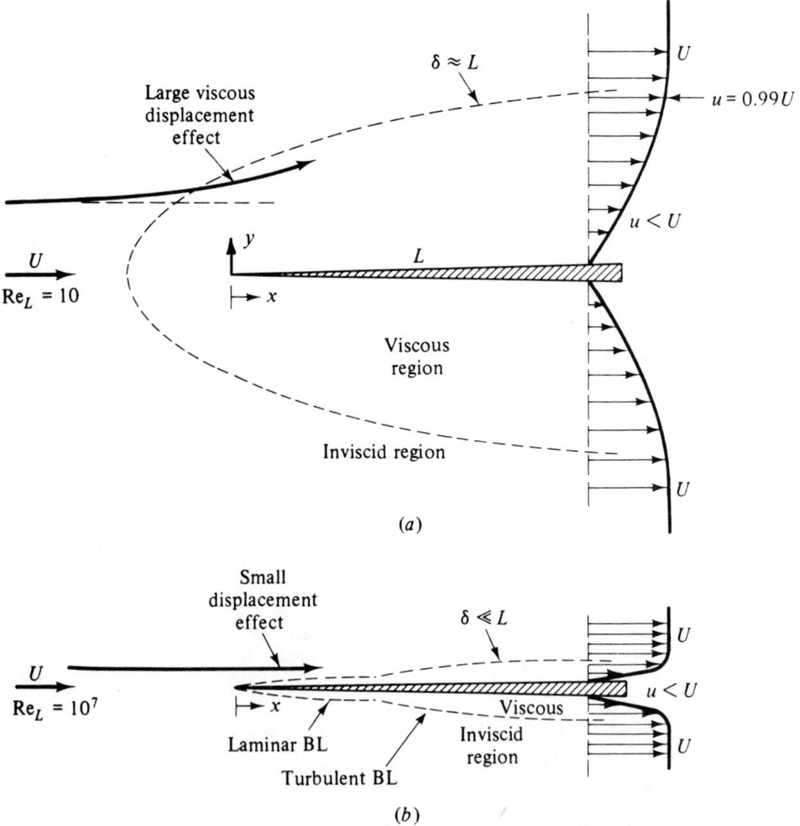

FIGURE 8.29 Comparison of flow past a sharp, flat plate at low and high Reynolds numbers: (*a*) laminar, low Reynolds number flow; (*b*) high Reynolds number flow. (*From White,*[2] *p. 401.*)

or

$$\tau_x = 0.332\mu \frac{U}{x}\sqrt{\mathrm{Re}_x} \tag{8.40}$$

The local *skin-friction coefficient* C_{f_x} is defined by

$$C_{f_x} = \frac{\tau_w}{\frac{1}{2}\rho U^2} \tag{8.41}$$

and for laminar boundary-layer flow over a flat plate, it is

$$C_{f_x} = \frac{0.664}{\sqrt{\mathrm{Re}_x}} \tag{8.42}$$

The wall shear stress varies with x, and in order to obtain the total force exerted by the fluid passing over the plate, we must integrate. The *total drag force* F_D on a plate of length L is

$$F_D = \int_0^L \tau_w(x) b\, dx = \frac{0.664\rho U^2 bL}{\sqrt{\mathrm{Re}_L}} \tag{8.43}$$

where $\mathrm{Re}_L = \rho UL/\nu$ and b is the width of the plate.

The skin-friction drag coefficient C_{D_f} is defined by

$$C_{D_f} = \frac{F_D}{\frac{1}{2}\rho U^2 bL} = \frac{1.328}{\sqrt{\mathrm{Re}_L}} \tag{8.44}$$

Note that C_{f_x} is a local skin-friction coefficient and C_{D_f} is a coefficient expressing the integrated drag over a length L of a surface.

2. Turbulent Flow. A critical Re_x for transition of laminar to turbulent boundary layer is approximately 5×10^5. The turbulent boundary thickness is

$$\frac{\delta}{x} \simeq \frac{0.37}{(\mathrm{Re}_x)^{1/5}} \tag{8.45}$$

The effect of turbulence is to bring about a more uniform profile than would be obtained for the laminar boundary layer (Fig. 8.29). It has been found that a profile based on

$$\frac{u}{U} = \left(\frac{y}{\delta}\right)^{1/7} \tag{8.46}$$

gives a very good correlation with experimental data over a wide range of turbulent Reynolds numbers.

The $\frac{1}{7}$-power law is not valid in the immediate vicinity of the wall, since in this region there exists the laminar sublayer in which turbulence is damped out by the wall. It is therefore necessary to rely on the experimental data to obtain an expression for τ_w. An equation that yields good agreement with data for

$$5 \times 10^5 < \mathrm{Re}_x < 10^7$$

is the Blasius resistance formula, that is,

$$\frac{\tau_w}{\rho U^2} = 0.0225\left(\frac{\nu}{U\delta}\right)^{1/4} \tag{8.47}$$

From Eqs. (8.45) and (8.47),

$$\tau_w = \frac{0.029\rho U^2}{(\mathrm{Re}_x)^{1/5}} \tag{8.48}$$

Also, analogously to Eqs. (8.42), (8.43), and (8.44),

$$C_{f_x} = \frac{0.058}{(\mathrm{Re}_x)^{1/5}} \tag{8.49}$$

$$F_D = \frac{0.036\rho U^2 bL}{(\mathrm{Re}_L)^{1/5}} \tag{8.50}$$

$$C_{D_f} = \frac{0.074}{(\mathrm{Re}_L)^{1/5}} \tag{8.51}$$

Figure 8.30 includes values for C_{D_f} as function of Re_L.

3. Flow Over Immersed Bodies. In analyzing uniform flow past a body surface, the engineer is most often interested in the resultant fluid forces acting on the body. Such forces can be divided into two components: *lift force*, acting upward and normal to the approach flow, and *drag force*, acting in the same direction as the approach flow, as shown in Fig. 8.31. For uniform flow past a finite body, such as an airfoil or cylinder, the boundary layer is subjected to both positive and negative pressure gradients (Fig. 8.18), and separation of the flow from

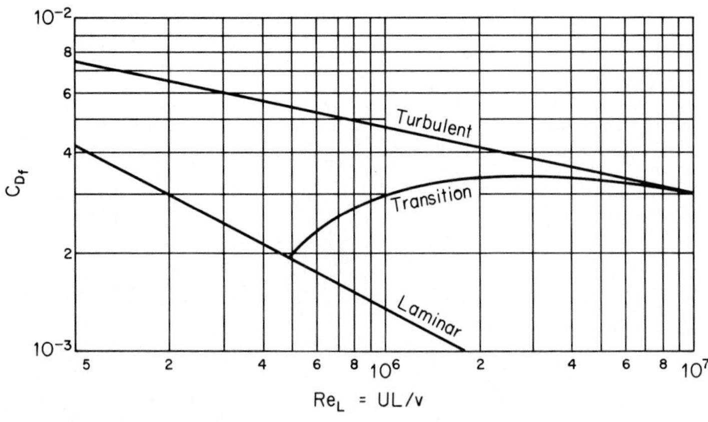

FIGURE 8.30 The drag law for smooth plates. (*From Streeter and Wylie,[1] p. 217.*)

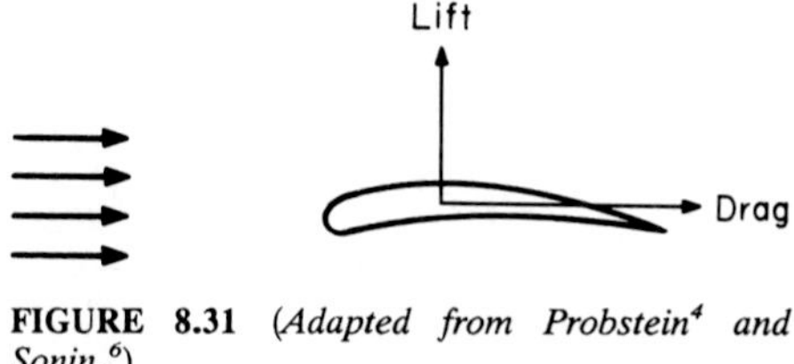

FIGURE 8.31 (*Adapted from Probstein*[4] *and Sonin.*[6])

the body surface is a result of a positive pressure gradient acting on the boundary layer.

When analyzing a flow over immersed bodies, it is assumed that the Reynolds numbers are large enough so that the flow can be divided into two parts: a viscous region, consisting of a thin boundary layer, separation region, and wake, and a nonviscous region, where the effects of viscosity can be neglected, as shown in Fig. 8.32.

Pressure distribution on the surface of a cylinder immersed in uniform steady flow is shown in Fig. 8.33. Owing to boundary-layer separation, the average pressure on the rear half of the cylinder is less than that on the front half. On the rear half of the cylinder, the pressure on the cylinder surface does not recover to the full stagnation pressure as in the case of potential flow (Fig. 8.33). Thus, by integrating the pressure forces over the cylinder surface for the real-fluid case, we obtain a pressure drag on the cylinder. It is obvious that pressure drag can be reduced by shaping or streamlining the body so as to place the point of boundary-layer separation in the vicinity of the trailing edge of the body. In the two pre-

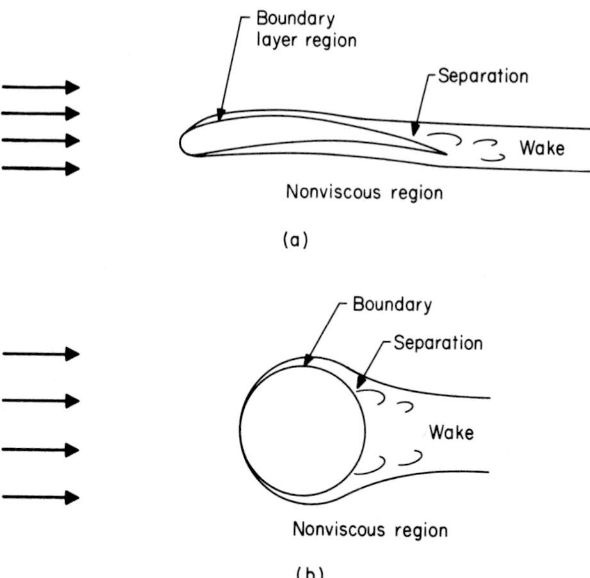

FIGURE 8.32 (*a*) Real fluid flow about an airfoil. (*b*) Real fluid flow about a circular cylinder. (*Adapted from Probstein*[4] *and Sonin.*[6])

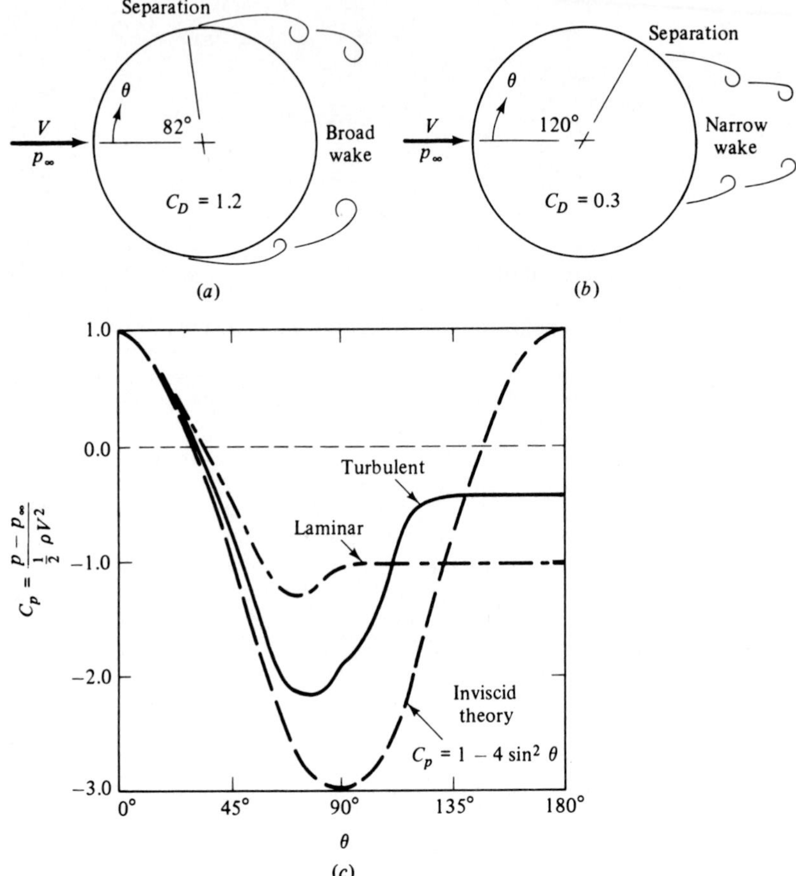

FIGURE 8.33 Flow past a circular cylinder: (*a*) laminar separation; (*b*) turbulent separation; (*c*) theoretical and actual surface-pressure distributions. (*From White,*[2] *p. 431.*)

ceding sections, relations for skin-friction drag are given. The total drag on a body is due to the sum of skin-friction drag ("surface" drag) and pressure drag ("shape" drag). For uniform flow over a flat plate aligned with the flow direction (Fig. 8.34*a*) with no pressure gradients, the entire drag is due to skin friction. For a flat plate normal to the flow direction (Fig. 8.34*b*), the entire drag is pressure drag. For a circular cylinder (Fig. 8.33), over 90 percent of the drag is pressure drag, with only a small fraction due to skin friction.

The drag of a bluff body is expressed in terms of a nondimensional parameter C_D, called the *drag coefficient*:

$$C_D = \frac{F_D}{\frac{1}{2}\rho U^2 A} \tag{8.52}$$

where A is the projected frontal area of the bluff body normal to the flow direc-

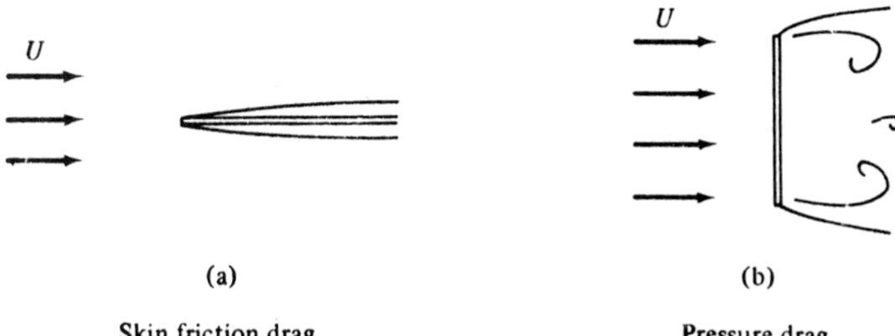

FIGURE 8.34 (*a*) Flat plate aligned with flow. (*b*) Flat plate normal to flow. (*Adapted from Probstein*[4] *and Sonin.*[6])

tion. Values of the drag coefficient for circular cylinders and spheres are given in Figs. 8.35 and 8.36. Table 8.1 includes C_D values for several shapes.

An analytical solution is available for flow about a sphere at Reynolds numbers less than 1 in which the inertial forces can be neglected in comparison to the viscous forces. This analysis, initially derived by Stokes, showed that for $\text{Re}_D < 1$,

$$C_D = \frac{24}{\text{Re}_D} \tag{8.53}$$

Equation (8.53) is useful for the design of settling basins for separating small solid particles from fluids ($\text{Re}_D = \rho UD/\mu$).

A typical pressure distribution for a symmetrical airfoil is given in Fig. 8.37. As the angle of attack α of the foil is increased from zero, the fluid moving over the top surface must accelerate more rapidly, yielding a more negative pressure coefficient C_p. The fluid traveling over the lower surface undergoes a much more gradual acceleration. The resultant difference in pressure between the upper and

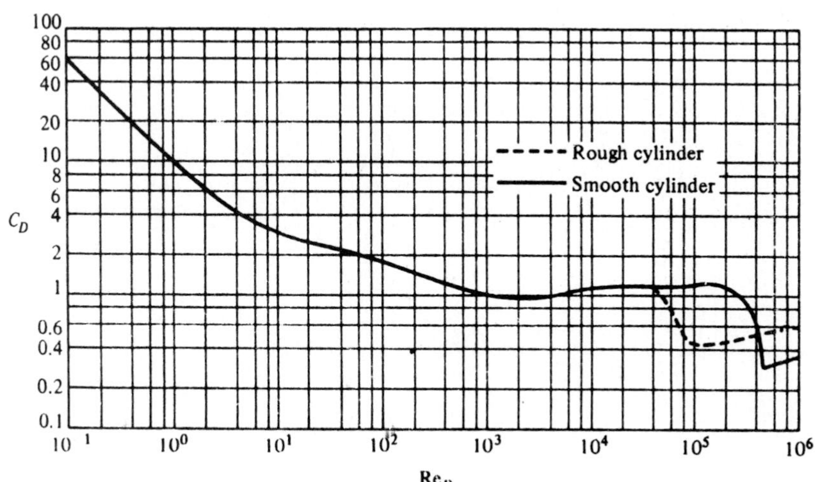

FIGURE 8.35 Drag coefficients of circular cylinders. (*Adapted from Schlichting,*[11] *p. 17.*)

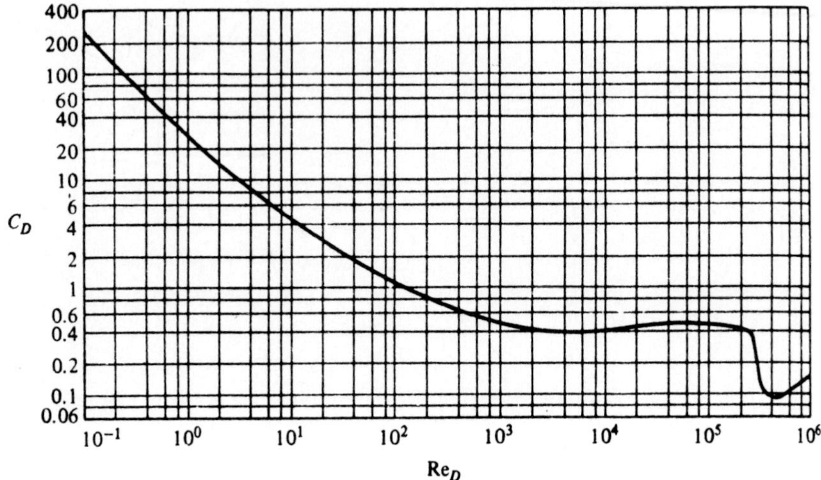

FIGURE 8.36 Drag coefficients of spheres. (*Adapted from Schlichting,*[11] *p. 17.*)

lower surfaces yields a positive lift force on the foil, generally expressed in terms of a lift coefficient C_L with

$$C_L = \frac{F_L}{\frac{1}{2}\,\rho U^2 A} \tag{8.54}$$

For small angles of attack, the lift coefficient varies linearly with the angle of attack, with $C_L = 0$ for $\alpha = 0$ for a symmetrical airfoil. Figure 8.38 gives typical values of lift and drag coefficients for an airfoil.

FLOW IN PIPES

An internal flow in a duct is constrained by the bounding walls, and the viscous effects will grow and meet and permeate the entire flow. Figure 8.17 shows an internal flow in a long duct. There is an *entrance region* where a nearly inviscid upstream flow converges and enters the tube. Viscous boundary layers grow downstream, retarding the axial flow $u(r, x)$ at the wall and thereby accelerating the center-core flow to maintain the incompressible continuity requirement

$$Q = \int_A u\,dA = \text{constant}$$

At a finite distance from the entrance, the boundary layers merge and the inviscid core disappears. The tube flow is then entirely viscous, and the axial velocity adjusts slightly further until at $x = L_e$ it no longer changes with x and is said to be *fully developed*, that is, $u \approx u(r)$ only.

Downstream of $x - L_e$ the velocity profile is constant, the wall shear is constant, and the pressure drops linearly with x for either laminar or turbulent flow.

TABLE 8.1 Drag Coefficients for Several Shapes

2-dimensional shape		Reynolds number	C_D
Circular cylinder	→ ○	10^4 to 10^5	1.2
Semitubular	→ (	4×10^4	1.2
Semitubular	→)	4×10^4	2.3
Square cylinder	→ □	3.5×10^4	2.0
Flat plate	→ ▯	$10^4 \times 10^6$	1.98
Elliptical cylinder	→ ⬭ 2:1	10^5	0.46
Elliptical cylinder	→ ⬭ 8:1	2×10^5	0.20

3-dimensional shape		Reynolds number	C_D
Sphere	→ ○	10^4 to 10^5	0.47
Hemisphere	→ ◖	10^4 to 10^5	0.42
Hemisphere	→ D	10^4 to 10^5	1.17
Cube	→ □	10^4 to 10^5	1.05
Cube	→ ◇	10^4 to 10^5	0.80
Rectangular plate with $\frac{\text{Length}}{\text{Width}} = 5$	→ ▯↕ width	10^3 to 10^5	1.20

Sources: Adapted from Refs. 1, 2, and 7 to 11.

Reynolds number is the only parameter affecting entrance length. The accepted correlations for L_e are

$$\frac{L_e}{D} \simeq 0.06\text{Re} \qquad \text{(laminar)} \tag{8.55}$$

$$\frac{L_e}{D} \simeq 4.4\text{Re}^{1/6} \qquad \text{(turbulent)} \tag{8.56}$$

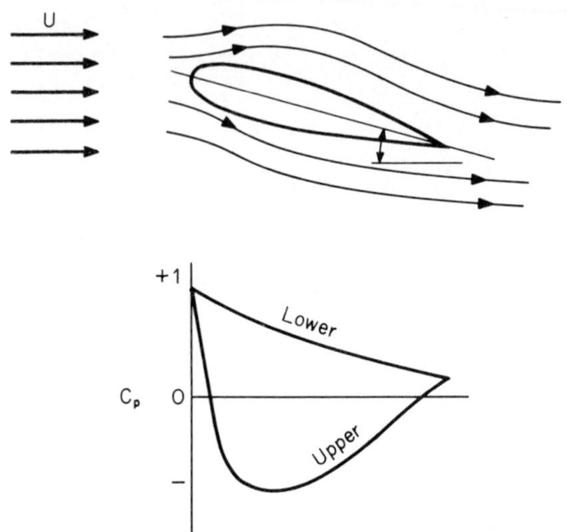

FIGURE 8.37 (*Adapted from Probstein*[4] *and Sonin.*[6])

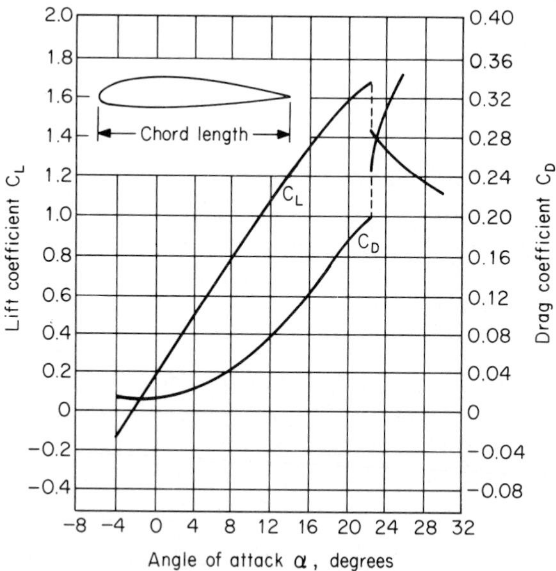

FIGURE 8.38 Typical lift and drag coefficients for an airfoil. (*From Streeter and Wylie,*[1] *p. 224.*)

An accepted design value for pipe-flow transition from laminar to turbulent is

$$\mathrm{Re}_{\mathrm{crit}} \approx 2300 \tag{8.57}$$

However, for the Reynolds-number range between 2000 to 4000, the flow is unstable and this zone is called the *transition zone*.

The pressure loss in a pipe flow may be expressed in terms of pipe-head loss [see Eq. (8.30), $h_{1-2} = h_f$]:

$$h_f = \frac{\Delta p}{\rho g} = f \frac{L}{D} \frac{V^2}{2g} \tag{8.58}$$

where the dimensionless parameter f is called the *Darcy friction factor*, given in terms of wall shear stress

$$\frac{f}{4} = \frac{\tau_w}{\frac{1}{2} \rho V^2} \tag{8.59}$$

where V is mean flow velocity and D is tube diameter.

If the duct is not circular, an equivalent hydraulic diameter D_h is used, defined as

$$D_h = \frac{4 \times \text{cross-sectional area of flow}}{\text{perimeter wetted by fluid}} \tag{8.60}$$

In *laminar flow*, the flow resistance is due to viscous forces only, so that it is independent of the pipe surface roughness. For *laminar flow*,

$$f = \frac{64}{\mathrm{Re}} \tag{8.61}$$

And for *turbulent flow*, the friction factor for Re > 4000 is computed using the Colebrook equation, that is,

$$\frac{1}{\sqrt{f}} = -2 \log \left(\frac{2.51}{\mathrm{Re} \sqrt{f}} + \frac{\epsilon/D}{3.7} \right) \tag{8.62}$$

Figure 8.39 (Moody diagram) is a graphical presentation of this equation. In order to determine a value of the friction factor f from the Moody diagram, a knowledge of relative roughness is necessary. Typical values of roughness for various types of pipe are shown in Table 8.2. For noncircular pipes, D is replaced with D_h.

1. Losses in Pipe Fittings and Valves. In addition to losses due to wall friction in a piping system, there are also losses of mechanical energy and pressure due to flow through valves or fittings (called *minor losses*). The minor losses can be treated as an equivalent frictional loss by expressing the pressure drop due to fittings as

$$\Delta p = -\tfrac{1}{2} K \rho V^2 \tag{8.63}$$

or in terms of heat loss,

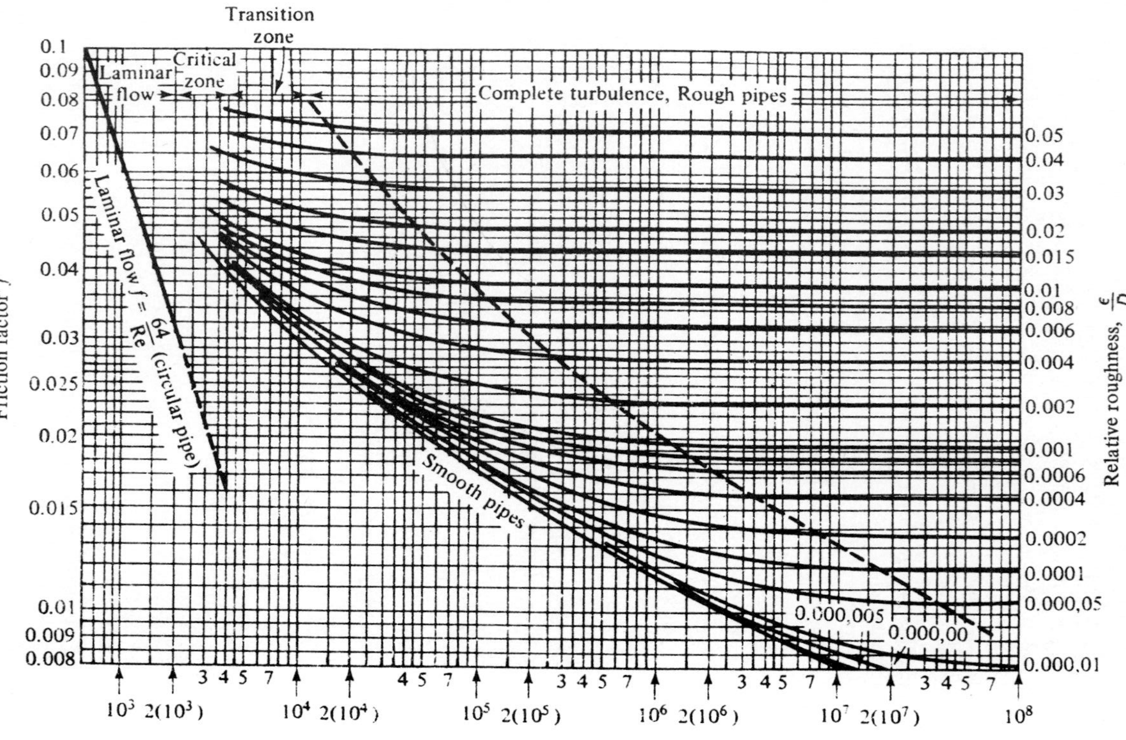

FIGURE 8.39 Friction factors for pipe flow. (*From Moody.*[12])

TABLE 8.2 Roughness of Various Types of Pipe

Type	ϵ (mm)
Glass	Smooth
Asphalted cast iron	0.12
Galvanized iron	0.15
Cast iron	0.26
Wood stave	0.18–0.90
Concrete	0.30–3.0
Riveted steel	1.0–10
Drawn tubing	0.0015

Sources: Adapted from Refs. 1, 2, and 7 to 11.

$$h_m = -K\frac{V^2}{2g} \tag{8.64}$$

where K is the loss (resistance) coefficient. Representative values of K are given in Tables 8.3 and 8.4.

In piping systems, three basic types of calculations are encountered:

1. For a given piping system and flow rate, the pressure drop might be required.
2. For a given piping system and pressure drop, the flow rate might be required.
3. For a given flow and pressure drop, it might be required to design the system, i.e., to determine the necessary pipe diameter.

Example 8.5. For the piping shown in Fig. 8.40, determine the pressure p_2. There are 100 m of 15-cm-I.D. cast-iron pipe and 30 m of 7.3-cm-I.D. cast-iron pipe. Water flow rate is 0.01 m^3/s, the water temperature is 20°C, and the gage pressure at 1 is 250 kPa.

TABLE 8.3 Values of K for Valves and Fittings

Fitting or valve	K
Standard 45° elbow	0.35
Standard 90° elbow	0.75
Long-radius 90° elbow	0.45
Coupling	0.04
Union	0.04
Gate valve	
Open	0.20
¾ Open	0.90
½ Open	4.5
¼ Open	24.0
Globe valve	
Open	6.4
½ Open	9.5
Tee (along run, line flow)	0.4
Tee (branch flow)	1.5

Sources: Adapted from Refs. 1, 2, and 7 to 11.

TABLE 8.4 Values of K for Representative Entrance and Exit Flow Conditions

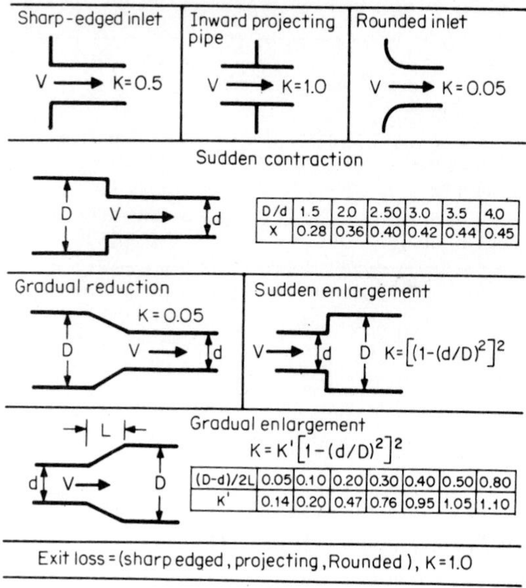

D/d	1.5	2.0	2.50	3.0	3.5	4.0
K	0.28	0.36	0.40	0.42	0.44	0.45

(D-d)/2L	0.05	0.10	0.20	0.30	0.40	0.50	0.80
K'	0.14	0.20	0.47	0.76	0.95	1.05	1.10

Source: From Avallone and Baumeister,[3] p. 3-58.

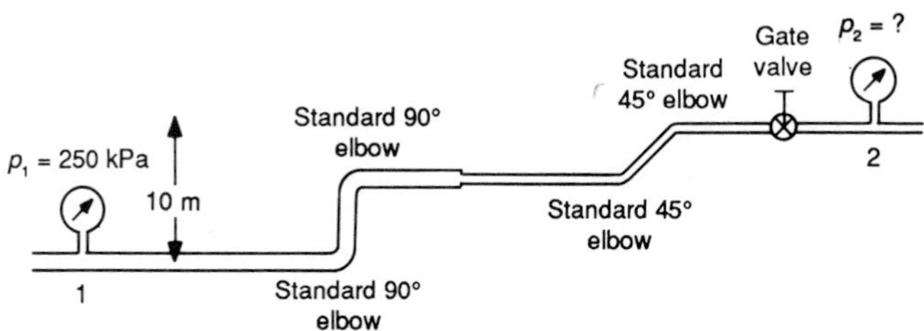

FIGURE 8.40 (*Adapted from Probstein*[4] *and Sonin.*[6])

Solution. First write the modified Bernoulli equation [Eq. (8.30)] between 1 and 2, including minor losses:

$$\frac{p_1}{\rho} + \frac{V_1^2}{2} + gz_1 = \left(\frac{fL}{D}\frac{V^2}{2}\right)_{\substack{\text{15-cm}\\\text{pipe}}} + \left(\Sigma K\frac{V^2}{2}\right)_{\substack{\text{15-cm}\\\text{pipe}}} + \left(\frac{fL}{D}\frac{V^2}{2}\right)_{\substack{\text{7.5-cm}\\\text{pipe}}} + \left(\Sigma K\frac{V^2}{2}\right)_{\substack{\text{7.5-cm}\\\text{pipe}}} + \frac{V_2^2}{2} + gz_2 + \frac{p_2}{\rho}$$

The velocity in the 15-cm pipe is

$$V_{15} = \frac{Q}{A} = \frac{0.01\ \text{m}^3/\text{s}}{(\pi/4)(0.15)^2\ \text{m}^2} = 0.5659\ \text{m/s}$$

and in the 7.5-cm pipe it is

$$V_{7.5} = \frac{Q}{A} = \frac{0.01\ \text{m}^3/\text{s}}{(\pi/4)(0.075)^2\ \text{m}^2} = 2.264\ \text{m/s}$$

At 20°C, for water, $\nu = 1.00 \times 10^{-6}\ \text{m}^2/\text{s}$, so

$$\text{Re}_{15} = \frac{(0.5659\ \text{m/s})(0.15\ \text{m})}{1.00 \times 10^{-6}\ \text{m}^2/\text{s}} = 8.489 \times 10^4$$

and

$$\text{Re}_{7.5} = \frac{(2.264\ \text{m/s})(0.075\ \text{m})}{1.00 \times 10^{-6}\ \text{m}^2/\text{s}} = 1.698 \times 10^5$$

Also,

$$\left(\frac{\epsilon}{D}\right)_{15} = \frac{0.26 \times 10^{-3}\ \text{m}}{0.15\ \text{m}} = 0.00173$$

$$\left(\frac{\epsilon}{D}\right)_{7.5} = \frac{0.26 \times 10^{-3}\ \text{m}}{0.075\ \text{m}} = 0.00346$$

From the Moody diagram (Fig. 8.39), we obtain

$$f_{15} = 0.025$$

$$f_{7.5} = 0.027$$

The minor losses in the 15-cm pipe are given by

$$(\Sigma K)_{15} = \underset{\substack{\text{standard}\\ \text{elbow}}}{0.75} + \underset{\substack{\text{standard}\\ \text{elbow}}}{0.75} = 1.50$$

For the 7.5-cm pipe,

$$(\Sigma K)_{7.5} = \underset{\text{contraction}}{0.33} + \underset{\text{elbow}}{0.35} + \underset{\text{elbow}}{0.35} + \underset{\text{gate value}}{0.20} = 1.23$$

Substituting into the Bernoulli equation, we have

$$\frac{p_2 - p_1}{\rho} = g(z_1 - z_2) + \frac{V_{15}^2}{2}\left(1 - \frac{fL}{D} - \Sigma K\right)_{15} - \frac{V_{7.5}^2}{2}\left(1 + \frac{fL}{D} + \Sigma K\right)_{7.5}$$

$$= 9.81\ \text{m/s}^2(-10\ \text{m}) + \left(\frac{0.5659^2}{2}\ \text{m}^2/\text{s}^2\right)\left(1 - \frac{0.025 \times 100}{0.15} - 1.50\right)$$

$$- \left(\frac{2.264^2}{2}\ m^2/s^2\right)\left(1 + \frac{0.027 \times 30}{0.075} + 1.23\right)$$

$$= -98.1\ \text{m}^2/\text{s}^2 + (0.1601\ \text{m}^2/\text{s}^2)(-17.17) - (2.563\ \text{m}^2/\text{s}^2)(13.03)$$

$$= -98.1\ \text{m}^2/\text{s}^2 - 2.749\ \text{m}^2/\text{s}^2 - 33.40\ \text{m}^2/\text{s}^2$$

$$= -134.2\ \text{m}^2/\text{s}^2$$

or

$$p_2 - p_1 = -(998.3\ \text{km/m}^3)(134.2\ \text{m}^2/\text{s}^2)$$

$$= -134\ \text{kPa}$$

or

$$p_2 = 250 - 134$$

$$= 116\ \text{kPa (gage)}$$

Example 8.6. For the piping system shown in Fig. 8.41, determine the water flow rate. The water temperature is 25°C, with all pipe asphalted cast iron. There are 250 m of 10-cm-I.D. pipe.

Solution. Write the modified Bernoulli equation between the reservoir surface and pipe outlet:

$$\frac{p_0}{\rho} + gz_0 + \frac{V_0^2}{2} = \frac{fL}{D}\frac{V^2}{2} + \Sigma K\frac{V^2}{2} + \frac{p_e}{\rho} + gz_e + \frac{V_e^2}{2}$$

where $p_0 = p_e$ = atmospheric pressure, $z_0 - z_e = 20$ m, and $V_0 = 0$. For this case, with flow velocity the unknown, Reynolds numbers cannot be directly determined; a trial-and-error solution is called for. As a first trial, assume $f = 0.02$. Then,

$$(9.81 \text{ m/s}^2)(20 \text{ m}) = \left[\frac{0.02 \times 250}{0.10} + (\underset{\substack{\text{edged}\\ \text{inlet}}}{0.5} + \underset{\text{4 elbows}}{3.0} + \underset{\substack{\text{globe}\\ \text{valve}}}{6.4}) + 1\right]\frac{V_e^2}{2}$$

$$\frac{V_e^2}{2} = \frac{9.81(20)}{50 + 9.9 + 1} \text{ m}^2/\text{s}^2$$

or

$$V_e = 2.53 \text{ m/s}$$

For this velocity,

$$\text{Re} = \frac{VD}{\nu} = \frac{(2.53 \text{ m/s})(0.10 \text{ m})}{1.00 \times 10^{-6} \text{ m}^2/\text{s}} = 2.53 \times 10^5$$

$$\frac{\epsilon}{D} = \frac{0.12 \times 10^{-3} \text{ m}}{0.10 \text{ m}} = 1.2 \times 10^{-3}$$

From the Moody diagram, we obtain $f = 0.022$. Since this value does not agree with our initial trial, we shall now make a second trial, starting with $f = 0.022$.

$$\frac{fL}{D} = \frac{0.022 \times 250}{0.10} = 55$$

$$\frac{V_e^2}{2} = \frac{9.81(20)}{55 + 9.9 + 1}$$

or

$$V = 2.44 \text{ m/s}$$

For this velocity,

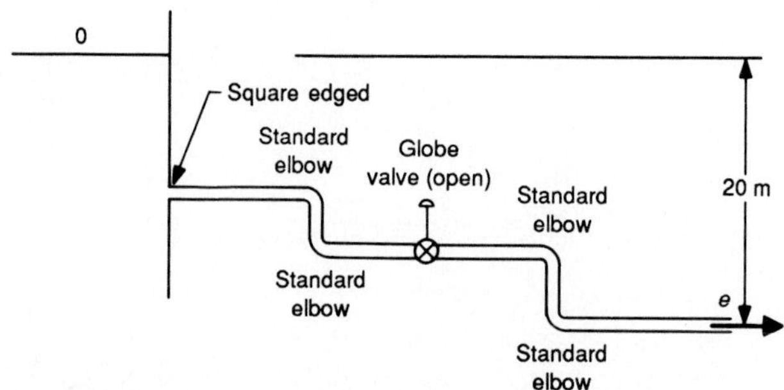

FIGURE 8.41

$$\text{Re} = \frac{VD}{\nu} = \frac{2.44 \times 0.10}{10^{-6}} = 2.44 \times 10^5$$

From the Moody diagram, we find $f = 0.022$. It can be seen that generally we are able to converge on the correct answer quite rapidly; no more than two trials are required as a rule. The water flow rate is

$$\begin{aligned} Q &= AV \\ &= \left[\frac{3.14}{4}(0.10)^2 \text{ m}^2\right] 2.44 \text{ m/s} \\ &= 0.01915 \text{ m}^3\text{/s} \\ &= 19.15 \text{ L/s} \end{aligned}$$

Example 8.7. A pump is to be used to supply 5 L/s of water from a reservoir to a point 400 m from the reservoir at the same level as the reservoir surface. Determine the minimum size cast-iron pipe required. The water temperature is 15°C; assume that minor losses can be neglected (see Fig. 8.42). The pump supplies 50 kW of power to the water flow.

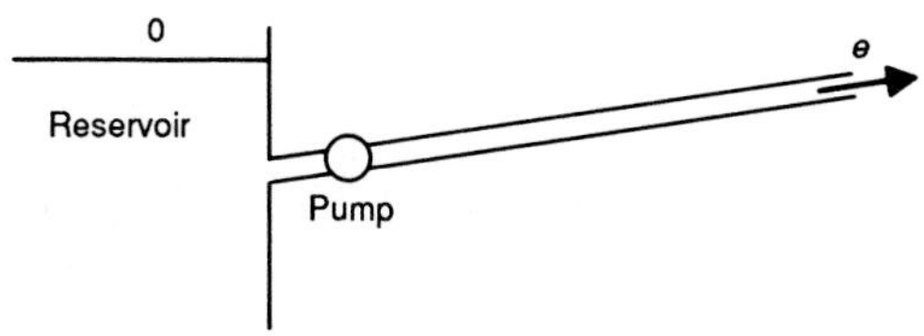

FIGURE 8.42

Solution. From Eqs. (8.31) and (8.32), we can express the pump work in terms of reservoir surface and outlet conditions:

$$-\frac{1}{\dot{m}}P = \frac{p_e - p_0}{\rho} + g(z_e - z_0) + \frac{fL}{D}\frac{V^2}{2} + \frac{V_e^2}{2}$$

For our case,

$$m = \rho Q = (999.2 \text{ kg/m}^3)(5 \times 10^{-3} \text{ m}^3\text{/s}) = 4.996 \text{ kg/s}$$

$$-\frac{1}{\dot{m}P} = -\frac{1}{4.996 \text{ kg/s}}(-50 \times 10^3 \text{ N} \cdot \text{m/s}) = +10{,}010 \text{ m}^2\text{/s}^2$$

Note. P is negative work in a thermodynamic sense, for it represents work done on the fluid.

We now have

$$10{,}010 = \left(\frac{fL}{D} + 1\right)\frac{V^2}{2}$$

where

$$\begin{aligned} V = \frac{Q}{A} &= \frac{5 \times 10^{-3} \text{ m}^3\text{/s}}{(\pi/4)D^2 \text{ m}^2} \\ &= \frac{6.366 \times 10^{-3}}{D^2} \text{ m/s} \qquad \text{with } D \text{ in meters} \end{aligned}$$

$$10{,}010 = \left(\frac{400f}{D} + 1\right)\frac{4.053 \times 10^{-5}}{2D^4}$$

For this case, with D the unknown, the Reynolds number and f cannot be immediately determined. A trial-and-error procedure is required. As a first trial, assume that $f = 0.025$. Then,

$$10{,}010 = \left[\frac{400(0.025)}{D} + 1\right]\frac{4.053 \times 10^{-5}}{2D^4}$$

$$= \frac{20.26 \times 10^{-5}}{D^5} + \frac{2.026 \times 10^{-5}}{D^4}$$

The second term on the right is small compared to the first, so, to a first approximation,

$$10{,}010 \simeq \frac{2.026 \times 10^{-4}}{D^5} \qquad \text{and} \qquad D = 0.0289 \text{ m}$$

For this first trial,

$$V = \frac{6.366 \times 10^{-3}}{(0.0289)^2} = 7.622 \text{ m/s}$$

$$\text{Re} = \frac{7.622 \times 0.0289}{1.15 \times 10^{-6}} = 1.915 \times 10^5$$

$$\frac{\epsilon}{D} = \frac{0.26 \times 10^{-3}}{0.0289} = 0.0090$$

From the Moody diagram, we obtain $f = 0.036$. To start a second iteration, let $f = 0.036$. Therefore,

$$10{,}010 = \left[\frac{400(0.036)}{D} + 1\right]\frac{2.026 \times 10^{-5}}{D^4}$$

or

$$10{,}010 \simeq \frac{0.0002917}{D^5} \qquad \text{and} \qquad D = 0.0311 \text{ m}$$

For this second trial,

$$V = \frac{6.366 \times 10^{-3}}{(0.0311)^2} = 6.582 \text{ m/s}$$

$$\text{Re} = \frac{6.582 \times 0.0311}{1.15 \times 10^{-6}} = 1.78 \times 10^5$$

$$\frac{\epsilon}{D} = \frac{0.26 \times 10^{-3}}{0.0311} = 0.0084$$

From the Moody diagram, we obtain $f = 0.036$. The agreement with the assumed value is good enough that $D = 0.0311$ m can be taken as the required diameter.

OPEN-CHANNEL FLOW

1. One-Dimensional Approximation. *Open-channel flow* is the flow of a liquid in a conduit with a free surface. Owing to the presence of the free surface, the pressure can be taken constant along the free surface and essentially equal to atmospheric pressure. Unlike flow in closed ducts, pressure gradient is not a direct factor in open-channel flow, where the balance of forces is confined to gravity and friction. Surface tension is rarely important because open channels are normally quite large and have very large Weber numbers. But the free surface complicates the analysis because its shape is a priori unknown: The depth profile changes with conditions and must be computed as part of the problem, especially in unsteady problems involving wave motion.

An open channel always has two sides and a bottom, where the flow satisfies the nonslip condition. Therefore, even a straight channel has a three-dimensional velocity distribution. Some measurements of straight-channel velocity contours are shown in Fig. 8.43. The profiles are quite complex, with maximum velocity typically occurring in the midplane about 20 percent below the surface. Little theoretical work has been done on velocity distributions like those in Fig. 8.43. Instead, the engineering approach is to assume one-dimensional flow with an average velocity $V(x)$ at each cross-sectional area $A(x)$, where x is distance along the channel. Since density variations are negligible in low-speed liquid flow, for steady flow the volume flux Q is constant along the channel from continuity:

$$Q = V(x)A(x) = \text{constant} \tag{8.65}$$

A second relation between velocity and depth is the Bernoulli equation includ-

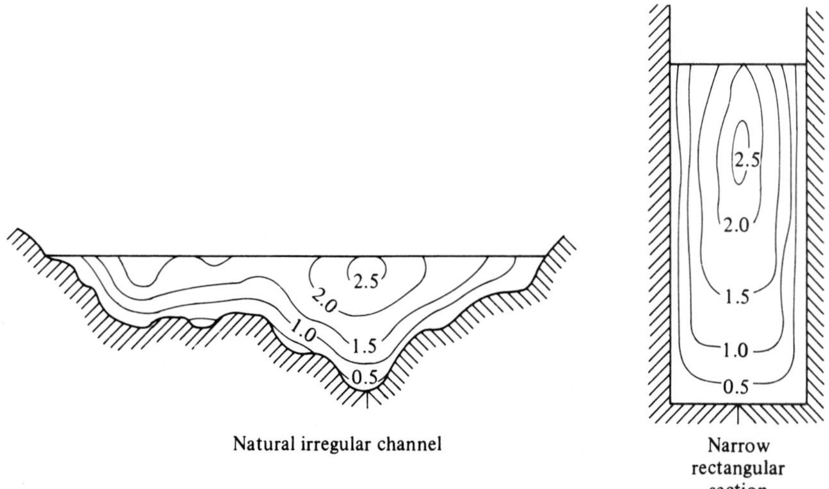

FIGURE 8.43 Measured isovelocity contours in typical straight open-channel flow. (*From White,[2] p. 598.*)

ing friction losses. If points 1 (upstream) and 2 (downstream) are on the free surface, $p_1 = p_2 = p_a$, and we have, for steady flow,

$$\frac{V_1^2}{2g} + z_1 = \frac{V_2^2}{2g} + z_2 + h_f \tag{8.66}$$

where h_f is the friction-head loss. The wall friction in channel flow is quite similar to that in steady duct flow and can be correlated adequately with Eq. (8.62) for turbulent flow with a rough surface where $D = D_h = 4A/P$. Most open-channel analyses use the hydraulic radius R_h instead, which is one-fourth the hydraulic diameter:

$$R_h = \tfrac{1}{4} D_h = \frac{A}{P} \tag{8.67}$$

where P is the wetted perimeter, which includes the sides and bottom of the channel but not the free surface and not the parts of the sides above the water lever.

Open channels are usually large and deep and contain low-viscosity water; hence they are almost always turbulent.

2. Falling Liquid Films. In the case of liquid film flowing down an inclined or vertical wall (Fig. 8.44*a*), three different flow regimes have been observed experimentally:

1. At Reynolds numbers $\text{Re}_\Gamma = 4\Gamma/\mu$ that do not exceed 20 to 30, there exists the usual viscous flow regime, and the film thickness is constant.
2. At Reynolds numbers $\text{Re}_\Gamma > 30$ to 50, a so-called wave regime appears, in which wave motion is superimposed on the forward motion of the film.
3. At $\text{Re}_\Gamma \approx 1600$, the laminar regime is replaced by turbulent motion.

In general, waves at the interface create good mixing within the film, especially in the vicinity of the interface. Experimental measurements suggest that the

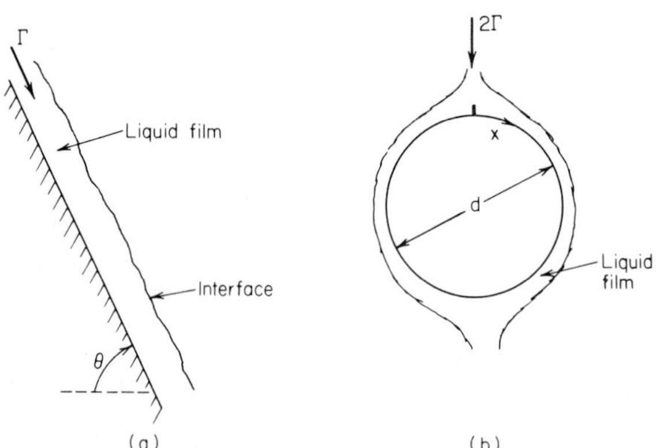

FIGURE 8.44 Falling film on (*a*) an inclined plate and (*b*) a horizontal tube. (*From Rohsenow, Hartnett, and Ganić,*[5] *p. 12-74.*)

turbulence influences the momentum transfer relatively little as compared to the wave effect.

Note that Γ is the mass flow rate of the film per unit width. For example, in the case of flow on the outside surface of a vertical tube with outer diameter d, Γ is equal to the mass flow rate divided by πd.

By equating the gravity forces with friction forces and satisfying the condition of no slip at the wall, the film thickness of a smooth laminar falling film is given as

$$\delta^+ = 0.866\text{Re}_\Gamma^{0.5} \tag{8.68}$$

where

$$\delta^+ = \frac{\delta(g\delta)^{0.5}}{\nu} \tag{8.69}$$

If the film surface is inclined at $\theta < 90°$ to the horizontal, then g in Eq. (8.69) should be replaced with $g \sin \theta$. For calculation of the mean film thicknesses of turbulent falling films, several empirical relations have been proposed in the form

$$\delta^+ = C\text{Re}_\Gamma^n \tag{8.70}$$

Recent comparisons[5] indicated that the specifications of $C = 0.095$, $n = 0.8$ are the best. The thickness of the liquid film on the horizontal tube (Fig. 8.44*b*) varies around the tube periphery because the gravity force component in the flow direction varies; in this case, g in Eq. (8.69) should be replaced with $g \sin (2x/d)$. Note that Γ for the case of a horizontal tube is the flow rate per axial unit length over one side of the horizontal tube (i.e., Γ is equal to the mass flow rate falling on a single tube divided by $2L$, where L is the tube length).

TWO-PHASE FLOW

1. General Relations. In general, when two phases flow in a pipe (Figs. 8.45 and 8.46), they do not move at the same velocity. The continuity equation for each phase fixes a relation between the phase velocities V_l and V_g, the void fraction α, and the flowing quality x. This relation is as follows. First, the definitions of the phase velocities:

$$V_l = \frac{Q_l}{A(1 - \alpha)} \tag{8.71}$$

$$V_g = \frac{Q_g}{A(\alpha)} \tag{8.72}$$

Using these definitions, it is possible to develop the following relationship between the velocity ratio, the void, and the quality:

$$\frac{1 - \alpha}{\alpha} = \left(\frac{V_g}{V_l}\right)\left(\frac{\rho_g}{\rho_l}\right)\left(\frac{1 - x}{x}\right) \tag{8.73}$$

2. Pressure-Drop Calculation by Homogeneous Method. The pressure drop in a straight pipe is the result of the action of three factors: the wall friction force,

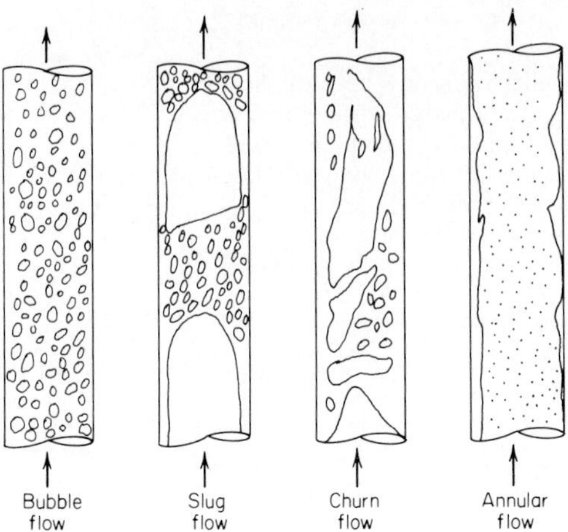

FIGURE 8.45 Flow patterns in vertical flow. (*From Rohsenow, Hartnett, and Ganić,*[5] *p. 13-13.*)

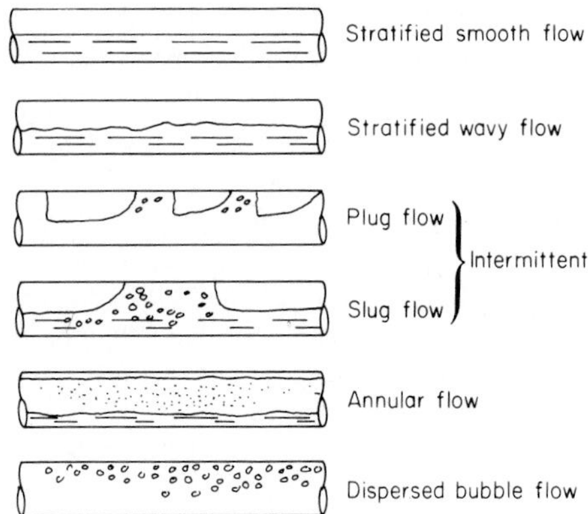

FIGURE 8.46 Flow patterns in horizontal flow. (*From Rohsenow, Hartnett, and Ganić,*[5] *p. 13-16.*)

the gravity force, and momentum changes. The general equation for pressure drop in a straight pipe is

$$\Delta p = \Delta p_f + \Delta p_G + \Delta p_m \tag{8.74}$$

The density of a homogeneous two-phase mixture is calculated assuming that the velocity of each phase is the same. That is,

$$V = \frac{1}{\rho} = V_l + xV_{lg} \tag{8.75}$$

The friction pressure drop is determined from

$$\Delta p_f = f\left(\frac{L}{D}\right)\frac{G^2V}{2} \tag{8.76}$$

The specific volume is calculated from Eq. (8.75). The mass velocity is calculated from the total flow rate and pipe cross-sectional area. Thus

$$G = \frac{\dot{m}}{A} \tag{8.77}$$

The friction factor is determined from the Moody diagram assuming that pure liquid is flowing in the pipe at the mixture mass velocity. The friction factor so determined will generally be low at qualities below 70 percent and will be too high at qualities above 70 percent.

The gravity pressure drop is calculated from

$$\Delta p_G = \rho g L \sin\theta \tag{8.78}$$

The density ρ is evaluated from Eq. (8.75). The angle θ is measured from the horizontal.

The momentum pressure drop is a result of either a pressure drop or heat transfer giving rise to a change in the volume flow rate. It is

$$\Delta p_m = G^2(V_2 - V_1) \tag{8.79}$$

The first term in the parentheses is the specific volume at the discharge, and the second term is the specific volume at the entrance. Both are evaluated using Eq. (8.75).

Note. Use this method where simplicity and an analytical espression are needed. Precision is low for low velocities or low pressure where the gravity contribution to the pressure drop is large. Precision improves for high quality and high pressure. The homogeneous method is used as the basis of the calculation of pressure drop for fittings.[5]

ACKNOWLEDGMENTS

The numerical examples in the text are based on material from Refs. 1, 2, 4, and 6 to 10.

NOMENCLATURE

Symbol = *definition, SI units (U.S. Customary units)*

A = area, m^2 (ft^2)

c = velocity of sound, m/s (ft/s)

C_D = drag coefficient

C_f = skin-friction coefficient

C_L = lift coefficient
D = inside tube diameter, m (ft); outside diameter (in "Flow over Immersed Bodies" section), m (ft)
D_h = hydraulic diameter, m (ft)
d = tube outer diameter, m (ft)
F, $\mathbf{F}$ = force (magnitude and vector), N (lb_f)
F_D = drag force, N (lb_f)
F_L = lift force, N (lb_f)
f = friction factor
G = flow rate per unit area ($= \dot{m}/A$), kg/(s · m²) [lb/(s · ft²)]
g, $\mathbf{g}$ = acceleration of gravity (magnitude and vector) m/s² (ft/s²)
h = vertical distance, m (ft)
h_f = friction head, m · N/N (ft · lb_f/lb_f)
K = loss coefficient
L = length, m (ft)
L_e = entrance length, m (ft)
$\dot{m}$ = mass flow rate, kg/s (lb/s)
p = pressure, N/m², Pa (lb_f/ft^2)
P = power, W (Btu/h)
Q = volume flow rate, m³/s (ft³/s)
Δp = pressure drop, N/m² (lb_f/ft^2)
r = radius, m (ft)
Re = Reynolds number
Re_Γ = film Reynolds number ($= 4\Gamma/\mu$)
R = gas constant, J/(kg · K) [Btu/(lb · °R)]
s = distance, m (ft)
T = temperature, K (°R)
T = torque (in "Moment-of-Momentum Equation" section), N · m (lb_f · ft)
t = time, s (s)
u = velocity in the axial direction (x direction), m/s (ft/s)
U = free-stream velocity (in "Boundary-Layer Flows" section), m/s (ft/s)
V = volume (in "Fluid Statics" section), m³ (ft³)
V, $\mathbf{V}$ = velocity (magnitude and vector), m/s (ft/s)
W = weight, N (lb_f)
w_{1-2} = work per unit mass, N · m/kg (lb_f · ft/lb)
v = specific volume, m³/kg (ft³/lb)
x = rectangular coordinate, m (ft)

y = rectangular coordinate, m (ft)
z = rectangular coordinate, m (ft)

Greek

α = void fraction (in "Two-Phase Flow" section)
δ = liquid film thickness (in "Falling Liquid Films" section)
δ^+ = dimensionless film thickness (in "Falling Liquid Films" section)
ϵ = surface roughness, m (ft)
μ = viscosity (dynamic viscosity), Pa · s [lb/(h · ft)]
τ = shear stress, N/m^2 (lb$_f$/ft^2)
ν = kinematic viscosity, m^2/s (ft^2/s)
ρ = density, kg/m^3 (lb/ft^3)
σ = surface tension, N/m (lb$_f$/ft)
γ = specific weight (= ρg), N/m^3 (lb$_f$/ft^3)
θ = angle, rad (degree)
η = eddy viscosity, Pa · s [lb/(h · ft)]
Γ = liquid film flow rate per perimeter, kg/(s · m) [lb/(s · ft)]
ω = angular velocity, 1/s

Subscripts

1 = cross section 1
2 = cross section 2
f = friction
l = liquid
g = gas
lg = g-l
m = momentum
G = gravity
x = x direction; function of x
w = wall
∞ = free-stream condition

Mathematical Operation Symbols

∇ = del operator ($\nabla = \mathbf{i}\, \partial/\partial x + \mathbf{j}\, \partial/\partial y + \mathbf{k}\, \partial/\partial z$),L/m(L/ft)
i, j, k = unit vectors
d/dx = derivative with respect to x, 1/m (1/ft)
$\partial/\partial x$ = partial derivative with respect to x, 1/m (1/ft)

REFERENCES*

1. V. L. Streeter and E. B. Wylie, *Fluid Mechanics*, 7th ed., McGraw-Hill, New York, 1979.
2. F. M. White, *Fluid Mechanics*, McGraw-Hill, New York, 1979.
3. E. A. Avallone and T. Baumeister, III (eds.), *Marks' Standard Handbook for Mechanical Engineering*, 9th ed., McGraw-Hill, New York, 1987.
4. R. F. Probstein, *Advanced Fluid Dynamics* 2.25, MIT class notes, 1974–1975.
5. W. M. Rohsenow, J. P. Hartnett, and E. N. Ganić (eds.), *Handbook of Heat Transfer Fundamentals*, 2d ed., McGraw-Hill, New York, 1985.
6. A. A. Sonin, *Fluid Mechanics* 2.20, MIT class notes, 1974–1975.
7. I. H. Shames, *Mechanics of Fluids*, McGraw-Hill, New York, 1962.
8. W.-H. Li and S.-H. Lam, *Principles of Fluid Mechanics*, Addison-Wesley, Reading, Mass., 1976.
9. R. H. F. Pao, *Fluid Dynamics*, Charles E. Merrill Books, Inc., Columbus, Ohio, 1967.
10. J. E. A. John and W. L. Haberman, *Introduction to Fluid Mechanics*, 2d ed., Prentice-Hall, Englewood Cliffs, N.J., 1980.
11. H. Schlichting, *Boundary Layer Theory*, 7th ed., McGraw-Hill, New York, 1979.
12. L. E. Moody, "Friction Factors for Pipe Flow," *Trans. ASME*, 68: 672, 1944.
13. R. J. Goldstein, *Fluid Mechanics Measurements*, Hemisphere, Washington, 1983.
14. R. H. Perry, D. W. Green, and J. O. Maloney, *Perry's Chemical Engineers' Handbook*, 6th ed., McGraw-Hill, New York, 1984.
15. L. Prandtl and O. G. Tietjens, *Fundamentals of Hydro- and Aeromechanics*, Dover Books, New York, 1934.
16. C. E. O'Rourke, *General Engineering Handbook*, 2d ed., McGraw-Hill, New York, 1940.
17. A. H. Shapiro, *Illustrated Experiments in Fluid Mechanics* (The NCFMF Book of Film Notes), MIT Press, Cambridge, Mass., 1972.

*Those references listed here but not cited in the text were used for comparison of different data sources, clarification, clarity of presentation, and, most important, reader's convenience when further interest in the subject exists.

CHAPTER 9
HEAT TRANSFER

Heat is defined as energy transferred by virtue of a temperature difference. It flows from regions of higher temperature to regions of lower temperature. It is customary to refer to different types of heat-transfer mechanisms as *modes*. The basic modes of heat transfer are *conduction* and *radiation*. In some textbooks on heat transfer, *convection* is also considered as a separate mechanism.

CONDUCTION

Conduction is the transfer of heat from one part of a body at a higher temperature to another part of the same body at a lower temperature or from one body at a higher temperature to another body at a lower temperature in physical contact with it. The conduction process takes place at the molecular level and involves the transfer of energy from the more energetic molecules to those with a lower energy level. This can be easily visualized within gases, where we note that the average kinetic energy of molecules in the higher-temperature regions is greater than those in the lower-temperature regions. The more energetic molecules, being in constant and random motion, periodically collide with molecules of a lower energy level and exchange energy and momentum. In this manner there is a continuous transport of energy from the high-temperature regions to those of lower temperature. In liquids, the molecules are more closely spaced than in gases, but the molecular energy exchange process is qualitatively similar to that in gases. In solids that are nonconductors of electricity (dielectrics), heat is conducted by lattice waves caused by atomic motion. In solids that are good conductors of electricity, this lattice vibration mechanism is only a small contribution to the energy-transfer process, the principal contribution being that due to the motion of free electrons, which move in a similar way to molecules in a gas.

At the macroscopic level the heat flux (i.e., the heat transfer rate per unit area normal to the direction of heat flow) q'' is proportional to the temperature gradient:

$$q'' = -k \frac{dT}{dx} \tag{9.1}$$

Reproduced in part from *Handbook of Heat Transfer Fundamentals*, chap. 1, by W. M. Rohsenow, J. P. Hartnett, and E. N. Ganić, eds. Copyright © 1985. Used by permission of McGraw-Hill, Inc. All rights reserved.

where the proportionality constant k is a *transport* property known as the *thermal conductivity* and is a characteristic of the material. The minus sign is a consequence of the fact that heat is transferred in the direction of decreasing temperature. Equation (9.1) is the one-dimensional form of *Fourier's law* of heat conduction. Recognizing that the heat flux is a vector quantity, we can write a more general statement of Fourier's law (i.e., the conduction rate equation) as

$$\mathbf{q}'' = -k \nabla T \tag{9.2}$$

where ∇ is the three-dimensional del operator, and T is the scalar temperature field. From Eq. (9.2) it is seen that the heat flux vector $\mathbf{q}''$ represents actually a current of heat (thermal energy) which flows in the direction of the steepest temperature gradient. Scalar components of the heat flux vector $\mathbf{q}''$ are given in Table 10.8.

1. Composite Walls

Flat Solid Wall. If we consider a one-dimensional heat flow along the x direction in the plane wall shown in Fig. 9.1*a*, direct application of Eq. (9.1) can be made, and then integration yields

$$q = \frac{kA}{\Delta x}(T_2 - T_1) \tag{9.3}$$

where the thermal conductivity is considered constant, Δx is the wall thickness, and T_1 and T_2 are the wall-face temperatures. Note that $q/A = q''$, where q is the heat transfer rate through an area A. Equation (9.3) can be written in the form

$$q = \frac{T_2 - T_1}{\Delta x/kA} = \frac{T_2 - T_1}{R_{th}} = \frac{\text{thermal potential difference}}{\text{thermal resistance}} \tag{9.4}$$

where $\Delta x/kA$ assumes the role of a *thermal resistance* R_{th}. The relation of Eq. (9.4) is quite like Ohm's law in electric circuit theory. The equivalent electric circuit for this case is shown in Fig. 9.1*b*.

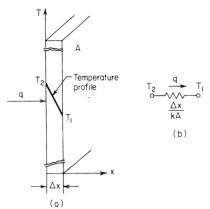

The electrical analogy may be used to solve more complex problems involving both series and parallel resistances. Typical problems and their analogous electric circuits are given in many heat transfer textbooks.[1–4]

In treating conduction problems it is often convenient to introduce another property which is related to the thermal conductivity, namely, the thermal diffusivity α,

$$\alpha = \frac{k}{\rho c} \tag{9.5}$$

where ρ is the density and c is the specific heat.

FIGURE 9.1 One-dimensional heat conduction through a plane wall (*a*) and electrical analogue (*b*). (*From Rohsenow, Hartnett, and Ganić,*[5] *p. 1-3.*)

Cylindrical Solid Wall. The rate of heat transfer through a cylindrical solid wall of length L is calculated from the equation

$$q = 2Lk \frac{T_1 - T_2}{\ln (r_2/r_1)} \tag{9.6}$$

where r_2 and r_1 are outside and inside radii, and ln is logarithm with base e. Similarly for the spherical solid wall

$$q = \frac{4kr_1r_2(T_1 - T_2)}{r_2 - r_1} \tag{9.7}$$

An important case of heat transfer is that from a hot fluid on one side of a solid wall to a cooler fluid on the other side. The wall may be a cylindrical or a flat wall of a single material or composite of different materials. The rate of heat transfer is calculated from

$$q = UA(T_h - T_c) \tag{9.8}$$

where T_h and T_c are the temperatures of the hot and cold fluids, respectively.

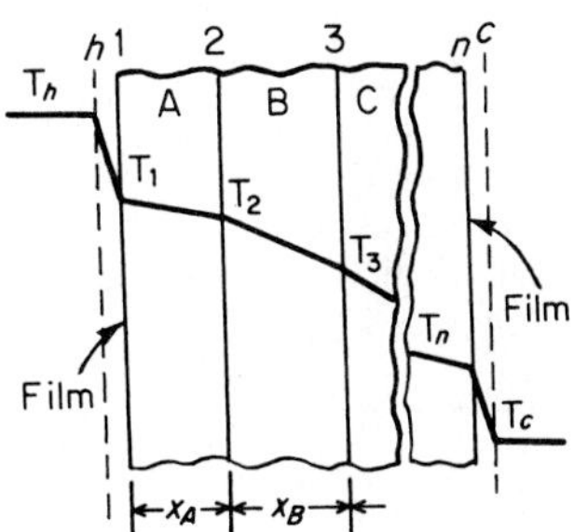

FIGURE 9.2 Composite wall.

Flat Composite Wall. Equation (9.8) is used to calculate the rate of heat transfer from a hot fluid successively through the hot-side film, layers of solid material of the wall, and cold-side film to the cold fluid (Fig. 9.2). For this case, U of Eq. (9.8) is

$$U = \frac{1}{(1/h_h) + (x_A/k_A) + (x_B/k_B) + \cdots + (1/h_c)} \tag{9.9}$$

where h_h and h_c are the hot- and cold-side film coefficients of heat transfer (see the section "Convection" for a definition of h).

Cylindrical Composite Wall. Equation (9.8) is also used to calculate the rate of heat transfer from a hot to a cold fluid through a composite cylindrical pipe wall (Fig. 9.3). For this case, the product UA of Eq. (9.8) is

$$UA = \frac{1}{\dfrac{1}{2r_1Lh_h} + \dfrac{\ln (r_2/r_1)}{2Lk_A} + \dfrac{\ln (r_3/r_2)}{2Lk_B} + \cdots + \dfrac{1}{2r_nLh_C}} \tag{9.10}$$

Spherical Composite Wall

$$UA = \frac{1}{\dfrac{1}{4r_1^2h_h} + \dfrac{r_2 - r_1}{4kr_1r_2} + \cdots + \dfrac{1}{4r_n^2h_c}} \tag{9.11}$$

General Conduction Equation. The differential equation for temperature distributions in solids is given by Eq. (10.23).

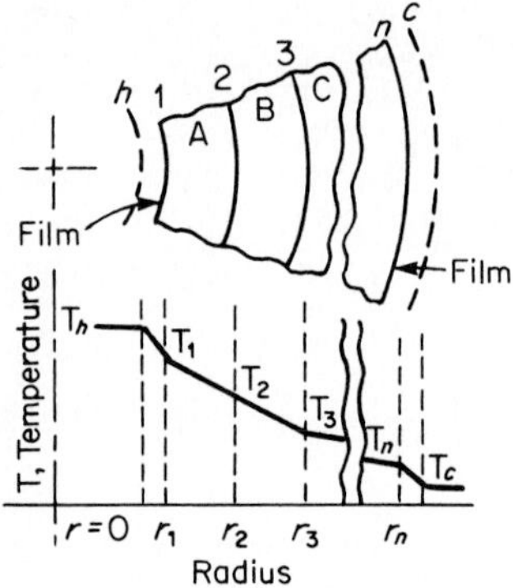

FIGURE 9.3 Composite cylinder or sphere.

2. Mass Transfer by Diffusion. As mentioned above, heat transfer will occur whenever there exists a temperature difference in a medium. Similarly, whenever there exists a difference in the concentration or density of some chemical species in a mixture, mass transfer must occur. Hence, just as a temperature gradient constitutes the driving potential for heat transfer, the existence of a concentration gradient for some species in a mixture provides the driving potential for transport of that species. Therefore, the term *mass transfer* describes the relative motion of species in a mixture due to the presence of concentration gradients.

Since the same physical mechanism is associated with heat transfer by conduction (i.e., heat diffusion) and mass transfer by diffusion, the corresponding rate equations are of the same form. The rate equation for mass diffusion is known as *Fick's law*, and for a transfer of species 1 in a binary mixture, it may be expressed as

$$j_1 = -\mathbf{D}\,\frac{dC_1}{dx} \tag{9.12}$$

where C_1 is a mass concentration of species 1 in units of mass per unit volume. This expression is analogous to Fourier's law, Eq. (9.1). Moreover, just as Fourier's law serves to define one important transport property, the thermal conductivity, Fick's law defines a second important transport property, namely the *binary diffusion coefficient* or mass diffusivity **D**. The quantity j_1 [mass/(time × surface area)] is defined as the mass flux of species 1, i.e., the amount of species 1 that is transferred per unit time and per unit area perpendicular to the direction of transfer. In vector form, Fick's law is given as

$$\mathbf{j}_1 = -\mathbf{D}\,\nabla C_1 \tag{9.13}$$

In general, the diffusion coefficient **D** for gases at low pressure is almost composition-independent; it increases with temperature and varies inversely with pressure. Liquid and solid diffusion coefficients are markedly concentration-dependent and generally increase with temperature. Differential equations for concentration distribution in solids and fluid medium are given in the section "The Conservation Equation for Species" in Chap. 10.

RADIATION

Radiation, or more correctly *thermal radiation*, is electromagnetic radiation emitted by a body by virtue of its temperature and at the expense of its internal energy. Thus thermal radiation is of the same nature as visible light, x-rays, and radio waves, the difference between them being in their wavelengths and the source of generation. The eye is sensitive to electromagnetic radiation in the region from 0.39 to 0.78 μm; this is identified as the visible region of the spectrum. Radio and hertzian waves have a wavelength of 1×10^3 to 2×10^{10} μm, and x-rays have wavelengths of 1×10^{-5} to 2×10^{-2} μm, while the bulk of thermal radiation occurs in rays from approximately 0.1 to 100 μm. All heated solids and liquids as well as some gases emit thermal radiation. The transfer of energy by conduction requires the presence of a material medium, while radiation does not. In fact, radiation transfer occurs most efficiently in a vacuum. On the macroscopic level, the calculation of thermal radiation is based on the *Stefan-Boltzmann law*, which relates the energy flux emitted by an ideal radiator (or *blackbody*) to the fourth power of the absolute temperature:

$$e_b = \sigma T^4 \tag{9.14}$$

Here σ is the Stefan-Boltzmann constant with a value of 5.669×10^{-8} W/(m$^2 \cdot$ K^4), or 1.714×10^{-9} Btu/(h $\cdot$ ft$^2 \cdot$ °R^4). Engineering surfaces in general do not perform as ideal radiators, and for real surfaces, the preceding law is modified to read

$$e = \epsilon\sigma T^4 \tag{9.15}$$

The term ϵ is called the *emissivity* of the surface and has a value between 0 and 1. When two blackbodies exchange heat by radiation, the net heat exchange is then proportional to the difference in T^4. If the first body "sees" only body 2, then the net heat exchange from body 1 to body 2 is given by

$$q = \sigma A_1(T_1^4 - T_2^4) \tag{9.16}$$

When, because of the geometrical arrangement, only a fraction of the energy leaving body 1 is intercepted by body 2,

$$q = \sigma A_1 F_{1\text{-}2}(T_1^4 - T_2^4) \tag{9.17}$$

where $F_{1\text{-}2}$ (usually called a *shape factor* or a *view factor*) is the fraction of energy leaving body 1 that is intercepted by body 2.

Since the shape factor is a purely geometrical quantity, consider the coordinate system of Fig. 9.4. The two infinitesimal surfaces dA_1 and dA_2 are separated by a distance r, and the line of connection forms the polar angles β_1 and β_2, respectively, with surface normals n_1 and n_2. For dA_1 and dA_2, the definition of a shape factor gives

$$F_{1\text{-}2} = \frac{1}{A_1}\int_{A_1}\int_{A_2} \frac{\cos\beta_1 \cos\beta_2}{\pi r^2}\, dA_1\, dA_2 \tag{9.18}$$

and

$$A_1 F_{1\text{-}2} = A_2 F_{2\text{-}1} \tag{9.19}$$

The preceding reciprocity relation for the shape factors of two finite surfaces is particularly useful in many engineering calculations. Values of shape factor for most common engineering applications (Fig. 9.5) are given in Table 9.1.

If the bodies are not black, then the view factor $F_{1\text{-}2}$ must be replaced by a new factor, $\mathcal{F}_{1\text{-}2}$, which depends on the emissivity ϵ of the surfaces involved as well as the geometrical view. Finally, if the bodies are separated by gases or liq-

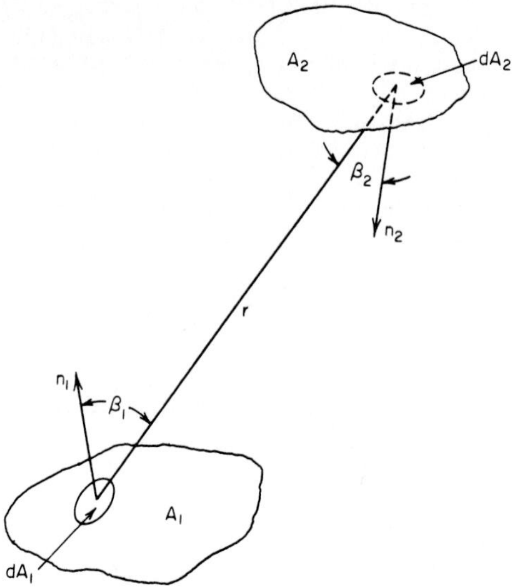

FIGURE 9.4 Coordinate system for shape factors. (*From Rohsenow, Hartnett, and Ganić,*[5] *p. 14-38.*)

uids that impede the radiation of heat through them, a formulation of the heat-exchange process becomes more involved (see Chap. 14 of Ref. 5).

CONVECTION

Convection, sometimes identified as a separate mode of heat transfer, relates to the transfer of heat from a bounding surface to a fluid in motion or to the heat transfer across a flow plane within the interior of the flowing fluid. If the fluid motion is induced by a pump, a blower, a fan, or some similar device, the process is called *forced convection*. If the fluid motion occurs as a result of the density difference produced by the heat transfer itself, the process is called *free* or *natural convection*.

Detailed inspection of the heat-transfer process in these cases reveals that although the bulk motion of the fluid gives rise to heat transfer, the basic heat-transfer mechanism is *conduction*; i.e., the energy transfer is in the form of heat transfer by conduction within the moving fluid. More specifically, it is not *heat* that is being convected but *internal energy*.

However, there are convection processes for which there is, in addition, *latent* heat exchange. This latent heat exchange is generally associated with a phase change between the liquid and vapor states of the fluid. Two special cases are *boiling* and *condensation*.

1. Heat-Transfer Coefficient. In convective processes involving heat transfer from a boundary surface exposed to a relatively low velocity fluid stream, it is

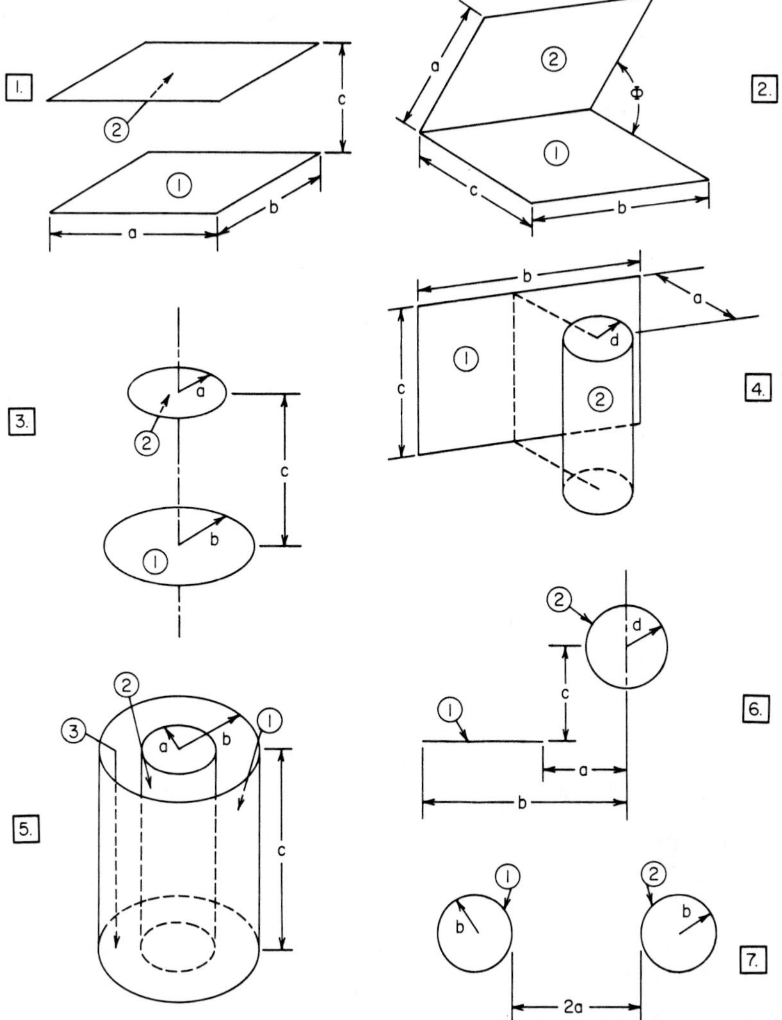

FIGURE 9.5 Schematic representations of configurations 1 through 7 (Table 9.1). (*From Rohsenow, Hartnett, and Ganić,*[5] *p. 14-39.*)

convenient to introduce a heat-transfer coefficient h, defined by Eq. (9.20), which is known as *Newton's law of cooling*:

$$q'' = h(T_w - T_f) \tag{9.20}$$

Here T_w is the surface temperature and T_f is a characteristic fluid temperature.

For surfaces in unbounded convection, such as plates, tubes, bodies of revolution, etc., immersed in a large body of fluid, it is customary to define h in Eq.

TABLE 9.1 Shape Factor Relations (Fig. 9.5)

Configuration 1: $X = a/c$, $Y = b/c$
$$F_{12}\left(\frac{\pi XY}{2}\right) = \ln\left[\frac{(1+X^2)(1+Y^2)}{1+X^2+Y^2}\right]^{1/2} + Y\sqrt{1+X^2}\tan^{-1}\frac{Y}{\sqrt{1+X^2}} + X\sqrt{1+Y^2}\tan^{-1}\frac{X}{\sqrt{1+Y^2}} - Y\tan^{-1}Y - X\tan^{-1}X$$
Configuration 2: $X = a/b$, $Y = c/b$, $Z = X^2 + Y^2 - 2XY\cos\Phi$
$$F_{12}(\pi Y) = -\frac{\sin 2\Phi}{4}\left[XY\sin\Phi + \left(\frac{\pi}{2}-\Phi\right)(X^2+Y^2) + Y^2\tan^{-1}\frac{X - Y\cos\Phi}{Y\sin\Phi} + X^2\tan^{-1}\frac{Y - X\cos\Phi}{X\sin\Phi}\right] + \frac{\sin^2\Phi}{4}\left[\left(\frac{2}{\sin^2\Phi}-1\right)\ln\frac{(1+X^2)(1+Y^2)}{1+Z} + Y^2\ln\frac{Y^2(1+Z)}{(1+Y^2)Z} + X^2\ln\frac{X^2(1+X^2)^{\cos 2\Phi}}{Z(1+Z)^{\cos 2\Phi}}\right] + Y\tan^{-1}\frac{1}{Y} + X\tan^{-1}\frac{1}{X} - \sqrt{Z}\tan^{-1}\frac{1}{\sqrt{Z}} + \frac{\sin\Phi\sin 2\Phi}{2}X\sqrt{1+X^2\sin^2\Phi} \times \left(\tan^{-1}\frac{X\cos\Phi}{\sqrt{1+X^2\sin^2\Phi}} + \tan^{-1}\frac{Y - X\cos\Phi}{\sqrt{1+X^2\sin^2\Phi}}\right] + \cos\Phi\int_0^Y \sqrt{1+\xi^2\sin^2\Phi}\left(\tan^{-1}\frac{X-\xi\cos\Phi}{\sqrt{1+\xi^2\sin^2\Phi}} + \tan^{-1}\frac{\xi\cos\Phi}{\sqrt{1+\xi^2\sin^2\Phi}}\right)d\xi$$
Configuration 3: $X = a/c$, $Y = c/b$, $Z = 1 + (1 + X^2)Y^2$
$$F_{12} = \tfrac{1}{2}\left(Z - \sqrt{Z^2 - 4X^2Y^2}\right)$$
Configuration 4: $X = a/d$, $Y = b/d$, $Z = c/d$
$$A = Z^2 + X^2 + \xi^2 - 1,\quad B = Z^2 - X^2 - \xi^2 + 1$$ $$F_{12} = \frac{2}{Y}\int_0^{Y/2} f(\xi)\,d\xi$$ $$f(\xi) = \frac{X}{X^2+\xi^2} - \frac{X}{\pi(X^2+\xi^2)} \times \left[\cos^{-1}\frac{B}{A} - \frac{1}{2Z}\left(\sqrt{A^2+4Z^2}\cos^{-1}\frac{B}{A\sqrt{X^2+\xi^2}} + B\sin^{-1}\frac{1}{\sqrt{X^2+\xi^2}} - \frac{\pi A}{2}\right)\right]$$

Source: From Rohsenow, Hartnett, and Ganić,[5] pp. 14-40 to 14-41.

TABLE 9.1 Shape Factor Relations (Fig. 9.5) (*Continued*)

Configuration 5: $X = b/a$, $Y = c/a$

$$A = Y^2 + X^2 - 1,\; B = Y^2 - X^2 + 1$$

$$F_{12} = \frac{1}{X} - \frac{1}{\pi X}\left\{\cos^{-1}\frac{B}{A} - \frac{1}{2Y}\left[\sqrt{(A+2)^2 - (2X)^2}\cos^{-1}\frac{B}{XA} + B\sin^{-1}\frac{1}{X} - \frac{\pi A}{2}\right]\right\}$$

$$F_{11} = 1 - \frac{1}{X} + \frac{2}{\pi X}\tan^{-1}\frac{2\sqrt{X^2-1}}{Y}$$

$$- \frac{Y}{2\pi X}\left[\frac{\sqrt{4X^2+Y^2}}{Y}\sin^{-1}\frac{4(X^2-1) + \frac{Y^2}{X^2}(X^2-2)}{Y^2 + 4(X^2-1)}\right.$$

$$\left. - \sin^{-1}\frac{X^2-2}{X^2} + \frac{\pi}{2}\left(\frac{\sqrt{4X^2+Y^2}}{Y} - 1\right)\right]$$

$$F_{13} = \tfrac{1}{2}(1 - F_{12} - F_{11})$$

Configuration 6: $X = c/d$, $Y = a/d$, $Z = b/d$

$$F_{12} = \frac{1}{Z - Y}\left(\tan^{-1}\frac{Z}{X} - \tan^{-1}\frac{Y}{X}\right)$$

Configuration 7: $X = 1 + (a/b)$

$$F_{12} = \frac{2}{\pi}\left(\sqrt{X^2-1} - X + \frac{\pi}{2} - \cos^{-1}\frac{1}{X}\right)$$

(9.20) with T_f as the temperature of the fluid far away from the surface, often identified as T_∞ (Fig. 9.6). For bounded convection, including such cases as fluids flowing in tubes or channels, across tubes in bundles, etc., T_f is usually taken as the enthalpy-mixed-mean temperature, customarily identified as T_m.

The heat-transfer coefficient defined by Eq. (9.20) is sensitive to the geometry, to the physical properties of the fluid, and to the fluid velocity. However, there are some special situations when h can depend on the temperature difference $T_w - T_f = \Delta T$. For example, if the surface is hot enough to boil a liquid surrounding it, h will typically vary as ΔT^2, or in the case of natural convection, h varies as some weak power of ΔT—typically as $\Delta T^{1/4}$ or $\Delta T^{1/3}$. It is important to

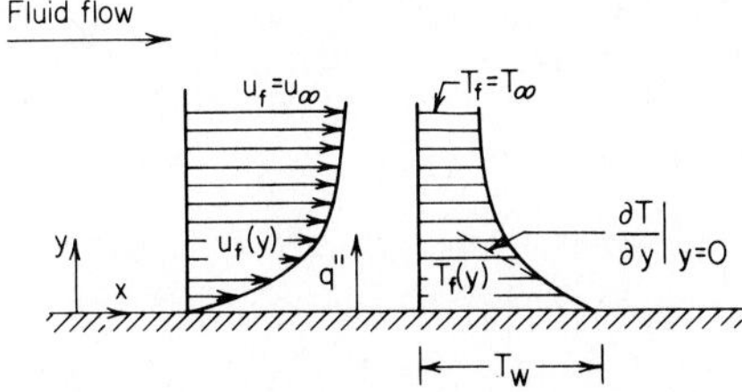

FIGURE 9.6 Velocity and temperature distributions in flow over a flat plate. (*From Rohsenow, Hartnett, and Ganić,*[5] *p. 1-5.*)

note that Eq. (9.20) as a definition of h is valid in these cases too, although its usefulness may well be reduced.

Since $q'' = q/A$, from Eq. (9.20) the thermal resistance in convection heat transfer is given by

$$R_{\text{th}} = \frac{1}{hA}$$

which is actually the resistance at a surface-to-fluid interface.

At the wall, the fluid velocity is zero, and the heat transfer takes place by conduction. Therefore, we may apply Fourier's law to the fluid at $y = 0$ (where y is the axis normal to the flow direction; Fig. 9.6):

$$q'' = -k \left.\frac{\partial T}{\partial y}\right|_{y=0} \tag{9.21}$$

where k is the thermal conductivity of fluid. By combining this equation with Newton's law of cooling [Eq. (9.20)], we then obtain

$$h = \frac{q''}{T_w - T_f} = -\frac{k(\partial T/\partial y)|_{y=0}}{T_w - T_f} \tag{9.22}$$

so that we need to find the temperature gradient at the wall in order to evaluate the heat-transfer coefficient.

The temperature gradient at the wall can be obtained by solving Eq. (10.22). The heat-transfer coefficient h can be expressed conveniently in terms of a nondimensional Nusselt number

$$\text{Nu} = \frac{hL}{k} \tag{9.23}$$

where k is the fluid thermal conductivity and L is the characteristic length (i.e., for the case of flow in a circular tube, $L = D$, where D is the tube diameter).

2. Convective Mass Transfer. Similar results may be obtained for *convective mass transfer*. If a fluid of species concentration $C_{1,\infty}$ flows over a surface at which the species concentration is maintained at some value $C_{1,w} \neq C_{1,\infty}$, transfer of the species by convection will occur. Species 1 is typically a vapor that is transferred into a gas stream by evaporation or sublimation at a liquid or solid surface, and we are interested in determining the rate at which this transfer occurs. As for the case of heat transfer, such a calculation may be based on the use of a convection coefficient.[3,5] In particular, we may relate the mass flux of species 1 to the product of a transfer coefficient and a concentration difference:

$$j_1 = h_{\text{D}}(C_{1,w} - C_{1,\infty}) \tag{9.24}$$

Here h_{D} is the convection mass-transfer coefficient, and it has a dimension of L/t.

At the wall, $y = 0$, the fluid velocity is zero, and species transfer is due only to diffusion; hence

$$j_1 = -\mathbf{D} \left.\frac{\partial C_1}{\partial y}\right|_{y=0} \tag{9.25}$$

Combining Eqs. (9.24) and (9.25), it follows that

$$h_D = -\frac{\mathbf{D}\,(\partial C_1/\partial y)|_{y=0}}{C_{1,w} - C_{1,\infty}} \tag{9.26}$$

Therefore, conditions that influence the surface concentration gradient, $(\partial C_1/\partial y)|_{y=0}$, will also influence the convection mass-transfer coefficient and the rate of species transfer across the fluid layer near the wall.

Conservation equations related to convective mass transfer are given in the section "The Conservation Equation for Species" in Chap. 10. Relations for estimating convective heat-transfer and mass-transfer coefficients for different flow geometries are given in Table 9.2. All symbols from Table 9.2 are defined in the list of nomenclature, see also Table 10.10 for physical interpretions of dimensionless groups.

Note. For convective processes involving high-velocity gas flows (high subsonic or supersonic flows), a more meaningful and useful definition of the heat-transfer coefficient is given by

$$q'' = h(T_w - T_{aw}) \tag{9.27}$$

Here T_{aw}, commonly called the *adiabatic wall temperature* or the *recovery temperature*, is the equilibrium temperature the surface would attain in the absence of any heat transfer to or from the surface and in the absence of radiation exchange between the surroundings and the surface. In general, the adiabatic wall temperature is dependent on the fluid properties and the properties of the bounding wall. Generally, the adiabatic wall temperature is reported in terms of a dimensionless recovery factor **r** defined as

$$T_{aw} = T_f + \mathbf{r}\frac{V^2}{2c_p} \tag{9.28}$$

The value of **r** for gases normally lies between 0.8 and 1.0. It can be seen that for low-velocity flows the recovery temperature is equal to the free-stream temperature T_f. In this case, Eq. (9.27) reduces to Eq. (9.20). From this point of view, Eq. (9.27) can be taken as the generalized definition of the heat-transfer coefficient.

COMBINED HEAT-TRANSFER MECHANISMS

In practice, heat transfer frequently occurs by two mechanisms in parallel. A typical example is shown in Fig. 9.7. In this case, the heat conducted through the plate is removed from the plate surface by a combination of convection and radiation. An energy balance in this case gives

$$-k_s A \left.\frac{dT}{dy}\right|_w = hA(T_w - T_\infty) + \sigma A \epsilon (T_w^4 - T_a^4) \tag{9.29}$$

where T_a is the temperature of the surroundings, k_s is the thermal conductivity of the solid plate, and ϵ is the emissivity of the plate (i.e., in this special case, $\mathcal{F}_{1-2} \equiv \epsilon$, since the area of the plate is much smaller than the area of the

TABLE 9.2 Illustrative Relations for Estimating Convective Transfer Coefficients

(Isothermal wall conditions, properties of fluids at reference temperature†; for mass transfer, replace Pr with Sc and Nu with Sh in flow geometries 1 to 9.)*

Flow geometry	Laminar flow	Transition	Turbulent flow
1. Flow parallel to a flat plate	$Nu_x = 0.332 Re_x^{1/2} Pr^{0.33} (Pr \geq 0.5)$ $Nu_x = 0.565 Re_x^{1/2} Pr^{1/2} (Pr \leq 0.025)$	$Re_x \doteq 5 \times 10^5$	$Nu_x = 0.0296 Re^{0.8} Pr^{0.6} (10.0 \geq Pr \geq 0.5)$ $Nu_L = 0.664 Re_{trans}^{1/2} Pr^{1/3} + \frac{5}{4}\left[1 - \left(\frac{Re_{trans}}{Re_L}\right)^{0.8}\right](0.0296) Re_L^{0.8} Pr^{0.6}$
2. Flow in a duct (a) A straight pipe	$Nu_D = 3.65 + \frac{0.065(D/L) Re_D Pr}{1 + 0.04[(D/L) Re_D Pr]^{2/3}}$	$Re_D \doteq 2300$	$Nu_D = 0.023 Re_D^{0.8} Pr^{0.33}; \; Pr > 0.5$ $Nu_D = 4.8 + 0.003 Re_D Pr; \; Pr < 0.1$
(b) Parallel plates	$Nu_{D_h} = 7.54 + \frac{0.03(D_h/L) Re_{D_h} Pr}{1 + 0.016[(D_h/L) Re_{D_h} Pr]^{2/3}}$	$Re_{D_h} \doteq 2800$	Nu_{D_h} same as 2a
3. Flow across a circular cylinder	$Nu_D = 0.3 + \frac{0.62 Re_D^{1/2} Pr^{1/3}}{[1 + (0.4/Pr)^{2/3}]^{1/4}}$ $Re_D Pr > 0.2$ $Re_D < 10{,}000$	$Re_D \doteq 10{,}000$	$Nu_D = Nu_D(\text{laminar}) + \left[1 + \left(\frac{Re_D}{282{,}000}\right)^{5/8}\right]^{4/5}$
4. Flow across a sphere	$Nu = 2 + 0.3 Re^{0.6} Pr^{0.33}; \; Pr \geq 0.6$ $Nu = 2 + 0.4(RePr)^{0.5}; \; Pr < 0.6$	$Re_D \doteq 150{,}000$	

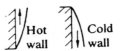

5. Free convection on a horizontal cylinder 	$\mathrm{Nu}_D = \dfrac{2}{\ln[1 + 4/(\mathrm{Gr_D Pr})^{1/4}]}$	$\mathrm{Gr}_D\mathrm{Pr} \doteq 10^9$	$\mathrm{Nu}_D^{1/2} = 0.60 + 0.387 \times \left(\dfrac{\mathrm{Gr}_D\mathrm{Pr}}{[1 + (0.559/\mathrm{Pr}^{9/16}]^{16/9}}\right)^{1/6}$
6. Free convection on a sphere 	$\mathrm{Nu}_D = 2 + K(\mathrm{Gr}_D\mathrm{Pr})^{1/4}$ $0 < \mathrm{Gr}_D\mathrm{Pr} < 50;\ K = 0.3$ $50 < \mathrm{Gr}_D\mathrm{Pr} < 200;\ K = 0.4$ $200 < \mathrm{Gr}_D\mathrm{Pr} < 10^6;\ K = 0.5$ $10^6 < \mathrm{Gr}_D\mathrm{Pr} < 10^8;\ K = 0.6$	$\mathrm{Gr}_D \doteq 10^8$	
7. Free convection on a vertical wall 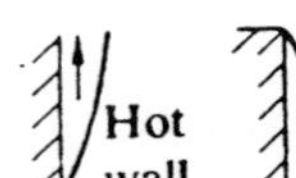	$\mathrm{Nu}_x = 0.508\dfrac{\mathrm{Gr}_x^{1/4}\mathrm{Pr}^{1/2}}{[0.952 + \mathrm{Pr}]^{1/4}}$ $\mathrm{Nu}_L = \dfrac{4}{3}\mathrm{Nu}_x(x = L)$	$\mathrm{Gr}_x \doteq 10^8$	$\mathrm{Nu}_x = 0.149(\mathrm{Pr}^{0.175} - 0.55)\mathrm{Gr}_x^{0.36}$ $\mathrm{Nu}_L = \left(\dfrac{4}{3}\right)(0.508)\dfrac{\mathrm{Gr}_{\mathrm{trans}}^{1/4}\mathrm{Pr}^{1/2}}{[0.952 + \mathrm{Pr}]^{1/4}} + \dfrac{1}{1.08}\left[1 - \left(\dfrac{\mathrm{Gr}_{\mathrm{trans}}}{\mathrm{Gr}_L}\right)^{0.92}\right](0.149) \times (\mathrm{Pr}^{0.175} - 0.55)\mathrm{Gr}_L^{0.36}$
8. Free convection on a horizontal square (a) Facing up 	$\mathrm{Nu}_L = 0.54(\mathrm{Gr}_t\mathrm{Pr})^{1/4}$ $10^3 < \mathrm{Gr}_L < 7 \times 10^7$ (For the rectangle use the shorter side for L)	$\mathrm{Gr}_L \doteq 7 \times 10^7$	$\mathrm{Nu}_L = 0.12(\mathrm{Gr}_L\mathrm{Pr})^{1/3}$ (Note that L is of no importance).
(b) Facing down	$\mathrm{Nu}_L = 0.27\ (\mathrm{Gr}_L\mathrm{Pr})^{1/4}$ $3 \times 10^5 < \mathrm{Gr}_L < 3 \times 10^{10}$		

TABLE 9.2 Illustrative Relations for Estimating Convective Transfer Coefficients (*Continued*)

Flow geometry	Laminar flow	Transition	Turbulent flow
9. Flow through a packed bed of spheres	$\text{StPr}^{2/3} = 1.625\text{Re}_D^{-1/2}$ $15 < \text{Re}_D < 120$ $\text{Re}_D = \dfrac{\dot{m}D}{A\mu}$ D: sphere diameter A: bed cross-sectional area	$\text{Re}_D = 120$ $\dfrac{\mathcal{P}}{A} = \dfrac{6(1 - \epsilon_v)}{D}$ ϵ_v: volume void fraction $\mathcal{P}$: perimeter (= mass-transfer surface area per unit bed length)	$\text{StPr}^{2/3} = 0.687\text{Re}^{-0.327}$ $120 < \text{Re} < 2000$
10. Falling film evaporation	Nusselt: $\text{Nu} = 1.10\text{Re}_F^{-1/3}; \text{Re}_F < 30$ Wavy laminar: $\text{Nu} = 0.82\text{Re}_F^{-0.22}; \text{Re}_F > 30$ $\text{Nu} = \dfrac{h(\nu^2/g)^{1/3}}{k}; \text{Re}_F = \dfrac{4\Gamma}{\mu}$	$\text{Re}_F = 1000 - 1800$ (Use the larger of the wavy laminar and turbulent film correlations.)	$\text{Nu} = 3.8 \times 10^{-3}\text{Re}_F^{0.4}\text{Pr}^{0.65}$
11. Film condensation on a vertical surface	Nusselt: $\text{Nu}_x = [\text{Gr}_x\text{Pr}_l/4\text{Ja}_l]^{1/4}; \text{Re}_F < 30$ Wavy laminar $\bar{h}_c = 0.57\text{Re}_{F,L}^{0.11}\bar{h}_{c,\text{Nu}}; \text{Re}_F > 30$	$\text{Re}_F = 1000 - 1800$ (See item 10.)	
12. Film condensation on tubes 	$\text{Nu}_D = [\text{Gr}_D\text{Pr}_l/3.6N\text{Ja}_l]^{1/4}$ N: number of tubes		

13. Nucleate, saturated, pool boiling 	$\mathrm{Nu}_{L_{cl}} = \dfrac{\mathrm{Ja}_L^2}{C^3 \mathrm{Pr}_l^m}$ $L_c = \sqrt{\sigma_l / g(\rho_l - \rho_v)}$ m = 2 for water m = 4.1 for other fluids C = 0.013 water-copper or stainless steel C = 0.006 water-nickel or brass C = 0.027 ethyl alcohol-chromium	$q \le q_{\max}$ $\mathrm{St}_{\max} = \dfrac{q_{\max}}{\rho_v V_{\max} i_{fg}} = 0.145$ $V_{\max} = [\sigma_l(\rho_l - \rho_v)g/\rho_v^2]^{1/4}$
14. Film boiling on a horizontal plate	$\mathrm{Nu}_{L_c,v} = 0.425\left[\mathrm{Gr}_{L_c,v}\mathrm{Pr}_v\left(\dfrac{1 + 0.4\mathrm{Ja}_v}{\mathrm{Ja}_v}\right)\right]^{1/4}$ $\mathrm{Gr}_{L_c,v} = \dfrac{g[(\rho_l - \rho_y)/\rho_c]L_c^3}{\nu_v^2}$ $L_c = \sqrt{\sigma_l / g(\rho_l - \rho_v)}$	$q \ge q_{\min}$ $\mathrm{St}_{\min} = \dfrac{q_{\min}}{\rho_v V_{\min} i_{fg}} = 0.09$ $V_{\min} = [\sigma_l(\rho_l - \rho_v)g/(\rho_l + \rho_v)^2]^{1/4}$
15. Film boiling on a horizontal cylinder 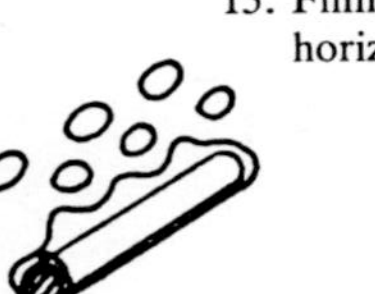	$\mathrm{Nu}_{D,v} = 0.62\left[\mathrm{Gr}_{D,v}\mathrm{Pr}_v\left(\dfrac{1 + 0.4\mathrm{Ja}_v}{\mathrm{Ja}_v}\right)\right]^{1/4}$ $\mathrm{Gr}_{D,v} = \dfrac{g[(\rho_l - \rho_v)/\rho_v]D^2}{\nu_v^2}$	
16. Film boiling on a sphere	$\mathrm{Nu}_{D,v} = 0.4\left[\mathrm{Gr}_{D,v}\mathrm{Pr}_v\left(\dfrac{1 + 0.4\mathrm{Ja}_v}{\mathrm{Ja}_v}\right)\right]^{1/3}$ $\mathrm{Gr}_{D,v} = \dfrac{g[(\rho_l - \rho_v)/\rho_v]D^3}{\nu_v^2}$	

*For external flows $\overline{Nu}$ same for uniform wall heat flux.

†$T_{ref} = T_a + \alpha(T_{a\,(\text{or } b)} - T_a)$ for low-speed flows. For general engineering use $\alpha \doteq \frac{1}{2}$, but often more precise specifications are available; e.g., for laminar natural convection of gases $\alpha = 0.38$ except for $\beta = 1/T_a$. Dimensionless parameters are also defined in Table 10.10.

Source: Adapted from Edwards, Denny, and Mills,[6] pp. 166–170.

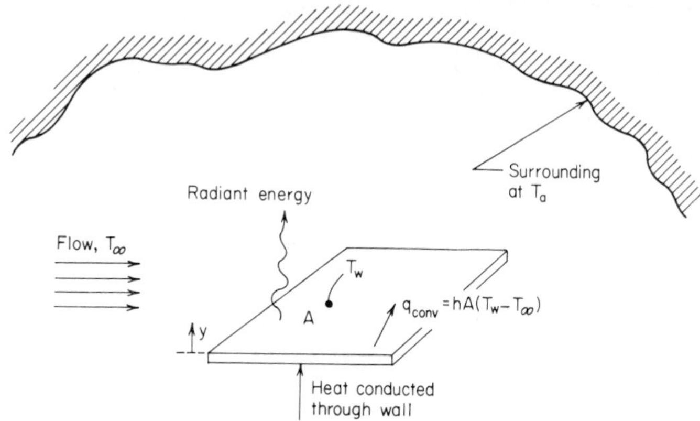

FIGURE 9.7 Combination of conduction, convection, and radiation heat transfer. *(From Rohsenow, Hartnett, and Ganić,[5] p. 1-12.)*

surroundings[3]). The plate and the surroundings are separated by a gas that has no effect on radiation.

There are many applications where radiation is combined with other modes of heat transfer, and the solution of such problems can often be simplified by using a thermal resistance R_{th} for radiation. The definition of R_{th} is similar to that of the thermal resistance for convection and conduction. If the heat transfer by radiation, for the example in Fig. 9.7, is written

$$q = \frac{T_w - T_a}{R_{th}} \tag{9.30}$$

the resistance is given by

$$R_{th} = \frac{T_w - T_a}{\sigma A \epsilon (T_w^4 - T_a^4)} \tag{9.31}$$

Also, a heat-transfer coefficient h_r can be defined for radiation:

$$h_r = \frac{1}{R_{th}A} = \frac{\sigma\epsilon(T_w^4 - T_a^4)}{T_w - T_a} = \sigma\epsilon(T_w + T_a)(T_w^2 + T_a^2) \tag{9.32}$$

Here we have linearized the radiation rate equation, making the heat rate proportional to a temperature difference rather than to the difference between two temperatures to the fourth power. Note that h_r depends strongly on temperature, while the temperature dependence of the convection heat-transfer coefficient h is generally weak.

RELATION OF HEAT TRANSFER TO THERMODYNAMICS

The subjects of thermodynamics and heat transfer are highly complementary, although some fundamental differences exist between them. Although thermodynamics is concerned with heat interaction and the vital role that it plays in the first and second laws, it considers neither the basic mechanisms that provide for heat exchange nor the methods that exist for computing the rate of heat exchange.[3,7] Thermodynamics is concerned with equilibrium states. More specifically, thermodynamics may be used to determine the amount of energy required in the form of heat for a system to pass from one equilibrium state to another. The discipline of heat transfer does what thermodynamics is inherently unable to do. It considers the rate at which heat transfer occurs in terms of thermal nonequilibrium. This is done through the introduction of a new set of physical principles. These principles are transport laws, which are not a part of the subject of thermodynamics. They include Fourier's law, the Stefan-Boltzmann law, and Newton's law of cooling, which were defined earlier. In summary, it is important to remember that a description of heat transfer requires that the additional principles be combined with the first law of thermodynamics.

NOMENCLATURE

Symbol = *Definition, SI Units (U.S. Customary Units)*

C = mass concentration of species, kg/m^3 (lb$_m$/ft^3)

c = specific heat, J/(kg · K) [Btu/(lb$_m$ · °F)]

c_p = specific heat at constant pressure, J/(kg · K) [Btu/(lb$_m$ · °F)]

c_v = specific heat at constant volume, J/(kg · K) [Btu/(lb$_m$ · °F)]

D = tube inside diameter; tube outside diameter (Table 9.2); diameter, m (ft)

D_h = hydraulic diameter (4 × flow area/wetted perimeter), m (ft)

$\mathbf{D}$ = diffusion coefficient, m^2/s (ft^2/s)

e = emissive power, W/m^2 [Btu/(h · ft^2)]

e_b = blackbody emissive power, W/m^2 [Btu/(h · ft^2)]

F_{1-2} = view factor (geometric shape factor for radiation from one blackbody to another)

$\mathcal{F}_{1-2}$ = real body view factor (geometric shape and emissivity factor for radiation from one gray body to another)

Gr = Grashof number (Table 9.2 and Table 10.10)

g = gravitational acceleration, m/s^2 (ft/s^2)

h = heat-transfer coefficient, W/(m^2 · K) [Btu/(h · ft^2 · °F)]

h_D = mass-transfer coefficient, m/s (ft/s)

i = enthalpy per unit mass, J/kg (Btu/lb$_m$)

i_{lg} = latent heat of evaporation, J/kg (Btu/lb$_m$)

j = mass diffusion flux of species, kg/(s · m^2) [lb$_m$/(h · ft^2)]

$\mathbf{j}$ = mass diffusion flux of species (vector), kg/(s · m^2) [lb$_m$/(h · ft^2)]

k = thermal conductivity, W/(m · K) [Btu/(h · ft · °F)]

L = length, m (ft)

m = mass flow rate, kg/s (lb/s)

Nu = Nusselt number (Table 9.2 and Table 10.10)

P = pressure, Pa, N/m^2 (lb$_f$/ft^2)
Pr = prandtl number (Table 9.2 and Table 10.10)
ΔP = pressure drop, Pa, N/m^2 (lb$_f$/ft^2)
q = heat-transfer rate, W (Btu/h)
$\mathbf{q}''$ = heat flux (vector), W/m^2 [Btu/(h · ft^2)]
q'' = heat flux, W/m^2 [Btu/(h · ft^2)]
q''' = volumetric heat generation, W/m^3 [Btu/(h · ft^3)]
R_{th} = thermal resistance, K/W (h · °F/Btu)
Re = Reynolds number
Re_x = local Reynolds number ($=\rho Vx/\mu$)
r = radial distance in cylindrical or spherical coordinate, m (ft)
$\mathbf{r}$ = recovery factor, Eq. (9.28)
Sc = Schmidt number (Table 9.2 and Table 10.10)
Sh = Sherwood number (Table 9.2 and Table 10.10)
St = Stanton number (Table 9.2 and Table 10.10)
T = temperatures °C, K (°F, °R)
ΔT = temperature difference, °C (°F)
t = time, s
u = velocity component in the axial direction (x direction) in rectangular coordinates, m/s (ft/s)
V = velocity, m/s (ft/s)
$\mathbf{V}$ = velocity (vector), m/s (ft/s)
v = velocity component in the y direction in rectangular coordinates, m/s (ft/s)
x = rectangular coordinate, m (ft)
y = rectangular coordinate, m (ft)
z = rectangular or cylindrical coordinate, m (ft)

Greek

α = thermal diffusivity, m^2/s (ft^2/s)
β = coefficient of thermal expansion, K^{-1} (°R^{-1})
δ = hydrodynamic boundary-layer thickness, m (ft)
δ_D = concentration boundary-layer thickness, m (ft)
δ_T = thermal boundary-layer thickness, m (ft)
ϵ = emissivity
Γ = film flow rate, kg/s · m (lb/s · ft)
μ = dynamic viscosity, Pa · s [lb/(h · ft)]
ν = kinematic viscosity, m^2/s (ft^2/s)
ρ = density, kg/m^3 (lb/ft^3)

σ = surface tension (Table 9.2), N/m (lb_f/ft); Stefan-Boltzmann radiation constant, W/($m^2 \cdot K^4$) [Btu/(h · $ft^2 \cdot °R^4$)]

σ_t = surface tension (Table 9.2), N/m (lb_f/ft)

Subscripts

a = surroundings
aw = adiabatic wall
cr = critical
f = fluid
g = gas (vapor)
i = species i
l = liquid
m = mean
r = radiation, Eq. (9.36)
s = solid
sat = saturation
t = total
w = wall
x = x component
y = y component
z = z component
θ = θ component
ϕ = ϕ component
v = vapor
max = maximum
1 = species 1 in binary mixture of 1 and 2
∞ = free-stream condition

Superscripts

$\bar{\ }$ = average (e.g., $\bar{X}$ is the average of X)

Mathematical Operation Symbols

d/dx = derivative with respect to x, m^{-1} (ft^{-1})
$\partial/\partial t$ = partial time derivative operator, s^{-1}
d/dt = total time derivative operator, s^{-1}
D/Dt = substantial time derivative operator, s^{-1}
∇ = "del" operator (vector), m^{-1} (ft^{-1})
∇^2 = Laplacian operator, m^{-2} (ft^{-2})

REFERENCES*

1. F. Kreith and V. Z. Black, *Basic Heat Transfer*, Harper & Row, New York, 1980.
2. J. P. Holman, *Heat Transfer*, 5th ed., McGraw-Hill, New York, 1981.
3. F. P. Incropera and D. P. DeWitt, *Fundamentals of Heat Transfer*, Wiley, New York, 1981.
4. M. N. Özisik, *Basic Heat Transfer*, McGraw-Hill, New York, 1977.
5. W. M. Rohsenow, J. P. Hartnett, and E. N. Ganić, eds., *Handbook of Heat Transfer Fundamentals*, McGraw-Hill, New York, 1985.
6. D. K. Edwards, V. E. Denny, and A. F. Mills, *Transfer Processes: An Introduction to Diffusion, Convection, and Radiation*, 2d ed., Hemisphere, Washington, and McGraw-Hill, New York, 1979.
7. J. H. Lienhard, *A Heat Transfer Textbook*, Prentice-Hall, Englewood Cliffs, N.J., 1981.
8. R. E. Treybal, *Mass-Transfer Operations*, 3d ed., McGraw-Hill, New York, 1980.

*Those references listed here but not cited in the text were used for comparison of different data sources, clarification, clarity of presentation, and, most important, reader's convenience when further interest in the subject exists.

CHAPTER 10
CONSERVATION EQUATIONS AND DIMENSIONLESS GROUPS

CONSERVATION EQUATIONS IN FLUID MECHANICS, HEAT TRANSFER, AND MASS TRANSFER

Each time we try to solve a new problem related to momentum and heat and mass transfer in a fluid, it is convenient to start with a set of equations based on basic laws of conservation for physical systems. These equations include

1. The continuity equation (conservation of mass)
2. The equation of motion (conservation of momentum)
3. The energy equation (conservation of energy, or the first law of thermodynamics)
4. The conservation equation for species (conservation of species)

These equations are sometimes called the *equations of change*, inasmuch as they describe the change of velocity, temperature, and concentration with respect to time and position in the system.

The first three equations are sufficient for problems involving a pure fluid (a *pure* substance is a single substance characterized by an unvarying chemical structure). The fourth equation is added for a mixture of chemical species, i.e., when mass diffusion with or without chemical reactions is present.

- *The control volume:* When deriving the conservation equations, it is necessary to select a control volume. The derivation can be performed for a volume element of any shape in a given coordinate system, although the most convenient shape is usually assumed for simplicity (e.g., a rectangular shape in a rectangular coordinate system). For illustration purposes, different coordinate systems are shown in Fig. 10.1. In selecting a control volume, we have the option of using a volume fixed in space, in which case the fluid flows through the

Adapted in part from *Handbook of Heat Transfer Fundamentals*, chap. 1, by W. M. Rohsenow, J. P. Hartnett, and E. N. Ganić, eds. Copyright © 1985. Used by permission of McGraw-Hill, Inc. All rights reserved.

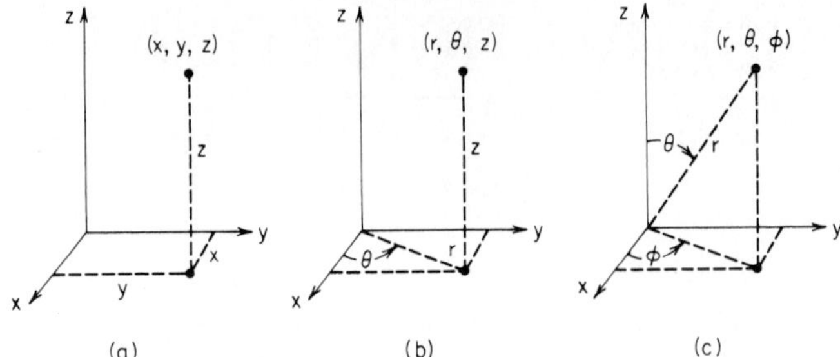

FIGURE 10.1 Coordinate systems: (*a*) rectangular, (*b*) cylindrical, (*c*) spherical.

boundaries, or a volume containing a fixed mass of fluid and moving with the fluid. The former is known as the *eulerian viewpoint*, and the latter is the *lagrangian viewpoint*. Both approaches yield equivalent results.

- *The partial time derivative* $\partial B/\partial t$: The partial time derivative of $B(x,y,z,t)$, where B is any continuum property (e.g., density, velocity, temperature, concentration, etc.), represents the change of B with time at a fixed position in space. In other words, $\partial B/\partial t$ is the change of B with t as seen by a *stationary observer*.
- *Total time derivative* dB/dt: The total time derivative is related to the partial time derivative as follows:

$$\frac{dB}{dt} = \frac{\partial B}{\partial t} + \frac{dx}{dt}\frac{\partial B}{\partial x} + \frac{dy}{dt}\frac{\partial B}{\partial y} + \frac{dz}{dt}\frac{\partial B}{\partial z} \tag{10.1}$$

where dx/dt, dy/dt, and dz/dt are the components of the velocity of the moving observer. Therefore, dB/dt is the change of B with time as seen by the moving observer.

- *Substantial time derivative* DB/Dt: This derivative is a special kind of total time derivative where now the velocity of the observer is just the same as the velocity of the stream; i.e., the observer drifts along with the current:

$$\frac{DB}{Dt} = \frac{\partial B}{\partial t} + u\frac{\partial B}{\partial x} + v\frac{\partial B}{\partial y} + w\frac{\partial B}{\partial z} \tag{10.2}$$

where u, v, and w are the components of the local fluid velocity $\mathbf{V}$. The substantial time derivative is also called the *derivative following the motion*. The sum of the last three terms on the right-hand side of Eq. (10.2) is called the *convective contribution* because it represents the change in B due to translation.

The use of the operator D/Dt is always made when rearranging various conservation equations related to the volume element fixed in space to an element following the fluid motion. The operator D/Dt also may be expressed in vector form:

$$\frac{D}{Dt} = \frac{\partial}{\partial t} + (\mathbf{V} \cdot \nabla) \tag{10.3}$$

Mathematical operations involving ∇ are given in many textbooks. Applications of ∇ in various operations involving the conservation equations are given in Refs. 1 and 2. Table 10.1 gives the expressions for D/Dt in different coordinate systems.

TABLE 10.1 Substantial Derivative in Different Coordinate Systems

Rectangular coordinates (x, y, z):

$$\frac{D}{Dt} = \frac{\partial}{\partial t} + u\frac{\partial}{\partial x} + v\frac{\partial}{\partial y} + w\frac{\partial}{\partial z}$$

Cylindrical coordinates (r, θ, z):

$$\frac{D}{Dt} = \frac{\partial}{\partial t} + v_r\frac{\partial}{\partial r} + \frac{v_\theta}{r}\frac{\partial}{\partial \theta} + v_z\frac{\partial}{\partial z}$$

Spherical coordinates (r, θ, ϕ):

$$\frac{D}{Dt} = \frac{\partial}{\partial t} + v_r\frac{\partial}{\partial r} + \frac{v_\theta}{r}\frac{\partial}{\partial \theta} + \frac{v_\phi}{r\sin\theta}\frac{\partial}{\partial \phi}$$

1. The Equation of Continuity. For a volume element fixed in space,

$$\frac{\partial \rho}{\partial t} = -\underbrace{(\nabla \cdot \rho \mathbf{V})}_{\substack{\text{net rate of mass}\\ \text{efflux per unit}\\ \text{volume}}} \qquad (10.4)$$

The continuity equation in this form describes the rate of change of density at a fixed point in the fluid. By performing the indicated differentiation on the right side of Eq. (10.4) and collecting all derivatives of ρ on the left side, we obtain an equivalent form of the equation of continuity:

$$\frac{D\rho}{Dt} = -\rho(\nabla \cdot \mathbf{V}) \qquad (10.5)$$

The continuity equation in this form describes the rate of change of density as seen by an observer "floating along" with the fluid.

For a fluid of constant density (incompressible fluid), the equation of continuity becomes

$$\nabla \cdot \mathbf{V} = 0 \qquad (10.6)$$

Table 10.2 gives the equation of continuity in different coordinate systems.

2. The Equation of Motion (Momentum Equation). The momentum equation for a stationary volume element (i.e., a balance over a volume element fixed in space) with gravity as the only body force is given by

$$\underbrace{\frac{\partial \rho \mathbf{V}}{\partial t}}_{\substack{\text{rate of increase}\\ \text{of momentum}\\ \text{per unit volume}}} = \underbrace{-(\nabla \cdot \rho \mathbf{V})\mathbf{V}}_{\substack{\text{rate of}\\ \text{momentum gain}\\ \text{by convection}\\ \text{per unit volume}}} - \underbrace{\nabla P}_{\substack{\text{pressure force}\\ \text{on element per}\\ \text{unit volume}}} + \underbrace{\nabla \cdot \tau}_{\substack{\text{rate of}\\ \text{momentum gain}\\ \text{by viscous}\\ \text{transfer per unit}\\ \text{volume}}} + \underbrace{\rho \mathbf{g}}_{\substack{\text{gravitational force}\\ \text{on element per}\\ \text{unit volume}}} \qquad (10.7)$$

TABLE 10.2 Equation of Continuity in Different Coordinate Systems

Rectangular coordinates (x, y, z):

$$\frac{\partial \rho}{\partial t} + \frac{\partial}{\partial x}(\rho u) + \frac{\partial}{\partial y}(\rho v) + \frac{\partial}{\partial z}(\rho w) = 0 \qquad (A)$$

Cylindrical coordinates (r, θ, z):

$$\frac{\partial \rho}{\partial t} + \frac{1}{r}\frac{\partial}{\partial r}(\rho r v_r) + \frac{1}{r}\frac{\partial}{\partial \theta}(\rho v_\theta) + \frac{\partial}{\partial z}(\rho v_z) = 0 \qquad (B)$$

Spherical coordinates (r, θ, ϕ):

$$\frac{\partial \rho}{\partial t} + \frac{1}{r^2}\frac{\partial}{\partial r}(\rho r^2 v_r) + \frac{1}{r \sin\theta}\frac{\partial}{\partial \theta}(\rho v_\theta \sin\theta) + \frac{1}{r \sin\theta}\frac{\partial}{\partial \phi}(\rho v_\phi) = 0 \qquad (C)$$

Incompressible flow

Rectangular coordinates (x, y, z):

$$\frac{\partial u}{\partial x} + \frac{\partial v}{\partial y} + \frac{\partial w}{\partial z} = 0 \qquad (D)$$

Cylindrical coordinates (r, θ, z):

$$\frac{1}{r}\frac{\partial}{\partial r}(r v_r) + \frac{1}{r}\frac{\partial v_\theta}{\partial \theta} + \frac{\partial v_z}{\partial z} = 0 \qquad (E)$$

Spherical coordinates (r, θ, ϕ):

$$\frac{1}{r^2}\frac{\partial}{\partial r}(r^2 v_r) + \frac{1}{r \sin\theta}\frac{\partial}{\partial \theta}(v_\theta \sin\theta) + \frac{1}{r \sin\theta}\frac{\partial v_\phi}{\partial \phi} = 0 \qquad (F)$$

Equation (10.7) may be rearranged, with the help of the equation of continuity, to give

$$\rho \frac{D\mathbf{V}}{Dt} = -\nabla P + \nabla \cdot \tau + \rho \mathbf{g} \qquad (10.8)$$

The last equation is a statement of Newton's second law of motion in the form *mass × acceleration = sum of forces*.

These two forms of the equation of motion, Eqs. (10.7) and (10.8), correspond to the two forms of the equation of continuity, Eqs. (10.4) and (10.5).

As indicated above, the only body force included in Eqs. (10.7) and (10.8) is gravity. In general, electromagnetic forces also may act on a fluid.[4]

The scalar components of Eq. (10.8) are listed in Table 10.3, and the components of the stress tensor τ are given in Table 10.4.

For the flow of a newtonian fluid with varying density but constant viscosity μ, Eq. (10.8) becomes

$$\rho \frac{D\mathbf{V}}{Dt} = -\nabla P + \frac{1}{3}\mu\nabla(\nabla \cdot V) + \mu\nabla^2\mathbf{V} + \rho\mathbf{g} \qquad (10.9)$$

TABLE 10.3 Equation of Motion in Terms of Viscous Stresses [Eq. (10.8)]*

Rectangular coordinates (x, y, z)

x direction

$$\rho\left(\frac{\partial u}{\partial t}+u\frac{\partial u}{\partial x}+v\frac{\partial u}{\partial y}+w\frac{\partial u}{\partial z}\right)=-\frac{\partial P}{\partial x}+\left(\frac{\partial \tau_{xx}}{\partial x}+\frac{\partial \tau_{yx}}{\partial y}+\frac{\partial \tau_{zx}}{\partial z}\right)+\rho g_x \qquad (A)$$

y direction

$$\rho\left(\frac{\partial v}{\partial t}+u\frac{\partial v}{\partial x}+v\frac{\partial v}{\partial y}+w\frac{\partial v}{\partial z}\right)=-\frac{\partial P}{\partial y}+\left(\frac{\partial \tau_{xy}}{\partial x}+\frac{\partial \tau_{yy}}{\partial y}+\frac{\partial \tau_{zy}}{\partial z}\right)+\rho g_y \qquad (B)$$

z direction

$$\rho\left(\frac{\partial w}{\partial t}+u\frac{\partial w}{\partial x}+v\frac{\partial w}{\partial y}+w\frac{\partial w}{\partial z}\right)=-\frac{\partial P}{\partial z}+\left(\frac{\partial \tau_{xz}}{\partial x}+\frac{\partial \tau_{yz}}{\partial y}+\frac{\partial \tau_{zz}}{\partial z}\right)+\rho g_z \qquad (C)$$

Cylindrical coordinates (r, θ, z)

r direction

$$\rho\left(\frac{\partial v_r}{\partial t}+v_r\frac{\partial v_r}{\partial r}+\frac{v_\theta}{r}\frac{\partial v_r}{\partial \theta}-\frac{v_\theta^2}{r}+v_z\frac{\partial v_r}{\partial z}\right)$$
$$=-\frac{\partial P}{\partial r}+\left[\frac{1}{r}\frac{\partial}{\partial r}(r\tau_{rr})+\frac{1}{r}\frac{\partial \tau_{\theta r}}{\partial \theta}-\frac{\tau_{\theta\theta}}{r}+\frac{\partial \tau_{zr}}{\partial z}\right]+\rho g_r \qquad (A)$$

θ direction

$$\rho\left(\frac{\partial v_\theta}{\partial t}+v_r\frac{\partial v_\theta}{\partial r}+\frac{v_\theta}{r}\frac{\partial v_\theta}{\partial \theta}+\frac{v_r v_\theta}{r}+v_z\frac{\partial v_\theta}{\partial z}\right)$$
$$=-\frac{1}{r}\frac{\partial P}{\partial \theta}+\left[\frac{1}{r^2}\frac{\partial}{\partial r}(r^2\tau_{r\theta})+\frac{1}{r}\frac{\partial \tau_{\theta\theta}}{\partial \theta}+\frac{\partial \tau_{z\theta}}{\partial z}\right]+\rho g_\theta \qquad (B)$$

z direction

$$\rho\left(\frac{\partial v_z}{\partial t}+v_r\frac{\partial v_z}{\partial r}+\frac{v_\theta}{r}\frac{\partial v_z}{\partial \theta}+v_z\frac{\partial v_z}{\partial z}\right)$$
$$=-\frac{\partial P}{\partial z}+\left[\frac{1}{r}\frac{\partial}{\partial r}(r\tau_{rz})+\frac{1}{r}\frac{\partial \tau_{\theta z}}{\partial \theta}+\frac{\partial \tau_{zz}}{\partial z}\right]+\rho g_z \qquad (C)$$

Spherical coordinates (r, θ, ϕ)

r direction

$$\rho\left(\frac{\partial v_r}{\partial t}+v_r\frac{\partial v_r}{\partial r}+\frac{v_\theta}{r}\frac{\partial v_r}{\partial \theta}+\frac{v_\phi}{r\sin\theta}\frac{\partial v_r}{\partial \phi}-\frac{v_\theta^2+v_\phi^2}{r}\right)=-\frac{\partial P}{\partial r}$$
$$+\left[\frac{1}{r^2}\frac{\partial}{\partial r}(r^2\tau_{rr})+\frac{1}{r\sin\theta}\frac{\partial}{\partial \theta}(\tau_{\theta r}\sin\theta)+\frac{1}{r\sin\theta}\frac{\partial \tau_{\phi r}}{\partial \phi}-\frac{\tau_{\theta\theta}+\tau_{\phi\phi}}{r}\right]+\rho g_r \qquad (A)$$

θ direction

$$\rho\left(\frac{\partial v_\theta}{\partial t}+v_r\frac{\partial v_\theta}{\partial r}+\frac{v_\theta}{r}\frac{\partial v_\theta}{\partial \theta}+\frac{v_\phi}{r\sin\theta}\frac{\partial v_\theta}{\partial \phi}+\frac{v_r v_\theta}{r}-\frac{v_\phi^2\cot\theta}{r}\right)=-\frac{1}{r}\frac{\partial P}{\partial \theta}$$
$$+\left[\frac{1}{r^2}\frac{\partial}{\partial r}(r^2\tau_{r\theta})+\frac{1}{r\sin\theta}\frac{\partial}{\partial \theta}(\tau_{\theta\theta}\sin\theta)+\frac{1}{r\sin\theta}\frac{\partial \tau_{\phi\theta}}{\partial \phi}+\frac{\tau_{r\theta}}{r}-\frac{\tau_{\phi\phi}\cot\theta}{r}\right]+\rho g_\theta \qquad (B)$$

ϕ direction

$$\rho\left(\frac{\partial v_\phi}{\partial t}+v_r\frac{\partial v_\phi}{\partial r}+\frac{v_\theta}{\partial r}\frac{\partial v_\phi}{\partial \theta}+\frac{v_\phi}{r\sin\theta}\frac{\partial v_\phi}{\partial \phi}+\frac{v_\phi v_r}{r}+\frac{v_\theta v_\phi\cot\theta}{r}\right)$$
$$=-\frac{1}{r\sin\theta}\frac{\partial P}{\partial \phi}+\left[\frac{1}{r^2}\frac{\partial}{\partial r}(r^2\tau_{r\phi})+\frac{1}{r}\frac{\partial \tau_{\theta\phi}}{\partial \theta}+\frac{1}{r\sin\theta}\frac{\partial \tau_{\phi\phi}}{\partial \phi}+\frac{\tau_{r\phi}}{r}+\frac{2\tau_{\theta\phi}\cot\theta}{r}\right]+\rho g_\phi \qquad (C)$$

*Components of the stress tensor τ for newtonian fluids are given in Table 10.4. This equation also may be used for describing nonnewtonian flow. However, we need relations between the components of τ and the various velocity gradients; in other words, we have to replace the expressions given in Table 10.4 by other relations appropriate for the nonnewtonian fluid of interest. The expressions for τ for some nonnewtonian fluid models are given in Ref. 2. See also Ref. 4.

TABLE 10.4 Components of the Stress Tensor τ for Newtonian Fluids*

Rectangular coordinates (x, y, z)	
$\tau_{xx} = \mu\left[2\frac{\partial u}{\partial x} - \frac{2}{3}(\nabla \cdot \mathbf{V})\right]$	(A)
$\tau_{yy} = \mu\left[2\frac{\partial v}{\partial y} - \frac{2}{3}(\nabla \cdot \mathbf{V})\right]$	(B)
$\tau_{zz} = \mu\left[2\frac{\partial w}{\partial z} - \frac{2}{3}(\nabla \cdot \mathbf{V})\right]$	(C)
$\tau_{xy} = \tau_{yx} = \mu\left[\frac{\partial u}{\partial y} + \frac{\partial v}{\partial x}\right]$	(D)
$\tau_{yz} = \tau_{zy} = \mu\left[\frac{\partial v}{\partial z} + \frac{\partial w}{\partial y}\right]$	(E)
$\tau_{zx} = \tau_{xz} = \mu\left[\frac{\partial w}{\partial x} + \frac{\partial u}{\partial z}\right]$	(F)
$(\nabla \cdot \mathbf{V}) = \frac{\partial u}{\partial x} + \frac{\partial v}{\partial y} + \frac{\partial w}{\partial z}$	(G)
Cylindrical coordinates (r, θ, z)	
$\tau_{rr} = \mu\left[2\frac{\partial v_r}{\partial r} - \frac{2}{3}(\nabla \cdot \mathbf{V})\right]$	(A)
$\tau_{\theta\theta} = \mu\left[2\left(\frac{1}{r}\frac{\partial v_\theta}{\partial \theta} + \frac{v_r}{r}\right) - \frac{2}{3}(\nabla \cdot \mathbf{V})\right]$	(B)
$\tau_{zz} = \mu\left[2\frac{\partial v_z}{\partial z} - \frac{2}{3}(\nabla \cdot \mathbf{V})\right]$	(C)
$\tau_{r\theta} = \tau_{\theta r} = \mu\left[r\frac{\partial}{\partial r}\left(\frac{v_\theta}{r}\right) + \frac{1}{r}\frac{\partial v_r}{\partial \theta}\right]$	(D)
$\tau_{\theta z} = \tau_{z\theta} = \mu\left[\frac{\partial v_\theta}{\partial z} + \frac{1}{r}\frac{\partial v_z}{\partial \theta}\right]$	(E)
$\tau_{zr} = \tau_{rz} = \mu\left[\frac{\partial v_z}{\partial r} + \frac{\partial v_r}{\partial z}\right]$	(F)
$(\nabla \cdot \mathbf{V}) = \frac{1}{r}\frac{\partial}{\partial r}(rv_r) + \frac{1}{r}\frac{\partial v_\theta}{\partial \theta} + \frac{\partial v_z}{\partial z}$	(G)
Spherical coordinates (r, θ, ϕ)	
$\tau_{rr} = \mu\left[2\frac{\partial v_r}{\partial r} - \frac{2}{3}(\nabla \cdot \mathbf{V})\right]$	(A)
$\tau_{\theta\theta} = \mu\left[2\left(\frac{1}{r}\frac{\partial v_\theta}{\partial \theta} + \frac{v_r}{r}\right) - \frac{2}{3}(\nabla \cdot \mathbf{V})\right]$	(B)
$\tau_{\phi\phi} = \mu\left[2\left(\frac{1}{r\sin\theta}\frac{\partial v_\phi}{\partial \phi} + \frac{v_r}{r} + \frac{v_\theta \cot\theta}{r}\right) - \frac{2}{3}(\nabla \cdot \mathbf{V})\right]$	(C)
$\tau_{r\theta} = \tau_{\theta r} = \mu\left[r\frac{\partial}{\partial r}\left(\frac{v_\theta}{r}\right) + \frac{1}{r}\frac{\partial v_r}{\partial \theta}\right]$	(D)

*It should be noted that the sign convention adopted here for components of the stress tensor is consistent with that found in many fluid mechanics and heat-transfer books; however, it is opposite to that found in some books on transport phenomena, e.g., Refs. 2, 3, and 5.

TABLE 10.4 Components of the Stress Tensor τ for Newtonian Fluids (*Continued*)

$$\tau_{\theta\phi} = \tau_{\phi\theta} = \mu \left[\frac{\sin\theta}{r} \frac{\partial}{\partial\theta} \left(\frac{v_\phi}{\sin\theta} \right) + \frac{1}{r\sin\theta} \frac{\partial v_\theta}{\partial\phi} \right] \quad (E)$$

$$\tau_{\phi r} = \tau_{r\phi} = \mu \left[\frac{1}{r\sin\theta} \frac{\partial v_r}{\partial\phi} + r \frac{\partial}{\partial r} \left(\frac{v_\phi}{r} \right) \right] \quad (F)$$

$$(\nabla \cdot \mathbf{V}) = \frac{1}{r^2} \frac{\partial}{\partial r} (r^2 v_r) + \frac{1}{r\sin\theta} \frac{\partial}{\partial\theta} (v_\theta \sin\theta) + \frac{1}{r\sin\theta} \frac{\partial v_\phi}{\partial\phi} \quad (G)$$

If ρ and μ are constant, Eq. (10.9) may be simplified by means of the equation of continuity ($\nabla \cdot \mathbf{V} = 0$) to give for a newtonian fluid

$$\rho \frac{D\mathbf{V}}{Dt} = -\nabla P + \mu\nabla^2\mathbf{V} + \rho\mathbf{g} \quad (10.10)$$

This is the famous Navier-Stokes equation in vector form. The scalar components of Eq. (10.10) are given in Table 10.5.

For $\nabla \cdot \tau = 0$, Eq. (10.8) reduces to Euler's equation:

$$\rho \frac{D\mathbf{V}}{Dt} = -\nabla P + \rho\mathbf{g} \quad (10.11)$$

which is applicable for describing flow systems in which viscous effects are relatively unimportant.

As mentioned before, there is a subset of flow problems, called *natural convection*, where the flow pattern is due to buoyant forces caused by temperature differences. Such buoyant forces are proportional to the coefficient of thermal expansion β, defined as

$$\beta = -\frac{1}{\rho} \left(\frac{\partial\rho}{\partial T} \right)_P \quad (10.12)$$

where T is absolute temperature. Using an approximation that applies to low fluid velocities and small temperature variations, it can be shown[2,3] that

$$\nabla P - \rho\mathbf{g} = \rho\beta\mathbf{g}(T - T_\infty) \quad (10.13)$$

Then Eq. (10.8) becomes

$$\rho \frac{D\mathbf{V}}{Dt} = \nabla \cdot \tau - \underbrace{\rho\beta\mathbf{g}(T - T_\infty)}_{\text{buoyant force on element per unit volume}} \quad (10.14)$$

The preceding equation of motion is used for setting up problems in natural convection when the ambient temperature T_∞ may be defined.

3. The Energy Equation. For a stationary volume element through which a pure fluid is flowing, the energy equation reads

$$\frac{\partial}{\partial t}\rho(\mathbf{u} + \tfrac{1}{2}V^2) = -\nabla \cdot \rho\mathbf{V}(\mathbf{u} + \tfrac{1}{2}V^2) - \nabla \cdot \mathbf{q}'' + \rho(\mathbf{V}\cdot\mathbf{g})$$

rate of gain of energy per unit volume; rate of energy input per unit volume by convection; rate of energy input per unit volume by conduction; rate of work done on fluid per unit volume by gravitational forces

$$- \nabla \cdot P\mathbf{V} + \nabla \cdot (\boldsymbol{\tau} \cdot \mathbf{V}) + q''' \tag{10.15}$$

rate of work done on fluid per unit volume by pressure forces; rate of work done on fluid per unit volume by viscous forces; rate of heat generation per unit volume ("source term")

The left side of the preceding equation, which represents the rate of accumulation of internal and kinetic energy, does not include the potential energy of the fluid, since this form of energy is included in the work term on the right side. Equation

TABLE 10.5 Equation of Motion in Terms of Velocity Gradients for a Newtonian Fluid with Constant ρ and μ, Eq. (10.10)

Rectangular coordinates (x, y, z)

x direction

$$\rho\left(\frac{\partial u}{\partial t} + u\frac{\partial u}{\partial x} + v\frac{\partial u}{\partial y} + w\frac{\partial u}{\partial z}\right) = -\frac{\partial P}{\partial x} + \mu\left(\frac{\partial^2 u}{\partial x^2} + \frac{\partial^2 u}{\partial y^2} + \frac{\partial^2 u}{\partial z^2}\right) + \rho g_x \tag{A}$$

y direction

$$\rho\left(\frac{\partial v}{\partial t} + u\frac{\partial v}{\partial x} + v\frac{\partial v}{\partial y} + w\frac{\partial v}{\partial z}\right) = -\frac{\partial P}{\partial y} + \mu\left(\frac{\partial^2 v}{\partial x^2} + \frac{\partial^2 v}{\partial y^2} + \frac{\partial^2 v}{\partial z^2}\right) + \rho g_y \tag{B}$$

z direction

$$\rho\left(\frac{\partial w}{\partial t} + u\frac{\partial w}{\partial x} + v\frac{\partial w}{\partial y} + w\frac{\partial w}{\partial z}\right) = -\frac{\partial P}{\partial z} + \mu\left(\frac{\partial^2 w}{\partial x^2} + \frac{\partial^2 w}{\partial y^2} + \frac{\partial^2 w}{\partial z^2}\right) + \rho g_z \tag{C}$$

Cylindrical coordinates (r, θ, z)

r direction

$$\rho\left(\frac{\partial v_r}{\partial t} + v_r\frac{\partial v_r}{\partial r} + \frac{v_\theta}{r}\frac{\partial v_r}{\partial \theta} - \frac{v_\theta^2}{r} + v_z\frac{\partial v_r}{\partial z}\right)$$

$$= -\frac{\partial P}{\partial r} + \mu\left[\frac{\partial}{\partial r}\left(\frac{1}{r}\frac{\partial}{\partial r}(rv_r)\right) + \frac{1}{r^2}\frac{\partial^2 v_r}{\partial \theta^2} - \frac{2}{r^2}\frac{\partial v_\theta}{\partial \theta} + \frac{\partial^2 v_r}{\partial z^2}\right] + \rho g_r \tag{A}$$

θ direction

$$\rho\left(\frac{\partial v_\theta}{\partial t} + v_r\frac{\partial v_\theta}{\partial r} + \frac{v_\theta}{r}\frac{\partial v_\theta}{\partial \theta} + \frac{v_r v_\theta}{r} + v_z\frac{\partial v_\theta}{\partial z}\right)$$

$$= -\frac{1}{r}\frac{\partial P}{\partial \theta} + \mu\left[\frac{\partial}{\partial r}\left(\frac{1}{r}\frac{\partial}{\partial r}(rv_\theta)\right) + \frac{1}{r^2}\frac{\partial^2 v_\theta}{\partial \theta^2} + \frac{2}{r^2}\frac{\partial v_r}{\partial \theta} + \frac{\partial^2 v_\theta}{\partial z^2}\right] + \rho g_\theta \tag{B}$$

z direction

$$\rho\left(\frac{\partial v_z}{\partial t} + v_r\frac{\partial v_z}{\partial r} + \frac{v_\theta}{r}\frac{\partial v_z}{\partial \theta} + v_z\frac{\partial v_z}{\partial z}\right)$$

$$= -\frac{\partial P}{\partial z} + \mu\left[\frac{1}{r}\frac{\partial}{\partial r}\left(r\frac{\partial v_z}{\partial r}\right) + \frac{1}{r^2}\frac{\partial^2 v_z}{\partial \theta^2} + \frac{\partial^2 v_z}{\partial z^2}\right] + \rho g_z \tag{C}$$

TABLE 10.5 Equation of Motion in Terms of Velocity Gradients for a Newtonian Fluid with Constant ρ and μ, Eq. (10.10) (*Continued*)

Spherical coordinates (r, θ, ϕ)†

r direction

$$\rho\left(\frac{\partial v_r}{\partial t} + v_r\frac{\partial v_r}{\partial r} + \frac{v_\theta}{r}\frac{\partial v_r}{\partial \theta} + \frac{v_\phi}{r\sin\theta}\frac{\partial v_r}{\partial \phi} - \frac{v_\theta^2 + v_\phi^2}{r}\right)$$

$$= -\frac{\partial P}{\partial r} + \mu\left(\nabla^2 v_r - \frac{2v_r}{r^2} - \frac{2}{r^2}\frac{\partial v_\theta}{\partial \theta} - \frac{2v_\theta\cot\theta}{r^2} - \frac{2}{r^2\sin\theta}\frac{\partial v_\phi}{\partial \phi}\right) + \rho g_r \qquad (A)$$

θ direction

$$\rho\left(\frac{\partial v_\theta}{\partial t} + v_r\frac{\partial v_\theta}{\partial r} + \frac{v_\theta}{r}\frac{\partial v_\theta}{\partial \theta} + \frac{v_\phi}{r\sin\theta}\frac{\partial v_\theta}{\partial \phi} + \frac{v_r v_\theta}{r} - \frac{v_\phi^2\cot\theta}{r}\right)$$

$$= -\frac{1}{r}\frac{\partial P}{\partial \theta} + \mu\left(\nabla^2 v_\theta + \frac{2}{r^2}\frac{\partial v_r}{\partial \theta} - \frac{v_\theta}{r^2\sin^2\theta} - \frac{2\cos\theta}{r^2\sin^2\theta}\frac{\partial v_\phi}{\partial \phi}\right) + \rho g_\theta \qquad (B)$$

ϕ direction

$$\rho\left(\frac{\partial v_\phi}{\partial t} + v_r\frac{\partial v_\phi}{\partial r} + \frac{v_\theta}{r}\frac{\partial v_\phi}{\partial \theta} + \frac{v_\phi}{r\sin\theta}\frac{\partial v_\phi}{\partial \phi} + \frac{v_\phi v_r}{r} + \frac{v_\theta v_\phi}{r}\cot\theta\right)$$

$$= -\frac{1}{r\sin\theta}\frac{\partial P}{\partial \phi} + \mu\left(\nabla^2 v_\phi - \frac{v_\phi}{r^2\sin^2\theta} + \frac{2}{r^2\sin\theta}\frac{\partial v_r}{\partial \phi} + \frac{2\cos\theta}{r^2\sin^2\theta}\frac{\partial v_\theta}{\partial \phi}\right) + \rho g_\phi \qquad (C)$$

†For spherical coordinates the Laplacian is

$$\nabla^2 = \frac{1}{r^2}\frac{\partial}{\partial r}\left(r^2\frac{\partial}{\partial r}\right) + \frac{1}{r^2\sin\theta}\frac{\partial}{\partial \theta}\left(\sin\theta\frac{\partial}{\partial \theta}\right) + \frac{1}{r^2\sin^2\theta}\left(\frac{\partial^2}{\partial \phi^2}\right)$$

(10.15) may be rearranged, with the aid of the equations of continuity and motion, to give[2,6]

$$\rho\frac{D\mathbf{u}}{Dt} = -\nabla\cdot\mathbf{q}'' - P(\nabla\cdot\mathbf{V}) + \nabla\mathbf{V}:\tau + q''' \qquad (10.16)$$

A summary of $\nabla\mathbf{V}:\tau$ in different coordinate systems is given in Table 10.6. For a newtonian fluid,

$$\nabla\mathbf{V}:\tau = \mu\Phi \qquad (10.17)$$

and values of the dissipation function Φ in different coordinate systems are given in Table 10.7. Components of the heat flux vector $\mathbf{q}'' = -k\nabla T$ are given in Table 10.8 for different coordinate systems.

Often it is more convenient to work with enthalpy rather than internal energy. Using the definition of enthalpy, $i = \mathbf{u} + P/\rho$, and the mass conservation equation [Eq. (10.5)], then Eq. (10.16) can be rearranged to give

$$\rho\frac{Di}{Dt} = \nabla\cdot k\nabla T + \frac{DP}{Dt} + \mu\Phi + q''' \qquad (10.18)$$

TABLE 10.6 Summary of Dissipation Term $\nabla\mathbf{V}:\tau$ in Different Coordinate Systems

Rectangular coordinates (x, y, z):

$$\nabla\mathbf{V}:\tau = \tau_{xx}\left(\frac{\partial u}{\partial x}\right) + \tau_{yy}\left(\frac{\partial v}{\partial y}\right) + \tau_{zz}\left(\frac{\partial w}{\partial z}\right) + \tau_{xy}\left(\frac{\partial u}{\partial y} + \frac{\partial v}{\partial x}\right) + \tau_{yz}\left(\frac{\partial v}{\partial z} + \frac{\partial w}{\partial y}\right) + \tau_{zx}\left(\frac{\partial w}{\partial x} + \frac{\partial u}{\partial z}\right) \quad (A)$$

Cylindrical coordinates (r, θ, z):

$$\nabla\mathbf{V}:\tau = \tau_{rr}\left(\frac{\partial v_r}{\partial r}\right) + \tau_{\theta\theta}\left(\frac{1}{r}\frac{\partial v_\theta}{\partial \theta} + \frac{v_r}{r}\right) + \tau_{zz}\left(\frac{\partial v_z}{\partial z}\right) + \tau_{r\theta}\left[r\frac{\partial}{\partial r}\left(\frac{v_\theta}{r}\right) + \frac{1}{r}\frac{\partial v_r}{\partial \theta}\right] + \tau_{\theta z}\left(\frac{1}{r}\frac{\partial v_z}{\partial \theta} + \frac{\partial v_\theta}{\partial z}\right) + \tau_{rz}\left(\frac{\partial v_z}{\partial r} + \frac{\partial v_r}{\partial z}\right) \quad (B)$$

Spherical coordinates (r, θ, ϕ):

$$\nabla\mathbf{V}:\tau = \tau_{rr}\left(\frac{\partial v_r}{\partial r}\right) + \tau_{\theta\theta}\left(\frac{1}{r}\frac{\partial v_\theta}{\partial \theta} + \frac{v_r}{r}\right) + \tau_{\phi\phi}\left(\frac{1}{r\sin\theta}\frac{\partial v_\phi}{\partial \phi} + \frac{v_r}{r} + \frac{v_\theta \cot\theta}{r}\right) + \tau_{r\theta}\left(\frac{\partial v_\theta}{\partial r} + \frac{1}{r}\frac{\partial v_r}{\partial \theta} - \frac{v_\theta}{r}\right) + \tau_{r\phi}\left(\frac{\partial v_\phi}{\partial r} + \frac{1}{r\sin\theta}\frac{\partial v_r}{\partial \phi} - \frac{v_\phi}{r}\right) + \tau_{\theta\phi}\left(\frac{1}{r}\frac{\partial v_\phi}{\partial \theta} + \frac{1}{r\sin\theta}\frac{\partial v_\theta}{\partial \phi} - \frac{v_\phi \cot\theta}{r}\right) \quad (C)$$

For most engineering applications, it is convenient to have the equation of thermal energy in terms of the fluid temperature and heat capacity rather than the internal energy or enthalpy. In general, for pure substances,[3]

$$\frac{Di}{Dt} = \left(\frac{\partial i}{\partial P}\right)_T \frac{DP}{Dt} + \left(\frac{\partial i}{\partial T}\right)_P \frac{DT}{Dt} = \frac{1}{\rho}(1 - \beta T)\frac{DP}{Dt} + c_p\frac{DT}{Dt} \quad (10.19)$$

where β is defined by Eq. (10.12). Substituting this into Eq. (10.18), we have the following general relation:

$$\rho c_p \frac{DT}{Dt} = \nabla \cdot k\nabla T + T\beta\frac{DP}{Dt} + \mu\Phi + q''' \quad (10.20)$$

For an ideal gas, $\beta = 1/T$, and then

$$\rho c_p \frac{DT}{Dt} = \nabla \cdot k\nabla T + \frac{DP}{Dt} + \mu\Phi + q''' \quad (10.21)$$

Note that c_p need not be constant.

We could have obtained Eq. (10.21) directly from Eq. (10.18) by noting that for an ideal gas, $di = c_p\, dT$, where c_p is constant, and thus

$$\frac{Di}{Dt} = c_p\frac{DT}{Dt}$$

TABLE 10.7 The Viscous Dissipation Function Φ

Rectangular coordinates (x, y, z):

$$\Phi = 2\left[\left(\frac{\partial u}{\partial x}\right)^2 + \left(\frac{\partial v}{\partial y}\right)^2 + \left(\frac{\partial w}{\partial z}\right)^2\right] + \left(\frac{\partial v}{\partial x} + \frac{\partial u}{\partial y}\right)^2 + \left(\frac{\partial w}{\partial y} + \frac{\partial v}{\partial z}\right)^2 + \left(\frac{\partial u}{\partial z} + \frac{\partial w}{\partial x}\right)^2 - \frac{2}{3}\left(\frac{\partial u}{\partial x} + \frac{\partial v}{\partial y} + \frac{\partial w}{\partial z}\right)^2 \qquad (A)$$

Cylindrical coordinates (r, θ, z):

$$\Phi = 2\left[\left(\frac{\partial v_r}{\partial r}\right)^2 + \left(\frac{1}{r}\frac{\partial v_\theta}{\partial \theta} + \frac{v_r}{r}\right)^2 + \left(\frac{\partial v_z}{\partial z}\right)^2\right] + \left[r\frac{\partial}{\partial r}\left(\frac{v_\theta}{r}\right) + \frac{1}{r}\frac{\partial v_r}{\partial \theta}\right]^2 + \left[\frac{1}{r}\frac{\partial v_z}{\partial \theta} + \frac{\partial v_\theta}{\partial z}\right]^2 + \left(\frac{\partial v_r}{\partial z} + \frac{\partial v_z}{\partial r}\right)^2 - \frac{2}{3}\left[\frac{1}{r}\frac{\partial}{\partial r}(rv_r) + \frac{1}{r}\frac{\partial v_\theta}{\partial \theta} + \frac{\partial v_z}{\partial z}\right]^2 \qquad (B)$$

Spherical coordinates (r, θ, ϕ):

$$\Phi = 2\left[\left(\frac{\partial v_r}{\partial r}\right)^2 + \left(\frac{1}{r}\frac{\partial v_\theta}{\partial \theta} + \frac{v_r}{r}\right)^2 + \left(\frac{1}{r\sin\theta}\frac{\partial v_\phi}{\partial \phi} + \frac{v_r}{r} + \frac{v_\theta\cot\theta}{r}\right)^2\right]$$
$$+ \left[r\frac{\partial}{\partial r}\left(\frac{v_\theta}{r}\right) + \frac{1}{r}\frac{\partial v_r}{\partial \theta}\right]^2 + \left[\frac{\sin\theta}{r}\frac{\partial}{\partial \theta}\left(\frac{v_\phi}{\sin\theta}\right) + \frac{1}{r\sin\theta}\frac{\partial v_\theta}{\partial \phi}\right]^2 + \left[\frac{1}{r\sin\theta}\frac{\partial v_r}{\partial \phi} + r\frac{\partial}{\partial r}\left(\frac{v_\phi}{r}\right)\right]^2$$
$$- \frac{2}{3}\left[\frac{1}{r^2}\frac{\partial}{\partial r}(r^2 v_r) + \frac{1}{r\sin\theta}\frac{\partial}{\partial \theta}(v_\theta\sin\theta) + \frac{1}{r\sin\theta}\frac{\partial v_\phi}{\partial \phi}\right]^2 \qquad (C)$$

TABLE 10.8 Scalar Components of the Heat Flux Vector **q**″

Rectangular (x, y, z)		Cylindrical (r, θ, z)		Spherical (r, θ, ϕ)	
$q''_x = -k\dfrac{\partial T}{\partial x}$	(A)	$q''_r = -k\dfrac{\partial T}{\partial r}$	(D)	$q''_r = -k\dfrac{\partial T}{\partial r}$	(G)
$q''_y = -k\dfrac{\partial T}{\partial y}$	(B)	$q''_\theta = -k\dfrac{1}{r}\dfrac{\partial T}{\partial \theta}$	(E)	$q''_\theta = -k\dfrac{1}{r}\dfrac{\partial T}{\partial \theta}$	(H)
$q''_z = -k\dfrac{\partial T}{\partial z}$	(C)	$q''_z = -k\dfrac{\partial T}{\partial z}$	(F)	$q''_\phi = -k\dfrac{1}{r\sin\theta}\dfrac{\partial T}{\partial \phi}$	(I)

For an incompressible fluid with specific heat $c = c_p = c_v$, we go back to Eq. (10.16) ($d\mathbf{u} = c\, dT$) to obtain

$$\rho c\frac{DT}{Dt} = \nabla \cdot k\nabla T + \mu\Phi + q''' \qquad (10.22)$$

Equations (10.16), (10.18), and (10.20) can be easily written in terms of energy (heat) and momentum fluxes using relations for fluxes given in Tables 10.4, 10.6, and 10.8. The energy equation given by Eq. (10.22) (with $q''' = 0$ for simplicity) is given in Table 10.9 in different coordinate systems.

TABLE 10.9 The Energy Equation (for Newtonian Fluids of Constant ρ and k)*

Rectangular coordinates (x, y, z):

$$\rho c_p\left(\frac{\partial T}{\partial t} + u\frac{\partial T}{\partial x} + v\frac{\partial T}{\partial y} + w\frac{\partial T}{\partial z}\right) = k\left(\frac{\partial^2 T}{\partial x^2} + \frac{\partial^2 T}{\partial y^2} + \frac{\partial^2 T}{\partial z^2}\right)$$

$$+ 2\mu\left\{\left(\frac{\partial u}{\partial x}\right)^2 + \left(\frac{\partial v}{\partial y}\right)^2 + \left(\frac{\partial w}{\partial z}\right)^2\right\} + \mu\left\{\left(\frac{\partial u}{\partial y} + \frac{\partial v}{\partial x}\right)^2 + \left(\frac{\partial u}{\partial z} + \frac{\partial w}{\partial x}\right)^2 + \left(\frac{\partial v}{\partial z} + \frac{\partial w}{\partial y}\right)^2\right\} \quad (A)$$

Cylindrical coordinates (r, θ, z):

$$\rho c_p\left(\frac{\partial T}{\partial t} + v_r\frac{\partial T}{\partial r} + \frac{v_\theta}{r}\frac{\partial T}{\partial \theta} + v_z\frac{\partial T}{\partial z}\right) = k\left[\frac{1}{r}\frac{\partial}{\partial r}\left(r\frac{\partial T}{\partial r}\right) + \frac{1}{r^2}\frac{\partial^2 T}{\partial \theta^2} + \frac{\partial^2 T}{\partial z^2}\right]$$

$$+ 2\mu\left\{\left(\frac{\partial v_r}{\partial r}\right)^2 + \left[\frac{1}{r}\left(\frac{\partial v_\theta}{\partial \theta} + v_r\right)\right]^2 + \left(\frac{\partial v_z}{\partial z}\right)^2\right\} + \mu\left\{\left(\frac{\partial v_\theta}{\partial z} + \frac{1}{r}\frac{\partial v_z}{\partial \theta}\right)^2 + \left(\frac{\partial v_z}{\partial r} + \frac{\partial v_r}{\partial z}\right)^2\right.$$

$$\left. + \left[\frac{1}{r}\frac{\partial v_r}{\partial \theta} + r\frac{\partial}{\partial r}\left(\frac{v_\theta}{r}\right)\right]^2\right\} \quad (B)$$

Spherical coordinates (r, θ, ϕ):

$$\rho c_p\left(\frac{\partial T}{\partial t} + v_r\frac{\partial T}{\partial r} + \frac{v_\theta}{r}\frac{\partial T}{\partial \theta} + \frac{v_\phi}{r\sin\theta}\frac{\partial T}{\partial \phi}\right) = k\left[\frac{1}{r^2}\frac{\partial}{\partial r}\left(r^2\frac{\partial T}{\partial r}\right) + \frac{1}{r^2\sin\theta}\frac{\partial}{\partial \theta}\left(\sin\theta\frac{\partial T}{\partial \theta}\right)\right.$$

$$\left. + \frac{1}{r^2\sin^2\theta}\frac{\partial^2 T}{\partial \phi^2}\right] + 2\mu\left\{\left(\frac{\partial v_r}{\partial r}\right)^2 + \left(\frac{1}{r}\frac{\partial v_\theta}{\partial \theta} + \frac{v_r}{r}\right)^2 + \left(\frac{1}{r\sin\theta}\frac{\partial v_\phi}{\partial \phi} + \frac{v_r}{r} + \frac{v_\theta\cot\theta}{r}\right)^2\right\}$$

$$+ \mu\left\{\left[r\frac{\partial}{\partial r}\left(\frac{v_\theta}{r}\right) + \frac{1}{r}\frac{\partial v_r}{\partial \theta}\right]^2 + \left[\frac{1}{r\sin\theta}\frac{\partial v_r}{\partial \phi} + r\frac{\partial}{\partial r}\left(\frac{v_\phi}{r}\right)\right]^2 + \left[\frac{\sin\theta}{r}\frac{\partial}{\partial \theta}\left(\frac{v_\phi}{\sin\theta}\right)\right.\right.$$

$$\left.\left. + \frac{1}{r\sin\theta}\frac{\partial v_\theta}{\partial \phi}\right]^2\right\} \quad (C)$$

*The terms contained in braces { } are associated with viscous dissipation and may usually be neglected except in systems with large velocity gradients.

For *solids*, the density may usually be considered constant, and we may set $\mathbf{V} = 0$, and Eq. (10.22) reduces to

$$\rho c\frac{\partial T}{\partial t} = \nabla \cdot k\nabla T + q''' \quad (10.23)$$

which is the starting point for most problems in heat conduction.

The Energy Equation for a Mixture. The energy equations in the preceding section are applicable for a pure fluid. A thermal energy equation valid for a mixture of chemical species is required for situations involving simultaneous heat and mass transfer. For a pure fluid, conduction is the only diffusive mechanism of heat flow; hence Fourier's law was used, which resulted in the term $\nabla \cdot k\nabla T$. More generally, this term may be written $-\nabla\mathbf{q}''$, where $\mathbf{q}''$ is the diffusive heat flux, i.e., the heat flux relative to the mass average velocity. More specifically, for a mixture, $\mathbf{q}''$ is now made from three contributions: (1) ordinary conduction, described by Fourier's law, $-k\nabla T$, where k is the *mixture* thermal conductivity; (2) the contribution due to interdiffusion of species, given by $\Sigma_i \mathbf{j}_i i_i$; and (3) diffusional conduction (also called the *diffusion-thermo effect* or *Dufour effect*[1,7]). The third contribution is of the second order and is usually negligible:

$$\mathbf{q}'' = -k\nabla T + \sum_i \mathbf{j}_i i_i \tag{10.24}$$

Here $\mathbf{j}_i$ is a diffusive mass flux of species i, with units of mass/(area × time), as mentioned before. Substituting Eq. (10.24) in, for example, Eq. (10.18), we obtain the energy equation for a mixture:

$$\rho \frac{Di}{Dt} = \frac{DP}{Dt} + \nabla \cdot k\nabla T - \nabla \cdot \left(\sum_i \mathbf{j}_i i_i \right) + \mu\Phi + q''' \tag{10.25}$$

For a nonreacting mixture the term $\nabla \cdot (\Sigma i \mathbf{j}_i i_i)$ is often of minor importance. But when endothermic or exothermic reactions occur, the term can play a dominant role. For reacting mixtures, the species enthalpies

$$i_i = i_i^0 + \int_{T^0}^{T} c_{pi}\, dT$$

must be written with a consistent set of heats of formation i_i^0 at T^0.[8]

4. The Conservation Equation for Species. For a stationary control volume, the conservation equation for species is

$$\underbrace{\frac{\partial C_i}{\partial t}}_{\substack{\text{rate of storage}\\ \text{of species } i \text{ per}\\ \text{unit volume}}} = \underbrace{-\nabla \cdot (C_i \mathbf{V})}_{\substack{\text{net rate of}\\ \text{convection of species}\\ i \text{ per unit volume}}} - \underbrace{\nabla \cdot \mathbf{j}_i}_{\substack{\text{net rate of diffusion}\\ \text{of species } i \text{ per unit}\\ \text{volume}}} + \underbrace{r_i'''}_{\substack{\text{production rate}\\ \text{of species } i \text{ per}\\ \text{unit volume}}} \tag{10.26}$$

Using the mass conservation equation, the preceding equation can be rearranged to obtain

$$\rho \frac{Dm_i}{Dt} = -\nabla \cdot \mathbf{j}_i + r_i''' \tag{10.27}$$

where m_i is mass fraction of species i, i.e., where $m_i = C_i/\rho$, where ρ is the density of the mixture, $\Sigma_i C_i = \rho$, and C_i is a partial density of species i (i.e., a mass concentration of species i).

The conservation equation of species also can be written in terms of mole concentration and mole fractions, as shown in Refs. 2, 3, and 7. The mole concentration of species i is $c_i = C_i/M_i$, where M_i is the molecular weight of the species. The mole fraction of species i is defined as $x_i = c_i/c$, where $c = \Sigma_i c_i$. As is obvious, $\Sigma_i m_i = 1$ and $\Sigma_i x_i = 1$.

Equations (10.26) and (10.27) written in different coordinate systems are given in Ref. 2.

5. Use of Conservation Equations to Set Up Problems. For a problem involving fluid flow and simultaneous heat and mass transfer, equations of continuity, momentum, energy, and chemical species, Eqs. (10.5), (10.8), (10.18), and

(10.27), are a formidable set of partial differential equations. There are four *independent variables*: three space coordinates (say, x, y, z) and a time coordinate t.

If we consider a pure fluid, then there are five equations: the continuity equation, three momentum equations, and the energy equation. The accompanying five *dependent variables* are pressure, three components of velocity, and temperature. Also, a thermodynamic equation of state serves to relate density to the pressure, temperature, and composition. (Notice that for natural-convection flows, the momentum and energy equations are coupled.)

For a mixture of n chemical species, there are n species conservation equations, but one is redundant, since the sum of mass fractions is equal to unity.

A complete mathematical statement of a problem requires specification of boundary and initial conditions. Boundary conditions are based on a physical statement or principle (for example, for viscous flow, the component of velocity parallel to a stationary surface is zero at the wall; for an insulated wall, the derivative of temperature normal to the wall is zero; etc.).

A general solution, even by numerical methods, of the full equations in the four independent variables is difficult to obtain. Fortunately, however, many problems of engineering interest are adequately described by simplified forms of the full conservation equations, and these forms can often be solved easily. The governing equations for simplified problems are obtained by deleting superfluous terms in the full conservation equations. This applies directly to laminar flows only. In the case of turbulent flows, some caution must be exercised. For example, on an average basis a flow may be two-dimensional and steady, but if it is unstable and as a result turbulent, fluctuations in the three components of velocity may be occurring with respect to time and the three spatial coordinates. Then the remarks about dropping terms apply only to the time-averaged equations.[7,8]

When simplifying the conservation equation given in a full form, we have to rely on physical intuition or on experimental evidence to judge which terms are negligibly small. Typical resulting classes of simplified problems are

- Constant transport properties
- Constant density
- Timewise steady flow (or quasi-steady flow)
- Two-dimensional flow
- One-dimensional flow
- Fully developed flow (no dependence on the streamwise coordinate)
- Stagnant fluid or rigid body

Terms also may be shown to be negligibly small by order-of-magnitude estimates.[7,8] Some classes of flow that result are

- Creeping flows: Inertia terms are negligible.
- Forced flows: Gravity forces are negligible.
- Natural convection: Gravity forces predominate.
- Low-speed gas flows: Viscous dissipation and compressibility terms are negligible.
- Boundary-layer flows: Streamwise diffusion terms are negligible.

DIMENSIONLESS GROUPS AND SIMILARITY IN FLUID MECHANICS AND HEAT TRANSFER

Modern engineering practice in the fields of fluid mechanics and heat transfer is based on a combination of theoretical analysis and experimental data. Often the engineer is faced with the necessity of obtaining practical results in situations where, for various reasons, physical phenomena cannot be described mathematically or the differential equations describing the problem are too difficult to solve. An experimental program must be considered in such cases. However, in carrying the experimental program, the engineer should know how to relate the experimental data (i.e., data obtained on the model under consideration) to the actual, usually larger, system (prototype). A determination of the relevant dimensionless parameters (groups) provides a powerful tool for that purpose.

The generation of such dimensionless groups in heat transfer and fluid mechanics (known generally as *dimensional analysis*) is basically done (1) by using differential equations and their boundary conditions (this method is sometimes called a *differential similarity*) and (2) by applying the dimensional analysis in the form of the Buckingham pi theorem.

The first method (differential similarity) is used when the governing equations and their boundary conditions describing the problem are known. The equations are first made dimensionless. For demonstration purposes, let us consider the relatively simple problem of a binary mixture with constant properties and density flowing at low speed, where body forces, heat source term, and chemical reactions are neglected. The conservation equations are, from Eqs. (10.6), (10.10), (10.16), and (10.27),

Mass

$$\nabla \cdot \mathbf{V} = 0 \tag{10.28}$$

Momentum

$$\rho \frac{D\mathbf{V}}{Dt} = -\nabla P + \mu \nabla^2 \mathbf{V} \tag{10.29}$$

Thermal energy

$$\rho c \frac{DT}{Dt} = k\nabla^2 T + \mu\Phi \tag{10.30}$$

Species

$$\frac{Dm_1}{Dt} = \mathbf{D}\nabla^2 m_1 \tag{10.31}$$

Using L and V as characteristic length and velocity, respectively, we define the dimensionless variables

$$x^* = \frac{x}{L} \qquad y^* = \frac{y}{L} \qquad z^* = \frac{z}{L} \tag{10.32}$$

$$\mathbf{V}^* = \frac{\mathbf{V}}{V} \tag{10.33}$$

$$t^* = \frac{t}{L/V} \tag{10.34}$$

$$P^* = \frac{P}{\rho V^2} \tag{10.35}$$

and also

$$T^* = \frac{T - T_w}{T_\infty - T_w} \tag{10.36}$$

$$m^* = \frac{m_1 - m_{1,w}}{m_{1,\infty} - m_{1,w}} \tag{10.37}$$

where the subscript ∞ refers to the external free-stream condition or some average condition and the subscript w refers to conditions adjacent to a bounding surface across which transfer of heat and mass occurs. If we introduce the dimensionless quantities, Eqs. (10.32) to (10.37), into Eqs. (10.28) to (10.31), we obtain, respectively,

$$\nabla^* \cdot \mathbf{V}^* = 0 \tag{10.38}$$

$$\frac{D\mathbf{V}^*}{Dt^*} = -\nabla^* P^* + \frac{1}{Re} \nabla^{*2} \mathbf{V}^* \tag{10.39}$$

$$\frac{DT^*}{Dt^*} = \frac{1}{\mathrm{Re\ Pr}} \nabla^{*2} T^* + \frac{2\ \mathrm{Ec}}{\mathrm{Re}} \Phi^* \tag{10.40}$$

$$\frac{Dm^*}{Dt^*} = \frac{1}{\mathrm{Re\ Sc}} \nabla^{*2} m^* \tag{10.41}$$

Obviously, the solutions of Eqs. (10.38) to (10.41) depend on the coefficients that appear in these equations. Solutions of Eqs. (10.38) to (10.41) are equally applicable to the model and prototype (where the model and prototype are geometrically similar systems of different linear dimensions in streams of different velocities, temperatures, and concentration), if the coefficients in these equations are the same for both model and prototype. These coefficients, Pr, Re, Sc, and Ec (called *dimensionless parameters* or *similarity parameters*), are defined in Table 10.10.

Focusing attention now on heat transfer, from Eq. (9.22), using the dimensionless quantities, the heat-transfer coefficient is given as

$$h = \left. \frac{k \partial T^*}{L \partial y^*} \right|_{y^*=0} \tag{10.42}$$

or in dimensionless form,

$$\frac{hL}{k} = \left. \frac{\partial T^*}{\partial y^*} \right|_{y^*=0} = \mathrm{Nu} \tag{10.43}$$

where the dimensionless group Nu is known as the *Nusselt number*. Since Nu is the dimensionless temperature gradient at the surface, according to Eq. (10.40) it must therefore depend on the dimensionless groups that appear in this equation; hence

$$\mathrm{Nu} = f_1(\mathrm{Re, Pr, Ec}) \tag{10.44}$$

TABLE 10.10 Summary of the Chief Dimensionless Groups*

Group	Symbol	Definition	Physical significance (interpretation)	Main area of use
Biot number	Bi	$\frac{hL}{k_s}$	Ratio of internal thermal resistance of solid to fluid thermal resistance	Heat transfer between fluid and solid
Biot number† (mass transfer)	$\mathrm{Bi}_{\mathbf{D}}$	$\frac{h_{\mathbf{D}}L}{\mathbf{D}}$	Ratio of the internal species transfer resistance to the boundary-layer species transfer resistance	Mass transfer between fluid and solid
Coefficient of friction (skin friction coefficient)	c_f	$\frac{\tau_w}{\rho V^2/2}$	Dimensionless surface shear stress	Flow resistance
Eckert number	Ec	$\frac{V_\infty^2}{c_p(T_w - T_\infty)}$	Kinetic energy of the flow relative to the boundary-layer enthalpy difference	Forced convection (compressible flow)
Euler number	Eu	$\frac{\Delta P}{\rho V^2}$	Ratio of friction to velocity head	Fluid friction
Fourier number	Fo	$\frac{\alpha t}{L^2}$	Ratio of the heat conduction rate to the rate of thermal energy storage in a solid	Unsteady-state heat transfer
Fourier number (mass transfer)	$\mathrm{Fo}_{\mathbf{D}}$	$\frac{\mathbf{D}t}{L^2}$	Ratio of the species diffusion rate to the rate of species storage	Unsteady-state mass transfer
Froude number	Fr	$\frac{V^2}{gL}$	Ratio of inertial to gravitational force	Wave and surface behavior (mixed natural and forced convection)
Graetz number	Gz	$\mathrm{Re}\,\mathrm{Pr}\,\frac{D}{L} = \frac{\rho c_p V D^2}{kL}$	Ratio of the fluid stream thermal capacity to convective heat transfer	Forced convection
Grashof number	Gr	$\frac{g\beta\,\Delta T L^3}{\nu^2}$	Ratio of buoyancy to viscous forces	Natural convection

TABLE 10.10 Summary of the Chief Dimensionless Groups (*Continued*)

Group	Symbol	Definition	Physical significance (interpretation)	Main area of use
Colburn *j* factor (heat transfer)	j_H	$\mathrm{St}\,\mathrm{Pr}^{2/3}$	Dimensionless heat transfer coefficient	Forced convection (heat, mass, and momentum transfer analogy)
Colburn *j* factor (mass transfer)	$j_{\mathbf{D}}$	$\mathrm{St}_{\mathbf{D}}\,\mathrm{Sc}^{2/3}$	Dimensionless mass transfer coefficient	Forced convection (heat, mass, and momentum transfer analogy)
Jakob number	Ja	$\dfrac{\rho_l c_{pl}(T_w - T_{sat})}{\rho_g i_{lg}}$	Ratio of sensible heat absorbed by the liquid to the latent heat absorbed	Boiling
Knudsen number	Kn	$\dfrac{\lambda}{L}$	Ratio of molecular mean free path to characteristic dimension	Low-pressure (low-density) gas flow
Lewis number	Le	$\dfrac{\alpha}{\mathbf{D}} = \dfrac{\mathrm{Sc}}{\mathrm{Pr}}$	Ratio of molecular thermal and mass diffusivities	Combined heat and mass transfer
Mach number	Ma	$\dfrac{V}{a}$	Ratio of the velocity of flow to the velocity of sound	Compressible flow
Nusselt number	Nu	$\dfrac{hL}{k}$	Basic dimensionless convective heat transfer coefficient (ratio of convection heat transfer to conduction in a fluid slab of thickness L)	Convective heat transfer
Péclet number	Pe	$\mathrm{Re}\,\mathrm{Pr} = \dfrac{\rho c_p V L}{k}$	Dimensionless independent heat transfer parameter (ratio of heat transfer by convection to conduction)	Forced convection

Péclet number (mass transfer)	$Pe_{\mathbf{D}}$	$\text{Re Sc} = \frac{VL}{\mathbf{D}}$	Dimensionless independent mass transfer coefficient (ratio of bulk mass transfer to diffusive mass transfer)	Mass transfer
Prandtl number	Pr	$\frac{\mu c_p}{k} = \frac{\nu}{\alpha}$	Ratio of molecular momentum and thermal diffusivities	Forced and natural convection
Rayleigh number	Ra	$\text{Gr Pr} = \frac{\rho g \beta \, \Delta T L^3}{\mu \alpha}$	Modified Grashof number (see interpretation for Gr and Pr)	Natural convection
Reynolds number	Re	$\frac{\rho V L}{\mu}$	Ratio of inertia to viscous forces	Forced convection; dynamic similarity
Schmidt number	Sc	$\frac{\nu}{\mathbf{D}}$	Ratio of molecular momentum and mass diffusivities	Mass transfer
Sherwood number	Sh	$\frac{h_{\mathbf{D}} L}{\mathbf{D}}$	Ratio of convection mass transfer to diffusion in a slab of thickness L	Convective mass transfer
Strouhal number‡	Sr	$\frac{Lf}{V}$	Ratio of the velocity of vibration Lf to the velocity of the fluid	Flow past tube (shedding of eddies)
Stanton number	St	$\frac{\text{Nu}}{\text{Re Pr}} = \frac{h}{\rho c_p V}$	Dimensionless heat transfer coefficient (ratio of heat transfer at the surface to that transported by fluid by its thermal capacity)	Forced convection
Stanton number (mass transfer)	$St_{\mathbf{D}}$	$\frac{\text{Sh}}{\text{Re Sc}} = \frac{h_{\mathbf{D}}}{V}$	Dimensionless mass transfer coefficient	Convective mass transfer
Weber number	We	$\frac{\rho V^2 L}{\sigma}$	Ratio of inertia force to surface tension force	Droplet breakup; thin-film flow

*In these dimensionless groups, L designates characteristic dimension (e.g., tube diameter, hydraulic diameter, length of the tube or plate, slab thickness, radius of a cylinder or sphere, droplet diameter, thin-film thickness, etc.). Physical properties are usually evaluated at mean temperature unless otherwise specified.

†*Note*: $\mathbf{D} = \mathbf{D}_{12}$ ($\mathbf{D}_{12}$ is also a commonly used symbol for binary diffusion coefficient; $\mathbf{D}_{ij}$ is the multicomponent diffusion coefficient). When species 1 is in very small concentration, the symbol $\mathbf{D}_{1m}$ is occasionally used,[7] representing an effective binary diffusion coefficient for species 1 diffusing through the mixture.

‡In some engineering texts, the symbol St is also used for this group.

For processes where viscous dissipation and compressibility are negligible, which is the case in many industrial applications, we have

$$\mathrm{Nu} = f_2(\mathrm{Re}, \mathrm{Pr}) \qquad \text{(forced convection)} \tag{10.45}$$

In the case of buoyancy-induced flow, Eq. (10.29) should be replaced with the simplified version[9] of Eq. (10.14), and following a similar procedure, we obtain

$$\mathrm{Nu} = f_3(\mathrm{Gr}, \mathrm{Pr}) \qquad \text{(natural convection)} \tag{10.46}$$

where Gr is the Grashof number, defined in Table 10.10. Also, using the relation of Eq. (9.26), and dimensionless quantities,

$$h_{\mathrm{D}} = \frac{\mathbf{D}}{\mathrm{L}} \frac{\partial \mathrm{m}^*}{\partial \mathrm{y}^*}\bigg|_{y^*=0} \tag{10.47}$$

or

$$h_{\mathrm{D}} \frac{L}{\mathbf{D}} = \frac{\partial m^*}{\partial y^*}\bigg|_{y^*=0} = \mathrm{Sh} \tag{10.48}$$

This parameter, termed the *Sherwood number*, is equal to the dimensionless mass fraction (i.e., concentration) gradient at the surface, and it provides a measure of the convection mass transfer occurring at the surface. Following the same argument as before [but now for Eq. (10.41)], we have

$$\mathrm{Sh} = f_4(\mathrm{Re}, \mathrm{Sc}) \qquad \text{(forced convection, mass transfer)} \tag{10.49}$$

The significance of expressions such as Eqs. (10.44) to (10.46) and (10.49) should be appreciated. For example, Eq. (10.45) states that convection heat-transfer results, whether obtained theoretically or experimentally, can be represented in terms of three dimensionless groups, instead of seven parameters (h, L, V, k, c_p, μ, and ρ). The convenience is evident. Once the form of the functional dependence of Eq. (10.45) is obtained for a particular surface geometry (e.g., from laboratory experiments on a small model), it is known to be universally applicable; i.e., it may be applied to different fluids, velocities, temperatures, and length scales, as long as the assumptions associated with the original equations are satisfied (e.g., negligible viscous dissipation and body forces). Note that the relations of Eqs. (10.44) and (10.49) are derived without actually solving the system of Eqs. (10.28) and (10.31). References 7 to 12 cover the preceding procedure with more details and also include many different cases.

It is important to mention here that once the conservation equations are put in dimensionless form, it is also convenient to make an order-of-magnitude assessment of all terms in the equations. Often a problem can be simplified by discovering that a term that would be very difficult to handle if large is in fact negligibly small.[7,8] Even if the primary thrust of the investigation is experimental, making the equations dimensionless and estimating the orders of magnitude of the terms are good practice. It is usually not possible for an experimental test to include (simulate) all conditions exactly; a good engineer will focus on the most important conditions. The same applies to performing an order-of-magnitude analysis. For example, for boundary-layer flows, allowance is made for the fact that lengths transverse to the main flow scale with a much shorter length than those measured in the direction of main flow. References 7, 11, and 13 cover many examples of the order-of-magnitude analysis.

When the governing equations of a problem are unknown, an alternative approach of deriving dimensionless groups is based on use of dimensional analysis in the form of the Buckingham pi theorem.[3,5,9,12,14] The Buckingham pi theorem proves that in a physical problem including n quantities in which there are m dimensions, the quantities can be arranged into $n-m$ independent dimensionless parameters. The success of this method depends on our ability to select, largely from intuition, the parameters that influence the problem. The procedure is best illustrated by an example.

Example 10.1. The discharge through a horizontal capillary tube is thought to depend on the pressure drop per unit length, the diameter, and the viscosity. Find the form of the equation. The quantities with their dimensions are as follows:

Quantity	Symbol	Dimensions
Discharge	Q	L^3t^{-1}
Pressure drop/length	$\Delta p/l$	$ML^{-2}t^{-2}$
Diameter	D	L
Viscosity	μ	$ML^{-1}t^{-1}$

Then

$$F\left(Q, \frac{\Delta p}{l}, D, \mu\right) = 0$$

Three dimensions are used, and with four quantities there will be one Π parameter:

$$\Pi = Q^{x_1}\left(\frac{\Delta p}{l}\right)^{y_1} D^{z_1}\mu$$

Substituting in the dimensions gives

$$\Pi = (L^3T^{-1})^{x_1}(ML^{-2}T^{-2})^{y_1}L^{z_1}ML^{-1}t^{-1} = M^0M^0L^0$$

The exponents of each dimension must be the same on both sides of the equation. With L first,

$$3x_1 - 2y_1 + z_1 - 1 = 0$$

and similarly for M and t,

$$y_1 + 1 = 0$$

$$-x_1 - 2y_1 - 1 = 0$$

from which $x_1 = 1, y_1 = -1, z_1 = -4$, and

$$\Pi = \frac{Q\mu}{D^4\,\Delta p/l}$$

After solving for Q,

$$Q = C\frac{\Delta pD^4}{l\mu}$$

from which dimensional analysis yields no information about the numerical value of the dimensionless constant C; experiment (or analysis) shows that it is $\pi/128$.

In the preceding example, if kinematic viscosity had been used in place of dynamic viscosity, an incorrect formula would have resulted.

Example 10.2. A fluid-flow situation depends on the velocity V; the density ρ; several linear dimensions l, l_1, and l_2; pressure drop Δp; gravity g, viscosity μ; surface tension σ; and bulk modulus of elasticity E. Apply dimensional analysis to these variables to find a set of parameters.

$$F(V, \rho, l, l_1, l_2, \Delta p, g, \mu\ \sigma, E) = 0$$

Since three dimensions are involved, three repeating variables are selected. For complex situations, V, p, and l are generally helpful. There are seven Π parameters:

$$\Pi_1 = V^{x1}\rho^{y1}l^{z1}\Delta p$$

$$\Pi_2 = V^{x2}\rho^{y2}l^{z2}g$$

$$\Pi_3 = V^{x3}\rho^{y3}l^{z3}$$

$$\Pi_4 = V^{x4}\rho^{y4}l^{z4}\sigma$$

$$\Pi_5 = V^{x5}\rho^{y5}l^{z5}E$$

$$\Pi_6 = \frac{l}{l_1}$$

$$\Pi_7 = \frac{l}{l_2}$$

By expanding the quantities into dimensions as in the first example, we have

$$\Pi_1 = \frac{\Delta p}{\rho V^2} \quad \Pi_2 = \frac{gl}{V^2} \quad \Pi_3 = \frac{\mu}{Vl\rho} \quad \Pi_4 = \frac{\sigma}{V^2\rho l}$$

$$\Pi_5 = \frac{E}{\rho V^2} \quad \Pi_6 = \frac{l}{l_1} \quad \Pi_7 = \frac{l}{l_2}$$

and
$$f\left(\frac{\Delta p}{\rho V^2}, \frac{gl}{V^2}, \frac{\mu}{Vl\rho}, \frac{\sigma}{V^2\rho l}, \frac{E}{\rho V^2}, \frac{l}{l_1}, \frac{l}{l_2}\right) = 0$$

It is convenient to invert some of the parameters and to take some square roots:

$$f_1\left(\frac{\Delta p}{\rho V^2}, \frac{V}{\sqrt{gl}}, \frac{Vl\rho}{\mu}, \frac{V^2 l\rho}{\sigma}, \frac{V}{\sqrt{E/\rho}}, \frac{l}{l_1}, \frac{l}{l_2}\right) = 0$$

The first parameter, usually written $\Delta p/(\rho V^2/2)$, is the pressure coefficient; the second parameter is the Froude number Fr; the third is the Reynolds number Re; the fourth is the Weber number We; and the fifth is the Mach number Ma. Hence

$$f_1\left(\frac{\Delta p}{\rho V^2}, \text{Fr}, \text{Re}, \text{We}, \text{Ma}, \frac{l}{l_1}, \frac{l}{l_2}\right) = 0$$

After solving for pressure drop,

$$\Delta p = \rho V^2 f_2\left(\text{Fr}, \text{Re}, \text{We}, \text{Ma}, \frac{l}{l_1}, \frac{l}{l_2}\right)$$

in which f_1, f_2 must be determined from analysis or experiment. By selecting other repeating variables, a different set of pi parameters could be obtained.

For example, knowing in advance that the heat-transfer coefficient in fully developed forced convection in a tube is a function of certain variables, that is, $h = f(V, \rho, \mu, c_p, k, D)$, we can use the Buckingham pi theorem to obtain Eq. (10.45), as shown in Ref. 3. However, this method is carried out without any consideration of the physical nature of the process in question; i.e., there is no way to ensure that all essential variables have been included. However, as shown above, starting with the differential form of the conservation equations, we have derived the similarity parameters (dimensionless groups) in rigorous fashion.

In Table 10.10 those dimensionless groups which appear frequently in fluid-flow, heat-, and mass-transfer literature have been listed. The list includes groups already mentioned above as well as those found in special fields of heat transfer. Note that although similar in form, the Nusselt and Biot numbers differ in both definition and interpretation. The Nusselt number is defined in terms of thermal conductivity of the fluid; the Biot number is based on the solid thermal conductivity.

NOMENCLATURE

Symbol = *Definition, SI Units (U.S. Customary Units)*

A = heat-transfer area, m^2 (ft^2)

$\mathbf{a}$ = acceleration, m/s^2 (ft/s^2)

a = speed of sound, m/s (ft/s)

C = mass concentration of species, kg/m^3 (lb_m/ft^3)

c = specific heat, J/(kg · K) [Btu/(lb_m · °F)]

c_p = specific heat at constant pressure, J/(kg · K) [Btu/(lb_m · °F)]

c_v = specific heat at constant volume, J/(kg · K) [Btu/(lb_m · °F)]

D = tube inside diameter, diameter, m (ft)

$\mathbf{D}$ = diffusion coefficient, m^2/s (ft^2/s)

Ec = Eckert number (Table 10.10)

e = emissive power, W/m^2 [Btu/(h · ft^2)]

e_b = blackbody emissive power, W/m^2 [Btu/(h · ft^2)]

F = force, N (lb_f)

f = frequency of vibration (Table 10.10), s^{-1}

f_1, f_2, f_3, f_4 = denotes function of Eqs. (10.44) to (10.46) and (10.49)

Gr = Grashof number (Table 10.10)

g, $\mathbf{g}$ = gravitational acceleration (magnitude and vector), m/s^2 (ft/s^2)

h = heat-transfer coefficient, W/(m^2 · K) [Btu/(h · ft^2 · °F)]

h_D = mass-transfer coefficient, m/s (ft/s)

i = enthalpy per unit mass, J/kg (Btu/lb_m)
i_{lg} = latent heat of evaporation, J/kg (Btu/lb_m)
i^0 = heat of formation, J/kg (Btu/lb_m)
j, $\mathbf{j}$ = mass diffusion flux of species (magnitude and vector), $kg/(s \cdot m^2)$ [$lb_m/(h \cdot ft^2)$]
k = thermal conductivity, W/(m · K) [Btu/(h · ft · °F)]
L = length, m (ft)
M = mass, kg (lb_m)
m = mass fraction of species [Eq. (10.27)]
Nu = Nusselt number (Table 10.10)
P = pressure: Pa, N/m^2 (lb_f/ft^2)
Pr = Prandtl number (Table 10.10)
ΔP = pressure drop, Pa, N/m^2 (lb_f/ft^2)
q = heat-transfer rate, W (Btu/h)
$\mathbf{q}''$ = heat flux (vector), W/m^2 [$Btu/(h \cdot ft^2)$]
q'' = heat flux, W/m^2 [$Btu/(h \cdot ft^2)$]
q''' = volumetric heat generation, W/m^3 [$Btu/(h \cdot ft^3)$]
Re = Reynolds number (Table 10.10)
r = radial distance in cylindrical or spherical coordinate, m (ft)
r''' = volumetric generation rate of species, $kg/(s \cdot m^3)$ [$lb_m/(h \cdot ft^3)$]
Sc = Schmidt number (Table 10.10)
Sh = Sherwood number (Table 10.10)
St = Stanton number (Table 10.10)
T = temperature: °C, K (°F, °R)
ΔT = temperature difference, °C (°F)
t = time, s
u = velocity component in the axial direction (x direction) in rectangular coordinates, m/s (ft/s)
$\mathbf{u}$ = internal energy per unit mass, J/kg (Btu/lb_m)
V, $\mathbf{V}$ = velocity (magnitude and vector), m/s (ft/s)
v = velocity component in the y direction in rectangular coordinates, m/s (ft/s)
v_r = velocity component in the r direction, m/s (ft/s)
v_z = velocity component in the z direction, m/s (ft/s)
v_θ = velocity component in the θ direction, m/s (ft/s)
v_ϕ = velocity component in the ϕ direction, m/s (ft/s)
w = velocity component in the z direction in rectangular coordinates, m/s (ft/s)
x = rectangular coordinate, m (ft)
y = rectangular coordinate, m (ft)
z = rectangular or cylindrical coordinate, m (ft)

Greek

α = thermal diffusivity, m^2/s (ft^2/s)
β = coefficient of thermal expansion, K^{-1} ($°R^{-1}$)
ϵ = emissivity
θ = angle in cylindrical and spherical coordinates, radians (degrees)
λ = molecular mean free path, m (ft)
μ = dynamic viscosity, Pa · s [$lb_m/(h \cdot ft)$]
υ = kinematic viscosity, m^2/s (ft^2/s)
ρ = density, kg/m^3 (lb_m/ft^3)
σ = surface tension (Table 10.10), N/m (lb_f/ft)
τ = shear stress, N/m^2 (lb_f/ft^2)
$\boldsymbol{\tau}$ = shear stress tensor, N/m^2 (lb_f/ft^2)
Φ = dissipation function (Table 10.7), s^{-2}
ϕ = angle in spherical coordinate system, rad (degrees)

Subscripts

a = surroundings
aw = adiabatic wall
cr = critical
f = fluid
g = gas (vapor)
i = species i
l = liquid
m = mean
s = solid
sat = saturation
t = total
w = wall
x = x component
y = y component
z = z component
θ = θ component
ϕ = ϕ component
1 = species 1 in binary mixture of 1 and 2
∞ = free-stream condition

Mathematical Operation Symbols

d/dx = derivative with respect to x, m^{-1} (ft^{-1})
$\partial/\partial t$ = partial time derivative operator, s^{-1}
d/dt = total time derivative operator, s^{-1}

D/Dt = substantial time derivative operator, s^{-1}
∇ = "del" operator (vector), m^{-1} (ft^{-1})
∇^2 = laplacian operator, m^{-2} (ft^{-2})

REFERENCES*

1. W. M. Kays and M. E. Crawford, *Convective Heat and Mass Transfer*, 2d ed., McGraw-Hill, New York, 1980.
2. R. B. Bird, W. E. Stewart, and E. N. Lightfoot, *Transport Phenomena*, Wiley, New York, 1960.
3. W. M. Rohsenow and H. Y. Choi, *Heat, Mass, and Momentum Transfer*, Prentice-Hall, Englewood Cliffs, N.J., 1961.
4. W. M. Rohsenow, J. P. Hartnett, and E. N. Ganić, eds., *Handbook of Heat Transfer Applications*, chap. 2, McGraw-Hill, New York, 1985.
5. A. S. Foust, L. A. Wenzel, C. W. Clump, L. Mans, and L. B. Andersen, *Principles of Unit Operations*, 2d ed., Wiley, New York, 1980.
6. S. Whitaker, *Elementary Heat Transfer Analysis*, Pergamon, New York, 1976.
7. D. K. Edwards, V. E. Denny, and A. F. Mills, *Transfer Processes: An Introduction to Diffusion, Convection, and Radiation*, 2d ed., Hemisphere, Washington, and McGraw-Hill, New York, 1979.
8. H. Schlichting, *Boundary-Layer Theory*, 7th ed., McGraw-Hill, New York, 1979.
9. B. Gebhart, *Heat Transfer*, 2d ed., McGraw-Hill, New York, 1971.
10. F. P. Incropera and D. P. DeWitt, *Fundamentals of Heat Transfer*, Wiley, New York, 1981.
11. F. M. White, *Viscous Fluid Flow*, McGraw-Hill, New York, 1974.
12. V. P. Isachenko, V. A. Osipova, and A. S. Sukomel, *Heat Transfer*, Mir Publishers, Moscow, 1977.
13. E. R. G. Eckert and R. M. Drake, Jr., *Analysis of Heat and Mass Transfer*, McGraw-Hill, New York, 1972.
14. J. H. Lienhard, *A Heat Transfer Textbook*, Prentice-Hall, Englewood Cliffs, N.J., 1981.
15. W. C. Reynolds and H. C. Perkins, *Engineering Thermodynamics*, 2d ed., McGraw-Hill, New York, 1977.
16. W. M. Rohsenow, J. P. Hartnett, and E. N. Ganić, eds, *Handbook of Heat Transfer Fundamentals*, 2d ed., chap. 1, McGraw-Hill, New York, 1985.

*Those references listed here but not cited in the text were used for comparison of different data sources, clarification, clarity of presentation, and, most important, reader's convenience when further interest in subject exists.

CHAPTER 11
TOPICS IN APPLIED PHYSICS

ELECTRIC FIELDS

1. Electric Charge. Any observed charge q, positive or negative, is in magnitude an integral multiple of

$$e = 1.6022 \times 10^{-19}\ \text{C} \tag{11.1}$$

where $-e$ is the charge on a single electron. Note that C (coulomb) is the derived unit for electric charge, since $1\ \text{C} = 1\ \text{A} \cdot \text{s}$. One coulomb is a large amount of charge; hence the microcoulomb (μC) is frequently used.

An electric current in a wire corresponds to the transport of a certain amount of electric charge through a fixed cross section in a certain time interval. Electric current is measured in amperes (A).

2. Coulomb's Law. If q_1 and q_2 are magnitudes of two charges, separated by a distance r, then the force between them is

$$F = b\,\frac{q_1 \cdot q_2}{r^2} \tag{11.2}$$

This force is repulsive when q_1 and q_2 are of the same sign and attractive when they are of opposite signs. By experiment, it is found that

$$b \simeq 9 \times 10^9\ \text{N} \cdot \text{m}^2/\text{C}^2$$

For mathematical reasons, it is convenient to replace b by another constant, ϵ_0, called the *permissivity of empty space* and defined by

$$\epsilon_0 = \frac{1}{4\pi b} = 8.85432 \times 10^{-12}\ \text{C}^2/(\text{N} \cdot \text{m}^2)$$

In terms of ϵ_0, Coulomb's law is written

$$F = \frac{q_1 \cdot q_2}{4\pi\epsilon_0 r^2} \tag{11.3}$$

Forces between point charges act independently. Suppose that charges $q_1, q_2, q_3, \ldots$ are fixed in an inertial frame. Then the force F_1 on a charge due to q_1 is computed by Coulomb's law as if $q_2, q_3, \ldots$ did not exist; etc. Finally, the total force on q is just the vector sum of the individual forces:

$$F_{\text{total}} = F_1 + F_2 + F_3 + \cdots \tag{11.4}$$

3. Electric Fields. If a charge q experiences a force F in the field, then the electric field strength is defined to be the force per unit charge, that is,

$$E = \frac{F}{q} \tag{11.5}$$

Equivalently, the force on a point charge q located at a point at which the electric field is E (regardless of what arrangement of charges may have established E) is given by

$$F = qE \tag{11.6}$$

Note that E is a vector in the direction of F and represents the force (in newtons) on a unit charge (1 C) at the point considered.

From Eqs. (11.4) and (11.5),

$$E_{\text{total}} = E_1 + E_2 + E_3 + \cdots \tag{11.7}$$

That is, the electric field at any point due to a distribution of charges (or charge elements) $q_1, q_2, q_3, \ldots$ is found by adding (or integrating) the fields independently established at that point by individual charges.

4. Electric Flux. The *electric flux* through the elementary area dS is defined as

$$d\psi = E \cdot dS \tag{11.8}$$

Note that $d\psi$ can be interpreted as the number of lines (the lines to which E is tangent at each point) cutting dS, provided that the lines are supposedly drawn with normal density E.

The total flux through S in the outward direction is given by integration of Eq. (11.8) as

$$\psi = \int_S E \cdot dS \tag{11.9}$$

5. Gauss's Law. Based on definition in previous sections, Grauss's law reads

$$\psi = \frac{Q}{\epsilon_0} \tag{11.10}$$

That is, the total electric flux ψ out of an arbitrary closed surface in free space is proportional to the net electric charge within the surface Q; the constant of proportionality is $1/\epsilon_0 = 4\pi b$. Gauss's law is valid even for enclosed charges in motion.

6. Electric Potential. The electric force F exerted on a test charge q by some stationary distribution of charges is a conservative force (i.e., it can be derived from a function of position). Therefore, the test charge possesses electric potential energy U. When U is known, then the components of the electric force F along the x, y, and z axes are given by

$$F_x = -\frac{\partial U}{\partial x} \qquad F_y = -\frac{\partial U}{\partial y} \qquad F_z = -\frac{\partial U}{\partial z}$$

or in vector form

$$\mathbf{F} = -\left(\frac{\partial U}{\partial x}\mathbf{i} + \frac{\partial U}{\partial y}\mathbf{j} + \frac{\partial U}{\partial z}\mathbf{k}\right) \tag{11.11}$$

If the only force acting on a point charge is the electric force **F**, the conservation of energy relation is

$$\Delta K_e + \Delta U = 0 \tag{11.12}$$

where K_e is the kinetic energy of the point charge.

The electric potential energy per unit test charge is called the *electric potential* ϕ or the *voltage* v. Note that ϕ and **E** enjoy the same relation as U and **F**, that is,

$$\mathbf{E} = -\left(\frac{\partial \phi}{\partial x}\mathbf{i} + \frac{\partial \phi}{\partial y}\mathbf{j} + \frac{\partial \phi}{\partial z}\mathbf{k}\right) \tag{11.13}$$

Notice that the unit of electric potential is the volt (V), where 1 V = 1 J/C; also, 1 V/m = 1 N/C. It follows also that

$$\phi_{\text{total}} = \phi_1 + \phi_2 + \phi_3 + \cdots \tag{11.14}$$

The Electronvolt. When an electron (charge $-e$) travels through a potential difference of 1 V (i.e., when it moves from a given location to a location where the electric potential is 1 V higher), the electron loses

$$e\ (1\text{ V}) = 1.602 \times 10^{-19}\text{ J}$$

of electric energy and gains [Eq. (11.12)] the same amount of kinetic energy. This quantity of energy is referred to as *one electronvolt* (1 eV). In general, any energy expressed in joules may be converted to electronvolts by dividing by 1.602×10^{-19} J/eV.

7. Electric Current, Resistance, and Power. The rate of flow of electric charge across a given conductor area is defined as the *electric current I* through that area, or

$$I = \frac{dq}{dt} \tag{11.15}$$

Also, in terms of J, the electric current through a surface S (e.g., a cross section of the conductor) is

$$I = \int_S J \cdot dS = \int_S J \cdot dA \tag{11.16}$$

where $dA = dS \cos\theta$ is the projection of dS perpendicular to the flow direction, and J is the electric current density at a point within a conductor.

Ohm's Law. Empirically, it is found that the current I in a resistor (long wire, conducting bar, etc.) is very nearly proportional to the difference v in electric potential between the ends of the resistor. This proportionality is known as *Ohm's law*:

$$v = I \cdot R \tag{11.17}$$

where the proportionality factor R is called the *resistance*. The unit of resistance, the ohm (Ω), is defined as $1\ \Omega = 1$ V/A. The resistance of a conductor of length L and uniform cross-sectional area A is

$$R = \rho \frac{L}{A} \tag{11.18}$$

where ρ, the *resistivity*, is a property of the material and is temperature-dependent. The units of ρ are ($\Omega \cdot$ m).

The difference in potential between the terminals of any source of electric energy such as a battery or a mechanically driven generator, when delivering no current, is a measure of the *electromotive force* v_e of the source. Note that v_e is often called the *open-circuit voltage* of a battery or generator. The actual source of electric energy is

$$v_t = v_e \pm IR_0 \tag{11.19}$$

where v_t is terminal voltage, I is the current passing through the source, and R_0 is the internal resistance of the source. The negative sign is used when the source is delivering current, and the positive sign is used when I has the opposite direction (e.g., charging a storage cell).

If an amount of charge dq moves through a difference in potential v, the change in electric energy is

$$dE = (dq)v \tag{11.20}$$

The electric power P is the time rate of change of electric energy, that is,

$$P = \frac{dE}{dt} = \left(\frac{dq}{dt}\right)v = I \cdot v \tag{11.21}$$

Note that (1 A)(1 V) = 1 W or 1 V = 1 W/A. The power output of an energy source having terminal voltage v_t and supplying current I is given as

$$P = I \cdot v_t = I(v_e - I \cdot R_0) \tag{11.22}$$

The power absorbed by a pure resistance R when the potential difference across it is v_{1-2} is given by

$$P = I \cdot v_{1-2} = I^2R = \frac{v_{1-2}^2}{R} \tag{11.23}$$

This latter power is dissipated as heat.

8. Kirchhoff's Laws. For any network composed of resistors and sources (batteries), given the values of a certain number of the electromotive forces v_e, currents I, and resistances R, the values of the remaining quantities can be found by an application of Kirchhoff's laws.

Kirchhoff's Junction Law. The algebraic sum of all currents into any junction is zero. That is, at any junction,

$$\Sigma I = 0 \tag{11.24}$$

In the last equation, a current out of the junction is counted as negative.

Kirchhoff's Loop Law. The algebraic sum of all electromotive forces around any closed circuit (loop) is equal to the algebraic sum of the voltage drops (the $R \cdot I_R$ and the $R_0 \cdot I$) around the loop, that is,

$$\Sigma v_e = \Sigma R \cdot I_R + \Sigma R_0 I_{R_0} \tag{11.25}$$

where the summations are taken along any continuously directed path from p_1 to p_2 in the network. If v_{1-2} comes out positive, this means that p_1 is at a positive potential relative to p_2.

Example 11.1. Referring to Fig. 11.1, find I_1, I_2, and I_3. Directions shown for the currents are arbitrary.

Junction Equations. At junction a,

$$I_1 - I_2 - I_3 = 0 \tag{1}$$

Another junction equation can be written at d, but it is the negative of Eq. (1).

Loop Equations. Traversing the loop *adcba* of the planar network, we obtain

$$100 = 4I_3 + 6I_1 + 10I_1 \tag{2}$$

and traversing *afeda*, we obtain

$$60 = 12I_2 + 4(-I_3) \tag{3}$$

(It is assumed that the batteries have no internal resistance.) Note that the loop *afedcba* would not give an independent equation, but rather the sum of Eq. (2) and Eq. (3).

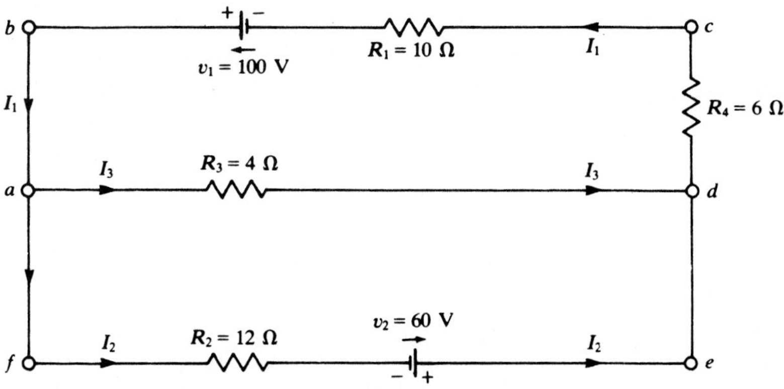

FIGURE 11.1 *(From Wells and Slusher,[2] p. 237.)*

The independent equations (1), (2), and (3) can now be solved simultaneously (applying usual methods) for I_1, I_2, and I_3, yielding

$$I_1 = 6.053 \text{ A} \qquad I_2 = 5.263 \text{ A} \qquad I_3 = 0.7895 \text{ A}$$

Note that each value of I is positive; thus the arbitrarily chosen directions indicated in Fig. 11.1 happen to be correct.

Example 11.2. Figure 11.2*a* and *b* shows two possibilities, the so-called series and parallel combinations of resistors. Using also Ohm's law, it can be shown that resistances in series add directly and resistances in parallel add as their reciprocals.

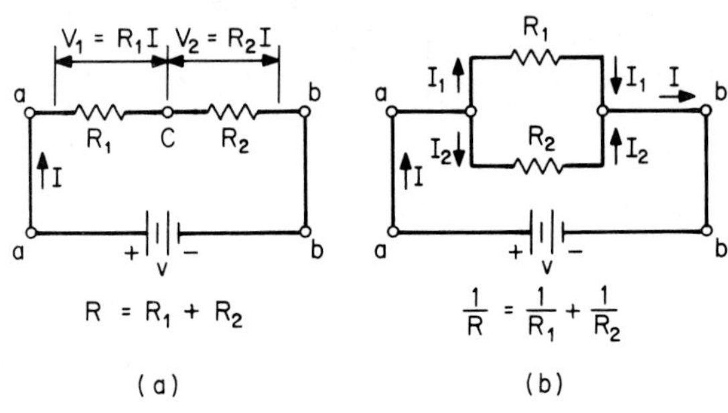

FIGURE 11.2

Example 11.3. Figure 11.3*a* and *b* shows two possibilities, the capacitors connected in parallel (*a*) and the capacitors connected in series (*b*). Where capacitors C_1, C_2, $C_3, \ldots$ are added in parallel (Fig. 11.3*a*), the total capacitance C increases, that is,

$$C = C_1 + C_2 + C_3 \tag{1}$$

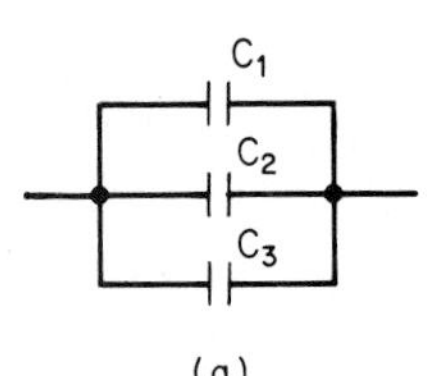

(a)

C1 C2 C3

(b)

FIGURE 11.3

Where capacitors C_1, C_2, $C_3, \ldots$ are added in series (Fig. 11.3*b*), the total capacitance C decreases, that is,

$$\frac{1}{C} = \frac{1}{C_1} + \frac{1}{C_2} + \frac{1}{C_3} \tag{2}$$

The *capacitance* C of a capacitor is the ratio of quantity of electricity q stored in it and voltage v across it, that is,

$$C = \frac{q}{v} \tag{3}$$

Where a capacitor requires a charge of 1 C to be charged to a voltage of 1 V, its capacitance is 1 F (farad), that is,

$$1 \text{ F} = 1 \frac{\text{C}}{\text{V}} = 1 \frac{\text{A} \cdot \text{s}}{\text{V}}$$

9. Direct and Alternating Currents. *Direct current* (dc) flows in one direction only through a circuit. The associated direct voltages, in contrast to alternating voltages, are of unchanging polarity. Direct current corresponds to a drift or displacement of electric charge in one unvarying direction around the closed loop or loops of an electric circuit. Direct currents and voltages may be of constant magnitude or may vary with time.

Alternating current (ac) reverses direction periodically, usually many times per second. One complete period, with current flow first in one direction and then in the other, is called a *cycle*, and 60 cycles per second (60 hertz, or Hz) is the customary frequency of alternation in the United States and 50 Hz in Europe.

Alternating current is shown diagrammatically in Fig. 11.4. Time is measured

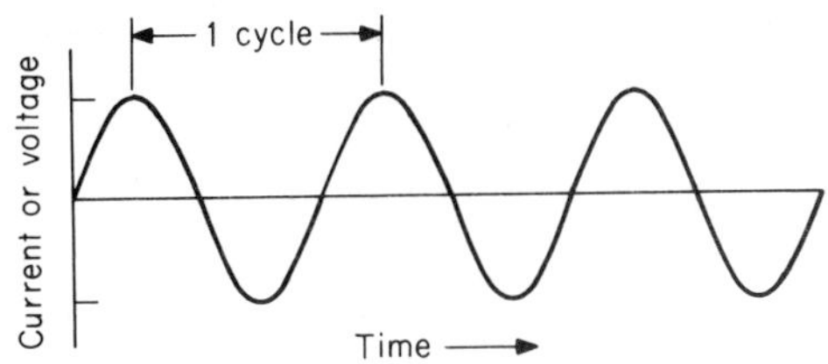

FIGURE 11.4 Diagram of sinusoidal alternating current.

horizontally (beginning at any arbitrary moment), and the current at each instant is measured vertically. In this diagram it is assumed that the current is alternating sinusoidally; that is, the current i is described by the following relation:

$$i = I_m \sin 2\pi ft \tag{11.26}$$

where I_m is the maximum instantaneous current, f is frequency per second (Hz), and t is time in seconds. Electric energy is ordinarily generated as alternating current.

MAGNETIC FIELDS

1. Magnetism. *Magnetism* comprises those physical phenomena involving magnetic fields and their effects on materials. Magnetic fields may be set up on a macroscopic scale by electric currents or by magnets. On an atomic scale, individual atoms cause magnetic fields when their electrons have a net magnetic moment as a result of their angular momentum. A magnetic moment arises whenever a charged particle has an angular momentum. It is the cooperative effect of the atomic magnetic moments that causes the macroscopic magnetic field of a permanent magnet.

2. Magnetic Field Forces. A region in space is the site of a magnetic field if a test charge q moving in the region experiences a force $\mathbf{F}$ by virtue of its motion with respect to an inertial frame. This force may be described in terms of a field vector $\mathbf{B}$, called the *magnetic induction* or *magnetic flux density*, or simply, the *magnetic field*. If u is the velocity of the charge, then the force is given by

$$\mathbf{F} = q(\mathbf{u} \times \mathbf{B}) \tag{11.27}$$

or in magnitude,

$$F = quB \sin \theta \tag{11.28}$$

where θ is the smaller angle between the vectors **u** and **B**. The unit of B is the tesla (T), where

$$1\ \mathrm{T} = 1\,\frac{\mathrm{N \cdot s}}{\mathrm{C \cdot m}} = 1\,\frac{\mathrm{N}}{\mathrm{A \cdot m}} = 1\,\frac{\mathrm{V \cdot s}}{\mathrm{m^2}} \tag{11.29}$$

When both a magnetic field **B** and an electric field **E** are present in the region, the force on a charge q is the vector sum of the electric and magnetic forces, that is,

$$\mathbf{F}_{\mathrm{tot}} = q[\mathbf{E} + (\mathbf{u} \times \mathbf{B})] \tag{11.30}$$

This relation is known as the *Lorentz equation*.

An element of length ds of a wire of arbitrary shape carrying current I in a magnetic field **B** experiences a force $d\mathbf{F}$ given by

$$d\mathbf{F} = I(ds \times \mathbf{B}) \tag{11.31}$$

Here the ds vector is of magnitude ds and in the direction of motion of positive charge. Net force on the wire (assuming a straight wire of length L) in a uniformed field **B** is, by integration,

$$\mathbf{F} = I(\mathbf{L} \times \mathbf{B}) \tag{11.32}$$

or

$$F = ILB \sin \theta \tag{11.33}$$

Magnetic Flux. Just as electric flux is associated with the electric field **E**, so magnetic flux Φ is associated with the magnetic field **B**. Thus the magnetic flux through an elementary area dS is defined as

$$d\Phi = \mathbf{B} \cdot dS \tag{11.34}$$

or

$$\Phi = \int_S \mathbf{B} \cdot \mathrm{dS} \tag{11.35}$$

The unit of magnetic flux is the weber (Wb), where

$$1\ \mathrm{Wb} = 1\ \mathrm{T \cdot m^2} = 1\ \mathrm{V \cdot s} = 1\ \mathrm{J/A}$$

Also, uncommonly, **B** (the magnetic flux density) is given in webers per square meter ($\mathrm{Wb/m^2}$).

3. Sources of the Magnetic Field. The *magnetic field of a moving charge* is given by

$$\mathbf{B} = \frac{\mu_0 q}{4\pi r^3}(\mathbf{u} \times \mathbf{r}) \tag{11.36}$$

where **u** is the velocity of a point charge q, **r** is the displacement vector, and μ_0 is the permeability of empty space ($= 4\pi \times 10^{-7}$ H/m). Here the henry (H) is the unit of inductance, where

$$1\ \mathrm{H} = 1\ \frac{\mathrm{Wb}}{\mathrm{A}}$$

In magnitude (in μT), we have

$$B = \frac{qu}{10r^2} \sin\theta \qquad (11.37)$$

where θ is the angle between **u** and **r**.

The *magnetic field of a current filament* is given by

$$dB = \frac{\mu_0 I}{4\pi r^3}(ds \times \mathbf{r}) \qquad (11.38)$$

Equation (11.38) is known as the *Biot-Savart law*.

Ampere's circuital law reads

$$\oint \mathbf{B} \cdot ds = \mu_0 I \qquad (11.39)$$

where the line integral on the left side is around an arbitrary closed path l, and I is the current through any open surface bounded by the path l.

4. Faraday's Law. *Faraday's law of induction* is related to an induced electromotive force (emf) to the change in magnetic flux that produces it. For any flux change that takes place in a circuit, Faraday's law states that the magnitude of the emf (v_i) induced in the circuit is proportional to the rate of change of flux as

$$v_i \approx -\frac{d\Phi}{dt} \qquad (11.40)$$

The time rate of change of flux in this expression may refer to any kind of flux change that takes place. If the change is motion of a conductor through a field, $d\Phi/dt$ refers to the rate of cutting flux. If the change is an increase or decrease in flux linking a coil, $d\Phi/dt$ refers to the rate of such change. For a coil of N turns,

$$v_i = -N\frac{d\Phi}{dt} \qquad (11.41)$$

5. Inductance. For an explanation of self-inductance of a coil refer to Fig. 11.5. In Fig. 11.5, current I establishes a magnetic field in and around coil C_0 as indicated. Each line of flux threads (or "links") some or all of the turns. The total linkage is $N\Phi$, where N is the total number of turns and Φ is the average flux linking a turn.

If Φ changes (as by sliding the contact on R), an electromotive force

$$v_i = -N\frac{d\Phi}{dt}$$

is induced in the coil. For an increase in Φ (decrease in R), $d\Phi/dt$ is positive and v_i is negative, that is, in the direction opposite to that of I. For Φ decreasing (increasing R), v_i is positive, that is, in the direction of I.

Assuming C_0 in empty space and no magnetic material nearby, the flux established, and thus the flux linkage, is directly proportional to I, that is,

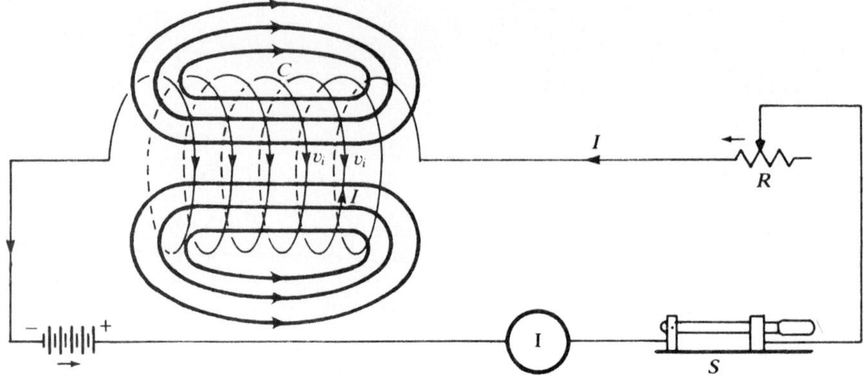

FIGURE 11.5 *(From Wells and Slusher,[2] p. 272.)*

$$N\Phi = LI \qquad \text{or} \qquad L = \frac{N\Phi}{I} \tag{11.42}$$

The proportionality factor L is referred to as the *self-inductance* (or just *inductance*). The SI unit of inductance, the henry (H), was defined in the section "Sources of the Magnetic Field."

Differentiating Eq. (11.42) with respect to time gives for the (back-) emf in the inductor

$$v_i = -N\frac{d\Phi}{dt} = -L\frac{dI}{dt} \tag{11.43}$$

This amount of energy stored in the final magnetic field around the inductor, i.e., principally in the space enclosed by the turns of C_0, is

$$E = \tfrac{1}{2}LI^2 \tag{11.44}$$

SIMPLE ELECTRIC CIRCUITS (EXAMPLES)

1. Series R-L-C Circuit. The type of circuit to be treated here is illustrated in Fig. 11.6. When switch S is closed, the source (a battery) begins to deliver charge q, thus establishing a current I in the circuit. Given the values of R, L, C, and emf v (assumed constant), we have to find expressions for q and I as functions of time. A more complete analysis also includes a determination of voltages v_{ab}, v_{bc}, and v_{cd}; energies stored in L and C; and the power delivered by the source—all as functions of time.

Applying Kirchhoff's voltage relation to the circuit, we have

$$v = v_{ab} + v_{bc} + v_{cd} = RI + L\frac{dI}{dt} + \frac{q}{C}$$

where the three terms on the right have been respectively obtained from relations given in previous sections. Writing $\dot{q}$ for I, we put this equation in the form

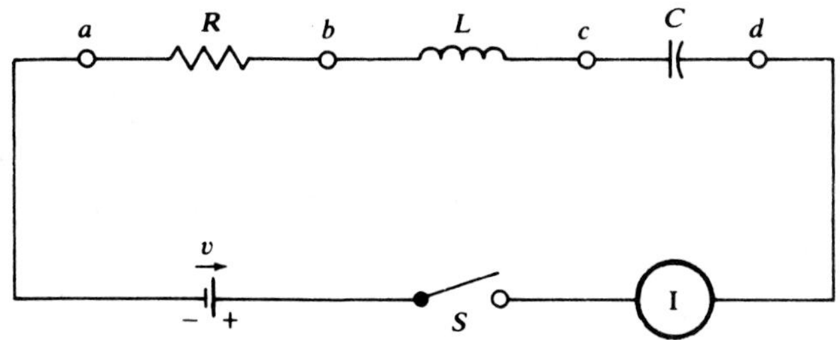

FIGURE 11.6 (*From Wells and Slusher,*[2] *p. 285.*)

$$L\ddot{q} + R\dot{q} + \frac{1}{C}q = v \tag{11.45}$$

A solution of this linear second-order differential equation, subject to appropriate initial conditions, gives q as a function of t, and from this all other results can be obtained.

2. Series AC Circuit. Referring to the circuit in Fig. 11.7, assume that a sinusoidal voltage, $u_a = v_a \sin \omega t$, is maintained between points a and b by an ac generator ("alternator"), where u_a is the instantaneous value of the voltage, v_a is the maximum value or amplitude of the voltage wave, and $\omega = 2\pi f$, where f is frequency in hertz (Hz).

The instantaneous values of voltage across resistor R, inductor L, and capacitor C are expressed in terms of instantaneous current i and instantaneous charge q by

$$u_R = Ri = R\dot{q} \qquad u_L = L\frac{di}{dt} = L\ddot{q} \qquad u_C = \frac{q}{C}$$

These must sum to u_a, which gives as the differential equation of the circuit

$$L\ddot{q} = R\dot{q} + \frac{1}{C}q = v_a \sin \omega t \tag{11.46}$$

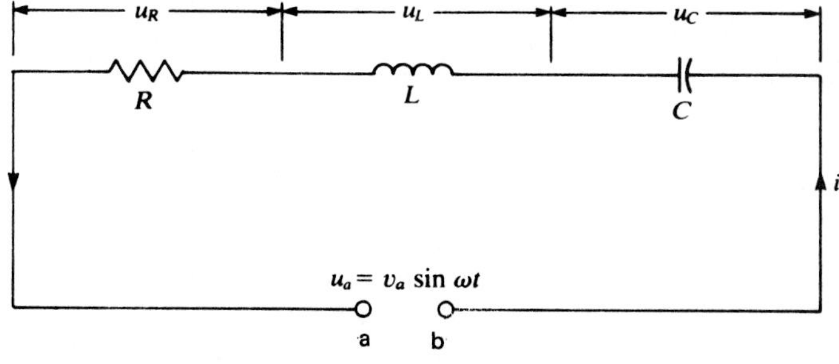

FIGURE 11.7 (*From Wells and Slusher,*[2] *p. 292.*)

After solving the last equation,

$$i = \frac{v_a}{Z} \sin(\omega t - \varphi) \equiv I \sin(\omega t - \varphi)$$

$$u_R = IR \sin(\omega t - \varphi) \equiv v_R \sin(\omega t - \varphi)$$

$$u_L = IX_L \cos(\omega t - \varphi) \equiv v_L \cos(\omega t - \varphi)$$

$$u_C = -IX_C \cos(\omega t - \varphi) \equiv -v_C \cos(\omega t - \varphi)$$

where the amplitudes of the four waves have been indicated as $I \equiv v_a/Z$, $v_R \equiv IR$, $v_L \equiv IX_L$, and $v_C \equiv IX_C$. Also,

$$X_C \equiv (\omega C)^{-1}$$

$$X_L \equiv \omega L$$

$$Z \equiv \sqrt{R^2 + (X_L - X_C)^2}$$

$$\tan \varphi \equiv \frac{X_L - X_C}{R}$$

WAVES

1. Wave Phenomena. A *wave* is a periodic disturbance that transports energy from one place to another. By a *periodic disturbance*, we mean certain changes that repeat themselves again and again in time. A wave $y(x, t)$ traveling along x with speed v is given by

$$y = f(x \pm vt) \tag{11.47}$$

where the minus and plus signs refer to wave propagation in the positive and negative x directions, respectively. This function satisfies the one-dimensional wave equation, that is,

$$\frac{\partial^2 y}{\partial x^2} = \frac{1}{v^2} \frac{\partial^2 y}{\partial t^2} \tag{11.48}$$

The speed of a wave traveling on a stretched uniform string, causing small transverse displacement $y(x, t)$, is given as

$$v = \sqrt{\frac{F}{\rho}} \tag{11.49}$$

where F is the tension in the string and ρ is the density of the string. The instantaneous power transmitted by the wave is given by

$$P = -F \frac{\partial y}{\partial x} \frac{\partial y}{\partial t} \tag{11.50}$$

The sinusoidal traveling wave

$$y = A \cos (kx - \omega t) \tag{11.51}$$

has a wave speed

$$v = \frac{\omega}{k}$$

where period T and wavelength λ are gives by

$$T = \frac{2\pi}{\omega} \tag{11.52}$$

$$\lambda = \frac{2\pi}{k} \tag{11.53}$$

or

$$v = \frac{\lambda}{T} = \lambda f \tag{11.54}$$

where f is frequency.

2. Longitudinal and Transverse Waves. There are two very different kinds of waves whose motion satisfies Eq. (11.54). The first of these is a *longitudinal wave* in which the periodic motion of the particles producing the wave is parallel to the direction of propagation of the wave (see Fig. 11.8*a*). For example, in a sound wave in air, the air molecules vibrate back and forth in the direction in which the wave is moving and pass on energy to other molecules by collisions. In this way energy is transported through the air. Since the result of this vibrational motion is to produce a series of high- and low-pressure regions of space in the

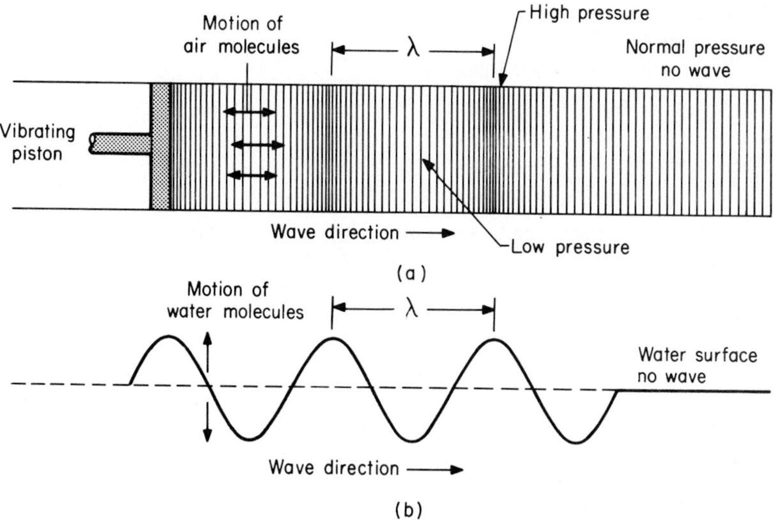

FIGURE 11.8 Longitudinal sound wave (*a*) and transverse water wave (*b*). *(From Mulligan,[3] p. 288.)*

direction in which the wave is traveling, a sound wave is sometimes called a *compressional wave*.

A second, very different kind of wave is one in which the particles involved in the wave move at right angles to the direction of propagation. This we call a *transverse wave* (see Fig. 11.8*b*). A good approximation of a transverse wave is a water wave on the surface of a pond. The water molecules execute a periodic motion up and down at their fixed positions on the surface of the pond. The wave, however, moves along the surface of the pond at right angles to the up-and-down motion of the water molecules.

3. Sound Waves. Sound waves are longitudinal waves in a medium. The longitudinal displacement l of the medium and the excess pressure p_e are related by

$$p_e = -E_0 \frac{\partial l}{\partial x} \tag{11.55}$$

where E_0 is the bulk modulus of the medium. Note that p_e and l satisfy the wave equation [Eq. (11.48)]. The wave speed is given by

$$v = \sqrt{\frac{E_0}{\rho_0}} \tag{11.56}$$

where ρ_0 is the density of the medium when it is in equilibrium. If the medium is a gas,

$$v = \sqrt{\frac{\gamma p_0}{\rho_0}} = \sqrt{\frac{\gamma KT}{m}} \tag{11.57}$$

where p_0 = undisturbed gas pressure
γ = specific heat ratio
K = Boltzmann's constant (1.38×10^{-23} J/K)
m = mass of the molecule of the gas

If γ is constant, then the speed of sound in a given gas varies as the square root of the absolute temperature.

Doppler Effect. Suppose that a source emitting sound waves of frequency f_s and an observer move along the same straight line. Then the observer will hear sound of frequency

$$f_o = f_s \frac{v + v_o}{v - v_s}$$

where v_o and v_s are the velocities of the observer and source relative to the transmitting medium, and v is the sound speed in this medium. Velocity v_o is positive if the observer is heading toward the source, and v_s is positive if the source is heading toward the observer.

Transverse waves (light) also exhibit a *Doppler effect*, but in the case of light, the effect depends only on the relative velocity between source and observer, there being no transmitting medium.

4. Electromagnetic Waves. In the case of mechanical waves, such as sound waves or water waves, there is always a material medium through which the wave passes, the molecules of a gas for a sound wave and the molecules of water for a water wave. Without these molecules, there would be no wave. However, there is another highly important class of waves that require no material medium for their propagation through space. These are the electromagnetic waves produced, for example, by electrons oscillating back and forth in the transmitting antenna of a radio station. These electromagnetic waves travel through a perfect vacuum just as easily as they move through air.

Electromagnetic waves are transverse waves consisting of periodic variations of electric and magnetic fields in space, and these fields can exist even where no matter is present. Sound waves in gases have different speeds depending on the properties of the gas, and these speeds often change with frequency. Electromagnetic waves in a vacuum (sometimes called *free space*), on the other hand, have speeds determined only by the properties of space. All electromagnetic waves in free space, therefore, travel with the same speed, no matter what the frequency or wavelength. This speed is found from experiments to be

$$c = 3.00 \times 10^8 \text{ m/s}$$

Since $c = \lambda f$, from Eq. (11.54), if the frequency of an electromagnetic wave is known, its wavelength is easily determined, and vice versa.

All electromagnetic waves (i.e., all waves produced by changing electric and magnetic fields) are of the same basic nature and travel at the same speed in a vacuum. They differ only in their frequency and wavelength. Waves of different frequencies interact differently with matter, however, and often appear to have very dissimilar properties. This is one example of a frequently occurring situation in physics in which quantitative changes (in this case, in wavelength and frequency) lead to important qualitative differences. Figure 11.9 shows the complete electromagnetic spectrum, with the wavelengths, frequencies, and energies corresponding to each part of that spectrum. It stretches all the way from long radio waves with wavelengths longer than 10^6 m to gamma rays with wavelengths shorter than 10^{-12} m.

5. Light. As we have seen from Fig. 11.9, visible light occupies that portion of the electromagnetic spectrum consisting of wavelengths from 0.4 μm (that is, 0.4×10^{-6} m) to 0.7 μm. From a physical point of view, the ultraviolet region, with wavelengths shorter than 0.4 μm, blends continuously into the visible region, which in turn blends continuously into the infrared region, with wavelengths longer than 0.7 μm. (Note that the *infra* in *infrared* and the *ultra* in *ultraviolet* refer to frequency, not wavelength.) The distinguishing feature of the visible portion of the spectrum is the human eye's sensitivity to wavelengths between 0.4 and 0.7 μm, in contrast to its insensitivity to infrared and ultraviolet radiation. The human eye sees each different wavelength of the visible region as a different color; violet corresponds to 0.4 μm and red to 0.7 μm, and there is a continuous change from violet to blue-green to orange to yellow to red between these two limits.

In free space, the speed of all electromagnetic waves is the same. This is not true in a material substance such as glass, where different wavelengths have different speeds.

Reflection and Refraction. Consider a ray of light. In an isotropic medium, rays are straight lines, along which energy travels at speed $v = c/n$, where n is the

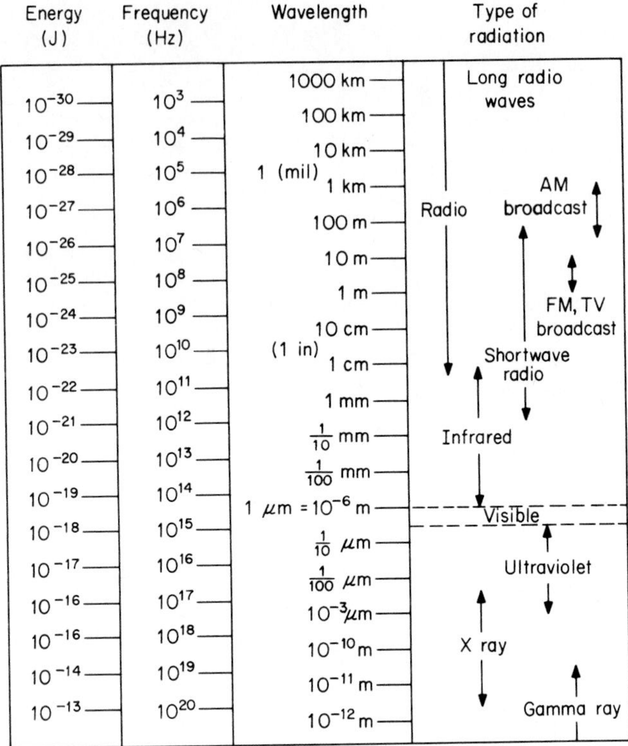

FIGURE 11.9 The complete electromagnetic spectrum. *(From Mulligan,*[3] *p. 291.)*

refractive index of the medium. Because $n > 1$, the speed is less than the speed in vacuum, $c = 3 \times 10^8$ m/s. The refractive index is a function of the wavelength (in vacuum) of the light.

Let a ray in medium 1 be incident on the interface with medium 2, making angle θ_1 with the normal to the interface (Fig. 11.10). Then, in general, there will be a reflected ray in medium 1 and a refracted ray in medium 2 such that

1. The three rays and the normal all lie in a common plane, the plane of incidence.
2. The angle of incidence equals the angle of reflection, that is, $\theta_1 = \theta_r$.
3. The directions of the incident and refracted rays are related by Snell's law:

$$n_1 \sin \theta_1 = n_2 \sin \theta_2$$

If $n_1 > n_2$ and θ_1 exceeds the critical angle θ_c, where

$$\sin \theta_c = \frac{n_2}{n_1}$$

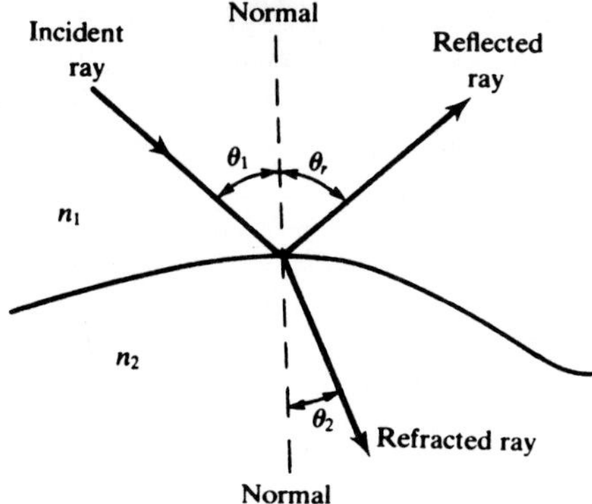

FIGURE 11.10 *(From Wells and Slusher,[2] p. 301.)*

then there will be no refracted ray—a phenomenon called *total reflection* or, from the point of view of medium 1, *total internal reflection*.

Geometrical Optics. Geometrical optics is concerned with the effects on light rays of mirrors (plane, concave, convex) and lenses (converging, diverging). It is assumed here that all lenses are thin (i.e., their thickness is small compared to their radii of curvature), and that all rays are paraxial (i.e., they make small angles with the axis of the optical system).

The two basic relations for both mirrors and lenses are

$$\frac{1}{d_o} + \frac{1}{d_i} = \frac{1}{f} \tag{11.58}$$

and

$$\frac{d_o}{d_i} = \frac{l_o}{l_i} \tag{11.59}$$

where f = focal length (cm)
d_o = distance of object (cm)
d_i = distance of image (cm)
l_o = linear size of object (cm)
l_i = linear size of image (cm)

The five quantities are conventionally given in centimeters. Each quantity carries an algebraic sign, as specified in Table 11.1. The light rays diverge from a real object and converge toward a virtual object. Contrarily, light rays converge toward a real image (which could not be projected directly on a screen).

Example 11.4. Describe the image formed by a concave mirror when a real object is situated outside the center of curvature C.

In Fig. 11.11, real rays *a, b, c* are drawn from a point P_1 on the real object P_1P_2. For convenience, *a* is drawn parallel to the optical axis, *b* is drawn through F, and *c* is drawn through C. From the geometry of the mirror and the law of reflection, reflected ray a' passes through F, b' is parallel to the optical axis, and c' returns through C along the path of *c*. The intersection of the real rays a', b', and c' at P_1'

TABLE 11.1

Quantity	Sign	
	+	−
f	converging lens, concave mirror	diverging lens, convex mirror
d_o	real object	virtual object
d_i	real image	virtual image
ℓ_o	erect object	inverted object
ℓ_i	erect image	inverted image

Source: From Wells and Slusher,[2] p. 309.

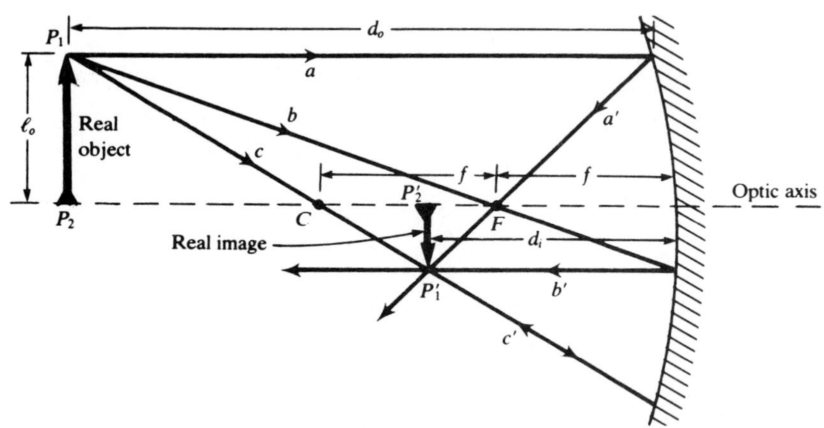

FIGURE 11.11 (*From Wells and Slusher,*[2] *p. 311.*)

locates one point on the real image $P_1'P_2'$. Rays a, b, and c may seem very special. However, any (paraxial) ray from P_1, after reflection, passes through P_1', as may be shown by an application of the law of reflection. Note that the image is inverted.

Let $f = -20$ cm, $d_o = +45$ cm, and $l_o = +5$ cm. Then Eq. (11.58) gives

$$\frac{1}{45} + \frac{1}{d_i} = \frac{1}{20} \qquad \text{or} \qquad d_i = +36 \text{ cm}$$

(a real image), and Eq. (11.59) gives

$$\frac{45}{36} = -\frac{5}{l_i} \qquad \text{or} \qquad l_i = -4 \text{ cm}$$

(an inverted, minified image).

NOMENCLATURE

Symbol = *Definition, SI Units (U.S. Customary Units)*

A = area, m^2 (ft^2)

B, $\mathbf{B}$ = magnetic flux density (magnitude and vector), T

C = capacitance, F

E, $\mathbf{E}$ = electric field (magnitude and vector), N/C

E = energy (in "Ohm's law" and "Inductance" sections), J (Btu)

e = electric charge, C

F, $\mathbf{F}$ = force (magnitude and vector), N (lb_f)

F = tension (in "Waves" section), N/m^2 (lb_f/ft^2)

I = electric current, A

i = alternating electric current, A

J = electric current density, A/m^2

K_e = kinetic energy, N · m (lb_f · ft)

L, $\mathbf{L}$ = length (magnitude and vector), m (ft)

L = inductance (in "Inductance" section), H

P = electric power, W

Q = electric charge, C

q = electric charge, C

q_1 = electric charge, C

q_2 = electric charge, C

R = resistance, Ω

R_0 = internal resistance of a source, Ω

r = distance, m (ft)

r, $\mathbf{r}$ = displacement (magnitude and vector), m (ft)

S = surface area, m^2 (ft^2)

s, $\mathbf{s}$ = length (magnitude and vector), m (ft)

T = temperature, K (°R)

t = time, s

U = electric potential energy, N · m (lb_f · ft)

u, $\mathbf{u}$ = velocity (magnitude and vector), m/s (ft/s)

v = voltage, V; velocity (in "Waves" section), m/s (ft/s)

Greek

θ = angle, rad, deg.

ψ = electric flux, N · m^2/C

ϕ = electric potential, V

Φ = magnetic flux, Wb

ρ = resistivity, $\Omega \cdot$ m; density (in "Waves" section), kg/m^3 (lb/ft^3)
μ_0 = permeability of empty space, H/m

ACKNOWLEDGMENTS

Material presented in this chapter is drawn mostly from Refs. 1 to 3. Reference 4 was used to compare certain definitions. All numerical examples are from Ref. 2. Several other texts in physics also were consulted for selection of material presented here.

REFERENCES

1. S. P. Parker, *McGraw-Hill Encyclopedia of Physics*, McGraw-Hill, New York, 1983.
2. D. A. Wells and H. S. Slusher, *Physics for Engineering and Science*, McGraw-Hill, New York, 1983.
3. J. F. Mulligan, *Practical Physics*, McGraw-Hill, New York, 1980.
4. E. U. Condon and H. Odishaw, *Handbook of Physics*, 2d ed., McGraw-Hill, New York, 1967.

CHAPTER 12
AUTOMATIC CONTROL

INTRODUCTION

The purpose of an automatic control on a system is to produce a desired output when inputs to the system are changed. Inputs are in the form of commands, which the output is expected to follow, and disturbances, which the automatic control is expected to minimize.

The various aspects of automatic control can best be described by use of an example. Consider a process such as that shown in Fig. 12.1. The flowing liquid

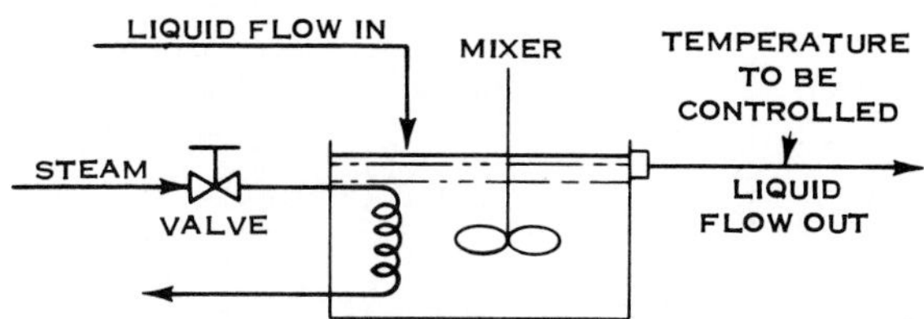

FIGURE 12.1 Simple heat-exchange process. (*From Perry et al.*[1])

is to be heated to a desired temperature by steam flowing through heating coils. The temperature of the exit flow is affected by factors (process variables) such as the temperature of the incoming liquid, the flow rate of the liquid, the temperature of steam, the flow rate of steam, heat capacities of the fluids, heat loss from the vessel, and mixer speed.

1. Open- and Closed-Loop Systems*. The system shown in Fig. 12.1 is normally classified as "open-loop." Open-loop control systems are those in which information about the controlled variable (in this case, temperature) is not used to adjust any of the system inputs to compensate for variations in the process variables. The term *open loop* is often encountered in discussions of control systems to indicate that the uncontrolled process dynamics are being studied.

*Sections 1 to 3 taken in part from *Perry's Chemical Engineers' Handbook*, 6th ed., by R. H. Perry, D. W. Green, and J. O. Maloney (eds.). Copyright © 1984. Used by permission of McGraw-Hill, Inc. All rights reserved.

A closed-loop control system implies that the controlled variable is measured and that the result of this measurement is used to manipulate one of the process variables, such as steam flow.

2. Feedback Control. In the closed-loop control system, information about the controlled variable is fed back as the basis for control of a process variable; hence the designation *closed-loop feedback control*. This feedback can be accomplished by a human operator (manual control) or by use of instruments (automatic control).

For manual control, referring to Fig. 12.1, an operator periodically measures the temperature; if this temperature, for example, is below the desired value, the operator increases the steam flow by opening the valve slightly. For automatic control, a temperature-sensitive device is used to produce a signal (electrical, pneumatic, etc.) proportional to the measured temperature. This signal is fed to a controller, which compares it with a preset desired value, or set point. If a difference exists, the controller changes the opening of the steam-control valve to correct the temperature, as in Fig. 12.2.

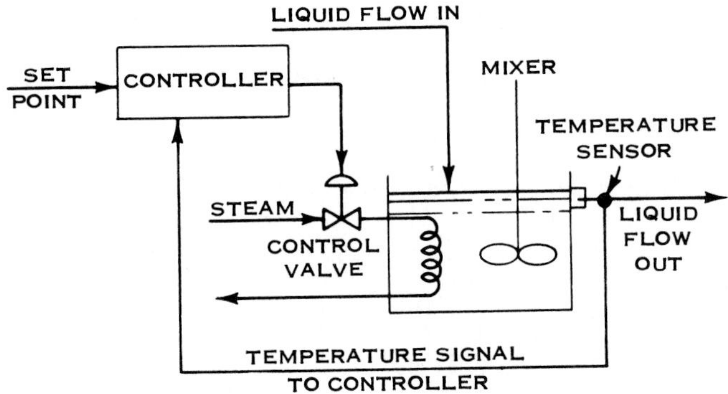

FIGURE 12.2 Automatic feedback control of heat-exchange process. (*From Perry et al.*[1])

3. Feedforward Control. Feedforward control is becoming widely used. Process disturbances are measured and compensated for without waiting for a change in the controlled variable to indicate that a disturbance has occurred. Feedforward control is also useful when the final controlled variable cannot be measured.

The general theories and definitions of automatic control have been developed to aid the designer to meet primarily three basic specifications for the performance of the control system, namely, stability, accuracy, and speed of response.

4. Nomenclature*. The nomenclature of automatic control has been reviewed by both the American Society of Mechanical Engineers (ASME) and the American Institute of Electrical Engineers (AIEE) in recent years. Many terms adopted by the ASME naturally tend somewhat toward the vocabulary of the process-

*This section taken in part from *Marks' Standard Handbook for Mechanical Engineers*, 9th ed., by E. A. Avallone and T. Baumeister III (eds.). Copyright © 1987. Used by permission of McGraw-Hill, Inc. All rights reserved.

control engineer and in many cases are not sufficiently broad for general control application. The following terms and definitions have been selected to assist the reader and to serve as reference to a complex area of technology whose breadth crosses several professional disciplines.

Automatic Regulator. An apparatus which measures the value of a quantity or condition which is subject to change with time, and operates to maintain within limits this measured value

Controlled Variable (Types). That quantity or condition of the controlled system which is directly measured or controlled

Throttling Range. That range of values through which the variable must change to cause the final control element to move from one extreme position to the other

Set Point. The value of the controlled variable that it is desired to maintain

Deviation. The difference at any time between the controlled variable and the set point

Corrective Action. A change in the flow of the control agent initiated by the measuring means of the automatic controller

Control Agent. Process energy whose flow is directly varied by the control element

Self-Regulation. That operating characteristic which inherently assists the establishment of equilibrium

Two-Position Controller Action. That in which the final control element is moved immediately, from one extreme to the other of its stroke, at predetermined values of the variable

Proportional-Position Controller Action. That in which there is a continuous linear relation between the values of the controlled variable and the rate of motion of a final control element

Floating Controller Action. That in which there is a predetermined relation between the values of the controlled variable and the rate of motion of a final control element

Proportional-Speed Floating Controller Action. That in which there is a continuous linear relation between the rate of motion of the final control element and the deviation of the controlled variable

Derivate Controller Action. That in which there is a predetermined relation between a derivative function of the controlled variable and the position of a final control element

Proportional Plus Floating Controller Action (Proportional + Reset). That in which proportional-position and proportional-speed floating actions are additively combined

Proportional Band. The range of scale values through which the controlled variable must pass in order that the final control element be moved through its entire range

Floating Speed. The rate of movement of a final control element corresponding to a specified deviation

Command Signal. The input which is established or varied by some means external to, and independent of, the feedback control system under consideration

Disturbance. A signal which tends to affect the value of the controlled variable

Response Time. The time required for the controlled variable to reach a specified value after the application of a step input of disturbance

Peak Time. The time required for the controlled variable to reach its first maximum following the application of a step input

Rise Time. The time required for the controlled variable to increase from one specified percentage of the final value to another, following the application of a step input

Settling Time. The time required for the absolute value of the difference between the controlled variable and its final value to become and remain less than a specified amount following the application of a step disturbance

Compensation. A method of changing or maintaining the state of a system by employing means to offset the effects of disturbances without causal relationship between the error in the state of the system and the action of the compensating means.

BASIC AUTOMATIC-CONTROL SYSTEM*

The general components of a basic automatic-control system are shown in Fig. 12.3. Each block in the diagram represents a function which must be performed

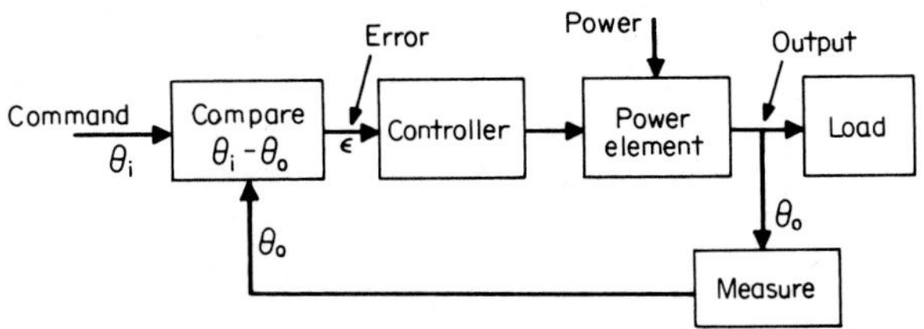

FIGURE 12.3 Functional diagram of an automatic-control system. (*From Baumeister et al.*[3])

by the control. The operation may be explained as follows: (1) A command signal θ_i is applied to the input and compared with the instantaneous position of the output θ_o. (2) The result of this comparison ϵ, representing an error, is amplified by a controller and used to position a power element. (3) The power device in turn further amplifies the error signal to supply large amounts of power to the output or load to reduce the difference between θ_i and θ_o.

ANALYSIS OF CONTROL SYSTEM

The stability, accuracy, and speed of response of a control system are determined by analyzing the steady-state and the transient performance. It is desirable to achieve the steady state in the shortest possible time, while maintaining the output within specified limits. Steady-state performance is evaluated in terms of the accuracy with which the output is controlled for a specified input. The transient performance, i.e., the behavior of the output variable as the system changes from one steady-state condition to another, is evaluated in terms of such quantities as maximum overshoot, rise time, and response time (Fig. 12.4).

*This section taken in part from *Marks' Standard Handbook for Mechanical Engineers*, 9th ed., by E. A. Avallone and T. Baumeister III (eds.). Copyright © 1987. Used by permission of McGraw-Hill, Inc. All rights reserved.

An automatic control normally has only two places where disturbances can be expected: at the input or at the load. For a purely mechanical system the input disturbance may take the form of a periodic oscillation, a displacement, a velocity, or an acceleration. Disturbances at the output are usually load changes expressed as a torque or force quantity. Nonmechanical systems have disturbances expressed in different quantities; however, they are directly analogous to the mechanical system.

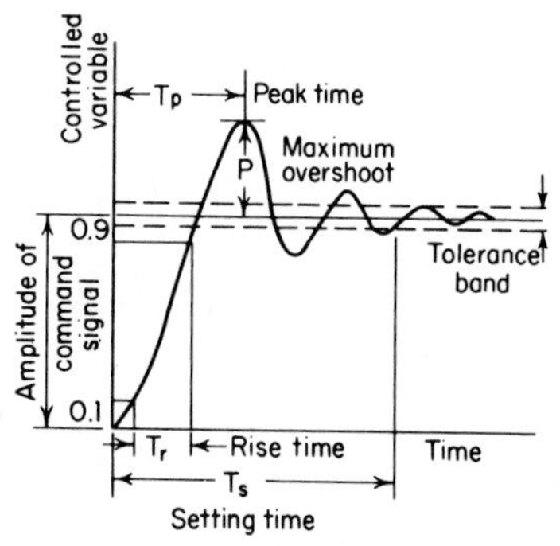

FIGURE 12.4 System response to a unit step-function command. (*From Avallone and Baumeister.*[2])

1. Laplace Transforms*. The steady-state and dynamic behavior of a system can be determined by solving the differential equation representing that system. This may be a long and tedious task, especially if there are many elements in the system. One technique for solving such differential equations uses the Laplace transformation. Here the problem is stated in terms of a second variable which allows the problem to be solved algebraically. Then, by transformation back to the original independent variable, the solution to the original differential equation is obtained (see Chap. 3).

Laplace transforms, as useful as they are, can be employed only for linear differential equations. Such equations describe a linear system, one in which the rules of superposition apply. That is, if the time response of the system is $y_1(t)$ when a forcing function $x_1(t)$ is applied and the response is $y_2(t)$ when $x_2(t)$ is applied, then if $x_1(t) + x_2(t)$ is applied, the response is $y_1(t) + y_2(t)$. A linear differential equation of this type is

$$p_n(t)\frac{d^n y(t)}{dt^n} + p_{n-1}(t)\frac{d^{n-1}y(t)}{dt^{n-1}} + \cdots + p_0(t)y(t) = x(t)$$

where the coefficients $p_i(t)$ are not functions of the dependent variable $y(t)$ or any of its derivatives. Generally, the solution of such an equation with time-varying coefficients is difficult. Most chemical operations are nonlinear over a wide range, but the assumption of linearity over a small region near the operating point can usually be justified (linearization). The coefficients of this linearized differential equation are usually independent of time. Therefore, for process control the basic process differential equation is usually of the form

$$m_n\frac{d^n y(t)}{dt^n} + m_{n-1}\frac{d^{n-1}y(t)}{dt^{n-1}} + \cdots + m_0 y(t) = x(t)$$

which can be solved routinely with Laplace transforms.

*This section taken in part from *Perry's Chemical Engineers' Handbook*, 6th ed., by R. H. Perry, D. W. Green, and J. O. Maloney (eds.). Copyright © 1984. Used by permission of McGraw-Hill, Inc. All rights reserved.

2. Transfer Functions*. To illustrate the convenience of the Laplace transformation, consider a differential equation given by

$$T\frac{dc(t)}{dt} + c(t) = RMu(t) \qquad c(0) = 0 \tag{12.1}$$

where T represents the time constant of the system, and $Mu(t)$ describes a step input to the system of magnitude M. We wish to find the response of the output $c(t)$. The Laplace transforms needed are taken from tables. We have

$$\mathcal{L}\left(\frac{dc(t)}{dt}\right) = sC(s) - c(0^+)$$

$c(0^+) = 0$ since the initial conditions were zero. Thus

$$\mathcal{L}\,[c(t)] = C(s)$$

$$\mathcal{L}\,[Mu(t)] = M\left(\frac{1}{s}\right)$$

The Laplace transform of Eq. (12.1) is then

$$TsC(s) + C(s) = R\left(\frac{M}{s}\right)$$

$$C(s) = \frac{RM}{s(Ts + 1)} \tag{12.2}$$

To transform this equation back into the time domain, we look up inverse transform $\mathcal{L}^{-1}$ and find that

$$c(t) = \mathcal{L}^{-1}\,[C(s)] = RM(1 - e^{-t/T})$$

In a system of many elements the transformed equation corresponding to Eq. (12.2) may be quite complicated. However, it can usually be manipulated into a form for which the inverse transform can be found in a table.

In process-control work Laplace transforms are used to determine responses to disturbances. Steady-state or constant terms will usually drop out of the solutions of the differential equations because initial conditions will usually be assumed to be zero.

The transfer function is defined as the ratio of the Laplace transform of the responding variable (output) to the Laplace transform of the disturbing variable (input). From Eq. (12.2), we get for the transfer function $KG(s)$

$$KG(s) = \frac{\text{output}}{\text{input}} = \frac{C(s)}{M(s)} = \frac{R}{Ts + 1} \tag{12.3}$$

The convention for designating transfer functions in a control diagram is the expression $KG(s)$. Capital letters are used when the functions are in the s domain, and small letters are used in the time domain. $G(s)$ represents the dynamic por-

*This section taken in part from *Perry's Chemical Engineers' Handbook*, 6th ed., by R. H. Perry, D. W. Green, and J. O. Maloney (eds.). Copyright © 1984. Used by permission of McGraw-Hill, Inc. All rights reserved.

tion of transfer function, and K is related to the steady-state gain through an element. In the transfer function, Eq. (12.3), $K = R$ and $G(s) = 1/(Ts + 1)$.

It is common practice in drawing block diagrams in the s domain to omit the (s) from the $F(s)$'s. Instead only the capital-letter designation is used to represent the s-domain transforms.

Combining Transfer Functions. Fluid- and thermal-process systems exhibit many different dynamic characteristics, but many systems may be described by combinations of five transfer functions:

Proportional element: K

Capacitance element: $\dfrac{1}{Ts}$

First-order element: $\dfrac{1}{Ts + 1}$

Second-order element: $\dfrac{1}{T^2s^2 + 2\xi Ts + 1}$

Dead-time element: e^{-Ls}

Transfer functions are important tools in the analysis of control systems. Each block or element of the control system has its own characteristic transfer function. If the s-domain transfer function notation $KG(s)$ is used for each block, the system elements can be combined by algebraic procedures into an overall expression for the entire control system.

3. Block Diagrams*. A useful representation of the mathematical relationships defining the flow of information and energy through the control system is by means of a block diagram. In the diagram the components of the control system are considered as functional blocks in series and parallel arrangements according to their position in the actual control system. Each component is represented by its transfer function, the ratio of the Laplace transform of the output variable to the input variable with all initial conditions taken as zero.

The block diagram of a single-loop feedback-control system subjected to a command input $R(s)$ and a disturbance $U(s)$ is shown in Fig. 12.5.

When $U(s) = 0$ and the input is a reference change, the system may be reduced as follows:

$$E(s) = \theta_i(s) - \theta_o(s)\,H(s)$$

$$\theta_o(s) = E(s)\,[G_1(s)G_2(s)]$$

Therefore

$$\frac{\theta_o(s)}{\theta_i(s)} = \frac{G_1(s)G(s)}{1 + G_1(s)G_2(s)H(s)} \tag{12.4}$$

*This section taken in part from *Marks' Standard Handbook for Mechanical Engineers*, 9th ed., by E. A. Avallone and T. Baumeister III (eds.). Copyright © 1987. Used by permission of McGraw-Hill, Inc. All rights reserved.

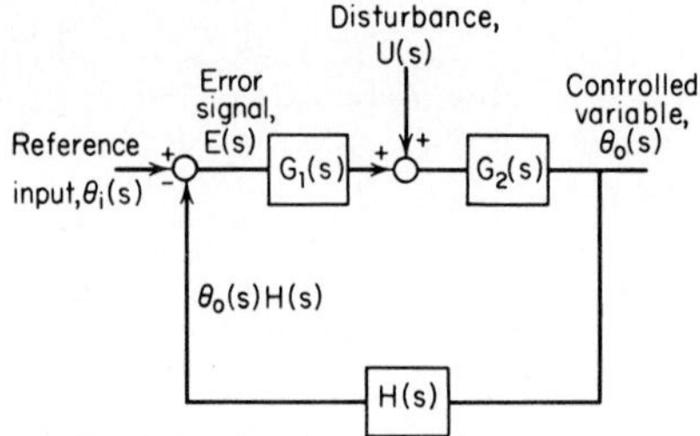

FIGURE 12.5 Single-loop feedback control system. *(From Avallone and Baumeister.*[2]*)*

When $\theta_i(s) = 0$ and the input is a disturbance, the system may be reduced as follows:

$$E(s) = -\theta_o(s)H(s)$$

$$[E(s)G_1(s) + U(s)]\,G_2(s) = \theta_o(s)$$

$$\frac{\theta_o(s)}{U(s)} = \frac{G_2(s)}{1 + G_1(s)G_2(s)H(s)} \qquad (12.5)$$

Equations (12.4) and (12.5) are in the form

$$\frac{\text{Response function}}{\text{Excitation function}} = \text{system function}$$

The system function is expressible as the ratio of two polynomials, $A(s)/B(s)$. The equation $B(s) = 0$ is the characteristic equation. When the excitation function is specified, inverse transformation of $\theta_o(s)$ yields $\theta_o(t)$, the transient response.

4. Examples*

1. *First-Order Lag Element (Time-Constant Element):* A first-order lag element may be illustrated by the liquid level in the tank shown in Fig. 12.6, where the flow out of the tank is assumed proportional to the level in the tank. The dynamics are described by

$$C_L \frac{dh}{dt} = f_1 - f_2 \qquad \text{and} \qquad f_2 = \frac{h}{R_L}$$

Therefore,

$$R_L C_L \frac{dh}{dt} + h = R_1 f_1 \qquad (12.6)$$

The product RC appears often in analyses of automatic control systems. By analogy from electric-circuit theory, it is called the *time constant* and is usually des-

*This section taken in part from *Perry's Chemical Engineers' Handbook*, 6th ed., by R. H. Perry, D. W. Green, and J. O. Maloney (eds.). Copyright © 1984. Used by permission of McGraw-Hill, Inc. All rights reserved.

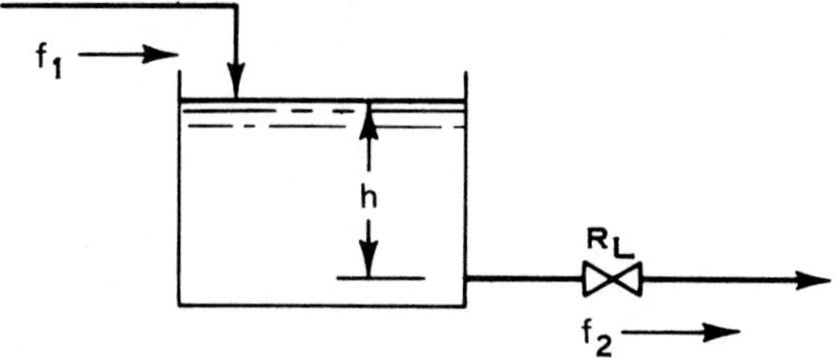

FIGURE 12.6 First-order lag, liquid-flow element. (*From Perry et al.*[1])

ignated as T. Time constants are probably the most used indicators of process-element characteristics. Substituting $T = R_L C_L$ Eq. (12.6) in the s domain gives

$$(Ts + 1)H(s) = R_L F_1(s)$$

$$H(s) = \frac{R_L}{Ts + 1} F_1(s) \tag{12.7}$$

The transfer function for this first-order lag element is $R_L/(Ts + 1)$; see Fig. 12.7. R_L is a magnitude factor arising here because head or potential was process-element output. First-order lag elements are characterized by the expression $1/(Ts + 1)$.

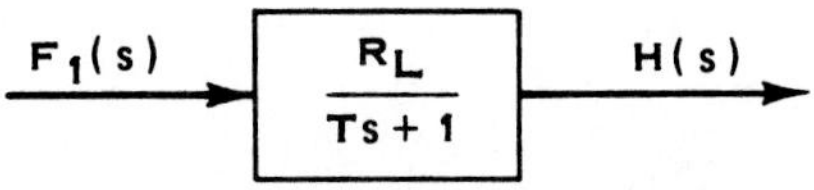

F1(s) → [$\frac{R_L}{Ts + 1}$] → H(s)

FIGURE 12.7 Block diagram of a liquid-flow, first-order lag process. (*From Perry et al.*[1])

The time response of a first-order lag can be obtained by solving Eq. (12.6) for a step change of inflow with the tank initially empty.

Rearranging terms,

$$\frac{dh}{R_L f_1 - h} = \frac{dt}{T} \qquad f_1 = \text{constant}$$

Integrating between the limits $h_0 = 0$ gives

$$- \ln (R_L f_1 - h) \Big|_{h}^{h_0} = \frac{t}{T}$$

$$h(t) = R_l f_1 (1 - e^{-t/T})$$

A plot of the process-element output for a unit-step increase in input in Fig. 12.8 illustrates that the time constant is the time required to reach $1 - e^{-1}$, 63.2 percent of the final value. The final value of the output will be R, where $R = R_L f_1$.

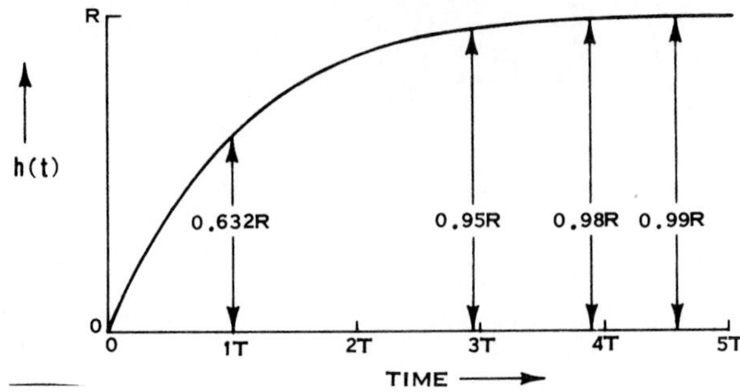

FIGURE 12.8 Time response of a first-order lag resulting from a unit-step increase in input at time equals zero. (*From Perry et al.*[1])

2. *Second-Order Lag Element (Quadratic or Oscillatory Element):* The second-order system is important in the study of automatic control systems because the actual response of a controlled system is often compared with that of an oscillatory second-order system. Few open-loop chemical processes exhibit oscillatory characteristics; however, the closed-loop control system is often tuned to have characteristics similar to the second-order oscillatory system.

Second-order-system characteristics are illustrated by the spring mass, and damper system shown in Fig. 12.9. The mass is acted on by four forces: (1) the spring-displacement force kx, (2) the velocity dependent damping force $p\ dx/dt$, (3) the acceleration-dependent inertial force $(W/g)\ (d^2x/dt^2)$ (g is gravitational acceleration), and (4) the applied force f. The output of this element is the displacement x. The equation relating the forces is

$$\frac{W}{g}\frac{d^2x}{dt^2} + p\frac{dx}{dt} + kx = f$$

If the initial displacement and velocity are zero, the Laplace transformations are

$$\mathcal{L}[x(t)] = X(s) \qquad \mathcal{L}\left(\frac{dx}{dt}\right) = sX(s) \qquad \mathcal{L}\left(\frac{d^2x}{dt^2}\right) = s^2X(s)$$

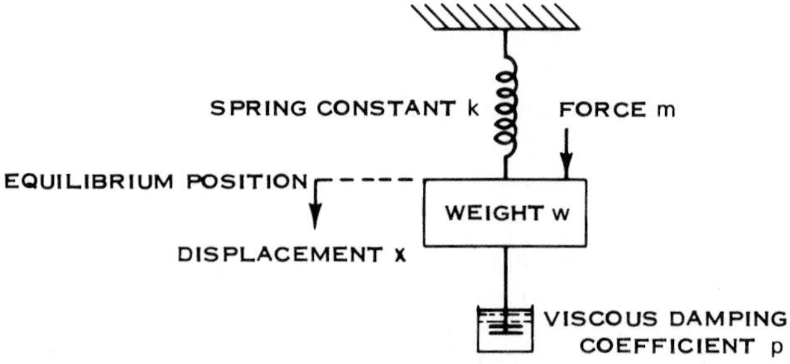

FIGURE 12.9 Mechanical representation of the oscillatory element. (*From Perry et al.*[1])

After rearranging, the transformed equation is

$$\frac{X(s)}{F(s)} = \frac{1/k}{(W/gk)s^2 + (p/k)s + 1}$$

where $F(s)$ and $X(s)$ are the system input and output, respectively.

This transfer function is characteristic of all quadratic lags. For discussion, it is convenient to write this transfer function in a standard form:

$$KG(s) = \frac{X(s)}{F(s)} = \frac{K}{T_c^2 s^2 + 2\xi T_c s + 1} \tag{12.8}$$

where K = system gain = $1/k$

T_c = characteristic time = $\sqrt{W/gk}$

ξ = damping factor = $(p/2k)(1/T_c)$

The significance of T_c and ξ can be seen from the response of the system to a unit-step change in the input shown in Fig. 12.10. The response depends on the value of ξ. For $\xi < 1$ it is oscillatory, while for $\xi > 1$ it is nonoscillatory. Systems for which $\xi > 1$ are termed *overdamped*; for $\xi < 1$, they are termed *underdamped*; and for $\xi = 1$, the systems are *critically damped*.

For $\xi < 1$:

$$x(t) = K\left[1 - \frac{1}{\sqrt{1 - \xi^2}} e^{-\xi(t/T_c)} \sin(\omega_r t + \phi)\right] \tag{12.9}$$

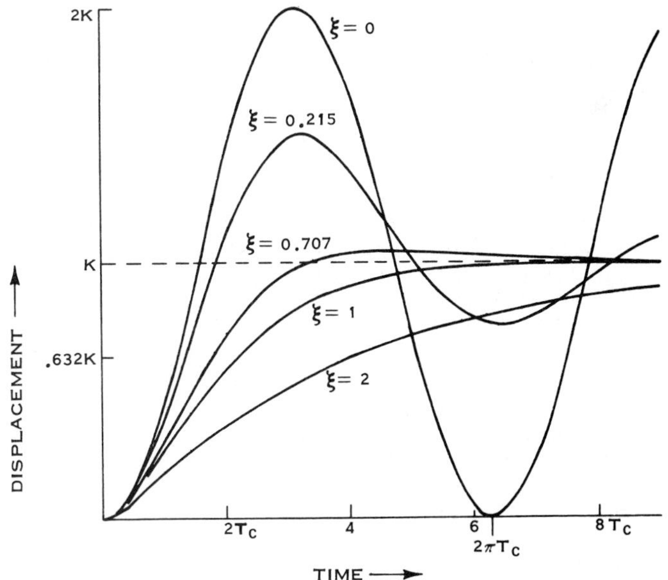

FIGURE 12.10 Time response of an oscillatory, or second-order, system to a unit-step disturbance for various values of damping factor. (*From Perry et al.*[1])

where $\omega_r = \dfrac{\sqrt{1 - \xi^2}}{T_c}$ and $\phi = \tan^{-1} \dfrac{\sqrt{1 - \xi^2}}{\xi} = \cos^{-1} \xi$

For $\xi = 1$:

$$x(t) = K\left[t - \left(1 + \frac{1}{T_c}\right) e^{-(t/T_c)}\right]$$

For $\xi > 1$:

$$x(t) = K\left[1 - \frac{1}{\sqrt{\xi^2 - 1}} e^{\xi(t/T_c)} \sinh(\omega_d t + \phi)\right]$$

where $\omega_d = \dfrac{\sqrt{\xi^2 - 1}}{T_c}$ and $\phi = \cosh^{-1} \xi$

The underdamped response $\xi < 1$ is commonly used for describing control-system performance. From Eq. (12.9) some specific relationships can be noted. The frequency of oscillation f is a function of both ξ and T_c:

$$f = \frac{\omega_n}{2\pi} = \sqrt{1 - \xi^2/2\pi T_c}$$

The undamped natural frequency f_n for $\xi = 0$,

$$f_n = \frac{\omega_n}{2\pi} = \frac{1}{2\pi T_c} \tag{12.10}$$

defines the characteristic time T_c.

Note that damping is a property of the system which opposes a change in the output variable.

The immediately apparent features of an observed transient performance are (1) the existence and magnitude of the maximum overshoot, (2) the frequency of the transient oscillation, and (3) the response time.

When an automatic-control system is underdamped, the output variable overshoots its desired steady-state condition and a transient oscillation occurs. The first overshoot is the greatest, and it is the effect of its amplitude which must concern the control designer. The primary considerations for limiting this maximum overshoot are (1) to avoid damage to the process due to excessive excursions of the controlled variable beyond that specified by the command signal, and (2) to avoid the excessive settling time associated with highly underdamped systems. Obviously, exact quantitative limits cannot generally be specified for the magnitude of this overshoot. However, experience indicates that satisfactory performance can generally be obtained if the overshoot is limited to 30 percent or less.

An undamped system oscillates about the final steady-state condition with a frequency of oscillation which should be as high as possible in order to minimize the response time. The designer must, however, avoid resonance conditions where the frequency of the transient oscillation is near the natural frequency of the system or its component parts.

Although these quantities ξ, T_c, ω_n, . . . are defined for a second-order system, they may be useful in the early design states of higher-order systems if the response of the higher-order system is dominated by roots of the characteristic equation near the imaginary axis.

3. *Distance-Velocity Lag (Dead-Time Element):* The dead-time element, commonly called *distance-velocity lag* or *true time delay*, is often encountered in process systems. For example, if a temperature-measuring element is located downstream from a heat exchanger, a time delay occurs before the heated fluid leaving the exchanger arrives at the temperature-measurement point. If some element of a system produces a dead time of L time units, then any input $f(t)$ to that element will be reproduced at the output as $f(t - L)$. Transforming to the s domain gives

$$\mathcal{L}[f(t)] = F(s) = \text{input}$$

$$\mathcal{L}[f(t - L)] = e^{-Ls}F(s) = \text{output}$$

and

$$KG(s) = \frac{\text{output}}{\text{input}} = \frac{e^{-Ls}F(s)}{F(s)} = e^{-Ls}$$

Putting this into block-diagram notation gives Fig. 12.11.

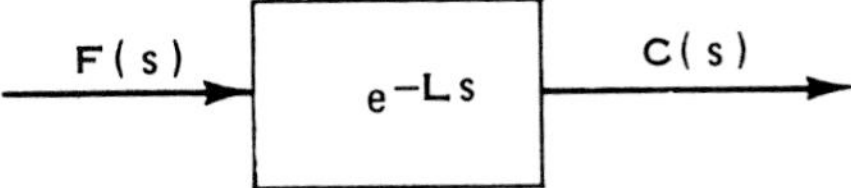

FIGURE 12.11 Block diagram of a dead-time process element. *(From Perry et al.[1])*

*FREQUENCY RESPONSE**

Although it is the time response of the control system that is of major importance, study of the effect on transient response of changes in system parameters, either in the process or controller, is more conveniently made from a frequency-response analysis of the system. The frequency response of a system is the steady-state output of the system to input sinusoids of varying frequency. The output for a linear system can be completely described in terms of the amplitude ratio of the output sinusoid to the input sinusoid to the phase of the output sinusoid to the input sinusoid. The amplitude ratio or gain, and phase, are functions of the frequency of the input sinusoid. For purposes of system analysis the frequency response function is more useful than that of the closed loop. Means for obtaining the closed-loop frequency response and evaluating transient performance from the open-loop frequency response are discussed in Ref. 2.

The frequency response can be obtained analytically from the transfer functions of the components, or system, by replacing s, the Laplace operator, with $j\omega$. Table 12.1 shows the frequency responses for some common control-system elements.

*This section taken in part from *Marks' Standard Handbook for Mechanical Engineers*, 9th ed., by E. A. Avallone and T. Baumeister III (eds.). Copyright © 1987. Used by permission of McGraw-Hill, Inc. All rights reserved.

TABLE 12.1. Frequency-Response Equations for Some Common Control-System Elements

Description	Transfer function $G(s)$	Frequency response $G(j\omega)$	Magnitude ratio	Phase angle
1. Dead time	$\epsilon^{-T_L s}$	$\epsilon^{-j\omega T_L}$	1	$-\omega T_L$ radians
2. First-order lag	$\frac{1}{Ts+1}$	$\frac{1}{j\omega T+1}$	$\frac{1}{\sqrt{\omega^2T^2+1}}$	$-\tan^{-1}(\omega T)$
3. Second-order lag	$\frac{1}{(Ts+1)(aTs+1)}$	$\frac{1}{-a\omega^2T^2+j(1+a)\omega T+1}$	$\frac{1}{\sqrt{(1-a\omega^2T^2)^2+(1+a)^2\omega^2T^2}}$	$-\tan^{-1}\left[\frac{(1+a)\omega T}{1-aT^2\omega^2}\right]$
4. Quadratic (underdamped)	$\frac{1}{\left(\frac{s}{\omega_n}\right)^2+\frac{2\zeta}{\omega_n}s+1}$	$\frac{1}{-\left(\frac{\omega}{\omega_n}\right)^2+j2\zeta\frac{\omega}{\omega_n}+1}$	$\frac{1}{\sqrt{\left(1-\frac{\omega^2}{\omega_n^2}\right)^2+4\zeta^2\left(\frac{\omega}{\omega_n}\right)^2}}$	$-\tan^{-1}\left[\frac{2\zeta\frac{\omega}{\omega_n}}{1-\left(\frac{\omega}{\omega_n}\right)^2}\right]$
5. Ideal proportional controller	K	K	K	0
6. Ideal proportional-plus-reset controller $T_i = \frac{1}{r}$ r = reset rate	$K\left(1+\frac{1}{T_is}\right)$ or $K\frac{T_is+1}{T_is}$	$K\left(1+\frac{1}{j\omega T_i}\right)$ or $K\frac{j\omega T_i+1}{j\omega T_i}$	$K\sqrt{1+\left(\frac{1}{\omega T_i}\right)^2}$	$-\tan^{-1}\left(\frac{1}{\omega T_i}\right)$
7. Ideal proportional-plus-rate controller	$K(1+T_ds)$	$K(1+j\omega T_d)$	$K\sqrt{1+\omega^2T_d^2}$	$\tan^{-1}(\omega T_d)$
8. Ideal proportional-plus-reset-plus-rate controller	$K\left(1+T_ds+\frac{1}{T_is}\right)$	$K\left(1+j\omega T_d+\frac{1}{j\omega T_i}\right)$ or $K\frac{j\omega T_i-\omega^2T_dT_i+1}{j\omega T_i}$	$K\sqrt{(\omega T_i)^2+(1-\omega^2T_dT_i)^2}$	$\tan^{-1}\left(\omega T_d-\frac{1}{\omega T_i}\right)$

Source: From Avallone and Baumeister.[2] Can also be found in Considine.[4]

The frequency response can also be obtained experimentally for systems not readily amenable to mathematical analysis by subjecting the system to input sinusoids of varying frequency.

STABILITY AND PERFORMANCE OF AN AUTOMATIC CONTROL

An automatic-control system is stable if the amplitude of transient oscillations decreases with time and the system reaches a steady state. The stability of a system can be evaluated by examining the roots of the differential equation describing the system. The presence of positive real roots or complex roots with positive real parts dictates an unstable system. Any stability test utilizing the open-loop transfer function or its plot must utilize this fact as the basis of the test.

1. The Nyquist Stability Criterion*. The $KG(j\omega)$ locus for a typical single-loop automatic-control system plotted for all positive and negative frequencies is shown in Fig. 12.12. The locus for negative values of ω is the mirror image of the positive ω locus in the real axis. To complete the diagram, a semicircle (or full circle if the locus approaches $-\infty$ on the real axis) of infinite radius is assumed to connect in a positive sense, the + locus at $\omega \rightarrow 0$ with the negative locus at $\omega \rightarrow -0$. If this locus is traced in a positive sense from $\omega \rightarrow \infty$, to $\omega \rightarrow 0$, around the circle at ∞, and then along the negative-frequency locus, the following may be concluded: (1) If the locus does not enclose the $-1 + j0$ point, the system is stable; (2) if the locus does enclose the $-1 + j0$ point, the system is unstable. The

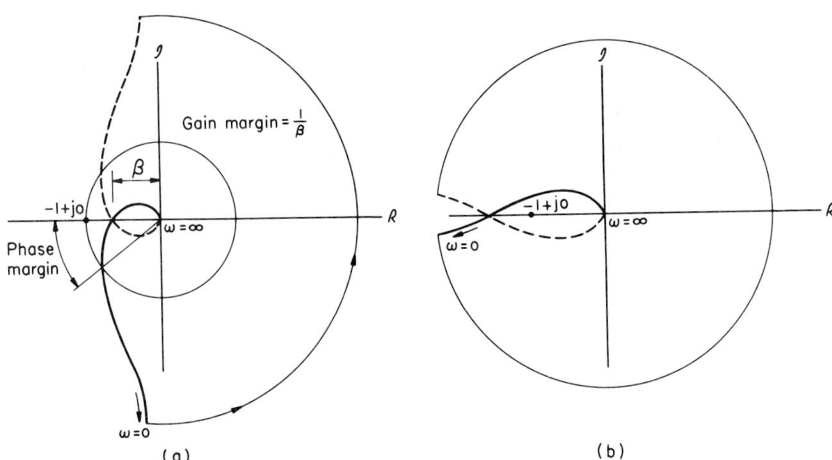

FIGURE 12.12 Typical $KG(j\omega)$ loci illustrating application of Nyquist's stability criterion: (*a*) stable; (*b*) unstable. (*From Avallone and Baumeister.*[2])

*This section taken in part from *Marks' Standard Handbook for Mechanical Engineers*, 9th ed., by E. A. Avallone and T. Baumeister III (eds.). Copyright © 1987. Used by permission of McGraw-Hill, Inc. All rights reserved.

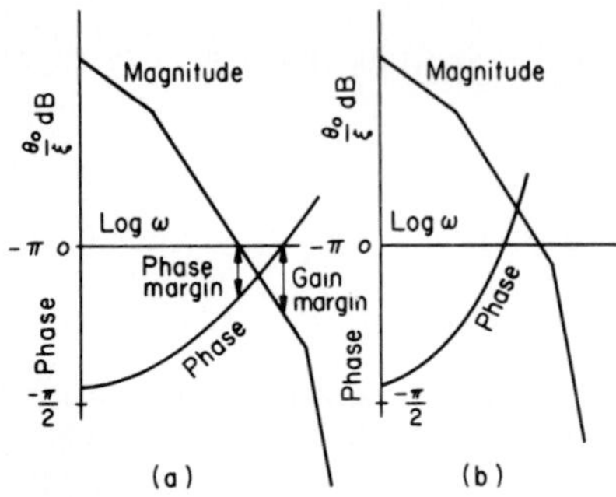

FIGURE 12.13 Nyquist stability criterion in terms of log magnitude $KG(j\omega)$ diagrams: (*a*) stable; (*b*) unstable. (*From Avallone and Baumeister.*[2])

Nyquist criterion can also be applied to the log magnitude of $KG(j\omega)$ and phase-vs.-log ω diagrams. In this method of display, the criterion for stability reduces to the requirement that the log magnitude of $KG(j\omega)$ must cross the 0-dB axis at a frequency less than the frequency at which the phase curve crosses the $-180°$ line. Two stability conditions are illustrated in Fig. 12.13. The Nyquist criterion not only provides a simple test for the stability of an automatic-control system but also indicates the degree of stability of the system by indicating the degree to which the $KG(j\omega)$ locus avoids the $-1 + j0$ point.

The concepts of phase margin and gain margin are employed to give this quantitative indication of the degree of stability of an automatic-control system. Phase margin is defined as the additional negative phase shift necessary to make the phase angle of the transfer function $-180°$ at the frequency where the magnitude of the $KG(j\omega)$ vector is unity. Physically, phase margin can be interpreted as the amount by which the unity KG vector has to be shifted to make a stable system unstable.

In a similar manner, gain margin is defined as the reciprocal of the magnitude of the KG vector at $-180°$. Physically, gain margin is the number by which the gain must be multiplied to put the system to the limit of stability. Satisfactory results can be obtained in most control applications if the phase margin is between 40 and 60° while the gain margin is between 3 and 10 (10 to 20 dB). These values will ensure a small transient overshoot with a single cycle in the transient. The margin concepts are qualitatively illustrated in Figs. 12.12 and 12.13.

2. Routh's Stability Criterion. The frequency-response equation of a closed-loop automatic control is

$$\frac{\theta_o}{\theta_i} = \frac{KG(j\omega)}{1 + KG(j\omega)}$$

The characteristic equation obtained therefrom has the algebraic form

$$A(j\omega)^n + B(j\omega)^{n-1} + C(j\omega)^{n-2} + \cdots = 0$$

The purpose of Routh's method is to determine the existence of roots of this equation which are positive or which are complex with positive real parts and thus identify the resulting instability. To apply the criterion, the coefficients are written alternately in two rows as

$$\begin{matrix} A & C & E & G \\ B & D & F & H \end{matrix}$$

This array is then expanded to

$$\begin{array}{l} A\ C\ E\ G \\ B\ D\ F\ H \\ \alpha_1\ \alpha_2\ \alpha_3 \\ \beta_1\ \beta_2\ \beta_3 \\ \gamma_1\ \gamma_2 \end{array}$$

where α_1, α_2, α_3, β_1, β_2, β_3, γ_1, and γ_2 are computed as

$$\alpha_1 = \frac{BC - AD}{B} \qquad \beta_1 = \frac{D\alpha_1 - B\alpha_2}{\alpha_1}$$

$$\alpha_2 = \frac{BE - AF}{B} \qquad \beta_2 = \frac{F\alpha_1 - B\alpha_3}{\alpha_1}$$

$$\alpha_3 = \frac{BG - AH}{B} \qquad \beta_3 = \frac{H\alpha_1 - B_0}{\alpha_1}$$

When the array has been computed, the left-hand column (A, B, α_1, β_1, γ_1) is examined. If the signs of all the numbers in the left-hand column are the same, there are no positive real roots. If there are changes in sign, the number of positive real roots is equal to the number of changes in sign. It should be recognized that this is a test for instability; the absence of sign changes does not guarantee stability.

3. Examples*

1. A servo system with a loop delay is represented by the block diagram in Fig. 12.14 and the function e^{-sT}. Find the maximum value of T for a stable system.

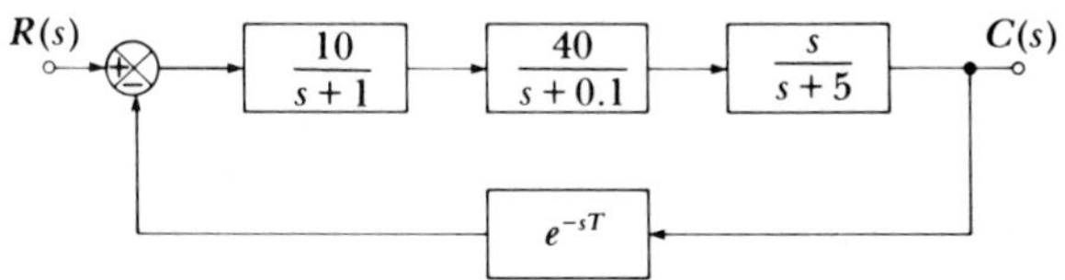

FIGURE 12.14 Block diagram of servo system.

Find the loop gain first. The loop gain is

$$GH(s) = G(s)H(s) = G_1(s)G_2(s)G_3(s)e^{-sT}$$

*From *Standard Handbook of Engineering Calculations*, 2nd ed., by T. G. Hicks. Copyright © 1985. Used by permission of McGraw-Hill, Inc. All rights reserved.

Neglecting the phase shift, we obtain $GH'(s)$:

$$GH'(s) = \frac{10}{s+1}\,\frac{40}{s+0.1}\,\frac{s}{s+5} = \frac{400s}{(s+0.1)(s+1)(s+5)}$$

$$GH'(j\omega) = \frac{400j\omega}{(0.1+j\omega)(1+j\omega)(5+j\omega)}$$

Plot the function next. For a starting point, find $G(j\omega)$ far from a breakpoint (for accuracy purposes), and plot the function. Pick $\omega = 0.01$ rad/s, and solve for $GH(j\omega)$:

$$GH(j0.01) = \frac{4j}{(0.1)(1)(5)} = 8j$$

$$|GH(j0.01)| = 20 \log_{10} 8 = 18.6 \text{ dB}$$

Now plot the straight-line approximation as shown in Fig. 12.15. (*Note*: One octave separates any two frequencies which are in ratio 2 or ½.)

Determine the phase at the crossover point: It appears as though the plot crosses 0 dB at 20 rad/s. The phase at this crossover point is

$$GH(j20) = \frac{400(j20)}{(0.1+j20)(1+j20)(5+j20)} \approx \frac{400}{(20\angle 87.1^\circ)(20.6\angle 78^\circ)}$$

$$GH(j20) = 0.97\angle -163.1^\circ$$

The gain checks since it is close to 1. The phase margin is only 16.9° (=180° − 163.1°) and is already potentially unstable. It will surely be unstable if another 16.9° is added. Therefore, at $\omega = 20$ rad/s

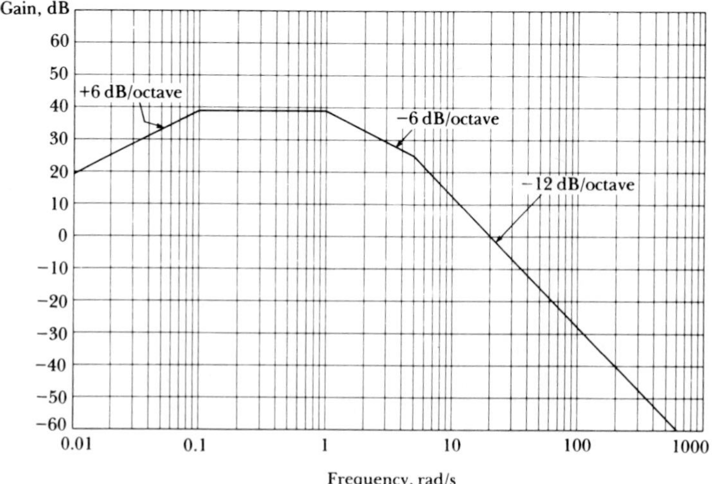

FIGURE 12.15 Bode plot straight-line approximation for $GH(s)$.

$$\omega T = 16.9° \times \frac{2\pi \text{ rad}}{360°}$$

$$\omega T = 0.294 \text{ rad}$$

$$T = \frac{0.294}{20} = 14.75 \text{ ms}$$

2. The servo system shown in Fig. 12.16 is to be used in an industrial application. Determine the range of K for the system to be stable.

To solve, use the Routh criterion. Using the Routh criterion gives

$$G(s) = G_1(s)G_2(s) = \frac{K}{(s + 2)(s^2 + 4s + 20)} = \frac{K}{D}$$

We set $1 + G(S)H(s) = 0$:

$$1 + \frac{K}{(s + 2)(s^2 + 4s + 20)} \frac{10}{s} = 0$$

$$s(s + 2)\,(s^2 + 4s + 20) + 10K = 0$$

$$s^4 + 6s^3 + 28s^2 + 40s + 10K = 0$$

Set up the Routh array:

s^4	1	28	$10K$
s^3	6	40	0
s^2	$\frac{(6)(28) - (1)(40)}{6} = 21.33$	$\frac{60K - 1(0)}{6} = 10K$	0
s^1	$\frac{(21.33)40 - 60K}{21:33} = 40 - 2.813K$	0	
s^0	$10K$		

Solve for the range of K:

$$K = 0 \quad \text{(for the } s^0 \text{ term)}$$

$$40 - 2.813K = 0 \quad \text{(for the } s^1 \text{ term)}$$

$$2.813K = 40$$

$$K \geq 14.22$$

Therefore K must lie between 0 and 14.22.

3. An angular-position servo system uses potentiometer feedback and has these gain-transfer functions:

Amplifier:

$$G_1(s) = \frac{100}{(s + 5)} \quad \frac{V}{V}$$

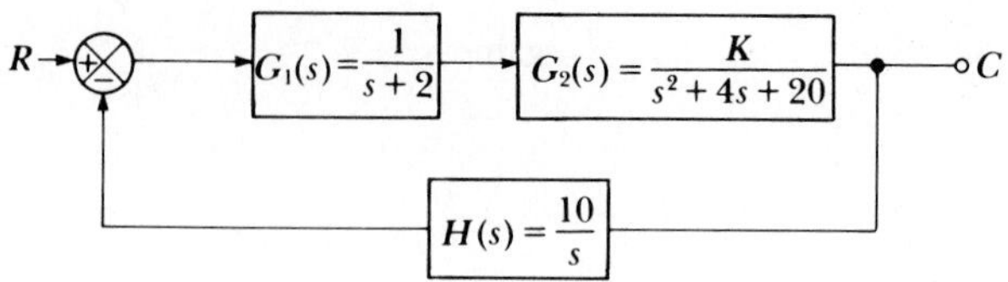

FIGURE 12.16 Servo system block diagram.

Motor mechanical transfer function:

$$G_2(s) = \frac{1}{40s} \quad \frac{\text{rad}}{A}$$

Motor electrical transfer function:

$$G_3(s) = \frac{1}{s + 4} \quad \frac{A}{V}$$

Potentiometer gain constant:

$$G_4(s) = K \quad \frac{V}{\text{rad}}$$

Draw a block diagram for this system, and determine the open-loop transfer function and the closed-loop transfer function. If $K = 100$, is the system stable?

Write the open-loop transfer function. The block diagram is drawn as shown in Fig. 12.17. The open-loop transfer function is

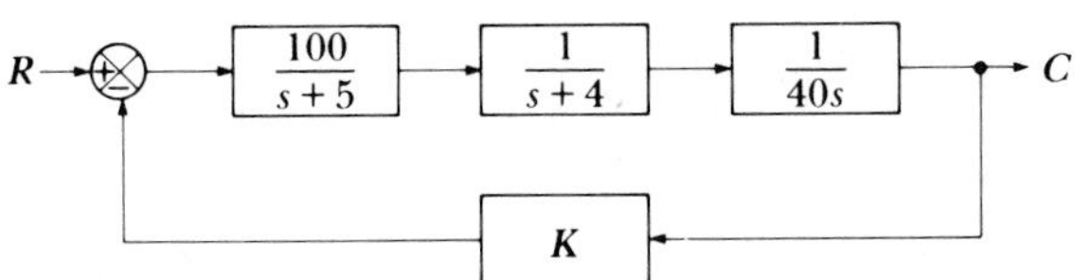

FIGURE 12.17 Angular-position servo system.

$$G(s) = G_1(s)G_2(s)G_3(s) = \frac{100}{40s(s + 5)(s + 4)} = \frac{2.5}{s(s + 5)(s + 4)}$$

Write the closed-loop transfer function. The closed-loop transfer function requirements suggest that the system be put in the canonical form as

$$\frac{C}{R} = \frac{G(s)}{1 + G(s)H(s)} = \frac{2.5}{(s)(s + 5)(s + 4) + 2.5K}$$

Determine whether the system is stable. The characteristic equation for this system is

$$(s)(s + 5)(s + 4) + 2.5K = 0$$

$$s^3 + 9s^2 + 20s + 2.5K = 0$$

Set up a Routh array to find the gain range of K for a stable system:

s^3	1	20	0
s^2	9	$2.5K$	0
s^1	$\frac{180 - 2.5K}{9}$	0	
s^0	$2.5K$	0	

Setting the s^1 and s^0 terms in the first column ≥ 0, solve for the range of K that makes the system stable. Thus,

For $180 - 2.5K \geq 0$: $K \leq 72$

For $2.5K \geq 0$: $K \geq 0$

So the range of K for a stable system is $0 \leq K \leq 72$ V/rad. So the system is not stable for $K = 100$.

SAMPLED-DATA CONTROL SYSTEMS*

Definition. Sampled-data control systems are those in which continuous information is transformed at one or more points of the control system into a series of pulses. This transformation may be performed intentionally, e.g., the flow of information over long distances is transformed to preserve the accuracy of the data during the transmission, or it may be inherent in the generation of the information flow, e.g., radiating energy from a radar antenna which is in the form of a train of pulses, or the signals developed by a digital computer during a direct digital control of machine-tool operation.

Methods of analysis analogous to those for continuous-data systems have been developed for the sampled-data systems.

Transformation. In the analysis of continuous-data systems, it has been shown that the Laplace transformation can be used to reduce ordinary differential equations to algebraic equations. For sampled-data systems, an operational calculus, the z transform, can be used to simplify the analysis of such systems (see Ref. 2).

STATE FUNCTIONS CONCEPT IN CONTROL*

State-space methods permit system analysis and design by study of a set of first-order differential equations rather than a single higher-order equation. This is convenient for solution by numerical methods, using the digital computer, and

*This section taken in part from *Marks' Standard Handbook for Mechanical Engineers*, 9th ed., by E. A. Avallone and T. Baumeister III (eds.). Copyright © 1987. Used by permission of McGraw-Hill, Inc. All rights reserved.

especially useful for systems with nonlinearities, time-varying characteristics, and multiple inputs and outputs.

For a system previously described as

$$A_n \frac{d^n c}{dt^n} + A_{n-1} \frac{d^{n-1} c}{dt^{n-1}} + \cdots + A_0 c = u(t)$$

where $c(t)$ is the output and $u(t)$ the input, allowing $x_1 = c, x_2 = \dot{c}, x_3 = \ddot{c}, \ldots, x_n = c^{(n-1)}$ yields the state-variable representation

$$\dot{\mathbf{x}} = \mathbf{P}\mathbf{x} + \mathbf{B}\mathbf{u}$$

$$\mathbf{c} = \mathbf{L}\mathbf{x}$$

where $\mathbf{x}$ is state vector and

$$\mathbf{x} = \begin{bmatrix} x_1 \\ x_2 \\ \cdot \\ \cdot \\ \cdot \\ x_{n-1} \\ x_n \end{bmatrix} \qquad \mathbf{P} = \begin{bmatrix} 0 & 1 & 0 & \cdots & 0 \\ 0 & 0 & 1 & \cdots & 0 \\ \cdots & \cdots & \cdots & \cdots & \cdots \\ 0 & 0 & 0 & \cdots & 1 \\ -A_0/A_n & -A_1/A_n & -A_2/A_n & \cdots & A_{n-1}/A_n \end{bmatrix}$$

$$\mathbf{x} = \begin{bmatrix} x_1 \\ x_2 \\ \cdot \\ \cdot \\ \cdot \\ x_n \end{bmatrix} \qquad \mathbf{B} = \begin{bmatrix} 0 \\ 0 \\ \cdot \\ \cdot \\ \cdot \\ 1/A_n \end{bmatrix} \qquad \mathbf{L} = [1, 0, 0, \ldots, 0]$$

The vector **P** is called the *companion matrix*.

Transition Matrix. The transition matrix relates the transition of the system state at time $t_0 = 0$ to the state at some later time t.

From $\dot{\mathbf{x}} = \mathbf{P}\mathbf{x} + \mathbf{B}\mathbf{u}$ can be derived

$$\mathbf{x}(t) = \phi(t)\mathbf{x}^{(0+)} + \int_0^t \phi(t - \tau)\, \mathbf{B}\mathbf{u}(\tau)\, d\tau$$

which is called the *state transition equation of the system*.

$\phi(t)$ is the transition matrix, calculable from

$$\phi(t) = \mathcal{L}^{-1}[(s\mathbf{I} - \mathbf{P})^{-1}]$$

$\phi(t)$ has the properties

$$\phi(0) = \mathbf{I}$$

$$\phi(t_2 - t_0) = \phi(t_2 - t_1)\phi(t_1 - t_0)$$

$$\phi(t + \tau) = \phi(t)\phi(\tau)$$

$$\phi^{-1}(t) = \phi(-t)$$

Since $\mathbf{c}(t) = \mathbf{Lx}(t)$, the following is the system output in terms of the transition matrix:

$$\mathbf{c}(t) = \mathbf{L}\phi(t)\mathbf{x}(0^+) + \int_0^t \mathbf{L}\phi(t - \tau)\mathbf{Bu}(\tau)\, d\tau$$

See Ref. 6 for digital-computer methods for determining $\phi(t)$, given **P**. *Note*: Ref. 6 has complete coverage of the state functions concept in control engineering.

GENERAL DESIGN PROCEDURE*

The initial performance specifications for an automatic control generally prescribe such quantities as the range of operation of the input variable and its derivatives, the maximum acceptable value of the steady-state error, and possibly other quantities, such as maximum settling time and peak overshoot. With preliminary knowledge of the nature of the input variable and the load, the designer integrates the components of the basic automatic-control system, develops the open-loop transfer function of this basic system, and examines its $G(s)$ locus. The system gain K is then adjusted to satisfy the steady-state error requirements and the resulting locus $KG(s)$ is again examined for stability. If instability exists at the required gain, the $KG(s)$ locus is reshaped through the use of derivative or integral compensation by means of a phase-lead or phase-lag component to display acceptable phase and gain margins. A detailed discussion of gain adjustment and phase compensation may be found in the references.

Note on Computer Control. Digital computers are being used with increasing frequency in the control of diverse processes. All but a few of the digital applications are supervisory or optimizing in nature; the computer, programmed to a model of the process, accepts measured data from conventional instruments, calculates optimum control settings for conventional controllers, and corrects them automatically. The computer need not be concerned only with optimizing the variables of a process for physical stability and quality but can also be used for economic optimization.

A second use of the digital computer is direct digital control (DDC), in which conventional automatic-control instruments are directly displaced by a special purpose digital computer time-shared among many control loops. The advantages of such a system are higher accuracy, more flexibility in incorporating advanced control techniques, and savings in control room costs because of compactness.

Applications of supervisory control computers are found in the electric utility industry, where they are applied to load frequency control and automatic dispatch as well as closed-loop control; in the steel industry, where they are applied

*This section taken in part from *Marks' Standard Handbook for Mechanical Engineers*, 9th ed., by E. A. Avallone and T. Baumeister III (eds.). Copyright © 1987. Used by permission of McGraw-Hill, Inc. All rights reserved.

to rolling mills; and in the chemical industry, where they are used for closed-loop process control.

MODELING OF PHYSICAL SYSTEMS

The behavior of real physical systems which engineers must design, analyze, and understand is controlled by the flow, storage, and interchange of various forms of energy. In almost all cases, real systems are extremely complex and may involve several interacting energy phenomena.

The analysis of a dynamic system always involves the formulation of a conceptual model made up of basic building blocks (lumped model elements) that are idealizations of the essential physical phenomena occurring in real systems. An adequate conceptual model of a particular physical device or system will behave approximately like the real system. The best system model is the simplest one which yields the information necessary for engineering decision making.

In modeling system elements it is convenient to determine certain functional relationships between variables of a system which are measurable characteristics of a system and may change with time. A lumped system element is usually described by a relationship between two physical variables, a *through-variable*, which has the same value at the two terminals or ends of the element, and an *across-variable*, which is specified in terms of a relative value or difference between the terminals.

Force, torque, current, fluid flow, and heat flux are through-variables. Velocity, angular velocity, voltage, pressure difference, and temperature difference are across-variables.

A convenient symbol for the relationship between through and across variables is the *linear graph* (a line segment). The two ends or terminals of this graph indicate the across-variables for the element, and the line between these terminals represents the continuity of the through-variables in the element. Linear graphs for ideal system elements are shown in Table 12.2. This table includes also the constitutive relationships and ideal elemental equations for the system elements. These system elements are for the purpose of notation classified as:

A type = the mass, inertia, and capacitance store energy by virtue of their across-variables—velocity and voltage

T type = springs and inductance store energy by virtue of their through-variables

D type = dampers and resistance dissipate energy elements

Equations for the automatic control analysis of a system made up of several system elements can be derived by combining linear graphs of elements and using their constitutive relationships (see Refs. 6 and 7).

An illustrative example of step-by-step modeling of a rather complex dynamic system (modeling of an automobile suspension system) is shown in Fig. 12.18. Note that all necessary relationships for the elements on Fig. 12.18 are given in Table 12.2.

TABLE 12.2 Summary of Ideal System Elements*

Type of element	Physical element	Linear graph	Diagram	Constitutive relationship	Energy or power function	Ideal elemental equation	Ideal energy or power
T-type energy storage $\mathscr{E} \geq 0$ v_2, x_2 f v_1, x_1	Translational spring	2, k, v_{21}, 1, F	F, F, v_2, x_2, v_1, v_1	$x_{21} = f(F)$	$\mathscr{E} = \int_0^F F\, dx_{21}$	$v_{21} = \frac{1}{k}\frac{dF}{dt}$	$\mathscr{E} = \frac{1}{2}\frac{F^2}{k}$
	Rotational spring	2, K, Ω_{21}, 1, T	T, T, Ω_2, Θ_2, Ω_1, Θ_1	$\Theta_{21} = f(T)$	$\mathscr{E} = \int_0^T T\, d\Theta_{21}$	$\Omega_{21} = \frac{1}{K}\frac{dT}{dt}$	$\mathscr{E} = \frac{1}{2}\frac{T^2}{K}$
Pure: $x_{21} = f(f)$; $\mathscr{E} = \int_0^f f\, dx_{21}$ Ideal: $x_{21} = Lf$; $\mathscr{E} = \frac{1}{2} Lf^2$	Inductance	2, L, v_{21}, 1, i	i, i, v_2, λ_2, v_1, λ_1	$\lambda_{21} = f(i)$	$\mathscr{E} = \int_0^i i\, d\lambda_{21}$	$v_{21} = L\frac{di}{dt}$	$\mathscr{E} = \frac{1}{2} Li^2$
	Fluid inertance	2, I, P_{21}, 1, Q	Q, Q, P_2, Γ_2, P_1, Γ_1	$\Gamma_{21} = f(Q)$	$\mathscr{E} = \int_0^Q Q\, d\Gamma_{21}$	$P_{21} = I\frac{dQ}{dt}$	$\mathscr{E} = \frac{1}{2} IQ^2$
A-type energy storage $\mathscr{E} \geq 0$ v_2 f,h v_1	Translational mass	2, m, v_{21}, 1, F	F,p, v_2, v_1 = const	$p = f(v_2)$	$\mathscr{E} = \int_0^{v} v_2\, dp$	$F = m\frac{dv_2}{dt}$	$\mathscr{E} = \frac{1}{2} mv_x^2$
	Inertia	2, J, Ω_{21}, 1, T	T,h, Ω_2, Ω_1 = const	$h = f(\Omega_2)$	$\mathscr{E} = \int_0^{\Omega_2} \Omega_2\, dh$	$T = J\frac{d\Omega_2}{dt}$	$\mathscr{E} = \frac{1}{2} J\Omega_2^2$
	Electrical capacitance	2, C, v_{21}, 1, i	i, q, i, v_2, v_1	$q = f(v_{21})$	$\mathscr{E} = \int_0^{v_{21}} v_{21}\, dq$	$i = C\frac{dv_{21}}{dt}$	$\mathscr{E} = \frac{1}{2} Cv_{21}^2$
Pure: $h = f(v_{21})$; $\mathscr{E} = \int_0^{v_{21}} v_{21} dh$ Ideal: $h = Cv_{21}$; $\mathscr{E} = \frac{1}{2} Cv_{21}^2$	Fluid capacitance	2, C_f, P_{21}, 1, Q	Q,V, P_2, P_1 = const	$V = f(P_2)$	$\mathscr{E} = \int_0^{P_2} P_2\, dV$	$Q = C_f\frac{dP_2}{dt}$	$\mathscr{E} = \frac{1}{2} C_f P_2^2$
	Thermal capacitance	2, C_t, θ_{21}, 1, q	q,H, θ_2, θ_1 = const	$\mathscr{H} = f(\theta_0)$	$\mathscr{E} = \int_0^{\theta_2} q dt = \mathscr{H}$	$q = C_t\frac{d\theta_2}{dt}$	$\mathscr{E} = C_t\theta_2$

TABLE 12.2 Summary of Ideal System Elements* (*Continued*)

Type of element		Physical element	Linear graph	Diagram	Constitutive relationship	Energy or power function	Ideal elemental equation	Ideal energy or power
D-type energy dissipators $\mathcal{P} \geq 0$ 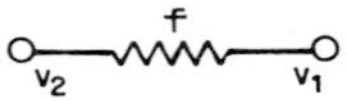		Translational damper	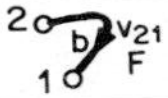	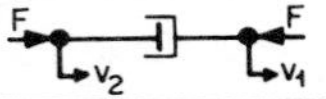	$F = f(v_{21})$	$\mathcal{P} = Fv_{21}$	$F = bv_{21}$	$\mathcal{P} = bv_{21}^2$
		Rotational damper	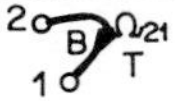	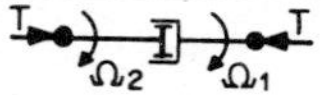	$T = f(\Omega_{21})$	$\mathcal{P} = T\Omega_{21}$	$T = B\Omega_{21}$	$\mathcal{P} = B\Omega_{21}^2$
		Electrical resistance		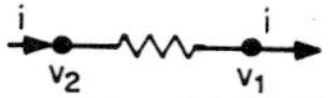	$i = f(v_{21})$	$\mathcal{P} = iv_{21}$	$i = \frac{1}{R} v_{21}$	$\mathcal{P} = \frac{1}{R} v_{21}^2$
Pure	Ideal	Fluid resistance	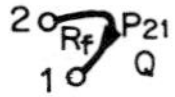	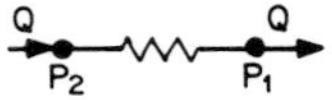	$Q = f(P_{21})$	$\mathcal{P} = QP_{21}$	$Q = \frac{1}{R_f} P_{21}$	$\mathcal{P} = \frac{1}{R_f} P_{21}^2$
$f = f(v_{21})$ $\mathcal{P} = v_{21} f(v_{21})$	$f = \frac{1}{R} v_{21}$ $\mathcal{P} = \frac{1}{R} v_{21}^2$ $= Rf^2$	Thermal resistance	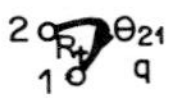	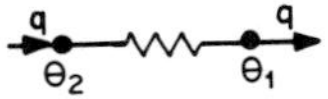	$q = f(\theta_{21})$	$\mathcal{P} = q$	$q = \frac{1}{R_t} \theta_{21}$	$\mathcal{P} = \frac{1}{R_t} \theta_{21}$
Energy sources $\mathcal{P} \gtrless 0$ $\mathcal{E} \gtrless 0$		*A*-type across-variable source	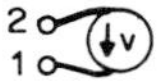	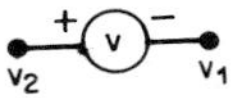	$v_{21} = f(t)$	$\mathcal{P} = fv_{21}$		
		T-type through-variable source			$f = f(t)$	$\mathcal{P} = fv_{21}$		

*Nomenclature: E = energy, $\mathcal{P}$ = power, f = generalized through-variable, F = force, T = torque, i = current, Q = fluid flow rate, q = heat flow rate, h = generalized integrated through-variable, p = translational momentum, h = angular momentum, q = charge, V = fluid volume displaced, $\mathcal{H}$ = heat, v = generalized across-variable, v = translational velocity, Ω = angular velocity, v = voltage, P = pressure, θ = temperature, x = generalized integrated across-variable, x = translational displacement, θ = angular displacement, λ = flux linkage, Γ = pressure-momentum, L = generalized ideal inductance, $1/k$ = reciprocal translational stiffness, $1/K$ = reciprocal rotational stiffness, L = inductance, I = fluid inertance, C = generalized ideal capacitance, m = mass, J = moment of inertia, C = capacitance, C_f = fluid capacitance, C_t = thermal capacitance, R = generalized ideal resistance, $1/b$ = reciprocal translational damping, $1/B$ = reciprocal rotational damping, R = electrical resistance, R_f = fluid resistance, R_t = thermal resistance.

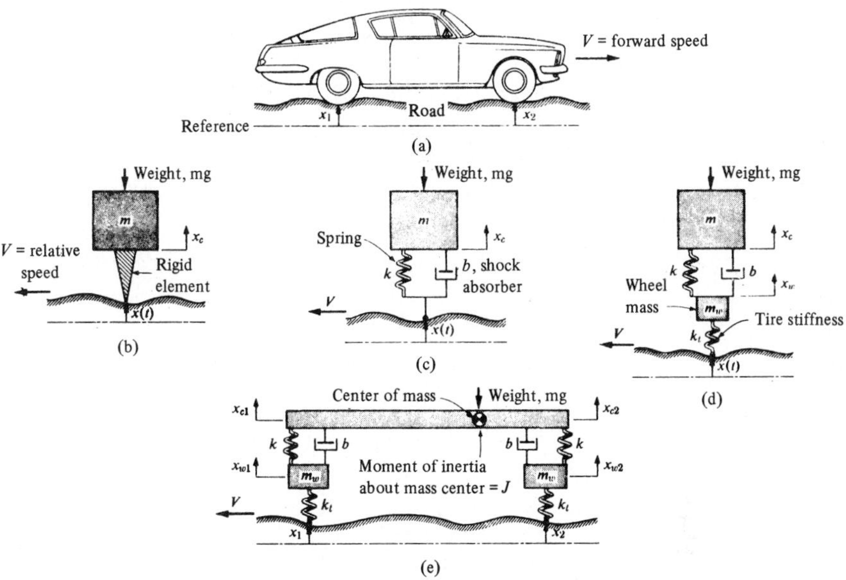

FIGURE 12.18 Successive models of automobile and suspension system. (*a*) Automobile and suspension system; (*b*) lumped-mass model; (*c*) mass, spring, and damper model; (*d*) model including tire stiffness and wheel and axle mass; (*e*) two-dimensional model. (*From Shearer et al.*[7])

NOMENCLATURE

Symbol = *Definition, SI units (U.S. Customary units)*

C = capacitance, F

C_f = fluid capacitance, m^5/N (ft^5/lb_f)

C_t = thermal capacitance, J/°C (Btu/°F)

E = energy, J ($lb_f \cdot ft$)

F = force, N (lb_f)

f = real variable function

H = heat, J (Btu)

I = fluid inertance, $N \cdot s^2/m^5$ ($lb_f \cdot s^2/ft^5$)

i = current, A

L = inductance, W/A

m = mass, kg (lb)

P = power, W ($lb_f \cdot ft/s$)

p = translational momentum, $N \cdot s$ ($lb_f \cdot s$)

p = pressure, N/m^2 (lb_f/ft^2)

Q = fluid flow rate, m^3/s (ft^3/s)

q = heat flow rate, J/s (Btu/s)

q = charge, C
R = resistance, V/A (or Ω for ohms)
R_f = fluid resistance, N · s/m^5 (lb$_f$ · s/ft^5)
R_t = thermal resistance, °C · s/J (°F · s/Btu)
s = complex variable
T = torque; N · m (lb$_f$ · ft)
t = time, s
V = fluid volume displacement, m^3 (ft^3)
v = translational velocity, m/s (ft/s)
v = voltage, V
x = translational displacement, m (ft)

Greek

θ = temperature, °C (°F)
ϕ = angular displacement, rad (deg)
λ = flux linkage, w
Γ = pressure momentum, N · s (lb$_f$ · s)
Ω = angular velocity, rad/s

Subscripts

1 = at point 1
2 = at point 2
21 = 2 − 1 (e.g., $x_{21} = x_2 - x_1$)

Note. Other symbols are defined in the text.

REFERENCES*

1. R. H. Perry, D. W. Green, and J. O. Maloney (eds.), *Perry's Chemical Engineers' Handbook*, 6th ed., McGraw-Hill, New York, 1984, Sec. 22.
2. E. A. Avallone and T. Baumeister III (eds.), *Marks' Standard Handbook for Mechanical Engineers*, 9th ed., McGraw-Hill, New York, 1987, Sec. 16.
3. H. A. Rothbart, *Mechanical Design and Systems Handbook*, 2d ed., McGraw-Hill, New York, 1985.
3. T. Baumeister, E. A. Avallone, and T. Baumeister III (eds.), *Marks' Standard Handbook for Mechanical Engineers*, 8th ed., McGraw-Hill, New York, 1987.
4. D. Considine, *Process Instruments and Controls Handbook*, McGraw-Hill, New York, 1974.

*Those references listed above but not cited in the text were used for comparison of different data sources, clarification, clarity of presentation, and, most important, reader's convenience when further interest in the subject exists.

5. T. G. Hicks, *Standard Handbook of Engineering Calculations*, 2d ed., McGraw-Hill, New York, 1985, Sec. 7.
6. D. G. Schultz and J. L. Melsa, *State Functions and Linear Control Systems*, McGraw-Hill, New York, 1967.
7. J. L. Shearer, A. T. Murphy, and H. H. Richardson, *Introduction to System Dynamics*, Addison-Wesley, Reading, Mass., 1971.

CHAPTER 13
MECHANICAL ENGINEERING

MECHANICAL DESIGN ENGINEERING*

PRINCIPLES OF MECHANISM

1. Definitions and Concepts. A *machine* is a combination of bodies which have definite form, and the strength to maintain that form. These bodies or *machine members* are so shaped and connected that they can move upon each other, but only in a particular way. Freedom for relative motion of parts differentiates the machine from the structure; constraint to a definite relative motion makes it more than a random assemblage of connected bodies.

Regularly one member (more extensive than the others) is fixed or basal as the *frame* of the machine, upon or within which the moving members function. Commonly one of these receives an input of driving force and acts as driver; this force, or the work of which it is a factor, is transmitted through the machine, with change in force and velocity and some loss due to friction; and by a final or driven member, at the place of output, force or work is applied against a useful resistance or to a useful effect.

2. Typical Examples. The members of the engine outlined in Fig. 13.1 are frame or bed 1, piston and slide 2, connecting rod 3, and crank and shaft or crank

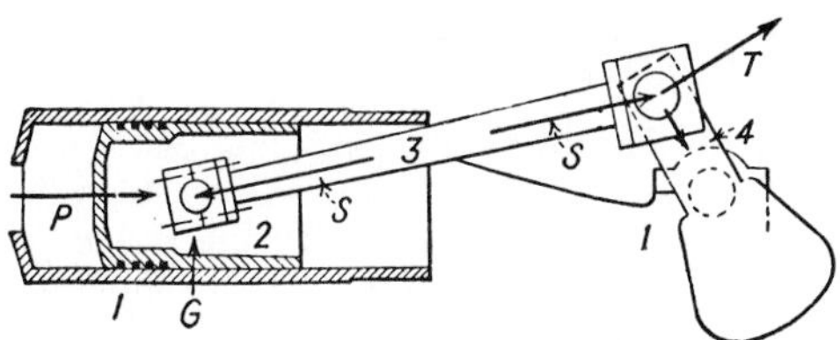

FIGURE 13.1 Outline of engine mechanism.

shaft 4. The piston is driver, receiving total gas pressure *P*. Through the connect ing rod, by means of its total lengthwise stress *S*, the driving effect is carried over to the crank pin, becoming active in tangential component *T*. The shaft is the output member, turning against whatever load is imposed upon it.

As a mechanism, or from the viewpoint of motion more than force action or work transmission, the chief effect of the engine is to change from the back-and-forth or reciprocating motion of the piston to the turning of the shaft.

The hoist gearing or hoisting-gear train of Fig. 13.2 starts with motor *M* and

*From *General Engineering Handbook*, 2d ed., by C. E. O'Rourke. Copyright © 1940. Used by permission of McGraw-Hill, Inc. All rights reserved. Updated and metricated by the editors.

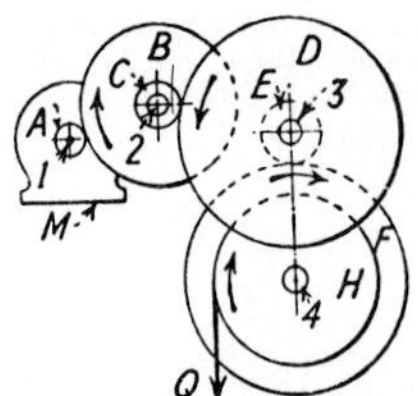

FIGURE 13.2 A hoisting-gear train.

ends with hoist drum H. It has four turning members, numbered at their shaft axes and connected by the gear pairs AB, CD, and EF. The effect of this machine is to change by successive steps from the input condition of a small force at high speed to the output condition of a big resistance at low speed.

3. Composition of Machines. A machine must have at least two members, definitely connected and with one fixed. A loose prying lever, such as a crowbar or canthook, is merely a tool or appliance, as is also a roller placed beneath a heavy body to help move it.

Examples of machines having but one moving member are:

An old-fashioned lever jack, for wagon axles

A simple windlass, without gearing

A waterwheel or turbine, a centrifugal pump or blower, an electric motor or generator; and, with greater completeness of function, such a combination as turbine and generator or motor and pump

A steam hammer or pile driver or the small direct-acting pump or compressor

Along with the extreme internal simplicity of such machines is likely to go a high complexity in the principles of the external driving or resisting force actions. Also, they commonly require complicated secondary control mechanisms or apparatus.

More complete machines, with separation of input and output members, must have at least three or four members and may have many more. With greater or less ease the more elaborate machines may be resolved into units of mechanism, or they are built up by combining such units.

4. Units of Mechanism. The distinctive meaning of the word *mechanism* is well shown by reducing the engine to its bare geometrical outline (Fig. 13.3); this *kinematic skeleton* shows all of form and dimension that is needed for motion analysis and determination. To this essential form for motion the name *mechanism* is given.

Figures 13.3 and 13.4 are typical *linkage mechanisms*, composed of cranks or arms, rods, and slides. The first has three pin or turning joints and one sliding joint, the second is the four-pin linkage. The regular linkage unit contains four members; but the special wedge mechanism in Fig. 13.5, with only sliding joints, has but three members. The *screw mechanism* (Fig. 13.6) is really a derivative of

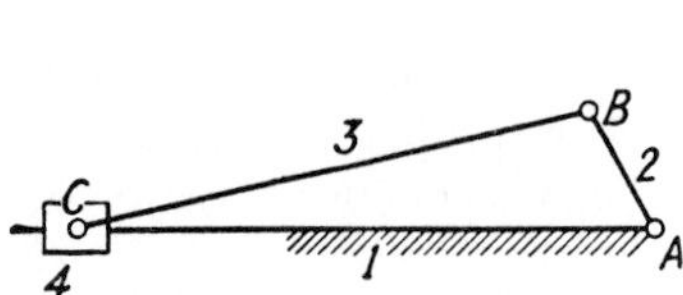

FIGURE 13.3 The engine linkage.

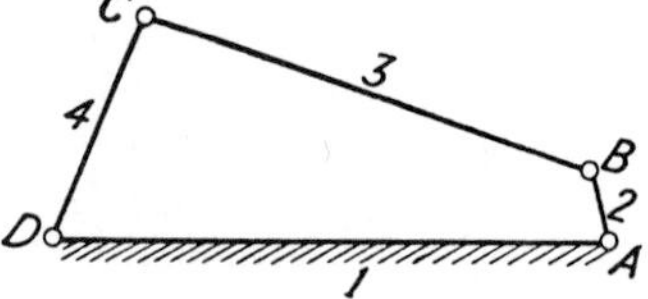

FIGURE 13.4 The four-pin linkage.

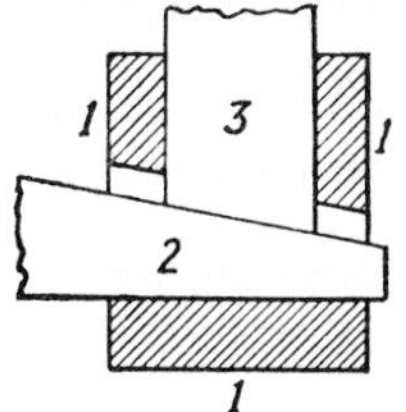

FIGURE 13.5 The wedge mechanism.

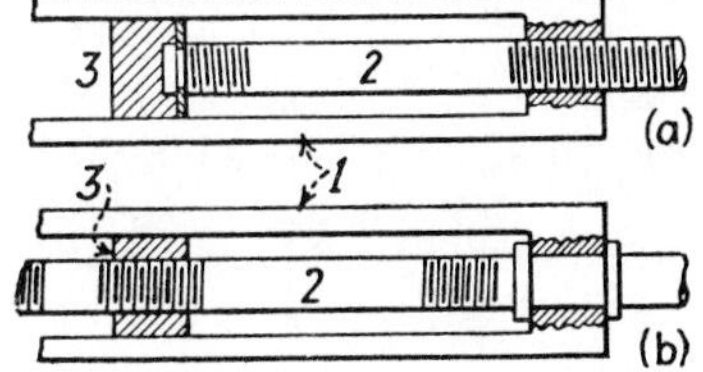

FIGURE 13.6 Screw mechanisms.

the wedge unit, substituting long-range for short-range action. Note the interchange of screw-and-nut joint and end-thrust turning joint as between the two cases.

A very common three-member unit, shown in three forms in Fig. 13.7, consists of a pair of gears and the frame carrying their shaft bearings. A solid disk or rim with teeth on the outside is called a *spur gear*. A pair of these, pair 2–3 in Fig. 13.7*a*, has opposite directions of turning. By the use of an *internal gear*, with teeth on the inside of the rim as indicated in Fig. 13.7*b*, like directions are obtained. The straight toothed bar 3′ in Fig. 13.7*a* is called a *rack*. A gear rim of large radius, made in segments to be bolted to the base of a gun mount or turret or of a turntable, may be called a *circular rack*. For tooth forms and dimensions see Sec. 60.

The profiles in Fig. 13.7 stand equally for friction wheels, which are serviceable in light transmission of power; and the wheel-and-rack pair 2–3′ in sketch *a* represents also the vehicle wheel on its road or track.

Another form of wheel unit contains a pair of pulleys or sprocket wheels with a belt or chain as flexible connector (Fig. 13.8). The dotted lines indicate the crossed belt which gives reversal of turning, but the chain can of course drive only directly.

The remaining type of unit mechanism is that containing a cam; but before considering this some matters regarding machine-member contacts must be made clear.

5. Kinematic Pairs. With emphasis upon machine motion, leaving questions of strength, construction, and functioning in the background, the fundamental kinematic unit is the pair of bodies in definite contact and connection for relative motion. The contact surfaces, with so much of the bodies as is needed to back up and support them, constitute a *kinematic pair*. Machine members carry and connect halves of two or more pairs, the machine being built up by the combination of such members.

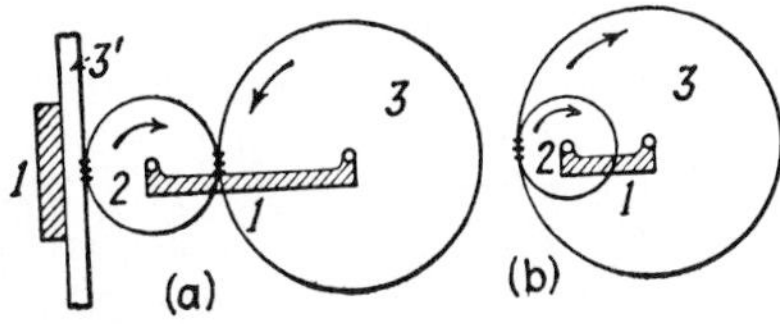

FIGURE 13.7 Gear pairs.

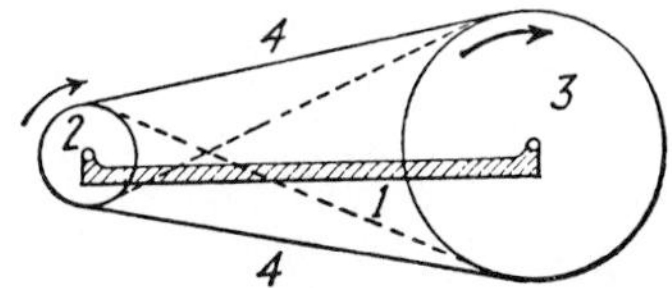

FIGURE 13.8 Belt-and-pulley pairs.

There are two types of moving contact for machine members, one distributed over complementary surfaces which have equal and uniform curvature, the other concentrated at the line or spot in which dissimilar surfaces touch each other. Kinematic pairs having contact of the first type are called *lower pairs*; those with contact of the second type are *higher pairs*.

There are four types of lower pairs, using the four kinds of possible contact surfaces. These are:

1. The journal-and-bearing or pin-and-eye pair, using the cylinder and other surfaces of revolution to permit relative turning about a common axis.
2. The sliding pair, using the prism of whatever cross-section, with zero curvature in the direction of motion. Examples are the piston and crosshead of an engine, the sliding table of a machine tool, and an elevator or hoist car running on its guide bars.
3. The screw and nut, with simultaneous turning about and sliding along an axis.
4. The sphere, which allows rotation in any direction about a point or center. Examples are found in the ball-and-socket joint used in linkages when cranks or arms move in different planes, and in the supporting pivot of a surveying instrument.

The three classes of higher pairs in common use are:

1. Rolling wheels in pairs, or a wheel, roller, or ball on its track. The principal motion is pure rolling, but there may be some secondary slip, incidental to force transmission through friction or due to form conditions, as when a ball rolls in a grooved race.
2. Gear teeth of the various kinds, as described in Secs. 60 to 63. In these pairs the relative surface motion is a combination of rolling and slip.
3. Cams and followers (see Sec. 6).

In general, higher pairs are less positive and complete in their effect to constrain two machine members to a definite relative motion; and they are also less adapted to heavy force action than are lower pairs with surface contact.

There is also a class of wrapping pairs, as of belt, rope, or chain, running over a pulley or winding on a drum. These stand by themselves, distinct from the pairs in which relative motion of surfaces occurs.

6. The Cam—Principle and Action. Consider the *eccentric*, a circular disk which can function as an enlarged crank pin (Fig. 13.9) or from which motion can be taken by a line-contact follower as in Figs. 13.10 to 13.12. In the latter use the eccentric has become a *cam*, a machine member which commonly serves to produce some form of reciprocating or oscillating motion. As to the form of this motion the cam mechanism has a freedom far greater than has the linkage mechanism. This freedom comes from the fact that the cam profile is not restricted to a simple circular form, but within certain limitations it may be given any desired shape. The gas-engine valve cam (Fig. 13.13) is a conspicuous example.

As to contact motion the variations of the follower in Figs. 13.10 to 13.12 show the three typical cases of pure sliding, of sliding combined with rolling, and of pure rolling. The last is decidedly the best condition in any service of continuous operation with high contact pressure. The motion produced by a cam depends

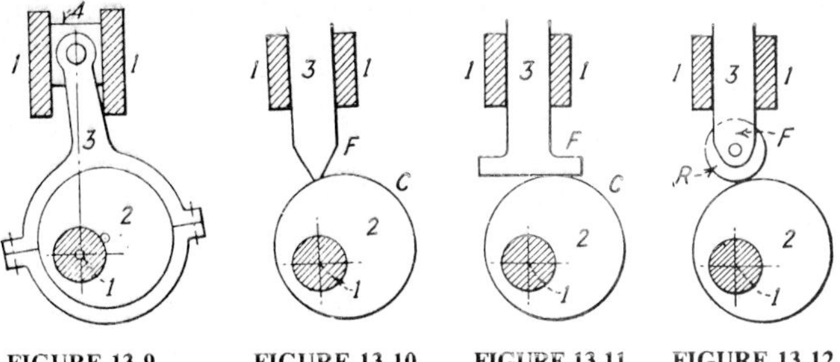

FIGURE 13.9 FIGURE 13.10 FIGURE 13.11 FIGURE 13.12

FIGURES 13.9 to 12 Various cams and followers.

somewhat upon the shape of the follower, being here slightly different as between Figs. 13.11 and 13.12. Note the oscillating follower shown as an alternative in Fig. 13.13.

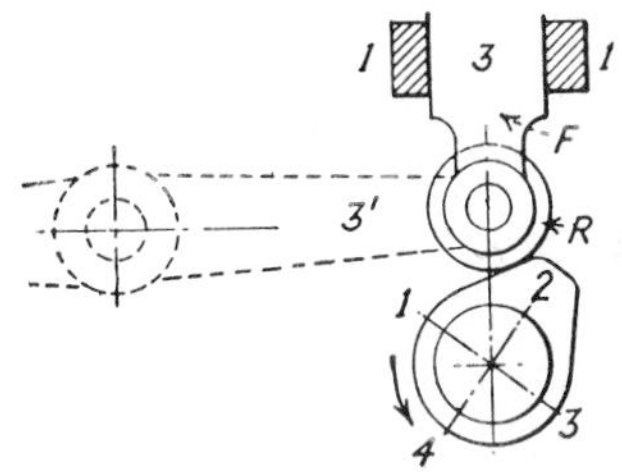

FIGURE 13.13 Gas-engine cam arrangements.

The cam in Fig. 13.13 is formed by building out a symmetrical extension from a circular disk. To the follower it gives a quick up-and-down movement, then a "dwell" or period of rest. This action is shown by the time-displacement curve in Fig. 13.14*b*, a type of diagram that depicts very clearly the performance of a rotating cam.

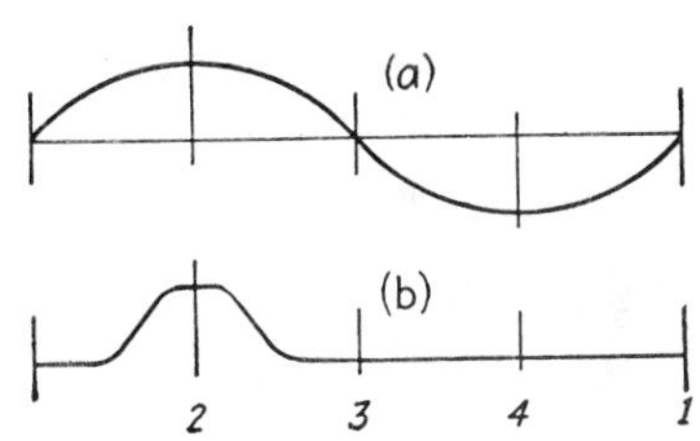

FIGURE 13.14 Curves of slide movement.

In Fig. 13.14 part *a* shows the movement of slide 4 in Fig. 13.9 and part *b* the movement of follower 3 in Fig. 13.13, with the lift of the cam taken equal to the radius or arm of the eccentric. Primarily the horizontal base is the angular position of the eccentric or cam in its rotation; but if, as is usual, shaft 2 turns at a uniform rate, this becomes also a time base. The quadrant ordinates in diagram *b* correspond to a vertical position of the like-numbered radii in Fig. 13.13.

7. Various Cams. In contrast with the symmetrical valve-gear cam, the powerful shear outlined in Fig. 13.15 is actuated by a cam so profiled from *A* to *B* as to give a slow cutting "stroke" at uniform velocity, followed by a quick return *BC* and a dwell *CA*. This is an unusually heavy service for a cam drive; but the trip cam in Fig. 13.16 also belongs to a machine of the rough-and-ready type: this is a lift-and-release mechanism regularly used in ore stamps for crushing a hard ore, such as gold-bearing quartz.

An oscillating cam of a steam-engine-valve gear is seen in Fig. 13.17, with range of swing from the dotted to the full-line position. For this mechanism the time base for a diagram like those in Fig. 13.14 would go back to the eccentric from which comes the oscillating movement of cam 2.

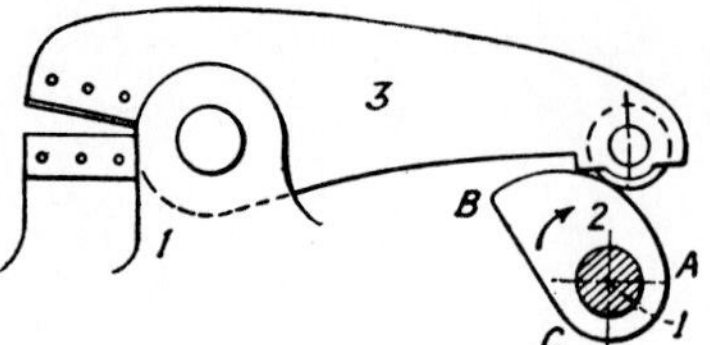

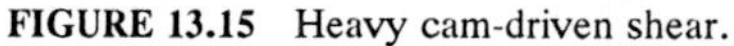

FIGURE 13.15 Heavy cam-driven shear.

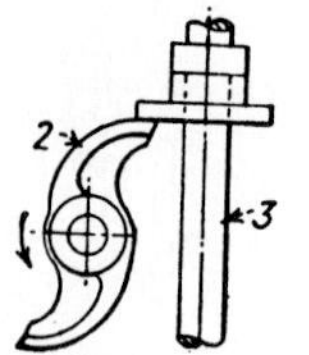
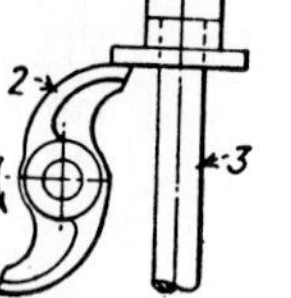
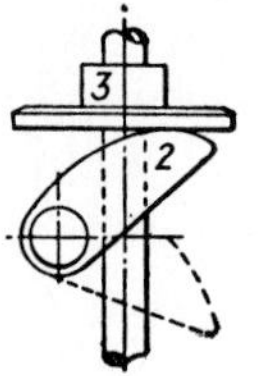

FIGURES 13.16 and 13.17 Trip and oscillating cams.

All of the examples thus far given are profiled disks or sectors (the latter in Fig. 13.17). The profile may also be on a cylinder as in Fig. 13.19, or even on a cone or sphere. The cylinder or drum cam has major use in the automatic turret lathe, where it produces the feed and return motions of the turret carriage. In this service the profile is adjustable, being made up of short helical strips screwed fast on the cylinder. During a cutting period the drum turns very slowly; for the return of the tool carriage it is turned rapidly—in either case through only a fraction of a revolution at any one time.

8. Positive Cams. One way to make a cam drive positive is to form a slot in which moves a roller or pin on the follower. This is seen in Figs. 13.18 and 13.19,

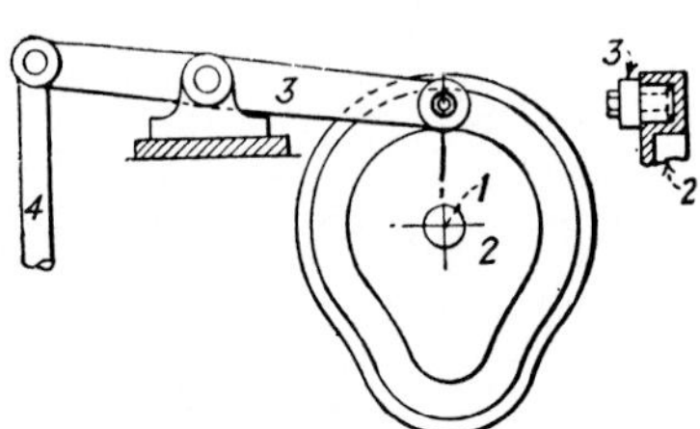

FIGURE 13.18 Positive disk cam.

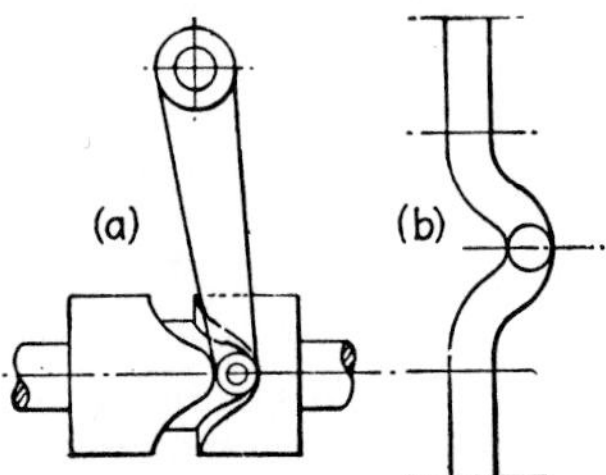

FIGURE 13.19 A drum cam and its profile.

both from machine-tool feed drives and producing the same motion. The "developed" slot profile in Fig. 13.19*b* is a diagram equivalent to Fig. 13.14*b*, which is made real in a more direct fashion when rolled on the cylinder than when replotted radially on the disk.

In general it takes two higher-pair contacts to give positive motion, as between two machine members. Such a distinct doubling of pairs is seen in Figs. 13.20 and 13.21. The first example is an equidistant or constant-breadth cam, necessarily rather restricted in its form possibilities; in the revolution or cycle there are two periods of movement and two of dwell of the follower, the mechanism being drawn in one of the latter.

Figure 13.21 shows a form of belt shifter for a metal planer; slide 2 is pushed a short distance by the planer table at each end of its stroke and there are two of these cam mechanisms, one for each belt.

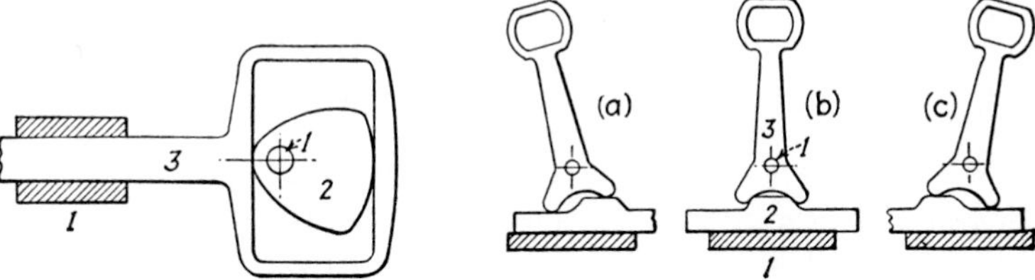

FIGURE 13.20 A double cam.

FIGURE 13.21 A belt-shifter cam.

9. Definition and Layout. A cam is a machine piece with a profiled edge of variable or irregular curvature. This working surface is formed either to a plane curve when on the rim of a disk or sector or on the side of a bar, or to a curve of helical type when wrapped on a cylinder. Driving through line contact and itself receiving either rotation (continuous or intermittent) or a simple oscillation derived from a linkage, the cam gives to its mating piece or follower some desired periodic motion.

In addition to uses suggested by the preceding examples, the cam has wide application in automatic machines where a number of arms or slides must successively come into working position according to some timing program.

The laying out of a cam is not a difficult problem, but calls for care to avoid too abrupt movements or too steep slopes. In the case of a roller on the follower the path of the center of the roller (on the plane of the cam) is the primary curve to which the actual curve is an equidistant. In high-speed service, as in automotive engines, questions of acceleration become important; shock is to be avoided and the spring must be strong enough to keep the follower in contact during its return movement.

10. Elaborate Mechanisms. More elaborate mechanisms or machines are built up by combining the units that have been described. Thus a multicylinder engine, of steam or internal-combustion type, is a mere aggregation of units, like Fig. 13.1 or Fig. 13.3, all having the frame and the crank shaft as members in common. The gear train in Fig. 13.2 is a series of units like 1–2–3 in Fig. 13.7*a*, a shaft with two wheels of different size being the moving member in common between any two successive units. More elaborate gear trains are used in the speed-changing drives of machine tools. Linkage mechanisms may become quite complex; thus a steam locomotive has two main engine mechanisms and two rather intricate valve gears, with their additional control rig. For a few linkage examples see Figs. 13.28 to 13.30.

11. Classification of Machines. Any classification of mechanisms that may be made has to do primarily with the internal constitution of the machine, between input and output points. Machines may also be classified as to service or function, or on the basis of external relationship. Important and obvious classes are as follows:

1. Prime movers or power-generating machines, which receive a natural driving force and deliver power to drive other machines; this class includes all kinds of engines and motors, driven by fluid pressure or by electromagnetic attraction.

2. Reversed prime movers, or pumps, compressors, and electric generators, which work against the same forces that drive prime movers.
3. Transmission machinery, between prime movers and working machines. Such an aggregation of shafting, pulleys, belts, etc., is an alternative to electrical transmission with a motor at and for each machine.
4. Transportation machinery of all kinds—locomotives, self-propelled cars, propelling equipment of ships and aircraft.
5. Hoisting and conveying machines or machinery, including various types of excavators.
6. Many groups of process machines, which work upon various materials to convert or shape or fabricate them.

FORCE AND WORK RELATIONS

12. Machine Forces. Machine forces are those which act upon or between machine members. External input forces and output resistances are accepted as to action and size, without question as to their causes or principles. The stresses within members, by virtue of which they sustain the action of impressed forces, are not considered in this section. The presence and effect of inertia forces must be recognized, but their determination belongs to a closer study of machine action. Classes of machine forces are:

1. The static or kinetic pressure of fluids
2. Electromagnetic attraction or repulsion
3. The contact pressures of solid bodies upon each other
4. Weight or gravity, of machine members and as a load force
5. Inertia, which becomes large in high-speed machines
6. Friction, wasteful in machine joints or moving contacts, useful in transportation, power transmission, brakes, and clutches

13. Working Force and Holding Force. Contact pressures between machine members may be either working forces or holding forces; the former interact at moving contacts, the latter at fixed contacts or bearings. On crank pin B in Fig. 13.22 acts a driving force P, applied by the connecting rod (see Fig. 13.1). So far as tendency to move the shaft axis is concerned, P is balanced by the parallel, equal, and opposite bearing reaction H as holding force, leaving the pure moment or torque Pa to cause turning. But even P is not wholly a working force; this function belongs to its tangential component T, along the path of the pin center, leaving the normal component N as a reaction against the constraint of that point to its circle path.

The pressure on a gear tooth, force P in Fig. 13.23, is not perpendicular to radius or tangent to pitch circle, but oblique at the angle σ from the tangent. Again it is held by an equal and opposite bearing reaction H and resolves into the components T and N.

Figure 13.24 brings out the fact that increase in normal component N involves a much smaller increase in total bearing pressure H; force P or AB is at the obliquity $\sigma = 15°$, while P' or AB' is at 30°. The second set of relative numbers shows

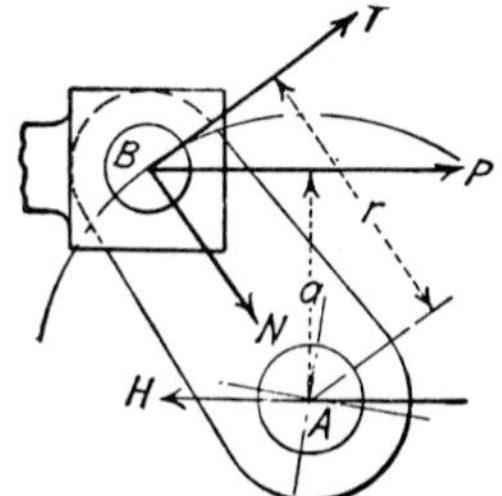

FIGURE 13.22 Forces on crank.

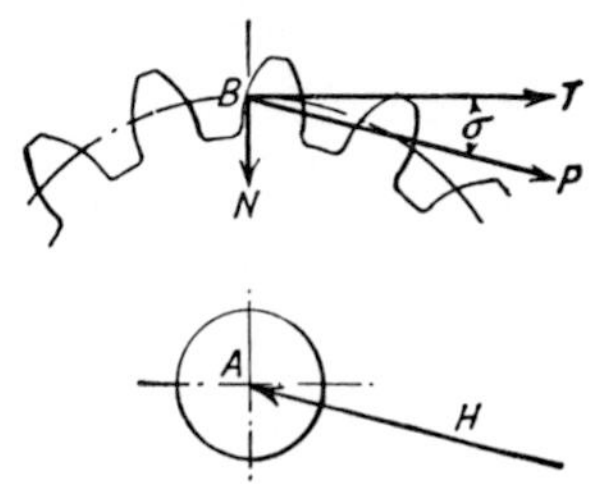

FIGURE 13.23 Forces on gear wheel.

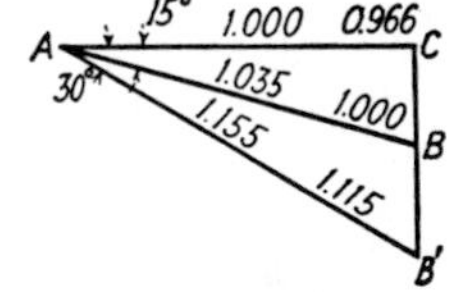

FIGURE 13.24 Variation of obliquity.

that while N is more than doubled from 15 to 30°, force H is increased by less than 12 percent (for the same effective force T).

The turning moment on the shaft may be taken as either Pa or Tr, with arms a and r as dimensioned on Fig. 13.22.

In the following discussions P will be used for the effective working force, which in the preceding closer view has appeared as tangential component T; that is, P will be the working force along the path of the point of application. Further, it will be convenient to use P for a driving force (in the direction of motion), and Q for a resistance (against motion but also tangent to path).

14. Work and Power. When a force P drives a body through a distance S against an equal and opposite resistance Q it performs the work $P \times S$ or $Q \times S$. With P in lb (N) and S in ft (m) the work is in ft · lb (N · m). If this is done in time t, so that the velocity (constant or mean) is $V = S/t$, the work rate is PV or QV. Velocity V is often that of a point in a circle path, due to the rotation of a radius R at N turns or r/min, so that $V = 2\pi RN$. With V in ft/min, the expressions for power are

$$\text{Power} = \frac{PV}{33{,}000} = \frac{P \times 2\pi RW}{33{,}000} = \frac{PRN}{5252.1} \text{ hp} \qquad (1 \text{ hp} = 746 \text{ W}) \tag{13.1}$$

In the closer study of machines it is convenient to express rate of turning as angular velocity in radians per second. The *radian* is the angle subtended by an arc equal in length to the radius; there are 2π or 6.2832 such angles, of the value 57.295°, in one turn or revolution. With $N/60 = n$ r/s, and ω for angular velocity,

$$\omega = 2\pi n = 2\pi \frac{N}{60} = 0.10472N \tag{13.2}$$

In Eq. (13.1) substitute ω for $2\pi n$. Then

$$\text{Power} = \frac{PRW}{550} \text{ hp} \qquad (1 \text{ hp} = 746 \text{ W}) \tag{13.3}$$

the expression $PR\omega$ giving ft · lb/s (N · m/s) as the product of moment in ft · lb (N · m) and angular velocity in radians/s. Emphasizing moment or torque PR and distinguishing PR' in lb · ft (N · m) and PR'' in lb · in (N · m)

$$PR' = 5252.1 \frac{\text{Power (hp)}}{N} \qquad (1 \text{ hp} = 746 \text{ W}) \tag{13.4}$$

$$PR'' = 63{,}025 \frac{\text{Power (hp)}}{N} \qquad (1 \text{ hp} = 746 \text{ W})$$

15. Pressure Work. Let a pressure p lb/in^2 (kPa) act upon a piston of area A in^2 (m^2), the piston making N strokes of length S ft/min (m/min). Then $pA = P$ is total pressure or driving force and NS is velocity V; therefore the power developed is

$$\text{Power} = \frac{pANS}{33{,}000} \text{ hp} \qquad (1 \text{ hp} = 746 \text{ W}) \tag{13.5}$$

This formula shows the method of calculating the power of an engine, pressure p or p_m being the *mean effective pressure* determined by the indicator diagram. If the crank shaft makes N r/min, this is also the number of working strokes of each piston face in the ordinary double-acting steam engine.

Consider pressure p on area A, pushing this area through S ft (m). The work done is

$$W = pA \times S = P \times S \tag{13.6}$$

Change the grouping of factors and convert area A to ft^2 (m^2). Then

$$W = 144p \times \frac{AS}{144} = 144p \times (v_2 - v_1) \tag{13.7}$$

in which $(v_2 - v_1)$ is the volume in ft^3 (m^3) displaced or "swept through" by the piston. This second definition or determination of work makes it the product of pressure in lb/ft^2 (kPa) by volume change in ft^3 (m^3).

16. Transmission of Work and Power. As the simplest type of example—because conditions stay constant at any one place in the mechanism—consider a gear train like Fig. 13.2. The transmission process starts with the input work rate P_iV_i and runs through to the output rate P_oV_o; and at each gear contact there is a change in the two factors, force P increasing as velocity V decreases. Also a progressive decrease occurs in the value of PV, due to machine friction.

In the case of turning member 1 in Fig. 13.2 there are no definite factors P_i and V_i, but rather a moment or torque PR on the armature of the motor and an angular velocity ω. With M as a single symbol for moment the alternative work rate is $M\omega$.

For the overall effect of a machine let P be the input force and Q the output resistance, with the work rates $U_p = PV_p$ and $U_q = QV_q$. The ideal or frictionless machine would have no internal loss, making

$$PV_p = QV_q \qquad Q_0 = P \times r_v \qquad P_0 = Q \div r_v \tag{13.8}$$

Here r_v is the velocity ratio V_p/V_q, taken from input toward output and expressing the kinematic or motional effect of the mechanism; Q_0 is the ideal load for a given driving force, greater than actual Q; P_0 is the ideal driving force for a given load, less than actual P.

17. Friction and Efficiency. Let U_f stand for the work rate of friction (work absorption) and e for mechanical efficiency. Then in the actual machine

$$PV_p = QV_q + U_f \qquad e = \frac{QV_q}{PV_p} \tag{13.9}$$

Efficiency is primarily a ratio of works or work rates; but by r_v for V_p/V_q and by Eq. (13.8) it becomes

$$e = \frac{Q}{Pr_v} = \frac{Q}{Q_0} = \frac{P_0}{P} \tag{13.10}$$

making efficiency a ratio of forces.

From the first form of Eq. (13.10) come the actual relations,

$$Q = P \times er_v \qquad P = Q \div er_v \tag{13.11}$$

in contrast with the ideal relations in Eq. (13.8). Velocity ratio r_v is often called the *mechanical advantage* of the mechanism. It is now seen that r_v is an ideal and er_v an actual or realized value of the ratio of Q to P.

18. Hoisting Work. Let the unit of load be one ton of 2000 lb (908 kg) and consider the height through which it can be lifted by one 1 hp (746 W) in 1 min. Without friction this would be

$$h_o = 33{,}000 \div 2000 = 16.5 \text{ ft } (5.0 \text{ m})$$

With an efficiency ranging from 0.6 to 0.8 the velocity of lift will be from 10 to 13 ft (3.0 to 3.9 m) by 1 input hp/ton (746 W/1016 kg).

19. The Wheel Pair. Consider the pair of gears A and B in Fig. 13.25, turning on fixed axes 1 and 2. Their relative speeds or turning rates are determined by the fact of a common linear, tangential velocity at the place of circle contact, at the ends of radii a and b, so that

$$aN_1 = bN_2 \qquad N_2 = N_1 \frac{a}{b} = \frac{N_1}{r_n} \tag{13.12}$$

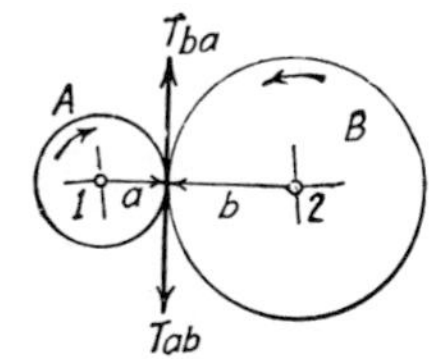

FIGURE 13.25 Gear relations.

Turning ratio r_n is taken in the same direction as r_v in Sec. 16, from driver to follower and usually from greater to less. Note that speeds N_1 and N_2 are inverse to radii a and b, the smaller wheel running faster, the large one more slowly.

At the place of tangency or wheel contact there is a pair of equal and opposite working forces: T_{ab} is the driving force which A exerts on B, and T_{ba} is the resistance which B offers to the motion of A. These forces have the unequal moments $M_a = Ta$ and $M_b = Tb$, directly as the radii; but with angular velocities inverse to these same radii the work rates $(M\omega)_a$ and $(M\omega)_b$ are equal.

Working force T is a moving force, despite the fact that in a diagram of the wheel pair it will remain fixed in position as the wheels turn. In detail, this force acts upon a part of either body (a tooth face) that is moving; but as one such body surface moves out of action another comes into play.

20. Gear-Train Relations. In the gear train outlined in Fig. 13.26 the first thing to be noted is the relative directions of turning. A single pair of ordinary spur gears (outside contact) reverses direction. With two or any even number of such pairs the follower and driver turn alike; with an odd number of pairs there is overall reversal. Second arrangement *b* in Fig. 13.26 shows the introduction of an idle wheel or idler X into pair AB; this does not affect velocity or force transmission, but merely reverses direction.

For the turning ratio of a train take first the successive single pairs, then combine them, thus,

$$\frac{N_1}{N_2} = \frac{b}{a} \qquad \frac{N_2}{N_3} = \frac{d}{c} \qquad \frac{N_1 N_2}{N_2 N_3} = \frac{N_1}{N_3} = r_n = \frac{bd}{ac} \tag{13.13}$$

In words, *the turning speed of the first driving wheel or shaft is to the speed of the last driven shaft as the continued product of (the radii of) all the followers is to the product of all the drivers.*

Figure 13.27 puts the gear train into power-transmission service, by applying

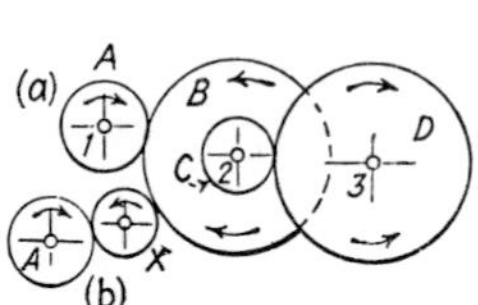

FIGURE 13.26 Directions of turning.

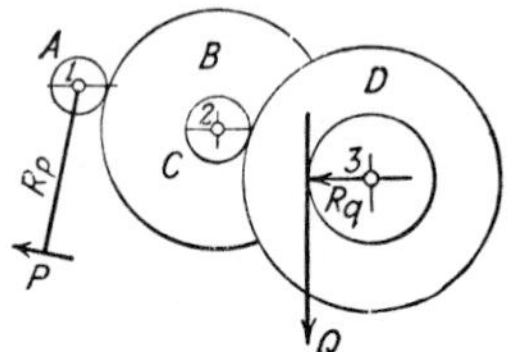

FIGURE 13.27 Force relations.

the driving moment PR_p to member 1 and the resisting moment QR_q to member 3. The no-friction relation would be

$$PR_p \times r_n = QR_q \tag{13.14}$$

while with friction it is

$$PR_p \times er_n = QR_q \tag{13.15}$$

These relationships apply definitely to gear and chain transmissions, but not so exactly to friction wheels and to belt and rope systems; in the latter there is some small slip in normal operation, which may become considerable or prohibitive of action under overload.

The primary requirement in the majority of many-pair gear trains is to produce a high velocity ratio. This is effected by combining changes of angular velocity (different diameters in a pair) with changes of linear velocity (different diameters on the same shaft).

21. High-Ratio Linkages. A few linkage mechanisms also have the function of producing a high velocity ratio. One example is the compound-lever system of a weighing scale for big loads: this is much like a wheel train in essence, but has only a very small range of oscillation instead of being formed for continuous rotation.

A simple type of linkage that can exert a very heavy pressure through a short distance is illustrated in Figs. 13.28 to 13.30. Unit mechanism 1–2–3–4 in each of

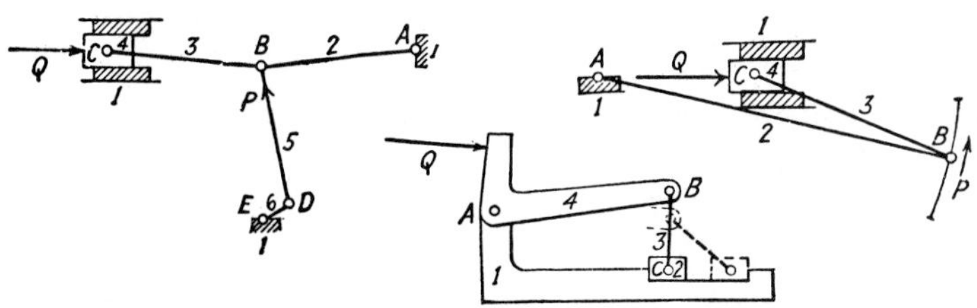

FIGURES 13.28 (left), 13.29 (right), and 13.30 (center) Various toggle mechanisms.

the first two cases contains the elements (if not the proportions) of the engine mechanism, with AB representing the crank and BC the connecting rod. An essential matter is that the range of motion of linkage ABC ends with AB and BC very nearly in line, so that a small force P at joint B will thrust against a large resistance Q. The first form is used in stone crushers, with a swing jaw in place of slide 4, and in coin presses; of course, short crank DE makes a complete rotation. Figure 13.29 is from a riveting machine, with a compressed-air cylinder to drive pin B upward. Figure 13.30 shows a common clamping device for friction clutches; in the full-line closed position the mechanism is self-locked, force Q having no ability to push block 2 to the right.

CONSTRUCTIVE ELEMENTS OF MACHINES

22. Foundations. The principal requirement that must be met by any foundation is to support the imposed load without settling, or without a local settling that will distort the superstructure. Heavy machines are generally set on separate foundations, especially when vibration or shock must be sustained; lighter machines, typically those for manufacturing processes, are frequently carried by the building.

Machine foundations, generally deeper and more massive than footings and foundations for buildings, are made of concrete. For this nothing weaker than the medium mixture 1:2½:5 should be used, while for more severe conditions a 1:2:4 mix (or richer) is preferable. A heavy spread-out machine particularly needs a rigid, monolithic support in order that the maintenance of geometrical form shall not depend on the machine frame itself. The most trying conditions are met when an extended machine of only moderate inherent stiffness is carried by a structure which is also low in stiffness—typically, a multicylinder engine of great fore-and-aft length (along its own shaft axis) in a ship's hull.

23. The Machine on Its Foundation. A compact machine with a complete bed or frame, forming a strong and rigid structure, is said to be *self-contained*. Having this quality in highest degree, the machine can be picked up bodily by a crane and shipped or moved in one piece. More extended machines will have the frame in sections, to be bolted together during erection. Then there is the machine with separate structural units, held in relation to each other wholly by the foundation. The extreme of extension is seen in transmission machinery, where each shaft bearing has its own support, commonly the building structure.

Sometimes support is flexible. If possibility of a distorting reaction is to be avoided completely a state of three-point support must be realized. This is done in the locomotive, where regularly the wheels (or their axle boxes) are "equalized" in three groups by levers and springs, each group having its definitely located reaction.

On a solid foundation even the most completely self-contained machine should be relieved of frame strain, at least by driving the leveling wedges with equal hammer blows. More extended machines may require skillful leveling and alignment before they are bolted fast.

In all kinds of heavy service the holding-down bolts are set into the foundation, being suspended in place from a locating template before the concrete is poured; only for light machines on a floor may the bolts be set into drilled holes. To avoid a difficult and costly degree of accuracy, both in locating holes in the machine frame and in laying out the bolt template, it is usual to form an enlarged hole about the bolt, of two or three times its diameter. This is done by placing a piece of pipe or a wooden box around the bolt, to be pulled out after the concrete begins to set but before it hardens. After the machine is placed this hole is usually filled with cement mortar so as to fix the bolt in position. For a foundation bolt see Fig. 13.45.

The actual fit of machine to foundation is made by *grouting*. In leveling up the machine it will be raised from ¼ to ¾ in (6.4 to 19.1 mm) from the foundation by the leveling wedges; if these are of cast iron the excess length can be broken off and the wedge need not be pulled out after the grout is set. A dam of sand or clay is laid all around the machine, then the clearance space is filled with a cream-like grout of cement, either neat or with some fine sand. Before it is quite hard the excess is trimmed and smoothed to a neat finish. After this grouting is fully set, the bolts are screwed down hard. Properly the machine is held in place by bolt friction and grout adhesion, not by a pin action of the bolts.

24. Machine Frames. For compact and self-contained machines it has been long-established practice to make the frame an iron or steel casting, or to assemble such castings into a framework, as for the locomotive. Now there is a rapidly growing tendency to build up frames and bodies by welding steel pieces together; this saves the cost of a pattern when only a few frames are needed and gives a lightness which in some cases is desirable. Only in extended machines that are really moving structures, such as the various kinds of cranes, is the riveted type of construction used.

25. Bearing Supports. These are the support elements of transmission machinery. Sometimes they rest on small separate foundations; oftener they are fastened to the ceiling (under side of floor), wall, or frame of the factory room or building; or they may be mounted on the framework of an extensive machine structure. As examples, Fig. 13.31 shows a rigid pillow block, on a special base plate with wedge adjustment, and Fig. 13.32 a shaft hanger with only the place indicated for the separate bearing unit.

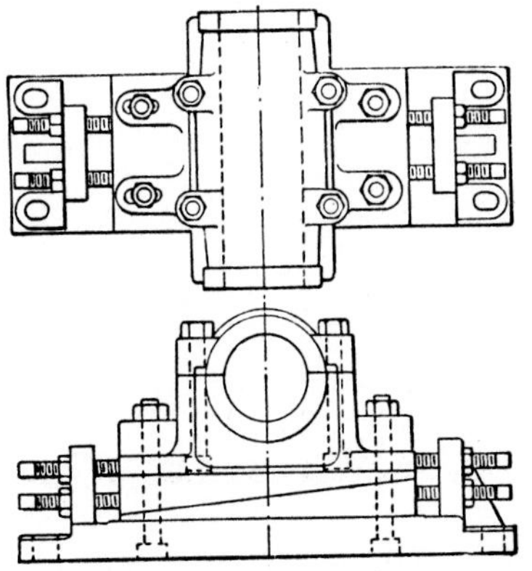

FIGURE 13.31 Adjustable pillow block.

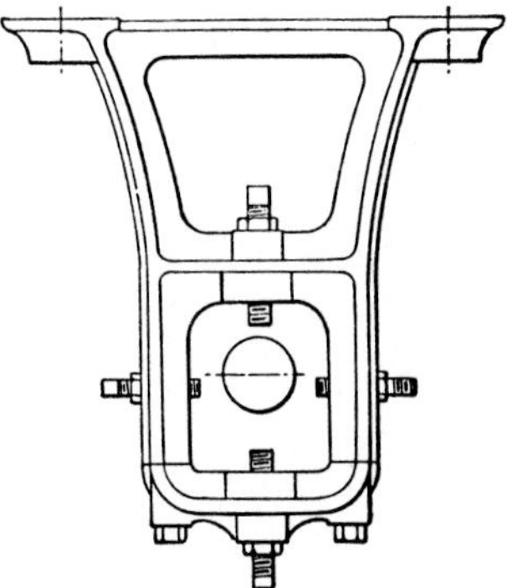

FIGURE 13.32 Hanger frame.

Any independent bearing consists of the bearing proper (see Secs. 50 to 58), and the supporting body or frame. The rigid bearing (see Fig. 13.31) is cheaper in light service and may be better in very heavy service if lined up with great care, but in most conditions of support, installation, and maintenance, a flexible and adjustable support is decidedly preferable.

The direct-support element of the hanger in Fig. 13.32 is a rectangular frame, enclosing a space within which the bearing is held and adjusted; and this enclosure can be opened for convenience in placing or removing the shaft. The footing portion of the structure may be above (as here) for bolting fast to a ceiling or floor beam; or it may be formed at the side, for support by wall or column, giving a post hanger; or the regular hanger (usually in the heavier designs) may be inverted and used as a floor stand for heavier shafting.

The shaft hanger, in its various forms, is most adaptable in wooden buildings or with heavy wooden floors. With the help of special clamps and supports a mechanical transmission system is readily carried by a steel building frame. But the concrete building—unless special fittings such as T-slot bars have been incorporated into its construction—offers very poor support for a shafting system; the difficulty and cost of such installation are a strong argument (in addition to others that may exist) for the driving of machines by individual electric motors.

26. Fastenings—Riveted Joints. Fastenings, such as rivets, bolts, screws, keys, and even forced or shrunk fits, are elements of machine structures, whether in the frame or in moving members.

Rivets and riveted joints find little place in actual machines, but in one class of mechanical structure, namely, existing older boilers and pressure tanks, they are very important. The button-head rivet shown in Fig. 13.33*a* is the form most used; but there is also the "pan" head (a truncated cone), and the countersunk head (Fig. 13.33*b*) which is used where a smooth surface is desired, especially in ship construction.

Lap joints, single and double riveted, are seen in Fig. 13.34, and with the first is sketched a distorted section, with the two plates "in line." The actual joint has a tendency to pull into this shape, which induces extra stress in the plates. For heavy service, in structures as well as pressure vessels, the butt joint in Fig.

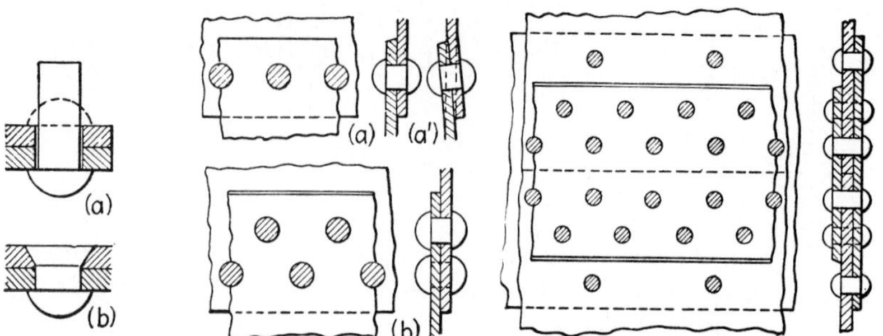

FIGURE 13.33 Rivets. **FIGURE 13.34** Lap joints. **FIGURE 13.35** High-type butt joint.

13.35 is better than the lap joint. Satisfactory working stresses for boilers are: tension in plate, 11,000 lb/in^2 (75.8 MPa) on the net section; shear in rivet, 8800 lb/in^2 (60.7 MPa); bearing of rivet on plate, 19,000 lb/in^2 (131 MPa).

The ratio of plate left between holes to unpunched plate is the efficiency of the joint. In the single-riveted lap joint this efficiency will average from 0.55 to 0.60; with double riveting it may exceed 0.65; and with a wide-spaced outer row (Fig. 13.35) the efficiency may reach 0.88.

In the sections shown in Figs. 13.34 and 13.35 the outside of the boiler is to the left. Note the bevel of each plate edge on this side, for ease in *calking* the joint to make it tight; a tool like a very blunt chisel drives the edge hard down upon the plate beneath.

27. Bolts and Screws. The three typical forms of bolts appear in Figs. 13.36, 13.37, and 13.38, namely:

Figure 13.36, bolt, through bolt, or machine bolt

Figure 13.37, stud bolt or stud

Figure 13.38, tap bolt or cap screw

The nominal length of the bolt is shown by dimension l in each case. Note that regularly the bolt is loose in its hole, with a diameter difference of 1/16 in (1.59 mm) for small sizes to 1/4 in (6.35 mm) for large sizes. For the exceptional use of bolts as pins see Sec. 41.

The stud bolt is supposed to be screwed in tightly and permanently; this is better (especially with cast iron) than cap-screw action when the joint has to be opened at intervals, as in the case of a cylinder head.

Screw-thread profiles for bolts and machine screws are drawn in Figs. 13.39, 13.40, and 13.41, as follows:

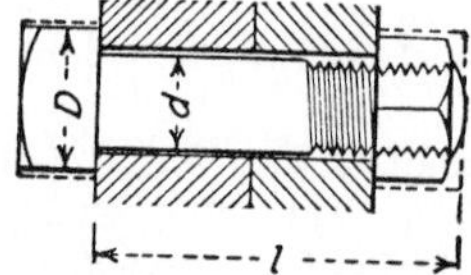

FIGURE 13.36 Machine bolt.

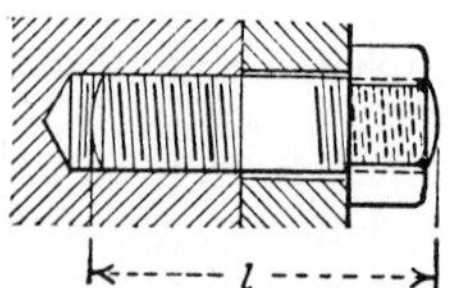

FIGURE 13.37 Stud bolt.

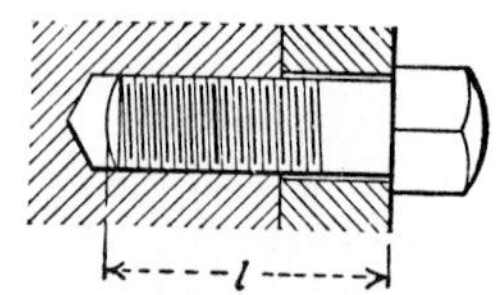

FIGURE 13.38 Cap screw.

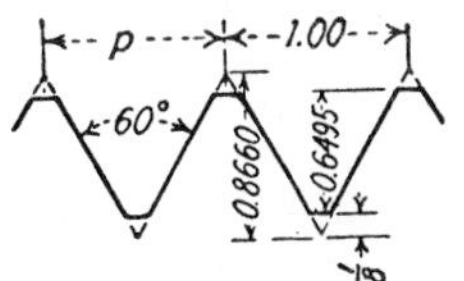

FIGURE 13.39 American thread.

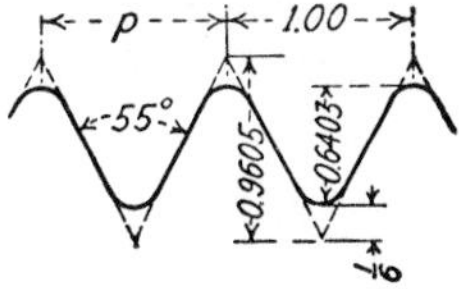

FIGURE 13.40 Whitworth thread.

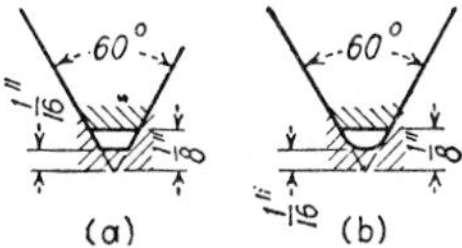

FIGURE 13.41 Groove details.

Figure 13.39, American, U.S. Standard or Sellers thread, standard also in the metric countries: 60° angle, ¾ depth of sharp V

Figure 13.40, Whitworth or British thread, 55° angle, ⅔ depth of sharp V

Figure 13.41*a*, American Society of Mechanical Engineers' (ASME) thread for machine sçrews, later recommended for all bolts; groove cut deeper, so that even when the tap or die is worn there will not be binding on top of thread

Figure 13.41*b*, similar International Standard of 1898, but with bottom of groove rounded

28. Bolt and Screw Proportions. The most important of these are established by the American Standards Association and the American Society of Automotive Engineers.

29. Various Bolts. The machine bolt and stud are sketched in flange joints in Fig. 13.42. An alternative to the stud, found in older practice and in rougher construction, is the T-head bolt in Fig. 13.43. The form of eyebolt in Fig. 13.44 is especially effective on chemical digesters or other pressure tanks that must be opened or closed frequently; the nuts need only be slacked off a few threads, then the bolts can be swung out of the way, making much easier the removal and replacement of the cover; and the screw threads are safe from being battered and spoiled.

Very long bolts are preferably double-ended, or are threaded for a nut at each end; this facilitates making the bolt and also putting it in place. Best practice is to cut the thread on an enlarged short portion of the bolt, as indicated in Fig. 13.54—thus making the bolt shank of the same diameter as the core inside of threads, or even a little smaller.

Foundation bolts, of which a highly developed example is shown in Fig. 13.45, are usually double-ended, with special large washers for bearing under the con-

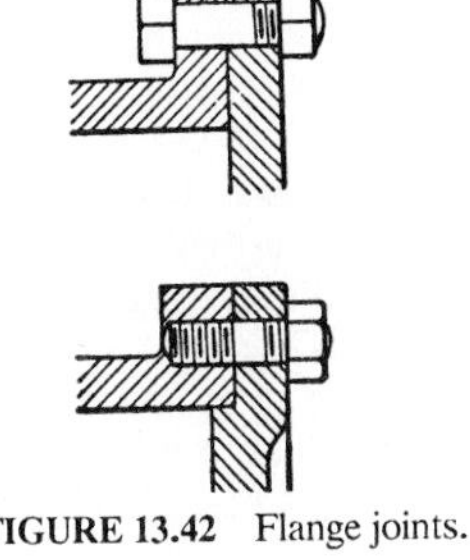
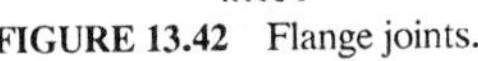

FIGURE 13.42 Flange joints.

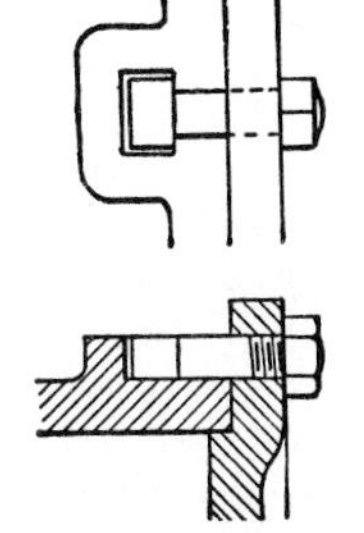

FIGURE 13.43 Tee bolt.

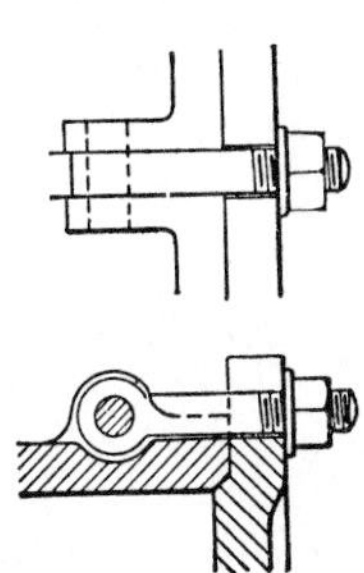

FIGURE 13.44 Eyebolt.

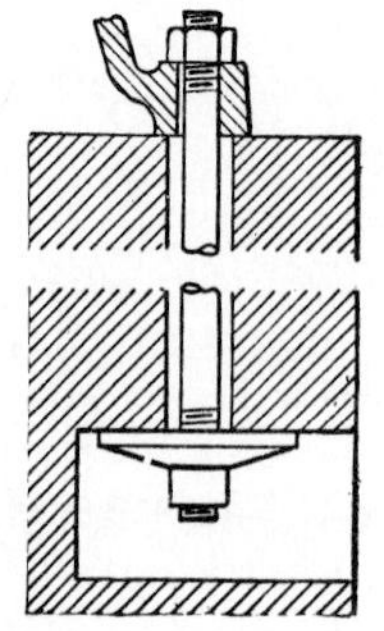
FIGURE 13.45 Foundation bolt.

crete. When bolts are near the edge of a foundation the formation of pockets around their lower ends gives better control of location and of replacement in case of possible breakage. Here the hole is not supposed to be grouted as described in Sec. 23, but that is the usual practice when heads are enclosed.

30. Screws. Machine screws are like cap screws in general form and service, but have a range of smaller sizes. The two ranges overlap, however, and the real distinction is that cap screws have diameters in binary fractions of the inch (mm), while machine-screw diameters are in a numbered series. Further, machine screws of a given number come with several numbers of threads per in (mm).

Forms of machine-screw heads are shown in Fig. 13.46, as follows:

1. *Fillister* or *cylindrical head*, here resting on flat surface
2. *Oval* or rounded head
3. *Flat fillister head*, sunk in counterbore, the preferable arrangement
4. *Countersunk head*, either flat or oval (dotted profile)

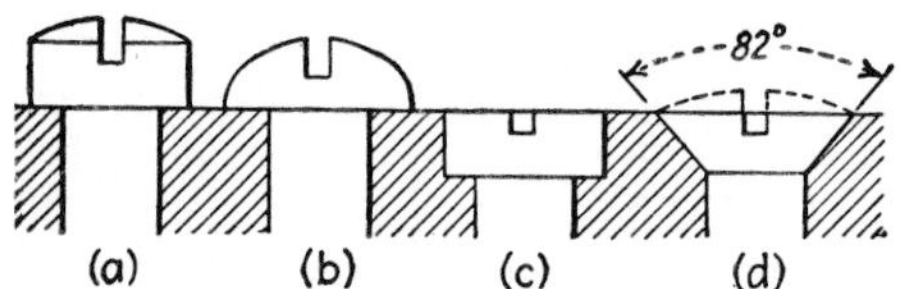

FIGURE 13.46 Machine-screw heads.

These forms are also used on cap screws, in addition to the hexagon head of Fig. 13.38. Round bolt heads are added in Fig. 13.47, of "button" and countersunk forms. The broad flat heads (full-line) are especially for use on wood; and to prevent turning, the shank is square under the head, as in case *b*, or has fins or ribs on it. The wide button head and short length of square shank characterize the *carriage bolt*. For use on metal the button head is higher; the track bolt dotted in Fig. 13.47*a* has this feature and is of oval cross section under the head, to fit a hole punched in the rail-joint fish plate. The dotted outline in *b* is a cap-screw head of smaller angle than in Fig. 13.46*d*.

For more complete information see manufacturers' catalogs of screws and of taps and dies.

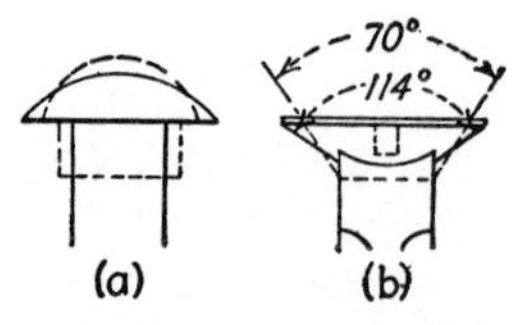

FIGURE 13.47 Round bolt heads.

The wood-screw thread has a much larger pitch than that of the machine screw; but the metal thread is thin, so as to leave a thick thread in the wood; note in Fig. 13.48*a* how the bottom of the space between threads is cut at a slant, so as to give a wider bearing surface on the pressure side. The wood-screw thread is sometimes called a *skein thread*. In large diameters, by sixteenths or eighths up to 1 in (25.4 mm), and with a square head, the wood screw becomes a *lag screw*. Very often this has the same gimlet point as the wood screw; but *b* and *c* of Fig. 13.48 show the cone point, the second with a different shape of thread, better for hammer driving of the screw.

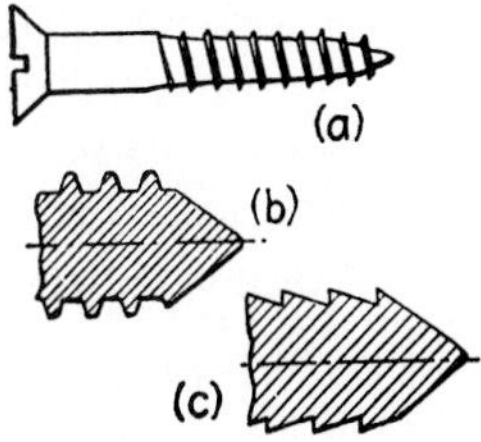

FIGURE 13.48 Threads for wood.

31. Set Screws. The function of the set screw is to exert pressure in the direction of its axis, with the major purpose of preventing sidewise motion of the piece beneath it. If the purpose is merely to hold that piece in a certain position the set screw becomes a clamping or adjusting screw, as in Figs. 13.50 to 13.52. Set screws are alike in being threaded over the whole length of shank, but they differ in heads and points.

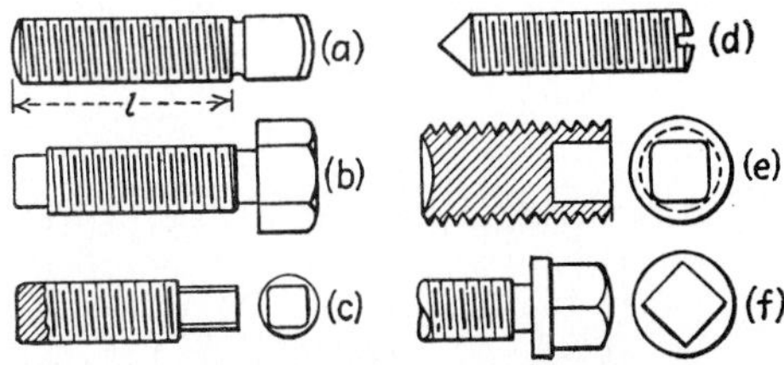

FIGURE 13.49 Set screws, points, and heads.

These are shown in Fig. 13.49, without implication that any particular point goes with a particular head. The heads are:

1. Regular, square, same diameter as shank, neck beneath head.
2. Hexagon, bolt size, special on some kinds of finished work.
3. Small square head, allows screw to be backed in from below, or a lock nut may be run down over head.
4. Headless, slotted screw, used to hold a small solid pulley on its shaft.
5. Allen safety or socket head; can be sunk down flush with surrounding surface, good where a projecting head would be unsightly or dangerous.
6. The collar head is good on a clamping screw in frequent use; holds wrench in place, makes turning of screw more easily a one-hand job.

For forms 3 and 5 special wrenches are required.
The set-screw points in Fig. 13.49 are:

1. Round or oval point, general service

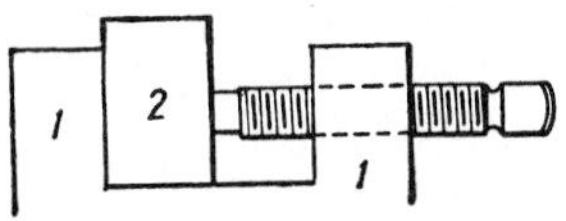

FIGURE 13.50 Clamp screw.

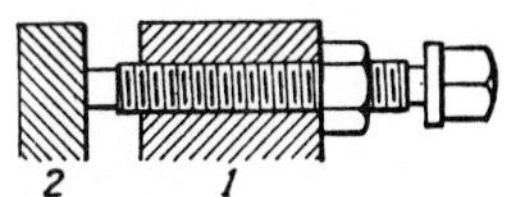

FIGURE 13.51 Screw with lock nut.

FIGURE 13.52 Loose nut.

2. Pivot point, more usual for clamping
3. Plain flat point, for clamping
4. Cone point, for light pulley service; hole should be countersunk into shaft
5. Cup point, best for a strong grip against sidewise movement; raises a burr on surface, which may well have a depressed strip where the screw rests

For set screws holding a wheel on shaft, see Fig. 13.68. Simple clamping is shown in Fig. 13.50. The adjusting screw in Fig. 13.51 does not necessarily grip piece 2; hence a lock nut is needed to hold it at a given setting. In order to avoid the need of tapping for such a screw, a square nut let into a pocket in the casting is advantageous, as shown in Fig. 13.52; the screws in Fig. 13.32 are of this type.

32. Various Screw Threads. The profiles of regular bolt and screw threads have been shown in Figs. 13.39 to 13.41 and wood-screw threads in Fig. 13.48. Before taking up other profiles consider general screw variations.

Bolt and screw threads (or fastening threads in general) are single, or have but one helical coil winding around a cylinder; contrast Fig. 13.53 with Fig. 13.55.

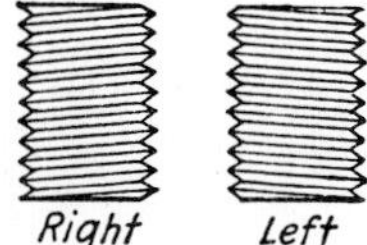

FIGURE 13.53 Right-and-left threads.

FIGURE 13.54 Turnbuckle.

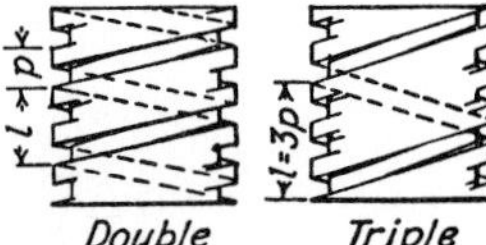

FIGURE 13.55 Multiple threads.

All ordinary threads are right hand (see Fig. 13.53); seldom is there need for a left-hand screw in a fastening, but the turnbuckle in Fig. 13.54 is a conspicuous example of the right-and-left screw.

A double and a triple thread are shown in Fig. 13.55 (motion screws). Here are distinguished pitch p of thread profile and lead l of the screw, or its advance per turn.

Profiles for motion screws are given in Fig. 13.56. The 29° Acme thread is now generally preferred to the square thread; it can be formed by a milling cutter, while the square thread must be cut on a lathe. Sometimes the screw will be moved under a light load, and then have to take a heavy standing load in one direction, as is the case with the big tension rods of a hydraulic press or a testing machine. In such service the buttress thread in Fig. 13.57 is most effective, with

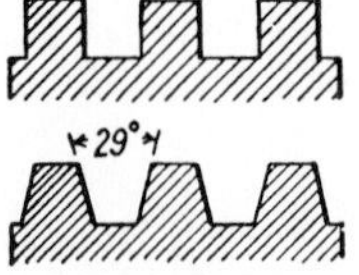

FIGURE 13.56

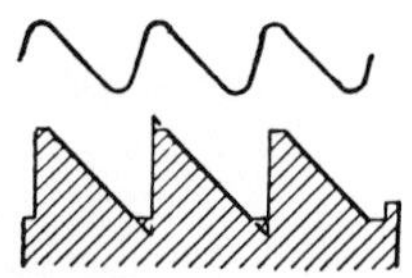

FIGURE 13.57

FIGURES 13.56 and 13.57 Threads for motion.

a flat surface for the heavy pressure. The upper profile in Fig. 13.57 shows an ordnance thread, for the breech block of a gun.

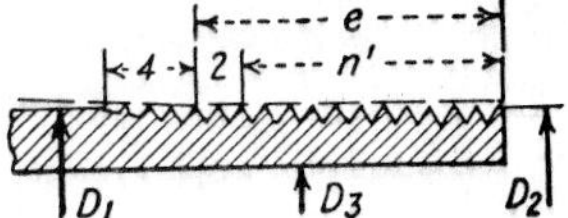

FIGURE 13.58 Pipe thread.

33. Pipe and Pipe Threads. The American standard pipe thread is profiled in Fig. 13.58. This thread is tapered or is cut on a long cone, the taper being 1 in 16 in the diameter or 1 in 32 on a side. Three diameters are of interest here, namely:

D_1, actual outside diameter of pipe as made, to be realized as accurately as possible with methods of manufacture.

D_2, overall diameter at end of thread when just fully cut, to be precise within 0.001 in (0.025 mm).

D_3, actual inside diameter, subject to variation; standard pipe may be light or full weight; extra heavy and double extra heavy pipe are also available, each with a smaller internal diameter.

For rough calculations the wall thickness of extra heavy (or extra strong) pipe up to and including 5 in (127 mm) may be taken as 1.42 times that of standard pipe. The thickness of extra heavy 6-in (152.4-mm) pipe is 0.432 in (10.9 mm), and for sizes from 7 to 12 in (177.8 to 304.8 mm) inclusive, the thickness is 0.500 in (12.7 mm). The thickness of double extra heavy (or double extra strong) pipe in sizes ½ to 6 in (12.7 to 152.4 mm) is twice that of extra heavy pipe; 7- and 8-in (177.8- and 203.2-mm) sizes are 0.875 in (22.2 mm) thick.

In Fig. 13.58 dimension n' is the number of full threads, reaching to the place where the cone tangent to thread tops intersects the pipe surface if of nominal diameter. This number is given by the formula

$$n' = 0.8D_1 + 4.8 \tag{13.16}$$

Beyond n' the standard shows two threads yet complete at the bottom, then a chamfer of the die at an angle of 12 or 15° with its axis. The threads have 60° angle and the depth is $0.8p$ or $0.8 \times 0.866 = 0.693$ of the full, sharp V.

34. Nut Holding. When a screwed joint is made up tight, or a nut is screwed down hard, a large primary holding friction is developed between the threads and between the nut and its seating. Under static conditions this prevents the nut from working loose, but under vibration and jarring it becomes necessary to employ some additional means or device to ensure that the nut will not shake loose. Several principles appear in the methods of solving this problem; the chief ones are to increase internal friction and to use an external locking device of some kind.

Holding pressure is due to elastic reaction, or to the compression of connected parts and tension of the bolt. Necessarily the dimensional range of this action is very short; while the turning moment to start a nut "off" is not much less than that used to tighten it, stress is released and resistance to turning disappears with only a little backward movement. The *spring washer* (Fig. 13.59) functions to increase greatly the range of elastic holding effect.

The *lock nut* or jam nut is a second nut screwed hard upon the first, as in Fig.

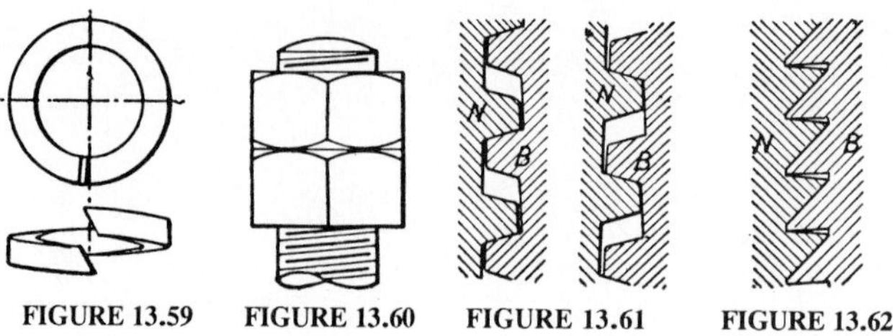

FIGURE 13.59 FIGURE 13.60 FIGURE 13.61 FIGURE 13.62

FIGURES 13.59 to 13.62 Various frictional nut locks.

13.60; or a nut upon the body into which a set screw is threaded, as in Fig. 13.51. It may increase primary friction, or it may act to produce heavy friction in the threads when there is no hard main gripping action to develop primary friction. Note in Fig. 13.60 that the tensile load on the bolt comes upon the top nut and the bolt threads within it.

The *Dardelet gripping thread* (Fig. 13.61) has a slant fit of top of nut thread on bottom of bolt groove, with a wide clearance between narrowed threads; the first sketch shows the loose or free-turning position of nut on bolt; and the second shows the position when screwed down hard. Proportions are such that the nut can be released and again tightened up, with renewed gripping effect.

The *Harvey grip thread*, for track or rail-joint bolts, is shown in Fig. 13.62. The thread is of buttress type, with a difference of angle on the "flat" or pressure side, as between bolt and nut. In a hard screw-up the nut thread is deformed beyond its elastic limit, giving a very intense local pressure and high internal friction.

The simplest external holding device is a split cotter pin in a hole through the bolt above the nut. This is a safety device rather than one to ensure that the joint will remain tight. Where the fastening is a permanent one a slender solid taper pin through nut and rod, held by slightly riveting its end, is an effective simple nut lock.

The three-grooved *castle nut* in Fig. 13.63, developed for the automobile, has come into wide use. It works better on rather long bolts and with fine screw threads, so that adjustment by 1/6 turn will not be too coarse. In older practice various forms of locking plates, held on the bolting surface by a small cap screw and fitting the side of the nut, have been used with similar effect.

A simple device that gives free adjustment in screw-up is a special thin washer of heavy sheet steel (Fig. 13.64). If a row or ring of bolts is near an edge or a shoulder, one wing of the washer is bent over or against this, then another is turned up against a side of the nut. An elaborate special nut with set-screw lock (the *Penn nut*) is shown in Fig. 13.65, used on a large engine connecting rod where conditions justify such construction.

35. Seating for Nuts. In rough work common plate washers are used, with larger cast-iron washers for bearing on wood. On a casting such as a machine frame there is usually a raised boss around each bolt hole, readily faced off for a true fit. Sometimes a flat surface or seating is depressed by counterboring perhaps 1/8 in (3.2 mm) deep. Finished washers are rather unusual, but nuts that are to rest on a finished surface are generally "washed faced" as indicated in Fig. 13.63; outside of the diameter across flats the bottom surface is cut back about 1/64 in (0.4 mm) so that the bolt "corners" will not scratch the seating surface.

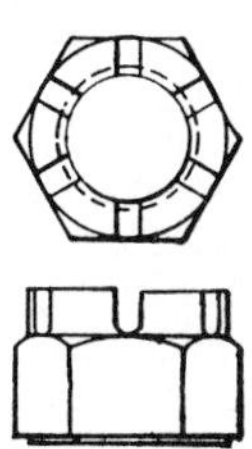

FIGURE 13.63 Castle nut.

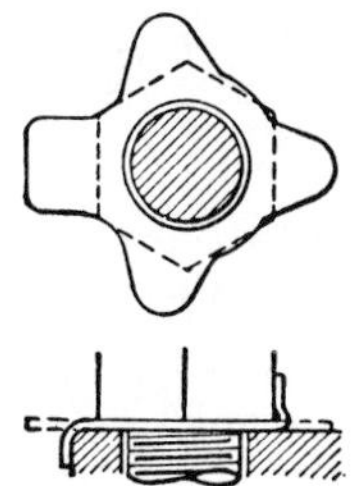
FIGURE 13.64 Lock washer.

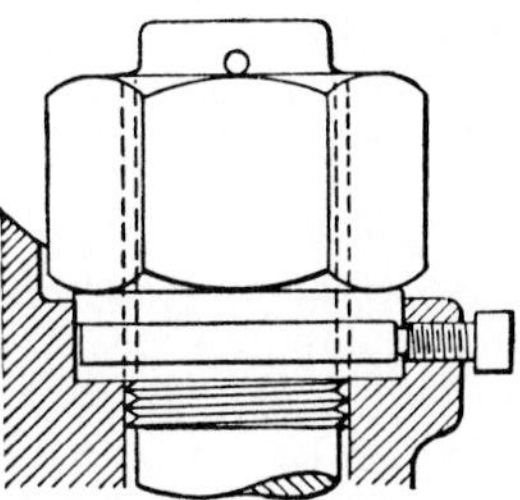
FIGURE 13.65 The Penn nut.

36. Screwing Up a Bolted Joint. When there are a number of bolts these are to be taken in turn and screwed up by stages. In a ring of bolts, for instance, first set them all up to an equal bearing by hand, then screw up moderately tight, finally set up hard. Work back and forth across diameters and swing by angles. The idea is to avoid severe local overstrain of the pieces bolted together.

37. Wheel on Shaft. The service of holding a wheel or pulley against turning on its shaft may be performed by friction alone, as in Figs. 13.66*a* and 13.67; by friction plus a slight degree of positive fit, as in Figs. 13.66*b* and 13.68; or by a positive fastening of varying strength, i.e., by a key or keys.

The *saddle key* in Fig. 13.66*a* is hollowed to the curvature of the shaft and tapered lengthwise (compare Fig. 13.71*a*) so as to function as a very flat wedge. Its advantage is that the wheel can be placed anywhere, without preparation of the shaft, but tightness of grip is somewhat hindered by frictional resistance to the driving of the wedge. The *flat key* in Fig. 13.66*b* requires that a seating be filed for it, but holds a little more strongly. Both are old forms, for light service only.

Figure 13.67 shows the hub of a *split pulley*, made to a large diameter, fitted to the shaft by a halved bushing and held by friction due to a strong bolted grip. This scheme is used with some cast-iron pulleys, with built-up pressed-steel pulleys and with wooden split pulleys—bushings for the last being of hard wood.

Set screws (Fig. 13.68) are largely used in pulleys for light and medium service, characterized by high speed and moderate torque. With either oval or cup points screwed hard into the shaft there is a small degree of positiveness of hold, but a slip due to overload will mar the shaft badly.

For very light wheels the small *taper pin* may be used as a key, in one of the three ways indicated in Fig. 13.69. For pulleys under heavy loading, and regularly

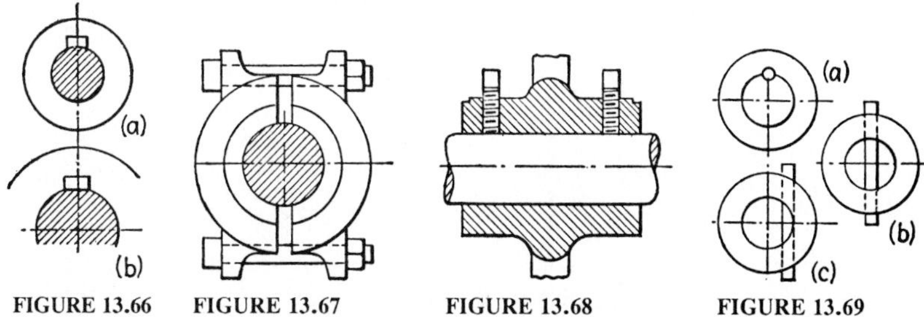

FIGURE 13.66 FIGURE 13.67 FIGURE 13.68 FIGURE 13.69

FIGURES 13.66 to 13.69 Various wheel-on-shaft fastenings.

for toothed gears, the *square* or *rectangular key* (Figs. 13.70 and 13.71) is the standard fastening. From the viewpoint of ease of placing and removal, the most convenient arrangement is the *straight key*, made to an easy slip fit and with clamping set screws above it, as in Fig. 13.70. The *taper* or *wedge key* in Fig. 13.71*a* may have a "gib" head as here, or may be headless; it requires a greater length of keyway and may be difficult to remove; taper is ⅛ in (3.2 mm) to the foot (0.3 m) or 1 in 96.

Often a gear has to slide along its shaft. The key in this service is called a *spline* or *feather*. The design in Fig. 13.71*b* implies that the keyway runs to the end of the shaft; if this is impossible, some more complicated method of holding the key in the wheel must be devised.

The *Woodruff key* (Fig. 13.72) is a convenient special form, much used on the short shafts of machines; the small dotted circle indicates the stem of the milling cutter that makes the key slot. This key is not good for transmission shafting, since its deeper cut (see comparison in Fig. 13.75) seriously diminishes the torsional strength of the shaft.

For powerful sliding-fit drives the *square shaft* (Fig. 13.73*a*) is an effective but old form, now rarely encountered. The solid *splined shaft* (Fig. 13.73*b*) is a standard automobile form. It is used for slip driving in the gear box and is effective on light hollow shafts for constant driving.

Key forms for heavy machinery, to fasten wheels and cranks, are grouped in Fig. 13.74; these are:

1. Double rectangular keys, set at quadrant
2. Kennedy keys, set at 45° angle
3. Lewis keys, same general scheme as preceding; different proportions
4. Nordberg heavy round key, slightly tapered, hole-drilled and reamed in assembled parts

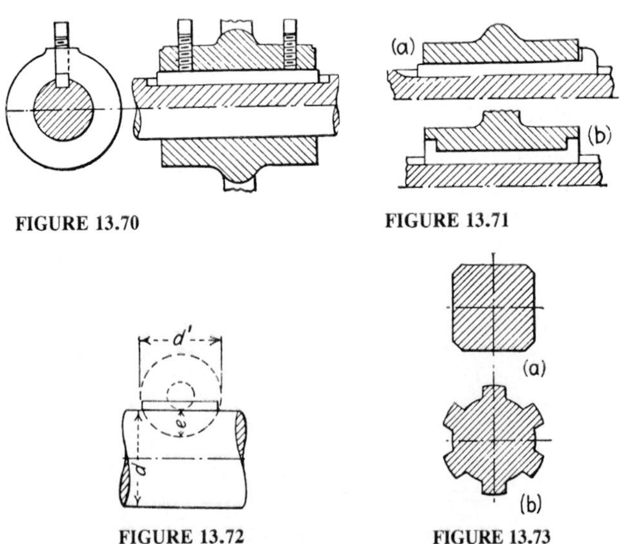

FIGURE 13.70

FIGURE 13.71

FIGURE 13.72

FIGURE 13.73

FIGURES 13.70 to 13.73 Key forms and arrangements.

Keys 2 and 3 are easier to fit, or to bring to equal fit and bearing, than are keys 1; in either case the key is tapered in one direction. These double keys are used in a wheel when it is at the end of the shaft—a crank arm is necessarily in that position.

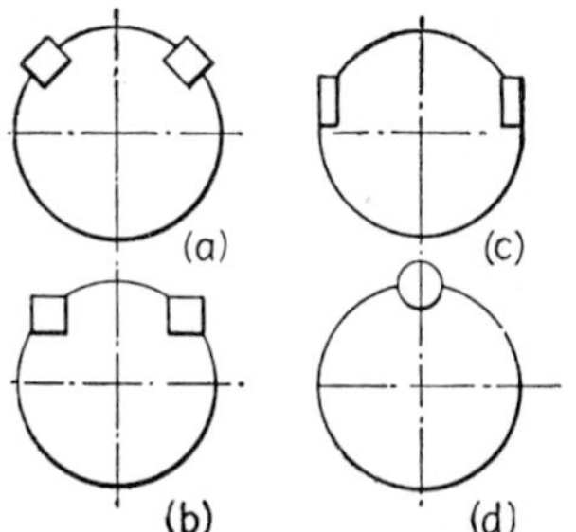

FIGURE 13.74 Heavy keys.

38. Key Dimensions. A standard series of stock key sizes is recommended by the American Engineering Standards Committee. This embodies the usual proportion that width *w* of the key shall be about ¼ diam. *d* of the shaft (see Fig. 13.75); but in dropping a number of odd-sixteenth sizes that have been in use it conforms less closely to this rule than do some older manufacturers' standards. For the "flat" or rectangular key height *h* is three-fourths to two-thirds of *w*. As shafts are larger, keys are relatively smaller, width *w* dropping to one-fifth or one-sixth of *d*.

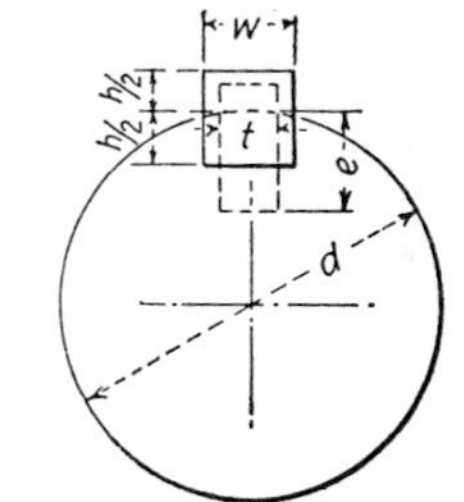

FIGURE 13.75 Key proportions.

Woodruff keys come in a number series of diameters, with supplementary letter sizes. For shafts up to 2½ in (63.5 mm) the ranges of recommended key sizes are such that the key thickness is one-fifth to one-sixth of the shaft diameter for smaller sizes and one-sixth to one-seventh for larger sizes. The key segment entering the shaft subtends an angle of 140 to 145° (of the key), or its depth *e* (Fig. 13.72) is from six-tenths to seven-tenths of the key radius.

39. Cotters or Cross-Keys. The shaft key is subject to shearing stress in a plane through its long axis. When a key is under shear across the long axis it is properly called a *cotter* (Figs. 13.76 to 13.79). The first three examples belong to older practice, but those in Fig. 13.79 continue in regular use.

A rod coupling is seen in Fig. 13.76, such as was used in a down-shaft mine pump. The side sketches show amount cut away to make room for cotter and compare sections of the rod and of its enlarged end at the cotter slot.

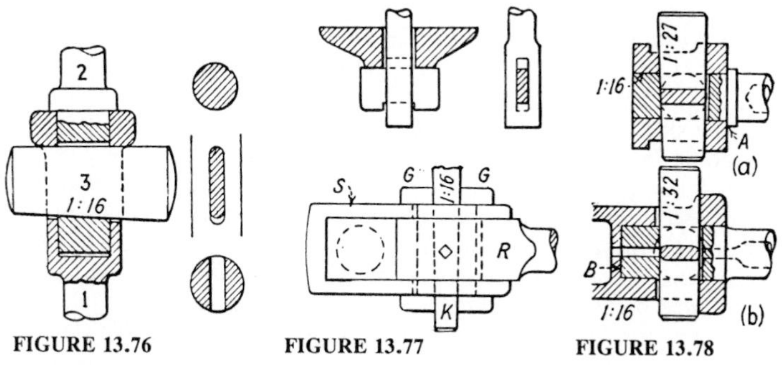

FIGURE 13.76 **FIGURE 13.77** **FIGURE 13.78**

FIGURES 13.76 to 13.79 Various uses of the cotter or cross-key.

Figure 13.77 shows a cotter at the lower end of a foundation bolt, instead of a nut. The bolt is squared at the end (strictly, is of rectangular cross section in a hole cored in the heavy cast-iron washer) and is punched for the cotter.

The old key-and-gib joint for a connecting-rod end is shown in Fig. 13.78. Strap *S* fits over rod *R*, is clamped to it by hooked "gibs" *G G*, and is held against end-wise movement by the combination of key and gibs into an adjustable cotter.

Present use of the cotter is seen in Fig. 13.79, which shows two examples of the holding of piston rod in crosshead on a locomotive. In Fig. 13.79*a* there is shoulder contact at *A* and the rod-end metal at sides of cotter slot is under the initial tension due to hard driving of the cotter. In case Fig. 13.79*b* the rod tip has end bearing on bottom of socket at *B*, so that driving of the cotter puts this end into compression; now the metal in the critical section, at sides of slot, receives only the tension due to working forces along the rod. Note how both of these piston rods are made hollow; this saves weight without sacrifice of strength, since possible bending under compression is the critical and determining condition.

40. Small Pins and Cotters. Small cotter pins and cotters are sketched in Fig. 13.80. Taper pins are rated by the big-end diameter, according to the following scale:

No.	00	0	1	2	3	4	5	6	7	8	9	10
d, in	0.141	0.156	0.172	0.193	0.219	0.250	0.289	0.341	0.409	0.492	0.591	0.706
d, mm	3.6	3.9	4.4	4.9	5.6	6.4	7.3	8.7	10.4	12.5	15.0	17.9

The taper is ¼ in (6.4 mm) to the foot (0.3 m); special reamers for these pins are available. Lengths range from ½ to 2 in (12.7 to 25.4 mm) for size 00 and from 1½ to 6 in (38.1 to 152.4 mm) for size 10.

Spring cotters or split pins, of half-round wire and with easy fit in a hole of their nominal size, come in the diameters 3/32 to 7/32 in (2.4 to 5.6 mm) by 64ths (0.4 mm), ¼ to ½ in (6.4 to 12.7 mm) by 16ths (1.6 mm), and ⅝ (15.9 mm), with a range of lengths like that for taper pins.

The flat spring key (Fig. 13.80*c*) is used to hold rigging together—as the brake rigging of a railway car—with the advantage over the round pin that it is thinner, hence does not so much weaken a rod or bolt.

41. Bolts as Pins. Sometimes bolts are under shear as a principal stress. The locomotive connecting-rod end in Fig. 13.81 is an example; the steamer-shaft coupling in Fig. 13.82 is another. Bolted joints are used also in assembling the sections of the side frame of a locomotive; the bolts are very slightly tapered and are driven home in reamed holes. Usually a portion (perhaps a large portion) of the shearing load is taken by rectangular keys, of the wheel-on-shaft type, set into the joint with an accurate driving fit on their sides. Straight bolts with driving fit have been used in a coupling like Fig. 13.82, but the taper form is decidedly preferable in original assembly and yet more in dismantling and reassembly. In the smaller keyed-on shaft coupling (Fig. 13.88), with relatively much larger bolt circle, the bolts have loose fit and the drive is by friction between flanges.

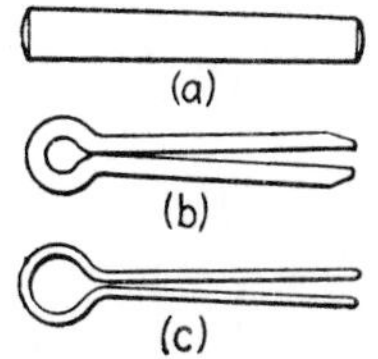

FIGURE 13.80 Cotter pins.

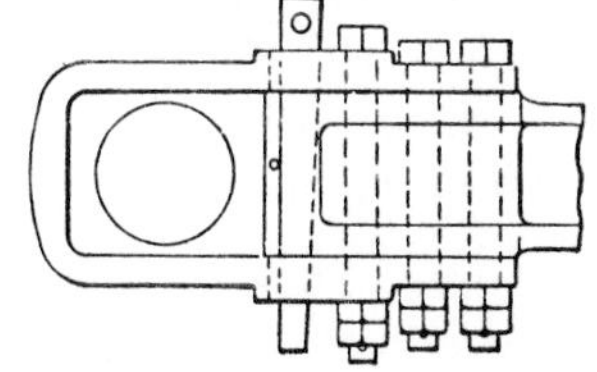
FIGURE 13.81 Bolted rod end.

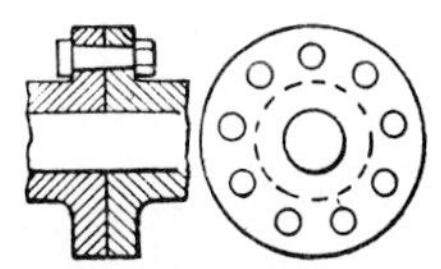
FIGURE 13.82 Heavy flange coupling.

42. Forced and Shrunk Fits. In what may be called *strained joints* the hole in a member, such as a wheel hub, a crank arm, or a steel hoop or tire, is smaller than the shaft or disk that is to go into it. By hammer drive or screw-press drive with small fits, or by means of the hydraulic press with large work, the parts are forced together; or the hole member may be expanded by heat so that the plug member can be slipped into place, then after cooling and at equal temperatures there will be a tight gripping. Out of a total of eight classes of fits, three are for tight and strong joints, namely:

Class 6. Tight Fit. Slight negative allowance, 1 in 4000; light pressure required to assemble and parts in more or less permanent assembly; gears on shafts in high-grade machines (plus a key) are good examples; this allowance is good for drive fits in thin sections or for very long fits in other sections.

Class 7. Medium Force Fit. Negative allowance, 1 in 2000; considerable pressure, permanent assembly; locomotive and car wheels on axles, armatures, and crank disks on shafts; shrink fits in light work, heavy force fits of great length. This allowance is the greatest to be used when hole is in cast iron.

Class 8. Heavy Force and Shrink Fit. Considerable negative allowance, 1 in 1000; for steel-hole members, where stress can be heavy without coming too close to elastic limit; likely to be excessive for force fits (developing excessive friction) unless these are short. Shrink fits are used where forcing would be impracticable, as for locomotive tires and the crank disks or arms of large engine shafts.

43. Fits in General. The several grades of loose fits in the ASME scheme are illustrated in Fig. 13.83; for the particular diameter [4 in (102 mm)], the sketches show allowance, tolerances, and tightest and loosest fit in each class, vertical dimensions being in mils or thousandths of an inch.

The heights of lines as lettered for case 1 in Fig. 13.83 are as follows:

Line A, minimum diameter of hole, equal to nominal diameter d of the fit

Line B, maximum diameter of hole, above A by hole tolerance t_h, which is always +

Line C, maximum diameter of shaft or plug, below A by allowance a; in general

$$a = d_{h\ \min} - d_{s\ \max} \tag{13.17}$$

Line D, minimum diameter of shaft, below C by shaft tolerance t_s, always −

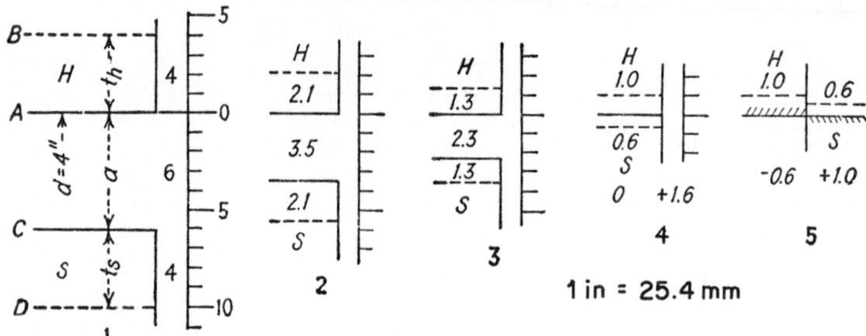

FIGURE 13.83 Allowances and tolerances.

The first four classes are interchangeable fits, the parts to be made within the range of tolerance and assembled at random. Then in any class the looseness ranges from a to $a + 2t$. The classes are as follows:

Class 1—Loose Fit. Large Allowance. Considerable freedom, for cases where accuracy is not essential; agricultural and mining machinery; textile, rubber, candy, and bread machinery; general machinery of similar grade.

Class 2—Free Fit, Liberal Allowance. Running fits with speeds of 600 r/min and over, journal pressures of 600 lb/in^2 (4.1 MPa) and over; dynamos and engines, many machine-tool parts, some automotive parts.

Class 3—Medium Fit, Medium Allowance. Running fits under 600 r/min, journal pressures less than 600 lb/in^2 (4.1 MPa), and sliding fits; the more accurate machine-tool and automotive parts.

Class 4—Snug Fit, Zero Allowance. Closest fit that can be assembled by hand, calls for high precision; used where no perceptible shake is permissible; motion not under load, as in instruments.

Class 5—Wringing Fit, Zero to Negative Allowance. Metal to metal fit, assembly should be selective, to avoid interference shown as one extreme of conditions, in Fig. 13.83, for this class.

The variation of allowance and tolerances with size or diameter is shown in Fig. 13.84, for class 1. This shows steps by 0.001 in (0.025 mm). These dimensions come from formulas which for all the classes of fits are given in Table 13.1.

44. Interchangeability and Selection. As already noted, the first four classes of fits are supposed to be assembled with random choice of parts, although some selection may improve results. Class 5 calls for selection, fitting larger plug into

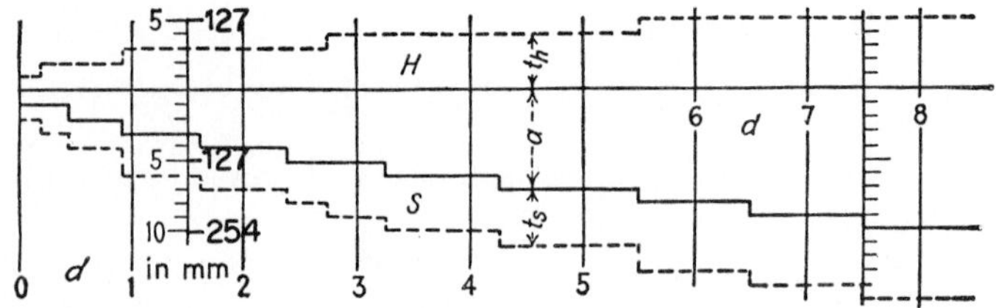

FIGURE 13.84 Variation with diameter.

TABLE 13.1 Recommended Allowances and Tolerances

Fit	Allowance	Hole tolerance	Shaft tolerance
1. Loose....................	$+0.0025\sqrt[3]{d^2}$	$+0.0025\sqrt[3]{d}$	$-0.0025\sqrt[3]{d}$
2. Free.....................	$+0.0014\sqrt[3]{d^2}$	$+0.0013\sqrt[3]{d}$	$-0.0013\sqrt[3]{d}$
3. Medium................	$+0.0009\sqrt[3]{d^2}$	$+0.0008\sqrt[3]{d}$	$-0.0008\sqrt[3]{d}$
4. Snug....................	none	$+0.0006\sqrt[3]{d}$	$-0.0004\sqrt[3]{d}$
5. Wringing...............	none	$+0.0006\sqrt[3]{d}$	$+0.0004\sqrt[3]{d}$
6. Tight....................	$-0.00025d$	$0.0006\sqrt[3]{d}$	$0.0006\sqrt[3]{d}$
7. Medium force...........	$-0.0005d$	$0.0006\sqrt[3]{d}$	$0.0006\sqrt[3]{d}$
8. Heavy force.............	$-0.001d$	$0.0006\sqrt[3]{d}$	$0.0006\sqrt[3]{d}$

larger hole, smaller plug into smaller hole. The three classes of forced fit described in Sec. 42 require close selection, so as to keep very near to the prescribed allowance.

A practical way to control forced fits, as in the case of car wheels on axles, is to specify a minimum and a maximum forcing pressure. Below the minimum the fit is too loose for assured holding; above the maximum there would be dangerous stress in the wheel hub.

Under reciprocating force action, as in the pin and shaft bearings of a high-speed engine, accurate manufacture must be supplemented by hand fitting and by adjustment. A scheme of adjustment for big bearings is to have a gap or clearance between cap and base and fill it with graduated *shims* or layers of sheet metal. Suppose that the clearance is ½ in (12.7 mm); then one shim of each of the following thicknesses, ¼, ⅛, 1/16, 1/32, and 1/64 in (6.4, 3.2, 1.6, 0.79, and 0.39 mm) and two of 1/128 (0.198 mm) will make possible adjustment to any nearest 128th (0.198 mm) of an inch. In a small bearing the last shims may be 0.001 (0.025 mm) or even 0.0005 in (0.013 mm) thick.

A high degree of control is exercised in gun assembly; for each shrunk fit the outer jacket or hoop is first bored and reamed and is micrometered at every foot of length with a "star gage," and the inner part is then finished to these dimensions plus the calculated allowance for shrinkage.

45. The Shrunk Fit. Simple average values of the two most important coefficients of linear expansion by heat, per degree Fahrenheit (1.8°C), are

For iron and steel, $$a = 0.0000060 \pm 3 \tag{13.18a}$$

For brass and bronze, $$a = 0.0000100 \pm 6 \tag{13.18b}$$

the range of variation being, of course, in the last figure or the seventh decimal place.

For a fit of steel parts the hole member will be heated from 300 to 500°F (149 to 260°C) above room temperature, the maximum of this range being well below the beginning of red heat. The proportional increase of diameter is then from 0.0018 to 0.003, plenty for easy slipping together.

The rough shrink fit in which the ring or hoop is heated red hot belongs to the cruder type of construction, for instance, the strengthening or repair of a wheel hub (rough casting) or the tire on a wooden wagon wheel.

MOTIVE ELEMENTS OF MACHINES

46. Shafting and Shafts. The word *shafting* is used for transmission lines and systems. A *shaft* is a single line or piece in a system or in a machine. Complete shafting systems were used before the electric motor became common, such usage culminating before 1900.

Modern practice has tended away from elaborate mechanical transmissions, putting an electric motor on each individual machine or using possibly a motor-driven line shaft for a group of smaller machines. But when a large amount of power is to be applied at low speed, as in a rubber mill or in a wire-drawing mill, a heavy line shaft from an engine or a big motor holds its own against the higher cost of individual motors.

47. Shaft Dimensions. Ordinary transmission shafting comes in diameters 1⁄16 in (1.59 mm) scant of the quarter (6.4 mm) or half-inch sizes (12.7-mm), from 13⁄16 to 215⁄16 in (30.2 to 74.6 mm) by quarters to 6 in (152.4 mm) by halves; anything bigger than 6 in (152.4 mm) is usually special. Machine shafts are made to eighths, quarters, or halves. Lengths up to 24 ft (7.3 m) are regular, but 16-, 18-, or 20-ft (4.9-, 5.5-, or 6.1-m) sections are common, to fit spacing of hangers as set by building construction.

The steel in common shafting has an ultimate tensile strength of 60,000 to 65,000 lb/in^2 (414 to 448 MPa). By cold rolling this is raised to approximately 80,000 lb/in^2 (552 MPa); and it is usual to say that cold-rolled is to turned (hot-rolled) shafting as 5:4 in transmitting capacity.

By Eq. (13.4) (Sec. 14) with M_t for turning moment PR'' in in · lb (N · m) and with S_s for maximum working shear stress, the torsion formula for a simple solid round shaft becomes

$$M_t = \frac{\pi}{16} d^3 S_s = \frac{1}{5.1} d^3 S_s = 63{,}025 \frac{H}{N} \qquad (13.19)$$

Using S_s = 10,000, and solving for d:

$$d = 0.08 \sqrt[3]{M_t} = 3.18 \sqrt[3]{H/N} \qquad (13.20)$$

With other values of S_s, the value of d as obtained from Eq. (13.20) must be multiplied by the proper factor F_1, as given below:

S_s	F_1	S_s	F_1	S_s	F_1	S_s	F_1
3500	1.419	5000	1.260	8,000	1.077	12,000	0.941
4000	1.357	6000	1.186	9,000	1.036	14,000	0.894
4500	1.305	7000	1.126	11,000	0.969	16,000	0.855

In shafting practice stress S_s is likely to range from 4000 to 8000 lb/in^2 (28 to 55 MPa) for the maximum load on common steel shafting, and from 5000 to 10,000 (34 to 69 MPa) on cold-rolled steel.

If in addition to turning moment or torque M_t the shaft is also subject to a bending moment M_b, with $M_b = kM_t$, a further factor F_2 must also be applied to d as computed from Eq. (13.20); its value is

$$F_2 = \sqrt[6]{k^2 + 1} \tag{13.21}$$

and values are as follows:

k	0.1	0.2	0.3	0.4	0.5	0.6	0.7	0.8	0.9	1.0
F_2	1.002	1.007	1.015	1.025	1.038	1.053	1.069	1.086	1.104	1.122
k	1.1	1.2	1.3	1.4	1.5	1.6	1.7	1.8	1.9	2.0
F_2	1.141	1.160	1.177	1.198	1.217	1.236	1.254	1.272	1.290	1.308

Machine shafts are a more highly developed type, running to better material (alloy steel, heat treated) and to more complex forms (as the crank shaft of an engine). They involve problems in machine design of a rather advanced type.

48. Shaft Couplings. The simplest coupling for shaft sections is the plain *muff* or *sleeve* in Fig. 13.85; one long key might be used, driven in from one end, but separate keys, as here, can more surely be driven to a tight fit in each shaft end.

A coupling cannot be put onto a shaft end with anything tighter than a very easy driving fit, whereas the strength and stiffness of a forced fit may be desirable. For this reason a number of compression or gripping couplings have come into general use. The simplest is the *ring-compression* or *friction-clip coupling* in Fig. 13.86: The double-tapered sleeve is split along one side, so that it can spring inward, and the rings are driven on hard with a sledge, or perhaps shrunk on with moderate heating; a key may or may not be used. This is a rough-and-ready coupling, good under exposure to dampness and more easily taken apart than one having screws to rust fast.

The *cone-wedge principle* is used in several more highly developed designs that are regularly manufactured, wedge drive being effected by screw action. One scheme is a solid outer sleeve, into which split cones are drawn by bolts parallel to axis of shaft; another has a sleeve like that in Fig. 13.86 (somewhat steeper cones) with rings threaded on its ends to push on the compression rings; yet another uses flanged rings (somewhat similar to the flanges in Fig. 13.88) on the same kind of a sleeve.

The split and bolted *compression coupling* in Fig. 13.87 is simple and powerful. Note how the bolt heads are protected by the "ribs" of the casting, and also how a flange projects over them in Fig. 13.88; it is dangerous to have projecting

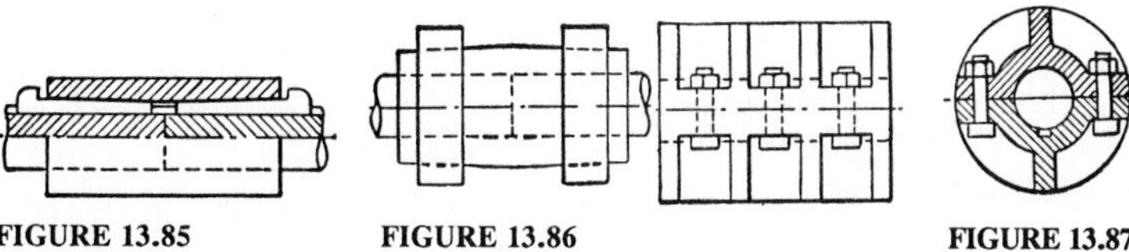

FIGURE 13.85 FIGURE 13.86 FIGURE 13.87

FIGURES 13.85 to 13.87 Simple shaft couplings.

bolt or set-screw heads on a shaft if men must work near it, or can get to it while running.

The *flange coupling* (Fig. 13.88) is a much used form. Made separate and keyed to the shaft it is much cheaper and more convenient than the solid-forged coupling on the shaft in Fig. 13.82. Note how one shaft end projects a little way into the other half-coupling, so as to assure alignment—which is not assured by the loose-fitting bolts.

The *Oldham coupling* (Fig. 13.89) may be used to couple shafts that are likely to be out of line but not out of parallel by more than a fraction of a degree. A tongue-and-groove connection between flange *A* and disk *B* and one between disk *B* and flange *C* (at right angles to the first) permits freedom but ensures an exact transmission of angular velocity.

To meet the condition of nonparallelism, but with little or no allowance for misalignment, a number of types of *flexible couplings* are in use. Generally a flange is keyed to each shaft end, as in Fig. 13.88; then these are coupled together by one of the following methods:

1. Pins on the flange faces are connected by heavy links of leather or by some such device.
2. One flange is large, the other small, and they are connected by a ring disk of leather or light plate metal, which is fastened to them at its respective edges.

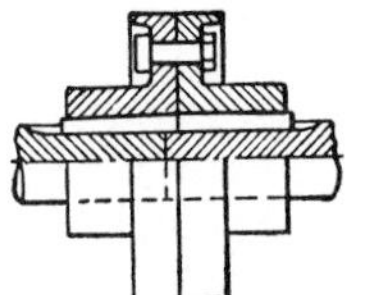

FIGURE 13.88 Flange coupling.

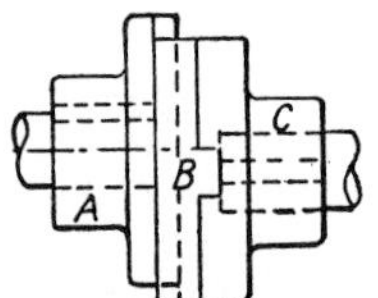

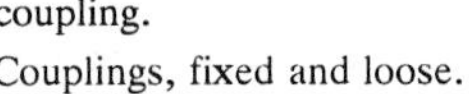

FIGURE 13.89 Oldham coupling.

Couplings, fixed and loose.

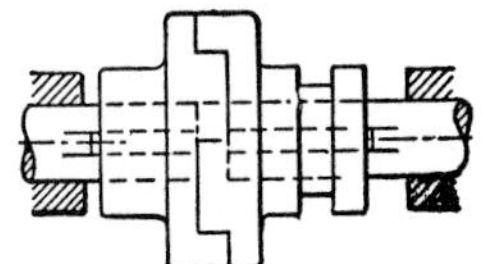

FIGURE 13.90 Jaw clutch (a separable coupling).

3. Flexible pins—in one case of blades of thin spring steel—are placed and held in holes in the disks.

The *separable coupling* in Fig. 13.90 introduces the class of positive or jaw clutches, but is here supposed to be used only for occasional connection or disconnection.

The *universal joint* (Fig. 13.91) has two forked shaft ends *A* and *C* engaging an intermediate cross *B*. It is highly flexible but does not transmit a uniform angular velocity. If shaft *A* (Fig. 13.92) turns uniformly, shaft *B* will oscillate back and

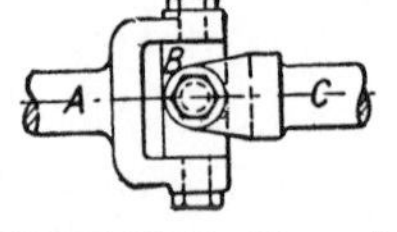

FIGURE 13.91 The universal joint.

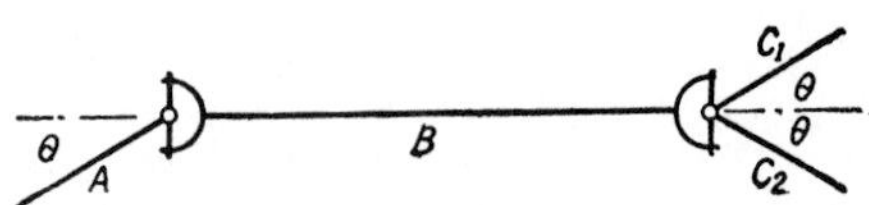

FIGURE 13.92 Diagram of action.

forth with respect to uniformity. But the three-shaft arrangement, all in one plane and with equal angles θ of A and C from B and with the two forks on shaft B in the same plane, does give true and smooth motion from A to C.

49. Clutches. The *jaw clutch* (Fig. 13.90) gives powerful and positive drive but abrupt engagement, and hence can be thrown in only at low speed. When there is to be but one direction of drive, engagement is made easier by slanting off the rear or noncontact sides of the teeth. Such clutches are used at the low-speed end of machine-tool gear trains, with friction clutches at the high-speed engagements. A class of special interest comprises the positive clutches used in punching and shearing machines and in presses; any of these has a control device that makes its release always occur when the tool head or ram is at top of stroke.

A *conical friction clutch* of heavier and rougher type is partly sketched in Fig. 13.93; its function is to connect hoist drum 3 to shaft 1 through disk 2. The way that gripping forces P, P are placed is intended to suggest that this is what may be called an *open clutch*. That is, the gripping action comes from outside of the clutch system, the forces being applied through moving contacts or end bearings. Such an arrangement may be justified when simplicity and low first cost are primary considerations; but in general the closed or self-contained arrangement represented by Fig. 13.96 is very much to be preferred.

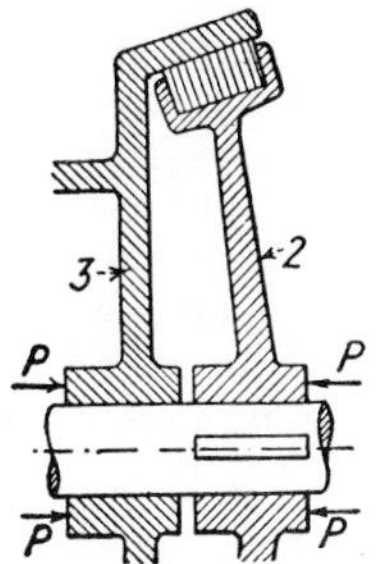

FIGURE 13.93 Cone clutch.

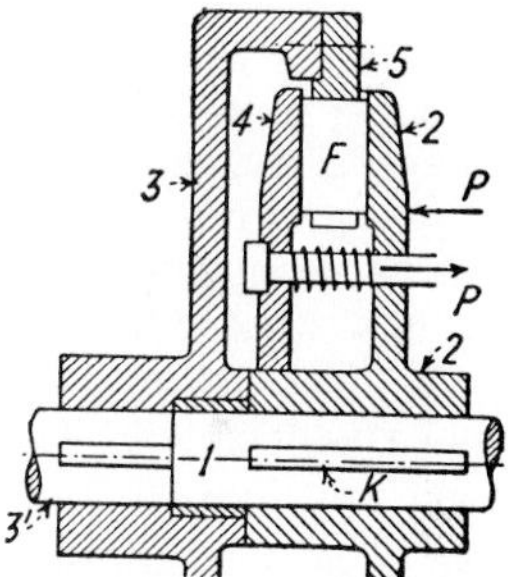

FIGURE 13.94 Disk clutch.

A *disk clutch*, in shaft-coupling service, is outlined in Fig. 13.94, with no more than a suggestion of the action of the gripping mechanism: there will be four or six units of this mechanism, at angles of 90° or 60° around disk 2. Note the brass-bushed bearing where the right-hand shaft end projects into disk 3. Here and in Fig. 13.93 one rubbing face is of hard wood. On smooth cast iron and with a very slight amount of greasiness of surface, the coefficient of friction is likely to range from 0.12 to 0.15.

The *multidisk* or *Weston clutch* in Fig. 13.95 is much used, with large modification in detail from this simple example. The essential fact is that one set of disks 2,2 is splined to shaft 1, the other set 4,4 to casing 3. With pressure P effective over a large number of contacts, ample driving force can be secured even with a good degree of lubrication; and this lubrication makes for easy running when loose and for a smooth and easy engagement.

Dry friction is strong, but it is very likely to be abrupt and "jerky" when "taking hold."

The example in Fig. 13.95 has metal disks and implies a powerful gripping ac-

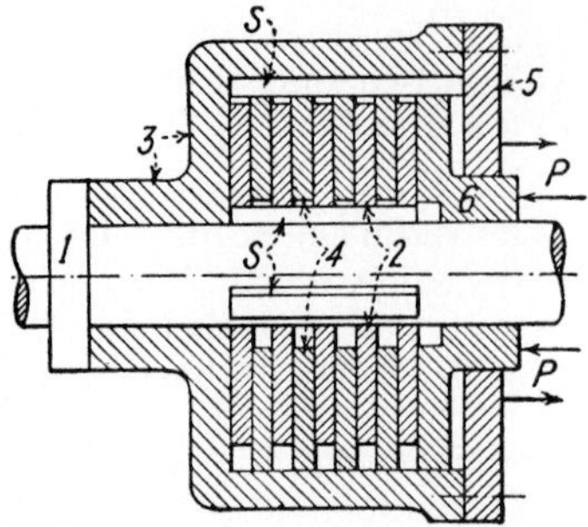

FIGURE 13.95 Multidisk or Weston clutch.

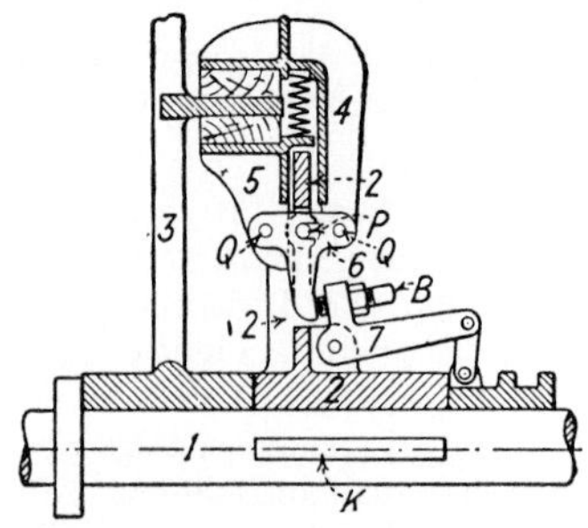

FIGURE 13.96 A ring clutch.

tion. The quite common automobile clutch of this type has a contact of asbestos on metal and gets its full driving effect with comparatively low gripping pressures.

A *ring clutch* of the shafting-transmission kind, in the service of connecting pulley 3 to shaft 1, is shown in Fig. 13.96. Arms 4 and 5 have a short radial movement in a slot in disk 2 and carry wooden blocks that bear on a ring, cast with the pulley arms. This drawing shows the use of the toggle mechanism of Fig. 13.30.

A very compact and powerful small clutch is shown in Fig. 13.97, out of machine-tool service. The parts are of strong steel, hardened at the contact surfaces; ring 4 is expanded by a pair of spreading levers actuated by a push cam fastened inside of sliding sleeve 7.

The most powerful type of friction clutch (within set limits of size and in the sense of getting a strong grip from a small gripping force) is the *coil clutch* of Fig. 13.98. Steel coil 4 is bored to an easy fit on drum 2, then one end is anchored to disk 3, then the other end is given a small movement by lever 5, which is pivoted on piece 3 and actuated by push cam 6; the function is to clutch disk 3 to shaft 1.

50. Bearings. The meaning of *journal* (inner, turning) and *bearing* (outer, standing) is made clear by Fig. 13.99, where the bearing might be of plain cast iron; also the simple lubricating system of oil hole and oil groove is shown. In Fig. 13.100 appears the more usual scheme of a lining or *bushing* of special ma-

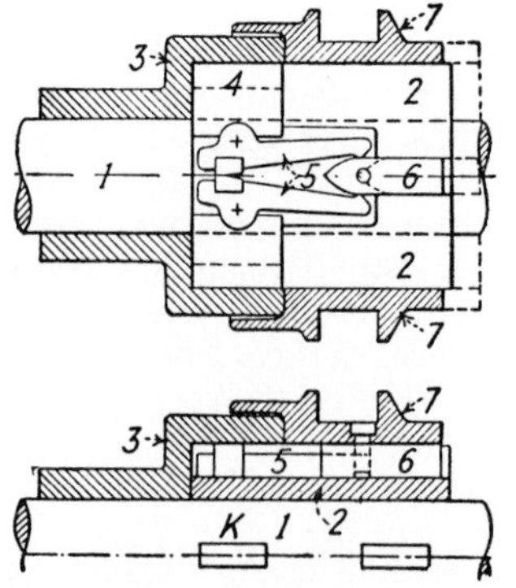

FIGURE 13.97 Small, powerful clutch.

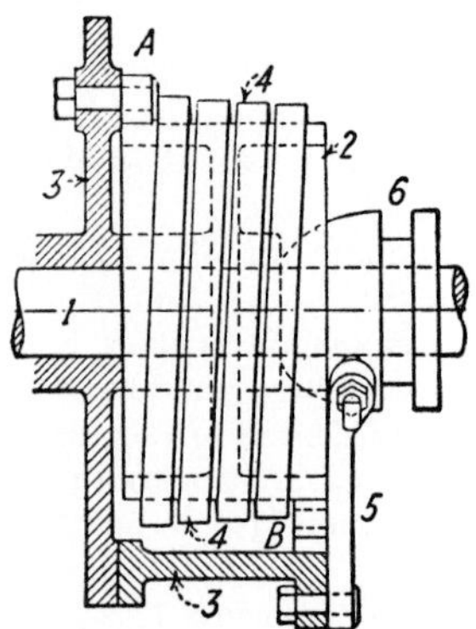

FIGURE 13.98 Coil clutch.

FIGURE 13.99 FIGURE 13.100 FIGURE 13.101

FIGURES 13.99 to 13.101 Forms of bearings, fixed and moving.

terial for the bearing. Sometimes this turning pair (see Sec. 5) is inverted, the inner or journal part being fixed and the hollow part turning; wheels on wagons and on small industrial cars (for short curves) are so arranged, and this inversion is fairly frequent in machine-tool gear trains. Figure 13.101 shows the case, out of linkage mechanism, in which both parts move.

Any shaft has to be held or constrained against endwise movement, even when forces along the axis are small and incidental. This may be done by collars at the ends of one bearing as in Fig. 13.102, or at opposite ends of any two bearings; these collars are set-screwed in place, preferably by sunk-head screws. The single concealed collar, also shown in Fig. 13.102 shrunk on the shaft and running in a groove in a special bearing, is a very neat device. For resistance to heavy end thrust, see Sec. 52.

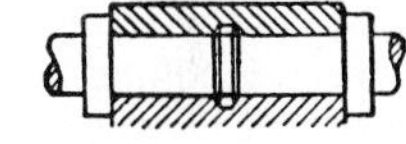

FIGURE 13.102 Collars.

Journal proportions, as defined by the ratio of length l to diameter d, are illustrated in Figs. 13.99 to 13.103; this ratio ranges from 4 (net and 5 overall) in Fig. 13.103 down to about 1 in Fig. 13.101. Large and heavy bearings are short, especially in engines, those for transmission shafting are long.

51. Bearings and Lubrication. The first step beyond intermittent squirt-can oiling, as suggested by Fig. 13.99, is to provide a scanty continuous lubrication from a small reservoir, by wick or pad functioning through capillarity. Grease-cup lubrication is similar, furnishing a continuous moderate lubrication with occasional attention. Lubricating greases are specially manufactured and are fed by spring pressure or through incipient melting by heat due to friction. Under neglect, grease is more persistent than oil; and a thick paste of grease is effective in excluding dirt and dust from the bearing.

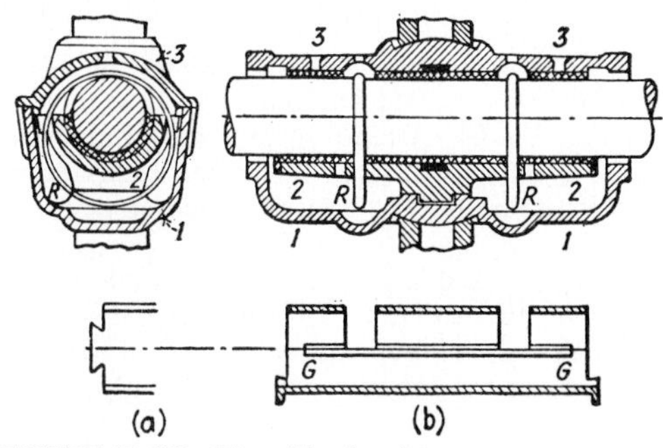

FIGURE 13.103 Ring-oiling bearing.

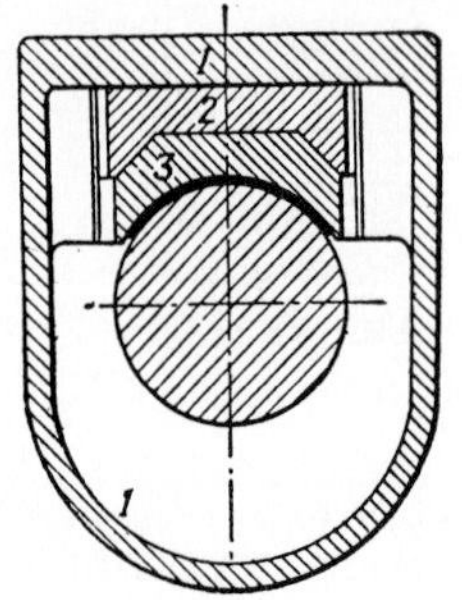

FIGURE 13.104 Car-journal bearing.

Next comes copious self-oiling from a large reservoir, as exemplified by Figs. 13.103 and 13.104. External oiling begins with individual sight-feed drip cups, is made far more certain and more convenient by the use of a central reservoir piped to the drips, and has fullest development in the pressure systems used in steam turbines and other high-speed machines.

An example of the *ring-oiling bearing*, much used for shafting and in stationary machines, is shown in detail in Fig. 13.103. This is a distinct unit with ball-and-socket support in a hanger or pillow-block frame, by large hollow cast-iron screws. It consists of outer lower shell 1, formed to serve as an oil reservoir; of lower bearing 2, with continuous surface; and of top bearing and cap 3, having its surface interrupted so as to make place for the oil rings *R,R*. These rings are of light half-round spring steel, bent to a mere contact joint so that they can be sprung over the shaft anywhere; in machine bearings the rings may be solid. In Fig. 13.103*b* is seen the oil groove *GG* into which the oil (carried up by the rings) feeds itself for distribution to the whole bearing surface. View *a* shows a simple oil scraper, to prevent oil in any quantity from working out along the shaft; this device or its equivalent takes a number of forms.

Figure 13.104 is a simplified section of the journal box of a railway car, which is the outstanding example of continuous and copious lubrication by an *oil pad*. The top of "wedge" 2 is shaped to a long curve in the direction of the axle or perpendicular to the plane of drawing, so that it functions as a rocker against body 1 and gives flexible support to "brass" 3. The last is usually of hard bronze and has a babbitt-metal lining, which is "sweated" fast or cast on a tinned surface so as to hold itself in place by adhesion alone; this type of construction is usual in high-class smaller machines such as automotive engines. The space below the journal is loosely "packed" with wool "waste" (much more open and spongy than cotton) and poured full of oil nearly to the bottom of the journal.

52. Thrust Bearings. This name denotes a bearing which has its heavy load in the direction of the shaft axis. Figure 13.105 shows one out of a number of forms of end or *pivot bearings*, with a brass shell or bushing for incidental side pressures and a hard steel "step" for end load; the shaft end should also be hardened to avoid wear.

The long-used *collar bearing*, necessary whenever the shaft must continue through the bearing and giving a large amount of surface for a big load, is shown in merest outline in Fig. 13.106. A plain thrust bearing does not lubricate itself nearly so well as does a journal bearing; hence unit pressure must be low and a large area of surface is required. The conspicuous use of this bearing is on the propeller shafts of ships.

Except by the high-pressure pumping in of oil, complete lubrication (see Sec. 56) is attained only when there is a wedge-shaped space into which oil can be drawn by the motion of one part upon the other. For a flat surface this requirement is met by the design of the *Kingsbury* or *Michel* type of bearing, the essential form of which is shown in Fig. 13.107. This drawing is a "developed" or

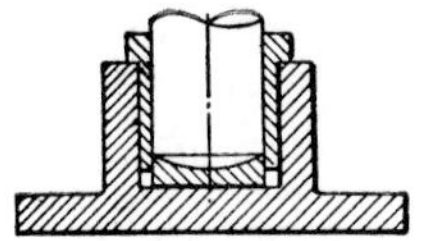

FIGURE 13.105 Pivot bearing.

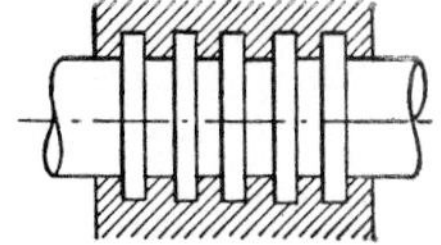

FIGURE 13.106 Collar bearing.

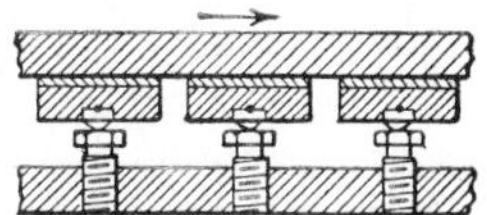

FIGURE 13.107 Kingsbury bearing.

unrolled section, really made by a cylinder on the shaft axis. Bearing surface is provided by a number of short shoes (sectors), each supported on a central pivot so as to be free to adjust itself to the natural shape of the oil film. A further refinement places a set of equalizing levers beneath the shoes, to insure equal division of the load.

53. Bearing Pressures. The term *bearing pressure*, as regularly used with regard to cylindrical bearings, means average pressure on the projected area, or on the rectangle of diameter d and length l, and is thus a simplified representation of something that may really be very complex. If P is the total pressure or load and p the unit pressure in lb/in^2 (kPa),

$$p = \frac{P}{dl} \tag{13.22}$$

For a thrust bearing product dl will be replaced by the actual area of a circle, a ring, or a number of rings.

Allowable bearing pressure depends on a number of considerations, which fall under three heads, as follows:

1. Kind and degree of lubrication and character of surfaces, which determine the strength or amount of frictional resistance, or the coefficient of friction μ.
2. Velocity V of slip or sliding, which has an influence upon μ but is chiefly the second factor of the work done in overcoming friction.
3. Freedom of escape or removal of heat. A bearing may be enclosed, may be open to a free circulation of air, may have a rapid circulation of oil (cooled outside), or may be watercooled; typically, the bearings of a marine engine are backed by hollow spaces through which water can be pumped.

A useful measure of the burden on a bearing is found in the product

$$q = pV \tag{13.23}$$

with pressure p by Eq. (13.22) and velocity V in ft/min (m/min). Multiplication by coefficient μ gives the friction work μV (ft · lb.) per in^2/min (N · m per cm^2/min). With varying degrees of ventilation, but no special cooling arrangements, the rate of escape of heat may range from 100 to 1500 ft · lb. per in^2/min. (21.0 to 315.3 N · m per cm^2/min) the maximum temperature being about 150° F (65.6° C) above the surrounding air. When the air is strongly agitated, as in the case of engine crank pins, the heat rate may be appreciably higher than the limit just stated.

For the cylindrical bearing, coefficient μ is based on the simple external load P, not on the larger total pressure between journal and bearing—compare Fig. 13.103. Note in Fig. 13.104 how the actual width of the bearing surface is only about 0.8 of the journal diameter.

54. Friction and Lubrication. This is a large and complex subject, containing elements of indeterminate or indefinite type. Conditions range from dry friction at one end to complete lubrication at the other, with all intermediate stages likely to be encountered.

With *dry surfaces* or the mere beginning of lubrication, friction is nearly proportional to load and independent of contact area, so that μ is approximately constant over a wide range of unit pressures. Recorded experiments on machine materials, under laboratory conditions, show an upper limit of 0.3 to 0.6 for μ, but in working conditions the actual range is from 0.10 to 0.30. Dry friction varies widely with slip velocity; certain classic experiments on railway brakes showed a drop from 0.33 static to 0.10 at 60 mi/h (96.5 km/h). Dry friction is a useful engineering force in traction, in power transmission, and in clutches and brakes.

One concept is that resistance to motion is due to the interlocking of minute irregularities of the two surfaces, strongest when the contact is static, weaker as rapid motion causes a "jumping" action. On the other hand, a smooth surface may give stronger friction, as in the case of belt on pulley; and for the general run of bearing materials there is a limit of pressure at which dry smooth surfaces begin to adhere or "seize." This limit lies in the range from 500 to 800 lb/in^2 (3.4 to 5.5 MPa); much below that which the same surfaces will carry with lubrication.

In *complete lubrication* the metallic surfaces do not touch each other, but are separated by an indrawn film of oil: the conditions for such action are described in Sec. 56. Resistance to motion is now within the oil film and is very low, μ approaching a value of 0.001 when the bearing is closely proportioned to its load. Subject to variation with a number of other conditions, this resistance is nearly independent of load; then ratio μ will vary inversely with the load on a given bearing, rising as the bearing is underloaded.

With *partial lubrication* there is metallic contact, mitigated in varying degree by the oil or grease that makes the surfaces slippery. In large classes of machinery the conditions for complete lubrication are unattainable, for either technical or economic reasons; therefore these machines must operate against more friction. With intermittent or scanty lubrication, μ is likely to range between 0.05 and 0.10; with more copious oil supply it will go below 0.05.

55. Working Pressures in Bearings. Practice in regard to the loading of bearings, evolved under the conditions enumerated in Sec. 53, is partially exemplified by the condensed statements in Table 13.2. Understand that, in general, stationary machines are under less severe restrictions as to weight and size than are machines in transportation service; therefore they can have bigger bearings and operate with lower pressures. They are expected to get along with much less attention than has to be given to the more heavily loaded machines, and are subject to much less rapid wear. The quantities given in Table 13.2 are pressure p by Eq. (13.22), slip velocity V [ft/min (m/min)], and their product q according to Eq. (13.23).

Line 4 of Table 13.2 expresses an old rule by Thurston that q or pV should not exceed 50,000 (105.2). When q rises to much higher values, lubrication must be strengthened (by a stronger or more viscous oil or by pressure feeding), or there will be rapid wear (as under locomotive conditions).

TABLE 13.2 Working Pressures in Bearings

Service	p		V		q		Lubrication
	lb/in²	MPa	ft/min	m/min	lb · ft/in² · min	MPa · m/min	
1. Light shafting	–50	0.34	–200	–60.9	10,000	63.1 –	Scanty
2. Medium shafting	–100	0.69	–300	–91.4	30,000 –	63.1 –	Ring-oiling
3. Heavy shafting	–150	1.03	–250	–76.2	40,000 –	78.5 –	Ring-oiling
4. Stationary machines	—	—	—	—	50,000 –	105.2 –	Copious
5. Railway axles	200–325	1.4–2.2	–1,000	–304.8	200,000 ±	420.4 ±	Oil-pad
6. Steam turbines	50–100	0.34–0.69	–10,000	–3,048	500,000 –	1,052.7 –	Forced, cooled
7. Engine stationary	150–250	1.03–1.7	200–500	76.2–152.4	80,000 –	168.4 –	Rapid drop or forced
8. Engine shafts, marine	250–400	1.7–2.8	300–600	91.4–182.9	120,000 –	252.7 –	Rapid drop or forced
9. Engine shafts, combustion	500–1000	3.4–6.9	—	—	100,000 –	210.5 –	Rapid drop or forced
10. Crank, stationary	400–1500	2.8–10.3	100–500	30.5–152.4	200,000 ±	420.3 ±	Rapid drop or forced
11. Crank pins, marine	400–500	2.8–3.4	300–600	91.4–182.9	200,000 ±	420.3 ±	Rapid drop or forced
12. Crank pins, locomotive	900–1500	6.2–10.3	300–500	91.4–152.4	300,000 ±	630.4 ±	Scanty, oil-cup
13. Crank pins, combustion	1,500–2000	10.3–13.8	—	—	200,000 ±	420.3 ±	Forced
14. Wrist pins	1,000–3000	6.9–20.7	Small	Small	—	—	—
15. Punch or press	3,000–4000	20.7–27.6	Small	Small	—	—	—
16. Thrust bearings	50–200	0.34–1.4	200–500	60.9–152.4	10,000	20.7 –	Weak

Note: A number such as –50 or –200, under p or V, means up to 50 or 200; 10,000 – under q means less than 10,000, setting a maximum; 200,000 ± sets a mean.

Lines 1 to 6 and 16 are cases of loads steady in direction and amount; the rest are pulsing or intermittent loads. In partial lubrication the latter are supposed to give better action. In combustion engines, and more in punches or presses, the very heavy pressures act but a short time (too short for the oil film to be squeezed out), and during the light-load periods the film is renewed. With irregular loads quantity q should be based on average pressures, not on the maximum; in most cases this is only roughly approximate.

56. Complete Lubrication. Prerequisites to the establishment and maintenance of a complete oil film are as follows:

1. A sufficient clearance or difference of diameter between journal and bearing (the amount of clearance has been indicated in Sec. 43); on the average it will be about 1-mil-per-in (0.025-mm) diam. at 2.5-in (63.5-mm) diam. and about 0.5 mil per in (0.13 mm) at 10-in (254-mm) diam.
2. The surfaces must be very true to geometrical form, whether cylinder or plane; the range of high and low should not exceed a few tenths of a mil, which means high-grade and costly crafting.
3. There must be a copious supply of oil, of sufficient viscosity at the highest temperature at which the bearing will run.

The shape of film space to which the two members of the pair adjust themselves is shown in Fig. 13.108, with clearance tremendously exaggerated. The bearing is only a half-cylinder, above in Fig. 13.108*a* and below in Fig. 13.108*b*. The place *D* of closest approach is at a considerable angle (here 40°) beyond the middle point *C*, which is on the line of (vertical) loading. The eccentricity *OQ* is an important dimension in the analysis of film action. The shape of the film space is shown further by the developed section in Fig. 13.108*c*, with the same exaggeration of clearance; the bearing is carried to a full circle, but in any case its active portion lies within range *ACB*.

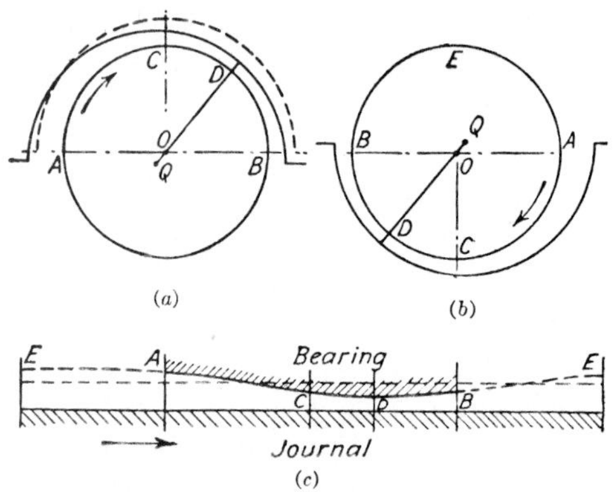

FIGURE 13.108 Action of oil film.

The variation of the oil pressure is shown in Fig. 13.109, first in a radial plot on the half-circle *ACB*, then on diameter *AB*; at points related by the vertical *FF*, ordinates *FH* are the same in the two plots. Roughly, the maximum pressure is about 1.5 times the average across the diameter.

If complete lubrication is intended and the needed copious supply of oil is provided, it is essential that there be no break in the bearing surface on the loaded side, as by an oil groove. To cut oil grooves is one way to ensure that the lubrication shall be only partial.

57. Bearing Surface and Material. If the conditions for complete lubrication (i.e., high accuracy of form and a smooth finish, with a plentiful supply of oil) can be provided initially and then surely maintained, the material of the bearing will be a matter of indifference. It is when such lubrication cannot be expected or maintained that special bearing metals of the brass-bronze and of the white-metal groups are required.

Hard steel on hard steel is effective where velocities are low with weak lubrication, as in the joints of a light linkage, such as an engine valve gear.

Cast iron may be used as a bearing for o. dinary steel shafting. It has been so used by a few makers of transmission machinery and is frequently found in light and cheap machinery. With a properly finished surface, smooth and burnished, it works well; but if a failure of lubrication starts abrasion, it will proceed rapidly, affecting journal as well as bearing.

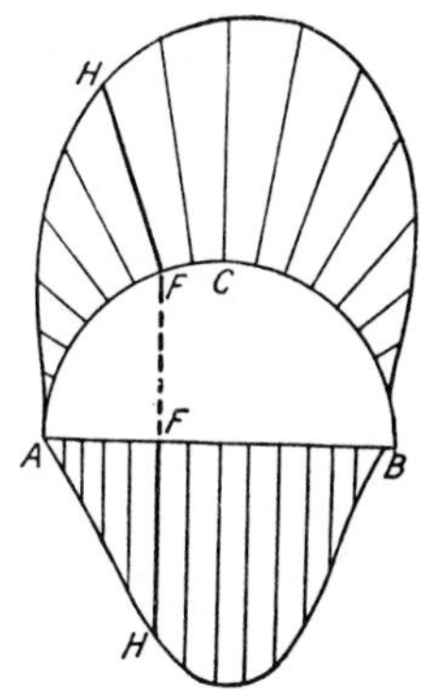

FIGURE 13.109 Pressure in oil film.

The worst combination is soft steel on soft steel; there is a strong tendency to start abrasion or "cutting," which will be rapid and destructive.

The advantage of the bearing metals is that they wear readily to a smooth surface and accurate form, and tend to have a low coefficient of greased friction; and if lubrication fails and the bearing overheats and cuts, damage will be largely confined to the bearing, much easier to replace than a spoiled journal.

Brass and bronze are used for the higher pressures, the white metals for lower pressures, with division at from 300 to 400 lb/in^2 (2.1 to 2.8 MPa) of pressure p. Also, a compound surface is sometimes used, with round or rectangular insets of babbitt in a brass "box."

The better grades of babbitt metal run from 80 to 90 percent tin, with copper and antimony for the rest, and are high priced. To some extent, zinc can replace tin, making the alloy harder but more brittle. Softer alloys, more readily fusible but good for lighter pressures, are made by using increasing proportions of lead. Generally the bearing metals are bought ready made, coming under trade names.

One type not yet mentioned is the *oilless bearing*, for light service as a matter of convenience and where possible dripping or smearing of oil might do damage, as in certain textile machines. One scheme is to impregnate wood with some hard waxy compound. Another is to press graphite into grooves in a bronze bushing. Yet another is to compress a mixture of graphite and powdered bronze into a bushing of adequate strength, under very high pressure.

Flake or powdered graphite, mixed with oil, makes a very effective lubricant under certain conditions.

58. Roller and Ball Bearings. Concerning these low-resistance bearings a large amount of detailed and quantitative information is available in the technical literature issued by the manufacturers. Regularly, these bearings are made up and purchased as assembled units, ready to be installed in the machine. Questions of load capacity and of mounting should in any case be taken up with the makers. In general, the mounting should support but not distort, permit and maintain alignment, retain lubricant, and exclude dirt. Lubrication is needed because of the slight friction of rolling, the incidental friction of guiding, and to prevent rusting.

In mechanical history the roller bearing is much older than the ball bearing; and it is much the more likely to be built for the particular job and in a general

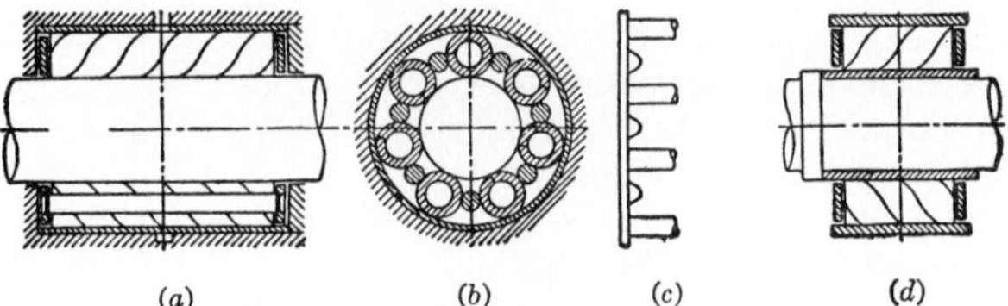

FIGURE 13.110 Hyatt bearing with flexible rollers.

shop, especially in the larger sizes, as for a crane footing or a turntable. Regularly the rollers are closely guided, whether they are cylindrical for a radial-thrust bearing or conical for end thrust; the guiding medium is some form of cage, which may become a fairly intricate structure of rings and rods. Of the small ready-made class the most prominent is probably the *Hyatt bearing* (Fig. 13.110). Note the flexible helical roller coiled from a rectangular bar of steel, the cage that guides the rollers by contact rods between them and riveted into the rings, and the type of race or track for the rollers. In Fig. 13.110*a* only an outer race is provided, as for insertion into a cast-iron frame, while Fig. 13.110*d* shows both inner and outer races; these bushings are so light that maintenance of true form is closely dependent upon the parts into or onto which they are lightly pressed or driven. The cage detail in Fig. 13.110*c* shows small teats pressed out from the rings, to hold the rollers in assembly.

Several forms of radial-thrust ball bearings are shown in Fig. 13.111, with the type shown in Fig. 13.111*a* as the typical design for heavy service. The groove in the race is only a little larger than the ball in its arc profile (one proportion being 1.04:1.00 in radii), so that with elastic compression under heavy load there may be line contact over an arc of 30 to 40°. The similarly proportioned *Hoffman roller bearing* in Fig. 13.111*b* is an evolution from the ball bearing rather than from the older form with long rollers; here the rollers are guided chiefly by their flat ends, and very little by the spacing cage. The double bearing shown in Fig. 13.111*c* is used for heavy service where alignment is strong and dependable; because of possible unequalization of load it is rated at only about 1.8 times the carrying capacity of the single row of balls. The self-aligning form shown in Fig. 13.111*d* is good for lighter service in machines that may lack rigidity; note how it has gone to spot contact on the outer race.

Figure 13.112 illustrates the way the balls are put into a bearing which has solid deep-grooved races. The maximum number of balls fills a little more than half the circle when in close contact as in Fig. 13.112*a*, then they are spread out as in Fig. 13.112*b* and held by a spacer with light guiding contact.

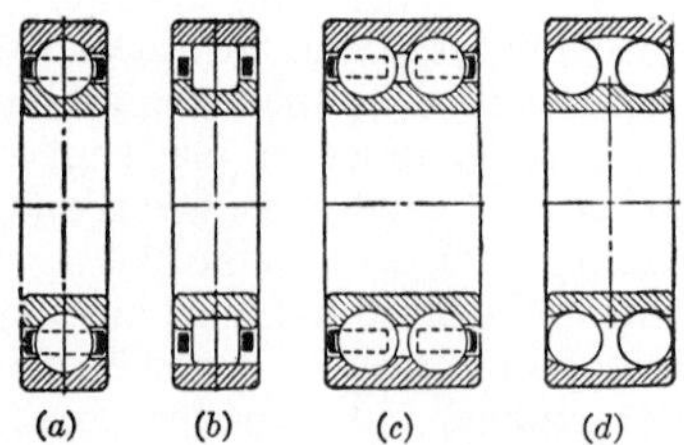

FIGURE 13.111 Various radial bearings.

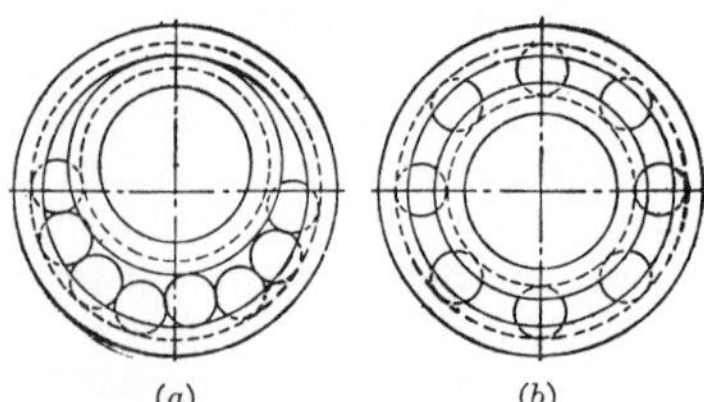

FIGURE 13.112 Filling a bearing.

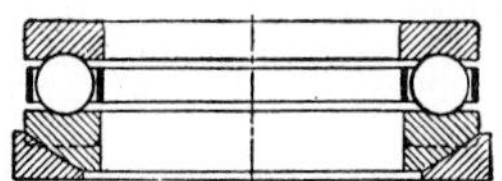

FIGURE 13.113 The round-groove thrust bearing.

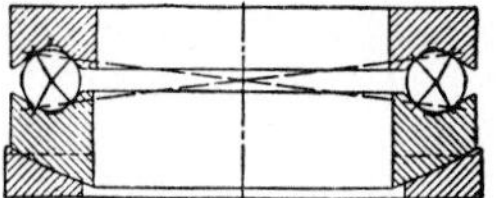

FIGURE 13.114 The V-groove thrust bearing.

Any of the radial bearings in Fig. 13.111 will hold against incidental end thrust, of the kind for which collars are used with common bearings as in Fig. 13.102. For heavy axial load the separate thrust bearing as in Fig. 13.113 or 13.114 is usual; the first example has round grooves, the second has V grooves so formed as to give true conic rolling. Note the spherical seating for one race in each case, to ensure equalization of pressure.

Oblique bearings are also used, regularly in pairs. Figure 13.115*a* shows the old cone-and-cup bearing, from the bicycle and the early automobile. Figure 13.115*b* is a current design, with smaller axial component in the average direction of contact pressure. The complete (double) *Timken roller bearing* in Fig. 13.116 has oblique conic rolling with plenty of line contact; to ensure such rolling, the two race-surface cones have a common apex. The rollers are held in alignment by a closely fitting cage, but all end thrust is taken by shoulder guiding on the inner race.

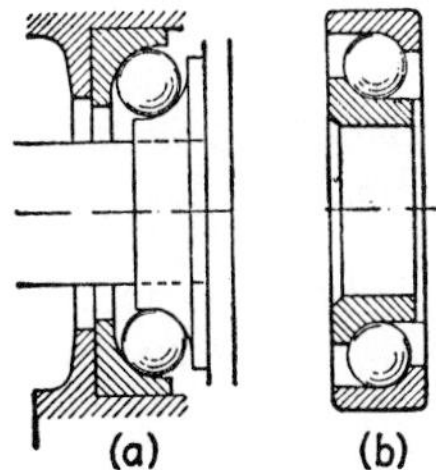

FIGURE 13.115 Thrust bearings.

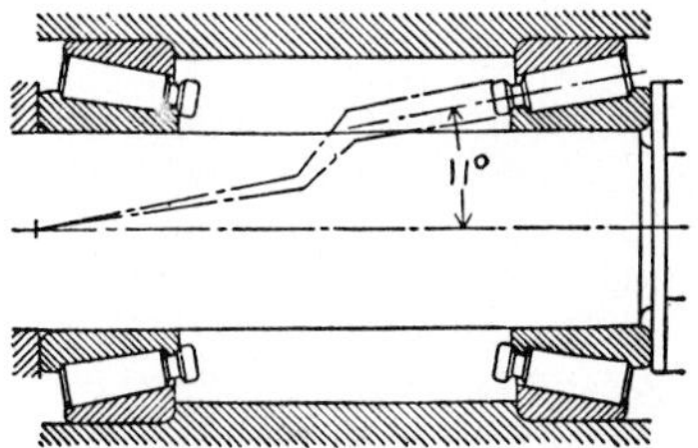

FIGURE 13.116 Complete Timken bearing.

In the matter of frictional resistance the ball bearing is likely to run a little better than the roller bearing. This resistance is made up of rolling friction proper and of secondary frictions due to guiding and to possible small misalignments. Reducing combined resistances to an equivalent friction tangent to the shaft, coefficient μ may run as low as 0.0002, as against a minimum of 0.005 for a journal bearing with complete lubrication and oil of little excess viscosity.

As compared with the lowest friction that a journal bearing can show under the best possible conditions, the ball or roller bearing has not a sufficient advantage to justify its much higher cost. But the ball or roller bearing ruggedly holds to its low resistance unless badly abused, while complete lubrication of a journal bearing is a sensitive condition, difficult to attain and to maintain in the general run of machines.

59. Load Capacity of Bearings. The crushing load on a ball is proportional to the square of its diam. d; and the working load P is either

$$P = cd^2 \qquad \text{or} \qquad P = D\,cd^2 \tag{13.24}$$

With race diam. D in fairly constant ratio to ball diam. d, the second expression is equivalent to the first. In $P = cd^2$, coefficient c will range from less than 1000 to more than 5000, depending upon material, accuracy of production methods, kind of contact (ranging from a flat race to a closely curved groove), and speed. As to speed, one rule makes P vary inversely as the cube root of the r/min or of N.

In a radial bearing, the question arises as to the number of balls, of the total number n, which are active in carrying the load. As angle θ from the line of direct thrust is greater, the ball receives less elastic compression. Hence its pressure or reaction P is less, and of this P, only a component is effective against load Q. It can be shown that this component pressure is proportional to the 5/2 power of cos θ. The effective number n_s of balls under full central load P is approximately

$$n_s = 0.23n \tag{13.25}$$

commonly assumed as $0.2n$; this gives

$$Q = \frac{Pn}{5} \tag{13.26}$$

for the total load Q in terms of safe ball load P.

The most used type of formula for roller bearings is

$$Q = kdln \tag{13.27}$$

with l for length and d, n, and Q as before. This probably underrates the influence of diam. d, which should have an exponent ranging from 1.3 to 1.4.

For definite load ratings, see makers' bulletins.

Balls and rollers and their races are made of special alloy steels, with chromium as the ingredient most effective in providing the needed combination of hardness and tough strength; requirements are most severe in the case of ball material. Very high precision is desired and attained, with such a tolerance as ± 0.1 mil or even ± 0.05 mil on diameter and sphericity.

60. Gear Teeth. In Fig. 13.7, Sec. 4, pairs of gears have been represented by their pitch circles, which are the profiles of kinematically equivalent friction wheels, assumed to roll upon each other without slip. To make this relative motion positive and to give capacity for transmitting force and work, teeth are formed by cutting into and building out from these cylinders. The primary requirement imposed upon tooth profiles is that their motion upon each other shall produce and ensure perfect rolling of the pitch circles. If the teeth are inaccurately spaced or are of incorrect profile, the driven gear or follower will receive an irregular motion and the pair will run noisily, especially at high speed.

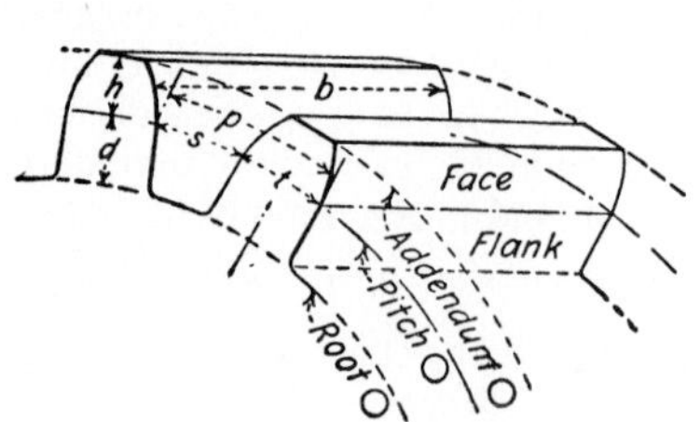

FIGURE 13.117 Parts and dimensions of teeth.

The parts and dimensions of gear teeth are shown in Fig. 13.117. The primary dimension is *circular pitch* p, the length of pitch-circle arc from a point on one tooth to the like point on the next. Strictly, pitch is measured or is made actual as an angle, since it is the function of the indexing mechanism of any gear-cutting machine to divide 360° of angle into any desired integral number of equal parts.

The size of teeth is regularly designated, however, not by the circular pitch p but by the *diametral pitch* p_d, which is the number of teeth per inch (mm or cm) of diameter. With pitch-circle diam. D and the number n of teeth in the wheel, the two pitches are

$$p = \frac{\pi D}{n} \tag{13.28a}$$

$$p_d = \frac{n}{D} \tag{13.28b}$$

whence $pp_d = \pi$. Being the number of teeth to 1-in (25.4-mm) diam., p_d is the same number per π in (mm) of circumference. As p_d is larger, the teeth are smaller.

Under the SI metric system this designating ratio is inverted to the *module*, which is the number of millimeters of diameter per tooth, or $m = D/n$. Between our diametral pitch and the metric module the relation is

$$mp_d = 25.4 \tag{13.29}$$

this being the number of millimeters to the inch. Using the module as a convenient linear unit in our system, its value is simply $1/p_d$ in.

The other dimensions shown on Fig. 13.117, with their values in a long-established standard system of gear teeth, are as follows:

h = *addendum*, equal to the module or $1/p_d$

d = *dedendum* or root depth, $1.157/p_d$

t = tooth *thickness*, at pitch circle

s = width of tooth *space*, also at pitch circle

b = *breadth* of gear rim

As appears in Fig. 13.118, the total *working depth*, or the depth of engagement or of *mesh*, is $2h$. The excess of d over h, here one-twentieth of p, is the bottom clearance of the tooth space. For accurately cut gears, space width s exceeds thickness t by a few mils/in (mm per mm) of pitch, or by just enough for an easy fit. With rough cast teeth the difference may be as much as one-tenth of the pitch. Such looseness or side clearance, which lets one gear swing back and forth while the other is held rigid, is called *backlash*.

61. Tooth Profiles. The once largely used but now nearly obsolete system of *cycloidal profiles* is represented by Fig. 13.118. The geometrical system that produces these curves adds the two auxiliary *describing* circles FF and GG, which roll with the pitch circles in common tangency at pitch point C. In this motion a particular point on either auxiliary circle "describes" a curve on the moving plane of each gear; and since the two curves (face of one tooth and flank of other) are traced simul-

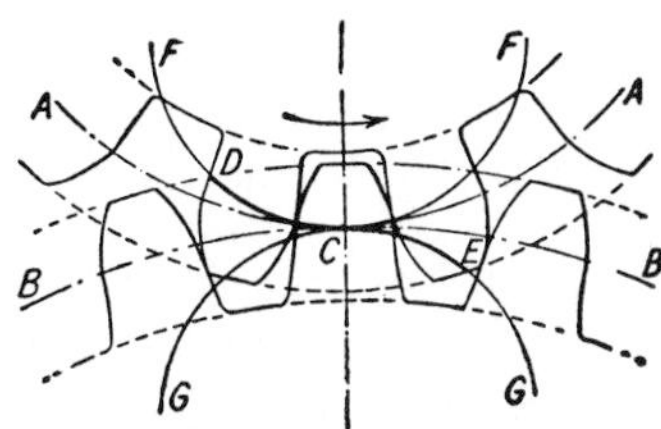

FIGURE 13.118 Tooth profiles.

taneously and in tangency at any instantaneous position of the describing point, the profiles work together in proper fashion for true pitch-circle rolling.

With gear A as driver and motion as indicated by the arrow, compound curve DCE is the *locus of contact* of the profiles of a pair of teeth as these pass through the range of contact action. This locus is limited by the addendum circle of the follower at D and by that of the driver at E. The range of contact must at least exceed one pitch; and with large tooth numbers it will go above two pitches, so that two pairs of teeth will always be in contact.

The now almost universally used *involute* system is presented by Fig. 13.119.

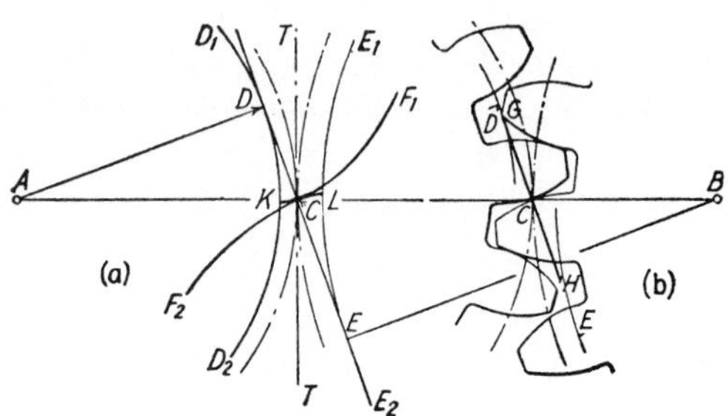

FIGURE 13.119 The layout of involute profiles.

The pitch circles, through point C, no longer enter directly into profile generation. First a line DE is drawn through the pitch point at an angle of obliquity from common tangent TT. Then the *base circles* D_1D_2 and E_1E_2 are drawn tangent to this line. As the pitch circles roll upon each other, line DE will roll upon the two base circles, also without slip; and a point of this line will trace simultaneously a pair of involutes, one to each base circle and on the moving plane of its gear.

In the displaced drawing shown in Fig. 13.119*b* these profiles are applied to teeth. Describing line DE is now the locus of tooth contact, with its active length limited at G and H by the respective addendum circles, as in the case of cycloidal profiles. It is important that limit points G and H shall lie inside of tangency points D and E; otherwise there will be a troublesome *interference* of profiles.

The old standard angle of obliquity is about 14.5°, exactly $\sin^{-1} 0.25$. With this angle, interference would occur at low tooth numbers, if not obviated by a small distortion from the exact involute which does no harm at low speeds. In Fig. 13.119 the obliquity is increased to 20° and the smaller gear, with 15 teeth, just keeps G inside of D.

In the Fellows system of *stub teeth*, with 20° obliquity, the teeth are made shorter by cutting down the addendum. These teeth have a double designation: thus, 4/5, 5/7, 6/8, 7/9, 8/10, 9/11, 10/12, 12/14. Here the first number is the actual diametral pitch of the tooth; the second is the pitch whose reciprocal is the addendum height h. Diminishing the addendum will shorten the range CG or CH in Fig. 13.119.

In Fig. 13.119 the involute curves run down to the base circles; then, below that, the profiles are radial straight lines down to the fillets that connect them to the root circle. This is the form used for milling cutters, but the tooth-generating methods produce curves that make the tooth root wider and stronger.

62. Kinds of Gears. Toothed gears may connect shafts with parallel axes, with axes that intersect, or with axes that cross but do not intersect.

With axes parallel, the wheels are called *spur gears*. The teeth on a gear may be straight (parallel to axis) as in Fig. 13.117, inclined or helical (after the manner of the gears in Fig. 13.123), or be double helical or of "herringbone" pattern. Helical teeth give smoother running at high speed, since all portions of the pair of profiles are in contact simultaneously, at successive cross sections; under this condition the minute imperfections that are unavoidable even by the best production process have least effect. With a double rim, or with two rows of teeth of opposite slant, end thrust is avoided.

When the axes intersect, the wheels become *bevel gears*, as in Figs. 13.120 and 13.121. Now the pitch surfaces are cones instead of cylinders and all tooth surfaces are conic, with the elements of straight teeth converging to the intersection of the axes; slant teeth can be used on bevel as well as on spur gears. The axes may be at right angles (Fig. 13.120) or at any oblique angle (Fig. 13.121).

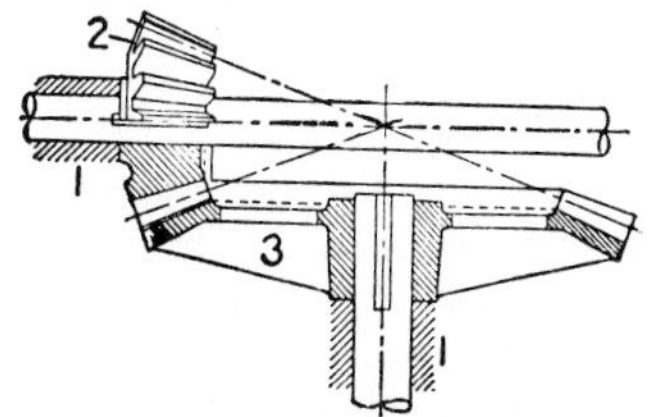

FIGURE 13.120 Bevel gear and pinion.

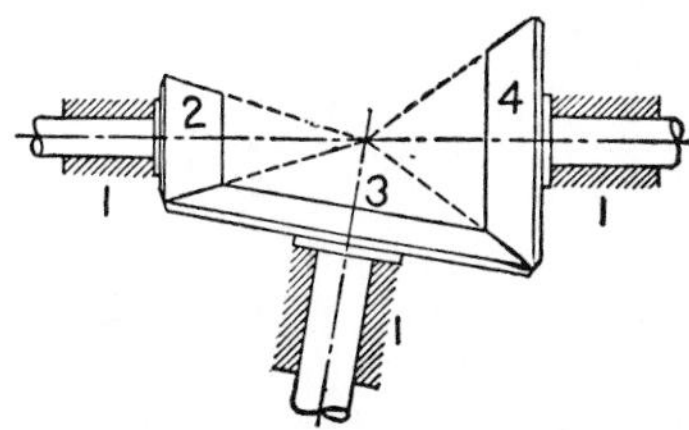

FIGURE 13.121 Gears with oblique axes.

With axes that cross, the *worm* and *worm wheel* (Fig. 13.122), the pair of *helical* or *spiral gears* (Fig. 13.123), and the geometrically much more complex hyperboloid or "skew-bevel" pair are used. The worm is essentially a coarse-thread screw; and when made into a cutting tool called the *hob*, by gashing its thread so as to make teeth, it generates a mating gear that has line contact with its own teeth (or those of a like worm). The one shown in Fig. 13.122 is a single-thread screw, and the pair has a high ratio of turning speeds but also a large proportion of work loss in friction. The more recent high-angle worm pair, with a much lower turning ratio, is more efficient in power transmission.

The *helical pair* in Fig. 13.123 is simple and convenient, but has the serious

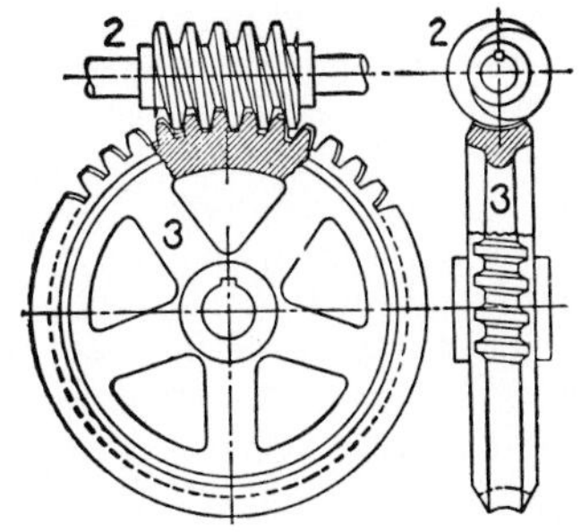

FIGURE 13.122 Worm and worm wheel.

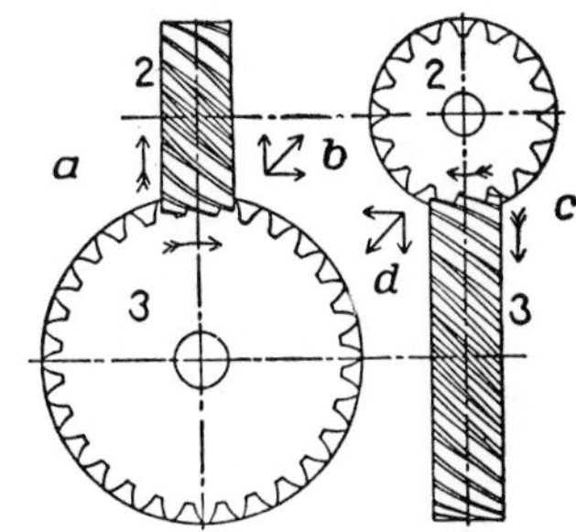

FIGURE 13.123 Helical or spiral gears.

disadvantage of only spot contact between the teeth; consequently, excessive wear can be avoided only by holding it down to light loads. Little diagrams *b* and *d* show a scheme for readily seeing the direction of relative turning. As to speed ratio, these gears are much more flexible than spur gears, since tooth angle as well as diameters enter into the question of respective tooth numbers.

63. Tooth Production. One general method of producing accurate teeth is to use a *formed cutter*, in most cases a milling cutter as in Fig. 13.124, but occasionally a planing tool. In an automatic gear-cutting machine the cutter is set to full depth of tooth, fed across to cut the groove or tooth space, and brought back; then during a pause the machine "indexes" the gear blank through a pitch angle and starts over again. This method can be made to give correct and accurate teeth, if the cutter is made for the particular wheel; but in the production of interchangeable wheels, having the same pitch but a range of all sizes or tooth numbers, there must be approximation if this is to be done with a limited number of cutters. The Brown and Sharpe system of involute cutters is listed in Table 13.3, with eight regular shapes and seven extra ones. The cutter is made correct for its lowest tooth number, which has greatest width of space between tips of teeth; it cuts the space a little too wide at the top for higher numbers.

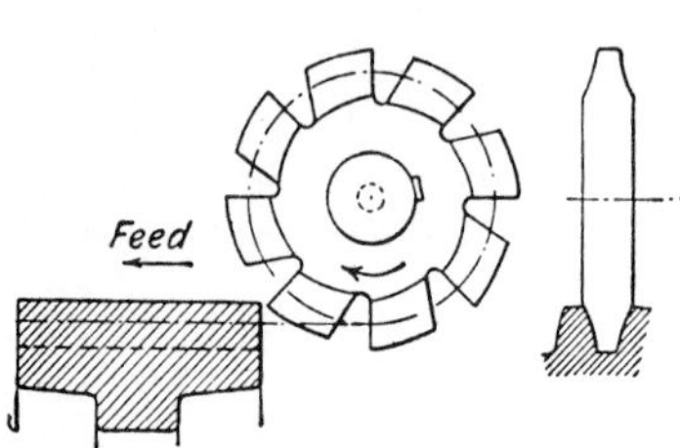

FIGURE 13.124 Milling teeth.

TABLE 13.3 Involute Tooth Cutters

Regular		Extra		Regular		Extra	
Number	Teeth	Number	Teeth	Number	Teeth	Number	Teeth
1	135–∞	1.5	80–134	5	21–25	5.5	19–20
2	55–134	2.5	42– 54	6	17–20	6.5	15–16
3	35–54	3.5	30– 34	7	14–16	7.5	13
4	26–34	4.5	23– 25	8	12–13		

The way in which a first wheel, of correct tooth profile and made into a cutting tool, can generate teeth on a mating blank has been suggested in the description of the worm-and-wheel pair. Figure 13.125 shows the hob in position to begin cutting, hob and blank being connected by an external gear train so that they turn together in precisely correct relative motion. The gear blank is fed slowly toward the hob (a very little for each revolution of the blank) until the desired depth of tooth is reached.

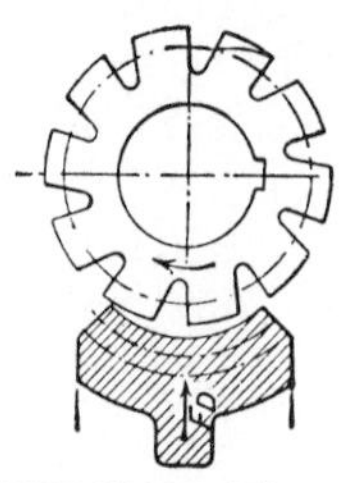

FIGURE 13.125 The worm.

One scheme for generating spur-gear teeth (Fellows') is to use a cutter like a correctly profiled wheel, giving it a planing motion in the direction of its own axis. At first the cutter is fed toward the blank somewhat as in Fig. 13.125, but with neither turning until full depth is reached; then cutter and blank are started on a slow relative turning (feed motion), which continues until all of the teeth have been formed.

A very important practical fact about the involute system is that its rack tooth has straight-line profiles, as shown in Fig. 13.126. This rack can be embodied readily into an accurately formed cutting tool for planing motion or into a hob for continuous cutting. A planing tool of rack form must necessarily be short; hence the generating motion must be intermittent. The blank and tool are rolled upon each other, back and forth through an angle of perhaps six or seven pitches, then the blank is indexed and the generating cycle repeated. The hob, set at such an angle that its helix fits into the desired direction of tooth space, is started as in Fig. 13.124 and fed slowly across the blank as the two turn together in continuous action.

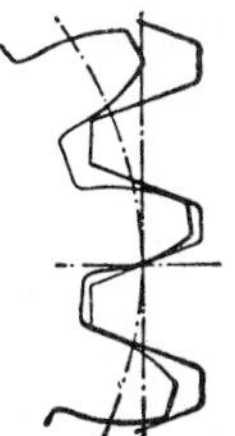

FIGURE 13.126 Rack generation.

The general method of rolling gear on rack is used also for generating bevel-gear teeth. Here the rack has become a gear with a cone angle of 180° (a flat circle), called a *crown gear*. The plane sides of one of its teeth are described by the cutting edges of two planing tools, moving with the necessary convergence.

For precise finish, teeth are ground, especially on automobile gears which may have been distorted slightly by heat treatment after cutting. One method is to use a profiled grinding wheel, similar to the milling cutter in Fig. 13.124, dressed at short intervals by a guided diamond tool to keep it true to form. Another scheme uses a wheel of straight, rack-tooth profile and the same kind of generating motion as for gear cutting.

64. Strength of Gear Teeth. Figure 13.127 indicates that the gear tooth functions as a short cantilever beam; the thrust and moment of pressure P are balanced by a combination of shearing and flexural stress at section BB. Total P may come on one tooth at its tip, especially if this tooth is on a small pinion; and in that location the tooth will have least width at root BB. Assuming different sizes of teeth to be similar in form, dimensions will be proportional to pitch p. Then with a and c as numerical coefficients, b for breadth of gear face as in Fig. 13.117, and S for the stress due to bending, the relation of load moment to resisting moment is

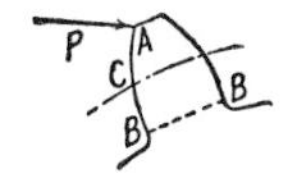

FIGURE 13.127

$$aPp = cSbp^2 \qquad \text{whence } P = kSbp \tag{13.30}$$

with k as a comprehensive constant or factor.

The much used *Lewis formula*, based on Eq. (13.30) and on a study of varying tooth proportions, uses two equations for k, namely,

$$k = 0.124 - \frac{0.684}{n} \qquad \text{(14.5° obliquity)} \tag{13.31a}$$

$$k = 0.154 - \frac{0.912}{n} \qquad \text{(20° obliquity)} \tag{13.31b}$$

These take account of the decrease of root width as tooth number n is smaller, holding down to $n = 12$. For stub teeth, k would be larger, but has not been formulated.

For pitch-circle velocities not greater than 100 ft/min (30.5 m/min), values of nominal stress S to be used in Eq. (13.30) are

Cast iron: 8000 lb/in^2 (55.2 MPa)

Cast steel: 20,000 lb/in^2 (137.8 MPa)

Forged machinery steel: 25,000 lb/in^2 (172.4 MPa)

Alloy steel, up to: 40,000 lb/in^2 (275.8 MPa)

With rising velocity, stress S diminishes, rapidly at first and then more slowly, falling to about one-fourth of the preceding values at $V = 2000$ ft/min (609.6 m/min).

Breadth b is commonly around $3p$, except on high-precision helical-tooth gears for high-speed transmissions, in which special precautions are taken to keep the axes truly parallel.

Besides the ferrous metals and brass or bronze, gears are made of several fibrous materials, such as rawhide, impregnated cloth, and several patented "compositions." Piled-up disks of leather or cloth are clamped between steel disks under heavy pressure and held by rivets, then teeth are cut in the usual manner. Regularly a steel pinion is mated with a soft gear, the purpose being to obtain quiet running at high speed with moderate pressures.

In the high-efficiency low-ratio worm pair, the worm is of steel and the wheel of hard phosphor bronze; here, slip velocities are much greater than in any other type of heavily loaded gear, and the same considerations prevail as in a journal-and-bearing pair.

All kinds of gears can be bought ready-made from a number of specializing manufacturers. Anyone not an experienced machine designer should consult with such a manufacturer as to gears for a proposed job.

POWER ENGINEERING*

PUMPS

1. Definitions. A *pump* is a machine or device for raising a liquid, which is a relatively incompressible fluid, to a higher level or to a higher pressure. A *compressor* is a machine or device for raising a gas, which is a compressible fluid, to a higher pressure. However, devices for exhausting air from closed vessels are called *air pumps*, though in reality they are air compressors working below atmospheric pressure.

A *blower*, as distinguished from a compressor, compresses a gas to a comparatively low pressure only. A *fan* is intended primarily to move large volumes of gas; the pressure developed by the fan is quite small and is secondary in importance.

*From *General Engineering Handbook*, 2d ed., by C. E. O'Rourke. Copyright © 1940. Used by permission of McGraw-Hill, Inc. All rights reserved. Updated and metricated by the editors.

2. Measurement of Head. The head which a pump has to develop or work against is the static lift plus all the friction losses in the piping. This value may be computed, but in actual operation it would be determined in a test by measuring the pressures in the piping adjacent to the pump on both suction and discharge sides. Let h = total head in ft (m), p = pressure expressed in ft (m) of the liquid, z = elevation of the center of the discharge gage above the point at which the suction pressure is measured, V = velocity in ft/s (m/s) at the section where the gage is attached, g = acceleration of gravity in ft/s^2 (m/s^2), the subscript d denotes discharge, and the subscript s denotes suction values. Then

$$h = p_d - p_s + z + \frac{V_d^2}{2g} - \frac{V_s^2}{2g} \tag{13.32}$$

If the pressure on the intake side is below atmospheric, and if gage pressures are used in the above equation, then p_s will be negative.

3. Power. If q = rate of discharge in ft^3/s (m^3/s), G = gal/min (L/min), w = density of the liquid in lb/ft^3 (kg/m^3), the horsepower (W) delivered in the liquid, called *water horsepower*, is: Water power = $wqh/550$ hp (1 hp = 0.75 kW). In the case of water of the customary density of 62.4 lb/ft^3 (8.0 kg/m^3), this may be reduced to

$$\text{Water power} = \frac{qh}{8.81} = \frac{Gh}{3960} \text{ hp (1 hp = 0.75 kW)} \tag{13.33}$$

For any other liquid of specific gravity s, the two expressions in the above equation, and in the one below, should be multiplied by s. If e is the overall efficiency of the pump, then the power input to the pump, often called *brake horsepower*, is

$$\text{Brake power} = \frac{qh}{e \times 8.81} = \frac{Gh}{e \times 3960} \text{ hp(1 hp = 0.75 kW)} \tag{13.34}$$

4. Efficiencies Defined. *Efficiency*, sometimes called *total* or *overall efficiency*, is the ratio of the power delivered in the liquid to the power input to the pump. That is

$$e = \frac{\text{water hp}}{\text{brake hp}} = \frac{\text{water kW}}{\text{brake kW}} \tag{13.35}$$

Hydraulic efficiency, e_h, is the ratio of the power actually delivered *in* the water to the power expended *on* the water or other liquid. These two quantities differ by the amount of the hydraulic-friction losses.

Mechanical efficiency, e_m, is the ratio of the power expended *on* the water to the power supplied to run the pump. These two differ by the amount of the mechanical-friction losses, such as friction of bearings, stuffing boxes, etc.

Volumetric efficiency, e_v, is the ratio of the amount of water actually delivered to that which would be delivered if there were no leakage losses, imperfect valve action, etc. *Slip*, in the case of a positive displacement pump, means the difference between the actual displacement and the volume of the fluid actually delivered, expressed as a percentage of the displacement. In certain combinations of a reciprocating pump and pipe line the inertia of the water causes flow to continue even while the pump is on dead center, and thus secures a discharge larger than the actual displacement volume. The relation between slip and volumetric efficiency is: slip = $100(1 - e_v)$.

The *total efficiency* is the product of the hydraulic, mechanical, and volumetric efficiencies, that is,

$$e = e_h \times e_m \times e_v \tag{13.36}$$

Duty is another means of expressing the efficiency of steam-driven pumping engines. It is usually expressed as the foot-pounds of work done per 1000 lb (J/454 kg) of steam supplied, but is more precisely defined as the foot-pounds of work done per million Btu (J/1.1 MJ) supplied.

5. Suction Lift. The theoretical suction lift may be computed as follows:

$$L = b - p_v - h_f - \frac{V_s^2}{2g} \tag{13.37}$$

where L = lift, b = barometer pressure in ft (m) of the liquid, p_v = vapor pressure of the liquid in ft (m), h_f = friction losses in foot valve, suction piping, etc., and V_s = velocity at intake of pump. For water it is desirable to maintain a pressure at least about 10 ft (3 m) more than the vapor pressure. Hence the maximum allowable lift is about 10 ft (3 m) less than given by the above equation. In practice the lift is usually about 20 ft (6 m) for cold water, decreasing as the water temperature increases; above 160°F (71°C), the water should be supplied under a positive pressure. For centrifugal pumps, see Sec. 42 below.

POSITIVE DISPLACEMENT PUMPS—RECIPROCATING TYPE

6. Direct-Acting Steam Pumps. The direct-acting steam-driven pump, such as shown in Fig. 13.128, is one of the simplest types of reciprocating pumps. If this consists of a steam cylinder and one water cylinder only, it is called a *simplex* pump. A pair of such pumps mounted together constitute a *duplex* pump. In the case of the latter, the movement of one piston operates the steam valve for the other one. In the simplex pump one steam piston is reduced to a small dummy for the sole purpose of operating the steam valve for the main cylinder, but its motion in turn is controlled by the movement of the main piston.

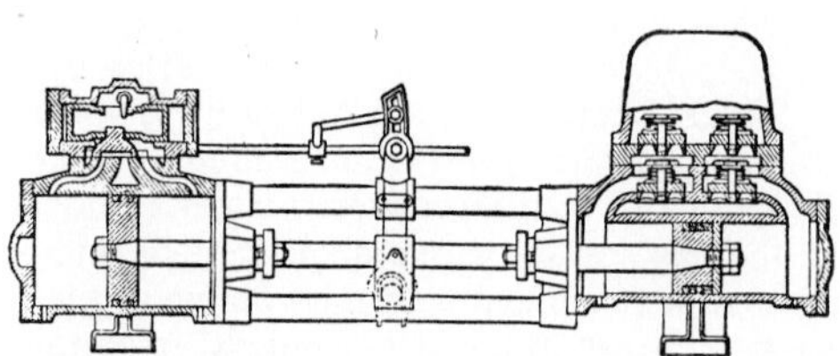

FIGURE 13.128 Single-cylinder steam pump, submerged-piston type.

Since the pressure on the water piston is substantially constant throughout the stroke, the pressure on the steam piston must also be constant, and hence the steam cannot be used expansively. For this reason direct-acting steam pumps are very uneconomical, requiring from 100 to 300 lb (45 to 136 kg) of steam per hp · h (0.75 kWh). They are therefore usually used only where the exhaust steam can be utilized for some heating purpose, or where economy is secondary to simplicity, ruggedness, and reliability.

If p_s and p_w denote steam and water pressures, respectively, while A_s and A_w denote the areas of the steam and water pistons, respectively, then

$$p_s A_s = m p_w A_w \tag{13.38}$$

where m is a factor greater than unity, in order to allow for friction losses, etc. For a boiler-feed pump, where the water pressure is but little greater than the steam pressure, m is given a value as high as 3 or 4 in order to provide an ample margin. In other cases it need be no larger than the reciprocal of the pump efficiency.

The dimensions of the water end may be computed from the equation

$$e_v d^2 L N' = 294G \tag{13.39}$$

where e_v is the volumetric efficiency which may range from 0.85 to 0.99, depending upon the condition of the pump; d is the diameter of the piston in in (mm); L the length of stroke in in (mm); N' the number of working strokes per min; and G is gal/min (L/min). If the pump is single-acting, then N' is the same as N, the number of "revolutions" per min. If it is double-acting, then $N' = 2N$. For accuracy, correction must be made for the area of the piston rod in the latter case. In general, the revolutions or cycles per min of a reciprocating pump should be less than 100 in large sizes and usually not more than 200 in small sizes. The piston speed, $LN/6$, should ordinarily be from 50 to 150 ft/min (15 to 46 m/min). Table 13.4 is a guide for average conditions. For minimum mechanical friction a large diameter and a short stroke should be employed. However, this means a rapid stroke and a greater number of cycles per min, attended with more slippage. Usually the length of stroke is slightly greater than the diameter.

In order to increase the economy, direct-acting steam pumps are sometimes compounded, having high- and low-pressure steam cylinders in tandem. Occasionally they are even made triple-expansion, with three steam cylinders in tandem.

TABLE 13.4 Piston Speeds for Direct-Acting Pumps

Stroke		General service		Boiler feed		High-pressure pumps	
ft/min	m/min	ft/min	m/min	ft/min	m/min	ft/min	m/min
3	7.6	28	8.5	18	5.5	20	0.1
4	10.2	33	10.1	22	6.7	24	0.3
5	12.7	38	11.6	24	7.3	26	0.9
6	15.2	40	12.2	26	7.9	29	0.8
8	20.3	47	14.3	30	9.1	33	0.1
10	25.4	58	17.7	38	11.6	42	12.8
12	30.5	60	18.3	39	11.9	43	13.1

7. Crank-and-Flywheel Pumps. In order to use steam expansively, it is necessary to use a flywheel or equivalent device so as to equalize between the steam and water cylinders. The addition of a flywheel necessitates a crank and connecting-rod mechanism, and thus a more complex construction than with the direct-acting type. Large triple-expansion pumping engines of the crank-and-flywheel type have given steam consumptions as low as about 10 lb (4.6 kg) of steam per hp · h (0.75 kW).

8. Power Pumps. A power pump is a positive displacement pump of the reciprocating type operated through cranks and connecting rods by power supplied to the crankshaft. The source of power is usually an electric motor which may be connected to the shaft through gears or by a belt and pulleys. The latter is more quiet.

The direct-acting steam pump is inherently a variable-speed pump. The power pump is inherently a constant-speed pump, and is thus not so well suited to a variable rate of discharge.

9. Pistons and Plungers. Reciprocating pumps may be provided with a piston, as in Fig. 13.129*a*, or a plunger, as in Fig. 13.129*b* and *c*. For the same displacement, the plunger pump requires a larger cylinder than the piston pump, but it is better for water carrying grit or other foreign substances. For high pressures the outside-packed plunger pump is best because the leakage can be detected and the packing tightened while the pump is running. Owing to the greater friction of the large stuffing boxes, the efficiency of outside-packed plunger pumps is about 5 percent less than that of other types.

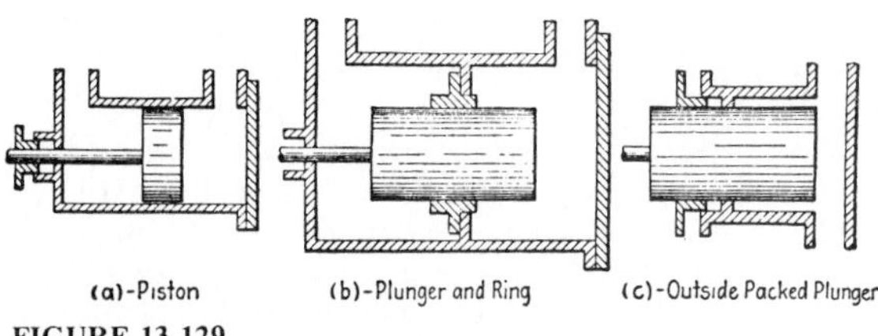

FIGURE 13.129

An outside-packed plunger pump is inherently single-acting, but it may be made double-acting by a duplication of cylinders, as shown in Fig. 13.130. Hence for a given capacity, this type of pump is the largest in size.

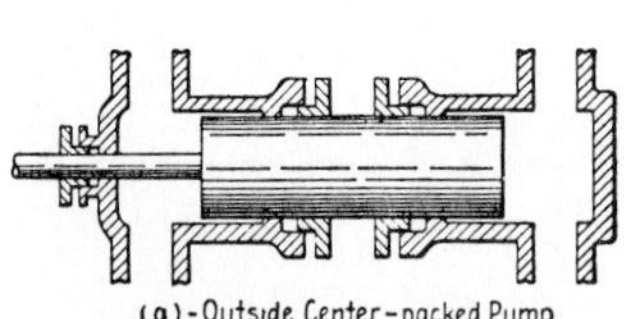

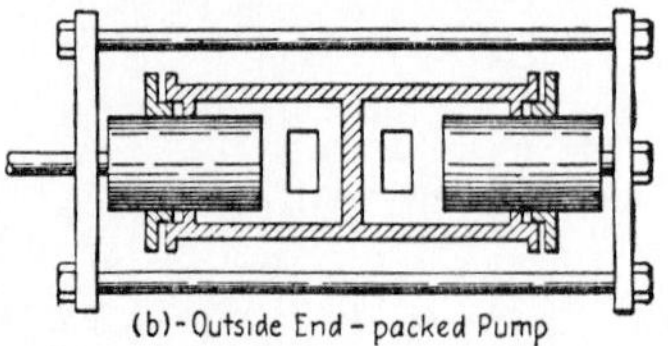

FIGURE 13.130

10. Bucket or Deep-Well Pumps. Certain vertical pumps may be of the bucket type, as shown in Fig. 13.131, in which case the suction valves are in the piston itself, which is then called a *bucket*. Reciprocating pumps in deep wells are necessarily of this type. Such pumps are single-acting.

11. Differential Pumps. By an enlargement of the piston rod to a size about half of that of the piston or plunger, a pump may be made to discharge on both strokes, while taking suction on one stroke only. The effect of this is to make the discharge more regular. Such an arrangement is shown in Fig. 13.132.

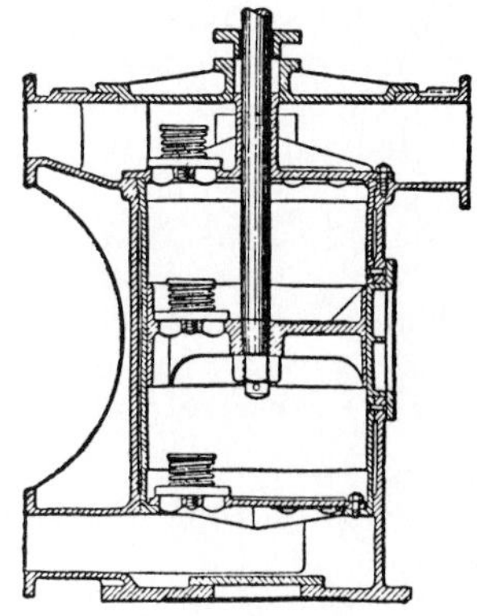

FIGURE 13.131 Bucket pump.

12. Pump Valves. Pump valves are usually of the disk type and range in diameter from 1½ to 9 in (38 to 229 mm), but about 4½ in (114 mm) is the usual practice. They are made of rubber or of composition according to the nature of the service. Soft rubber may be used for low pressures, medium-soft for pressures of about 75 lb/in^2 (517 kPa), medium-hard for pressures of about 150 lb · ft (1,034 kPa), and hard for higher pressures. Metal valve disks may also be used for the higher pressures. Special composition is necessary where hot water is to be handled. (See Fig. 13.133.)

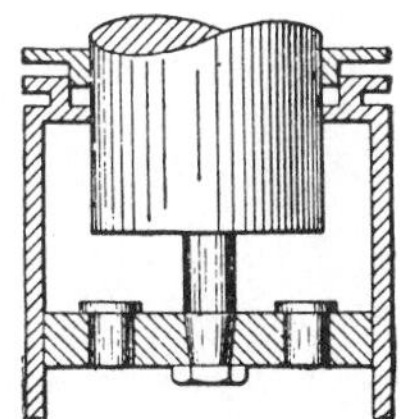

FIGURE 13.132 Differential bucket pump.

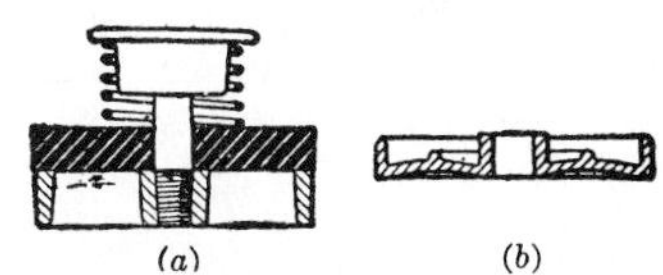

FIGURE 13.133 (*a*) Disk valve. (*b*) Metal valve disk.

Valves may also be leather-faced and may be of either the flat-disk type or the conical type, also known as *wing valves*. (See Fig. 13.134.) Metal conical wing valves without any facing are used for very high pressures because they can be ground to a tight seat. The disadvantage is that they must lift higher to give the same net opening as a flat valve.

The net valve area required is from 45 to 60 percent of the piston area, assuming the piston speed to be 100 ft/min (30 m/min). The resulting water velocity through the valve opening then varies from 22 to 167 ft/min (6.7 to 50.9 m/min).

An approximate rule for the spring pressure on disk valves in pounds per square inch (kPa) of inside seat area is that, for discharge valves, it should vary

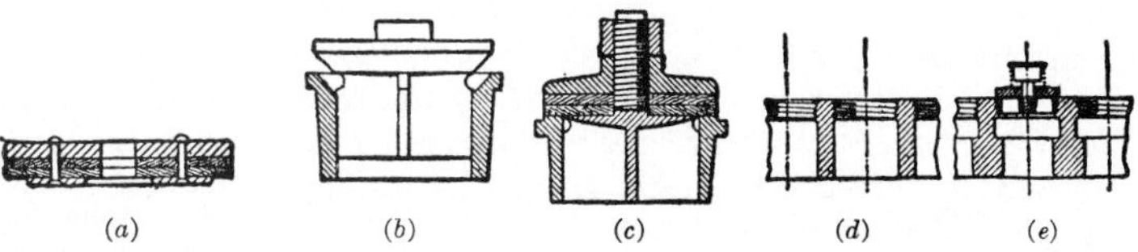

FIGURE 13.134 (*a*) Leather-faced valve. (*b*) Conical wing valve. (*c*) Leather-faced wing valve. (*d*) and (*e*) Valve decks.

from 0.005 to 0.01 × water pressure, with a maximum of 5 lb/in^2 (34.5 kPa). For suction valves under suction lift, it may be from 0.25 to 0.50 lb/in^2 (1.7 to 3.4 kPa) and under a positive pressure twice these values.

Ball valves are used for pumps handling liquids with foreign matter. The balls are usually of bronze or of rubber with metal cores.

Pump valves are almost invariably automatically operated, but, since such a valve lags in operation behind the piston or plunger, attempts have been made to produce a valve which will be closed mechanically at the exact end of the stroke. So far such designs have not been successful, with the possible exception of the Riedler valves used abroad.

13. Steam Valves. For the direct-acting pump, the steam valves are made without any lap and are of either the D or the B type, as shown in Fig. 13.135. With the duplex pump, the D valve is generally used. For large pumps the plain slide valve may be replaced by piston valves. In order to secure a desired pause at the end of the pump stroke, the valve is not rigidly connected to the mechanism, but instead a certain amount of lost motion is allowed.

The steam cylinders of direct-acting pumps are double-ported, in order that when the piston closes the exhaust port a certain amount of cushion steam will be trapped and bring the piston quietly to rest. The inlet port, on the other hand, should extend beyond the extreme end of the stroke, so that there can be no possibility of the pump failing to start the stroke.

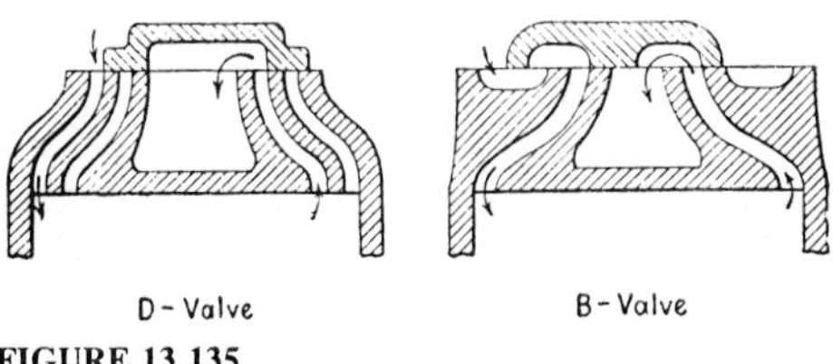

FIGURE 13.135

14. Air Chambers. The rate of discharge from a reciprocating pump continually varies, as may be seen in Fig. 13.136. This variation in the flow produces shocks and water hammer in the piping, unless air chambers are provided. While ample-sized air chambers are desirable, they may be dispensed with in the case of direct-acting duplex steam pumps, but are practically necessary for single-cylinder pumps and all crank-and-flywheel or power pumps. The size of the air chamber ranges from one to eight times the displacement volume of one plunger per stroke. The larger-sized air chambers are required for higher-speed pumps or for those with the fewer number of cylinders.

Air chambers are desirable on the suction side as well as on the discharge side of the pump. Although the inertia forces in the relatively short suction piping are much less in magnitude than on the discharge side, the effects may be quite as important so far as the smooth running of the pump is concerned. If it is desired to have the same percentage variation in pressure on both sides, the air chamber on the suction side should be of the same size as that on the discharge side.

Since water under pressure will absorb air in time, it is necessary that the air be replenished periodically in the discharge air chamber. The air may be provided

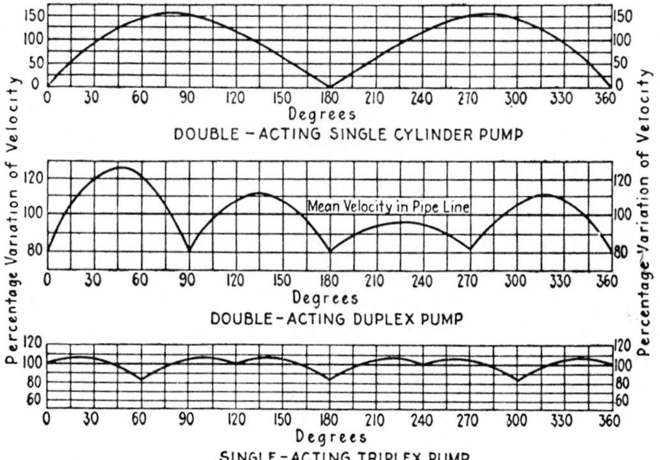

FIGURE 13.136 Variation of pump discharge.

by a small air compressor. A simple way is to admit small quantities of air into the suction piping, and this air will then tend to collect in the air chambers.

15. Efficiencies of Reciprocating Pumps. The overall efficiencies of reciprocating pumps, whether direct-acting or power pumps, range from about 45 percent to about 90 percent or more, depending upon the size and the condition of the pump. The higher values are usually found only in large pumps and the lower values in very small pumps. Crank-and-flywheel or power pumps are usually a few percent more efficient than direct-acting pumps of the same size. Table 13.5 indicates probable values for the efficiencies of direct-acting steam pumps in good operating condition.

16. Calculations for Reciprocating Pump. To obtain the essential dimensions of a reciprocating pump, proceed as follows: Assume values of piston speed and efficiency, being guided by the values in Tables 13.4 and 13.5. Also allow about 3 percent for slip; i.e., assume the volumetric efficiency to be 97 percent. Then the

TABLE 13.5 Efficiencies of Direct-Acting Steam Pumps in Percent

Stroke				High-pressure pumps	
in	cm	Piston to 250 lb/in² (1.7 MPa)	Plunger to 300 lb/in² (2.1 MPa)	To 1000 lb/in² (6.9 MPa)	Above 1000 lb/in² (6.9 MPa)
3	7.6	50	47	45	39
4	10.2	55	52	50	43
5	12.7	60	57	54	47
6	15.2	65	61	58	51
8	20.3	70	66	63	55
10	25.4	75	71	67	58
12	30.5	77	73	69	60

length of stroke and the diameter of piston may be determined so as to satisfy the equation $e_v d^2 LN' = 294G$ (see Sec. 6). The power required to drive the pump may be computed as:

$$\text{Brake power} = \frac{Ghs}{(e \times 3960)} \text{ hp}(1 \text{ hp} = 0.75 \text{ kW}) \qquad (13.40)$$

(see Sec. 3).

17. Characteristics of Reciprocating Pumps. A distinguishing characteristic of the positive displacement pump is that the capacity—neglecting slip—is equal to the displacement. Thus the capacity is directly proportional to the speed of the pump, and for a constant speed the capacity is constant. If a variable quantity is desired, it is necessary to vary the speed of the pump, or to simply bypass some of the fluid, which is wasteful of power.

The slip or leakage increases as the pressure on a given pump is increased. It tends to become less in percentage value as the pump speed is increased.

The head or pressure which the positive displacement pump will develop is determined by the resistance against which the pump works, and is not fixed by any property of the pump. Its maximum or limiting value is determined by the power available to drive the pump and the strength of the various parts. Obviously such a pump should be protected against excessive pressures by some form of relief valve. The positive displacement pump is of especial value, compared with the centrifugal pump, for small capacities, and especially for very high heads.

POSITIVE DISPLACEMENT PUMPS—ROTARY TYPE

18. Types of Rotary Pumps. The principal types of rotary pumps are shown in Fig. 13.137. These are the lobe type, eccentric type, and gear type. In each of them the fluid is trapped within the spaces bounded by the casing and the vanes, teeth, or lobes, and is delivered to the discharge side. Such pumps have no valves. Since the leakage loss is kept within a reasonable value by nothing more than the small clearance between the various parts, such pumps deteriorate rapidly with wear. Hence they should not be used for a liquid containing grit. They are of especial value for such liquids as oils, whose physical properties are such as to produce less leakage loss than water and to provide better lubrication to such rubbing surfaces as are in contact.

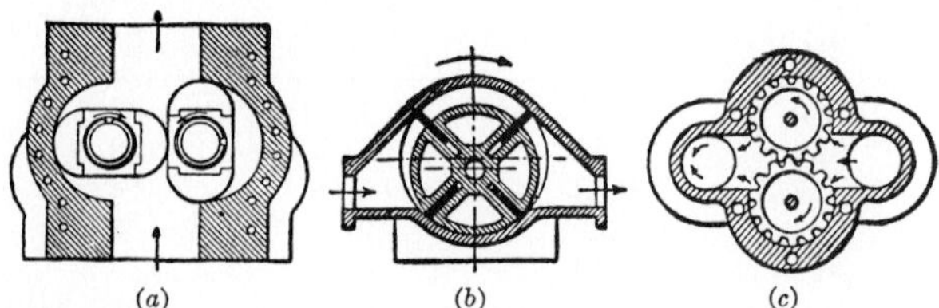

FIGURE 13.137 Rotary pumps. (*a*) Lobe type. (*b*) Eccentric type. (*c*) Gear-wheel pump.

The *lobe type* may be used for moderate rotative speeds. It produces a continuous but nonuniform rate of discharge and hence requires air chambers. Efficiencies as high as 85 percent are claimed for it. The *eccentric type* has blades which slide in and out and thus produce inertia forces due to their reciprocating action, which limits the speed of such pumps. The blades can never wear to a tight fit because of the varying curvature and, furthermore, they produce considerable friction loss. There is considerable slip around the ends of the blades and the rotor, which is hard to check satisfactorily. The *gear type* is best suited to high rotative speeds, but is not quite so efficient as the lobe type. In addition to these three there is a *twin-screw type*, which consists of two parallel screws within a casing in which the thread of one fits into that of the other.

19. Characteristics of Rotary Pumps. Rotary pumps, being positive displacement pumps, are very similar to reciprocating pumps in most respects. The principal advantage is the substitution of rotation for reciprocation; this permits higher speeds, with a reduction in the size of the pump for the same capacity. Rotary pumps tend to be inefficient, however, especially with wear, which permits the leakage loss to become excessive. Even when such pumps are new, they tend to suffer from too much slip, on the one hand, or too high a mechanical friction, on the other. While they may be used for heads up to several hundred feet, they are best adapted to moderate or low pressures.

CENTRIFUGAL PUMPS

20. Classifications. Centrifugal pumps may be divided into turbine pumps and volute pumps. In the former the impeller is surrounded by diffusion vanes which are so formed as to reduce the velocity of the water and efficiently convert the kinetic energy into pressure. In the latter the impeller is merely surrounded by a spiral casing, called the *volute*, which is also so designed as to convert efficiently the kinetic energy into pressure (see Figs. 13.138 and 13.139).

They may also be classified as single-stage and multistage, according to whether only one impeller is used, or whether there are more than one in series. In the latter case each stage adds to the pressure produced by the preceding

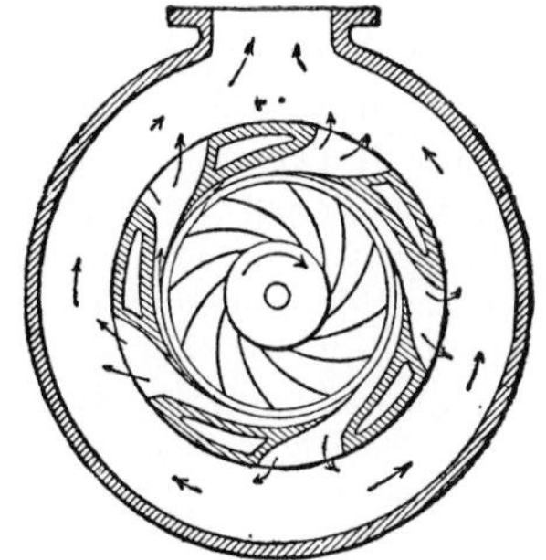

FIGURE 13.138 Turbine pump with circular case.

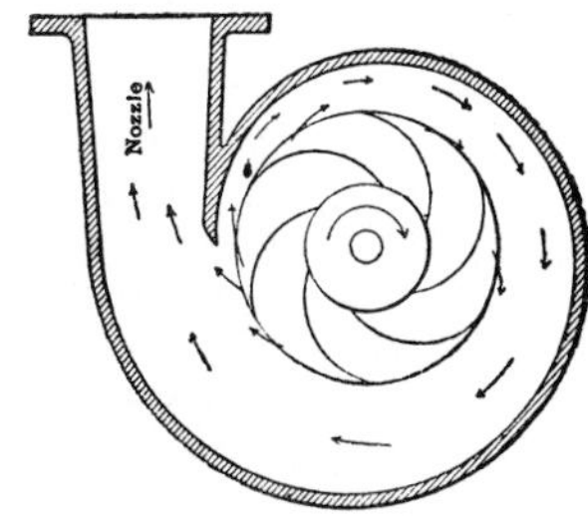

FIGURE 13.139 Volute pump.

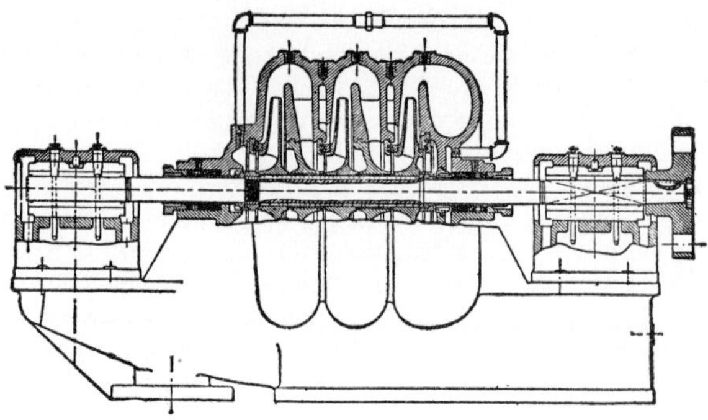

FIGURE 13.140 Multistage pump.

stage, so that the total pressure is equal to that developed by one impeller alone multiplied by the number of stages (see Fig. 13.140).

If the impeller takes water from one side only, the pump is called a *single-suction pump*. If it takes water from both sides, it is a *double-suction pump*. A single-suction impeller produces an end thrust which must be provided for, while a double-suction impeller is in hydraulic balance.

A deep-well turbine or centrifugal pump is merely one that is so proportioned that it can be placed within a well casing, the diameter of which is usually from 12 to 24 in (30.5 to 60.9 cm) at most. Thus, such a pump is necessarily a vertical-shaft type with impellers and case of relatively small diameter. In order that it may develop the head required, despite the small diameter of the impeller, it is necessary that it be built with many stages. Sometimes deep-well pumps are the axial-flow type, in which case the impellers become very similar in appearance to screw propellers as used for ships. They are, in fact, a form of propeller pump. Propeller pumps are also used in large sizes for pumping large quantities of water against relatively low heads.

21. Nominal Pump Size. The size of a centrifugal pump is usually designated by the diameter in inches of the discharge-pipe connection. As the velocity of the water at that section is usually about 10 ft/s (3.0 m/s), one may thus obtain an approximate value of the pump capacity by multiplying the area so obtained by 10. But this is subject to considerable deviation.

22. Theory. The centrifugal pump is so called because of the fact that centrifugal action is an important factor in its operation. If a body of water is rotated about a vertical axis through its own center, the surface will assume the shape of a paraboloid, and at any point where the velocity is u the surface will be at a height $u^2/2g$ above the level at the axis of rotation. Thus centrifugal action has lifted a particle of water at that point to that height above the center. But the particle is likewise moving with a velocity u, and thus possesses kinetic energy which is measured by $u^2/2g$ also. In any efficient pump it is necessary to convert into pressure as large a fraction of this kinetic energy as possible. To do that is the function of the diffusion vanes in the turbine pump or the volute casing in the volute pump.

Referring to Fig. 13.141, where V = absolute velocity of the water, v = water velocity relative to the impeller, and u = velocity of a point on the impeller, all in ft/s (m/s), while the subscript 2 denotes values at the point of discharge from the impeller, the head (or energy per unit weight) imparted to the water by the impeller is

$$H = \frac{u_2 V_2 \cos \alpha_2}{g} = \frac{u_2(u_2 - v_2 \cos \beta_2)}{g} \tag{13.41}$$

This may also be expressed in another form, which is equivalent to Eq. (13.41), as follows:

$$H = \frac{u_2^2}{2g} - \frac{v_2^2}{2g} + \frac{V_2^2}{2g} \tag{13.42}$$

The actual head h developed by the pump is less than this by the amount of head lost in hydraulic friction. Or it may be obtained by multiplying by the hydraulic efficiency. Thus

$$h = e_h \times H \tag{13.43}$$

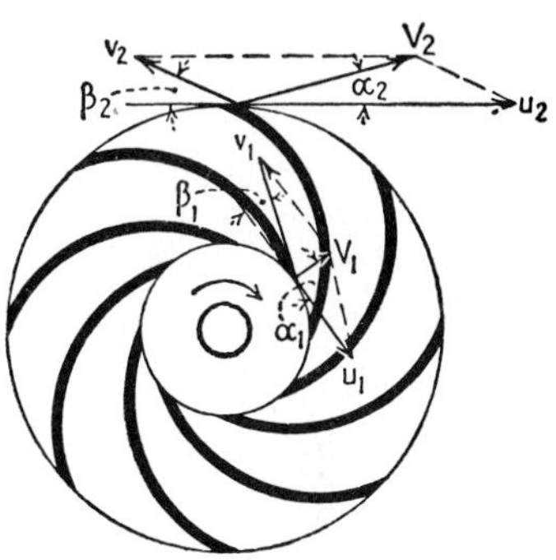

FIGURE 13.141

The hydraulic-friction losses consist of: (1) friction losses due to eddies or turbulence in flow throughout the entire pump, including the eye of the impeller, the impeller passages, and the case; (2) additional turbulence loss at the entrance to the impeller, in case the water velocity and the impeller vane velocity and shape at that point are not suitable to each other; (3) large turbulence loss at discharge from the impeller, resulting in a loss in converting kinetic energy into pressure. This last is probably the principal source of hydraulic loss in the centrifugal pump.

If all of the normal hydraulic-friction losses throughout the pump are assumed to be proportional to the square of the quantity of water pumped, they may all be covered by a term of the form $ku_2^2/2g$. But the turbulence loss at discharge from the impeller will not follow this same law, for it will be a maximum at a flow approaching zero and will decrease as the discharge increases, reaching a minimum value at somewhere near the normal capacity of the pump and then increasing again. If the proportion of the kinetic energy of the water leaving the impeller that is converted into pressure be given by $mV_2^2/2g$, the actual head developed by the pump may be expressed as

$$h = \frac{u_2^2}{2g} - (1 + k)\frac{v_2^2}{2g} + \frac{mV_2^2}{2g} \tag{13.44}$$

In this equation the factor m is a variable, its value ranging from 0 at shutoff up to 0.75 or more at normal flow.

23. Head at Shutoff. When a centrifugal pump is run with the discharge valve closed so that no flow can take place, there is obtained a simple case of centrifugal action. Theoretically the shutoff head would be $h_0 = u_2^2/2g$, where u_2 = the peripheral velocity of the impeller in ft/s (m/s). In reality, there is some rotation of the water in the casing surrounding the impeller, and there is also some slight flow through the impeller because of the leakage through the clearance spaces from the pressure to the suction side. Owing to these and similar factors, the head at shutoff is actually

$$h_0 = 0.85 \text{ to } 1.10 \frac{u_2^2}{2g}$$

24. Pump Characteristics. Owing to the practical difficulty of determining the proper values of certain velocities, angles, and factors to use in the equations presented in Sec. 22, the relation between head, discharge, and pump speed is best determined by test rather than by calculation. Test curves for pumps running at constant speed are shown in Fig. 13.142. Centrifugal pumps are said to have rising or falling characteristics, according to whether the head increases somewhat as the discharge increases from zero or whether it continuously decreases. The latter may also be divided into flat and steep characteristics. The propeller pump has an even steeper characteristic curve than the normal centrifugal pump.

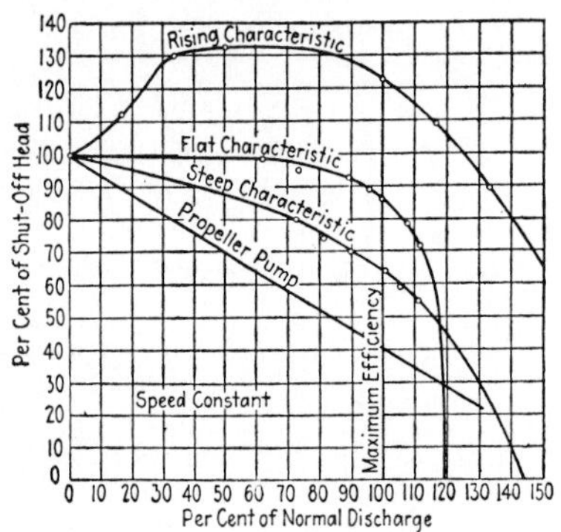

FIGURE 13.142 Head-discharge characteristics.

In Fig. 13.143, horsepower and efficiency curves are included with the head-discharge curve for a pump running at constant speed. The head and capacity at which the pump is rated at this speed are those values obtained when the efficiency is a maximum. The value of the head at the point of maximum efficiency is usually less than at shutoff, but may be higher in exceptional cases. The usual range is such that the head for maximum efficiency is $h = 0.60 \text{ to } 1.2\ u_2^2/2g$.

The power curve is seen to rise to a maximum value and then diminish. It is desirable that the maximum power should not be much greater than that at the normal rate of discharge; otherwise the motor is not run at its most efficient load

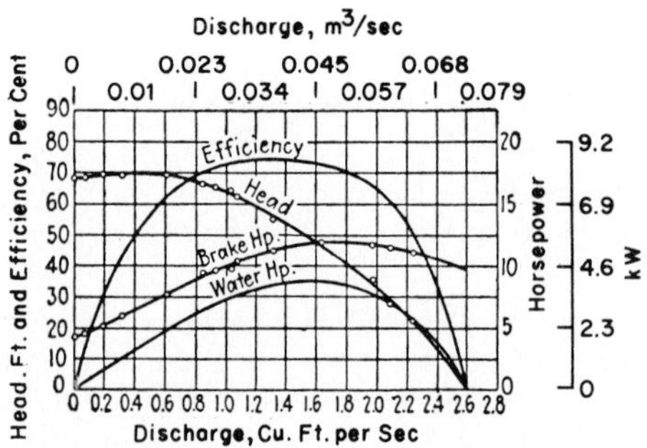

FIGURE 13.143 Characteristics of a centrifugal pump at constant speed.

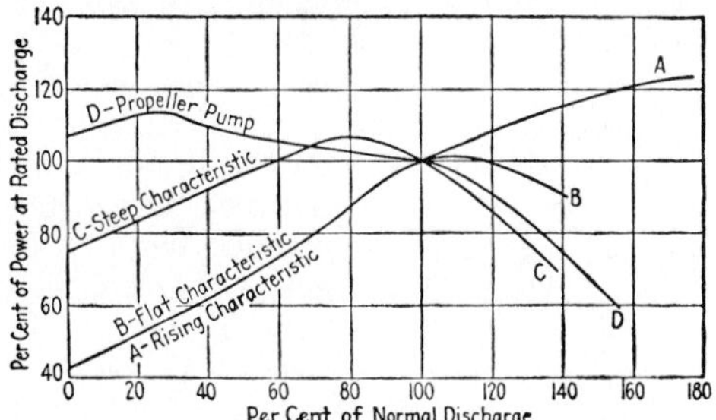

FIGURE 13.144 Horsepower at constant speed.

most of the time, or else is in danger of being overloaded owing to any abnormal increase in the rate of discharge, such as that due to the rupture of the pipe.

From this standpoint the rising-characteristic pump is the least desirable, since the power curve tends to rise with flows above normal, while the steep-characteristic pump has a flatter power curve. The propeller pump may actually have a power curve which is a maximum at shutoff head (see Fig. 13.144).

25. Centrifugal Pump at Different Speeds. In Fig. 13.145 is shown a series of curves for the same pump at different speeds. If a series of parabolas are drawn through the origin of coordinates, points on the various curves intersected by any one parabola are related to each other in a very simple way, but there is no simple relationship between two points in general at two different speeds. Corresponding points along any parabola are related as follows: q varies as N, h varies as N^2, power varies as N^3, while the efficiency is very nearly constant. q = ft³/s (m³/s), N = r/min, and h = head in ft (m).

For very large changes in speed the foregoing ratios should be applied with caution. Cavitation, which is a function of the quantity of water pumped, since the suction pressure at entrance to the impeller continuously diminishes as the velocity of flow increases, may make the head-discharge curve much steeper than normal. Aside from this the hydraulic efficiency is constant. But the mechanical friction losses do not follow the same laws as the hydraulic-friction losses, and consequently the overall efficiency is not quite the same at different speeds. But, for a reasonable range, the variation is negligible.

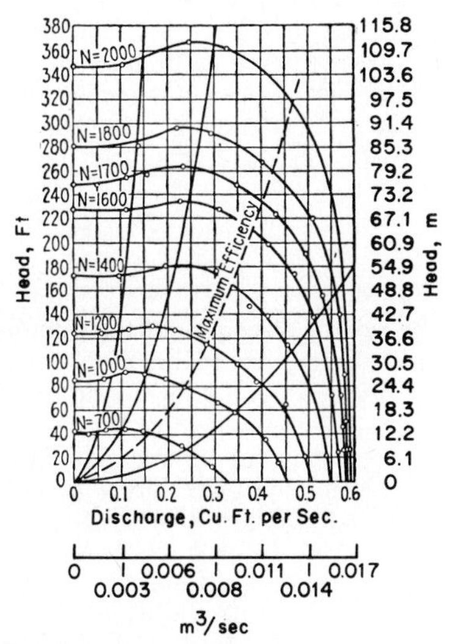

FIGURE 13.145 Relation between head and discharge at various speeds.

26. Centrifugal Pump and Pipe Line. In Fig. 13.146 is shown a curve representing the static lift plus the friction losses, which vary as the square of the

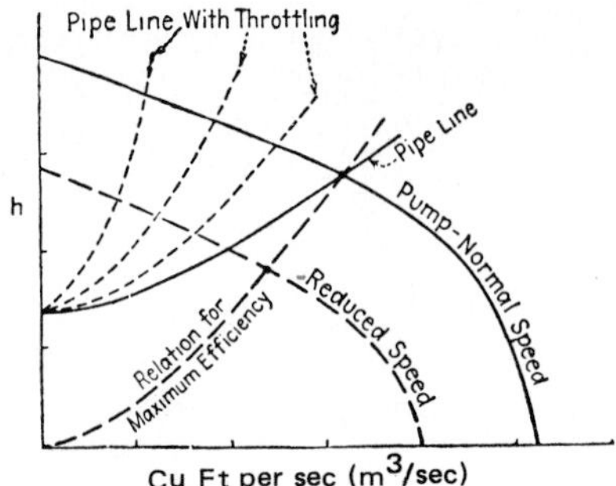

FIGURE 13.146 Centrifugal pump and pipe-line head-capacity curves.

quantity of flow, in a given pipe line. Also there is shown the head-capacity curve for a centrifugal pump at constant speed. The intersection of the two determines the amount of water that will be pumped and the head accompanying it. It is impossible to secure any greater flow than this without changing the pipe line or increasing the speed of the pump. But if additional resistance is added to the line, as by closing a throttle valve, the dotted curves show that the flow will be decreased. Decreasing the discharge by throttling is inefficient, but it is commonly done because it is the simplest solution.

The pump speed may be reduced, but that is not so readily accomplished, unless the pump is driven by a variable-speed motor. For maximum efficiency at all flows it would be necessary for the static lift to be zero for the pipe-line curve, and for it and the parabola for constant efficiency at different speeds to be identical, and this is seldom the case. Hence, there is usually some loss in efficiency in pumping a reduced quantity of water by reducing the pump speed, but it is not so great a loss as is occasioned by throttling.

Whether a variable rate of discharge should be obtained by throttling or by varying the pump speed is solely a question of economics. The constant-speed motor costs less than the variable-speed motor, but the annual cost of power with it may be more if the pump discharge is throttled much of the time.

27. General Laws and Factors. It is sometimes convenient to use factors ϕ and c such that $u_2 = \phi\sqrt{2gh}$ and $v_2 = c\sqrt{2gh}$, where u_2 = the peripheral velocity of the impeller, v_2 = the relative velocity of water at exit from impeller, both in ft/s (m/s), while h is the head in ft (m) of the fluid.

While for a given pump, values of ϕ and c may vary over a wide range, the values of especial interest are those for the head and capacity at which maximum efficiency is obtained. These values for maximum efficiency are

$$\phi = 0.90 \text{ to } 1.30$$

$$c = 0.10 \text{ to } 0.30$$

The rate of discharge of the pump in ft²/s (m³/s) is $q = a_2 v_2$, where a_2 = the cross-sectional area of the streams of water at exit from the impeller, measured normal to the direction of v_2 and expressed in ft² (m²).

The radial component of the water velocity is $V_2 \sin \alpha_2$ or $v_2 \sin \beta_2$. If B and D represent impeller width and diameter in inches (mm), respectively, n the number of vanes, and t their thickness measured along the circumference, the net circumferential area in ft² (m²) is $B(\pi D - nt)/144$. Hence

$$q = v_2 \sin \beta_2 \frac{B(\pi D \quad nt)}{144} \quad (0.028 \text{ m}^{\cdot}/\text{s})\ (q) \tag{13.45}$$

The factor giving the radial component of the velocity is $c \sin \beta_2$. The vane angle β_2 may be anything from 10° to 80° or more, but it is usually between 20° and 30°.

Thus the factor for the radial component of velocity may range from about 0.04 to 0.15 in ordinary practice.

Along any one parabola in Fig. 13.145, the factor ϕ is constant. When ϕ is constant, then c and, likewise, the hydraulic efficiency e_h are also constant. And only when ϕ and c are constant may one apply the simple ratios previously given, that

$$q \text{ varies as } N$$

$$h \text{ varies as } N^2$$

$$\text{hp (W) varies as } N^3$$

The overall efficiency will not be quite constant, but may be assumed so for a reasonable speed range.

28. Variation with Diameter. For a series of homologous impellers, in which all angles are the same, and all ratios of dimensions are the same, but which differ from each other in actual dimensions only so that each one is simply an enlargement or a reduction of another, all areas will vary as the square of the diameter. And, for the same rotative speed in all cases, all velocities will vary as the first power of the diameter. Thus for such a series at the same rotative speed, the following will hold:

$$q \text{ varies as } D^3$$

$$h \text{ varies as } D^2$$

$$\text{hp (W) varies as } D^5$$

But if the impellers were not homologous but differed in diameter only, while the impeller width remained unchanged, it would be found that

$$q \text{ varies as } D^2$$

$$h \text{ varies as } D^2$$

$$\text{hp (W) varies as } D^4$$

This assumes that each impeller is a separate construction such that the diameter is the only dimension that is changed, except that each diameter of impeller is in a case of suitable proportions for it. If the diameter of a given impeller is changed, as by turning it down in a lathe, and if this reduced size of impeller is used in the same case as the original, no exact ratios are known except as they are determined experimentally. But for a moderate change in diameter, it will be found that q varies very nearly as D, though the exponent appears to be somewhat more than 1 but less than 2. The head h varies approximately as D^2 but not exactly.

The diameter of impeller, speed, and head may all be combined in the following useful expression:

$$DN = 1840\phi\sqrt{h} \tag{13.46}$$

For a series of homologous impellers, ϕ is constant. Thus, for that series $DN/\sqrt{h}$ is constant. In practice, for different designs of centrifugal pumps this combination may range in value from 1660 to 2400, according to the value of ϕ for maximum efficiency.

29. Specific Speed. A most useful factor is the specific speed, which may be expressed as*

$$N_s = \frac{N\sqrt{gal/min}}{h^{3/4}} \tag{13.47}$$

If preferred, cubic feet per second may be used instead of gallons per minute, but the latter will give values 21.2 times that of the former. In this expression the values for speed, capacity, and head should be those at which the maximum efficiency is obtained. [Note that this differs from specific speed as used for hydraulic turbines in which horsepower (kW) instead of capacity is employed.] The specific speed of any one pump is a constant, and thus the above may be used to obtain possible combinations of the three factors involved. But it is also a constant for a series of homologous centrifugal pumps. The value of the head to be used to obtain N_s is the head per stage in the case of a multistage pump.

The preceding expression for specific speed applies directly to single-suction pumps. For double-suction pumps the specific speed should be computed by employing one-half of the actual pump capacity, since such an impeller is equivalent to two single-suction impellers placed back to back.

Small values of N_s are found with pump impellers whose width is small compared to the diameter, while large values are found for large values of that ratio. Usually the ratio of B/D will be found to range from 2 to 70.

For the centrifugal pump, values of specific speed are usually between the limits of 500 to 9000. With propeller pumps, this value may be increased up to 16,000.

30. Efficiency of Centrifugal Pumps. The efficiency obtained with centrifugal pumps depends among other things upon the absolute size, as large pumps are more efficient than small ones. In Figs. 13.147 and 13.148 are shown probable optimum values of efficiency as a function of the pump capacity.

Efficiency is also a function of specific speed, as shown by Fig. 13.149. In fact the actual efficiency depends upon both the capacity and the specific speed; and to get the highest values, the pump must be of reasonably large capacity and also of a favorable specific speed.

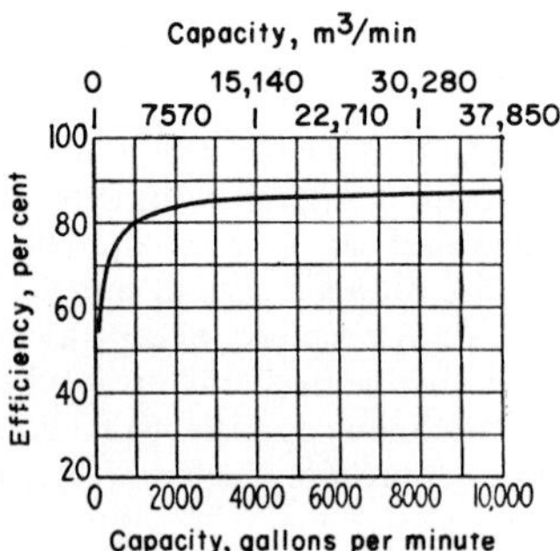

FIGURE 13.147 Optimum efficiency of turbine pumps as a function of capacity.

31. Head per Stage. Usual practice is to limit the head per stage to about 100 to 150 ft (30.5 to 45.7 m). But heads of several hundred feet (m) per stage have been employed success-

*The SI specific speed = 0.613 times the U.S. Customary System N_s when flow is expressed in L/s and head in m.

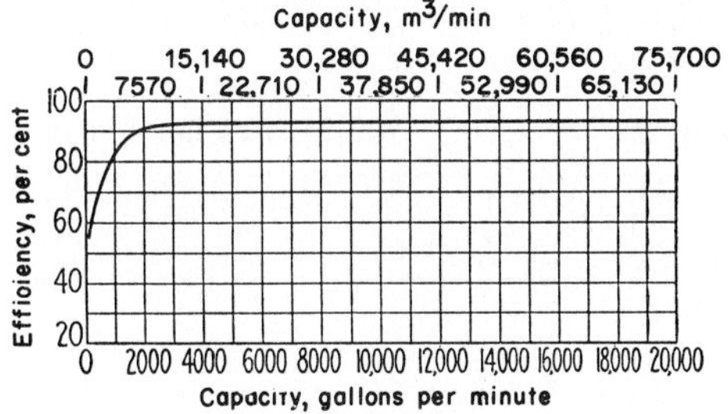

FIGURE 13.148 Optimum efficiency of volute pumps as a function of capacity.

fully. The principal criterion is whether a favorable or even a possible value of specific speed may be obtained. Provided a favorable value for N_s is obtained, the efficiency does not seem to be materially affected by the value of the head used.

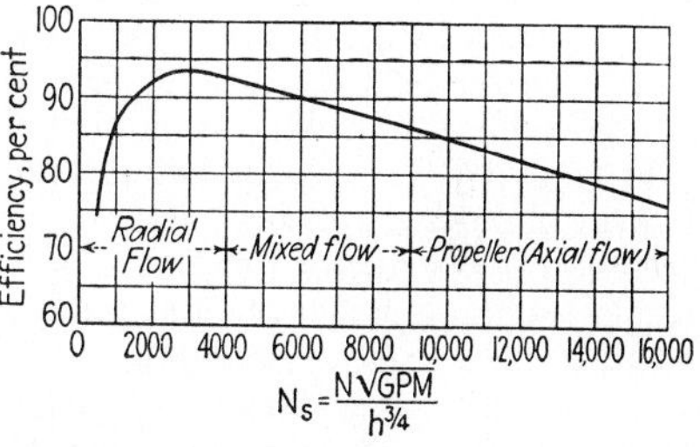

FIGURE 13.149 Optimum efficiency as a function of specific speed.

32. Disk Friction. An important source of power loss in the centrifugal pump is the disk friction or *rotation loss* due to the friction between the sides of the impeller and the fluid in the space between it and the casing. This may amount to from 1 to 20 percent of the total power, and more in extreme cases, even with water as the fluid. If a viscous liquid such as oil is being pumped, the loss may be even greater.

The power consumed by a disk rotating in a fluid is given by the equation

$$\text{Power} = KD^5N_s^3\left(\frac{U}{D^2N}\right) \quad \text{hp (1 hp} = 0.75 \text{ kW)} \tag{13.48}$$

where D = the diameter in in (mm), N = r/min, s = specific gravity (for water = 1), and U = kinematic viscosity, here taken as absolute viscosity in centipoises divided by density. For water at 68.4°F (20.2°C), $U = 1$.

The value of n is 0.5 if the flow of the fluid in contact with the disk is all laminar, and is 0.1 if it is all turbulent. But since the velocity of a point on the disk, and hence the velocity of the fluid, varies with the radius, it is possible to have laminar flow near the axis and turbulent flow near the periphery. Hence the transition from one type of flow to the other is very gradual, and n may therefore have values intermediate between 0.5 and 0.1. In fact, both n and K are functions of Reynolds' number, which for convenience may here be arbitrarily expressed as $R = D^2N/U$. If R is 1200 or less, the flow is all laminar and $n = 0.5$. If R is 70,000 or more, the flow is all turbulent and $n = 0.1$.

The equation for disk friction may also be written as

$$\text{Power} = KD^{5-2n}N^{3-n}sU^n \quad \text{hp(1 hp} = 0.75 \text{ kW)} \tag{13.49}$$

which for certain values of n may be reduced to

$$\text{Power} = \frac{D^4 N^{2.5} s U^{0.5}}{(436 \times 10^{10})} \text{ hp} \qquad \text{for laminar flow} \tag{13.50a}$$

$$\text{Power} = \frac{D^{4.6} N^{2.8} s U^{0.2}}{(477 \times 10^{11})} \text{ hp} \qquad \text{for } n = 0.2 \tag{13.50b}$$

$$\text{Power} = \frac{D^{4.8} N^{2.9} s U^{0.1}}{(240 \times 10^{12})} \text{ hp} \qquad \text{for turbulent flow} \tag{13.50c}$$

The constant in the last equation applies to smooth brass disks. For rough cast iron the power would be about 15 percent more.

The preceding equations apply to a solid disk, whereas an impeller has an "eye" in one or both sides. A deduction may be made for this in the case of a double-suction impeller by multiplying the above equations by $(1 - y^{5-2n})$ or practically by $(1 - y^5)$ for turbulent flow, where $y = D_1/D$ or the ratio of the diameter of the eye to the diameter of the impeller. For a ratio of 0.5, which is quite common, the value of this factor is 0.97, and for a ratio of 0.7 it is 0.83. For a single-suction impeller obviously this should be applied to only half of the power as given by the above equations.

If the disk has a thickness b, which is appreciable, the power should be increased by multiplying by the factor $[1 + (5 - 2n)b/D]$. The correction factor for both of these cases together is $[1 + (5 - 2n)b/D - y^{5-2n}]$.

These formulas for disk friction may be applied to shrouded impellers only, i.e., those which have the vanes enclosed. For the open type of impeller the friction loss would be very much higher. For a multistage pump, the disk friction given by these formulas should be multiplied by the number of stages.

The head developed by an impeller depends upon the peripheral velocity, and a given peripheral velocity may be obtained by various combinations of D and N. Since the disk friction always varies as a higher power of D than of N, it is seen that for a given head, disk-friction loss will be less for a small diameter of impeller at high rotative speed rather than the reverse. In other words, the disk-friction loss is relatively larger with low-specific-speed pumps for the same head and capacity. In fact, this is one reason why the efficiencies of very low specific-speed pumps are not so good as for those of higher specific speeds.

Strictly speaking, the values of K in the preceding equations apply to a disk rotating in a stationary fluid. In most pumps there is a circulation of fluid in the pump casing adjacent to the sides of the impeller which has the effect of reducing the relative velocity between impeller and fluid and thus making the disk friction less. However, the equations do give the true law by which the disk friction varies with diameter and rotative speed.

33. End Thrust and Balancing Pistons. In every single-suction pump there will be end thrust along the shaft, since the forces on the two sides of the impeller are not symmetrical. Even in double-suction pumps there may occasionally be some end thrust in case there is not exact similarity of flow in the two halves. This end thrust may be taken care of by a mechanical thrust bearing or by some form of hydraulic balancing. The former consumes power in mechanical friction;

the latter consumes power in pumping the small quantity of water that must be permitted to pass through it.

A typical example of a balancing piston is shown in Fig. 13.150. Water from the pressure side of the pump flows down between the back side of the impeller and the casing, through the small but constant area *c* into space *s*, then through the small but variable area *v* into the suction. If the impeller and shaft drift toward the right, the variable area *v* diminishes, and the pressure builds up in space *s* and stops the movement. On the other hand, if the impeller and shaft move to the left, the area *v* increases, the pressure in space *s* drops, and the movement ceases when the net forces on the impeller are in balance with the forces on the piston. Thus any hydraulic balancing device consists essentially of an area on which water pressure may act, and two fissures, the area of one of which remains unchanged while that of the other varies. The impeller and shaft must be permitted a very slight end movement; some leakage of water must always be had.

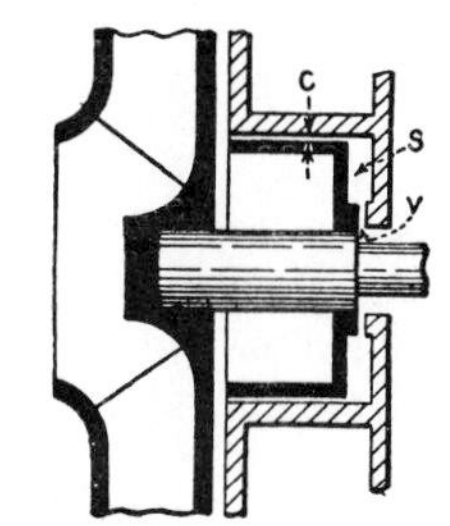

FIGURE 13.150 Balancing piston.

34. Wearing Rings or Clearance Rings. In order to reduce the leakage of water from the discharge to the suction side and yet avoid any actual contact between the impeller and the case, close running fits are employed. Since this small clearance area will become enlarged in time owing to wear even with clear water alone, or more rapid wear if any grit is present, it is desirable that these parts be easily replaceable. Hence both the case and the impeller are provided with renewable rings for this purpose.

These rings may be very plain and simple and provide nothing more than a straight passageway of small area, or, in order to diminish still further the flow of water past them, they may be made of the labyrinth type, as shown in Fig. 13.151. The rings are placed as near to the eye of the impeller as possible. They are usually made of bronze.

35. Gland Seals. Packing glands are provided where the shaft passes out through the casing. Since leakage of air on the suction side is very undesirable in most cases, and since it may be necessary to have the packing unduly tight to prevent it, it is customary to use a gland seal. This consists of a space that is left open about in the middle of the packing rings and is kept open by means of a gland cage, which is a skeleton structure of metal that slips on over the shaft in the place of a ring of packing. A small pipe from the pressure side of the case conducts water to this space. Thus, if the packing is loose enough to leak, water will be drawn back into the pump on the suction side instead of air. It is merely necessary to keep the packing tight enough so that water does not leak out in the outside in any appreciable quantity.

FIGURE 13.151 Wearing rings.

36. Impellers and Cases. Pump impellers may be made of cast iron, brass, or bronze. They are usually cast in one piece.

Casings are usually made of cast iron, though for high pressures they may be made of cast steel.

Cases may be split along the shaft, which makes it easier to open up the pump and to make repairs. Also it is not necessary to disconnect the piping when so doing. The so-called side-plate type is often a more economical form of construction, but it is not so convenient to get at the interior of the pump when it is employed.

With multistage pumps the impellers may all be arranged in the same way along the shaft, and the water may be led from one to the next in order from one end of the shaft to the other. This arrangement leads to the simplest type of case construction but requires the end thrust of all the impellers to be taken care of by either a thrust bearing or a balancing piston, if all the impellers are single-suction. They may all be double-suction, which complicates the case construction, but does minimize end thrust.

On the other hand, single-suction impellers of multistage pumps may be arranged back-to-back in various combinations, and the water may be led from the discharge of one to the suction of the other through a passageway cast in the casing or by means of a curved pipe, which is quite separate from the case. Such arrangements obviously require an even number of impellers.

37. Priming Centrifugal Pumps. In case the pump is placed above the water level so that the water does not flow to it under pressure, it will be necessary to prime the pump before it is started. If a small water supply is available for the purpose, the pump and suction piping may simply be filled with water from this source. A foot valve on the end of the suction pipe is necessary, and also pet cocks for the escape of air should be placed at various high points on the pump, where air pockets could form.

On the other hand, the air may be exhausted from the casing by means of a vacuum pump operated by hand, or by means of an ejector operated by either compressed air, steam, or water. With this arrangement it is necessary to have the valve closed in the discharge pipe, but a foot valve on the suction pipe is not essential. However a foot valve is often a convenience, as the pump does not have to be primed if it is shut down for only a short time.

38. Centrifugal Pumps with Viscous Liquids. Centrifugal pumps may also be used for pumping oils and other liquids with viscosities much greater than that of water. Owing to the increased friction of the fluid as well as the increased disk friction, the efficiency continually decreases as the viscosity increases.

Also, for a given pump at constant speed the head-capacity curve becomes continually steeper with increasing viscosity. Thus both the head and the capacity diminish. See Figs. 13.152 and 13.153. In one case the oil pumped had an absolute viscosity 3200 times that of water, which indicates the range covered.

The larger the size of the pump, the less is the effect of the viscosity of the liquid pumped. It is also found that the centrifugal-pump design, which is most efficient for high viscosity, is not the most efficient for water.

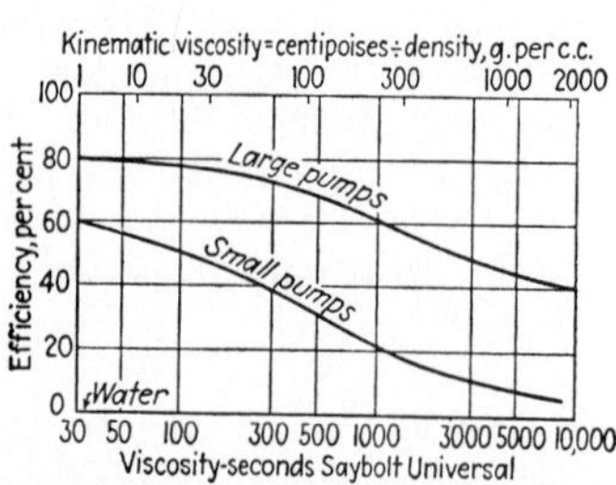

FIGURE 13.152 *(Adapted from Bulletin 130, Goulds Pumps, Inc.)*

39. Calculations for Centrifugal Pump. The quantity of water to be pumped and the total head may be assumed to be fixed by certain physical conditions, but the rotative speed and the number of stages or the head per stage might be determined by a consideration of the specific speed. The type of characteristic curve desired may then be decided upon, and values of ϕ and c and vane angle β_2 selected in accordance with experience.

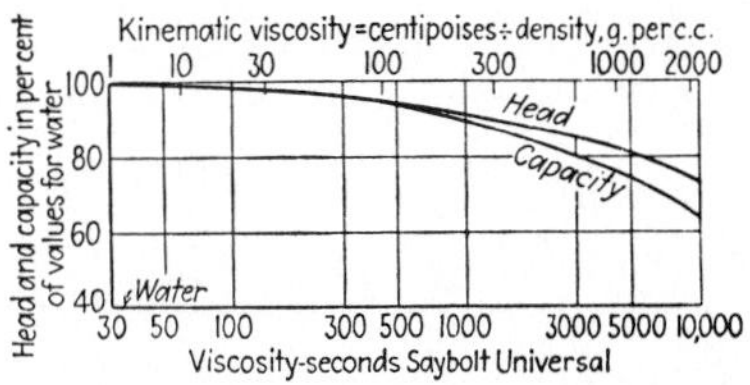

FIGURE 13.153 *(Adapted from Bulletin 130, Goulds Pumps, Inc.)*

Then, the impeller diameter will be given by $D = 1840\phi\sqrt{h}/N$. The impeller width will be given by

$$B = \frac{0.04 \times \text{gal/min}}{(\pi D - nt)c \sin \beta_2 \sqrt{h}} \tag{13.51}$$

the notation being as given in Sec. 27.

At entrance the diameter D_1 is usually from 0.3 to $0.6D$ and the width B_1 is from 1 to $2B$. The diameter of the "eye" should be so chosen that the water velocity will not be too high and cause cavitation.

The vane angle at entrance should be so determined that at the normal rate of discharge the absolute velocity of the water is radial. The value of the absolute velocity at this point is not very far different from $v_2 \sin \beta_2$. In fact, the radial component of the velocity does not change much between the inlet and outlet of the impeller. If β_1 = the vane angle at entrance, $\tan \beta_1 = V_1/u_1$.

If it is a turbine pump, the angle of the diffuser vanes is found from

$$\tan \alpha_2 = \frac{c \sin \beta_2}{(\phi - c \cos \beta_2)} \tag{13.52}$$

If it is a volute pump, the area of the volute at its point of maximum cross section may be determined as $a_2 \times c/(\phi - c \cos \beta_2)$. The volute should normally be so designed that the velocity is constant around the circumference of the impeller. If the case has a circular cross section, the outer boundary curve is a parabolic spiral, whose equation is $r = \sqrt{c\theta} + K$, where θ is the angle measured from the "cutwater," which is where the spiral starts; K is the radius to the cutwater; and c is a constant.

Only the directions of the vanes at the two ends enter into the theory. There is no theory at all that determines the shape of the vane throughout its length. But if losses of energy are to be avoided, the vane should be without any abrupt changes in shape and still give the required directions at the two ends.

In reality, the theory is concerned with the directions and the velocities of the water, and the directions of the stream lines are not always the same as the directions of the vanes. The values of the velocities are not always obtained by dividing the discharge by what appears to be the cross-sectional areas. Hence any theory should be used with caution. The successful design of centrifugal pumps requires much experience and empirical data to go along with any theory.

40. Operating Characteristics of Centrifugal Pumps. Unlike the positive displacement pump, the head and capacity are not independent of each other but are mutually related. Thus the head cannot be varied without changing the rate of discharge. The head developed is a function not only of the rate of discharge but also of the pump speed. On the other hand, the pump may be operated at a constant speed and the rate of discharge may be varied; this is impossible with the positive displacement pump.

The centrifugal pump has the advantage of being suited to the high rotative speeds that are found also with electric motors and steam turbines.

Since the friction losses of a positive displacement pump are about the same regardless of the head, such a pump becomes less efficient as the head becomes less. The centrifugal pump is not so affected and thus makes its most favorable comparison with the piston pump under low heads. For very high heads and very small capacities, it is not so well adapted, though its use under such conditions is becoming more common.

41. Cavitation. If in the operation of a centrifugal pump the pressure at the entrance to the impeller is continuously decreased while all other values remain the same, eventually the pressure, which is a function of the suction lift, friction losses, and the local velocity head at the point in question, will be reduced to the vapor pressure of the liquid. In that case bubbles of vapor and dissolved air will develop. This tends to reduce the flow, and that in turn permits a slight rise in pressure, whereupon the bubbles immediately collapse, and then the cycle is repeated. This rapid pulsation of pressure causes the pump to vibrate and is evidenced by noise as if rocks were passing through the system. The phenomenon is called *cavitation*.

The immediate effect of cavitation is to produce a drop in pump capacity and also in the efficiency, as shown by Fig. 13.154. Continued operation of a pump under cavitation conditions produces a rapid wearing away of portions of the impeller and thus shortens materially the life of the pump.

42. Maximum Suction Lift. For a centrifugal pump the determination of the limiting suction lift at which cavitation will just begin is somewhat more complicated than is indicated in Sec. 5, which applies more directly to positive displacement pumps. The difficulty is due to the fact that it is not feasible to calculate the local velocity head at a given point on the impeller vane, since the velocity distribution is far from uniform. It is therefore necessary to rely upon an empirical factor σ in the equation

$$L_{max} = b - p_v - h_f - \sigma h$$

where L_{max} = maximum suction lift, b = barometer pressure, p_v = vapor pressure of the liquid in question, h_f = friction losses in suction piping, and h = total head developed by the pump at the point of maximum efficiency. All values are to be expressed in feet (m) of the liquid in question. If L as computed by this equation has a positive value, it indicates that the fluid can be lifted to that height from a level below the pump, but if the value is negative, as it may be with large values of σh, then the liquid must flow to the pump by gravity.

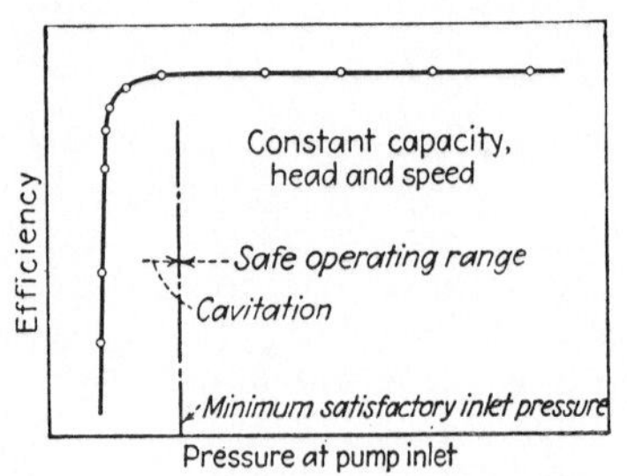

FIGURE 13.154 Drop in efficiency as a function of inlet pressure.

43. Values of Cavitation Factor. In order to determine the value of σ at which cavitation is impending, it is necessary to obtain a curve for the pump such as that shown in Fig. 13.154, and that requires a high degree of accuracy in testing. For this reason very few data are available aside from some values for a limited range of specific speed determined at the California Institute of Technology. Using these values as a basis and developing a theory as to the possible variation as a function of specific speed, G. F. Wislicenus, R. M. Watson, and I. J. Karassik have developed a tentative formula from which Table 13.6 has been computed. The values in this table are reliable for specific speeds around 2000 but should be used with caution for those from 8000 up.

For a given specific speed different values of σ may be found, depending upon the actual design of the impeller and other factors. In Table 13.6 the values labeled maximum are for ordinary centrifugal pumps, while those labeled minimum are for special pumps in which great care has been taken in design and construction.

It can be seen that the higher the head on the pump, the less is the allowable suction lift, and for high heads it is necessary that the pump be submerged. Also, the higher the specific speed, the less is the allowable suction lift, and for very high-specific-speed pumps (i.e., propeller pumps) it might be necessary to submerge them to a very considerable depth in comparison with the total pumping head.

44. Selection of Type of Pump. The selection of the type of centrifugal pump is primarily a matter of choice of specific speed. For a given installation the total head is usually fixed by the physical setting and is not subject to change. However, it may be developed in one stage or divided among a number of stages of a multistage pump. The total quantity to be pumped is also fixed but, if large, may be divided among several pumps. The rotative speed may also be given different values, but it is usually confined to a limited number of synchronous speeds if the pump is directly connected to an alternating-current motor. Hence by varying any one or all three of the terms entering into the specific speed, the latter may be given a series of values, often over a very wide range.

The factors that affect the selection of the most suitable specific speed are as follows: The higher the specific speed, the smaller the pump and electric motor and the less the cost. But Fig. 13.149 shows that after a specific speed of about

TABLE 13.6 Values of σ as a Function of Specific Speed

Specific speed	σ	
	Maximum	Minimum
500	0.030	0.018
1,000	0.075	0.046
1,500	0.13	0.080
2,000	0.19	0.11
3,000	0.33	0.20
4,000	0.48	0.29
5,000	0.64	0.31
6,000	0.82	0.52
7,000	1.0	0.62
8,000	1.2	0.74
9,000	1.5	0.90
10,000	1.6	1.0
12,000	2.1	1.2
14,000	2.5	1.5
16,000	3.0	1.8

3000 is exceeded, the possible efficiency to be obtained becomes less. Also Table 13.6 shows that the higher the specific speed, the less is the allowable suction lift. In fact, unless the head per stage is low, the required pump submergence may be impracticable. Thus this last consideration may impose a limit upon the maximum specific speed that can be employed in a given case.

Figures 13.147 and 13.148 show that the volute pump is slightly more efficient than the turbine pump, and hence there seems small justification for the use of the latter, since its construction is more complicated and thus its cost is usually greater.

Figure 13.149 shows that very low specific-speed pumps are not desirable because of their lower efficiency and because their first cost is often more, but they are sometimes necessary for very low capacities combined with high heads per stage. To avoid them it is necessary to employ more stages or use still higher rotative speeds, but often either one of these is impracticable, and hence the lower efficiency must be tolerated.

45. Determination of Pump Size. For manufacturing reasons it is customary to employ a series of impellers of different diameters and different widths at discharge in the same pump case. Theoretically only one diameter and width will be most suitable, and hence such a practice results in a slight loss in efficiency, which increases with increasing departure from the optimum values. When the loss of efficiency becomes too great, then another case of a different size becomes necessary. It is thus apparent that a single case is suitable for a certain range of capacity and head for a single rotative speed, and by varying the latter, the same case may be made to cover a still greater range of operating conditions. Since the size of a centrifugal pump is specified by the diameter of the discharge pipe in inches (mm), as stated in Sec. 21, it is seen that any given size of pump may be made to fulfill a variety of requirements.

It is customary for manufacturers to express these relations by a series of charts, each one for a different synchronous speed. The coordinates on each chart would be head and capacity, and a certain area on such a chart would be covered by one nominal pump size. Also for the same rotative speed a different chart would be used for different numbers of stages. It is thus seen that a large number of charts would be necessary to cover the entire field. Furthermore, the practice of different manufacturers would not be identical, and hence such charts, if presented, would not be of universal application. Consequently no attempt will be made to present a series here.

However, a fair approximation of the pump size may be made by applying Sec. 21, though the velocity at discharge may actually range from 5 to 30 ft/s (1.5 to 9.1 m/s). The impeller diameter may be calculated by the principles in Sec. 28; i.e., $D = 1660 \text{ to } 2400\sqrt{h}/N$.

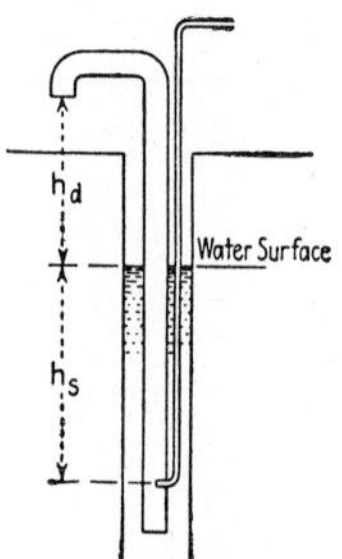

FIGURE 13.155 Air lift.

MISCELLANEOUS TYPES OF PUMPS

46. The Air Lift. The air lift shown in Fig. 13.155 consists of a long pipe, of which a considerable portion is submerged below the surface, into which air is admitted near the lower end. This air bubbles up through the water in the pipe and, when water is flowing out at the top, the weight of the mixture of air and water within the pipe produces a unit pressure at the lower end that is slightly less than that produced by the water surrounding the pipe.

Increasing h_s will reduce the quantity of air that is required, but will increase the pressure necessary, for the air pressure must be slightly greater than that due to the depth of water h_s. In practice h_s is from 1.0 to $4.0h_d$.

The volume of air required, expressed as cubic feet of free air, or air at atmospheric pressure, will be given by

$$v = \frac{Q(h_d + h_f + h_v)}{34 \log_e (p_1/p_a)} \quad (0.028v \text{ m}^3) \tag{13.53}$$

where v = ft³ (m³) of free air, Q = ft³ (m³) of water, h_d = lift, h_f = friction head, h_v = velocity head, all in ft (m); while p_1 is pressure at depth h_s, and p_a is atmospheric pressure.

The efficiency of the air lift varies from about 25 to about 45 percent, considering efficiency as the ratio of the useful power in lifting the water the height h_d to the indicated horsepower (W) in the air-compressor cylinder.

The air lift is of value in obtaining large flows from deep wells of small diameter. It is of especial value if the water contains grit or other material which would damage any working parts of a pump. This type of pump is also very useful in pumping oil from very deep wells. Instead of air, a gas obtained from the oil is usually used, and it is then known as a *gas lift*.

47. Jet Pumps. Jet pumps are of two types: the *injector* and the *ejector*. The injector (Fig. 13.156) uses steam as the working medium and is employed to pump feed water into the boiler. Steam, discharging at high velocity through a nozzle, mixes with cold water, which condenses the steam and, at the same time, warms the water, and the mixture then passes on into the boiler. The heat-balance equation for this is

$$0.98(H - h_2) = W(h_2 - h_1) \tag{13.54}$$

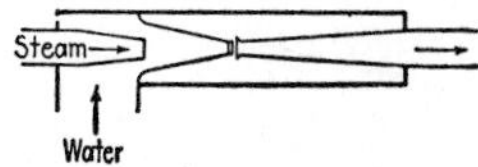

FIGURE 13.156 Injector.

where W = lb of water per lb of steam (kg/kg), H = total heat per lb (kg) of steam supplied, h_2 = heat of liquid of water going to boiler, h_1 = heat of liquid of cold water supplied.

The law of momentum, with a correction factor introduced to take care of losses, is

$$0.5(V_1 + WV_2) = (1 + W)V_3 \tag{13.55}$$

where V_1 = velocity of steam jet, V_2 = velocity of cold water, usually negligible in practice, and V_3 = velocity of the mixture. In accordance with Bernoulli's theorem, the velocity V_3 is largely transformed into pressure before entrance into the boiler by means of the diverging tube as shown.

The efficiency of the injector, considered solely as a pump, is very low, being only about 2 percent. But since all the heat of the steam used is returned to the boiler, save for a small radiation loss, its thermal efficiency approaches 100 percent. The steam consumption of an injector is about 400 lb of steam per hp · h (182 kg per 0.75 kWh). The injector cannot be used to lift hot water, and it is not an economical water heater. Its principal merits are its small size and light weight.

The ejector (Fig. 13.157) is operated with either water or steam or compressed

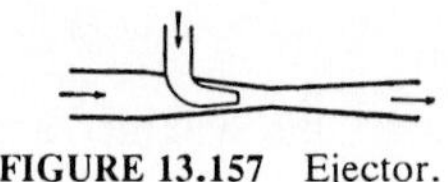
FIGURE 13.157 Ejector.

air as the working medium, and it may pump either a liquid or a gas. In any event it usually uses a small quantity of the working medium at a high pressure and pumps a large quantity at a low pressure. The equations for the injector apply to it also, except that if steam is not used, the heat-balance equation cannot be employed. The efficiency of the ejector is very low, being from 15 to 30 percent. It may be used for capacities up to 700 gal/min (2650 L/min).

48. The Hydraulic Ram. The hydraulic ram (Fig. 13.158) requires a large volume of water under low head to pump a small volume to a high head. It consists of a drive pipe of length L, in which there is a periodic water-hammer action which forces small quantities of water up a delivery pipe. The length of the drive pipe L is given by $L = (h_d + 2)(h_d - h_s)/h_s$, but should be not less than either $0.75h_d$ or $5h_s$.

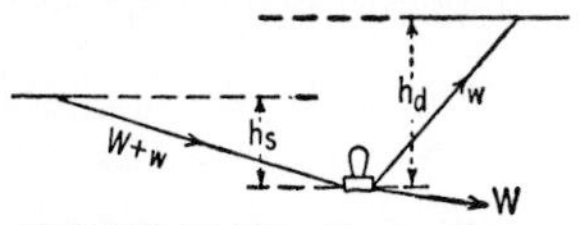

FIGURE 13.158 Hydraulic ram.

The capacity of hydraulic rams ranges from 1 to 30 gal/min (3.8 to 113.6 L/min). The efficiency, $e = w(h_d - h_s)/Wh_s$, is from 40 to 60 percent. In this equation, w = weight of water pumped, while W is the corresponding weight that is wasted. The quantity of water flowing in the drive pipe is $w + W$.

The efficiency decreases very rapidly as the ratio of h_d/h_s increases. With ratios of 4, 8, 16, and 24, values of efficiency are 72, 52, 25, and 4 percent, respectively.

COMPRESSORS

49. Density of Air. The weight of dry air, lb/ft^3 (kg/m^3), is given by $w_a = 2.7p_a/T_a$, where p_a is the air pressure, in lb/in^2 abs. (kPa abs), and T_a the temperature, in °F abs (°C abs).

The true pressure of the air is $p_a = B - hp_v$, where B is the barometer pressure (in same units as p), h the relative humidity, and p_v the vapor pressure for saturation at T_a. This is the value given in steam tables. The weight of water vapor, lb/ft^3 (kg/m^3), is $w_v = h/V_s$, where V_s is the specific volume of dry saturated steam at T_a.

The weight of the moist air is $w = w_a + w_v$. The quantity w is the actual weight per cubic foot (m^3) that is handled by the compressor. At 65°F (18.3°C), 14.7 lb/in^2 (101.3 kPa) barometer, and 70 percent humidity, $w = 0.0752$.

50. Free Air. By free air is meant air at the pressure and temperature of the place from which the compressor draws its supply. Its volume may be obtained from the volume at any other set of conditions by the equation pV/T = constant.

51. Gas Laws. If a perfect gas is compressed or expanded, it follows the law pV^n = constant. If the process is isothermal, $n = 1$. If it is adiabatic, $n = k$, which for air is approximately 1.4. In a piston compressor, n is usually from 1.3 to 1.35, and in a centrifugal compressor very nearly 1.4.

The temperature change during compression or expansion is determined by the relation

$$\frac{T_2}{T_1} = \left(\frac{V_1}{V_2}\right)^{n-1} = \left(\frac{p_2}{p_1}\right)^{(n-1)/n} \tag{13.56}$$

52. Volumetric Efficiency. This term is really a capacity factor, as it represents the relationship between the actual capacity of a machine and its physical size, and in an air compressor has but small effect upon true efficiency. It is defined as the ratio of the volume of free air actually delivered by the compressor to its displacement volume.

In Fig. 13.159, the compressed air in the clearance space at the end of the stroke expands down to 4 before the suction valve can open. The volume of air drawn into the machine is only V_1, while the displacement volume is V_p. If $c = V_c/V_p$, or the percent of clearance divided by 100, then

$$V_1 = V_p\left[1 + c - c\left(\frac{p_2}{p_1}\right)^{1/n}\right] \tag{13.57}$$

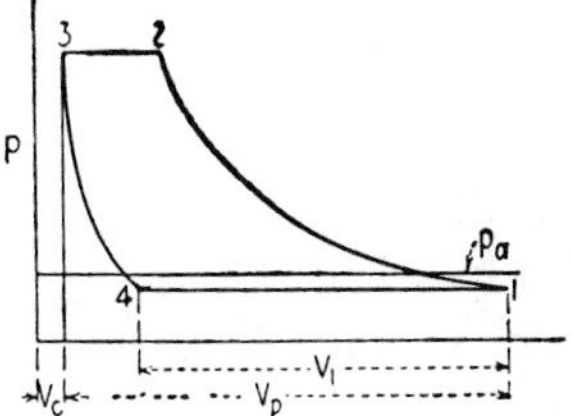

FIGURE 13.159

To reduce the volume V_1 to that of free air, multiply it by (p_1/p_a) (T_a/T_1). The effect of the clearance and the value of p_1/p_a can be obtained from an indicator card, but the temperature of the air in the cylinder at the end of the suction stroke is not known. It is only certain that it is much higher than T_a but less than T_2. In addition to the effect of clearance, pressure drop, and temperature rise, there must also be included leakage and losses due to imperfect valve action. Thus, by calculation, it is possible to determine only the clearance and pressure factors on volumetric efficiency, and it would be necessary to assume temperature and leakage factors. The correct value of the volumetric efficiency can only be determined by a test in which the actual volume of the air delivered is measured.

For piston compressors, clearance is usually from 2 to 6 percent, the larger values being found with smaller machines. The actual volumetric efficiencies of single-stage compressors are from 50 to 85 percent, depending upon the clearance and the pressure. Figure 13.160 gives actual values with all factors considered.

53. Effect of Clearance. For a given pressure, the effect of clearance on volumetric efficiency is usually greater than that of any other factors. The larger the clearance, the less is the capacity of the compressor. But the area of the indicator card in Fig. 13.159 is decreased in the same ratio. Hence the efficiency of the machine is apparently unaffected by the clearance. Actually, a decreased volumetric efficiency means that a larger size of machine will be required for the same capacity, so

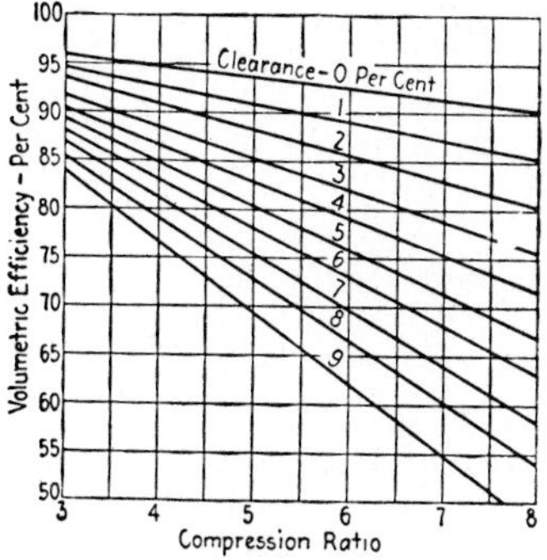

FIGURE 13.160 Volumetric efficiency.

that the mechanical friction losses will cause the real efficiency of the compressor to be lower. The other factors in volumetric efficiency, such as pressure drop and leakage, also decrease the real efficiency.

54. Operation under Different Discharge Pressures. In Fig. 13.161 is shown the operation of the same compressor under various discharge pressures. As the discharge pressure increases, the capacity continuously decreases and becomes zero at some definite pressure, which cannot be exceeded. This maximum pressure is $p_m = p_1[(1 + c)/c]^n$.

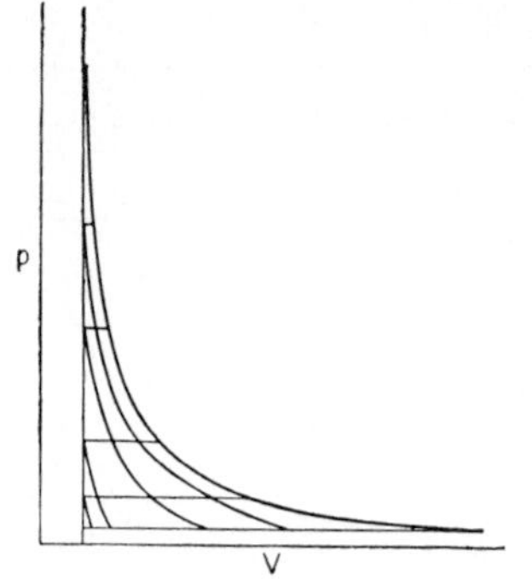

FIGURE 13.161 Compressor at different discharge pressures.

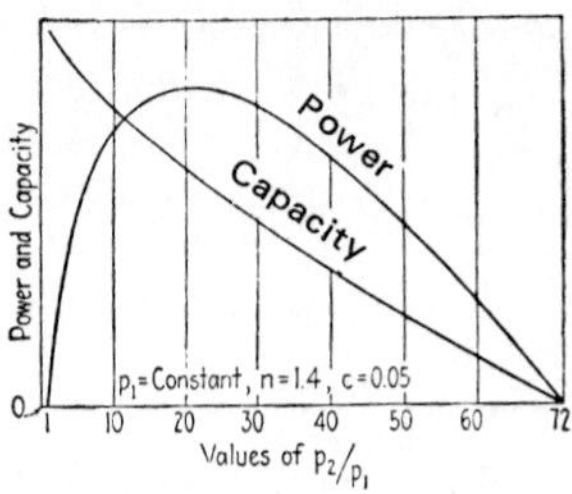

FIGURE 13.162 Effect of discharge pressure on power and capacity.

The power to run the compressor increases to some maximum value and then decreases to zero, theoretically, when the maximum pressure is reached. Figure 13.162 shows a specific case.

The practical significance is that with a given compressor it is impossible to exceed a certain pressure even by closing a throttle valve, and that a certain power cannot be exceeded no matter how high the pressure may be carried.

55. Multistaging. For high pressures, compressors are built in two or more stages, thus reducing the pressure range within each cylinder and hence maintaining a reasonable volumetric efficiency. If the air is also cooled, between stages, to the initial temperature, there is a saving of power as shown in Fig. 13.163, where *ABC* is the compression line for a single cylinder and *AD* is an isothermal.

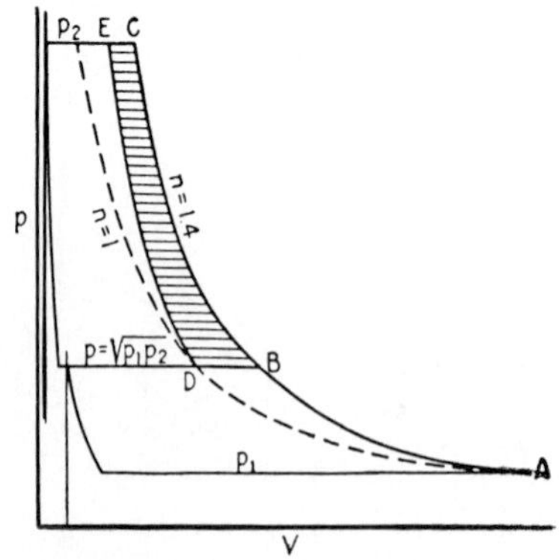

FIGURE 13.163 Two-stage compression with intercooling.

For minimum work with perfect intercooling, the intermediate pressures are given by $p = \sqrt{p_1 p_2}$ for a two-stage compressor; $p' = \sqrt[3]{p_1^2 p_2}$ and $p'' = \sqrt[3]{p_1 p_2^2}$ for a three-stage machine.

If there were no pressure drops in the intercoolers, the cylinder volumes would be inversely proportional to these pressures and to p_1. Owing to the higher volumetric efficiency with the lower pressure, the volume of the low-pressure cylinder will be less than that of a single-stage compressor of the same capacity.

56. Compressor Efficiencies. The *mechanical* efficiency is the work done in the air cylinders as given by the actual indicator card divided by the indicated work in the engine cylinders, if driven by a direct-connected engine, or by the work delivered by a belt, if it is belt-driven. For piston compressors, mechanical efficiencies are about 90 percent for direct-connected machines and a few percent higher for the belt-driven types.

The *compression* efficiency is the ratio of the work required to compress and deliver the gas, as given by an ideal cycle with isothermal (or adiabatic) compression, to the work done on the air, as shown by actual indicator cards. Values of *isothermal compression* efficiency are usually from 65 to 70 percent and nearly independent of the pressure, while *adiabatic compression* efficiencies are: 64 percent for $p_2/p_1 = 1$; 76 percent for $p_2/p_1 = 2$; 87 percent for $p_2/p_1 = 3$; and about 90 percent for all higher ratios.

57. Power Required for Compressor. The power required to compress and deliver to the receiver any gas is given theoretically for adiabatic compression by

$$\text{Power} = 0.01525 p_1 V_1 \left[\left(\frac{p_2}{p_1}\right)^{0.286} - 1\right] \text{ hp (1 hp = 0.75 kW)} \qquad (13.58)$$

where p_1 is initial pressure and p_2 final pressure, in lb/in^2 abs (kPa abs), and V_1 is ft^3 of gas per min (m^3/min) at p_1. This is for a single stage only. For a multistage compressor with S stages and perfect intercooling, multiply 0.01525 by S and divide 0.286 by S.

For isothermal compression the equation for power becomes

$$\text{Power} = 0.01 p_1 V_1 \log_{10}\left(\frac{p_2}{p_1}\right) \text{ hp (1 hp = 0.75 kW)} \qquad (13.59)$$

The actual power required to drive the compressor may be obtained from either of the above by dividing by the mechanical efficiency and the proper compression efficiency as given in Sec. 56.

The theoretical power for other values of n may be obtained by the equation

$$\text{Power} = 0.00437\left(\frac{n}{(n-1)}\right) p_1 V_1 \left[\left(\frac{p_2}{p_1}\right)^{(n-1)/n} - 1\right] \text{ hp (1 hp = 0.75 kW)} \qquad (13.60)$$

For small pressure differences, where $(p_2 - p_1)/p_1$ is less than 0.1, such as with fans and blowers, it is sufficiently accurate to use the "hydraulic" formula for a fluid of constant density, which is power = $0.00437\,(p_2 - p_1)V_1$ hp (1 hp = 0.75 kW) in place of any of the preceding equations.

Values of horsepower for both adiabatic and isothermal compression are shown in the following table for 100 ft^3/min (2.8 m^3/min) at 14.7 lb/in^2 (101.3 kPa) compressed to various gage pressures:

Final pressure, hp (kPa)	10 (68.9)	20 (137.9)	40 (275.8)	60 (413.6)	80 (551.5)
Adiabatic hp (kW)	3.59 (2.7)	6.24 (4.7)	10.25 (7.6)	13.3 (9.9)	15.8 (11.8)
Isothermal hp (kW)	3.32 (2.5)	5.48 (4.1)	8.40 (6.3)	10.4 (7.6)	11.9 (8.9)

58. Piston Compressors. Reciprocating piston compressors are built for pressures from 1 to 100 lb/in^2 or more in single stages, up to 500 lb/in^2 (3.4 MPa) in two stages, 1200 to 2500 lb/in^2 (8.3 to 17.2 MPa) in three stages, and up to 5000 lb/in^2 (34.5 MPa) in four stages. For moderate pressures double-acting pistons are used, but for very high pressures it is customary to use a single-acting plunger. Compressor cylinders and cylinder heads should be water-jacketed. Very small sizes may be air-cooled by metal fins.

Speeds recommended by the Compressed Air Society are

Stroke, in (cm)	6 (15.2)	10 (25.4)	14 (35.6)	18 (45.7)	24 (60.9)
R/min	350	300	257	214	167

59. Compressor Valves. Compressor valves are generally automatic, and for small machines or for high-pressure stages poppet valves of the single-beat type may be employed. For other services, valves are made of very light-weight plate or strips which are held in suitable cages.

60. Unloading Devices. Since compressors usually run at constant speed and the demand for air is variable, some form of unloading device is necessary. This is actuated by the air pressure in the receiver, so that when the pressure rises to a predetermined amount the device stops further delivery. This may be accomplished by holding the intake valve open, by holding it closed, by closing an auxiliary valve in the air-intake line, or by varying the clearance in the compressor by opening the normal clearance space to special clearance pockets.

61. Rotary Blowers. Rotary blowers are also of the positive displacement type, and are usually limited to pressures of from 0.5 to 12 lb/in^2 gage (3.4 to 82.7 kPa). They are very similar to rotary pumps for liquids, as shown in Fig. 13.137*a* and most of the statements made in Secs. 18 and 19 apply here equally well. For small sizes and high pressure, the slip tends to be excessive; but, under favorable conditions and for low pressures, efficiencies as high as 80 to 90 percent are claimed.

62. Centrifugal Compressors. Centrifugal compressors are very similar to centrifugal pumps in all respects, the difference being that they handle a gas of variable density instead of a liquid of constant density. In addition to its increase in density, the temperature of the gas rises during flow through the machine, and thus introduces thermodynamic effects. Because of the increase in density, all impeller dimensions should be decreased with successive stages, but for manufacturing reasons impellers either are made all alike if the number of stages is small, or are arranged in groups if many stages are employed. Another difference from the centrifugal pump is that the machine is usually water-jacketed. (See Sec. 20 and following.)

Single-stage compressors are built for pressures of from 1 to 5 lb/in^2 gage (6.9

to 34.4 kPa) and even up to 15 lb/in^2 (103.4 kPa). Capacities range from 100 to 100,000 ft^3/min (2.8 to 2831 m^3/min). Multistage machines are built for pressures up to 125 lb/in^2 (861.8 kPa) and even up to 200 lb/in^2 (1379 kPa) in a few cases. Peripheral velocities up to 1000 ft/s (304.8 m/s) are employed, and efficiencies are as high as 75 percent.

63. Theory. The energy input *to* the gas per pound (kilogram) is given by H, which is identical with that of Sec. 22 for the centrifugal pump, while the energy delivered *in* the gas is $e_h \times H$. The hydraulic losses, mechanical losses, disk-friction losses, and leakage losses are all similar to the same quantities for centrifugal pumps.

The power delivered in the gas is given by

$$\text{hp} = \frac{Sd_1V_1e_hH}{33{,}000} \quad (1\text{hp} = 0.75\text{k}W) \qquad (13.61)$$

where S = number of stages, H = head per stage, d_1 = density, in lb/ft^3 (kg/m^3), and V_1 = ft^3/min (m^3/min), at p_1. It is quite customary for the outlet of the vanes of air compressors to be radial, or $\beta_2 = 90°$. For that special case, $H = u_2{}^2/g$, but for other vane angles the appropriate expression for H should be used. This equation for power should give values identical with those obtained by the equations in Sec. 58. See Fig. 13.164.

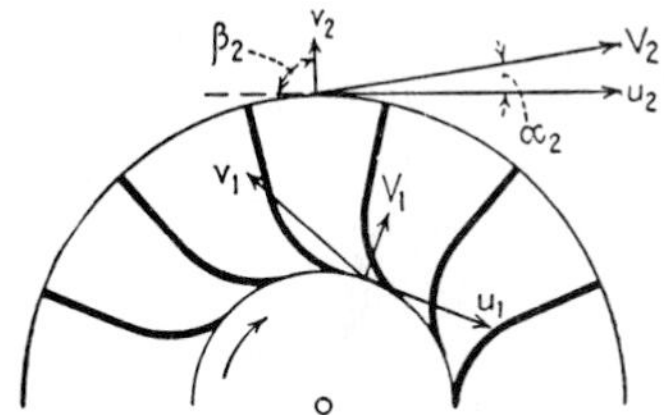

FIGURE 13.164

The pressure rise in the compressor with perfect intercooling between stages may be found from

$$p_2 = p_1\left(1 + \frac{d_1e_hH(n-1)}{144np_1}\right)^{Sn/(n-1)} \qquad (13.62a)$$

while, if the cooling is continuous throughout, the pressure is

$$p_2 = p_1\left(1 + \frac{Sd_1e_hH(n-1)}{144np_1}\right)^{n/(n-1)} \qquad (13.62b)$$

The heat produced in a centrifugal compressor owing to internal friction of the air may be such that the temperature rise may be more than that due to an adiabatic process with $n = 1.4$. With the cooling assumed to be such that $n = 1.4$, and with the further substitution of the special case for $\beta_2 = 90°$, $p_1 = 14.7$, $d_1 = 0.0752$, $e_h = 0.70$, the above reduce to

$$p_2 = 14.7\left(1 + \frac{u_2^2}{4{,}530{,}000}\right)^{3.5^S}$$

and

$$p_2 = 14.7\left(1 + \frac{Su_2^2}{4{,}530{,}000}\right)^{3.5}$$

The characteristics of centrifugal compressors are very similar to those of centrifugal pumps, except that with radial vanes, the pressure-quantity curve tends to be flatter than for the usual pump. The variation with D and N would be the same as stated in Secs. 27 and 28.

64. Centrifugal Fans. The centrifugal fan is practically identical with the centrifugal pump so far as its theory is concerned, because the variation of the density of the gas within it is negligible (see Secs. 22 through 28). Fans are built to produce pressures up to 0.5 lb/in^2 (3.4 kPa) or 28 in (711.2 mm) of water, but in usual practice the figure is very much less than that. Capacities range as high as 400,000 ft^3/min (189 m^3/s).

65. Fan Pressures. Because of their small values, fan pressures are usually stated in terms of inches (mm) of water or ounces per square inch (kPa). One inch (25.4 mm) of water is equivalent to 0.577 oz/in^2 or 0.0362 lb/in^2 (0.25 kPa) or 69.3 ft (21.1 m) of air whose density is 0.0752 lb/ft^3 (1.2 kg/m^3). Three pressures are considered: static, velocity, and impact, which is the sum of the first two. The last is usually used for computing fan efficiency, but the static pressure may also be used. Hence it is necessary to specify which pressure is used.

If the velocity head of a gas is represented by h_v in (mm) of water, the velocity in feet per minute (m/min) is $V = (1100/\sqrt{w})\sqrt{h_v}$, where w = density in lb/ft^3 (kg/m^3).

66. Air Horsepower. The power in the air delivered by the fan is $5.2Qh/33{,}000$, where Q = ft^3/min (m^3/min), and h = in (mm) of water. It may be either static pressure or impact pressure.

67. Fan Efficiency. The overall efficiency of a centrifugal fan is e = ahp/bhp, where ahp is the air horsepower and bhp the brake horsepower or the power necessary to run the fan (e = akW/bkW). Its value is usually about 60 to 70 percent.

68. Construction Features. Owing to the light pressures, certain structural features of centrifugal fans, such as the use of built-up fan wheels and sheet-plate construction of wheels and case, are quite different from corresponding parts in the centrifugal pump; but despite this the similarity is very close.

The use of radial blades in fan wheels is very common, but vanes are also curved backward, as in centrifugal pumps, and also forward. Radial blades are generally used where the width of the fan wheel is not too large as compared with its diameter, the inner diameter is not large as compared to the outer diameter, and the vanes are few. It is a comparatively low-speed type.

The so-called multiblade fan is one with curved blades, generally curved forward in the direction of rotation; the width of the wheel is large compared to its diameter; the inner diameter is a large proportion of the outer diameter; and there are many vanes placed close together. Such a fan is of a high capacity and a high rotative speed.

69. Fan Characteristics. The characteristics of centrifugal fans are identical with those of the centrifugal pump (Sec. 24), but instead of one curve for h there are often two, the static-pressure and the impact-pressure curves, and likewise there are two efficiency curves to correspond. The various curves may be plotted

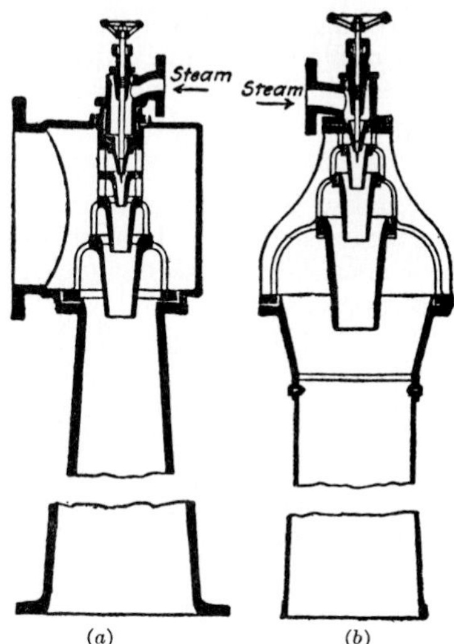

FIGURE 13.165 (*a*) Exhauster. (*b*) Steam-jet air compressor.

with Q as a base or, since the static head is often of greater interest and more readily measured, all values may be plotted against it.

70. Steam-Jet Blowers and Vacuum Pumps. For many purposes the steam-jet compressor or vacuum pump (Fig. 13.165) is very useful. It is simple and compact, but is inefficient unless exhaust steam is used for it. A form of this is widely used for air pumps on steam condensers. They use from 0.19 to 0.37 lb of steam per 1000 ft^3 (0.09 to 0.17 kg per 28.3 m^3) of air.

ENGINEERING THERMODYNAMICS*

FUELS AND COMBUSTION

1. Fuels. In a commercial sense, a fuel is any material of suitable cost that will combine, in part, with the oxygen of the air to liberate heat. Fuels may be classified as *solid, liquid*, and *gaseous. Natural* solid fuels are coal, lignite, peat, and wood; *prepared* solid fuels include coke, pulverized coal, and charcoal; *waste* solid fuels include sawmill refuse and bagasse (crushed sugar cane).

2. Coal. Coal, the most important and most abundant solid fuel, is of vegetable origin and exists in a variety of forms having different chemical and physical characteristics. It contains carbon, hydrogen, oxygen, nitrogen, sulfur, and ash.

*From General Engineering Handbook, 2d ed., by C. E. O'Rourke. Copyright © 1940. Used by permission of McGraw-Hill, Inc. All rights reserved. Updated and metricated by the editors.

Sampling Coal. In collecting a sample of coal for analysis or for heating-value determination, it is of primary importance that the sample be truly representative. For the standard method of sampling coal, adopted by the American Society of Mechanical Engineers, see ASME "Test Code for Solid Fuels."

2a. Coal Analysis. Two kinds of analyses are commonly used to ascertain the composition of coal: the ultimate analysis and the proximate analysis.

Ultimate Analysis. The chemical analysis of a fuel, giving the percentages of carbon, hydrogen, oxygen, nitrogen, sulfur, and ash, is known as the *ultimate analysis*. For the standard method of making these determinations, see ASME "Test Code for Solid Fuels."

Proximate Analysis. This analysis determines the percentages by weight of moisture, volatile matter, fixed carbon, and ash in the fuel. Usually a statement of the sulfur content and the calorific value of the fuel is included in the commercial proximate analysis. *Moisture* is considered to be the loss in weight of a sample (1 g) of coal when dried at a temperature of 104 to 110°C. for 1 h. *Combustible* is arbitrarily defined as that portion of the coal remaining after subtraction of ash and moisture. *Volatile matter* is the total combustible minus the fixed carbon; it includes the carbon which is combined with hydrogen, together with free oxygen, nitrogen, and other gas-forming constituents of the dry fuel which are driven off by heat. *Fixed carbon*, or uncombined carbon, is the combustible remaining after the volatile matter has been driven off. It is determined by subtracting from the weight of the original sample the weight of the moisture, ash, and volatile matter. *Ash* is the incombustible residue from the complete burning of the coal. For the standard method of making these determinations, see ASME "Test Code for Solid Fuels."

3. Heating Value. The *calorific* or *heating value* of a fuel is the amount of heat recovered when the products of complete combustion of unit quantity of the fuel are cooled to the initial temperature of the air and fuel. The heating values of solid and liquid fuels are expressed on the weight basis in Btu per pound (kJ/kg). For gaseous fuels, the heating values are expressed in Btu per cubic foot (kJ/m^3) of the gas; common standard conditions for the measurement of the gas volume are a temperature of 60°F (15.6°C), a total pressure of 30 in Hg (76.2 cm Hg), and "saturation" of the gas with water vapor. When a fuel contains hydrogen, there are a number of possible heating values depending upon the fractional part of water vapor formed during combustion that condenses when the products of combustion are cooled to the initial temperature. If none of the water vapor condenses, the resultant heating value is the *lower heating value*; if all of the water vapor condenses, the heating value is the *higher heating value*. The numerical difference between the higher and lower heating value of any hydrogen-containing fuel is the product of the weight of water vapor formed from the complete combustion of unit quantity of the fuel and the latent heat of condensation of that water vapor. The heating value of fuels may be determined by calorimeter tests or may be estimated, with a fair degree of accuracy, from a knowledge of their chemical and physical properties. The approximate higher heating value (HHV) of coal, in Btu per pound (kJ/kg), may be calculated from the ultimate analysis by a formula of the *Dulong* type:

$$\text{HHV} = 14{,}500C + 62{,}000\left(H - \frac{O}{8}\right) + 4000S\ (\times\ 2.3 = \text{kJ/kg}) \tag{13.63}$$

where C, H, O, and S are the fractions by weight of carbon, hydrogen, oxygen, and sulfur in the coal.

4. Classification of Coal. A classification of coals based on fixed carbon, volatile matter, and moisture is given in Fig. 13.166.

Although there is no universal standard of coal screening in practice, coal sizes are approximately as given in Table 13.7.

5. Liquid Fuels. Petroleum in its unrefined state, frequently called *crude oil*, is a viscous, dark brown or greenish liquid occurring in natural reservoirs in the earth's crust in many parts of the world. Regardless of their source, petroleum and petroleum products have ultimate analyses that generally fall within the following limits:

	Carbon	Hydrogen	Sulfur	O + N	Moisture and sediment
Percentage by weight	80 to 87	10 to 15	0.3 to 2.5	1 to 7	0.1 to 1.5

In the purchase and sale of petroleum and its distillates, the specific gravity is of importance. In general, the higher the specific gravity, the lower is the content of lighter hydrocarbons and the lower the heating value of the oil. Commercially the specific gravity is ascertained by means of a hydrometer. The reading of the hydrometer is related to the true specific gravity as indicated by the expression: Specific gravity = 141.5/(131.5 + B), in which B = deg. API (American Petroleum Institute) as read from the scale of the hydrometer. Table 13.8 gives the relation of API readings to density and volume. Fuel oil is generally sold by the barrel, each barrel containing 42 gal of 231 in^3, at a temperature of 60°F (159 L of 3785 cm^3 at 15.6°C).

Crude oil has highly inflammable volatile components. When the oil is refined by fractional distillation, these are removed and such valuable products as gasoline and kerosene are obtained, leaving a residual oil commonly called *fuel oil*.

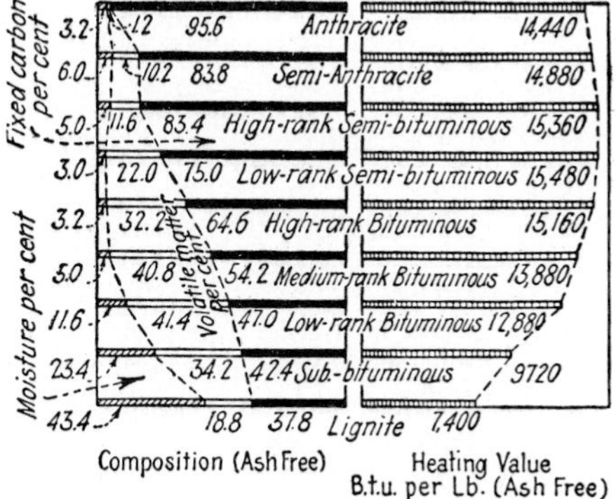

FIGURE 13.166 U.S. Geological Survey's classification of coals. (1 Btu/lb = 2.33 kJ/kg.)

TABLE 13.7 Coal Sizes

Trade name	Diameter of opening through or over which coal passes, in (cm)	
	Through	Over
Anthracite		
Egg	3¼ (8.3)	2⅜ (6.0)
Stove	2⅝ (6.7)	1¼ (3.2)
Chestnut	1¼ (3.2)	¾ (1.9)
Pea	¾ (1.9)	½ (1.3)
No. 1 buckwheat (buckwheat)	½ (1.3)	¼ (0.6)
Eastern bituminous		
Lump	...	1¼ (3.2)
Nut	1¼ (3.2)	¾ (1.9)
Slack	¾ (1.9)	
Western coals		
Nut, 3 in	3 (7.6)	1¼ (3.2)
Nut, 1¼ in	1¼ (3.2)	¾ (1.9)
Nut, ¾	¾ (1.9)	⅝ (1.6)
Screenings	1¼ (3.2)	

The higher heating value of fuel oil, in Btu per pound (kJ/kg), is given approximately by the Sherman and Kropf empirical equation:

$$\text{HHV} = 18{,}650 + 40(B - 10) = 13{,}050 + \frac{5600}{\text{Sp. gr.}} \quad (\times\ 2.3 = \text{kJ/kg}) \tag{13.64}$$

The suitability of a fuel oil for *power-plant use* depends on the cost on the basis of heating value, and freedom from grit, acids, sulfur, and other objectionable constituents; the oil should have a flash point over 150°F (65.6°C) and a suitable viscosity for pumping.

Fuel oil for use in *engines* must be clean and noncorrosive, and must have a viscosity that permits its flow through small passages. *Gasoline* is the best known and most widely used engine fuel and is obtained largely by distillation of petroleum. *Casing-head gasoline* is obtained from natural gas by compression and absorption methods; it is too volatile for commercial sale and is blended to produce a satisfactory engine fuel. Gasoline is also produced by a *cracking* process: The heavy oils are subjected to high pressure and temperature and break up to form

TABLE 13.8 Relation of API Readings to Density and Volume

Deg. API	Sp. gr. 60°/60°F (15.6°C/15.6°C)	Lb per U.S. gal (kg/L)	Weight per bbl, lb (kg)
10.0	1.0000	8.328 (0.998)	349.8 (158.7)
15.0	0.9659	8.044 (0.964)	337.8 (153.2)
20.0	0.9340	7.778 (0.932)	326.7 (148.2)
25.0	0.9042	7.529 (0.902)	316.2 (143.4)
30.0	0.8762	7.296 (0.874)	306.4 (138.9)
35.0	0.8498	7.076 (0.848)	297.2 (134.8)
40.0	0.8251	6.870 (0.823)	288.5 (130.9)

compounds of greater volatility. The higher heating value of gasoline in Btu per pound (kJ/kg), is closely given by the following empirical equation:

$$\text{HHV} = 12{,}720 + \frac{5600}{\text{Sp. gr.}} \;(\times\ 2.3 = \text{kJ/kg}) \qquad (13.65a)$$

Octane (C_8H_{18}) is about the average of the mixture of several hydrocarbons that constitute gasoline.

Kerosene is the distillation product that lies between gasoline and the distillate fuel oils; it is used in tractor and similar engines. Kerosene is composed of a number of hydrocarbons of which $C_{12}H_{26}$ is about the average. The higher heating value of kerosene, in Btu per pound (kJ/kg), is closely given by the following empirical equation:

$$\text{HHV} = 12{,}840 + \frac{5600}{\text{Sp.gr}} \;(\times\ 2.3 = \text{kJ/kg}) \qquad (13.65b)$$

The so-called *engine distillates* consist of the fractions distilled off between kerosene and the lubricating oils. The lighter distillates may be thought of as low-grade kerosene and are used in semi-Diesel and high-speed Diesel engines. The specific gravity of these fuels may vary from 28° to 44° on the API scale. In Table 13.9 are given representative data that apply to crude oil, fuel oil, distillates, kerosene, and gasoline.

Benzol consists principally of benzene, C_6H_6, and is obtained from byproduct coke ovens. The specific gravity of benzol is 0.88, the higher heating value about 18,000 Btu per lb (41,868 kJ/kg), and the lower about 17,300 Btu per lb (40,240 kJ/kg). Commercial benzol, or 90 percent benzol, a satisfactory motor fuel, contains about 70 percent of benzol, 24 percent of toluol, and 6 percent of heavy hydrocarbons.

Alcohols, both ethyl (grain) and methyl (wood), are suitable engine fuels. Pure ethyl alcohol, C_2H_6O, has higher and lower heating values of about 12,820 and 11,580 Btu per lb (29,819 and 26,935 kJ/kg); pure methyl alcohol, CH_4O, has higher and lower heating values of 9600 and 8410 Btu per lb (22,330 and 19,562

TABLE 13.9 Analyses and Heating Values of American Crude Petroleum, Fuel Oils, and Typical Distillates

Product	API	Specific gravity at 60°F (15.6°C)	Weight per gal		Higher heating value		Ultimate analysis				
			lb	kg	Btu per lb	kJ/kg	C	H	S	N	O
California crude	22.8	0.917	7.836	3.55	18,910	43,985	84.00	12.70	0.75	1.70	1.20
Mexican crude	13.6	0.975	8.120	3.68	18,755	43,624	83.70	10.20	4.15		
Oklahoma crude	31.3	0.869	7.236	3.28	19,502	45,362	85.70	13.11	0.40	0.30	
Pennsylvania crude	42.6	0.813	6.769	3.07	19,505	45,369	86.06	13.88	0.06	0.00	0.00
Texas crude	30.2	0.875	7.286	3.30	19,460	45,264	85.05	12.30	1.75	0.70	0.00
Gasoline	60.0	0.739	6.152	2.79	20,750	48,265	84.90	14.76	0.08		
Kerosene	41.2	0.819	6.822	3.09	19,810	46,078					
Gas oil	32.5	0.863	7.184	3.26	19,200	44,659					
Furnace oil	26.4	0.896	7.462	3.39	19,025	44,252					
Fuel oil (Mex.)	11.8	0.987	8.223	3.73	18,510	43,054	84.02	10.06	4.93		
Fuel oil (mid-continent)	27.1	0.892	7.428	3.37	19,376	45,069	85.62	11.98	0.35	0.50	0.60
Fuel oil (Calif.)	16.6	0.9554	7.956	3.61	18,835	43,810	84.67	12.36	1.16		

Note: A barrel of oil contains 42 U.S. gal (159 L).
Source: From Avallone and Baumeister.[2]

kJ/kg), respectively. Commercial alcohols, containing water and denaturants, have higher and lower heating values of about 11,500 and 10,500 Btu per lb (26,749 and 24,423 kJ/kg). Benzol is usually added to commercial alcohols to prevent excessive corrosion from the formation of acetic acid. Alcohol is used in tropical countries as an engine fuel.

6. Gaseous Fuels. Natural gas, blast-furnace gas, producer gas, and byproduct coke-oven gas are used as fuels for the generation of steam and directly for the generation of power in large internal-combustion engines. In Table 13.10 are given representative data on manufactured gaseous fuels; these fuels differ widely in physical and chemical properties, and, for accurate data, tests of a particular fuel must be made.

Natural gas is of organic origin and is usually found along with petroleum. The composition of natural gas depends upon the location of the gas field, but representative limits are given in Table 13.10*a*.

Coal gas results from the destructive distillation of coal and is popularly known as *city gas* or *illuminating gas*. In general, coal gas is too costly for engine use. It is, however, a highly satisfactory fuel. *Coke-oven gas* is a byproduct of coke ovens. It has a good heating value and, when freed from dust, sulfur compounds, and ammonia, makes a very good engine fuel. Sulfur compounds along with moisture form sulfuric acid (H_2SO_4), and ammonia likewise forms nitric acid (HNO_3), both of which corrode metal engine parts. Engines of large rating are operated on coke-oven gas. Difficulties from tar and naphthalene will arise unless these constituents are removed from the gas. *Blue water gas* is a mixture of permanent gases, chiefly hydrogen and carbon monoxide, and is produced by the interaction of steam and incandescent carbon. Small amounts of carbon dioxide and methane are also present. This gas is not of importance as an engine fuel. *Carbureted water gas* is blue water gas into which, during the manufacturing process, petroleum oil is sprayed; this process forms a composite gas that has a greater heating value per cubic foot and can be burned with a luminous flame. Much city gas is in whole or part carbureted water gas. It is a fine engine fuel, but like coal gas its cost is usually too high for use in engines. *Oil gas* is produced by the decomposition (cracking) of petroleum oils by the application of high temperatures. It is similar in composition to coal gas. Its uses are similar to those of coal gas. It is an excellent engine fuel but, like that of coal gas, its cost is ordinarily too high to allow extensive use. *Producer gas* is formed when a mixture of air and steam is blown up through a thick, hot mass of coal or coke. The greater portion of the original heating value of the coal or coke is contained in the carbon monoxide (CO) and the hydrogen (H_2) which are formed. It is the cheapest gas for industrial purposes that can be manufactured. Low-grade fuels and crude oil may be used in gas producers. Producer gas from anthracite coal has been satisfactory as an engine fuel. Producer gas from bituminous coal has been used, but usually without satisfaction, because of the difficulties in removing tar and dust. Producer gas is not of large importance at the present time as an internal-combustion-engine fuel. *Blast-furnace gas* is an important engine fuel and is generally used in large engine units. It is also burned in large quantities directly under boilers. The volume of gas produced per ton of iron made is from 130,000 to 150,000 ft^3 (3680 to 4247 m^3). Allowing for the gas used in stoves, loss when tapping, and irregular gas supply, about 83,000 ft^3 (2350 m^3) of blast-furnace gas per ton of iron is available for power purposes. When the dust has been removed,

TABLE 13.10 Volumetric Composition and Heating Value of Gases

Manufactured gases	Methane (CH_4), %	Ethylene C_2H_4), %	Carbon monoxide (CO), %	Carbon dioxide (CO_2), %	Hydrogen (H_2), %	Oxygen (O_2), %	Nitrogen (N_2), %	Heating value, Btu/ft^3 at 60°F, 30 in Hg (MJ/kg at 15.6°C, 76.2 cmHg), sat. with H_2O: Higher	Heating value: Lower	Air required per ft^3/gas, ft^3(m^3)
Coal gas	34.0	6.6	9.0	1.1	47.0	...	2.3	634 (23.6)	560 (20.9)	5.50 (0.16)
Coke-oven gas	28.5	2.9	5.1	1.4	57.4	0.5	4.2	536 (19.9)	476 (17.7)	4.65 (0.13)
Coke-oven gas	33.9	5.2	6.1	2.6	47.9	0.6	3.7	600 (22.4)	538 (20.0)	5.28 (0.15)
Blue water gas	...	...	43.4	3.5	51.8	...	1.3	310 (11.6)	285 (10.6)	2.28 (0.07)
Carbureted water gas	14.8	12.8	33.9	1.5	35.2	...	1.8	578 (21.5)	529 (19.7)	4.85 (0.14)
Oil gas	27.0	2.7	10.6	2.8	53.5	...	3.4	516 (19.2)	461 (17.2)	4.25 (0.12)
Producer gas	2.6	0.4	22.0	5.7	10.5	...	58.8	136 (5.1)	128 (4.8)	1.08 (0.03)
Blast-furnace gas	...	...	26.2	13.0	3.2	...	57.6	93 (3.5)	91.6 (3.4)	0.70 (0.02)

TABLE 13.10*a* Approximate Limits of Composition and Heating Value of Natural Gas

Higher heating value, Btu/ft³ at 60°F, 30 inHg, sat (MJ/kg at 15.6°C, 76.2 cmHg)	Methane, CH_4, percent by vol.	Ethane, C_2H_6, percent by vol.	Nitrogen, N_2, percent by vol.	Carbon dioxide, CO_2, percent by vol.	Ratio $\frac{\text{Dens. of gas}}{\text{Dens. of air}}$
700 to 1500 (26.1 to 55.9)	30 to 99	0 to 70	0 to 40	0 to 25	0.6 to 0.9

blast-furnace gas is a very good gas-engine fuel, because it is low in hydrogen and relatively free from impurities.

7. Minimum Amount of Air Required for Combustion. The air required for the complete combustion of a fuel depends upon the chemical composition of that fuel. The atomic weights of elements important in combustion are: carbon, 12.00; hydrogen, 1.00; sulfur, 32.06; nitrogen, 14.01. For combustion calculations, atmospheric air may be treated as a mechanical mixture of nitrogen and oxygen in the proportion of 79 parts N_2 to 21 parts O_2, by volume, and 77 parts N_2 to 23 parts O_2 by weight. By writing the equation of the chemical reaction, the weight of oxygen (and of air) required to burn completely each combustible element of the fuel may be determined. For example, for the complete combustion of 1 lb (0.45 kg) of carbon, $C + O_2 = CO_2$. By substitution of atomic weights, 12 lb (5.4 kg) of carbon + 32 lb (14.5 kg) of oxygen yields 44 lb (19.9 kg) of CO_2. In other words, to burn completely 1 lb (0.45 kg) lb of carbon requires 32/12(0.23) or 11.6 lb (5.3 kg) of air. The weight of air required to burn completely 1 lb (0.45 kg) of hydrogen is 34.8 lb (15.8 kg), and to burn 1 lb (0.45 kg) of sulfur, 4.35 lb (1.97 kg).

Excess Air. If the minimum amount of air required for complete combustion were supplied to an actual furnace, the fuel would not burn completely, largely because of imperfect mixing of the oxygen and combustibles. It is necessary to supply excess air in order to prevent the loss of heat due to incompleteness of combustion. An increase in excess air causes a greater loss of heat in the exit gases, however, and there is an optimum amount of excess air for any one fuel and set of operating conditions where the total loss of heat will be a minimum. For any one fuel, the CO_2 content of the products of combustion may be used as an index of the excess air. The coefficients in the chemical-reaction equations may be used to represent volumes of gaseous constituents. For example, in the complete combustion of carbon in air, one volume of CO_2 results from each volume of oxygen consumed. Since each volume of oxygen is mixed with $^{79}/_{21}$ or 3.76 volumes of nitrogen in air, the products of combustion will consist of 1 ft³ (0.028 m³) of CO_2 and 3.76 ft³ (0.106 m³) of the inert N_2 per cubic foot of oxygen consumed. By volume, then, the products of combustion of carbon will consist of 21 percent CO_2 and 79 percent of N_2 when there is no excess of air and no deficiency. When excess air is supplied, excess O_2 and additional N_2 appear in the products of combustion and decrease the percent by volume of CO_2. For fuels containing hydrogen, the maximum CO_2 content of the products of combustion is less than 21 percent, decreasing as the hydrogen increases. For example, in the combustion of methane, CH_4, in air, the chemical-reaction equation (with minimum air for complete combustion) is

$$CH_4 + 2O_2 + 7.52N_2 = CO_2 + 2H_2O + 7.52N_2$$

The fraction by volume of CO_2 in the *dry* products of combustion is 1/8.52 or 11.7 percent.

The heat lost in the dry products of combustion leaving an engine or a boiler setting may be found from the relation $Q = 0.24w(t_2 - t_1)$, where Q = heat lost per lb (kg) of fuel, in Btu (MJ); w = weight of dry products of combustion per lb (kg) of fuel, in lb (kg); t_2 = temperature of the exit products, in °F (°C); and t_1 = temperature of the air supplied for combustion, in °F (°C) (× 2.3 = kJ/kg). If the fuel contains much hydrogen, the latent heat of water vapor formed during combustion must also be included as a heat loss; any incomplete combustion is an additional source of heat loss.

INTERNAL-COMBUSTION ENGINES

8. Characteristics. Important characteristics distinguishing internal-combustion engines from steam engines and other heat engines are the combustion of the fuel with the necessary air directly in the engine cylinder and the use of the resulting products of combustion as the working substance. For all ordinary considerations, the working substances may be considered as having the properties of air. In internal-combustion engines the maximum gas temperature existing in the cylinder occurs during the explosion or combustion process; 3300°R may be taken as a representative average. At full load, the temperature of the gases at the end of expansion will average about 1800°R, and the temperature of the exhaust gases will average about 1200°R. The internal-combustion engine operates over an average range of temperature of 1500°F (816°C), while the steam plant operates over an average range of about 700°F (371°C). Consequently, the theoretical efficiencies of practical internal-combustion-engine cycles are much higher than those of practical steam cycles. In general, any first-class internal-combustion engine of any given rating commercially available at the present time will operate with a much higher actual overall thermal efficiency than any first-class commercially available steam plant of the same rating.

9. Heat and Thermodynamics. Chemical potential energy is the real energy which is supplied to any internal-combustion engine and is associated with the fuel supplied to the engine. No internal-combustion engine can utilize this chemical energy directly, and it must therefore be converted into heat energy. This conversion is effected by the explosion or combustion process within the cylinder. It is the function of any internal-combustion engine to convert as much of the available heat energy as possible into useful mechanical work. Theoretically it is immaterial as to the kind of fuel which is used to supply heat by the combustion process to the compressed charge of air in the engine cylinder; actually, however, there are other considerations which make certain fuels better adapted for use than others.

10. Cycles. All commercial engines operate on either (1) the *Otto cycle*, (2) the *Diesel cycle*, or (3) a cycle which embodies characteristics of both the Otto and the Diesel and is variously termed the *semi-Diesel cycle, mixed cycle*, or *Sabathe cycle*.

Otto Cycle. In this cycle combustion occurs at constant volume, and the fuel and air are mixed in the proper ratio before entering the cylinder. Consequently, the compression pressure is limited by preignition. This cycle has by far the widest use. Its most extensive applications are to automobile, aircraft, tractor, marine, and miscellaneous small engines. In these applications gasoline is the principal fuel, although many tractor and industrial engines use kerosene. This cycle is also extensively used in large stationary engines operating on such fuels as natural gas, blast-furnace gas, producer gas, coal gas, coke-oven gas, etc. The Otto cycle is not well adapted to the use of low-grade fuels or heavy oils. Engines operating on this cycle are low in first cost, long-lived, and very satisfactory in operation.

Diesel Cycle. In this cycle, combustion occurs at constant pressure. The outstanding features of the cycle are: (1) air only is drawn into the cylinder on the suction stroke; (2) compression is carried to a pressure of 375 to 550lb/in^2 gage (2585 to 3792 kPa), thus heating the air to 900 to 1100°F (482 to 593°C); (3) air or solid injection of so-called nonvolatile fuel oils or crude oils. The temperature of the air in the cylinder at the time of injection of the fuel is well above the ignition point of the fuel, and burning occurs. It is ordinarily intended that there shall be no rise in pressure over that existing at the end of the compression stroke, the rate of injection being adjusted so that any tendency toward a rise in pressure is just offset by the losses and the rate at which work is being done by the engine. This condition is not always realized in practice.

Diesel engines are built in sizes from a 6-bhp (4.5-kW) farm engine to a 35,000-bhp (26,110-kW) eight-cylinder engine for central-station service. The overall thermal efficiency of Diesel engines is very high, even in small sizes, and the thermal efficiency is well maintained over a large load range. Until quite recently the Diesel engine was rather costly and was made only in fairly large sizes. It is now being manufactured in small sizes, and costs have been much reduced. The ability of the Diesel engine to use practically any fuel, the absence of carburetors and ignition equipment, and excellent torque characteristics are outstanding advantages.

Semi-Diesel or Mixed Cycle. In general, any cycle wherein some of the fuel is burned at constant volume and some at constant pressure falls under this classification. All internal-combustion-engine cycles in commercial use today that are not classifiable as Otto or Diesel cycles are properly and best classified as *mixed cycles*.

Almost coincident with the commercial application of engines operating on the Otto cycle were the development and application of engines that use the less volatile and cheaper fuel oils and operate on some form of mixed cycle. Only air is compressed in the cylinder, but the resulting temperature is not sufficiently high to ignite the fuel. Ignition is usually accomplished by leaving some portion of the combustion chamber uncooled, this portion usually being in the form of a bulb or pocket. The fuel is injected against this uncooled surface and is ignited by contact with this surface and the heated air. Semi-Diesel engines may operate with either constant-volume or constant-pressure combustion, but in general operate somewhere between these two types. The more nearly an engine operates with constant-volume combustion, the higher will be its efficiency.

In engines of this class the compression pressure rarely goes over 350 lb/in^2 (2413 kPa). Many engines of this type in small and medium sizes are in various classes of service. Some outstanding advantages of this class of engine are low first cost, reliability in the hands of unskilled operators, absence of carbureting devices and electrical ignition equipment, and ability to operate with a large variety of low-grade nonvolatile fuels. This type is well adapted to two-cycle operation, which makes for reasonably low first cost per brake-horsepower (kW) rating. The fuel consumption of engines of this class is, in general, 20 to 30 percent greater than that of Diesel engines for the same service. This type of engine also requires external heating at starting or must be equipped with electrically heated devices for starting. In view of these conditions, and also because in the last few years Diesel engines have become cheaper and very reliable and have become available in much smaller sizes, the semi-Diesel engine has rapidly been losing ground; and some former manufacturers of this type have abandoned it altogether in favor of the Diesel engine.

Four-Cycle and Two-Cycle Operation. The preceding discussion of cycles is largely concerned with their theoretical and thermodynamic features. In actual engines all of the above cycles may be completed in four strokes of the piston and two revolutions of the crankshaft, in which case there is a *four-stroke-cycle* operation or, briefly, a *four-cycle* operation; or the cycles may be completed in two strokes of the piston and one revolution of the crankshaft, in which there is a *two-stroke-cycle* operation or, briefly, a *two-cycle* operation. Compared to two-cycle operation, we have the following: Four-cycle operation is employed to a far greater extent than two-cycle operation. It is almost exclusively used in automobile, truck, tractor, motorcycle, and aircraft engines. It is widely used in all other types of engines for all classes of service. Some of the important advantages of four-cycle operation are reliability, flexibility in operation, good mechanical efficiency, and low fuel consumption. Four-cycle engines are built to be single-acting or double-acting, and have various numbers of cylinders and cylinder arrangements. They operate with any fuels that may be used in internal-combustion engines.

Compared to four-cycle operation, we have the following: Two-cycle operation will give about 1.8 times as much power output from a given cylinder as can be obtained from a four-cycle engine of the same cylinder size and running at the same speed. This may give a lower weight per unit of output. Two-cycle engines may be built entirely without valves. Precompression of the mixture may be carried out in the crankcase or in a separate cylinder. When volatile fuels are used, two-cycle engines are usually built in relatively small sizes for marine and general utility service. When using nonvolatile fuels such as fuel oil, etc., these engines are built in very large sizes for stationary and marine service. In general, the specific fuel consumption per brake horsepower per hour is not quite so good as for four-cycle engines. In engines using nonvolatile fuels, the two-cycle principle is freed from many of its limitations and is continually finding more extensive application.

OTTO-CYCLE ENGINES

11. General. By far the greater number of internal-combustion engines in use operate on the *Otto four-stroke cycle*. This class of engine is used in stationary service for power generation and in portable equipment such as tractors, power shovels, excavators, etc. They are used in marine service and for aircraft propulsion. Automobile engines are almost exclusively of this type. Many engines operate on the *Otto two-stroke cycle*. In the stationary field this type is confined to engines of rather small size, operating usually on gasoline. They are also used in very large sizes and operate usually with gaseous fuels such as natural gas, blast-furnace gas, producer gas, etc. A great many small and medium-sized marine engines are of the two-cycle type. The two-cycle type has no commercial importance in automobile or aircraft service. It is characteristic of engines operating on the Otto cycle that (1) the fuel and air are mixed in the proper ratio by a suitable measuring and mixing device (a carburetor in the case of an engine operating on a liquid fuel, such as gasoline, and a mixing valve in the case of a gas engine) *outside of the cylinder*; (2) the *fuel-air mixture undergoes compression*; and (3) the mixture is ignited in the cylinder by an electrical jump spark or make-and-break ignition system. Stationary engines are commonly built with any number of cylinders from one to eight and, in exceptional cases, more. They are built with horizontal and with vertical cylinders and may be single- or double-acting, single-acting engines being much more common and universally used in small stationary engines and in automotive and aircraft engines. The great majority of all internal-combustion engines are water-cooled. Exceptions are very small stationary engines [of about ¼ to 1½ hp (1.7 to 10.3 kW)], motorcycle engines and aircraft engines, and a few small miscellaneous types, which are air-cooled. With the exception of air-cooled aircraft and motorcycle engines, all are provided with forced air circulation by means of a fan or blower driven by the engine itself. It is worth noting that all engines are ultimately air-cooled, since the heat taken up by the cooling water is finally transferred through the radiator to the air or else by direct contact of the water with the air. The essential elements of an Otto single-acting, four-cycle engine are shown in Fig. 13.167. A small, three-port, two-cycle engine is shown in Fig. 13.168. This represents the simplest type of internal-combustion engine. There are no valves, and the crankcase is used to compress slightly the charge of the explosive mixture before it enters the cylinder. Ignition occurs by the usual jump-spark system. Large two-cycle engines frequently use a separate cylinder instead of the crankcase for precompression of the charge.

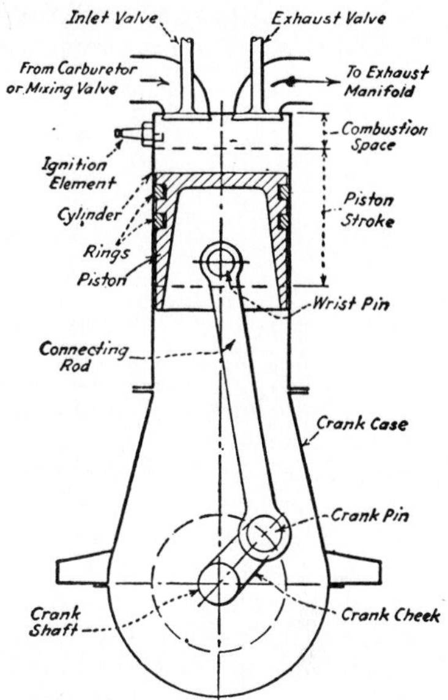

FIGURE 13.167 Elementary, four-cycle, internal-combustion engine.

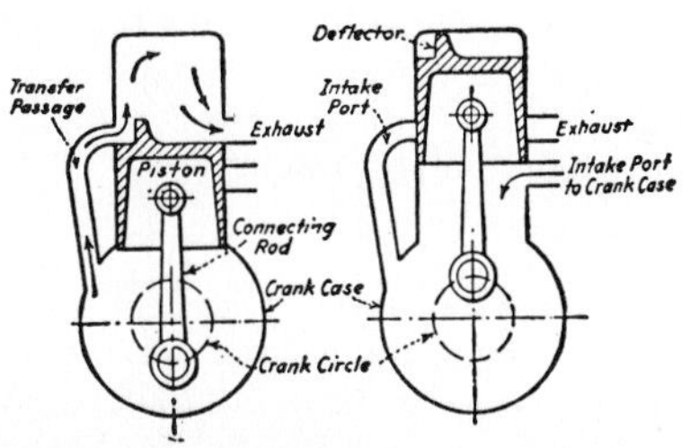

FIGURE 13.168 Three-port, two-cycle internal-combustion engine.

12. Engine Classification and Performance. Engines for *stationary* uses include many sizes and types, from the small gasoline farm engines to the larger gas engines driving gas compressors, electric generators, pumps, or blowers. The development of the oil industry has made available large quantities of natural gas which has been piped to the large industrial centers and used in many gas engines.

OIL ENGINES

13. Low- and Moderate-Compression Engines. An oil engine is an internal-combustion engine that uses oil for fuel. Air, only, is drawn into the cylinder and compressed, and the fuel is metered, injected, and atomized at the proper time in the stroke by means of a fuel pump driven from the engine. Oil engines may be arbitrarily classified as low-compression, moderate-compression, or high-compression (Diesel) engines

Low-compression oil engines have compression pressures that are usually less than 150 lb/in^2 (1034 kPa). The temperature of the air at the end of compression is not high enough for autoignition, and some uncooled portion of the cylinder, not bulb, hot bolt, or similar device must aid in producing ignition of the fuel. The igniting device must be externally heated during starting or during long periods of operation at light loads. This type of engine can successfully use some grades of Diesel fuel oils, and for this reason is sometimes called a *semi-Diesel engine*. The combustion is substantially at constant volume, however, and the engine usually operates on the Otto cycle. Many low-compression oil engines are in use, but only a few are being manufactured, as the moderate-compression and high-compression (Diesel) engines are displacing them.

Moderate-compression oil engines employ compression pressures between 200 and 400 lb/in^2 (1379 to 2758 kPa) and operate on a mixed cycle with part of the combustion at constant volume and part at substantially constant pressure. In general, these engines now have complete cylinder cooling and will start cold under ordinary conditions, the heat of compression producing the necessary ignition temperature. Under low-temperature conditions and with heavy fuels, an electrically heated resistance glow plug is sometimes used to assist in starting, or some provision is made for heating the air before it enters the cylinder. The line of demarcation is not sharp between this class of engine and Diesel engines. In general, they operate with lower compression pressures than Diesels and with somewhat higher fuel consumption.

14. Diesel Engines. Diesel or high-compression engines are used in stationary service, ships, trains, tractors, busses, trucks, and aircraft. The Diesel engine is characterized by its ability to burn low-grade and cheap fuels; by its high thermal efficiency over a wide range of loads; by the absence of all electrical ignition equipment and carbureting devices.

Fuel injection is an important feature in engine performance and may be accomplished by *air injection* or *solid injection* of the fuel. For air injection a multistage air compressor is employed, and it delivers air at about 800 to 1500 lb/in^2 (5515 to 10,341 kPa). This compressor is usually built integrally with the engine but may be an independent unit. A fuel pump is also required to deliver the proper amount of fuel to the injection nozzle at the proper time in the stroke. This pump is driven directly by the engine. At the proper time an air-admission valve opens, and the fuel is atomized and forced into the cylinder. One method of solid injection employs a pump unit for each cylinder which is capable of producing very high pressures, frequently in excess of 10,000 lb/in^2 (68.9 MPa). This high pressure together with suitable injection nozzles delivers the fuel in a highly atomized condition. This is the method most commonly used. Another method uses one pump to maintain a uniformly high pressure in the oil-supply lines to all the nozzles. Each injection nozzle is controlled by a fuel valve operated by the engine. The fuel pump is usually driven by the engine but may be independently driven. The great advantages of a solid-injection system are the absence of an air compressor and its general simplicity.

In small and moderate-sized engines the *Hvid* or *Brons* principle of injecting the fuel is employed. The advantage of this principle is that no air compressor or fuel pump is required, as the fuel can flow to the engine by gravity and the gas pressure generated in the device atomizes, ignites, and forces the fuel into the main cylinder. The device consists of an uncooled cup in the engine cylinder. Fuel flows by gravity into the cup during the suction stroke, and the amount passed is controlled by a governor-controlled fuel valve. During compression, air is forced through very small holes close to the bottom of the cup. At the end of compression the air is hot enough to ignite the more volatile fuel elements and create a pressure in the cup above the pressure in the cylinder and atomize and force the remaining fuel into the main combustion space.

Stationary or *heavy-duty Diesel engines* are built for two-cycle and four-cycle operation, and are of the single- and double-acting, single- and multi-cylinder, and horizontal and vertical types. Dominant characteristics of recent heavy-duty American Diesel engines are given in Table 13.11. The weights of these engines per brake horsepower range from 25 to 300 lb (15 to 182 kg/kW), with an average of 100 lb (61 kg).

Marine engines do not differ markedly from stationary engines, except that their piston speeds are a little higher and they are fitted with the necessary auxiliaries; fuel consumption is about the same.

High-speed Diesel engines find extensive use in applications where rotative speeds from 600 to 1200 r/min are desired. High-speed Diesel engines run at speeds up to 2500 r/min with piston speeds of 2000 ft/min (610 m/min), and they weigh 10 to 15 lb/hp (6 to 9 kg/kW). Characteristics of high-speed Diesel engines recently installed in this country are given in Table 13.12.

From the standpoint of safety and economy, the high-speed Diesel engine has many advantages that suit it for use in aircraft, and many Diesel aircraft engines are being developed. For transportation, in general, the advantages of the Diesel engine over the gasoline engine are higher thermal efficiency, the use of a safer and probably cheaper fuel, and the absence of electrical ignition equipment. Where these advantages are not so important as very high speeds, flexibility, quietness, and lightness, the gasoline engine will still continue to be used.

15. Testing Internal-Combustion Engines. Codes for testing general types of internal-combustion engines are issued by the ASME, and additional codes for testing high-speed automotive and aircraft engines are issued by the SAE (Society of Automotive Engineers). These codes should be followed in the conducting and reporting of tests.

TABLE 13.11 Characteristics of Heavy-Duty Diesel Engines

Service	Bhp (kW)	R/min	Cylinder data		Cycle	Type of fuel injection	Piston speed, ft/min (m/min)	Fuel consumption, lb. per bhp.-hr. (kg/kWh)
			No.	Size, in (cm)				
Generator	7000 (5227)	167	8	24 × 36* (61 × 91)	2	Solid	1002 (305)	0.385 (0.234)
Generator	4100 (3059)	124	6	30 × 42 (76 × 107)	2	Solid	868 (265)	0.416 (0.253)
Generator	1000 (746)	257	8	16¼ × 24½ (41 × 62)	4	Air	1050 (320)	0.41 (0.249)
Generator	600 (448)	225	4	16¼ × 23 (41 × 58)	4	Air	863 (263)	0.42 (0.255)
Generator	600 (448)	400	6	12 × 15 (30 × 38)	2	Solid	1000 (305)	0.40 (0.243)
Generator	300 (224)	300	4	14 × 17 (36 × 43)	2	Solid	850 (259)	0.38 (0.231)
Generator	150 (112)	360	6	8½ × 12 (22 × 30)	4	Solid	720 (219)	0.45 (0.274)
Pump	1000 (746)	300	8	16¼ × 21 (41 × 53)	4	Solid	1050 (320)	0.38 (0.231)

*Double-acting.

TABLE 13.12 Characteristics of High-Speed Diesel Engines

Service	Bhp (kW)	R/min	Cylinder data		Cycle	Type of fuel injection	Piston speed, ft/min (m/min)	Fuel consumption, lb. per bhp.-hr. (kg/kWh)
			No.	Size, in (cm)				
Drilling	125 (93)	850	6	5¾ × 8 (15 × 20)	4	Solid	1133 (345)	0.43 (0.26)
Generator	100 (75)	600	6	6½ × 8½ (17 × 22)	4	Solid	850 (259)	0.40 (0.24)
Generator	75 (60)	1200	4	3½ × 4¾ (9 × 12)	2*	Solid	950 (290)	
Generator	18 (13)	1200	4	3⅝ × 4½ (9 × 11)	4	Solid	900 (274)	
Compressor	70 (52)	900	6	4⅞ × 6 (12 × 15)	4	Solid	900 (274)	

*Opposed pistons.

STEAM-POWER PLANT EQUIPMENT

16. Furnaces. In steam generation, correct furnace design and operation are essential to the efficient combustion of fuels. The furnace must be adapted to the fuel; it must be constructed and operated so that the combustible gases will be thoroughly mixed with the proper amount of air to support combustion, and the gases must be maintained at a temperature above their ignition point until they have been completely burned.

Coals high in volatile matter, when burned in bulk form, require a furnace design which will cause the distillation of the volatile matter to take place at low temperatures. The resulting light hydrocarbons are more likely to burn completely without depositing soot than the heavier compounds resulting from high-temperature distillation.

For maximum thermal efficiency, the furnace temperature should be the highest that can be maintained, but the brickwork will deteriorate rapidly if subjected to the full temperature available from most fuels. Either a compromise must be made between thermal efficiency and furnace-wall maintenance, or special features must be introduced to keep the wall temperature below the danger point. One method is to bring in some of the air for combustion through ducts in the walls; another is to place a portion of the boiler heating surface in the wall where it will absorb radiant heat from the furnace and reduce the temperature at the wall.

Air-cooled walls require a greater furnace volume for a given rate of energy release or output than do *water-cooled walls*. With hand or stoker firing, the cross section of the furnace is usually fixed by the area of the grate; the height of the furnace must then be great enough to give the required furnace volume. Water-cooled walls may be of the (1) bare-plate, (2) bare-tube, or (3) covered-tube type. A water-cooled wall that consists of tubes covered with protective, refractory-lined metal blocks is shown in Fig. 13.169.

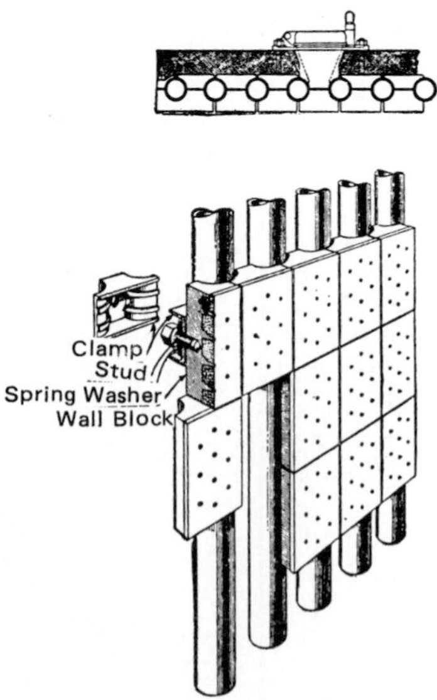

FIGURE 13.169 Bailey furnace wall.

Refractories. Refractory brick for lining the furnace of a steam-power plant should possess the following properties: (1) high fusion point, (2) low thermal conductivity, (3) low thermal expansion, and (4) high resistance to abrasion. Failure in service may be due to: (1) fusion, (2) plastic deformation, (3) activity—expansion, shrinkage, and spalling—(4) slagging. Spalling means any breaking or cracking of bricks, whether due to thermal shock, pinching because of expansion, or changes in structure. Slag action results in the erosion of the brick due to fluxing by the ash of the fuel, or to the building up of layers of solid slag upon the surface.

17. Hand-Fired Grates. Not widely used today, stationary grates for hand-fired furnaces are constructed of cast-iron sections. The common forms are: the straight bar, the tupper or half-herringbone, and the herringbone. Tupper and herringbone bars are less liable to warp than the straight bar, but they are not so convenient for use with coals which clinker badly. Pinhole grates are used in burning sawdust, tanbark, and very small sizes of coal.

Stationary grates have the advantage of low first cost, but the fire is not easily cleaned. If the air spaces are allowed to become clogged, combustion is hindered. Frequent cleaning is wasteful of fuel, permits the entrance of large amounts of cold air through the open fire doors, and thus lowers the efficiency. The width of air space in the grate is selected for the size of coal to be used, and it should be as large as possible without permitting loss of coal to the ashpit.

The ratio of grate surface to heating surface ranges from 1:25 to 1:60. For No. 1 buckwheat it may be taken as 1:40; for No. 2, 1:35; for No. 3, 1:30; for culm, 1:25. For bituminous coals, the ratio of the grate surface to the heating surface for economy is from 1:45 to 1:60, and for capacity, from 1:40 to 1:55.

Sectional shaking grates are frequently selected for coals that clinker and have high ash content. They provide means of manipulating the fuel bed without opening the fire doors.

Combustion rates depend on the available draft, thickness of the fuel bed, characteristics of the coal and ash, total grate surface, and area available for air passage. A given rate of combustion may be obtained by using a large grate surface and light draft, or a smaller surface, thicker fuel bed, and heavier draft.

18. Mechanical Stokers. Mechanical stoking is always superior to hand firing when thermal efficiency and smokeless combustion are alone considered. Overall economy, however, measured in dollars, must give consideration to fixed charges and maintenance, as well as to operating expense. Automatic stokers are high in first cost and, if they are applied to too small units, the fixed charges and maintenance may offset the saving in fuel. This is very likely to be the case when no reduction in the firing force can be made and where the plant operates only a fraction of the time.

Commercial stokers are classified as (1) *overfeed* and (2) *underfeed*, depending upon whether the coal is fed to the fuel bed above or below the level of the supply of primary air. The *overfeed* type may be divided into three classes: (a) front-feed inclined-grate stokers in which the fuel is received at the front and fed down an incline to the ash dump at the bottom, (b) double-inclined side-feed stokers with the fuel fed from both sides, and (c) traveling grates with continuous horizontal feed. The underfeed type may be *single-retort* or *multiple-retort*. Draft for mechanical stokers may be natural or forced.

19. Overfeed Stokers. *Inclined overfeed stokers* may be used with all grades of coal and also with sawdust, hogged fuel, etc., but they are particularly well adapted to the burning of high-volatile, high-ash, coking coals. At high rates of combustion with some fuels, there is difficulty in getting rid of the ash and clinker; the maximum rate of combustion with natural draft and a coking coal is about 35 lb/ft^2 (171 kg/m^2) of grate surface per hour. The use of these stokers is restricted to moderate- and small-sized boiler units [under 10,000 ft^2 (929 m^2) of heating surface].

Traveling-grate stokers consist of an endless grate that passes into and out of the furnace like a chain between sprockets at the front and rear. In *chain-grate* stokers, the grate bars form the links of the chain; in *bar-grate* stokers, the grates are carried on crossbars attached at their ends to steel drive chains. Coal is fed by gravity from a hopper at the front and travels through the furnace at a uniform slow rate; there is no agitation of the fuel bed, but the links of the chain-grate stoker shear off ash that sticks to the grates at the rear sprocket. The chain-grate stoker is well adapted to the burning of noncoking bituminous coal but will also burn coke breeze, anthracite, and lignite. Bar-grate stokers are adapted to the burning of small sizes of anthracite (buckwheat, river coal, and culm) but will also burn noncoking bituminous coals if the ash does not stick to the grate. For either type of stoker, an ash content of the coal of at least 7 percent is necessary to insulate the grate.

For burning bituminous coal with forced draft, efficient combustion rates are between 30 and 40 lb/ft^2 (146 and 195 kg/m^2) per hour, with a maximum rate of about 70 (341); for low-volatile anthracite or coke breeze with forced draft, efficient combustion rates are between 30 and 38 (146 and 186); for free-burning coals with natural draft, efficient rates are between 20 and 30 (98 and 146). Traveling-grate stokers show low maintenance costs, but for meeting sudden and large variations in load, they are not so suitable as underfeed stokers.

20. Underfeed Stokers. Underfeed stokers are particularly suitable for burning bituminous and semibituminous coals at high rates of combustion when the ash content is low and has a high fusion point. Low-volatile fuels, such as anthracite and coke breeze, may be burned when mixed with bituminous coals. Usually, the fuel is fed from front hoppers by reciprocating rams into the bottom of troughlike retorts with air admitted through rows of tuyeres along the upper edge of the retort. Forced draft is required, and ignition arches are unnecessary. The green coal is gradually pushed up from below, and as it becomes heated, the volatile gases rise and pass through the incandescent fuel bed without the formation of much soot or smoke.

The most efficient rates of combustion with multiple-retort underfeed stokers are between 35 and 45 lb/ft^2 (171 and 220 kg/m^2) per hour; rates as high as 89 (435) have been carried efficiently, however. From ¾ to 1 hp (0.56 to 0.75 kW) is required per retort burning from 700 to 1100 lb of coal (318 to 499 kg) per hour to feed and distribute the coal, operate the ash dumps, and drive the fans. Modern stokers require about 7 kWh of energy per ton (907 kg) of coal burned for the operation of the fans and driving motors. The average wind-box pressure required is from ¾ to 1 in (1.9 to 2.5 cm) of water for each 10 lb of coal burned per ft^2 (49 kg/m^2) per hour; higher pressures are required for low-volatile than for high-volatile coals. In Fig. 13.170 is shown an underfeed stoker installed under a cross-drum boiler.

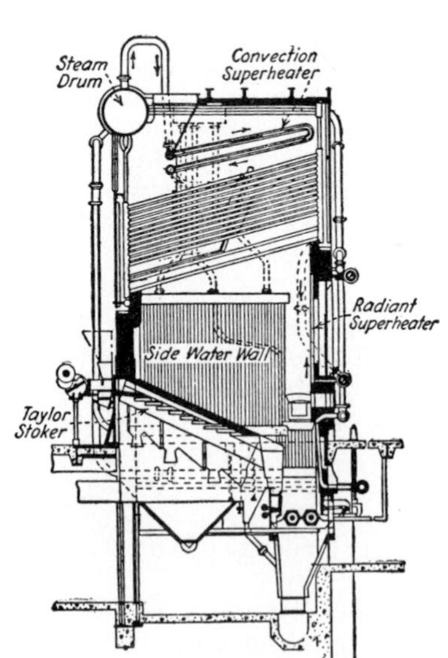

FIGURE 13.170 Taylor underfeed stoker installed under Babcock & Wilcox cross-drum boiler.

21. Pulverized Coal. Coal that has been crushed, dried, and ground to a powder has been used as a power-plant fuel since about 1918. The advantages are complete combustion with minimum excess air; high furnace efficiency and capacity; and the fact that cheaper and widely varying grades of coal may be burned, and fuel and air supply can readily be controlled to conform with variations in load. Other factors which must be given consideration are first cost, size of plant, space requirements, preparation costs, maintenance; and ash and dust collection and disposal.

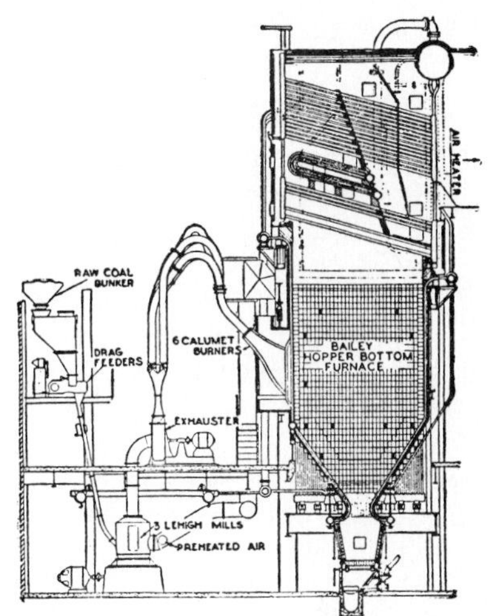

FIGURE 13.171 Pulverized-coal-fired boiler.

There are two types of pulverized-coal plants: the central or storage system and the unit or direct-firing system. In the *storage* system, the pulverized coal is delivered from a central preparation plant to storage bins and then, as needed, to the various burners. Pulverizing may be done at a constant rate and during off-peak hours; failure of the coal mill will not cause an immediate shutdown, and good operating conditions may be maintained with wide variations of load. The cost of the preparation plant and bins is high, however. The energy required for the coal-preparation apparatus is between 15 and 30 kWh per ton (907 kg).

In the *unit* system of *direct firing*, the coal is piped directly to the burner, without storage, from a single piece of apparatus that contains the tramp-iron separator, feeder, pulverizer, and fan. The separating air used in the mill serves as the carrying agent and enters the furnace through the burner (see Fig. 13.171). This system is lower in first cost than the storage system because it is simpler and can be arranged compactly in a small space close to the steam-generating unit. This system is commonly chosen for industrial plants, and its selection in preference to the storage system is becoming more frequent in the public-utility field.

Combinations of the two systems may be installed with a unit system handling the larger and steadier loads and a storage system of sufficient capacity to assist in meeting maximum and minimum demands.

The *pulverized-coal burner* must direct the stream of coal and primary air into the furnace and secure proper mixture with the secondary air. Burners are of the *straight-shot* or *turbulent* type. Straight-shot burners (Fig. 13.172) discharge the coal and primary air in a straight line and may be horizontal, vertical, or inclined. Turbulent

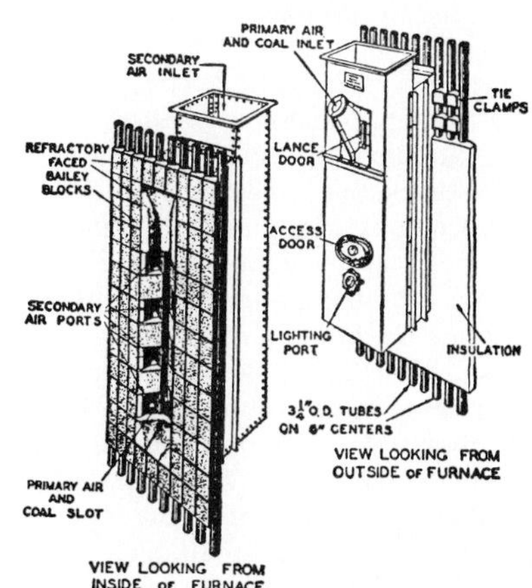

FIGURE 13.172 Calumet burner designed for firing through water-cooled furnace walls.

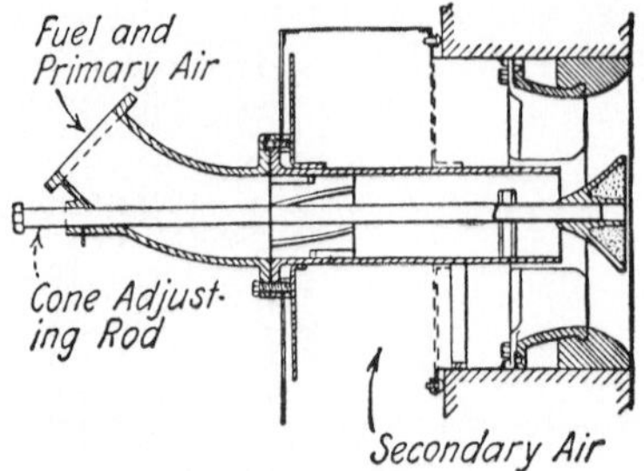

FIGURE 13.173 Circular horizontal turbulent burner.

burners (Fig. 13.173) provide for intimate mixing of the fuel with all of the air (primary and secondary) immediately upon leaving the burner, and they effect complete combustion in a minimum of time and space. The arrangement of the burner, as well as the design, may promote turbulence. Tangential firing, where a burner is placed in each of the four corners of the furnace to direct a stream of coal and air horizontally and tangent to a circle, gives very intensive mixing. Steam-generating units with a wide range of load are often equipped with an auxiliary low-load burner.

The design of the modern pulverized-fuel furnace is controlled largely by the ash problem. Where the fusion point of the ash is high and the combustion rate is low, ash may be removed dry. Slag screens of widely spaced water tubes may be placed above the furnace bottom to cool the ash. If the fusion point of the ash is low, the slag may be kept molten at high rates of combustion; a pool of molten slag then fills the bottom of the furnace and may be tapped off intermittently or continuously (such furnaces are *slag-tap* furnaces). Pulverized-coal furnaces require larger volumes or lower energy-release rates than do furnaces using other fuels. The limit of the rate of heat liberation in refractory furnaces is the excessive erosion of the refractories; in water-cooled furnaces, the limit is generally the deposition of slag in boiler tubes and the clogging of gas passages.

A pulverized-fuel furnace is well adapted to the burning of a variety of coals; also, oil and gas may be burned in the same furnace if fuel costs warrant a change-over.

22. Oil Burners. The use of oil as a fuel for steam-generating units in industrial and central-station power plants is widespread. The oil commonly used is that meeting the Bunker C specifications of the U.S. Navy. This oil has a viscosity of not over 300 sec. Saybolt Furol at 122°F (50°C) and a density of 5 to 14 API. Oil burners are classified as (1) mechanical, (2) steam- or air-atomizing, and (3) rotary-cup. Mechanical burners use mechanical means of atomizing the oil, and are supplied with oil at pressures of from 200 to 300 lb/in^2 (1379 to 2068 kPa); burners of this type are provided with registers for combustion air. Capacity is governed by changing the oil pressure or by cutting burners in and out; between 15 and 1100 gal of oil (57 to 4164 L) per hour are usually handled per burner. Most steam-atomizing burners use dry steam at pressures above 30 lb/in^2 (207 kPa) to cut across the oil stream and atomize the fuel; steam required for atomization is usually between 2 and 3 percent of that generated. The capacity of a single steam-atomizing burner typically is 250 gal/h (946 L/h). In the rotary-cup burner, oil is thrown off the rim of a rotating cup and atomized; burners of this type typically burn 150 gal per hr (568 L/h). Mechanical burners are selected for the larger installations because of better economy; steam-atomizing burners are well suited to the burning of heavy sludges, and rotary-cup burners are widely used in large heating boilers. Energy-release rates of 60,000 Btu/ft^3 · h (2236 MJ/m^3 · h) of furnace volume may be efficiently attained in oil-fired furnaces.

Gas Burners. In a gas burner, either the air, the gas, or both air and gas may be supplied under pressure. Many burners are of the venturi type in which the gas is supplied under a pressure that is adjusted to the load, while the air is supplied under atmospheric pressure; the rate of induced air flow is substantially proportional to the rate of gas flow, and the mixture ratio is maintained nearly constant. Other burners may depend upon furnace draft flow to draw in air around the gas jet.

Boilers and Superheaters

23. Boilers. A complete steam-generating unit consists at least of a boiler and furnace and may include some or all of the following heat-transfer elements: superheater, economizer, air preheater, steam reheater, and auxiliaries.

Boilers are divided into two general classes: (1) fire-tube, and (2) water-tube. *Fire-tube boilers* are made in sizes up to about 6000 ft^2 (557 m^2) of heating surface. Fire-tube boilers are commonly selected for small installations where low first cost is more important than operating cost. For construction use, small, vertical, tubular boilers are built in sizes from 21 to 66 in (0.5 to 1.7 m) in diameter, with 40 to 1000 ft^2 (3.7 to 93 m^2) of surface, for pressures not exceeding 150 lb/in^2 (1034 kPa); large, vertical, tubular boilers are built in sizes ranging from 500 to 6000 ft^2 (46 to 557 m^2) for pressures up to 200 lb/in^2 (1379 kPa); stationary locomotive-type boilers are commonly built in sizes ranging from 150 to 2500 ft^2 (232 m^2), and for pressures from 50 to 150 lb/in^2 (345 to 1034 kPa) the stationary Scotch marine boilers have outer shells that range from 6 to 16 ft (1.8 to 4.9 m) in diameter and have been built for pressure up to 300 lb/in^2 (2068 kPa). For power-plant use, the horizontal return-tubular boiler has been widely selected in sizes from 150 to 3750 ft^2 (14 to 348 m^2) of heating surface, with corresponding shell sizes from 36 in (diameter) × 8 ft (length) to 90 in × 20 ft (0.9 × 2.4 to 2.3 × 6.1 m); steam pressures seldom exceed 150 lb/in^2 (1034 kPa).

Water-tube boilers are made with a number of different arrangements of tubes and drums with steam-generating capacities from 10,000 to over 1,000,000 lb/h (4536 to 453,590 kg/h). They are used in central-station service and in industrial plants requiring more than about 10,000 ft^2 (929 m^2) of heating surface. The materials and construction of stationary boilers should conform to the ASME Boiler Construction Code and also to any inspection regulations to which the installation may be subject.

24. Capacity and Load. The *capacity* of a steam-generating unit may be expressed in terms of: (1) the maximum rate of heat absorption by the unit, stated in kilo Btu (kB) per hour (W) or in mega Btu (mB) per hour (W) where mB = 1000 kB = 1,000,000 Btu; or (2) the maximum rate of steam generation, in pounds (kg) per hour, together with the pressure and temperature of the steam and the temperature of the feed water. The *load* of a steam-generating unit is the actual rate of heat absorption or steam generation at any time, expressed in the same units as capacity. The older practice of measuring the boiler output in terms of boiler horsepower, where 1 boiler hp (0.75 W) is equivalent to a rate of heat absorption of 33,479 Btu/h (9813 W); of assigning a "builder's rating" of 1 boiler hp per 10 ft^2 (0.8 W/m^2) of heating surface; and of calculating the "percent of rating" by the quotient of the output to the "builder's rating," is rapidly becoming obsolete.

25. Performance. The performance of a steam-generating unit may be expressed in terms of its efficiency, an energy balance, or a combination of both. The *efficiency* of a steam-generating unit may be expressed as the ratio of the quantity of heat actually absorbed by the water and steam passing through the unit in a given time to the quantity of heat supplied by the fuel used in that time. In the form of an equation, the net efficiency is

$$\text{Eff.} = \frac{W_n(h_2 - h_1)}{W_f(\text{HHV})} \tag{13.66}$$

where W_n = net weight of steam delivered by the unit, in lb/h (kg/h)
= gross weight minus auxiliary steam
h_2 = enthalpy of steam at outlet from unit, in Btu/lb (kJ/kg)
h_1 = enthalpy of feed water at inlet to first heating element of unit, in Btu/lb (kJ/kg)
W_f = weight of fuel fired, in lb/h (kg/h)
HHV = higher heating value of fuel as fired, in Btu/lb (kJ/kg)

Auxiliaries may consist of steam engines, steam turbines, steam jets, or electric motors. When the auxiliary is steam-driven, the steam consumed is subtracted directly from the gross weight delivered. If the auxiliary is not steam-driven, the equivalent weight of steam necessary to supply the auxiliary energy must be found and subtracted from the gross weight. Steam-generating units, in large sizes and with high load factors, may be expected to show monthly average net efficiencies of around 85 percent. Large industrial plants may show monthly average efficiencies of around 75 percent with lower values for the smaller plants.

Energy balances may be prepared from complete tests of steam-generating units, and the distribution of the various losses may be studied. The energy balance should include the following items:

1. Energy absorbed by the water and steam
2. Loss due to incomplete combustion of carbon
3. Loss due to unconsumed carbon in refuse
4. Loss due to heat carried away in dry exit gas
5. Loss due to moisture in fuel
6. Loss due to moisture formed from combustion of hydrogen in fuel
7. Loss due to water vapor in air
8. Loss due to external heat transfer, and unaccounted for

When expressed in terms of Btu/lb (kJ/kg) of fuel fired, these items must add to equal the heating value of the fuel as fired; when expressed as a percent of the heating value of the fuel, these items must add to 100.

26. Superheaters. Superheaters increase the temperature of the steam without increasing its pressure. *Separately fired superheaters*, requiring separate settings, may be used for superheating steam exhausted from engines before use in low-pressure turbines, or for superheating a portion of the steam to a higher temperature than is required for the remainder. *Integral superheaters* are installed within the setting of the steam generator; they may be classified as *radiant* or *convection* superheaters, according to the predominant mode of heat transfer.

The flame temperature to which a radiant superheater is exposed increases less rapidly with increasing load than does the weight of steam generated, and the temperature of the steam falls. With convection superheaters, the steam temperature will rise with increased load until a maximum is reached at extremely high loads. Uniform superheat is desired, and by placing part of the superheating surface in the path of the gases and part to receive radiant heat, the rising characteristic of the convection type and the falling characteristic of the radiant type may be combined to give a nearly constant steam temperature at all loads. (See Fig. 13.170.)

DRAFT AND DRAFT EQUIPMENT

27. Draft. In steam-power-plant practice, draft usually means the pressure difference available to overcome the various resistances to the flow of air into and through the fuel bed and to the flow of products of combustion through the furnace. Draft is usually measured in terms of the height, in inches (cm), of a column of water. If the water temperature is 80°F (26.7°C), 1 in (2.54 cm) of water is equivalent to a pressure of 0.036 lb/in^2 (0.25 kPa).

The resistance of the fuel bed generally causes the greatest loss of draft; this resistance varies widely with different grades of coal and different rates of combustion, and increases with (1) the thickness of the fuel bed, (2) the ash content of the coal, (3) decreasing volume of the interstices. The resistance offered by the boiler depends upon the type of boiler and setting, the arrangement of the tubes and baffles, and the load; manufacturers will furnish information on such resistance. Breechings, or flues, are usually proportioned to have velocities of flow about 20 percent lower than in the chimney. The draft loss through the breeching may be estimated at 0.05 in (0.13 cm) for each right-angle bend and 0.10 in per 100 ft (0.25 cm per 30.5 m) of length.

28. Natural Draft. Draft produced by a chimney or stack is known as *natural* draft; draft produced by means of fans, blowers, or steam jets is known as *mechanical* draft. Some small and medium-sized power plants use natural draft. Such draft is produced by the difference in weight between the column of atmospheric air outside the chimney and the column of heated gases inside the chimney. This difference in weight for a given set of temperature conditions varies directly with the height of the stack. For a barometric pressure of 14.7 lb/in^2 (101 kPa), the static draft produced by a chimney Z ft in height is about $7.64Z[(1/T_a) - (1/T_g)]$ in of water (× 2.54 = cm of water), where T_a and T_g are the absolute temperatures of the cool air and hot gases, respectively. The static draft is the maximum one that can be produced by a chimney of given height when filled with stationary hot gases. When a flow of gases occurs, the chimney offers frictional resistance to flow which must be overcome, and the discharge velocity head must also be produced by the draft. The friction loss of head in the chimney may be estimated from Eq. (13.62*a*) by using a friction factor of between 0.004 and 0.005 for the common case. The draft available at the flue connection to overcome the losses through the fuel bed and furnace is the difference between the static draft and the sum of the frictional loss of head in the chimney and the gas-velocity head.

Large plants having steam-generating equipment which offers considerable resistance to the flow of gases and produces low exit-gas temperatures commonly use mechanical draft, but a chimney is then necessary to discharge the products of combustion at a suitable level.

29. Chimneys. Chimneys, or stacks, may be constructed of brick, reinforced concrete, or steel. The tallest chimneys are made of radial brick or reinforced concrete. *Self-supporting* steel stacks may be mounted on girders located near the roof level and supported by columns between the boilers. These stacks can be erected more rapidly, occupy less space, and cost less than masonry chimneys; they must be painted frequently to protect the steel from the air and from the corrosive action of the stack gases. *Guyed* steel stacks are seldom built in sizes larger than 72 in (183 cm) in diameter and 150 ft (45.7 m) high.

The proportions of the chimney should be such as will meet the requirements of capacity and draft with the least cost. Reducing the chimney diameter will increase the gas velocity and the frictional resistance to flow and necessitate an increase in the height of the chimney to produce the draft needed at the flue connection. Many radial-brick chimneys have a height that is about 22 times the diameter; velocities of gas flow are commonly between 20 and 30 ft/s (6.1 and 9.1 m/s).

30. Mechanical Draft. Large steam-power plants use both forced-draft and induced-draft fans. *Forced-draft* fans force air from a closed duct into the furnace through the fuel bed. *Induced-draft* fans are placed between the boiler and the stack, and must handle gases that are at a high temperature and are frequently dust-laden. Although this duty is more severe than that of a forced-draft fan, the use of an induced-draft fan is often necessary in order to maintain a suction throughout the boiler setting. *Steam jets* are sometimes used to produce draft in small plants; forced draft may be produced when the jets are in the ashpit, and induced draft when the jets are in the flue or stack. They are low in first cost but have a large steam consumption and a limited capacity.

FEEDWATER, ACCESSORIES, AND PIPING

31. Feedwater Heaters and Economizers. Feedwater heaters are of two general classes: (1) *open* heaters, in which the water comes into direct contact and mixes with the heating steam; (2) *closed* heaters, in which the steam and water are separated by metal surfaces through which heat is transferred. The *deaerating* heater is a special form of open heater which is designed to remove dissolved gases from the feedwater. The steam used in feedwater heating may be exhaust steam from engines or processes or steam extracted from turbines. The purpose of feedwater heating is to improve the thermal performance of the plant; a fuel saving of approximately 1 percent will result from each 12°F (6.7°C) rise in temperature of the feedwater if the steam used in the heater would otherwise be wasted.

Other heat exchangers may be incorporated in the feedwater system to supplement feedwater heaters. *Economizers* are closed heaters that utilize the heat in the exit gases to heat the feedwater. Economizers may be *independent* and located above or behind the boiler; or *integral*, i.e., located within the boiler setting and connected directly to the boiler circulating system.

32. Boiler-Water Conditioning. Untreated water contains solids and dissolved gases that may cause scale formation, corrosion, caustic embrittlement, foaming, or priming. The presence or formation of insoluble solids in feedwater causes *scale formation*; calcium sulfate and silica form extremely hard scale of low thermal conductivity which may cause overheating and failure of boiler tubes. *Corrosion* of steam-generating and feedwater equipment may be hastened by the presence of dissolved oxygen, acids, surface deposits, electrolytes, and unlike metal couples. *Caustic embrittlement* results in the formation of intercrystalline cracks below the water line of drums, around rivet heads, at tube ends, and at plate seams when the ratio of sodium sulfate to sodium hydroxide in the boiler water is low. The sodium hydroxide and carbonate alkalinity of the water is obtained by determining the amount of acid required to neutralize a water sample, and then calculating the weight of sodium carbonate necessary to neutralize this acid. The ASME Boiler Code recommends the following minimum values of the ratio of the sodium sulfate to equivalent sodium carbonate alkalinity:

Working pressure of boiler		Minimum safe ratio of sulfate to carbonate
lb/in^2	kPa	
0–150	0–1034	1
150–250	1034–1724	2
250 and over	1724 and over	3

Foaming refers to the formation of foam in the boiler drum due to failure of the steam bubbles to break. The water films of the bubbles are carried over with the steam, and a high moisture content results. *Priming* means the carrying of water into the steam space as a result of violent ebullition. The water carried over by priming is generally in larger and more compact masses or slugs than are present in foaming. Foaming and priming are apparently increased by the presence of dissolved and suspended solids.

Water conditioning may consist of sedimentation or filtration to reduce suspended solids, softening by chemical treatment to reduce formation of hard scale, evaporation of makeup water to reduce the concentration of suspended solids and dissolved gases, blowdown to reduce the concentration of solids, and deaeration to reduce the concentration of dissolved or mechanically entrained gases to minimize corrosion.

33. Boiler Accessories. Boiler accessories are those devices which are directly connected to the boiler for the purpose of giving safe and convenient operation. Certain accessories are required by law, and others are used for improving performance. Accessories commonly provided are a steam-pressure gage, water column, safety valve, stop-and-check valve, fusible plug, feedwater regulator, blowoffs, soot blowers, and tube cleaners.

34. Piping. Standard pipe is supplied in sizes from ⅛ to 12 in (3.2 to 304.8 mm); in random lengths of 18 to 20 ft (5.5 to 6.1 m), both ends threaded with coupling on one end. Extra-strong pipe is heavy pipe used for high steam, gas, or hydraulic pressures; sizes are ⅛ to 12 in (3.2 to 304.8 mm); and random lengths are 12 to 22 ft (3.7 to 6.7 m).

STEAM TURBINES

35. General Characteristics. Steam turbines are built in sizes from less than 1 kW to over 200,000 kW. The small requirements of space per unit of capacity, the enormous capacities possible, the high rotative speed, and close regulation make turbines particularly suitable as prime movers for central-station generation of electrical energy. The large specific volume of steam at low pressures may be handled more satisfactorily by a turbine than by a reciprocating engine, and back pressures of under 1 in Hg (2.5 cm Hg) abs are successfully used. Highly superheated steam may be used without lubrication difficulties, and the exhaust steam is free from oil.

36. Types. Steam turbines are divided into two general classes: (1) *impulse* turbines, in which the expansion of the steam occurs only in the stationary blades or nozzles; (2) *reaction* turbines, in which the steam expands in both the stationary and the moving blades.

In *impulse* turbines, the energy made available by the drop in pressure of the steam as it passes through the nozzles or stationary blades imparts velocity to the steam. The steam jet strikes the blades on a rotating disk and transfers its velocity energy to them. To utilize the energy efficiently in one passage of the steam over a set of blades involves very high rotative speeds. Lower speeds are obtained by passing the steam at constant pressure through two and sometimes three sets of rotating blades, each set absorbing its share of the energy. This is known as *velocity staging* or velocity compounding; the principle is illustrated in Fig. 13.174, which shows diagrammatically how the pressure and velocity of the steam change during its passage through the nozzles and blades of a turbine having one pressure stage and two velocity stages.

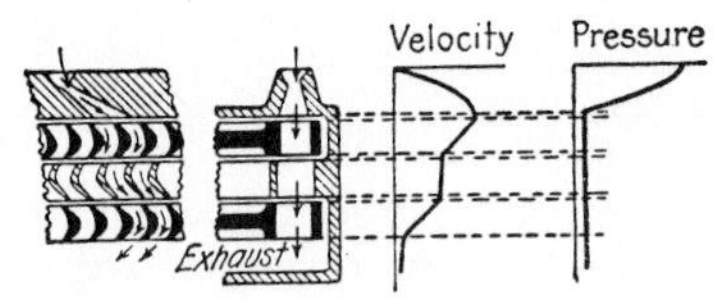

FIGURE 13.174 Pressure and velocity changes in an impulse turbine having one pressure stage and two velocity stages.

Figure 13.175 shows a method of velocity staging with a single row of moving blades. This is known as the *axial-flow, reëntry* impulse turbine. The reëntry principle is also applied to *tangential* flow of steam as in the Terry single-wheel impulse turbine; steam discharged from the nozzle enters pocketlike buckets on the rim of the wheel, and reversing stationary guide passages attached to the cas-

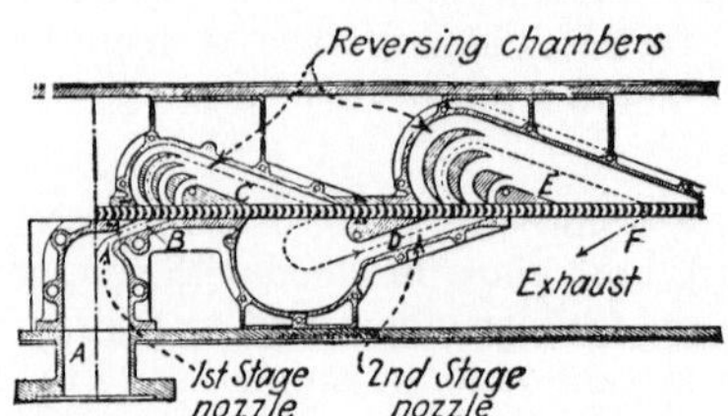

FIGURE 13.175 Reentry impulse turbine.

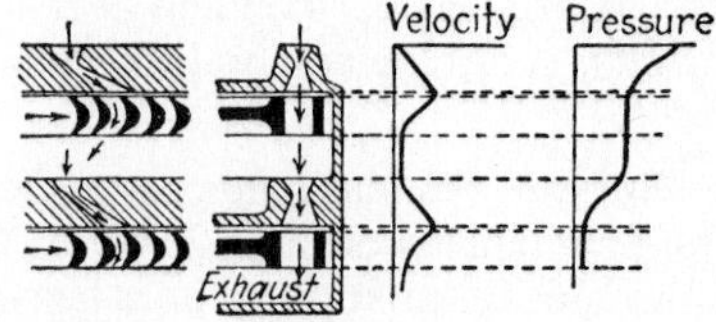

FIGURE 13.176 Pressure and velocity changes in an impulse turbine.

ing opposite the buckets redirect the steam into the buckets. Velocity staging is used only in turbines of the impulse type.

Low peripheral speed with high efficiency can also be obtained by *pressure staging* as in the *Rateau* turbine (see Fig. 13.176). The available pressure drop is divided into steps or stages, the rotating blades in each successive stage utilizing a part of the total available energy. This class of turbine may be thought of as consisting of a series of single-velocity-stage impulse turbines all on one shaft and each supplied with steam at a lower pressure than the one preceding.

A combination of velocity staging and pressure staging is illustrated in Fig. 13.177.

37. Reaction Turbine. The *reaction turbine*, frequently called the *Parsons type*, is a multipressure-stage machine each stage consisting of a set of fixed blades or nozzles followed by a set of rotating blades, as indicated in Fig. 13.178. The pressure drop is continuous and gradual from supply to exhaust; in any one stage it is small, seldom over 3 lb/in^2 (21 kPa), and consequently the number of stages is large. The stationary blades extend around the entire circumference because partial peripheral admission is impossible. The clearances must be small to minimize steam leakage around the blades to a lower pressure region. The leakage loss diminishes toward the exhaust end.

All the blades of a reaction turbine, both stationary and moving, have the same general form, but the passage areas increase toward the exhaust end of the machine to provide for the growth in volume of the steam resulting from reduced pressure. Each row of blades acts as a convergent nozzle and causes an increase in velocity with a small drop in pressure. The steam enters each set of moving blades with the velocity developed in the stationary blades, and it acquires more velocity by the nozzle action of the moving blades; but these blades recede at such a rate that the velocity is reduced to almost zero as the steam leaves to enter the next set of stationary blades.

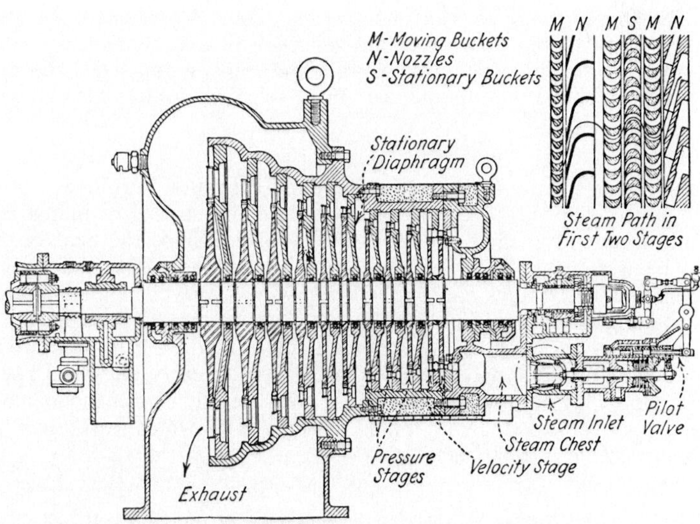

FIGURE 13.177 Axial-flow impulse turbine.

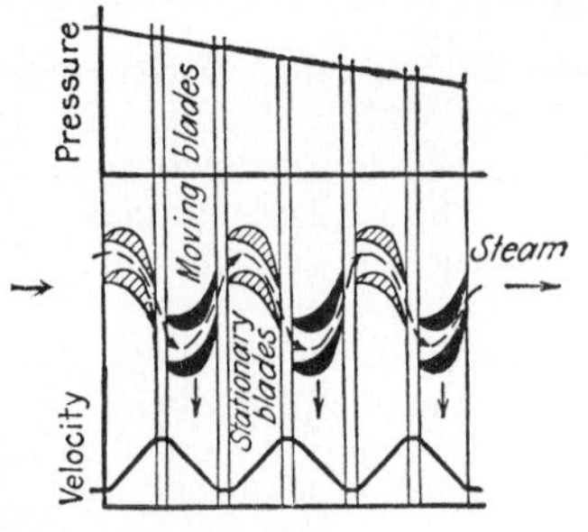

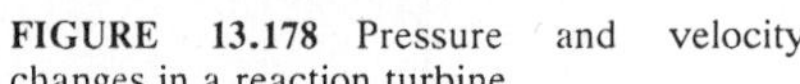
FIGURE 13.178 Pressure and velocity changes in a reaction turbine.

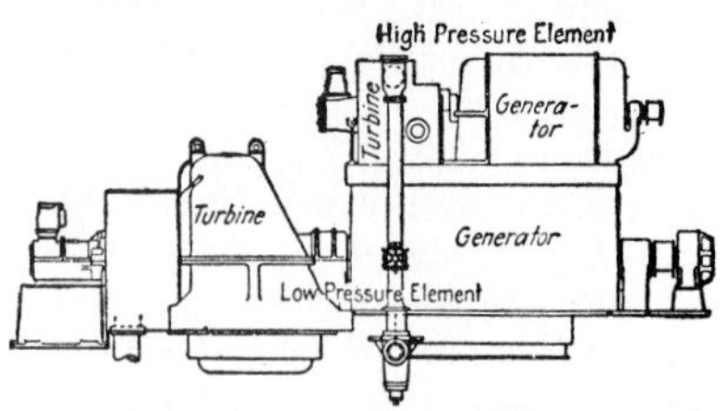

FIGURE 13.179 A 50-MW 1200-lb/in^2 (8.3-MPa) vertical compound turbine-generator set.

38. Combined Impulse and Reaction Turbines. Owing to the small pressure drop per stage, the shaft of the reaction turbine becomes quite long when a large total pressure drop is used. To shorten the machine and also to reduce the loss due to blade-tip leakage, which is greatest at the high-pressure end, impulse elements are frequently used at the high-pressure end, followed by reaction elements continued to the low-pressure end, so that a *combination type* is obtained.

39. Multicylinder or Compound Turbines. The increase in capacity of steam-turbine units has led to the development of compound machines having two or three cylinders. After partial expansion of the steam in the high-pressure cylinder, the steam passes to the next cylinder, where the process is continued. The cylinders may be arranged in tandem with a common spindle, or each unit may have its own shaft, this being the cross-compound arrangement. With multicylinder construction, the temperature range per cylinder and the consequent stresses in the rotor and casing can be reduced; more flexibility in design is also possible.

A *vertical compound* turbine-generator developed by the General Electric Company for use with 1200-lb/in^2 (8.3-MPa) pressure steam is shown in Fig. 13.179. The high-pressure element, comprising about 25 percent of the total capacity of the unit, is built on top of the generator of the low-pressure element. This compact construction saves in cost of foundations, requires less piping, conserves floor space, and permits the use of one set of air coolers for both generators.

40. Back-Pressure, Low-Pressure, and Mixed-Pressure Turbines. Turbines operated noncondensing, in connection with building heating or industrial processes, are frequently called *back-pressure* turbines. Since the heat in the exhaust can be usefully applied, the power from the machine is obtained at very low cost. *Low-pressure* turbines are supplied only with exhaust steam from other apparatus and should exhaust into high vacuum. *Mixed-pressure* turbines are designed to operate on low-pressure steam, supplemented by high-pressure steam when the low-pressure supply becomes insufficient to carry the load. The high-pressure supply may be throttled down to a pressure suitable for the machine, or the turbine may be provided with a high-pressure stage which runs idle when no high-pressure steam is being used.

41. Extraction or Bleeder Turbines. *Extraction* or *bleeder* turbines are supplied with high-pressure steam and operated condensing, but they have the distinctive feature of permitting the withdrawal of steam from the machine at one or more intermediate points between supply and exhaust, for use in building heating, feedwater heating, or manufacturing processes.

42. Turbine Performance. The best measure of comparative performance of steam turbines is the engine efficiency; brake engine efficiency is the ratio of the brake thermal efficiency of the turbine-generator to the thermal efficiency of the corresponding ideal engine (Rankine engine if there is no reheating or regenerative feedwater heating). The steam consumption may readily be estimated, when the engine efficiency, condition of supply steam, and exhaust pressure are known. The engine efficiency of turbines is primarily dependent upon the size. There are other factors that affect the efficiency, and turbine manufacturers can supply different machines of the same capacity that will show different efficiencies; however, when the first cost of the machine is considered in relation to the annual load factors and fuel costs, the size alone is a good index of the probable engine efficiency of the usual installation. Typical values of the brake engine efficiency of steam turbines are given in Table 13.13. These values are not exact, and manufacturers' guarantees should be obtained for new units.

TABLE 13.13 Brake Engine Efficiencies of Steam Turbines

Brake hp	kW	Brake engine efficiency, %
100	74.6	50
500	373.0	58
1,000	746.0	63
5,000	3730.0	72
10,000	7460.0	76
20,000–100,000	14,920–74,600	80–83

The *first cost* of steam prime movers of a fixed size varies almost directly as the engine efficiency. Engines or turbines of large output cost less per unit of capacity than small ones.

CONDENSING EQUIPMENT

43. General. The object in applying condensing equipment to steam prime movers is to improve their economy and increase their capacity by lowering the back pressure against which they exhaust. For a given condition of steam supply, a reduction in back pressure increases the available energy per pound (kg) of steam. When the back pressure is reduced to 1 in Hg (2.54 cm Hg) [a 29-in (73.7-cm) vacuum referred to a 30-in (76.2 cm) barometer], the temperature of the exhaust steam is only about 80°F (26.7°C).

Condensers are of two general types: (1) *surface* and (2) *direct contact*. Both types use cooling water to change the exhaust vapor to a liquid of much smaller volume, but the surface condenser keeps the two fluids separate while the direct-contact condenser brings them together in a mixture. The selection of type of condenser depends upon: (1) water quality and abundance, (2) fixed charges on condenser and auxiliaries, (3) power cost for condenser pumps, (4) maintenance expense, and (5) space requirements.

44. Surface Condensers. This type of condenser consists of a shell surrounding a compact nest of tubes. Cooling water is circulated through the tubes, and exhaust steam fills the shell surrounding them. Tube sizes are usually ¾ in (1.9 cm), ⅞ in (2.2 cm), or 1 in (2.5 cm), and 0.049 in (0.12 cm) (No. 18 B.W.G.) thick; ⅝-in (1.59-cm) tubes are sometimes used in small condensers. Small tubes are desirable from the standpoint of heat transfer but are often difficult to keep clean. The overall coefficient of heat transfer in surface condensers depends primarily upon the water velocity, cleanliness of the tubes, and the amount of noncondensible gases in the steam. Economic water velocities are usually between 6 and 8 ft/s (1.8 and 2.4 m/s). The heat balance for a surface condenser is given as follows:

$$W_s(h_s - h_c) = W_w(t_2 - t_1) = UA\theta_m \tag{13.67}$$

where W_s = weight of steam condensed, in lb/h (kg/h)
h_s = enthalpy of exhaust steam entering the condenser, in Btu/lb (kJ/kg)
h_c = enthalpy of condensate leaving condenser, in Btu/lb (kJ/kg)
W_w = weight of cooling water flowing through condenser, in lb/h (kg/h)
t_2 = temperature of exit cooling water, in °F (°C)
t_1 = temperature of entering cooling water, in °F (°C)
U = overall coefficient of heat transfer, in Btu/h · ft^2 · °F (W/m^2 · °C) (for rough approximations, a value of 350 may be used for USCS)
A = area of tube surface, in ft^2 (m^2)
θ_m = mean temperature difference between steam and water, °F (°C)

With single-pass condensers, the condensing surface provided for turbine installations is usually between 0.6 and 1.0 ft^2/kW (0.06 and 0.09 m^2/kW), while for two-pass condensers between 0.7 and 1.4 ft^2/kW (0.07 and 0.13 m^2/kW) is common; the larger figure in each case is for small turbines. The water circulated is commonly between 1.0 and 2.0 gal/min · ft^2 (0.7 and 1.4 L/s · m^2).

45. Direct-Contact Condensers. This group includes (1) jet, (2) barometric, and (3) ejector condensers. In jet and barometric condensers the exhaust steam and cooling water enter near the top, the water being discharged into the steam in a fine spray. The resulting mixture of condensed steam, warmed cooling water, and air must be pumped from the condenser unless it is of the barometric type. An independent air-removal pump is necessary for high vacuum. The barometric condenser is elevated sufficiently above the hot well to cause the water to flow out by gravity against atmospheric pressure. A barometric condenser requires a circulating pump and sometimes a dry air pump. Ejector condensers operate on the same principle as a steam ejector. No removal pump is needed. High velocity of the fluids is utilized to discharge the mixture of condensed steam, cooling water, and air against atmospheric pressure.

46. Air Removal. Air, entering the boilers with the feedwater, passes through the prime movers and is discharged into the condensers with the exhaust steam. Unless it is promptly removed, it will accumulate, impair the vacuum, and cause the back pressure to rise. Air may also enter the condensers owing to leakage. Pumps for removing air from condensers are commonly of the steam-jet type.

47. Recooling Condensing Water. When cooling water is deficient in quantity or quality, cooling ponds, spray cooling systems, or cooling towers may be used to cool the water after it has passed through the condenser. In this way the same cooling water may be used repeatedly with a makeup of from 2 to 8 percent to replace that lost by evaporation.

Cooling ponds are so constructed as to expose a large area to the atmosphere. The water is cooled by surface evaporation. The necessary pond area may be estimated at 8 ft^2 for each lb (1.64 m^2/kg) of steam condensed per hour. A cooling effect of about 4 Btu/h per °F (22.7 W/m^2 per °C) temperature difference between the water and the air will be obtained, per square foot of pond area, in the summer, and about 2 Btu/h · °F (11.4 W/m^2 · °C) in the winter.

Spray cooling systems accomplish the same results as cooling ponds but require much less area. The cooling water is piped to a system of nozzles through which the water is sprayed into the air. It falls back into a basin below. The amount of water surface exposed to the atmosphere is thus greatly increased. A typical spray cooling system with a capacity of 4200 gal/min at 7 lb/in^2 pressure (265 L/s at 48 kPa) discharged through 105 nozzles attached to 2-in (50.8-mm) pipes requires a pond or basin 138 ft long and 100 ft wide (42 × 30 m). The nozzles are in 21 standard clusters of 5 each; the clusters are in 3 rows; the rows are 25 ft (7.6 m) apart; each row has 7 clusters on 13-ft (3.9-m) centers. The reduction in temperature by spraying depends on the atmospheric conditions but is usually 15 to 20°F (8.3 to 11.1°C).

A *cooling tower* is a structural assemblage of cooling surface which may conveniently be placed on the top of a building or in an open field. Where space is ample, the atmospheric type, depending on natural draft, is more economical than the forced-draft type, which uses fans to pass the air upward through the structure. A casing made of wood, steel, or concrete surrounds a network of wooden partitions, sheet-iron trays, wire mats, or tile. Water introduced at the top falls downward in such a manner as to expose a large surface to the air passing upward.

The water may be cooled, as an ideal limit, to the wet-bulb temperature of the air, but cooling to within 5°F (2.8°C) of this temperature represents good practice. A cooling tower is commonly guaranteed by the manufacturer to cool a prescribed weight of water per unit of time through a definite temperature range at a stated atmospheric temperature and relative humidity.

REFRIGERATION ENGINEERING*

REFRIGERATION MACHINES AND PROCESSES

Refrigeration is a special aspect of heat transfer and involves the production and utilization of below-atmospheric temperatures by a number of practical pro-

*Excerpted from *Marks' Standard Handbook for Mechanical Engineers*, by E. A. Avallone and T. Baumeister III (eds.). Copyright © 1987. Used by permission of McGraw-Hill, Inc. All rights reserved. Updated and metricated by the editors where necessary.

cesses. Substances are cooled when their heat is transferred, via a temperature drop, to solid, liquid, or gaseous media which are naturally or artificially colder, their lower temperature stemming from radiation, sensible- or latent-heat physical effects, or endothermic chemical, thermoelectric, or even magnetic effects. Effects, such as cold streams, melting ice, and sublimating solid carbon dioxide, are included.

Of particular practical use is the achievement of lowered temperature for a circulating fluid (*refrigerant*) on a continuous basis, with heat-absorption capability, by (1) expansive flow of a fluid through a pressure drop in a restriction ("expansion valve") *without* production of external work, i.e., throttling (Joule-Thomson) effect, and (2) expansive flow of a fluid (usually a gas) through a pressure drop in a machine (expander) *with* production of external work.

The integrated systems for continuous cold effect are best recognized through the similarity of format in their closed flow circuits, particularly in (1) vapor-compression and (2) gas-compression systems. Both these systems contain a high-level and a low-level heat exchanger and two pressure difference devices, a pressure generator and a pressure-reducing element, typical of energy-transport systems. In the vapor-compression system the heat exchangers are characterized by latent-heat effects; in the gas compression, by sensible-heat effects primarily. Although some cases may entail discharge of some circulant at low pressure and low temperature with subsequent fresh make-up, the closed systems are more usual. Generally, the low-level-exchanger refrigeration effect is achieved through energy input, such as work input to a compressor.

With continuous recirculation of the cycle fluid, the low-temperature heat pickup must be pumped, via energy input, to a higher level so as to discharge it to atmospheric temperature. In this sense the system acts as an *energy-transport device* or *heat pump* or, from another viewpoint, as a temperature transformer. This philosophy of energy transport may equally well be applied to more remote systems, such as thermoelectric. It is, of course, a manifestation of the principles of the Second Law, and the effectiveness of the system's operation may be evaluated numerically.

Coefficient of performance is the ratio of useful refrigeration to the work supplied, either directly or indirectly, as energy capable of doing work. The ratio of the coefficient of performance of any actual cycle to that of the Carnot cycle operating between the same two temperature levels is called the *relative efficiency* of the cycle. Relative efficiencies must always be less than unity, but coefficients of performance can be severalfold greater than one.

Refrigeration capacity is defined in terms of the "ton."

1. Units of Refrigeration. In the United States, a *standard ton of refrigeration* corresponds to a heat absorption at a rate of 288,000 Btu/day or 200 Btu/min (3.5168 kW). The heat absorption per day is approximately the heat of fusion of 1 ton (907.19 kg) of ice at 32°F (0°C). The *standard rating* of a refrigerating machine, using a condensable vapor, is the number of standard tons of refrigeration it can produce under the following conditions: (1) liquid only enters the expansion valve and vapor only enters the compressor or the absorber of an absorption system; (2) the liquid entering the expansion valve is subcooled 9°F (5°C) and the vapor entering the compressor or absorber is superheated 9°F (5°C), these temperatures to be measured within 10 ft (3.05 m) of the compressor cylinder or absorber; (3) the pressure at the compressor or absorber inlet corresponds to a saturation temperature of 5°F (− 15°C); and (4) the pressure at the compressor or absorber outlet corresponds to a saturation temperature of 86°F (30°C).

The *British unit of refrigeration* corresponds to a heat-absorption rate of 237.6 Btu/min (4.175 kW) with inlet and outlet pressures corresponding to saturation temperatures of 23°F (−5°C) and 59°F (15°C), respectively. On the European continent a unit of refrigeration capacity called the *frigorie* is used. A frigorie is approximately equivalent to 50 Btu/min (0.8786 kW), or ¼ of a standard ton of refrigeration.

PROPERTIES OF REFRIGERANTS

Refrigerants are the transport fluids which convey the heat energy from the low-temperature level to the high-temperature level, where it can, in terms of heat transfer, give up its heat. In the broad sense, gases involved in liquefaction processes or in gas-compression cycles go through low-temperature phases and hence may be termed *refrigerants*, in a way similar to the more conventional vapor-compression fluids.

Refrigerants are *designated by number*. The identifying number of the refrigerant, or the word *Refrigerant* or both, may be used in conjunction with the trade name. For example, Refrigerant 12 can also be referred to as Freon 12, Isotron 12, Freon Refrigerant 12, Freon 12 Refrigerant, or F-12 Refrigerant. The number designation of refrigerants is covered in greater detail in ANSI/ASHRAE Standard 34–78.

The desirable thermal properties of a refrigerant are (1) convenient evaporation and condensation pressures, (2) high critical and low freezing temperatures, (3) high latent heat of evaporation and high vapor specific heat, (4) low viscosity and high film heat conductivity. Desirable practical properties include (1) low cost, (2) chemical and physical inertness under operating conditions, (3) noncorrosiveness toward ordinary construction materials, and (4) low explosive hazard both alone and mixed with air. The refrigerant should be nonpoisonous and nonirritating and should not cause deterioration in the lubricant used. Leakage should be detectable by simple tests, easily performed.

The *specific volume of refrigerant* to be handled is important with reciprocating compressors, as it determines the size of the compressor; but with centrifugal compression a large volume is not objectionable and may be a positive advantage for small units. A large compression ratio is undesirable in reciprocating compressors from the standpoint of clearance losses and may make the use of compound compression necessary.

A comparison of various refrigerants based on *ideal performance* is given in Table 13.14. The results are for the *standard temperature range* of 5 to 86°F (−15 to 30°C) and also for certain other ranges. For these calculations, zero piston clearance and expansion through a throttle valve were assumed. The use of an expansion cylinder instead of a valve would have yielded work (available for supplying some of the work of compression), but this is always a negligible quantity, except when the compression pressure approaches the critical pressure, as for carbon dioxide. Theoretical *coefficients of performance* may be obtained by dividing 4.72 by the theoretical horsepower given in the table.

Ammonia is used extensively in large installations, industrial and commercial. It is toxic and because of corrosive action must be kept out of contact with copper or copper-bearing alloys. It has a high latent heat of evaporation and convenient pressure-specific volume relations. It is not miscible to any large extent with lubricating oil. *Carbon dioxide* was used for a long time as a safety refrigerant; exposure to it in a confined space is not dangerous unless concentrations are high. With cooling water at 70°F (21°C), the condensing pressure of CO_2 is high,

TABLE 13.14 Ideal Performance of Refrigerants for Various Temperature Ranges

Refrigerant (number)	Operating temperature range, °F	Suction pressure, lb/in² abs	Head pressure, lb/in² abs	Ratio of head to suction pressure	With dry and saturated suction vapor, per ton: Weight of vapor, lb/min	Piston displacement, ft³/min	Theoretical hp	Temperature at end of compression, °F	Type of compressor
Air (729)	5–86	14.7	73.5	5.0	7.02	82.3	2.82	277.0	Recip. and exp. cyl
Water (718)	32–86	0.0885	0.6152	6.95	0.1957	647.0	0.618	332	Centrif. or ejector
	32–100	0.0885	0.9492	10.73	0.1985	656.2	0.819	420	
	40–100	0.1217	0.9492	7.80	0.1978	483.4	0.687	366	
Carbon dioxide (CO_2) (744)	5–86	332.0	1,043.0	3.14	3.61	0.960	1.827	160.3	Recip.
Ammonia (NH_3) (717)	5–86	34.27	169.2	4.94	0.421	3.44	0.99	209.8	Recip.
	20–100	48.21	211.9	4.40	0.421	2.49	0.94	212.8	
	40–100	73.32	211.9	2.89	0.427	1.70	0.65	176.0	
Freon 11 (CCl_3F) (11)	5–86	2.931	18.28	6.20	3.058	37.0	0.94	112.7	Centrif.
	20–100	4.342	23.60	5.44	3.086	26.2	0.89	122.1	
	40–100	7.032	23.60	3.36	2.945	16.08	0.63	114.2	
Freon 12 (CCl_2F_2) (12)	5–86	26.51	107.9	4.07	3.916	5.82	1.00	100.2	Recip. or centrif.
	20–100	35.75	131.6	3.68	4.054	4.54	0.97	112.5	
	40–100	51.68	131.6	2.55	3.880	3.07	0.67	108.0	
Freon 21 ($CHCl_2F$) (21)	5–86	5.243	31.23	5.96	2.364	20.87	0.94*	99.0*	Rotary
	20–100	7.699	40.04	5.20	2.404	15.61	0.94*	110.4*	
	40–100	12.32	40.04	3.25	2.315	10.19	0.68*	105.9*	
Freon 22 ($CHClF_2$) (22)	5–86	43.02	174.5	4.06	2.926	3.65	1.03	131.7	Recip.
	20–100	57.98	212.6	3.67	3.023	2.83	0.99	152.7	
	40–100	83.72	212.6	2.54	2.936	1.93	0.68	131.0	
Methylene chloride (CH_2Cl_2) (30)	5–86	1.28	10.07	8.56	1.485	74.0	0.96*	205.1*	Centrif.
	20–100	1.92	13.25	6.90	1.520	47.72	0.91*	157.1*	
	40–100	3.38	13.25	3.92	1.493	27.76	0.63*	167.7*	

Methyl chloride (CH_3Cl)	5–86	21.15	94.70	4.48	1.331	5.95	0.96	178.1	Recip.
(40)	20–100	29.16	116.7	4.00	1.363	4.51	0.90	184.4	
	40–100	43.25	116.7	2.69	1.342	3.07	0.62	157.0	
Sulfur dioxide (SO_2)	5–86	11.81	66.45	5.63	1.415	9.08	0.97	191.4	Recip.
(764)	20–100	17.18	84.52	4.92	1.453	6.52	0.92	193.4	
	40–100	27.10	84.52	3.12	1.444	4.17	0.63	162.6	
Propane (C_3H_8) (290)	5–86	42.1	155.3	3.69	1.653	4.10	1.35*	92.9*	Recip.
	20–100	55.5	187.0	3.37	1.730	3.29	1.32*	103.9*	
	40–100	78.0	187.0	2.40	1.646	2.26	0.90*	101.5*	
Ethane (C_2H_5) (170)	5–86	236.0	675.9	2.87	3.41	1.82	2.180	105	Recip.
Ethyl chloride (C_2H_5Cl)	5–86	4.65	27.10	5.83	1.405	24.0	0.95*	106.3*	Rotary
(160)	20–100	6.80	34.79	5.12	1.425	17.19	0.92*	116.1*	
	40–100	10.79	34.79	3.22	1.375	10.73	0.63*	109.9*	

*Values may be slightly in error.

and since its critical temperature is 87.8°F (31°C) condensation will not occur at water temperatures above this. Power consumption is high on CO_2 machines. *Sulfur dioxide* has rapidly gone out of use even in the small household units where it was previously widely employed. It is extremely corrosive unless absolutely anhydrous and although not dangerous in the quantities used in a household unit [2 to 3 lb (0.9 to 1.4 kg)], it may be dangerous in larger installations. It is extremely irritating even in small amounts. *Methyl chloride* (CH_3Cl), an anesthetic in amounts of 5 to 10 percent by volume, has been used in air-cooled units of either moderate or small sizes. It is miscible with mineral oils; small amounts of moisture in a methyl chloride system will cause trouble by freezing in expansion valves. The series of fluoro-chloro hydrocarbons known as *Freons*, or *Genetrons*, are widely used in both small- and medium-size household and commercial units, as well as large industrial units. Many compounds of this type are commercially produced; those which are nontoxic,nonirritating, and nonflammable are commonly employed. They include F-11, trichloromonofluoromethane (CCl_3F); F-23, dichlorodifluoromethane (CCl_2F_2); F-13, monochlorotrifluoromethane ($CClF_3$); F-21, dichloromonofluoroethane ($CHCl_2F$); F-22, monochlorodifluoromethane ($CHClF_2$); F-113, trichlorotrifluoroethane (CCl_2FCClF_2); and F-114, dichlorotetrafluoroethane ($C_2Cl_2F_4$). F-12 is widely used for air conditioning and general commercial cooling. Methyl chloride, having similar physical properties, has been largely supplanted by F-12 in this field. Refrigerants containing chlorine as part of the molecule are readily detected in minute amounts by passing them through a copper gauze kept hot by the essentially colorless flame of burning methyl alcohol. Even traces of such refrigerants in air give a readily detected test due to the intense green color imparted to the flame by their presence.

For industrial work, *butane* (C_4H_{10}), *propane* (C_3H_8), and, at low temperatures, *ethane* (C_2H_6) are used. *Dieline* (dichloroethylene, $C_2H_2Cl_2$) and *Carrene* (dichloromethane, CH_2Cl_2) have been used to some extent, usually in centrifugal compressors.

The pressure-temperature relations for the saturated vapors of many of the commercially important refrigerants are given in Fig. 13.180. If the temperature at which the refrigeration is desired and the temperature at which heat can be discarded (condenser temperature) are known, the chart is convenient for determining for any chosen refrigerant the pressures which must be maintained. For instance, if the use of methyl chloride is contemplated in an air-cooled cabinet food freezer located in a room where the air temperature may rise to 90°F (32°C), the compressor discharge must be at least 100 lb/in^2 abs (690 kPa absolute), and if the cabinet cooling coil is to be held at −30°F (−34°C), compressor suction will be 9 lb/in^2 abs (62 kPa absolute).

The large volumes required for F-11, dieline, and water vapor can be handled satisfactorily by centrifugal compressors. When the evaporating pressure is below atmospheric, as in the case of dieline, ethyl chloride, sulfur dioxide, water vapor, and butane, air leaks are likely to be excessive; where the refrigerant is nearly odorless (Freon 12 and methyl chloride), leaks into the atmosphere are difficult to detect, and loss of the refrigerant may be excessive.

OVERALL CYCLES

Figure 13.181 represents the simple closed-circuit *vapor-compression* system. The upper heat exchanger is a vapor condenser (with some superheat), and after

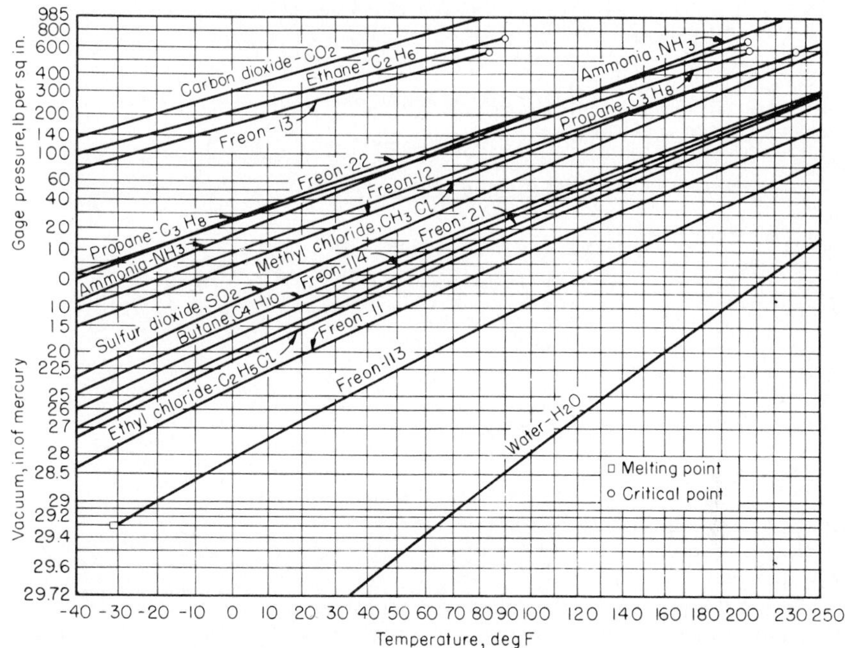

FIGURE 13.180 Pressure-temperature relations for saturated vapors of refrigerants.

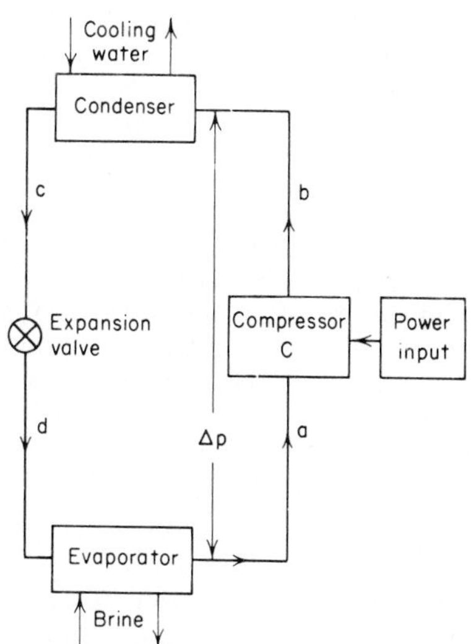

FIGURE 13.181 Simple closed-circuit vapor-compression system.

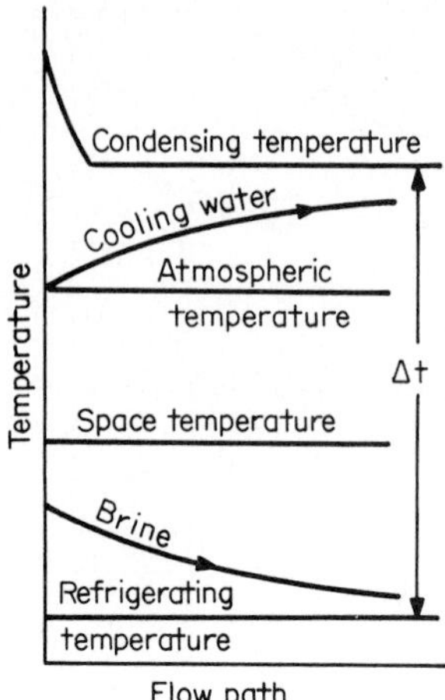

FIGURE 13.182 Comparative temperature relations for fluids of a simple system.

the throttling expansion valve, the lower one is the evaporator in which the refrigerant liquid at reduced pressure and temperature evaporates with the inward refrigeration heat flow. The refrigerant vapor is elevated by the compressor to a higher pressure and condensing temperature so that it will liquefy in its transfer of heat to the atmospheric level. Figure 13.182 shows comparative temperatures for the cycle.

COMPONENTS OF COMPRESSION SYSTEMS

A refrigeration system is an energy-transport complex of assorted components. It may be a conventional system in which the same fluid, the refrigerant, recirculates continuously in a closed circuit, or it may entail a partially open system with discharge of some of the processed fluid either as a liquefied gas or as a solidified gas, with replacement makeup, e.g., by air liquefaction and solid carbon dioxide production. Components include vapor and gas compressors; liquid pumps; heat-transfer equipment (gas coolers, intercoolers, aftercoolers, exchangers, economizers); vapor condensers and the counterpart evaporators; liquid coolers and receivers; expanders; control valves and pressure-drop throttling devices (capillaries, refrigerant-mixture separating chambers, stream-mixing chambers); and connecting piping and insulation.

Compression may be single stage (Fig. 13.181) or multiple stage. The *compressors* may be displacement or kinetic. The former include piston *reciprocating* machines in which the refrigerant, as vapor, is raised in pressure, usually with some simultaneous cooling. These machines may be horizontal or vertical, single acting or double acting; the vertical arrangement, with multiple cylinders in automotive arrangement and running at high rotative speeds, is common. Of utmost importance in the realized performance of a displacement compressor is the real volumetric effectiveness, or "efficiency," which measures the extent of utilization of the machine displacement volume. It is influenced by cylinder clearance volume, pressure ratio, nature of the gas, pressure drops through the valves and suction heating, and the equivalent compression exponent n. [The numerical value of n expresses the effect of heat transfer to or from the gas during compression or expansion in a cylinder. Normally it lies between $n = 1.0$ and $n = k$ (adiabatic), such as for the ideal (isentropic).]

Air-cooled compressors are used where discharge temperatures are low, such as with Refrigerant 12; *water cooling*, where discharge temperatures are high, as with ammonia. *Oil separators* return to the compressor the lubricating oil carried over by the refrigerant vapors. *Rotary compressors* of the vane, eccentric, gear, and screw types are in use.

Kinetic compressors include high-speed centrifugal and axial flow machines, usually multistaged, and jet-entrainment devices. Centrifugal machines are especially adapted to high-volume flow (> 500 ft^3/min) (14.2 m^3/min).

The entire machine-compression operation may be replaced by a secondary absorber-pump-generator system, in which the complex is known as an *absorption refrigeration system*.

Dual (or *multiple-effect*) *compression* may be used when refrigeration at two temperatures is desired. The compressor takes vapor from a lower temperature expansion coil during the first part of its intake stroke, and from a higher temperature expansion coil at or near the end of the stroke. The mixture is then compressed and condensed.

In *wet compression*, cooling is obtained during compression by spraying liquid refrigerant into the compressor cylinder. The desuperheating of the compressed vapors results in better heat transfer in condensers and more nearly isothermal compression; the disadvantages are reduced compressor capacity and the problem of control of the amount of injection. Wet compression is not widely used.

Condensers are usually shell-and-tube type, with the refrigerant passing outside the tubes. Older industrial installments still use *double-pipe condensers*. In double-pipe condensers, gas flows between the two pipes while water passes through the inner. The outside pipe diameter may be 2 in (5 cm); the inner, 1¼ in (3.2 cm). In some instances, the pipes are exposed to the atmosphere either with or without water drip. In an *evaporative condenser* the refrigerant vapor is condensed as it passes through tubes over which water is sprayed; the water is then evaporated by air flowing over the wet tubes. In this way, cooling-water requirements are reduced to from 5 to 15 percent of the water requirements of a nonevaporative condenser with no water reuse. The evaporative condenser combines in a single unit a refrigerant condenser and the atmospheric cooling tower or spray pond which is required if the cooling water is to be reused. Air-cooled finned-tube condensers with forced ventilation are widely used on small units, and shell-and-coil or double-tube water-cooled condensers on medium units.

Evaporators may have finned or plain surfaces. Defrosting may be automatic or manual. *Flooded* evaporators are operated practically full of liquid refrigerant, the level being controlled by a float valve. *Wet-expansion* evaporators are operated with a level approaching that for flooded operation; *dry-expansion* (once-through) units operate with an indefinite amount of liquid in the evaporator. Flooded or wet operation gives high heat-transfer rates, but requires larger amounts of refrigerant than dry operation. Pumped recirculation of refrigerant is used on some flooded evaporator systems to promote heat transfer, and also on shell-and-tube evaporators (liquid chillers), e.g., in water-vapor systems. For a low-cost unit operating in limited space under conditions of motion (as on boats, trains, etc.) dry operation is preferred.

Controls are required on the liquid level of the refrigerant and on the temperature of the refrigerated space. The liquid control regulates the flow of refrigerant into the evaporator and also serves as the pressure barrier between the high operating pressure of the condenser and the lower operating pressure of the evaporator. It may take the form of a *capillary tube* between the condenser (high side) and the evaporator (low side), in which case it is nonadjustable. Plugging due to dirt is a common difficulty. Capillary-tube liquid control is largely confined to relatively small units assembled and charged at the factory and particularly for hermetically sealed systems. The *constant-pressure expansion valve*, maintaining a constant evaporator pressure, and the *thermal expansion valve*, maintaining a constant superheat leaving the evaporator, are standard liquid controls for most commercial applications. A *low-side float* liquid control, used with a flooded evaporator operating at evaporator (low) pressure, consists of a float-operated valve to admit liquid refrigerant to the evaporator in accordance with demand so that a constant liquid level is held in it. A *high-side float* liquid control is often used with a single flooded evaporator; the float operating the valve between the evaporator and the condenser is in a float chamber containing liquid refrigerant at the condenser (high-side) pressure. As the liquid level in the float chamber falls, the valve closes, thus preventing hot gaseous refrigerant from passing from the high to the low side as is possible when using a capillary or a low-side float.

Steam-jet refrigeration systems are used where cooling to temperatures above 32°F (0°C) is desired. Applications include industrial air conditioning and cooling of city gas to condense out tar and other objectionable impurities, of gas absorbers to increase efficiency of absorption, of reaction units where heat removal and temperature control are important during chemical transformations, and of wort and mash in the brewing and other fermentation industries particularly in the summer months. Jet refrigeration has been used for cooling passenger trains and is coming into popularity for marine installations, e.g., cooling of banana boats and large passenger vessels. The system is a compression-type refrigerator: it uses water as a refrigerant, a part of which is evaporated to produce cooling of the remainder, steam-jet ejectors to compress the water vapor resulting from evaporation, and a condenser, either the surface type, where refrigerant water is to be reused, or the barometric type, where refrigerant water can be discarded. The advantages of the system are: low installation cost; the absence of moving parts except for small liquid pumps; and safety, since no noxious or toxic refrigerants are used.

For considerable differences in temperatures between condensing-water temperature and refrigerating temperatures (10°C or more), cooling of the refrigerant water by evaporation in two stages operating at different pressures (and temperatures) will usually give better operating economy, but at somewhat higher installation cost.

Figure 13.183 shows the variation of refrigerating capacity with variation of the chilled-water temperature, capacity increasing as chilled-water temperature rises.

ABSORPTION SYSTEMS

Absorption refrigeration machines are essentially vapor-compression plants (Fig. 13.181) in which the mechanical compressor has been replaced by a thermally activated arrangement (Fig. 13.184). The basic elements are the *absorber, pump, heat exchanger, throttle valve*, and *generator*.

In an absorption cycle, a secondary fluid, the *absorbent*, is used to absorb the primary fluid, the *refrigerant*, after the refrigerant has left the evaporator. The vaporized refrigerant is converted back to liquid phase in the absorber. Heat released in the absorption process must be rejected to cooling water. The solution of absorbent and refrigerant is then pumped to the generator. A heat exchanger may be used for heat recovery and a corresponding increase in efficiency. In the generator, heat is added, and the more volatile refrigerant is separated from the absorbent through distillation. The refrigerant continues to the condenser, expansion valve, and evaporator, while the absorbent returns to the absorber.

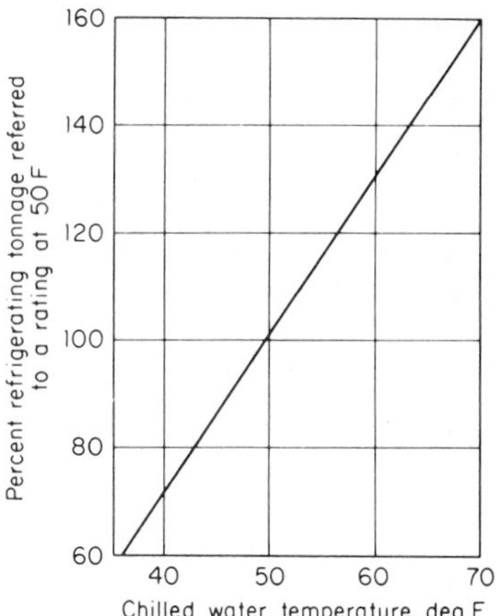

FIGURE 13.183 Variation of refrigerating capacity with the jet-chilled water temperature in steam refrigeration.

Although there are many refrigerant-absorbent combinations, the most common systems are *ammonia-water* and *water-lithium bromide*. A common variation of the absorption system is the *Electrolux-Servel* process (Platen-Munters Patent). It has had wide application in gas-fired household refrigeration machines and air-conditioning systems. The use of hydrogen in a hermetically sealed system eliminates the use of pumps and other moving parts.

Absorption machines represent only a small percentage of refrigeration systems in operation. This is because they are generally bulky and difficult to operate efficiently. There has been renewed interest in absorption machines with the advent of high energy costs. This is because an absorption machine is an excellent way to utilize waste heat. Examples of potential applications are varied. One example may be on an exothermic process, or where extensive use of steam turbine with limited turndown results in vented steam especially during warm weather when air conditioning is required. A site where peak electrical demand is set by air-conditioning load and ratcheted demand charges are high may justify absorption refrigeration if adequate steam-generation facilities are already present.

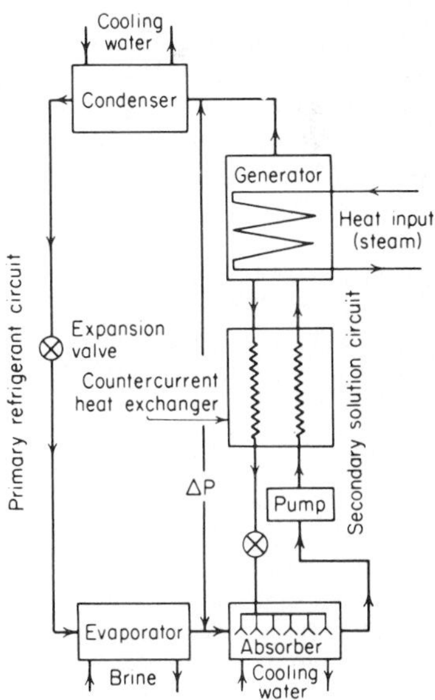

FIGURE 13.184 Elemental absorption system circuit.

A typical absorption machine requires about 20 lb (9 kg) of steam for 1 ton (3.5 kW) of refrigeration effect at 45°F (7°C).

A more detailed explanation of absorption systems is found in Ref. 3, pp. 1.20 to 1.25.

THERMOELECTRIC COOLING

Thermoelectric cooling utilizes the *Peltier effect* whereby a temperature differential will occur between two junctions of a closed loop of dissimilar metallic conductors when an electric current is imposed. It is variously called *Peltier effect cooling, thermoelectric refrigeration*, and *electronic cooling*. It is the inverse of the thermocouple operating principle whereby a temperature differential across the junctions of two dissimilar metals produces an electric current.

The capacity of a thermoelectric couple is small. System capacity can be increased by increasing the number of elements used. The temperature produced can be lowered by cascading the systems. Capacities approaching 25 tons (87.9 kW) cooling effect have been achieved, and temperatures as low as −266°F (−166°C) have been attained.

Aside from limited practical capacities, the major disadvantage of a thermoelectric cooling system is its poor coefficient of performance. The absence of moving parts, silent operation, ability to function in zero gravity, and lack of pressurized vessels, have given thermoelectric cooling wide space-program application. Other applications include use on submarines, electronic equipment cooling, and small units such as drinking water coolers or units for use in recreational vehicles.

The elements of a thermoelectric cooler are quite simple and are equivalent to those of a thermoelectric generator. For development of various output capacities, the number of basic elements is proportionately increased; they all act in parallel, and their outputs add together. *Semiconductor* materials, such as bismuth-telluride-selenide and bismuth-antimony-telluride alloys, are employed.

With an input to the circuit of low-voltage direct current and with heat continuously abstracted at one junction at room temperature, the other junction will become cold. Essentially each of the two junctions becomes an "activity cell," in which at the lower refrigeration temperature level, heat is converted into electrical effect, and at the higher atmospheric temperature, the electrical effect is converted into heat. This establishes the basis of "heat-pump" operation: the lower-temperature heat requirement must be supplied, essentially as in the evaporator of a vapor-compression system, by heat withdrawal via heat transfer, from all connected surroundings. The energy-transport circuit thus parallels the machine-activated circuit of Fig. 13.181. The amount of heat for a single-component circuit depends on the circulating electric current, on the properties and dimensions of the two conductors (p and n types), and on the resultant temperature difference between the high side room temperature and the low side input heat that is divided into two groups, useless and useful. The *useless heat* is the heat conduction between the high and low temperature through the thermoelements plus the current-generated resistance-loss heat within the elements that flows to the colder junction. The *useful heat* is the refrigeration effect in a regular evaporator. All the heat must in turn be discharged via the Peltier effect at the higher room temperature, as in the condenser of a vapor-compression system. Ultimate performance of the circuit depends on the voltage-generating nature of the two dissimilar conductors and on their electrical resistance, thermal conductivity, and physical dimensions. The different properties are mathematically combined into a *figure of merit, Z*.

METHODS OF APPLYING REFRIGERATION

In *direct expansion systems* the evaporator is placed in the space which is to be cooled; in *indirect systems* a secondary fluid (*brine*) is cooled by contact with the evaporator surface, and the cooled brine goes to the space which is to be refrigerated. Brine systems require 40 to 60 percent more surface than do direct expansion; they have an equalizing effect due to the large heat capacity of cold brine, they are safer (particularly if the refrigerating effect must be carried considerable distance or widely distributed), and they permit closer temperature regulation than is possible with direct expansion. If two temperatures are to be held, the lower may be by direct expansion, the higher by brine cooling. Development of better controls and newer piping methods has made direct expansion more attractive than previously.

Brines used for industrial refrigeration are usually aqueous solutions of calcium chloride, ethylene glycol, propylene glycol, or undiluted methylene chloride, and Freon 11. Calcium chloride should not contain over 0.2 percent magnesia, calculated as magnesium chloride. Calcium chloride brines are recommended down to −45°F (−43°C). Brines should be chemically neutral: acidic brines attack ferrous materials, alkaline brines attack zinc, ammonia in brine (resulting from leaks in the ammonia system) is especially harmful to most nonferrous metals. Corrosion by brine is increased by the presence of oxygen, air, or carbon dioxide and by galvanic action between dissimilar metals. Corrosion inhibitors are widely used, a satisfactory one being about 1600 g of sodium chromate or dichromate per 1 m^3 of calcium chloride brine or 3200 g per 1 m^3 of sodium chloride brine.

Brine coolers may be of three types: shell-and-tube, shell-and-coil, and double pipe. The shell-and-tube type most widely used is brine flowing through the tubes which are surrounded by the evaporating refrigerant. Tubes may be arranged for multipass operation. Effective heat-transfer surface varies from 8 to 15 ft^2/ton (0.21 to 0.4 m^2/kW), varying with temperature and brine velocity. A submerged coil in an open brine tank is used for ice making by the can process.

The *double-pipe cooler* is usually of 2-in (50-mm) inner or brine-flow pipe and 3-in (75-mm) outer pipe. The commercial rating is 15 to 20 ft (5 to 6 m) length of coil per ton of refrigeration.

The *shell-and-tube cooler* is used with closed heads and is erected both vertically and horizontally; brine flows through the tubes and ammonia is in the shell. It is made in sizes from 1 to 350 tons with ratings of 8 to 15 ft^2 (0.7 to 1.4 m^2) effective surface per ton, varying with the temperature and brine velocities; tubes 1 to 2½ in (25 to 63 mm) arranged multipass. This type of cooler has largely displaced all other types in recent installations.

REFRIGERANT PIPING

It is important to the proper operation of a refrigeration system that the piping or mains interconnecting the compressors, condensers, evaporators, and receivers be properly sized. This piping must be considered in three categories, viz., liquid lines, suction lines, and discharge lines. Essentially pipeline sizing is governed by pressure drop, first cost, operating cost, noise, and oil entrainment. Excessive pressure drops penalize compressor efficiencies and may affect control-valve operation adversely. Liquid-line velocities for most refrigerants are in the order of 60 to 400 ft/min (18 to 122 m/min), suction lines from 700 to 4600 ft/min (38 to 250 m/s), and discharge lines from 1000 to 5000 ft/min (213 to 1402 m/min). Pressure drops vary approximately as the square of the velocity (or tonnage) and directly as the length of the piping. For liquid lines, when evaporator is located above condenser, the following pressure drops in pounds per square inch per foot (kilopascals per meter) of static lift should be allowed: ammonia, 0.26 (5.9); Freon 12, 0.57 (12.9); Freon 22, 0.51 (11.5); Freon 11, 0.64 (14.5).

2. Ammonia Mains. Standard-weight (schedule 40) steel pipe is used for ammonia mains, except for liquid lines 1½ (38 mm) and smaller where extra-strong (schedule 80) pipe is used. Joints may be either screwed, flanged, or welded, but welding is preferred.

3. Freon Mains. Freon 22 piping will, in general, be smaller for the same tonnage than Freon 12 piping. Suction lines will handle about one-third more tonnage for the same pressure drop. Liquid and discharge lines can be the same as for the equivalent tonnage of Freon 12, or slightly smaller.

Freon piping may be standard-weight steel pipe, but copper tubing is used in most cases. Medium-weight type L copper tubing is normally used for land installations, and the heavier type K copper tubing is used for marine work. Joints in the copper tubing may be flared compression fittings for tubing ¾ in (18 mm) OD and smaller. However, hard solder (silver-base alloy melting above 1000°F) (540°C) is preferred for most tubing connections. Copper tubing is almost always used for liquid lines. Steel pipe may be used for the larger suction and discharge lines but should be sandblasted on the inside and carefully cleaned before installation.

4. Mains for Other Refrigerants. For sizing piping for methyl chloride, sulfur dioxide, carbon dioxide, and other refrigerants, see Ref. 3, Chap. 34.

COLD STORAGE

5. Insulation. In recent years newer insulating materials have largely supplanted the use of plaster and corkboard in the cold-storage room construction. Insulation has generally been used in three forms for this application: board form, panelized, and poured, sprayed, or foamed in place.

Board form insulation includes corkboard, rigid polystyrene, polyisocyanurate, and polyurethane, foamed, and fibrous glass.

Panelized insulation has gained acceptance, particularly in prefabricated installations. The outer metal skin can serve as a vapor barrier, and the panels are factory insulated with polystyrene, polyisocyanurate, polyurethane, or fibrous glass. *U* values as low as 0.035 have been attained in 4-in (100-mm) thick panels.

Poured, sprayed, or foamed in place insulation has its greatest applications on existing installations. It is primarily a urethane material. The application requires special equipment, and precautionary measures must be taken.

6. Cold-Storage Temperatures. A great deal of research and experience are required to obtain authoritative information on optimum temperatures and humidities for various products in cold storage. The safe storage period depends upon the product and the storage temperature, and operational techniques vary greatly. Modern cold-storage warehouses of the larger concerns are cooled by brine which is furnished at two different temperatures only. The higher temperature for the mild-temperature warehouses is 10 to 12°F (−12 to −11°C), and the low-temperature brine for the freezers is −10 to −12°F (−23 to −24°C). All temperatures above that of the brine are obtained by regulating the amount of brine circulated in any particular set of coils. In the low-temperature warehouses, the piping is arranged for two classes of service: (1) *sharp freezers*, where the goods which are to be frozen are kept while their temperature is brought down quickly to the holding temperature (say, in from 6 to 10 h) after which they are stored in (2) *holding rooms* where the desired temperature is maintained.

The system of cooling which is now being installed in the highest type of warehouses consists of a coil room containing the necessary brine coils, through which the air from the different rooms is circulated by a pressure blower. The inlet and outlet of each room are so arranged that the cooled circulating air will cover the entire room in its transit; this is usually accomplished by having the cold-air inlet in the center of the room and two return outlets—one at each end of the room. The piping ratio for the coil rooms in this system, assuming a high-grade insulation with 2 to 4 Btu transmission per ft^2 per 24 h per °F (14.2 W/m^{2}°C) temp diff, should be 1 ft^2 (0.09 cm^2) of external pipe surface to 15 ft^3 (0.42 m^3) of space to be cooled [with brine at −10 to −12°F (−23 to −24°C)] for warehouses carrying temperatures of zero and below, and 1 ft^2 (0.09 m^2) of pipe surface to 24 ft^3 (0.68 m^3) of space to be cooled [with brine at 10 to 12°F (−12 to −11°C)] in warehouses carrying mild temperatures of from 30 to 40°F (−1 to 4°C). Another system is used in which the blower and coils are supplemented by coils in the rooms. With this arrangement, it is possible to reduce temperatures quickly and hold them with the coils.

The average coil transmission in cold-storage rooms without forced circulation of the air is about 2 Btu per ft^2 of outside metal surface per h per °F (11.4 $W/m^2 \cdot$ °C) temp diff with horizontal piping and 2.5 Btu (14.2 $W/m^2 \cdot$ °C) with vertical piping. When forced air circulation is used, the transmission rate will increase to 20 Btu (114 $W/m^2 \cdot$ °C) or more. In *brine circulation*, the brine, at same compressor back pressure, has a higher temperature than the ammonia, and consequently 1½ times as much pipe or more is used in brine circulation as in direct expansion for a given back pressure.

7. Piping of Rooms. The size of pipe usually employed for piping rooms varies from 1 to 2 in (25 to 50 mm) with either brine circulation or direct expansion.

The extra cost of liberal piping allowance will often be offset by the consequent improvement in the efficiency of operation of the compressor. An expansion valve should be provided for every 500-ft (150-m) length of 1-in (25-mm) pipe, every 650 ft (200 m) of 1¼-in (30-mm) pipe, and every 1000 ft (300 m) of 2-in (50-mm) pipe when direct expansion is used.

Brine circulation is generally preferred to direct expansion so as to avoid danger from escaping ammonia or other refrigerant in case the pipes should leak. An advantage of the brine system is that there is always a considerable mass of refrigerated brine which can be drawn on in case the machinery should have to be stopped for any reason. In small plants, the general machinery may be stopped at night and only the brine pump be kept going to distribute the surplus refrigeration which has been accumulated in the brine during the day. Brine piping must consist of two lines, a flow and a return, usually of the same size. Brine storage is seldom used in large plants because of its bulk, its first cost, and the practical inability to store much refrigeration.

The *brine coils* in each cooled space are in parallel across the supply and return pipes. It has been common practice to allow 100 to 120 running ft (30 to 36 m) of pipe per circuit for low temperatures, and 400 to 440 (120 to 135 m) for high temperatures. The *tons refrigeration produced by brine* at various temperature differences and rates of pumping may be calculated approximately as follows: tons refrigeration = gal/min × °F range/28 (=0.91 × tons refrigeration for SI tonnage).

CRYOGENICS

Cryogenics is the study, production, and utilization of low temperatures. Cryogenic temperatures have been defined so ambiguously that "upper limits" to the cryogenic range from 216 to 396°R (120 K to 220 K) may be found in the literature. In this section the cryogenic range for a given property is considered to embrace the scale between absolute zero and the temperature above which the property has the expected or normal behavior. Cryogenics thus embraces the unusual and unexpected variations which appear at low temperatures and make extrapolations of properties from ambient to low temperatures unreliable.

Progressively lower temperatures become increasingly difficult to attain in practice. As the working temperature of a refrigerator is lowered, the work required to transfer a given amount of heat increases as demonstrated by the Carnot limitation, to wit, $W = Q[(T_1 - T_2)/T_2]$, where W is the work required to extract the heat Q at a low temperature T_2 and reject it at a higher temperature T_1. The actual work is always greater than this because of inefficiencies of mechanical equipment, thermal losses associated with finite temperature differences in heat exchangers, and heat leaks from the surroundings to the cold equipment.

8. Refrigeration Methods. *Cryogenic refrigerators* may be classified by (1) the functions they perform (e.g., the delivery of liquid cryogens, the separation of mixtures of gases, and the maintenance of spaces at cryogenic temperatures, (2) their refrigerating capacities and (3) the temperatures they reach. Large industrial-sized plants (a) deliver LNG (~120 K), LO_2 (~83 K), LN_2 (~77 K), LH_2 (~20 K), and LHe (~4 K); (b) separate gaseous mixtures, e.g., the constituents of the atmosphere, H_2 from petroleum refinery gases, H_2 and CO from coke oven and coal-water gas reactors, and He from natural gas, and (c) provide refrigeration to maintain spaces at low temperatures. For the latter, i.e., (c), a unit has been designed, manufactured, and is awaiting testing that will deliver 25 kW of refrigeration at 3.8 K.

An important area of refrigerator development that is being commercially exploited now is laboratory-sized cryogenic refrigerators for laboratory research and development at LHe temperatures (<5 K). These refrigerators are used for many different purposes, refrigerating for example: high-field superconducting electromagnets, LHe bubble chambers for high-energy (nuclear) particle research, experimental superconducting power transmission lines, experimental superconducting electric generators and motors and superconducting magnets for levitation of railroad trains. Another area of commercial exploitation that is important for the progress of cryogenic physics research is the development of refrigerators for the continuous production of refrigeration at temperatures below 1 K. These include L^3He evaporation refrigerators and L^3He- L^4He- dilution refrigerators that reach temperatures of 0.4 and 0.003 K, respectively.

Of various methods of refrigeration, the most commonly used to produce temperatures as low as 1 K are (1) the evaporation of a volatile liquid (referred to as the *cascade* method when applied in several successive stages using progressively lower-boiling liquids), (2) Joule-Thomson (isenthalpic) expansion of a compressed gas, and (3) an adiabatic (isentropic) expansion of a compressed gas in an engine (reciprocating or turbine) or from a bomb or cylinder through a throttling valve (Simon expansion). Using L^3He, method 1 is capable of reaching temperatures as low as ~ 0.4 K. Numerous refrigeration cycles have been devised utilizing various combinations of the above three methods. The final stage of refrigeration in plants that deliver liquid cryogens is generally a Joule-Thomson (isenthalpic) expansion. *Expansion engines* (reciprocating and turbine) are commonly used for producing the refrigeration needed to reach the final stage. Turbine expanders are preferred for the large refrigerators because their efficiencies (70 to 85 percent) are higher than for reciprocating expanders. The efficiencies of the heat exchangers for counterflowing "cold" and "warm" gases are an important factor in the overall refrigerator efficiency. In industrial plants heat exchanger efficiencies reach 98 percent. Their design represents a compromise between the attainment of a high heat-transfer coefficient, on the one hand, and a low resistance to the flow on the other.

The *Stirling refrigerating engine* and modifications of it are used in a number of commercial makes of laboratory and miniature-sized refrigerators. The Stirling-cycle refrigerator is well suited to (1) the liquefaction of air on a laboratory scale (~7 L/h), (2) the recondensation of evaporated liquid cryogens, and (3) the refrigeration of closed spaces where the refrigeration load is not very large. These laboratory scale refrigerators supply ~1 kW of refrigeration at ~80 K and ~2 kW at 160 K. They are used with liquid air fractionating columns for the production of LN_2 (~7 L/h) and LO_2 (~5 L/h).

The Stirling-cycle refrigerator consists of a piston for compressing isothermally the working fluid (usually He), and a displacer which can operate in the same cylinder with the piston. The displacer and piston are connected to the same electrically driven shaft but displaced in phase by 90°. The displacer pushes compressed gas isochorically from the warm region where it was compressed, through the regenerator into the cold region where the compressed gas is expanded isothermally doing work on the piston and producing refrigeration. The regenerator consists of a porous mass (packed metal wool of high heat capacity) in which a steep temperature gradient is established between the warm region of compression and the cold region of expansion. The displacer returns the gas after expansion to the region for compression through the regenerator in which it is warmed to the temperature of the isothermal compression. The refrigerant (He) is recycled. The regenerators in the larger refrigerators are usually placed around the outside of the engine cylinders, but in small capacity, lower temperature refrigerators they are put inside the displacers.

In Stirling refrigerators that reach temperatures of 17 K and produce 1 W of refrigeration at 25 K, the compressed He is expanded in two or three stages at temperatures intermediate between room temperature and the lowest temperature reached. There is only a single piston for compression. The expansion in stages allows part of the heat that leaks into the coldest region of a single stage engine to be absorbed and removed at a higher (intermediate) temperature where the efficiency for transferring heat is greater.

The *Vuilleumier refrigerating engine* is a modification of the Stirling refrigerator. It *resembles* two single Stirling engines placed back to back. One of the engines operates as a heat engine and the other as a refrigerator. The lower temperature at which the heat engine discharges its waste heat is also the top temperature at which the refrigerator engine discharges its waste heat. Hence, the Vuilleumier refrigerator operates at elevated, intermediate and low temperature levels which may be 800, 300 (ambient), and 90 K, respectively. Each engine has a cylinder, a displacer, and a regenerator, and the two engines are connected to a common crankshaft but displaced by 90°. Very little external power is needed to operate the two displacers which are operated in sinusoidal motion, displaced 90° in phase. The working fluid (He) is recycled.

INDUSTRIAL AND MANAGEMENT ENGINEERING*

The American Institute of Industrial Engineers defines industrial engineering as a branch of engineering "concerned with the design, improvement, and installation of integrated systems of people, material, equipment, and energy. It draws upon the specialized knowledge and skills in the mathematical, physical and social sciences together with the principles and methods of engineering analysis and design to specify, predict, and evaluate the results to be obtained from such systems."

BACKGROUND

The advent of industrial engineering is frequently associated with the Industrial Revolution in the 19th century. It was during this period that household production was replaced by production in factories. Frederick W. Taylor is usually recognized as the father of industrial engineering and scientific management. His pioneering work in the design, measurement, planning, and scheduling of work from 1880 to the time of his death in 1915 was the impetus for the conceptualization and growth of industrial engineering. Taylor introduced the concepts of time study, methods engineering, tool standardization, costing methods, routing, employee job selection, and incentives.

The term classical has been applied to the traditional industrial engineering activities of Taylor and F. B. Gilbreth. These techniques have had great practical value, and for decades have been considered the hallmarks of industrial engineering. Modern industrial engineering techniques address the more quantitative computer-systems approach to the solution of industrial engineering problems.

The classical activities of methods, job standards, plant layout, and costs are vital to effective management; in fact, many of the quantitative systems models need data which are obtained by time study, synthetic time standards, or work sampling. For an industrial engineer to be effective necessarily demands the capability of evaluating and utilizing the appropriate classical and the more sophis-

*Excerpted from the *McGraw-Hill Encyclopedia of Science and Technology*, 6th ed. Copyright © 1987. Used by permission of McGraw-Hill, Inc. All rights reserved.

ticated management science–operations research techniques in solving unstructured, real-life problems.

ACTIVITIES

Industrial engineering is similar to the other major engineering specializations (civil, mechanical, electrical, and chemical) in that it is concerned with analysis and design, and applying the laws and materials of nature to useful and constructive purposes. It is different from these other fields of engineering in that it is specifically concerned with equipment and systems in which people are an integral part. The industrial engineer must be able to use mathematics, materials, machinery, devices, chemistry, electricity, electronics, and so on, just as all of the other types of engineers. But, unlike them, the industrial engineer must also understand and be able to integrate people into his or her designs, and must know their physical, physiological, psychological, and other characteristics—singly and in groups.

The early, and still major, activities of industrial engineers include work methods analysis and improvement, work measurement and the establishment of standards, job and workplace design, plant layout, materials handling, wage rates and incentives, cost reduction, suggestion evaluation, production planning and scheduling, inventory control, maintenance scheduling, equipment evaluation, assembly line balancing, systems and procedures, overall productivity improvement, and special studies—all done, almost exclusively in the early days, in manufacturing industries. As the technology evolved, additional activities were added to this list. These include machine tool analysis; numerically controlled machine installation and programming; computer analysis, installation, and programming; linear programming; queueing and other operations research techniques; simulations; management information systems; value analysis; human factors engineering; human/machine systems design; ergonomics; biomechanics; and the use of robots and automation. Moreover, it was found that the industrial engineering techniques used successfully in the factory could also be applied in the office, laboratory, classroom, hospital (including the operating room), the government, the military, and other nonindustrial areas.

The industrial engineer of today is using computers more in his or her production and test equipment, controls and systems, and for analyses and special studies. Minicomputers and microprocessors are in wide and growing use.

Computers are already being used to eliminate much of the calculation drudgery of work measurement. The power of the computer also facilitates the synthesis of various methods configurations.

Computer-aided manufacturing (CAM) makes use of the computer in an attempt to develop a total systems approach to the manufacturing process. It links together computer-aided design (CAD), automatically programmed tools, work measurement, sequencing, materials handling, and inventory control. Computer interactive graphics are also being introduced in the design and analysis functions.

Numerical control (N/C) is the term usually associated with automatically programmed equipment. An N/C machine consists of a reader, a control unit, and the machine tool. Preprogrammed machine instructions on tape are transmitted to the control unit, which interprets the instructions and causes the machine tool to execute. While N/C has been used primarily for metal removal machinery, it has also been applied successfully in material storage, assembly, and packaging.

The capability of N/C can be enhanced by direct numerical control (DNC), which uses an on-line computer to provide the instructions directly to the machine. Further extensions are in the area of adaptive control systems and robots.

The production function is dynamic by its inherent activities, and traditional industrial engineers are frequently called upon to determine the impact of worker power changes, product mix variances, and equipment additions or deletions on facilities arrangement and line balancing in the manufacturing area. This task becomes increasingly more complex today with the additional energy computation and pollution level considerations. This complicated system of interrelated activities is almost impossible to evaluate manually; however, through the use of simulation on a computer, answers can be given to "what if" types of questions which provide significant input to the decision maker.

The age-long inventory control function can now be monitored constantly by a real-time computerized system. The computer greatly facilitates the storage and retrieval of historical and current information on the status of inventory and concomitant purchase orders. Computers can be programmed to evaluate the tradeoffs between the savings in inventory investment and the activity required in the planning and control phases.

In the early days of industrial engineering, safety engineering was considered an important element in the educational programs. However, in the 1950s and 1960s safety courses were dropped in many of the curricula as more sophisticated analytical techniques were added. However, the huge costs of industrial accidents, and the regulatory measures established by the Occupational Safety and Health Administration (OSHA) demand that safety be considered in the initial facilities design in a more in-depth manner than previously. It should be recognized that the criteria utilized for plant layout, such as minimizing material handling, minimizing congestion and hazards, and providing for good housekeeping, are compatible with the objectives associated with safety engineering.

Plant layout is considered a classical industrial engineering activity; however, the traditional approach to facilities design has been supplemented by the use of computers in the design and calculation process of a layout, and also by the application of mathematical modeling and optimization to those problems. It must be emphasized that the basic concepts of plant layout are still important, and the computer and operations research methods must be skillfully blended with the traditional methods in order to obtain better answers to facilities problems which become increasingly more complex—for example, by inclusion of accident risk factors.

SYSTEMS CONCEPT

The trend in industrial engineering is from micro- to macrolevel structures, such as a computer-aided manufacturing system, a total information system design of a company, or the complex materials-handling function involving many interrelated departments.

Management of industrial, business, or service activities is growing increasingly complicated as the sciences and humanities become more interdependent. These organizations need not only management personnel, but individuals capable of designing and installing new integrated systems of people, materials, and equipment which will function effectively in a society made more complex by the technological explosion. Today's industrial engineer has a foundation in the basic physical and social sciences, engineering, and computers, and a skill in systems analysis and design which crosses traditional disciplinary lines. The result is a capability to meet the demands imposed by a dynamic social system.

The logical structure of systems study may be outlined as follows:

1. Identification of the components of a complex system. Examples include a production scheduling system, a computer software system, an educational scheduling system, a project planning and control system for underdeveloped countries, a hospital information system, and a space information system.
2. Development of the topological properties of the system, and the generation of mathematical models and analogs, or the adaptation of a heuristic approach when desirable.
3. Establishment of the functional relationships between the variables of the system, together with the necessary feedback required to control the operational system.
4. Selection and evaluation of optimization criteria. Examples of criteria are profit, costs, idle capacity, energy consumption, and information entropy.
5. Analysis and design of the nondeterministic functions to make them amenable to solution of the total system structure.
6. Integration of behavioral patterns in different environments, as required, within the physical framework of the system. Examples are the sociopolitical environment of a country for which the system is designed, the behavioral attitudes of nurses in a hospital, and those of production line workers in a fabrication shop.
7. Design of simulation models which permit a rigorous critique of the parameters of a dynamic system.
8. Economic analysis of the total system and concomitant subsystems with extensions into alternative structure.

OPERATIONS RESEARCH AND INDUSTRIAL ENGINEERING DESIGN

The real-world problems are complex systems configurations which require the use of more sophisticated methods than previously to present meaningful results to the decision maker. In the development of modern systems technology, the pacing factor is management. This is the function which directs, coordinates, and controls the many facets of a system. Technological advancement depends on optimal use of all resources and requires that new discoveries be translated into new products in minimal time. Productivity must be increased. The terms *optimal* and *minimal* are characteristics of management science and operations research.

Management science and operations research are often used synonymously since the tenor of their objectives is compatible. Operations research has been referred to as a sharper kind of industrial engineering. Even though no two definitions of operations research are exactly the same, the following four common denominators can be abstracted from most definitions: (1) formulation of objectives—seek to attain goals established by management; (2) systems perspective—this is frequently concerned with the interrelationship among many components; (3) alternatives—choices exist among decision variables; (4) optimization—at-

TABLE 13.15 Applications of Operations Research Techniques in Production*

Application areas	Linear programming	Dynamic programming	Network models	Simulation	Queueing theory	Game theory	Regression analysis	Others
Production scheduling	30 (41.1)	7 (9.6)	6 (8.2)	26 (35.6)	9 (12.3)	0 (0.0)	5 (6.8)	10 (13.7)
Production planning and control	19 (26.0)	3 (4.1)	7 (9.6)	18 (24.7)	4 (5.5)	0 (0.0)	3 (4.1)	3 (4.1)
Project planning and control	10 (13.7)	1 (1.4)	28 (38.4)	9 (12.3)	2 (2.7)	0 (0.0)	0 (0.0)	3 (4.1)
Inventory analysis and control	11 (15.1)	3 (4.1)	3 (4.1)	27 (37.0)	4 (5.5)	1 (1.4)	12 (16.4)	7 (9.6)
Quality control	2 (2.7)	0 (0.0)	1 (4.1)	2 (2.7)	0 (0.0)	0 (0.0)	15 (20.5)	9 (12.3)
Maintenance and repair	0 (0.0)	1 (1.4)	3 (4.1)	8 (11.0)	3 (4.1)	1 (1.4)	4 (5.5)	3 (4.1)
Plant layout	13 (17.8)	0 (0.0)	5 (6.8)	19 (26.0)	5 (6.8)	1 (1.4)	2 (2.7)	3 (4.1)
Equipment acquisition and replacement	4 (5.5)	0 (0.0)	1 (1.4)	11 (15.1)	1 (1.4)	0 (0.0)	0 (0.0)	7 (9.6)
Blending	32 (43.8)	0 (0.0)	1 (1.4)	6 (8.2)	1 (1.4)	0 (0.0)	3 (4.1)	1 (1.4)
Logistics	27 (37.0)	1 (1.4)	8 (11.0)	24 (32.9)	3 (4.1)	2 (2.7)	6 (8.2)	2 (2.7)
Plant location	32 (43.8)	2 (2.7)	8 (11.0)	23 (31.5)	1 (1.4)	0 (0.0)	5 (6.8)	4 (5.5)
Other	7 (9.6)	1 (1.4)	2 (2.7)	7 (9.6)	1 (1.4)	1 (1.14)	3 (4.1)	4 (5.5)

*All data are expressed in terms of numbers (percent). The percentages do not total 100% because many respondents indicated they used more than one technique in a given application area.

Source: From A. N. Ledbetter and J. F. Cox, "Are OR Techniques Being Used?" *Ind. Eng.*, p. 21, February 1977.

tempt to make best decision relative to cited objectives, recognizing that in the real world this is not usually achieved in the strict sense.

In the current identification of engineering design and operations research, two apparently diverse disciplines, their components, and processes are found to be very similar. Industrial engineers are involved with both the design and analysis aspects of engineering in solving many of their problems. Engineering design is the process of devising a system, which includes its components and processes, to meet desired needs. It is a decision-making process (often iterative and interactive) in which the basic sciences, mathematics, and engineering sciences are applied to convert resources optimally to meet a stated objective. Among the fundamental elements of the design process are the establishment of objectives and criteria, synthesis, analysis, construction, testing, and evaluation.

The criteria established for engineering design evaluation provide a powerful impetus to utilize operations research foundations in engineering problems. It is important to note that in industrial engineering operations research is currently being used in solving the problems. The data in Table 13.15 provide insights relative to the use of operations research techniques in the production function, which is a major area of activity for industrial engineers. The techniques used most frequently are linear programming and simulation. Network models have been used extensively in project planning and control, and regression analysis in inventory and quality control.

REFERENCES

1. C. E. O'Rourke, *General Engineering Handbook*, 2d ed., McGraw-Hill, New York, 1940.
2. E. A. Avallone and T. Baumeister III (eds.), *Marks' Standard Handbook for Mechanical Engineers*, 9th ed., McGraw-Hill, New York, 1987.
3. *ASHRAE Handbook, 1985 Fundamentals*, American Society of Heating, Refrigerating and Air-Conditioning Engineers, Atlanta, Ga., 1985.
4. *McGraw-Hill Encyclopedia of Science and Technology*, 6th ed., McGraw-Hill, New York, 1987.

CHAPTER 14

CIVIL ENGINEERING AND HYDRAULIC ENGINEERING

CIVIL ENGINEERING*

SURVEYING

1. Measurement of Distance

Units of Measurement. Distances are usually measured in feet and tenths, hundredths, and (for accurate work) thousandths of feet (meters, centimeters, and millimeters). For many older surveys, distances were measured in chains and links. A chain is 66 ft (20.1 m) in length and is divided into 100 links, each 7.92 in (201.2 mm) long. The metric system, in which the unit of distance is the meter and its decimal fractions or multiples, is in use in most countries and is expected to replace the English system of measurement in the United States. See Chap. 1 for conversion factors.

Four methods used for the direct measurement of distance are pacing, stadia reading, taping, and electronic distance recording.

Pacing. Pacing is a rapid means of checking more accurate measurements of distance. The precision of pacing under average conditions is from 1:100 to 1:200.

Stadia Reading. The use of stadia furnishes a rapid method of determining distances with a fair degree of accuracy. Under average conditions, a precision of from 1:300 to 1:1000 can be obtained.

Measurement with Tape. The most commonly used method of determining distance is by measurement with a tape. Steel tapes, ranging in length from 50 to 300 ft (15.2 to 91.4 m), are generally used, but tapes of other materials may be used where accuracy is not essential. The precision of a tape measurement depends on the degree of refinement with which the measurement is made. The precision of taping ordinarily used in surveys is from 1:3000 to 1:5000.

For ordinary taping, a tape accurate to 0.01 ft (0.003 m) should be used. The tension of the tape should be about 15 lb (66.7 N). The temperature should be determined within 10°F (5.6°C), and the slope of the ground within 2 percent, and the proper corrections applied. The correction to be applied for temperature when using a steel tape is

$$C_t = 0.0000065s(T - T_0) \tag{14.1}$$

The correction to be made to measurements on a slope is

$$C_h = s(1 - \cos\theta) \quad \text{exact} \tag{14.2}$$

or

$$= 0.00015s\theta^2 \quad \text{approximate} \tag{14.2a}$$

---

*From *Engineering Manual*, 3d ed., by R. H. Perry (ed.). Copyright © 1976. Used by permission of McGraw-Hill, Inc. All rights reserved. Updated and metricated by the editors.

or
$$C_h = \frac{h^s}{2s} \qquad \text{approximate} \tag{14.2b}$$

where C_t = temperature correction to measured length, ft (× 0.3048 for meters)
C_h = correction to be subtracted from slope distance, ft (m)
s = measured length, ft (m)
T = temperature at which measurements are made, °F (°C)
T_0 = temperature at which tape is standardized, °F (°C)
h = difference in elevation at ends of measured length, ft (m)
θ = slope angle, deg

In more accurate taping, using a tape standardized when fully supported throughout, corrections should also be made for tension and for support conditions. The correction for tension is

$$C_p = \frac{(P_m - P_s)s}{SE} \tag{14.3}$$

The correction for sag when not fully supported is

$$C_s = \frac{w^2L^3}{24P_m^2} \tag{14.4}$$

where C_p = tension correction to measured length, ft (m)
C_s = sag correction to measured length for each section of unsupported tape, ft (m)
P_m = actual tension, lb (N)
P_s = tension at which tape is standardized, lb (N) [usually 10 lb (44.5 N)]
S = cross-sectional area of tape, in^2 (mm^2)
E = modulus of elasticity of tape, lb/in^2 (29 million lb/in^2 for steel)
w = weight of tape, lb/ft (N/m)
L = unsupported length, ft (m)

Electronic Distance Measurement (EDM). Electronic measuring devices, utilizing principles similar to radar, are now in general use. Distances are measured by determination of the time required for a wave (infrared, radio, or laser beam) to travel at the speed of light to and from a point. For some types, readings of distance are obtained directly from a counter, eliminating the need for calculations. Advantages include ability to take readings across bodies of water, rugged terrain, and brush much more rapidly than with tape measurement. The precision of EDM ranges up to 1:300,000.

2. Measurement of Difference in Elevation. Difference in elevation may be measured by three methods: barometric leveling, stadia leveling, and direct leveling. Barometric methods are used for rough or preliminary work. Stadia reading is a rapid method and will give results having an error, in feet (meters), of $1.0\sqrt{\text{distance in miles}}$ (Sec. 4). Direct leveling is the most accurate and most commonly used method for determining difference in elevation. EDM instruments (Sec. 1) make it possible to determine differences in elevation by measurement of slope distances and vertical angles.

Rough Leveling. Rough leveling is practiced on preliminary or reconnaissance surveys. Sights are permitted up to 1000 ft (304.8 m) in length, and rod readings are made to 0.1 ft (0.004 m). Precision in feet is $0.4\sqrt{\text{distance in miles}}$.

Ordinary Leveling. Ordinary leveling is used in the construction and location of highways, railroads, and the like. Sights are permitted up to 500 ft (152 m) in length, and rod readings are made to 0.01 ft (0.003 m). Precision in feet (meters) is $0.1\sqrt{\text{distance in miles}}$.

Accurate Leveling. Accurate leveling is used for establishing important bench marks. Sights are limited to 300 ft (91.4 m) in length, and rod readings are made to 0.001 ft (0.0003 m). Precision in feet (meters) is $0.05\sqrt{\text{distance in miles}}$.

Precise Leveling. Precise leveling is used for establishing bench marks at widely separated locations. Sights are limited to 300 ft (91.4 m) in length, and rod readings are made to 0.001 ft (0.0003 m). Special equipment and extreme care are used, and several runs are usually made. Precision in feet (meters) is $0.02\sqrt{\text{distance in miles}}$.

3. Measurement of Angles. Angles may be measured with either a compass or a transit. The precision of compass measurement is from 30′ to 1°. The precision of transit measurements is from 1″ to 2′, depending on the type of instrument used and the care exercised. Angle measurements should have a precision consistent with distance measurements. Surveys in which distances are measured to 0.01 ft (0.003 m) should have angles measured to 15″, with a resulting accuracy of better than 1:10,000, and surveys with distances measured to 0.1 ft (0.03 m) and angles measured to 1′ should have an accuracy of about 1:5000.

4. Stadia Surveying. In stadia surveying, a transit having horizontal stadia cross hairs above and below the central horizontal cross hair is used. The difference in the rod readings at the stadia cross hairs is termed the rod intercept. The intercept may be converted to the horizontal and vertical distances between the instrument and the rod by the following formulas:

$$H = Ki(\cos a)^2 + (f + c)\cos a \tag{14.5}$$

$$V = \tfrac{1}{2}Ki(\sin 2a) + (f + c)\sin a \tag{14.6}$$

where H = horizontal distance between center of transit and rod, ft (m)

V = vertical distance between center of transit and point on rod intersected by middle horizontal cross hair, ft (m)

K = stadia factor (usually 100)

i = rod intercept, ft (m)

a = vertical inclination of line of sight, measured from the horizontal, deg

$f + c$ = instrument constant, ft (m) [usually taken as 1 ft (0.3 m)]

In the use of these formulas, distances are usually calculated to feet (meters) and differences in elevation to tenths of feet (meters).

5. Latitudes and Departures. The latitude of a line is the projection of the line upon a true or assumed north-south meridian. The latitude of a line of length s is $s \cos \beta$ where β is the bearing of the line (the angle between the direction of the line and the direction of the north-south axis). If the line is considered as running from the southerly end to the northerly end, the latitude is positive; if from the northerly end to the southerly end, the latitude is negative.

The departure of a line is the projection of the line upon a parallel at right angles to the meridian. The departure of a line of length s is $s \sin \beta$. If the line is considered as running from the westerly end to the easterly end, the departure is positive; if from the easterly end to the westerly end, the departure is negative.

6. Balancing a Closed Traverse. The geometry of a closed traverse requires that the algebraic sum of the latitudes and of the departures be zero. If these sums are not zero, but are small enough to indicate that the discrepancy is not the result of an actual error, the traverse may be balanced by adjusting each line. The compass rule is usually used in balancing. The rule states: The correction to be applied to the latitude (or departure) of any line in the traverse is to the total error in latitude (or departure) as the length of the line is to the length of the traverse.

7. Calculation of Areas of Land. Areas of land within the limits of a balanced traverse are usually calculated by the method of double meridian distances. For convenience in the use of this method, a meridian is assumed to pass through one corner (usually the most westerly corner) of the traverse. The meridian distance of any point in the traverse is then the departure of the point, departures to the east of the meridian being considered positive and departures to the west, negative. The double meridian distance of a line in the traverse is the sum of the departures (or meridian distances) of the two ends of the line. Double meridian distances (DMD) may be calculated by the following rules:

1. The DMD of the first line in the traverse (the line one end of which is on the reference meridian) is equal to the departure of that line.
2. The DMD of any other line is equal to the DMD of the preceding line, plus the departure of the preceding line, plus the departure of the line itself.
3. The DMD of the last line is the same as the departure of that course with opposite sign.

Algebraic values should be used with due regard for signs. The area of the traverse is equal to one-half the algebraic sum of the products of the DMD and the latitude of each line.

The area of irregular tracts of land may be determined by the trapezoidal method or by Simpson's method. Both of these methods require that a straight line enclose one side of the area and that offsets from this line be measured at regular intervals to the irregular boundary. In addition, Simpson's method requires that the number of offsets be odd (or the number of regular intervals be even). The trapezoidal method states: The area is equal to the product of the interval between offsets and the sum of the intermediate offsets and one-half each end offset. Simpson's method states: The area is equal to one-third the product of the interval between offsets and the sum of the end offsets, twice each odd intermediate offset, and four times each even intermediate offset.

8. Circular Curves. Circular curves are the most common type of horizontal curve used to connect intersecting tangent (or straight) sections of highways or railroads. In the United States, two methods of defining circular curves are in use: the first, in general use in railroad work, defines the degree of curve as the central angle subtended by a *chord* of 100 ft (30.4 m) in length; the second, used in highway work, defines the degree of curve as the central angle subtended by an *arc* of 100 ft (30.4 m) in length. In the metric system, the degree of curve is sometimes expressed as the number of degrees subtended by an arc or chord 20 m long.

The terms and symbols generally used in reference to circular curves are listed below and shown in Figs. 14.1 and 14.2.

PC = point of curvature, beginning of curve
PI = point of intersection of tangents
PT = point of tangency, end of curve
R = radius of curve, ft (m)
D = degree of curve (see above)
I = deflection angle between tangents at PI, also central angle of curve
T = tangent distance, distance from PI to PC or PT, ft (m)
L = length of curve from PC to PT measured on 100-ft (30.4-m) chord for chord definition, on arc for arc definition, ft
C = length of long chord from PC to PT, ft (m)
E = external distance, distance from PI to midpoint of curve, ft (m)
M = midordinate, distance from midpoint of curve to midpoint of long chord, ft (m)
d = central angle for portion of curve ($d < D$)
l = length of curve (arc) determined by central angle d, ft (m)
c = length of curve (chord) determined by central angle d, ft (m)
a = tangent offset for chord of length c, ft (m)
b = chord offset for chord of length c, ft (m)

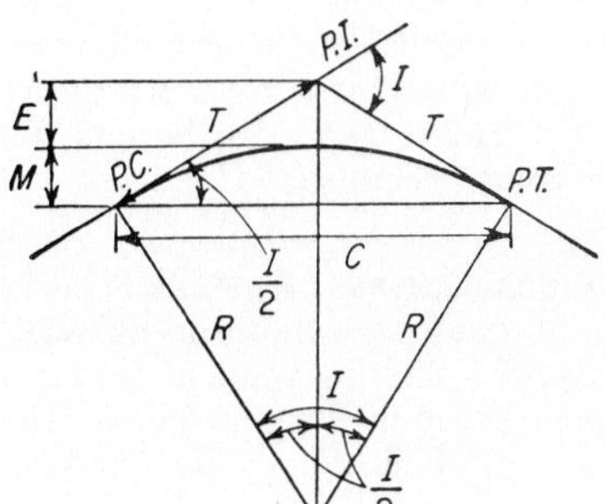

FIGURE 14.1 Circular curve.

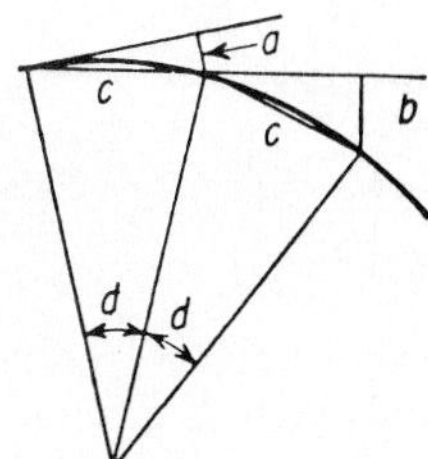

FIGURE 14.2 Offsets to circular curve.

Equations of Circular Curves

$$R = 5729.578/D \quad \text{exact for arc definition, approximate for chord definition} \tag{14.7}$$

$$= 50/\sin \tfrac{1}{2}D \quad \text{exact for chord definition} \tag{14.8}$$

$$T = R \tan \tfrac{1}{2}I \quad \text{exact} \tag{14.9}$$

$$E = R \operatorname{exsec} \tfrac{1}{2}I = R(\sec \tfrac{1}{2}I - 1) \quad \text{exact} \tag{14.10}$$

$$M = R \operatorname{vers} \tfrac{1}{2}I = R(1 - \cos \tfrac{1}{2}I) \quad \text{exact} \tag{14.11}$$

$$C = 2R \sin \tfrac{1}{2}I \quad \text{exact} \tag{14.12}$$

$$L = 100I/D \quad \text{exact} \tag{14.13}$$

$$L - C = L^3/24R^2 = C^3/24R^2 \quad \text{approximate} \tag{14.14}$$

$$d = Dl/100 \quad \text{exact for arc definition} \tag{14.15}$$

$$Dc/100 \quad \text{approximate for chord definition} \tag{14.16}$$

$$\sin \tfrac{1}{2}d = c/2R \quad \text{exact for chord definition} \tag{14.17}$$

$$a = c^2/2R \quad \text{approximate} \tag{14.18}$$

$$b = c^2/R \quad \text{approximate} \tag{14.19}$$

Layout of Circular Curve. The field layout of a circular curve depends on the geometric property of a circle that the angle between a tangent and a chord is one-half the included angle. The procedure is shown in Fig. 14.3, where the length of the first chord (or arc) is so chosen that point 1 is at an even 100-ft (30.4-m) station. Point 1 is located by measurement of the chord distance c from the PC and by the deflection angle $\tfrac{1}{2}d$ from the tangent. Point 2 is then located by measurement of the 100-ft (30.4-m) chord [or the chord corresponding to the 100-ft (30.4-m) arc] from point 1 and by the total deflection angle ($\tfrac{1}{2}D + \tfrac{1}{2}d$) from the tangent. Succeeding points are similarly located. The entire curve can be laid out with the transit set at the PC.

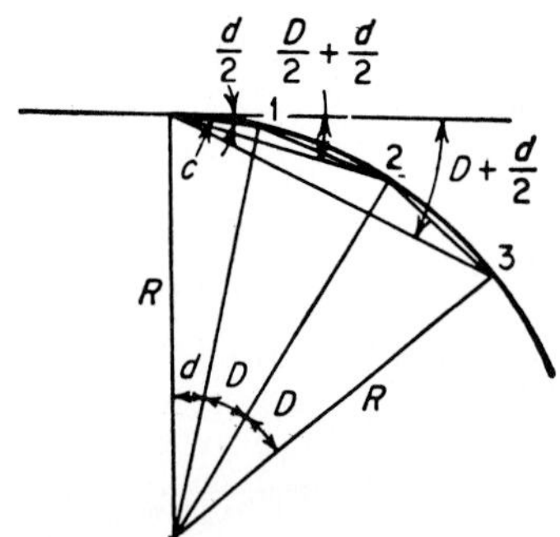

FIGURE 14.3 Layout of a circular curve.

9. Parabolic Curves. Parabolic curves are used to connect sections of highways or railroads of differing gradient. The use of a parabolic curve provides a gradual change in direction along the curve. The terms and symbols generally used in reference to parabolic curves are listed below and shown in Fig. 14.4.

PVC = point of vertical curvature, beginning of curve
PVI = point of vertical intersection of grades on either side of curve
PVT = point of vertical tangency, end of curve
G_1 = grade at beginning of curve, ft/ft (m/m)
G_2 = grade at end of curve, ft/ft (m/m)
L = length of curve, ft
R = rate of change of grade, ft/ft^2 (m/m^2)
V = elevation of PVI, ft (m)
E_0 = elevation of PVC, ft (m)
E_t = elevation of PVT, ft (m)
x = distance of any point on the curve from the PVC, ft (m)
E_x = elevation of point x distant from PVC, ft (m)
x_s = distance from PVC to lowest point on a sag curve or highest point on a summit curve, ft (m)
E_s = elevation of lowest point on a sag curve or highest point on a summit curve, ft (m)

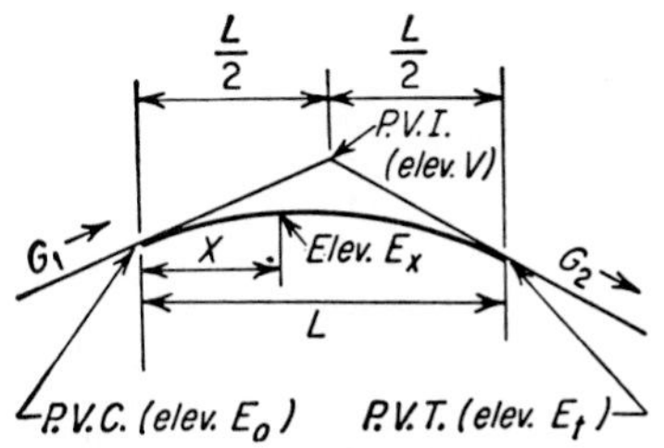

FIGURE 14.4 Vertical parabolic curve (summit curve).

Equations of Parabolic Curves. In the parabolic-curve equations given below, algebraic quantities should always be used. Upward grades are positive and downward grades are negative.

$$R = (G_2 - G_1)/L \tag{14.20}$$

Note. K as used in Figs. 14.10 and 14.11 is equal to $1/100R$.

$$E_0 = V - \tfrac{1}{2}LG_1 \tag{14.21}$$

$$E_x = E_0 + G_1x + \tfrac{1}{2}Rx^2 \tag{14.22}$$

$$x_s = -G_1/R \tag{14.23}$$

Note. If x_s is negative or if $x_s > L$, the curve does not have a high point or a low point.

$$E_s = E_0 - G_1^2/2R \tag{14.24}$$

10. Photogrammetry. Photogrammetry is a method of obtaining measurements through use of ground or aerial photography. For large-scale projects, aerial photogrammetry techniques permit substantial savings in mapping time, but establishment of the location and elevation of control points by conventional ground survey are necessary. Through photoanalysis and interpretation, a trained interpreter can obtain reliable qualitative information concerning the type and characteristics of the soils, surface and ground waters, and manufactured features such as roads and bridges. Matched groups of overlapping photographs are viewed through a stereoscope and deductive and inductive methods used for evaluation. A selective field check is an important part of the photo-interpretative methodology.

The scale of aerial photography is given as a ratio, such as 1:6000, equivalent to 1 in = 500 ft. For a standard camera with a 6-in focal length lens and using 9-in square film, the following relationships apply:

Flight height above ground, ft (m)	Photographic scale	Coverage per single photograph, mi^2 (m^2)	Map scale (5 × enlargement of photographs), in = ft (mm = m)	Contour interval, ft (m)
1,200 (365.8)	1:2,400	0.11 (284,889)	1 = 40 (25.4 = 12.2)	1 (0.3)
3,000 (914.4)	1:6,000	0.72 (1,864,730)	1 = 100 (25.4 = 30.5)	2.5 (0.76)
4,000 (1219.2)	1:8,000	1.29 (3,340,975)	1 = 133 (25.4 = 40.5)	(1)
6,000 (1828.8)	1:12,000	2.90 (7,510,719)	1 = 200 (25.4 = 60.9)	5 (1.5)
8,000 (2438.4)	1:16,000	5.16 (13,363,901)	1 = 267 (25.4 = 81.4)	(2)
12,000 (3657.6)	1:24,000	11.62 (30,094,677)	1 = 400 (25.4 = 121.9)	10 (3.05)
18,000 (5486.4)	1:36,000	26.15 (67,725,973)	1 = 600 (25.4 = 182.9)	15 (4.6)
20,000 (6096.0)	1:40,000	32.28 (83,602,080)	1 = 667 (25.4 = 203.3)	(5)

For mapping, the photographed area is usually covered in series of parallel strips, with photographs of the same strip overlapping 60 percent and photographs of adjacent strips overlapping 30 percent. Horizontal and vertical ground control is required for the preparation of maps.

SOIL MECHANICS AND FOUNDATIONS

11. Grain Size. The grain size classification of soils used by the U.S. Department of Agriculture is given as follows:

Soil type	Particle diam, mm	Soil type	Particle diam, mm
Gravel..........	>2.0	Sand, very fine....	0.10–0.05
Gravel, fine.......	2.0–1.0	Silt.............	0.05–0.005
Sand, coarse......	1.0–0.5	Clay............	0.005–0.0002
Sand, medium.....	0.5–0.25	Colloids.........	<0.0002
Sand, fine.........	0.25–0.10		

12. Bureau of Public Roads Soil Classification. The U.S. Bureau of Public Roads (now the Federal Highway Administration of the U.S. Department of Transportation) developed a detailed method for classifying soils for use as highway subgrades which was subsequently expanded and adopted by the American Association of State Highway and Transportation Officials (AASHTO). Soils are classified in seven major groups as shown in Table 14.1 and described below.

Granular Materials

Group A-1. This group includes granular materials with or without nonplastic or feebly plastic soil binders. Subgroup A-1-a includes materials consisting predominantly of stone fragments or gravel, either with or without a well-graded binder of fine material. Subgroup A-1-b includes materials consisting predominantly of coarse sand either with or without a well-graded soil binder.

Group A-3. This group includes fine beach sand or fine desert blow sand without silty or clay fines or with a very small amount of nonplastic silt and stream-deposited mixtures of poorly graded fine sand with limited amounts of coarse sand and gravel.

Group A-2. This group includes a wide variety of "granular" materials which are at the border line between materials falling in groups A-1 and A-3 and the silt-clay materials of groups A-4, A-5, A-6, and A-7. Subgroups A-2-4 and A-2-5 include such materials as gravel and coarse sand with silt content or plasticity index in excess of the limitations of group A-1 and fine sand with nonplastic silt content in excess of the limitations of group A-3. Subgroups A-2-6 and A-2-7 include materials similar to those described under subgroups A-2-4 and A-2-5, except that the fine portion contains plastic clay having the characteristics of the A-6 or A-7 group.

Silt-Clay Materials

Group A-4. The typical material of this group is a nonplastic or moderately plastic silty soil. The group also includes mixtures of fine silty soil and sand and gravel.

Group A-5. The typical material of this group is similar to that described under group A-4, except that it is usually of diatomaceous or micaceous character and may be highly elastic as indicated by the high liquid limit.

Group A-6. The typical material of this group is a plastic clay soil. The group also includes mixtures of fine clayey soil and sand and gravel. Materials of this group usually have a high volume change between wet and dry states.

Group A-7. The typical material of this group is similar to that described under group A-6, except that it has the high-liquid-limit characteristics of the A-5 group and may be elastic as well as subject to high volume change. Subgroup A-7-5 includes those materials with moderate plasticity indexes which may be highly elastic as well as subject to considerable volume change. Subgroup A-7-6 includes those materials with high plasticity indexes in relation to liquid limit which are subject to extremely high volume change.

TABLE 14.1 Classification of Highway Subgrade Materials*

General classification	Granular materials (35% or less passing No. 200 sieve—0.075 mm)							passing No. 200 sieve—0.075 mm)			
Group classification	A-1		A-3	A-2				A-4	A-5	A-6	A-7**
	A-1-a	A-1-b		A-2-4	A-2-5	A-2-6	A-2-7				A-7-5, A-7-6
Sieve analysis, % passing:											
No. 10 (2.00 mm)	50 max										
No. 40 (0.425 mm)	30 max	50 max	51 min								
No. 200 (0.075 mm)	15 max	25 max	10 max	35 max	35 max	35 max	35 max	36 min	36 min	36 min	36 min
Characteristics of fraction passing No. 40:											
Liquid limit	......	......	......	40 max	41 min	40 max	41 min	40 max	41 min	40 max	41 min
Plasticity index	6 max	6 max	NP	10 max	10 max	11 min	11 min	10 max	10 max	11 min	11 min
Group index	0	0	0	0	0	4 max	4 max	8 max	12 max	16 max	20 max
Usual types of significant constituent materials	Stone fragments, gravel, and sand		Fine sand	Silty or clayey gravel and sand				Silty soils		Clayey soils	
General rating as subgrade	Excellent to good					Fair to poor					

*Classification procedure: With required test data available, proceed from left to right on above chart and correct group will be found by process of elimination. The first group from the left into which the test data will fit is the correct classification.

**Plasticity index of A-7-5 subgroup is equal to or less than LL minus 30. Plasticity index of A-7-6 subgroup is greater than LL minus 30.

Group Index. The group index is used as an approximate within-group evaluation of the materials of the A-2-6, A-2-7, A-4, A-5, A-6, and A-7 groups.

$$\text{Group index} = 0.2a + 0.005ac + 0.01bd$$

where a = that portion of the percentage passing the No. 200 sieve greater than 35 and not exceeding 75 percent, expressed as a positive whole number (1 to 40)

b = that portion of the percentage passing the No. 200 sieve greater than 15 and not exceeding 55 percent, expressed as a positive whole number (1 to 40)

c = that portion of the numerical liquid limit greater than 40 and not exceeding 60, expressed as a positive whole number (1 to 20)

d = that portion of the numerical plasticity index greater than 10 and not exceeding 30, expressed as a positive whole number (1 to 20)

Under average conditions of good drainage and thorough compaction, the supporting value of a material as a subgrade is in inverse ratio to its group index; that is, a group index of 0 indicates a good subgrade material and a group index of 20 indicates a very poor subgrade material.

13. Relationship among Soil Classifications. Other important soil classifications and measures of supporting strength include the following:

1. California Bearing Ratio, the ratio (expressed as a percentage) of the load required to cause a specified penetration in a given soil to the load required to cause the same penetration in a compacted gravel
2. Casagrande soil classification
3. Civil Aeronautics Administration soil classification
4. Resistance value *R*
5. Bearing value

The approximate relationships among these classifications are shown in Fig. 14.5.

In the Casagrande soil classification, the following symbols are used:

G = gravel, gravelly soil
S = sand, sandy soil
O = organic silt or clay
C = clay
M = silt or very fine sand
F = fine
P = poorly graded
W = well graded
L = low to medium compressibility
H = high compressibility

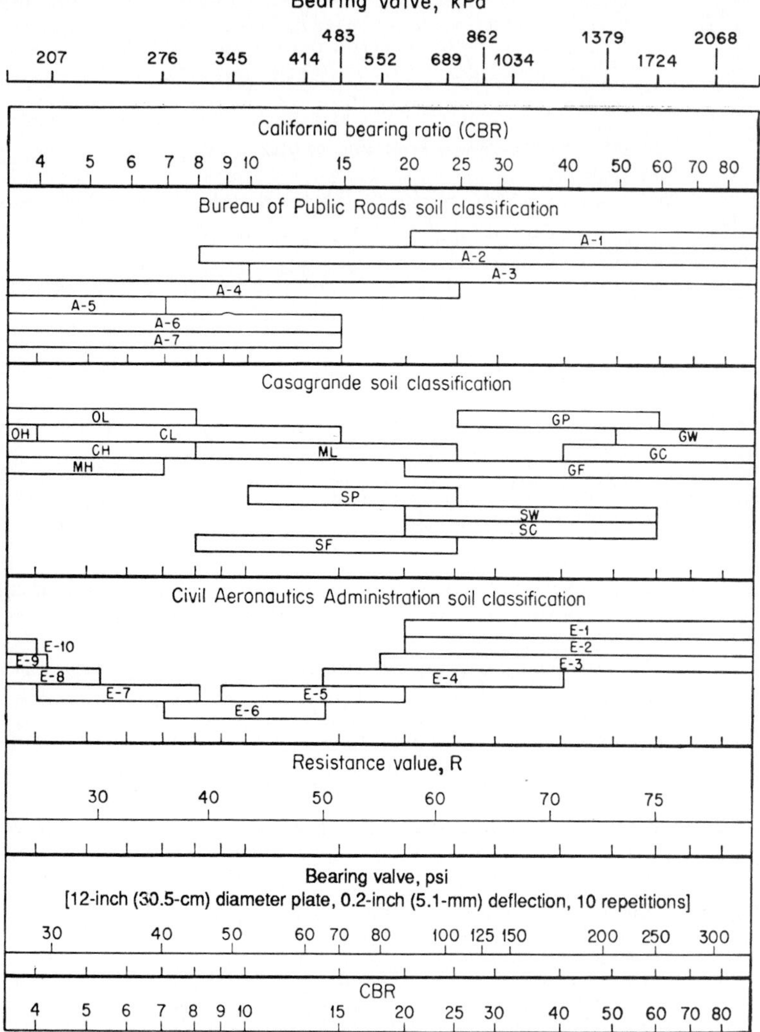

FIGURE 14.5 Approximate relationship among soil classifications.

14. Relationship of Weights and Volumes in Soil. The unit weight of soil varies, depending on the amount of water contained in the soil. Three unit weights are in general use: the saturated unit weight γ_{sat}, the dry unit weight γ_{dry}, and the buoyant unit weight γ_b.

$$\gamma_{sat} = \frac{(G + e)\gamma_0}{1 + e} = \frac{(1 + w)G\gamma_0}{1 + e} \qquad S = 100\% \tag{14.25}$$

$$\gamma_{dry} = \frac{G\gamma_0}{1 + e} \qquad S = 0\% \tag{14.26}$$

$$\gamma_b = \frac{(G - 1)\gamma_0}{1 + e} \qquad S = 100\% \tag{14.27}$$

Unit weights are generally expressed in pounds per cubic foot or grams per cubic centimeter. Representative values of unit weights for a soil with a specific gravity of 2.73 and a void ratio of 0.80 are

$$\gamma_{sat} = 122 \text{ lb/ft}^3 = 1.96 \text{ g/cm}^3$$

$$\gamma_{dry} = 95 \text{ lb/ft}^3 = 1.52 \text{ g/cm}^3$$

$$\gamma_b = 60 \text{ lb/ft}^3 = 0.96 \text{ g/cm}^3$$

The symbols used in Eqs. (14.25) to (14.27) and in Fig. 14.6 are

- G = specific gravity of soil solids (specific gravity of quartz is 2.67; for majority of soils specific gravity ranges between 2.65 and 2.85; organic soils would have lower specific gravities)
- γ_0 = unit weight of water (62.4 lb/ft^3 or 1.0 g/cm^3)
- e = voids ratio, volume of voids in mass of soil divided by volume of solids in same mass [also equal to $n/(1 - n)$, where n is porosity—volume of voids in mass of soil divided by total volume of same mass]
- S = degree of saturation, volume of water in mass of soil divided by volume of voids in same mass
- w = water content, weight of water in mass of soil divided by weight of solids in same mass (also equal to Se/G)

15. Atterberg Limits. The Atterberg limits are used to define the change in the strength properties of fine-grained soils with a change in water content. The *liquid limit* w_l is the highest water content at which the soil has a small but definite shear resistance. At the liquid limit, the cohesion of the soil is practically zero. The *plastic limit* w_p is the lowest water content at which the soil is plastic. The

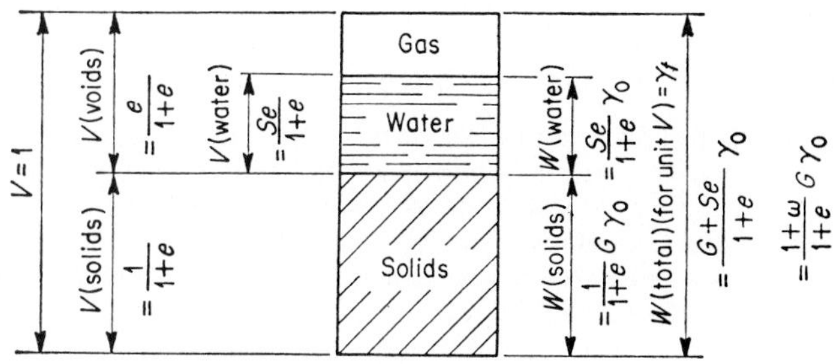

Total volume (solids + water + gas) = 1

FIGURE 14.6 Relationship of weights and volumes in soil.

shrinkage limit w_s is the lowest water content that can occur in a soil when it is completely saturated.

The *plasticity index* I_p is the liquid limit minus the plastic limit and is the range of water content throughout which the soil is plastic. When the plastic limit is equal to the liquid limit, the plasticity index is zero and the soil is entirely lacking in plasticity.

16. Permeability. The coefficient of permeability of a soil is the volume of water which would be forced through a mass of soil having a unit cross-sectional area and a unit length by a unit head of water. The permeability of sand usually ranges from 20×10^{-4} to 3000×10^{-4} cm/sec (5 to 850 ft/day). The permeability of clays is usually less than 10×10^{-4} cm/sec (2.8 ft/day).

Natural soils occurring in stratified formations have a permeability in the direction of stratification much greater than in the direction perpendicular to the stratification.

17. Internal Friction and Cohesion. The angle of *internal friction* for a soil is expressed by

$$\tan \phi = \frac{\tau}{\sigma} \tag{14.28}$$

where ϕ = angle of internal friction
$\tan \phi$ = coefficient of internal friction
σ = normal force on given plane in cohesionless soil mass
τ = shearing force on same plane when sliding on plane is impending

For medium and coarse sands, the angle of internal friction is about 30 to 35°. The angle of internal friction for clays ranges from practically 0 to 20°.

The *cohesion* of a soil is the shearing strength which the soil possesses by virtue of its intrinsic pressure. The value of the ultimate cohesive resistance of a soil is usually designated by c. Average values for c are given below:

General soil type	Cohesion c		General soil type	Cohesion c	
	lb/ft²	N/m²		lb/ft²	N/m²
Almost liquid clay	100	4,788	Medium clay	1000	47,880
Very soft clay	200	9,576	Damp, muddy sand	400	19,152
Soft clay	400	19,152			

18. Vertical Pressures in Soils. The vertical stress in a soil caused by a vertical, concentrated surface load may be determined with a fair degree of accuracy by the use of elastic theory. Two equations are in common use, the Boussinesq and the Westergaard. The Boussinesq equation applies to an elastic, isotropic, homogeneous mass which extends infinitely in all directions from a level surface. The vertical stress at a point in the mass is

$$\sigma_z = \frac{3P}{2\pi z^2[1 + (r/z)^2]^{5/2}} \tag{14.29}$$

The Westergaard equation applies to an elastic material laterally reinforced with horizontal sheets of negligible thickness and infinite rigidity, which prevent the mass from undergoing lateral strain. The vertical stress at a point in the mass, assuming a Poisson's ratio of zero, is

$$\sigma_z = \frac{P}{\pi z^2[1 + 2(r/z)^2]^{3/2}} \tag{14.30}$$

where σ_z = vertical stress at a point, lb/ft^2 (N/m^2)
P = total concentrated surface load, lb (N)
z = depth of point at which σ_z acts, measured vertically downward from surface, ft (m)
r = horizontal distance from projection of surface load P to point at which σ_z acts, ft (m)

For values of r/z between 0 and 1, the Westergaard equation gives stresses appreciably lower than those given by the Boussinesq equation. For values of r/z greater than 2.2, both equations give stresses less than $P/100z^2$.

The Westergaard equation is somewhat preferable for use in analyses in sedimentary soils because the assumptions on which it is based are probably nearer to the conditions existing in stratified soils.

Equations (14.29) and (14.30) may be used for loads spread over an area, provided that the area of loading has a maximum dimension less than one-third the depth z at which the stress is to be computed. Areas having greater dimensions should be subdivided for purposes of the computation, and the resulting stresses added.

19. Lateral Pressures in Soils, Forces on Retaining Walls. The Rankine theory of lateral earth pressures, used for estimating approximate values for lateral pressures on retaining walls, assumes that the pressure on the back of a vertical wall is the same as the pressure that would exist on a vertical plane in an infinite soil mass. Friction between the wall and the soil is neglected. The pressure on a wall consists of (1) the lateral pressure of the soil held by the wall, (2) the pressure of the water, if any, behind the wall, and (3) the lateral pressure from any surcharge on the soil behind the wall.

Symbols used in this section are as follows:

γ = unit weight of soil, lb/ft^3 (kg/m^3) (saturated unit weight, dry unit weight, or buoyant unit weight, depending on conditions)
P = total thrust of soil, lb/linear ft of wall (kg/m)
H = total height of wall, ft (m)
ϕ = angle of internal friction of soil, deg
i = angle of inclination of ground surface behind wall with horizontal; also angle of inclination of line of action of total thrust P and pressures on wall with horizontal
K_A = coefficient of active pressure
K_P = coefficient of passive pressure
c = cohesion, lb/ft^2 (N/m^2)

Lateral Pressure of Cohesionless Soils. For walls that retain cohesionless soils and are free to move an appreciable amount, the total thrust from the soil is

$$P = \frac{1}{2}\gamma H^2 \cos i \frac{\cos i - \sqrt{(\cos i)^2 - (\cos \phi)^2}}{\cos i + \sqrt{(\cos i)^2 - (\cos \phi)^2}} \tag{14.31}$$

When the surface behind the wall is level, the thrust is

$$P = \tfrac{1}{2}\gamma H^2 K_A \tag{14.32}$$

$$K_A = [\tan (45° - \phi/2]^2 \tag{14.33}$$

The thrust is applied at a point $H/3$ above the bottom of the wall, and the pressure distribution is triangular, with the maximum pressure of $2P/H$ occurring at the bottom of the wall.

For walls that retain cohesionless soils and are free to move only a slight amount, the total thrust is $1.12P$, where P is as given above. The thrust is applied at the midpoint of the wall and the pressure distribution is trapezoidal with the maximum pressure of $1.4P/H$ extending over the middle six-tenth of the height of the wall.

For walls that retain cohesionless soils and are completely restrained (very rare), the total thrust from the soil is

$$P = \frac{1}{2}\gamma H^2 \cos i \frac{\cos i + \sqrt{(\cos i)^2 - (\cos \phi)^2}}{(\cos i)^2 - \sqrt{(\cos \phi)^2}} \tag{14.34}$$

When the surface behind the wall is level, the thrust is

$$P = \tfrac{1}{2}\gamma H^2 K_P \tag{14.35}$$

$$K_P = [\tan (45° + \phi/2)]^2 \tag{14.36}$$

The thrust is applied at a point $H/3$ above the bottom of the wall, and the pressure distribution is triangular, with the maximum pressure of $2P/H$ occurring at the bottom of the wall.

Lateral Pressure of Cohesive Soils. For walls that retain cohesive soils and are free to move a considerable amount over a long period of time, the total thrust from the soil (assuming a level surface) is

$$P = \tfrac{1}{2}\gamma H^2 K_A - 2cH\sqrt{K_A} \tag{14.37}$$

or, since highly cohesive soils generally have small angles of internal friction,

$$P = \tfrac{1}{2}\gamma H^2 - 2cH \tag{14.38}$$

The thrust is applied at a point somewhat below $H/3$ from the bottom of the wall, and the pressure distribution is approximately triangular.

For walls that retain cohesive soils and are free to move only a small amount or not at all, the total thrust from the soil is

$$P = \tfrac{1}{2}\gamma H^2 K_P \tag{14.39}$$

since the cohesion would be lost through plastic flow.

Water Pressure. The total thrust from water retained behind a wall is

$$P = \tfrac{1}{2}\gamma_0 H^2 \tag{14.40}$$

where H = height of water above bottom of wall, ft (m)
γ_0 = unit weight of water, lb/ft^3 (kg/m^3) [62.4 lb/ft^3 (999.02 kg/m^3) for fresh water and 64 lb/ft^3 (1024.6 kg/m^3) for salt]

The thrust is applied at a point $H/3$ above the bottom of the wall, and the pressure distribution is triangular, with the maximum pressure of $2P/H$ occurring at the bottom of the wall. Regardless of the slope of the surface behind the wall, the thrust from water is always horizontal.

Lateral Pressure from Surcharge. The effect of a surcharge on a wall retaining a cohesionless soil or an unsaturated cohesive soil can be accounted for by applying a uniform horizontal load of magnitude K_{Ap} over the entire height of the wall, where p is the surcharge in pounds per square foot (kilopascals). For saturated cohesive soils the full value of the surcharge p should be considered as acting over the entire height of the wall as a uniform horizontal load. K_A is defined in list of nomenclature, above.

20. Stability of Slopes

Cohesionless Soil. A slope in a cohesionless soil without seepage of water is stable if

$$i < \phi \tag{14.41}$$

With seepage of water parallel to the slope, and assuming the soil to be saturated, an infinite slope in a cohesionless soil is stable if

$$\tan i < \left(\frac{\gamma_b}{\gamma_{sat}}\right) \tan \phi \tag{14.42}$$

where i = slope of ground surface
ϕ = angle of internal friction of soil
γ_b, γ_{sat} = unit weights, lb/ft^3 (kg/m^3) (Sec. 14)

Cohesive Soils. A slope in a cohesive soil is stable if

$$H < \frac{C}{\gamma N} \tag{14.43}$$

where H = height of slope, ft (m)
C = cohesion, lb/ft^2 (kPa)
γ = unit weight, lb/ft^3 (kg/m^3)
N = stability number, dimensionless

For failure on the slope itself, without seepage water,

$$N = (\cos i)^2(\tan i - \tan \phi) \tag{14.44}$$

Similarly, with seepage of water,

$$N = (\cos i)^2\left[\tan i - \left(\frac{\gamma_b}{\gamma_{sat}}\right)\tan \phi\right] \tag{14.44a}$$

where terms are as defined for Eq. (14.42).

For failure encompassing all or part of the slope, together with soil at the top or toe of the slope, approximate values of the stability number N are given in Table 14.2. In the use of formula (14.44a) and Table 14.2, appropriate values

TABLE 14.2 Stability Numbers for Simple Slopes

i	N for various values of ϕ			
	0°	5°	15°	25°
90°	0.261	0.239	0.199	0.165
75°	.219	.196	.154	.118
60°	.191	.165	.120	.082
45°	.170	.141	.085	.048
30°	.156	.114	.048	.012
15°	.145	.072		

must be used for ϕ and γ. When the slope is submerged, ϕ is the angle of internal friction of the soil and γ is equal to γ_b. When the surrounding water is removed from a submerged slope in a short time (sudden drawdown), ϕ is the weighted angle of internal friction [equal to $(\gamma_b/\gamma_{sat})\phi$] and γ is equal to γ_{sat}.

21. Bearing Capacity of Soils. The approximate ultimate bearing capacity under a long footing at the surface of a soil is given by Prandtl's equation as

$$q_u = \left(\frac{c}{\tan \phi} + \frac{1}{2}\gamma_{dry}b\sqrt{K_p}\right)(K_p e^{\pi \tan \phi} - 1) \tag{14.45}$$

where q_u = ultimate bearing capacity of soil, lb/ft² (kPa)
c = cohesion, lb/ft² (kPa)
ϕ = angle of internal friction, deg
γ_{dry} = unit weight of dry soil, lb/ft³ (kg/m³) (Sec. 14)
b = width of footing, ft (m)
d = depth of footing below surface, ft (m)
K_p = coefficient of passive pressure = $[\tan (45 + \phi/2)]^2$
e = 2.718...

For footings below the surface, the ultimate bearing capacity of the soil may be modified by the factor $1 + Cd/b$. The coefficient C is about 2 for cohesionless soils and about 0.3 for cohesive soils. The increase in bearing capacity with depth for cohesive soils is often neglected.

Typical values of the allowable bearing capacity of various soils as given in the National Building Code of the National Board of Fire Underwriters are shown in

TABLE 14.3 Allowable Bearing Capacity of Soils

Soil	Allowable bearing capacity	
	tons/ft^2	kPa
Medium soft clay	1.5	1,436
Medium stiff clay	2.5	2,394
Sand, fine, loose	2	1,915
Sand, coarse, loose; compact fine sand; loose sand-gravel mixture	3	2,873
Gravel, loose; compact coarse sand	4	3,830
Sand-gravel mixture, compact	6	5,746
Hardpan and exceptionally compacted or partially cemented gravels or sands	10	9,576
Sedimentary rocks, such as hard shales, sandstones, limestones, and silt stones, in sound condition	15	14,364
Foliated rocks, such as schist or slate, in sound condition	40	38,304
Massive bedrock, such as granite, diorite, gneiss, and trap rock, in sound condition	100	95,760

Table 14.3. These values represent the ultimate bearing capacity divided by an appropriate safety factor.

22. Settlement under Foundations. The approximate relationship between loads on foundations and settlement is

$$\frac{q}{P} = C_1\left(1 + \frac{2d}{b}\right) + \frac{C_2}{b} \tag{14.46}$$

where q = load intensity, lb/ft^2 (kPa)
P = settlement, in (mm)
d = depth of foundation below ground surface, ft (m)
b = width of foundation, ft (m)
C_1 = coefficient dependent on internal friction
C_2 = coefficient dependent on cohesion

The coefficients C_1 and C_2 are usually determined by bearing-plate loading tests.

23. Allowable Loads on Piles. (See also Sec. 54, Pile Driving.) A dynamic formula extensively used in the United States to determine the allowable static load on a pile is the *Engineering News* formula. For piles driven by a drop hammer, the allowable load is

$$P_a = \frac{2HW}{p + 1} \tag{14.47}$$

For piles driven by a single-acting hammer, the allowable load is

$$P_a = \frac{2WH}{p + 0.1} \tag{14.48}$$

For piles driven by a double-acting hammer, the allowable load is

$$P_a = \frac{2E}{p + 0.1} \tag{14.48a}$$

where P_a = allowable pile load, lb (kg)
W = weight of hammer, lb (kg)
H = height of drop or stroke, ft (m)
E = actual energy delivered per blow, ft · lb (N · m)
p = penetration of pile per blow, in (cm)

For a group of piles penetrating a soil stratum of good bearing characteristics and transferring their loads to the soil by point bearing on the ends of the piles, the total allowable load would be the sum of the individual allowable loads for each pile. For piles transferring their loads to the soil by skin friction on the sides of the piles, the total allowable load would be less than the sum of the individual allowable loads for each pile, because of the interaction of the shearing stresses and strains caused in the soil by each pile.

24. Types of Piles. Foundation piles used to carry structure loads may be timber, concrete, composite (timber with concrete upper section), or steel. Piles which distribute the load throughout their length to the soil are called *friction piles*. Those which carry the load to firm substrata are *end-bearing piles*. Sheet piles used to retain soil or water may be wood planking, steel sheeting, or precast concrete sheets. Waterproofing is obtained by interlocking or overlapping of sections.

HIGHWAY AND TRAFFIC ENGINEERING

(See also Sec. 55, Pavements.) For detailed data on the geometric design of highways, see Ref. 2. Information on the capacity of highways will be found in Ref. 3. Much of the material herein is taken from these publications.

25. Highway Design Controls

Vehicle Characteristics. Dimensions of the four design vehicles recommended for use by the American Association of State Highway and Transportation Officials (AASHTO) as controls for geometric design are shown in Table 14.4. Minimum turning paths for these vehicles are shown in Fig. 14.7. The vehicle which should be used in design is the largest one which represents a significant percentage of the traffic. For design of most highways accommodating truck traffic, one of the design semitrailer combinations should be used. A design check should be made for the largest vehicle expected, in order to ensure that such a vehicle can negotiate the designated turns, particularly if pavements are curbed.

Design Speed. The design speed of a highway is the maximum safe speed that can be maintained over a specified section when conditions are favorable, so that the design features of the highway govern the speed.

TABLE 14.4 Dimensions of Design Vehicles

Design vehicle type and symbol	Dimensions, ft (m)					
	Wheel-base	Overhang Front	Overhang Rear	Overall length	Overall width	Height
Passenger car, P	11 (3.35)	3 (0.91)	5 (1.52)	19 (5.79)	7 (2.13)	
Single-unit truck, SU	20 (6.09)	4 (1.22)	6 (1.83)	30 (9.14)	8.5 (2.59)	13.5 (4.1)
Semitrailer combination, intermediate, WB-40	13 + 27 = 40 (12.2)	4 (1.22)	6 (1.83)	50 (15.2)	8.5 (2.59)	13.5 (4.1)
Semitrailer combination, large, WB-50	20 + 30 = 50 (15.2)	3 (0.91)	2 (0.61)	55 (16.8)	8.5 (2.59)	13.5 (4.1)

Source: From AASHTO.[2]

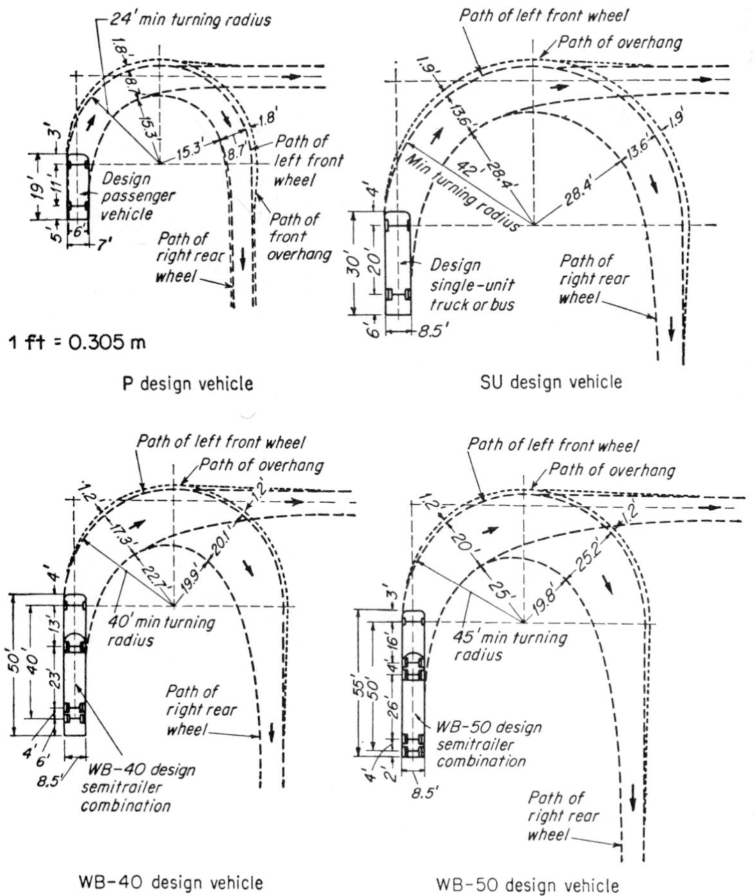

FIGURE 14.7 Minimum turning paths for design vehicles.

TABLE 14.5 Traffic Elements and Their Relation for Rural Highways

Traffic element	Explanation and nationwide percentage or factor
Average daily traffic, ADT	Average 24-hr volume for a given year; total for both directions of travel, unless otherwise specified
DHV	Design hour volume (two-way unless otherwise specified), usually the thirtieth highest hourly volume of the design year (30HV)
K	DHV expressed as a percentage of ADT, both two-way; normal range 12 to 18%
D	Directional distribution of DHV, one-way volume in predominant direction of travel expressed as a percentage of two-way DHV; general range 55 to 80%, average 67%
T	Trucks (exclusive of light delivery trucks) expressed as a percentage of DHV; normal range 5 to 12%, averagae 8%

Traffic. The principal measures of traffic volume and character and the relationship between the various elements for rural highways are shown in Table 14.5. Determination of the relationship between the traffic elements for urban highways usually requires special study.

Types of Arterial Highways. A major street is an arterial highway with intersections at grade and direct access to abutting property and on which geometric design and traffic control measures are used to expedite the safe movement of through traffic. An expressway is a divided arterial highway with full or partial control of access and generally with grade separations at intersections. A freeway is an expressway with full control of access. A parkway is a type of arterial highway provided for noncommercial traffic, with full or partial control of access and usually located within a park or ribbon of parklike development.

26. Elements of Geometric Design

Stopping Sight Distance. Design stopping sight distance is the minimum distance required for a vehicle traveling at or near the design speed to stop before reaching an object in its path. It is the sum of the distances traveled during perception and brake reaction time and the distance traveled while braking to a stop. Stopping sight distance is measured from a point 3.75 ft (1.14 m) above the road surface to a point 6 in (15.2 cm) above the road surface. The sight distance at every point on a highway should be at least as great as the minimum distances shown in Table 14.6.

Passing Sight Distance. Design passing sight distance is the minimum distance required to make safely a normal passing maneuver on two- and three-lane highways at passing speeds representative of nearly all drivers, commensurate with design speed. Passing sight distance is measured from a point 3.75 ft (1.14 m) above the road surface to a second point 4.5 ft (1.37 m) above the road surface. The minimum passing sight distance is shown in Table 14.7.

TABLE 14.6 Stopping Sight Distance

Design speed		Min. stopping sight distance	
mi/h	km/h	ft	m
15	24.1	80	24.4
20	32.2	120	36.6
25	40.2	160	48.8
30	48.3	200	60.9
40	64.4	275	83.8
50	80.5	350	106.7
60	96.5	475	144.8
70	112.6	600	182.9
80	128.7	750	228.6

TABLE 14.7 Passing Sight Distance for Two-Lane Highways

Design speed		Min. passing sight distance	
mi/h	km/h	ft	m
30	48.3	1100	335.3
40	64.4	1500	457.2
50	80.5	1800	548.6
60	96.5	2100	640.1
70	12.6	2500	762.0
80	28.7	2700	822.9

Maximum Horizontal Curvature and Superelevation. The maximum horizontal curvature for a given design speed is limited by the maximum rate of superelevation and the allowable side friction. The maximum superelevation that is considered generally desirable is 0.10 ft/ft (0.098 m/m) pavement width. Values from 0.06 to 0.12 are used for maximum superelevation rates, depending on local conditions such as ice formation, frequency of intersection, and similar factors. For a maximum superelevation rate of 0.10 ft/ft (0.098 m/m), the design superelevation rates recommended by AASHTO for various speeds are shown on Fig. 14.8. The figure also indicates the maximum curvature for various design speeds at the maximum superelevation rate of 0.10.

Maximum Grades. The maximum grades recommended by AASHTO for main highways are shown in Table 14.8. Maximum grades for secondary highways may be about 2 percent steeper than those shown in the table.

Sight Distance on Horizontal Curves. Stopping sight distance must be provided on all horizontal curves. Figure 14.9 shows the required clearance from the center line of the inside lane to provide the minimum stopping sight distance. Design of two-lane highways for passing sight distance must in general be confined to tangent or very flat alignment conditions because of the excessive clearances that would be required on curves.

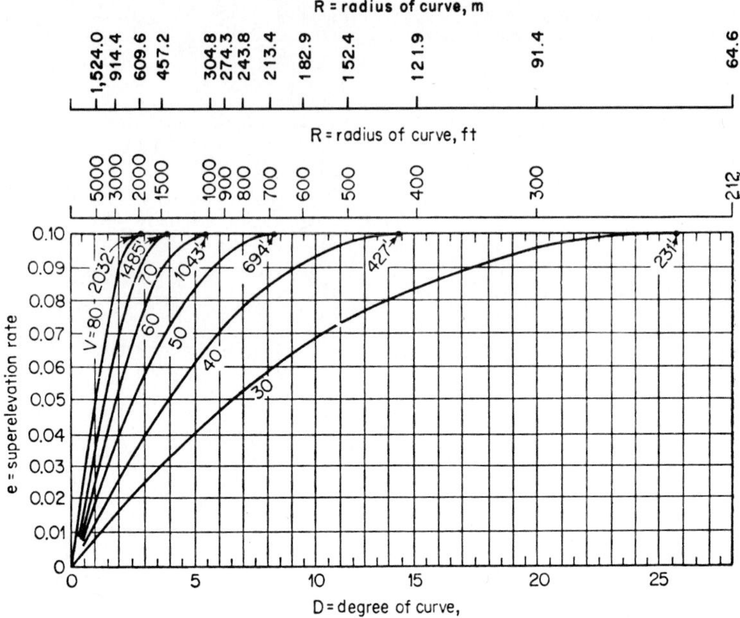

FIGURE 14.8 Design superelevation rates for maximum superelevation rate of 0.10 ft/ft (0.03 m/m).

TABLE 14.8 Maximum Grades, Percent

	Design speed, mi/h (km/h)					
Topography	30 (48.3)	40 (64.4)	50 (80.5)	60 (96.5)	70 (112.6)	80 (128.7)
Flat	6	5	4	3	3	3
Rolling	7	6	5	4	4	4
Mountainous	9	8	7	6	5	

Sight Distance on Vertical Curves. Vertical curves must be designed to provide stopping sight distance. Other factors that enter into the determination of the length of a vertical curve are rider comfort and drainage control. It is generally impractical to design vertical curves for passing sight distance. The minimum length of vertical curve L for various algebraic differences in grade A is shown in Fig. 14.10 for crest vertical curves and in Fig. 14.11 for sag vertical curves.

27. Highway Cross Sections

Pavement Type and Cross Slope. The type of pavement is determined by the volume and composition of traffic, the availability of materials, the initial cost, and the extent and cost of maintenance. *High-type* pavements have smooth riding

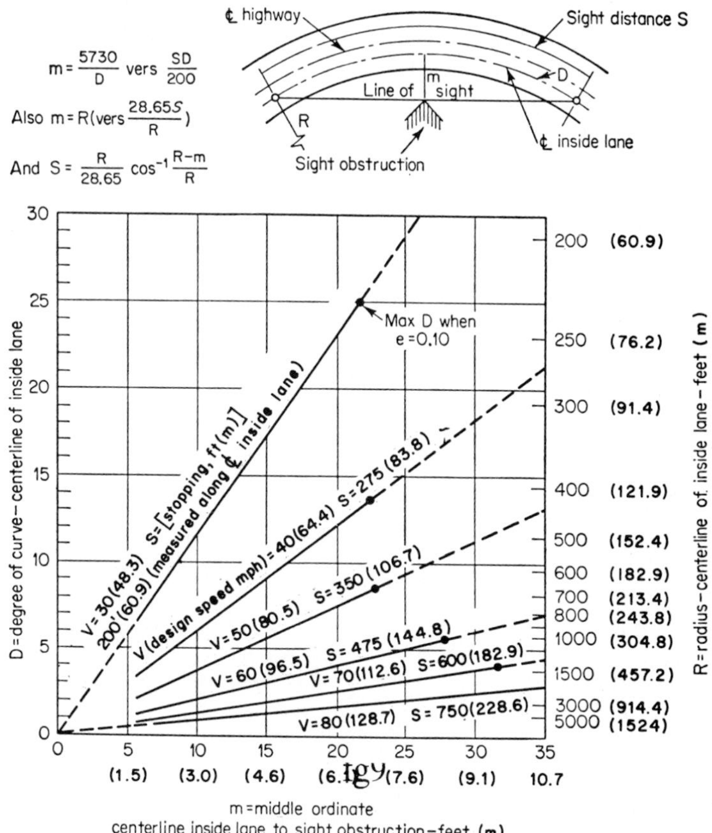

FIGURE 14.9 Stopping sight distance on horizontal curves (open road conditions).

qualities and good antiskid properties in all weather, and should support adequately the expected volume and weight of vehicles without fatigue. *Intermediate-type* pavements vary from those only slightly less costly than the high type to surface treatments. *Low-type* surfaces range from surface-treated earth to loose surfaces such as earth, shell, or gravel.

The range of cross slopes applicable to each type of pavement for adequate drainage is as follows:

Surface type	Cross slope	
	ft/ft	m/m
High	0.01–0.02	0.003–0.006
Intermediate	0.015–0.03	0.0046–0.0091
Low	0.02–0.04	0.006–0.012

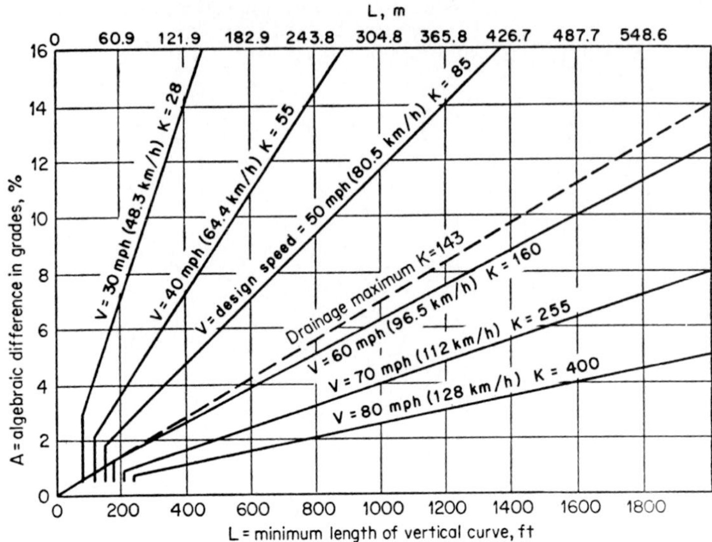

FIGURE 14.10 Design controls for crest vertical curves (stopping sight distance).

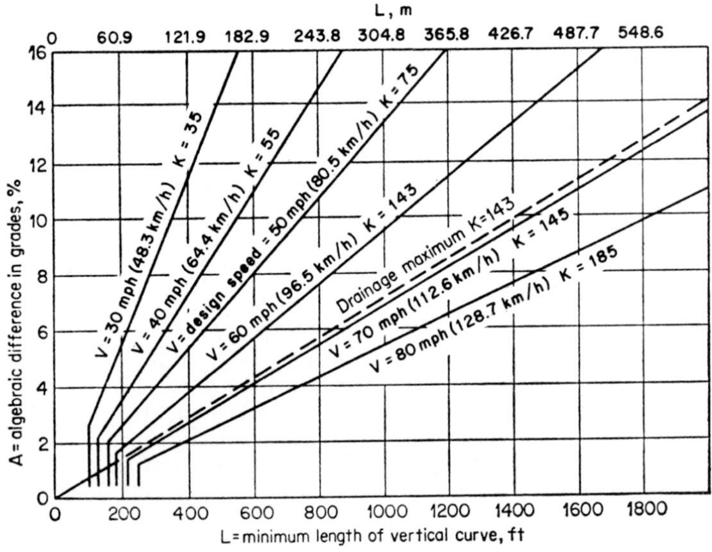

FIGURE 14.11 Design controls for sag vertical curves.

When curbs are located at the pavement edge, the above values should be increased slightly.

Vertical Clearance. Clear heights of 14 ft (4.3 m) should be provided over all highways, except for routes limited to noncommercial traffic, where 12.5 ft (3.8 m) is adequate. In many states, 16 ft (4.9 m) is required on interstate highway routes. These clearances are often increased by 4 to 6 in (10 to 15 cm) to provide for future resurfacing.

Pavement Width. The desirable width of pavement on a highway is 12 ft (3.66 m) per lane. This width may be reduced where traffic is light or speed is low.

Shoulder Width. The usable width of shoulder is that which can be used when a driver makes an emergency stop. The overall width of shoulder, the dimension between the edge of pavement and the intersection of the shoulder and side-slope planes, is 1 to 3 ft (0.3 to 0.9 m) greater than the usable shoulder width, except where the side slopes are 4:1 or flatter. Usable shoulders 10 to 12 ft (3.0 to 3.7 m) wide are desirable on all highways, but narrower shoulders may be used in low-volume highways.

Design Guides. Design guides for cross-section elements for two-lane rural highways are shown in Table 14.9.

Four-lane highways should, if possible, be designed as divided highways. Design guides for cross-section elements for various types of four-lane divided rural highways are shown in Table 14.10. These cross-section elements should provide a balanced total section. Where restrictions are necessary, the border width should be reduced before decreasing the median, and both should be cut to a minimum before considering a reduction in shoulder or lane width.

Design guides for major streets in urban areas are shown in Table 14.11. Design speeds for major streets are 30 mi/h (48.3 km/h) in built-up districts and as high as 50 mi/h (80.5 km/h) in outlying areas.

An expressway-at-grade is intermediate between a major street and a freeway with respect to design features. The expressway-at-grade is a surface facility practically free from roadside interference and on which crossing or entering traf-

TABLE 14.9 Design Guides for Two-Lane Rural Highways

Cross-section element	Element dimension, ft (m)		
	Low	Intermediate	High
Surfacing	18–20 (5.5–6.1)	20–24 (6.1–7.3)	24 (7.3)
Usable shoulder	4–8 (1.2–2.4)	8 (2.4)	10 (3.0)
Roadway	26–36 (7.9–10.9)	36–40 (10.9–12.2)	44 (13.4)
Border, each side	18–25* (5.5–7.6)	20–30* (6.1–9.1)	25–35* (7.6–10.7)
Right of way	66–80* (18.3–24.4)	80–100* (24.4–30.5)	100–120* (30.5–36.6)

*Preferably more.

TABLE 14.10 Design Guides for Four-Lane Divided Rural Highways

Cross-section element	Element dimension, ft (m)		
	Restricted	Intermediate	Desirable
Pavement, each	24 (7.3)	24 (7.3)	24 (7.3)
Usable shoulder	8–10 (2.4–3.0)	10 (3.0)	10–12 (3.0–3.7)
Median	4–15 (1.2–4.6)	20* (6.1)	40* (12.2)
Border, each side	12–15* (3.7–4.6)	25–40* (7.6–12.2)	50–80* (15.2–24.4)
Right of way	90–110* (27.4–33.5)	140–180* (42.7–54.9)	210–310* (64.0–94.5)

*Preferably more.

fic from minor streets is eliminated. Design speeds range from 40 mi/h (64.4 km/h) in built-up districts to 60 mi/h (96.5 km/h) in outlying areas. Design guides for expressways-at-grade are shown in Table 14.12.

Design guides for freeways are similar to those for expressways-at-grade. Design speeds range from 50 mi/h (80.5 km/h) in built-up districts to 60 or 70 mi/h (96.5 or 112.6 km/h) in outlying areas. Frontage roads are often required for depressed freeways to provide continuity in the local street system. Grade separations for cross streets may occur at intervals of one to two blocks in downtown areas, three to five blocks in intermediate areas, and at greater distances in outlying areas. Interchanges are generally located about 2 mi (3.2 km) apart in urban areas, 4 mi (6.4 km) apart in suburban areas, and 8 mi (12.9 km) apart in rural areas.

28. Highway Capacity and Levels of Service. The capacity of a highway is the maximum number of vehicles which has a reasonable expectation of passing over a given section of a lane or a roadway in one direction (or in both directions for a two-lane or a three-lane highway) during a given time period under prevailing roadway and traffic conditions. Capacity is equivalent to level of service *E*, defined below with other levels of service:

A = free flow, low volumes, high speeds, little or no restriction in maneuverability

B = stable flow, operating speeds somewhat restricted by traffic conditions (level *B* suitable for design of rural highways)

C = stable flow, operating speeds satisfactory but closely controlled by traffic conditions (level *C* suitable for urban design)

D = approaching unstable flow, tolerable operating speed, little freedom to maneuver

E = unstable flow, momentary stoppages (level *E* is capacity)

F = forced flow, congestion (volumes below capacity)

Service volumes for two-lane highways and for freeways and expressways under uninterrupted flow conditions are shown in Tables 14.13 and 14.14. These

TABLE 14.11 Design Guides for Major Urban Streets

Section	Type of urban area	Through traffic lanes					Median—width				Shoulders, pavement widening at curbs, or parking lanes—width				Border—width				Right of way—width			
			Width																			
			ft*		m		ft		m		ft		m		ft		m		ft		m	
Section	Type of urban area	No.	A	B	A	B	A	B	A	B	A	B	A	B	A	B	A	B	A	B	A	B
Shoulders—no curbs	Res.	2	11	20	3.4	3.7	0	0	0	0	10	10	3.0	3.0	12	12	3.7	6.1	66	84	20.1	25.6
	Res	4	11	12	3.4	3.7	0	14	0	4.3	10	10	3.0	3.0	8	12	2.4	3.7	80	106	24.4	32.3
Curbed—no parking	Com.	4	11	12	3.4	3.7	0	4	0	1.2	1	2	0.3	0.6	8	12	2.4	3.7	62	80	18.9	24.4
	Res.	4	11	12	3.4	3.7	0	4	0	1.2	1	2	0.3	0.6	12	16	3.7	4.9	70	88	21.3	26.8
	Com.	6	11	12	3.4	3.7	0	4	0	1.2	1	2	0.3	0.6	8	12	2.4	3.7	84	104	25.6	31.7
	Res.	6	11	12	3.4	3.7	0	4	0	1.2	1	2	0.3	0.6	12	16	3.7	4.9	92	112	28.0	34.1
Curbed with parking lanes	Com.	4	11	12	3.4	3.7	0	4	0	1.2	10	11	3.0	3.4	8	12	2.4	3.7	80	98	24.4	29.9
	Res.	4	11	12	3.4	3.7	0	4	0	1.2	10	10	3.0	3.0	12	16	3.7	4.9	88	104	26.8	31.7
	Com.	6	11	12	3.4	3.7	0	4	0	1.2	10	11	3.0	3.4	8	12	2.4	3.7	102	122	31.1	37.2
	Res.	6	11	12	3.4	3.7	0	4	0	1.2	10	10	3.0	3.0	12	16	3.7	4.9	110	128	33.5	39.0
Divided with parking lanes†	Com.	4	11	12	3.4	3.7	4	14	1.2	4.3	10	12	3.0	3.7	8	12	2.4	3.7	84	110	25.6	33.5
	Res.	4	11	12	3.4	3.7	4	14	1.2	4.3	10	11	3.0	3.4	12	16	3.7	4.9	92	116	28.0	35.4
	Com.	6	11	12	3.4	3.7	4	14	1.2	4.3	10	12	3.0	3.7	8	12	2.4	3.7	106	134	32.3	40.8
	Res.	6	11	12	3.4	3.7	4	14	1.2	4.3	10	11	3.0	3.4	12	16	3.7	4.9	114	140	34.7	42.7

Note: A = acceptable minimum, B = desirable minimum, Res. = residential, Com. = commercial.
*Ten-foot widths may be considered in special cases, but not on two-lane streets.
†Without parking lanes, deduct 20 ft (6 m) from right of way.

TABLE 14.12 Design Guides for Expressways-at-Grade

Cross-section element	Element dimension, ft (m)		
	Restricted	Intermediate	Desirable
Pavement, each:			
Four-lane	24 (7.3)	24 (7.3)	24 (7.3)
Six-lane	36 (10.9)	36 (10.9)	36 (10.9)
Shoulder	10 (3.0)	10 (3.0)	10 (3.0)
Median	4 (1.2)	14–25 (4.3–7.6)	40* (12.2)
Border, each side	12 (3.7)	20 (6.1)	30* (9.1)
Right of way:			
Four-lane	96 (29.3)	120–130 (36.6–39.6)	170* (51.8)
Six-lane	120 (36.6)	145–155 (44.2–47.2)	195* (59.4)

*Preferably more.

TABLE 14.13 Operating Speeds and Service Volumes, Two-Lane Highway
(Uninterrupted flow conditions; rural)

Level of service	Operating speed, mi/h (km/h)	% of length with passing sight distance of 1500 ft (457.2 m) or more	Maximum service volume under ideal conditions (total passenger cars per hour, both directions)			
			For 70-mi/h design speed (112.6 km/h)	For 60-mi/h design speed (96.5 km/h)	For 50-mi/h design speed (80.5 km/h)	For 40-mi/h design speed (64.4 km/h)
A	60 or more (96.5)	100	400	*	*	*
		50	270	*	*	*
		0	80	*	*	*
B	50 or more (80.5)	100	900	800	*	*
		50	720	540	*	*
		0	480	240	*	*
C	40 or more (64.4)	100	1400	1320	1120	*
		50	1270	1070	850	*
		0	1080	760	360	*
D	35 or more (56.3)	100	1700	1660	1500	1160
		50	1650	1550	1350	960
		0	1600	1320	1020	380
E†	30 ±	n.p.‡	2000	2000	2000	2000
F	Less than 30 (48.3)	n.p.‡	Variable	Variable	Variable	Variable

*This level of service not attainable at this design speed.
†Capacity.
‡No passing at this level.

TABLE 14.14 Operating Speeds and Service Volumes, Freeways and Expressways *(Uninterrupted flow conditions; rural or small metropolitan areas)*

		Maximum service volume under ideal conditions (total passenger cars per hour, one direction)			
		For 70-mi/h (112.6-km/h) design speed		For 60-mi/h (96.5-km/h) design speed per lane in one direction	For 50-mi/h (80.5-km/h) design speed per lane in one direction
Level of service	Operating speed, mi/h (km/h)	For two lanes in one direction	Each additional lane in one direction		
A	60 or more (96.5)	1400	1000	*	*
B	55 or more (88.5)	2000	1500	500	*
C	50 or more (80.5)	2300	1400	700	*
D	40 or more (64.4)	2800	1400	1200	700
E†	30–35 (48.3–56.3)	4000	2000	2000	2000
F	Less than 30 (less than 48.3)	Variable	Variable	Variable	Variable

*This level of service not attainable at this design speed.
†Capacity.

service volumes must be adjusted for roadway and traffic conditions. Adjustments for lane widths and lateral clearances are given in Table 14.15. Adjustments for trucks may be made by converting trucks to equivalent passenger cars using the factors from Table 14.16. The service volumes shown do not apply at intersections or in the vicinity of ramp termini.

Typical design service volumes per lane for urban arterial routes with allowances for the factors discussed above and for roadside and intersection interferences are shown in Fig. 14.12.

29. Intersection Capacity and Levels of Service. The capacity of a signalized intersection approach is the maximum number of vehicles that the approach can reasonably accommodate under the existing geometric, environmental, and traffic characteristics and controls. Capacity is equivalent to level of service *E*, defined below with other levels of service:

A = free operation, no vehicle waits longer than one red indication: load factor = 0.0

B = stable operation, occasional approach cycle fully utilized, many drivers somewhat restricted by traffic conditions: load factor = 0.1 or less (level *B* suitable for design of rural intersections)

C = stable operation, intermittent loading, most drivers somewhat restricted

TABLE 14.15 Effects of Narrow Traffic Lanes and Restricted Lateral Clearances

Clearance from pavement edge to obstruction	Percentage of service volume*							
	Lanes with obstruction on one side, ft (m)				Lanes with obstruction on both sides, ft (m)			
	12 (3.7)	11 (3.4)	10 (3.0)	9 (2.7)	12 (3.7)	11 (3.4)	10 (3.0)	9 (2.7)
Two-lane highways								
6	100	86	77	70	100	86	77	70
4	96	83	74	68	92	79	71	65
2	91	78	70	64	81	70	63	57
0	85	73	66	60	70	60	54	49
Freeways and expressways, two lanes each direction								
6	100	97	91	81	100	97	91	81
4	99	96	90	80	98	95	89	79
2	97	94	88	79	94	91	86	76
0	90	87	82	73	81	79	74	66

*Applies to level of service *B* for two-lane highways and to all levels for freeways and expressways.

TABLE 14.16 Average Passenger Car Equivalent of Trucks over Extended Lengths of Highways

Type of route	Level of service	Level terrain	Rolling terrain	Mountainous terrain
Freeway and expressways	All	2	4	8
Two-lane highways	*A*	3	4	7
	B and *C*	2.5	5	10
	D and *E*	2	5	12

by traffic conditions: load factor = 0.3 or less (level *C* suitable for design of urban intersections)

D = approaching instability, substantial delays during short peaks within peak period: load factor = 0.7 or less

E = delays of several cycles, queues developing: load factor = 0.7 to 1.0, depending on conditions, with an average of 0.85 (level *E* is capacity)

F = jammed conditions, traffic flow controlled by downstream conditions, volumes unpredictable

Load factor is the proportion of green-signal intervals that are fully utilized.

Service volumes at intersections on two-way and one-way urban streets with parking are shown on Figs. 14.13 and 14.14. Service volumes of intersections are expressed as vehicles per hour of green signal time, and the service volume for an approach must be adjusted for the proportion of total time allocated to the approach. Adjustments must also be made for metropolitan-area size, the peak-hour factor (for intersections, the ratio of the volume during the peak hour to four times the volume during the peak 15 min: range 0.25 to 1.00), the location within

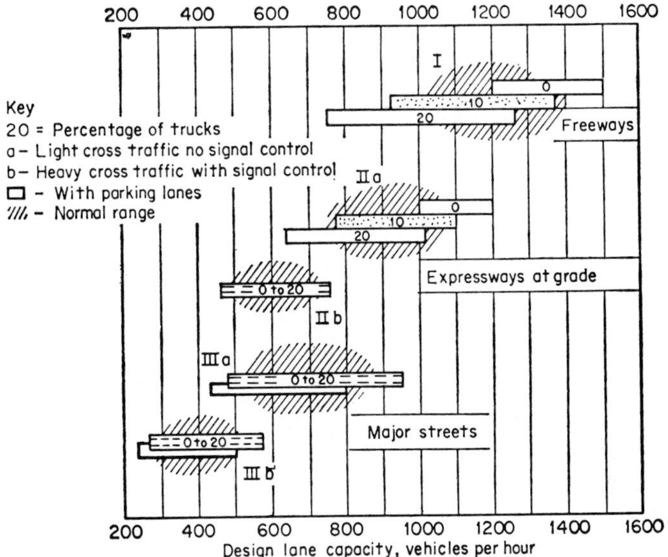

FIGURE 14.12 Lane capacities for urban routes.

the metropolitan area, and the effects of commercial-vehicle traffic, turning movements, and parking prohibitions.

Peak-hour factors of 1.00 are rarely found. Where long lines of waiting vehicles are typically present, a peak-hour factor of 0.90 or 0.95 may be used. The usual conditions in a metropolitan area are equivalent to a peak-hour factor of 0.85, but where a high rate of flow occurs over a period shorter than an hour, factors of 0.75, 0.70, or less should be used. Adjustments to service volumes for various peak hours are combined in Figs. 14.13 and 14.14 with adjustments for metropolitan-area size. The figures also indicate adjustments for the location of the intersection within the metropolitan area.

For streets with approach widths of 21 to 29 ft (6.4 to 8.8 m) where turning movements differ from 10 percent right and 10 percent left, multiply the service volume by

$$[1.00 - (0.005)(R - 10)][1.00 - (0.010)(L - 10)] \qquad \text{[for two-way streets]}$$

$$[1.00 - (0.005)(R - 10)][1.00 - (0.005)(L - 10)] \qquad \text{[for one-way streets]}$$

where R = percentage of right turns (maximum $R = 30$)
L = percentage of left turns (maximum $L = 30$)

Special adjustments must be made for intersections having separate lanes and/or separate signal indications for turning movements.

For intersections where the percentage of trucks and through buses differs from 5 percent, multiply the service volume by $[1.00 - (0.010)(T - 5)]$ where T is the percentage of trucks and through buses. Adjustments for buses which stop in the vicinity of the intersection require special procedures.

Prohibition of parking can result in increases of 30 to 50 percent over the service volumes shown in Fig. 14.14 for two-way streets and 20 percent to as much as 100 percent over the volumes shown in Fig. 14.13 for one-way streets.

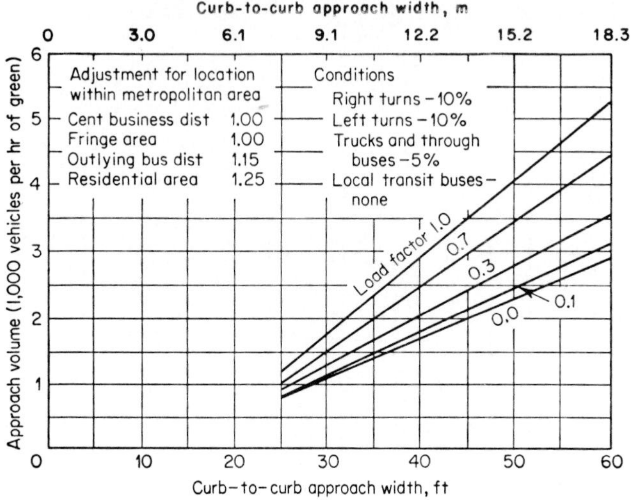

Adjustment for Peak-hour Factor and Metropolitan-area Size

Metropolitan-area pop. (1,000's)	Peak-hour factor						
	0.70	0.75	0.80	0.85	0.90	0.95	1.00
Over 1,000	1.00	1.05	1.09	1.14	1.19	1.24	1.29
1,000	0.97	1.02	1.07	1.11	1.16	1.21	1.26
750	0.94	0.99	1.04	1.09	1.14	1.18	1.23
500	0.91	0.96	1.01	1.06	1.11	1.16	1.21
375	0.88	0.93	0.98	1.03	1.08	1.13	1.18
250	0.85	0.90	0.95	1.00	1.05	1.10	1.15
175	0.82	0.87	0.92	0.97	1.02	1.07	1.12
100	0.80	0.85	0.89	0.94	0.99	1.04	1.09
75	0.77	0.82	0.87	0.92	0.96	1.01	1.06

FIGURE 14.13 Urban intersection approach service volume, in vehicles per hour of green-signal time, for one-way streets with parking both sides.

30. Parking Requirements

Automobile Parking. The space used for automobile parking is shown in Fig. 14.15 and Table 14.17.

Truck Parking. Truck loading/unloading platform heights range from 48 to 52 in (1.2 to 1.3 m) to match truck floors. Parking stall should be 10 to 14 ft (3.0 to 4.3 m) in width. Other dimensions are shown in Fig. 14.15.

RAILROADS

31. Track Gage. Standard railroad gage in North America is 4 ft 8½ in (1435 mm) between inside of rails, measured ⅝ in (15.9 mm) below top of rail. The gage may be varied from 4 ft 8⅜ in (1432 mm) on some high-speed tangent track to 4

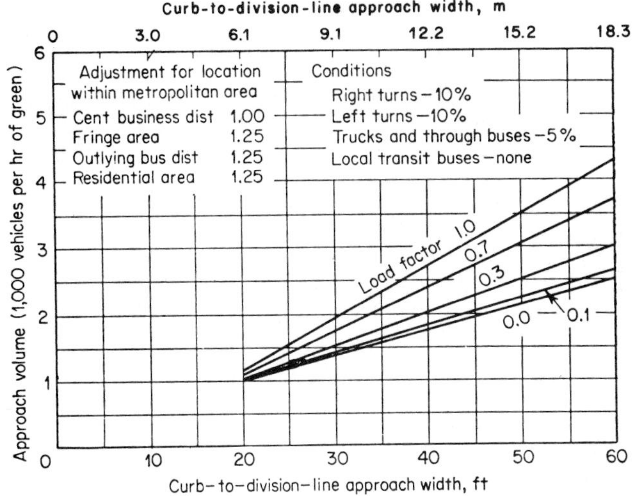

Adjustment for Peak-hour Factor and Metropolitan-area Size

Metropolitan-area pop. (1,000's)	Peak-hour factor 0.70	0.75	0.80	0.85	0.90	0.95	1.00
Over 1,000	1.00	1.05	1.10	1.14	1.19	1.24	1.29
1,000	0.97	1.02	1.07	1.11	1.16	1.21	1.27
750	0.94	0.99	1.04	1.09	1.13	1.18	1.23
500	0.91	0.96	1.01	1.06	1.11	1.15	1.20
375	0.89	0.93	0.98	1.03	1.08	1.12	1.17
250	0.86	0.91	0.95	1.00	1.05	1.10	1.14
175	0.83	0.88	0.92	0.97	1.02	1.07	1.11
100	0.80	0.85	0.90	0.94	0.99	1.04	1.09
75	0.77	0.82	0.87	0.91	0.96	1.01	1.06

FIGURE 14.14 Urban intersection approach service volume, in vehicles per hour of green-signal time, for two-way streets with parking.

TABLE 14.17 Street Space Used for Parking

Angle of parking at curb	Width of street used when parked ft	m	Width needed for parking plus maneuvering ft	m	Length of curb per car ft	m
Parallel	7	2.1	19	5.8	22.0	6.7
45°	17	5.2	29	8.8	11.3	3.4
60°	18	5.5	36	10.9	9.2	2.8
90°	17	5.2	40	12.2	8.0	2.4

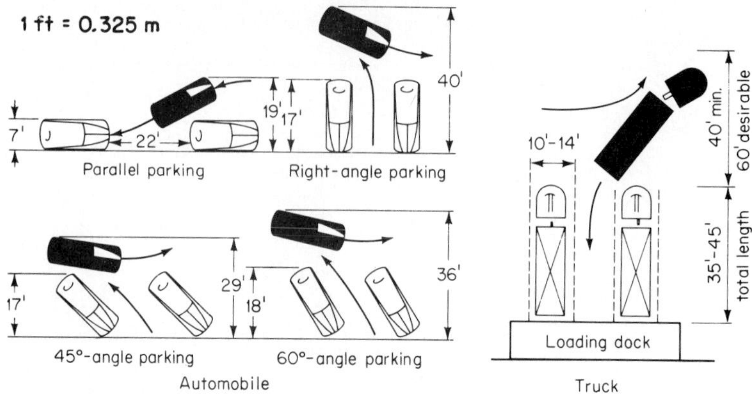

FIGURE 14.15 Storage and maneuvering space used for various parking positions.

ft 9⅛ in (1451 mm) on curves of small radius. Table 14.18 lists major railroad gages in use throughout the world.

32. Track Materials. Ties are usually oak, pine, or fir, but concrete ties are in use. Ties are generally 8 ft 6 in (2591 mm) or 9 ft (2743 mm) (recommended) in length for standard gage and range from 6 in (152 mm) deep and 6 in (152 mm) wide for yard and sidetracks to 7 in (178 mm) deep and 10 in (254 mm) wide for principal main lines. Ties spacing varies from 19½ in (495 mm) on centers for main lines to 24 in (610 mm) for secondary tracks.

Rail is designated by weight per yard and section. Weights up to 155 lb/yd (76.9 kg/m) are used for high-speed track, but rail ranging from 100 to 130 lb/yd (49.6 to 64.5 kg/m) is in more common use. Rail is usually rolled in 39-ft (11.9-m) lengths, but 78-ft (23.8-m) lengths are often provided to decrease maintenance costs. The use of continuous welded rail with lengths of 5000 to 6000 ft (1524 to 1829 m) is increasing. Length is governed by turnout and signal locations since ends of signal circuits require insulated joints.

Rails are connected by means of joint bars which hold adjoining rails in horizontal and vertical alignment. Tie plates are placed under rails to distribute rail loads and reduce wear. Both rail and tie plates are held to the ties by means of spikes. Longitudinal rail movement is prevented by rail anchors.

33. Curvature and Superelevation. Maximum curvature for railroads is usually determined by train operating speeds and allowable superelevation. Maximum superelevation is 6 to 7 in (152 to 178 mm) but may be less on a particular railroad. The AREA (American Railway Engineering Association) formula for equilibrium superelevation is

$$e = 0.0007DV^2 \tag{14.49}$$

where D = degree of curve, deg
V = speed, mi/h (km/h)

For high-speed trains, as much as 3 in (76 mm) of unbalanced superelevation may be permitted, so that

TABLE 14.18 Major Railway Gages of the World

36 in. (914 mm)	$39\frac{3}{8}$ in. (1,000 mm)	42 in. (1,067 mm)	$56\frac{1}{2}$ in. (1,435 mm)	60 in. (1,524 mm)
Colombia	Algeria*	Angola	Algeria*	Czechoslovakia*
El Salvador	Argentina*	Australia*	Australia*	Finland
Guatemala	Austria*	Costa Rica	Austria*	Panama*
Haiti	Bangladesh	Ecuador	Argentina*	USSR*
Honduras*	Belgium*	Gabon	Belgium*	
Panama*	Bolivia	Honduras*	Brazil*	
Venezuela*	Brazil*	Indonesia	Bulgaria	
	Chile*	Malawi	Canada	**63 in. (1,600 mm)**
	Czechoslovakia*	Mozambique	China	
	Egypt*	New Zealand	Cuba	Brazil*
	Ethiopia	Nicaragua	Czechoslovakia*	
	France*	Nigeria	Denmark	
	Germany, West*	Norway*	Egypt*	
	Germany, East*	Philippines	France*	
	Kenya	Rhodesia	Germany, West*	
	Malaysia	Sudan	Germany, East*	**$65\frac{5}{8}$ in. (1,668 mm)**
	Peru*	Union of South Africa	Great Britain	
	Puerto Rico	Venezuela*	Greece	Spain*
	Spain*	Zaire*	Hungary	
	Switzerland*	Zambia	Iran	
	Tanzania		Iraq	
	Thailand		Ireland	
	Tunisia		Italy	
	USSR*		Japan	**66 in. (1,676 mm)**
	Vietnam		Mexico	
	Zaire*		Morocco	Argentina*
			Netherlands	Chile*
			Norway*	India
			Paraguay	Pakistan
			Peru*	Portugal
			Poland	Spain*
			Romania	
			Sweden	
			Switzerland*	
			Syria	
			Turkey	
			United States	
			Uruguay	

* Countries having more than one standard gage.

$$e = 0.0007DV^2 - 3 \tag{14.49a}$$

Curvature may also be limited because of coupling difficulties on curves of more than 6°. Some railroads have established an absolute maximum curvature of 16° because of dimensions of the rigid wheelbase of cars and engines and permissible swing of couplers. Reverse curves should be separated by at least two car lengths.

Vertical curves should be of sufficient length to limit gradient changes to 0.05%/100 ft (30.5 m) for sags and 0.10%/100 ft (30.5 m) for crests on main lines; secondary lines may have vertical curves one-half the lengths required for main lines.

34. Clearances. Track clearances are established from center line of track for horizontal dimensions and from top of rail for vertical dimensions. Track spacings vary among railroads, but the following are common centerline spacings:

Main track–main track	13 ft 6 in (4.1 m)
Main track–yard or passing tracks	15 ft 0 in (4.6 m)
Yard track–yard track	13 ft 6 in (4.1 m)
Ladder track–yard track	18 ft 0 in (5.5 m)
Ladder track–ladder track	19 ft 0 in (5.8 m)

Track spacings are increased on curves at a rate of 2 in/deg (51 mm). Where adjacent tracks have different superelevation, additional clearance is provided of 3 in/in (76 mm/mm) of difference in superelevation.

Clearances from track to fixed structures vary by railroad and by state. Generally, vertical clearances of 22 ft (6.7 m) are required, except at building entrances, where 17 ft (5.2 m) may be permitted. Horizontal clearances of 8 ft (2.4 m) and 8 ft 6 in (2.6 m) are standard for tangent track, with additional allowances of 1 in/deg (25.4 mm/deg) of curve and 3 in/in (76 mm/mm) of superelevation for curved track.

35. Turnouts and Crossovers. Turnouts, used to divert trains from one track to another, consist of a switch, a frog to carry the wheel flanges over crossing rails, closure rails connecting the switch rails and the frog, and guard rails to guide the flanges at the frog. Control points for turnouts are the actual or ⅛-in (3.2 mm) point of the switch [ground to a width of ⅛ in (3.2 mm)] and the actual or ½-in (12.7-mm) point of frog. Turnouts with No. 16 to No. 20 frogs are used for high-speed main-line movements, No. 10 to No. 12 for slow-speed main-line movements, and No. 8 for yards and sidings. Crossovers, used to transfer trains between parallel tracks, consist of two turnouts and connecting rails.

WATER SUPPLY, SEWERAGE, AND DRAINAGE

36. Water Supply and Treatment

Quantity of Water. Average annual water requirements in metropolitan areas with metered systems generally range from 100 gal (378.5 L) per capita per day (gpcpd) to 200 gpcpd (757 Lpcpd), with a median value of about 150 gpcpd (567.8 Lpcpd). Unmetered supply systems have considerably higher consumption, and large water-using industries require special determination of demand.

Seasonal variations in water demand occur largely because of irrigation, lawn sprinkling, and air-conditioning loads, and maximum monthly consumption is generally about 125 percent of average annual demand but may range up to 200 percent of average annual demand. Maximum daily demands of 150 percent of average annual demand and maximum hourly demands of from 200 to 250 percent of annual average demand are commonly used for design.

Fire demand is often the determining factor in the design of mains, distribution storage tanks, and pumps, even though the total quantity of water required for fire fighting is small during a long period. For communities of less than 200,000 population, the fire demand is given by the National Board of Fire Underwriters as

$$Q = 1020\sqrt{P}(1 - 0.01\sqrt{P}) \qquad (14.50)$$

where Q = fire demand, gal/min
P = population in thousands

The fire demand is added to the normal demand on the maximum day to determine the total maximum demand.

Design Period. Pipes less than 12 in (304.8 mm) in diameter are generally designed to be adequate for the full development of the area served; pipes more than 12 in (304.8 mm) in diameter and wells, distribution systems, and filtration and treatment plants are generally designed for the flow expected 15 to 25 years in the future; and large dams and conduits are generally designed to be adequate for 25 to 50 years.

Quality of Water. The outstanding requirement for a domestic water supply is freedom from pathogenic bacteria. In addition, there are reasonable limits for certain impurities, as listed below:

Impurity	Limit, ppm	Impurity	Limit, ppm
Turbidity	10	Iron plus manganese	0.3
Color	20	Magnesium	125
Lead	0.1	Total solids	500
Fluoride	1.5	Total hardness (calcium plus magnesium salts)	100
Copper	3.0		

Water Treatment. Water treatment usually consists of filtration through either a slow or a rapid sand filter and disinfection with chlorine. In addition, water may be softened to remove hardness and aerated to remove iron and manganese.

The slow sand filter operates at a rate of 2 to 10 million gal/acre · day (1858 to 9291 L/m^2 · day) and is effective in removing tastes and odors from raw water. About 99 percent of the bacterial content is also removed.

The rapid sand filter operates at a rate of 125 to 250 million gal/acre · day (125,000 to 250,000 L/m^2 · day). However, preliminary treatment of the raw water is required, including chemical coagulation and sedimentation. The entire treatment process is effective in removing about 99.98 percent of the bacterial content, but removal of the color and turbidity is less dependable than for the slow sand filter and requires particular attention to the coagulation process.

Softening is accomplished by the addition of lime, or lime and soda ash, and sedimentation. The addition of lime and passage through a zeolite softener is also used. Iron and manganese may be removed by aeration and sedimentation.

Disinfection by the addition of chlorine is the final stage of any treatment process. Common practice is to add sufficient chlorine so that a small free chlorine residual is maintained.

37. Water Distribution. Transmission mains connecting the source of supply to the distribution system must be large enough to supply at least the maximum daily demand plus fire flow. If the distribution system does not include storage, supply mains must also be adequate to deliver maximum hourly demands. Both transmission and distribution mains are usually designed by using the Hazen-Williams formula.

38. Sewage Collection and Treatment

Quantity of Sewage. The average flow of sewage from a metropolitan area is about 100 gpcpd (gallons per capita per day) (378.5 Lpcpd). This rate may vary from 240 gpcpd (908 Lpcpd) in a maximum hour, 160 gpcpd (606 Lpcpd) on a maximum day, 70 gpcpd (265 Lpcpd) on a minimum day, and 40 gpcpd (151 Lpcpd) in a minimum hour. In addition, infiltration of ground water into sewers may be taken at about 600 gal/day · in diam. · mi.

Design Period. Laterals and submains less than 15 in (381 mm) in diameter are generally designed to be adequate for the full development of the area served; main sewers, outfalls, and intercepter sewers are generally designed for the flow expected from 40 to 50 years in the future; and treatment works are generally designed for the flow expected from 10 to 25 years in the future.

Sewer Design. Sewers should be at least 8 in (203 mm) in diameter and should be laid on a grade sufficient to produce a velocity of 2 ft/s (0.6 m/s) when flowing full, to prevent the deposition of suspended solids. Sewer lines should be designed with straight alignment and uniform grade between manholes, which should have a maximum spacing of 400 ft (122 m).

Quality of Sewage. Sewage is approximately 99.92 percent water, with the remaining 0.08 percent (800 ppm by weight) composed of organic and mineral matter, as shown below:

	Organic matter, ppm	Mineral matter, ppm
Suspended solids....	100	50
Colloidal solids......	140	60
Dissolved solids.....	160	290

The "biochemical oxygen demand" of sewage, BOD, is the quantity of oxygen which must be supplied during the aerobic stabilization of sewage, and thus is a direct measure of the pollutional effect. Residential sewage has an average BOD of 0.24 lb (0.11 kg) oxygen per capita.

Sewage Treatment. The degree of sewage treatment required should be based on the size, characteristics, and usage of the receiving body of water and upon the amount and quality of sewage to be treated. Complete sewage treatment might include preliminary treatment, such as screening to remove large suspended solids, grit removal, and grease removal; primary treatment, such as plain sedimentation or chemical precipitation; secondary treatment of a biological nature, such as the trickling filter or the activated sludge process; final treatment by chlorination; and finally disposal by dilution in a body of water. Approximate values of the BOD and suspended-solids removal of primary and secondary treatment are shown below:

Treatment process	Percentage removal		Treatment process	Percentage removal	
	BOD	Suspended solids		BOD	Suspended solids
Plain sedimentation	25–40	40–70	Trickling filter	80–95	80–90
Chemical precipitation	50–75	70–90	Activated sludge	85–95	85–95

39. Sizes and Slopes of Sewers. Sewer sizes and slopes are usually designed by using the Manning formula

$$v = \frac{1.486}{n} R^{2/3} S^{1/2} \tag{14.51}$$

where v = average velocity of flow, ft/s (m/s)
n = coefficient of roughness
R = hydraulic radius, ft (m) = A/P = $D/4$ for circular conduit flowing full
S = hydraulic gradient, ft head loss/ft length (m/m)
A = cross-sectional area of flow, ft^2 (m^2)
P = wetted perimeter, ft (m)
D = diameter of circular conduit, ft (m)

For circular sewers flowing full, the Manning formula can be written as

$$Q = \frac{0.4632}{n} D^{2/3} S^{1/2} = \text{conveyance factor} \times S^{1/2} \tag{14.52}$$

where Q = quantity of flow, ft^3/s (m^3/s).

40. Quantity of Runoff. The rational method for the determination of the quantity of storm water which appears as runoff involves the use of

$$Q = ciA \tag{14.53}$$

where Q = runoff from rainfall, ft^3/s (m^3/s)
c = coefficient of runoff, dimensionless
i = rainfall intensity, expressed as a rate, in rain/h (mm/h)
A = tributary area, acres (m^2)

These factors are discussed below.

Coefficient of Runoff. The coefficient of runoff for a particular area depends on the character of the surface, the type and extent of vegetation, the slope of the surface, and other less important factors. Approximate values of the coefficient of runoff c are given in Table 14.19.

Rainfall Intensity. The rainfall intensity is dependent on the recurrence interval and the time of concentration. The recurrence interval is the period of time within which, on the average, a rainfall of a given intensity will be equaled or exceeded

TABLE 14.19 Runoff Coefficients for Rational Formula

Type of area	Flat:slope <2%	Rolling: slope -10%	Hilly: slope >10%
Pavements, roofs, etc.	0.90	0.90	0.90
City business areas	.80	.85	.85
Suburban residential areas	.45	.50	.55
Dense residential areas	.60	.65	.70
Grassed areas	.25	.30	.30
Earth areas	.60	.65	.70
Cultivated land:			
Impermeable (clay, loam)	.50	.55	.60
Permeable (sand)	.25	.30	.35
Meadows and pasture lands	.25	.30	.35
Forests and wooded areas	.10	.15	.20

only once. Recurrence intervals of from 5 to 25 years are generally used, but for important structures periods of 100 years have been used.

For a particular area and a given recurrence interval, a study of rainfall records will permit the determination of an intensity-duration curve, which gives the rainfall intensity [in inches (mm) per hour] as a function of the duration of rainfall. The rainfall intensity is greatest for short periods and decreases sharply as the duration of rainfall becomes greater. The intensity to use for a particular design is that for which the duration is equal to the time of concentration.

Time of Concentration. The time of concentration for a particular inlet to a drainage system is the time required for rainfall falling on the most remote part of the tributary area drained by the inlet to reach the inlet. At this time, the entire area tributary to the inlet will be contributing to the runoff and the total runoff will be a maximum. The time for water to flow overland from the most remote part of the tributary area to the inlet may be approximated by

$$t = C\left(\frac{L}{Si^2}\right)^{1/2} \tag{14.54}$$

where t = time of overland flow, min
L = distance of overland flow, ft (m)
S = slope of land, ft/ft (m/m)
i = rainfall intensity, in/h (mm/h)
C = coefficient: 0.5 for paved areas, 1.0 for bare earth, 2.5 for turf

For any portions of the flow carried in ditches, the time of flow to the inlet may be computed by means of the Manning formula.

41. Flow in Drainage Channels. Drainage channels are usually of such lengths that head losses other than those due to friction are negligible. Design of drainage channels is generally by the Manning formula:

$$v = \frac{1.486}{n} R^{2/3} S^{1/2} \tag{14.55}$$

where $S = h_f/l$ = hydraulic gradient, ft head loss/ft length (m/m)
n = coefficient of roughness, dimensionless
R = hydraulic radius, ft (m) = A/P
A = cross-sectional area of flow, ft^2 (m^2)
P = wetted perimeter, ft (m)
h_f = head loss due to friction, ft (m)
l = length of channel or conduit, ft (m)
v = average velocity of flow, ft/s (m/s)

42. Steel Design

Working Stresses. Structural steel has a weight of 490 lb/ft³ (7845 kg/m³), a modulus of elasticity of 29 million lb/in² (199.9 GPa), and a shearing modulus of 12 million lb/in² (82.7 GPa). Structural steels used in buildings and the minimum yield points are listed in Table 14.20.

The allowable unit working stresses for structural steel as given in Ref. 4 are shown below. Somewhat lower stresses than those shown are used for bridges, in recognition of the more severe service and the greater possibility of overloading such structures.

Tension. Tension on net section, except at pinholes:

$$F_t = 0.60F_y \tag{14.56}$$

Tension on net section at pinholes:

$$F_t = 0.45F_y \tag{14.57}$$

where F_t = allowable tensile stress, lb/in² (kPa)

The slenderness ratio Kl/r [defined following Eq. (14.63)] preferably should not exceed 240 for main members or 300 for bracing or other secondary members, other than rods.

TABLE 14.20 Designations and Yield Points for Structural Steels

ASTM Specification	Designation	Specified minimum yield point F_y lb/in²*	Specified minimum yield point F_y MPa
"Structural Steel"	A36	36,000	248.1
"Structural Steel"	A529	42,000	289.5
"High-strength Low-alloy Structural Steel"	A242	42,000	289.5
"High-strength Structural Steel"	A440	42,000	289.5
"High-strength Low-alloy Structural Magnesium Vanadium Steel"	A441	42,000	289.5
"High-strength Low-alloy Structural Columbium Vanadium Steel, Grade 42"	A572	42,000	289.5
"High-strength Low-alloy Structural Steel with 50,000 psi Minimum Yield Point"	A588	50,000†	344.7
"High-Yield Strength, Quenched and Tempered Alloy Steel Plate"	A514	90,000	620.5

*Values given are for heavy sections and for plates 1½ to 4 in (102 mm) thick; higher values may be allowed for light sections and thinner plates and lower values for thicker plates.

†For sections weighing more than 600 lb/ft, F_y = 42,000 lb/in² (289.5 MPa).

Shear. Shear on gross section:

$$F_v = 0.40F_y \tag{14.58}$$

where F_v = allowable shear stress, lb/in (kPa)

Compression. Compression on the gross section of axially loaded compression members when Kl/r is less than C_c:

$$F_a = \frac{\{1 - [(Kl/r)^2/2C_c^2]\}\, F_y}{SF} \tag{14.59}$$

Compression on the gross section of axially loaded columns when Kl/r exceeds C_c:

$$F_a = \frac{149{,}000{,}000}{(Kl/r)^2} \tag{14.60}$$

Compression on the gross section of axially loaded bracing and secondary members when l/r exceeds 120:

$$F_{as} = \frac{F_a \text{ [from Eq. (14.59) or (14.60), depending on } C_c]}{1.6 - l/200r} \tag{14.61}$$

Compression on the gross area of plate girder stiffeners:

$$F_a = 0.60F_y \tag{14.62}$$

Compression on the web of rolled shapes at the toe of the fillet:

$$F_a = 0.75F_y \tag{14.63}$$

where F_a = allowable comprehensive stress permitted in absence of bending moment, lb/in^2 (kPa)
F_{as} = allowable comprehensive stress permitted in absence of bending moment for bracing and other secondary members, lb/in^2 (kPa)
F_y = minimum yield point, lb/in^2 (kPa)
K = effective-length factor (suggested design values shown in Fig. 14.16)
l = actual unbraced length, in (mm)
r = radius of gyration corresponding to K and l, in (mm) ($=\sqrt{I/A}$)
I = moment of inertia, in^4 (mm^4)
A = gross cross-sectional area, in^2 (mm^2)
C_c = slenderness ratio separating elastic and inelastic buckling

$$C_c = \sqrt{\frac{2\pi^2 E}{F_y}} \tag{14.64}$$

SF = factor of safety

$$SF = 1.67 + \frac{3(Kl/r)}{8C_c} - \frac{(Kl/r)^3}{8C_c^3} \tag{14.65}$$

The slenderness ratio Kl/r of compression members must not exceed 200.

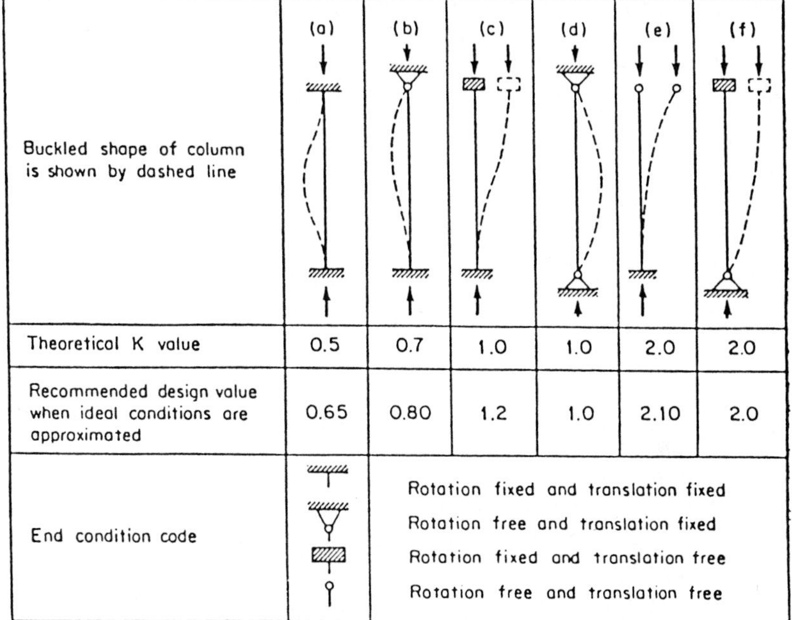

Buckled shape of column is shown by dashed line	(a)	(b)	(c)	(d)	(e)	(f)
Theoretical K value	0.5	0.7	1.0	1.0	2.0	2.0
Recommended design value when ideal conditions are approximated	0.65	0.80	1.2	1.0	2.10	2.0
End condition code		Rotation fixed and translation fixed Rotation free and translation fixed Rotation fixed and translation free Rotation free and translation free				

FIGURE 14.16 Effective-length factors for members subject to axial load.

Bending. Tension and compression on extreme fibers of laterally supported compact shapes* having an axis of symmetry in the plane of loading:

$$F_b = 0.66F_y \tag{14.66}$$

where F_b = allowable bending stress in absence of axial load, lb/in^2 (kPa)

Laterally supported members have transverse movement of the compression flange prevented at points of support not more than $2400b_f/\sqrt{F_y}$, or $20{,}000{,}000\ A_f/dF_y$ in (mm) apart, where

b_f = compression flange width, in (mm)

A_f = cross-sectional area of compression flange, in^2 (mm^2)

d = depth of member, in (mm)

*A compact shape has the flanges continuously connected to the web or webs; the width of unstiffened projecting elements of the compression flange does not exceed $2050/\sqrt{F_y}$ times the flange thickness; the width of flange plates does not exceed $6000/\sqrt{F_y}$ times the flange-plate thickness; and the depth of the web does not exceed $20{,}200[1 - 3.74(f_a/F_a)]/\sqrt{F_y}$ times the web thickness where (f_a/F_a) is the ratio of computed axial stress to allowable axial stress in the absence of bending moment, except that it need not be less than $8100/\sqrt{F_y}$.

Tension and compression on extreme fibers of laterally supported unsymmetrical members (except channels) or box-type members, and tension on other rolled shapes or built-up members:

$$F_b = 0.60F_y \tag{14.67}$$

Compression on extreme fibers of other rolled shapes and built-up members (except box-type members), the larger value from Eqs. (14.68) and (14.69), but not more than $0.60F_y$:

$$F_b = \left[1.0 - \frac{(l/r)^2}{2C_c^2}\right] 0.60F_y \tag{14.68}$$

$$F_b = \frac{12{,}000{,}000}{ld/A_f} \tag{14.69}$$

where l = unsupported length of compression flange, in (mm)
r = radius of gyration of compression flange plus one-sixth web about an axis in plane of web, in (mm)
d = depth of member, in (mm)
A_f = cross-sectional area of compression flange, in² (mm²)
$C^c = \sqrt{2\pi^2 E/F_y}$

Equation (14.68) may be further modified in certain cases by consideration of the moments at each end of the unsupported length.

Compression on extreme fibers of channels, the value from Eq. (14.69) but not more than $0.60F_y$.

Tension and compression on extreme fibers of pins:

$$F_b = 0.90F_y \tag{14.70}$$

Tension and compression on extreme fibers of rectangular bearing plates:

$$F_b = 0.75F_y \tag{14.71}$$

Bearing. Bearing on milled surfaces and pins in reamed, drilled, or bored holes:

$$F_p = 0.90F_y \tag{14.72}$$

Bearing on bolts or rivets:

$$F_p = 1.35F_y \tag{14.73}$$

where F_p = allowable bearing stress, lb/in² (kPa)

Columns and Tension Members. Columns are designed on the basis of the gross area of the section used. Tension members, however, are designed on the basis of net area, with deductions made for rivet and other holes. In determining net area, net width is obtained by deducting from the gross width the sum of the diameters of all the holes in any chain of holes in any diagonal or zigzag direction and adding for each gauge space in the chain the quantity $s^2/4g$, where s is the longitudinal spacing (pitch) in inches (millimeters) of any two successive holes, and g is the transverse spacing (gage) in inches (millimeters) of the same two holes. Several chains of holes should be tried until the one giving the least net width is found. The net area of a tension member taken through a hole is limited to 85 percent of the gross area. The diameter of a rivet or bolt hole is taken as ⅛ in (3.2 mm) greater than the nominal diameter of the rivet or bolt.

Beams. The extreme fiber stress in bending for a steel beam is computed as

$$f_b = \frac{M}{S} \tag{14.74}$$

where f_b = maximum fiber stress, lb/in^2(kPa)
M = bending moment, lb · in (N · m)
S = section modulus (= I/c), in^3 (mm^3)
I = moment of inertia of cross-sectional area, in^4 (mm^4)
c = distance from extreme fiber to neutral axis, in (mm) (c = one-half the depth for a symmetrical cross section)

The section moduli for selected standard rolled-steel sections are shown in Table 14.21. For beams requiring greater section moduli, built-up sections or plate girders are generally used.

The shearing stress in a steel beam with flanges is relatively constant over the depth of the web, and may be computed as

$$f_v = \frac{V}{d_w t} \tag{14.75}$$

where f_v = shearing stress in web, lb/in^2 (kPa)
V = total shear at section, lb (N)
d_w = depth of web, in (mm)
t = thickness of web, in (mm)

Built-up sections will usually require stiffeners to prevent web buckling.

TABLE 14.21 Elastic Section Modulus S_x for Shapes Used as Beams

in^3 (cm^3)	Shape	in^3 (cm^3)	Shape	in^3 (cm^3	Shape
1110 (18,190)	W 36×300	252 (4,130)	S 24×120	110 (1,803)	W 21×55
		250 (4,097)	W 24×100	108 (1,770)	W 18×60
1030 (16,879)	W 36×280	250 (4,097)	W 21×112	107 (1,753)	W 12×79
				104 (1,704)	W 16×64
952 (15,600)	W 36×260	243 (3,982)	W 27×94	103 (1,688)	S 18×70
		236 (3,867)	S 24×105.9	103 (1,688)	W 14×68
894 (14,650)	W 36×245				
		221 (3,622)	W 24×94	98 (1,612)	W 18×55
837 (13,716)	W 36×230	220 (3,605)	W 18×114		

Source: *Manual of Steel Construction*, American Institute of Steel Construction.

43. Structural Shapes. Designations for structural shapes are related to the profiles of the shapes. W shapes have essentially parallel flange surfaces. HP bearing pile shapes have essentially parallel flange surfaces and equal web and flange thicknesses. S shapes and American Standard channels (C) have a slope of approximately 2 in 12 on their inner flange surfaces. The letter M designates shapes not classified as W, HP, or S. MC designates channels not American Standard. Current and former designations for selected structural shapes are shown in Table 14.22.

44. Steel Connections

Rivets. Rivets vary in size from ⅜ to 1¼ in (9.5 to 31.8 mm) in diameter, with the ¾- and ⅞-in (1.91- and 22.2-mm) sizes most commonly used. The standard sizes and cross-sectional areas of rivets are shown in Table 14.23. Rivet holes are considered to be ⅛ in (3.2 mm) larger in diameter than the rivet.

Bolts. Both turned and unfinished bolts are available in the same sizes as rivets. Larger sizes are also available. Cross-sectional areas are as shown in Table 14.23.

TABLE 14.22 Hot-Rolled Structural Steel Shape Designations

Current designation	Type of shape	Former designation
W 24 × 76 W 14 × 26	W shape	24 WF 76 14 B 26
S 24 × 100	S shape	24 I 100
M 8 × 18.5 M 10 × 9 M 8 × 34.3	M shape	8 M 18.5 10 JR 9.0 8 × 8 M 34.3
C 12 × 20.7	American Standard Channel	12 ⊏ 20.7
MC 12 × 45 MC 12 × 10.6	Miscellaneous Channel	12 × 4 ⊏ 45.0 12 JR ⊏ 10.6
HP 14 × 73	HP shape	14 BP 73
L 6 × 6 × ¾ L 6 × 4 × ⅝	Equal Leg Angle Unequal Leg Angle	∠ 6 × 6 × ¾ ∠ 6 × 4 × ⅝
WT 12 × 38 WT 7 × 13	Structural Tee cut from W shape	ST 12 WF 38 ST 7 B 13
ST 12 × 50	Structural Tee cut from S shape	ST 12 I 50
MT 4 × 9.25 MT 5 × 4.5 MT 4 × 17.15	Structural Tee cut from M shape	ST 4 M 9.25 ST 5 JR 4.5 ST 4 M 17.15
PL ½ × 18	Plate	PL 18 × ½
Bar 1 ⌷ Bar 1¼ ϕ Bar 2½ × ½	Square Bar Round Bar Flat Bar	Bar 1 ⌷ Bar 1¼ ϕ Bar 2½ × ½
Pipe 4 Std. Pipe 4 X - Strong Pipe 4 XX - Strong	Pipe	Pipe 4 Std. Pipe 4 X-Strong Pipe 4 XX-Strong
TS 4 × 4 × .375 TS 5 × 3 × .375 TS 3 OD × .250	Structural Tubing: Square Structural Tubing: Rectangular Structural Tubing: Circular	Tube 4 × 4 × .375 Tube 5 × 3 × .375 Tube 3 OD × .250

Source: *Manual of Steel Construction*, American Institute of Steel Construction.

TABLE 14.23 Rivet and Bolt Diameters and Areas

Rivets and bolts		Bolts only		Rivets and bolts		Bolts only	
Nominal diam., in (mm)	Cross-sectional area, in^2 (mm^2)	Cross-sectional area, thread root, in^2 (mm^2)	Threads/in (threads/mm)	Nominal diam., in (mm)	Cross-sectional area, in^2 (mm^2)	Cross-sectional area, thread root, in^2 (mm^2)	Threads/in (threads/mm)
⅜	0.110	0.068	16	⅞	0.601	0.419	9
(9.5)	(70.9)	(43.9)	(0.63)	(22.2)	(387.7)	(270.3)	(0.35)
½	0.196	0.126	13	1	0.785	0.551	8
(12.7)	(126.5)	(81.3)	(0.51)	(25.4)	(506.5)	(355.5)	(0.32)
⅝	0.307	0.202	11	1⅛	0.994	0.693	7
(15.9)	(198.1)	(130.3)	(0.43)	(28.6)	(641.3)	(447.1)	(0.28)
¾	0.442	0.302	10	1¼	1.227	0.890	7
(19.1)	(285.2)	(194.8)	(0.39)	(31.8)	(791.6)	(574.2)	(0.28)

Welds. The allowable loads on butt welds of the same size as the connected members are the same as for the members. The allowable load per inch (millimeters) of fillet weld is determined on the minimum cross section; for an equal leg weld, the minimum section at the throat is 0.707 times the dimension of the weld leg.

Working Stresses

Rivets. Allowable stresses for A502, Grade 1 hot-driven rivets are 20,000 lb/in^2 (137.9 MPa) in tension and 15,000 lb/in^2 (103.4 MPa) in shear; for A502, Grade 2 hot-driven rivets, stresses are 27,000 lb/in^2 (186.1 MPa) in tension and 20,000 lb/in^2 (137.9 MPa) in shear.

Bolts. Allowable stresses for A307 bolts are 20,000 lb/in^2 (137.9 MPa) in tension and 10,000 lb/in^2 (68.9 MPa) in shear; for other threaded parts of other steels, stresses are $0.60F_y$ in tension and $0.30F_y$ in shear.

Allowable stresses for A325 and A490 bolts are shown below:

Allowable Bolt Stresses, lb/in^2 (MPa) (Buildings)

Bolts	Tension	Shear		
		Friction-type connections	Bearing-type connections: Threading excluded from shear planes	Bearing-type connections: Threading not excluded from shear planes
A325	40,000 (275.8)	15,000 (103.4)	22,000 (151.7)	15,000 (103.4)
A490	54,000 (372.3)	20,000 (137.9)	32,000 (220.6)	22,500 (155.1)

Welds. The allowable stress for welds on A36, A242, and A441 steels is 21,000 lb/in^2 (144.8 MPa); except that complete-penetration groove welds with any type of loading and partial-penetration groove welds loaded in compression, bearing, or tension parallel to the axis of the weld may be stressed to the full allowable stress of the connected material.

45. Reinforced-Concrete Design

Concrete Mixes. The proportioning of concrete ingredients is by weight or volume. Weight measures are considered more reliable. Concrete mixes are designated by the proportion of each ingredient in the order: cement, sand, coarse aggregate. For example, a 1:2:3 mix is one part cement, two parts sand, and three parts stone or gravel. A bag of cement [94 lb (42.6 kg)] is equivalent to 1 ft^3 (0.28 m^3). Water for concrete should be free of injurious amounts of oils, acids, alkalis, salts, or organic matter.

Strength and Durability of Concrete. The most important factor affecting the strength and durability of concrete is the water-cement ratio. For concrete made from average materials, compressive strengths to be used for design are shown in Table 14.24. Strengths greater than those shown may be used, based on compressive-strength tests. Water-cement ratios for various types of construction and exposure conditions are shown in Table 14.25, but any concrete subject to freezing temperatures while wet should have a water-cement ratio not more than 6 gal (22.7 L) per bag and should contain entrained air.

Design Methods. Reinforced concrete may be designed by either one of two methods: working-stress design or ultimate-strength design. Both methods are permitted under current codes, and the selection of a method is left to the designer. Both methods are discussed herein; reference should be made to the American Concrete Institute (ACI) Building Codes, which are the bases for the following discussion.

Reinforcing Bars. Steel bars for concrete reinforcement are available in the sizes shown in Table 14.26.

TABLE 14.24 Maximum Permissible Water-Cement Ratios for Concrete

	Maximum permissible water-cement ratio*			
	Non-air-entrained concrete		Air-entrained concrete	
Specified compressive strength at 28 days f_c, lb/in^2 (MPa)	U.S. gal per 94-lb bag of cement (L/42.6 kg)	Absolute ratio by weight	U.S. gal per 94-lb bag of cement (L/42.6 kg)	Absolute ratio by weight
2500 (17.2)	7.3 (27.6)	0.65	6.1 (23.1)	0.54
3000 (20.7)	6.6 (24.9)	0.58	5.2 (19.7)	0.46
3500 (24.1)	5.8 (21.9)	0.51	4.5 (17.0)	0.40
4000 (27.6)	5.0 (18.9)	0.44	4.0 (15.0)	0.35

*Including free surface moisture on aggregates.
Source: *Building Code Requirements for Reinforced Concrete* American Concrete Institute.

TABLE 14.25 Maximum Permissible Water-Cement Ratios [Gallons per Bag (Liters per Bag)] for Different Types of Structures and Degrees of Exposure

Type of structure	Exposure conditions*					
	Severe wide range in temperature or frequent alternations of freezing and thawing (air-entrained concrete only)			Mild temperature, rarely below freezing, or rainy, or arid		
		At the water line or within the range of fluctuating water level or spray			At the water line or within the range of fluctuating water level or spray	
	In air	In fresh water	In sea water or in contact with sulfates†	In air	In fresh water	In sea water or in contact with sulfates†
Thin sections, such as railings, curbs, sills, ledges, ornamental or architectural concrete, reinforced piles, pipe, and all sections with less than 1-in. (25.4-mm) concrete cover over reinforcing	5.5 (20.8)	5.0 (18.9)	4.5‡ (17.0)	6 (22.7)	5.5 (20.8)	4.5‡ (17.1)
Moderate sections, such as retaining walls, abutments, piers, girders, beams	6.0 (22.7)	5.5 (20.8)	5.0‡ (18.9)	§	6.0 (22.7)	5.0‡ (18.9)
Exterior portions of heavy (mass) sections	6.5 (24.6)	5.5 (20.8)	5.0‡ (18.9)	§	6.0 (22.7)	5.0‡ (48.9)
Concrete deposited by tremie under water	—	5.0 (18.9)	5.0 (18.9)	—	5.0 (18.9)	5.0 (18.9)
Concrete slabs laid on the ground	6.0 (22.7)	—	—	§	—	—
Concrete protected from the weather, interiors of buildings, concrete below ground	§	—	—	§	—	—
Concrete which will later be protected by enclosure or backfill but which may be exposed to freezing and thawing for several years before such protection is offered	6.0 22.7	—	—	§	—	—

*Air-entrained concrete should be used under all conditions involving severe exposure, and may be used under mild exposure conditions to improve workability of the mixture.

†Soil or ground water containing sulfate concentrations of more than 0.2 percent.

‡When sulfate-resisting cement is used, maximum water-cement ratio may be increased by 0.5 gal (1.9 L) per bag.

§Water-cement ratio should be selected on the basis of strength and workability requirements.

Source: *Recommended Practice for Selecting Proportions for Concrete*, ACI publication 614-59, American Concrete Institute.

TABLE 14.26 ASTM Standard Reinforcing Bars

Bar size no.	Weight, lb/ft (N/mm)	Diameter, in (mm)	Cross-sectional area, in^2 (mm^2)	Perimeter, in (mm)
3	0.376 (0.005)	0.375 (9.5)	0.11 (70.9)	1.178 (29.9)
4	0.668 (0.010)	0.500 (12.7)	0.20 (129.0)	1.571 (39.9)
5	1.043 (0.015)	0.625 (15.9)	0.31 (200.0)	1.963 (49.9)
6	1.502 (0.022)	0.750 (19.1)	0.44 (283.9)	2.356 (59.8)
7	2.044 (0.030)	0.875 (22.2)	0.60 (387.1)	2.749 (69.8)
8	2.670 (0.039)	1.000 (25.4)	0.79 (509.7)	3.142 (79.8)
9	3.400 (0.050)	1.128 (28.7)	1.00 (645.2)	3.544 (90.0)
10	4.303 (0.063)	1.270 (32.3)	1.27 (819.4)	3.990 (101.3)
11	5.313 (0.078)	1.410 (35.8)	1.56 (1006.5)	4.430 (12.5)
14	7.650 (0.112)	1.693 (43.0)	2.25 (1451.6)	5.319 (135.1)
18	13.600 (0.199)	2.257 (57.3)	4.00 (2580.6)	7.091 (180.1)

Working-Stress Design

Design Loadings. In working-stress design, members should be designed to withstand actual service loads, consisting of dead loads, live loads, wind loads, and earthquake loads in any combination. Members subject to stress produced by wind or earthquake may be proportioned for stresses one-third greater than those given in Tables 14.27 and 14.28, provided that the section thus required is not less than required for dead plus live loads.

Working Stresses. Allowable unit stresses for concrete and reinforcing steel are given in Tables 14.27 and 14.28.

Beams. Concrete beams may be considered to be of three principal types: rectangular beams with tensile reinforcing only, T beams with tensile reinforcing only, and beams with tensile and compressive reinforcing.

Rectangular Beams with Tensile Reinforcing Only. This type of beam includes slabs [for which b = 12 in (304.8 mm) when the moment and shear are expressed per foot (meter) of width]. The stresses in the concrete and steel are

$$f_c = \frac{2M}{kjbd^2} \tag{14.76}$$

$$f_s = \frac{M}{A_s jd} = \frac{M}{pjbd^2} \tag{14.77}$$

TABLE 14.27 Allowable Stresses in Concrete

Description	Allowable stresses				
		For strength of concrete shown below			
	For any strength of concrete	$f_c' = 2500$ lb/in^2 (=17.2 MPa)	$f_c' = 3000$ lb/in^2 (=20.7 MPa)	$f_c' = 4000$ lb/in^2 (=27.0 MPa)	$f_c' = 5000$ lb/in^2 (≐34.5 MPa)
Modulus of elasticity ratio: n	$\frac{29{,}000{,}000}{w^{1}33\sqrt{f_c'}}$				
For concrete weighing 145 lb/ft^3 (7321 kg/m^3, n	—	10	9	8	7
Flexure: f_c					
Extreme fiber stress in compression, f_c	$0.45f_c'$	1125 (7.6)	1350 (9.3)	1800 (12.4)	2250 (15.5)
Extreme fiber stress, in tension in plain concrete footings and walls, f_c	$1.6\sqrt{f_c'}$	80 (0.55)	88 (0.61)	102 (0.70)	113 (0.78)
Shear: v (as a measure of diagonal tension at a distance d from the face of the support)					
Beams with no web reinforcement, v_c	$1.1\sqrt{f_c'}$	55 (0.38)	60 (0.41)	70 (0.48)	78 (0.54)
Joists with no web reinforcement, v_c	$1.2\sqrt{f_c'}$	61 (0.42)	66 (0.46)	77 (0.53)	86 (0.59)
Members with vertical or inclined web reinforcement or properly combined bent bars and vertical stirrups, v	$5\sqrt{f_c'}$	250 (1.7)	274 (1.91)	316 (2.2)	354 (2.4)
Slabs and footings (peripheral shear), v_c	$2\sqrt{f_c'}$	100 (0.69)	110 (0.76)	126 (0.87)	141 (0.97)
Bearing: f_c					
On full area	$0.25f_c'$	625 (4.3)	750 (5.2)	1000 (6.9)	1250 (0.6)
On one-third area or less*	$0.375f_c'$	938 (6.5)	1125 (7.8)	1500 (10.3)	1875 (12.9)

*This increase is permitted only when the least distance between the edges of the loaded and unloaded areas is a minimum of one-fourth of the parallel side dimension of the loaded area. The allowable bearing stress on a reasonably concentric area greater than one-third but less than the full area is to be interpolated between the values given.

Note: f_c' = compressive strength of concrete, lb/in^2 (kPa), n = ratio of modulus of elasticity of steel to that of concrete; w = weight of concrete, lb/ft^3 (kg/m^3).

Source: *Building Code Requirements for Reinforced Concrete*, American Concrete Institute.

TABLE 14.28 Allowable Stresses in Steel for Concrete Reinforcement

Types	lb/in^2	MPa
In tension		
For billet-steel or axle-steel concrete-reinforcing bars of structural grade	18,000	124.1
For main reinforcement, ⅜ in (9.5 mm) or less in diameter, in one-way slabs of not more than 12-ft (3.7-m) span, 50% of the minimum yield strength specified by the American Society for Testing Materials for the reinforcement used, but not to exceed	30,000	206.8
For deformed bars with a yield strength of 60,000 lb/in^2 (413.6 MPa) or more and in sizes No. 11 and smaller	24,000	165.5
For all other reinforcement	20,000	137.9
In compression, vertical column reinforcement		
Spiral columns, 40% of the minimum yield strength, but not to exceed	30,000	206.8
Tied columns, 85% of the value for spiral columns, but not to exceed	25,500	175.8
Composite and combination columns:		
Structural steel sections:		
For ASTM A36 steel	18,000	124.1
Cast-iron sections	10,000	68.9
Spirals [yield strength for use in Eq. (14.91)]		
Hot-rolled rods, intermediate grade	40,000	275.8
Hot-rolled rods, hard grade	50,000	344.7
Hot-rolled rods, ASTM A432 grade and cold-drawn wire	60,000	413.6

Source: Building Code Requirements for Reinforced Concrete, American Concrete Institute.

where b = width of beam [equals 12 in (304.8 mm) for slab], in (mm)
d = effective depth of beam, measured from compressive face of beam to centroid of tensile reinforcing (Fig. 14.17), in (mm)
M = bending moment, lb · in (N · m)
f_c = compressive stress in extreme fiber of concrete, lb/in^2 (kPa)
f_s = stress in reinforcement, lb/in^2 (kPa)
A_s = cross-sectional area of tensile reinforcing, in^2 (mm^2)
j = ratio of distance between centroid of compression and centroid of tension to depth d
k = ratio of depth of compression area to depth d
p = ratio of cross-sectional area of tensile reinforcing to area of the beam ($= A_s/bd$)

For approximate design purposes, j may be assumed to be ⅞ and k ⅜. For average structures, the following guides to the depth d of a reinforced-concrete beam may be used:

Member	d
Roof and floor slabs	$l/25$
Light beams	$l/15$
Heavy beams and girders	$l/12$–$l/10$

where l is the span of the beam or slab in inches (millimeters). The width of a beam should be at least $l/32$.

For a balanced design, one in which both the concrete and the steel are stressed to the maximum allowable stress, the following formulas may be used:

$$bd^2 = \frac{M}{K} \tag{14.78}$$

$$K = \tfrac{1}{2} f_c kj = pf_s j \tag{14.79}$$

Values of K, k, j, and p for commonly used stresses are given in Table 14.29.

T Beams with Tensile Reinforcing Only. When a concrete slab is constructed monolithically with the supporting concrete beams, a portion of the slab acts as the upper flange of the beam. The effective flange width should not exceed (1) one-fourth the span of the beam, (2) the width of the web portion of the beam plus 16 times the thickness of the slab, or (3) the center-to-center distance between beams. T beams where the upper flange is not a portion of a slab should have a flange thickness not less than one-half the width of the web and a flange width not more than 4 times the width of the web. For preliminary designs, the formulas given above for rectangular beams with tensile reinforcing only can be used, since the neutral axis is usually in or near the flange. The area of tensile reinforcing will usually be critical.

Beams with Tensile and Compressive Reinforcing. Beams with compressive reinforcing are generally used when the size of the beam is limited. The allowable beam dimensions are used in the formulas given above to determine the moment

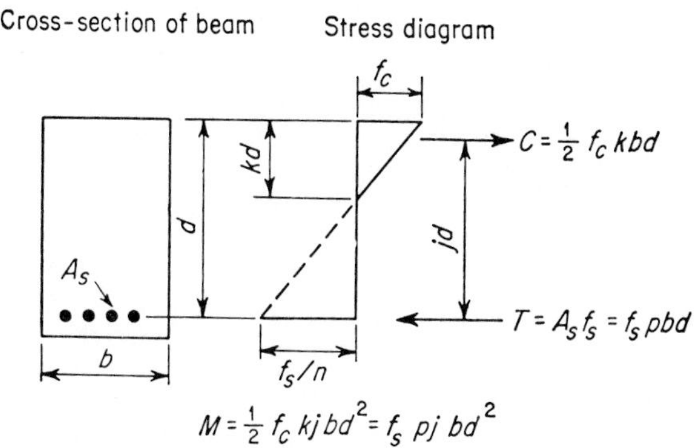

FIGURE 14.17 Rectangular concrete beam with tensile reinforcing only.

TABLE 14.29 Values of *K*, *k*, *j*, *p* for Rectangular Sections (Balanced Conditions)

f_c'	f_c	K	k	j	p
	f_s = 18,000 lb/in² (124.1 MPa)				
2500	1125	207	0.429	0.857	0.0134
3000	1350	248	0.429	0.857	0.0161
4000	1800	340	0.444	0.852	0.0222
5000	2250	444	0.467	0.845	0.0292
	f_s = 20,000 lb/in² (137.9 MPa)				
2500	1125	196	0.403	0.866	0.0113
3000	1350	236	0.403	0.866	0.0136
4000	1800	325	0.419	0.861	0.0188
5000	2250	422	0.440	0.853	0.0247
	f_s = 24,000 lb/in² (165.5 MPa)				
2500	1125	177	0.359	0.880	0.0084
3000	1350	213	0.359	0.880	0.0101
4000	1800	295	0.375	0.875	0.0141
5000	2250	387	0.396	0.868	0.0186

which could be carried by a beam without compressive reinforcement. The reinforcing requirements may then be approximately determined from

$$A_s = \frac{8M}{7f_s d} \tag{14.80}$$

$$A_{sc} = \frac{M - M'}{nf_c d} \tag{14.81}$$

where A_s = total cross-sectional area of tensile reinforcing, in² (mm²)
A_{sc} = cross-sectional area of compressive reinforcing, in² (mm²)
M = total bending moment, lb · in (N · m)
M' = bending moment which would be carried by beam of balanced design and same dimensions with tensile reinforcing only, lb · in (N · m)
n = ratio of modulus of elasticity of steel to that of concrete

Check of Stresses in Beam. Beams designed by the above approximate formulas should be checked to ensure that the actual stresses do not exceed the allowable, and that the reinforcing is not excessive. This can be accomplished by determining the moment of inertia of the beam. In this determination, the concrete below the neutral axis should not be considered as stressed, while the reinforcing steel should be transformed into an equivalent concrete section. For tensile reinforcing, this transformation is made by multiplying the area A_s by n, the ratio of the modulus of elasticity of steel to that of concrete. For compressive reinforcing, the area A_{sc} is multiplied by $(2n - 1)$. This factor includes allowances for the concrete in compression replaced by the compressive reinforcing and for the plastic flow of concrete. The neutral axis is then located by solving

$$\tfrac{1}{2}bc_c^2 + (2n - 1)A_{sc}c_{sc} = nA_s c_s \tag{14.82}$$

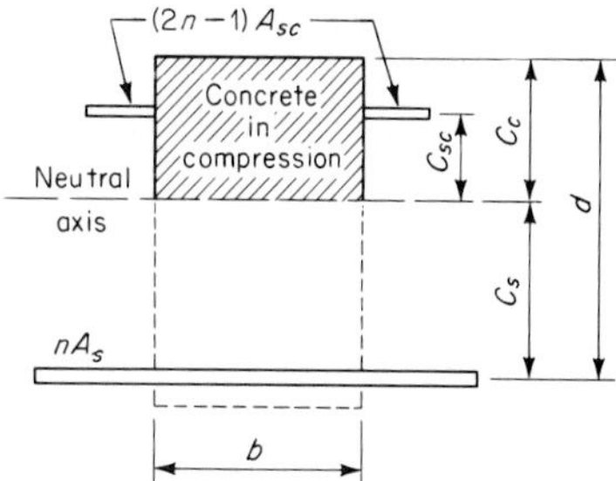

FIGURE 14.18 Transformed section of concrete beam.

for the unknowns c_c, c_{sc}, and c_s (Fig. 14.18). The moment of inertia of the transformed beam section is

$$I = \tfrac{1}{3}bc_c^3 + (2n - 1)A_{sc}c_{sc}^2 + nA_sc^2 \tag{14.83}$$

and the stresses are

$$f_c = \frac{Mc_c}{I} \tag{14.84}$$

$$f_{sc} = \frac{2nMc_{sc}}{I} \tag{14.85}$$

$$f_s = \frac{nMc_s}{I} \tag{14.86}$$

where f_c, f_{sc}, f_s = actual unit stresses in extreme fiber of concrete, in compressive reinforcing steel, and in tensile reinforcing steel, respectively, lb/in² (kPa)
c_c, c_{sc}, c_s = distances from neutral axis to face of concrete, to compressive reinforcing steel, and to tensile reinforcing steel, respectively, in (mm)
I = moment of inertia of transformed beam section, in⁴ (mm⁴)
b = beam width, in (mm)

and A_s, A_{sc}, M, and n are as defined for Eqs. (14.80) and (14.81).

Shear and Diagonal Tension in Beams. The shearing unit stress, as a measure of diagonal tension, in a reinforced-concrete beam is

$$v = \frac{V}{bd} \tag{14.87}$$

where v = shearing unit stress, lb/in² (kPa)
V = total shear, lb (N)
b = width of beam (for T beam use width of stem), in (mm)
d = effective depth of beam

If the value of the shearing unit stress as computed above exceeds the allowable shearing unit stress (v_c in Table 14.27) web reinforcement should be provided. Such reinforcement will usually consist of stirrups. The cross-sectional area required for a stirrup placed perpendicular to the longitudinal reinforcement is

$$A_v = \frac{(V - V')s}{f_v d} \tag{14.88}$$

where A_v = cross-sectional area of web reinforcement in distance s (measured parallel to longitudinal reinforcement), in^2 (mm^2)
f_v = allowable unit stress in web reinforcement, lb/in^2 (kPa)
V = total shear, lb (N)
V' = shear which concrete alone could carry ($=v_c bd$), lb (N)
s = spacing of stirrups in direction parallel to that of longitudinal reinforcing, in (mm)
d = effective depth, in (mm)

Stirrups should be so spaced that every 45° line extending from the middepth of the beam to the longitudinal tension bars is crossed by at least one stirrup. If the total shearing unit stress is in excess of $3\sqrt{f_c'}$ lb/in^2 (kPa), every such line should be crossed by at least two stirrups. The shear stress at any section should not exceed $5\sqrt{f_c'}$ lb/in^2 (kPa).

Bond and Anchorage for Reinforcing Bars. In beams in which the tensile reinforcing is parallel to the compression face, the bond stress on the bars is

$$u = \frac{V}{jd \sum_0} \tag{14.89}$$

where u = bond stress on surface of bar, lb/in^2 (kPa)
V = total shear, lb (N)
d = effective depth of beam, in (mm)
Σ_0 = sum of perimeters of tensile reinforcing bars, in (mm)

For preliminary design, the ratio j may be assumed to be ⅞. Bond stresses may not exceed the values shown in Table 14.30. To provide sufficient anchorage to develop the strength of reinforcing steel, tensile bars should be extended beyond the point at which they are needed to resist stress and should be terminated in a compression region, over a support, or with a hook.

Columns. The principal columns in a structure should have a minimum diameter of 10 in (254 mm) or, for rectangular columns, a minimum thickness of 8 in (203.2 mm) and a minimum gross cross-sectional area of 96 in^2 (619.4 cm^2).

Short Columns, Spiral Reinforcing. For short columns with closely spaced spiral reinforcing enclosing a circular concrete core reinforced with vertical bars, the maximum allowable load is

$$P = A_g(0.25f_c' + f_s p_g) \tag{14.90}$$

where P = total allowable axial load, lb (N)
A_g = gross cross-sectional area of column, in^2 (mm^2)
f_c' = compressive strength of concrete, lb/in^2 (kPa)

TABLE 14.30 Allowable Bond Stresses, lb/in^2 (kPa)

	Horizontal bars with more than 12 in (304.8 mm) of concrete cast below the bar	Other bars
Tension bars with sizes and deformations conforming to ASTM A305	$\frac{3.4\sqrt{f_c'}}{D}$ or 350, whichever is less	$\frac{4.8\sqrt{f_c'}}{D}$ or 500, whichever is less
Tension bars with sizes and deformations conforming to ASTM A408	$2.1\sqrt{f_c'}$	$3\sqrt{f_c'}$
Deformed compression bars	$6.5\sqrt{f_c'}$ or 400, whichever is less	$6.5\sqrt{f_c'}$ or 400, whichever is less
Plain bars	$1.7\sqrt{f_c'}$ or 160, whichever is less	$2.4\sqrt{f_c'}$ or 160, whichever is less

Note: f_c' = compressive strength of concrete, lb/in^2 (kPa); D = nominal diameter of bar, in (mm)

f_s = allowable stress in vertical concrete reinforcing, lb/in^2 (kPa), equal to 40 percent of the minimum yield strength, but not to exceed 30,000 lb/in^2 (206.8 MPa)

p_g = ratio of cross-sectional area of vertical reinforcing steel to gross area of column A_g

The ratio p_g should not be less than 0.01 nor more than 0.08. The minimum number of bars to be used is six, and the minimum size is No. 5. The spiral reinforcing to be used in a spirally reinforced column is

$$p_s = \frac{0.45(A_g/A_c - 1)f_c'}{f_y} \tag{14.91}$$

where p_s = ratio of spiral volume to concrete-core volume (out-to-out spiral)

A_c = cross-sectional area of column core (out-to-out spiral), in^2 (mm^2)

f_y = yield strength of spiral reinforcement, lb/in^2 (kPa), but not to exceed 60,000 lb/in^2 (413.6 MPa)

The center-to-center spacing of the spirals should not exceed one-sixth of the core diameter. The clear spacing between spirals should not exceed one-sixth the core diameter, or 3 in (76.2 mm), nor be less than 1⅜ in (34.9 mm), or 1½ times the maximum size of coarse aggregate used.

Short Columns with Ties. The maximum allowable load on short columns reinforced with longitudinal bars and separate lateral ties is 85 percent of that given in Eq. (14.90) for spirally reinforced columns. The ratio p_g for a tied column should not be less than 0.01 nor more than 0.08. The longitudinal reinforcing should consist of at least four bars, and the minimum size is No. 5.

Ties should be at least ¼ in (6.4 mm) in diameter, and should be spaced apart not over 16 bar diameters, 48 tie diameters, or the least dimension of the column.

Long Columns. Allowable column loads where compression governs design must be adjusted for column length, as follows:

1. If the ends of the column are fixed so that a point of contraflexure occurs between the ends, applied axial loads and moments should be divided by R from Eq. (14.92) (R cannot exceed 1.0):

$$R = \frac{1.32 - 0.006h}{r} \tag{14.92}$$

2. If relative lateral displacement of the ends of the column is prevented and the member is bent in single curvature, applied axial loads and moments should be divided by R from Eq. (14.93) (R cannot exceed 1.0):

$$R = \frac{1.07 - 0.008h}{r} \tag{14.93}$$

where h = unsupported length of column, in (mm)
r = radius of gyration of gross concrete area, in (mm)
= 0.30 times depth for rectangular column
= 0.25 times diameter for circular column
R = long-column load reduction factor

Applied axial load and moment when tension governs design should be similarly adjusted, except that the factor R varies linearly with the axial load from the values given by Eqs. (14.92) and (14.93) at the balanced condition, as defined by Eq. (14.99), to a value of 1.0 when the axial load is 0.

Combined Bending and Compression. The strength of a symmetrical column is controlled by compression if the equivalent axial load N has an eccentricity e in each principal direction no greater than given by Eq. (14.94) or (14.95) and by tension if e exceeds these values in either principal direction.

For spiral columns:

$$e_b = 0.43p_g mD_s + 0.14t \tag{14.94}$$

For tied columns:

$$e_b = (0.67p_g m + 0.17)d \tag{14.95}$$

where e = eccentricity, in (mm)
e_b = maximum permissible eccentricity, in (mm)
N = eccentric load normal to cross section of column
p_g = ratio of area of vertical reinforcement to gross concrete area
m = $f_y/0.85f_c'$
D_s = diameter of circle through centers of longitudinal reinforcement, in (mm)
t = diameter of column or overall depth of column, in (mm)
d = distance from extreme compression fiber to centroid of tension reinforcement, in (mm)
f_y = yield point of reinforcement, lb/in^2 (kPa)

Design of columns controlled by compression is based on Eq. (14.96), except that the allowable load N may not exceed the allowable load P [Eq. (14.90)] permitted when the column supports axial load only.

$$\frac{f_a}{F_a} + \frac{f_{bx}}{F_b} + \frac{f_{by}}{F_b} \leq 1.0 \tag{14.96}$$

where f_a = axial load divided by gross concrete area, lb/in² (kPa)
f_{bx}, f_{by} = bending moment about x and y axes, divided by section modulus of corresponding transformed uncracked section, lb/in² (kPa)
F_b = allowable bending stress permitted for bending alone, lb/in² (kPa)
F_a = $0.34(1 + p_g m)f_c'$

The allowable bending moment on columns controlled by tension varies linearly with the axial load from M_0 when the section is in pure bending to M_b when the axial load is N_b.

For spiral columns:

$$M_0 = 0.12 A_{st} f_y D_s \tag{14.97}$$

For tied columns:

$$M_0 = 0.40 A_s f_y (d - d') \tag{14.98}$$

where A_{st} = total area of longitudinal reinforcement, in² (mm²)
f_y = yield strength of reinforcement, lb/in² (kPa)
D_s = diameter of circle through centers of longitudinal reinforcement, in (mm)
A_s = area of tension reinforcement, in² (mm²)
d = distance from extreme compression fiber to centroid of tension reinforcement, in (mm)
d' = distance from extreme compression fiber to centroid of compression reinforcement, in (mm)

N_b and M_b are the axial load and moment at the balanced condition, i.e., when the eccentricity e equals e_b as determined from Eq. (14.94) or (14.95). At this condition, N_b and M_b should be determined from Eq. (14.96) so that

$$M_b = N_b e_b \tag{14.99}$$

When bending is about two axes,

$$\frac{M_x}{M_{0x}} + \frac{M_y}{M_{0y}} \leq 1 \tag{14.100}$$

where M_x and M_y are bending moments about the x and y axes, and M_{0x} and M_{0y} are the values of M_0 for bending about these axes.

Ultimate-Strength Design

Loadings. In ultimate-strength design, proportioning of members is based upon design loads determined from appropriate combinations of dead loads, live loads, wind loads, and earthquake loads. The design loads are determined by multiplying the actual loads by various safety factors. Design loads are

$$U = 1.4D + 1.7L \tag{14.101}$$

or, with wind load a factor,

$$U = 0.75(1.4D + 1.7L + 1.7W) \tag{14.102}$$

or, with L absent,

$$U = 0.9D + 1.3W \tag{14.103}$$

where U = design load, lb (N) (use maximum value from above equations)
D = dead loads, lb (N)
L = live loads plus impact, lb (N)
W = wind loads, lb (N)

For earthquake loading, substitute $1.1E$ for W in Eqs. (14.102) and (14.103), where E = earthquake loads, lb (N).

Capacity Reduction Factors. To compensate for variations in dimensions, crafting skill, and materials, the ultimate-strength equations include a capacity reduction factor ϕ. Typical values of ϕ are shown in Table 14.31 and should be applied to computed resisting forces and moments.

TABLE 14.31 Capacity Reduction Factors—Reinforced Concrete

Condition	ϕ Value
Bending, with or without axial tension	0.90
Axial tension	0.90
Axial compression, with or without combined bending	0.70*
Shear and torsion	0.85
Bearing on concrete	0.70

*May be increased to 0.75 when spiral reinforcement given by Eq. (14.91) is provided; values up to 0.90 are allowable under special conditions when axial compressive loading is small.

Assumptions in Design. (See Fig. 14.19.) In ultimate-strength design, basic assumptions include (1) maximum strain occurs at extreme compression fiber and is 0.003 in/in (mm/mm), (2) stress in reinforcing bars below the yield strength, f_y, is 29,000,000 lb/in^2 (199.9 GPa) times the steel strain, (3) strain in concrete is directly proportional to the distance from neutral axis, (4) tensile strength of concrete is neglected in flexural calculations, and (5) concrete stress intensity of $0.85f_c'$ is uniformly distributed over a depth $a = kc$, where c is the distance from the extreme compression fiber to the neutral axis. For $f_c' \leq 4000$ lb/in^2 (27.6 MPa), use $k = 0.85$; for each 1000 lb/in^2 (6.9 MPa) in excess of 4000 lb/in^2 (27.6 MPa), k is reduced by 0.05.

Design Values. For conventional structures, values normally used are compressive strength of concrete $f_c' = 4000$ lb/in^2 (27.6 MPa) and yield strength of reinforcement $f_y = 60{,}000$ lb/in^2 (413.6 MPa).

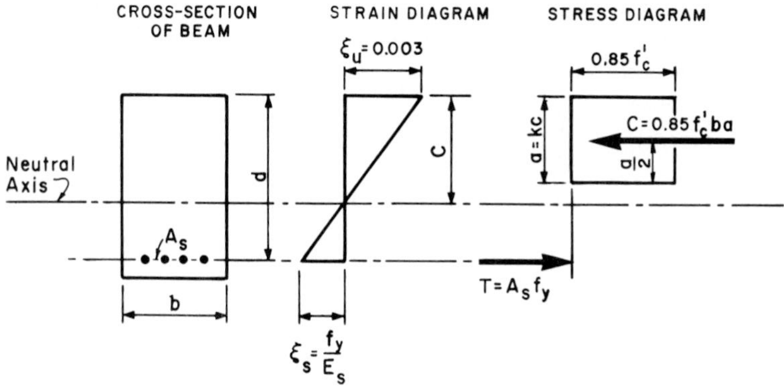

FIGURE 14.19 Assumptions in ultimate-strength design.

Rectangular Beams with Tensile Reinforcing Only. This type of beam includes slabs [for which $b = 12$ in (304.8 mm) when the moment and shear are expressed per foot of width]. The ultimate resisting moment is

$$M_u = \phi[bd^2 f_c' q(1 - 0.59q)] = \phi\left[A_s f_y\left(d - \frac{a}{2}\right)\right] \quad (14.104)$$

where A_s = area of tensile reinforcement, in^2 (mm^2)

a = depth of rectangular stress block = $A_s f_y / 0.85 f_c' b$, in (mm)

b = width of compressive face of flexural member, in (mm)

d = distance from extreme compression fiber to centroid of tension reinforcement, in (mm)

f_c' = compressive strength of concrete, lb/in^2 (kPa)

f_y = yield strength of reinforcement, lb/in^2 (kPa)

M_u = ultimate resisting moment, in · lb (N · m)

p = A_s/bd, $< 0.75 p_b$

p_b = reinforcement ratio producing balanced conditions*

q = $A_s f_y / bd f_c' = p f_y / f_c'$

ϕ = capacity reduction factor, Table 14.31

k = concrete stress intensity factor = 0.85 when f_c' = ≤ 4000 lb/in^2 (27.6 MPa)

T Beams with Tensile Reinforcing Only. For preliminary designs, the formulas given above for rectangular beams with tensile reinforcing only can be used, since the neutral axis is usually in or near the flange. The area of tensile reinforcing will usually be critical.

---

*Balanced conditions occur when the tension reinforcement is at its yield strength f_y and the concrete in compression is at its assumed ultimate strain of 0.003; balanced conditions exist when

$$p_b = \left(\frac{0.85 k f_c'}{f_y}\right)\left(\frac{87{,}000}{87{,}000 + f_y}\right) \quad (14.104a)$$

Beams with Tensile and Compressive Reinforcing. Beams with compression reinforcing are generally used when the size of the beam is limited, or where $p > 0.75p_b$. The ultimate resisting moment is then

$$M_u = \phi\left[(A_s - A_{sc})f_y\left(d - \frac{a}{2}\right) + A_{sc}f_y(d - d')\right] \tag{14.105}$$

where A_s = area of tensile reinforcement, in² (mm²)
A_{sc} = area of compressive reinforcement, in² (mm²)
a = depth of equivalent rectangular stress block = $(A_s - A_{sc})f_y/0.85f_c'b$, in (mm)
b = width of compressive face of flexural member, in (mm)
d = distance from extreme compression fiber to centroid of tension reinforcement, in (mm)
d' = distance from extreme compression fiber to centroid of compression steel, in (mm)
f_y = yield strength of reinforcement, lb/in² (kPa)
M_u = ultimate resisting moment, in · lb (N · m)
p = A_s/bd
p' = A_{sc}/bd $(p - p' < 0.75p_b)$
p_b = reinforcement ratio producing balanced conditions, Eq. (14.104*a*)
ϕ = capacity reduction factor, Table 14.30

Shear and Diagonal Tension in Beams. The ultimate shearing unit stress, as a measure of diagonal tension, in a reinforced-concrete beam is

$$v_u = \frac{V_u}{bd} \tag{14.106}$$

where v_u = ultimate shearing unit stress, lb/in² (kPa)
V_u = ultimate total shear, lb (N)
b = width of beam (for T beam use width of stem), in (mm)
d = effective depth of beam, in (mm)

For design, the maximum shear is considered to occur at a distance d from the face of the support. The shear stress carried by concrete v_c should not exceed $2\phi\sqrt{f_c'}$. Wherever the ultimate shear stress v_u exceeds the shear stress v_c, web reinforcement is mandatory. Such reinforcement will usually consist of stirrups. The cross-sectional area required for a stirrup placed perpendicular to the longitudinal reinforcement is

$$A_u = \frac{V_u s}{\phi f_y d} \tag{14.107}$$

where A_u = total area of web reinforcement in tension within a distance measured in a direction parallel to the longitudinal reinforcements, in² (mm²)
d = effective depth, in (mm)
f_y = yield strength of reinforcement, lb/in² (kPa)
s = spacing of stirrups, in (mm)
V_u = total ultimate shear, lb (N)
ϕ = capacity reduction factor, Table 14.31

The shear stress v_u should not exceed $10\phi\sqrt{f_c'}$ and f_y should not exceed 60,000 lb/in^2 (413.6 MPa). Stirrups should be anchored at both ends to be considered effective and so spaced that every 45° line extending from the middepth of the beam to the longitudinal tension bars is crossed by at least one stirrup.

Bond and Anchorage for Reinforcing Bars. In beams in which the tensile reinforcing is parallel to the compression face, the bond stress on the bars is

$$u_u = \frac{V_u \phi}{jd \sum_0} \tag{14.108}$$

where u_u = ultimate bond stress on surface of bar, lb/in^2 (kPa)
V_u = total ultimate shear, lb (N)
jd = distance between centroid of compression and centroid of tension, in (mm)
Σ_0 = sum of perimeters of tensile reinforcing bars, in (mm)
ϕ = capacity reduction factor, Table 14.31
d = effective depth of beam, in (mm)

For preliminary design, the ratio j may be assumed to be ⅞. Bond stresses should not exceed the values shown in Table 14.32. To provide sufficient anchorage to develop the strength of reinforcing steel, tensile bars should be extended beyond the point at which they are needed to resist stress and should be terminated in a compression region, over a support, or with a hook.

Columns—General. Columns should be designed for the axial load computed by using Eqs. (14.101), (14.102), and (14.103) and for the actual eccentricity e of the applied loading which should not be less than $0.05t$ for spirally reinforced columns or $0.10t$ for tied columns where t is the overall depth of a rectangular section or diameter of a circular section. The maximum load capacities given by Eqs. (14.109) through (14.114) apply only to short columns; adjustment factors for long columns are given by Eqs. (14.92) and (14.93).

TABLE 14.32 Maximum Bond Stresses, lb/in^2 (kPa)

	Horizontal bars with more than 12 in (304.8 mm) of concrete cast below the bar	Other bars
Tension bars with sizes and deformations conforming to ASTM A305	$\frac{6.7\sqrt{f_c'}}{D}$ or 560 lb/in^2 (413.6 MPa)	$\frac{9.5\sqrt{f_c'}}{D}$ or 800 lb/in^2 (5.5 MPa)*
Tension bars with sizes and deformations conforming to ASTM A408	$4.2\sqrt{f_c'}$	$6\sqrt{f_c'}$
Deformed compression bars	$13\sqrt{f_c'}$ or 800 lb/in^2 (5.5 MPa)*	$13\sqrt{f_c'}$ or 800 lb/in^2 (5.5 MPa)*
Plain bars	250 (1.7 MPa)	250 lb/in^2 (1.7 MPa)

*Use lower of two values.
Note: f_c' = compressive strength of concrete, lb/in^2 (kPa); D = nominal diameter of bar, in (mm).

Short Columns—Rectangular. The ultimate strength of short rectangular columns where the reinforcement is in one or two faces, each parallel to the axis of bending and all reinforcement in any one face is located at approximately the same distance from the axis of bending, is computed by the empirical formulas

$$P_u = \phi(0.85f_c'ba + A_{sc}f_y - A_sf_s) \tag{14.109}$$

$$P_ue' = \phi\left[0.85f_c'ba\left(d - \frac{a}{2}\right) + A_{sc}f_y(d - d')\right] \tag{14.110}$$

where a = depth of equivalent rectangular stress block, in (mm)
A_g = gross area of section, in^2 (mm^2)
A_s = area of tension reinforcement, in^2 (mm^2)
A_{sc} = area of compression reinforcement, in^2 (mm^2)
A_{st} = total area of longitudinal reinforcement, in^2 (mm^2)
b = width of compression face of flexural member, in (mm)
d = distance from extreme compression fiber to centroid of tension reinforcement, in (mm)
d' = distance from extreme compression fiber to centroid of compression reinforcement, in (mm)
D = overall diameter of circular section, in (mm)
D_s = diameter of the circle through centers of reinforcement arranged in a circular pattern, in (mm)
e = eccentricity of axial load at end of member measured from plastic centroid* of the section, calculated by conventional methods of frame analysis, in (mm)
e' = eccentricity of axial load at end of member measured from the centroid of the tension reinforcement, calculated by conventional methods of frame analysis, in (mm)
f_c' = compressive strength of concrete, lb/in^2 (kPa)
f_s = calculated stress in reinforcement when less than the yield strength f_y, lb/in^2 (kPa)
f_y = yield strength of reinforcement, lb/in^2 (kPa)
m = $f_y/0.85f_c'$
P_u = axial load capacity under combined axial load and bending, lb (N)
p_t = A_{st}/A_g
t = overall depth of a rectangular section or diameter of a circular section, in (mm)
ϕ = capacity reduction factor, Table 14.31

Short Columns—Circular. The ultimate strength of short circular columns with reinforcing bars circularly arranged is computed by the following empirical formulas.

When tension controls:

$$P_u = \phi\left\{0.85f_c'D^2\left[\sqrt{\left(\frac{0.85e}{D} - 0.38\right)^2 + \frac{p_tmD_s}{2.5D}} - \left(\frac{0.85e}{D} - 0.38\right)\right]\right\} \tag{14.111}$$

*Centroid of the resistance to load computed for assumptions that concrete is stressed uniformly to $0.25f_c'$ and steel is stressed uniformly to f_y.

When compression controls:

$$P_u = \phi\left\{\frac{A_{st}f_y}{(3e/D_s) + 1} + \frac{A_g f_c'}{[9.6De/(0.8D + 0.67D_s)^2] + 1.18}\right\} \tag{14.112}$$

Short Columns—Square. The ultimate strength of short square columns with reinforcing bars circularly arranged is computed by the following empirical formulas.

When tension controls:

$$P_u = \phi\left\{0.85btf_c'\left[\sqrt{\left(\frac{e}{t} - 0.5\right)^2 + 0.67\frac{D_s}{t}p_t m} - \left(\frac{e}{t} - 0.5\right)\right]\right\} \tag{14.113}$$

When compression controls

$$P_u = \phi\left\{\frac{A_{st}f_y}{(3e/D_s) + 1} + \frac{A_g f_c'}{[12te/(t + 0.67D_s)^2] + 1.18}\right\} \tag{14.114}$$

Equations (14.109) through (14.114) for the ultimate strength of columns subjected to combined axial compression and bending require assumption of a concrete cross section and reinforcement, computation of an allowable ultimate load and eccentricity, and comparison of the allowable loading with the design loading. This procedure usually requires iteration and may involve several assumptions regarding member size and amount of reinforcement. In practice, columns are usually designed by reference to extensive design tables, such as appear in *Concrete Reinforcing Steel Handbook* (Ref. 5), which lists allowable axial loads and eccentricities for a large selection of cross sections and reinforcement arrangements. An alternative method involves the development of a strain diagram for an assumed member with the concrete strain in the extreme compressive fiber at 0.003 in/in (0.08 mm/mm); use of the compressive stress in concrete as $0.85f_c'$ over an area determined by using the k factor described earlier under Assumptions in Design; computation of the tensile and compressive stresses in the reinforcement using strains compatible with the strains in the concrete; converting reinforcement strains to stresses using E_s = 29,000,000 lb/in^2 (413.6 MPa); and determination of resisting forces using appropriate stresses and areas. The axial load and eccentricity determined from the summation of the resisting forces represent the allowable ultimate loading for the assumed member and strain diagram.

Concrete Protection for Reinforcing Steel. The concrete protection for reinforcing steel should not be less than that given in Table 14.33.

CONSTRUCTION ENGINEERING

The construction of civil engineering projects involves the scheduling and planning of personnel, materials, and equipment. Consideration is given to means of access to the project area, delivery schedules for materials, availability of skilled and unskilled labor, and construction procedures suited to the project schedule and climate.

TABLE 14.33 Minimum Concrete Protection for Reinforcement

	Minimum concrete protection	
	in	mm
Concrete placed directly against the ground	3	76.2
Concrete exposed to weather:		
Bars less than no. 5 in size	1½	38.1
Bars no. 5 in size or larger	2	50.8
Concrete not exposed to ground or weather:		
Slabs and walls	¾	19.1
Beams, girders, and columns	1½	38.1
Concrete exposed to sea water	4	101.6

46. Earthmoving. The volume and density of earth changes when it is excavated, hauled, and compacted as fill. Payment for earthmoving may be based on bank-measure volume (in a borrow pit or cut prior to excavation) or fill volume (after compaction). Loose measure volume (after excavation but before compaction) is used to determine haul payloads. The change from the bank-measure volume to the loose volume in fill after compaction is termed *shrinkage*; the change from the bank-measure volume to the loose volume before compaction is termed-*swell*. Moisture content of soils directly affects shrinkage and swell allowances. Typical volume relationships for various materials are shown below:

Typical soil type	Volume factors		
	Bank measure	Loose	Compacted fill
Sand	1.00	1.10	0.95
Loam	1.00	1.25	0.90
Clay	1.00	1.45	0.90
Rock (blasted)	1.00	1.50	1.30

47. Pile Driving. There are three major types of pile hammers, classified by power source and action:

1. Single-acting hammers, raised by compressed air, steam, or other means, and dropped
2. Double-acting hammers, similar to single-acting but with power-assisted downward stroke
3. Vibratory drivers, transmitting a vibration to the pile through clamps

Variables governing the selection of a pile hammer include type and weight of pile, ground conditions, penetration requirements, and soil characteristics. For pile driving, soils can be separated into two groups: cohesive and noncohesive. Table 14.34 is a guide to pile hammer selection.

48. Pavements. The two basic types of pavements are rigid and flexible. The principal rigid type is concrete with or without a bituminous wearing sur-

TABLE 14.34 Guide to Selection of Pile Hammers

Pile type	Cohesive soils		Noncohesive soils	
	Soft	Stiff	Soft	Stiff
Wood	DA	SA	DA	SA
Open pipe	V-DA	DA-SA	V-DA	V-DA-SA
Closed pipe	DA-SA	SA	DA	SA
H-pile	V-DA	DA-SA	V-DA	V-DA-SA
Sheet	V-DA	DA-SA	V	V-DA
Concrete	DA-SA	SA	DA	SA

V = vibratory, DA = double-acting, SA = single-acting.

face. Flexible types consist of a variable depth of bituminous material mixed with aggregate with or without a separate bituminous surface course. Both types require a base course of granular material. Soil-cement mixtures may be used to improve the bearing value of marginal or substandard granular base or subbase materials.

Rigid Types. Concrete slabs are designed to distribute wheel loads over large areas of subgrade. Typical concrete thicknesses for primary roads range from 8 to 12 in (203 to 305 mm). Transverse joints are provided to relieve temperature stresses, and longitudinal joints are formed between lanes to control irregular cracking and mark traffic lanes. Pavement concrete is usually reinforced with wire mesh. A properly placed concrete pavement has a probable life of 25 years or more.

Flexible Types. Types of bituminous construction most common for pavements are surface treatment, penetration macadam, mixed-in-place, and plant mix. The usual thicknesses of bituminous surface courses and their probable life expectancies are

Type	Thickness		Probable life, years
	in	mm	
Surface treatment	½–1	12.7–25.4	5–8
Penetration macadam	2½–3	63.5–76.2	15–17
Mixed-in-place	1½–3	38.1–76.2	10–12
Plant mix (asphaltic concrete)	2–3 per course	50.8–76.2	15–20

Bituminous paving materials are solutions of asphalt cement in a solvent (rapid-curing, RC, and medium-curing, MC) or straight-run distillations or blends (slow-curing, SC). Within these three grades, additional classifications are made on the basis of viscosity. The relative hardness of the asphalt cement varies from high for the RC grades, to medium for MC, to low for the SC types. The type of bituminous construction selected is influenced by thickness of the required surface course, by the climate, materials, and equipment available, and by the relative costs.

ECONOMIC, SOCIAL, AND ENVIRONMENTAL CONSIDERATIONS

The scope of civil engineering includes many types of capital projects where the expenditure of funds must be justified by economic and financial analyses. Major civil engineering projects usually require analysis to determine their effect on community and regional social structures and on the environment.

49. Economic Analyses. Economic analyses compare the economic benefits of a project with its economic costs, while financial analyses compare the monetary return from a project with its financial costs. Economic analyses may be summarized in terms of a benefit-cost ratio using the formula

$$B/C = \frac{\Delta t + \Delta u + \Delta a}{\Delta c + \Delta m} \tag{14.115}$$

where B/C = benefit-cost ratio

Δt = present worth of reduction in user time costs (computed on annual basis over the life of a project, discounted to present worth)

Δu = present worth of reduction in user operating costs (computed on annual basis over the life of a project, discounted to present worth)

Δa = present worth of reduction in user accident and damage costs (computed on annual basis over the life of a project, discounted to present worth)

Δc = present worth of incremental costs of initial and recurring capital investments over the life of a project

Δm = present worth of increase in maintenance and operating costs (computed on annual basis over the life of a project, discounted to present worth)

In some cases the benefits of an engineering project are determined as the net value of increased farm or industrial output attributable to the project.

In Eq. (14.115) the interest cost to be used is the opportunity cost of capital or market rate of interest; the opportunity cost of capital generally ranges from 8 to 15 percent. The life of a project is determined from consideration of the inherent durability of project components but should not exceed the period in which technological obsolescence may occur. For most construction projects, the life will range up to 30 years, but longer lives may be used for major projects. The benefit-cost ratio (B/C) indicates the economic justification of a project. A project with a benefit-cost ratio of one is marginal; values higher than one indicate greater justification.

An internal rate of return calculation is often used instead of a benefit-cost ratio. The internal rate of return is determined by setting B/C in Eq. (14.115) equal to unity and determining the interest rate used for discounting at which the numerator equals the denominator. The interest rate is then compared with the opportunity cost of capital.

Benefit-cost or internal rate of return calculations should be made on an incremental basis for comparison among alternatives. One alternative which must be included is the "do nothing" alternative. Incremental analyses among alternatives permit the determination of the incremental return obtained from successively greater capital investments.

Economic analyses usually involve comparison of benefits which accrue to the public with costs which are incurred by governments. Financial analyses involve the same type of computation but consideration is limited to revenues and costs received by and incurred by a private or quasi-public organization.

50. Social and Environmental Factors. Civil engineering projects may have significant effects on the social structure of a community, a region, or a nation. Projects may affect population growth, quality of life, community relationships, and distribution of income. Analyses of project effects involve consideration of changes in demographic or cultural patterns to determine the positive or negative benefits of project development.

Civil engineering projects normally cause some environmental changes. Project planning should maximize positive environmental effects and minimize negative effects. Such effects may be short term, generally limited to the construction period, or long term, occurring through the life of the project.

For social and environmental analyses, the impact of a project normally involves consideration of the following:

1. Biological resources—wildlife, vegetation, fish, etc.
2. Air quality and pollution
3. Noise control
4. Water quality and flow
   a. Ground water
   b. Streams and rivers
   c. Lakes, beaches and shores, and estuaries and other larger bodies of water
5. Commitment of natural resources—types, approximate amount, and probable source of needed construction materials
6. Aesthetics
   a. Visual quality of the facility
   b. Special architectural or engineering features
7. Open space and recreational opportunities
8. Archaeological or paleontological resources, with consideration of salvage operations during construction
9. Historical areas or sites
10. Future land uses of surrounding area, including neighborhood compatibility
11. Potential for joint development of public or private facilities
12. Comprehensive plans for the development of the region or area
13. Programs of public agencies
14. Displacements of persons, businesses, and farms and proposed relocation treatment
15. Residential and business property values
16. Local area tax base
17. Local churches, clubs, and other social activities
18. Community services, such as police and fire protection, health services, public and private utilities, and mail
19. Educational institutions, including modification of attendance boundaries or the actual school environment
20. Area employment in terms of number and location of jobs

Consideration should be given to the interactions among these factors within a complete system.

Environmental and social considerations are weighed in conjunction with economic and financial factors in the assessment of project justification and selection.

HYDRAULIC ENGINEERING*

HYDRAULIC TURBINES

1. Types. There are two types of turbine in general use, the *tangential water wheel* and the *reaction turbine*. The reaction type is further subdivided into the *mixed-flow reaction* and the *propeller* types. While the power and r/min are also involved in choice of type (see Sec. 3 below) the tangential water wheel is essentially a high-head machine, the mixed-flow reaction turbine is a medium-head machine, and the propeller type is a low-head machine. The *Kaplan turbine* is a propeller type in which the blades may be rotated to give variable pitch.

2. Homologous Machines. When two machines are identical except as to size, they are said to be *homologous*. In a line of homologous turbines all of that line have the same proportions for their respective parts, and the parts are similarly located. For such a line of machines, the performance of one having been found by test, the performance of the others at any head may be found as follows: Let N = r/min, q = discharge in ft^3/s (m^3/s), P = brake hp (kW), D = diameter in in (cm), h = head in ft (m). Then

$$N = \frac{1840\phi\sqrt{h}}{D} \tag{14.116}$$

$$q = k_1 D^2 \sqrt{h} \tag{14.117}$$

$$P = k_2 D^2 h^{1.5} \tag{14.118}$$

The constants ϕ, k_1, and k_2 are typical of that line of machines, and are obtained from the test of the first machine.

3. Specific Speed. Another turbine constant of great utility is variously called characteristic speed, type characteristic, or specific speed, N_s:

$$N_s = \frac{N\sqrt{P}}{h^{1.25}} \tag{14.119}$$

The value of N_s is a criterion of type, as follows:

Tangential water wheel, N_s up to 6
Mixed-flow turbine, N_s = 20 to 120
Propeller turbine, N_s = 120 to 200 or higher

Experience has shown that, for satisfactory operation of the mixed-flow turbine, the specific speed should not exceed the value

$$N_s = \frac{5050}{h + 32} + 19 \tag{14.120}$$

*From *General Engineering Handbook*, 2d ed., by C. E. O'Rourke. Copyright © 1940. Used by permission of McGraw-Hill, Inc. All rights reserved. Updated and metricated by the editors.

With values of N_s greater than that obtained from this empirical equation, the metal of the runner may be eaten. This phenomenon, called *pitting*, is not very well understood. In addition to being a function of the head, it appears to be influenced by draft head or the height above tail water level at which the machine is placed. High velocities and low pressures are apparently conducive to pitting. The propeller-type turbine has been found to be practically free from this trouble, if the machine is not set too high.

4. Efficiency. A well-designed tangential water wheel will operate with an efficiency of 85 to 90 percent, in the usual sizes found in power plants. The mixed-flow reaction turbine, with a specific speed between 40 and 80, may be expected to have an efficiency of 88 to 94 percent. The lower value has been obtained even with small-size laboratory machines. The propeller-type machines, with specific speeds between 100 and 150, have shown efficiencies as high as 90 percent. For any type, the higher efficiency will be associated with the larger machine operating under moderate head for its class. In order to estimate the efficiency of a large machine when a test on a small homologous model is available, the following empirical formula by L. F. Moody is extensively used:

$$E = 1 - (1 - e)\left(\frac{d}{D}\right)^{1/4}\left(\frac{h}{H}\right)^{1/10} \tag{14.121}$$

where E is maximum efficiency of large turbine; e is maximum efficiency of small turbine; d and D are diameters of small and large turbines, respectively; h is the head under which the model is tested; and H is the head under which the large machine will operate. Another formula sometimes used is the Cammerer formula:

$$E = 1 - (1 - e)\frac{1.4 + (1/D)}{1.4 + (1/d)} \tag{14.122}$$

In those cases where the two formulas do not agree, the Moody formula is likely to give results which are too high, while the Cammerer results will be too low.

5. Performance Curves at Constant Head. The complete performance of a turbine is obtained by operating it at or near constant head and at all speeds from standstill to runaway. Consideration of Eq. (14.116) shows that for a given machine, an increase in speed is equivalent to a decrease in head. Hence, for convenience in laboratory work, performance under variable head is determined from tests at approximately constant head but with variable speed.

In general, the relations between torque, power, speed, discharge, etc., are of the same nature for any type of turbine. The shape of the curve expressing the relation between any two of these quantities will change with change of type, but will be of the same general appearance.

The curves usually plotted from a complete turbine test will be shown. The machine is a small laboratory model, with 15-in (38.1-cm) diameter runner. By means of Eqs. (14.116), (14.117), and (14.118) all data have been reduced to 1-ft (0.3-m) head for the purpose of getting a convenient constant value. The actual tests were made at 19- to 21-ft (5.8- to 6.4-m) head. Figure 14.20 shows the relation between torque and speed, each curve being for a separate gate opening. Figure 14.21 shows efficiency-speed relations, also for constant gate openings. Figure 14.22 is a plot of all turbine characteristics. This is called the *characteristic curve*. Ordinates are r/min and abscissas are discharge in s · ft (s · m). The dash lines show the speed-discharge relation at constant gate opening. The solid lines of contour appearance are curves of equal efficiency, iso-efficiency curves. From this curve, the operation of the machine at any head, speed, and load may be determined.

To illustrate its use, find the best performance of this machine at a head of 25 ft (7.6 m). From Fig. 14.21 the maximum efficiency comes at a speed of 106 r/min at 1-ft (0.3-m) head. Hence the best speed at 25-ft (7.6-m) head is 106 × $\sqrt{25}$ = 530 r/min. At full gate opening for this speed, the discharge is 2.2 s · ft (0.67 s · m); efficiency is 87 percent. For 25-ft (7.6-m) head, then, discharge is 2.2 × $\sqrt{25}$ = 11.0 s · ft (3.4 s · m); efficiency is the same; brake horse-

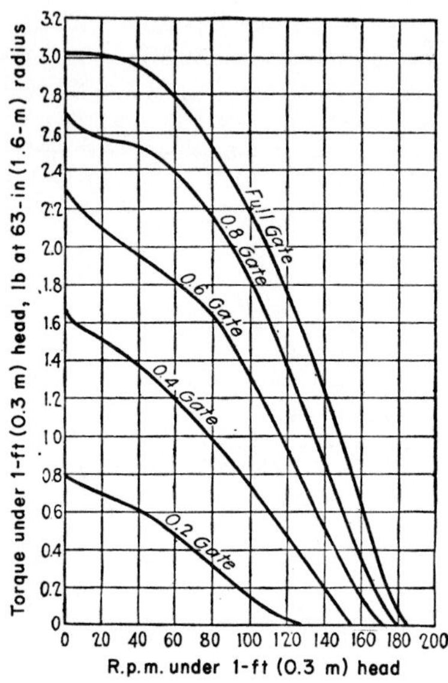

FIGURE 14.20 Torque-speed curves at constant gate opening.

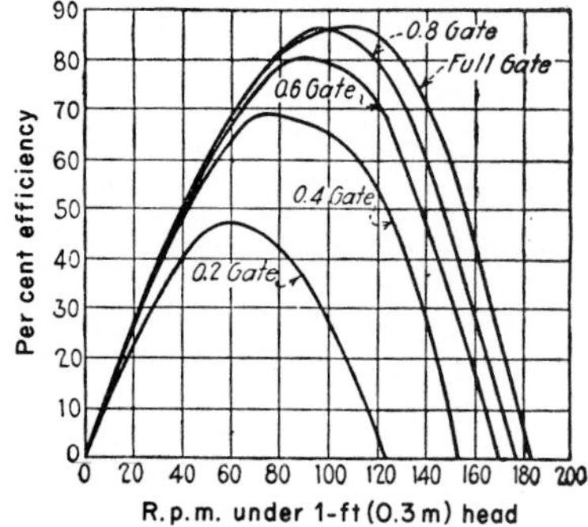

FIGURE 14.21 Efficiency curves at constant gate opening.

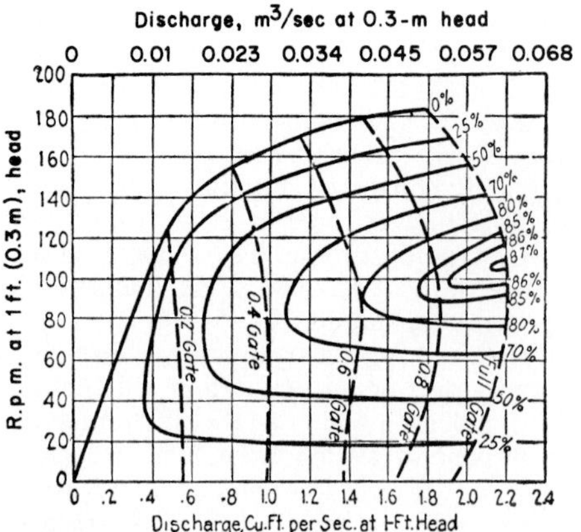

FIGURE 14.22 Characteristic curve for reaction turbine.

power is discharge times head times efficiency, or 11.0 × 25 × 0.87 ÷ 8.8 = 27.2 hp (20.3 kW). At other gate openings similar computations give results which are plotted in Fig. 14.23, with load as abscissa. The efficiency curve there shown gives the data usually sought for in a power plant operating at constant head. For plants operating with variable head, similar curves can be plotted each for a different head, as in Fig. 14.24.

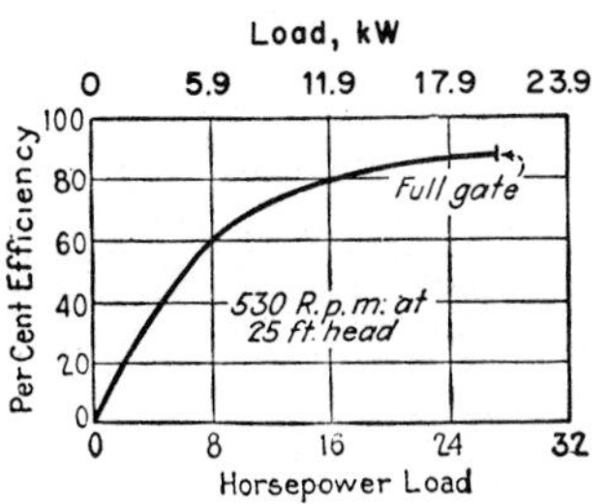

FIGURE 14.23 Efficiency at constant speed and head.

In Figs. 14.25 and 14.26 are shown portions of the characteristic curves for an impulse type and propeller type, respectively.

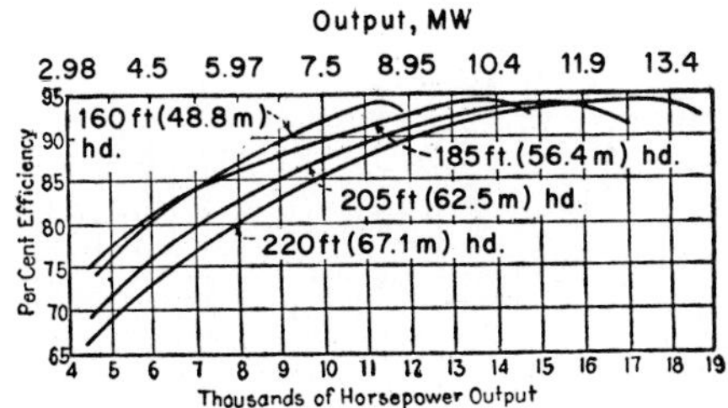

FIGURE 14.24 Efficiency at varying head. *(Courtesy S. Morgan Smith Co.)*

6. Diameter of Turbine. In order to determine the size of machine required in any reaction-turbine installation, Fig. 14.27 may be used. For the given head, the specific speed to be used may be determined from Eq. (14.120). Reference to Fig. 14.26 will then give the principal dimensions required to determine the plan dimensions of the installation. Since the figure shows the diameters of the machine which will deliver 1 hp (0.75 kW) at 1-ft (0.3-m) head, the actual diameters may be found by the use of Eq. (14.118), which for this case may be rearranged as follows: $D = \sqrt{P}/\sqrt[4]{h^3}$ multiplied by the diameter given in Fig. 14.27.

For the impulse wheel an approximate diameter of the impulse circle can be found from $D = 860\sqrt{P}/N_s\sqrt[4]{h^3}$. The impulse circle diameter is twice the perpendicular distance from the center of the shaft to the jet from the nozzle. The maximum overall diameter of the wheel will be about 15 percent greater, but varies somewhat with the design. The diameter of the jet from the nozzle is, then,

$$d \text{ (in)} = \frac{DN_s}{55} \qquad [= 2.5d \text{ (cm)}]$$

In order to find the speed at which any of these machines will run, Eq. (14.119) may be used.

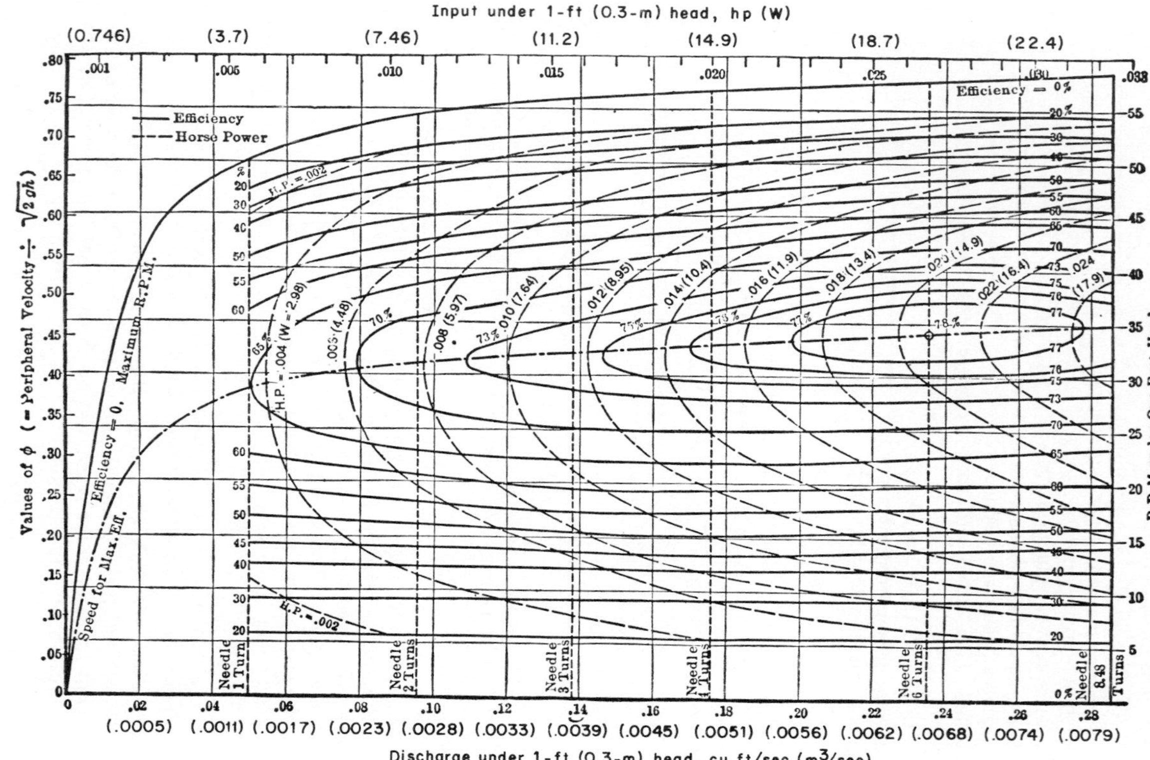

FIGURE 14.25 Characteristic curve for a 24-in (60-cm) tangential water wheel.

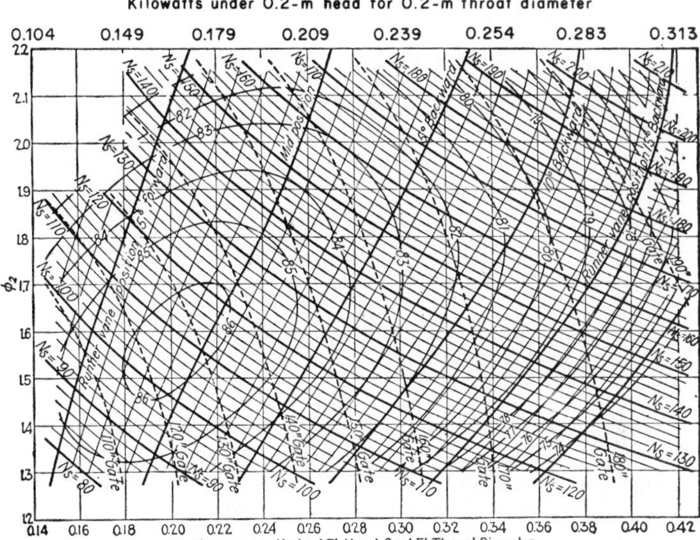

FIGURE 14.26 Characteristic curve for propeller turbine. *(Courtesy I. P. Morris & De La Vergne, Inc.)*

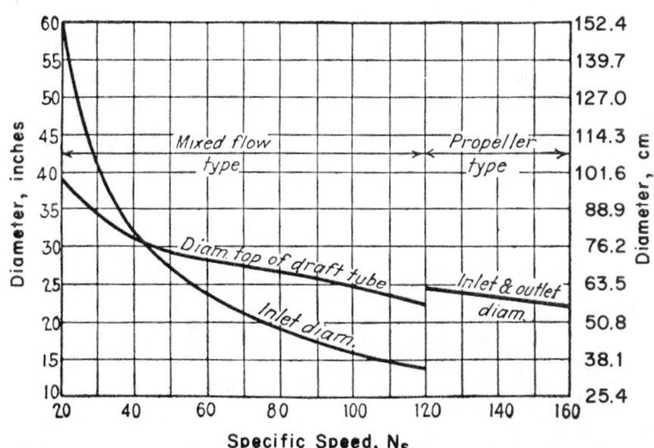

FIGURE 14.27 Turbine dimensions for 1 hp at 1-ft head (0.75 kW at 0.3 m).

7. Turbine Setting. The purpose of the turbine setting is to conduct the water to the machine and get it away from the machine, back to the stream, with as little loss as possible.

For the impulse wheel this means a pipe line attached to the base of the nozzle, the nozzle, and a case which is large enough so that water after leaving the buckets does not splash back into the wheel but falls down into the tailrace (see Fig. 14.28).

For the reaction types, there are two general arrangements for leading the wa-

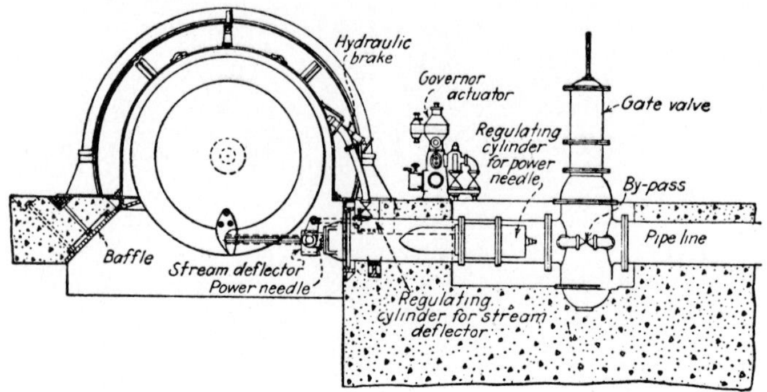

FIGURE 14.28 Typical setting of impulse turbine. *(Courtesy of Allis-Chalmers Manufacturing Co.)*

ter to the machine. One of these is the *open-flume setting* and the other is the *encased setting.*

The *open-flume setting* is illustrated in Fig. 14.29. Here the *speed ring* for the support of loads from above, and the *guide vanes* which direct the water into the runner, are placed in an open flume. More generally the generator load is carried by the powerhouse structure and the speed ring is omitted, as is the case in Fig. 14.29. This type of setting is common for low heads but is rarely economical for heads greater than about 30 ft (9.1 m), with machines of over 2000 hp (1.49 MW). Water-passage areas should be ample to ensure low water velocities if good efficiency is to be expected. Velocities should not exceed 3 ft/s (0.9 m/s), and for heads of 10 ft (3.0 m) should be only 2 ft/s (0.6 m/s). Distance between the centers of units should not be less than about three times the diameter of the guide-vane circle. With this type of setting the turbine shaft is practically always vertical. Almost all reaction-type turbines are set with a vertical shaft because of the higher efficiency of this arrangement.

The *encased setting* consists of a formed passageway of minimum dimensions for efficient performance. For low heads, the concrete of the powerhouse substructure is often made to serve as a casing for the turbines. For heads above about 100 ft (30 m), the amount of steel required for reinforcing a casing becomes so great, and the water velocities so high, that plate-steel cases are usually more economical. Both riveted and welded cases are found in this class. For the higher heads, cast iron or cast steel is used. Manufacturers are now prepared to build reaction turbines for heads as high as 1000 ft (304 m). Figure 14.30 shows a section of an installation where the case is formed in the concrete of the powerhouse substructure.

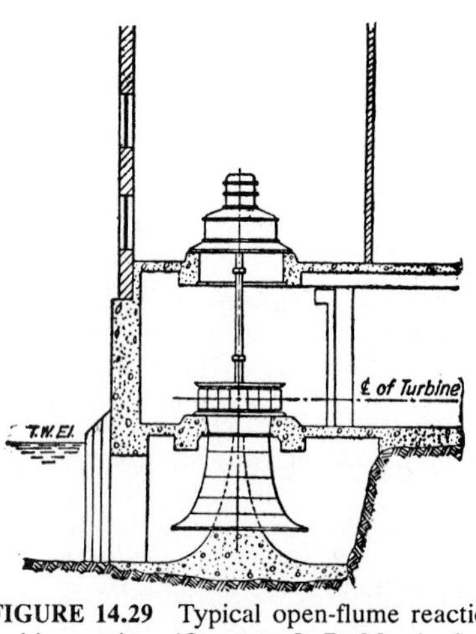

FIGURE 14.29 Typical open-flume reaction-turbine setting. *(Courtesy I. P. Morris & De La Vergne, Inc.)*

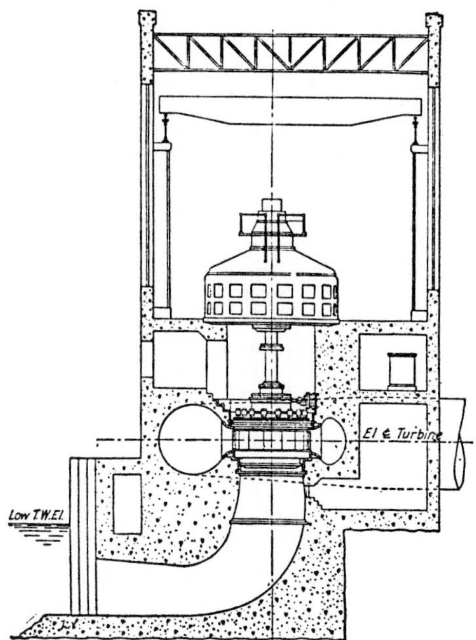

FIGURE 14.30 Typical concrete spiral setting of reaction turbine. *(Courtesy S. Morgan Smith Co.)*

Figure 14.31 shows a similar installation for higher head with a steel case for the turbine.

8. Draft Tubes. With the advent of the high-speed mixed-flow turbine, the velocities of water leaving the machine were high enough to carry away a very appreciable amount of energy and thus reduce the overall plant efficiency. The then common draft tubes were not very effective because of high losses in the passages. Research received an added impetus when the propeller type of runner was introduced. In this type there is not only a high axial component of velocity at discharge but also a high tangential component; these together give spiral flow in the draft tube. Two types of draft tube were found which increased markedly the efficiency and power capacity of the high-speed turbine. These are known as the *Moody spreading draft tube* (Fig. 14.32) and the *White hydraucone* (Fig. 14.33). Further research resulted in new forms of the old elbow type of draft tube (Fig. 14.34) which have in many cases given results equally as good as those from any other type. The use of tests on models of turbine settings is becoming quite common where a small increase in efficiency or power will justify the cost of the research. In these model tests various types of casing or draft tube can be compared and modified to suit the given conditions.

9. Selection of Turbines. The specific speed, as defined by Eq. (14.119), is all that is necessary to determine the type of turbine to be used. Since both the *speed* and *power* of a unit can be varied for any given installation, it is the choice

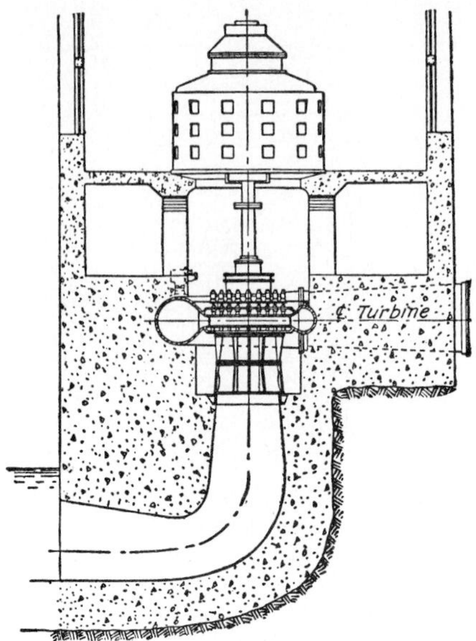

FIGURE 14.31 Typical steel case setting of reaction turbine. *(Courtesy S. Morgan Smith Co.)*

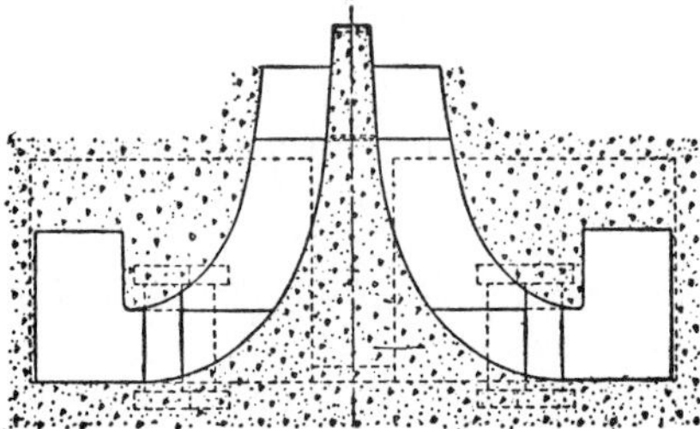

FIGURE 14.32 Moody spreading draft tube. *(Courtesy I. P. Morris & De La Vergne, Inc.)*

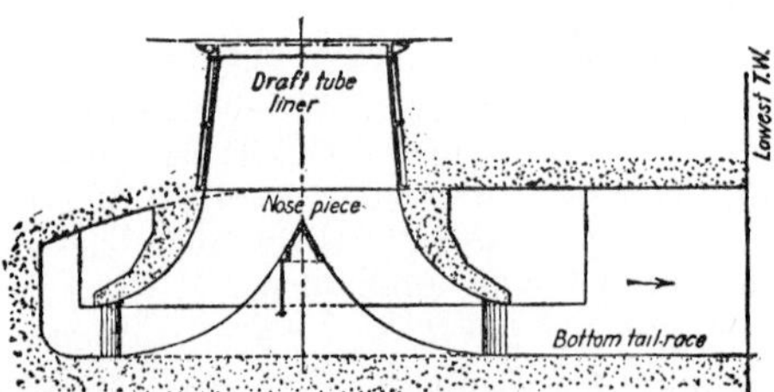

FIGURE 14.33 White hydraucone draft tube. *(Courtesy Allis-Chalmers Manufacturing Co.)*

of these which will determine the type of turbine. The *power* of a unit will be selected from considerations of size of the development and the load to be served. From the standpoint of stream flow, the size of unit should be small enough so as to run at or near full load or best efficiency during the low-water season when the flow is a minimum. Having determined from the stream-flow record the minimum load which the plant is to carry, the size of unit should be but little larger than this. It should be small enough, on the other hand, so that not less than about three machines will be required altogether unless the plant is one in a large system in which breakdown of one unit will not disturb the ability to carry the load. With an isolated water-power plant, one extra machine should be provided for standby in case of damage or repairs to any one machine. Thus, for a 100,000-hp (74.6-MW) plant, install 5 units at 25,000 hp (18.7 MW), or 6 units at 20,000 hp (14.9 MW) or 11 units at 10,000 hp (7.5 MW). Unless the smallest size is required on account of minimum stream flow, as noted above, that combination which gives the cheapest installation should be chosen. It should be noted that the larger sizes are ordinarily more efficient, and therefore a somewhat greater investment may be warranted in order to secure a greater plant capacity with the given stream flow.

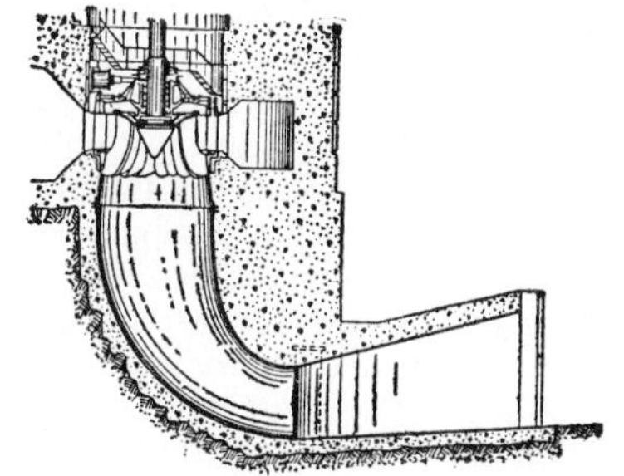

FIGURE 14.34 Elbow-type draft tube. *(Courtesy Allied Engineers, Inc.)*

The size of unit having been chosen, the *speed* is chosen either to accommodate the load to be driven or to require a type of turbine best suited for the installation. For a generator directly connected, as is the case in a great majority of hydraulic plants, the higher the speed the cheaper will be the generator for heads under 60 ft (18.3 m). In these cases the speed is normally selected for hydraulic reasons. For the mixed-flow reaction turbine the specific speed should not exceed the value given by Eq. (14.120); for the propeller-type, present practice limits its specific speed to a maximum value of about 175, although higher values can be had. Where a definite choice can not be made on this basis alone, cost and efficiency must be estimated and used as a basis for a final decision. For heads above 60 ft (18.3 m), only the mixed-flow type is now being used, and installations have been made under heads as high as 850 ft (259 m). Within this range, the speed is normally selected by means of Eq. (14.120), except for the higher heads, with which lower speeds are necessary because of generator limitations. When the tangential water wheel is used, the limited economical range of specific speed fixes the revolutions per minute within rather narrow limits.

In the very large sizes of reaction turbine, *transportation facilities* have sometimes limited the size of machine. Engineers have been rather hesitant about accepting runners built in more than one piece, so that railroad clearances have been the limiting factor in some cases. Propeller-type runners are better adapted mechanically to construction in several parts, and these have been used in sizes which will pass 10,000 ft^3 of water/s (283 m^3/s). The mixed-flow type is rarely economical in sizes larger than 7000 ft^3/s (198 m^3/s). The maximum power available in a single machine is about 75,000 hp (56 MW), although manufacturers are prepared to build machines of 100,000 hp (74.6 MW), where the head makes a machine of this size safe from pitting.

10. Testing of Turbines. The usual purpose for which tests are made is to determine the efficiency and power of the machine for comparison with the guarantees given by the manufacturer. Owing to the cost and difficulty of measuring the rate of discharge, in the past only the larger machines were tested in place. In some cases a small model of the big machine has been built and tested in a laboratory. The results are then stepped up in accordance with the data of Secs. 2 and 4. With the advent of the salt-velocity and pressure-time methods, the cost of measuring rate of discharge has been so reduced that tests of completed machines in place have been quite commonly made. Besides rate of discharge, head, speed, and power must be measured and efficiency computed from these. Details of test methods are available in the *Test Code for Hydraulic Power Plants and Their Equipment*, published by the American Society of Mechanical Engineers. The results of such a test may be plotted as shown in Figs. 14.23 and 14.24.

11. Speed Regulation of Hydraulic Turbines. For practically all installations it is necessary to maintain approximately constant speed for the hydraulic turbines. With the advent of the electric clock controlled by frequency regulation of an alternating-current system, the method of regulation has been changed radically. Before this, regulation was done by maintaining a constant rate of rotation of the turbine through the agency of the usual flyball mechanism driven from the turbine shaft either mechanically or electrically. This method is not sensitive enough for clock drive. The newer system uses, in addition, a master-clock pendulum which controls the machine speed to maintain correct time of the electric clocks.

In general, there are three functions which speed-regulation devices may be called upon to perform: first, to hold speed as nearly constant as possible immediately following a change of load; second, to regulate the speeds of the units so that the generators connected in parallel will divide the load in proportion to their capacities without continual shifting of the load between machines; and third, frequency regulation to maintain constant frequency at all times.

The first of these functions is performed by a flyball-actuated system. Following a load change comes a speed change which moves the collar on the flyballs. This actuates a relay valve which admits oil under pressure to the cylinders attached to the turbine-gate mechanism of a reaction-type machine or to the nozzle of a tangential water wheel. In this way the flow of water to the turbine is adjusted to that required to carry the load. In order that the parallel-connected generators may divide the total load in proportion to their ratings, the governors are made to decrease slightly the speed of the machines with increasing load. If one machine tries to take too much of the load, its governor tries to lower its speed slightly by reducing the input until it again comes back to the required amount. The ordinary adjustments give a full-load speed from 1½ to 5 percent lower than at no load. Without in any way disturbing the above two functions of the governors, a master clock may be utilized to shift the range of speed up or down as may be required in order to maintain a constant average speed. Referring to Fig. 14.35, when the load is 50 percent of capacity, the governor action follows the line AB, while with 100 percent load the master clock will change the governor action to follow line CD, thus giving a constant frequency which may be utilized for time service.

Quick-acting governors, capable of completely shutting off the water supply of a turbine operating at full load in 2 to 3 s, have been much used. In the isolated power plant this is still often necessary in order to maintain a reasonably constant speed with large and sudden load changes. This practice has been costly because of increased governor cost and also because of additional expenditures in pipe line or its protection to prevent damage due to water hammer. In the larger power systems of today this practice is being abandoned. Slow-acting governors are found to be satisfactory where the flywheel effect (Wr^2) of the load, added to that of the generators and turbines, is large. Even in smaller systems, their use is increasing because close regulation on large load changes does not seem to be so important as it used to be. The increased sensitiveness of present equipment has cut down the lag between speed change and governor action so that, with a quicker start, the governor does not have to travel so fast.

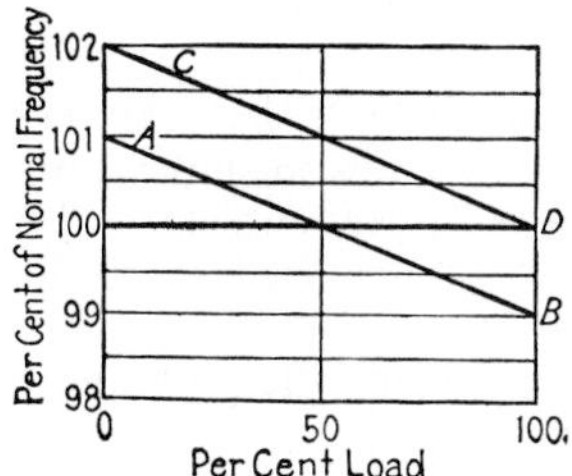

FIGURE 14.35 Speed-regulation curves.

12. Governors. As yet, nothing has been found more satisfactory for detecting and correcting speed changes than the flyball element. In hydraulic turbine installations, this may be belt-driven from the machine shaft or mounted directly on the shaft. When the turbine drives an alternating-current generator, the flyballs are sometimes electrically driven by a small synchronous motor, connected to the generator. Owing to the higher speeds obtained this way and to the smaller weights thus made possible, the cost may be reduced by the electric drive. Since the forces required to operate the needle or deflecting nozzle of a tangential water wheel or the gates of a reaction turbine are larger than can be exerted by the collar of the flyball element, an indirect system is used. On the turbine is mounted a cylinder with its piston connected to the nozzle or gates. This is called the *servomotor*. Oil, under pressure, is admitted to either end of the cylinder to move the piston either way, as required, to increase or decrease the flow of water. The oil flow is controlled by a small relay valve, the piston of which is connected to the flyball collar. In this way, small forces exerted by the flyball element on the relay valve adjust the flow of oil at 200 to 300 lb/in^2 (1.4 to 2.1 MPa) so that any size of machine can be controlled easily. Mechanical linkages between the servomotor and relay valve permit the adjustment of the speed change from no load to full load and operate a compensation device which prevents hunting and continual motion of the gates. An adjustment of the relay-valve ports permits variation of quickness of motion, and a synchronizing device permits a manual or clock-controlled adjustment of speed at any load. The compensation device stops gate movement before return to normal speed. Its adjustment depends upon the Wr^2 of the machine and load and on the water-hammer effect. A device for limiting the load to be taken by the machine is also a part of most modern governors. In some cases, frequency control is connected to this rather than to the synchronizing device. Overspeed of turbine-generator is prevented by an automatic overspeed device which, when tripped, shuts down the machine.

13. Water Hammer. If a column of water is to be accelerated, some force is required, over and above the static pressure, to cause this acceleration. When a hydraulic turbine is started up, there is a pressure drop at the machine. This causes an unbalance of forces, and the water is speeded up. Similarly, when the turbine is shut down, there is an increase of pressure at the machine, and this unbalance causes the water to slow down or stop. This phenomenon is called *water hammer*. It sometimes happens that when the turbine is partly shut down, the increase in pressure due to the reduction in velocity may be great enough to cause a momentary increase in input to the machine where a decrease is desired. It is for this reason that governor compensation is needed as noted above. In order to compute the pressure changes due to water hammer, Allievi's charts are usually used. Figure 14.36 is his chart for finding the maximum pressure increase due to complete gate closure, and Fig. 14.37 is for finding the pressure decrease due to gate opening from the closed position. In these charts θ is the conduit characteristic and ρ the time characteristic.

$$\theta = \frac{at}{2L} \tag{14.123}$$

$$\rho = \frac{av_0}{2gy_0} \tag{14.124}$$

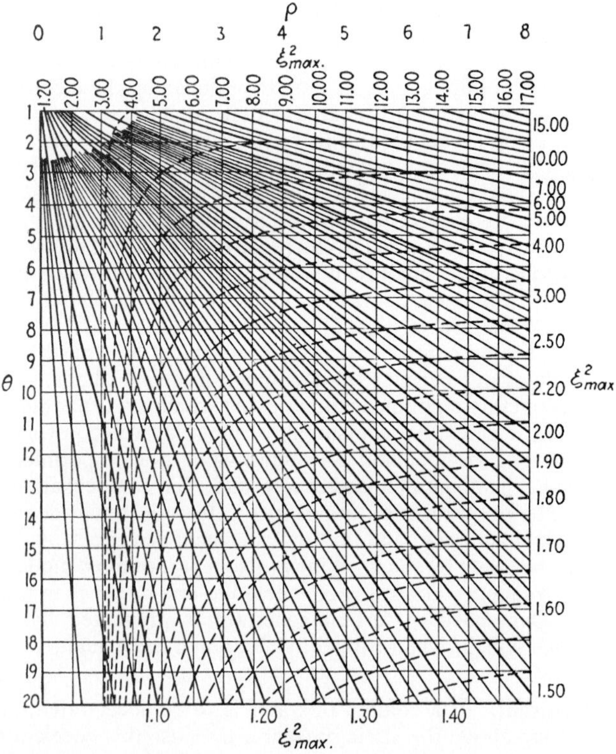

FIGURE 14.36 Allievi's water-hammer chart for complete closure.

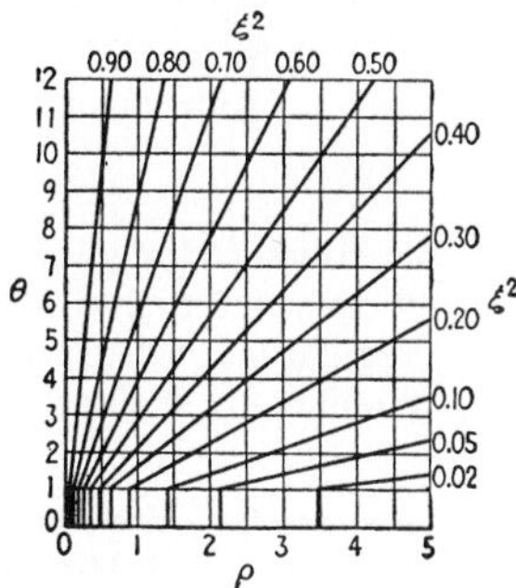

FIGURE 14.37 Allievi's water-hammer chart for gate opening.

$$a = \frac{4675}{\sqrt{1 + MN}} \tag{14.125}$$

where t = time, in s, that the gates are in motion

L = length of the pipe line, in ft (m), from forebay (or surge tank if there is one) to the tailrace

v_0 = velocity of flow before beginning closure or after opening when steady conditions are reached, ft/s (m/s)

g = acceleration of gravity, 32.2 ft/s^2 (9.8 m/s^2)

y_0 = pressure head at gates just before gate motion starts, ft (m)

a = velocity of propagation of pressure wave along pipeline, ft/s (m/s)

M = ratio of modulus of elasticity of water to that of pipe material

N = ratio of pipe diameter to pipe-wall thickness, which is zero for tunnels and passages through powerhouse substructure in concrete

Owing to the fact that governors do not move at a uniform rate, but start and stop more slowly than in the middle of travel, the time t should be reduced to 50 to 85 percent of the time required for gate closure, the actual value depending on the speed of closure, the higher value for very slow closure and the lower value for rapid closure. No correction in time is necessary for gate opening. Where diameters, and therefore velocities, change along the length of the pipe, v_0 may be found from

$$v_0 = \frac{v_1 L_1 + v_2 L_2 + \text{etc.}}{L} \tag{14.126}$$

where the subscripts refer to successive lengths of the total length L, and their respective velocities. Similarly,

$$a = \frac{a_1 L_1 + a_2 L_2 + \text{etc.}}{L} \tag{14.127}$$

for a corrected value of velocity of pressure-wave propagation for a pipeline of varying size, wall thickness, or material.

Having determined the two constants θ and ρ, refer to Fig. 14.36 for pressure increase due to complete closure, or to Fig. 14.37 for opening from complete closure. At the intersection of the two values, obtain the value of the constant ξ^2 by interpolating between the radiating lines so marked. The total maximum pressure head, equal to the sum of static head and water hammer, is this value of ξ^2 times the static head.

Where the pipeline is long, the conduit characteristic θ is small and the pressure increase is large. This would cause excessive cost of a pipeline designed for this maximum pressure, and the regulation of the turbines would be difficult. It is therefore usual to protect long pipelines by the use of surge tanks.

14. Surge Tanks. Two types of surge tank are in general use, the simple surge tank and the Johnson differential surge tank.* The purpose of either is to put a point of relief as close as possible to the point where the water flow is controlled. Each consists of a vertical tank, of diameter several times the pipe diameter, connected to the pipeline as near the powerhouse as possible and rising to a height somewhat above the water level in the forebay. The differential type differs from the simple tank in that it has a central riser pipe extending from the bottom of the tank to within a short distance of its top. The function of either one is to accept water rejected by the turbine while the velocity in the pipeline is slowly decreasing, or to supply water to the turbine while the pipeline velocity is slowly increasing to meet an increased demand. The theory and design of either of these surge tanks are too complicated for inclusion here.

15. Increasing Head by Use of Excess Flow. During periods of high water, the head on most hydraulic power plants is reduced and the plant capacity is therefore less than normal. At the same time it may be necessary to waste water because of insufficient pondage or storage. Several ways have been devised whereby this excess stream flow may be utilized to increase the head on the turbines and thus help to maintain plant capacity. Several of these schemes introduce this excess flow into the turbine draft tube for the purpose of increasing the suction head at the turbine discharge; in these the Venturi principle is employed. Several of these are employed in existing power plants. These are the *Repogle discharge accelerator*, the *Tefft tube spillway*, and the *Moody ejector turbine*. Another type, the *Thurlow backwater suppressor*, utilizes the flow over the dam to create a hydraulic jump so that the discharge from the turbine draft tube may take place at low elevation, before the jump returns the water level to the higher tailrace elevation.

16. Venting Reaction Turbines. At light loads, and more particularly at no load when the connected generator is used for power-factor correction or voltage regulation, considerable water may be saved by introducing air into the top of the draft tube. In some cases, an air valve is controlled by the turbine-gate mechanism, which automatically admits the correct amount of air at small gate openings to give maximum efficiency. At no load, the gates should not be shut, as the water passing through the machine is normally used to prevent heating of the bearings and of the runner where it may rub in the clearances. The function of the air is to prevent recirculation of water through the turbine runner and upper section of the draft tube which, in effect, gives the action of a pump with the consequent power loss.

*This is a patented design.

REFERENCES

1. R. H. Perry (ed.), *Engineering Manual*, 3d ed., McGraw-Hill, New York, 1976.
2. *A Policy on Geometric Design of Rural Highways*, AASHTO, Washington, D.C.; *A Policy on Arterial Highways in Urban Areas*, AASHTO, Washington, D.C.
3. *Highway Capacity Manual*, Transportation Research Board—National Academy of Sciences, Washington, D.C.
4. *Specification for the Design, Fabrication and Erection of Structural Steel for Buildings*, American Institute of Steel Construction, Chicago, Ill.
5. *Concrete Reinforcing Steel Handbook*, Concrete Reinforcing Steel Institute, Detroit, Mich.
6. C. E. O'Rourke, *General Engineering Handbook*, 2d ed., McGraw-Hill, New York, 1940.

CHAPTER 15

CHEMICAL ENGINEERING, ENVIRONMENTAL ENGINEERING, AND PETROLEUM AND GAS ENGINEERING

CHEMICAL ENGINEERING*

DIFFUSIONAL OPERATIONS

1. Mass-Transfer Fundamentals. The transfer of material from one phase to another is a primary means of separating multicomponent solutions. In general, two equilibrium phases of a multicomponent mixture will have different chemical compositions, and this difference offers a means for separating a mixture into its individual components. Repetitive phase changes can provide increasingly pure solutions and in the limiting case can produce pure individual components. Analysis of these phase-change separation processes depends on three factors: thermodynamic equilibrium, mass-transfer rates, and pattern of contact between phases.

Equilibria. When two phases are brought into contact, their temperatures, pressures, and compositions will adjust until thermodynamic equilibrium is reached. The compositions of the phases are functions of temperature, pressure, and overall composition. The number of such variables that may be specified independently is fixed according to the Gibbs phase rule of thermodynamics. Data relating equilibrium phase compositions are ultimately dependent on experimental measurement. Phase equilibrium data for many systems have been tabulated in the literature.[2–4] Estimating techniques are available for predicting phase equilibria, as are equations correlating experimental data. Since proper design is dependent on the accuracy of the equilibrium relationships used, the use of accurate experimental data is preferable.

Mass-Transfer Rates. The amount of contacting required to bring two phases into equilibrium is dependent on the rate of mass transfer. The rate at which mass

*This section is from *Engineering Manual*, 3d ed., by R. H. Perry (ed.). Copyright © 1976. Used by permission of McGraw-Hill, Inc. All rights reserved. Updating and SI units were added by the editors.

is transferred between phases is controlled by the *driving force* for mass transfer, the *resistance* to mass transfer, and the *interfacial area* between phases, according to

$$N_{M_i} = K_L a V(x_i - x_i^*) = K_G a V(y_i - y_i^*) \tag{15.1}$$

The overall mass-transfer coefficients are dependent on resistance to mass transfer in interfacial films (in a manner analogous to film resistances in convective heat transfer), which depend on molecular parameters, fluid turbulence near the interface, and the equilibrium relationship between phases. For equilibrium relationships which are essentially the straight lines $y = mx$, mass-transfer coefficients are given by

$$\frac{1}{K_G} = \frac{1}{k_G} + \frac{m}{k_L} \tag{15.2}$$

$$\frac{1}{K_L} = \frac{1}{k_L} + \frac{1}{mk_G} \tag{15.3}$$

The coefficients are ordinarily determined experimentally as volumetric coefficients $K_G a$ and $K_L a$, since interfacial areas are difficult to determine.

Continuous Contacting. Continuous-contact processes may be run with cocurrent, crosscurrent, or countercurrent flow patterns. In terms of efficient mass transfer, countercurrent contacting, which allows the greatest amount of mass transfer between phases for a given initial composition difference between feed streams, is preferred. Other flow patterns are generally used only in special cases where countercurrent flow is impractical.

The *transfer unit* is a standard degree of separation used to describe the performance of a contacting device. The greater the number of transfer units, the more thorough is the separation. The number of transfer units required to accomplish a specified degree of separation between components is dependent on the product compositions and relative flow rates of the two phases and may be based on either phase, according to

$$N_{OG} = \int_{y_2}^{y_1} \frac{(1-y)_{lm}}{(1-y)(y-y^*)}\,dy \cong \int_{y_2}^{y_1} \frac{dy}{y-y^*} \tag{15.4}$$

$$N_{OL} = \int_{x_1}^{x_2} \frac{(1-x)_{lm}}{(1-x)(x-x^*)}\,dx \cong \int_{x_2}^{x_1} \frac{dx}{x-x^*} \tag{15.5}$$

The approximation holds for dilute solutions, where transfer of material does not change the overall stream molar flow rates, and is a true equality for equimolar counterdiffusion.

The *height of a transfer unit* (*HTU*) is an indication of the amount of contacting required to accomplish the standard separation of one transfer unit. It is dependent on mass-transfer coefficients, packing type, and specific surface area, flow patterns of the contacting phases, and flow rates. The heights of the overall transfer units defined by Eqs. (15.4) and (15.5) are

$$H_{OG} = \frac{G_M}{K_G a(1-y)_{lm}} \tag{15.6}$$

$$H_{OL} = \frac{L_M}{K_L a(1 - x)_{lm}} \tag{15.7}$$

The height of a transfer unit is often much more constant than the mass-transfer coefficient for a given column and packing at various flow rates and is therefore more commonly used. HTUs are determined empirically by measuring mass transfer in a given column, calculating N_{OG} or N_{OL} by Eq. (15.4) or (15.5) and dividing into column height, that is,

$$H_{OG} = \frac{Z}{N_{OG}} \tag{15.8}$$

$$H_{OL} = \frac{Z}{N_{OL}} \tag{15.9}$$

In design applications, the number of transfer units required for a specified separation is calculated by Eq. (15.4) or (15.5). This is combined with HTU data for similar (or pilot plant) operations to obtain total column height.

Nomenclature for Eqs. (15.1) through (15.9) corresponds to that used for gas absorption. The equivalent equations for other continuous contacting operations have slightly different forms appropriate to the parameters commonly encountered. A thorough treatment of mass transfer is found in Treybal's comprehensive text.[5]

Staged Operations. In staged contacting, two phases initially not in equilibrium are held in contact for a length of time assumed sufficient to attain equilibrium. The two phases are then separated, and each is fed to an adjacent stage where it is again held in contact with a nonequilibrium mixture of the opposite phase. The two phases flow from stage to stage in opposite directions. As one phase advances through the contactor it becomes progressively more concentrated in a particular component or group of components. As the other phase advances in the opposite direction it becomes progressively less concentrated in that same component or group of components.

Design of staged contacting systems (or performance analysis of existing systems) requires repetitive computation using equations based on three basic principles: material balances on the various independent components, equilibrium relationships between streams, and degree of approach to equilibrium (efficiency) of each contacting stage. The approach to equilibrium depends on the same factors as control mass-transfer rates in continuous contacting. For practical calculations, efficiencies are always estimated on the basis of experience with similar contacting equipment and multiphase systems.

2. Distillation

Definitions. *Simple distillation* is the partial vaporization of a solution of liquid components with separate recovery of vapor and liquid residue. The concentration of the more volatile components (sometimes termed the *lighter components*) is greatest in the condensed vapor, while the concentration of the less volatile components (the *heavier components*) is greatest in the liquid residue. The degree of separation in a simple distillation process depends on the relative volatility of the components, which, in turn, depends on the thermodynamic properties of the mixture. Ordinarily, single-stage distillation does not provide adequate separation of components. (The major exception is evaporation, which is treated separately in the section "Evaporation.") Simple distillation may be carried out as a batch operation, in which case the instantaneous compositions of vapor product and liquid residue change continuously. Alternatively, it may be carried out as a continuous operation, with liquid feed to the still pot to replenish the vapor removed. In this case, the vapor product composition remains constant once steady state is achieved.

Rectification is continuous countercurrent contact of the vapor resulting from a simple distillation with a condensed portion of the vapor product. This countercurrent contact results in greater enrichment of the vapor (often termed *overhead*) product with the more volatile components than is possible with a single stage of simple distillation. The condensed vapor returned to accomplish this is termed *reflux*. In rectification, the feed is to the simple distillation stage (bottom) and the more important product is removed as a vapor (top).

Stripping is continuous countercurrent contacting of the liquid feed with the vapor resulting from a simple distillation. This countercurrent contacting results in a more complete removal of the more volatile components from the liquid product than could be accomplished by a single stage of simple distillation. In stripping, the liquid feed is introduced at the top of the column and the vapor is supplied by partial vaporization of the liquid stream at the bottom of the column in a *reboiler*.

Commonly, rectification is carried out on the vapor product leaving a stripping operation. This is termed *fractional distillation*, or *fractionation*. In fractional distillation, column feed is introduced at a *feed stage* within the column. Above the feed stage is the rectification section, and below it is the stripping section. Overhead product is condensed, and a portion is returned as reflux. Bottom product is partially vaporized, and the vapor is returned to the column as *boilup*. The stripping section is for recovery of overhead product and purification of bottom product, while the rectification section is for purification of top product and recovery of bottom product. The important factors controlling separation are the number of stages in each section, the vapor-liquid flow ratios, and the relative volatilities of the components.

Equilibrium Data. Vapor-liquid equilibrium data are required for the design of stills. In general, these must be observed experimentally. For binary systems, they are reported usually as tables or graphs of corresponding x and y values. Many such data are summarized by Hala et al.,[6] Smith, Block, and Hickman,[7] and by Hala et al.[8]

If equilibrium data are not available, they may be estimated for a binary system by a modification of Raoult's law:

$$y_1 = \frac{P_1}{\Pi} = \frac{\gamma_1 x_1 P_1^0}{\Pi} = \frac{\alpha_{12} x_1}{1 + (\alpha_{12} - 1)x_1} \tag{15.10}$$

For ideal mixtures, $\gamma_1 = \gamma_2 = 1.0$; as a system departs from ideality, use of Eq. (15.10) becomes less reliable. Values of γ may be estimated for a number of binaries by a method summarized in Smith, Block, and Hickman.[7]

Simple Batch Distillation (Rayleigh or Differential Distillation). In this case, a batch of material is charged to a still pot, boiling is initiated, and the vapors are continuously removed, condensed, and collected until their average composition has reached a desired value. As the distillation proceeds, the concentration of less volatile components in the vapor continually increases. Sometimes successive portions of the vapor product (termed *cuts*) are collected separately; these decrease in purity (with respect to the more volatile component) as the distillation proceeds. Ordinarily, simple distillation is used only when there are large differences in relative volatility among the components.

Simple distillation is analyzed by assuming that at any instant the vapors are in equilibrium with the average liquid composition in the still. Though not strictly correct, this is usually a good assumption. In this case, the composition of the liquid in the still is related to the total amount vaporized by

$$\ln \frac{S_2}{S_1} = \int_{x_1}^{x_2} \frac{dx}{y - x} \tag{15.11}$$

where y is related to x by the equilibrium relationship $y = y(x, \Pi, t)$. The average composition of the vapors collected during this period is related to the initial and final liquid compositions and quantities by a simple material balance:

$$x_{\text{ave}} = \frac{x_1 S_1 - x_2 S_2}{S_1 - S_2} \tag{15.12}$$

Equilibrium Flash Distillation. Liquid at an elevated temperature and pressure is throttled into a *flash chamber* maintained at a pressure below the vapor pressure of the liquid. A certain fraction of the feed vaporizes, reducing the system temperature until the vapor pressure of the remaining liquid at the new temperature matches the flash chamber pressure. Vapor and liquid fractions are led separately from the still. If vapor and liquid streams are in equilibrium, this is equivalent to a single-stage simple distillation.

Compositions and flow rates of vapor and liquid streams are obtained by simultaneous solution of material balances

$$(1 - f)x_i + fy_i = x_{Fi} \tag{15.13}$$

and energy balance

$$(t_F - t)C_F = F\lambda_m \tag{15.14}$$

together with the equilibrium relationship

$$y_i = y(x_i, \Pi, t) \tag{15.15}$$

and the relationship between temperature, vapor pressure, and total pressure

$$P_i = P_i(t) \tag{15.16}$$

$$\Pi = \Sigma P_i(t) \tag{15.17}$$

where f is the fraction vaporized, and C_F and λ_m are the heat capacity and latent heat of vaporization of the feed (assumed constant). Equations (15.13) through (15.17) are solved simultaneously by trial and error.

Continuous Binary Rectification

Plate Columns—Plate-to-Plate Calculations. The simultaneous solution of material balance, energy balance, and equilibrium relationships between each successive two stages in a column permits the exact computation of the column behavior from one terminal stream to another. Until recently a generally impractical procedure, it now may be utilized economically if the need for repeated designs is sufficient to justify its being programmed for a digital computer. This method is beyond the scope of this book (see Smith, Block, and Hickman[7]).

McCabe-Thiele Graphical Method. If the molar latent heat of vaporization is relatively independent of composition, and if heat losses are negligible, liquid and vapor molar flow rates within the column will be constant and the following graphical procedure is acceptable:

1. Plot the vapor-liquid equilibrium data as y versus x.

2. Write and plot the operating-line (material balance) equations for each section of the column which relate passing streams within the column to feed or product streams. With reference to Fig. 15.1, the equations are, for section II (enriching section),

$$y_n = x_{n+1}\frac{L_{n+1}}{V_n} + x_d\frac{D}{V_n} = \frac{R}{R+1}x_{n+1} + \frac{x_d}{R+1} \tag{15.18}$$

where $R = L_E/D$ = *reflux ratio* and for section III (stripping section),

$$y_m = x_{m+1}\frac{L_{m+1}}{V_m} - x_w\frac{W}{V_m} \tag{15.19}$$

The liquid rates in the two sections of the column are constant but generally different because of the introduction of the feed, which may be liquid, vapor, or a mixture (see below).

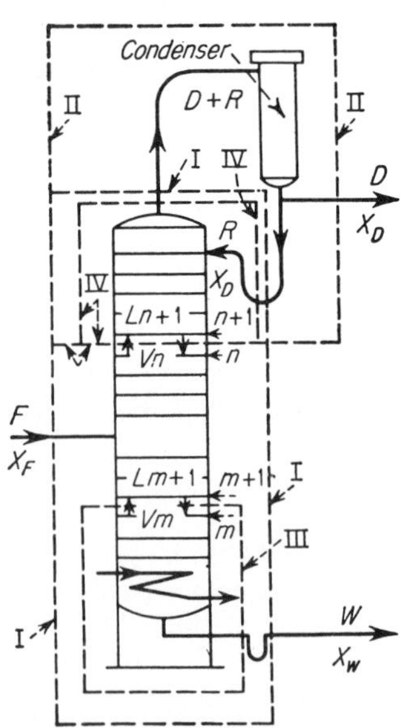

FIGURE 15.1 Schematic of continuous distillation column.

3. Determine the number of stages by stepping from operating line vertically to equilibrium line, then horizontally to operating line. Begin with bottom product and continue in this fashion until the feed composition is reached; then use upper operating line until upper product concentration is reached.

4. Correct the theoretical number of stages to the actual number of stages using stage efficiencies (estimated from previous experience) according to

$$N_{\text{act}} = \frac{N_{\text{theor}}}{\text{efficiency}} \tag{15.20}$$

The upper and lower operating lines intersect on a *feed line* described by Eq. (15.13), where f is the moles of vapor introduced to the rectifying section per mole of feed introduced to the feed stage. Note that f is fractional between zero and unity if the feed is introduced as a mixture of liquid and vapor; $f = 0$

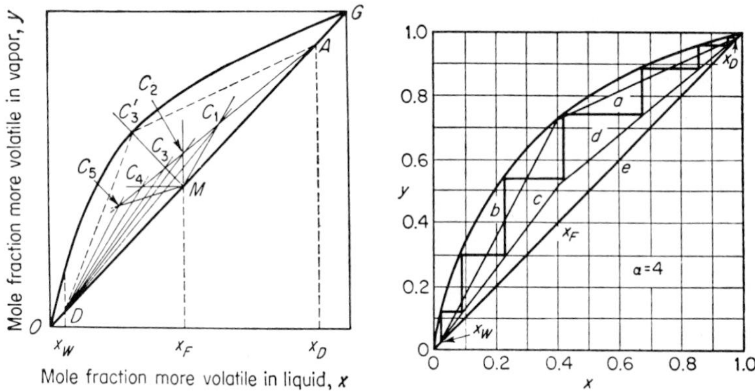

FIGURE 15.2 (*a*) Effect of thermal condition of feed on operating lines and minimum reflux ratio. (*b*) McCabe-Thiele graphical method for determining number of theoretical stages.

if feed is a saturated liquid; $f < 0$ if feed is a cold liquid; $f = 1$ if feed is saturated vapor; and $f > 1$ if feed is a superheated vapor. In Fig. 15.2*a*, five different feed lines are shown with the same upper operating line and five lower operating lines leading to the same bottom product composition. Intersections between operating lines for feeds of different qualities are represented by C_1 through C_5:

$C_1(f < 0)$ $C_2(f = 0)$ $C_3(0 < f < 1)$ $C_4(f = 1)$ and $C_5(f > 1)$

Three sets of operating lines are shown on Fig. 15.2*b*. Lines *a* and *b*, intersecting the feed line on the equilibrium curve, represent minimum reflux and require an infinite number of theoretical stages to separate the feed with composition x_F into products with compositions x_d and x_w. Line *e*, where both operating lines coincide with the diagonal, represents total reflux, which requires the smallest number of stages to achieve the desired separation, but produces no product. Lines *c* and *d* represent practical operating lines between the limits of minimum and total reflux.

Analytical Method for Mixtures of Constant Relative Volatility

1. Solve for the number of theoretical plates necessary at "total" reflux (the condition when vapor and liquid rates within the column are infinitely large compared to feed, overhead, and bottom drawoff rates) using the Fenske-Underwood equation

$$N_T + 1 = \log (x_1/x_2)_d(x_2/x_1)_w/\log \alpha \tag{15.21}$$

2. Estimate the minimum reflux ratio (the ratio of liquid to distillate rates if the column were infinitely tall) by using

$$\frac{R}{R_{\min}} = \frac{x_d[1 + (\alpha - 1)x_F] - \alpha x_F}{(\alpha - 1)x_F(1 - x_F)} \tag{15.22}$$

3. By use of Fig. 15.3, estimate the number of theoretical trays necessary for the reflux ratio to be employed.

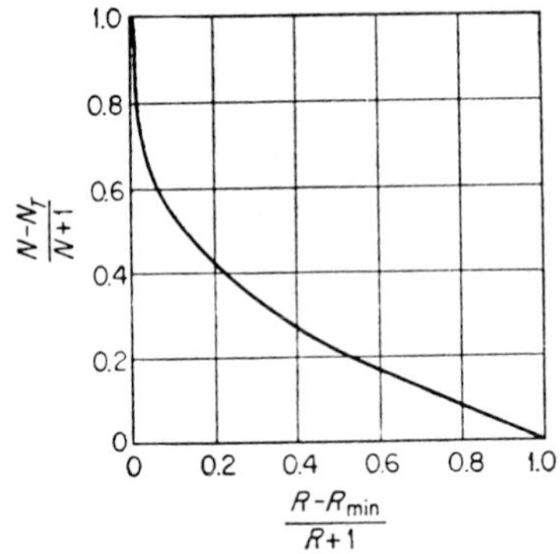

FIGURE 15.3 Correlation of Gilliland for number of theoretical stages.

Batch Binary Rectification. In batch distillations, generally three "products" are withdrawn from the still. These are an initial product high in purity with regard to the more volatile or light-key component, an intermediate product that will usually be recycled for redistillation, and finally, a product high in purity with regard to the heavy or less volatile component. Obviously a desirable separation is one which will minimize the middle or recycle cut. For most batch distillations, the following rule of thumb will hold and is useful in setting the reflux to be used:

$$\frac{L}{D}(\alpha - 1) \geq 10 \qquad (15.23)$$

If a constant-reflux ratio is maintained, the product purity of the more volatile component will drop off as the distillation proceeds. The speed at which this decline in purity occurs will be a function of the particular reflux ratio employed, the relative volatility, and the amount of volatile component originally present.

It must be remembered that the practicing engineer is most often confronted with a different problem in designing for continuous operation than for batch distillation: In the former, a column is usually designed and built for a given separation; in the latter, the usual problem is how a given piece of equipment should be operated to effect a desired separation. We shall therefore assume that the equipment and the number of plates are specified.

Of importance in planning for the operation of any batch-distillation apparatus are answers to the following questions:

1. What is the overhead-product composition as a function of still-pot composition?
2. How many moles of steam (as a heating medium) will be required to effect the separation?

For the constant-reflux case, the vapor requirement, and thus the steam requirement, is obtained from

$$V = (L/D + 1)D \qquad (15.24)$$

The relation between still-pot and overhead compositions is obtained by plotting on a $y - x$ diagram (as shown in Fig. 15.4) lines of constant slope equal to

$$L/V = (L/D)/(L/D + 1) \qquad (15.25)$$

and stepping off the number of theoretical plates in the column. To obtain the relation between the amount distilled and the still-pot composition, plot x_p versus $1/(x_p - x_d)$. The area under the curve is equal then to P/D, where P represents the amount of liquid in the still pot.

A second method of operating batch-distillation columns is to maintain product purity over a period of time by constantly increasing the reflux ratio. The relation between amount distilled and still-pot composition is now found by a simple material balance. The steam requirement is obtained by finding the area under a curve of $[P_0(x_0 - x_p)]/[(x_p - x_d)^2(1 - L/V)]$ versus x_p. Appropriate values of x_p are read as a function of L/V from a $y - x$ plot.

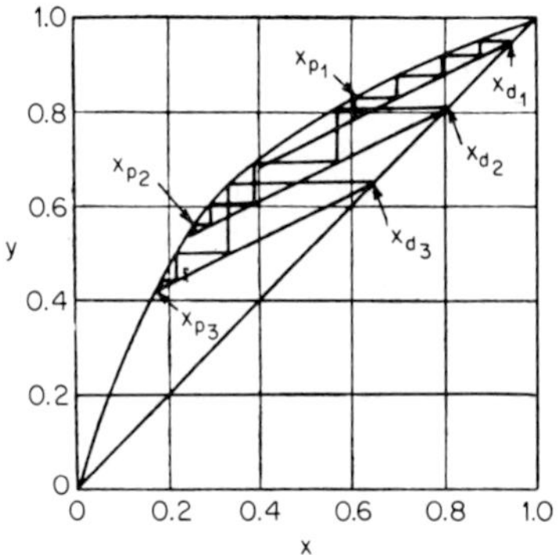

FIGURE 15.4 Diagram for batch distillation at constant reflux ratio.

Multicomponent Rectification. This subject is too complicated for treatment here. The interested reader should refer to a more comprehensive source (e.g., Smith, Block, and Hickman[7]).

Plate Efficiency. The estimation of plate efficiencies is empirical. In a properly designed column, the value should exceed 0.60 and may exceed 0.95. Fair et al.[9] present estimation methods and typical values.

Limiting Factors. Proper operation of a column requires gas velocities within a narrow range of values. Velocities must be high enough to give good gas-liquid contacting, yet low enough to prevent entrainment or excessive pressure drop with resultant flooding. These factors control column diameter for a given vapor rate.

Tray spacing must be large enough to prevent carryover of liquid from the tray below, either by entrainment or because of froth height. Some liquid-vapor systems cause severe operating problems because of formation of a relatively stable foam with considerable liquid carryover to the tray above.

A detailed discussion of design procedures with regard to these problems is beyond the scope of this work. The interested reader is referred to Treybal.[5]

Design Optimization. When designing a distillation process, the designer usually must meet specified product purities for both products or one specified purity plus a percentage recovery of a component. Choice of a given reflux ratio specified the number of stages required to meet product specifications. Higher reflux ratios lead to smaller numbers of stages (hence lower column heights and column costs). Normal optimal design procedure is to determine minimum reflux ratio and then to incrementally increase reflux ratio toward total reflux. For each reflux ratio capital and operating costs are calculated, and an acceptable design is determined by the economic balance. Final details of design can then be determined using this reflux ratio. In practice, the economical reflux ratio is often between 1.2 and 2.0 times the minimum.

3. Solvent Extraction

Definitions. Solvent extraction consists of the transfer of a component dissolved in a liquid (called the *feed solution*) to a second liquid (called the *solvent*) to form an *extract* solution of the transferred component and to leave a *raffinate* solution relatively lean in the transferred component. Solvent extraction is used when distillation is impractical, as with close-boiling or temperature-sensitive mixtures.

There is a strong analogy between extraction and distillation, solubility being the counterpart of volatility and the solvent that of heat (Fig. 15.5).

Equilibrium Data. Phase equilibria for liquids are so specific that it is best to refer to laboratory data for the system in question. Maddox[10] cites many such data.

For some systems, the equilibrium is well approximated by the ideal-distribution law

$$K' = w/w' \tag{15.26}$$

with consequent simplification of design procedures. In solvent extraction, fewer theoretical plates and much lower plate efficiencies are encountered than in distillation or absorption, with corresponding aggravation of inaccuracies implicit in simplified methods. For this reason, shortcut approximations should be used with caution.

Countercurrent Extraction. A feed mixture of two completely miscible components A and B is to be separated into its components by extraction with a solvent. If the solvent is partially miscible with each component of the feed, countercurrent extraction with reflux may be used to separate the feed components completely (in the limiting case). The method of operation in this case is

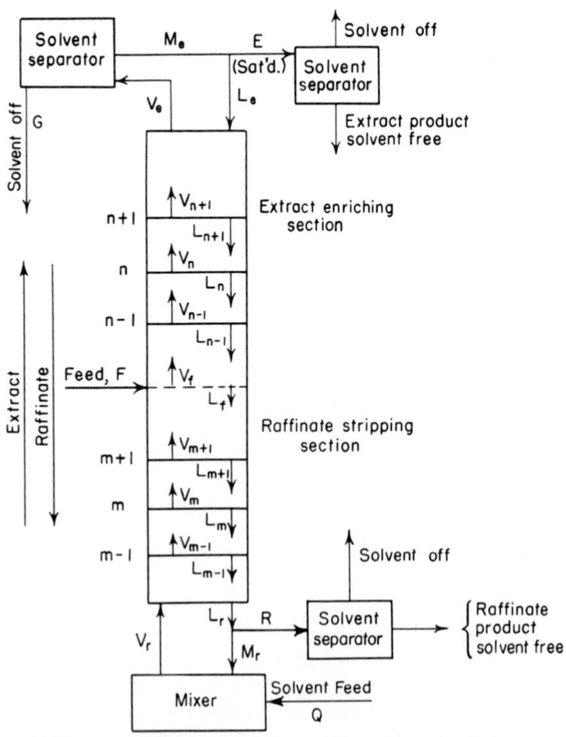

FIGURE 15.5 Flow diagram for countercurrent multistage extraction with reflux.

presented schematically in Fig. 15.5. If the solvent is partially miscible with one feed component A and totally miscible with the other component B, complete separation cannot be achieved. In this case, essentially pure component A can be obtained at the solvent-feed end of the column, but the ratio of components A and B in the overhead product is limited by the shape of the equilibrium diagram. Where purity of component B is limited, the extract-enriching section is not employed; i.e., the feed is at the top of the column, there is no extract reflux, and the top product after solvent recovery is a solution of A and B, although richer in component B than the feed stream.

Assume that the solvent is only partially miscible with each feed component, and the equipment is as shown in Fig. 15.5. Reflux is furnished to the top of the column by removing sufficient solvent from the extract phase (e.g., by distillation) to make it miscible with the raffinate phase. (Note that this does not change the ratio of A to B.) A portion of this raffinate phase is returned to the column as reflux; the remainder is purified further to remove any remaining solvent and is removed as product.

Raffinate phase at the bottom of the column is withdrawn and purified to remove any solvent present, e.g., by distillation. Raffinate reflux is sometimes supplied at the bottom of the column by adding a portion of the raffinate before purification to the solvent feed in a mixer. This is done primarily to ensure adequate dispersion of solvent in raffinate for good interphase contact and does not increase recovery of component A as extract reflux increases recovery of component B.

All recovered solvent is recycled to the solvent-feed end of the column and is supplemented by makeup solvent if product purification is not complete.

Contacting may be continuous, with one phase dispersed in the other, using a packed column or spray column, or it may be stagewise, using a tray column as in distillation or a countercurrent system of mixer-settler stages. Design procedures are presented for staged extraction systems. These are applicable for continuous systems by determining empirically the column-height equivalent to a theoretical stage (HETS).

A simplified design procedure similar to McCabe-Thiele graphical method (see earlier subsection) may be used, assuming that extract-layer and raffinate-layer flow rates are constant throughout the column.

1. Equilibrium data are plotted as mass fraction of A in the extract layer (*ordinate*) against mass fraction of A in the raffinate layer (*abscissa*), *both fractions being on a solvent-free basis*.
2. Extract and raffinate products are located on the $Y = X$ line.
3. Operating lines through these points and of slope L_c/V_c and L_r/V_r are drawn.
4. Plates are stepped off as in the McCabe-Thiele method previously described.

A more precise design procedure is illustrated in Fig. 15.6 and is outlined below. All flow rates and concentrations are on a solvent-free basis, unless otherwise noted.

1. From known equilibrium relationships, construct the S versus Y (extract layer) and s versus X (raffinate layer) lines on a working diagram.
2. Locate the operating point K at an abscissa of X_c (extract-product composition) and an ordinate of $G/E + S_c$.
3. Locate the operating point N at an abscissa of X_r (the A content of the raffinate product) and an ordinate of $(s_r - Q/R)$. A line joining K and N will intersect the s versus X line at Z_f (solvent content of the feed).

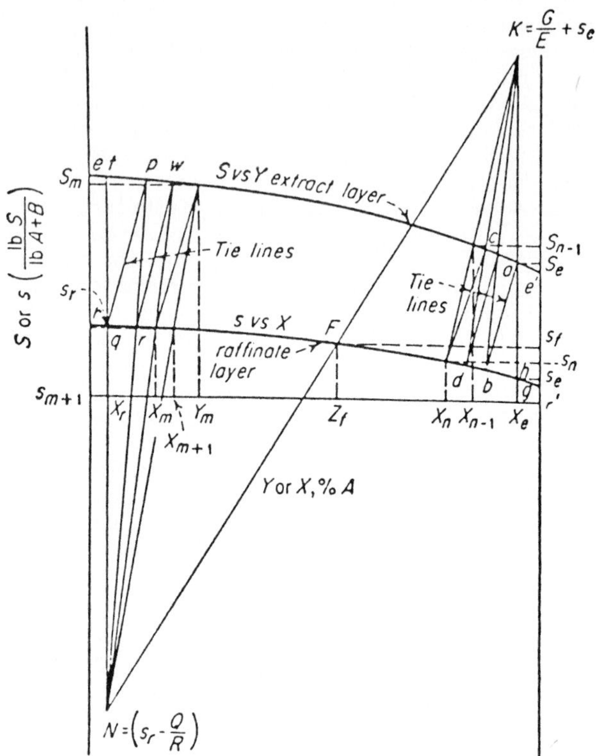

FIGURE 15.6 Graphical stepwise calculation of equilibrium stages on a solvent content-concentration diagram for operation with reflux.

It is now possible to "walk" across the diagram to determine the number of theoretical stages necessary to effect the separation. A line is drawn from K to X_c intersecting the S versus Y curve at a. Line ab is an equilibrium tie line wherein the composition represented by b is that in equilibrium with the composition represented by a. Another line from K can then be drawn to the point b so established, and another equilibrium tie line cd is drawn. The procedure is repeated until a ray from K coincides with the line joining K and N. To the left of this dividing line the same procedure is followed, using point N where point K was used before. The number of theoretical stages is then obtained by counting the total number of rays drawn from the two operating points.

Minimum reflux on this type of diagram is obtained by moving K vertically downward and N vertically upward (at such a relative rate that F always lies on a line joining them) until the line between them coincides with a tie line through F. The ordinate of K then corresponds to a point of minimum reflux. Economic balances of column size versus heat loads in solvent recovery of course determine the optimum degree of departure from this minimum-reflux point in actual operation.

Column Efficiency. For perforated-plate columns, the overall efficiency may be estimated as the fraction E_c:

$$E_c = \frac{89{,}500 Z_t^{0.5}}{\sigma g_c}\left(\frac{V_D}{V_c}\right)^{0.42} = \frac{0.9 Z'^{0.5}_t}{\sigma'}\left(\frac{V_D}{V_c}\right)^{0.42} \tag{15.27}$$

where Z_t is tray spacing in feet, and Z'_t is tray spacing in inches.

For packed columns, the efficiency is expressed in the height assigned to a transfer unit (HTU) or theoretical stage (HETS). Ellis[11] shows that for *rough estimates*, the following empirical relationships are useful for towers packed with Raschig rings larger than ⅜ in (9.5 mm):

1. Transfer of solute from continuous aqueous to dispersed organic phase:

$$\text{HETS} = \frac{94.5\mu_C(12d_F)^b(V_c/V_D)^{0.5}}{10^{0.0683s}\Delta\rho} \tag{15.28}$$

2. Transfer of solute from dispersed organic to continuous aqueous phase:

$$\text{HETS} = \frac{69\mu_C(12d_F)^b}{10^{0.0535s}\Delta\rho} \tag{15.29}$$

where $b = 2.15/10^{0.096s}$, d_F is in inches (millimeters), μ is in lb/ft · s (kg/m · s), and ρ is in lb/ft^3 (kg/m^3).

Here s is the average of the mutual solubilities of the solute-free contacted liquids in each other, expressed as weight percent, and provides a rough measure of interfacial tension. For liquid pairs as insoluble as toluene and water, s may be taken as zero.

Extraction-Tower Diameter. Limiting flows, and hence minimum allowable diameters, for liquid-liquid extraction columns may be calculated using Fig. 15.7.

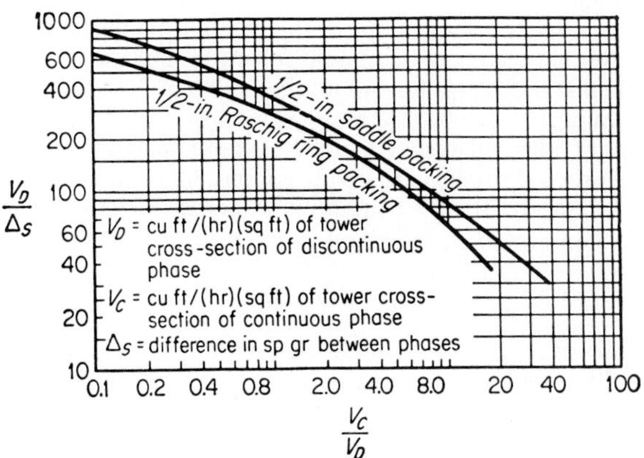

FIGURE 15.7 Colburn correlation of flooding data for packed extraction columns.

While strictly applicable only to packed towers, the figure may be used with extreme caution for tray columns in the absence of better data. Column diameters must be larger (i.e., liquid velocities lower) than those specified in the flooding correlation.

4. Gas Absorption

Definitions. *Gas absorption* consists of the transfer of a component from a gas phase to a liquid phase. The liquid phase is called the *solvent*, or *absorbent*; the transferred gas is called the *solute*, or *absorbate*. Usually, the solute is selectively absorbed from a carrier gas. Fundamental considerations and design methods that apply to absorption are useful generally for the reverse operation of *desorption*, or *stripping*.

Equilibrium Data. Gas solubility in a liquid is measured as a function of partial pressure or concentration of the gas in the equilibrium vapor phase. Solubilities sometimes are reported in the form of Henry's law constants. Equilibrium data for many systems may be found in standard reference sources.[2,12]

Equipment. Gas absorption or stripping is accomplished in three principal types of equipment: *absorption columns*, packed or plate; *spray chambers* or towers; and *bubble-sparged tanks*, frequently agitated. Only absorption columns, by far the most important, will be treated here.

Column Height. The height of a *packed column* is determined by the degree of separation to be achieved and by a characteristic contacting effectiveness of the packing. The former may be expressed by stream compositions or by number of transfer units [Eq. (15.4)]; the latter, by the appropriate transfer coefficient [Eq. (15.2)] or HTU [Eq. (15.6)]. The height of a transfer unit varies with application and should be determined experimentally for the gas-liquid system, column packing, and column loadings employed.

The number of transfer units required may be calculated from Eq. (15.4) or, if operating and equilibrium lines are approximately straight, by Fig. 15.8, use of which requires knowledge of the slope of the equilibrium line m and stipulation of G_M/L_M. The latter should be an economic selection. For most columns, 0.7 is an acceptable value for mG_M/L_M, but something less may be used if the solute is of low economic value.

The number of ideal plates in a *plate column* may be determined by the McCabe-Thiele procedure (see earlier subsection). Once the values of K_L and K_G are estimated for the particular conditions under consideration and then combined to give a value for H_{OG} by use of Eqs. (15.6) and (15.7),

$$Z = H_{OG}N_{OG} \tag{15.30}$$

The estimation of N_{OG} may be obtained from Fig. 15.8, assuming that the operating and equilibrium lines are both straight or, at worst, only slightly curved.

5. Humidification. The most frequently encountered gas-vapor system is that of air and water vapor. A distinct terminology has been developed for this system, but the principles and mechanisms involved apply to any gas-vapor system.

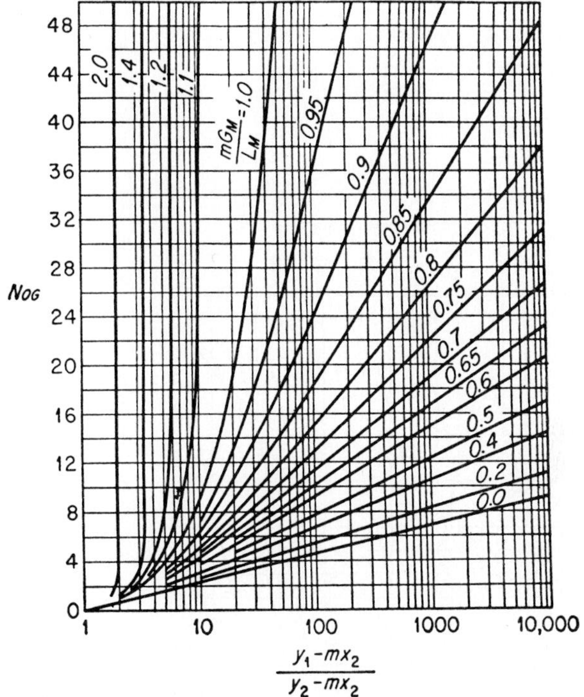

FIGURE 15.8 Number of transfer units in an absorption column. Subscripts 1 and 2 refer to the concentrated and dilute ends, respectively.

Humidification is the process by which the moisture content of air is increased; *dehumidification* is the reverse process.

If dry air and liquid water are held in intimate contact, some of the water will vaporize into the air in an effort to establish equilibrium between the phases. The concentration and the partial pressure of water vapor in the air will increase until the air becomes *saturated*; that is, until the partial pressure of the water vapor equals the vapor pressure of water at the equilibrium temperature.

The concentration of water vapor in the air is expressed as the *absolute humidity* (pounds of water vapor per pound of dry air), the *molal humidity* (moles of water vapor per mole of dry air), *relative humidity* (ratio of the actual partial pressure of vapor to the partial pressure if saturated at the existing temperature), or *percentage humidity* (ratio of actual humidity to the humidity if saturated at the existing temperature). The moisture level in air is indicated by the *dew point* (the temperature to which an air sample must be cooled to reach saturation) and by the difference between the air temperature and the *wet-bulb temperature* (the temperature assumed by a water-wet body held in a fast-flowing stream of the air). The preceding properties for the air–water vapor system are commonly presented graphically on a humidity or psychrometric chart as shown in Fig. 15.9.

Humidification and dehumidification are often the unavoidable secondary re-

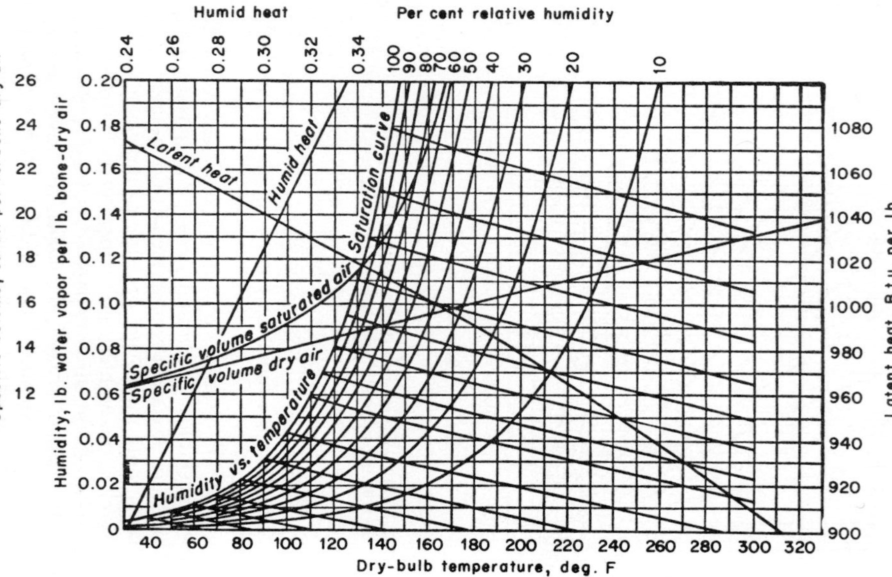

FIGURE 15.9*a* Humidity chart for air–water vapor mixtures, U.S. Customary units.

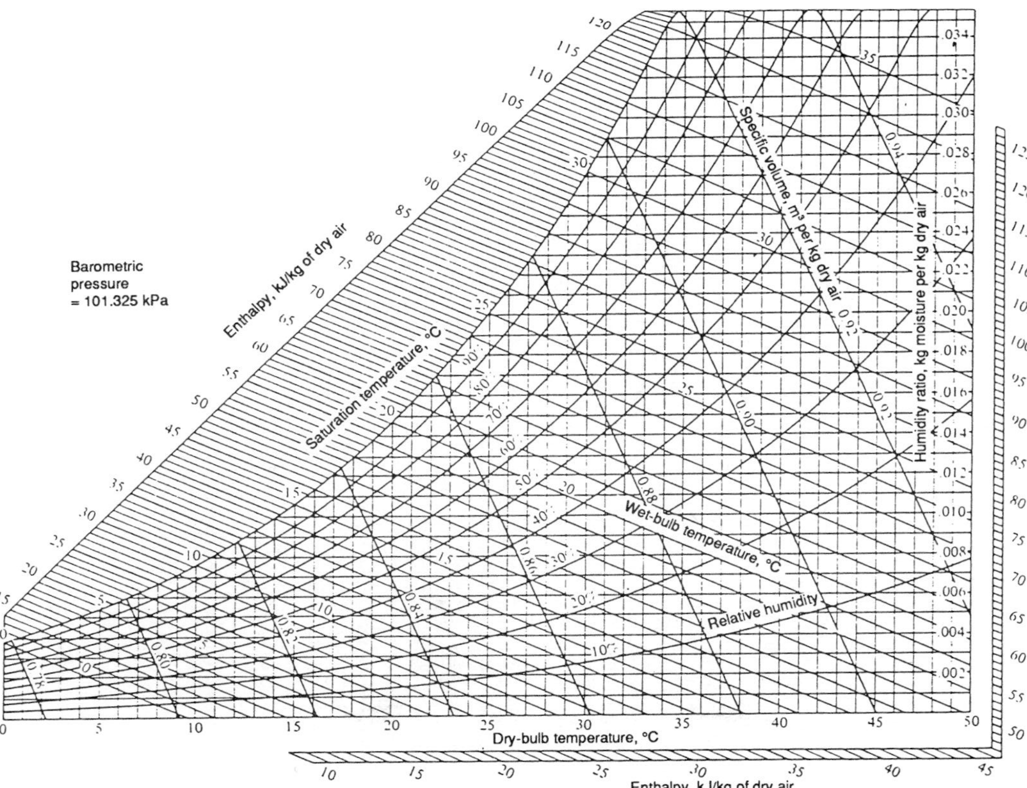

FIGURE 15.9*b* Humidity chart for air–water vapor mixtures, SI units. *(From W. F. Stoecker and J. W. Jones, Refrigeration and Air Conditioning, 2d ed., McGraw-Hill, New York.)*

sults of other operations such as drying, compression, absorption, and water cooling. The evaporative cooling of water is accomplished by adiabatically contacting it with a relatively large amount of unsaturated air; the resultant vaporization of some of the water cools the remainder, which approaches wet-bulb or adiabatic saturation temperature corresponding to the condition of the air.

The contacting of air and water is accomplished with spray ponds or cooling towers. Spray ponds depend on natural forces for air movement, but power must be supplied to atomize and spray the water into the air. Spray ponds require large land areas, and water loss through entrainment is often high. Spray-nozzle performance is critical in the design, since it controls power demand, air-water contact surface, and water loss by drift. Table 15.1 presents guidelines for the design of spray ponds.

Cooling towers are preferred for most large industrial installations because they conserve land area, are less sensitive to atmospheric changes, and offer greater flexibility in design and operation. They are of two types: atmospheric, which depends on the wind for air cross-flow, and mechanical, in which fans pull the air upward through the tower. In both the water is pumped to the top of the structure, where it is distributed uniformly across the tower and then cascades from grid deck to grid deck.

The design of induced-draft towers is based on the allowable air velocity, the cooling load and range, the approach to wet-bulb temperature, and the transfer coefficients attained.

TABLE 15.1 Spray-Pond Engineering Data and Design

Recommendations	Usual	Minimum	Maximum
Nozzle capacity, gpm each (L/s)	35–50 (132–189)	10 (38)	60 (227)
Nozzles per 12-ft (3.7-m) length of pipe	5–6 (1.5–1.8)	4 (1.2)	8 (2.4)
Height of nozzles above sides of basin, ft (m)	7–8 (2.1–2.4)	2 (0.6)	10 (3.0)
Nozzle pressure, lb/in^2 (kPa)	5–7 (34–48)	4 (28)	10 (69)
Size of nozzles and nozzle arms, in (mm)	2 (51)	1.25 (32)	2.5 (64)
Distance between spray lateral piping, ft (m)	25 (7.6)	13 (3.9)	38 (11.6)
Distance of nozzles from side of pond, unfenced, ft (m)	25–35 (7.6–10.7)	20 (6.1)	50 (15.2)
Distance of nozzles from side of pond, fenced, ft (m)	12–18 (3.7–5.5)	10 (3.0)	25 (7.6)
Height of louver fence, ft (m)	12 (3.7)	6 (1.8)	18 (5.5)
Depth of pond basin, ft (m)	4–5 (1.2–1.5)	2 (0.6)	7 (2.1)
Friction loss per 100 ft (30.5 m) pipe, in (mm) of water	1–3 (25–76)	— —	6 (152)
Design wind velocity, mi/h (km/h)	5 (8)	3 (4.8)	10 (16.1)

Source: From *Spray Pond Bull.* SP-51, p. 3, Marley Co. (Metricated by the editors.)

Norris[13] gives a design procedure, but the services of a reputable tower supplier are recommended. As a very rough estimate, 1 ft (0.3 m) of tower height will yield 1°F (0.56°C) of cooling when the water rate is 2 to 3 gpm/ft^2 (81.5 to 122.2 L/m^2) of tower cross section, the upward air velocity is 300 ft/min (91.4 m/min), and the approach to the wet-bulb temperature is 10 to 15°F (5.6 to 8.3°C).

6. Drying. The unit operation *drying* refers to the removal by vaporization of liquid from a solid; the liquid usually constitutes a relatively small fraction of the wet solid. Beyond this, the term is applied to the removal of traces of vapor from a gas and of small amounts of water from another liquid. Only drying of a solid is discussed here, and water and air will be used as examples of wetting liquid and surrounding gas, respectively.

The equipment for drying is classified according to the means by which the necessary heat is brought to the evaporating liquid and also according to the form and disposition of the wet solid. *Direct dryers* deliver heat to the liquid by direct contact with the hot gas stream into which the liquid vaporizes; *indirect dryers* supply the heat through a wall that separates the wet solid from a heat source, such as condensing steam.

Simultaneous heat transfer and mass transfer occur at equivalent rates during drying; the wet solid assumes the temperature needed to maintain this balance. The rate of drying depends on (1) the temperature of the heat source, (2) the resistance to transfer of heat to the vaporization site, (3) the resistance to transfer of mass from the vaporization site, and (4) the concentration (or partial pressure) of vapor in the gas in contact with the wet solid. In turn, (2) and (3) depend on the physical form and characteristics of the solid and its wetness (liquid concentration); thus the rate of drying can vary greatly as the process proceeds. It is very common for some of the liquid wetting a solid to be dispersed in such a way that it does not exert its normal vapor pressure; in such cases, the final extent of drying is limited by the temperature and vapor content of air in contact with the wet solid.

Typical of indirect dryers are drum dryers (for slurries), can dryers (for textiles), and cylinder dryers (for paper), all of which are rotating cylindrical vessels usually internally heated with steam. The material being dried is continuously applied to the outside surface, carried around for a part of a revolution, and then removed. Sufficient drying may be accomplished in one pass, as on a slow-turning drum dryer, or it may require repeated passes over a series of similar dryers, as in paper making. Generally, the drying rates can only be established experimentally; rates can range from the equivalent of 1000 to 4000 Btu/(h · ft^2) (3152 to 12,608 W/m^2) of dryer surface.

A wide variety of direct dryers are available, which differ primarily in the method of contacting the wet solid with the drying gas. The simplest and lowest cost is the batch tray drier in which the wet solid is spread on trays held in an enclosure and hot air is blown across the surface of the solid until it is sufficiently dry. The drying rate is usually low; depends greatly on the temperature, humidity, and velocity of the air; and can decrease substantially as the drying proceeds. If the wet solid can be granulated or pelletized and held on a perforated tray, improved rates can be obtained by passing the gas through the solid bed rather than across it.

A modification of the tray dryer is the truck dryer, in which the trays are loaded on trucks and rolled into the dryer compartment. Since the loading and unloading are done outside the dryer proper, the dryer is used much more efficiently.

Rotary dryers offer continuous operation and are particularly adaptable to finely divided, nonsticking, and nonagglomerating solids. The wet solid is fed into the upper end of a slightly sloped, rotating cylinder through which heated air or hot combustion gas is passed. Internal flights lift the solid and shower it through the hot gas while also advancing it to the discharge end. If the gas velocity is high, fine particles may be carried out with the gas and must be recovered in cyclones or bag filters.

The steam-tube dryer is a variation of the rotary dryer in which a number of tubes are supported longitudinally within the rotating cylinder and supply the required heat. Only a small flow of air through the cylinder is needed to sweep out the vapor, and the loss of fines is minimized.

Spray dryers are particularly applicable when the feed is a solution or a very dilute suspension that can be atomized into fine droplets. The dryer consists of an atomizing device, a hot-gas source, and a drying chamber in which the droplets contact the hot gases. The atomizing device may be a pressure spray nozzle or a rotating disk. The hot gases rise vertically counter to the falling droplets. The product of the spray dryer is usually a spherical particle, often hollow, which is advantageous in certain cases.

In pneumatic conveyor or flash dryers, the wet solid is dispersed in a high-velocity stream of hot gas, and drying occurs as the gas carries the solids to a separating device. Solids disintegration and solids size classification are often carried out in conjunction with flash drying.

Empirical relationships for estimating drying rates under very restricted conditions and performance data for several types of dryers are given by McCormick.[14] However, experimentally determined rates for the specific material and particular dryer type are much preferred when preparing dryer specifications.

7. Evaporation. *Evaporation* is the operation by which a volatile liquid is separated from a solution or suspension by vaporization. The separation is usually not complete, and either or both the vapor and concentrated liquor may be the desired product.

Direct-fired or steam-jacketed kettles may serve as evaporators, but the most common types use a tubular heating surface and are steam heated. The tubes may be horizontal or vertical, and the heating steam may be inside or surrounding the tubes. The tube bundle is placed within or external to the body of the evaporator and is so arranged that the boiling liquor can be circulated through or around the tubes. The body also serves to allow disengagement of the vapor from the liquor as it is formed. Figure 15.10 illustrates several types of tubular evaporators.

An evaporator is basically a heat-transfer device. Its evaporative capacity is related to the temperature difference between heating steam and the boiling solution, the area of the tubular heat-transfer surface, the heat-transfer coefficient and the latent heat of vaporization of the evaporating liquid:

$$\text{Capacity} = \frac{q}{\lambda} = \frac{(UA)(\Delta t)}{\lambda} \tag{15.31}$$

The coefficient U is chiefly a function of the velocity of the solution past the heating surface and its viscosity, the extent of fouling of the heating surface, the temperature difference between steam and liquor, and the height of the liquor level relative to the tubes. The fouling of the tubes can become significant or even the controlling factor when scale-forming materials are present in the liquor. The values of U are nearly always derived from experimentally determined rates of evap-

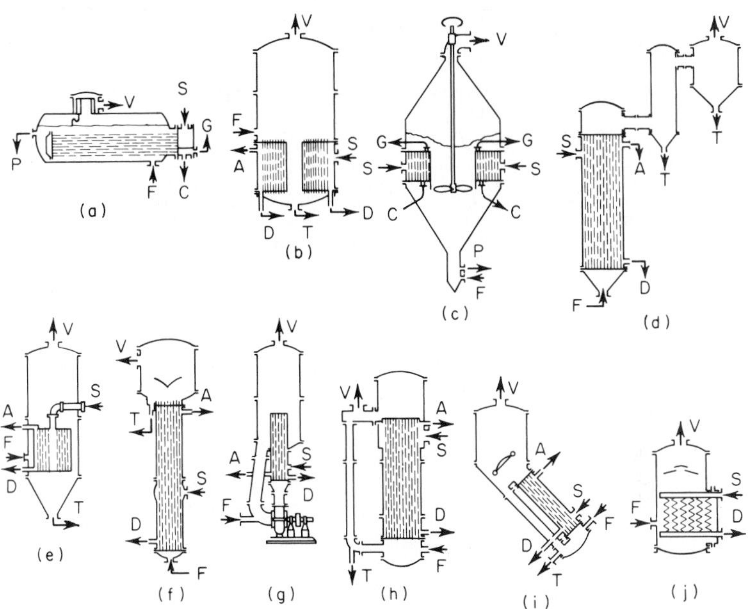

FIGURE 15.10 Typical evaporator designs: (*a*) horizontal tube, (*b*) short tube vertical, (*c*) propeller calandria, (*d*) long-tube vertical without vapor head, (*e*) basket type, (*f*) long-tube vertical, (*g*) forced circulation, (*h*) long-tube vertical with downtake, (*i*) buflovac inclined tube, (*j*) coiled tube.

oration as a function of Δt. The temperature difference observed, however, is more often apparent than real because it is inferred from pressure measurements, which correspond to temperature values that do not reflect boiling-point rise of the liquid due to dissolved solute or temperature shifts on the steam side due to vapor superheat or condensate subcooling. Such Δt values are known as *apparent temperature differences*, and the values of U corresponding to them are known as *apparent coefficients*. The error on the steam side usually is small. When the boiling-point rise is known and can be used to adjust the temperature difference, nearly correct values of Δt and U result; such values are known as *temperature difference* and *coefficient corrected for boiling-point rise*. When not otherwise stipulated, the values reported for evaporators usually are these.

Illustrative overall coefficients are given in Figs. 15.11 to 15.14. These coefficients are intended only for preliminary estimation; they are unreliable for accurate design.

Multiple-Effect Evaporation. A *multiple-effect evaporator* is merely a series of similar evaporators so connected that the vapor from one body is the heating medium for the next. Passing from single to multiple effect does not alter the major features of body construction; it merely affects the interconnecting piping and the operation.

The purpose of multiple-effect evaporation is to improve thermal efficiency. One pound (0.45 kg) of steam supplied to the first effect will evaporate approximately one pound (0.45 kg) of water in that effect. This pound (0.45 kg) of water vapor will then pass to the steam space of the second effect and, in condensing,

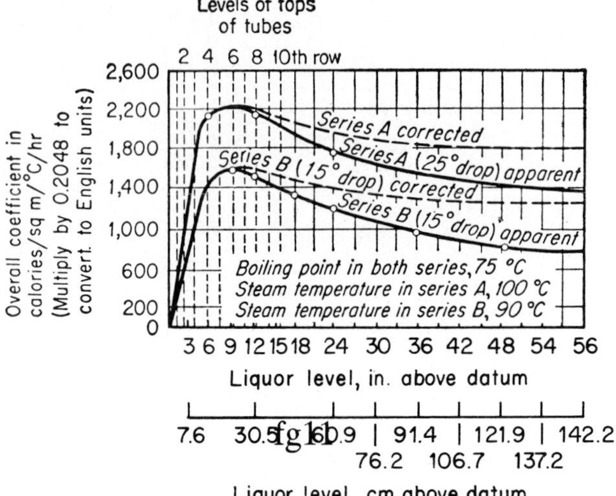

FIGURE 15.11 Heat-transfer coefficients in a horizontal-tube evaporator.

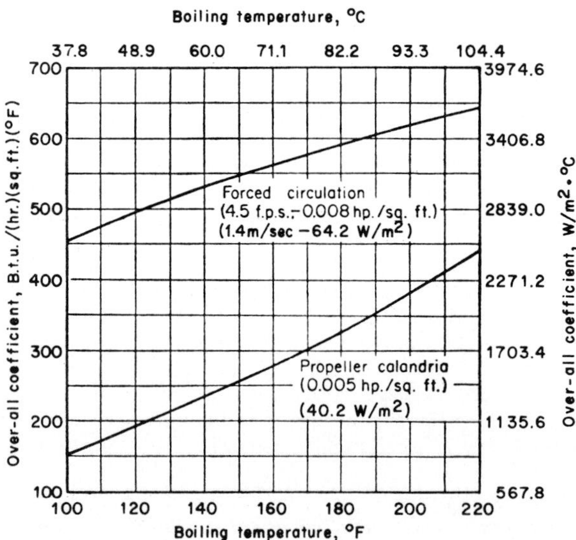

FIGURE 15.12 Heat-transfer coefficients in salt evaporators.

will evaporate approximately another pound (0.45 kg) of water, and so on, so that in N effects, 1 lb (0.45 kg) of steam will evaporate approximately (but somewhat less than) N lb (kg) of water. The pressure must be progressively reduced from effect to effect in order to produce a temperature difference between the boiling liquid of that effect and the condensing vapor from the preceding effect.

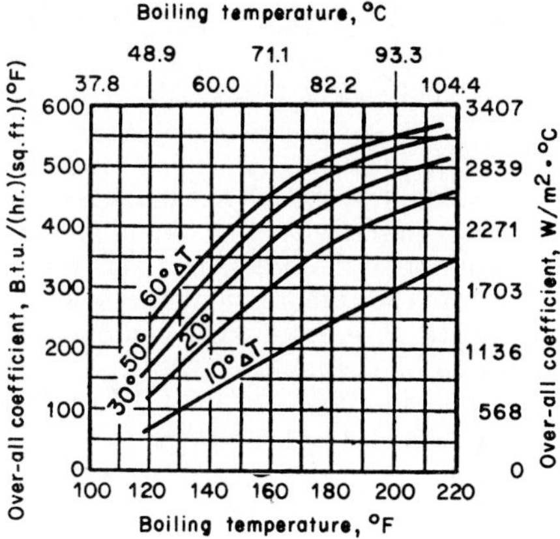

FIGURE 15.13 Heat-transfer coefficients for water in short tube.

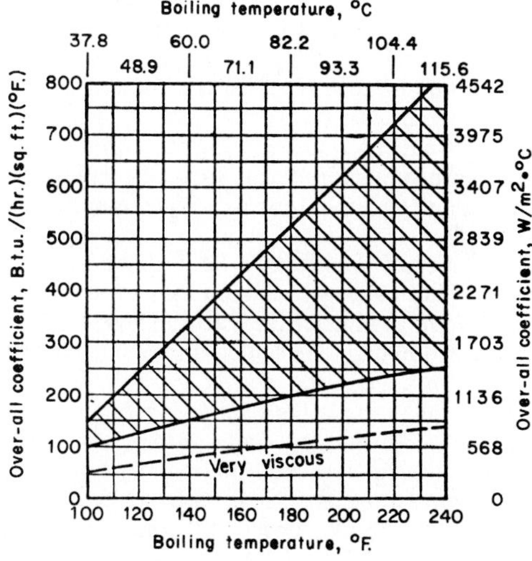

FIGURE 15.14 General range of LTV coefficients.

If it is assumed that the terminal temperatures (temperature of heating steam available and temperature corresponding to the vacuum that can be produced in the condenser) are fixed, then passing from a single to a multiple effect does not increase the capacity of an evaporator. If a single-effect evaporator is operating between these terminal conditions and requires A ft^2 (m^2) heating surface to accomplish the desired evaporation, an N-effect evaporator to be used between the same terminal conditions for the same weight of water evaporated will require N bodies of approximately A ft^2 (m^2) each to accomplish the same result. In short, passing from single- to multiple-effect operation decreases steam cost but increases apparatus cost.

The effects are commonly made the same size. The total area required may be calculated from the simultaneous solution of heat and material balance statements written for the several effects and the entire evaporator (e.g., see McCabe and Smith[15]).

Thermocompression. The simplest, though not the least expensive, means of reducing the thermal energy requirements of evaporation is to compress the vapor from a single-effect evaporator so that the vapor can be used as the heating medium in the same evaporator. The compression may be accomplished by mechanical means or by a steam jet. In order to keep the compressor cost and power requirements within reason, the evaporator must work with a fairly narrow temperature difference, usually from about 10 to 20°F (5.6 to 11.1°C). This means that a large evaporator heating surface is needed, partially offsetting the advantages of thermocompression.

8. Adsorption and Ion Exchange

Definitions. *Adsorption* is separation of the components of a fluid phase brought about by contacting the fluid phase with a fixed solid phase. The solid phase, termed the *sorbent*, consists generally of highly porous particles with greatly extended surface area, most of which is internal. One or more components of the fluid phase, termed *sorbates*, enter the sorbent particles and bond to the surface of the solid. In *physical adsorption*, weak forces such as ionic or Van der Waals forces bond sorbate molecules to the surface. In *chemisorption*, true chemical bonds form between sorbate and sorbent surface. *Molecular sieves* possess a large number of adsorption sites of uniform type and essentially molecular dimensions and can be extremely selective toward a given component or class of components.

Ion exchange is transfer of ions from a fluid phase to a solid phase with simultaneous transfer of other similarly charged ions in the reverse direction. The solid medium is a synthetic organic resin with functional groups having a given charge bonded to the resin matrix and hence immobile. Mobile ions of opposite charge are distributed throughout the matrix to maintain electrical neutrality. *Cation-exchange resins* have negatively charged functional groups (such as sulfonic groups) bound to the matrix, while *anion-exchange resins* have positively charged bound groups (quaternary ammonium groups in *strongly basic resins* and other amine groups in *weakly basic resins*). If the mobile ion X is predominant, the resin is said to be in the *X form* (e.g., if Cl^- ions are the predominant mobile ion, the resin is in *chloride form*).

Once the sorbent has become saturated, the sorbate is stripped from the sorbent and recovered, returning the sorbent to its original condition for reuse. This is termed *regeneration* and may be accomplished by heating, chemical reaction, or contacting with a *regenerant*.

Mechanisms. Design or operation of an ion-exchange or adsorption system depends on two factors: sorption equilibrium and rate of mass transfer. The relationship between solute concentration in the fluid phase and equilibrium amounts of solute retained in the solid phase is generally a strong function of temperature, and the curve describing this relationship is known as an *isotherm*. Many theoretical descriptions of sorption equilibrium have been developed and are summarized by Vermeulen et al.[16] Isotherms for a particular sorbent-sorbate system must be determined experimentally.

The important rate processes for adsorption are as follows: turbulent eddy transport within the fluid phase to the boundary of the sorbent particle, molecular diffusion through pores to the interior of the particle, reaction at the phase boundary to establish equilibrium between fluid and solid phases, and diffusion in the sorbed state (especially important for molecular sieves). In ion exchange, there is also equimolar counterdiffusion of the exchanged ion. Ordinarily, transport to the particle and reaction at the surface are relatively rapid; thus diffusion within pores or in the sorbed state is the rate-controlling step.

Continuous Operation. For simplification, assume adsorption (or ion exchange) of a single solute (or ion) on a fixed bed, packed initially with adsorbent (or resin) free of the desired sorbate. The fluid phase is assumed to move slowly relative to the rate at which equilibrium is attained on the particle surface. At the outset of flow through the bed, all the solute is adsorbed on the packing within a short distance of the inlet end. As the sorbed-phase concentration approaches equilibrium with the inlet solute concentration, solute must flow further through the bed before being adsorbed. During this time, the effluent concentration of sorbate is essentially zero, but eventually some of the solute passes through the entire depth of the bed without being adsorbed and appears in the effluent. This is termed *breakthrough*, and thereafter the effluent concentration rises until finally it equals the inlet concentration. At this time, the entire bed has a sorbed-phase concentration in equilibrium with the effluent concentration, and the bed is said to be *exhausted*. Concentration versus time profiles illustrating this sequence are presented in Fig. 15.15.

At some point in the operation, usually when the effluent reaches some specified maximum allowable concentration, the bed must be taken off stream and regenerated to restore its adsorptive capacity. The maximum allowable effluent concentration is based on product purity requirements, emissions limitations to protect environmental quality, or economic balances between the value of recovered solute and the cost of regeneration.

Regeneration usually is accomplished by contacting the exhausted bed with a regenerant, often at elevated temperature. The solute is desorbed into a much smaller volume of regenerant than in the initial carrier stream, thus giving a much increased concentration of solute than initially fed to the system. (If regeneration is by heat alone, the regeneration yields pure sorbate.) In ion exchange, the sorbate is removed in a relatively small quantity of a regenerant solution which is highly concentrated in the alternate ion (e.g., in a concentrated NaCl or HCl solution if the resin is regenerated in the chloride form). After repeated regenerations, adsorbents usually lose adsorptive capacity, requiring addition of fresh makeup adsorbent.

Because bed-type adsorption is basically a batch operation, multiple beds in series must be used to attain continuous operation. When the first bed in the series is exhausted, the feed is switched by a valving system to the second bed, thus making that the first bed, and the exhausted bed is regenerated and, again by means of valves, put on-stream as the last bed of the series. Thus each bed in the series successively becomes the first bed.

Design. Design for continuous operation generally depends on scale-up from an experimentally determined breakthrough curve for a particular sorbent-sorbate

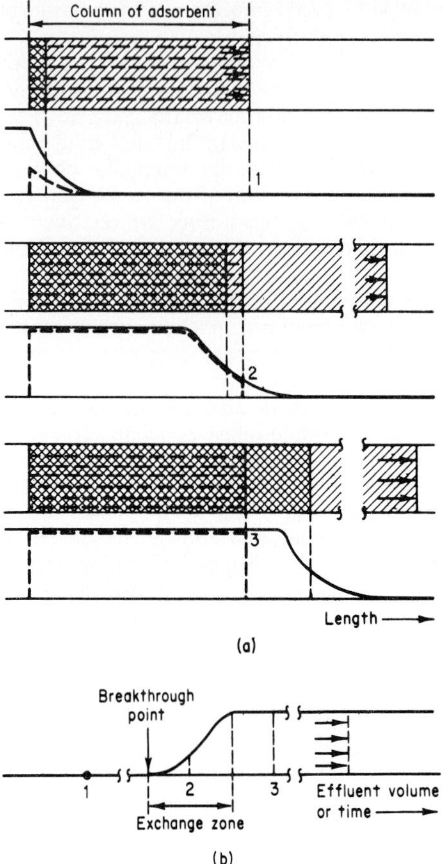

FIGURE 15.15 (*a*) Typical column-concentration profiles. (*b*) Effluent-concentration history for fixed beds. In (*a*), dashes show span of packed bed. Crosshatching, single, indicates penetration by carrier fluid; double, by solute. Length scale to right of bed is contracted to show entire exchange zone.

fluid system. It is generally found that for a given system the ratio of effluent concentration is a function of the quantity of liquid treated (expressed as volumes of liquid per volume of solid, dimensionless), residence time, and effective mass-transfer rate. For a given system at a given set of equilibrium conditions, the same functional relationship must hold:

$$'X = f\left(\frac{F^a V}{D_p^b F'}\right)\left(\frac{v}{V}\right) \tag{15.32}$$

where $F^a/D_p{}^b$ is a dimensional group controlling mass-transfer rate, V/F' is the residence time, and (v/V) is the total throughput in bed volumes. If mass transfer is rapid enough to ensure instantaneous equilibrium at all points, the effects of the mass-transfer parameter and residence time are not significant, and

$$X = f\left(\frac{v}{V}\right) \tag{15.33}$$

The shape of this curve is dependent on equilibrium considerations only (i.e., the feed concentration and the adsorption isotherm).

The shape of the breakthrough curve can be used to calculate the length of bed which has reached equilibrium with the feed (i.e., exhausted) and the bed loading, or amount of solute adsorbed.[3,17] From this information one can calculate the required time between regenerations as a function of bed volume and feed flow rate, as well as the appropriate number of bed volumes treated to ensure complete exhaustion before regeneration for multiple beds in series.

9. Membrane Processes. Several diffusional separation processes involve transfer of one or more components from one fluid phase through a porous solid medium or membrane to a second fluid phase. The porous medium, generally very thin, does not permit flow of the fluid but does allow the transfer of material between phases by molecular diffusion through pores. Membrane processes are classified according to the driving force across the membrane causing this diffusion; in *membrane permeation*, the driving force is a difference in partial pressure; in *dialysis*, the driving force is a concentration difference; in *ultrafiltration* and *reverse osmosis*, the driving force is a difference in pressure; and in *electrodialysis*, the driving force is a voltage gradient.

Owing to its physical, chemical, or electrochemical properties, a particular membrane material offers different resistances to the diffusional flow of different materials. This allows the separation of the components of solutions from one another.

Membrane Permeation. For steady-state processes, the rate of permeation of a component through the membrane is given by

$$N_M = \frac{P}{L}(\Delta p) \tag{15.34}$$

where Δp is the difference in partial pressure of that component across the membrane, and P is the *permeability*. Permeability is a function of the membrane composition and its structure, diffusing component, and system temperature and pressure. Correlations are available to predict permeabilities for various permeate-membrane combinations and the effects of temperature and pressure.[3]

Membrane permeation is most useful for those separations where a more conventional process is impractical and where the product can be recovered as a vapor. Some examples are separation of isomeric mixtures, thermally unstable compounds, and fluorides of different isotopes of uranium. Process design considerations include tradeoffs between membrane selectivity, permeation rates, membrane strength, and durability. Membrane separation is a capital-intensive process rather than energy-intensive and is only practical for difficult separations involving components with considerable economic value.

Dialysis and Osmosis. If two liquid solutions with different compositions are separated by a permeable membrane, the concentration gradient causes molecular diffusion through the membrane. For relatively dilute solutions, where no significant volume change occurs as the diffusion proceeds, mass-transfer rates are described by

$$N_{M_i} = U_i A \, \Delta C_{i,lm} \tag{15.35}$$

The overall dialysis coefficient U_i is determined by three film coefficients, analogous to heat transfer

$$\frac{1}{U_i} = \frac{1}{U_{i_{L1}}} + \frac{1}{U_{i_m}} + \frac{1}{U_{i_{L2}}} \tag{15.36}$$

where the individual coefficients are for the membrane and the two liquid films. If adequate mixing takes place in the two liquids, the membrane coefficient is controlling, $U_i = U_{i_m}$.

If the membrane resistance to diffusion differs significantly among solute components, mass transfer proceeds at different rates for the different components. In such cases, membranes may be used to recover a single component preferentially from a complex solution. In such cases, the complex solution is separated from a solvent by a selective membrane through which the desired component diffuses at a higher rate than other components. The resulting dilute solution of recovered solute is then concentrated by addition of makeup solute and returned to the plant for reuse. This is the basis for caustic recovery processes in the viscose-rayon industry and acid recovery from metallurgical liquors.

Design of a dialysis process depends on choice of membrane, concentration driving forces employed, and physical arrangement of the contacting streams. It is desirable to use a membrane that is thin and which has a high selectivity for the desired components. Thin membranes offer higher transfer rates but greater problems with mechanical stability. When large transfer areas are involved, the most effective and preferred arrangement is countercurrent staging of modules, with crosscurrent contacting within modules. Typical module designs have membranes separated by corrugated spacers; with corrugation at right angles on alternating layers, providing mechanical support for the membranes.

Design tradeoffs are total membrane area (capital cost) versus value of recovered solute and membrane type and thickness (maintenance costs) versus transfer rate (size and thus capital costs).

Reverse Osmosis and Ultrafiltration. Reverse osmosis and ultrafiltration processes recover the pure solvent from a solution. If a liquid solution is separated from the pure solvent by a membrane permeable to the solvent only, the osmotic pressure causes the solvent to diffuse from the pure liquid side to the solution side. A mechanical pressure applied to the solution side slows the diffusion process. If sufficient pressure is applied, the flow is reversed; i.e., solvent is transferred from the solution to the pure side, counter to the concentration gradient.

For dilute solutions, large solute molecules or high pressure differentials, Eq. (15.34) applies for reverse osmosis processes. For concentrated solutions of solutes with low molecular weight, a correction for osmotic pressure is required.

In most applications, the limiting factor is concentration polarization of the membrane. As solvent is removed from the solution, solutes greatly increase in concentration in a thin layer near the membrane. This causes an osmotic pressure that opposes the applied pressure gradient and reduces the rate of transfer. Another significant problem is buildup of deposits on the membrane surface.

Normal design tradeoffs involve membrane thickness, pressure differential, membrane surface area, and percentage of solvent recovered.

Electrodialysis. If an electric field is imposed on an electrolytic solution, negatively charged ions are attracted toward the positive pole of the field and positively charged ions are attracted to the negative pole. If the solution is partitioned by semipermeable membranes (alternately anion-permeable and cation-permeable), the flow of ions under the influence of the electric field will create alternately dilute and concentrated compartments (see Fig. 15.16). The feed solution is fed to the diluting compartments and a salt-recovery solution is fed to the concentrating compartments. Electrodialysis may be carried out on either a continuous or a batch basis.

Important parameters in electrodialysis are membrane selectivity, electric field strength, current efficiency, and desired degree of deionization of the feed. The current efficiency, or amount of solute transferred per coulomb of electric charge, and the total voltage drop required to maintain a constant field strength both increase with the number of compartments. The rate of transfer increases

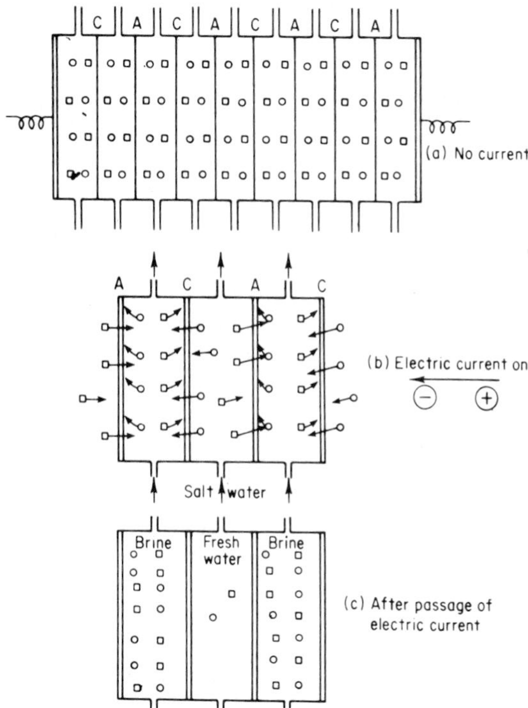

FIGURE 15.16 Principles of electrodialysis. ○ Positive ion (e.g., sodium). □ Negative ion (e.g., chloride). C and A, cation- and anion-permeable membrane, respectively. Ion migration under action of electric current causes salt depletion in alternate compartments and salt enrichment in adjacent ones. *(From Speigler, Salt Water Purification, Wiley, New York, 1962.)*

with increasing field strength, but for a given membrane, selectivity decreases as electric field strength is increased.

Operating problems include concentration polarization of membranes, scaling or fouling of membrane surfaces, mechanical instability of membranes, and current control. Typical design tradeoffs are larger number of parallel cells (increased current efficiency) versus higher membrane costs and higher operating voltages, larger membrane area versus electric power costs, higher concentrate cell concentration versus greater concentration polarization and membrane scaling, and lower diluent cell concentrations versus increased electrical resistance and power costs.

MULTIPHASE CONTACTING AND PHASE DISTRIBUTION

10. Agitation. *Agitation* is motion imparted to material to promote heat or mass transfer to, from, or within the material or to distribute another phase through the material. Examples include mixing of miscible liquids, formation of uniform solids suspensions, promotion of heat transfer between a process fluid and a heat-exchange surface, promoting dissolution of a gas in a liquid. The most important process applications of agitation involve a freely fluid liquid as the primary phase. The key equipment is a rotating element (sometimes more than one) called the *agitator* or *impeller*. Turbines, paddles, propellers, and special shapes are used, the choice depending on the properties of the agitated material and the type of agitation desired.

Agitation Power. Over a wide range of operating conditions, the parameter that most completely determines the performance of an agitator is the power delivered to the fluid—the greater the power delivered, the more effective is the agitation. The power required by a rotating impeller cannot be computed directly; rather it must be measured for a geometrically similar model of the impeller and its surroundings and then be scaled up or down. The most useful scaling correlations involve the impeller Reynolds number $D_a^2N\rho/\mu$, the power number $Pg_c/D_a^5\rho N^3$, and the Froude number g/N^2D_a. Examples of this correlation are presented in Fig. 15.17. Since agitator geometry (including impeller dimensions, impeller position, vessel dimensions, baffles, etc.) affects the relationship between Reynolds number and power number, it is preferable to use experimental data rather than published correlations.

Functional Performance of Agitators. The degree of effectiveness of agitation is related to the intensity of shear, the level of turbulence, and the circulation rate produced by the impeller. For an impeller of given design, these quantities are determined by the agitator speed and power delivery. The following equations relate velocity head, power, and circulation rate for geometrically similar impellers under turbulent flow conditions:

$$Q = N_Q N D_a^3 \tag{15.37}$$

$$H = \frac{N_P N^2 D_a^2}{N_Q g_c} \tag{15.38}$$

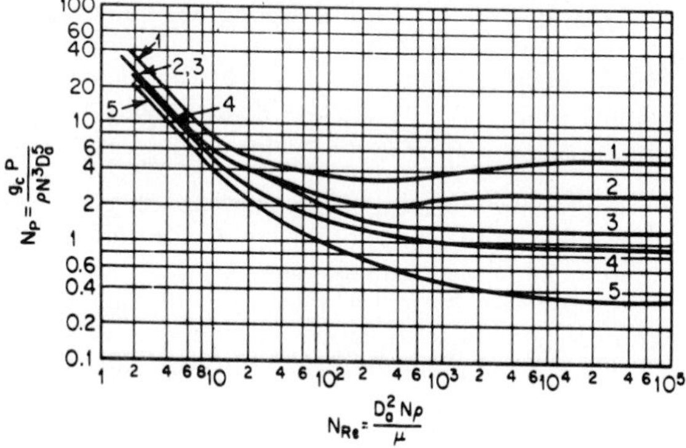

FIGURE 15.17 Impeller power correlations. (1) Six-blade turbine (curved blades), $D_a/W_i = 5$, with four baffles of height equal to 1/12 agitator tank diameter. (2) Six-blade turbine (straight vertical blades), $D_a/W_i = 8$; same baffle arrangement as (1). (3) Six-blade turbine (straight blades with 45° pitch), $D_a/W_i = 8$; same baffle arrangement as (1). (4) Propeller, pitch equal to $2D_a$; four baffles of height equal to 1/10 agitator tank diameter. (5) Propeller (pitch equal to D_a); same baffle arrangement as (4). *(Curves 4 and 5 from Rushton, Costich, and Everett, Chemical Engineering Progress 46: 395, 467, 1950, by permission; curves 2 and 3 from Bates, Fondy, and Corpstein, Industrial Engineering and Chemical Process Design Development 2: 310, 1963, by permission of the copyright owner, the American Chemical Society.)*

$$P = \rho QH = N_P \rho N^3 \frac{D_a^5}{g_c} \tag{15.39}$$

where Q is impeller discharge rate, ft^3/s (m^3/s); H is velocity head, $ft \cdot lb_f/lb_m$ ($N \cdot m/N$); N_P is power number; and N_Q is discharge coefficient, dimensionless.

The intensity of turbulence is primarily a function of the velocity head induced by the impeller. To achieve a higher discharge velocity head at a given rate of power consumption requires smaller impeller diameter and higher rotational speeds. The high-intensity turbulence of such an impeller is appropriate for (1) production of large interfacial area and small droplets in gas-liquid and immiscible liquid systems, (2) promotion of mass transfer between phases, (3) solids deagglomeration by shear forces, (4) rapid mixing in the impeller discharge system, and (5) suspension of particles with relatively high settling rates.

To achieve a larger impeller discharge rate at constant power consumption requires larger impeller diameters and lower rotational speeds. A large-diameter impeller with its associated high discharge rate is appropriate for (1) short times to complete mixing of a miscible liquid throughout the entire agitator volume, (2) promotion of heat transfer at a heat-exchange surface within the vessel, (3) maintaining uniform temperatures and concentrations throughout a backmix reactor, and (4) suspension of particles with relatively low settling rates.

11. Spray Generation. A *spray* is a mechanically produced, unstable suspension of liquid droplets in a gas. Sprays are generated from a continuous liquid by nozzles, of which there are three principal types: *pressure nozzles, rotating noz-*

zles, and *gas-atomizing nozzles* (or *two-fluid nozzles*). The nozzle converts mechanical energy into the surface free energy required to form liquid droplets, either by means of a pressure-head loss (pressure nozzles and gas-atomizing nozzles) or a spinning shaft (rotating nozzles). The smaller the average droplet diameter, the more energy is required per pound of liquid dispersed.

Pressure Nozzles. In pressure nozzles, classified as *hollow-cone, solid-cone, fan*, or *impact* types depending on the shape of the spray, the fluid is throttled through small openings, thereby attaining high velocities. Droplets are formed because of shear forces at the nozzle exit, by its instability at such high levels of inertia, or by impact with another jet or a solid surface. Discharge rates and droplet sizes are functions of orifice opening and applied pressure drop. Typical values are given by Tables 15.2 and 15.3.

Gas-Atomizing Nozzles. These nozzles disintegrate a stream of liquid by contact with a high-velocity stream of gas. The liquid may be preatomized by a pressure nozzle or injected as a generally continuous sheet. In either case, the primary energy source for atomization is the gas pressure drop, not the liquid pressure drop. Two-fluid nozzles produce very fine droplets at the price of high energy consumption. Typical drop-size distribution for a gas-atomizing nozzle is given in Table 15.4.

12. Gas Sparging. A *sparger* is a distributor that disperses gas into the body of a liquid by emitting bubbles or jets of gas through an individual orifice, an array of orifices, or a porous structure. Spargers are used to promote gas-liquid mass transfer or to produce dispersions; sometimes they are used as gentle agitators.

Simple Bubblers. Open-ended pipes or perforated tubes or plates with orifices ⅓ to ½ in (3.2 to 12.7 mm) in diameter are used as spargers. A perforated tube should be designed so that the pressure drop across the individual orifice is large compared to the pressure drop for flow through the tube. At practical operating rates, simple bubblers produce jets rather than bubbles; the jets disintegrate, but the resulting cloud of bubbles may include some as large as 0.5 in (12.7 mm). Their effectiveness as mass-transfer promoters is orders of magnitude below that of vigorously agitated tanks or packed columns, and they are used only for very easy transport operations (e.g., air humidification) or for gentle mechanical agitation.

Porous Septa. Porous plates, tubes, or disks are made by bonding or sintering carefully sized particles of carbon, ceramic, metal, or polymer. The resulting septa may be used as spargers to produce much smaller bubbles than will result from a simple bubbler.

The gas flux through a porous septum is limited on the lower side by the requirement that, for good performance, the whole sparger surface should bubble uniformly, and on the higher side by the onset of serious bubble coalescence. In a practical range of fluxes, the size of bubbles produced is a direct function of both pore size and pressure drop. Figure 15.18 shows the recommended limit of flux density for a typical porous medium. The working pressure drop across typical porous media is larger than the dry permeability. Wet-permeability values should be used in all design calculation.

Porous spargers are used generally to promote gas absorption. They are of the same order of effectiveness as packed or tray columns or agitated vessels, but no

TABLE 15.2 Discharge Rates and Included Angle of Spray of Typical Pressure Nozzles

		Discharge, gpm (L/min), and included angle of spray							
		10 lb/in² (69 kPa)		25 lb/in² (172 kPa)		50 lb/in² (345 kPa)		100 lb/in² (689 kPa)	
Nozzle type	Orifice diameter, in	Discharge	Angle, deg	Discharge	Angle, deg	Discharge	Angle, deg	Discharge	Angle, deg
Hollow cone	0.046 (1.2)	—	—	0.10 (0.4)	65	0.135 (0.51)	68	0.183 (0.69)	75
	0.140 (3.6)	0.535 (2.0)	82	0.81 (3.1)	88	1.10 (4.2)	90	1.50 (5.7)	93
	0.218 (5.5)	1.25 (4.7)	83	1.88 (7.1)	86	2.55 (9.7)	89	3.45 (13.1)	92
	0.375 (9.5)	7.2 (27.3)	62	11.8 (44.7)	70	16.5 (62.5)	70		
Solid cone	0.047 (1.2)	—	—	0.167 (0.63)	65	0.235 (0.89)	70	0.34 (1.3)	70
	0.188 (4.8)	1.60 (6.1)	55	2.46 (9.3)	58	3.42 (12.9)	60	4.78 (18.1)	60
	0.250 (6.4)	3.35 (12.7)	65	5.40 (20.4)	70	7.50 (28.4)	70	10.4 (39.4)	75
	0.500 (12.7)	17.5 (66.2)	86	27.5 (104.1)	84	38.7 (146.5)	73		
Fan	0.031 (0.79)	0.085 (0.32)	40	0.132 (0.50)	90	0.182 (0.69)	110	0.252 (0.95)	110
	0.093 (2.4)	0.70 (2.7)	70	1.12 (4.2)	76	1.57 (5.9)	80	2.25 (8.5)	80
	0.187 (4.8)	2.25 (8.5)	50	3.70 (14.0)	59	5.35 (20.3)	65	7.70 (29.1)	65
	0.375 (9.5)	9.50 (35.9)	66	15.40 (58.3)	74	22.10 (83.6)	75	30.75 (116.4)	75

Source: Data furnished through the courtesy of the Spray Engineering Co., Cambridge, Mass. (Metricated by the editors.)

TABLE 15.3 Drop-Size Distributions Produced by Three Hollow-Cone Nozzles of the Same Design

Nominal drop diameter, μm	Number of drops in each size group					
	0.063-in (1.6-mm) Orifice diameter			0.086-in (2.2-mm) Orifice diameter		0.128-in (3.3-mm) Orifice diameter,
	50 lb/in² (345 kPa)	100 lb/in² (689 kPa)	200 lb/in² (1379 kPa)	100 lb/in² (689 kPa)	200 lb/in² (1379 kPa)	200 lb/in² (1379 kPa)
10	375	800	1700	100	300	100
25	200	280	580	60	150	50
50	160	180	260	41	100	45
100	50	60	70	26	34	27
150	27	31	35	14	18	15
200	19	23	27	9	12	11
300	8	9	11	5	8	6
400	2	4	4	4	7	3
500	1	1	—	2	1	2
600	1	—	—	1	—	1

Note: 1 μm = 10^{-4} cm = 0.0000394 in (0.001 mm). The nominal diameter is the middiameter of a drop group which includes a finite range of sizes. The 25 group includes drops from 17.5 to 37.5 μm; the 50 group contains drops from 37.5 to 75 μm; etc. The number of drops has been adjusted in each case so that the total amount of fluid sprayed is the same for each size distribution.

TABLE 15.4 Drop-Size Distribution of a Small Atomizing Nozzle

Drop diameter, μm	Number of drops	Drop diameter, μm	Number of drops
2	390,000	35	1730
5	340,000	40	1080
10	165,000	45	650
15	40,200	50	430
20	11,680	60	350
25	4,970	70	220
30	2,160		

Note: The fluid pressure and the gas pressure were each 15 lb/in² (103.4 kPa). The total quantity of fluid represented by this size distribution is the same as that in Table 15.8, so that the numbers of drops are directly comparable.

generalized data or methods are available for their specification as mass-transfer devices. Their advantages of simplicity and inexpensiveness are balanced by their susceptibility to plugging and their awkwardness for countercurrent operations.

13. Fluidization

Definitions. If a gas is passed upward through an unrestrained and unconsolidated bed of granular solids with ever-increasing velocity, the pressure drop across the bed due to friction will increase until it becomes equivalent to the

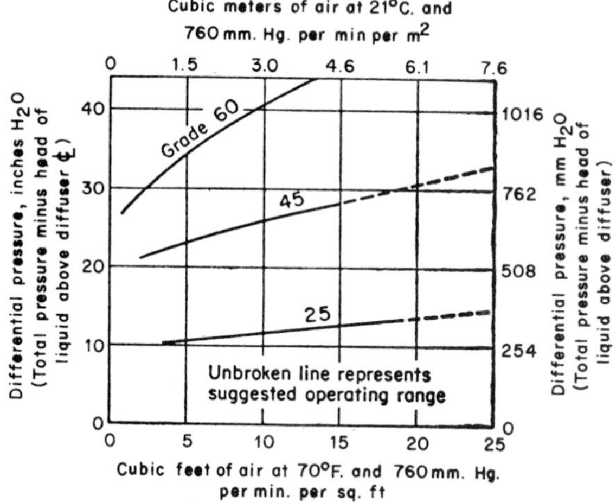

FIGURE 15.18 Pressure drop across porous carbon diffusers submerged in water at 70°F (21°C). *(National Carbon Co.)*

weight of the bed plus the friction between fluid and walls. With further increase in the gas velocity, the bed tends to rise as a unit, but its unconsolidated character causes it instead to expand until the increased porosity allows the friction again just to balance the pressure drop. As the bed becomes more expanded, individual particles achieve freedom to interchange position, and the bed can circulate. As the gas velocity is further increased, the drag force on individual particles eventually becomes large enough to balance gravitational forces, and individual particles can be entrained and carried out of the bed. The latter conditions are termed *particulate* or *continuous fluidization*, the former are termed *aggregative* or *batch fluidization*. The designation *fluidized bed* usually refers to aggregative fluidization.

Continuous Fluidization. One primary application of continuous fluidization is *pneumatic conveying* of a dispersed solid by means of compressed air. The solid phase is introduced into the carrier gas (e.g., by aspiration or by a screw conveyor), is immediately fluidized, and is transported to the destination where it is removed from the carrier gas by means of a cyclone or other collection device. The gas-pressure drop is greater than would occur if no solids were present, increasing with increased solids concentration. Empirical correlations between pressure drop and solids loading are imprecise, and experimental data should be used where possible.

Two operating problems associated with pneumatic conveying are *saltation*, or deposition of particles in the pipeline with restriction of gas flow to the unblocked section, and *erosion* of the pipe wall caused primarily by inertial impact of particles at elbows, etc. Saltation can be prevented by employing superficial gas velocities in excess of a certain minimum velocity, termed the *saltation velocity*, which is a function of particle type and mass loading in the gas and which must generally be determined experimentally. Erosion is best combated by protection at the inertial impact point, either by an erosion plate, reinforced pipe walls, or a saltation layer resulting from the flow pattern.

Batch Fluidization. Fluidized beds sometimes are called *boiling beds*. Indeed, the expanded suspended mass of the bed does resemble a boiling liquid. This mass has a zero angle of repose, seeks its own level, and assumes the shape of the containing vessel. Just as in a vessel designed for boiling a liquid, space must be provided for vertical expansion of the solids and for disengaging splashed and entrained material.

Conditions for Fluidization. The size of solid particles that can be fluidized varies greatly, from less than 1 μm to 2½ in (6.4 cm). It is generally concluded that particles distributed in size between 65 mesh and 10 μm are the best for smooth fluidization (least formation of large bubbles). Large particles cause instability and result in slugging or massive surges. Small particles (less than 10μm) frequently, even though dry, act as if damp, forming agglomerates or fissures in the bed or spouting. Adding finer-sized particles to a coarse bed or coarser-sized particles to a bed of fines often results in better fluidization.

The upward velocity of the gas is usually between 0.5 and 10 ft/s (0.15 and 3.0 m/s). This velocity is based on the flow through the empty vessel and is frequently referred to as the *superficial velocity*. Its upper limit is fixed by the terminal free-settling velocity of the smallest particles in the bed that should not be carried over. The velocity used is best determined by test in equipment where visual observations of the action of the bed can be made. The flow required to maintain a completely homogeneous bed of solids, whereby coarse or heavy particles will not segregate from the fluidized portion, is very different from the minimum fluidizing velocity discussed in most studies of fluidization.

Bed height is determined by a number of factors, either individually or collectively, such as

1. Space-time yield
2. Gas-contact time
3. *L*/*D* ratio required to provide staging
4. Space required for internal heat exchangers
5. Solids-retention time

Generally, bed heights are not less than 12 in (30.5 cm) or more than 50 ft (15.2 m).

Heat Transfer and Mixing in Fluidized Beds. Heat-exchange surfaces have been used to provide means of removing or adding heat to fluidized beds. Usually, these surfaces are provided in the form of vertical tubes manifolded at top and bottom. Other shapes have been used such as horizontal bayonets. In any such installations, adequate provision must be made for abrasion of the exchanger surface by the bed. Normally, the transfer rate is 5 to 25 times that for solids-free gas.

Heat transfer from solids to gas and gas to solids usually results in a coefficient of about 3 to 10 Btu/(h · ft^2 · °F) [17.0 to 56.8 W/(h · m^2 · °C)]. However, the large area of the solids per cubic foot of bed [15,000 ft^2/ft^3 (49,240 m^2/m^3)] for 60-μm particles of 40 lb/ft^3 (640 kg/m^3) bulk density] results in the rapid approach of gas and solids temperatures. With a fairly good distributor, essential equalization of temperatures occurs within 1 to 3 in (2.5 to 7.6 cm) of the top of the distributor.

Bed thermal conductivities in the vertical direction have been measured in the laboratory in the range of 20,000 to 30,000 Btu/(h · ft · °F/ft) [373 to 559 kW/(m · h · °C/m)]. Horizontal conductivities for ⅛-in (3.2-mm) particles in the range of 1000 Btu/(h · ft^2 · °F/ft) [18.6 kW/(h · m^2 · °C/m)] have been measured in large-scale experiments.

Except at large *L*/*D* ratios, the temperature in the fluidized bed is uniform, the temperature at any point being, generally, within 10°F (5.6°C) of any other point. The solids, too, will be well mixed. For all practical purposes, beds with *L*/*D* ratios of from 1.0 to 4 can be considered to be completely mixed continuous-reaction vessels as far as the solids are concerned.

Equipment. The use of the fluidization technique requires in almost all cases the employment of a fluidized-bed system rather than an isolated piece of equipment. Figure 15.19 illustrates the arrangement of components of a system used in cases

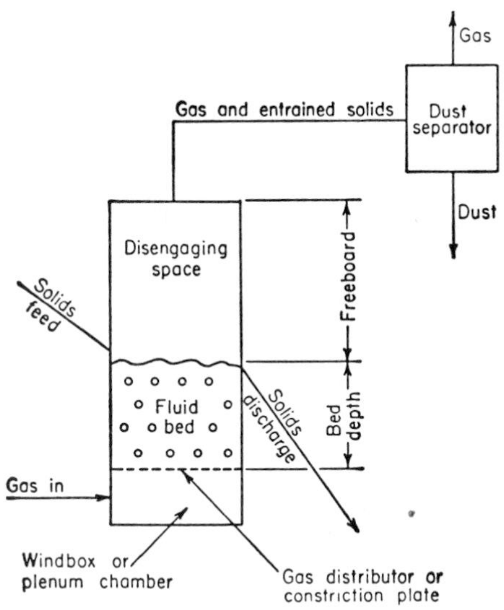

FIGURE 15.19 Noncatalytic fluidized-bed system.

where the flow of solids is small, such as is generally encountered in noncatalytic usages of the fluidized beds or in catalytic units where there is little or no deactivation of the catalyst. Figure 15.20 illustrates a catalytic-type unit such as is used for petroleum cracking where large quantities of solids flow into and out of the reactor and to and from the catalyst regenerator, which also is usually a fluidized bed. It is obvious that, in the simplified form, the only difference between a fluidized catalytic-cracking unit and the fluidized-bed units used in most other cases is the method and point of solids feed.

The major parts of a fluidized-bed system can be listed as follows:

1. Reaction vessel
   a. Fluidized-bed portion
   b. Disengaging space or freeboard
   c. Gas distributor
2. Solids feeder or flow control
3. Solids discharge

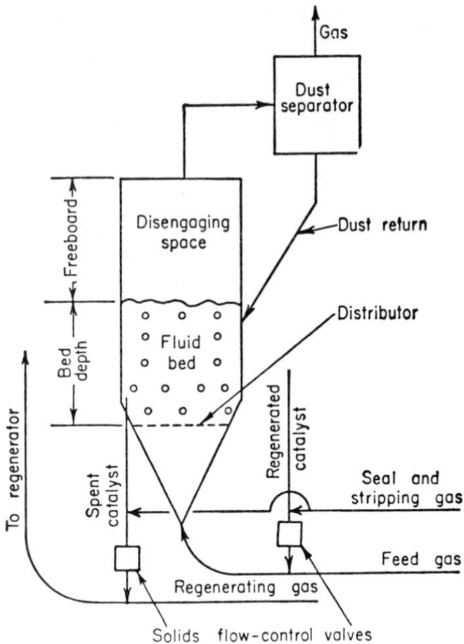

FIGURE 15.20 Catalytic fluidized-bed system.

4. Dust separator for the exit gases
5. Instrumentation
6. Gas supply

The reactor is usually a vertical cylinder; however, there is no real limitation on shape. The specific design features vary with operating conditions, available space, and use. The lack of moving parts tends toward simple, clean design.

The freeboard or disengaging height is frequently chosen rather arbitrarily or based on experience. It has been established that carry-over of solids entrained by the gases is reduced as the vertical distance between the top of the dense-phase fluidized bed and gas-outlet port is increased. Small-scale experiments also have shown that the size distribution of the solids entrained by the gases is reduced as the freeboard height or cross-sectional area is increased. However, for some distance (from a few inches to a number of feet) the size distribution of the solids in the dilute suspension just above the fluid bed is the same as the size distribution of the solids in the fluid bed.

The gas distributor has a considerable effect on proper operation of the fluidized bed. Basically, there are two types: (1) for use where the inlet gas contains solids, and (2) for use where the inlet gas is clean. In most cases, the distributor is designed to prevent backflow of solids during normal operation, and in many cases it is designed to prevent backflow during shutdown. In order to provide distribution, it is necessary to restrict the gas or gas and solids flow so that pressure drops across the restriction amount to from a few inches of water to a few pounds per square inch. As a general rule, pressure drops in excess of 2 lb/in^2 (13.8 kPa) are not used.

In cases where both solids and gases pass through the distributor, such as in catalytic-cracking units, a number of variations are or have been used, such as concentric rings in the same plane, with the annuli open, concentric rings in the form of a cone, grids of T bars or other structural shapes, flat metal perforated plates supported or reinforced with structural members, and dished and perforated plates concave both upward and downward. The last two forms are generally more economical.

Fluidized-bed reactors usually are designed by scaling up a laboratory or pilot unit. Considerable difficulty has been encountered in such scale-up because of the staging effect achieved in high L/D ratio units used in the laboratory or semiworks as compared with the lower L/D ratios used in commercial units.

MECHANICAL SEPARATIONS AND PHASE COLLECTION

14. Filtration. *Filtration* is the mechanical phase separation of a fluid-solid suspension or slurry by passage of the liquid through a porous septum, or filter medium, which retains the solids. The clarified liquid product is called the *filtrate*, and the retained solids are called the *filter cake*. As the filtration proceeds, the liquid in the feed slurry must flow through the cake that is being formed as well as the filter medium; therefore, the resistance to flow continually increases. The driving force for this flow can be gravity, a pump-developed pressure on the upstream side of the filter medium, or a vacuum applied to the downstream side. Filtering devices are categorized on the basis of the method for creating the driving force and the mode of operation.

Types of Filters. Gravity filters are the simplest and are always intermittent in operation. They are usually in the form of a false-bottomed tank; the false bottom is perforated and supports the filter medium. The slurry is pumped into the tank, and the liquid drains by gravity through the medium into the lower portion of the tank. The rate of filtration decreases continually as the filter cake builds up, and the unit must eventually be shut down and the cake removed.

The *plate-and-frame filter press* is an intermittent-pressure filter. The plates are solid with channeled or ribbed faces; the filter medium (e.g., filter cloth) is laid over the faces of each plate. The frames are hollow and of the same outer dimensions as the plates but are made in a variety of thicknesses. The plates carrying the medium and the frames are arranged alternately in a clamping device which, when tightened, forms a leak-tight assembly of cavities into which the slurry is pumped and paths for the escape of the filtrate which has passed through the filter medium. Channels and ports in each plate and frame permit the simultaneous flow of slurry into all frames, the flow of filtrate from all plates, and a flow for washing of the filter cake collected in each frame. When the frames become full, or when the resistance to flow becomes excessive, the clamping device is opened and the cake is removed from each frame.

Leaf filters are also intermittent-pressure filters. The leaf is a hollow frame covered with a wire support screen and the filter medium. A series of leaves is mounted in a pressure housing into which the slurry is pumped. The cake builds up on the leaves as the filtrate passes through the medium into the interior of the leaf and then through a conduit to the exterior of the housing. When the cake has built up to a predetermined thickness or the back pressure becomes excessive, the housing is drained and opened and the filter cake is removed.

There are several *continuous-vacuum filter* designs, but all are similar in basic principle. The filter medium is carried on a rotating drum or disk mounted on a hollow horizontal shaft. Vacuum is applied to the plenum behind the medium and the cake is formed during that part of the revolution when the medium is immersed in the slurry. The vacuum is maintained and the cake is dewatered after emerging from the slurry. At some later point in the revolution the vacuum is released and the cake is scraped off, leaving the medium ready for reimmersion in the slurry. The filtrate drains from the plenum through the central shaft to vacuum receivers. Provision also can be made for washing and air drying the cake before removing it.

Rate of Filtration. The rate of filtration is obviously the rate at which the filtrate can be forced through the cake and filter medium. The flow resistance of the cake is a function of its particle-size distribution, compressibility, thickness, and the viscosity of the filtrate. The volumetric flow rate through a hard, granular, noncompressible cake can be related to the operating conditions and cake properties:

$$\frac{dV}{d\theta} = \frac{A\Delta P_c}{\mu\alpha g_c(W/A)} \tag{15.40}$$

where α is the specific cake resistance (constant). For compressible cakes, α is a function of pressure drop, often correlated by

$$\alpha = \alpha'\Delta P_c^S \tag{15.41}$$

where α' and S are empirical constants depending on the particular solids being filtered. The compressibility S varies from zero for a hard, granular, noncompressible cake to unity for a soft, readily deformed, compressible cake; for most industrial slurry solids, S ranges from 0.1 to 0.8. For compressible cakes, the filtration rate is given by

$$\frac{dV}{d\theta} = \frac{A\Delta P_c^{(1-s)}}{\mu\alpha' g_c(W/A)} \tag{15.42}$$

The preceding rate expressions do not consider the flow resistance offered by the filter medium. This can be significant and even controlling particularly if the medium becomes blinded by deposition of solids in its interstices. A wide variety of filter-medium materials and weaves is available, and the choice involves a tradeoff between the resistance to flow and the clarity of the filtrate but must also consider resistance to wear and chemical attack, tendency to blind, mechanical strength, and cost.

The sizing of a filter for a particular slurry starts with laboratory or pilot-plant-scale tests followed by scale-up to the desired capacity. The choice of the type of filter involves a variety of factors and is best made with the advice of filter manufacturers. Table 15.5 may be used as a preliminary guide to filter selection.

15. Settling and Sedimentation. The separation of suspended solids from a fluid by gravitational forces is termed *settling* or *sedimentation*; differentiation between the two terms is not precise. *Settling* usually refers to very dilute suspensions of discrete rigid particles or discrete liquid droplets. *Sedimentation* is

TABLE 15.5 Filter Selection: Slurry Characteristics

	Fast filtering	Medium filtering	Slow filtering	Dilute	Very dilute
Slurry characteristics					
Cake-formation rate	in/s (mm/s)	in/min (mm/min)	0.05–0.25 in/min (1.3–6.4 mm/min)	< 0.05 in/min (< 1.3 mm/min)	No cake
Usual solids concentration	> 20%	10–20%	1–10%	< 5%	< 0.1%
Settling rate	Very rapid	Rapid	Slow	Slow	
Leaf test rate, $lb/(h \cdot ft^2)$ [$kg/(h \cdot m^2)$]	> 500 (> 2443)	50–500 (244–2443)	5–50 (24.4–244)	< 5 (< 24.4)	
Filtrate rate, $gal/(min \cdot ft^2)$ [$L/(min \cdot m^2)$]	> 5 (> 24.4)	0.2–5 (0.97–24.4)	0.01–0.02 (0.05–0.098)	0.01–2 (0.05–9.8)	0.01–2 (0.05–9.8)
Filters:					
Continuous vacuum filters:					
Multicompartment drum	X	X	X		
Single-compartment drum	X				
Dorrco	X				
Top-feed	X				
Horizontal-table	X				
Tilting-pan	X	X			
Horizontal-belt	X	X			
Disk	—	X	X		
Precoat	—	—	—	X	X
Batch vacuum leaf	—	X	X	X	X
Batch nutsche	X	X	X	X	X
Batch pressure filters:					
Plate-and-frame	—	X	X	X	X
Vertical leaf	—	X	X	X	X
Tubular	—	X	X	X	X
Horizontal plate	X	X	X	X	X
Cartridge, edge	—	—	—	—	X
Continuous pressure filters:					
Drum	—	X	X		
Fest	—	X	X	X	
Precoat	—	—	—	X	X

Source: Adapted from Porter et al., *Chemical Engineering* 78(4): 40, 1971.

applied to higher concentrations of solids and to solids that *flocculate* (i.e., agglomerate to form rigid lattices).

In the free or unhindered settling of discrete particles, the particle will accelerate until drag forces offered by the liquid exactly balance the net gravitational forces (i.e., particle weight minus buoyant forces) and thereafter will settle at a constant *terminal velocity* expressed by

$$v_t = \sqrt{\frac{4gD_p(\rho_p - \rho_f)}{3\rho_f C_D}} \tag{15.43}$$

where the drag coefficient C_D is a function of particle shape and the Reynolds number based on particle diameter, $N_{\text{Re}} = D_p v_t \rho_f/\mu$. At very low N_{Re} (<0.3), Stokes' law is valid for spherical or near-spherical rigid particles, and $C_D = 24/N_{\text{Re}}$; at higher N_{Re}, the coefficient decreases with increasing N_{Re}, but becomes less dependent on N_{Re}, becoming essentially constant; as the sphericity of a particle decreases, the drag coefficient increases. Nonrigid particles exhibit similar behavior, but droplet deformation and internal circulation affect the relationship between C_D and N_{Re}.

If the concentration of particles is sufficiently high that the particles collide or interact in any way, *hindered settling* results. In some cases Eq. (15.43) can be modified to give approximate values for hindered-settling velocities, but normally data must be obtained experimentally for each specific case. Since particle interactions increase drag forces, hindered-settling velocities are always lower than free-settling velocities. As the concentration of particles increases, interactions increase, leading to reduced terminal velocities.

When the suspended particles are flocculent, either naturally or by the addition of promoters termed *flocculating agents*, the settling behavior can vary greatly, depending on the concentration. At very low solids concentrations, flocs can settle freely and unhindered in a manner similar to single rigid particles. At higher concentrations, the flocs continue to grow as they settle, either by coalescence of flocs or by continued flocculation of fine particles. As a result, the settling velocity changes as the flocs settle. At still higher concentrations, the flocs coalesce completely and tend to settle as a single porous mass. In this case, fluid flow phenomena more closely resemble flow through a porous bed of solids than flow around a submerged particle.

In typical settling operations, the solids concentration (after a period of settling) becomes a function of depth. Distinct zones appear in which different mechanisms control settling velocities (i.e., free settling, hindered settling, flow in a porous bed), leading to the term *zone settling*. Since no mathematical analysis of settling is available other than for free settling of discrete particles, laboratory settling data must be obtained to evaluate design parameters.

16. Centrifugation. Centrifugal force can be applied to enhance the separation of liquid-solid or liquid-liquid suspensions by either settling or filtration. Centrifugal force is exerted on a mass that is following a curved path; the extent of the force depends on the radius of curvature and the angular velocity, increasing as velocity is increased or curvature is decreased. Thus a centrifugal force field thousands of times stronger than normal gravity can be generated in properly designed devices called *centrifuges*. A centrifugal force field can produce the same effects as a gravitational field but at rates proportional to the relative strengths of the fields. For example, the separation of a liquid-solids suspension by settling, which may be very slow when brought about by gravity, can be speeded up greatly by placing the suspension in a strong centrifugal field.

A centrifuge for the separation of phases consists of a rotatable vessel (the bowl or basket) in which the centrifugal force is generated, a drive mechanism for rotating the vessel, and means for introducing the separable mixture and removing, individually, the separated phases. The treatment vessel may be a bowl with solid walls or a basket with perforated walls. The operation may be batchwise or continuous; the former usually gives a more complete separation but the latter requires less operating labor. A summary of types and characteristics of centrifuges is given in Table 15.6.

Centrifuges may be classified by the operation carried out, namely, sedimentation or filtration. *Sedimentation centrifuges* are usually continuous-operation, solid-bowl machines yielding a dense phase of thickened sludge and a light phase of clear liquid. The concentration of solids in the sludge can be controlled by the throughput rate (or residence time) but often is limited by the flow (by the method of discharge) characteristics of the sludge. In the case of liquid-liquid suspensions, the degree of separation of the light and heavy phases is limited only by the residence time at any particular force field.

Filtration centrifuges usually have a perforated-wall cylindrical basket lined with a suitable filter medium, such as fine woven cloth or wire screen, which will retain the solids and allow the liquid to pass through. The centrifugal force acts directly on the liquid to push it through a cake of solids during buildup of the cake and also to drive it out of the pores of the cake during the dry spin. The latter action is much more effective than forcing air through the cake as in many standard filters. Both batch and continuous designs are available. The method of discharging the finished cake is a principal distinction among specific designs of both types of centrifuge.

The operating mode of batch-filtering centrifuges can be varied readily to provide the best conditions for the particular liquid-solids system. Continuous machines are much less flexible and must be much more closely specified for a particular application.

The design or specification of centrifuges, depending on the operation carried out, calls for knowledge of the type and rate of settling, volume of liquid retained in the bowl and the throughput rate, filtration or drainage rate, compressibility and porosity of the cake, and the particle-size range of the solids. Generally, predictions based on theory are very risky, and the specification of centrifuges should be based on scale-up from tests on laboratory machines of similar type and geometry. Most centrifuge manufacturers can provide testing services and are well versed in scale-up methods.

17. Screening. *Screening* is the mechanical separation of a mixture of particles into two or more fractions by means of a surface with multiple uniform perforations, termed a *screen*. Material retained on the first screen in a series (i.e., largest openings) is termed *oversize*; that passing through the last in a series (i.e., smallest openings) is termed *undersize*, or *fines*. A screen may consist of cloth woven from various fibers, a perforated plate, or uniformly spaced parallel bars. Screens are specified by the number of openings per linear inch (*mesh count*) or by the dimension of the openings, measured between and perpendicular to adjacent wires or bars (*aperture* or *clear opening*).

Commercial screening machines vary in specific design but have common principles. The material to be screened is fed to a screen surface which slants downward from inlet to outlet. The screen surface is rotated, vibrated, shaken,

TABLE 15.6 Characteristics of Commercial Centrifuges

Method of separation	Rotor type	Centrifuge type	Manner of liquid discharge	Manner of solids discharge or removal	Centrifuge speed for solids discharge	Capacity
Sedimentation	Batch	Ultracentrifuge	—	—	—	1 ml
		Laboratory, clinical	Batch	Batch manual	Zero	To 6 L (1.6 gal)
	Tubular	Supercentrifuge	Continuous*	Batch manual	Zero	To 1200 gal/h (4542 L/h)
		Multipass clarifier	Continuous*	Batch manual	Zero	To 3000 gal/h (11,355 L/h)
	Disk	Solid wall	Continuous*	Batch manual	Zero	To 30,000 gal/h (113,550 L/h)
		Light-phase skimmer	Continuous	Continuous for light-phase solids	Full	To 1200 gal/h (4542 L/h)
		Peripheral nozzles	Continuous	Continuous	Full	To 24,000 gal/h (90,840 L/h)
		Peripheral valves	Continuous	Intermittent	Full	To 3000 gal/h (11,355 L/h)
		Peripheral annulus	Continuous	Intermittent	Full	To 12,000 gal/h (45,420 L/h)
	Solid bowl	Constant-speed horizontal	Continuous*	Cyclic	Full (usually)	To 60 ft^3
			Continuous*	Cyclic	Zero or reduced	To 16 ft^3
		Variable-speed vertical	Continuous	Continuous screw conveyor	Full	To 18,000 gal/h (68,130 L/h) To 75 tons/hr. solids
		Continuous decanter				
Sedimentation and filtration	—	Screen bowl decanter	Continuous	Continuous	Full	To 16,000 gal/h (60,560 L/h) To 75 tons/h solids (76,200 kg/h)
Filtration	Conical screen	Wide-angle screen	Continuous	Continuous	Full	To 40 tons/h solids (40,640 kg/h)
		Differential conveyor	Continuous	Continuous	Full	
		Vibrating, oscillating, and tumbling screens	Continuous	Essentially continuous	Full	To 40 tons/h solids (40,640 kg/h) To 100 tons/h solids (101,600 kg/h)
		Reciprocating pusher	Continuous	Essentially continuous	Full	Limited data
	Cylindrical screen	Reciprocating pusher, single and multistage	Continuous	Essentially continuous	Full	To 30 tons/h solids (30,480 kg/h)
		Horizontal	Cyclic	Intermittent, automatic	Full (usually)	To 25 tons/h solids (25,400 kg/h)
		Vertical, underdriven	Cyclic	Intermittent, automatic or manual	Zero or reduced	To 6 tons/h solids (6096 kg/h)
		Vertical, suspended	Cyclic	Intermittent, automatic or manual	Zero or reduced	To 10 tons/h solids (10,160 kg/h)

*Interrupted during solids unloading.

oscillated, or otherwise mechanically energized to bring the material into contact with the screen and to help impel oversize particles through to the outlet. Undersize particles coming in contact with the screen fall through and are conveyed to a different outlet.

The efficiency of a screening device is dependent on the thickness of the particle bed above the screen, the length of time the particles are in contact with the screen, the amount of mechanical vibration (or rotation, etc.) to bring undersize particles into contact with the screen, and the characteristics of the particle mixture (agglomeration tendency, shape, etc.). Efficiency is defined in terms of percent of desirable fraction retained by the screening device divided by percent of that fraction in the feed. For a given screen, increased throughput generally results in reduced efficiency. The primary design variable is screen area required, which depends on the throughput to be handled and the desired efficiency.

18. Wet Classification. *Wet classification* is the separation of a mixture of particles into two or more fractions according to particle size or particle density by contact with a fluidizing medium, often water. It is used as a unit operation in the chemical process industry primarily for raw materials treatment (i.e., ore beneficiation, coal washing, etc.). Most types of wet classification utilize the different settling velocities of the different particles to remove coarse or dense particles (termed *sand*) from fine particles. Zones of settling are created (either hindered or free settling) in which the sands are retained behind a weir and removed in the underflow, and fines are carried along with the overflow. Sequential settling zones may be used to give several fractions as products. Mechanical agitation is often used to create zones of increased liquid velocities and to keep coarser particles in suspension, as well as water jets. Wet classification seldom produces sharp divisions of particle size or density between fractions, and it is most useful when there is a large difference in settling velocities between desired product and wastes. The application and operating characteristics of several types of wet classifiers are given in Table 15.7.

Jigging. *Jigging* is the separation of materials of different specific gravities by the pulsation of a stream of liquid flowing through a bed of the materials. The liquid pulsates, or "jigs," up and down, causing the heavy material to work down to the bottom of the bed and the lighter material to rise to the top. Each product is then drawn off separately.

The throughput capacity and power requirements of jigs depend on the character of the feed, the separation required, and the type of equipment used. The water consumption is high, 1200 to 2500 gal water per ton (4.5 to 9.3 L/kg) of solids processed.

Tabling. *Tabling* is the classification of particulate solids by means of an inclined, riffled, shaking surface (called a *table*) across which water or air is flowed. The particles are classified principally on the basis of density difference.

Wet tables require finer feed (dense ore, 6 to 150 mesh; light material, such as coal, < 1 in [25.4 mm]) than air tables [which handle ore up to ¼ in (6.4 mm) and coal up to 3 in (76.2 mm)].

Froth Flotation. *Froth flotation* is the fractionation of particulate solids based on differences in interfacial tensions between the solids, water, and air. It has been an important process in the beneficiation of ore. The ore particles are suspended in a liquid at a pulp density of 15 to 35 percent solids by mechanical or air agita-

TABLE 15.7 Sizes, Limitations, and Major Applications of Wet-Classification Machines

Type of classifier	Normal size range, ft (m)			Normal mesh of separation range*	Normal feed tonnage range	Maximum oversize in feed (mm)	Normal overflow, % solids range	Normal sand product, % solids range	Motor range, hp (kW)	Typical applications
	Width	Diameter	Maximum length							
Nonmechanical:										
Cone classifier	—	2–12 (0.6–3.1)	—	28–325	2–100 tons/h (2032–101,600 kg/h)	1/4 in (6.4)	5–30	35–60	None	For desliming and primary dewatering
Liquid cyclone	—	10 mm to 1 ft (10 mm to 1.2 m)	9 (2.7)	48 mesh to 5 μm	1/2–1500 gpm (1.9–5678 L/min)	14–325 mesh	5–30	55–70	Power for pressure head 5–60 lb/in^2 (34.5–414 kPa)	For medium or fine separations and closed-circuit grinding
Mechanical:										
Drag classifier	1–10 (0.3–3.0)	—	Not critical	28–200	5–350 tons/h (5080–355,600 kg/h)	1½ in (38.1)	5–30	70–83	1–10 (0.75–7.5)	For desliming, conveying, and closed-circuit grinding
Rake and spiral classifiers	1–20 (0.3–6.1)	—	40 (12.2)	20–200	5–350 tons/h (5080–355,600 kg/h)	1 in (25.4)	5–30	75–83	1/2–25 (0.4–18.7)	Closed-circuit grinding, washing and dewatering, desliming, process feed control
Bowl classifier	1½–20 (0.46–6.1)	4–28 (1.2–8.5)	40 (12.2)	100–325	5–200 tons/h (5080–203,200 kg/h)	1/2 in (12.7)	5–25	75–80	Bowl: 1–7½ (0.75–5.6) Rake: 1–25 (0.75–18.7)	Closed-circuit grinding, usually in secondary circuits
Bowl desiltor	4–16 (1.2–4.9)	20–50 (6.1–15.2)	40 (12.2)	100–325	5–250 tons/h (5080–254,000 kg/h)	1/2 in (12.7)	1–15	75–83	Bowl: 1–10 (0.75–7.5) Rake: 5–25 (3.7–18.7)	Recovery of fine sand, limestone, coal, and fine phosphate rock from large flow volumes

Hydroseparator	—	10–150 (3.0–45.7)	—	100–325	5–700 tons/h (5080–711,200 kg/h)	1/4 in (6.4)	1–20	30–50	1–15 (0.75–11.2)	For fine separation where large feed volumes are involved and drainage not critical
Solid-bowl centrifuge	—	18–54 in (0.5–1.4)	70 in (2.0)	200 mesh to 1 μm	10–600 gpm (37.9–2271 L/min)	1/4 in (6.4)	1–40	10–70	15–150 (11.2–112)	For fine-size fractionating
Sand washer	—	7–12 (2.1–3.7)	—	28–65	25–125 tons/h (25,400–127,000 kg/h)	1 in (25.4)	5–15	75–80	5–10 (3.7–7.5)	For desliming and dewatering large tonnages of solids
Countercurrent classifier	—	1 1/2–10 (0.46–3.0)	40 (12.2)	35–100	1–600 tons/h (1016–609,600 kg/h)	3 in (76.2)	5–30	75–83	1/4–25 (0.19–18.7)	Sand-slime separations, washing, closed-circuit grinding
Hydraulic:										
Sizer	1 1/2–20 (0.46–6.1)	—	5–20 (1.5–6.1)	8–150	2–100 tons/h (2022–101,600 kg/h)	3/16 in (4.8)	1–10	40–60	1–2 for air pressure (0.75–1.5)	Multiproduct unit for exceptionally clean sands fractionated into narrow size ranges: minimum of 3 tons hydraulic water per ton (3048 kg/kg) sand
Super sorter†	6 (1.8)	—	40 (12.2)	8–150	40–150 tons/h (40,640–152,400 kg/h)	3/8 in (9.5)	1–10	40–60	1 to operate pincer valves (0.75)	Multiproduct unit for exceptionally clean sands fractionated into narrow size ranges: minimum of 3 tons hydraulic water per ton (3048 kg/kg) sand
Siphon sizer‡	—	3–30 (0.9–9.1)	—	14–150	1–100 tons/h (1016–101,600 kg/h)	1 in (25.4)	1–10	40–60	None	Two-product unit efficient for desliming and exceptionally clean sands, washing, closed-circuit grinding; minimum of 2 tons hydraulic water per ton (2032 kg/kg) sand
Hydroscillator‡	4–12 (1.2–3.7)	4–14 (1.2–4.3)	40 (12.2)	20–150	5–250 tons/h (5080–254,000 kg/h)	1/2 in (12.7)	5–30	75–83	Oscillator: 3–10 (2.2–7.5) Rakes: 5–20 (3.7–14.9)	Two-product unit for exceptionally clean sand having low moisture content; closed-circuit grinding, washing; minimum of 0.5 ton hydraulic water per ton (508 kg/kg) sand

*Size of screen retaining 1½ percent of the overflow solids.
†Trademark of Deister Concentrator Co., Inc.
‡Trademark of Dorr-Oliver, Inc.

tion. The slurry is treated with chemicals, called *promoters*, which render the surfaces of specific minerals air-avid and water-repellent. Air bubbles are then introduced by direct aeration (see the section "Gas Sparging"), agitation, or injection of water saturated with air at a much higher pressure. The air bubbles adhere to the treated particles and carry them to the surface froth, which is skimmed off.

The valuable concentrates from froth flotation may be either the froth product which collects at the top or the underflow product. In the case of metallic sulfide ores of copper, lead, zinc, nickel, mercury, and molybdenum and native gold and silver, the values collect in the froth. In glass-sand flotation, iron-bearing minerals are floated off in the froth, while high-grade silica values appear as underflow.

19. Crystallization. *Crystallization* is the production and recovery of solid material from a solution brought about by reduction of solubility due to temperature change or by evaporation of solvent. A saturated solution is fed to a vessel, often an agitated vessel, and heat is added or removed so that the concentration in the solution is above the solubility at the operating conditions, i.e., supersaturated. The circulating slurry of crystals in supersaturated liquor is continuously withdrawn, and crystals are removed, e.g., by settling, centrifugation, or filtration. The mother liquor, saturated at the operating conditions, is returned to the process.

In the crystallization vessel, two processes take place simultaneously: crystal nucleation and crystal growth. For a given solvent-solute system, the rate of nucleation depends almost entirely on the degree of supersaturation in the crystallizer, increasing rapidly as supersaturation increases. The rate of crystal growth depends on number of growth sites (i.e., total surface area), resistance to diffusional transport to growth site, resistance to incorporation into crystal, and degree of supersaturation. In general, the greater the degree of supersaturation maintained in the crystallizer, the greater is the amount of nucleation, the faster is the rate of crystallization, and the smaller are the product crystals.

CHEMICAL KINETICS AND REACTOR DESIGN

20. Introduction. Nearly all industrial chemical processes involve one or more steps where the process steam undergoes a chemical transformation. Analysis and design of *chemical reactors* to bring about such transformations are a most important facet of chemical engineering practice, requiring understanding of many other phases of chemical engineering, notably fluid flow, heat and mass transfer, thermodynamics, and process control, in addition to a fundamental understanding of the chemical reactions involved.

The principles governing the mechanisms of chemical reactions and the rates at which they proceed comprise the field of *chemical kinetics*. Although the theory of chemical kinetics is imperfect, it is a useful guide for analyzing the results of experimental investigations and is thus the basis for design of practical industrial reactor systems.

The rate of a chemical reaction is best described in terms of number of moles of reactant converted (or moles of product produced) per unit of reactor volume per unit time. Thus for the reaction with the stoichiometric equation

$$aA + bB + cC \rightleftharpoons pP + qQ \tag{15.44}$$

the rate of reaction may be described by

$$rV = \frac{-1}{a}\frac{dN_A}{d\theta} = \frac{-1}{b}\frac{dN_B}{d\theta} = \frac{-1}{c}\frac{dN_C}{d\theta} = \frac{1}{p}\frac{dN_P}{d\theta} = \frac{1}{q}\frac{dN_Q}{d\theta} \tag{15.45}$$

The rate of appearance (or disappearance) of any product (or reactant) can be obtained in terms of the rate of change of any other participant by multiplication of Eq. 45 by the appropriate stoichiometric coefficient. If the reaction volume is constant, the concentration of reactants may be introduced:

$$r = \frac{1}{a}\frac{d(N_A/V)}{d\theta} = -\frac{1}{a}\frac{dC_A}{d\theta} \tag{15.46}$$

This is approximately true for most liquid-phase reactions taking place in a tank where material is neither added nor removed (*batch reaction*) or where no significant change in liquid level occurs in a flow reaction. It is exact for gas-phase reactions (flow or batch) confined in a rigid vessel.

For steady-state flow reactions with no longitudinal mixing, the composition at any point is constant with time, and the reaction rate may be defined in terms of change in composition with position as

$$r = \frac{F}{a}\frac{dX_A}{dV} \tag{15.47}$$

where F is volumetric feed rate, X_A is moles of A converted per unit volume of total feed, and V is the reactor volume upstream of that point in the reactor.

Reactions are classified as homogeneous or heterogeneous depending on whether they occur in a single phase or involve contact of several phases, as exothermic or endothermic according to whether they liberate energy or absorb energy from the surroundings, and as simple or complex depending on whether or not the rate law follows the stoichiometric coefficients directly.

21. Homogeneous Reactions. Reactions in which both reactants and products are in the same phase throughout the reaction are termed *homogeneous reactions*. Some reactions involving phase change may be treated as homogeneous even if there is a phase change between the initial reactant state and final product state provided the phase change occurs rapidly enough so as not to affect the overall reaction rate. (A reaction which results in a precipitated product, for example, may be treated as a homogeneous reaction during which the product concentration is constant at its solubility level.) Practically, homogeneous reactions must occur within a liquid or gas phase.

Reaction Order. In general, the rates of reactions whose mechanisms are simple have been found to be proportional to integral powers of the concentrations of some or all of the reacting components. That is, for a reaction involving reactants A, B, and C,

$$r = kC_A^{\alpha}C_B^{\beta}C_C^{\gamma} \tag{15.48}$$

The exponents are experimentally determined and are not necessarily equal to the stoichiometric coefficients of the reaction equation. In general, they have a value between 0 and 3. For the special case of single-step reactions involving a small number of reacting molecules, the exponents α, β, and γ will be integers.

Rate laws for more complex reactions having several steps, involving catalyzed reactions, chain reactions, etc., will be much more complicated than Eq.

(15.48). Such reactions may sometimes be described by Eq. (15.48) in order to obtain a correlation between rates and reactant concentrations. If this is done, the exponents α, β, and γ will often have noninteger values. When the reaction mechanism is unknown, this is often the only available procedure. It should be used with caution, however, and usually the values of α, β, and γ which are experimentally determined apply only to the conditions (reactant concentration, catalyst concentration, temperature, etc.) for which the data were taken.

The reaction kinetically described by Eq. (15.48) is said to be αth order with respect to A, βth order with respect to B, and γth order with respect to C; as a whole, its order is $(\alpha + \beta + \gamma)$th. Theory suggests that order may be related to molecularity of the reaction mechanism; if so, the order of simple homogeneous reactions with respect to each component should be finite and represented by an integer. Most, in fact, are of first, second, or third order, the latter being rare. Under certain conditions, however, reactions appear to be of zeroth order with respect to others, especially in complex reaction mechanisms.

To postulate a reaction mechanism and order without reference to actual rate data is speculative and dangerous if a reactor design is to be based on the postulate. The order assigned a reaction must rationalize reliable experimental kinetic data. On the other hand, data that indicate a homogeneous reaction to be of exotic order should be critically examined, or the method of their treatment should be questioned, or both.

Equation (15.48) gives the rate of a reaction proceeding irreversibly among three components. If the reaction of interest were, instead, a reversible one (as, strictly speaking, all reactions are), such as

$$A + B + C \leftrightharpoons P + Q \tag{15.49}$$

Eq. (15.49) would describe only the rate of the forward half-reaction. The reverse might be expected to exhibit a rate proportional to simple powers of the concentrations of the products P and Q:

$$r' = k' C_P^{\rho} C_Q^{\sigma} \tag{15.50}$$

Thus the reverse half-reaction would be ρth order with respect to P, σth order with respect to Q, and $(\rho + \sigma)$th order overall.

It should be noted that the net rate of a reversible reaction is the algebraic sum of its forward and reverse rates. For the reaction described by Eqs. (15.48) and (15.50),

$$r_{\text{net}} = r - r' \tag{15.51}$$

The coefficients k and k' of Eqs. (15.49) and (15.50) are known as specific rate constants, peculiar to a particular reaction and temperature but independent of concentrations of reactants.

Integrated Rate Equations. If one substitutes the appropriate rate law [e.g., Eq. (15.48)] into Eq. (15.46) and relates the concentrations by means of a material balance, it is possible to integrate Eq. (15.46) to give reactant concentrations as a function of time. For irreversible reactions of integral order, these integrated equations have a simple form. Table 15.8 presents integrated equations for simple irreversible, constant-volume reactions.

Reaction rate constants [k in Eq. (15.49)] may be determined from experimental data by plotting concentration versus time in a manner determined by the form

TABLE 15.8 Rate Equations for Reactions of Simple Order

Order	Differential equation	Constant-volume process
Zero	$-\frac{dN_A}{V d\theta} = k$	$k(\theta - \theta_0) = C_A{}^0 - C_A$
One-half	$-\frac{dN_A}{V d\theta} = kC_A{}^{1/2}$	$k(\theta - \theta_0) = 2(C_A{}^{01/2} - C_A{}^{1/2})$
First	$-\frac{dN_A}{V d\theta} = kC_A$	$k(\theta - \theta_0) = \ln \frac{C_A{}^0}{C_A}$
Second	$-\frac{dN_A}{V d\theta} = kC_A{}^2$	$k(\theta - \theta_0) = \frac{1}{C_A} - \frac{1}{C_A{}^0}$
	$-\frac{dN_A}{V d\theta} = kC_A C_B$	$k(\theta - \theta_0) = \frac{1}{C_B{}^0 - C_A{}^0} \ln \frac{C_A C_A{}^0 + C_A{}^0 C_B{}^0 - C_A{}^{02}}{C_A C_B{}^0}$ $\quad C_A{}^0 \neq C_B{}^0$*
Third	$-\frac{dN_A}{V d\theta} = kC_A{}^3$	$2k(\theta - \theta_0) = \frac{1}{C_A{}^2} - \frac{1}{C_A{}^{02}}$
	$-\frac{dN_A}{V d\theta} = kC_A C_B C_C$	$k(\theta - \theta_0) = \frac{1}{(C_B{}^0 - C_A{}^0)(C_C{}^0 - C_A{}^0} \ln \frac{C_A{}^0}{C_A} + \frac{1}{(C_B{}^0 - C_C{}^0)(C_B{}^0 - C_A{}^0} \ln \left(\frac{C_B{}^0}{C_A + C_B{}^0 - C_A{}^0}\right)$ $+ \frac{1}{(C_C{}^0 - C_B{}^0)(C_C{}^0 - C_A{}^0} \ln \left(\frac{C_C{}^0}{C_A + C_C{}^0 - C_A{}^0}\right)$ $\quad C_B{}^0 \neq C_C{}^0 \neq C_A{}^0$†

NOTE: C^0 and θ_0 are initial conditions for time and concentration, respectively.
* If $C_A{}^0 = C_B{}^0$, use expression for $-dN_A/V d\theta = kC_A{}^2$.
† If $C_A{}^0 = C_B{}^0 = C_C{}^0$, use expression for $-dN_A/V d\theta = kC_A{}^3$.

of the integrated rate law [e.g., for a first-order reaction plot $\ln(C_{A0}/C_A)$ versus time; the reaction rate constant is the slope of the line].

Reversible reactions, consecutive reactions ($A \rightarrow B \rightarrow D$), and parallel reactions

$$\left(A \begin{smallmatrix} \nearrow B \\ \searrow C \end{smallmatrix}\right)$$

require much more complicated rate equations for their description. Whereas the formulation of an appropriate rate statement is often simple, the solution of the resulting differential equations is likely to be difficult and is beyond the scope of this section. Many rate equations previously considered too difficult to solve because of their demand for awkward or tedious numerical approximation methods can now be solved by means of computers.

Equilibrium and Kinetics. Inasmuch as all chemical reactions are limited by a chemical equilibrium, reaction kinetics really describes the rate of approach to that equilibrium rather than to a stoichiometric completeness of the reaction. At equilibrium, the net rate of reaction Eq. (15.51) is zero, whence it follows that for a reaction whose rate law is described by the molecularity (i.e., stoichiometric equation), the equilibrium constant for the reaction is related to the forward and reverse specific rate constants. Thus,

$$\frac{C_A^a C_B^l C_C^c}{C_P^p C_Q^q} = K_c = \frac{k}{k'} \tag{15.52}$$

It is clear that the larger the value of K_c, the larger is the magnitude of the forward rate constant relative to the reverse and the closer to stoichiometric completeness is the equilibrium conversion. Also, the greater the concentration of reactant above the equilibrium concentration, the faster is the progress of the reaction toward equilibrium. All other things being equal, conditions that increase the value of the equilibrium constant are favorable to the net kinetics of the reaction.

Effect of Temperature. Homogeneous reactions are strongly temperature-dependent. Their specific reaction rate always increases with increasing temperature. The effect of temperature is described by the semitheoretical relation of Arrhenius:

$$k = Ae^{-E/RT} \tag{15.53}$$

The coefficient A (called the *frequency factor*) and the exponent E (called the *energy of activation*) have theoretical interpretations, but they are best regarded by the process designer as empirical quantities peculiar to a particular chemical reaction and evaluable from experimental rate data. Thus a plot of k against $1/T$ should be linear and should have a slope of $-E/R$ and an intercept of $\ln A$, provided A and E are independent of temperature. In fact, both the frequency factor and the energy of activation vary slightly with temperature, but over the temperature ranges normally encountered, they may be assigned constant average values without serious error. Figure 15.21 shows an Arrhenius plot for a second-order reaction. The activation energy represented by the line drawn through the data is 10,000 cal/(g · mol).

The failure of rate data to fit Eq. (15.53) may be accepted as evidence that

1. A reversible reaction has been treated as if it were irreversible, and the effect of temperature on the equilibrium is significant.

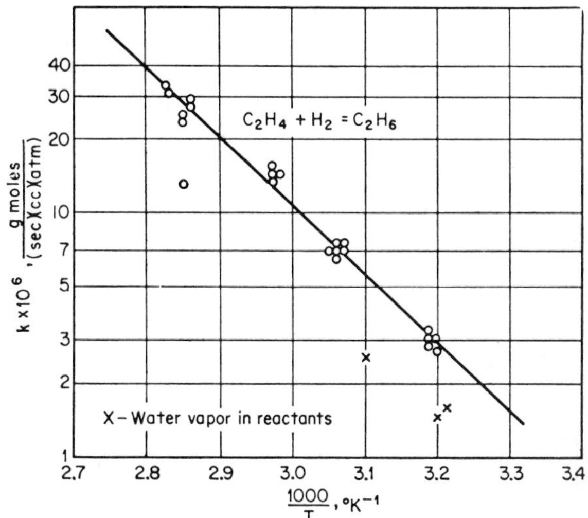

FIGURE 15.21 Arrhenius plot for hydrogenation of ethylene. *(From Smith, Chemical Engineering Kinetics, McGraw-Hill, New York, 1956.)*

2. An otherwise incorrect mechanism has been assigned to the reaction.
3. The specific rate constant has been evaluated from the experimental data incorrectly.
4. The reaction is heterogeneous, and its rate is influenced by adsorption or by some other physical process.

Energies of activation range from less than 1000 to greater than 100,000 cal/(g · mol). For most reactions, the value will be between 10,000 and 70,000 cal/(g · mol).

A long popular rule of thumb states that the rate of a reaction approximately doubles for each 10°C rise in temperature. Inspection of Eq. (15.53) shows that this can be true only for particular combinations of activation energy and temperature. At high temperatures especially, typical activation energies are such that considerably more than 10°C is required to double the reaction rate.

Effect of Concentration. At constant temperature, the specific rate constant is assumed to be independent of the concentration of reactants and products, so that equations like (15.48) and (15.50) show explicitly the effect of concentration on the progress of the reaction. In general, this is a valid assumption for homogeneous uncatalyzed reactions. If there seems to be a dependency of k on concentration, the most likely reasons are that the wrong order (or mechanism) has been postulated, that a catalyst is influencing the reaction, or that the temperature has not remained constant. Heterogeneous reactions may yield apparent rate constants that reflect complex combinations of physical and chemical processes and hence may vary with concentration.

Homogeneous Catalyzed Reactions. A *catalyst* is a substance that affects the rate of a chemical reaction without entering the reaction in any stoichiometric sense.

The catalyst may undergo net physical or chemical change in the course of the reaction, but often it does neither. Trace amounts of a catalyst can greatly influence the reaction rate and mechanism. A catalyst can be positive (increase the rate) or negative (decrease the rate); if not otherwise stipulated, a positive effect is implied. Negative catalysts are called *inhibitors*.

Some homogeneous gas-phase and liquid-phase (most commonly the latter) reactions can be catalyzed by materials that are soluble in the reacting mass and therefore do not destroy the homogeneity of the system. Sometimes homogeneous catalysis occurs without the prior knowledge of the kinetic experimenter. In such a case, grossly incorrect conclusions can be drawn about the kinetics of the reaction. Usually, catalytic behavior will be signaled by one or more of the following phenomena:

1. Irrationally rapid or accelerating rate of reaction
2. Irrationally slow rate of reaction
3. Apparent zero or fractional order with respect to known reactants
4. Abnormal temperature dependency of the rate

Although homogeneous catalysts are likely to be effective in very small amounts, most catalyzed reactions will exhibit a definite order with respect to the catalyst below a particular concentration. This order should be determined experimentally for the most reliable statement of the kinetics of the reaction. In some instances, however, the catalyst concentration is kept constant, and the reaction may be described satisfactorily for design purposes by the assignment of apparent orders to the reactants and products, with no explicit ordering of the catalyst.

22. Heterogeneous Reactions. A chemical reaction is said to be *heterogeneous* if more than one phase is an active participant and if transfer of materials to phase boundaries has an effect on the rate of reaction. Heterogeneous reactions commonly involve fluid-solid mass transfer (e.g., catalysis of a fluid reaction mixture on the surface of solid catalyst pellets, combustion of a solid fuel in air, acid leaching of metals from ores) or mass transfer between two fluid phases (e.g., absorption of gaseous sulfur dioxide by weak aqueous sodium hydroxide, nitration of toluene by nitric acid). In the limiting cases of extremely high reaction rates, heterogeneous reactions are analyzed as mass-transfer problems. Reactions where mass-transfer rates are sufficiently rapid can often be analyzed as homogeneous reactions taking place in one of the fluid phases.

Uncatalyzed Heterogeneous Reactions. In an uncatalyzed heterogeneous reaction, chemical action occurs among components that are simultaneously being transferred physically from phase to phase. The apparent rate of the reaction is in fact the rate of a more complicated process. It will be influenced not only by factors affecting chemical kinetics, but also by those affecting the rate of interphase mass transfer. Among the latter are

1. Amount of interfacial surface
2. Concentration of reactants in each phase
3. Concentration of products in each phase
4. Relative velocity at the interface
5. Temperature (in its effects on both phase equilibrium and diffusivities)

6. Presence of a solid resistance at the interface (e.g., an ash layer formed on a reacting solid)

Sometimes the conditions of a heterogeneous process are such that the chemical reaction is relatively rapid, whence the rate of physical transport becomes effectively that of the overall process. Sometimes the reverse is true. More generally, the rates of physical transfer and chemical reaction are of the same order of magnitude, in which case each contributes significantly to the kinetics of the overall process.

Inasmuch as temperature effects in chemical and physical processes are quite different, the selection of operating temperature in a heterogeneous reaction may determine which component process is controlling. Figure 15.22 shows the rate of a gas-solid reaction, in which an ash film is formed, as a function of temperature. Over section *AB* the process rate is essentially that of the chemical reaction; over section *CD*, that of the diffusion of reactant through the gas film; and over *EF*, that of diffusion through the ash layer. Over sections *BC* and *DE*, more than one phenomenon controls. *It should be noted that the overall rate never can be greater than that of the slowest component process*.

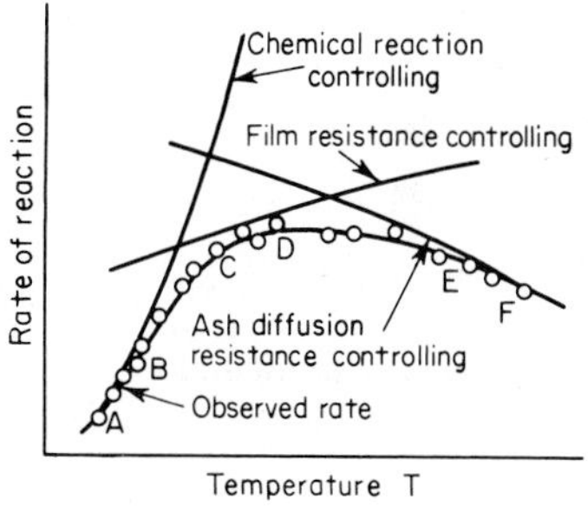

FIGURE 15.22 Rate of reaction as affected by combined resistances. *(From Levenspiel,*[18] *p. 355.)*

Whenever more than one phenomenon determines the effective reaction rate, the overall rate equation becomes very difficult to solve. Design equations combining reaction effects with mass-transfer effects have been derived for a number of special cases. These are still largely not supported by extensive experimental data. Such systems are often treated as modified mass-transfer problems.

A number of techniques exist which may be used to scale up reactor systems from empirical data. These must be applied carefully and generally are restricted by the conditions of the experiment. Such scale-up methods are beyond the scope of this text.

Catalyzed Heterogeneous Reactions. Although heterogeneous reactions responsive to catalysis may involve any combination of phases, the examples most common and industrially most important are solid-fluid systems in which the catalyst is the solid phase. The reactants and products may be gaseous, liquid, or both. The solid catalyst may be a container wall, a metal gauze, or a granular mass. Usually it is the latter, in either fixed- or fluidized-bed form, with particles seldom larger than 0.25 in (6.4 mm). A reaction catalyzed by a solid is believed to take place at the surface of the solid (the surface may be internal, i.e., interstitial within a porous particle) and to involve activated adsorption or chemisorption on that surface.

The mechanism of a fluid-phase reaction catalyzed by a solid is extremely complex and may comprise as many as seven sequential steps:

1. Diffusion of reactants to the outside catalyst surface from the body of the fluid phase

2. Diffusion of reactants into the catalyst pores (or through an inert deposit to active surface regions)
3. Adsorption of reactants
4. Chemical reaction in the adsorbed state
5. Desorption of products
6. Diffusion of products from the catalyst pores
7. Diffusion of products from the outside catalyst surface into the body of the fluid phase

Any one or more of these steps may be slow enough to control the rate of the entire sequence. Often control can be ascribed to a single step, and an adequate design procedure can be based on this premise.

Analysis of a catalyzed heterogeneous process, then, consists in examining the data for evidence of the rate-controlling step (the experimental program must have been planned to yield such evidence) and applying the data to evaluate coefficients and indices of whatever equations appropriately describe the rate.

The procedure is outlined as follows:

1. If the degree of conversion depends on the linear velocity of reacting fluid with respect to the catalyst (at constant space velocity), mass transfer between the body of the fluid and the external surface of the catalyst is controlling. The rate of the process is then dependent on the rate of diffusion to or from the catalyst, and the transfer coefficient may be calculated from established correlations for mass-transfer coefficients. The estimated value of the mass-transfer coefficient in the fluid phase can be used to calculate the concentration of reactants at the surface of the catalyst particle.

2. If the degree of conversion is independent of fluid velocity but depends on size of the catalyst pellets, diffusion of reactants or products within the pores of the catalyst is a rate-controlling step. In such a case, the data should be treated to evaluate an *effectiveness factor*, defined as the ratio of observed reaction rate to that which would obtain if pore diffusional resistance were negligible.

Figure 15.23 is a typical plot of effectiveness factor against a modulus

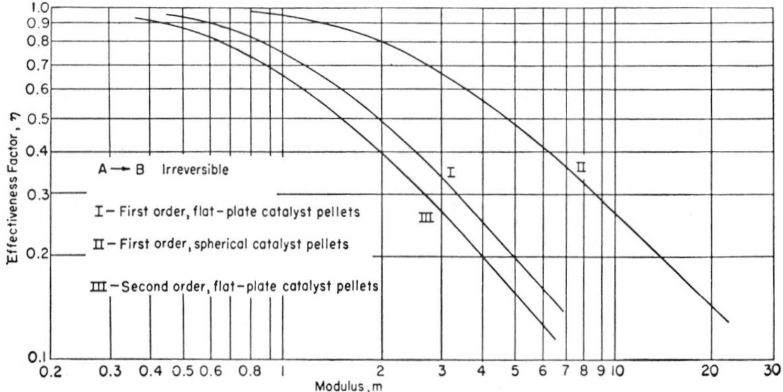

FIGURE 15.23 Effectiveness factor for equations of simple order.

$$m = 1 \sqrt{2k(C_i^0)^{n-1}(\bar{r} D_i)} \tag{15.54}$$

in which C^0 is the concentration of reactant at the external catalyst surface, and $\bar{r}$, the average pore radius, is calculated as $2V_g/S_g$ (examples are given in Table 15.9).

3. If absorption-desorption or chemical reaction at the catalyst surface is the controlling process, a mechanism must be found that will identify which of the possibilities are rate-controlling and which are equilibrium steps. For example, for the stoichiometric reaction

$$A \leftrightharpoons R \tag{15.55}$$

the following mechanistic steps involving the participants A and R and a catalyst site s may be postulated:

$$A + s \leftrightharpoons As \tag{15.55a}$$

$$As \leftrightharpoons Rs \tag{15.55b}$$

$$Rs \leftrightharpoons R + s \tag{15.55c}$$

Assumption of each of these in order as the controlling step results in a different rate equation, which may be validated against experimental kinetic data. If none meets the test, a new mechanism must be tried. Rate equations for a number of simple examples are summarized in Table 15.10.

The discovery of a suitable rate equation by the methods described does not constitute establishment of the true mechanism by which the heterogeneous cat-

TABLE 15.9 Values of Internal Surface Area, Pore Volume, and Average Pore Radius for Typical Catalysts

Catalyst	S_g, m²/g	V_g, cc/g	$\bar{r} = 2V_g/S_g$, A
Activated carbons	500–1500	0.6–0.8	10–20
Silica gels	200–700	0.4	15–100
Silica-alumina cracking catalysts ~10–20% Al_2O_3	200–700	0.2–0.7	15–150
Silica-alumina (steam-deactivated)	67	0.519	155
Silica-magnesia microsphere:			
Nalco, 25% MgO	630	0.451	14.3
Nalco, steam treated, 621°C, 400 psig for 24 hr	322	0.283	17.6
Da-5 silica-magnesia	656	0.365	11.1
Activated clays	150–225	0.4–0.52	~100
TCC clay pellets (MgO, CaO, Fe_2O_3, SO_4) = ~10%	276	0.363	26.3
Clays:			
Montmorillonite (raw)	214	0.297–0.306	~28
Montmorillonite (heated 550°C)	212	0.268	25.2
Vermiculite	35	0.063–0.057	~314
Activated alumina (Alorico)	175	0.388	45
CoMo on alumina	168–251	0.261–0.331	20–40
Kieselguhr (Celite 296)	4.2	1.14	11,000
Fe-synthetic NH_3 catalyst	4–13	0.12	200–1000
Co-ThO_2-Kieselguhr 100:18:100 (reduced) pellets	42.3	0.73	345
Co-ThO_2-MgO (100:6:12) (reduced) granular	84.1	0.80	190
Co-Kieselguhr 100:200 (reduced) granular	22.8	2.31	2030
Porous plate (Coors No. 760)	1.6	0.172	2150
Pumice	0.38		
Fused copper catalyst	0.23		
Ni film	8.4		
Ni on pumice, 91.8% pumice	1.27		

S_g = catalyst surface area
V_c = catalyst pore volume
$\bar{r}$ = average radius of pore
A = angstrom unit = 1×10^{-8} cm

TABLE 15.10 Mechanisms and Their Corresponding Rate Equations

Chemical equation	Catalytic steps	Rate equation*
$A \rightleftharpoons R$	$A + s \rightleftharpoons As$	$r = \dfrac{k(C_A - C_R/K)}{1 + K_R C_R}$
	$As \rightleftharpoons Rs$	$r = \dfrac{k(C_A - C_R/K)}{1 + K_A C_A + K_R C_R}$
	$Rs \rightleftharpoons R + s$	$r = \dfrac{k(C_A - C_R/K)}{1 + K_A C_A}$
$A \rightleftharpoons R$	$2A + s \rightleftharpoons A_2s$	$r = \dfrac{k(C_A^2 - C_R^2/K^2)}{1 + K_R C_R + K_R C_R^2}$
	$A_2s + s \rightleftharpoons 2As$	$r = \dfrac{k(C_A^2 - C_R^2/K^2)}{(1 + K_R C_R + K_A C_A^2)^2}$
	$As \rightleftharpoons Rs$	$r = \dfrac{k(C_A - C_R/K)}{1 + K_A C_A^2 + K_A' C_A + K_R C_R}$
	$Rs \rightleftharpoons R + s$	$r = \dfrac{k(C_A - C_R/K)}{1 + K_A C_A^2 + K_A' C_A}$
$A \rightleftharpoons R$	$A + 2s \rightleftharpoons 2A_{1/2}s$	$r = \dfrac{k(C_A - C_R/K)}{(1 + \sqrt{K_R C_R} + K_R' C_R)^2}$
	$2A_{1/2}s \rightleftharpoons Rs + s$	$r = \dfrac{k(C_A - C_R/K)}{(1 + \sqrt{K_A C_A} + K_R C_R)^2}$
	$Rs \rightleftharpoons R + S$	$r = \dfrac{k(C_A - C_R/K)}{1 + \sqrt{K_A C_A} + K_A' C_A}$
$A \rightleftharpoons R + S$	$A + s \rightleftharpoons As$	$r = \dfrac{k(C_A - C_R C_S/K)}{1 + K_{RS} C_R C_S + K_R C_R + K_S C_S}$
	$As + s \rightleftharpoons Rs + Ss$	$r = \dfrac{k(C_A - C_R C_S/K)}{(1 + K_A C_A + K_R C_R + K_S C_S)^2}$
	$Rs \rightleftharpoons R + s$; $Ss \rightleftharpoons S + s$	$r = \dfrac{k(C_A - C_R C_S/K)}{C_S(1 + K_A C_A + (K_{AS} C_A/C_S) + K_S C_S)}$
$A \rightleftharpoons R + S$	$A + s \rightleftharpoons As$	$r = \dfrac{k(C_A - C_R C_S/K)}{1 + K_R C_R + K_{RS} C_R C_S}$
	$As \rightleftharpoons Rs + s$	$r = \dfrac{k(C_A - C_R C_S/K)}{1 + K_A C_A + K_R C_R}$
	$Rs \rightleftharpoons R + s$	$r = \dfrac{k(C_A - C_R C_S/K)}{C_S(1 + K_A C_A + K_{AS} C_A/C_S)}$
$A + B \rightleftharpoons R$	$A + s \rightleftharpoons As$	$r = \dfrac{k(C_A - C_R/K C_B)}{1 + (K_{RB} C_R/C_B) + K_B C_B + K_R C_R}$
	$B + s \rightleftharpoons Bs$	$r = \dfrac{k(C_B - C_R/K C_A)}{1 + K_A C_A + (K_{RA} C_R/C_A) + K_R C_R}$
	$As + Bs \rightleftharpoons Rs + s$	$r = \dfrac{k(C_A C_B - C_R/K)}{(1 + K_A C_A + K_B C_B + K_R C_R)^2}$
	$Rs \rightleftharpoons R + s$	$r = \dfrac{k(C_A C_B - C_R/K)}{1 + K_A C_A + K_B C_B + K_{AB} C_A C_B}$
$A + B \rightleftharpoons R + S$	$A + s \rightleftharpoons As$	$r = \dfrac{k(C_A - C_R C_S/K C_B)}{1 + (K_{RS} C_R C_S/C_B) + K_B C_B + K_R C_R + K_S C_S}$
	$B + s \rightleftharpoons Bs$	$r = \dfrac{k(C_B - C_R C_S/K C_A)}{1 + (K_{RS} C_R C_S/C_A)^{1/2} + K_A C_A + K_R C_R + K_S C_S}$
	$As + Bs \rightleftharpoons Rs + Ss$	$r = \dfrac{k(C_A C_B - C_R C_S/K)}{(1 + K_A C_A + K_B C_B + K_R C_R + K_S C_S)^2}$
	$Rs \rightleftharpoons R + s$; $Ss \rightleftharpoons S + s$	$r = \dfrac{k[(C_A C_B/C_S) - C_R/K]}{1 + K_A C_A + K_B C_B + K_S C_S + K_{AB} C_A C_B/C_S}$

TABLE 15.10 Mechanisms and Their Corresponding Rate Equations (*Continued*)

Chemical equation	Catalytic steps	Rate equation*
$A + B \rightleftharpoons R + S$	$A + 2s \rightleftharpoons 2A_{1/2}s$	$r = \dfrac{k(C_A - C_R C_S/KC_B)}{[1 + K_{RS}C_R C_S/C_B + K_B C_B + K_R C_R + K_S C_S]^2}$
	$B + s \rightleftharpoons Bs$	$r = \dfrac{k(C_B - C_R C_S/KC_A)}{1 + \sqrt{K_A C_A} + (K_{RS}C_R C_S/C_A) + K_R C_R + K_S C_S}$
	$2A_{1/2}s + Bs \rightleftharpoons Rs + Ss + s$	$r = \dfrac{k(C_A C_B - C_R C_S/K)}{(1 + \sqrt{K_A C_A} + K_B C_B + K_R C_R + K_S C_S)^3}$
	$Rs \rightleftharpoons R + s$	$r = \dfrac{k(C_A C_B/C_S - C_R/K)}{1 + K_A\sqrt{C_A} + K_B C_B + (K_{AB}C_A C_B/C_S) + K_S C_S}$
	$Ss \rightleftharpoons S + s$	$r = \dfrac{k(C_A C_B/C_R - C_S/K)}{1 + \sqrt{K_A C_A} + K_B C_B + K_R C_R + K_{AB}C_A C_B/C_R}$
$A + B \rightleftharpoons R + S$	$B + s \rightleftharpoons Bs$	$r = \dfrac{k(C_B - C_S C_R/KC_A)}{1 + K_R C_R + K_{RS}C_R C_S/C_A}$
	$A + Bs \rightleftharpoons Rs + S$	$r = \dfrac{k(C_A C_B - C_R C_S/K)}{1 + K_R C_R + K_B C_B}$
	$Rs \rightleftharpoons R + s$	$r = \dfrac{k[(C_A C_B/C_S) - C_R/K]}{1 + (K_{AB}C_A C_B/C_S) + K_B C_B}$

NOTE: $K_{AB}\ldots$ = combined equilibrium constants; K = over-all equilibrium constant for the chemical equation; k = constant.
* The rate equation is opposite the catalytic step assumed to be rate-controlling.

alytic reaction is occurring. Nevertheless, whenever the true mechanism is unknown—and it usually is—a rate-equation formulation by the kind of semitheoretical approach outlined offers the most reliable device for rationalizing kinetic data and extending them to conditions not exactly covered in the experiment that yielded them.

23. Interpretation of Kinetic Data. The design of a chemical reactor should be based on properly collected and analyzed laboratory data. These data will, in general, be one of three types:

1. Measurements of composition as a function of time in a batch reactor of constant volume operated at various temperatures and pressures
2. Measurements of outlet composition as a function of feed rate to a flow reactor of constant volume operated at various pressure and temperature levels
3. Measurements of composition as a function of time in a variable-volume batch reactor operated at constant temperature and substantially constant pressure

The third type of data is much less common than the other two, and the experimental technique is more difficult. Data of the second type are generally the most dependable and simple to obtain. This method has the advantage of direct applicability to flow-type reactors. Data of the first type should not be used for the design of flow reactors unless it is certain that the extent of mixing is the same in both the batch and flow systems. In all cases it is important that the temperature does not vary with time in the batch reactor or with position in the flow reactor.

Data should be taken and analyzed to determine reaction order with respect to all participants, reaction rate constants, and temperature dependence. The first step is to determine the reaction order. This treatment consists, generally, of testing the validity of the rate equations, differential or integrated, suspected to be appropriate until one is found that fits the data. Many techniques exist for testing the validity of proposed rate equations. The best use some integrated form of the rate equation, though preliminary assessment can often be made by determining the differential reaction rate as a function of concentration. The use of differential rate equations requires the direct measurement of reaction rates in the laboratory (usually not feasible) or the differentiation of composition-time data (a procedure highly susceptible to error). Differential equations are useful for a quick, tentative assessment of a reaction, but integrated equations are preferred for final evaluation.

For simple reactions involving more than one reactant, the order with respect to any one component can be determined by holding all other reactant concentrations constant. This may be accomplished by using large excesses of these other reactants. For the specific reaction $A + B \rightarrow C$, with reactant B present in great excess, assumed to be substantially irreversible, several analysis techniques are presented below as examples of the general approach.

1. *Differential rate equation:* A logarithmic plot of $dC_A/d\theta$ (or, by approximation, $\Delta C_A/\Delta\theta$) against C_A yields a straight line, of which the slope is n, the order of the reaction (Fig. 15.24).

2. *Integrated rate equation, first order:* A plot of $-\log (C_A/C_{AO})$ against θ yields a straight line through the origin (Fig. 15.25).

3. *Integrated rate equation, other than first order:* When $n > 1$, the general rate equation

$$-\frac{dC_A}{d\theta} = kC_A{}^n \tag{15.56}$$

is integrated between the limits C_{AO} and C_A to give

$$\left(\frac{1}{C_A}\right)^{n-1} - \left(\frac{1}{C_{AO}}\right)^{n-1} = (n - 1)k\theta \tag{15.57}$$

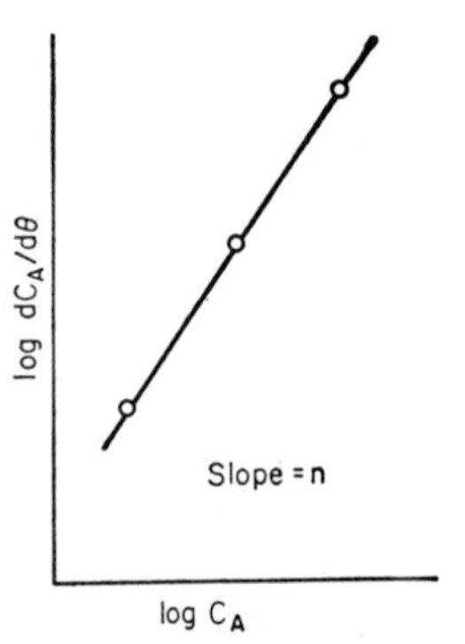

FIGURE 15.24 Determining reaction order: differentiation. *(From Walas,[19] Fig. 2.1.)*

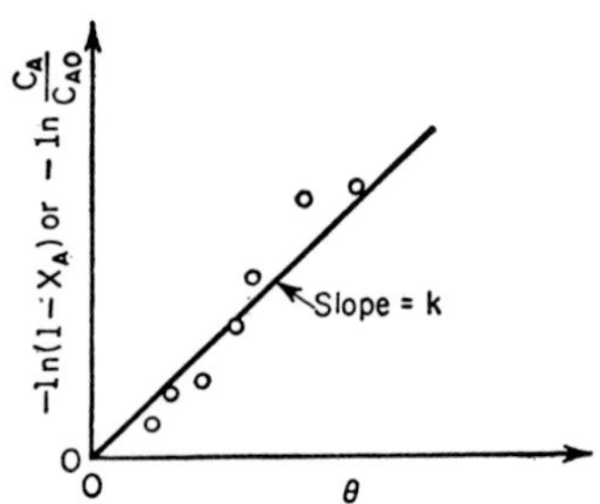

FIGURE 15.25 Test for first-order reaction. *(From Levenspiel,[18] p. 48.)*

From Eq. (15.57), a plot of $(1/C_A)^{n-1}$ against θ yields a straight line with the intercept $(1/C_{AO})^{n-1}$ and the slope $k(n - 1)$ (Fig. 15.26).

4. *Integrated rate equation: Method of half-life:* The *half-life*, or time for 50 percent conversion, is a useful criterion for order. Integration of Eq. (15.56) between the limits C_{AO} and $0.5C_{AO}$ yields the following values for half-life $\theta_{1/2}$:

$$\theta_{1/2} = \begin{cases} \dfrac{1}{2k} C_{AO} & n = 0 \\[2ex] \dfrac{0.69}{k} & n = 1 \\[2ex] \dfrac{1}{kC_{AO}} & n = 2 \\[2ex] \dfrac{2^{n-1} - 1}{k(n - 1)(C_{AO})^{n-1}} & n = 3 \end{cases}$$

Hence, if experiments are run at different initial concentrations and log $\theta_{1/2}$ is plotted against log C_{AO}, a straight line of slope $1 - n$ results (Fig. 15.27).

5. *Integrated rate equation: Method of reference curves:* Inspection of the integrated rate equations indicates that the ratio of times required for any two degrees of conversion is dependent only on those conversion fractions and on the reaction order. Walas[19] has calculated such ratios, taking 90 percent conversion as the arbitrary convenient reference, to give the useful curves of Fig. 15.28. If one plots percent conversion $100C_A/C_{AO}$ against $\theta/\theta_{0.9}$, to the same scale as in Fig. 15.28, comparison of the two graphs will reveal the order of the reaction.

The reaction rate constant k determined by these techniques includes the concentration of the excess component. The order with respect to other components must be determined using the same techniques. The true reaction rate constant can be determined from the pseudo rate constants by dividing out the (constant) concentrations of reactants raised to the appropriate order.

Failure of the experimental data to fit the form of the equations indicated in

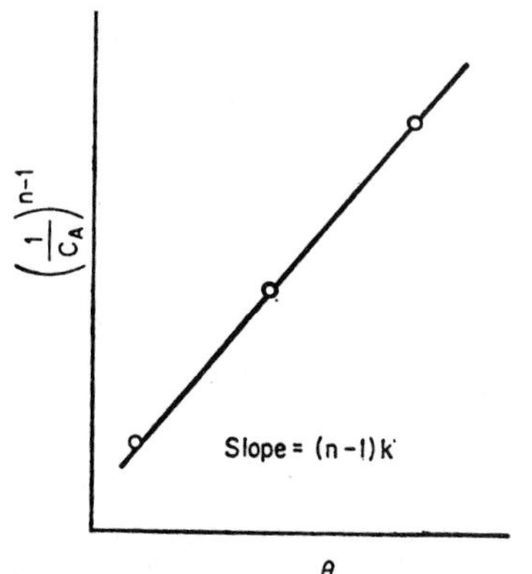

FIGURE 15.26 Determining order of action: integrated equation. *(From Walas,[19] Fig. 2-2.)*

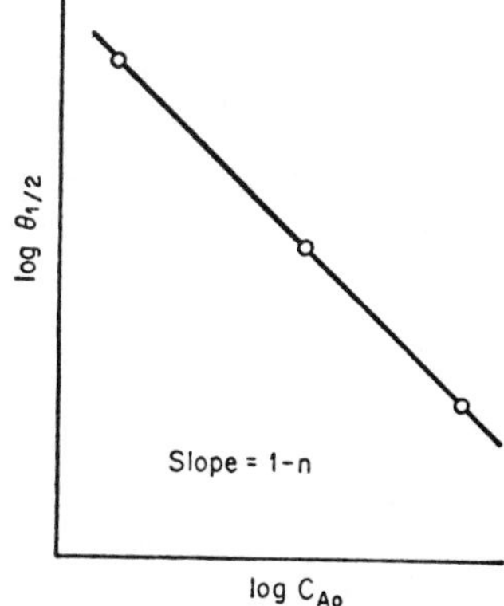

FIGURE 15.27 Determining order of reaction: half-lives. *(From Walas,[19] Fig. 2-3.)*

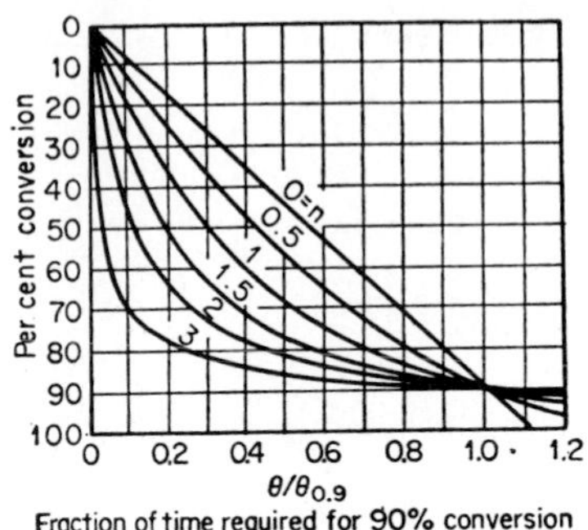

FIGURE 15.28 Generalized curves for determining order of reaction. *(From Walas,[19] Fig. 2-4.)*

steps 1 through 4 above is evidence that the postulated simple rate equation is in error. For example, the reaction may be reversible, may involve chain steps, etc.

Sometimes complicated reactions can be simplified in their analysis by approximations permitted in the way experimental data are taken. For example, the early stages of reversible reaction among initially pure reactants will act substantially as if the reaction were irreversible, and the data may be so treated. Again, the order of each of several reactants sometimes can be determined individually from experiments in which all but the reactant of interest is present in large stoichiometric excess. Extreme care must be taken in such approximations, however, inasmuch as sampling and analysis problems can induce large experimental error.

24. Reactor Design. Once a suitable rate equation that fits the experimental kinetic data has been discovered, design of the plant reactor can proceed. Five steps are involved:

1. Selection of the type of reactor
2. Selection of the shape or proportions of reactor
3. Sizing the reactor
4. Selection of materials of construction
5. Design of reactor auxiliaries

Type of Reactor. Reactors generally are of four basic types: batch, semicontinuous stirred tank, continuous stirred tank, and tubular (plug-flow). The choice of type depends on the state of the reactants and products, the nature of the reaction, the rate of production, and the character of the rest of the process of which the reaction is a part. The choice is often determined by economic factors. Table 15.11 indicates some of the conditions for which each type may be suitable.

Shape of Reactor. Except for the simplest of reactions, it is desirable that a plant reactor be of the same type and generally of the same shape as the laboratory unit from which the kinetic data for the reaction were obtained. In extremely complicated cases, a pilot prototype of the plant unit should be operated. Even in the latter instance, however, rational scale-up of the pilot unit will be required and may lead to a plant reactor of different proportions from the prototype.

For perfect scale-up, complete similarity (geometric, kinematic, dynamic, thermodynamic, and chemical) should be preserved between small and large models. If one is to operate with the same process stream in both models, as one must, complete similarity is impossible. A compromise must then be made, frequently requiring longer reaction times or lower reactor productivity (rate of pro-

TABLE 15.11 Types of Reactors and Their Applications

Type	*Conditions Suitable*
Batch	Intermittent operation Holdup of charge for testing required Individual fine adjustment necessary Small production rate Liquid-liquid or liquid-solid reactions Long induction period involved
Semicontinuous	Gas-liquid reactions Large excess of one reagent desired, for liquid-liquid or liquid-solid reactions
Continuous stirred tank	Homogeneous liquid-phase reactions best Steady availability of all reactants Steady demand for product Liquid-solid reactions satisfactory if solids are easily suspended Gas-solid reactions satisfactory if fluidization is feasible
Tubular (plug-flow)	Solid-catalyzed gas-phase reactions appropriate Homogeneous fluid-phase reactions best Steady availability of reactants Steady demand for product

duction per unit volume) in the large unit than in the small, but meeting the most critical demands of the system—good mixing, catalyst distribution, or temperature control, for example.

Sizing the Reactor. The volume of a reactor is calculated directly from the rate equation, such as Eq. (15.49), which may be rewritten as

$$\int_0^\theta V d\theta = \int_{N_{A_2}}^{N_{A_1}} \frac{dN_A}{r_A} \tag{15.58}$$

the correct-order expression being inserted for r_A. For flow reactors, a more useful form of Eq. (15.58) is

$$\int_0^V \frac{dV}{F} = \int_0^{X_A} \frac{dX_A}{r_A} \tag{15.59}$$

written in terms of the molal flow rate F and degree of conversion X_A, moles of A converted per mole of F. For steady-state continuous stirred tanks, the rate expression is simply

$$r_A = (F/V)(C_{A1} - C_{A2}) \tag{15.60}$$

where the subscripts 1 and 2 refer to the concentration of reactant A in the entering and leaving streams, respectively.

A plug-flow reactor (no longitudinal mixing) and a batch reactor (no concentration or temperature gradients) require the shortest residence time possible for a given reaction of finite order to proceed to the desired extent. The residence time is identical for these two types. The productivity (average rate of product availability per unit of reactor volume) is reduced for the batch reactor by the outage time (time required for emptying, cleaning, and refilling between batches).

Single-stage continuous stirred-tank reactors, which provide a reaction environment of the constant composition of the effluent or completed-reaction stream, have the lowest productivity of all reactors. As the total volume of the continuous stirred-tank reactor is subdivided into a series of equal-volume stages, the productivity increases, approaching that of a plug-flow reactor as the number of stages approaches infinity. Figure 15.29 shows a comparison of residence times in stirred-tank reactors of j equal stages and in plug-flow reactors for second-order irreversible reactions of various specific reaction constants and initial concentrations.

In actual practice, the assumptions of perfect mixing in continuous stirred tanks and of no longitudinal mixing in tubular reactors are only approximations. For stirred tanks of proper design with vigorous agitation and for small-diameter, high-velocity tubular reactors, the approximations are well within the limits of design accuracy. For large-diameter tubes and for packed beds, the effect of longitudinal mixing can be appreciable, resulting in a larger requirement of reactor volume. Figure 15.30 gives an idea of the effect of longitudinal mixing on the required reactor volume for irreversible second-order reactions. The axial dispersion coefficient D must be determined experimentally for a given reactor.

Materials of Construction. Materials for the fabrication of all process equipment are selected first for their ability to withstand chemical attack and thus avoid process-stream contamination and second for their economic life. Resistance to corrosion is especially important in reactors because (1) the combination of composition, temperature, and mechanical conditions is likely to be more severe there than in most other pieces of process equipment and (2) trace contamination due to dissolved metal can be disastrous in its catalytic effect. The high pressures, high or low temperatures, and rapid temperature changes that obtain in

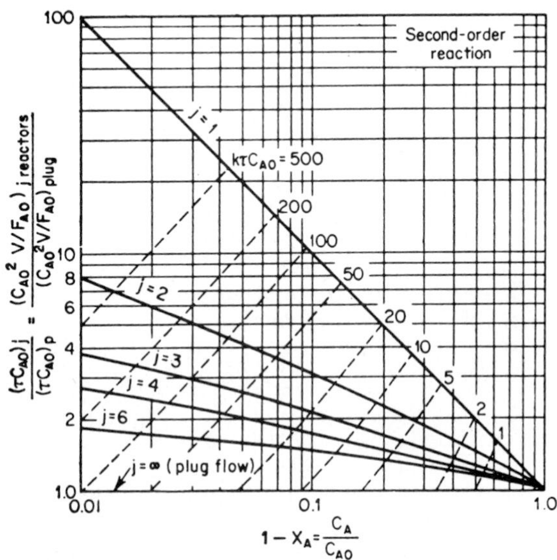

FIGURE 15.29 Comparison of plug-flow and a series of j equal-size backmix reactors. [$2A \rightarrow R$; $A + B \rightarrow R$; $C_{A_0} = C_{B_0}$ with negligible expansion. For the same processing rate of identical feed, the ordinate measures the volume ratio V_j/V_p or space-time ratio τ_j/τ_p directly.] *(From Levenspiel,[18] p. 141.)*

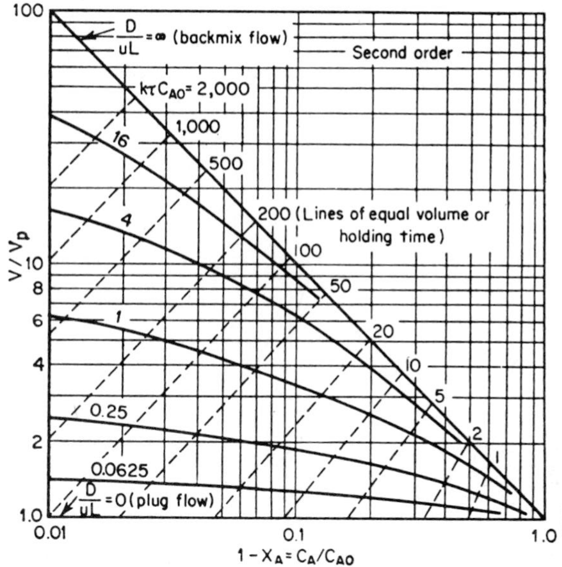

FIGURE 15.30 Comparison of real and ideal (plug-flow) reactors. *(From Levenspiel,[18] p. 280.)*

many reactors also make mechanical and structural integrity difficult to obtain. Detailed tables summarizing the chemical and mechanical characteristics of materials are available in a number of review sources.

Optimal Design. The optimization of reaction conditions and of the reactor design is highly complicated, because of the many variables that are involved, and is seldom achieved or even attempted. Certain aspects of the optimization must be considered, however, to arrive at a reasonable, if not optimal, operation.

The temperature of the reaction is chosen with regard to the following considerations: (1) reaction rates increase with temperature, and high temperatures favor low residence time in a reactor; (2) undesirable side reactions may be minimized by the proper choice of temperature; (3) equilibrium of exothermic reactions is less favorable, the higher the temperature; (4) maintaining high reactor temperatures may require high thermal costs; (5) temperature must be chosen with regard to its effect on catalysts; and (6) high temperatures are identified with high corrosion rates and with costly materials of construction. To achieve a compromise between kinetic and equilibrium effects in an exothermic reaction, a programmed temperature change during the course of the reaction may be used, as in sulfuric acid converters. Figure 15.31 shows how such programming can achieve the maximum average rate of reaction.

Reactant concentrations are selected in such a way as to allow maximum conversion of the most expensive or critical reactant, with due regard to product isolation and reactant recovery costs. Pressure is generally the equivalent of concentration in a gas-phase reaction and must be selected with additional regard for the equipment costs associated with high-pressure reactors. Pressure does not affect liquid-phase reactions.

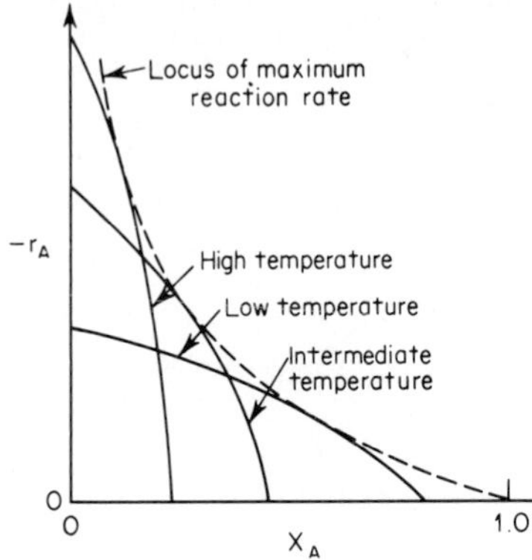

FIGURE 15.31 Reaction rate as a function of conversion and temperature for reversible exothermic reactions using a given feed material. Dashed line shows the temperature to use at each composition for optimum operations. *(From Levenspiel,*[18] *p. 217.)*

Degree of conversion must be chosen, keeping separation, recycle, and subsequent processing steps and their costs in mind. In general, these costs are balanced against the cost of the reactor and its operation; the former are high for low conversion, whereas the latter may be high for high conversion.

The optimization of reactor type, proportion, size, and materials of construction is extremely complicated and cannot be treated here. The interested reader is referred to special literature on the subject.[20]

PROCESS CONTROL

25. Introduction. Chemical processing of materials is generally carried out as a sequence of unit operations, either on a batch or continuous-flow basis. The output of any one process step is usually the input to one or more subsequent steps. For each subprocess there are a number of independently adjustable variables which must be specified by the engineer or process operator. The values specified for these *process variables* determine how the subprocess operates on the feed to produce the output and determine such factors as yield, purity, reaction time, etc. Examples of such independently adjustable variables are process-stream flow rates, energy input rates, and cooling-water flow rates.

Design procedures generally specify approximate values of these variables for a given input to the subprocess and a given desired output. However, a process must generally be "fine-tuned" to determine the appropriate relationship between input parameters and process variables required to achieve a given output or product.

The field of process control deals with the use of automatic devices to set values of process variables in order to achieve a given process output. Inputs to a given subprocess are often variable, depending on previous process steps, feed material compositions, temperatures of available cooling water, etc. Since changes in inputs to a process generally require changes in process variables in order to maintain a given output condition, the dynamic response of processes and their control systems is of primary importance.

Simple Control Systems. In the simplest control system, no information about either process input or output is used to determine the setting of process variables. Upsets in process inputs lead to changes in output, with no compensation by the control system to restore output to the desired condition. Obviously, this type of control is useful only for processes where the output response to input changes is either small or tolerable within its range of variance.

Feedback Control. Information about some aspect of the output, called the *controlled variable*, is used to determine the adjustment to be made in some process variables. Such control systems, in which measured values of a controlled variable are used to adjust a process variable upstream of the measurement point, are termed *closed-loop systems*. A simple feedback control process is illustrated by Fig. 15.32. Systems that do not make use of such feedback from downstream points to upstream points are termed *open-loop systems*.

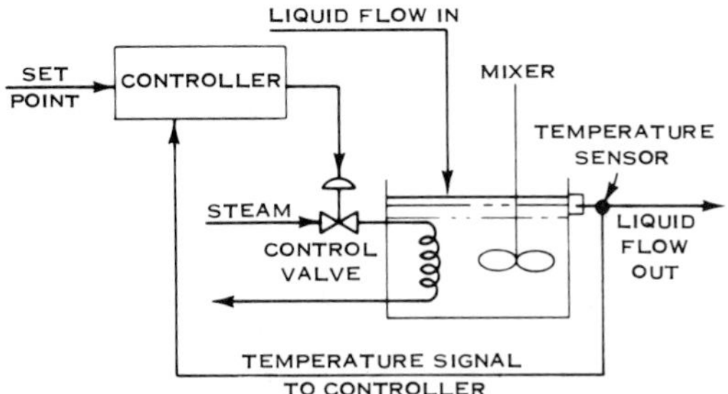

FIGURE 15.32 Automatic feedback control of heat-exchange process.

Feedforward Control. An input variable to the process is measured and transmitted to a feedforward controller. The controller uses a system model to determine values of process variables that will produce the desired value of the output (controlled variable). The effectiveness of pure feedforward control depends on whether the system model accurately predicts the response of the process to input and process variable changes. Feedforward control is most useful for those situations where there is a long time lag between upsets in input variables and the associated changes in the controlled variable. A simple feedforward control system is illustrated in Fig. 15.33.

One significant drawback to feedforward control is that few system models are accurate in predicting controlled variable responses. This is overcome by utilizing a feedback controller based on measurement of the controlled variable to modify the signal from the feedforward controller. This is most often accomplished by simply adding the signal from the feedback controller to that of the feedforward controller. Such control schemes (termed *combined feedforward and feedback systems*) are often the only effective method for controlling processes with long time lags between changes in process variables and the response of the controlled variable. The feedforward control adjusts the process to these

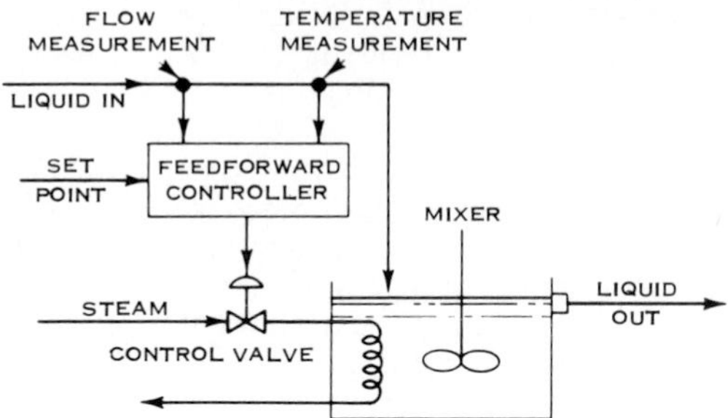

FIGURE 15.33 Feedforward control of heat-exchange process.

changes, and the feedback signal "trims" the signal of the feedforward controller to compensate for inadequacies of the system model. An example of combined feedforward and feedback control is shown in Fig. 15.34.

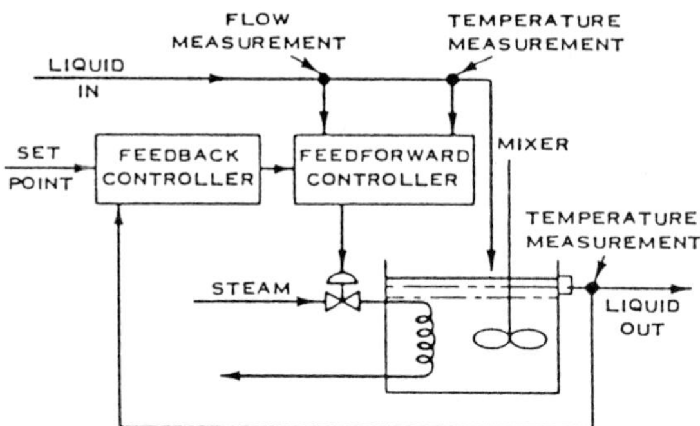

FIGURE 15.34 Combined feedforward and feedback control of heat exchange.

Cascade Control. Another approach to controlling processes with a long time lag between input variable change and controlled variable response is *cascade control*. The primary control loop measures the controlled variable and adjusts the set point on a secondary control loop. The secondary control loop measures some intermediate process variable and compares this to the output of the primary-loop controller to determine adjustments in the manipulated variable. The secondary loop can be either feedforward or feedback, but it must have a shorter time constant than the primary loop. Cascade control is valuable when the manipulated variable is subject to fluctuations outside the process being controlled, for example the pressure of a steam line being used to supply heat to a reboiler. The secondary loop, with its short time lag, prevents fluctuations in the input condition of the manipulated variable from affecting the process. The primary loop ensures that the adjustments made in the manipulated variable will give the desired output of the controlled variable. A simple cascade control system is illustrated in Fig. 15.35.

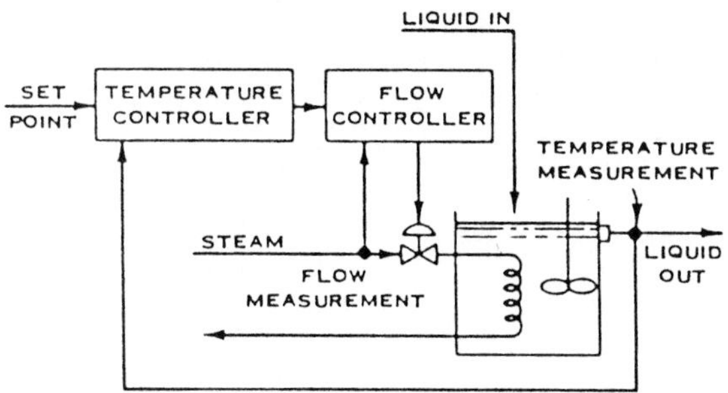

FIGURE 15.35 Cascade control system in which disturbances originating in the steam supply are prevented from entering the heat-exchange process.

26. Control System Analysis. Analysis of a process (either controlled or uncontrolled) may be carried out by deriving and solving the differential equations describing the dynamic (time-dependent) behavior of the system. Each independent element of the system is analyzed in terms of its governing physical, chemical, and electrical principles to obtain a series of simultaneous algebraic and time-dependent differential equations.

A generalized block diagram for such a system is shown in Fig. 15.36. The output *signal* of the controller is a function of the *error*, which is the difference between the measured value of the controlled variable and the desired value, which is represented by the controller *set point*. Equations describing this system are

$$f = f(\theta) \tag{15.61}$$

$$c = \phi(f, m, \theta) \tag{15.62}$$

$$m = \psi(e) \tag{15.63}$$

$$e = c - s \tag{15.64}$$

where c is the signal from an instrument measuring the controlled variable, s is the controller set point, m is the controller output signal, and f is the feed to the system. The function represents the physical, chemical, and thermodynamic

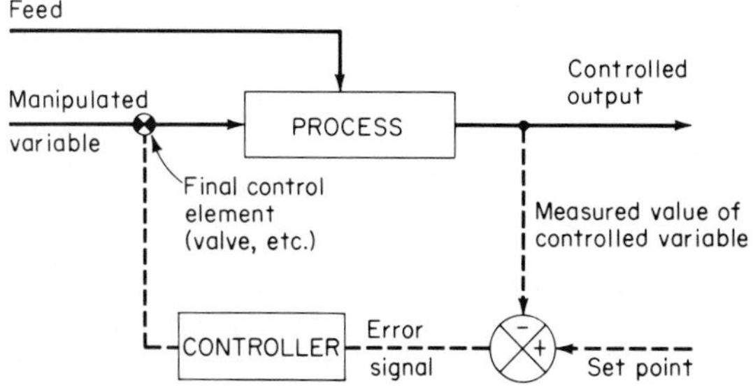

FIGURE 15.36 Generalized block diagram of feedback control system.

equations describing the process. The form of the function ψ depends on the nature of the controller.

More sophisticated system models include additional equations describing the dynamic behavior of measuring and transmission elements, as well as the characteristics of the control valve.

To design or analyze a control system for a given process, the parameters of the controller function ψ are adjusted and the responses of the controlled process to various changes in f are evaluated. There are several approaches to this procedure. The first is to linearize all equations in terms of a very narrow range of values near the normal steady-state operating conditions and then solve these linear simultaneous equations by means of Laplace transforms. This procedure is widely used and quite useful, since elements of the more standard controllers have simple Laplace transforms; the response of many systems has been analyzed in terms of transfer functions in the Laplace domain. An explanation of the theory of Laplace transforms is given in Chap. 12 of this book.

One major disadvantage of Laplace transforms is the approximation introduced by linearization of the process model. This is acceptable for small variations about the steady-state operating values but is likely to be significantly in error for large-scale variations such as at startup, in major process upsets, or in deliberate shifts to significantly different steady-state operating conditions.

An alternative procedure to the use of Laplace transforms is to solve the system model equations on either an analog or digital computer. This has the advantage of simulating system response to any input function, both simple pulse, step-change, or sine-wave inputs (such as are used in Laplace transform analysis) and more complex inputs which are not amenable to Laplace transforms. Furthermore, it may not be necessary to linearize the system model equations, making them applicable for large-scale process changes. This procedure has the disadvantage of requiring far more effort to design and tune those simple control systems where Laplace transforms are applicable.

Yet another procedure for analysis of control systems is the frequency-response approach. Deliberate sine-wave manipulations are made in the process input variables, and the associated response of the process output variable is measured. The relationships between amplitudes of input variable changes and output responses for different frequencies of sine-wave input are used to design controller systems. One method of frequency-response analysis (the Bode plot) is described in the section "Tuning Control Systems."

27. Feedback Control Systems

On-Off Control. The simplest and most common control system is the on-off controller. This is exemplified by the thermostat used in home heating. The controller puts out either a maximum or a minimum signal depending on the sign (but not the magnitude) of the error signal.

On-off control is simple, inexpensive, and stable control. If the controller is a perfect pure relay, instantaneously assuming either the maximum or minimum output condition as the error becomes different from zero, and if there is no time lag in the system or controller transmission lines, then on-off control is optimal control, responding to system upsets in the minimum time. In practice, time lags invariably occur in both the control system elements (transmission lines, control valves) and the process itself. Because of this, the control system tends to overcompensate, delivering a corrective control signal longer than necessary to drive the error to zero. This causes successive overshoot and undershoot, with the controlled variable oscillating around the desired value, as shown in Fig. 15.37. Typically, response dynamics of the controlled system will be such that the time-averaged system output will differ from the set point, even though the oscillation is around the set point. This difference is termed *offset* and generally is undesirable.

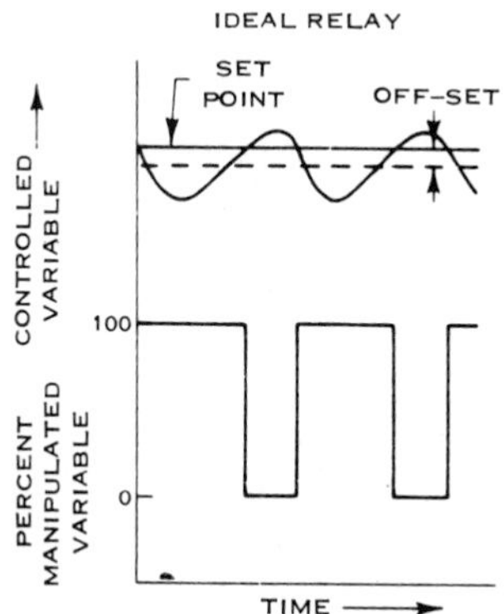

FIGURE 15.37 Relationship between controlled variables and the manipulated variable for an ideal relay.

The magnitude and period of the oscillations are of importance in on-off control and depend on both the process system dynamics and the controller characteristics. It is desirable to reduce the amplitude of the oscillations and eliminate offset while keeping the period of the oscillations as long as possible. Rapid oscillations can induce maintenance problems with the mechanical elements of the controller system, for example, through too rapid cycling of valves. High-amplitude oscillations can cause poor product quality or unstable operations.

Rapid cycling of the controller is often reduced by using a controller with hysteresis or dead band, which eliminates chattering of the controller system when at steady state but increases the amplitude of oscillations. The magnitude of oscillations can be reduced by using a smaller adjustment in the manipulated variable. This, however, slows the rate of return to steady state after a process upset.

For processes where small-scale deviations from steady-state conditions can be tolerated and time lags are relatively small, on-off control (usually with hysteresis or dead band) is often suitable. It is the least sophisticated and therefore least expensive control system. In practice, feasibility of on-off control is usually determined by whether the process can stand the rapid and frequent transition from fully-on to fully-off conditions. For example, the water-hammer effect resulting from rapid closure of a valve may not be acceptable. Additionally, rapid fluctuation of the final control element (valve, etc.) may cause frequent failure and maintenance problems.

Proportional Control. One approach to eliminating overshooting which occurs in on-off control is to reduce the adjustment to the manipulated variable as the magnitude of the error signal approaches zero. In *proportional control*, the controller output is proportional to the magnitude of the error signal, up to the maximum output of the controller, after which point it is a constant value. The proportionality constant by which the error signal is multiplied to obtain controller output is the *controller gain*. The inverse of the controller gain (times 100 percent) is the *proportional band*, which is similar to the dead band or hysteresis band in on-off control. The proportional controller is tuned to the process by adjusting the gain (or the width of the proportional band.)

The proportional controller suffers some of the disadvantages of the on-off controller, with certain exceptions. When the gain is high, a system with lag exhibits overshoot, oscillating around the set point. However, the amplitude of the oscillations can be made to decrease with time by using a low enough gain. As a result, the system can be made to seek a stable steady state with no oscillations. The higher the gain, the greater will be the initial overshoot as the system responds to an upset, the more rapid will be the oscillations, and the slower will be the decay to a new steady state. In the limiting case, a proportional control with very high gain becomes almost equivalent to an on-off controller, exhibiting a limit cycle around a steady-state value which does not decay.

If the gain is lowered to achieve stability, the amplitude of fluctuations will die out to produce a steady-state value. However, the new steady-state value may not be the desired output value. This offset is one of the most serious deficiencies of simple proportional control. It arises because a gain setting sufficiently low to achieve stability may not provide enough adjustment to the manipulated variable to correct for a changed process input.

For systems with no lag (or small lag), response to a process upset is slower than an on-off controller, becoming faster as gain is increased. For this type of system, on-off control is preferable, except for technical considerations previously mentioned. For systems with lag, a controller having proportional gain is preferable, often mandatory.

Proportional Plus Integral Control (PI or Two-Mode Control). The most significant drawback of proportional control is the offset that occurs in optimally tuned stable control systems. This problem can be reduced by adding *integral* or *reset action* to the controller. The output of a two-mode controller is proportional to the sum of the error (as in proportional control) plus the integral of the error with respect to time, up to the maximum controller output. A small error signal integrated over a long enough period of time can provide enough controller output to drive the process to the set point, thus eliminating offset.

The PI controller has two degrees of freedom which are adjusted to tune the controller to the process: proportional band or gain and reset time. *Reset time* is the time required for the integral-mode output of the controller to equal the proportional-mode output, assuming a constant error signal. Some controllers are calibrated in resets per minute, which is the inverse of reset time. Increasing the reset time decreases the rate at which the controller drives a small error to zero. In the limiting case, long reset time gives straight proportional control.

Proportional Plus Integral Plus Rate Control (PID or Three-Mode Control). The main disadvantages of PI control are slow response to a sudden change and a tendency to overshoot the desired value when tuned to give optimal response. Both these disadvantages can be reduced by addition of *derivative* or *rate action* to the controller. In three-mode control, the controller output is proportional to the sum of error signal, the integral of the error signal, and the time derivative of the error signal. This allows the controller to handle a rapidly changing error signal, since the faster the error increases, the greater will be the controller output to correct that error. On the other hand, as the corrective action of the controller catches up to the process disturbance and the magnitude of the error becomes smaller, the time derivative of the error becomes negative. This reduces the corrective action of the controller and thus limits the magnitude of the overshoot. This allows the use of somewhat higher gain for rapid controller action plus shorter reset times for rapid controller action and elimination of offset. Generally, high rate action is desirable; practically, this is limited by noise in the system. Noisy signals involve high time derivatives and create rapid fluctuation of the rate action of the controller, which reduces controller stability.

Three-mode controllers are the most sophisticated and expensive of simple feedback control systems. They are best suited for systems where precise control is required and no special problems, such as large-scale process time lags, occur. They are especially useful for systems where inertial effects are large.

28. Tuning Control Systems.

On-Off Systems. Tuning on-off control systems consists primarily of adjusting the relationship between amplitude and frequency of oscillations around the set point. This is done primarily by adjusting the dead band or hysteresis band of the controller. It is generally a tradeoff (based on experience) between allowable variation in the controlled variable and increased maintenance cost due to wear in the control equipment. Offset is reduced by adjusting the relationship between maximum and minimum manipulated variable corresponding to maximum and minimum controller output. If the offset is such that more corrective action is needed, the ratio of maximum value to minimum value is increased; if the offset indicates less corrective action, the ratio is reduced.

PID Controllers. The criteria for tuning three-mode controllers vary but generally include stability of the closed-loop system, the rate at which overshoot decays, and minimization of the error integral. It is obvious that the process must be designed so that the manipulated variables have sufficient range to maintain the desired steady-state output condition over the anticipated range of inputs to the system. This point is sometimes overlooked and cannot be compensated for by any control system.

Specific procedures for tuning controllers are beyond the scope of this text (see ref. 3). General procedures are presented below. The first step consists of determining the response of the open-loop (uncontrolled) system to a sine-wave variation of the manipulated variable. This may be done experimentally by manipulating the system or analytically by using Laplace transform techniques on the system model equations. In either case, a plot is made of the phase lag between the system response and the manipulated variable versus the logarithm of the frequency of oscillation of the manipulated variable. A second plot is made, on log-log coordinates, of amplitude ratio (ratio of system response to maximum input value of manipulated variable) versus frequency.

If the frequency-response characteristics of the control system are superimposed on these plots (note that the shape of the control-system plots depends on the controller gain, reset time, and rate-action time constants of the controller), the characteristics of the closed-loop system can be obtained by graphical addition of the two plots. The tuning procedure then involves adjusting the controller parameters to ensure that the controlled system is stable and to minimize the error integral.

Instability occurs when the phase lag approaches 180° and the amplitude ratio approaches unity. In this situation, the controller continually adjusts the process in the wrong direction, and the system gain is high enough to cause increasingly large oscillations around the set point. This is analogous to pushing a pendulum at the limit of its swing. If the controlled system has zero phase lag, the controller corrects the process at the time its effect is strongest. This is analogous to retarding a pendulum at the center of its swing.

In some cases, disturbances enter the system at some point other than the manipulated input variable. If the major sources of disturbance are not in the manipulated variable, the formalized technique above will not give adequate control. In such cases, the control system parameters obtained above are used as a starting point to tune the controller using a trial-and-error search technique. The controller parameters are adjusted slightly and changes in the total system performance are analyzed. This is a difficult procedure and calls for experience and good judgment.

There are many other methods for tuning controllers. One frequently used for tuning existing controllers is to determine the *ultimate gain* and *ultimate period*. The integral and derivative modes are turned off and the proportional gain is adjusted to obtain a metastable oscillation in the controlled system (i.e., an oscillation whose amplitude neither decays nor increases with time). The proportional gain at which this occurs is the *ultimate gain*, and the period of the oscillation is the *ultimate period*. The settings of the three controller parameters are then determined from the ultimate gain and period by one of various tuning-criteria rules. For example, if K_u is the ultimate gain and P_u is the ultimate period, then

$$K_c = 0.6K_u \tag{15.65}$$

$$\tau_I = 0.5P_u \tag{15.66}$$

$$\tau_D = 0.125P_u \tag{15.67}$$

will produce a decay of successive oscillations to one-fourth the amplitude of the previous oscillation, where K_c is the controller gain, τ_I is the reset time, and τ_D is the rate-action derivative time.

29. Design of Control Systems. Design of control systems is a complex subject that cannot be presented in great detail in any condensed text. The best procedure is generally to design the control system along with the process, with close interaction between design teams. Control systems "added on" after the fact seldom perform as well as those designed for the system from initial stages. Several critically important principles often overlooked in the design of a control system are presented below.

One independent process input variable must be manipulated for each controlled variable. It is best to select combinations of measured variable/manipulated variable that give the shortest possible time lag between input adjustment and response; this usually means as close together in the process flowsheet as possible. The paired variables should interact strongly rather than weakly.

For control loops where the controlled variable can be allowed to oscillate within a certain range, on-off controllers are the best choice, provided that there are no restrictions due to mechanical limitations of the control system or dynamic limitations on allowable rate of manipulation of input variables. If on-off control is unacceptable, the next choice would be proportional control. If the resulting offset is also unacceptable, it can be eliminated by adding integral mode (i.e., PI controller). Derivative control action is added to eliminate overshoot (PID control). If there is a significant time lag in the system, especially "dead time," some form of feedforward control is indicated. The measuring device used must be adequately sensitive to changes in the controlled variable and preferably fast acting.

One problem that is less obvious is control system incompatibility. It is possible that manipulation of one input variable to control a given output variable will cause upsets in a second output variable. It is desirable to reduce interaction between control loops whenever possible. The subject of interacting loops is complex and beyond the scope of this text.

Proper design of a process control system can eliminate many later operating problems and can improve the process, increasing yields, improving product purity, and shortening processing times. See Chap. 12 for additional data on control systems.

NOMENCLATURE

Basic Dimensions. Basic dimensions of parameters are given in terms of the following symbols and may be found with various compatible combinations of the units shown.

L = length, ft, cm
θ = time, s, h
T = temperature, °R or K
M = mass, lb mass, grams, tons
moles = mass/molecular weight, gram-moles, lb-moles
F = force, lb force, newtons
conc. = concentration, moles per unit volume
E = energy, cal, Btu, joules

Nomenclature

A = filter area, L^2
= collector plate area, L^2
= heat-transfer surface area, L^2
= membrane area, L^2
= chemical reactant species
= frequency factor, moles/($\theta \cdot$ conc.$^{n-1}$)
a = surface area of packing per unit volume of contactor, L^2/L^3
= stoichiometric coefficient for reactant A
B = chemical reactant species
b = stoichiometric coefficient for reactant B
C = chemical reactant species
= concentration, when subscripted, moles/L^3
C_D = drag coefficient
C_F = heat capacity of feed, $E/(M \cdot T)$
c = stoichiometric coefficient of reactant C
= controlled variable
D = distillate product rate, moles/θ
= chemical reactant species
D_a = impeller diameter, L
D_p = particle diameter, L
= diameter of ion-exchange resin bead, L
D_i = fluid-phase diffusion coefficient, L^2/θ
d = indicates differentiation
d_F = packing characteristic length, L
E = energy of activation, E

= extract flow rate, M/θ

E_c = column plate efficiency

e = error signal

exp = exponential function (natural antilogarithm)

F = feed flow rate, moles/θ, M/θ, L^3/θ

f = amount of vapor introduced to flash still or feed tray in rectifying column, moles/mole

= collected feed stream variables in process control

G = gas volumetric flow rate, L^3/θ

= rate of solute-free solvent removed in solvent separator, M/θ

G_M = gas molar flow rate, moles/θ

g = gravitational acceleration, L/θ^2

g_c = gravitational constant, $LM/(\theta^2 F)$

H = velocity head of agitated fluid, E/M

H_{OG} = height of transfer unit (HTU) based on overall (gas-phase) driving force, L

H_{OL} = height of transfer unit (HTU) based on overall (liquid-phase) driving force, L

K' = distribution coefficient for liquid-liquid equilibrium

K_c = equilibrium constant

= optimal controller gain

K_G = overall gas-phase mass-transfer coefficient, moles/($L^2\theta$ conc.)

K_L = overall liquid-phase mass-transfer coefficient, moles/(θL^2 conc.)

K_u = ultimate controller gain

k = reaction rate constant, forward reaction

k' = reaction rate constant, reverse reaction

k_G = gas-phase mass-transfer coefficient, moles/(θL^2 conc.)

k_L = liquid-phase mass-transfer coefficient, moles/(θL^2 conc.)

L = membrane thickness, L

= liquid molar flow rate in distillation column, moles/θ

L_M = liquid molar flow rate, moles/θ

L_e = extract-layer flow rate from last stage, M/θ

L_r = raffinate-layer flow rate from last stage, M/θ

l = nominal pore length in catalyst particle (radius of sphere or cylinder, half-thickness of slab), L

m = controller output signal

= slope of vapor-liquid equilibrium line

N = rotation speed of impeller, rpm

= number of equilibrium stages

= number of moles of subscripted species

N_{OG} = number of transfer units, based on over-all (gas-phase) driving force

N_{OL} = number of transfer units, based on over-all (liquid-phase) driving force

N_M = molar mass-transfer rate, moles/θ

N_P = impeller power number

N_Q = impeller discharge coefficient

N_T = number of theoretical stages at total reflux

N_{act} = actual number of stages in column

N_{theor} = number of equilibrium stages (theoretical stages) in column

n = order of chemical reaction

p = permeability of membrane

P = pressure of liquid, F/L^2

= agitator shaft power, E/θ

= chemical reactant species

= vapor pressure of subscripted component, F/L^2

= number of moles in still-pot (batch distillation)

P_u = ultimate period, θ

p = partial pressure of subscripted component, F/L^2

= stoichiometric coefficient of reactant P

Q = impeller discharge rate, L^3/θ

= rate of feed of solute-free solvent to raffinate end of extractor, M/θ

= chemical reactant species

q = rate of heat transfer, E/θ

= stoichiometric coefficient of reactant Q

R = ideal gas constant, $E/(T \cdot \text{moles})$

= raffinate-layer flow rate, M/θ

= reflux ratio

r = specific reaction rate, or specific reaction rate of subscripted species (forward reaction), moles/(θL^3)

r' = specific reaction rate (reverse reaction), moles/(θL^3)

$\bar{r}$ = average pore radius, L

τ_{net} = net forward rate of a reversible reaction, moles/(θL^3)

S = amount of material distilled in Rayleigh distillation, moles

= solvent content of extract layer, $M_{\text{solvent}}/M_{\text{dissolved material}}$

s = controller set point

= empirical constant for filter-cake compressibility

= average of mutual solubilities of solute-free contacted liquids

= solvent content of raffinate layer, $M_{\text{solvent}}/M_{\text{dissolved solids}}$

T = absolute temperature, °R or K

t = temperature, °C or °F

U = overall heat transfer coefficient, $E/(L^2\theta T)$

U_i = dialysis coefficient in subscripted phase, moles/$(L^2\theta)$ conc.

V = total volume upstream of a given point in a plug flow reactor, L^3
= volume of contactor or reactor, L^3
= molar vapor rate in column, M/θ
V_C = superficial velocity of continuous phase, $L^3/(L^2\theta)$
V_D = superficial velocity of dispersed phase, $L^3/(L^2\theta)$
V_e = extract-layer flow rate from last stage, M/θ
V_r = raffinate-layer flow rate from last stage, M/θ
v = total volume of feed to ion-exchange process, L^3
v_t = terminal settling velocity, L/θ
W = mass of solids retained in filter cake, M
= bottom product rate, moles/θ
w = equilibrium composition of raffinate phase, $M_{\text{solute}}/M_{\text{solute-free phase}}$
w' = equilibrium composition of extract phase, $M_{\text{solute}}/M_{\text{solute-free phase}}$
X = mass fraction solute in raffinate, solvent-free basis
= moles of subscripted component converted per mole of feed to continuous reactor
x = mole fraction (of subscripted component) in liquid phase
x^* = mole fraction in liquid phase which would be in equilibrium with actual gas phase at that point in contactor
Y = mass fraction of solute in extract, solvent-free basis
y = mole fraction (of subscripted component) in vapor or gas phase
y^* = mole fraction in vapor phase which would be in equilibrium with actual liquid phase at that point in contactor
Z = total column height, L
Z_t = tray spacing, L
Z_f = solvent content of feed, $M_{\text{solvent}}/M_{\text{solvent-free stream}}$
α, α_{12} = relative volatility
α = reaction order with respect to reactant A
= specific cake resistance, L/M
= empirical constant for wet scrubber efficiencies
α' = empirical constant for filter cake resistance
β = reaction order with respect to reactant B
γ = activity coefficient
= empirical constant for wet scrubber efficiencies
= reaction order with respect to reactant C
Δ = difference or change
ΔP_c = pressure drop across filter cake, F/L^2
θ = time
$\theta_{1/2}$ = reaction half-life
λ_m = molar latent heat of vaporization, E/moles
λ = latent heat of vaporization, E/M

μ = viscosity $M/(L\theta)$
Π = total pressure, F/L^2
ρ = density, M/L^3
= reaction order with respect to reactant P
ρ_f = density of fluid, M/L^3
ρ_p = density of particle, M/L^3
σ = reaction order with respect to reactant Q
σ = interfacial tension, lb mass/hr^2
σ' = interfacial tension, dynes/cm
$\phi(\ )$ = arbitrary function of ( )
$\psi(\ )$ = arbitrary function of ( )
τ_I = reset time, θ
τ_D = rate action derivative time, θ

Subscripts

A = reactant A
B = reactant B
C = reactant C
= continuous phase
D = disperse phase
d = distillate product
F = feed stream
i = component i
= at interface
L_1 = liquid phase 1
L_2 = liquid phase 2
lm = logarithmic mean
M = molar property
m = molar property
O = initial value
p = in still pot
P = reactant P
Q = reactant Q
w = bottom product
in = entering
out = leaving
1 = component 1
= at position 1 (upstream) or initial time
2 = component 2
= at position 2 (downstream) or final time

ENVIRONMENTAL ENGINEERING*

INTRODUCTION

The concept of environmental engineering is relatively new. Historically, governments have concerned themselves with only the most flagrant forms of pollution, such as a public water supply contaminated with sewage. In general, the world's atmosphere and bodies of water were considered a limitless garbage dump, and natural purification processes were sufficient to handle human wastes for thousands of years. But now the combined impacts of continued industrial development and population growth make environmental engineering a necessity.

The field spans an enormous range of activities and embraces all the traditional engineering disciplines. This section will concentrate on the legal and technological aspects of wastewater treatment and control of air pollution from stationary sources.

The world's industrialized nations did not give serious attention to the environment until after World War II. In the United States, federal water laws were passed in 1948, 1956, 1965, and 1970. The first federal air pollution act became law in 1955, with amendments in 1967.

Federal legislation was generally ineffective because enforcement remained in the hands of state and local governments. Few communities were willing to spend time and money on pollution control, which had not been recognized as a serious problem. (Exceptions were cities such as Pittsburgh and Los Angeles that began fighting pollution in the 1950s.)

By 1970 it was obvious that pollution would not go away by itself, and the environmental movement became popular in Congress. Among the results were passage of stringent federal laws and consolidation of all federal pollution control activities under a single agency, the Environmental Protection Agency (EPA).

1. Clean Air Act of 1970. The Clean Air Act of 1970 was the first law to require national ambient air quality standards based on geographic regions. Air quality is regulated by two sets of standards determined by EPA (Table 15.12). The primary standards are intended to be the minimum level of air quality that is necessary to preserve human health, while the secondary standards are aimed at preventing damage to animals, plant life, and property.

State governments retain the authority to determine how ambient air quality standards are to be met within their borders. However, the state implementation plans are subject to approval by EPA, which can revise any state plan it considers unsatisfactory.

EPA also has the power to set performance standards for new stationary sources of air pollution. It has published standards for a number of major sources, including fossil-fuel-fired steam generators, incinerators, cement plants, sulfuric and nitric acid manufacturing, petroleum refineries, asphalt plants, steel mills, secondary lead smelters, and various other nonferrous metal operations.

Air pollutants that are particularly hazardous to human health are controlled on an individual basis under emission standards set by EPA. Asbestos, beryllium, and mercury were among the first materials to be regulated.

---

*This section is from *Engineering Manual*, 3d ed., by R. H. Perry (ed.). Copyright © 1976. Used by permission of McGraw-Hill, Inc. All rights reserved. Updating and SI units were added by the editors.

TABLE 15.12 National Ambient Air Quality Standards*

	Primary standard		Secondary standard	
	$\mu g/m^3$	ppm	$\mu g/m^3$	ppm
Sulfur oxides:				
Annual arithmetic mean	80	0.03		
24-hr concentration	365†	0.14†		
3-hr concentration	.....	.....	1.300†	0.5†
Suspended particulate matter:				
Annual geometric mean	75	.....	60‡	
24-hr concentration	260†	.....	150†	
Carbon monoxide:				
8-hr concentration	.....	9.0†	Same as primary	
1-hr concentration	.....	35.0†		
Photochemical oxidants:				
1-hr concentration	160†	0.08†	Same as primary	
Hydrocarbons (corrected for methane):				
3-hr concentration (6–9 am)	160†	0.24†	Same as primary	
Nitrogen oxides:				
Annual arithmetic mean	100	0.05	Same as primary	

* ppm = parts per million, $\mu g/m^3$ = micrograms per cubic meter.
† Not to be exceeded more than once a year.
‡ A guide for assessing achievement of the 24-hr standard.

An early problem that developed under the 1970 act was how to treat regions where air quality was better than the ambient standards. Environmentalists argued that EPA had a duty to protect pristine areas, while others said that a certain amount of development (and pollution) was necessary for the good of the people. After several conflicting court decisions, EPA has set guidelines that allow some degradation of clean air up to the limits of the secondary air quality standards. However, the matter is still subject to further review by the courts and Congress.

As written in 1970, the Clean Air Act required compliance with the national primary standards by 1975, although the deadline could be extended for 2 years in those cases where compliance was "technologically impossible." At the beginning of 1975, there were still two major gaps in the clean air program: automobile emissions and emissions of sulfur oxides from power plants. Both problems involved not only questions of control technology, but also whether the clean-air timetable should be revised because of energy considerations.

2. Water Act of 1972. The U.S. Congress tackled water pollution with the Water Pollution Control Act Amendments of 1972. This bill is considerably more stringent than any previous water legislation and provides penalties up to $25,000 per day for willful or negligent violations. All discharges to navigable waters must have been treated by best practicable technology by July 1, 1977 and best available technology by July 1, 1983. The EPA holds responsibility for determining exactly what is required under the two levels of treatment technology.

The principal control mechanism of the 1972 amendments is the National Pollutant Discharge Elimination System (NPDES). This means that every wastewater discharger must have a permit. The permit lists the quantities of pollutants that the source may discharge in terms of the usual parameters such as biochemical oxygen demand, dissolved solids, etc. EPA has the basic authority for issuing permits, although it is required to turn over this authority to any state that has established an acceptable permit program.

During 1973 and 1974, EPA set effluent guidelines for 28 industry categories. These serve as the basis for pollutant limitations in the discharge permits. The guidelines sparked much controversy (and a number of legal suits). Further action by the Congress may be required. The objections to the guidelines involve complex legal questions, but in general they are being challenged on the grounds that EPA exceeded its statutory authority under the 1972 law.

3. Information Sources. The engineer who needs to determine the specific pollution-control requirements for a particular operation in a certain location has many sources of information available—too numerous to be listed here. In general, the starting point for legal information should be at the lowest applicable level of government. Usually this is a state body, although in metropolitan areas a county or city agency may have jurisdiction.

When in doubt, the EPA has vast information resources. Established in 1970 to consolidate all federal environmental efforts under a single agency, EPA maintains 10 regional offices that are familiar with local problems and requirements.

WASTEWATER TREATMENT

4. Preliminary Treatment

Equalization. Wastewater flow rates are erratic. Municipal flow rates peak during the day and reach a minimum during the night; within these periods, the flow rises and falls with the time of day and with the weather (rainfall). Industrial wastewater flows are equally variable, being a function of production rate, frequency of batch operations, etc. Thus many industries, although surprisingly few municipalities, have installed equalization tanks (holding basins) to smooth out process flows and allow the wastewater treatment plant to operate at a steady rate.

Neutralization. Biological wastewater treatment proceeds optimally at pH values near 7 (neutrality). Slight deviations from this value will reduce efficiency, while large differences may result in total inactivation of the bacteria.

While municipal sewage normally needs no pH adjustment, industrial waste quite often does. Using pH recorder-controllers, some incorporating analog computers, acids and bases can be added in optimum fashion. Less control equipment is needed if pH is adjusted manually in the equalization tank. Sodium hydroxide or ammonia raises the pH; sulfuric or phosphoric acid lowers the pH. Ammonia and phosphoric acid may serve the dual purpose of pH control and a source of nitrogen or phosphorus, which are important nutrients.

Oil Removal. Generally, oil concentrations greater than 50 mg/L inhibit biological action. Thus gravity settling basins are usually installed to separate oil and water when they are combined in a nonemulsified form.

In general, oil-water separators are rectangular, multichannel structures that produce low flow velocity and a minimum of turbulence while allowing time for the oil globules to float to the surface and be removed. Gravity treatment, of course, only separates free oil. When emulsified oil is present in industrial wastes, aeration followed by gravity separation is usually sufficient to reduce oil concentrations for subsequent biological treatment.

Discharge to a receiving water directly from oil separators generally is not allowed by effluent standards. Secondary oil removal must follow the oil-separation stage. Typical processes include dissolved air flotation, granular-media filtration, and activated carbon adsorption.

Dissolved air flotation serves the dual purpose of oil removal and clarification (removal of suspended solids). Air is dissolved in the wastewater under pressure [30 to 70 lb/in_g^2 (207 to 483 kPa_g)], and then nucleated as fine bubbles by a rapid pressure reduction when the pressurized water stream enters a flotation chamber at atmospheric pressure. The micron-size bubbles attach themselves to suspended matter and carry it to the surface for removal by skimmers. Though more expensive than separators, flotation devices produce a higher-quality effluent.

Oil emulsions can be broken by chemical pretreatment with acids or aluminum and iron salts. Subsequent addition of polyelectrolytes aids in floc formation and leads to more efficient operation.

Metallic-Ion Removal. Ions of the heavy metals (Cu, Zn, Al, Fe, Cr, etc.) in concentrations greater than 1 to 10 mg/L, depending on their nature, must be removed because they inhibit subsequent treatment processes. Adding lime, raising the pH to about 10.5, normally results in a reasonable reduction of metallic-ion concentration, because many of the metals form insoluble hydroxides at this pH level. It may even be possible to go below the level predicted by the solubility product because of adsorption of the metallic ions on the chemical floc. For cadmium, lead, and mercury, precipitation with lime may be incomplete, and addition of soda ash (for lead) and sodium sulfide (for cadmium and mercury) may be needed.

5. Primary Treatment. Screens, comminutors, and/or grit chambers usually precede primary sedimentation in municipal plants. Screens, normally composed of iron bars or grates with ¾- to 3-in (1.9- to 7.6-cm) openings, remove the largest suspended particles, while comminutors grind large particles into smaller ones. Grit chambers are designed to remove the heavier solids (i.e., inorganic matter such as sand whose specific gravity is greater than 2) through a slight reduction in flow velocity.

Considerable advantage is gained by removing suspended particulate matter in the primary basin rather than in the secondary. Normally, the initial sedimentation process removes 50 percent of the suspended solids, carrying with it 25 to 40 percent of the biochemical oxygen demand (BOD); reduction of the BOD load to the secondary system is an obvious benefit in reducing air-supply requirements and production of microbial solids.

Simplistic design of sedimentation systems is accomplished by choosing either a residence detention time (generally 90 to 150 min) or an overflow rate [the normal range is 600 to 1200 gal/ft^2 per day (24,446 to 48,891 L/m^2 per day)]. From these data, the tank depth can easily be selected.

Chemical Addition. Chemicals are being used increasingly in the initial stages of wastewater treatment for (1) phosphorus removal, (2) increased efficiency of BOD removal in the primaries, and (3) proper conditioning of the wastewater for filtration, carbon adsorption, and reverse osmosis. Hence the proper selection of chemicals that assist in the aggregation of colloidal particles (0.001 to 10 μm) is crucial to the process operation.

A number of suitable coagulating agents can be used: lime, alum, iron salts, and polyelectrolytes (Table 15.13). The first step is to test each in the laboratory. Besides determining the best additive, the laboratory study should investigate the

TABLE 15.13 Chemical Coagulant Selection*

Chemical process	Dosage range, mg/L	Resulting pH	Sludge dewatering	Conditions favoring use
High lime	150–600	11.5–12.0	Easy	1. For colloid coagulation and P removal 2. High and variable influent P 3. Large plants >10 mgd, where recalcination is feasible 4. Wastewater with high calcium content and low noncalcium dissolved solids content
Low lime	75–250	9.5–10.5	Easy	1. For colloid coagulation and P removal 2. Wastewater with low alkalinity and high and variable P 3. Medium-size plants, 1–20 mgd, no recalcination
Very low lime	50–100	8 0–8.5	Fair	1. For P removal 2. Use in combination with the activated sludge process where a 6–8-hr detention time is available for the precipitation reaction 3. Wastewater with low alkalinity and high and variable P
Solid alum	75–250	4.5–7.0	Difficult	1. For colloid coagulation and P removal 2. Wastewater with high alkalinity and low and stable P 3. Where the difficult alum sludge can be handled
Liquid alum	75–250	4.5–7.0	Difficult	1. Same as for solid alum with the following differences: a. It is cheaper than solid alum when you are located within 100–150 mi of an alum manufacturing plant b. It is easy to feed and handle and is, therefore, well suited to small (<1 0 mgd) plants c. These differences between solid and liquid forms also apply to $FeCl_3$
$FeCl_3$, $FeCl_2$ $FeSO_4$, $7H_2O$	35–150 70–200	4.0–7.0 4.0–7.0	Fair Fair	1. For colloid coagulation and P removal 2. Wastewater with high alkalinity, and low and stable P 3. Where leaching of iron in the effluent is allowable or can be controlled 4. Where economical source of waste iron is available (steel mills, etc.)
Cationic polymers	2–5	No change	May improve	1. For colloid coagulation or to aid coagulation with a metal 2. Where the buildup of an inert chemical is to be avoided
Anionic and some nonionic polymers	0.25–1.0	No change	May improve	1. Used as a flocculating aid to speed flocculation and settling and to toughen floc for filtration
Weighting aids and clays	3–20	No change	May improve	1. Used for very dilute colloidal suspensions for weighting

*mg/L = milligrams per liter, mgd = millions of gallons per day, (1 mgd = 3.8 millions of liters per day), P = phosphorus.

Source: From T. J. Tofflemire and L. J. Hetling, "Chemicals and Clarifiers, Which Are Best," *Water and Wastes Engineering* 10: F24–F26, November 1972.

sequence of chemical addition, as well as blending, flocculation, and detention times. Additional data sought should be estimates of effluent quality, volumetric sludge production rate, and a measure of the sludge dewatering characteristics.

Lime, ferric chloride, ferric sulfate, sodium aluminate, and alum are the most widely used inorganic coagulants. *Polyelectrolytes* are high-molecular-weight (15,000 to 10,000,000) polymers with multiple ionic charges along the chain;

nonionic polymers also have been termed *polyelectrolytes*. These polymers can be linear, branched, or cross-linked and are normally more effective as molecular weight increases.

The performance of chemical additives in wastewater treatment is affected by a number of factors: dosage and molecular weight of the coagulant, temperature, agitation rate and duration, order of addition of chemicals, nature of suspended matter, pH or amount of alkali present, and time between chemical additions. Coagulant dosage is important because the optimum concentration lies in a narrow range. Low dosages fail to destabilize while excess dosages restabilize colloids.

6. Bioxidation. Simple sedimentation, as described in the previous section, removes a major fraction of the suspended solids but none of the dissolved contaminants. Microorganisms can utilize dissolved organic material as nutrients for their growth and metabolism. The process, shown mechanistically in Fig. 15.38,

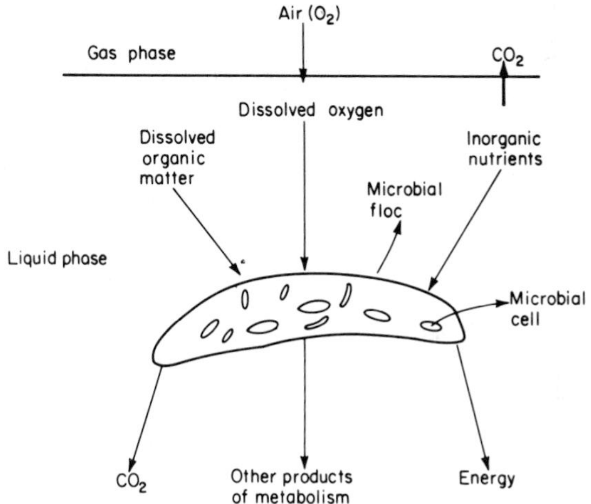

FIGURE 15.38 Schematic for bioxidation of dissolved organic matter.

could be termed *flameless oxidation* in the liquid state because bacteria convert carbon to carbon dioxide via metabolic pathways, therein consuming oxygen and yielding energy.

In nature, the rate of supply of oxygen may be low. Engineers circumvent this limitation by designing waste treatment systems that optimize the contact between the microorganisms (catalysts), food (dissolved organic matter), and air (oxygen source). In essence, the system is balanced so that the rates of oxygen supply and consumption are equal.

There are two different biological oxidation, or bioxidation, processes, with the distinction between them based on microbial mobility:

1. Suspended—the microorganisms are suspended in the liquid; the system operates as a continuous biological reactor, called an *activated sludge plant* or *oxidation pond*.

2. Fixed bed—the microorganisms are embedded in a gelatinous mass attached to a solid support medium. This is referred to as a *trickling filter* or *bioxidation tower*.

The role of the bioxidation system is to remove the dissolved organic matter from the wastewater. Because excess cells are produced, they are usually removed by sedimentation in a secondary clarifier, which is similar to the primary clarifier.

Three types of reactions take place in the bioxidation process: (1) assimilative respiration, (2) synthesis of cells, and (3) endogenous respiration. Assimilative respiration provides the necessary energy for the life processes of cells and can be represented by the following equation:

$$C_xH_yO_z + (x + \tfrac{1}{4}y - \tfrac{1}{2}z)O_2 \rightarrow CO_2 + \tfrac{1}{2}yH_2O + \text{energy} \qquad (15.68)$$

Besides the energy released, the cells derive energy for the oxidation process to use in building new cells:

$$n(C_xH_yO_z) + nNH_3 + n(x - \tfrac{1}{4}y - \tfrac{1}{2}z - 5)O_2 \rightarrow (C_5H_7NO_2)n + n(x - 5)CO_2 + \tfrac{1}{2}n(y - 4)H_2O \qquad (15.69)$$

This process is termed *endogenous respiration*; it is actually autoxidation or self-destruction of cellular material and can be represented by

$$(C_5H_7NO_2)n + 5nO_2 \rightarrow 5nCO_2 + 2nH_2O + nNH_3 + \text{energy} \qquad (15.70)$$

The formula representative of bacterial cells is $C_5H_7NO_2$. The need for nitrogen in order to grow cells is obvious by its inclusion in the formula. Not shown, however, is another important nutrient, phosphorus, which is needed in lesser amounts. Phosphorus, nitrogen, and BOD should be present in a ratio of about 1:5:100 to support growth and oxidation. Municipal sewage normally contains sufficient amounts of nitrogen and phosphorus, but many industrial wastes are deficient in one or the other and must be supplemented chemically.

Activated Sludge System. The conventional activated sludge process is analogous to a continuous biological fermentor because nutrients in the untreated wastewater are continually added, oxygen is supplied through an aeration system, and cells are produced and removed in the effluent stream (Fig. 15.39). The product is a combination of treated wastewater and suspended cells, which are removed by sedimentation and partially recycled to the reactor. The treated effluent overflows to discharge or reuse systems.

The activated sludge process provides a smooth-running, highly efficient operation, with excellent removal of organic matter; BOD removals of 90 percent are easily achieved. However, many problems can develop to upset the process. Shock loading, through changes in concentration and volumetric flow rate, abrupt variations in pH, temperature, or other environmental factors, or the presence of toxic materials are among the major causes of trouble.

The keys to the process are the active bacteria embedded in gelatinous clumps in the reactor (activated sludge basin) that oxidize biodegradable organic material. Also present are a host of other microorganisms, fungi, protozoa, rotifers, and sometimes nematods that are necessary for a healthy, balanced population. Important design and operating variables for the process are the food-to-

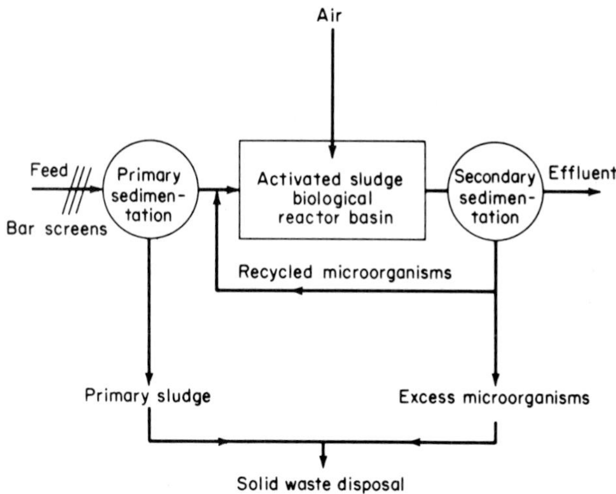

FIGURE 15.39 Simplified flowsheet for activated sludge plant.

microorganism ratio [the *F/M* ratio in pounds (kilograms) of food added per pound (kilogram) of cells per day], the amount of BOD supplied per unit volume of reactor [lb BOD/1000 ft^3 of reactor volume (kg BOD/1000 m^3)] and oxygen consumption rate [lb O_2/lb BOD treated (kg O_2/kg BOD)].

The conventional activated sludge process can be divided into a series of steps:

1. Mixing of the activated sludge and the sewage
2. Aeration and agitation of the mixed liquor for the prescribed period of time
3. Separation of the sludge cells from the water
4. Recycle of the proper amount of cells (to step 1)
5. Disposal of the excess sludge

Modification of these basic steps has shown the versatility of this process and has led to many variations: tapered aeration, step aeration, contact stabilization, and extended aeration to name just a few (Fig. 15.40).

Tapered Aeration. As the organic contaminants are consumed by the bacteria in a long reactor, the *F/M* ratio and oxygen demand decrease with length. The process modification from conventional activated sludge is a variation of the airflow rate according to oxygen needs achieved by putting in decreasing numbers of diffuser tubes with increasing distance from the inlet. The obvious advantage is lower power consumption than uniform (and excess) aeration would require.

Step Aeration. Another way to balance air requirements is to add the wastewater at different points along the aeration tank. This modification keeps the *F/M* ratio constant as a function of tank length and leads to uniform oxygen demand by the microorganisms. With these changes, BOD loading may be increased and detention time and consumption rate may be lowered.

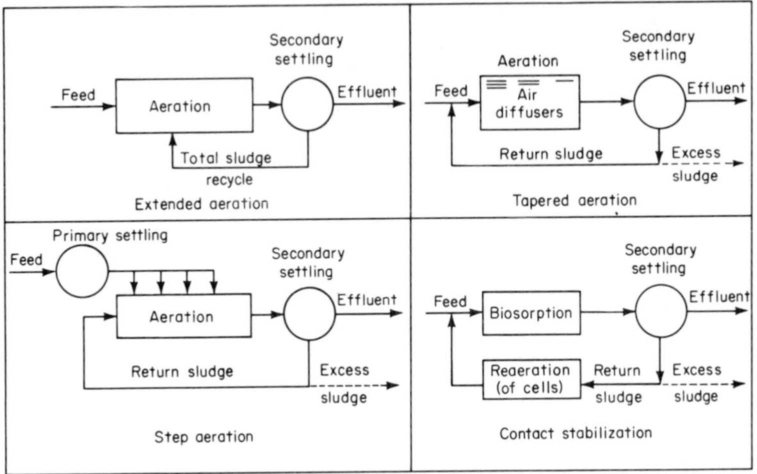

FIGURE 15.40 Four variations of the activated sludge process.

Contact Stabilization. The adsorptive properties of activated sludge particles are advantageous when a large part of the BOD is present in suspended form. Sludge that has been conditioned by reaeration is brought into short (15 to 30 min) but unaerated contact with the wastewater. During this stabilization period, the activated sludge adsorbs and absorbs a high percentage of the colloidal, suspended, and even dissolved organic matter. The mixture is then discharged to a settling tank where solids are removed and directed to a regenerator (reactor) tank for active aeration for approximately 3 h before reentering the cycle.

Extended Aeration. Extended aeration systems operate essentially without any sludge production because they include neither a primary sedimentation tank nor any allowance for continuous sludge withdrawal from the bioxidation system. However, a penalty for eliminating these units must be paid: a high oxygen supply rate is needed to oxidize all organic matter to carbon dioxide at low BOD loading values.

Aeration tanks usually are sized to provide 24- to 30-h retention with air added mechanically or by submerged diffusers. A final settling basin where the detention is approximately 4 h removes most solids for total recycle—although normally provision is made for intermittent sludge withdrawal to holding tanks for later off-site disposal. In general, the process functions best at BOD loadings of less than 15 lb (6.8 kg) per day per 1000 ft^3 (28.3 m^3) of aeration tank and less than 0.1 lb BOD/lb (6.05 kg BOD/kg) of cells.

Advantages include simple operation, availability of prefabricated units, and adaptability to a variety of topographies and climate conditions, and generally substantial BOD reduction (an average of 87 percent) has been reported. Disadvantages include high power cost, long retention times, and periodic loss of fine floc in the effluent.

Oxygen Demand and Supply. Generally, 2 mg/L of dissolved oxygen should be maintained at all points of the activated sludge basin to keep the cells working at their maximum rate. Hence the mixed liquor in the reactor is continuously aerated by either a diffused air system or by mechanical surface aerators. The aeration process serves three functions: (1) it mixes the sludge and the sewage, (2) it keeps the sludge in suspension, and (3) it supplies the oxygen needed for bioxidation.

In diffused air systems, air is supplied by blowers and forced through diverse types of porous materials mounted in plates or tubes; the air is released as fine bubbles several feet below the surface of the liquid. In mechanical systems, surface aerators cause liquid turbulence, at which point the oxygen is dissolved and dispersed downward into the fluid.

Oxygen must be supplied at a rate sufficient to meet the demand of the cells. The amount needed varies with the waste being treated, the activity of the microbial cells, and length of aeration period, but generally it is in the range of 1 to 2 lb O_2/lb BOD (0.45 to 0.9 kg O_2/kg BOD) removed.

Oxygenation economy is extremely important in sewage plant design and operation. A critical factor is the amount of oxygen dissolved in the water per unit of power, expressed in lb O_2 dissolved/(hp · h). Surface aerators yield an average of 3.5 lb O_2/(hp · h) (2.1 kg/kWh) and submerged diffusers about 2.5 (1.5).

Lagoons. The term *lagoon* or *stabilization pond* generally is applied to all bodies of water artificially created with the intention of retaining sewage or organically contaminated wastewaters until the wastes are rendered stable or unobjectionable through biological decomposition and the waters are suitable for disposition, either by discharge into receiving waters or by way of ground seepage and evaporation.

Lagooning is a common and cheap means of eliminating dissolved or suspended organics. The feed enters at one end of the lagoon. Sedimentation occurs first, with settlable solids being deposited near the entrance. Additional sedimentation occurs throughout the pond area, with the action of soluble salts aiding precipitation of colloid matter. Added to the bottom (benthic) deposits through sedimentation are the bacteria and algae produced by biological activity.

Algae are a significant part of the purification cycle in lagoons. They use carbon dioxide, sulfates, nitrates, phosphates, water, and sunlight to synthesize their own cellular material, with free oxygen as a waste product. This process reverses the bacterial degradation sequence to generate an organic material that, if discharged, is itself a pollutant. Treatment efficiencies vary, with indication that well-designed and well-operated ponds produce effluents equivalent to or better than conventional secondary plants.

In design of a pond, several physical factors must be considered:

1. Depth—experience indicates that a depth of 3 to 4 ft (0.9 to 1.2 m) with flexibility to permit 5 to 6 ft (1.5 to 1.8 m) is desirable.
2. Sidewall—sloped at a ratio of 3:1, freeboard of 2 to 5 ft (0.6 to 1.5 m).
3. Bottom—level and clear of vegetation and debris.
4. Inlet—feed should be mixed quickly and distributed throughout the pond; short-circuiting is to be eliminated.
5. Retention time—usually 2 to 30 days.

Trickling Filters. *Trickling*, or *biological, filter* is the name given to the process whereby microorganisms attached to stones (or plastic media) strip wastewater of its organic components as it flows (trickles) down through a packed tower. The attached growths adsorb, absorb, and oxidize suspended and colloidal matter in the wastewater. Crushed stone [1½ to 4 in (3.8 to 10.2 cm)] normally is used as the packing medium for the bed, although recently plastic packing has found substantial application. The wastewater is applied to the filter through a central rotating distributor, which intermittently applies water to the various segments of the bed.

Approximately 60 percent of the dissolved organic matter removed from the waste stream by the bioxidation process is utilized for growth of new microorganisms, with the remaining 40 percent of the material being oxidized to terminal metabolic products, carbon dioxide, and water while providing energy. Growth is most active at the liquid-slime surface where the concentration of organic material is highest. Oxygen diffuses from the liquid into the slime, but progressively lower oxygen concentrations are found at greater depths because oxygen is consumed by the cells. At some limiting slime depth, the oxygen is gone and anaerobiosis occurs.

The growing slime layer is constantly being scoured by the downward flow of liquid. After the microbial layer reaches a critical thickness, outer portions slough off (break off) and are discharged in the effluent leaving at the bottom of the tower. Sloughing will occur either intermittently or more or less continuously, depending on the hydraulic load. Various biological processes are compared in Table 15.14.

Design Equations. The rate of removal of BOD in a filter is a function of the BOD concentration of the wastewater and the adsorptive capacity of the biological growth. The rate of stabilization controls the adsorption capacity of the biological growth.

Eckenfelder[21] has proposed equations for design of rock filters with and without recirculation:

TABLE 15.14 Comparison of Biological Processes

Process	Area, acres (m^2)	Biological loading, lb BOD/1000 ft^3 (kg BOD/m^3)	BOD removal, %
Stabilization pond	57* (32,218)	0.09–0.23 (0.001–0.004)	70–90
Aerated lagoon	5.75† (23,426)	1.15–1.60 (0.018–0.026)	80–90
Activated sludge:			
Extended	0.23 (937)	11.0–30.0 (0.176–0.481)	95+
Conventional	0.08 (326)	33.0–400 (0.529–6.41)	90
High rate	0.046 (187)	57.0–150 (0.913–2.4)	70
Trickling filter:			
Rock	0.2–0.5 (815–2037)	0.7–50 (0.011–0.801)	40–70
Plastic media	0.02–0.08 (81–326)	20–200 (0.32–3.21)	50–70

*5ft (1.5m) deep.
†10ft (3.0 m) deep.
Source: *Chemical Engineering* 76; 63–70, 1970.

No recirculation: $$\frac{L_c}{L_0} = \frac{100}{1 + 2.5(D^{0.67}/Q^{0.50})} \tag{15.71}$$

Recirculation: $$\frac{L_c}{L_0} = \frac{1}{(1 + N)[1 + 2.5(D^{0.67}/Q^{0.5})] - N} \tag{15.72}$$

when $$L_0 = \frac{L_a + NL_c}{N + 1}$$

where L_c = BOD of effluent, mg/L
L_a = influent BOD, mg/L
L_0 = BOD of mixed influent plus recirculated streams, mg/L
N = recirculation ratio
D = filter depth, ft (m)
Q = hydraulic application rate, million gal/(acre · day) [million L/(m^2 · day)]

Recirculation has several benefits:

1. Smooths out diurnal flows
2. Stops long retention times in settling basins at low flows
3. Provides a microbial seed
4. Brings back organic matter that escapes removal in the first pass

There is, however, a limit to the amount of organic matter that can be removed through successive treatment by increasing the recirculation rate. Each succeeding pass removes a smaller amount of the unoxidized BOD, because the remaining BOD is more refractory (i.e., resists biodegradation). As the slime is forced to use more dilute waste, it dissolves itself as endogenous respiration occurs. Additionally, as the recirculation ratio increases, the hydraulic load on the filter increases.

Temperature Effect. The effect of temperature on treatment efficiency is given by the formula

$$E = E_{20}1.035^{(T-20)} \tag{15.73}$$

when E is the efficiency of BOD removal at any temperature T (°C), and E_{20} is the efficiency of removal at 20°C.

Plastic Packing. Plastic media have found wide acceptance in municipal and industrial systems. The medium consists of corrugated plastic sheets running parallel to each other and bonded together to form a bundle or a pack. The uniform but open pattern of the plastic grid largely eliminates clogging problems that exist with rock and allows the grid to operate at higher organic loadings.

The basic design formula for plastic-media trickling filters was developed by Germain:[22]

$$\frac{L_c}{L_0} = e^{-kD/Q^{1/2}} \tag{15.74}$$

where D = the depth of the filter, ft (m)

Q = the hydraulic dosage rate of primary effluent, gal/(min · ft^2) [L/(min · m^2)]

k = the reaction rate coefficient (treatability factor) in consistent units to render the exponent dimensionless

The treatability factor is important because its magnitude determines rate of BOD removal. For municipal sewage, Germain determined a value of 0.088.

Because plastic media can handle high BOD loads, it is an ideal packing for a roughing filter, i.e., to serve as a first and important step in the BOD reduction process. In a plant that has become overloaded, installation of a device of this type prior to the activated sludge tanks has proved economical and efficient.

7. Sludge Treatment. *Sludge*, a suspension of solids in water, must be processed further to reduce its high water content and to reduce the concentration of organic matter. If untreated and used as landfill, it will decay and produce an offensive odor. Concentration and stabilization are the primary functions of the sludge-processing system that can constitute as much as 25 to 50 percent of the operating and capital cost of a wastewater treatment plant.

According to Burd,[23] whose FWPCA report on sludge handling and disposal has become the standard handbook in the field, the following general observations can be made about sludge handling and disposal practices: anaerobic digestion followed by sand-bed dewatering is the most common method presently employed because of its simplicity and low cost; lagooning is the most common method employed by industry; for coastal and near-coastal cities, anaerobic digestion followed by pipeline or barge transportation to an open dumping site is cheapest; marketing of dried waste sludge has been a failure; almost all the common methods now employed were used in the 1930s.

Burd has been able to note trends in sludge treatment, among which are (1) increasing adoption of mechanical dewatering with the centrifuge is of increasing importance over vacuum filters; (2) sludge incineration is the process with the brightest future; (3) raw sludge incineration is replacing anaerobic digestion at medium and large plants; (4) the popularity of composting sludge in foreign countries is declining; and (5) sludge volumes are rising and becoming more difficult to dewater.

In municipal plants, the two main sources of sludge are the primary clarifiers, which produce a sludge having 1 to 4 percent solids, and the secondary clarifiers, which yield biological sludge with a solids concentration of less than 1 percent. Of the two, primary sludge is simpler to process, dewater, stabilize, and dispose of. Sludge production characteristics of typical municipal wastewater treatment plants are shown in Table 15.15.

Conditioning. Sludge is most often conditioned chemically. Coalescence can be enhanced, in theory, by chemicals that neutralize charges on suspended particles, causing them to agglomerate and simultaneously lessening the particle's tendency to bind water.

Synthetic polymers have taken on a role of increasing importance in flocculation, but inorganic compounds such as alum and ferric chloride are still widely used. Other conditioning agents include fly ash and diatomaceous earth.

Elutriation has been defined by Burd[23] as "a washing operation which removes sludge constituents that interfere with thickening and dewatering pro-

TABLE 15.15 Normal Quantities of Sludge Produced by Different Treatment Processes*

Treatment process	Normal quantity of sludge: Gal/million gal of sewage (L/million L)	Normal quantity of sludge: Tons/million gal of sewage (kg/million L)	Normal quantity of sludge: Ft^3/1000 persons daily (m^3/1000 persons)	Moisture, %	Specific gravity of sludge solids	Specific gravity of sludge	Dry solids: Lb/million gal of sewage (kg/million L)	Dry solids: Lb/1000 persons daily (kg/1000 persons)
Primary sedimentation:								
Undigested	2950	12.5 (3355)	39.0 (1.1)	95	1.40	1.02	1250 (149.8)	125 (56.7)
Digested in separate tanks	1450	6.25 (1678)	19.0 (0.54)	94	—	1.03	750 (89.9)	75 (34.0)
Digested and dewatered on sand beds	—	0.94 (252)	5.7 (0.16)	60	—	—	750 (89.9)	75 (34.0)
Digested and dewatered on vacuum filters	—	1.36 (365)	4.3 (0.12)	72.5	—	1.00	750 (89.9)	75 (34.0)
Trickling filter	745	3.17 (851)	9.9 (0.28)	92.5	1.33	1.025	476 (57.0)	48 (21.8)
Chemical precipitation	5120	22.0 (5905)	68.5 (1.93)	92.5	1.93	1.03	3300 (395.4)	330 (149.7)
Dewatered on vacuum filters	—	6.0 (1611)	19.3 (0.55)	72.5	—	—	3300 (395.4)	330 (149.7)
Primary sedimentation and activated sludge:								
Undigested	6900	29.25 (7852)	92.0 (2.61)	96	—	1.02	2340 (280.4)	234 (106.1)
Undigested and dewatered on vacuum filters	1480	5.85 (1570)	20.0 (0.57)	80	—	0.95	2340 (280.4)	234 (106.1)
Digested in separate tanks	2700	11.67 (3127)	36.0 (1.02)	94	—	1.03	1400 (167.7)	140 (63.5)
Digested and dewatered on sand beds	—	1.75 (470)	18.0 (0.51)	60	—	—	1400 (167.7)	140 (63.5)
Digested and dewatered on vacuum filters	—	3.5 (939)	11.7 (0.33)	80	—	0.95	1400 (167.7)	140 (63.5)
Activated sludge:								
Wet sludge	19,400	75.0 (20,132)	258.0 (7.30)	98.5	1.25	1.005	2250 (269.6)	225 (102.0)
Dewatered on vacuum filters	—	5.62 (1509)	19.0 (0.54)	80	—	0.95	2250 (269.6)	225 (102.0)
Dried by heat dryers	—	1.17 (314)	3.0 (0.09)	4	—	1.25	2250 (269.6)	225 (102.0)
Septic tanks, digested	900	—	12.0 (0.34)	90	1.40	1.04	810 (97.1)	81 (36.7)
Imhoff tanks, digested	500	—	6.7 (0.19)	85	1.27	1.04	690 (82.7)	69 (31.3)

*Based on a sewage flow of 100 gallon (378.5 L) per capita per day and 300 mg/L, or 0.25 lb per capita daily, of suspended solids in sewage.

Source: Metcalf and Eddy, *Wastewater Engineering: Collection, Treatment, Disposal*, McGraw-Hill, New York, 1972.

cesses." The simplest method of elutriation occurs in a single-stage batch process by a fill and draw procedure, with sedimentation and decantation performed in a single step.

Gravity Thickening. Gravity thickening basins are usually circular and are provided with a raking mechanism to convey the solids to the point of discharge at the center of the tank. For municipal sewage sludges, the following loading rates, given in lb/ft^2 per day (kg/m^2 per day), are recommended for thickeners: (1) 22 (107.4) for primary sludge, (2) 15 (73.2) for primary plus trickling-filter biofloc, (3) 8 to 12 (39.1 to 58.6) for a blend of primary and waste-activated sludge; and (4) 4 (19.5) for waste-activated sludge alone.

It is obvious that activated sludge alone releases its water very slowly, while blends of primary and secondary sludge respond better. Thus blending is recommended when both sludges are available.

Flotation. Flotation is attractive for sludge thickening because of a faster dewatering rate that is not affected adversely by decomposing solids, which evolve gases and cause sludge bulking. In flotation units, air is dissolved under pressure in the liquid being treated. When the pressurized air-wastewater stream is discharged into an aeration tank, maintained at atmospheric pressure, the supersaturated air nucleates as small bubbles in the 10- to 100-μm range.

The air bubbles, normally carrying a charge, collide with the solid particles and form agglomerates whose specific gravity is less than that of the water. The "lightened" solids rise to the surface, forming a blanket that is skimmed off by mechanical scrapers.

Processes to prepare sludge for dewatering and disposal are as diverse as any unit operation in waste treatment. They range from those employed for many years (anaerobic digestion) to those having only research interest at present (ultrasound), from aerobic to anaerobic digestion by bacteria, and from heating to freezing.

Aerobic Digestion. In *aerobic digestion*, waste sludge is simultaneously the food and the oxidation system; i.e., in autoxidation or endogenous systems the living cells utilize nutrients released when other cells die and dissolve. Thus the microbial population and the amount of biodegradable organic matter are continually decreasing. The process is carried out until the volatile suspended solids are reduced to about 50 percent, a level at which the sludge product will not cause a nuisance. The residence time should be about 15 days to produce a good sludge that forms a filter cake of 20 to 25 percent solids.

For medium- to small-sized plants, aerobic digestion has real promise and offers several advantages over anaerobic digestion, including lower BOD of the supernatant fluid, production of a better dewatering sludge, and lower capital cost. Offsetting these advantages is the major added cost of power to supply air to oxidize the cellular matter.

Anaerobic Digestion. *Anaerobic digestion* is the most popular method of sludge stabilization. The term applies to the process in which organic material is decomposed biologically in an environment devoid of oxygen. This decomposition results from the action of two major groups of bacteria: acid formers, consisting of facultative bacteria that convert carbohydrates, fats, and proteins to organic acids and alcohols; and methane bacteria, which are strict anaerobes that convert the organic acids and alcohols produced by the acid formers into carbon dioxide and methane. Small amounts of hydrogen and hydrogen sulfide are also formed.

The gaseous mixture produced by these symbiotic (mutually benefiting) reactions contains approximately 60 percent methane and 35 percent carbon dioxide, with the remainder being water vapor, hydrogen sulfide, hydrogen, ammonia, and nitrogen. Its heat value is in the range of 600 to 650 Btu/ft^3 (22.4 to 24.2 MJ/m^3) as contrasted to 960 Btu/ft^3 (35.8 MJ/m^3) for pure methane. This gas, which can be burned to produce energy, is one of the major advantages of anaerobic digestion. Many investigators are examining ways of optimizing methane production by anaerobic processes. Other advantages of this process include a high degree of waste stabilization, combined with production of a low amount of sludge that is inoffensive, has low hydrophilic properties, and is easy to dewater.

The standard-rate digestor, shown schematically in Fig. 15.41, has wide application in sewage treatment. To obtain significant stabilization of the sludge, retention times are long—on the order of 30 to 60 days with loadings of 30 to 100 lb volatile solids per 1000 ft^3 (13.6 to 45.4 kg per 28.3 m^3). This results in large vessels. Further, only about one-third of the volume is used for actual digestion. The remainder of the space is occupied by stabilized solids, supernatant fluid, scum, and gas.

The feed to the digestor and withdrawal of byproducts are intermittent. Upon entering the digestor, the new sludge rises to the scum zone and undergoes initial decomposition, thereby producing much of the gas. As decomposition proceeds, the partially decomposed solids fall to the bottom of the tank and build up a layer of digested and digesting solids. The intermediate layer is a supernatant liquid, which must be returned to the liquid treatment system because it contains a high concentration of dissolved and suspended solids.

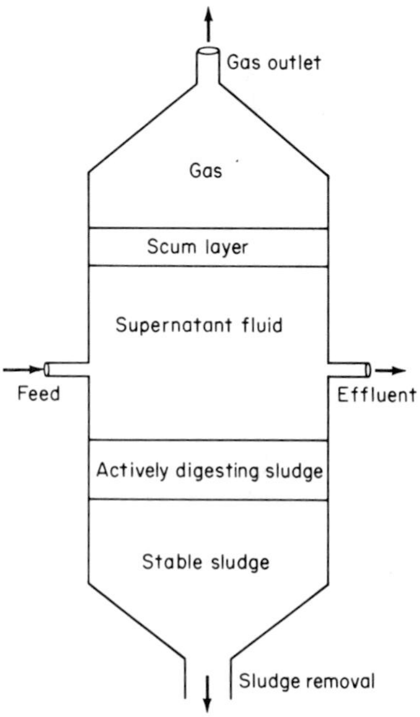

FIGURE 15.41 Active regions in conventional anaerobic digestor.

The size of the tank is based on population. For a facility treating primary sludge, design values are 2 to 3 ft^3 (0.06 to 0.09 m^3) of volume per capita, 4 to 5 ft^3 (0.11 to 0.14 m^3) if the plant is a trickling-filter, and 4 to 6 ft^3 (0.11 to 0.17 m^3) if it is activated sludge. The higher values should be used in each case for installations handling the waste of populations of 5000 or less.

Temperature exerts a significant effect on anaerobic digestion. As the temperature increases, the time required to attain a specific degree of stabilization decreases. The optimum temperature for digestor operation in the mesophilic range is 98°F (36.7°C). Thermophilic temperatures, it has been shown, cause digestion to occur more efficiently, but temperature control is difficult and more thermal energy is required.

Reduction of residence time through continuous mixing is the basis of the high-rate digestion process. Continuous feed and withdrawal and heating of the entire tank contents result in stable operation with a detention time of 10 to 15 days at loadings of 100 to 500 lb volatile solids per 1000 ft^3 (45.4 to 227.0 kg per 28.3 m^3). Since there is no scum layer and no stratification, the entire tank is actively used for sludge digestion.

Although anaerobic digestion is mainly used for the stabilization of sludges, it is employed to a limited extent for the treatment of high-strength wastes. Anaerobic digestion becomes economical at effluent concentrations in excess of 4000 mg/L and has increased economic advantages as the strength of the organic waste rises. At the upper limit (above 50,000 mg/L), evaporation and solid recovery begin to look attractive.

Thermal Conditioning. Heat treatment of sludge under pressure for about 30 min produces a sludge that is easily dewatered without the use of chemicals. Essentially, the process consists of grinding the sludge to eliminate large particles, and sparging the sludge with live steam in a reactor for 30 to 45 min at 250 lb/in_g^2 (1.7 MPa_g).

Sludge is withdrawn to a decanter where the solid material settles rapidly to about one-third of its original volume. Vacuum filters handling heat-treated sludge average 12 lb/ft^3 per day (58.6 kg/m^3 per day). The filter cake, which prior to heat treatment averaged 20 to 25 percent solids, is increased to 40 to 45 percent.

Dewatering. Used for many years, applicable to a wide variety of waste, and relatively cheap, vacuum filtration is a popular method of mechanically dewatering sludge. A typical vacuum filter consists of a rotating cylindrical drum partially submerged in sludge. A vacuum of 10 to 20 inHg (254 to 508 mmHg) is applied to the center of the drum, drawing the water through the filter medium but leaving the solids deposited on the periphery. Continual application of vacuum as the drum rotates out of the liquid removes more water. Just before the drum reenters the sludge, the cake is removed.

Several factors affect the design of the vacuum filter and dictate its performance. The major ones are

1. Sludge type, age, temperature, and compressibility
2. Solids concentration in the feed
3. Flocculant type, dosage, and procedure

These considerations affect vacuum level and vacuum pump capacity, drum size and degree of submergence, and filter medium. The vacuum level is important in determining the final solids level. Doubling the vacuum from 10 to 20 in (254 to 508 mm) will increase by several percentage points the solids content of the cake. In addition, the filtration rate varies approximately as the 0.3 power of the vacuum. To size the vacuum pump, one generally designs for a flow rate of 2 $ft^3/(min \cdot ft^2)$ [6.01 $m^3/(s \cdot m^2)$] at 20 inHg (508 mmHg). Typical performance of a vacuum filter is shown in Table 15.16.

Another parameter affecting the filtration rate is the solids concentration. There is a direct relation between the two, with the cake solids concentration increasing as the feed solids increase. Practically, however, there is a limit of 8 to

TABLE 15.16 Expected Performance of Vacuum Filters Handling Properly Conditioned Sludge

Type of sludge	Yield, lb/(ft^2 · h) [kg/m^2 · h)]
Fresh solids:	
Primary	4–12 (19.5–58.6)
Primary + trickling filter	4–8 (19.5–39.1)
Primary + activated	4–5 (19.5–24.4)
Activated (alone)	2.5–3.5 (12.2–17.1)
Digested solids (with or without elutriation):	
Primary	4–8 (19.5–39.1)
Primary + trickling filter	4–5 (19.5–24.4)
Primary + activated	4–5 (19.5–24.4)

Source: Sewage Treatment Plant Design, *Manual of Practice*, vol. 8. Water Pollution Control Federation, Washington, D.C., 1959.

10 percent to the feed solids concentration; if this concentration level is exceeded, chemical conditioning and sludge distribution become difficult.

Pressure Filtration. This essentially batch operation is relatively labor-intensive, but the product is so dry (50 to 60 percent solids) that in many cases the advantages in disposal of the resultant solid waste outweigh the cost of labor. Besides a well-dewatered product, the filtrate generally contains a low concentration (20 mg/L or less) of suspended solids.

The filter press is an alternating series of plates and frames. The feed is pumped into the interior of a frame and forced through the filter medium by the pressure applied by the pump. Filtrate leaves through channels in the plates, while the solids are retained in the frame. When full, continuous hydraulic pressure applied by the feed pump forces (presses) more liquid out. Then the press is opened and the sludge removed. Normal cycle times are in the range of 2 to 4 h.

Centrifugation. In recent years, centrifuges have gained increasing acceptance in the handling of municipal and industrial sludges. Among the reasons are the development of centrifuges requiring less maintenance than in the past while operating at higher feed rates and greater sludge recovery. Disadvantages still include the maintenance problems and noise that are inherent with high-speed equipment.

There are basically three types of centrifuges: disk, solid-bowl conveyor, and solid-bowl baskets. Each has its own operating characteristics and special advantages, requiring a study of the potential application to choose the proper design.

In the *disk centrifuge*, the feed stream is divided among many narrow channels in the bowl and discharges from it through peripheral orifices. A *solid-bowl centrifuge* essentially consists of a cylindrical bowl in which solids are carried by an internal helical conveyor to one end of the bowl, while the clarified effluent is discharged at the other end. In the *basket centrifuge*, solids are retained in a perforated basket while the liquid passes through. Performance data for centrifuges on a wide variety of sludges are found in Table 15.17.

Sand Drying Beds. Although mechanical methods of sludge dewatering have received the greatest research attention recently, operators, especially those of

TABLE 15.17 Performance of Centrifuges on Various Sludges

Sludge source	Sludge type	Cake, % solids	Recovery of solids, %	Chemical needs, lb/ton (kg/908 kg)
Primary	Raw	30–40	70–80	None (none)
Secondary activated	Undigested	15–20	85–100	10–15 (4.5–6.8)
Secondary—extended aeration	Activated	5–15	90–100	5–10 (2.3–4.5)
Biological	Heated	40–45	90–95	None (none)
Chemical/antibiotic	—	20–30	High	—
Refinery	Blended	15–35	82–86	None (none)

small plants, still use sand beds effectively. Drying beds are constructed by placing 4 to 6 in (102 to 152 mm) of sand over graded layers of gravel or crushed stone underlaid by tile drains. The sides of the beds are made of concrete or wood; internally the beds are subdivided into cells, usually less than 20 ft (6.1 m) wide and 100 ft (30.5 m) long.

The digested sludge (raw sludge cannot be dried this way because it will cause an odor problem) flows into the bed to a depth of 6 to 12 in (152 to 305 mm) and is allowed to dewater. The process occurs by both drainage and evaporation. Although drainage is the predominant mechanism, evaporation is also important.

Burd[23] has recommended as design criteria the following sizes for open beds in 40 to 45°N latitude.

Primary sludge	1.00 ft^2/capita (0.093 m^2/capita)
High-rate trickling filter sludge	1.50 ft^2/capita (0.14 m^2/capita)
Activated sludge	1.35 ft^2/capita (0.13 m^2/capita)

After drying, the sludge normally is removed by hand, although mechanical handling equipment has been developed. The requirement for hand labor is one of the major disadvantages of sand beds. Another is the large land area required. But for small plants, sand beds following anaerobic digestion are the most economical treatment sequence.

8. Sludge Disposal. It should be evident that waste treatment does not destroy pollution, but merely converts the pollutant to another form. Hence a large fraction of wastewater contaminants, both suspended and dissolved, are converted to a solid waste whose ultimate disposal location is on the land, in the atmosphere, or in the ocean. Economics, location, and environmental impact are the three factors normally controlling process selection.

Incineration. With suitable sanitary landfill areas rapidly disappearing and ocean dumping being closely scrutinized, incineration is becoming an increasingly popular method of sludge disposal. Burning accomplishes a significant reduction in sludge volume and produces a sterile ash. The fundamental concern is the supplementary heat required (if any) to sustain combustion. Dried solids contain considerable heat value. Based on the ultimate analysis, the heat value is

$$Q = 14{,}600\ \mathrm{C} + 62{,}000\ (\mathrm{H} - \mathrm{O}/8) \tag{15.75}$$

where Q = Btu/lb (J/kg) dried sludge
C = percent carbon
H = percent hydrogen
O = percent oxygen

A typical sewage sludge will have a heat value of 7000 Btu/lb (16.3 MJ/kg) dry solids. If inorganic coagulants are used in the thickening process, this value is reduced.

Factors affecting the amount of supplementary heat required for combustion include

1. Heat value of the dried sludge.
2. Percent moisture of the feed. This is an extremely important variable since approximately 1000 Btu (2.3 MJ) is needed to evaporate each pound (kilogram) of water accompanying the solids.
3. Concentration of inert chemicals combined with the sludge as a result of their use in conditioning or phosphate removal.
4. Temperature of combustion, which determines the heat losses from the furnace by radiation and convection in addition to the heat lost in the ash and gases.
5. Excess air required and degree of preheating.

Multiple-hearth incinerators are quite popular. The furnace consists of circular hearths, one above the other, enclosed in a refractory-lined steel shell. Entering at the top of the furnace, the sludge is raked by blades that are attached to a central vertical rotating shaft. Solids follow a spiral path across the hearth to a drop hole where they fall to the next hearth. The sludge passes through three successive zones: drying, for removal of water to the point at which combustion can occur; combustion; and subsequent cooling to recover heat from the solids. The ash, freed of all organic matter, is a sterile mixture of the oxides of silicon, aluminum, magnesium, sulfur, sodium, calcium, and iron.

The prime requisite for ultimate disposal of sewage sludge, processed or unprocessed, is that it not cause a secondary pollution problem. For example, minimization of air pollution from incinerators is accomplished with wet scrubbers. Comparable steps must be taken to ensure the environmental soundness of any disposal technique for solid matter on land or into water.

Land Spreading. There are two principal methods of disposing of sludge on land, a process that has the advantage of recovering the fertilizer value of the sludge. Drying and sale of the material for fertilizer have had limited success. The city of Milwaukee for many years has produced Milorganite, a fertilizer that gained prominence largely due to marketing techniques. However, synthetic fertilizers have provided severe competition, because they deliver more nitrogen and phosphorus at lower cost.

The sludge also can be spread wet, thus avoiding the cost of drying. Many farmers seek this material if available locally, although there may be problems with odor and possible accumulation of heavy metals in the soil.

Ocean Disposal. For cities near the ocean, barging or pipeline transport of sludge to deep water has been popular because ocean disposal is cheap and simple. Although all traces of sludge disappear within ½ h of the discharge and studies by large cities using the method claim no detectable degradation of the sea, there are serious concerns over the environmental impact of this method. Thus ocean dumping of sludge will probably not last.

9. Tertiary Treatment. *Tertiary treatment* is defined as any process that follows secondary biological systems. For the purpose of this section, it has been divided into four areas: removal of nutrients, fine solids, organic material, and dissolved solids. None of the steps to remove these pollutants is cheap in view of the incremental cost. Microscreening, for example, can reduce BOD by a further 5 percent at a cost of 1.5 cents per 1000 gal (1.5 cents per 3785 L)—compared to a 90 percent reduction in secondary treatment for 11 cents. For dissolved solids removal, the cost increases dramatically.

Phosphorus. Although phosphorus can be removed biologically and chemically (ion exchange included), physical processes appear to be better understood and more controllable. It is generally agreed that primary sedimentation will remove, on average, 10 percent of the influent phosphorus. Conventional activated sludge plants, including primary sedimentation, achieve about 46 percent removal. There are, however, cases where significantly improved phosphorus removal has been attained biologically. Even so, chemical precipitation is the favored method.

The two most common types of chemical treatment are alkaline removal with lime and adsorption or precipitation with a metallic hydroxide such as alum. Costs vary widely, but are usually less than 10 cents per 1000 gal (10 cents per 3785 L).

Lime, if added in proper concentration, not only precipitates phosphorus, but through increased flocculation and enhanced sedimentation removes BOD and suspended solids better than iron and aluminum salts. Moreover, lime systems are less susceptible to shock loading and corrosion in the sludge-handling system and present less trouble in removing oil, grease, and scum. They also put fewer dissolved solids into the effluent (sulfate and chloride concentrations are increased by the other two chemicals mentioned). Finally, lime systems offer cost savings because the lime can be recycled by recalcination.

Lime is cheap but must be added in large doses (approximately 200 to 400 mg/L) to raise the pH to 9.5 or above. At this level of alkalinity, a good settling floc is obtained and the phosphorus is reduced to 0.5 mg/L or less.

Iron and aluminum salts have advantages, especially in the lesser amount of sludge generated and its treatability. Aluminum is generally supplied by alum (aluminum sulfate) or sodium aluminate; removal of phosphorus by these salts is optimum at pH 6. Unlike lime, which must be used at high bactericidal pHs, aluminum salts can be added directly to the biological system. Although the cost for this additive is higher, its lower dosage is a balancing factor.

Ferric salts are also employed in phosphorus removal, especially with a suitable polymer. The required dosage is generally low, and the sludge is readily settleable. Waste pickle liquor, if available locally, should be considered as an iron source.

Nitrogen. Although conventional sewage plants remove some nitrogen, there are still varying amounts released in diverse forms, such as ammonia, organic nitrogen, and nitrates. A process for nitrogen removal receiving much attention at the present time is nitrification-denitrification. Nitrogen is first oxidized to its highest oxidation state; the nitrates are then reduced to nitrogen, which is air-stripped. Nitrification occurs in two steps. The bacteria capable of oxidizing ammonia to nitrate are believed to be primarily autotrophic nitrifiers, among which *Nitrosomonas* are most common:

$$NH_4^+ + 1.50_2 \rightarrow 2H^+ + H_2O + NO_2^- \qquad (15.76)$$

The second phase of the reaction, nitrite to nitrate, is carried out by *Nitrobacter*:

$$NO_2^- + 0.50_2 \rightarrow NO_2^- \qquad (15.77)$$

One factor of great importance in nitrification is the large amount of oxygen required per unit of ammonia—theoretically 4.56 mg O_2/mg NH_3.

The first step can be carried out in activated sludge basins if proper conditions are maintained: optimum pH of 8 to 8.5 and temperature 30 to 35°C (the reactor will work in the pH range 7 to 9 and temperature range 15 to 35°C, but at lower rates and efficiency), dissolved oxygen concentration 1 to 2 mg/L or higher, and low loading equivalent to extended aeration systems [0.2 lb BOD per day per lb MLSS* (0.09 kg BOD per day per kg MLSS)].

Dentrification is also a biological process in which bacteria, in the absence of free molecular oxygen, reduce the nitrates to nitrogen. In order to achieve a reasonable efficiency, a supplementary source of carbon, commonly methanol, is added to accomplish the following:

$$5CH_3OH + 6NO_3^- \rightarrow 5CO_2 + 7H_2O + 3N_2 + 6OH^- \qquad (15.78)$$

The air-stripping process for ammonia removal consists in raising the pH to 10.5 to 11.5 and degassing the ammonia through air-water contact. Cooling towers have been used for this step. Important design variables for the process are pH control, the rate of interphase ammonia transfer, and the air-liquid flow rate. The process can be controlled to attain selected ammonia removals, but there are several disadvantages, including low efficiency at low temperatures, deposits of calcium in the tower, air pollution problems, and deterioration of wood packing of the tower at high pH.

Almost all soluble nitrogen compounds in wastewater are in ionic form and theoretically can be removed by deionizing or desalting processes such as reverse osmosis, electrodialysis, or distillation. However, none of these processes has a favorable selectivity for either the ammonium or nitrate ion, and all would require 85 to 90 percent removal of all salts to remove 90 percent of the nitrogen. Distillation is not useful because ammonia would be in the distillate unless the liquid was acidic, in which case nitrous acid would form.

A potentially promising process is ion exchange using a naturally occurring material called *clinoptilolite*, a natural zeolite that has a strong affinity for the ammonium ion. Removals of 93 to 99 percent of the ammonia nitrogen can be obtained.

Breakpoint chlorination has been used for many years in disinfection of drinking water and wastewater. Applied to the latter in sufficient concentration, it has been shown that essentially all the ammonia can be oxidized to nitrogen gas. Side products such as nitrates and nitrogen trichloride can be produced if the pH is uncontrolled, but effective mixing and pH management can limit their concentration to satisfactory levels.

*MLSS = mixed liquor suspended solids.

Ammonia removal proceeds according to the reaction

$$3Cl_2 + 2NH_3 \rightarrow N_2 + 6HCl \tag{15.79}$$

The *breakpoint* is defined as the point where the nitrogen in the form of ammonia is eliminated and free available chlorine is detected.

Removal of Trace Suspended Solids. The characterization of trace suspended solids is difficult. Particle size, concentration, and physical and chemical properties (specific gravity, toxicity, stickiness, etc.) are all highly variable and dependent on the pollution source. Sewage plant effluents, especially those from biological plants, contain small amounts of suspended solids that can have a serious impact on the receiving body of water.

A prime variable that must be considered in selecting tertiary solids-removal equipment, aside from chemical composition, is the fineness of the material. The spectrum ranges from coarse particles that rapidly sink or float to colloids and macroscopic, microscopic, and submicroscopic organisms dispersed in the discharge. Of concern in this section is the removal of small amounts (generally less than 30 mg/L of organic solids) in effluents emanating from biological waste treatment.

Microstraining. Essentially, a *microstrainer* consists of a rotating drum with a fine screen around its periphery. Feedwater enters the drum through an open end and passes radially through the screen while solids are deposited on the inner surface of the screen. At the top of the drum, pressure jets of effluent water, amounting to 1 to 5 percent of the treated flow, are directed onto the screen to remove the mat of deposited solids. The portion of backwash water that permeates the screen and dislodged solids are captured in a waste hopper and are removed through the hollow axle of the unit.

The effectiveness of removal depends on the screen pore size. The screens used in microstraining are made of a variety of plastics and stainless steel and have extremely small openings. They have high porosity to effectuate high flow rate at low pressure drops [6 in (152 mm) due to the fabric and 12 to 15 in (305 to 381 mm) overall].

Sand Filtration. The most common filter medium is a graded bed of silica sand. Developed in Great Britain and used for water clarification, these filters were operated at rates of 0.04 to 0.12 gpm/ft^2 [0.027 to 0.081 L/(s · m^2)]. In the United States, precoagulation has increased flow rates to 1 to 4 gpm/ft^2 [0.68 to 2.7 L/(s · m^2)]. The two processes have thus been called *slow* and *rapid sand filtration*.

A slow sand filter consists of a watertight basin containing a layer of sand 3 to 5 ft (0.9 to 1.5 m) thick supported on a layer of gravel 6 to 12 in (152 to 305 mm) thick. The gravel is overlaid by a system of open-joint underdrains placed 10 to 20 ft (3 to 6.1 m) apart, which leads the filtered water to a single outlet where the rate of flow through the filter is controlled. The most common effective size of the sand is 0.35 mm.

The filter is operated with a water depth of 3 to 5 ft (0.9 to 1.5 m) above the sand surface. The solids removed from the water in a properly functioning slow sand filter are found mainly in the mat of previously trapped solids in the upper inch of sand. When the pressure drop equals the head of fluid, the filter is removed from service, drained, and cleaned.

Cleaning is traditionally done by scraping an inch of sand from the surface before returning the filter to service. The dirty sand is cleaned hydraulically and stored for future replacement in the filter. Slow sand filters remove most suspended solids except for fine clays and other colloidal solids.

With the exception of gravity sedimentation, deep-bed filtration is the most widely used unit operation for solid-liquid separation. At the present time, virtually all deep-bed filters utilized for waste treatment are the rapid sand type. This type consists of a layer of sand or other granular medium 18 to 30 in (457 to 762 mm) thick, supported on an underdrain system. It can be an open-air gravity system or an enclosed pressure filter.

Rapid sand filters differ from slow sand filters in three major respects: thinner layers of granular media are used, higher loading rates are possible, and cleaning is accomplished by backwashing. Preceded by coagulation, rapid filters are more effective than slow filters and are better suited to cope with fine particles. Slow sand filters, on the other hand, can remove 99 percent of the bacteria even without pretreatment as opposed to 80 percent for rapid sand filters. The feedwater is filtered by forcing it through the sand layer by either pressure or gravity. When the quantity of suspended solids in the effluent reaches an acceptable level or the pressure drop exceeds the set point, the filter medium is washed. Washing is normally done by reversing the flow of the water through the filter at a rate adequate to lift the grains of filter medium into suspension. The deposited material, dislodged from the grains by the hydraulic shearing action of the rising water and by the abrasion of grains of filter medium rubbing against each other, is thus flushed up through the expanded bed.

Mixed-Media Filter. One major disadvantage of a single-medium filter is that the particles tend to pack themselves so that the largest, having the fastest settling velocities, fall to the bottom of the filter after backwashing. Thus the particle size of the medium decreases with distance from the bottom of the bed—just the opposite of what is desired.

The mixed-media filter solves this problem. Combinations of anthracite coal, sand, garnet, and ilmenite, with average specific gravities of 1.5, 2.6, 4.2, and 4.8, respectively, have been used. Properly formed after washing, the bed has the largest particles on top and the smallest on the bottom (if a downflow filter). The feed stream first meets and deposits its heaviest load of solids in the pores between the largest particles. The smaller pores, near the bottom, remove the last of the solids.

Organic Removal by Activated Carbon. Activated carbon selectively adsorbs dissolved organics (phenol, xylenes, etc.) and has a high capacity because of its large surface area, typically 500 to 1400 m^2/g.

Activated carbon is used to produce high-quality effluents. Two processes involve tertiary treatment: passing biologically treated wastewater through carbon columns or adding powdered activated carbon to the activated sludge basin. Another method of using activated carbon is part of physical-chemical treatment that eliminates biological oxidation entirely; this process is described in the next subsection.

The engineer has a great variety of configurations to choose from to attain contact between the wastewater and the carbon: upflow or downflow, pressurized gravity flow, packed or expanded beds, series or parallel column configuration, and flat or conical bottom shapes. Key variables are contact time between the waste and the carbon, flow rate, and pressure drop. See Table 15.18 for typical design criteria.

As pollutants are adsorbed, the pores of the activated carbon become filled

TABLE 15.18 Design Criteria for Activated Carbon Adsorption Columns

Design Parameter	Design Range
Flow rate	4–8 gpm/ft^2 [2.7–5.4 L/(s · m^2)]
Bed depth	8–12 ft (2.4–3.7 m)
Contact time	20–50 min (longer for some industrial wastes)
Media size	8 × 30 mesh
Backwash frequency	Daily
Backwash rate:	
Air scour	3–5 ft^2/(min · ft^2) [0.015–0.025 m^3/(s · m^2)]
Water wash	14–18 ft^3/(min · ft^2) [0.07–0.09 m^3/(s · m^2)]
Number of stages:	
Low-quality effluent	1
High-quality effluent	2
Carbon loss	2–10% per cycle
Cost (tertiary sewage treatment)	10–30¢/1000 gal (10–30¢/3785 L)

Source: Eimco Processing Machinery Div., Envirotech Corp., Salt Lake City, Utah.

and finally breakthrough occurs, i.e., the point where no more contaminants can be removed from the solution. Then the exhausted carbon must be replaced with fresh material. Normally, thermal regeneration is used to destroy adsorbed organics so the carbon can be recycled. The thermal process has three steps: drying, baking to pyrolize the adsorbates, and reactivation by oxidation at the surface to increase the number of active sites. Regeneration, like initial activation, is accomplished by heating at 910 to 940°C in an atmosphere of steam.

Reverse Osmosis. Filtration through semipermeable membranes under pressure removes biological and colloidal matter, as well as most dissolved organics that affect color, odor, and taste. Additionally, iron and hardness are removed as is done in conventional water treatment processes; however, reverse osmosis also reduces the concentration of all dissolved solids (chlorides, sulfates, nitrates, fluorides, trace metals, etc.) and can even extract pesticides and radioactive material.

Presently, the municipal application for reverse osmosis is the desalting of brackish water having total dissolved solids between 1000 and 15,000 mg/L. Industrially, there are a wide variety of uses. The important design parameter is the rate of transport of water across the membrane, called the *flux J*, which is

$$J = C(\Delta P_g - \Delta P_o) \qquad (15.80)$$

where ΔP_g = physical (gage) pressure (feedside-productside)
ΔP_o = osmotic pressure (feed-product)

Osmosis is a phenomenon that occurs naturally. It results whenever a dilute liquid (such as freshwater) and a concentrated liquid (such as saltwater or sugar) are separated by a semipermeable, selective material that permits one kind of molecule to pass through but not the other. The osmotic pressure is a function of the concentration of the dissolved ions. For every 100 mg/L of dissolved inorganic solids, an osmotic pressure of about 1 lb/in^2 (6.9 kPa) is generated.

If pressure is applied to impure water in excess of the osmotic pressure, the reverse process takes place: Water molecules from the saline solution are forced through the membrane into the freshwater with the selective, semipermeable membrane acting as a barrier to passage of the contaminant.

The most popular membranes developed so far are made of cellulose acetate. They are solvent-cast like photographic films and are characterized by extremely thin, dense skins on one side of highly porous substructures. They are designed for flow from the dense side only of about 20 gal/ft^2 per day (814.9 L/m^2 per day).

These membranes do, however, have problems: temperatures greater than 140°F (60°C) render them impermeable; they are vulnerable to attack by enzymes and microorganisms; and if not kept wet, they crack.

Du Pont and Dow are developing hollow-fiber technology to take the place of flat or rolled membranes. Dow has used cellulose acetate; Du Pont has developed a nylon fiber they say is more resistant to bacterial attack and degradation by acidic and alkaline solutions. Flow rates are a modest 0.10 gal/ft^2 per day (4.1 L/m^2 per day) over an estimated lifetime of 5 years, but the packing density is high; 30×10^4 fibers with an area of 85,000 ft^2 (7897 m^2) is practical. These fibers typically have an outside diameter of 0.002 in (0.05 mm) and a thickness of 0.0005 in (0.013 mm).

The primary factors that determine the performance of these systems are applied pressure, concentration of the dissolved solids in the feed, feed temperature, and porosity or permeability of the membrane. To ensure optimum performance over long periods of time, it is important in most cases to pretreat the feed by filtration and chemical addition.

Ion Exchange. *Ion exchange* is a process in which ions, held by electrostatic force to charged functional groups on the surface of a solid, are exchanged for ions of a similar charge in a solution in which the solid is immersed. Because the charged functional groups at which exchange occurs are on the surface of the solid, and because the exchanging ions must undergo a phase transfer from liquid phase to residence on a solid phase, ion exchange is a sorption process.

Although synthetic resins are used for most ion-exchange applications, the phenomenon of exchange is known to occur with a number of natural solids, including soil, humus, cellulose, wood, active carbon, coal, lignin, metallic oxides, and living cells such as algae and bacteria.

Two important terms are used to define resin capacity. The first is the total amount of exchangeable ions per unit weight of dry resin. More important, however, is the breakthrough capacity. This is an operational term that is a function of flow rate, concentration of ions, grain size, temperature, etc. Breakthrough occurs when the ion being removed from the solution appears in the column effluent.

Ion-exchange resins adsorb different ions selectively, based on the relative strength of the ion's charge and its radius of hydration. Multicharged ions are more readily adsorbed than single-charged ions, and smaller ions more readily than larger ones. As with activated carbon, ion exchange takes place in the capillaries of the solid particles. The size and number of these porous structures are important. Most silicate-based ion exchangers can be used over only a very small pH range, whereas synthetic resins are generally quite stable. The selectivity of a resin decreases with increasing temperatures but attains equilibrium faster.

A typical application of ion exchange is in preparing high-purity waters for boilers. Recent advances have shown potential for control and recovery of metallic ions in the plating industry; nickel, gold, and silver wastes can be treated. Radioactive wastes, too, are amenable to ion-exchange treatment.

Until recently, ion exchange was thought to be impractical for municipal wastewater treatment because extensive pretreatment would be needed to remove organic matter to prevent fouling of the resins. The development of macroporous resins that can adsorb organic matter without fouling and require less expensive regenerates makes the process worth considering.

10. Physical-Chemical Processes. Physical-chemical treatment (involving coagulation, carbon adsorption, and filtration) is an alternative to biological wastewater treatment methods. It replaces microbial stabilization of the dissolved organic matter with a combination of physical treatment processes that do not rely on bacterial action. A typical flow diagram for physical-chemical treatment is shown in Fig. 15.42.

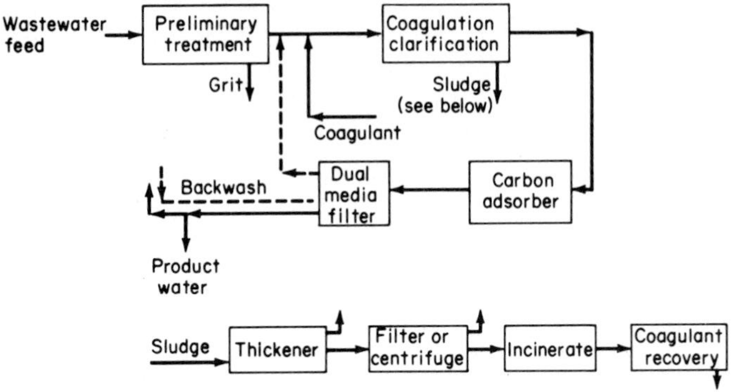

FIGURE 15.42 Principal operations in physical-chemical treatment.

The primary goal of pretreatment is to remove most of the suspended solids from the raw waste. In the case of physical-chemical treatment of municipal sewage, lime and high-molecular-weight water-soluble polymers and other flocculants are combined with primary sedimentation and filtration through mixed-media filters to produce as clear an effluent as possible. An additional benefit of chemical treatment is that it takes out almost all the phosphorus in the wastewater.

After removing the major portion of pollutants, the clarified wastewater is given a final polishing by activated carbon. Overall, physical-chemical treatment produces an effluent essentially free of suspended solids, and with TOC (total organic carbon) in the range of 5 to 10 mg/L.

11. Disinfection. Wastewater treatment processes remove some but not nearly all of the pathogens (Table 15.19). Hence the discharge from biological treatment plants, as well as natural waters contaminated by human sources, can carry diseases caused by bacteria and viruses. Thus, disinfection of sewage plant effluents is widely practiced.

Disinfection, it must be emphasized, is not the same as sterilization. The latter is total inactivation of all microbial life, while the former is selective destruction of most of the disease-causing pathogens.

Many chemical agents kill bacteria. The bactericidal effect of heavy metals (such as silver and mercury) has long been recognized, and phenols and alcohols have been used in hospitals for many years. However, the most popular chemical

TABLE 15.19 Removal of Bacteria During Wastewater Treatment

Operation	*Bacterial Removal,* %
Coarse screen	0–5
Fine screen	10–20
Grit chamber	10–25
Sedimentation	25–75
Chemical precipitation	40–80
Trickling filtration	90–95
Activated sludge	90–98

Source: Metcalf and Eddy, *Wastewater Engineering: Collection, Treatment, Disposal.* McGraw-Hill, New York, 1972.

disinfectants are oxidizing agents such as ozone, the halogens (chlorine, iodine, bromine), hydrogen peroxide, and dyes.

Chlorination. Of all the chemical disinfectants, chlorine is perhaps most commonly used. It is reasonably economical, toxic to microorganisms that carry waterborne diseases, tasteless and nonpoisonous to humans at low concentrations, and can be detected in residual amounts some time after application.

Chlorine is quite soluble in water—7160 mg/L at 20°C and at atmospheric pressure, forming hypochlorite according to the following equation:

$$\begin{array}{l} Cl_2 + H_2O \rightleftharpoons HOCl + H^+ + Cl^- \\ \qquad\qquad\qquad\quad \uparrow\downarrow \\ \qquad\qquad\qquad\quad H^+ + OCl^- \end{array} \tag{15.81}$$

When ammonia is present, it reacts to form chloramines, which are slow-acting disinfectants:

$$NH_3 + HOCl \rightarrow NH_2Cl + H_2O \tag{15.82}$$

$$NH_3 + 2HOCl \rightarrow NHCl_2 + 2H_2O \tag{15.83}$$

$$NH_3 + 3HOCl \rightarrow NCl_3 + 3H_2O \tag{15.84}$$

Chlorine, present as NH_2Cl or $NHCl_2$, is called *combined available chlorine.* When all reducing agents have been satisfied, then free chlorine is detected. This is called the *breakpoint*, which occurs only after the chloramines have been oxidized to N_2O and N_2.

Chemical oxidation by chlorine depends on the usual factors affecting a chemical reaction: temperature, time of contact, concentration of chlorine, pH, and surface tension. Because it is a microbial system, other items of importance are the species of microorganisms, the nature of the suspending fluid, the number of microorganisms, and the way in which they are suspended in the fluid, i.e., as single, discrete cells, in flocs, or attached to particulates. Table 15.20 lists dosages for several types of sewage.

Ozonization. A relatively new disinfection method utilizes ozone, a molecule that consists of three atoms of elemental oxygen (O_3). In water, these atoms break down rapidly, so they are free to oxidize organic matter and inactivate bacteria and viruses. Additionally, ozonization adds oxygen to the system but not dissolved solids. Further, ozone will deodorize gases in a few seconds of contact time at low levels of concentration.

TABLE 15.20 Chlorine Dosages Required to Yield 0.2 ppm Residual after 10 to 15 Minutes of Contact Time

Sewage Type	*Dosage, mg/l*
Raw:	
Fresh to stale	6–12
Septic	12–25
Settled:	
Fresh to stale	5–10
Septic	12–40
Effluent chemical precipitation	3–6
Trickling filter:	
Normal	3–5
Poor	5–10
Activated sludge:	
Normal	2–4
Poor	3–8
Intermittent sand filter:	
Normal	1–3
Poor	3–5

Source: W. W. Eckenfelder, Jr., P. A. Krenkel, and C. A. Adams, *Advanced Wastewater Treatment* (AIChE Today Series), AIChE, New York.

Ozone has a half-life of about 20 min in water. It is approximately 13 times more soluble than oxygen. The amount of ozone needed for disinfection of water or wastewater depends on the temperature, physical properties, and contaminants in the water being treated. For control purposes, one should be able to detect some residual ozone at the end of 5 min.

AIR POLLUTION CONTROL

12. Definitions. *Air pollution* is defined as the "presence in the atmosphere of one or more contaminants of such quantity and duration as may be injurious to human, plant, or animal life, or property, or which may unreasonably interfere with comfortable enjoyment of life, property, or conduct of business." Air pollution results from a two-part phenomenon, a time-concentration relationship. This relationship is recognized in the National Ambient Air Quality Standards (Table 15.12).

Classes of Pollutants. Atmospheric pollutants may exist as gases or particulates. Frequently, a third category, odors, is recognized. Chemically, odors fall into one of the previous categories. They are frequently separated because of their objectionable olfactory sensation at concentrations where they are neither toxic nor hazardous. Table 15.21 lists typical gaseous pollutants emitted to the atmosphere.

The term *particulates* denotes the release of both solid and liquid particles. The term should apply only to materials that are particles at their point of release and that could be captured by particulate-control equipment. However, regulatory definitions frequently include as particulates substances that may be vapors at the point of release but that can produce particulates by subsequent cooling and condensation to a standard temperature such as 70°F (21°C). Figure 15.43 compares sizes of typical particulates in the atmosphere and effective ranges for particle sizing and particle collection devices.

Several terms are used to describe particulates, which causes some confusion. *Aerosol* applies to a mixture of suspended particulates, either liquid or solid, or a

TABLE 15.21 Significant Gaseous Pollutants

Major class	Subclass	Typical pollutants
Organic gases	Hydrocarbons	Hexane, benzene, methane, butane, ethylene, butadiene
	Aldehydes and ketones	Formaldehyde, acetone
	Other organics	Alcohols, chlorinated hydrocarbons
Inorganic gases	Oxides of N_2	NO, NO_2
	Oxides of S	SO_2, SO_3
	Oxides of C	CO, CO_2
	Other inorganics	H_2S, HF, HCl, NH_3

combination of both. It usually applies to particles that settle slowly, if at all, due to gravity and includes particles from submicron to 10 to 20 μm in size. *Dusts* are coarser solid particles generated by entrainment from handling, crushing, grinding, and similar operations. They do not flocculate spontaneously, and generally they settle due to gravity. Sizes range from 1.0 to 2000 μm. *Smoke* generally applies to carbon and soot particles resulting from incomplete combustion, although occasionally the term may be applied to other finely divided solid particles such as ammonium chloride smoke. Smoke particles are 0.01 to 1.0 μm in size and do not settle by gravity. *Mists* are suspended liquid droplets usually resulting from condensation. *Fumes* are fine solid suspensions generated by condensation from the gaseous state, often in smelting and metallurgical operations.

Table 15.22 lists the sources of pollutants by general classes. Source emissions are best evaluated by sampling and reporting the contaminants released per unit of production. These results are called *emission factors*. Typical factors for many uncontrolled operations are given in *U.S. Public Health Serv. Publ.* 999-AP-42.

Effects. SO_2 and other acidic pollutants corrode architecture and sculpture (both metals and masonry, especially limestone and marble). Deterioration of historic works has increased significantly in the last 50 years. Carbonaceous deposits discolor buildings. Particulates and condensation nuclei reduce atmospheric visibility, hindering aircraft and ground transportation, block off distant scenic views, and decrease the quantity of solar radiation reaching the earth. Pollutants damage vegetation and reduce crop yields.

Pollutants also affect the health of animals and humans. High concentrations of pollutants retard recovery from many diseases. They contribute to respiratory diseases, and sufficiently high concentrations can lead to death. Effects of various contaminants on humans have been discussed in the Air Quality Criteria Documents.[24,25]

Studies on animals have shown a synergistic effect between particulate matter and SO_2. When SO_2 is combined with low concentrations of certain particulates, the SO_2 needed to produce harmful effects need be only 25 percent of the amount required in pure air to produce the same result.

13. Transport and Meteorological Effects. Pollutants are transported through the atmosphere by wind currents from their point of release to downwind receptors. They are dispersed and diluted so that an emission, toxic at its release point,

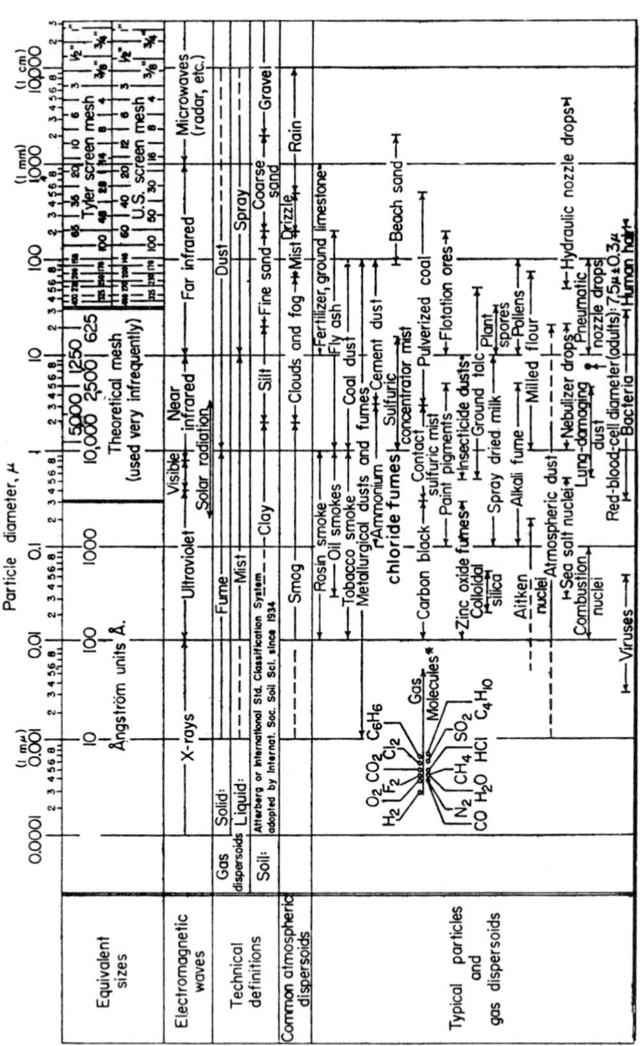

Particle diameter, μ
0.0001
0.001 (1 mμ)
0.01
0.1
1
10
100
1000 (1 mm)
10,000 (1 cm)
Equivalent sizes
Ångström units Å
10
100
1000
10,000
5000
2500
1250
625
Theoretical mesh (used very infrequently)
Tyler screen mesh
U.S. screen mesh
Electromagnetic waves
X-rays
Ultraviolet
Visible
Near infrared
Far infrared
Microwaves (radar, etc.)
Solar radiation
Technical definitions
Gas dispersoids
Solid:
Liquid:
Fume
Dust
Mist
Spray
Soil:
Atterberg or International Std. Classification System adopted by Internat. Soc. Soil Sci. since 1934
Clay
Silt
Fine sand
Coarse sand
Gravel
Common atmospheric dispersoids
Smog
Clouds and fog
Mist
Drizzle
Rain
Typical particles and gas dispersoids
Gas Molecules*
Rosin smoke
Oil smokes
Tobacco smoke
Metallurgical dusts and fumes
Ammonium chloride fumes
Carbon black
Contact sulfuric mist
Paint pigments
Zinc oxide fumes
Colloidal silica
Insecticide dusts
Ground talc
Spray dried milk
Alkali fume
Milled flour
Aitken nuclei
Atmospheric dust
Sea salt nuclei
Nebulizer drops
Combustion nuclei
Lung-damaging dust
Red-blood-cell diameter (adults): 7.5μ ± 0.3μ
Viruses
Bacteria
Human hair
Fertilizer, ground limestone
Fly ash
Coal dust
Cement dust
Sulfuric concentrator mist
Pulverized coal
Flotation ores
Beach sand
Plant spores
Pollens
Pneumatic nozzle drops
Hydraulic nozzle drops

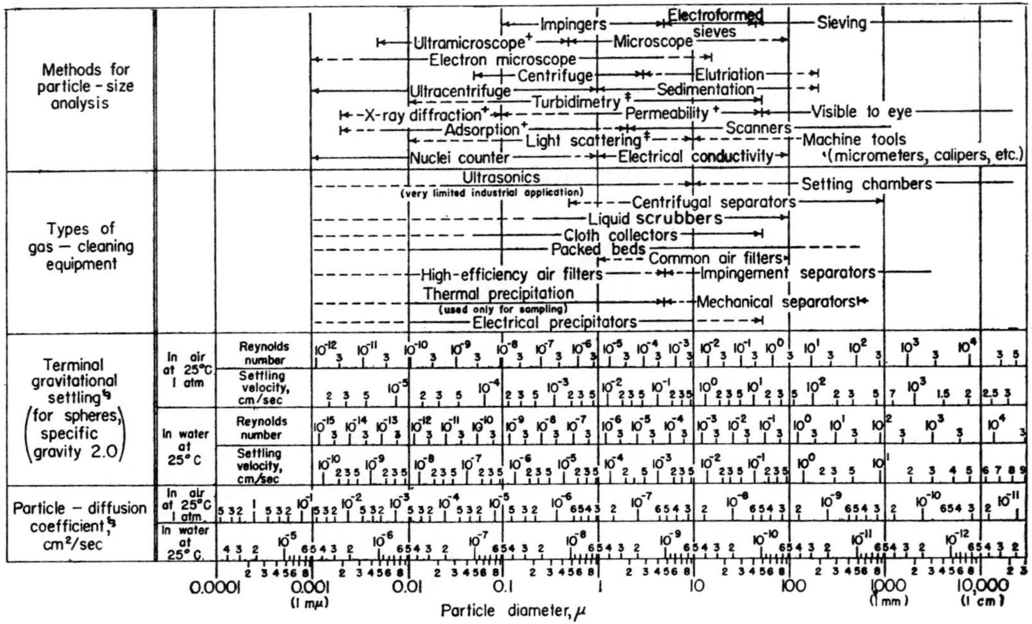

Methods for particle-size analysis

Impingers
Electroformed sieves
Sieving
Ultramicroscope+
Microscope
Electron microscope
Centrifuge
Elutriation
Ultracentrifuge
Sedimentation
Turbidimetry‡
X-ray diffraction+
Permeability+
Visible to eye
Adsorption+
Scanners
Light scattering‡
Machine tools (micrometers, calipers, etc.)
Nuclei counter
Electrical conductivity

Types of gas-cleaning equipment

Ultrasonics (very limited industrial application)
Setting chambers
Centrifugal separators
Liquid scrubbers
Cloth collectors
Packed beds
Common air filters
High-efficiency air filters
Impingement separators
Thermal precipitation (used only for sampling)
Mechanical separators
Electrical precipitators

Terminal gravitational settling§ (for spheres, specific gravity 2.0)

In air at 25°C 1 atm: Reynolds number; Settling velocity, cm/sec
In water at 25°C: Reynolds number; Settling velocity, cm/sec

Particle-diffusion coefficient,§ cm²/sec

In air at 25°C 1 atm
In water at 25°C

0.0001 0.001 (1 mμ) 0.01 0.1 1 10 100 1000 (1 mm) 10,000 (1 cm)

Particle diameter, μ

* Molecular diameters calculated from viscosity data at 0°C.

+ Furnishes average particle diameter but no size distribution.

‡ Size distribution may be obtained by special calibration.

§ Stokes-Cunningham factor included in values given for air but not included for water.

FIGURE 15.43 Characteristics of particles and particle dispersoids.

TABLE 15.22 Estimated Sources of Air Pollutants

Source	CO	Particulates	SO_x	HC*	NO_x	Total
Transportation	111.0	0.7	1.0	19.5	11.7	143.9
Stationary source fuel combustion	0.8	6.8	26.5	0.6	10.0	44.7
Industrial processes	11.4	13.1	6.0	5.5	0.2	36.2
Solid-waste disposal	7.2	1.4	0.1	2.0	0.4	11.1
Miscellaneous	16.8	3.4	0.3	7.1	0.4	28.0
Total	147.2	25.4	33.9	34.7	22.7	263.9

* Hydrocarbons.
SOURCE: U.S. Environmental Protection Agency.

may be harmless at ground level downwind. The higher the release point above the surroundings and the more buoyant is the plume, the greater is the dilution.

The major meteorological parameters controlling atmospheric dispersion are atmospheric stability and wind velocity. Atmospheric stability is affected by solar radiation and the vertical temperature gradient of the atmosphere called *lapse rate*. Stability can be characterized by comparing the actual lapse rate to the dry adiabatic lapse rate (DALR). An atmosphere at the DALR decreases in temperature 5.4°F (3.0°C) for each 1000-ft (305-m) increase in altitude (−0.98°C/100 m). An atmosphere that decreases in temperature faster than the DALR is *superadiabatic* (unstable atmosphere). One that decreases at the DALR rate is *neutral*. An atmosphere that decreases less rapidly than the DALR is *subadiabatic* and is more stable. An atmosphere at constant temperature is *isothermal* and is more stable than subadiabatic. When temperature increases with elevation, an *inversion* condition exists—an extremely stable atmosphere.

Best conditions for dispersion are high wind speeds and a highly unstable temperature gradient (superadiabatic). As the atmosphere becomes progressively more stable, wind speeds tend to decrease, and plume dispersion becomes progressively poorer. An indication of atmospheric stability can be gained from smoke plume behavior (Fig. 15.44).

Under superadiabatic conditions, high ground concentrations may occur close to the stack as the looping plume actually strikes the ground at times, but concentrations considerably downwind will be reduced. Under neutral conditions, dispersion is still good but less rapid. Ground concentrations close to the stack will be less than under unstable conditions, but they will be higher further downwind. Under inversion conditions, dilution occurs very slowly, and the plume meanders and spreads out horizontally. Ground concentrations will often be low, but the pollution accumulates aloft waiting to sink to ground level in large undiluted puffs at the moment of inversion breakup or fumigation. At these times, often of 0.5- to 1-h duration, ground concentrations up to 20 times normal neutral atmosphere ground concentrations may be experienced.

Increasing wind speed aids dispersion. The plume is blown over more rapidly and rises less in the atmosphere. Because of this, higher ground concentrations may be experienced closer to the stack, but as dilution occurs, concentrations farther downwind will be reduced.

Dispersion frequently follows a daily cycle. In late evening with clear sky, the earth radiates heat to space, cooling more rapidly than the air above. This creates an inversion that lasts throughout the night. Winds die down and pollutants accumulate at the base of the inversion layer. When the sun rises, the earth is

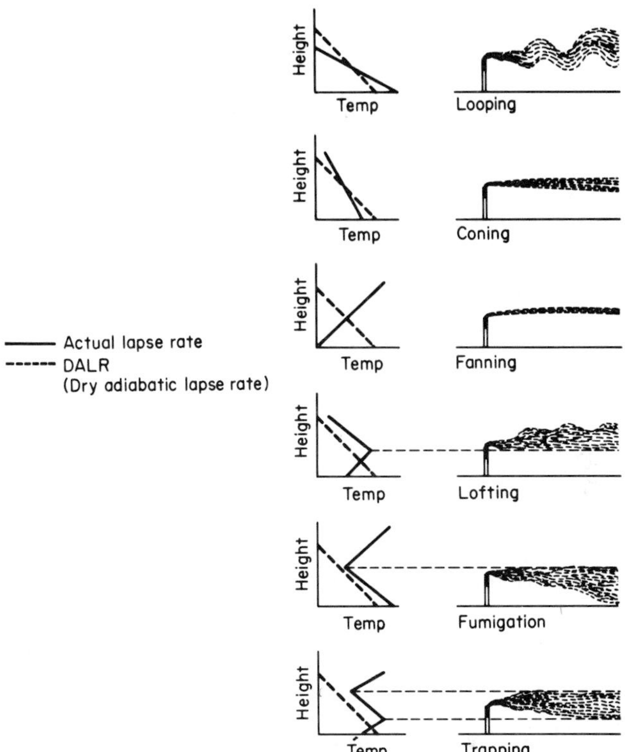

FIGURE 15.44 Smoke plume behavior as a function of temperature. *(From H. L. Perkins, Air Pollution, McGraw-Hill, New York, 1974, p. 169.)*

heated and a decreasing temperature lapse rate builds up from the earth's surface. When this reaches the inversion ceiling, the accumulated pollutants begin to sink to ground level, producing a fumigation. As further warming occurs, the DALR is restored, wind speeds increase, and the pollutants are blown away and dispersed. Conditions for good dispersal remain until cooling sets in again near sunset.

Dispersion Equations. Standard gaussian distribution can predict downwind concentrations with fair accuracy. Models currently most in use are described by Turner.[26] Figure 15.45 illustrates the model. The plume is discharged from a stack of height h_s. A buoyant plume continues to rise some distance above the stack, Δh. The overall effective stack height H is the sum of h_s and Δh. The plume is assumed to be released from a point source upwind from the stack such that it spreads out in the shape of a cone as it travels downwind. The plume is free to spread horizontally about its axis and upward. However, at the point where the plume hits the ground, the assumption is made that the pollutants are reflected upward. This is handled mathematically by assuming a virtual image source beneath the ground that adds to the concentration downwind beyond the

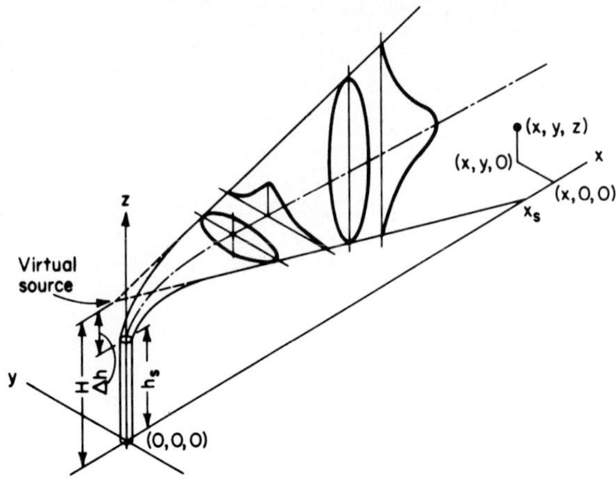

FIGURE 15.45 Gaussian model for dispersion equations.

point x_s. Coordinates are x, the downwind direction with wind speed of $\overline{U}$; y, the crosswind direction measured from the plume centerline; and z, the vertical height above the ground.

The downwind concentration C at any point (x, y, z) can be computed from Eqs. (15.85) to (15.89). For downwind distances up to x_s,

$$\frac{C_{(x,y,z)}\overline{U}}{Q} = \frac{1}{2\pi\sigma_y\sigma_z} \exp\left\{-\frac{1}{2}\left[\left(\frac{y}{\sigma_y}\right)^2 + \left(\frac{z-H}{\sigma_z}\right)^2\right]\right\} \tag{15.85}$$

Beyond x_s,

$$\frac{C_{(x,y,z)}\overline{U}}{Q} = \frac{1}{2\pi\sigma_y\sigma_z} \exp\left[-\frac{1}{2}\left(\frac{y}{\sigma_y}\right)^2\right]\left\{\exp\left[-\frac{1}{2}\left(\frac{z-H}{\sigma_z}\right)^2\right] + \exp\left[-\frac{1}{2}\left(\frac{z+H}{\sigma_z}\right)^2\right]\right\} \tag{15.86}$$

Beyond x_s, where ground-level concentrations are desired ($z = 0$),

$$\frac{C_{(x,y,0)}\overline{U}}{Q} = \frac{1}{\pi\sigma_y\sigma_z} \exp\left\{-\frac{1}{2}\left[\left(\frac{y}{\sigma_y}\right)^2 + \left(\frac{H}{\sigma_z}\right)^2\right]\right\} \tag{15.87}$$

If plume centerline concentrations only are desired ($y = 0$),

$$\frac{C_{(x,0,0)}\overline{U}}{Q} = \frac{1}{\pi\sigma_y\sigma_z} \exp\left[-\frac{1}{2}\left(\frac{H}{\sigma_z}\right)^2\right] \tag{15.88}$$

If the effluent is released at ground level ($H = 0$), the downwind plume centerline concentration is obtained by

$$\frac{C_{(x,0,0)}\overline{U}}{Q} = \frac{1}{\pi\sigma_y\sigma_z} \tag{15.89}$$

Any consistent set of units may be used such as C, concentration in g/m^3; $\overline{U}$, wind speed, m/s; Q, the pollutant release, g/s; and distances H, x, y, and z and diffusion coefficients σ_y and σ_z in meters.

For very accurate predictions of ground concentration, the values of σ_y and σ_z should be determined experimentally for the meteorological conditions and terrain under consideration. The model represents the real situation best in a neutral atmosphere (lapse rate close to the DALR).

There are six stability categories as defined in Table 15.23 with experimental

TABLE 15.23 Stability Categories

Surface wind speed, m/sec	Day: Incoming solar radiation			Night	
	Strong	Moderate	Slight	Thin overcast or $\geq 4/8$ cloudiness	$\leq 3/8$ cloudiness
<2	A	A–B	B		
2	A–B	B	C	E	F
4	B	B–C	C	D	E
6	C	C–D	D	D	D
>6	C	D	D	D	D

The neutral class, D, should be assumed for overcast conditions during day or night. A—extremely unstable conditions; B—moderately unstable conditions; C—slightly unstable conditions; D—neutral conditions; E—slightly stable conditions; F—moderately stable conditions.

Source: From Turner.[26]

values for σ_y and σ_z as shown in Figs. 15.46 and 15.47. These values of σ_y and σ_z are most accurate (± 100 percent) for level farm land. Ground structures, wooded areas, and rolling or mountainous terrain will result in larger actual departures from the calculated results. Figure 15.48 may be used with the stability categories of Table 15.23 to determine downwind plume centerline distance at which the maximum ground concentration will occur and its concentration.

The calculated concentrations are typical of those which would be obtained while sampling for 10 min. A sample averaged over a longer time period will give lower values due to normal fluctuations in wind direction. Equation (15.90) may be used to predict the average ground concentrations to be expected up to periods of several hours.

$$C_t = C_0\left(\frac{t_0}{t_t}\right)^p \tag{15.90}$$

where C_t = the concentration for a longer time period
C_0 = the concentration estimated for a 10-min period
t_0 and t_t = the respective time periods
p = value between 0.17 and 0.20

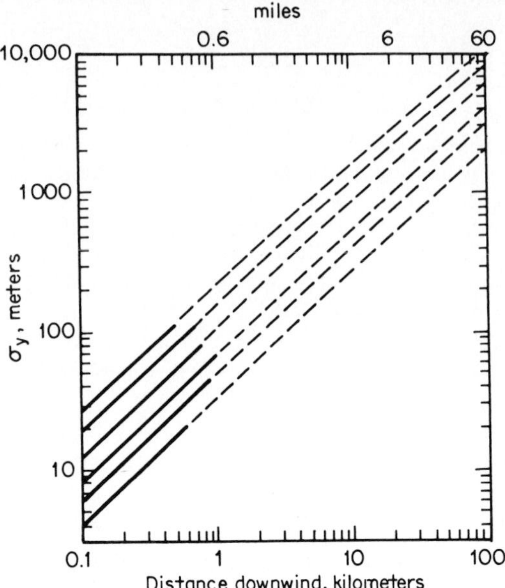

FIGURE 15.46 Horizontal dispersion coefficient as a function of downwind distance from the source. Letters refer to stability category; see Table 15.23. *(From Turner.[26])*

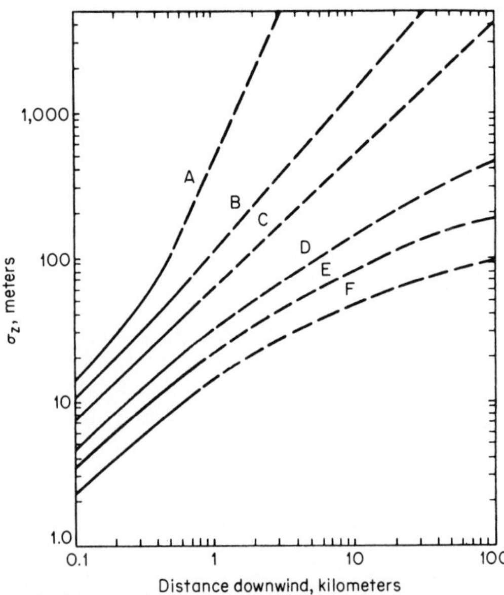

FIGURE 15.47 Vertical dispersion coefficient as a function of downwind distance from the source. Letters refer to stability category; see Table 15.23. *(From Turner.[26])*

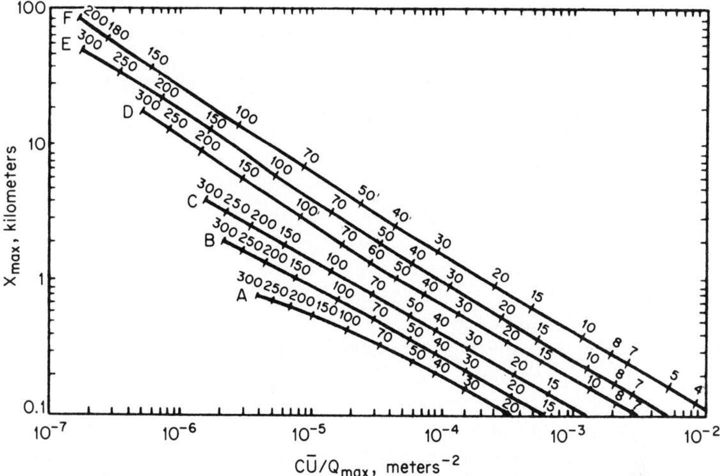

FIGURE 15.48 Distance of maximum concentration as a function of stability and effective height of emission. Letters indicate stability class; numbers, the effective stack height; see Table 15.23. *(From Turner.[26])*

The dispersion equations apply specifically to gases and suspended particulates. Owing to gravity fallout close to the stack, they should not be applied to particulates larger than 20 μm.

Plume Rise. The preceding dispersion equations require an estimate of the distance the plume continues to rise above the stack:

$$\Delta h = \frac{V_s d}{\overline{U}}\left[1.5 + 2.68 \times 10^{-3} pd \left(\frac{T_s - T_a}{T_a}\right)\right] \tag{15.91}$$

where Δh = the plume rise, m
V_s = stack discharge velocity, m/s
d = stack diameter, m
$\overline{U}$ = wind speed, m/s
p = atmospheric pressure, mbars
T_s = stack discharge temperature, K
T_a = ambient temperature, K

To correct plume rise for atmospheric stability, it is recommended that the value of Eq. (15.91) be multiplied by 1.15 for unstable to 0.85 for very stable conditions (1.0 for neutral).

Pollution Potential and Climatology. When considering a location for an operation with a major air pollution potential, locations well suited to atmospheric dispersion should be considered. Flat, open spaces are preferred. Narrow mountain valleys should be avoided. Locations adjacent to large lakes and oceans can present special problems. A study by Hosler[27] gives indications of geographic areas with high and low potential for air pollution problems. It is desirable to pick an area with a mean mixing depth as great as possible.

14. Air Pollution Measurements. Sampling may be occasional, intermittent, or infrequent to check on specific problems, or it may be conducted at regular intervals or even continuously to provide frequent monitoring. Positioning of samplers must be carefully considered. For a representative area sample, the intake should not be at a busy intersection, along a dusty road, or directly downwind of a major pollution source. Sampling for gases (and some particulates) is practiced by drawing a measured sample through a series of midget impingers. Water or a suitable solution absorbs the gaseous component.

For routine sampling, there are a number of devices of varying complexity and cost. Primitive methods still in use consist of the "dust-fall" bucket and the lead peroxide planchette. These measure, respectively, dust that settles by gravity and the SO_2 or sulfation rate, usually on a monthly basis.

The "hi-vol" sampler is the specified method for checking ambient-air compliance with national standards for particulates. It pulls about 50 ft^3/min (1.4 m^3/min) of ambient air through a glass-fiber filter for a 24-h period. Tape samplers that filter ambient air through a 1-in- (25.4-mm) diameter spot on a cellulose tape for a 2-h period are also used. After the sampling period, the tape is advanced automatically and the soil is measured photometrically by reduction in light transmittance. Tape-sampler results are usually reported as *coefficient of haze* (coh/1000 ft) (coh/305 m). Unfortunately, coh is not comparable with "hi-vol" gravimetric measurements ($\mu g/m^3$). Tape samplers specific for certain gases (H_2S and fluorides) have been developed in which two chemically treated tapes are brought together. Their contact activates the tapes for absorption of the gaseous component. Absorption causes discoloration of the tape which can be measured photometrically.

Atmospheric monitoring for SO_2 and other gases is frequently performed by taking successive 2-h composite samples with midget impingers in a sequential sampler. The absorbing liquid is analyzed using the West-Gaeke technique. (Ozone and oxidants are determined by their oxidizing effect on KI solution.) Nitrogen oxide analysis is usually based on a chemiluminescence reaction using a continuous instrumental analyzer.

Many communities use continuous monitors for SO_2 measurement. The coulometric instruments are probably the preferred choice, but colorimetric, conductivity, and flame photometry instruments also have been used. The standard method for measuring CO is an instrument using nondispersive infrared.

Source Sampling. There is a current tendency to use the U.S. EPA Method V sampler,[28] with modifications if needed, for many types of source sampling.

Process location for sample withdrawal must be carefully selected to obtain a representative sample. The EPA test method V specifies, "a composite sample shall be taken from at least 12 different locations in the same plane in a duct with each point the centroid of an equal area section." Under this circumstance, there should be at least 8 duct diameters of straight pipe upstream of the sampling point and 2 duct diameters downstream. When a location having a greater amount of flow disturbance is the only suitable location available, the effects of poor flow distribution may be compensated for by increasing the number of traverse points (Fig. 15.49).

Velocity Measurement. A measure of the desirability of a particular sampling location can be obtained from the velocity distribution across the duct. The more uniform the velocity distribution, the better is the chance that the pollutants will be somewhat uniformly distributed. Also, a knowledge of the velocity distribu-

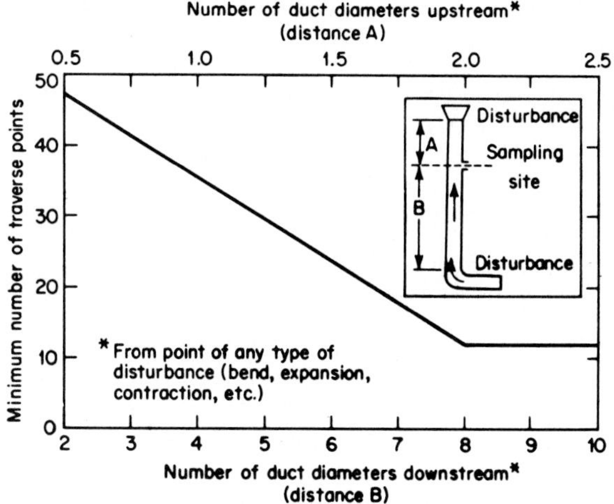

FIGURE 15.49 Number of sampling points when there are duct disturbances. *(From U.S. EPA.[28])*

tion is necessary in selection of the sample probe orifice, sampling flow rate, and total sampling time. The velocity distribution is usually determined in a round duct on two different diameters at right angles to each other using an S-type pitot tube. The velocity at each point is calculated with Eq. (15.92):

$$V_0 = 2.90(F)\sqrt{\left(\frac{29.92}{P}\right)\left(\frac{1.00}{G_d}\right)(\Delta h_p T_R)} \qquad (15.92)$$

where V_0 = local velocity, ft/s (m/s)
P = absolute pressure in duct, inHg (mmHg)
G_d = carrier gas specific gravity (air = 1)
Δh_p = pitot differential pressure, inH_2O (mmH_2O)
T_R = absolute temperature of carrier gas, °R

The value F in Eq. (15.92) is the discharge coefficient for the S-type pitot tube. This should be obtained for the velocity range being measured by calibration in a duct with clean air against a standard pitot tube having a discharge coefficient of unity. Recalibration of the S-type pitot should be performed at intervals, since the value of the discharge coefficient can change with abrasion and wear.

In a circular duct, it is usual to divide the area into one-fourth the number of concentric circles as there are sample traverse points or velocity determination points. The circles are proportioned so that each has equal area. A measurement point is then located at the intersection of the traverse diameter and the centroid of area for each annulus.

Table 15.24 locates measurement points for circular ducts. This method of locating velocity measurement points weights the individual velocities so that the average velocity for the entire duct is equal to the arithmetic average of all velocities. (If the pitot readings are used, the arithmetic average of the square roots of each pitot tube reading must be used.)

TABLE 15.24 Location of Traverse Points in Circular Stacks

Traverse point number on a diameter	Number of traverse points on a diameter											
	2	4	6	8	10	12	14	16	18	20	22	24
1	14.6	6.7	4.4	3.3	2.5	2.1	1.8	1.6	1.4	1.3	1.1	1.1
2	85.4	25.0	14.7	10.5	8.2	6.7	5.7	4.9	4.4	3.9	3.5	3.2
3	....	75.0	29.5	19.4	14.6	11.8	9.9	8.5	7.5	6.7	6.0	5.5
4	....	93.3	70.5	32.3	22.6	17.7	14.6	12.5	10.9	9.7	8.7	7.9
5	....	....	85.3	67.7	34.2	25.0	20.1	16.9	14.6	12.9	11.6	10.5
6	....	....	95.6	80.6	65.8	35.5	26.9	22.0	18.8	16.5	14.6	13.2
7	....	....	....	89.5	77.4	64.5	36.6	28.3	23.6	20.4	18.0	16.1
8	....	....	....	96.7	85.4	65.0	63.4	37.5	29.6	25.0	21.8	19.4
9	....	....	....	....	91.8	82.3	73.1	62.5	38.2	30.6	26.1	23.0
10	....	....	....	....	97.5	88.2	79.9	71.7	61.8	38.8	31.5	27.2
11	....	....	....	....	....	93.3	85.4	78.0	70.4	61.2	39.3	32.3
12	....	....	....	....	....	97.9	90.1	83.1	76.4	69.4	60.7	39.8
13	....	....	....	....	....	....	94.3	87.5	81.2	75.0	68.5	60.2
14	....	....	....	....	....	....	98.2	91.5	85.4	79.6	73.9	67.7
15	....	....	....	....	....	....	....	95.1	89.1	83.5	78.2	72.8
16	....	....	....	....	....	....	....	98.4	92.5	87.1	82.0	77.0
17	....	....	....	....	....	....	....	....	95.6	90.3	85.4	80.6
18	....	....	....	....	....	....	....	....	98.6	93.3	88.4	83.9
19	....	....	....	....	....	....	....	....	....	96.1	91.3	86.8
20	....	....	....	....	....	....	....	....	....	98.7	94.0	89.5
21	....	....	....	....	....	....	....	....	....	....	96.5	92.1
22	....	....	....	....	....	....	....	....	....	....	98.9	94.5
23	....	....	....	....	....	....	....	....	....	....	....	96.8
24	....	....	....	....	....	....	....	....	....	....	....	98.9

Source: From U.S. EPA.[28]

Sampling Technique. After determining from the velocity traverse that the sampling location is suitable, the diameter of the sampling probe tip and the desired sampling rate must be selected. *Isokinetic sampling* means sampling with the same inlet velocity at the probe tip as the flowing gas stream has at the same site. Sampling isokinetically is necessary to obtain a representative sample of large particles. (In sampling gaseous pollutants, isokinetic sampling is unnecessary.) The error from anisokinetic sampling is zero for particles smaller than 3 μm and less than 5 percent for 5-μm particles. For larger particles, the error can be very substantial.

Isokinetic sampling requirements set the velocity at the probe inlet. Efficient collection of pollutants in the sampling train may require a given flow rate through the train. The availability of a number of different size probe inlet tips is usually needed to make these two requirements compatible.

Sampling-train characteristics will usually make the pressure at which the train flow rate is measured different from that in the stack. Often the temperature is also different and the sample may be dried. Relating train flow rate to tip velocity for isokinetic sampling requires advance planning for quick computation.

The pollutant concentration in the sample must be related to its concentration and total mass in the stack. This means comparing measured sample volume to its original stack volume. With dry filtering equipment, this correction is made using the ideal-gas law. Where wet collection occurs, water condensed or evaporated from the sample train also must be accounted for. This is usually handled by running a material balance on the train covering the sampling period.

Particle-Size Measurement. Particle-size measurements are usually made either by laboratory analysis or by classification into various size ranges in a flowing gas sample. Where an unclassified sample is collected, particle size is usually measured by microscopic measurement or resuspension and classification in either an air stream or a liquid. With microscopic methods, a thin deposit of solids may be collected on a dry membrane filter or a glass slide using an electrostatic or thermal precipitator sampler. The sizes of particles are measured visually, and the number in each size range is counted. The particles are usually measured by equivalent area or by a predominant dimension. To obtain particle-size distribution on a mass basis, it is necessary to make assumptions about particle volume and density.

For determining particle size for cyclone performance, the American Society of Mechanical Engineers (ASME) has approved the Bahco dry classifier. A 10-g dust sample is required. The Bahco classifier will size the sample into a number of fractions between 44 and 1 μm. A number of questions arise about this technique. In collection on the filter, particles may agglomerate and the Bahco classifier may not resuspend them in the same size range they had initially. There also can be a tendency for finer particles to be lost from the sample by becoming embedded in the pores of the filter.

Particles collected wet or dry may be sized in a liquid-classifying device by measuring settling rate due to gravity or centrifugal force (Whitby centrifuge). The Coulter counter is also used to measure particles in a liquid. It measures the effect of a particle on an electric field, and the size distribution resulting is based on the particle volume.

Airborne particles often flocculate, producing chains and agglomerates held together by electrostatic charges. When the particles are dispersed in a liquid, these charges can be markedly different than in the original airstream. Because of this dispersion problem and the fact that different measurement techniques really measure different properties of the particle, it is seldom found that a single sample will give the same particle-size distribution when measured by different techniques. For comparative tests, it is essential that similar measuring techniques be used.

Probably the best way to measure particle sizes is to classify the particles by size while they are still suspended in the original gas. This is done using a classifying sampler such as a cascade impactor or an Andersen sampler. A very short sampling period should be used when dust loadings are high; otherwise, coarser particles collected in the first stages can be reentrained into the finer particle-size stages as the first stages become overloaded.

Monitoring. For monitoring pollutant gases, a number of instruments have been developed using infrared and ultraviolet absorption. These are adaptable to SO_2, NO, CO, CO_2, hydrocarbons, and other specific gases. Membrane-type cells involving oxidation-reduction reactions with the pollutant also have been developed.

For monitoring particulates from source emissions, tape samplers have recently been developed that filter out the particulates and determine their mass by β-ray attenuation.

In many situations, regulations on plume visibility are more restrictive than mass emission regulations. In-stack instruments that measure light attenuation across the stack can be calibrated in terms of emission mass when the particle-size distribution does not change with plant operation.

There is a need to predict plume opacity for compliance with regulations when certain particulate collection efficiencies are anticipated or specified. Ensor and Pilat[29] present a theoretical equation for relating plume opacity and particle concentration and properties:

$$W = -\frac{K\rho \ln (I/I_0)}{L} \qquad (15.93)$$

where W = the particle mass concentration, g/m^3
ρ = true particle density, g/cm^3
I/I_0 = ratio of light transmitted through plume to quantity transmitted if there were no emission [opacity = $1 - (I/I_0)$]
L = length of light path, m
K = a proportionality constant, cm^3/m^2, and dependent on particle properties and light wavelength

This equation is most accurate for a process in which particle-size distribution is constant and when the value of K has been determined experimentally by measuring opacity and particle concentration simultaneously. However, predictions of plume opacity can be made with fair accuracy from fundamental data on the mass-median particle size, the standard particle-size deviation, the density of the particles, and their refractive index. Theoretical values of K are given in Ensor and Pilat.[29]

15. Pollution Control Techniques

Minimizing Need for Collection Devices. Table 15.25 lists means of reducing or eliminating pollutants without specific removal devices. Examples are substitution of low-sulfur fuels, nonvolatile solvents, operation at lower temperatures to reduce NO_x formation or volatilization of a processed material; use of indirect rather than direct-contact heat transfer; heat transfer from radiant panels; pretreatment of natural raw materials to remove easily airborne fines or volatiles; and wetting or agglomeration of solids to reduce airborne releases during handling.

TABLE 15.25 Fundamental Means of Reducing or Eliminating Pollutant Emissions to the Atmosphere

I. Eliminate the source of the pollutant
  - Seal the system to prevent interchanges between system and atmosphere
    - Use pressure vessels
    - Interconnect vents on receiving and discharging containers
    - Provide seals on rotating shafts and other necessary openings
  - Change raw materials, fuels, etc., to eliminate the pollutant from the process
  - Change the manner of process operation to prevent or reduce formation of, or air entrainment of, a pollutant
  - Change the type of process step to eliminate the pollutant
  - Use a recycle gas or recycle the pollutants rather than using fresh air or venting

II. Reduce the quantity of pollutant released or the quantity of carrier gas to be treated
  - Minimize entrainment of pollutants into a gas stream
  - Reduce number of points in system in which materials can become airborne
  - Recycle a portion of process gas
  - Design hoods to exhaust the minimum quantity of air necessary to ensure pollutant capture

III. Use equipment for dual purposes, such as a fuel combustion furnace to serve as a pollutant incinerator

Application of Control Devices. Table 15.26 lists equipment types useful for reducing emission of gases, odors, and particulates. Some odors can be canceled by other antagonistic odor molecules. This provides a means of eliminating an odor without actually preventing its release.

Equipment for control of gaseous pollutants is often not suitable for collection of solid particulates, especially when the loading is heavy. A moderate loading of solid particulates can be handled in gas absorption equipment if the particulates are readily soluble in the absorbing liquid and all impacting surfaces are well flushed. A typical upper limit for handling solids in packed gas absorption devices is 5 gr/ft^3 (176.6 gr/m^3).

Even though it is difficult to handle gas and solid pollutant mixtures, many situations occur where this must be done. Three alternatives are generally available:

1. Use absorption equipment such as cross-flow packed towers when the particulate solid loading is light.
2. Use wet scrubbing equipment such as Venturi scrubbers, impingement tray towers, and fluidized-bed impaction spheres.
3. Use dry particulate collectors such as cyclones, bag filters, and electrostatic precipitators, followed by equipment for efficient gas collection.

In equipment selection, consider the physical form of the collected material most amenable to reuse or disposal. Collection in dry form may permit recycling to the process or blending into product. It might be the preferred form for landfill disposal, but care must be taken in handling to prevent redispersal into the atmosphere. Wetting and pugging of the dry dust may be necessary. Wet collection of

TABLE 15.26 Types of Equipment Applicable to the Control of Various Classes of Air Pollutants

Equipment type	Pollutant classification			
	Gas	Odor	Particulate	
			Liquid	Solid
Absorption	×	×		
Adsorption	×	×		
Air dispersion (stacks)	×	×	×	×
Condensation	×	×		
Centrifugal (dry)			×	×
Filtration, bags				×
beds			×	×
fine fibers			×	×
Gravitational settling				×
Impingement (dry)			×	×
Incineration	×	×	×	×
Precipitation, electric			×	×
thermal			×	×
Wet collection			×	×

insoluble particulates followed by settling and disposal of the sludge to an impoundment area by pipeline can be an inexpensive disposal technique. However, if the solids are water-soluble, disposal without causing groundwater pollution can be a major problem.

16. Collection and Removal of Gases

Absorption. This is one of the most frequently used methods for removal of water-soluble gases. Acidic gases (such as HCl, HF, and SiF_4) can readily be absorbed in water efficiently, especially if contact is with water having an alkaline pH. Less-soluble acidic gases (such as SO_2, Cl_2, and H_2S) can be absorbed more readily in a dilute caustic solution such as 5 to 10 percent NaOH. Scrubbing with an ammonium salt solution is also employed; the gas is often contacted with the more alkaline solution first and a neutral or slightly acid solution last to prevent loss of NH_3 to the atmosphere. Lime is an inexpensive alkali but often leads to plugging problems in absorption equipment if the calcium salts have only limited solubility. A better technique is to absorb with an NaOH solution, which is then limed external to the absorption tower. The calcium salts are settled out and the regenerated NaOH returned to the absorption system.

When flue gases containing CO_2 are being scrubbed with an alkaline solution, CO_2 also can be absorbed, resulting in use of an inordinate amount of caustic. However, if the pH of the scrubbing liquid is kept below 9, the amount of CO_2 absorbed can be reduced to a very small value.

Alkaline gases such as NH_3 can be removed by scrubbing with acidic solutions such as dilute H_2SO_4, H_3PO_4, or HNO_3. The resulting mixtures can often be disposed of as fertilizer ingredients.

Absorption equipment is frequently favored when the pollution concentration is quite low (chemical costs can become sizable for large-volume gas streams or high concentrations). The most common devices are packed towers and spray towers. Packed towers are smaller because of the better gas-liquid contact provided by the packing, but open spray chambers are more resistant to plugging. Bubble-cap columns, baffle towers, and sieve-plate towers are also used occasionally.

While there is theoretically no limit to the number of transfer units that can be built into a packed tower if it is made tall enough, there is a limit to spray towers due to spray entrainment, which results in loss of countercurrentcy. Vertical spray towers with upward gas flow parallel to the tower walls have been demonstrated to be capable of at least 5.8 transfer units. A spray tower with cyclonic gas flow can have up to 7 transfer units. Horizontal spray chambers have been reported with up to 3.5 transfer units.

Venturi scrubbers are sometimes used for gas absorption. Their capability appears to be limited to about 3 transfer units.

While water is the most common scrubbing liquid, organics such as dimethylaniline and various amines (mono-, di-, and triethanol amines, methyl diethanolamine) have been used with acidic gases. In these cases, the absorbing liquid is regenerated by stripping solute off in a concentrated form at a higher temperature or by treatment with lime to produce a precipitate. The volatility of the organic solvent and its possible loss to the atmosphere (or its oxidation) are possible problems that must be considered.

Adsorption. Adsorption processes consist of contacting a gas with a solid. The solids are essentially porous with an affinity for certain substances. Typical materials are activated carbon, activated alumina, silica gels, and molecular sieves. An adsorbent can hold from 8 to 25 percent of its own weight in adsorbed vapors. Generally, the adsorbate is held in liquid phase, even though physical principles would predict that its physical state should be a vapor. The adsorbate may be held in the pore structure by direct physical attraction or by the formation of chemical bonds. Since the adsorbate is held in the liquid phase, heat of condensation is released in the bed. To keep the bed from heating up, it may be necessary to precool the inlet gas. Capacity and efficiency of adsorption decrease with an increase in temperature.

Some adsorbents exhibit a selectivity for particular vapors; for instance, aluminas and silica gels have a special affinity for water vapor. Therefore, it may be necessary to dry the gas first to prevent saturating the bed with condensed water. Carbon does not selectively adsorb water vapor, which makes it useful for treating moist gases and for bed regeneration with steam.

The adsorption process is practically complete regardless of inlet gas concentration as long as the bed is not saturated. Once the bed becomes saturated, the exit concentration of the adsorbate increases exponentially. This fact makes adsorption processes especially attractive for control in those situations where extremely low exit gas concentrations must be reached.

The bed is regenerated by raising its temperature above the boiling point of the adsorbate and stripping the adsorbate from the bed with a hot gas which is recirculated to the bed after condensation of a portion of the adsorbate. The bed is then cooled and returned to service. Occasionally, the bed may be regenerated by heating and evacuating without use of a stripping gas.

Condensation. A number of vapors, especially hydrocarbons with low volatility, can be recovered by condensation. A tubular, water-cooled heat exchanger is adequate for many high-molecular-weight organic vapors. Where the volatility is greater, a refrigerated condenser following the water-cooled condenser may be necessary. In condensing many vapors, where heat transfer is more rapid than mass transfer, fog particles of the condensate (0.5- to 1.5-μm diameter) are apt to form in the bulk gas stream. These particles can result in plume opacity violations as well as a recovery from cooling that is not as great as predicted from vapor pressure-temperature data. Therefore, it is often necessary to follow the condenser with a mist eliminator having high efficiency on fine particles. A small-diameter indepth fiber-bed filter or an electrostatic precipitator are good for fog control. However, electrostatic precipitators should not be used where combustible mixtures are present.

When the vapor is too volatile for efficient removal by condensation alone, compressing the gas before condensation will result in adequate recovery at an economical cost in situations where the total gas volume is small and the quantity of vapor to be recovered is an appreciable portion of the total. Condensation is attractive where the gas stream is already at an elevated pressure and can be cooled prior to pressure release.

17. Special Gaseous Pollutant Control Methods. Two widely released gaseous pollutants are SO_2 and NO_x. Major sources of these pollutants are flue gases from combustion operations. Control of these two gases has received wide study, and a number of specialized techniques are available.

Sulfur Dioxide. One control technique is to substitute a low-sulfur or desulfurized fuel. Desulfurization has been commercialized for petroleum fuels.

Several processes have been demonstrated for coal, but it may be several years before they are commercial.

A number of processes have been developed for removal of SO_2 from flue gases. Among the most economical are those which react SO_2 with an inexpensive alkali such as limestone or quicklime. Dry injection of ground limestone through the boiler burner also has received extensive study. The limestone is calcined to quicklime, which reacts with SO_2. $CaSO_4$ is recovered in the unit's electrostatic precipitator. Removal of SO_2 is low at stoichiometric proportions of limestone, but as excess limestone is used, recovery efficiencies increase. However, only 50 percent recovery has been achieved with 100 percent excess limestone.

Reaction of SO_2 with lime or limestone slurry in a wet scrubbing system has been demonstrated to give 70 to 90 percent SO_2 removal. Difficulties with pluggage from $CaSO_4$ deposits have been experienced in some units. Scrubbing with a sodium alkali or an organic absorbent is a more pluggage-resistant method, but considerably more expensive in initial cost.

In smaller boilers, scrubbing with NaOH or $NaCO_3$ has been utilized to give efficient SO_2 removal, but disposal of a Na_2SO_3-Na_2SO_4 solution or solid may be troublesome unless the unit is located near a sulfite pulp mill.

The Cominco NH_3 scrubbing system also has been used. This is a two-stage scrubber in which the gas is contacted first with a $(NH_4)_2SO_4$ solution. The second stage, which prevents loss of NH_3, contains NH_4HSO_4 solution. The makeup NH_3 is added to the second stage, and a portion of the NH_4HSO_4 solution is fed to the first stage. The $(NH_4)_2SO_4$ produced is used in fertilizer manufacture.

Two other processes that recover SO_2 in the form of H_2SO_4 for sale are the Monsanto Cat-OX and the Chemico MgO scrubbing systems. In the Cat-OX process, particulates are removed with an electrostatic precipitator. The cleaned flue gas containing SO_2 is mixed with heated air and passed through a vanadium catalyst where the SO_2 is oxidized to SO_3 and absorbed in H_2SO_4. A fiber mist eliminator provides final cleanup, and the gas is reheated by interchange with heat from the gases leaving the catalyst bed.

In the Chemico process, MgO, suspended in a saturated solution of $MgSO_4$, contacts the flue gas in a Venturi scrubber. The SO_2 is absorbed, reacting with MgO to produce $MgSO_3$ and further $MgSO_4$ through air oxidation. Precipitated solids are removed in a centrifuge and charged to a kiln where $MgSO_3$ is decomposed. Carbon is added to decompose $MgSO_4$. The regenerated MgO is recycled to the process and the kiln off-gas, containing 15 to 16 percent SO_2, is sent to a sulfuric acid plant.

Both processes are capable of recovering 80 to 90 percent of the SO_2. The sales price of the H_2SO_4 produced is insufficient to carry the cost of operation when using flue gases from 3 percent sulfur coal. The cost of these plants is essentially a function of the flue-gas quantity rather than the H_2SO_4 produced. There is definite opportunity to improve the economics by increasing the H_2SO_4 production by using a high-sulfur (5 to 6 percent) coal.

Nitrogen Oxides. Combustion operations are a major source of NO_x pollutants. O_2 and N_2 from air react at high temperatures in the flame to produce NO. NO reacts more slowly at lower temperatures with O_2 to produce NO_2. In most furnaces, reaction rates to form NO are too slow to produce equilibrium amounts of NO corresponding to flame temperature, but it is not unusual for the flue gases from oil and coal combustion to contain 1000 to 2000 ppm by volume of NO when using 5 to 10 percent excess air.

The usual control techniques consist of modifying the combustion process to minimize the formation of NO. Common methods are (1) use of low excess air, (2) two-stage combustion, (3) flue-gas recirculation, and (4) combustion chamber modification.

Reducing the O_2 in the combustion products to 0.3 to 0.5 percent can produce a two- to fivefold reduction in the NO produced. Reducing the flame temperature lowers the quantity of NO produced by making the equilibrium less favorable and decreasing the rate of reaction. Recirculation of flue gas can be used as a means to reduce flame temperature. Formation of NO becomes extremely slow at temperatures below 1600°F (871°C).

Two-stage combustion involves burning the majority of the fuel at a high temperature for good heat transfer with a deficiency of O_2 (50 to 60 percent of theoretical) to discourage NO formation. Combustion of CO and other products is completed downstream at a lower temperature where the remaining combustion air is added. Flue-gas recirculation also may be used to provide flame cooling. A combination of two-stage combustion and flue-gas recirculation can result in a 90 percent reduction in NO formed.

Combustion chamber modifications speed up combustion rate and reduce flame temperature more quickly. One approach is to arrange burners so they burn tangentially along radiating refractory walls. The refractory surface speeds the reaction catalytically and absorbs heat from the flame providing rapid quenching.

Other techniques can be used where NO formation cannot be reduced adequately. For example, the gases may be passed through a combustion catalyst maintained at 1000 to 1400°F (538 to 760°C), where NO decomposes to O_2 and N_2. If the gases are too cold to keep the catalyst hot, some additional fuel is added.

NO can be scrubbed from gases with an alkaline scrubbing liquid as long as the mole ratio of NO_2 to NO is above unity. Unfortunately, the reaction rate of oxidizing NO to NO_2 is fairly slow, and equilibrium becomes more favorable as room temperature is approached. Thus scrubbing is usually practical only if the flue gases are cooled, mixed with additional air, and held in a large reaction chamber for several seconds before scrubbing.

Control of Pollutants by Incineration. Incineration is used to destroy combustible vapors such as hydrocarbons (especially unsaturated and aromatic compounds that are photochemically reactive), CO, H_2, H_2S, and mercaptans.

Consideration should be given to collecting valuable hydrocarbons and organic solvents by other means, such as condensation, rather than destroying them by combustion. If the quantities are appreciable, recovery for fuel value may be worthwhile. Gases containing sufficient combustibles to support combustion are burned in flares or waste-heat recovery boilers or are used for process heat.

Incineration is used to destroy odors in those cases where the odor substance can be oxidized. It is also possible to use an incinerator, when properly designed, to burn combustible airborne liquid and solid particles using a burner much like one designed to burn pulverized coal. In such an incinerator, it is necessary to provide rapid ignition of the particles by heating them above the kindling temperature and providing adequate residence time and flame space for complete combustion. The presence of noncombustible residue that can produce a solid or molten ash also must be considered.

Two types of gas incinerators are in use: direct flame and catalytic. In the direct-flame type, gases are heated in a fuel-fired refractory chamber to their autoignition temperature, where oxidation occurs with or without a visible flame. Autoignition temperatures vary with chemical structure but are generally in the range of 1000 to 1400°F (538 to 760°C). Gases can be incinerated by indirect heating, but a higher temperature is generally required than when a direct flame is present.

For continuous use, a direct-flame incinerator is usually equipped with heat interchangers to preheat the incoming gas with the exhaust gases to save fuel. The required residence time in the combustion chamber is that needed to provide 100 percent oxidation of the combustible materials. This varies with the substances to be oxidized and the temperature. It is specified that incinerators shall provide 0.3-s residence time at 1300°F (704°C) or above in many installations.

Catalytic incinerators oxidize substances at temperatures below which they would burn in air, usually around 500°F (260°C). Catalysts are from the platinum family of metals, but certain other metallic oxides are occasionally used. In catalytic oxidation, fuel must be provided initially to start the reaction. Once the catalyst bed is heated, frequently no further fuel is needed. Hence the advantages of catalytic oxidation are less fuel, no NO formation, and less bulk. However, the catalyst may be poisoned by heavy metals, phosphates, and arsenic. Its activity may be decreased temporarily by halogens and sulfur compounds. Further, it may be rendered inactive by surface coatings of soot and inorganic dust. There is also danger of destroying a catalytic incinerator by overheating if the gas composition is highly variable.

Flares for burning concentrated combustible gases are generally located at high elevations away from other structures to provide protection from the radiant heat of the flame. The design must provide against flashback; a water seal or gas purge is frequently used. To prevent air pollution, the flare should burn smokelessly. Clean burning can be achieved with steam injection [0.05 to 0.3 lb steam/lb (0.02 to 0.14 kg/kg) combustibles] or with multijet flares having a radiant refractory. Combustible particulates must not be fed to flares, since falling burning particles can create a fire hazard. Centrifugal knockout drums are often used to protect against entrained particulates.

18. Collection and Removal of Particulates. Six basic principles are used alone or in combination in particulate collectors:

1. Gravity settling
2. Flowline interception
3. Inertial deposition
4. Diffusional deposition
5. Electrostatic deposition
6. Thermal precipitation

Table 15.27 lists these mechanisms and their basic parameters. In addition, sonic agglomeration has been considered but has seldom become commercially practical.

Particle Dynamics. Larger particles encountered in air pollution obey Stokes' law (see ref. 3). Terminal settling velocities of particles as calculated from Stokes' law are shown in Fig. 15.50. Below 16 μm in size, the Stokes-Cunningham correction factor becomes important, and this correction has been applied. Below about 5 μm, particles tend to become suspended by Brownian motion because of the impact of gas molecules on the particle, and gravity settling ceases as an important influence.

TABLE 15.27 Summary of Mechanisms and Parameters in Aerosol Deposition

Deposition mechanism	Origin of force field	Deposition mechanism measurable in terms of: Basic parameter	Deposition mechanism measurable in terms of: Specific modifying parameters	System parameters
Flow-line interception*	Physical gradient*	$N_{sf} = \left(\frac{D_p}{D_b}\right)$	$N_{sc} = \left(\frac{N^2_{sf}}{N_{st}N_{sd}}\right) = \left(\frac{18\mu}{K_m\rho_p D_v}\right)$‡	Geometry: (D_{b1}/D_b), (D_{b2}/D_b), etc. ϵ_v α
Inertial deposition	Velocity gradient	$N_{si} = \left(\frac{K_m\rho_s D_p^2 V_o}{18\mu D_b}\right)$		
Diffusional deposition	Concentration gradient	$N_{sd} = \left(\frac{D_v}{V_o D_b}\right)$		
Gravity settling	Elevation gradient	$N_{sg} = \left(\frac{u_t}{V_o}\right)$		Flow pattern: NRe¶ NMa NKn Surface accommodation
Electrostatic precipitation	Electric-field gradient† Attraction	$N_{sec} = \left(\frac{K_m Q_p \epsilon b}{\mu D_p V_o}\right)$	δ_p, δ_b§	
	Induction	$N_{sei} = \left(\frac{\delta_p - 1}{\delta_p + 2}\right)\left(\frac{K_m D_p^2 \delta_o \epsilon b^2}{\mu D_b V_o}\right)$		
Thermal precipitation	Temperature gradient	$N_{st} = \left(\frac{T - T_b}{T}\right)\left(\frac{\mu}{K_m\rho D_b V_o}\right)\left(\frac{k_t}{2k_t + k_{tp}}\right)$	(T_b/T), (T_p/T),§ (N_{Pr}) (k_{tp}/k_t), (k_{tb}/k_t),§ (c_{hp}/c_h), (c_{hb}/c_h)§	

§Not likely to be significant contributors.

*This has also commonly been termed "direct interception" and in conventional analysis would constitute a physical boundary condition imposed upon particle path induced by action of other forces. By itself it reflects deposition that might result with a hypothetical particle having finite size but no mass or elasticity.

†In cases where the body charge distribution is fixed and known, ϵ_b may be replaced with Q_{bs}/δ_o.

‡This parameter is an alternative to N_{sf}, N_{sl}, or N_{sd} and is useful as a measure of the interactive effect of one of these on the other two. It is comparable with the Schmidt number.

¶When applied to the inertial deposition mechanism, a convenient alternative is $(K_m\rho_s/18\rho) = N_{sl}/(N_{sf}^2 N_{Re})$.

Source: Perry and Chilton (eds.), *Chemical Engineers' Handbook*, 5th ed., McGraw-Hill, New York, 1973, p. 20–80.

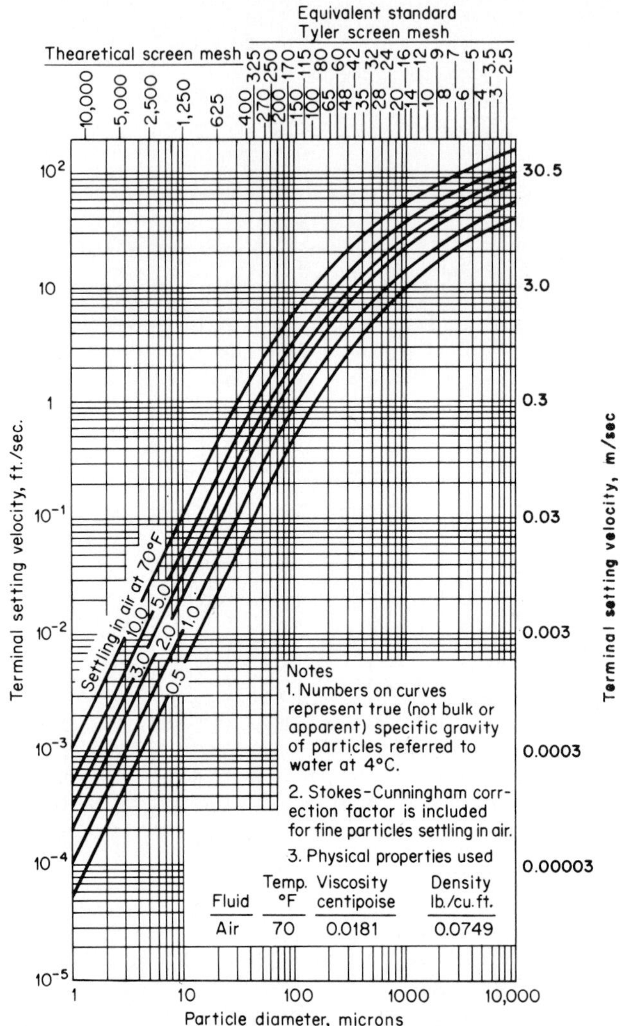

FIGURE 15.50 Terminal settling velocity of spherical particles.

Gravity Settling Chambers. Owing to low collection efficiency on smaller particles, gravity settling is seldom useful today as a sole collection device. However, such chambers may be used as preclassifiers to remove large abrasive particles ahead of another type of collector. Settling chambers are seldom efficient on particles smaller than 50 μm.

Basically, a *settling chamber* is a large horizontal enlargement in the duct where the gas slows down. Turbulence should be low to prevent reentrainment. The collection efficiency for any size particle can be calculated from the terminal settling velocity of the particle as determined in Fig. 15.50, the distance the particle has to fall to settle out, and the residence time in the chamber. Equation

(15.94) can be applied to each individual size particle to calculate collection efficiency for that size. Equation (15.95) gives the smallest size particle that can be collected with 100 percent efficiency.

$$\eta = \frac{u_t L_s}{H_s V_s} \tag{15.94}$$

$$D_{p,\min} = \sqrt{\frac{18\mu H_s V_s}{g_L L_s(\rho_s - \rho_g)}} \tag{15.95}$$

where η = fractional collection efficiency
u_t = particle terminal settling velocity, ft/s (m/s)
V_s = bulk gas velocity, ft/s (m/s)
L_s and H_s = the length and height of the settling path, respectively, ft (m)
μ = gas viscosity, lb/(s · ft) [kg/(s · m)]
g_L = local acceleration due to gravity, ft/s^2 (m/s^2)
$(\rho_s - \rho_g)$ = difference in particle and gas density, lb/ft^3 (kg/m^3)

Cyclonic Collectors. In cyclonic collectors, centrifugal force separates particles from the gas stream. Since centrifugal force can equal many times that of gravity, much smaller particles can be collected. Figure 15.51 shows a typical dust-collection cyclone in which the gas enters tangentially, spirals downward, reverses direction in the cone, and exits through the top in smaller spirals. The dust particles spiral downward along the wall and discharge at the bottom.

Cyclones are reasonably effective for collecting solid and liquid particles down to 5 to 10 μm. The smaller the diameter of the cyclone, the higher its efficiency on small particles. When cyclones are used to collect liquids, special modifications must be made to prevent reentrainment of droplets at the outlet tube.

The principles of centrifugal force are applied in many wet scrubbers in which the particles are centrifuged into a water film on a wall. Wet scrubbers have been built using a cyclone like that shown in Fig. 15.51 in which spray nozzles are placed in the annular space at the top or a film of water is allowed to flow down the walls.

The efficiency of a cyclone is computed by integration of the collection efficiency for each individual size particle as obtained from the manufacturer's efficiency curve. It is frequent practice to calculate the particle cut size D_{pc} (the diameter particle that is collected with 50 percent efficiency). Equation (15.96) gives the cut size for a cyclone with dimensions as given in Fig. 15.51.

$$D_{pc} = \sqrt{\frac{9\mu B_c}{2\pi N_c V_c(\rho_s - \rho_g)}} \tag{15.96}$$

where N_c = the number of turns the gas makes in the cyclone, often 5 to 10
B_c = the width of the gas inlet, ft (m)
V_c = inlet gas velocity, ft/s (m/s)

All other symbols are as defined for Eq. (15.95). Figure 15.52 relates the collection efficiency for other size particles to the cut size for the Fig. 15.51 cyclone.

Pressure drop through a cyclone depends on cyclone geometry. The manufacturer's calibration curve should be used. The pressure drop of the Fig. 15.51 cyclone in terms of the number of inlet velocity heads, F_{cv}, is given by Eq. (15.97).

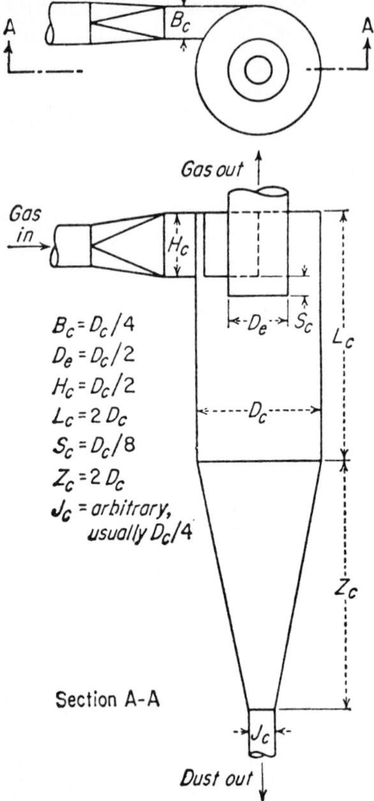

FIGURE 15.51 Cyclone separator proportions.

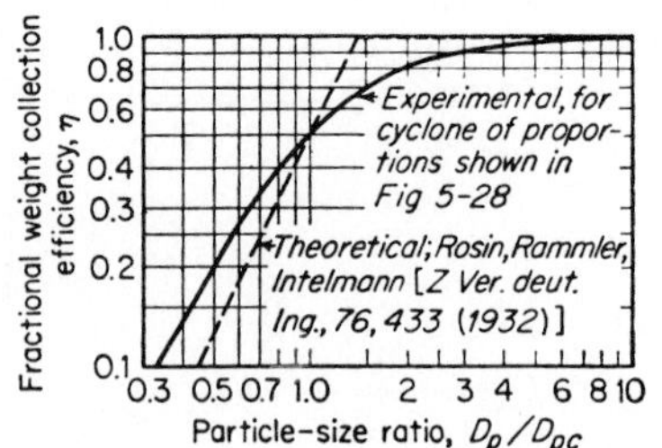

FIGURE 15.52 Separation efficiency of cyclones.

The value of K for this cyclone is 16. Other symbols are defined by Fig. 15.51.

$$F_{cv} = KB_cH_c/D_c^2 \qquad (15.97)$$

Impingement Collectors. Figure 15.53 illustrates several collection principles involved in impingement collection. If D_b represents a target, all particles contained upstream within the streamlines *A-B* will be collected on the target, unless the particles have sufficient mobility so that they can flow around the target with the gas. These collected particles are caught by flowline interception. Large particles just outside these streamlines, owing to their diameter, will have to move farther away from the target to prevent impact. However, the larger particles may have too much momentum to be deflected and also will be collected. This is known as *inertial deposition*. Other particles still farther away, but carrying an electric charge, may be attracted to the target, especially if the target develops an opposite charge.

Some dry impingement separators have been built in which the particulates are directed first at one target and then at another. Figure 15.54 gives the collection efficiency for a single row of targets of various shapes for collection by both direct and inertial impaction but does not include electrostatic attraction. Many wet scrubbers use the impaction principle to collect a particle in a liquid film on a target. The higher the upstream velocity and the larger the particles, the greater is the collection efficiency of a single stage. Overall efficiency is improved by using a number of stages, such as in a fiber-bed filter.

Granular-Bed Filters. In these particulate-removal devices, a gas is forced through a granular bed. Reported efficiencies are as high as 97 to 98 percent on particles of 2 to 5 μm. Often the primary collection technique is direct and inertial impaction. The high efficiency on small particles results from the many successive targets.

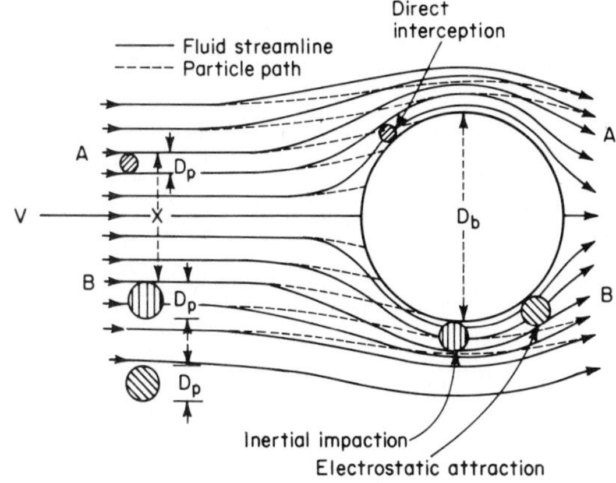

FIGURE 15.53 Mechanisms involved in impingement separators.

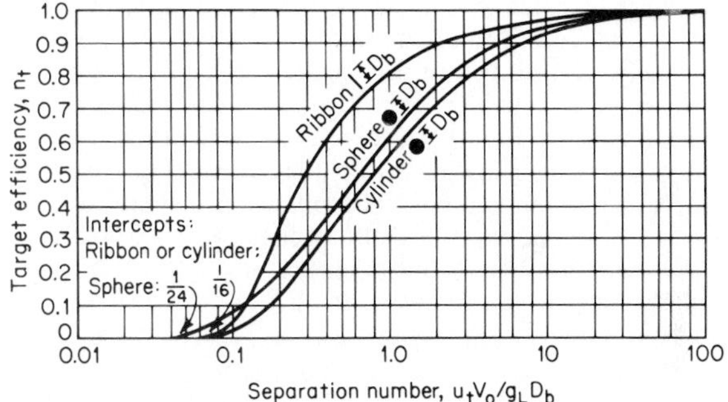

FIGURE 15.54 Target efficiency for conditions where Stokes' law applies to the motion of the particle. (*From Langmuir and Blodgett, U.S. Army Air Force Technical Report 541S, U.S. Department of Commerce, OTS, PB 27565.*)

In some beds, plastic granules that develop an electrostatic charge have been used to increase the collection efficiency. Some problems with granular-bed filters are adequate removal of the collected particles from the bed and particle reentrainment. (See Perry, Green, and Maloney[3] for further discussion on granular-bed filters.)

Bag Filters. Bag filters for the collection of particulates use a woven fabric or a nonwoven felt bag. Either can be specified for collection efficiencies of 99 percent or better. The collection process is not merely a screening or filtration of the dust, since the openings in the cloth are many times the size of the dust particles collected. Efficiency of a new bag may be fairly low for a few moments until it develops a precoat of dust that serves as the filtering layer. Once precoated, the bag usually retains sufficient solids in its pores so that it does not return to its original efficiency.

Pressure drop through the fabric is usually negligible compared to that through the layer of collected dust. Pressure drop through the dust layer can be expressed by

$$\Delta p_i = K_d \mu C_d V_f^2 \theta \tag{15.98}$$

where C_d = the concentration of dust in the gas stream
V_f = the gas velocity through the bag
μ = the gas viscosity
θ = the time since the last bag cleaning cycle

The resistance factor K_d is a function of the dust size, shape, packing density, etc. and is best determined experimentally. Usual practice is to specify a maximum desired pressure drop across the bag and pick a cleaning cycle such that this pressure drop is not exceeded. Typical operating pressure drops are 2 to 6 inH_2O (50.8 to 152 mmH_2O).

With woven fabrics, cleaning must be done by mechanically shaking the bag. The cycle may be controlled manually or automatically. More frequent shaking permits a smaller bag area, but bag life is shortened by the mechanical strains of shaking. Filters are usually sized on superficial velocity through the bag and will be in the range of 1 to 8 $ft^3/(min \cdot ft^2)$ [0.005 to 0.041 $m^3/(s \cdot m^2)$] of bag area. Longer bag life results from more generous sizing. For fine dusts, velocity should not exceed 3 $ft^3/(min \cdot ft^2)$ [0.015 $m^3/(s \cdot m^2)$]. Airflow through a compartment is shut off while the bags are shaken. On a continuous process this means providing one or two spare compartments for use while others are being cleaned.

With felt cloth, somewhat higher flow rates can be used, up to 15 $ft^3/(min \cdot ft^2)$ [0.075 $m^3/(s \cdot m^2)$]. Cleaning can be accomplished in several ways, but a reverse-flow air blast is one of the best. Table 15.28 lists the maximum desirable operating temperatures for a number of filter fabrics, while Table 15.29 lists the chemical compatibility of different fabrics.

Care must be taken to operate bag filters well above the dew point to prevent caking on the bags. To avoid condensation, baghouses are often insulated and occasionally steam traced. A few steam-jacketed baghouse designs are available.

When handling very fine dusts, problems with dust leakage through the bag are sometimes encountered. Leakage can be reduced by (1) a bag made of staple rather than monofilament fibers, (2) a napped-fiber bag, (3) a finer weave, (4) less frequent and less vigorous cleaning, (5) a bag fiber better suited to attract and retain the dust particles, and (6) a precoat material following cleaning.

TABLE 15.28 Maximum Desirable Operating Temperatures* for Filter Bags

Fiber	*Temperature, °C*
Cotton	80
Wool	100
Nylon 66 and 6	105
Dacron	140
Orlon	120
Dynel	85
Saran	70
Polyethylene	70
Glass	290
Teflon	260
Nomex nylon	230

*°F = 9/5 °C + 32.

TABLE 15.29 Chemical Compatibility of Fibers for Dust Collector Bags

Resistance	Acid media	Alkaline media
Excellent	Polyethylene Saran Teflon Nomex nylon	Dynel Nylon 66 Polyethylene Teflon
Good	Dacron Dynel Glass Wool	Cotton Nylon 6 Saran Saran
Unsuitable	Cotton Nylon 66 Nylon 6	Wool

Textile fibers tend to develop static charges as gases pass through the bags. If a leaking bag does not develop a strong charge, a bag developing a stronger charge or the opposite charge might be tested.

Occasionally, a bag retains its cake too tenaciously during cleaning and the bag is said to be subject to *blinding*. This problem may be improved by taking the reverse action to that used for leakage.

Electrostatic Precipitators. An electrostatic precipitator collects solid or liquid particulates with high efficiency and low energy utilization. The pressure drop is composed almost entirely of inlet and exit losses. Initial cost of precipitators is high, however, and their on-stream reliability may be less than that of other collection devices.

Collection is based on imparting a charge to particles in an electric field, which then causes them to migrate and deposit on a collection plate of the opposite charge. Liquids agglomerate on the collection plate and drain off, but solids must be removed by rapping. Reentrainment may occur during rapping.

When power is applied to a precipitator, no current flows until a sufficient voltage is achieved to create a corona discharge (gas ionization) at the discharge electrodes. Then particle charging and deposition begin. As voltage is increased further, the field strength becomes more intense and precipitator efficiency increases until the point where sparkover or arcing from the discharge electrode to the collecting plate occurs.

For maximum collection efficiency, the voltage should be maintained just short of sparkover. Unfortunately, sparkover potential is an operating variable affected by changes in temperature, dust resistivity, moisture content, and frequency and efficiency of removal of collected dust. For maximum operability, the difference in potential between start of corona and sparkover should be as large as possible. This difference decreases as gas temperature increases or system absolute pressure decreases. The difference is greater for negative-polarity current than for positive. Thus most industrial precipitators are operated with negative polarity.

Because of sparkover, precipitators should not be used to collect combustible or explosive mixtures. A two-stage precipitator in which the particle charging is done in one step and the deposition in a second step (developed primarily for air-conditioning) is somewhat safer because lower voltages are employed. Such units have been used with hydrocarbon mists in air. However, the safety of this application is still questionable.

Equation (15.99) gives the theoretical efficiency for a precipitator:

$$\eta = 1 - e^{-[(u_e A_e)/q]} = 1 - e^{-K_e u_e} \tag{15.99}$$

where η = the weight fractional efficiency
u_e = the velocity of migration of particle toward collecting electrode
A_e = the area of collecting electrodes
q = the total gas flow rate

Any consistent set of units that makes the exponent dimensionless can be used. The efficiency factor K_e is basically a geometric design factor for the precipitator. For a plate-type precipitator, $K_e = L/B_e V_e$. For a tube type, $K_e = 4L/D_t V_e$, where L is the length of a flow passage in the electric field, B_e is the distance between adjacent plates, D_t is the tube diameter, and V_e is the gas velocity through an individual flow passage. The migrational velocity u_c is a function of the particle size, shape, composition, and average electric field strength. Each different size particle has a different migration velocity. For a dust of a given particle-size distribution, an average migrational velocity is often used.

With a given precipitator, it is often desired to know the effect of changes in operating conditions. Using Eq. (15.99) with geometric values for K_e, the effects of changes in gas velocity or residence time can be determined without any knowledge of the value of u_e. However, changing gas temperature or particle size will change the value of u_e. For a constant field strength, the relative changes in u_e can be estimated by Eqs. (15.100) and (15.101).

For a change in temperature,

$$(u_e)_{t_2} = \frac{(u_e)_{t_1}(\mu_g)_{t_1}}{(\mu_g)_{t_2}} \tag{15.100}$$

where μ_g is the gas viscosity at temperatures t_1 and t_2.

For a change in median particle size at constant particle standard **deviation**,

$$u_{e_2} = \frac{u_{e_1} r_1}{r_2} \tag{15.101}$$

where r_1 and r_2 are the radius of the median particle sizes.

Mist Filters. Mists are created by condensation. They consist of small particles, from submicrometer size up to 10 μm. For coarser mists, 5 μm and up, knitted-wire-mesh separator pads, 4 to 6 in (10.2 to 15.2 cm) thick, have collection efficiencies of 98 percent and above with 1 to 2 in (2.5 to 5.1 cm) of water-pressure drop. These separators are most efficient when installed horizontally with upflow of gas, and gravity drainage of liquid. Collection is by impingement.

Impingement collectors consisting of thin mats, 1 to 2 in (2.5 to 5.1 cm) thick, of compressed glass or plastic fibers give 92 to 94 percent collection efficiency on 1-μm particles with 4 to 8 in (10.2 to 20.3 cm) of water-pressure drop. Similar fiber-bed filters, 4 to 6 in (10.2 to 15.2 cm) thick, and employing Brownian diffusion as the collection principle, are used to collect submicrometer mists with collection efficiencies up to 99 percent and pressure drops from 15 to 30 in (38.1 to 76.2 cm) of water. These filters, when irrigated, are also used to collect submicrometer particles that are soluble in the irrigating liquid.

Wet Scrubbers. A great variety of liquid scrubbing devices are available for collection of particulates. The major collection mechanism is inertial impaction with direct interception and Brownian diffusion of lesser importance. Some units contact the gas with atomized liquids using either gravity or centrifugal force. Others impinge the gas at high velocity against sheets and films of water, causing the water to shatter into droplets. In either case, the major portion of the collection occurs by collision between small liquid droplets and the particulate.

Figure 15.55 shows data on target efficiency for collection in a gravity spray tower between dust particles of specific gravity 2.0 and spray droplet size. It can be seen that the most efficient droplet sizes are from 500 to 1000 μm. The data also indicate that it would be difficult to obtain high efficiency in a gravity spray tower for particulates smaller than 3 μm.

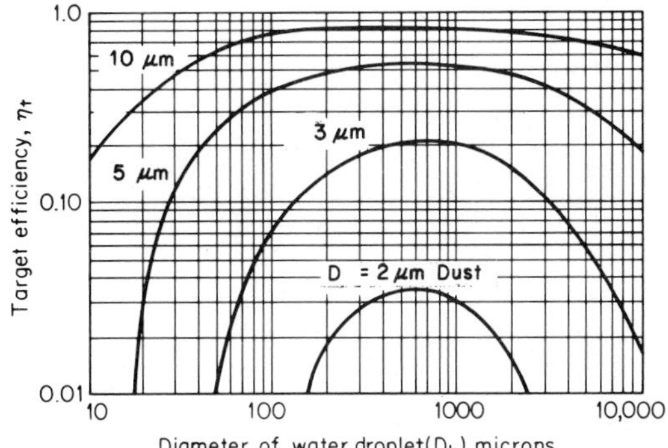

FIGURE 15.55 Target efficiency in a spray tower. *(From Stairmand, Journal of the Institute of Fuel 29: 58, 1956.)*

Collection efficiency can be improved greatly by the addition of centrifugal force. Figure 15.56 shows target efficiencies for water droplets in a centrifugal field of 100*g*. [A centrifugal force of 100*g* is equivalent to a tangential velocity of 57 ft/s (17.4 m/s) at a radius of 1 ft (0.3 m).] It can be seen that the optimum scrubbing droplet size has been reduced to 40 to 200 μm and that sizable collection efficiencies on particles down to 1 μm can now be obtained.

High collection efficiencies on submicrometer particles in wet scrubbers can only be obtained with Venturi scrubbers and wetted fibrous-bed scrubbers. The latter are subject to plugging if the dust loading is high or insoluble. The former requires high energy expenditure, in the order of 30 to 50 in (76.2 to 127.0 cm) of water pressure drop for high efficiency on 0.5- to 1.0-μm particles.

Table 15.30 indicates typical pressure drop and minimum particle size that can generally be collected at 80 percent collection efficiency in various types of wet scrubbers.

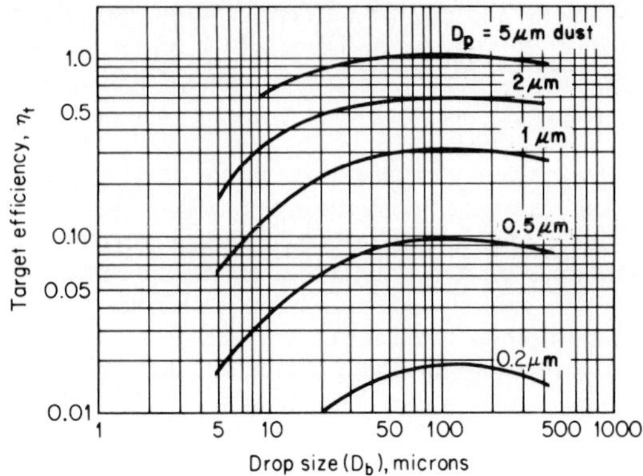

FIGURE 15.56 Target efficiency in 100 times gravitational field. *(From Johnstone and Roberts, Industrial Engineering and Chemistry 46: 1601, 1954.)*

TABLE 15.30 Minimum Particle Size for Various Types of Scrubbers

	Pressure drop,		Min. particle size, μm
	in water	cm water	
Spray towers	0.5–1.5	1.27–3.8	10
Cyclone spray scrubbers	2–10	5.1–25.4	2–10
Impingement scrubbers	2–50	5.1–127.0	1–5
Packed- and fluidized-bed scrubbers	2–50	5.1–127.0	1–10
Orifice scrubbers	5–100	12.7–254.0	1
Venturi scrubbers	5–100	12.7–254.0	0.8
Fibrous-bed scrubbers	5–110	12.7–279.4	0.5

Source: Perry, Green, and Maloney.[3]

Petroleum and Gas Engineering*

Petroleum engineering is a broad-based discipline comprising the technologies used for the exploitation of crude oil and natural gas reservoirs. It is usually subdivided into the branches of petrophysical, geological, reservoir drilling, production, and construction engineering. After an oil or gas accumulation is discovered, technical supervision of the reservoir is transferred to the petroleum engineering group, although in the exploration phase the drilling and petrophysical engineers have played a role in the completion and evaluation of the discovery.

*Excerpted from the *McGraw-Hill Encyclopedia of Science and Technology*, 6th ed. Copyright © 1987. Used by permission of McGraw-Hill, Inc. All rights reserved.

PETROPHYSICAL ENGINEERING

The petrophysical engineer is perhaps the first of the petroleum engineering group to become involved in the exploitation of the new discovery. By the use of down-hole logging tools and of laboratory analysis of cores made during the drilling operation, the petrophysicist estimates the porosity, permeability, and oil content of the reservoir rock that has been sampled at the drill site.

GEOLOGICAL ENGINEERING

The geological engineer, using the petrophysical data, the seismic surveys conducted during the exploration operations, and an analysis of the regional and environmental geology, develops inferences concerning the lateral continuity and extent of the reservoir. However, this assessment usually cannot be verified until additional wells are drilled and the geological and petrophysical analyses are combined to produce a firm diagnostic concept of the size of the reservoir, the distribution of fluids therein, and the nature of the natural producing mechanism. As the understanding of the reservoir develops with continued drilling and production, the geological engineer, working with the reservoir engineer, selects additional drill sites to further develop and optimize the economic production of oil and gas.

RESERVOIR ENGINEERING

The reservoir engineer, using the initial studies of the petrophysicist and geological engineers together with the early performance of the wells drilled into the reservoir, attempts to assess the producing rates [barrels of oil or millions of cubic feet (cubic meters) of gas per day] that individual wells and the entire reservoir are capable of sustaining. One of the major assignments of the reservoir engineer is to estimate the ultimate production that can be anticipated from both primary and enhanced recovery from the reservoir. The *ultimate production* is the total amount of oil and gas that can be secured from the reservoir until the economic limit is reached. The *economic limit* represents that production rate which is just capable of generating sufficient revenue to offset the cost of operating the reservoir. The proved reserves of a reservoir are calculated by subtracting from the ultimate recovery of the reservoir (which can be anticipated using available technology and current economics) the amount of oil or gas that has already been produced.

Primary recovery operations are those which produce oil and gas without the use of external energy except for that required to drill and complete the wells and lift the fluids to the surface (pumping). *Enhanced recovery*, or *supplemental recovery*, is the amount of oil that can be recovered over and above that producible by primary operation by the implementation of schemes that require the input of significant quantities of energy. In modern times, waterflooding has been almost exclusively the supplementary method used to recover additional quantities of crude oil. However, with the realization that the discovery of new petroleum resources will become an increasingly difficult achievement in the future, the reservoir engineer has been concerned with other enhanced oil recovery processes that promise to increase the recovery efficiency above the average 33 percent experienced in the United States (which is somewhat above that achieved in the rest of the world). The restrictive factor on such processes is the economic cost of their implementation.

DRILLING ENGINEERING

The drilling engineer has the responsibility for the efficient penetration of the earth by a well bore and for cementing of the steel casing from the surface to a depth usually just above the target reservoir. The drilling engineer or another specialist, the mud engineer, is in charge of the fluid that is continuously circulated through the drill pipe and back up to surface in the annulus between the drill pipe and the bore hole. This mud must be formulated so that it can do the following: carry the drill cuttings to the surface, where they are separated on vibrating screens; gel and hold cuttings in suspension if circulation stops; form a filter cake over porous low-pressure intervals of the earth, thus preventing undue fluid loss; and exert sufficient pressure on any gas- or oil-bearing formation so that the fluids do not flow into the well bore prematurely, blowing out at the surface. As drilling has gone deeper and deeper into the earth in the search for additional supplies of oil and gas, higher and higher pressure formations have been encountered. This has required the use of positive-acting blowout preventers that can firmly and quickly shut off uncontrolled flow due to inadvertent imbalances in the mud system.

PRODUCTION ENGINEERING

The production engineer, upon consultation with the petrophysical and reservoir engineers, plans the completion procedure for the well. This involves a choice of setting a liner across the formation or perforating a casing that has been extended and cemented across the reservoir, selecting appropriate pumping techniques, and choosing the surface collection, dehydration, and storage facilities. The production engineer also compares the productivity index of the well [barrels per day per pounds per square inch (cubic meters per pascal) of drawdown around the well bore] with that anticipated from the measured and inferred values of permeability, porosity, and reservoir pressure to determine whether the well has been damaged by the completion procedure. Such comparisons can be supplemented by a knowledge of the rate at which the pressure builds up at the well bore when the well is abruptly shut-in. Using the principles of unsteady-state flow, the reservoir engineer can evaluate such a buildup to assess quantitatively the nature and extent of well-bore damage. Damaged wells, like wells of low innate productivity, can be stimulated by acidization, hydraulic fracturing, additional performation, or washing with selective solvents and aqueous fluids.

CONSTRUCTION ENGINEERING

Major construction projects, such as the design and erection of offshore platforms, require the addition of civil engineers to the staff of petroleum engineering departments, and the design and implementation of natural gasoline and gas processing plants require the addition of chemical engineers.

GAS FIELD AND GAS WELL

Petroleum gas, one form of naturally occurring hydrocarbons of petroleum, is produced from wells that penetrate subterranean petroleum reservoirs of several kinds. Oil and gas production are commonly intimately related, and about one-third of gross gas production is reported as derived from wells classed as oil wells. If gas is produced without oil, production is generally simplified, in part at least because the gas flows naturally without lifting, and also because of fewer complications in reservoir problems. As for all petroleum hydrocarbons, the term *field* designates an area underlain with little interruption by one or more reservoirs of commercially valuable gas.

PETROLEUM ENHANCED RECOVERY

Novel technology has been designed to enhance the fraction of the original oil in place in a reservoir. Heightened interest in developing enhanced recovery technology has developed as it has become more certain that over two-thirds of the oil discovered in the United States, and a still greater percentage in the rest of the world, will remain unrecovered through the application of conventional primary and secondary (waterflood) operations. Thus there is a strong incentive for the development and implementation of advanced technology to recover some of this oil.

The problems encountered in developing such technology are very great because of the nature of fluid flow within a subsurface reservoir, and because of the inability of engineers to exercise any intimate degree of control over the flow and distribution of fluids within the reservoir. The only points of contact with the reservoir are at the surface of the producing and injection wells that lead down to the reservoir sands.

REFERENCES

1. R. H. Perry (ed.), *Engineering Manual*, 3d ed., McGraw-Hill, New York. 1976.
2. *International Critical Tables*.
3. R. H. Perry, D. W. Green, and J. O. Maloney (eds.), *Perry's Chemical Engineers' Handbook*, 6th ed., McGraw-Hill, New York, 1984.
4. *Journal of Chemical and Engineering Data*.

5. R. E. Treybal, *Mass Transfer Operations*, 2d ed., McGraw-Hill, New York, 1968.
6. Hala, Wichterle, et al., *Vapor Liquid Equilibrium Data at Normal Pressures*, Pergamon Press, New York, 1968.
7. Smith, Block, and Hickman, in R. H. Perry, D. W. Green, and J. O. Maloney (eds.), *Perry's Chemical Engineers' Handbook*, 6th ed., McGraw-Hill, New York, 1984.
8. Hala, Pick, Fried, and Vilim, *Vapor-Liquid Equilibrium*, 2d ed., Pergamon Press, New York, 1967.
9. Fair, et al., in R. H. Perry, D. W. Green, and J. O. Maloney (eds.), *Perry's Chemical Engineers' Handbook*, 6th ed., McGraw-Hill, New York, 1984.
10. Maddox, in R. H. Perry, D. W. Green, and J. O. Maloney (eds.), *Perry's Chemical Engineers' Handbook*, 6th ed., McGraw-Hill, New York, 1984.
11. Ellis, *Industrial Chemistry* 28: 483, 1952.
12. Sherwood and Pigford, *Absorption and Extraction*, McGraw-Hill, New York, 1952.
13. Norris, in R. H. Perry, D. W. Green, and J. O. Maloney (eds.), *Perry's Chemical Engineers' Handbook*, 6th ed., McGraw-Hill, New York, 1984.
14. McCormick, in R. H. Perry, D. W. Green, and J. O. Maloney (eds.), *Perry's Chemical Engineers' Handbook*, 6th ed., McGraw-Hill, New York, 1984.
15. McCabe and Smith, *Unit Operations of Chemical Engineering*, 3d ed., McGraw-Hill, New York, 1976, pp. 454–459.
16. Vermeulen et al., in R. H. Perry, D. W. Green, and J. O. Maloney (eds.), *Perry's Chemical Engineers' Handbook*, 6th ed., McGraw-Hill, New York, 1984.
17. J. J. Collins, *Chemical Engineers' Progress Symposium Service* 63(74): 31, 1967.
18. Levenspiel, *Chemical Reaction Engineering*, Wiley, New York, 1962.
19. Walas, *Reaction Kinetics for Chemical Engineers*, McGraw-Hill, New York, 1959.
20. Aris, *The Optimal Design of Chemical Reactors*, Academic Press, New York, 1961.
21. Eckenfelder, "Trickling Filtration Design and Performance," *Journal of the Sanitation Engineering Division, American Society of Civil Engineers* 87(SA4): 33–45, 1961.
22. Germain, "Economic Treatment of Domestic Waste by Plastic-Medium Trickling Filters," *Journal of Water Pollution Control Federation* 38: 192–203, 1966.
23. R. S. Burd, *A Study of Sludge Disposal and Handling* (Publication No. WP-20-4), U.S. Department of the Interior, Federal Water Pollution Control Administration, Washington, D.C., May 1968.
24. U.S. Public Health Service, Publications AP-49, *Particulates*, 1969; AP-50, SO_x, 1969; AP-62, *CO*, 1970, Washington, D.C.
25. U.S. Environmental Protection Agency, Publications AP-63, *Oxidants*, 1970; AP-64, *Hydrocarbons*, 1970; AP-84, NO_x, 1971.
26. D. B. Turner, *Workbook of Atmospheric Dispersion Estimates* (Publication No. AP-26), U.S. Environmental Protection Agency, Washington, D.C., 1970.
27. Hosler, "Low Level Inversion Frequency in the Contiguous US," *Monthly Weather Review* 89: 319–339, 1961.
28. U.S. Environmental Protection Agency, "Method V Sampler," *Federal Register* 36: 24893, 1971.
29. Ensor and Pilat, *Journal of the Air Pollution Control Association* 21: 496–501, 1971.
30. *McGraw-Hill Encyclopedia of Science and Technology*, 6th ed., McGraw-Hill, New York, 1987.

CHAPTER 16
ELECTRICAL ENGINEERING AND ELECTRONICS ENGINEERING

ELECTRICAL ENGINEERING*

ELECTRIC CIRCUITS AND THEIR CHARACTERISTICS

1. Electric Charge *Q* and Current *I*. The entities of electricity are called *electrons*. Each electron carries a *charge* $Q = 1.59 \times 10^{-19}$ coulomb. When electrons are set in motion, they produce an *electric current* denoted by the letter I or i. Electric current is measured in *amperes* and is equal to the *charge* in coulombs passing a given point per second, or

$$I = \frac{dQ}{dt} \tag{16.1}$$

The symbolic notation for current is an arrow →.

2. Electromotive Force emf or *E*. The agency which can set electrons in motion is called *electromotive force* (abbreviated emf and denoted by the letter E or e). The unit of electromotive force is the volt. There are several sources of emf, of which the following may be mentioned:

1. The motion of a metallic body in a magnetic field. (This is the agency which creates emf in rotating electrical machines.)
2. The change in the value of a magnetic field in the neighborhood of a metallic body. (This is the agency which creates emf in transformers.)
3. Chemical reactions. (These are the cause of emf in storage batteries.)
4. Light. (This is the cause of emf in photoelectric cells.)
5. Heat. (This is the cause of emf in a junction of two unequally heated metals.)
6. Pressure. (This is the cause of emf in certain crystals subject to mechanical pressure.)

In electrical engineering, it is customary to distinguish between two types of emf sources, as follows:

Direct (*continuous* or *unidirectional*) emf, which continues in the same direction. A symbolic notation for direct emf is ⊣ ⊢ with the light vertical line representing the positive terminal and the small block representing the negative terminal.

*This section is from *General Engineering Handbook*, 2d ed., by C. E. O'Rourke. Copyright © 1940. Used by permission of McGraw-Hill, Inc. All rights reserved. Updated by the editors.

Alternating emf, which varies periodically, and the mean value of which during each period is zero. A symbolic notation for alternating emf is a short wavy line: ~.

3. Circuits and Their Characteristics. Any arrangement in which electric current may flow is called a *circuit*. By definition, a circuit may contain an emf source to set the electrons in motion and a body through which the current may flow. The body through which current flows is known as the *characteristic* (or *parameter*) of the circuit.

4. Types of Circuit Characteristics. There are three fundamental types of circuit characteristics: (1) resistance, (2) inductance, and (3) capacitance.

5. Electric Resistance (*R* or *r*). With the present meager knowledge of the nature of electricity, it is not possible to explain the physical nature of electric resistance. It can be defined, however, through certain empirical relations. These are (1) Ohm's law and (2) Joule's law.

Ohm's Law. If an unvarying emf (Fig. 16.1) is impressed on a homogeneous metallic conductor, and if the temperature of that conductor is held constant, then the ratio of the current I in the conductor to the impressed emf E is constant. This constant ratio may be expressed in two ways:

$$R = \frac{E}{I} \tag{16.2a}$$

$$G = \frac{I}{E} \tag{16.2b}$$

R is known as the *resistance* of the conductor, and $G = 1/R$ is called its *conductance*. The unit of resistance is the ohm. The unit of conductance is the mho.

Equations (16.2) give a mathematical statement of Ohm's law and incidentally define resistance and conductance. The following limitations of Ohm's law must be emphasized:

1. Ohm's law holds true only for homogeneous metallic conductors. It fails to be applicable, for example, in the case of liquid and gaseous conductors, of conducting dielectric materials, or of two dissimilar metals joined or welded together.

2. Ohm's law applies only when the source of emf is constant or when no local emf's exist in the circuit.

3. Experiment shows that the resistance of a metallic conductor generally increases with an increase in temperature. Thus, if R_t is the resistance at a temperature t°C, and if R_c is the resistance at 0°C, then

$$R_t = R_c(1 + \alpha) \tag{16.3}$$

where α is a positive constant which depends upon the metal and the units used and is known as the *temperature coefficient* of the metal. Temperature coefficients for various materials are given in Tables 16.1 and 16.2.

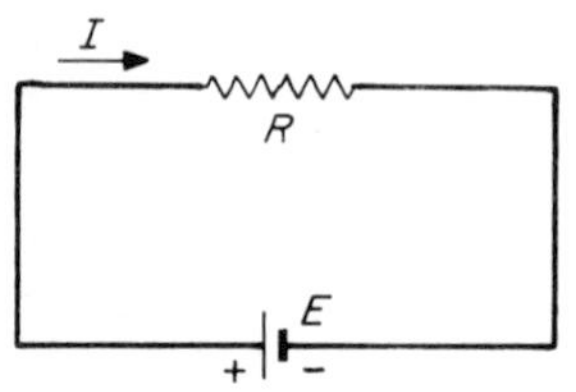

FIGURE 16.1

TABLE 16.1 Resistivities and Temperature Coefficients of Metals
To obtain the resistivity in microhms for 1 cm² divide by 6.01

Material	Resistivity at 0°C		Temperature coefficient per °C, at 20°C
	Circular mil-ft	$\mu\Omega/cm^3$	
Aluminum	17.1	2.85	0.00390
Copper, annealed	9.35	1.56	0.00393
Iron, pure	53.00	8.82	0.00600
Gold	12.36	2.06	0.00365
Lead	115.0	19.14	0.00390
Magnesium	30.0	4.99	0.00381
Mercury	564.00	93.84	0.00072
Nickel	41.60	6.92	0.00500
Platinum	66.00	10.98	0.00370
Silver	8.85	1.47	0.00400
Tantalum	87.60	14.58	0.00330
Tin	78.00	12.98	0.00365
Tungsten (hard drawn)	33.00	5.49	0.00320
Zinc	34.50	5.74	0.00400

TABLE 16.2 Resistivities and Temperature Coefficients of Low Temperature-Coefficient Alloys

Material	Resistivity at 20°C		Temperature coefficient per °C, at 20°C
	Circular mil-ft	$\mu\Omega/cm^3$	
Copper-nickel	100–250	16.64–41.59	0.000005–0.0004
Copper-nickel-zinc (nickel silver)	200–290	33.28–48.25	0.0002–0.00027
Iron-nickel	200–700	33.28–116.47	0.00034–0.001
Iron-nickel-chromium	520–720	86.52–119.80	0.00016–0.00072
Copper-manganese-nickel	249–270	41.43–44.93	$0.000025–3 \times 10^{-5}$

Joule's Law. The flow of current through a resistance is always accompanied by heat. Joule established the experimental fact that the amount of energy W dissipated in a metallic wire is proportional to the product of the square of the current I and the time t during which that current flows. Moreover, the coefficient of proportionality between W and I^2t is the resistance R of the circuit, or

$$W = RI^2t \tag{16.4}$$

This gives a definition of resistance from an energy viewpoint. Resistance is that characteristic of a circuit which accounts for the existence of heat. An electric circuit is devoid of resistance when the flow of current is not accompanied by loss of energy through heat. Some metals at a temperature 0° abs. offer practically no resistance to the flow of electricity.

According to Ohm's law, $E = RI$. Therefore Eq. (16.4) becomes

$$W = (RI)It = EIt \tag{16.4'}$$

which is the general expression for the energy supplied by a source of emf to a circuit obeying Ohm's law.

Resistance as an Intrinsic Property of Metals. Although resistance is computed and defined through the current and voltage relations, or through the energy dissipated in an electric circuit, it must not be inferred that it is a function of either E, I, or W. Resistance is independent of all. It is a function of the material of the conductor and its physical dimensions. Experiment shows that the resistance of a straight conductor having a constant length h in the direction of current flow and a constant cross-sectional area A perpendicular to the flow is

$$R = \rho \frac{h}{A} \tag{16.5a}$$

whence

$$G = \frac{1}{R} = \frac{1}{\rho}\frac{A}{h} = \gamma \frac{A}{h} \tag{16.5b}$$

where ρ is known as the *resistivity*, and $\gamma = 1/\rho$ is called the *conductivity* of the metal. ρ and γ are defined as the *resistance* and *conductance*, respectively, of a cube having sides of unit length (a centimeter cube* or an inch cube).

The resistance of a conductor may be computed by the use of Eq. (16.5) only if that conductor satisfies the following conditions:

1. Its length in the direction of current flow must be constant.
2. Its cross section must be uniform and constant.

When these two conditions are not satisfied, there exist two alternative methods of determining resistance: experimentally, through the use of Ohm's or Joule's law, and mathematically, through the use of calculus. The latter method is limited in its scope to conductors having shapes that can be expressed by simple mathematical relations.

For engineering work, resistivity is usually expressed as the resistance of a circular mil-foot.† For rectangular conductors, the square mil-foot is preferable. For scientific work, a centimeter cube is one of the units used, but for good conductors this gives values inconveniently small. Circular-mil-foot resistivities for a number of materials are given in Tables 16.1 and 16.2.

6. Inductance *L*. Inductance (similar in effect to mechanical inertia) is that characteristic of an electric circuit by virtue of which a sudden increase or decrease of the current is checked. Viewed from the standpoint of energy, inductance is that property which causes an electric circuit to store up energy while the current increases and to deliver energy while the current decreases.

*In defining resistivity and conductivity, distinction must be made between centimeter cube and cubic centimeter. The former specifies not only the volume but the dimensions as well. The latter gives the volume irrespective of dimensions. Resistivity is the resistance of a *centimeter cube* and not of a cubic centimeter. The resistance of a cubic centimeter is an indefinite quantity.

†A circular mil is the area of a circle 1/1000 inch in diameter. This is a convenient unit for measuring the areas of round wires because the area, expressed in circular mils, is equal to the square of the diameter. The area of 1 circular mil, in square mils, is $(1)^2\pi/4 = \pi/4$. The area of a circle of diameter D mils, in square mils, is $D^2\pi/4$. Hence if $\pi/4$ is used as the unit of area, the area of the circle becomes D^2 circular mils. A conductor of 1 circular mil area and of 1 ft length is 1 circular mil-ft. 1 mil = 0.001 in.

The Magnetic Field Accompanying a Current and the Nature of Inductance. Experiment shows that a circuit carrying an electric current is encircled by a magnetic field in a plane perpendicular to the direction of current flow. The magnetic field may be imagined to consist of lines or tubes of force (Fig. 16.2) which form closed paths around the current. Since current is a directed quantity, these tubes of force (Fig. 16.2) will assume reverse directions when they surround oppositely flowing currents. It is assumed that the direction of current flow and that of the magnetic field correspond respectively to the direction of the progress of a right-handed screw and that of its rotation. Thus imagine a screw to be turned clockwise so that it will progress away from the observer. A counterclockwise rotation makes a screw progress toward the observer. Similarly, if a current is directed into the plane of the paper (see X, Fig. 16.2) then the flux surrounding it is clockwise, and vice versa. These facts are illustrated in Fig. 16.2, which shows cross sections of two conductors carrying oppositely directed currents. Current flowing away from the reader is shown by X, while current flowing toward the reader is indicated by a dot. Circles representing the lines of force are drawn around the current. The direction of these lines of force follows the convention stated above. The same relation holds in the relative direction of the flux through a solenoid and its surrounding exciting current.

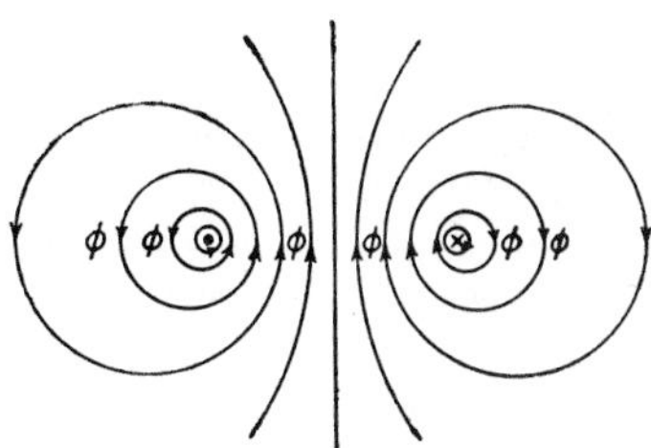

FIGURE 16.2

If the magnetic permeability of the circuit and that of the ambient medium are constant (i.e., if the phenomenon of magnetic saturation does not exist), then the flux linking with the circuit (Fig. 16.2) is proportional to the current, or

$$\phi = \frac{L}{N} i \tag{16.6}$$

where ϕ = magnetic flux linking with the circuit

i = current in the circuit

L/N = coefficient of proportionality between ϕ and i (the value of L/N depends on the physical dimensions of the circuit and the units used)

Experiment further shows that if the flux linking a circuit be changed, then an emf e_L is induced in the circuit which is proportional to the time rate of change of flux. In other words,

$$e_L = -N \frac{d\phi}{dt} \tag{16.7}$$

where N is the number of turns in the circuit which are completely linked by the flux ϕ.

There are, therefore, these two experimental facts:

1. In media of constant permeability, a change in the current produces a proportional change in the flux encircling it [Eq. (16.6)].
2. A change in the flux linking with a circuit induces an emf in that circuit [Eq. (16.7)].

The inevitable conclusion from these two premises is that a change in the current flowing in a circuit produces an emf in that circuit. Experiment shows that such is the case. Mathematically, this deduction may be easily verified by substituting Eq. (16.6) in (16.7). This gives

$$e_L = -L\frac{di}{dt} \tag{16.8}$$

L is known as the *inductance, self-inductance*, or *coefficient of self-induction* of the circuit. It is the coefficient of proportionality between the voltage e_L induced in a circuit and the time rate of change of current in it. The symbol for inductance is a helix (see Fig. 16.3).

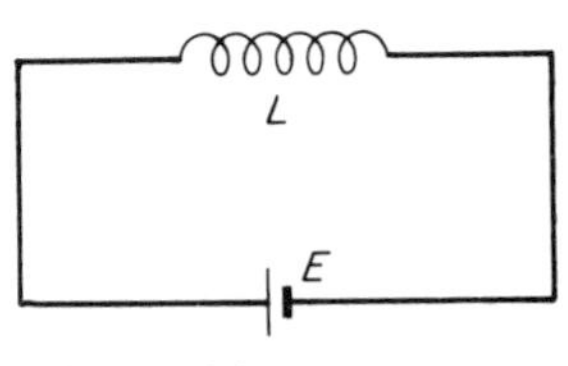

FIGURE 16.3

The negative sign in Eq. (16.8) indicates that the voltage induced in a circuit, due to a change in current, is such as to oppose the change. Thus, if the current increases, a voltage is induced which tends to check that increase, and vice versa. Lenz was the first to observe this fact. Its validity can be easily seen from the following reasoning.

Suppose that a continuous emf is suddenly impressed on a series circuit having resistance and inductance. The current, heretofore zero, will in time assume a certain definite steady value I. In the interim, the current must increase from zero to I. If each successive increase induces a voltage which tends to increase the current further, then, instead of assuming a finite value I, the current will ultimately become infinite—and this is contrary to physical facts.

Inductance Defined in Terms of Flux Linkages. Solving Eq. (16.6) for L,

$$L = \frac{N\phi}{i} \qquad \left(\text{or } N\frac{d\phi}{di}\right) \tag{16.9}$$

Equation (16.9) states that the inductance of a circuit is equal to the flux per unit current which links with the turns of N of the circuit. The product ($N\phi$) is often called the *flux linkages*. Hence inductance may be defined as the flux linkages per unit current.

The Limitations of the Concept of Inductance. In defining inductance by Eq. (16.8), certain restrictions were imposed. These are given below:

1. The concept of inductance applies only to conductors made of materials of constant permeability and surrounded by media of constant permeability. One cannot, for example, speak of the inductance of an iron conductor, nor can one refer to the inductance of a copper wire wound around an iron core unless it is agreed that the term is to be used only where the permeability is constant. This is due to the fact that Eq. (16.6) must hold true.

2. The emf induced in a circuit by virtue of its inductance is invariably due to the change of current carried by that circuit itself and is not due to any other cause. Thus the emf induced in the armature winding of a generator due to its rotation in a magnetic field is not an emf of self-induction. This is why the term *self-inductance* is preferable. It is more descriptive of the phenomenon.

3. Self-inductance disappears if, by any means whatever, the magnetic field due to a current is eliminated. Under these conditions, a change in current does not produce a corresponding change in the flux because ϕ is nonexistent. Although this is not fully realizable, an approach to it may be had if the circuit is wound noninductively.

Inductance as a Storehouse of Electromagnetic Energy. From an energy viewpoint, *inductance* may be defined as that property of an electric circuit by virtue of which energy is stored in a circuit while the current increases and is given up while the current decreases. This statement will be amplified by describing an experiment and actually computing the stored energy.

Figure 16.4*a* shows a circuit with a continuous emf source E, a variable resis-

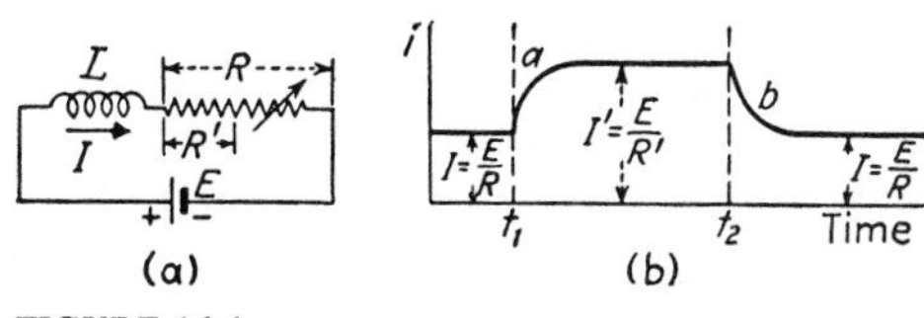

FIGURE 16.4

tance R which may be reduced to R', and a constant inductance L. Let the resistance be maintained at a value R, and let steady conditions be established. The steady current, according to Ohm's law, is (see Fig. 16.4*b*) $I = E/R$. Next let the resistance suddenly be reduced to R' at the instant t_1. The final steady current is (see Fig. 16.4*b*) $I' = E/R'$.

The current does not increase instantly from I to I' but rather rises gradually, following curve *a* (Fig. 16.4*b*). During the time of increase, a voltage of self-induction exists of such a direction as to oppose the impressed voltage. In other words, one part of the impressed voltage is devoted to counterbalance the emf of self-induction and the other part is available to send current through the circuit. That part of the impressed voltage which overcomes e_L is

$$e_L' = -e_L = +L\frac{di}{dt} \tag{16.10}$$

If the instantaneous value of the current is i, then the energy supplied by the emf source in time dt at an emf e_L' is

$$dW = e_L' i\,dt = Li\,di \tag{16.11}$$

and the total energy supplied by the source and stored in the circuit by virtue of its inductance is

$$W = \int_I^{I'} Li\,di = \tfrac{1}{2}L(I'^2 - I^2) \tag{16.12}$$

It can be shown similarly that during the decrease of the current (see curve *b*, Fig. 16.4*b*) from I' to I, the magnitude of the energy returned from the circuit to the continuous emf source is also expressed by Eq. (16.12).*

*The expression for returned energy is the negative of Eq. (16.12). The negative sign means that the emf source is receiving rather than supplying energy.

Inductance as an Intrinsic Property of a Circuit. The reader would ask whether L could possibly be computed from the dimensions of a circuit. The answer is yes only in very simple and special cases. It is outside the scope of this text to deal with this phase of inductance. It will therefore be assumed that the inductance of a circuit is either known or can be determined experimentally.

7. Capacitance *C*. When a dielectric separates two conductors, there results a *capacitor* or *condenser*. The name *condenser* or *capacitor* generally signifies a device to store electricity and which is made up of parallel metallic sheets with insulating sheets interleaved between them (see also Secs. 48 to 52). The dielectric may be air or any other nonconducting material. The conductors which the dielectric separates may be two metallic plates, two wires, a wire and the earth, metallic spheres, or a combination of any two or more of the above-mentioned conductors. The symbol for capacitance is two parallel lines (see d and f, Fig. 16.5a). For simplicity of treatment, it will be assumed that the dielectric separating the plates is perfect (for example, vacuum or air) and that the emf is kept low enough to eliminate corona and other disturbing influences.

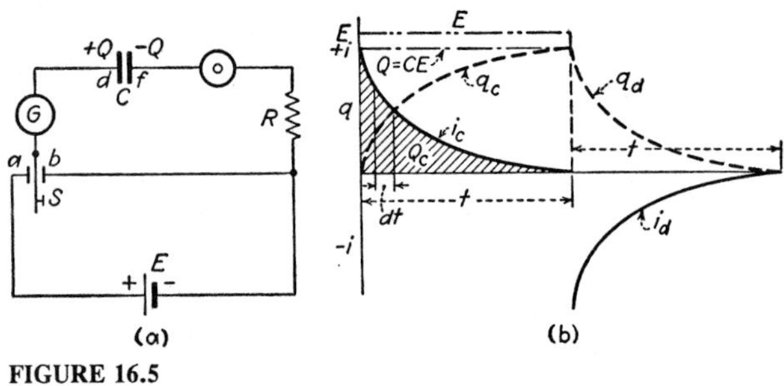

FIGURE 16.5

Relation Between the Charge and the Charging Current. Figure 16.5a shows a circuit consisting of a resistance R, a parallel-plate capacitor C, a ballistic galvanometer G, an oscillograph O, a source of continuous emf E, and a double-throw switch S. Upon moving S to point a, the emf E is impressed on the capacitor C. The ballistic galvanometer shows a deflection indicating that a quantity of electricity Q has passed from the source to the capacitor. This charge of electricity is nothing but a momentary current which continues to flow for a short time and then ceases altogether as shown by the curve i_c (Fig. 16.5b).

Electric current may be regarded as consisting of a transfer of electricity along a conductor. Consequently, the amount (or quantity) of electricity which passes through any cross section of a conductor in time t is proportional to the magnitude of the current and the time of flow. Thus the readings of the ballistic galvanometer and the oscillograph may be coordinated through the following equations:

$$Q = \int_0^t i\,dt \tag{16.13}$$

or

$$i = \frac{dq}{dt} \tag{16.14}$$

Equation (16.13) states that the total charge acquired by the condenser is the area under the curve i_c (Fig. 16.5*b*). Equation (16.14) states that the magnitude of the current flowing in the circuit (Fig. 16.5*a*) is at every instant equal to the rate at which the quantity of electricity is carried along the wires. This is graphically the slope of the curve q_c and is represented by the curve i_c (Fig. 16.5*b*).

The Charging EMF Is Equal to the Potential across the Plates of a Charged Capacitor. Let the switch S be open after a time t (Fig. 16.5*b*) when the current has become zero. No spark is observed when the switch is open, because no current has been interrupted. Upon examining the plates d and f, it will be found that the plate d has a positive charge $+Q$ and f has an equal negative charge $-Q$. Moreover, measurement of the potential difference E_c between the plates shows that the voltage is actually equal to the impressed emf E, or

$$E = E_c \tag{16.15}$$

Under these conditions the capacitor is said to be fully charged.

Equality of the Charge and Discharge. Let the switch S now be moved to the point b. The capacitor begins to discharge through the resistance R. The galvanometer shows a deflection equal in magnitude to that previously indicated, but opposite in direction. This indicates that the quantity of electricity Q_d obtained during the discharge is equal to the quantity Q_c supplied during the charge, or

$$Q_c = Q_d \tag{16.16}$$

A capacitor may therefore be defined as a device for storing electricity—not in the form of electromagnetic energy, as in the case of inductance, but as electrostatic energy (the electric charge).

Relation Between the Charging EMF and the Charge. Experiment shows that the charge acquired by a capacitor whose conducting terminals are separated by a perfect dielectric is proportional to the emf across its plates. This proportionality between E_c and Q may be expressed in the following two ways:

$$C = \frac{Q}{E_c} \tag{16.17a}$$

or

$$S = \frac{E_c}{Q} \tag{16.17b}$$

C is known as the *capacitance* of a condenser (or capacitor), and $S = 1/C$ is known as its *elastance*. Equations (16.17) are similar to Eqs. (16.2) and express what may be termed *Ohm's law for the electrostatic circuit*. Equation (16.17*a*) gives a definition of capacitance. It is simply the ratio Q/E_c.

Capacitance as an Intrinsic Property of a Condenser. The capacitance C for perfect dielectrics is independent of both Q and E and is simply a function of the dimensions and material of the capacitor. Experiment shows that for capacitors made of plates having equal areas placed parallel to each other and separated by a homogeneous perfect dielectric of area A and thickness h, the capacitance C is*

$$C = Kk_e \frac{A}{h} \tag{16.18}$$

where K = a constant known as the *relative permittivity of the material*, which is the ratio of the permittivity k of a material to the permittivity k_e of vacuum, or $K = k/k_a$

k_a = the *permittivity of vacuum* and may be taken as 8.842×10^{-14} F/cm^3

A = area of one plate or of a cross section of the dielectric in square centimeters

h = distance between plates in centimeters

Capacitance Viewed from an Energy Standpoint. The energy supplied to a capacitor is

$$dW = ei\,dt = e\,dq \tag{16.19}$$

But by Eq. (16.17)

$$dq = C\,de \tag{16.20}$$

$$\therefore W = \int_0^{E_c} Ce\,de = ½CE_c^2 = ½CE^2 \tag{16.21}$$

whence

$$C = \frac{2W}{E^2} \tag{16.22}$$

Thus the capacitance of a condenser is a measure of its capacity to store electric energy at a given applied potential.

8. Types of Circuits. There are three general types of circuits: (1) series circuits, (2) parallel circuits, and (3) series-parallel circuits.

1. *Series Circuit:* A *series circuit* (Fig. 16.6) is one which has the same current I in all its parts. Thus Fig. 16.6*a* consists of four resistances connected in series across the same emf E. They are in series because one and the same current I flows through all the resistances. Similarly, Fig. 16.6*b* comprises four inductances connected in series. Figure 16.6*c* consists of four capacitances in series, while Fig. 16.6*d* has two resistances, two inductances, and two capacitances in series.

2. *Parallel Circuit:* A *parallel circuit* (Fig. 16.7) is one which has the same emf across all its parts. Thus the circuit in Fig. 16.7*a* has the same emf E across all the characteristics R_1, R_2, R_3, R_4. The currents in these characteristics i_1, i_2, i_3, and i_4, respectively, may or may not be the same. Figure 16.7*b* is a circuit consisting of four inductances in parallel, while Fig. 16.7*c* consists of four capacitances in parallel. Figure 16.7*d* consists of two resistances, two inductances, and two capacitances, all connected in parallel.

3. *Series-Parallel Circuit (or Network):* A *series-parallel circuit* (or *net-*

*Equation (16.18) is true under the following limitations: (1) it neglects fringing of the electrostatic flux at the edges of the plates, (2) it assumes a perfect contact between the plates and the dielectric, and (3) it assumes a nonanomalous perfect dielectric in which no hysteroviscosity or conduction effects can be observed.

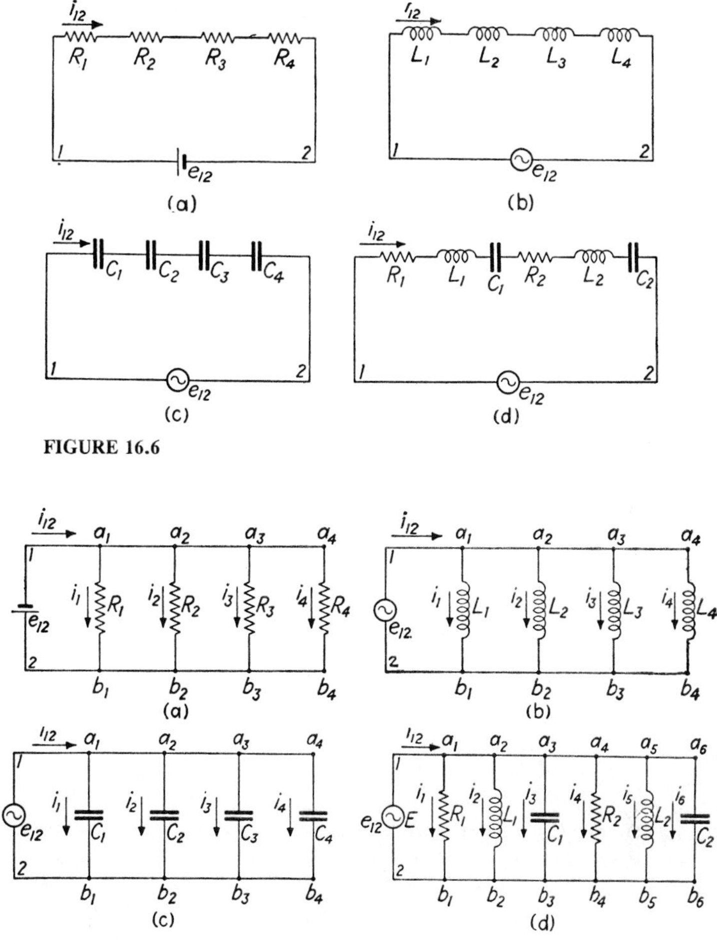

FIGURE 16.6

FIGURE 16.7

work) is a combination of circuit characteristics which includes both series and parallel (Fig. 16.8).

9. Kirchhoff's Laws

Branch. A *branch* is a part of a circuit which carries one current. A branch may consist of one characteristic or of several characteristics connected in series. Thus all the circuits in Fig. 16.6 have but one branch each. Figure 16.7*a* to *c* has four branches each: a_1b_1, a_2b_2, a_3b_3, and a_4b_4. Figure 16.7*d* has six branches: a_1b_1, a_2b_2, a_3b_3, a_4b_4, a_5b_5, and a_6b_6. Figure 16.8 has two branches: a_1b_1, and a_2b_2.

Junction. A *junction* is a point at which two or more branches meet. Thus Fig. 16.7*a* to *c* has two junctions each: $(a_1a_2a_3a_4)$ and $(b_1b_2b_3b_4)$. Figure 16.8 has two

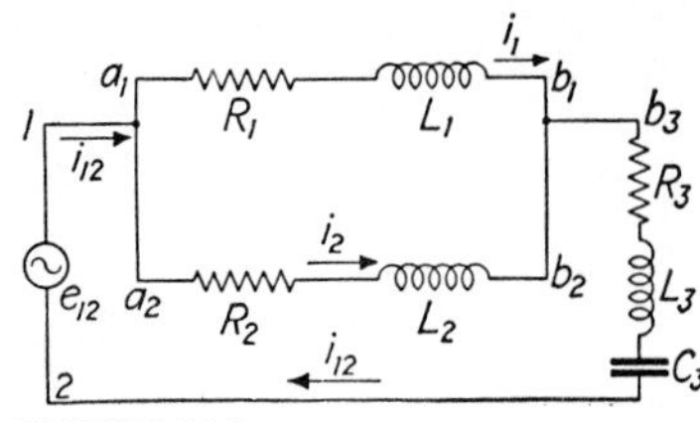

FIGURE 16.8

junctions: (a_1a_2) and (b_1b_2). Observe that any two or more points between which no circuit characteristic or emf exists are electrically identical and form one junction. Thus in Fig. 16.7 the points a_1, a_2, a_3, a_4 are electrically identical and form one junction a.

Loop. A *loop* is any closed path in a circuit. A loop may be *simple* or *compound*. A *simple loop* is one which encloses no branches. A compound loop is one enclosing branches. Thus Fig. 16.6*a* to *d* has one simple loop each. Figure 16.7*a* to *c* has the following simple loops: $1a_1b_12$, $a_1a_2b_2b_1$, $a_2a_3b_3b_2$, and $a_3a_4b_4b_3$. Figure 16.7*a* to *c* also has the following compound loops: $1a_2b_22$, $a_1a_2a_3b_3b_2b_1$, $a_1a_2a_3a_4b_4b_3b_2b_1$, $1a_1a_2a_3b_3b_2b_12$, $1a_1a_2a_3a_4b_4b_3b_2b_12$.

Potential Drop. The potential drop in any characteristic of a circuit is equal to the emf which is required to make a current flow in that characteristic. Thus if e_r, e_L, and e_c are the potential drops in a resistance, inductance, and capacitance, respectively (see Secs. 5, 6, and 7),

$$e_r = Ri \qquad (16.23a)$$

$$e_L = L\frac{di}{dt} \qquad (16.23b)$$

$$e_c = \frac{q}{C} = \frac{1}{C}\int i\,dt \qquad (16.23c)$$

EMF-Current Notations. The direction of the emf and current which it causes will be as follows: If E_{12} is an emf applied to the points 1 and 2 of a circuit, that emf will cause a current to flow from point 1 to 2 through the characteristics of that circuit. The direction of flow of the current is indicated by an arrow (see Figs. 16.6 to 16.8) pointing from point 1 to point 2 of the characteristics.

Kirchhoff's EMF Law. The sum of the potential drops around any loop is equal to the sum of the emf's in that loop, provided the notations are as in the preceding subsection. Any current which flows in a direction opposite to that specified in the preceding subsection causes a negative drop. Moreover, in any one loop, oppositely directed currents cause drops of opposite signs.

Kirchhoff's emf law for some of the loops in Figs. 16.6 to 16.8 is expressed in Eqs. (16.24) to (16.29):

Fig. 16.6*a*:

$$e_{12} = i_{12}R_1 + i_{12}R_2 + i_{12}R_3 + i_{12}R_4 = i_{12}(R_1 + R_2 + R_3 + R_4) \qquad (16.24)$$

Fig. 16.6*b*:

$$e_{12} = L_1\frac{di_{12}}{dt} + L_2\frac{di_{12}}{dt} + L_3\frac{di_{12}}{dt} + L_4\frac{di_{12}}{dt} \qquad (16.25a)$$

$$= \frac{di_{12}}{dt}(L_1 + L_2 + L_3 + L_4) \qquad (16.25b)$$

Fig. 16.6*c*:

$$e_{12} = \frac{q}{C_1} + \frac{q}{C_2} + \frac{q}{C_3} + \frac{q}{C_4} = q\left(\frac{1}{C_1} + \frac{1}{C_2} + \frac{1}{C_3} + \frac{1}{C_4}\right) \tag{16.26a}$$

$$= \left(\int i_{12}\,dt\right)\left(\frac{1}{C_1} + \frac{1}{C_2} + \frac{1}{C_3} + \frac{1}{C_4}\right) \tag{16.26b}$$

Fig. 16.6*d*:

$$e_{12} = i_{12}R_1 + L_1\left(\frac{di_{12}}{dt}\right) + \left(\int i_{12}\frac{dt}{C_1}\right) + i_{12}R_2 + L_2\left(\frac{di_{12}}{dt}\right) + \left(\int i_{12}\frac{dt}{C_2}\right) \tag{16.27a}$$

$$= (R_1 + R_2)i_{12} + (L_1 + L_2)\left(\frac{di_{12}}{dt}\right) + \left(\frac{1}{C_1} + \frac{1}{C_2}\right)\int i_{12}\,dt \tag{16.27b}$$

Fig. 16.7*a*:

Loop $1a_1b_12$: $$e_{12} = i_1R_1 \tag{16.28a}$$

Loop $a_1a_2b_2b_1a_1$: $$0 = i_1R_1 - i_2R_2 \tag{16.28b}$$

Loop $a_2a_3b_3b_2a_2$: $$0 = i_2R_2 - i_3R_3 \tag{16.28c}$$

Loop $a_3a_4b_4b_3a_3$: $$0 = i_3R_3 - i_4R_4 \tag{16.28d}$$

Fig. 16.8:

Loop $1a_1b_1b_22$:

$$e_{12} = i_1R_1 + L_1\left(\frac{di_1}{dt}\right) + R_3i_{12} + L_3\left(\frac{di_{12}}{dt}\right) + \int\frac{i_{12}\,dt}{C_3} \tag{16.29a}$$

Loop $a_1b_1b_2a_2$:

$$0 = i_1R_1 + L_1\left(\frac{di_1}{dt}\right) - i_2R_2 - L_2\left(\frac{di_2}{dt}\right) \tag{16.29b}$$

Kirchhoff's Current Law. The sum of currents entering a junction is equal to the sum of currents leaving that junction.

Note: Entrance to a junction is denoted by an arrow directed toward the junction; exit is indicated by an arrow directed away from that junction.

Kirchhoff's current law for the junction *a*, Figs. 16.7 and 16.8, is expressed in Eqs. (16.30).

Fig. 16.7*a* to *c*: $$i_{12} = i_1 + i_2 + i_3 + i_4 \tag{16.30a}$$

Fig. 16.7*d*: $$i_{12} = i_1 + i_2 + i_3 + i_4 + i_5 + i_6 \tag{16.30b}$$

Fig. 16.8: $$i_{12} = i_1 + i_2 \tag{16.30c}$$

10. Use of Kirchhoff's Laws. Kirchhoff's emf and current laws are used to determine the current flowing in each branch of a given circuit when the emf and characteristics of that circuit are specified. It can be shown that in order to solve for the currents, the number of Kirchhoff's emf equations must be equal to the number of simple loops in a circuit, while the number of Kirchhoff's current equations must be equal to the number of junctions less one.

DIRECT-CURRENT CIRCUITS

11. Circuit with One Resistance. Consider a circuit (Fig. 16.1) in which direct voltage E is impressed on a resistance R. In order to determine the current I, use is made of Ohm's law. Thus $E = IR$, and

$$I = \frac{E}{R} \tag{16.31}$$

12. Circuit with One Inductance. The circuit in Fig. 16.3 consists of a direct voltage E impressed on an inductance L. Equation (16.10) is used to determine the current i:

$$E = L\frac{di}{dt} \tag{16.32}$$

Solving Eq. (16.32) for i,

$$i = \frac{1}{L}\int E\,dt + k \tag{16.33a}$$

$$i = \frac{E}{L}t + k \tag{16.33b}$$

Equation (16.33b) shows that the current i increases indefinitely when a direct voltage E is impressed on an inductance. Indeed, an inductance acts virtually as a short circuit for a constant direct emf.

13. Circuit with One Capacitance. The circuit in Fig. 16.9 consists of a direct voltage E impressed on a capacitance C. Here, by Eq. (16.17a),

$$q = CE \tag{16.34a}$$

$$i = \frac{dq}{dt} = \frac{C\,dE}{dt} \tag{16.34b}$$

$$i = 0 \tag{16.34c}$$

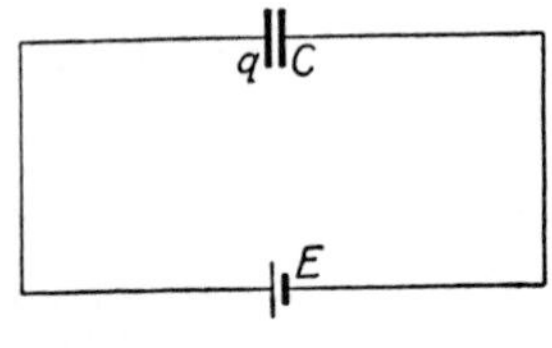

FIGURE 16.9

Equation (16.34c) states that the *steady* current in Fig. 16.9 is zero. In other words, a condenser acts virtually as an open circuit when a direct emf is impressed on it.

Since an inductance acts as a short circuit and a condenser acts as an open circuit whenever a direct emf is impressed on them, they will be considered as nonadmissible characteristics in this section. The section will be devoted, therefore, to series, parallel, and series-parallel circuits of resistances only.

14. Series Circuit with Several Resistances. In the circuit in Fig. 16.10*a*, there are n resistances $R_1, R_2, \ldots, R_n$ connected in series across a direct voltage E. By Kirchhoff's law,

$$E = IR_1 + IR_2 + \cdots + IR_n \tag{16.35a}$$

$$= I(R_1 + R_2 + \cdots + R_n) \tag{16.35b}$$

$$= IR \tag{16.35c}$$

where

$$R = R_1 + R_2 + \cdots + R_n \tag{16.35d}$$

Equations (16.35*c*) and (16.35*d*) state that the series circuit (Fig. 16.10*a*) may be replaced by a simple circuit comprising one resistance R (Fig. 16.10*b*) whose value is equal to the sum of the series resistances.

15. Parallel Circuits with Several Resistances. In the circuit in Fig. 16.11*a*, there are n resistances in parallel across the same voltage E. By Ohm's law,

$$I_1 = \frac{E}{R_1} \quad I_2 = \frac{E}{R_2} \quad I_n = \frac{E}{R_n} \tag{16.36}$$

Applying Kirchhoff's current law to the junction a,

$$I = I_1 + I_2 + \cdots + I_n \tag{16.37a}$$

$$= E\frac{1}{R_1} + \frac{1}{R_2} + \cdots + \frac{1}{R_n} \tag{16.37b}$$

$$= \frac{E}{R} \tag{16.37c}$$

where

$$\frac{1}{R} = \frac{1}{R_1} + \frac{1}{R_2} + \cdots + \frac{1}{R_n} \tag{16.37d}$$

Equations (16.36) give the currents in the various branches of the parallel circuit. Equation (16.37*a*) states that the total current I supplied by the emf source is the sum of these currents. Equations (16.37*c*) and (16.37*d*) state that the parallel circuit (Fig. 16.11*a*) may be replaced by a simple circuit (Fig. 16.11*b*) comprising one resistance whose value is expressed by Eq. (16.37*d*).

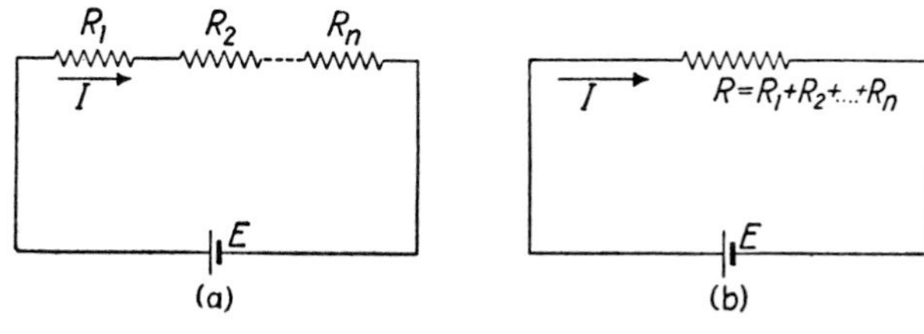

FIGURE 16.10

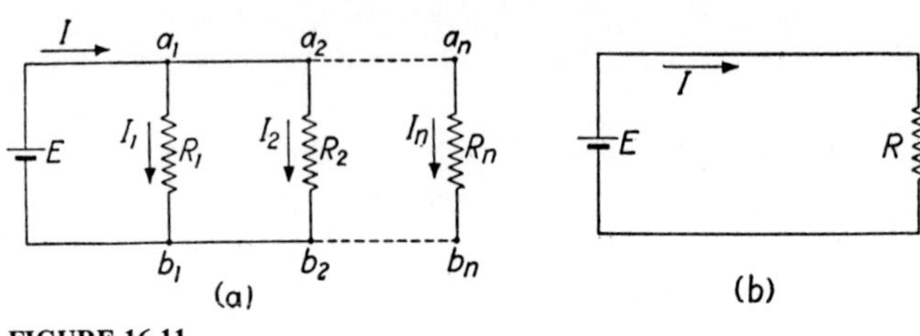

FIGURE 16.11

16. Series-Parallel Circuits (Networks). The determination of the currents in a network of resistances (Fig. 16.12) is accomplished by the application of Kirchhoff's laws. Thus, applying Kirchhoff's emf law to the loops $1a_1b_12$ and $a_1b_1b_2a_2$,

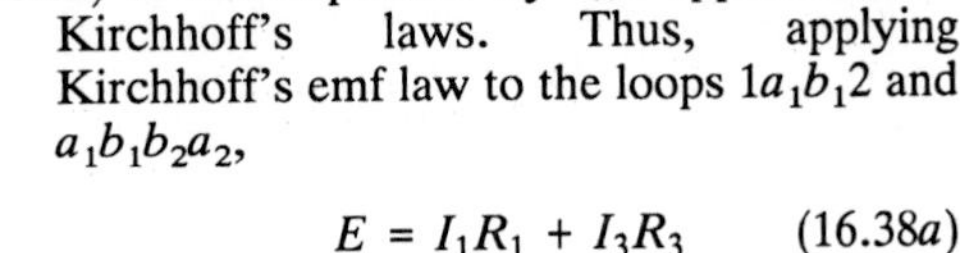

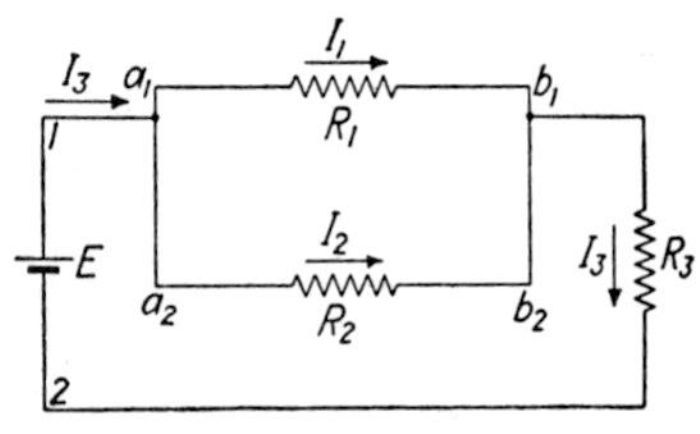

FIGURE 16.12

$$E = I_1R_1 + I_3R_3 \qquad (16.38a)$$

$$0 = I_1R_1 - I_2R_2 \qquad (16.38b)$$

Applying Kirchhoff's current law to the junction a,

$$I_3 = I_1 + I_2 \qquad (16.39)$$

Equations (16.38) and (16.39) give three independent relations among the currents I_1, I_2, and I_3 from which these currents can be determined. Thus write these equations as follows:

$$R_1I_1 + 0I_2 + R_3I_3 = E \qquad (16.40a)$$

$$R_1I_1 - R_2I_2 + 0I_3 = 0 \qquad (16.40b)$$

$$I_1 + I_2 - I_3 = 0 \qquad (16.40c)$$

Using determinants (see Sec. 13), let

$$\Delta = \begin{vmatrix} R_1 & 0 & R_3 \\ R_1 & -R_2 & 0 \\ 1 & 1 & -1 \end{vmatrix} = (R_1R_2 + R_1R_3 + R_2R_3 \qquad (16.41)$$

$$\therefore I_1 = \frac{\begin{vmatrix} E & 0 & R_3 \\ 0 & -R_2 & 0 \\ 0 & 1 & -1 \end{vmatrix}}{\Delta} = \frac{R_2E}{R_1R_2 + R_1R_2 + R_2R_3} \qquad (16.42a)$$

$$I_2 = \frac{\begin{vmatrix} R_1 & E & R_2 \\ R_1 & 0 & 0 \\ 1 & 0 & -1 \end{vmatrix}}{\Delta} = \frac{-ER_1}{R_1R_2 + R_1R_3 + R_2R_3} \qquad (16.42b)$$

$$I_3 = \frac{\begin{vmatrix} R_1 & 0 & E \\ R_1 & -R_2 & 0 \\ 1 & 1 & 0 \end{vmatrix}}{\Delta} = \frac{E(R_2 - R_1)}{R_1R_2 + R_1R_3 + R_2R_3} \tag{16.42c}$$

17. Power and Energy in Direct-Current Circuits. Power in a dc circuit is defined as follows:

$$P = EI \tag{16.43}$$

where E is the emf impressed on the circuit, and I is the total current furnished by that emf. If E is expressed in volts and I in amperes, P is in watts.

The power in any branch k of the circuit is defined through Joule's law as follows:

$$P_k = I_k^2 R_k \tag{16.44}$$

where P_k is the power in branch k, I_k is the current flowing through branch k, and, R_k is the resistance of branch k.

The energy supplied to a dc circuit in an interval of time of length t is

$$W = Pt = EIt \tag{16.45}$$

If P is in watts and t is in seconds, W is expressed in watt-seconds or joules. The energy supplied to each branch in an interval t is

$$W_k = P_k t = I_k^2 R_k t \tag{16.46}$$

But, by the principle of conservation of energy, the total energy supplied must be equal to the sum of the energies supplied to all the branches. Hence, if a circuit has n branches,

$$W = EIt = \sum_{k=1}^{k=n} I_k^2 R_k t \tag{16.47a}$$

or

$$EI = \sum_{k=1}^{k=n} I_k^2 R_k \tag{16.47b}$$

Equation (16.47*a*) is the law of conservation of *energy* in a dc circuit, and Eq. (16.47*b*) is the law of conservation of *power* in a dc circuit.

SINGLE-PHASE ALTERNATING-CURRENT CIRCUITS

18. Sine Functions

Definition. A sine function of x is of the form

$$e = E_m \sin (x + \alpha) \tag{16.48}$$

Graph. A graph of the sine function given in Eq. (16.48) is shown in Fig. 16.13, wherein E_m is the *maximum value* or *amplitude* of the sine function, e_0 is the

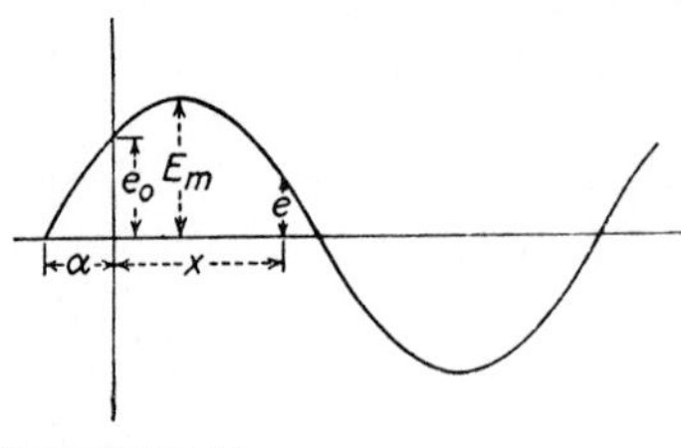

FIGURE 16.13

initial value of the sine function or its value corresponding to $x = 0$, and α is the initial angle of the sine function.

Periodicity. Sine functions are periodic and have a period of 2π. In other words, a sine function repeats itself every 2π radians, or

$$e = E_m \sin (x + \alpha) = E_m \sin (x + \alpha + 2n\pi) \tag{16.49}$$

where n is any integer.

Cycle. The set of values which a sine function assumes in the course of a period constitute a *cycle*. The term *cycle* is often used in ac circuits and is denoted by the symbol ~.

19. Average and Effective Values. Frequent reference is made in the text to the average and effective values of a sine function. These are now defined.

Average Value E_a. In general, the average value of any function between two limits θ and β is the value of the mean ordinate between these limits. The mathematical expression for this statement is

$$E_a = \left(\frac{1}{\theta - \beta}\right) \int_\beta^\theta e\, dx \tag{16.50}$$

Applying Eq. (16.50) to a sine function, Eq. (16.48),

$$E_a = \left(\frac{1}{\theta - \beta}\right) \int_\beta^\theta E_m \sin (x + \alpha)\, dx \tag{16.51a}$$

$$= \left(\frac{E_m}{\theta - \beta}\right) [\cos (\beta + \alpha) - \cos (\theta + \alpha)] \tag{16.51b}$$

It is evident from an inspection of Eq. (16.51) and Fig. 16.14 that the average value of a sine wave varies depending on the limits θ and β. For many purposes, however, the sine wave is integrated from its zero value to an immediately succeeding π value. Thus in Fig. 16.14 the limits are taken from $x = -\alpha$ to $x = (\pi - \alpha)$. Substituting these limits in Eq. (16.51) and simplifying,

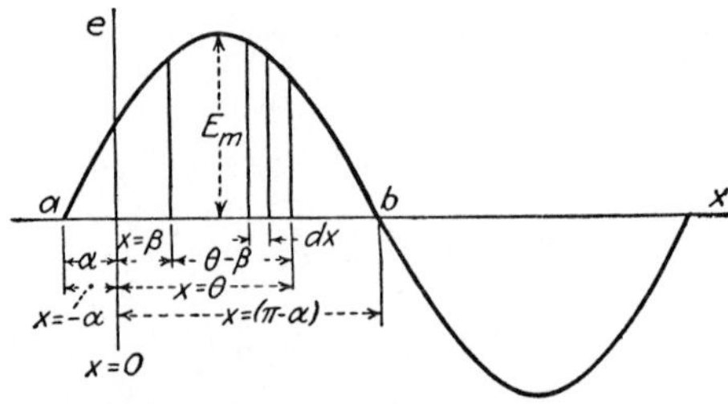

FIGURE 16.14

$$E_a = \frac{E_m}{\pi}(\cos 0 - \cos \pi) = 2\frac{E_m}{\pi} \tag{16.52}$$

Effective Value E. The effective value E of a periodic function between the limits β and θ is the square root of the area under the squared curve divided by the base.

Thus let the curve in Fig. 16.15 be obtained by squaring the ordinates of the curve in Fig. 16.14. The effective value of the sine wave between the limits θ and β is the square root of the quotient obtained by dividing the value of the shaded area by the base $\theta - \beta$. The mathematical expression for this operation is

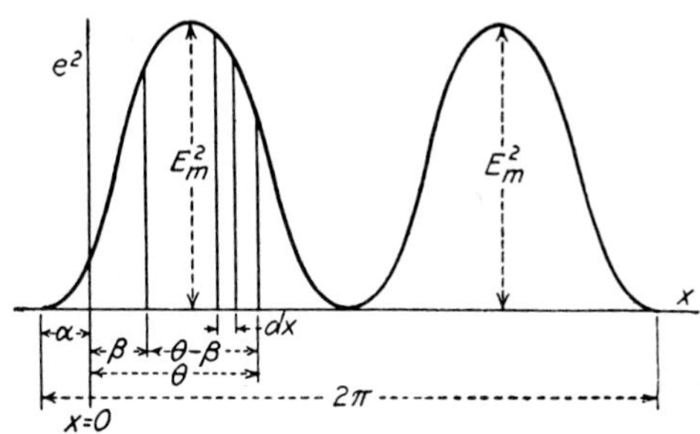

FIGURE 16.15

$$E = \left(\int_\beta^\theta \frac{e^2\,dx}{\theta - \beta}\right)^{1/2} \tag{16.53}$$

Applying Eq. (16.53) to a sine function having the form of Eq. (16.48),

$$E = \left(\int_\beta^\theta \frac{E_m^2 \sin^2(x + \alpha)\,dx}{\theta - \beta}\right)^{1/2} \tag{16.54}$$

Here again the effective value of a sine function will vary depending on the limits β and θ. For most purposes, however, the limits are taken as 0 and 2π. Substituting these limits in Eq. (16.54),

$$E = E_m\left[\frac{1}{2\pi}\int_0^{2\pi} \sin^2(x + \alpha)\,dx\right]^{1/2} = \frac{E_m}{\sqrt{2}} \tag{16.55}$$

Amplitude and Form Factors. The two values defined above have given rise to two factors known as the *amplitude* and *form factors*.

The amplitude factor k_a is the ratio of the maximum to the effective value of a periodic wave, or

$$k_a = \frac{E_m}{E} \tag{16.56}$$

For sine waves,

$$k_a = \sqrt{2} \tag{16.57}$$

The form factor k_f is the ratio of the effective value to the average value of a periodic wave, or

$$k_f = \frac{E}{E_a} \tag{16.58}$$

For sine waves,

$$k_f = \frac{\pi}{2\sqrt{2}} \cong 1.11 \tag{16.59}$$

20. Lead, Lag, and Phase. *Lead* means to be ahead, and *lag* means to be behind. Graphically (see curve a, Fig. 16.16), a sine wave leads the origin if its zero value (which is nearest to the origin) occurs *before* the point $x = 0$. A curve lags the origin if its zero value (which is nearest to the origin) occurs *after* the point $x = 0$ (see curve c, Fig. 16.16). Finally, a sine function is in phase with the origin if its zero value *coincides* with $x = 0$ (see curve b, Fig. 16.16). The *angle of lead or lag* (the *phase angle* or the *initial angle*) is measured from the point $x = 0$ to the nearest zero value of the sine wave. A positive angle is measured to the left, and a negative angle to the right.

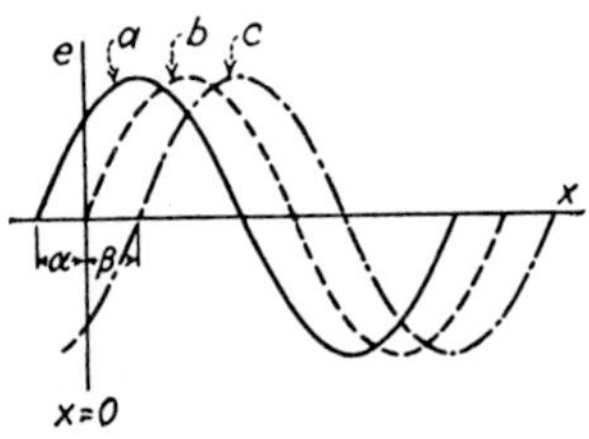

FIGURE 16.16

The terms *lead* and *lag* are relative. Thus if curve a in Fig. 16.16 leads the origin, then the origin lags curve a. These two statements have the same meaning and are interchangeable. Again, if curve a leads the origin by an angle α and curve c lags it by $-\beta$, then curve a leads c by an angle $(\alpha + \beta)$. When it is desired to give the phase angle without any reference to lead or lag, the term *out of phase* is used. Thus in Fig. 16.16 the curves a and b are $\alpha°$ out of phase and the curves a and c are $(\alpha + \beta)°$ out of phase.

It is very convenient to be able to state, by a mere inspection of the function and without actually plotting it, whether a sine function leads, lags, or is in phase with the origin. The value of the initial angle α [Eq. (16.48)] determines whether the nearest zero value of a sine function occurs before or after the point $x = 0$. In general, a sine function leads the origin (curve a, Fig. 16.16) if the initial angle is such that $0 < \alpha < \pi$ and lags (curve c, Fig. 16.16) if the initial angle is such that $\pi < \alpha < 2\pi$. If $\alpha = 0$ or $\alpha = 2n\pi$, the function is in phase with the origin. The use of either term, *lead* or *lag*, is justifiable when $\alpha = (2n + 1)\pi$.

21. Sine Functions and Revolving Vectors. Figure 16.17*a* shows a vector of length E_m and initial argument α. Let this vector revolve at a uniform angular velocity ω such that in time t it moves through an angle x. Then by definition

$$\omega = \frac{x}{t} \qquad \text{or} \qquad x = \omega t \tag{16.60}$$

If the period (time required to make one complete revolution, or 2π radians) is denoted by T, then

$$\omega = \frac{2\pi}{T} \tag{16.61}$$

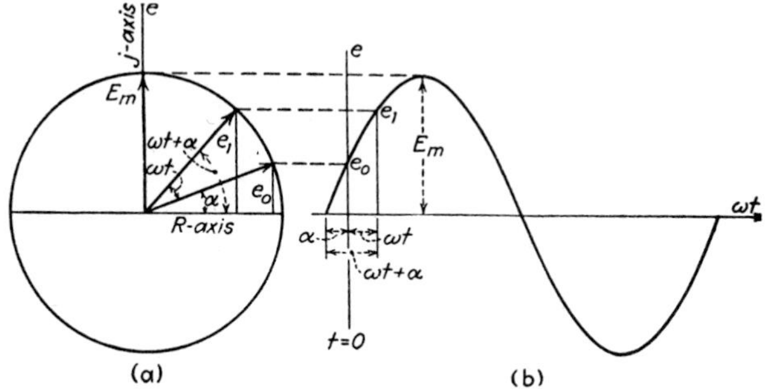

FIGURE 16.17

Consequently,*

$$x = \omega t = \frac{2\pi t}{T} = 2\pi f t \qquad (16.62)$$

where

$$f = \frac{1}{T} \qquad (16.63)$$

The quantity f is known as the *frequency of the revolving vector* and is the number of revolutions or cycles performed per second.

Now consider the projections of the revolving vector on the vertical axis. From the geometry of Fig. 16.17*a*, these are

$$e = E_m \sin (x + \alpha) = E_m \sin (\omega t + \alpha) = E_m \sin (2\pi f t + \alpha) \qquad (16.64)$$

But Eq. (16.64) is simply a sine function which may be represented by the sine wave (Fig. 16.17*b*). It can be concluded, therefore, that a revolving vector of modulus E_m and initial argument α moving at a uniform angular velocity ω has projections along the vertical axis which correspond exactly with the instantaneous values e of a sine function of amplitude E_m, argument ωt, and initial angle α.

22. The Use of Stationary Vectors in Alternating-Current Circuits. Let two sine waves (Fig. 16.18) be generated by two revolving vectors having the same angular velocity ω. Let the initial angles of the first and second be α and $-\beta$, respectively. It is evident that as long as the vectors move at the same velocity, the angle between them will always be $(\alpha + \beta)$. Now in many applications of sine functions to ac circuits this phase angle is sought. Since this angle is the same no matter whether the vectors revolve or are stationary, and since stationary vectors lend themselves easily to analytical treatments through complex numbers, the revolving vector is replaced by a stationary vector. Again, under steady conditions, many of the phenomena that occur in alternating electric circuits are such that what is true of any one instant is also true of all other instants. This furnishes a further justification for replacing revolving vectors by stationary ones which may be drawn for any arbitrary instant.

In representing sine functions by stationary vectors, the following facts should be remembered:

*Note that x may be expressed in radians or degrees depending on whether π is taken as 3.1416 rad or 180°.

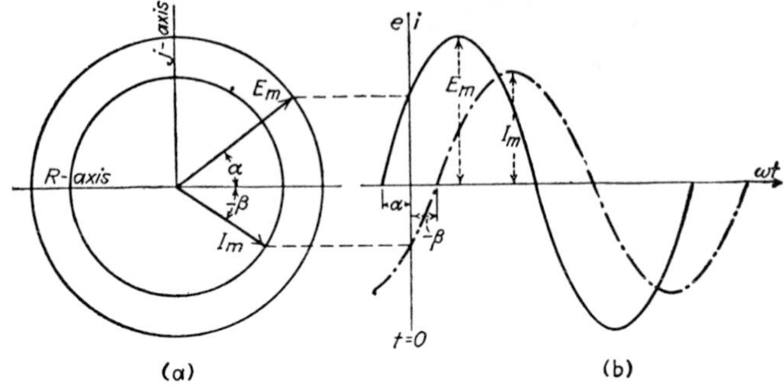

FIGURE 16.18

1. Stationary vectors may be used if, and only if, all the sine functions entering into the problem are of the same frequency f.

2. The phenomena must be such that what is true of one instant is also true of all instants (steady-state conditions must prevail).

3. The vector is so plotted that it makes an angle, with the reference axis, equal to the initial angle of the sine function. This angle is the argument of the stationary vector. It is, hence, the direction angle between the stationary vector and the reference axis. Thus a vector leads the reference axis if it has an argument α such that $0 < \alpha < \pi$. On the other hand, a vector lags the reference axis if its argument is such that $-\pi < \alpha < 0$. A vector is in phase with the R axis if $\alpha = 0$.

4. The angle that the vector makes with the reference axis is measured counterclockwise from the R axis to the vector. If the angle is measured clockwise from the R axis to the vector, it is considered a negative angle. Thus the angle β (Fig. 16.18a) is a negative angle.

5. The length of the vector is made proportional to the effective value E, instead of the amplitude E_m, of the sine function. This departure from the fundamental representation described above may be, at first, confusing. Its justification, however, for practical purposes becomes apparent when it is noted that meters read effective values rather than amplitudes, and most of the computations for electric circuits are based on effective values. Again, for sine waves, the relation between the effective value and the amplitude is definite [see Eq. (16.55)]. Hence a change of scale makes the same vector represent either E or E_m.

6. The vectors representing sine functions possess all the properties of vectors. Since all vectors used in this text are coplanar, complex numbers can be used to define their moduli and arguments.

In order to make these facts clear, the instantaneous expressions for the sine curves drawn in Fig. 16.18b, and the complex expressions for the stationary vectors representing them, are given below:*

$$e = E_m \sin (\omega t + \alpha) \tag{16.65a}$$

*Throughout this section, the symbol ϵ is used to represent the base of the Napierian or natural logarithms in order to avoid confusion between this term and instantaneous voltage, which is represented by the letter e.

$$i = I_m \sin(\omega t - \beta) \tag{16.65b}$$

$$\hat{E} = E\epsilon^{j\alpha} = \frac{E_m}{\sqrt{2}}\epsilon^{j\alpha} \tag{16.66a}$$

$$\hat{I} = I\alpha^{-j\beta} = \frac{I_m}{\sqrt{2}}\epsilon^{-j\beta} \tag{16.66b}$$

It should be noted that Eqs. (16.65) and (16.66) are reversible. In other words, given Eqs. (16.65), Eqs. (16.66) can easily be written, and vice versa.

23. Some Advantages of Vector Representation. The representation of sine functions by vectors possesses advantages which simplify the algebraic operations and render the physical phenomena clearer to the student. The demonstration of the latter aspect is reserved to succeeding pages. The following example shows how the addition and subtraction of sine functions are simplified through the use of vectors.

Let it be required to add the three sine waves

$$a = A_m \sin(\omega t + \alpha) \tag{16.67a}$$

$$b = B_m \sin(\omega t + \beta) \tag{16.67b}$$

$$c = C_m \sin(\omega t + \gamma) \tag{16.67c}$$

Vectorially these sine waves become

$$\hat{A} = A\epsilon^{j\alpha} = a' + ja'' \tag{16.68a}$$

$$\hat{B} = B\epsilon^{j\beta} = b' + jb'' \tag{16.68b}$$

$$\hat{C} = C\epsilon^{j\gamma} = c' + jc'' \tag{16.68c}$$

whence

$$\hat{D} = \hat{A} + \hat{B} + \hat{C} = (a' + b' + c') + j(a'' + b'' + c'') = d' + j'' = D\epsilon^{j\theta} \tag{16.69}$$

Then the sum of the three sine waves is a sine wave having an amplitude $D_m = D\sqrt{2}$ and an initial angle θ. The instantaneous value of this wave is

$$d = D_m \sin(\omega t + \theta) \tag{16.70}$$

Caution. Since the product of two sine functions yields a constant term and a double-frequency sine term, and since the frequency is entirely omitted in the representation of a sine function by a stationary vector, the student is hereby warned to avoid the use of vectors in multiplying two sine functions.

24. Cosine Functions. Cosine functions are encountered rather frequently. A separate treatment of these functions becomes superfluous when it is remembered that

$$\cos(x + \alpha) = \sin\left(x + \alpha + \frac{\pi}{2}\right) \tag{16.71}$$

Thus a cosine function can always be converted to a sine function by adding $\pi/2$ to the initial angle of the cosine function.

25. Circuit Containing Resistance. Consider a circuit (Fig. 16.19) which has a resistance R and a sine emf

$$e = E_m \sin(\omega t + \alpha) \tag{16.72}$$

The drop across the resistance is equal to the applied emf, or

$$e_r = Ri = E_m \sin(\omega t + \alpha) \tag{16.73}$$

Hence the current i is

$$i = \frac{E_m}{R}\sin(\omega t + \alpha) = I_m \sin(\omega t + \alpha) \tag{16.74a}$$

where $$I_m = \frac{E_m}{R} \tag{16.74b}$$

Equation (16.74) states that the current in a circuit with resistance is in phase with the voltage and has an amplitude $I_m = E_m/R$.

The vectors representing the voltage and current in a circuit are shown in Fig. 16.19 and may be expressed in complex form as follows:

$$\hat{I} = I\epsilon^{j\alpha} \qquad \hat{E} = \hat{I}R = E\epsilon^{j\alpha} \tag{16.75}$$

It may thus be concluded that in a circuit subjected to a sine emf the vector potential drop across the resistance is

$$\hat{E}_r = \hat{I}R \qquad \text{or} \qquad \hat{I} = \hat{E}_r G \tag{16.76}$$

where $G = 1/R$ is called the *conductance* of the circuit.

26. Circuits Containing Inductance. In the circuit in Fig. 16.20 let a current be

$$i = I_m \sin(\omega t - \beta) \tag{16.77}$$

Then by Eq. (16.10) the drop across the inductance which is equal to the applied emf is

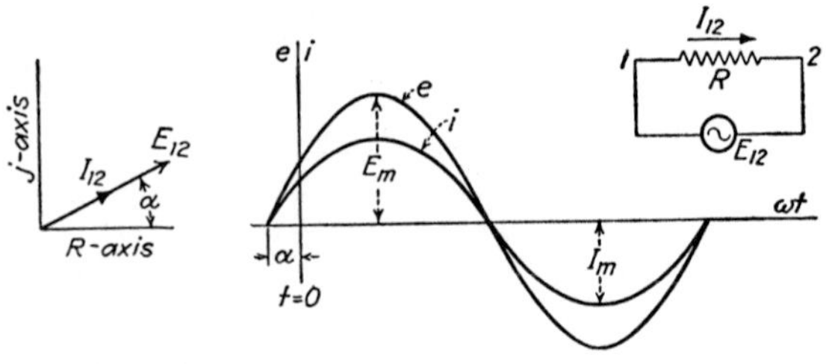

FIGURE 16.19

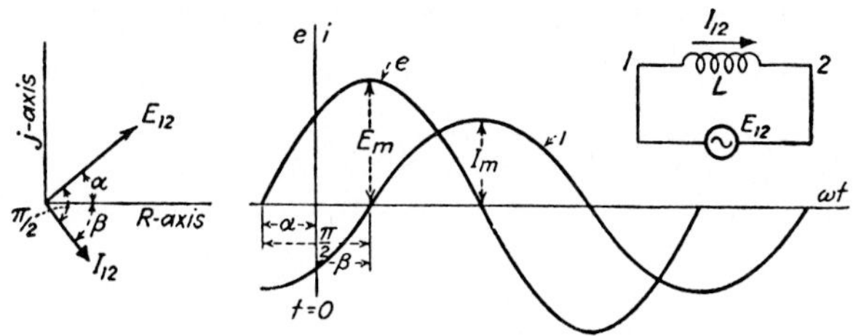

FIGURE 16.20

$$e_L = L\frac{di}{dt} = L\omega I_m \cos(\omega t - \beta) = X_L I_m \sin\left(\omega t - \beta + \frac{\pi}{2}\right) \tag{16.78a}$$

$$= E_m \sin\left(\omega t - \beta + \frac{\pi}{2}\right) = E_m \sin(\omega t + \alpha) \tag{16.78b}$$

where $$E_m = X_L I_m = \omega L I_m \quad \text{and} \quad \alpha = \frac{\pi}{2} - \beta \tag{16.78c}$$

Observe that in this case the voltage wave [compare Eqs. (16.77) and (16.78*a*)] leads the current wave by $\pi/2$ radians. The current is said to be in *lagging quadrature* with the voltage. The vectors representing the current and voltage, respectively, are shown in Fig. 16.20. Their vector expressions are

$$\hat{I} = I\epsilon^{-j\beta} \tag{16.79a}$$

$$\hat{E} = E\epsilon^{j[(\pi/2)-\beta]} = X_L I \epsilon^{j(\pi/2)}\epsilon^{-j\beta} = jX_L I\epsilon^{-j\beta} = jX_L\hat{I} \tag{16.79b}$$

The angle $\phi = \alpha - \beta = \pi/2$ between the voltage and current waves, as well as between the voltage and current vectors (see Fig. 16.20), is called the *power-factor angle* of the circuit. X_L is called the *inductive reactance* of the circuit and is defined by Eq. (16.78*c*) as follows:

$$X_L = \omega L = 2\pi f L \tag{16.80}$$

The vector drop across an inductance may therefore be defined, in magnitude and phase position, as follows:

$$\hat{E}_L = jX_L\hat{I} \tag{16.81a}$$

$$= \frac{\hat{I}}{-jB_L} \tag{16.81b}$$

B_L is called the *inductance susceptance* of the circuit and is defined as follows:

$$B_L = \frac{1}{X_L} = \frac{1}{2\pi f L} \tag{16.82}$$

27. Circuits Containing Capacitance. Let the emf impressed on the capacitance C of Fig. 16.21 be

$$e = E_m \sin(\omega t + \alpha) \tag{16.83}$$

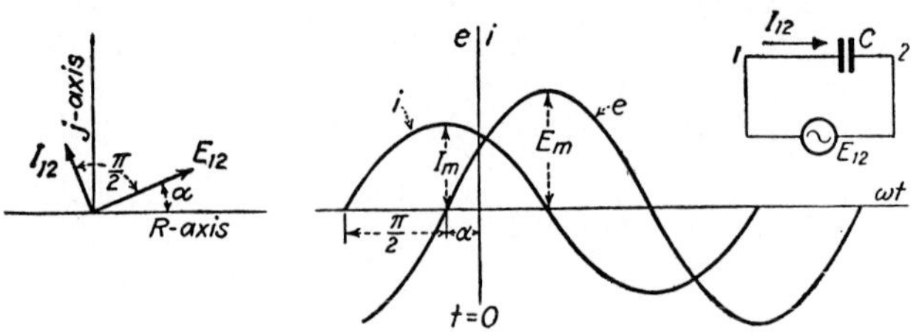

FIGURE 16.21

The charge accumulating on the capacitor (condenser) C at any instant is, by Eq. (16.17*a*),

$$q = Ce = CE_m \sin(\omega t + \alpha) \tag{16.84}$$

Hence the current flowing through the circuit is

$$i = \frac{dq}{dt} = C\omega E_m \cos(\omega t + \alpha) = B_c E_m \sin\left(\omega t + \alpha + \frac{\pi}{2}\right) \tag{16.85a}$$

$$= I_m \sin\left(\omega t + \alpha + \frac{\pi}{2}\right) = I_m \sin(\omega t + \beta) \tag{16.85b}$$

where $$I_m = B_c E_m = \omega C E_m \quad \text{and} \quad \beta = \left(\alpha + \frac{\pi}{2}\right) \tag{16.85c}$$

Observe that in this case the current i leads the voltage by $\pi/2$ radians or is in *leading quadrature* with the voltage. The vectors representing the current and voltage, respectively, are given in Fig. 16.21 and may be expressed as follows:

$$\hat{I} = I\epsilon^{j\beta} = I\epsilon^{j(\alpha+\pi/2)} = B_c E\epsilon^{j\alpha}\epsilon^{j(\pi/2)} = jB_c E\epsilon^{j\alpha} = jB_c\hat{E} \tag{16.86a}$$

$$\hat{E} = E\epsilon^{j\alpha} \tag{16.86b}$$

The angle $(\alpha - \beta) = -\pi/2$, between the voltage and current waves, as well as between the voltage and current vectors (see Fig. 16.21), is the *power-factor angle* of the circuit. $B_c = \omega C = 2\pi f C$ is called the *capacitive susceptance* of the circuit, and $jB_c = j\omega C$ is called the *complex capacitive susceptance* of the circuit.

The potential drop across a capacitance may be defined as follows:

$$\hat{E}_c = \frac{\hat{I}_c}{jB_c} = -jX_c\hat{I}_c \tag{16.87}$$

where X_c and B_c are known, respectively, as the capacitive reactance and capacitive susceptance of the circuit and are defined as

$$X_c = \frac{1}{2\pi f C} \tag{16.88a}$$

$$B_c = 2\pi f C \tag{16.88b}$$

28. Series Circuits with *R, L,* and *C*. An emf e is impressed on a series circuit (Fig. 16.22). Let

$$e = E_m \sin(\omega t + \alpha) \tag{16.89}$$

A current i flows in the circuit whose value is

$$i = I_m \sin(\omega t + \beta) \tag{16.90}$$

It is required to ascertain the relation between the effective values E and I and to determine the phase angle ϕ between the voltage and current waves or vectors.

Solution. The voltage drops across R, L, and C are, by Eqs. (16.76), (16.81), and (16.87),

$$\hat{E}_r = \hat{I}R \tag{16.91a}$$

$$\hat{E}_L = j\hat{I}X_L \tag{16.91b}$$

$$\hat{E}_c = -j\hat{I}X_c \tag{16.91c}$$

By Kirchhoff's emf law,

$$\hat{E} = \hat{E}_r + \hat{E}_L + \hat{E}_c = \hat{I}[R + j(X_L - X_c)] \tag{16.92a}$$

$$\hat{I} = \frac{\hat{E}}{R + j(X_L - X_c)} = \frac{\hat{E}}{\hat{Z}} = \frac{E\epsilon^{j\alpha}}{Z\epsilon^{j\phi}} = \frac{E}{Z}\epsilon^{j(\alpha - \phi)} = I\epsilon^{j\beta} \tag{16.92b}$$

where

$$\hat{Z} = R + j(X_L - X_c) \tag{16.93a}$$

$$Z = \sqrt{R^2 + (X_L - X_c)^2} \tag{16.93b}$$

$$\phi = \tan^{-1}\left(\frac{X_L - X_c}{R}\right) \tag{16.93c}$$

Z is known as the *impedance* of the series circuit (Fig. 16.22).

Equation (16.92*b*) states that the effective value of the current and its phase angle in the series circuit in Fig. 16.22 are, respectively,

$$I = \frac{E}{Z} \tag{16.94a}$$

$$\beta = (\alpha - \phi) \tag{16.94b}$$

The power-factor angle is, by Eq. (16.92*b*),

$$\phi = \alpha - \beta \tag{16.95}$$

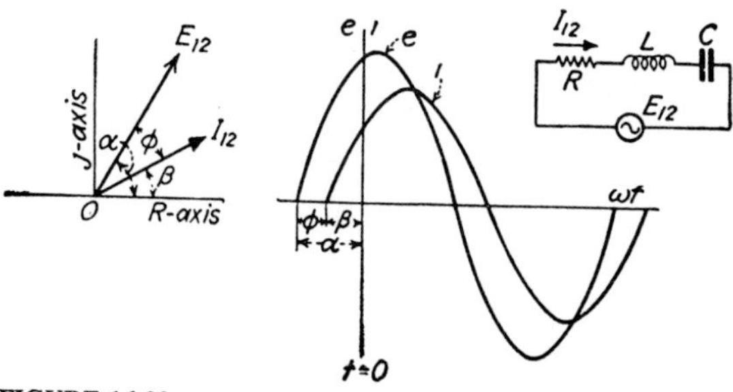

FIGURE 16.22

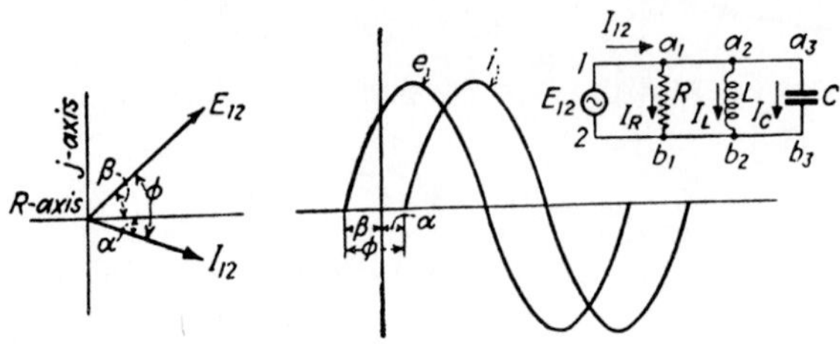

FIGURE 16.23

28a. Parallel Circuits with *R, L,* and *C*. An emf e is impressed on a parallel circuit (Fig. 16.23). Let

$$e = E_m \sin(\omega t + \alpha) \tag{16.96}$$

Since all the characteristics are across the same emf, from Eqs. (16.76), (16.81), and (16.87),

$$\hat{I}_r = \frac{\hat{E}}{R} = \hat{E}G \tag{16.97a}$$

$$\hat{I}_L = \frac{\hat{E}}{jX_L} = -j\hat{E}B_L \tag{16.97b}$$

$$\hat{I}_c = j\hat{E}B_c \tag{16.97c}$$

Applying Kirchhoff's current law to the junction a (Fig. 16.23),

$$\begin{aligned}\hat{I} &= \hat{I}_r + \hat{I}_L + \hat{I}_c \\ &= \hat{E}[G + j(B_c - B_L)] = \hat{E}\hat{Y} = E\epsilon^{j\alpha}Y\epsilon^{j\phi} = EY\epsilon^{j(\alpha+\phi)} = I\epsilon^{j\beta}\end{aligned} \tag{16.98}$$

where

$$\hat{Y} = G + j(B_c - B_L) \tag{16.99a}$$

$$Y = \sqrt{G^2 + (B_c - B_L)^2} \tag{16.99b}$$

$$\phi = \tan^{-1}\left(\frac{B_c - B_L}{G}\right) \tag{16.99c}$$

Y is called the *admittance* of the parallel circuit in Fig. 16.23.

Equation (16.98) states that the effective value of the current and its phase angle in the parallel circuit (Fig. 16.23) are, respectively,

$$I = EY \tag{16.100a}$$

$$\beta = \alpha + \phi \tag{16.100b}$$

The power-factor angle ϕ is, by Eq. (16.99c) and by Fig. 16.23,

$$\phi = \beta - \alpha \tag{16.101}$$

29. Voltage and Current Resonance

Voltage Resonance. Refer to the series circuit (Fig. 16.22) and observe by Eqs. (16.91) that

$$\hat{E}_L + \hat{E}_c = j\hat{I}(X_L - X_c) \tag{16.102}$$

When the sum of the voltage drops across the inductance and capacitance is nil, the series circuit in Fig. 16.22 is said to be *in resonance*. Obviously, this occurs when

$$X_L - X_c = 0 \qquad \text{or} \qquad X_L = X_c \tag{16.103}$$

In order to determine the resonant frequency f_r, substitute in Eq. (16.103) for X_L and X_c their values as given by Eqs. (16.80) and (16.88), and obtain

$$2\pi f_r L = \frac{1}{2\pi f_r C} \tag{16.104a}$$

or

$$f_r = \frac{1}{2\pi\sqrt{LC}} \tag{16.104b}$$

Current Resonance. Refer to the parallel circuit (Fig. 16.23) and observe, by Eq. (16.97), that

$$\hat{I}_L + \hat{I}_c = j\hat{E}(B_c - B_L) \tag{16.105}$$

When the sum of the currents in the inductance and capacitance of a parallel circuit is nil, that circuit is said to be *in resonance*. This occurs when

$$B_c - B_L = 0 \qquad \text{or} \qquad B_L = B_c \tag{16.106}$$

In order to determine the resonant frequency f_r, substitute in Eq. (16.106), for B_L and B_c, their values as given in Eqs. (16.82) and (16.88), and obtain

$$2\pi f_r C = \frac{1}{2\pi f_r L} \tag{16.107a}$$

or

$$f_r = \frac{1}{2\pi\sqrt{LC}} \tag{16.107b}$$

Comparing Eqs. (16.104*b*) and (16.107*b*), it may be concluded that the frequency at which resonance occurs in a series or parallel circuit is the same.

POLYPHASE ALTERNATING-CURRENT CIRCUITS

30. Three-Phase EMF Sources. A three-phase emf source is one having three distinct emf's. Such a three-phase source is said to be *balanced* when (1) the three emf's have the same effective values, and (2) the vectors representing these emf's are equally displaced from each other by $2\pi/3$ radians or 120 degrees. The three emf sources, their waves, and the vectors representing them are shown in Fig. 16.24. Commercial three-phase systems are approximately balanced, and this condition is assumed in the following sections.

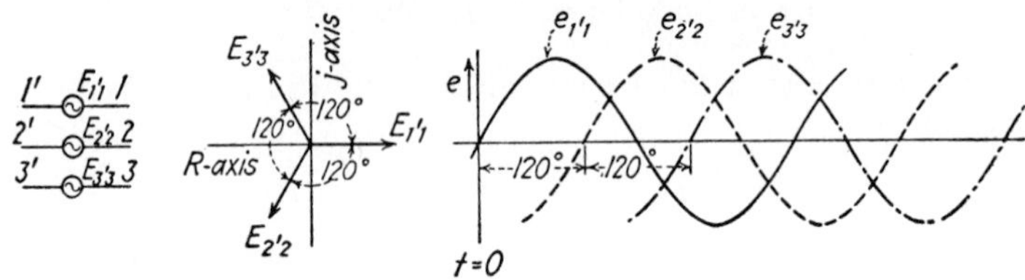

FIGURE 16.24

30a. Methods of Connecting Three-Phase Sources and Three-Phase Loads. Three-phase sources may be connected either in star (Y) or in mesh (Δ). Each of these types of connection is defined below.

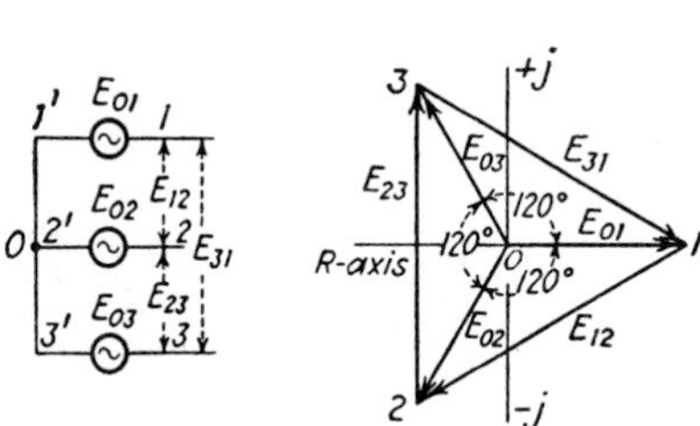

FIGURE 16.25

Star or Y Connection. Let the terminals 1′, 2′, and 3′ of the emf sources be connected together as in Fig. 16.25. Call the common terminal O. The voltages then become

$$\hat{E}_{1'1} = \hat{E}_{01} \tag{16.108a}$$

$$\hat{E}_{2'2} = \hat{E}_{02} \tag{16.108b}$$

$$\hat{E}_{3'3} = \hat{E}_{03} \tag{16.108c}$$

The vectors representing these emf's are also shown in Fig. 16.25. The Y connection is so named because of the shape of this vector diagram.

The emf's given in Eq. (16.108) are known as *phase emf's.* If a voltmeter is connected across the terminals 12, 23, and 31 (Fig. 16.25), respectively, it will read what is known as the *line emf's* E_{12}, E_{23}, and E_{31}. The vector relation between the line and phase emf's is obtained directly from Fig. 16.25. Thus

$$\hat{E}_{12} = \hat{E}_{02} - \hat{E}_{01} \tag{16.109a}$$

$$\hat{E}_{23} = \hat{E}_{03} - \hat{E}_{02} \tag{16.109b}$$

$$\hat{E}_{31} = \hat{E}_{01} - \hat{E}_{03} \tag{16.109c}$$

Adding Eqs. (16.109), it is seen that the sum of the line voltages is zero, or

$$\hat{E}_{12} + \hat{E}_{23} + \hat{E}_{31} = 0 \tag{16.110}$$

Again, in the vector diagram (Fig. 16.25), drop a perpendicular from O to any of the line voltages. Then, designating the magnitudes of the line and phase voltages by E_L and E_p, respectively,

$$0.5E_L = E_p \sin 60° = E_p\frac{\sqrt{3}}{2} \tag{16.111a}$$

or

$$E_L = \sqrt{3}E_p \tag{16.111b}$$

This is an extremely important relation between the effective values of the phase and line voltages.

The Mesh or Delta (Δ) Connection. Since the vectors representing the emf's in Fig. 16.24 are all equal and equally displaced from each other by 120°, their sum is zero. Hence the terminals of the emf's may be joined in series (Fig. 16.26) and there will be no residual voltage in the combination. Such a connection is known as the Δ (delta) connection. It is evident, by reference to Fig. 16.26, that the line voltages are

$$\hat{E}_{12} = \hat{E}_{2'2} \tag{16.112a}$$

$$\hat{E}_{23} = \hat{E}_{3'3} \tag{16.112b}$$

$$\hat{E}_{31} = \hat{E}_{1'1} \tag{16.112c}$$

Figure 16.26 also shows the vector diagram of the voltages in Eqs. (16.112).

Star and Mesh Loads. Consider three equal impedances of value $Z\epsilon^{j\phi}$. These impedances also may be connected in star (Y) or mesh (Δ), as shown in Fig. 16.27*a* and *b*.

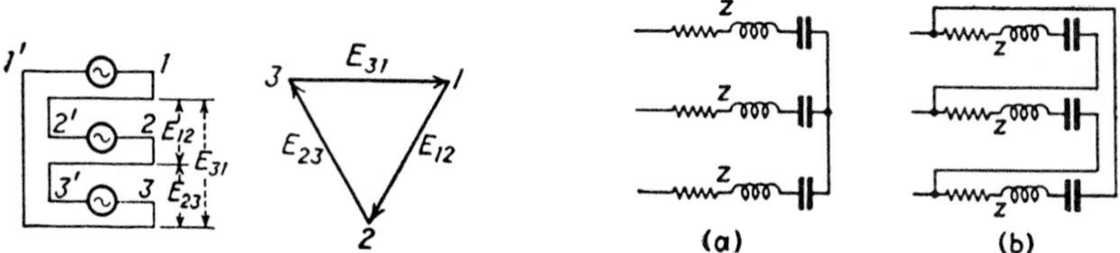

FIGURE 16.26

FIGURE 16.27 (*a*) Impedance connected in star; (*b*) impedance connected in Δ

31. Types of Polyphase Circuits. There are four types of three-phase circuits designated by the type of connection of source and load, as shown in Fig. 16.28 and described in Table 16.3.

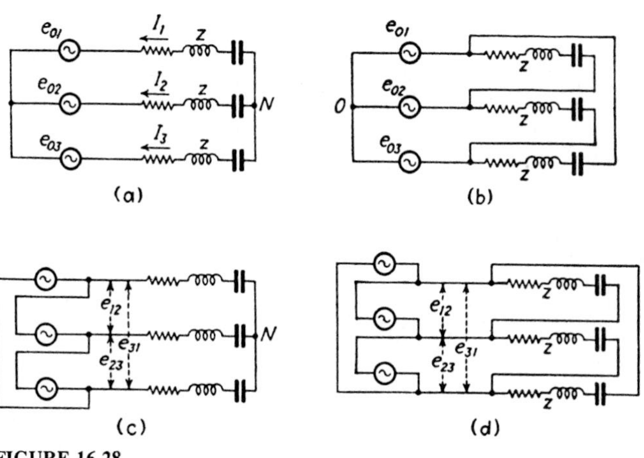

FIGURE 16.28

TABLE 16.3 Types of Polyphase Circuits in Fig. 16.28

Source	Load	Name	Figure
Star	Star	Star-star	16.28*a*
Star	Mesh	Star-mesh	16.28*b*
Mesh	Star	Mesh-star	16.28*c*
Mesh	Mesh	Mesh-mesh	16.28*d*

32. Current-EMF Relations. It can be shown that all the types of three-phase circuits can be converted to a star-star circuit. Hence the emf-current relations for a star-star balanced circuit only are given by

$$\hat{I}_1 = \frac{E_{01}}{\hat{Z}} \tag{16.113a}$$

$$\hat{I}_2 = \frac{\hat{E}_{02}}{\hat{Z}} \tag{16.113b}$$

$$\hat{I}_3 = \frac{\hat{E}_{03}}{\hat{Z}} \tag{16.113c}$$

33. Power in AC Circuits. The instantaneous power in an ac circuit is defined as

$$p = ei = E_m \sin(\omega t + \alpha) I_m \sin(\omega t + \alpha \pm \phi) \tag{16.114}$$

where e is the emf applied to the circuit, and i is the total current furnished by this emf source. Using the fact that $\sin x \sin y = ½[\cos(x - y) - \cos(x + y)]$, Eq. (16.114) becomes

$$p = \frac{E_m I_m}{2}[\cos(\pm\phi) - \cos(2\omega t + 2\alpha \pm \phi)] \tag{16.115a}$$

$$= EI[\cos(\phi) - \cos(2\omega t + 2\alpha \pm \phi) \tag{16.115b}$$

The average power in an electric circuit is defined as

$$P = \frac{1}{T}\int_0^T p\, dt \tag{16.116}$$

Substituting Eq. (16.115*b*) in (16.116) and integrating,

$$P = \frac{EI}{T}\int_0^T \cos\phi\, dt - \int_0^T \cos(2\omega t + 2\alpha \pm \phi)\, dt \tag{16.117a}$$

$$= EI\cos\phi + 0. \tag{16.117b}$$

Thus the *average power* in an ac circuit is

$$P = EI\cos\phi \tag{16.118}$$

The term $\cos\phi$ is known as the *power factor* (P.f.) of the circuit; the product EI is called the *apparent power* (P_a) and is measured in voltamperes (VA) or kilovoltamperes (kVA); the product $EI \sin \phi$ is called the *reactive power* (P_r), also measured in voltamperes or kilovoltamperes. Thus

$$\text{P.f.} = \cos \phi \tag{16.119}$$

$$P_a = EI \tag{16.120}$$

$$P_r = EI \sin \phi \tag{16.121}$$

THE MAGNETIC CIRCUIT

34. Fundamental Concepts. A piece of iron is said to be *magnetized*, to possess *magnetism*, to be in a *magnetic state*, or to be a *magnet* if it is capable of attracting or repelling other pieces of iron placed near it. Magnetism appears to be concentrated at the ends of a magnet. Two magnets placed side by side (Fig. 16.29) may either attract or repel each other. Magnets placed as in Fig. 16.29*a*

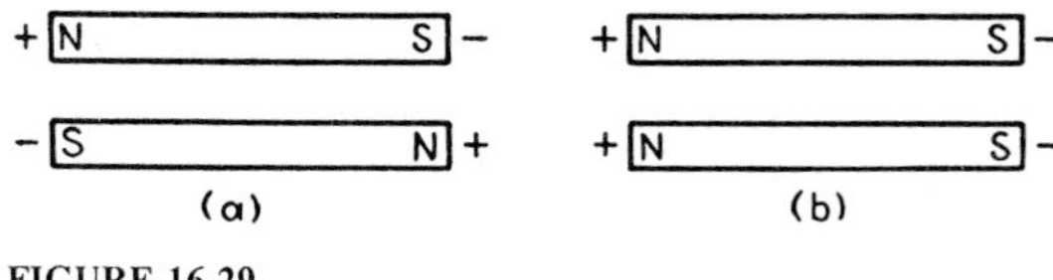

FIGURE 16.29

will attract each other. The reversal of one of them (Fig. 16.29*b*) results in repulsion between them. It is necessary, therefore, to distinguish between the positive (north) and negative (south) poles of a magnet.

A *magnetic needle* is a magnetized piece of steel having small mass and so suspended that it can move freely in any of three mutually perpendicular directions. Such a needle is used to detect the existence of magnetism, because a magnetic needle orients itself in a certain direction whenever it is placed near a magnetized object. The magnetic state is not confined to iron; it may exist in air. Thus the space (Fig. 16.30) surrounding a coil carrying a current I is in a *magnetic state*, because a magnetic needle placed near such a coil orients itself at each point in the direction tangent to the lines shown in Fig. 16.30. The earth itself is endowed with magnetism and is a huge magnet, which accounts for the fact that a magnetic needle orients itself according to north and south. The terms *north* and *south* as applied to the ends of a magnet originated because the north pole of a magnetic needle points northward. According to this terminology, the geographic north pole of the earth, since it attracts the north pole of a magnetic needle, is in reality the earth's magnetic south pole.

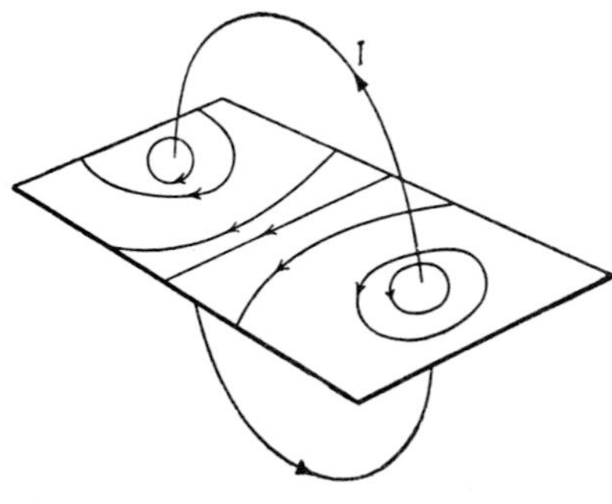

FIGURE 16.30

The region in which a magnetic state exists is a *magnetic field*. The magnetism at every point in a magnetic field is characterized by two properties, magnitude and direction. These two properties are defined by one term, *induction*, which is a vector. The magnitude of magnetism at any point is proportional to the force exerted on a magnetic needle placed at that point. The direction is the direction in which the needle is oriented. Curves (Fig. 16.30) whose tangents at every point correspond with the direction of magnetism are known as *lines of force* or *lines of flux*. These lines always form *closed paths*.

35. Graphical Representation of a Field. The induction of a magnetic field varies from point to point in space. Hence it may be represented by a vector whose length is proportional to the magnitude of the magnetism and whose direction is the direction of the lines of force. Such representation, while describing the induction at any one point, does not give a picture of the field simultaneously at several points. Two-dimensional figures are therefore generally used, giving a map of the field in a plane (Fig. 16.31). The *direction* of induction is here represented by the lines of force (flux), while its *magnitude* is indicated by the density of the lines of flux. Thus the number of lines of force drawn across a unit length is a measure of the magnitude of the induction at any point in the plane. This idea of replacing the magnitude of induction by flux density permeates all engineering literature and is a very useful concept. *Induction* may now be defined as a vector having at every point a magnitude equal to the flux density and a direction corresponding to that of a magnetic needle placed at this point.

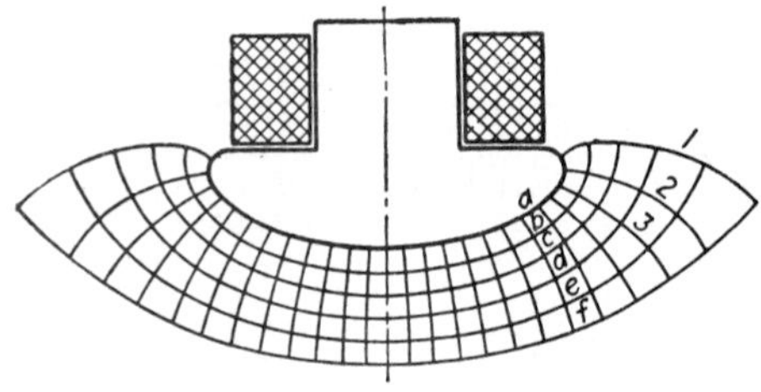

FIGURE 16.31

36. Magnetomotive Force. So far attention has been paid to the magnetic field without inquiring into its cause. The cause of such a field is, or may always be traced to, an electric current. The question as to why there is a field in a magnetized piece of iron or in magnetite when apparently no current exists will be taken up subsequently. Suffice it to say here that an electric current (Fig. 16.30) is always accompanied by a magnetic field whose direction is perpendicular to the direction of the current and whose flux density at any point varies with the strength of the current and the configuration of the current path.

Figure 16.30 shows that the current and field which it produces form links like the links of a chain. Moreover, the directions of the current and of the lines of flux are such that if a right-hand screw is rotated in the direction of current flow, the progress of the screw corresponds to the direction of the flux lines.

A few simple experiments will now be outlined in an effort to determine the relation between cause and effect in a magnetic field. Consider two currents I_1

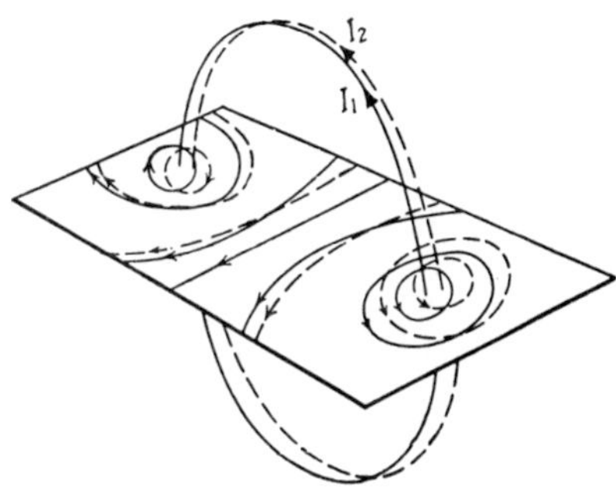

FIGURE 16.32

and I_2 (Fig. 16.32) in two separate wires of identical shape. The fields due to I_1 and I_2 are represented by solid and dotted lines, respectively. If the loops are large and the wires are thin and placed very close to each other, the two fields, for all practical purposes, will merge and may be considered as one.

Let $I_1 = I, I_2 = 0$, and the flux density at a point $= \mathcal{B}_x$. If now $I_1 = 2I$ and $I_2 = 0$, the flux density becomes $2\ \mathcal{B}_x$. It may be concluded, therefore, that the flux density at any point is directly proportional to the current. Next, let $I_1 = 0.5I$ and $I_2 = 0.5I$. The flux density due to I_1 is $0.5\mathcal{B}_x$, and that due to I_2 is also $0.5\mathcal{B}_x$. Hence the total density is $0.5\mathcal{B}_x + 0.5\mathcal{B}_x = \mathcal{B}_x$. But these two currents flowing through two separate loops are electrically equivalent to one current $0.5I$ flowing through *two turns*. Hence it may be concluded that flux density is proportional not only to the current but also to the number of turns this current makes. In other words, the cause of magnetic flux is the *ampere-turns* linking it.

By analogy to electromotive force, which is the cause of electric flow, the term *magnetomotive force* is used to describe the cause of magnetic flow or flux. Magnetomotive force is measured in ampere-turns in the m.k.s. (meter-kilogram-second) system of units.*

37. Magnetic Circuit. Just as an electric circuit consists of an emf and a current, so a magnetic circuit consists of a magnetomotive force (mmf or $\mathcal{M}$) and a flux. The study of the numerical relations between cause and effect in an electric circuit constitutes the subject of *electric-circuit analysis*. Similarly, the study of the numerical relations between cause and effect in a magnetic circuit is known as *magnetic-circuit analysis*. It must here be emphasized that while in most practical problems of electric-circuit analysis the flow of electricity is uniform and uniformly distributed over the path of flow, such is not the case in the majority of problems of magnetism. In general, a magnetic field has different values of induction at different points. The study of the magnetic circuit, therefore, presents extremely complicated problems. The following discussion is restricted generally to very simple circuits. Where complications do arise, simplifying assumptions will be made.

*The m.k.s. system of units was originally proposed by G. Giorgi in 1902. It is based on the meter, the kilogram, and the second as fundamental units. The simplicity of the system for engineering work can be seen from the following table, which gives the values of seven practical electrical units in terms of the c.g.s. units and the m.k.s. units:

	Length	Mass	Time	Coulomb	Ampere	Volt	Ohm	Henry	Farad	Weber
C.g.s.	1 cm	1 g	1 s	10^{-1}	10^{-1}	10^8	10^9	10^{-9}	10^{-9}	10^8
M.k.s.	1 m	1 kg	1 s	1	1	1	1	1	1	

37a. Magnetic Circuit with Uniform Circular Field. The lines of flux form closed paths. Hence the direction of a line of force can never be constant over all its path. The flux density, however, may be constant throughout the whole field. Thus the simplest field is one having constant flux density and variable direction. The magnetic circuit (Fig. 16.33) in which the flux density has constant magnitude (induction) and circular direction will be studied, with attention only to numerical relations, regardless of the direction of flux.

The mmf producing the field in Fig. 16.33 consists of one layer having N turns of very thin wire wound over a circular form of rectangular cross section and carrying a continuous current I. The mmf is, therefore,

$$\mathcal{M} = kNI \tag{16.122}$$

If the diameter of the ring is assumed to be large in relation to its radial thickness a_1, this mmf produces a uniform field of constant flux density $\mathcal{B}$. Moreover, since $\mathcal{B}$ is proportional to $\mathcal{M}$,

$$\mathcal{M} = k\mathcal{B} \tag{16.123a}$$

when

$$k = k_1 \nu \pi D \tag{16.123b}$$

k_1 is a factor affected only by the units used and has nothing to do with the physical dimensions or the material of the magnetic ring.

Substituting (16.123*b*) in (16.123*a*),

$$\mathcal{M} = k_1 \nu \pi D \mathcal{B} \tag{16.124}$$

Now let ϕ be the total flux, and let $A = a_1 a_2$ be the cross-sectional area of the toroid. Since the flux density $\mathcal{B}$ is constant,

$$\phi = A\mathcal{B} \qquad \therefore \qquad \mathcal{B} = \frac{\phi}{A} \tag{16.125}$$

hence

$$\mathcal{M} = \frac{k_1 \nu \pi D}{A} \phi = \mathcal{R}\phi \tag{16.126a}$$

where

$$\mathcal{R} = \frac{k_1 \nu \pi D}{A} \tag{16.126b}$$

$\mathcal{R}$ in Eq. (16.126) is known as the *reluctance* of the flux path and is analogous to resistance in the case of electric circuits, and ν is called the *reluctivity* of the flux path, or the reluctance of a path having unit length and unit sectional area.

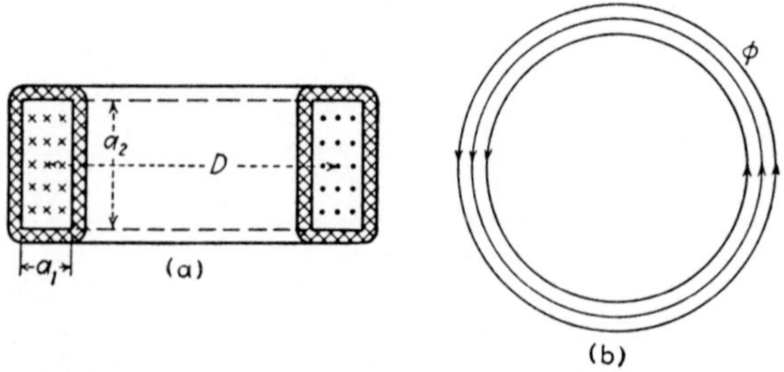

FIGURE 16.33

The analogy, between $\mathcal{M} = \mathcal{R}\phi$ in a magnetic circuit with uniform circular field and $E = RI$ in an electric circuit is complete. Hence Eq. (16.126*a*) is often referred to as *Ohm's law for the magnetic circuit*. Reference will hereafter be made to the term $\mathcal{R}\phi$ as the *mmf drop*.

Equation (16.126*a*) may be written in another form:

$$\phi = \frac{1}{\mathcal{R}}\mathcal{M} = \mathcal{P}\mathcal{M} \tag{16.127a}$$

where

$$\mathcal{P} = \frac{1}{\mathcal{R}} = \frac{1}{k_1}\mu\frac{A}{\pi D} \tag{16.127b}$$

$\mathcal{P}$ is known as the *permeance* of the flux path, and $\mu = 1/\nu$ is its *permeability*. By definition, μ is the permeance of a flux path of unit length and unit sectional area.

38. Point Relations. The mmf (Fig. 16.33) is entirely consumed in forcing the flux against the reluctance of the magnetic path. The flux density is uniform, and the magnetic path has a uniform reluctance. Since the mmf drop $\mathcal{R}\phi$ is uniformly distributed throughout the whole length of the lines of force, the mmf drop per unit length may be defined as

$$\mathcal{H} = \frac{\mathcal{R}\phi}{\pi D} \tag{16.128a}$$

$$= \left(\frac{k_1\nu\pi D}{A}\right)\left(\frac{\mathcal{B}A}{\pi D}\right) = k_1\nu\mathcal{B} \tag{16.128b}$$

$\mathcal{H}$ is known as the *mmf drop per unit length*, the *field gradient*, or the *field intensity*. It is a vector whose magnitude is proportional to that of $\mathcal{B}$ and whose direction is along $\mathcal{B}$. Both $\mathcal{B}$ and $\mathcal{H}$ are quantities defined at every point of a magnetic field. Hence Eq. (16.128*b*) expresses a *point relation* and holds true at every point of a field, irrespective of whether the field is uniform or not.

39. Units. The relations developed above are general and hold true independently of the units used. For purposes of numerical computations, however, it is necessary to use a consistent set of units, as listed in Table 16.4.

With this choice of units, the constants k_1, μ, and ν in Sec. 37a, for air or vacuum, become

$$k_1 = 1$$

$$\mu_0 = (4\pi)10^{-9} = 1.257 \times 10^{-8} \text{ henry per cm}^3 = 3.195 \times 10^{-8} \text{ henry per in}^3 \tag{16.129}$$

$$\nu_0 = (1/4\pi)10^{9} = 0.795 \times 10^{3} \text{ yrneh per cm}^3 = 0.313 \times 10^{3} \text{ yrneh per in}^3$$

and the relations of Secs. 37 and 38, for an air core, become

$$\begin{aligned} &\mathcal{M} = \mathcal{R}_0\phi \quad (a) && \mathcal{H} = \nu_0\mathcal{B} \quad (b) && \mathcal{R}_0 = \nu_0\frac{l}{A} \quad (c) \\ &\phi = \mathcal{M}\mathcal{P}_0 \quad (a') && \mathcal{B} = \mu_0\mathcal{H} \quad (b') && \mathcal{P}_0 = \mu_0\frac{A}{l} \quad (c') \end{aligned} \tag{16.130}$$

TABLE 16.4 Physical Quantities, Symbols, and Units

Symbol	Physical quantity	Equation	Dimension	Name of unit
l	Length	Fund. unit	[L]	Centimeter
t or T	Time	Fund. unit	[T]	Second
μ	Permeability	$\mu = \dfrac{\mathcal{B}}{\mathcal{H}}$	$[RTL^{-1}]$	Henry per cm. cube
I, i	Current	Fund. unit	[I]	Ampere
R, r	Resistance	Fund. unit	[R]	Ohm
E, e	E.m.f.	$E = RI$	[RI]	Volt
A	Area	$A = l_1l_2$	$[L^2]$	Sq. cm.
V	Volume	$V = l_1l_2l_3$	$[L^3]$	Cu. cm.
L	Inductance	$e = -L\dfrac{di}{dt}$	[RT]	Henry
ν	Reluctivity	$\nu = \dfrac{1}{\mu}$	$[R^{-1}T^{-1}L]$	Yrneh per cm. cube
ϕ	Magnetic flux	$e = \dfrac{d\phi}{dt}$	[RIT]	Weber
$\mathcal{B}$	Flux density	$\mathcal{B} = \dfrac{\phi}{A}$	$[RITL^{-2}]$	Weber per sq. cm.
$\mathcal{M}$	M.m.f.	$\mathcal{M} = NI$	[I]	Ampere-turn
$\mathcal{H}$	Magnetic intensity	$\mathcal{H} = \dfrac{\mathcal{M}}{l}$	$[IL^{-1}]$	Amp.-turns per cm.
P	Power	$P = EI$	$[RI^2]$	Watt
W	Energy	$W = PT$	$[RI^2T]$	Joule
$\mathcal{R}$	Reluctance	$\mathcal{R} = \dfrac{\nu l}{A}$	$[R^{-1}T^{-1}L^{-1}]$	Yrneh
$\mathcal{P}$	Permeance	$\mathcal{P} = \dfrac{\mu A}{l}$	[RTL]	Henry

The relations in Eqs. (16.130b) and (16.130b') are point relations and hold true irrespective of whether the field is uniform or not. The remaining equations in Eqs. (16.130) apply only to a uniform air field wherein l is the length of the flux path and A is the cross-sectional area perpendicular to the direction of flux.

40. Series and Parallel Reluctances and Permeances. Two permeances are in parallel if the same mmf acts on them. Two reluctances are in series if the same flux path passes through them. Expressions for the equivalent permeance of a parallel circuit and the equivalent reluctance of a series circuit are developed as follows.

Equivalent Permeance. Consider a set of n permeances in parallel. Let the flux through each of them be

$$\phi_1 = \mathcal{M}\mathcal{P}_1 \qquad \phi_2 = \mathcal{M}\mathcal{P}_2 \qquad \phi_n = \mathcal{M}\mathcal{P}_n \tag{16.131}$$

The total flux in the circuit is

$$\phi = \phi_1 + \phi_2 + \cdots + \phi_n = \mathcal{M}(\mathcal{P}_1 + \mathcal{P}_2 + \cdots \mathcal{P}_n) = \mathcal{M}\mathcal{P} \tag{16.132a}$$

where $$\mathcal{P} = \mathcal{P}_1 + \mathcal{P}_2 + \cdots + \mathcal{P}_n \tag{16.132b}$$

Thus in a parallel circuit the equivalent permeance is the sum of the parallel permeances.

Equivalent Reluctance. Consider next a set of reluctances in series. Let the mmf drop across each of them be

$$\mathcal{M}_1 = \mathcal{R}_1\phi \qquad \mathcal{M}_2 = \mathcal{R}_2\phi \qquad \mathcal{M}_n = \mathcal{R}_n\phi \tag{16.133}$$

Then the total mmf of the circuit is

$$\mathcal{M} = \mathcal{M}_1 + \mathcal{M}_2 + \cdots + \mathcal{M}_n = (\mathcal{R}_1 + \mathcal{R}_2 + \cdots + \mathcal{R}_n)\phi = \mathcal{R}\phi \qquad (16.134a)$$

where

$$\mathcal{R} = \mathcal{R}_1 + \mathcal{R}_2 + \cdots + \mathcal{R}_n \qquad (16.134b)$$

Thus in a series circuit the equivalent reluctance is the sum of the reluctances.

41. Magnetic Properties of Matter. According to its magnetic properties, matter falls into either one of the following classes:

1. Diamagnetic matter
2. Paramagnetic matter
   a. Ferromagnetic
   b. Nonferromagnetic

Diamagnetic matter includes substances whose magnetic permeability is less than μ_0. Among these may be mentioned antimony and bismuth. The permeability of bismuth, which is one of the most highly diamagnetic substances known, is 1.254×10^{-8} H/cm^3, which is not very different from $\mu_0 = 1.257 \times 10^{-8}$.

Paramagnetic matter includes substances whose magnetic permeability is equal to or greater than μ_0. Air, wood, copper, brass, etc. have a permeability which is approximately μ_0. Iron and iron alloys and nickel and cobalt and their alloys have permeabilities varying from μ_0 to $10^5\mu_0$, depending on the flux density used and the composition of the alloy. This latter class of paramagnetic matter which possesses high permeability is referred to as *ferromagnetic* in order to distinguish it from *nonferromagnetic* matter, whose permeability is very close to that of vacuum.

The absence of materials of very high reluctivity accounts for the fact that there is no known magnetic insulator. Such an insulator, if it existed, would have a very low permeability or very high reluctivity. The reluctance of a path made of this material would be high, and hence ϕ in the relation $\mathcal{M} = \mathcal{R}\phi$ would be low. In order to appreciate this fact, compare the ratio of the resistivity of insulators to that of conductors with the ratio of the reluctivity of paramagnetic or diamagnetic materials to that of ferromagnetic materials as follows:

Resistivity of silver: $\rho_1 = 1.63 \times 10^{-6}$
Resistivity of gutta-percha: $\rho_2 = 3.7 \times 10^{14}$
Reluctivity of permalloy: $\nu_1 = 2 \times 10^{-5}$
Reluctivity of bismuth: $\nu_2 = 8 \times 10^7$

$$\frac{\rho_2}{\rho_1} = \frac{3.7 \times 10^{14}}{1.63 \times 10^{-6}} = 2.27 \times 10^{20}$$

$$\frac{\nu_2}{\nu_1} = \frac{8 \times 10^7}{2 \times 10^{-5}} = 4 \times 10^{12}$$

From this comparison it is evident that while the variation in resistivity covers a very wide range, this is not true of reluctivity.

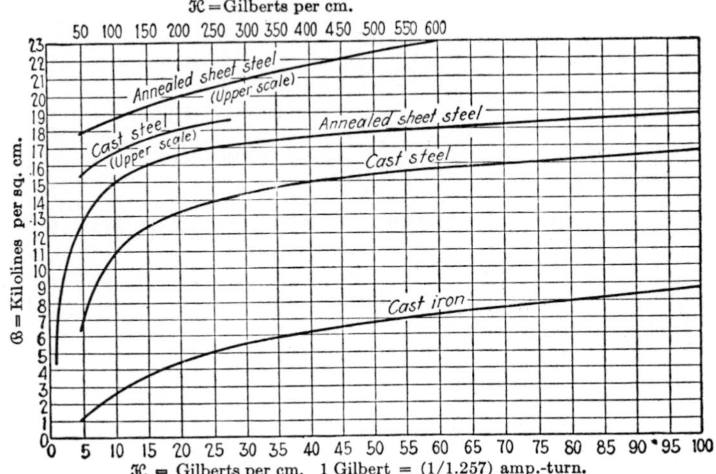

FIGURE 16.34 Magnetization or $\mathcal{B}$-$\mathcal{H}$ curves.

42. Magnetization or B-H Curves of Steel and Its Alloys. Some magnetization curves are shown in Fig. 16.34. Some observations relative to these curves should be pointed out because of their significance in the solution of magnetic-circuit problems.

1. The curves are not straight lines. This means that the permeability of iron μ is not constant but is a function of the flux density $\mathcal{B}$ or the magnetic intensity $\mathcal{H}$.

2. On account of (1), the flux ϕ in a magnetic circuit does not obey the principle of superposition. In other words, if $\mathcal{M}_1$ and $\mathcal{M}_2$ produce fluxes ϕ_1 and ϕ_2, respectively, in an iron path, and if $\mathcal{M}_1 = a\mathcal{M}_2$, then $\phi_1 \neq a\phi_2$. This fact is of extreme importance and has far-reaching consequences, some of which will be pointed out in the succeeding section.

43. Magnetic Circuits with Iron. Three types of magnetic circuits may be recognized: (1) series magnetic circuit (Fig. 16.35*a*), (2) parallel circuit (Fig. 16.35*b*), and (3) series-parallel circuit (Fig. 16.35*c*). Two types of problems occur in each of these types of circuit: (*A*) given the flux ϕ; one must find the mmf $\mathcal{M}$, and (*B*) given the mmf $\mathcal{M}$, one must find the flux ϕ.

The solutions of these two problems are discussed below for each type of circuit, with, in each case, the following assumptions: (1) the flux path follows the path marked b_k (Fig. 16.35) and consists of arcs of circles and straight lines, (2) there exists no leakage flux, so that in any series circuit the total flux is the same in each part of the path, (3) there exists no fringing of flux at the air gaps where two iron paths meet, and (4) the flux density is uniform in any one element of the circuit.

Case A_1: Given ϕ, Required $\mathcal{M}$ (Series Circuit). Referring to Fig. 16.35*a*, let the various drops in the series circuit be $\mathcal{H}_1 b_1$, $\mathcal{H}_2 b_2, \ldots, \mathcal{H}_6 b_6$. Then the required mmf is

$$\mathcal{M} = \mathcal{H}_1 b_1 + \mathcal{H}_2 b_2 + \cdots + \mathcal{H}_6 b_6 \tag{16.135}$$

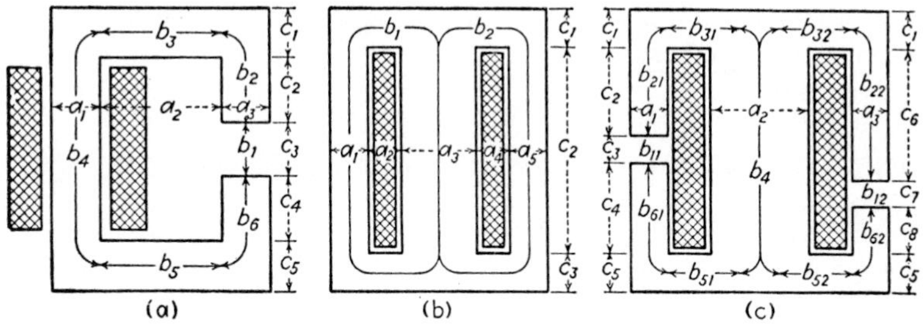

FIGURE 16.35

Since ϕ is given, the flux density in the various sections of the series circuit may be found from the relation $\mathcal{B}_k = \phi/A_k$. Knowing $\mathcal{B}_k$, the value of $\mathcal{H}_k$ may be found from the magnetization curves (Fig. 16.34) for the iron sections and from the relation $\mathcal{B} = \mu_0\mathcal{H}$ for the air sections. Thus all the $\mathcal{H}$'s in Eq. (16.135) are known, and $\mathcal{M}$ can be computed.

Case A_2: Given ϕ, Required $\mathcal{M}$(Parallel Circuit). Referring to Fig. 16.35*b*, let it be assumed that the total flux ϕ divides into two parts ϕ_1 and ϕ_2 which traverse the paths b_1 and b_2, respectively. Then the drops through the two parallel branches of the magnetic circuit are equal, or

$$\mathcal{M} = \mathcal{H}_1 b_1 = \mathcal{H}_2 b_2 \tag{16.136}$$

Two subcases may be considered:

1. ϕ, ϕ_1, ϕ_2 are all known. Here $\mathcal{B}_1$ and $\mathcal{B}_2$, and therefore $\mathcal{H}_1$ and $\mathcal{H}_2$, are known. Thus $\mathcal{M}$ in Eq. (16.136) becomes known and the problem is solved.

2. ϕ is known, but ϕ_1 and ϕ_2 are unknown. Here $\mathcal{M}$ cannot be found in a straightforward manner. It has to be determined by trial. The work is somewhat simplified through the following considerations: Since $\mathcal{M} = \mathcal{H}_1 b_1 = \mathcal{H}_2 b_2$,

$$\mathcal{H}_1 = \frac{b_2}{b_1}\mathcal{H}_2 \tag{16.137a}$$

also

$$\mathcal{B}_1 = \frac{\phi_1}{A_1} \quad \text{and} \quad \mathcal{B}_2 = \frac{\phi_2}{A_2} = \frac{\phi - \phi_1}{A_2} \tag{16.137b}$$

In order to solve the problem, assume values of ϕ_1 and compute $\mathcal{B}_1$ and $\mathcal{B}_2$. Find $\mathcal{H}_1$ and $\mathcal{H}_2$ from the magnetization curves (Fig. 16.34). If these satisfy Eq. (16.137*a*), the problem is solved. Otherwise try again.

Case A_3: Given ϕ, Required $\mathcal{M}$ (Series-Parallel Circuit). This case (Fig. 16.35*c*) also divides itself into two parts.

1. The flux in each of the branches (Fig. 16.35*c*) is known. Here the flux densities $\mathcal{B}_k$ and therefore the intensities $\mathcal{H}_k$ become known, and the problem simply reduces to adding the various drops $\mathcal{H}_k b_k$.

2. Only the total flux ϕ is known, and the branch fluxes ϕ_k are not known. Here a straightforward solution becomes impossible. The problem can be solved only by assuming values of ϕ_1 and ϕ_2, computing the various drops in the series-parallel branches, and testing the result by the following relation:

$$\mathcal{M} = \Sigma\mathcal{H}_{k1}b_{k1} = \Sigma\mathcal{H}_{k2}b_{k2} \tag{16.138}$$

where $\mathcal{H}_{k1}$ are the various intensities in branch 1, and $\mathcal{H}_{k2}$ are the various intensities in branch 2.

Case B_1: Given $\mathcal{M}$, Required ϕ (Series Circuit). This case can be solved only by trial. Thus, referring to Fig. 16.35*a*,

$$\mathcal{M} = \Sigma\mathcal{H}_k b_k \tag{16.139}$$

The solution consists of assuming values of ϕ and finding the flux densities $\mathcal{B}_k$. From the magnetization curves (Fig. 16.34), the corresponding $\mathcal{H}_k$ can be obtained. The substitution of these values in Eq. (16.139) should yield the given $\mathcal{M}$ if the correct value of ϕ has been assumed. Various values of ϕ are assumed until Eq. (16.139) is satisfied.

A special case arises when the series circuit has just two elements, one of which is iron with uniform flux density $\mathcal{B}_i$ and the other of which is an air gap having the same flux density ($\mathcal{B}_a = \mathcal{B}_i$). Here Eq. (16.139) reduces to

$$\mathcal{M} = \mathcal{H}_i b_i + \mathcal{H}_a b_a \tag{16.140a}$$

$$= \mathcal{H}_i b_i + \mu_0\mathcal{B}_a b_a = \mathcal{H}_i b_i + \mu_0\mathcal{B}_i b_a \tag{16.140b}$$

Equation (16.140*b*) is the equation of a straight line in the two unknowns $\mathcal{H}_i$ and $\mathcal{B}_i$. The other equation involving these two unknowns is the magnetization curve of the iron. Hence the required values of $\mathcal{B}_i$ and $\mathcal{H}_i$ are the coordinates of the point of intersection of the straight line [Eq. (16.140*b*)] and the magnetization curve (Fig. 16.34). Having determined $\mathcal{B}_i$, the value of ϕ can be obtained from the expression $\phi = \mathcal{B}_i A_i$, where A_i is the cross-sectional area of the iron.

Case B_2: Given $\mathcal{M}$, Required ϕ (Parallel Circuit). Referring to Fig. 16.35*b*,

$$\mathcal{M} = \mathcal{H}_1 b_1 = \mathcal{H}_2 b_2 = \cdots = \mathcal{H}_k b_k \tag{16.141a}$$

whence

$$\mathcal{H}_k = \frac{\mathcal{M}}{b_k} \tag{16.141b}$$

Having obtained $\mathcal{H}_k$, determine $\mathcal{B}_k$ from the magnetization curve and obtain ϕ from the relation $\phi = \mathcal{B}_k A_k$.

Case B_3: Given $\mathcal{M}$, Required ϕ (Series-Parallel Circuit). The solution of this problem is accomplished by trial. Thus various values of ϕ are assumed, and the corresponding values of the flux densities $\mathcal{B}_k$ are found. Then from the magnetization curves the values $\mathcal{H}_k$ are determined. The relation which must be satisfied by each branch of the series-parallel circuit is $\mathcal{M} = \mathcal{H}_k b_k$.

Conclusion. It will be observed that the only magnetic circuits which possess a straightforward solution are those discussed under cases A_1 and B_2. The remaining circuits are solvable only by trial. This emphasizes the difference between electric and magnetic circuits—a difference which is, in a great measure, due to the fact that magnetic materials do not obey the law of flux superposition. In other words, the magnetization curves (Fig. 16.34) are not straight lines.

44. Hysteresis. If the magnetizing force (Fig. 16.36) is gradually lowered from its maximum value, the $\mathcal{B}$-$\mathcal{H}$ curve will not be retraced. When the magnetizing force is reduced to zero, there will still be a considerable value of magnetic density $\mathcal{B}$. This is known as *residual magnetism*. The negative magnetizing force necessary to bring the density to zero is the *coercive force*. By continuing to build up the magnetomotive force in the negative direction and then again reversing it, as in the curve of Fig. 16.36, the *hysteresis loop* or *hysteresis curve* is produced. The area of this loop is a measure of the energy consumed in carrying the iron through the cycle of magnetization.

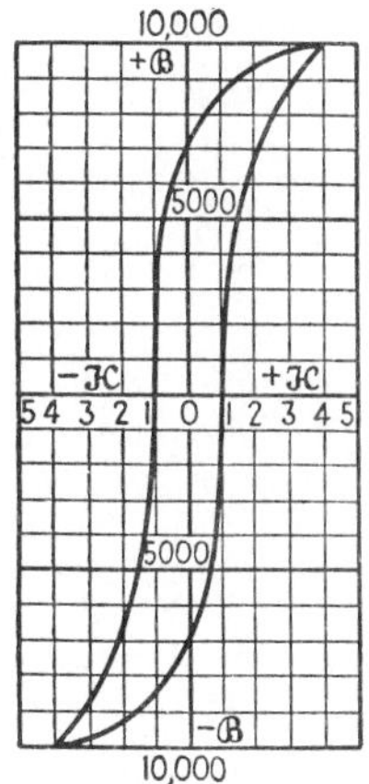

FIGURE 16.36 Hysteresis loop.

When a piece of iron is put through a cycle of positive and negative magnetization, a certain amount of energy is lost in the form of heat. This loss is measured by the area of the hysteresis curve (Fig. 16.36). Whenever iron is subjected to the magnetization effect of an alternating current, this cycle of hysteresis loss occurs during every period of the alternating current. Experimentally, it can be shown that the hysteresis loss P_h, in watts, varies as a power h (less than 2) of the magnetic density $\mathcal{B}_m$, as the volume V of the iron, as its magnetic quality k, and as the frequency f. Hence the loss may be expressed by the following relation:

$$P_h = kfV\mathcal{B}_m^h 10^{-7} \tag{16.142}$$

The exact data on all iron losses for the steel to be used should be obtained from the maker. The loss is lowered by alloying steel with silicon, thus forming what is known as silicon steel.

45. Eddy Currents. Whenever a conductor is placed in a varying magnetic field, emf's are set up in the conductor. These cause local currents known as *eddy currents* which produce heat and are a source of loss. To keep these losses down, the iron of a magnetic circuit, which carries the variable flux, is made up of thin laminations. The eddy-current loss P_{ee} in thin laminations is

$$P_{ee} = kf^2V\mathcal{B}_m^2 t^2 10^{-6} \tag{16.143}$$

where P_{ee} is the eddy-current loss in watts, t is the thickness of the laminations, and the other terms are as defined in Eq. (16.142).

Equation (16.143) shows the way in which the eddy-current loss varies with the various factors involved. Uncertainty with regard to the value of k, however, makes it impracticable to use the formula for quantitative data; hence test results are generally furnished by the manufacturer.

46. Magnetic Pull or Traction; Magnetic Mechanisms. When there is a cut at right angles across the iron of a magnetic circuit and the two faces are in close contact, the magnetic pull which is exerted between the faces is equal to $\mathcal{B}^2A/(8\pi \times 981)$ g or $\mathcal{B}^2A/11{,}183{,}000$ lb (lb $\times$ 4.45 = N) ($\mathcal{B}$ in gausses or lines per square centimeter, A in square centimeters). In such devices as magnetic

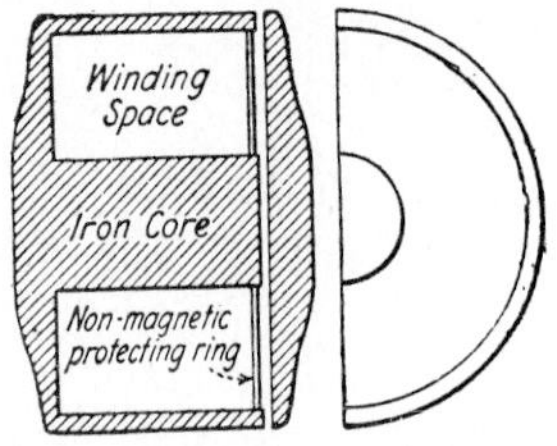

Figure 16.37 Iron-clad magnet.

clutches, faceplates, brakes, and hoists, this expression for traction makes the design comparatively simple. Where, however, the pull is called on to move part of the mechanism, the presence of leakage around the necessary gap complicates the calculation. The shorter the gap, the more nearly will the pull come up to that given by the formula. It is, therefore, desirable to make use of levers and other devices to shorten the required distance of travel of the magnet armature. One method of doing this is to use a conical pole piece, with an armature through which a corresponding conical hole is bored. A desirable form is the iron-clad magnet, in which one pole consists of an iron cylinder surrounding the coil, while the other pole constitutes the core of the coil. The armature is then a circular disk (see Fig. 16.37).

47. Magnetic Force Produced by Current-Carrying Conductors. Since whenever magnetic fields come in contact with each other a force is exerted between them, and since all current-carrying conductors are surrounded by magnetic fields, it follows that a force will be exerted by every current-carrying conductor on any other similar conductor and also on any magnetic pole in its vicinity. Conversely, any conductor carrying current and situated in a magnetic field will experience a pull at right angles to the conductor and the field. This pull is equal to $10.2\mathcal{B}I \times 10^{-8}$ kg/cm length when the conductor carries a current I and is in a field of $\mathcal{B}$ gauss. The operation of the electric motor is based on this fact. Two conductors of length l lying parallel to each other at a distance of d cm and carrying currents I_1 and I_2, respectively, will experience a pull of

$$\frac{(2.04 \times 10^{-8} I_1 I_2 l)}{d} \qquad (\text{kg} \times 9.81 = N)$$

When the conductors are surrounded by iron, the pull will, of course, be much greater. These magnetic forces may be great enough to do serious damage to electrical equipment, especially in the event of short circuits.

ELECTROSTATIC CIRCUIT

48. The Electrostatic Charge. When an emf E (Fig. 16.5*a*) is applied to the plates of a condenser (see Sec. 7) with a capacity C, an electrostatic field is established between these plates owing to the accumulation of negative and positive charges on their surfaces. The amount of charge Q that accumulates on the plates depends on (1) the potential E impressed on the plates, (2) the area A of the plates, (3) the distance d between the plates, (4) the kind of dielectric occupying the space between the plates, and (5) the time that elapses from the instant of impressing E to the instant of measuring the charge Q.

In most commercial dielectrics, the variation of charge with each of these factors can be ascertained only by experiment. Indeed, there seems to be no general law which all dielectrics do obey. In order to simplify this discussion, assume a perfect dielectric, i.e., one for which there exists a linear relation between Q and

E, and Q and A, as well as between Q and $1/h$. Moreover, such a perfect dielectric will have an infinite resistance so that no conduction current will flow through it. Finally, in such a dielectric, Q is independent of time, so that the dielectric will be assumed to take its full charge instantly. With the foregoing assumptions (which, except for time, hold true for air and vacuum), the following relation can be set forth as an empirical result for a two-plate condenser:

$$Q = CE = (kk_a)\left(\frac{A}{h}\right)E \tag{16.144}$$

where k = dielectric constant (permittivity) of the material
k_a = dielectric constant of air
A = area of one plate
h = distance between the plates
E = applied potential

From Eq. (16.144) the constant C, which is known as the *capacitance* of a condenser, may be determined from the dimensions of the condenser. Thus

$$C = kk_a \frac{A}{h} \tag{16.145}$$

49. Elastance as the Reciprocal of Capacitance. Another form of writing Eq. (16.144) is as follows:

$$E = \frac{Q}{C} = SQ = \left(\frac{1}{kk_a}\right)\frac{h}{A}Q \tag{16.146}$$

where

$$S = \frac{1}{C} = \left(\frac{1}{kk_a}\right)\frac{h}{A} = \sigma\sigma_a \frac{h}{A} \tag{16.147}$$

The symbol S stands for the elastance of a condenser and corresponds to resistance in the electrodynamic circuit. From Eq. (16.147) one observes that S is directly proportional to the distance between the plates of a condenser and inversely proportional to the area of any one plate.

50. Units. The preceding laws are true irrespective of the choice of units. In order to apply them to numerical problems, however, a definite system of units is necessary. The practical system of units, as given in Table 16.4, are recommended. The constants σ_a and k_a for this system of units have the following values:

$$\sigma_a = 11.3 \times 10^{12} \text{ darafs per cm}^3 = 4.45 \times 10^{12} \text{ darafs per in}^3$$

$$k_a = 0.08842 \times 10^{-12} \text{ farad per cm}^3 = 0.2244 \times 10^{-12} \text{ farad per in}^3$$

51. Charge Density and Potential Gradient. In the electrodynamic circuit, the concept of current density is helpful in the solution of problems, so here the concept of the charge density D is introduced, which will be defined as follows:

$$D = \lim_{\Delta A \to 0}\left(\frac{\Delta Q}{\Delta A}\right) \tag{16.148a}$$

When the charge Q is equally distributed over the area of the dielectric, Eq. (16.148a) becomes

$$D = \frac{Q}{A} \tag{16.148b}$$

The voltage gradient G is defined as

$$G = \lim_{\Delta h \to 0} \left(\frac{\Delta E}{\Delta h}\right) \tag{16.149a}$$

and for a linear variation of E with h, Eq. (16.149a) becomes

$$G = \frac{E}{h} \tag{16.149b}$$

The following interesting and useful relationship exists between G and D:

$$D = \frac{Q}{A} = \frac{CE}{A} = \left(\frac{kk_aA}{h}\right)\left(\frac{E}{A}\right) = kk_aG \tag{16.150a}$$

$$G = \left(\frac{1}{kk_a}\right)D = (\sigma\sigma_a)D \tag{16.150b}$$

52. Dielectric Strength. The *strength* of a dielectric is defined as the voltage it can stand per unit thickness (without breaking down) under certain specified conditions of temperature, humidity, time of application of potential, etc. The mechanism of breakdown in dielectrics is still unknown and offers a great field for research. But whatever the mechanism may be, it is a commonly observed fact that dielectrics do break down when the potential applied to them exceeds a certain fixed value known as the *dielectric strength* of the material. This phenomenon finds its analogue in the breakdown of materials such as steel due to the application of a stress exceeding their elastic limit.

Values of dielectric constants and dielectric strengths of various insulating materials are given in Tables 16.5 and 16.6. These values are to be used with great

TABLE 16.5 Dielectric Constants

Material	Dielectric constant k	Material	Dielectric constant k
Glass (easily fusible)	2.0 to 5.0	Air and other gases	1.0
Glass (difficult to fuse)	5.0 to 10.0	Alcohol, amyl	15.0
Gutta-percha	3.0 to 5.0	Alcohol, ethyl	24.3 to 27.4
Ice	3.0	Alcohol, methyl	32.7
Marble	8.3	Bakelite	4.5 to 5.5
Mica	2.5 to 6.6	Benzine	1.9
Paper with turpentine	2.4	Benzol	2.2 to 2.4
Paper or jute impregnated	4.3	Micarta	4.1
Porcelain	5.7 to 6.8	Olive oil	3.0 to 3.2
Rubber	2.4	Paraffin	2.3
Rubber, vulcanized	2.5 to 3.5	Paraffin oil	1.9
Shellac	2.7 to 4.1	Petroleum	2.0
Silk	1.6	Turpentine	2.2
Sulphur	4.0	Water	81

Source: From C. E. Magnusson, *Alternating Currents*, McGraw-Hill, New York.

TABLE 16.6 Dielectric Strengths

Material	Strength, kv. per mm.	Material	Strength, kv. per mm.
Air at atmospheric pressure (76 cm.)	3.0	Melted paraffin	7.5 to 20.0
Boiled linseed oil	8 to 19	Mica	25.0 to 220.0
Dry wood	0.4 to 0.6	Micanite	33 to 40
Fiber, vulcanized	8.0 to 18.0	Paraffined paper	40.0 to 60.0
Fish paper	10.0 to 15.0	Paraffin oil	16 to 21
Kraft paper	4.0 to 6.0	Turpentine	3.5 to 16
Marble	2.0 to 4.0	Varnished cambric	45.0 to 70.0
Maple, oiled or paraffined	3.0 to 45.0	Varnished silk	45.0 to 70.0
		Vulcanized rubber	9.0 to 10.0

Source: From C. E. Magnusson, *Alternating Currents*, McGraw-Hill, New York.

caution, however, because the conditions under which they were obtained may vary materially from the conditions of a given problem.

SOURCES OF EMF: GENERATORS

53. Electromagnetic Induction. Faraday discovered that when relative motion exists between a metallic conductor and a magnetic field, an emf is induced in the conductor. The magnitude of this emf is

$$e = \mathcal{B}vl \tag{16.151}$$

where e = voltage induced between the terminals of the conductor
$\mathcal{B}$ = flux density of the magnetic field in a direction perpendicular to the axis of the conductor
l = length of the conductor
v = relative velocity of the conductor and the magnetic field

The essential limitations of Eq. (16.151) are as follows:

1. $\mathcal{B}$, v, and l shall be directed along three mutually perpendicular axes, as shown in Fig. 16.38.
2. $\mathcal{B}$ shall be uniform along the entire length of the conductor, but not necessarily uniform along the direction of v.
3. Since the velocity v is relative, it is immaterial whether the conductor or the flux performs the motion. As a matter of fact, in most commercial alternators the conductor is stationary while the field revolves. In continuous-current machines, however, the flux is stationary and the conductor revolves.

54. Single-Phase Alternators. Sine voltages are produced by machines called *alternators* which utilize Faraday's law of electromagnetic induction. In order to make this statement more real, the following brief description of such a machine is given.

Figure 16.39 shows a cross section of four magnetic poles marked N (north pole) and S (south pole). These poles are mounted on a shaft and are free to revolve inside an iron cylinder A, known as the *armature* of the machine.

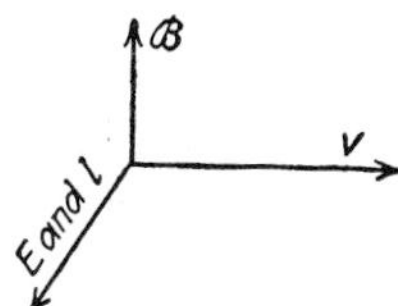

FIGURE 16.38

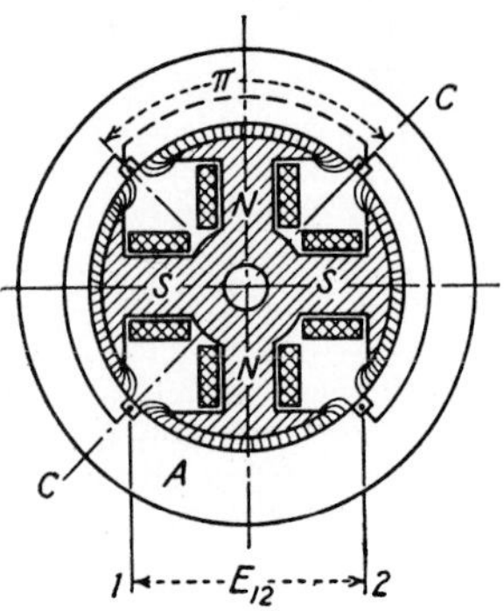

FIGURE 16.39

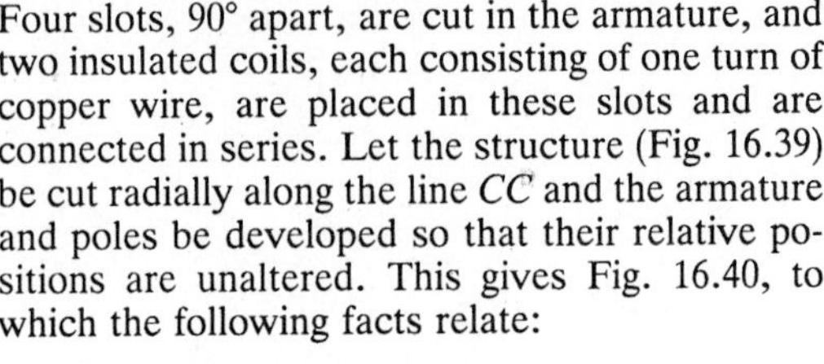

Four slots, 90° apart, are cut in the armature, and two insulated coils, each consisting of one turn of copper wire, are placed in these slots and are connected in series. Let the structure (Fig. 16.39) be cut radially along the line CC and the armature and poles be developed so that their relative positions are unaltered. This gives Fig. 16.40, to which the following facts relate:

1. The flux emanating from a north pole is considered positive, and the flux entering into a south pole is negative. This is indicated by drawing the flux-density wave (sine wave with amplitude $\mathcal{B}_m$) so that its positive values occur opposite a north pole and its negative values opposite a south pole.

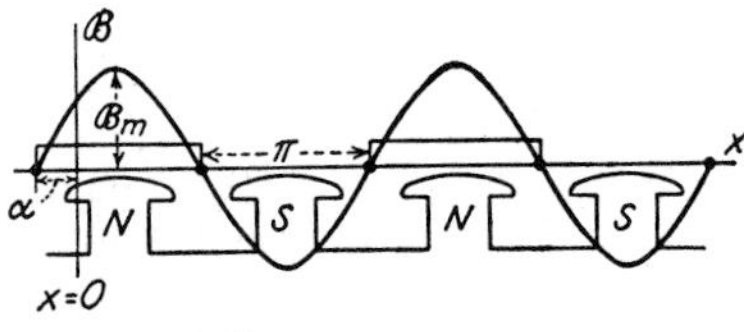

FIGURE 16.40

2. The flux-density curve is assumed to be a sine function of space along the air gap. Therefore, its positive maximum occurs opposite the midpoint of a north pole, and its negative amplitude is opposite the middle of a south pole. The mathematical expression for this wave is

$$\mathcal{B} = \mathcal{B}_m \sin (x + \alpha) \qquad (16.152)$$

3. From paragraph (2), it follows that a pole pitch (the distance from one point on a north pole to a similar point on an adjacent south pole) is equal to π radians measured along the abscissa of the flux wave.*

Now let v be the relative velocity of the pole structure and l the length per conductor. Then the emf between the terminals of each conductor is, according to Eq. (16.151),

$$e = \mathcal{B}_m v l \sin (x + \alpha) = E_m \sin (x + \alpha) \qquad (16.153)$$

where

$$E_m = \mathcal{B}_m v l \qquad (16.154)$$

Let the field rotate through an electrical angle x in time t, and let its velocity be such that it takes T seconds to move over a distance of 2π electrical radians. Then

$$\frac{x}{2\pi} = \frac{t}{T}$$

*When distance along the periphery of the armature is expressed in radians, measured along the abscissa of the flux curve (Fig. 16.40), the term *electrical radians* is used. Distinction should clearly be drawn between an electrical radian and a mechanical radian. Thus there always exist π electrical radians (Figs. 16.39 and 16.40) between any two corresponding points of a north pole and an adjacent south pole, irrespective of the number of poles in a machine. On the other hand, considering that there are 2π mechanical radians in a circle, the number of mechanical radians between any two adjacent pole centers is $2\pi/p$, where p is the number of poles in a machine.

$$\therefore x = \frac{2\pi t}{T} = 2\pi f t = wt \tag{16.155}$$

Substituting this value of x in Eq. (16.153),

$$\begin{aligned} e &= E_m \sin (wt + \alpha) = E_m \sin (2\pi f t + \alpha) \\ &= E_m \sin \left(\frac{2\pi t}{T} + \alpha\right) \end{aligned} \tag{16.156}$$

Thus there is a sine voltage of amplitude E_m and frequency f.*

55. Polyphase Alternators

Two-Phase Alternator. A *two-phase alternator* (Fig. 16.41a) differs from a single-phase machine in that it comprises two independent windings instead of just one. Thus let the armature in Fig. 16.41a have eight equidistant slots. Let these slots

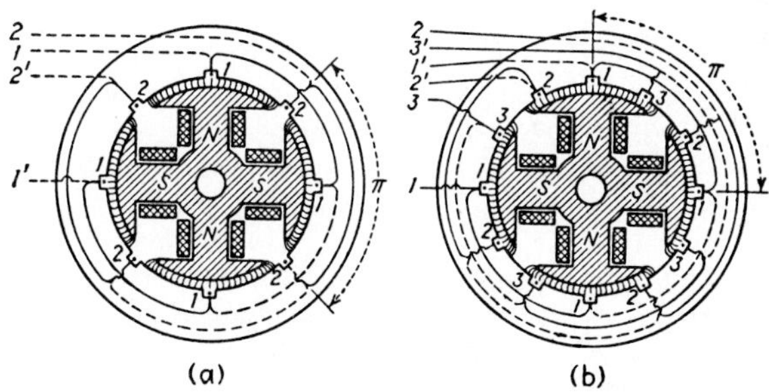

FIGURE 16.41

be divided into two groups by marking them alternately 1 and 2. Let insulated conductors be placed in these slots and be so connected that all conductors occupying slots 1 are connected in series and all those occupying slots 2 are also connected in series, so as to form two independent windings with terminals 11′ and 22′, respectively. If the flux distribution along the air gap is sinusoidal, then the voltage induced in any one conductor also will be sinusoidal. Furthermore, the sum of the emf's induced in the first group of conductors is equal to the terminal voltage $E_{1'1}$, and the sum of the emf's induced in the second group of conductors is equal to the terminal voltage $E_{2'2}$.

An inspection of Fig. 16.41a shows that at the instant when the first group of conductors is under the centers of the poles, the second group of conductors falls in the neutral spaces between two adjacent poles. In other words, when the flux cutting the first group of conductors is a maximum, the flux cutting the second group is zero, and vice versa. Since the induced voltage in a conductor is proportional to the density of the flux which cuts that conductor ($e = \mathcal{B}vl$), when the

*It is desirable to develop a relation between the frequency of the emf generated and the number of revolutions that the pole structure makes. Consider an alternator having p poles and making S revolutions per second. Each conductor passes through $p/2$ cycles of flux per revolution because it takes two poles to make a complete cycle of flux (see Fig. 16.40). Thus, in 1 s it passes through $pS/2$ cycles. Consequently $f = pS/2$ cycles per second.

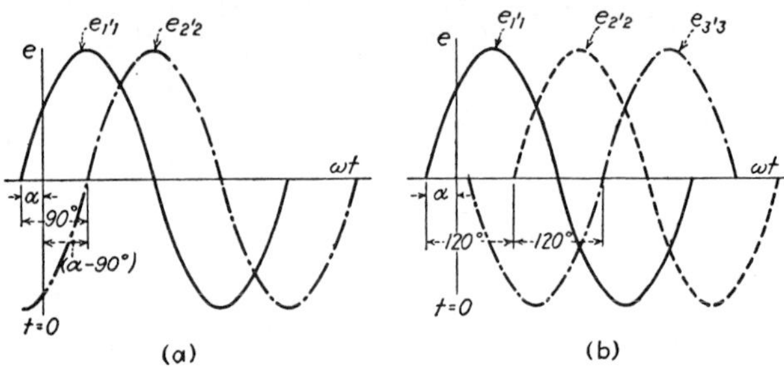

FIGURE 16.42

voltage induced in the first group of conductors is a maximum, that induced in the second group is zero, and vice versa. There exists, therefore, a phase displacement in time of 90° (Fig. 16.42*a*) between the induced voltage $e_{1'1}$ and $e_{2'2}$. The first group of conductors (Fig. 16.41*a*) forms a winding which is called *phase 1*, and the second group forms another winding which is known as *phase 2*. The machine is a two-phase alternator.

Three-Phase Alternator. The *three-phase system* has come to be almost universally used. Let each pole pitch (covering π electrical radians; Fig. 16.41*b*) be divided into three equal parts such that the points of division are 60 electrical degrees apart. Let slots be cut at these points of division, and let insulated conductors be placed in these slots. Figure 16.41*b* thus has 12 equidistant slots. Divide these slots into three groups by marking them consecutively 1, 3, 2 (not 1, 2, 3). Connect in series all the conductors occupying slots 1 and obtain a winding with the terminals 1′1. Do the same with the conductors occupying slots 2 and slots 3, respectively, and obtain two windings with terminals 2′2 and 3′3. The alternator is then said to be a *three-phase alternator* having three separate windings interconnected to give three terminals. The induced voltages $e_{1'1}$, $e_{2'2}$, and $e_{3'3}$, measured at these terminals, are displaced from each other (in time) by 120°, as shown in Fig. 16.42*b*.

Although alternators are generally constructed three-phase, the reader can see how an electric machine may be wound for any number m of phases by having an independent winding for each phase. The conductors constituting these separate windings are generally so displaced in space that there exist $2\pi/m$ electrical radians between any two corresponding conductors of two consecutive phases. Rotary converters furnish a practical illustration of electric machines built for six or more phases.

56. Direct-Current Generators

Construction. In Sec. 53 the statement was made that the relative velocity of the conductor and flux is the cause of emf generation. In Sec. 54 a machine was described in which the emf is generated by the rotation of the field structure while the armature (conductor structure) is held stationary. Let it now be assumed that the armature structure will revolve while the field will remain stationary. Such a machine is shown in Fig. 16.43*a*; the voltage induced in the conductors is a line voltage.

Commutator. Assume that in addition to the rotating armature there is a structure which contains pieces of copper C which are insulated from each other and

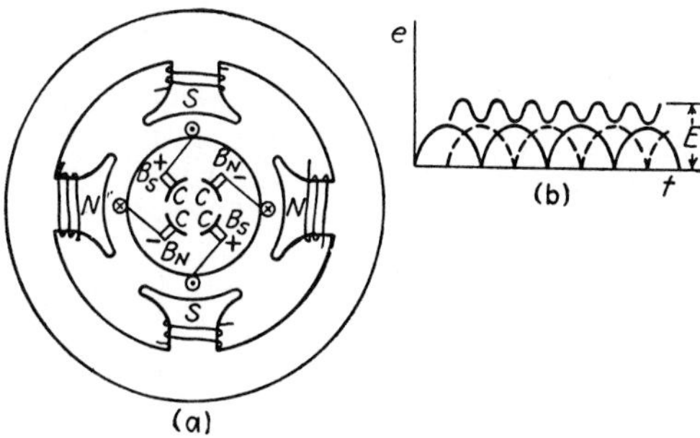

FIGURE 16.43 Direct-current generator.

to which the terminals of the armature coils are connected. Such a structure is called a *commutator*. Let stationary brushes be in contact with the commutator. As the armature revolves, the emf induced in any one coil side changes direction, but since each brush makes contact only with that coil side which is under a definite pole, the emf at the brush remains constant. The net result of such an arrangement is to produce, at any one brush, an emf which is unidirectional, as shown in Fig. 16.43*b*. Thus the actual emf produced by such a machine is (except for the ripples, which become less pronounced as more conductors are utilized) a dc voltage.

57. Types of DC Generators. Alternators require a separate source of direct emf to excite their field windings. This is not true of dc generators, whose field excitation may be taken directly from the emf generated in the dc machine. The methods of connecting the field winding or windings to the terminal voltage of a machine vary, and this variation gives rise to differing operating characteristics. Accordingly, dc generators are divided into the following classes:

1. A *shunt generator* (Fig. 16.44*a*) is one that has its field winding connected across the armature. Its field strength, therefore, varies with the armature volts. Since the *IR* drop through the generator varies with the load, for a fixed position of the field rheostat the generator will have a drooping voltage curve (see Fig.

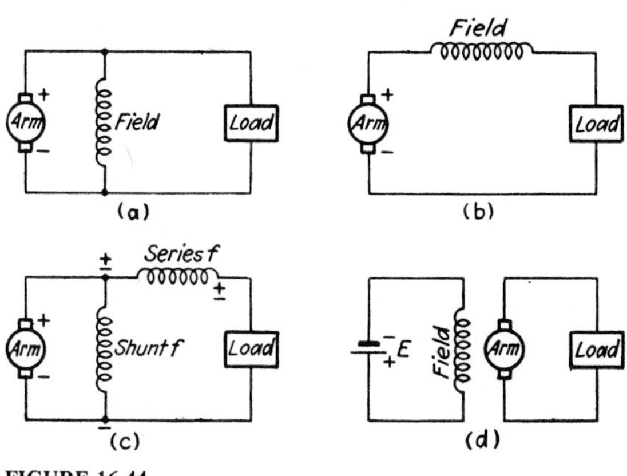

FIGURE 16.44

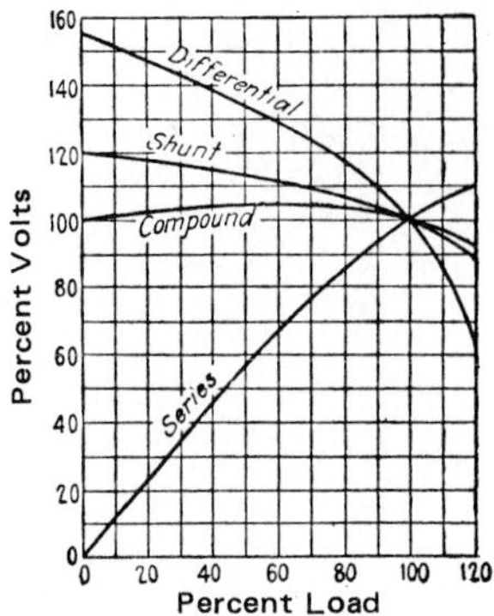

FIGURE 16.45 Characteristic curves of dc generators.

16.45). The amount of the drop in voltage from no load to full load depends on the magnetic densities, the amount of armature reaction, and the *IR* drop. For commutating-pole generators, the drop is about 15 to 20 percent of the line voltage. The shunt type of generator is not used to any great extent.

2. A *series-wound generator* (Fig. 16.44*b*) is one which has its field in series with the armature. Its excitation, therefore, varies with the load. The voltage rises rapidly with the load (see Fig. 16.45). Generators of this type are rarely built.

3. A *compound-wound generator* (Fig. 16.44*c*) is a combination of a shunt and series winding. The no-load voltage is obtained by the shunt winding. The series winding simply takes care of the *IR* drop and armature reaction. The voltage curve is called a *compound characteristic* (see Fig. 16.45). For slow-speed generators this curve (see Fig. 16.45) has a large hump; for moderate- and high-speed machines the curve is very flat. For some classes of service it is desired to *overcompound* the generator, i.e., to have a full-load voltage higher than the no-load voltage. This is to compensate for line drop under load.

4. A *differential generator* (Fig. 16.44*c*) has the series field in opposition to the shunt field, and thus there is more droop to the voltage curve (Fig. 16.45). This type is used in electric-shovel work, arc welding, and electric traction. By varying the percentage of differential and shunt fields, the load characteristic can be given any desired shape.

5. A *separately excited generator* (Fig. 16.44*d*) is usually a shunt-wound machine but with the field excited from an external source. This is necessary when the generator is operating over a variable voltage that starts from zero, especially if the generator has to carry a heavy load at low voltage. Under certain conditions, the series fields may be separately excited to obtain definite voltage curves.

The behavior of these various types of dc generators with load is shown in Fig. 16.45, where the terminal voltage of the machine is plotted as a function of the load current.

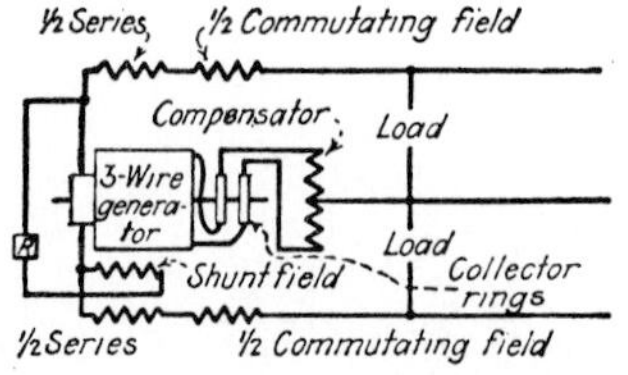

FIGURE 16.46 Three-wire generator connections.

58. Three-Wire Generators. Three-wire generators are built for use on the Edison three-wire system. They are usually 250-V and fundamentally standard dc generators with additions. There are two methods of connecting the third or neutral wire. One method is to build in a revolving compensator whose ends are connected to opposite sides of the dc armature winding, usually on the back end, and connect the center of its winding to a collector ring. The second and usual method is to connect two collector rings to the opposite sides of the armature winding and have an external compensator connected between them. The neutral wire is connected to the center of the compensator winding. In order to have the correct *IR* drop and field ampere-turns, alternate series and commutating fields are connected on the opposite sides of the armature (see Fig. 16.46).

SOURCES OF EMF: ELECTRIC BATTERIES

59. Definitions. *Electric battery cells* are devices for generating or storing electric energy by means of electrochemical action. Primary cells are those in which, by chemically transforming one or another of a group of materials, electric energy is generated. The *secondary cell* is one which, by means of a reversible electrochemical action, converts electric into chemical energy, which is then available to be returned again as electric energy. Such a cell is also called a *storage cell*. Two or more cells connected, usually in series, constitute a *battery*. The terminals of a cell are also called *poles*. The *positive terminal* or pole is that from which the current flows to the external circuit. The active chemical portions of the cell, connected to the terminals and immersed in the electrolyte, are the *electrodes*. The electrode connected to the negative terminal of the cell is called the *anode* because current flows from it into the electrolyte. The electrode into which the current flows in the cell is the *cathode*. In many primary cells, the anode is metallic zinc. *Polarization* is a secondary reaction occurring at the surface of one or both of the electrodes, which introduces a back emf and reduces that available at the terminals. A polarized cell, if allowed to stand on open circuit, will regain its normal emf. *Depolarizers* are means, usually chemical, for preventing polarization.

60. Wet Primary Cells. The development of storage cells and dry cells has greatly decreased the importance of primary cells using a fluid as an electrolyte. Before the development of the dynamo generator, such cells formed the only practicable means for the generation of electric energy. Tables 16.7 and 16.8 give the data for several of the most important wet cells.

61. Dry Cells. The only commercial dry cell makes use of the chemicals of the Leclanché wet cell, the ammonium chloride solution being held in an absorbent material which separates the electrodes. This cell is moist rather than dry and is useless when completely dried out. In the usual form, a cylindrical container of

TABLE 16.7 Wet Cells

Name	Electrolyte	Positive	Negative	Depolarizer
Chromic acid	H_2SO_4 or NaCl	Carbon	Zinc	CrO_3
Daniell	H_2SO_4 or $ZnSO_4$	Copper	Zinc	$CuSO_4$
Gravity	$ZnSO_4$ or H_2SO_4	Copper	Zinc	$CuSO_4$
Edison-Lalande	NaOH	CuO + Cu	Zinc	CuO
Leclanché	NH_4Cl	Carbon	Zinc	MnO_2

TABLE 16.8 Wet Cells

Name	E.m.f., volts	Resistance, ohms	End result
Chromic acid	2.00	5 to 4	$Cr_2(SO_4)_3 + 3ZnSO_4$
Daniell	1.07 to 1.14	0.3 to 30	$Cu + ZnSO_4$
Gravity	1.00	0.1 to 6	$Cu + ZnSO_4$
Edison-Lalande	0.75	0.02 to 0.1	$Cu + Na_2ZnO_2$
Leclanché	1.50	1 to 5	$Mn_2O_3 + 2NH_3 + ZnCl_2$

zinc is connected to the negative terminal, and a central rod of carbon to the positive. Surrounding the carbon rod is a mixture of carbon and manganese oxide used as a depolarizer. Between this and the inside of the zinc cylinder is the absorbent material, saturated with ammonium chloride and the zinc chloride which is formed during discharge. The open-circuit voltage, 1.5 V, is reduced to about 1 V on closed circuit, principally by polarization. Even a good dry cell, when unused, becomes worthless, owing to drying out and local action, after about a year. The most common size of large cell has a 2½-in (6.4-cm) diameter and a 6-in (15-cm) length. On short-circuit tests such a cell should give about 20 A or more. The cylindrical unit cell most used in flashlights is 1⁵⁄₁₆ × 2¼ in (3.3 × 5.7 cm); two or three such cells are generally used in series. The data in Table 16.9 show typical dry-cell capacities. For an end voltage of 1 V, the values were about two-thirds of the above. For 0.6-V end voltage, they were about 50 percent greater. If the service is intermittent, the capacity may be increased as much as two or three times, but if the test extends over many days, the total capacity is decreased by local action.

TABLE 16.9 Capacity of Dry Cells [6 in (15 cm)]

(End voltage, 0.8 volt)

Amp	Hr	Amp-hr
½	16	8
¼	51	12
⅛	143	18
1/16	414	26
1/32	1078	34

62. Storage Cells. These are cells which are reversible as to their electrochemical action. That is, charging a discharged cell with electric current will bring it back to its original chemical conditions and thus make it available for another discharge. In order that the plates may keep their original form, it is necessary that the active material should not be soluble in the electrolyte. Such cells are also called *secondary cells* or *accumulators*, as distinguished from primary cells. Two types of storage cells are in general use: the lead cell and the alkaline cell.

63. The Lead Cell. The electrodes of this cell consist of lead grids or frames which carry the active material. When the battery is completely charged, the active material on the positive plate is principally lead peroxide (PbO_2), and on the negative plate spongy metallic lead. During discharge both materials are converted into lead sulfate ($PbSO_4$). The positive plate is reddish brown and the negative plate is grayish in color. There is always one more negative than positive plate, because the negative is more robust and withstands better the effect of the reactions on one side only. The electrolyte is chemically pure sulfuric acid and water. For stationary batteries, the specific gravity varies from 1.150 to 1.160 on discharge and rises to 1.200 or 1.215 when fully charged. In portable and vehicle batteries, the specific gravity may be 1.300 when charged. The smaller cells are usually carried in hard rubber or glass jars, sealed at the top except for small ventilating holes. Large cells have lead-lined wooden tanks. During the life of the cell, the active material slowly disintegrates and falls from the plates and thus eventually terminates the usefulness of the cell. This sediment collects in the lower part of the container, so that it is necessary to support the plate a considerable distance from the bottom to prevent the sediment from coming in contact with, and short-circuiting, the plates. Thin separators of wood, plastic, or rubber are generally used between the plates.

There are, in general, two methods of applying the active material to the plate. For the *pasted* plate, the lead oxide is made into a paste and forced into the interstices of a suitable grid of pure lead which forms the plate. Such plates are assembled and placed in sulfuric acid. A forming current is then used to convert the positive plate into lead peroxide and the negative plate into spongy lead. In the *formed* or Planté plate, the active material is formed from pure lead, sometimes from the supporting frame itself, by chemical means. The Planté type is usually heavier, stronger, and more durable. The pasted type is generally used for portable and vehicle batteries.

64. Operation of Lead Storage Cell. During the charge, the emf rises from about 2.2 to 2.6 V, though this latter figure may range in extreme cases from 2.4 to 2.8. The discharge starts at about 2 V and is usually carried to 1.8 V or, with heavy current, as low as 1.7 V. The state of charge or discharge can be determined by the voltage only when normal current has been passing through the battery for a period of several minutes. Nothing concerning the condition of the cell is shown by the open-circuit voltage. The state of charge can be determined accurately by a hydrometer reading, and the completion of charging is indicated by the violent emission of gas, due to decomposition of the water.

The capacity rating of a cell is usually based on a discharge continued during a certain number of hours. For stationary batteries, this is usually 8 h, but for vehicle batteries it may be 3 or even as low as 2 h. A battery is not injured by a wide variation of discharge rates. When discharged (or charged) at a high rate, its ampere-hour capacity for the discharge in question is greatly decreased. For a 5-h discharge, the capacity of a cell is about ⅞ of the 8-h-rate capacity, and for a 1-h discharge, this becomes ½.

While heavy discharge rates can generally be handled without injury to the battery, frequent discharge below 1.8 V, and especially complete discharge, causes the formation of an insoluble lead sulfate, which seriously damages the battery. This sulfate may, to some extent, be reconverted into the normal active material by long-continued slow charging. Since, when the battery is discharged, the sulfuric acid is weakened, there is danger that the battery will freeze under such conditions. For a specific gravity of 1.15, normal when a battery is discharged, freezing occurs at about 5°F (−15°C); for 1.2, at −16°F (−26.7°C); and for 1.22, at −30°F (−34.4°C). The efficiency of a lead cell at a given discharge rate is the watthours received on discharge, divided by the watthours used on the previous charge. The charge must be at a specified rate, and the starting and stopping states of the battery must be the same.

65. Alkaline Storage Cells. The electrolyte of the Edison cell is a 21 percent solution of potassium hydrate in water; it also contains a little lithium hydrate. The active electrode materials for the charged cell are nickel peroxide for the positive plate and finely divided iron for the negative plate. Nickel flakes are added to the nickel peroxide to increase its conductivity, and small amounts of other chemicals are present to promote the normal reactions. When discharged, the active material of both plates becomes a mixture of iron and nickel hydrates and oxides. The active materials on both plates are held in their perforated containers of nickel steel, which form parts of the plates. Alternate positive and negative plates are mounted on insulated steel rods, and the elements so combined are placed in a nickel-steel container. These containers must, therefore, be insulated from each other in mounting. The chemical reactions at the two plates balance each other, so that there is no change in the electrolyte except some loss of water by decomposition and evaporation. In charging, the emf, starting at about 1.5 V, rises rapidly to 1.7 V and then gradually to 1.8 V when fully charged. The discharge, starting at 1.4 V, falls rapidly to 1.3 V, then gradually to 1.1 V on the completion of the discharge. The preceding voltages are considerably decreased at high discharge rates. With four times normal discharge rates, the operating range of voltage will be from 1.0 to 0.8 V. Seven hours is the normal charging rate for most Edison cells. High discharge rates can be used without injury. Overcharging or overdischarging does not seriously injure the cells, which are much more robust than the lead cells. The specified capacity of a cell is normally based on a 5-h discharge rate. The capacity of a cell increases during its early use, sometimes as much as 30 percent above the rating. A 40°F (22°C) rise in temperature decreases the voltage about 0.1 V, and the capacity of the cell is greatly decreased by low temperature.

At normal discharge rate, the watthour efficiency is about 72 percent, and at a 1-h discharge rate, about 58 percent. The internal resistance of the alkaline cells is considerably higher than that of the lead cells, so that there is an immediate drop of about 8 percent with normal discharge currents.

The Hubbell cell differs from the Edison cell principally in the use of cadmium instead of iron for the negative plate. It is used for miners' lamps.

66. Battery-Regulating Equipment. Owing to the large change in voltage between charged and discharged conditions, voltage-regulating equipment is needed. An *end-cell switch* is similar to the faceplate of a rheostat, but the contacts are connected in between the cells, so that a successively increasing number can be used, as the voltage per cell drops during discharge. This switch should be designed so that in passing from cell to cell, while the circuit is not opened, no cell is short-circuited through the contacts. Instead of cutting in regular cells, this switch may cut out *counter-emf cells*. These are cells of low capacity, the voltage of which is always used in opposition to the voltage of the main battery, so that the end cells are always charged during use. A booster generator may be used to obtain increased voltage required by the battery during charge.

TRANSFORMERS*

67. General Considerations. Figure 16.47 shows an iron core on which two windings are placed. The *primary winding* has n_1 turns; the *secondary winding* has n_2 turns. If an alternating voltage e is impressed on the primary, a current i_1 will flow. This current causes flux ϕ in the core of the transformer. Moreover,

*The remaining portion of this section has been adapted from the section on Electrical Machinery of the 1st edition of the *General Engineering Handbook*, which was prepared by Henry W. Chadbourne, Industrial Engineer, General Electric Company.

since the voltage and hence the current in the primary are alternating, the flux ϕ will alternate. But the flux threads both the primary and secondary windings. Hence in both primary and secondary windings a voltage e_2 is induced of such magnitude that

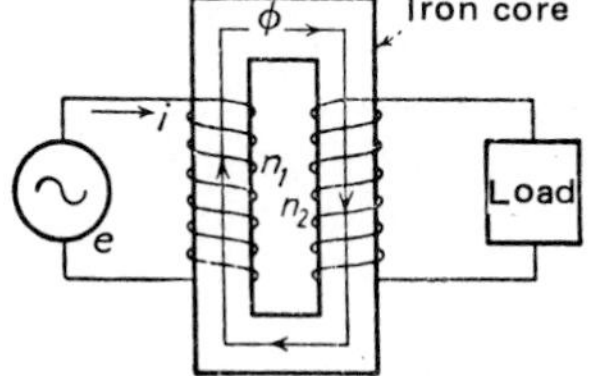

FIGURE 16.47 Transformer.

$$e_1 = n_1 \frac{d\phi}{dt} \qquad (16.157a)$$

$$e_2 = n_2 \frac{d\phi}{dt} \qquad (16.157b)$$

Dividing Eq. (16.157*a*) by Eq. (16.157*b*),

$$\frac{e_1}{e_2} = \frac{n_1}{n_2} \qquad (16.158)$$

Virtually, e_1 is equal to the applied voltage e. Moreover, Eq. (16.158) states that any desired ratio of primary to secondary voltage can be obtained merely by varying the ratio of the number of turns n_1 and n_2 in the primary and secondary. A *transformer* is therefore a device which is used to transform the voltage from high to low, or vice versa.

According to the law of conservation of power,

$$E_1 I_1 = E_2 I_2 \qquad (16.159a)$$

$$\therefore \frac{I_2}{I_1} = \frac{E_1}{E_2} = \frac{n_1}{n_2} \qquad (16.159b)$$

Thus while the ratio of primary to secondary voltage varies directly as the number of turns, the ratio of primary to secondary current varies inversely as the number of turns.

Generally speaking, in a powerhouse where electricity is generated, transformers are used to *step up* the voltage to a higher value for transmission purposes. At the other end of the transmission line, in substations and industrial plants, *step-down* transformers are used to reduce the voltage. Therefore, either the high-voltage winding or the low-voltage winding may be the primary or the secondary depending on the work the transformer is doing. *The power is supplied to the primary and given out by the secondary.*

68. Classes of Transformers. All transformers for power or distribution service are divided into two main classes: power transformers and distribution transformers. *Power transformers* are above 500 kVA and are used to step up the generated voltage to the transmission voltage; they are also used in substations to step down the transmission voltage to the distributing voltage or to a standard-service voltage. *Distribution transformers* are 500 kVA and under, being used to step down from a transmission voltage to a distribution voltage or to a standard-service voltage or from a distribution voltage to a service voltage.

A large percentage of all transformers are built *single-phase*; a small percentage of both power and distribution transformers are built *three-phase*. It is usually slightly cheaper to build one three-phase transformer than three single-phase transformers of equal total power. The reason for using single-phase transformers is flexibility. Where a system is supplied from a three-phase transformer, a short circuit or open circuit in this unit usually cuts off the whole system. In case of trouble with one single-phase transformer in a three-phase bank, this transformer can be cut out and the other two operated in open-delta, or a spare transformer can be substituted. With large sizes, the question of size and weight also must be considered on account of difficulties that may be encountered in shipping and handling.

Three single-phase transformers may be connected in either delta or Y, and these connections may be changed as desired. A three-phase transformer has the connections made permanently at the factory. It is the general practice in the United States to use single-phase transformers and in Europe to use three-phase transformers.

69. Cooling. Transformers may be *air-cooled*, by the natural circulation of the air around the core and coils, or they may be cooled by an *air blast*, the air being forced through the transformer. They may be *oil-immersed*, the cooling being obtained by natural circulation of oil within the tank and circulation of air around the tank, or *oil-immersed, forced oil-cooled*, where the oil is circulated by a pump through external coolers. They are *water-cooled, oil-immersed*, where the core and coils are immersed in oil but cooling water is forced through pipes within the transformer tank to carry off the heat. For the larger sizes, various devices—such as corrugated tanks, tanks with pipes welded to the outside, and radiators—are used to increase the radiating surface and improve the cooling process. Many large transformers are now being equipped with blowers and the necessary ducts to direct jets of air against the radiating surfaces, so that the cooling efficiency is greatly increased. Because the loss varies as the volume, and hence as the cube of the dimension, while the heat-radiating surface varies only as the square, the difficulty of cooling increases with the size of the transformer.

70. Mechanical Construction. There are several types of construction, each of which gives maximum service, best operating characteristics, and highest efficiency under certain conditions. A change in kilovoltampere capacity or operating voltage emphasizes different heat-radiating requirements, mechanical forces, or voltage stresses which may be best handled in the design by different windings, insulation, or tank construction. Transformers which are operating 24 h per day, at low-load factors, should have small exciting and iron losses with comparatively large load losses, while large power transformers operating at a high constant load may have low load losses and high exciting and iron losses, provided the total loss is a minimum. In either case, a high all-day efficiency is obtained.

Mechanically, transformers are of the *core type* or *shell type*, as shown in Figs. 16.48. As the stresses set up by heavy currents tend to make the coil cir-

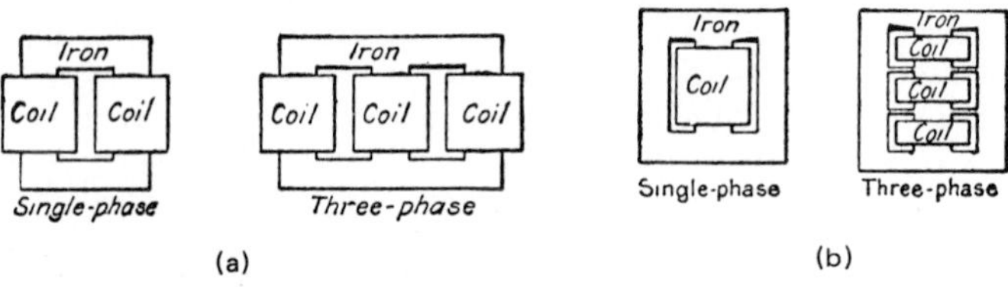

FIGURE 16.48 (*a*) Core-type transformers. (*b*) Shell-type transformers.

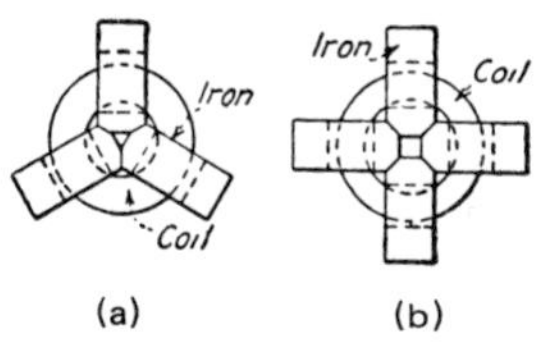

FIGURE 16.49

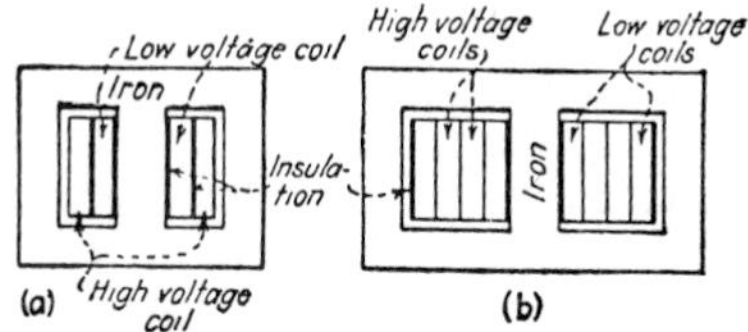

FIGURE 16.50

cular, most transformer coils are now built in that shape. A large percentage of the small single-phase transformers are built as a modified core type (Fig. 16.48*b*), but may have three or four outside legs (Fig. 16.49). Three-phase transformers are usually of the core type shown in Fig. 16.48*a*.

The *coils* may be wound in concentric cylinders (Fig. 16.50). The low-voltage winding may be all next to the core (Fig. 16.50*a*), or both windings may be split and wound (Fig. 16.50*b*). The coils may be wound in pancake form (Fig. 16.51*a*), or may be a combination (Fig. 16.51*b*). The windings and core are so arranged that the oil or air may circulate in ducts through them and thus maintain uniform cooling.

The *tanks* are of welded-steel construction. Nearly all tanks are provided with lifting lugs and, with the exception of pole-type units, have jack bosses. Since the great majority of transformers are used outdoors, the covers and bushings must be built for this service.

The *insulation* must be of a very high quality because of the high voltages involved and the compact arrangement of the transformers necessary for proper operating characteristics. High-class materials, such as mica, special thin paper built up to the required thickness and cemented together with synthetic resin under heat and pressure, and hard fiber, are used. All completed coils are vacuum-dried and treated with a moisture- and oil-resisting varnish. Oil is used as an insulating as well as a cooling medium. Oil introduces two difficulties: sludging or carbonization of the oil and the danger of explosion of oil vapor and air. Sludging, caused by oxygen coming into contact with hot oil, reduces the insulating and cooling value of the oil. Both these difficulties can largely be eliminated by providing some means of preventing the air from coming into contact with the hot oil. There are two methods in commercial use. Since the volume of oil changes with the temperature, it is not possible to seal the transformer tank. One method uses tanks with the space above the oil level filled with an inert gas to prevent air from coming into contact with the oil. Another method uses the transformers with oiltight covers and then adds a second, smaller tank slightly above to keep a constant head of oil on the transformer and to allow for expansion and contrac-

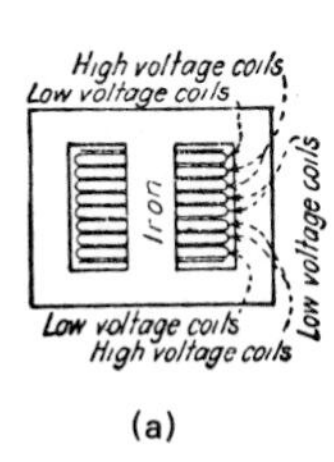

Low voltage circular disc coils

High voltage coils

Insulating cylinders between windings

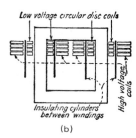

(b)

FIGURE 16.51

tion. As means are taken to ensure a very slow passage of oil in and out of this extra tank, its oil is cool, and contact with the air does not cause it to sludge. These methods also prevent explosions due to a mixture of oil vapor and air. Noncombustible insulating fluids are now being introduced as substitutes for oils.

71. Subtractive and Additive Polarity. Consider a single-phase transformer. Connect one high-voltage terminal to the adjacent low-voltage terminal and apply voltage across the two high-voltage terminals. Then if the voltage across the unconnected high-voltage and low-voltage terminals is less than the applied voltage, the polarity is *subtractive*, while if it is greater than the applied voltage, it is *additive* (see National Electrical Manufacturers' Association *Standards*). Additive polarity is standard for all single-phase transformers 200 kVA and smaller, whose high-voltage ratings are 2500 V and below. Subtractive polarity is standard for all others (see NEMA *Standards*).

72. Delta-Delta Connection. For three-phase circuits, transformers may have their primary or secondary windings connected in either delta or Y. Figure 16.52 shows a delta-delta connection.

Advantages. Any three similar single-phase transformers can be so connected. The bank can operate open delta if one unit is disabled (see Sec. 75). For low voltages and high currents, the delta connection gives a more economical design than the Y connection. This connection is free from all third-harmonic troubles. The primary and secondary delta carry the third-harmonic magnetizing current which does not appear on the lines.

Disadvantages. The neutral cannot be derived. Differences in the voltage ratios cause a circulating current in both primary and secondary windings, limited only by their impedances. Differences in impedances cause unequal load division among the units. For very high voltages, the delta connection costs more than the Y connection. The connection of Fig. 16.52*a* cannot be connected in multiple with that of Fig. 16.52*b*.

73. Y-Y Connection. This is shown in Fig. 16.53.

Advantages. The neutral can be brought out for grounding. Differences in impedances and ratios of the units do not cause any circulating currents or appreciably unequal load division. For relatively high voltages and small currents, the

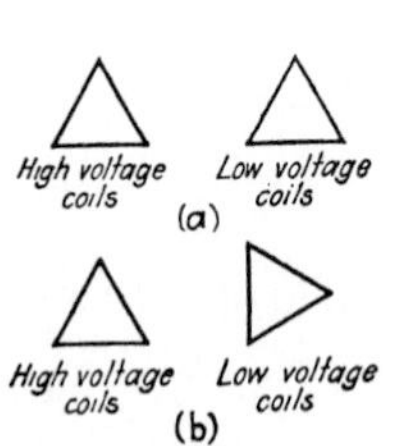

FIGURE 16.52 Delta-delta connection.

High Voltage Coils Low Voltage Coils
(a)
High Voltage Coils Low Voltage Coils
(b)

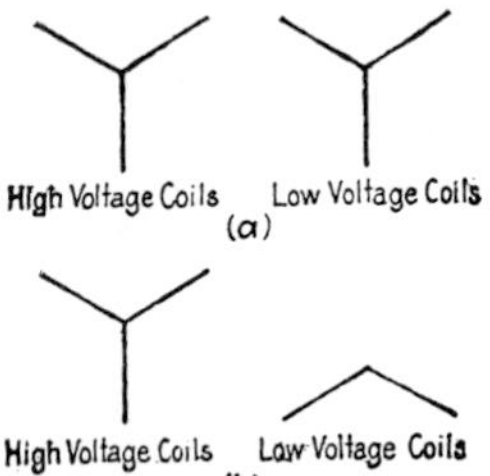

FIGURE 16.53 Y-Y connection.

Y connection is generally more economical than the delta. A short circuit in or on one unit does not cause a power short circuit, but does cause a very large magnetizing current because of the overexcitation of the remaining units at 1.73 times rated voltage.

Disadvantages. There is a third-harmonic voltage from line to neutral amounting to as much as 50 percent in some types of single-phase transformers. The neutral is unstable unless grounded. The units cannot be loaded, single-phase, line to neutral, unless the neutral of the primary is connected to that of the generator. If the neutral is grounded, it may cause telephone interference. The Y-Y bank cannot operate with two units when one is disabled. A short circuit in or on one unit raises the voltage on the other units to 1.73 times normal. Figure 16.53*a* cannot be connected in multiple with Fig. 16.53*b*.

74. Delta-Y Connection. This is generally considered to be the most satisfactory three-phase connection.

Advantages. The neutral can be brought out either for grounding or for loading. The neutral is stable, being locked by the delta. The connection is practically free from third-harmonic voltages, as third-harmonic magnetizing current circulates through the delta winding. Differences of magnetizing current, voltage ratio, and impedance in the different units are adjusted by a small magnetizing current circulating in the delta. A short circuit in one leg of the Y does not affect the voltages of the secondary lines. A single-phase short circuit on the secondary lines causes a smaller short-circuit stress on a delta-Y-setup bank than on a delta-delta bank. Figure 16.54*a* can be multipled with Fig. 16.54*b* by properly selecting the leads.

Disadvantages. The delta-Y bank cannot operate temporarily with two units when one is disabled. A short circuit on or in one unit is extended to all three units. If the delta is on the primary side and is accidentally opened, the unexcited leg on the Y side may resonate with the line capacitance and cause damage.

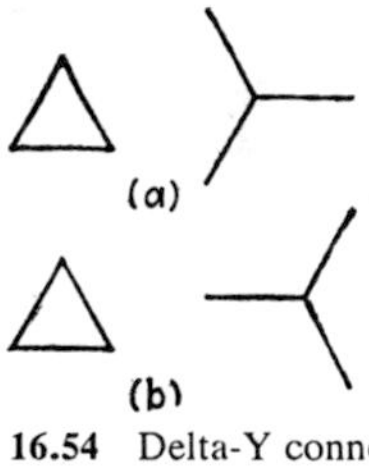

(a)

(b)

FIGURE 16.54 Delta-Y connection.

75. Open-Delta or V-V Connection (see Fig. 16.55). This connection requires only two units and is useful in an emergency; it is used in small control systems. Since the internal power factor is 0.866 (assuming a unity power-factor load), it can deliver only 0.866 of its rated kilovoltampere capacity or 58 percent of the capacity of three units. Load voltages become unbalanced under load, even with balanced three-phase load, the amount of unbalancing depending on the impedance of the units and the power factor of the load.

FIGURE 16.55 V-V connection.

76. Three-Phase, Two-Phase Connections. The *Scott connection* shown in Fig. 16.56 has the following *advantages*: It requires only two single-phase units or one two-phase unit. It can be used with either three-wire or four-wire two-phase service. Both two- and three-phase voltages can be obtained on the pri-

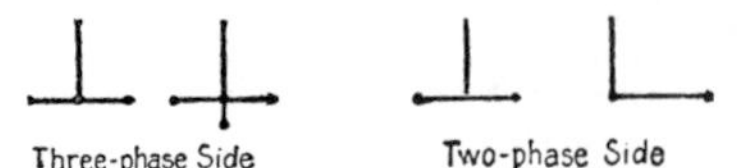

FIGURE 16.56 Three-phase two-phase transformer connection.

mary, with only four wires if the 86.6 percent tap of one winding is connected to the center of the other.

Disadvantages. If the two transformers are duplicates, there must be a 50 percent tap and an 86.6 percent tap on each unit. The three-phase side carries 15 percent more current than that corresponding to the two-phase side. It therefore requires 15 percent more copper, or the two-phase side must be operated at 86.6 percent of its capacity. The decreased use of two-phase systems limits the application of the Scott connection. It is seldom, if ever, used for transformation of voltage without change in number of phases.

77. Autotransformers. An *autotransformer* differs from a regular transformer in that it has only one winding instead of a primary and secondary winding. The change in voltage is obtained by connecting the winding across the impressed voltage and tapping off another circuit with more or fewer turns. The connection for single-phase is shown in Fig. 16.57. Autotransformers may be wound single-phase but usually are three-phase or two-phase.

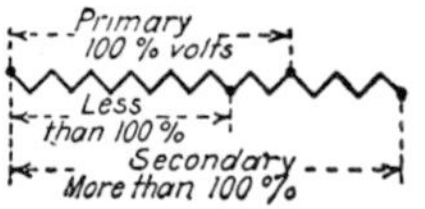

FIGURE 16.57

Advantages. For a given output, autotransformers are cheaper than transformers. This economy is greater the nearer the ratio comes to unity. Autotransformers also have better efficiency and regulation than transformers. This advantage also increases as the ratio approaches unity.

Disadvantages. There are some serious disadvantages in the use of autotransformers. The high- and low-voltage windings being continuous, the low-voltage circuit and connected apparatus are subjected to high voltage to ground and may be subjected to abnormal voltages due to disturbances and grounds on the high-voltage circuit. This is particularly true when there is a large difference between voltages. Short-circuit currents are larger with autotransformers, and this is worse as the voltage ratio approaches unity or the high and low voltages approach each other. Except for starting compensators, autotransformers are seldom used where the ratio exceeds 2:1.

Autotransformers are recommended for isolated systems and for grounded systems, provided the neutral of the autotransformer is also grounded. But they are used most extensively for supplying a low voltage for starting synchronous or induction motors.

The *Y connection* shown in Fig. 16.58 is the most economical and therefore the most used autotransformer connection. The ratio of rating to output is $(E_1 - E_2)/E_1$, where E_1 is the high voltage and E_2 is the low voltage. The neutral may be grounded if the generator neutral is also grounded. The disadvantages are the same as given for Y-Y transformers (see Sec. 73).

The *V* or *open-delta connection* (Fig. 16.59*a*) is free from third-voltage harmonics, and the ratio of rating to output is 15 percent more than for the Y connection.

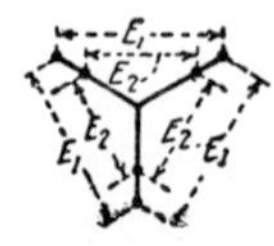

FIGURE 16.58 Y connection.

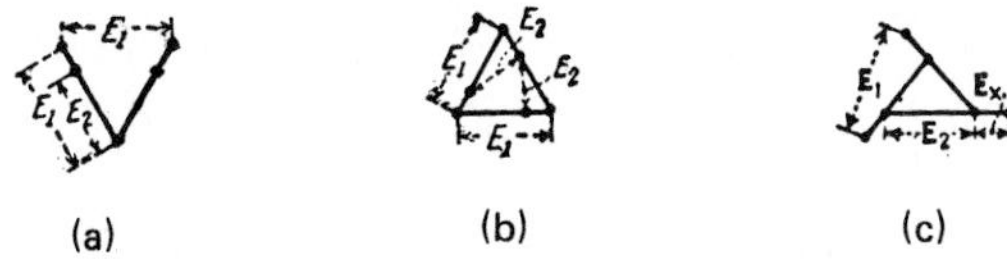

(a) (b) (c)

FIGURE 16.59 (*a*) Open-delta connection. (*b*) Delta connection. (*c*) Extended-delta connection.

The *delta connection* (Fig. 16.59*b*) has characteristics similar to those for transformers connected delta-delta (Sec. 72). The rating to output is $(E_1^2 - E_2^2)/1.73E_1E_2$.

The *extended-delta connection* (Fig. 16.59*c*) gives a ratio of rating to output of $\sqrt{1 - 0.25(E_2/E_1)^2} - 0.866E_2/E_1$ or $1.73E_x/E_1$, where E_x is the voltage of the extended portion. When the Y connection is undesirable, owing to the third harmonic, the V or extended-delta connection is more economical than the delta connection. When the low voltage is less than 92 percent of the high voltage, the extended-delta connection requires a smaller rating than either the straight delta or V connection. When this ratio is over 92 percent, the V connection requires the smallest rating of the three.

MOTORS

78. Introduction. Direct-current motors and ac synchronous motors are identical in construction with dc generators and ac generators, respectively. Motors and generators differ only in their functions. A *generator* converts mechanical energy into electric energy; a *motor* converts electric into mechanical energy. Like generators, motors are divided into two fundamental classes: ac motors and dc motors.

ALTERNATING-CURRENT MOTORS

79. Types. Alternating-current motors may be divided into *induction* and *synchronous* motors. A synchronous motor runs always at synchronous speed independently of the load; an induction motor runs at a speed below the synchronous speed (subsynchronous speed). The synchronous speed is determined by the following equation:

$$\text{rpm} = \frac{120f}{p}$$

where rpm is the number of revolutions per minute of the motor, f is the frequency of the supply source of emf, and p is the number of poles of the machine.

Both synchronous and induction motors may be single-phase or polyphase. Polyphase motors are, however, more used commercially because of their better operating characteristics. Single-phase motors are used where polyphase power is not available, as in homes. In all cases in which reference to cost or size is given, the squirrel-cage-type three-phase induction motor is used as a reference standard. Polyphase motors are usually three-phase. The different kinds of polyphase motors are described in the following paragraphs.

80. Squirrel-Cage Induction Motors. A *squirrel-cage induction motor* consists of a distributed wound stator and a short-circuited or "squirrel-cage" rotor. The power is applied to the stator winding only; this sets up a magnetic field which revolves at synchronous speed. The rotor acts as the secondary of a transformer. With the rotor at standstill, the transformer action generates a low voltage in its rotor at the same frequency that is impressed on the primary or stator windings. As the rotor is short-circuited, this voltage forces a high current through it. The value of this current depends on the resistance and reactance of the rotor windings. The magnetic field set up by this current tends to travel with the primary revolving field and thus develops torque. This torque varies with the current. As the rotor increases in speed, its voltage, current, and frequency decrease, so that theoretically at synchronous speed they are zero and no torque is developed. The rotor must "slip" behind the revolving field enough to develop sufficient secondary voltage to force the required current through the rotor. For this reason, the slip varies directly with the load. This is the simplest and cheapest type of motor. It is used for any application requiring approximately constant-speed operation.

1. The *standard-type squirrel-cage motor* has low starting, high pullout torque, and draws a high starting current; these vary with the capacity and speed. Since the starting torque varies with the rotor resistance and reactance, there are several variations from the standard; these are known as *high-resistance squirrel-cage motors, high-reactance low-torque squirrel-cage motors*, and *high-reactance high-torque* or *double-winding squirrel-cage motors*.

2. The *high-resistance squirrel-cage motor* may have a very high starting torque and a high starting current. Since this motor has a large slip, it is especially suitable for applications where a flywheel is used, such as punch presses, and for frequent-starting duty. It is also used for elevator service. Its efficiency and power factor are lower than those for a standard squirrel-cage motor, and therefore, it is not often used for continuous operation.

3. The *high-reactance low-torque motor* has a starting torque about the same as that of the standard-type motor, but the starting current is lower. The efficiency and power factor are usually slightly lower than for a similar standard-type motor, but in some ratings they are being used instead of the standard type.

4. The *high-reactance high-torque motor* may be designed for high starting torque with a moderate starting current. This motor has still lower efficiency and power factor. Average speed-torque and current-speed, efficiency, and power-factor curves are given in Figs. 16.60 to 16.62.

5. *Starting.* Any of the squirrel-cage-type motors can be designed for across-the-line starting. It is standard practice to start motors at 5 hp (3.7 kW) and below in this way. In most applications, the high-reactance and high-resistance motors up to 200 hp (150 kW) are thrown directly across the line. The limiting feature is the power system. For frequent starting or for applications where slow starting is desired, a reduced-voltage starter should be used.

81. Slip-Ring Induction Motor. This type has a stator winding similar to that of the squirrel-cage motor, but the rotor is not short-circuited; it is polar-wound,

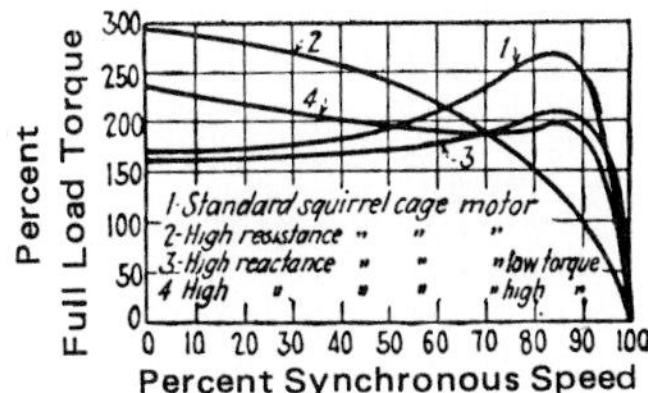

FIGURE 16.60 Torque characteristics of squirrel-cage motors.

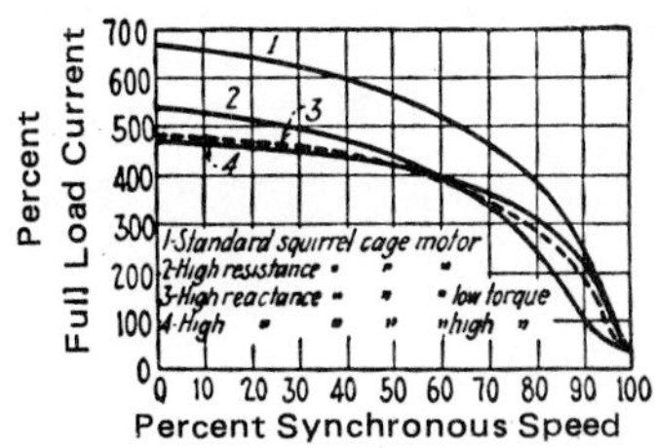

FIGURE 16.61 Current characteristics of squirrel-cage motors.

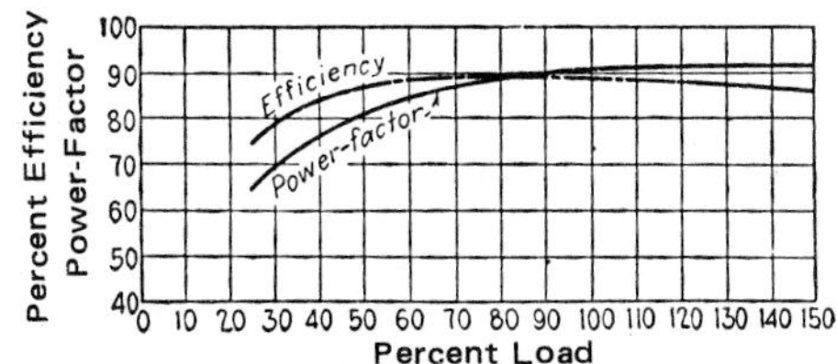

FIGURE 16.62

with the ends brought out to collector rings. There are usually three rings, although some special motors have six rings. The collector rings are connected, through brushes, to an external resistance. The speed at which any torque is developed may be controlled by varying the rotor or secondary resistance. That is, the higher the total rotor resistance, the greater must be the slip, below synchronism, to develop a given torque. A series of torque-speed curves for different values of rotor resistance is shown in Fig. 16.63, and the corresponding current curves are given in Fig. 16.64. The dotted curve *M* at the right in Fig. 16.64 gives the current of the motor with the brushes short-circuited. Curve 8 in Fig. 16.63 shows the actual slip with the resistance all cut out, but allows for the resistance of brushes, leads, etc. From these curves it may be seen that if a motor is operating at any speed and the power is reversed, it will develop torque in the opposite direction. This is called *countertorque* or *plugging* and is often used to

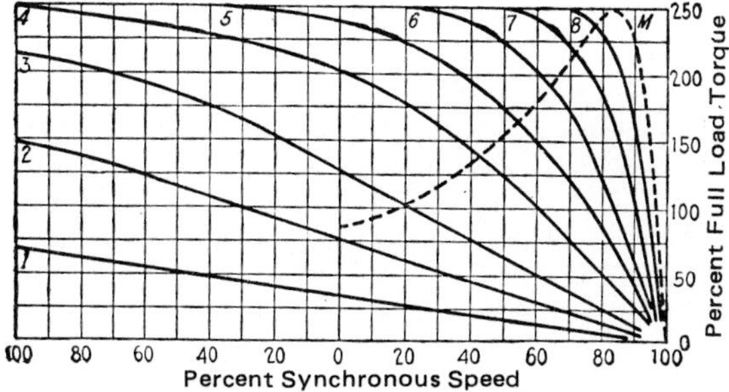

FIGURE 16.63 Torque-speed curves of slip-ring motors.

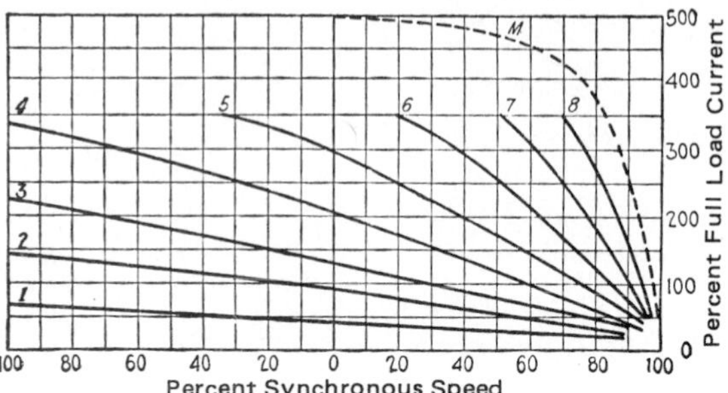

FIGURE 16.64 Current-speed curves of slip-ring motors.

stop the motor quickly. When a motor is plugged at normal speed, the secondary voltage is double the standstill voltage, and for values of torque up to 150 or 175 percent, it is nearly double what it would be at standstill with any given resistor value. In plugging, the rotor resistance also should be increased to correspond approximately to curve 3 in Fig. 16.63.

Regenerative braking can be obtained at speeds above synchronism when the load changes and drives the motor as a generator. With no secondary resistance, the motor, operating as a generator, will develop normal torque with normal slip above synchronism. Cutting in resistance increases the speed at which normal torque will be developed.

Dynamic braking is obtained by disconnecting the ac source of power and exciting the stator from a dc source, usually two phases in series, the third being open. For any given torque requirement, the speed can be changed by changing the resistance in the secondary circuit. The more resistance, the higher the speed. Below about 15 or 20 percent of full-load speed, the motor becomes unstable and may speed up and run away. The maximum torque which a given motor can develop when acting as a dynamic brake is about 75 percent of its breakdown torque as a motor.

Slip-ring motors are used for any applications requiring high starting torque, low starting current, speed control, or reversing. With secondary resistance, full-load torque may be obtained with approximately full-load current at any speed. This motor is used on such heavy duty applications as mine hoists, steel-mill main-roll drives, etc. Its cost is about 15 to 50 percent more than that of a squirrel-cage motor, depending on the size. Small motors have the greater cost difference. For slip-ring motors it is always convenient to start with some resistance in the rotor circuit. This resistance may be cut out in one or more steps during acceleration.

82. Concatenated Motors. By connecting two similar slip-ring motors in series, half speed is obtained at normal current and normal full-load torque. The rotors are mounted on the same shaft and thus are rigidly connected together. The stator of one motor is connected to the line, and its rotor is connected to the stator of the second motor (see Fig. 16.65). The rotor of the second motor may be short-circuited or may have a resistance in series for starting or speed control. At half speed the secondary of the first motor develops one-half normal frequency. Thus the second motor has half normal frequency applied, and its primary acts as a resistance for the secondary of the first motor. For this connection the open-circuit slip-ring voltage must equal the primary voltage. By suitable switching the same two motors may be connected in multiple to give normal speed. This combination gives a 2:1 speed ratio at constant torque and therefore gives similar results to a pole-changing motor. Owing to the heating at normal-load current, at one-half speed, the motors must be larger than would otherwise be required. This same type of connection can be used with two dissimilar motors to get a lesser speed change. For instance, if a large motor of 24 poles were used and a smaller one of 4 poles, the speed of the main motor would be reduced in the ratio of (24 + 4):24, or for a 60-cycle circuit, from 300 to 257 rpm. If normal speed is desired, the smaller motor is simply cut out of the circuit. The smaller motor may be 4, 6, 8, or 10 poles, as required. These motors have been built up to 500-hp capacity for steel-mill main-roll drive.

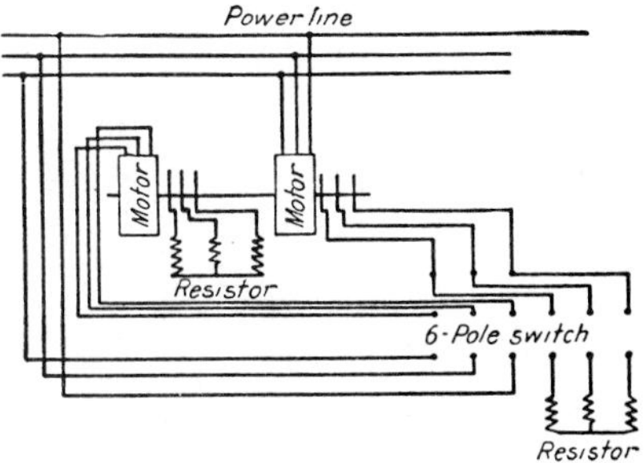

FIGURE 16.65 Concatenated motors.

83. Two- or Three-Speed Motors. Either the squirrel-cage or the slip-ring type of motor can be arranged with one or more sets of windings so as to have different synchronous speeds. This is known as *pole changing*. That is, a motor may be rated 6/12 poles, 100/50 hp (75/38 kW), 1200/600 rpm. One method of stator connection is shown in Fig. 16.66. This connection is for variable torque, two-speed, three-phase. There are a number of other connections. This type of motor is considerably more expensive than a single-speed motor. Pole changing may be accomplished by a knife switch or by contactors.

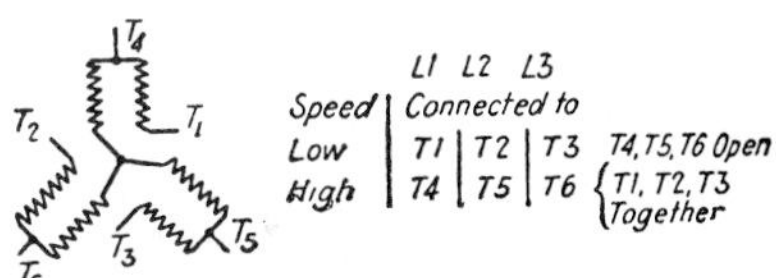

FIGURE 16.66 Connections for two-speed motors.

84. Adjustable-Speed Brush-Shifting AC Motors. These motors may be of either the series or shunt speed characteristic type. They are used where an adjustable-speed ac motor is required and the slip-ring type does not suit or is of too low efficiency at low speeds.

Series Type. The motor with a series speed characteristic consists of a stator winding like that of a squirrel-cage motor. The rotor is wound like a dc armature with a commutator, except that it is wound for three or more phases. If the power line is three-phase, the rotor may be wound for an increased number of phases, the only limit being the complexity of brush gear and transformer. The number of brush-holder stud locations equals the number of pairs of poles times the rotor phases, but with a large number of phases, part of the studs for each phase are sometimes omitted. Power is applied to the primary or stator leads, each phase of the stator winding being in series with one phase of the primary of a rotor transformer, the secondary of this transformer being in series with the rotor windings. Figure 16.67*a* shows this connection.

If the flux set up by the primary and that set up by the secondary winding have their axes in the same line, they oppose each other and no torque will be developed. For this reason, it is necessary to shift the brushes so that there will be a

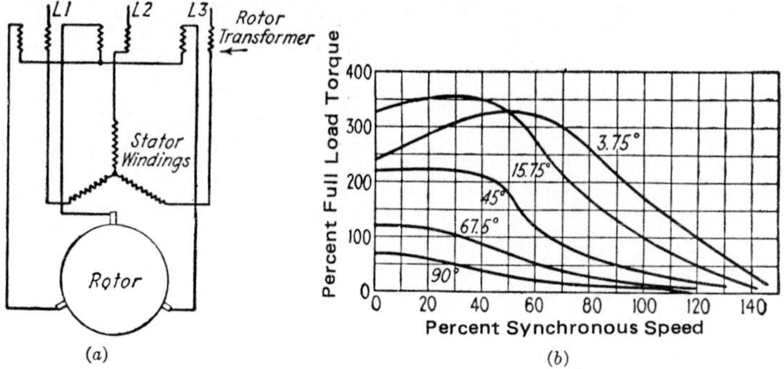

FIGURE 16.67 Connections (*a*) and speed-torque curves (*b*) of series-type motor.

resultant component of these fluxes or magnetic field in a direction to produce torque. This torque is caused by the repulsion between the current in the rotor coils and the magnetic field. A set of speed-torque curves is shown in Fig. 16.67*b*, each curve being marked with the degrees of brush shift. In this motor, the maximum torque is developed with a brush shift of about 15°.

With the brushes set to give full-load starting torque, the starting current is about 150 to 175 percent. The efficiency at full load, full-load speed is slightly less than that of a slip-ring motor. At lower speeds, the efficiency is better than that of the slip-ring motor with secondary resistance. The cost is higher. The power factor is better than that of a slip-ring motor. These motors are to a great extent superseded by the shunt-type motors for constant-torque loads.

For small sizes, the motors may be thrown across the line, with the brushes in the low-speed position. For larger sizes, a Y-delta switch is sometimes required to give easier starting conditions. Therefore, the only controls besides the rotor transformer are a line switch and possibly a Y-delta starting switch.

Shunt Type. The stator winding (the secondary) is constructed like the stator (primary) winding of an induction motor, except that phases are electrically independent and both ends of each phase are brought out for connection to the commutator brushes. The rotor has two windings. One rotor winding (primary) is identical in construction with the stator (primary) winding of a normal induction motor and is connected to the collector rings, to which the power is applied. The adjusting winding, which is in the top of the slots occupied by the primary winding, is connected to the commutator in the same manner as in a dc motor (Fig. 16.68).

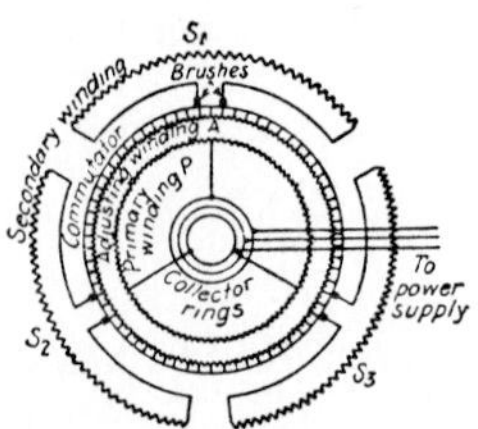

FIGURE 16.68 Connections for shunt-type adjustable-speed ac motor.

The motor may be compared with a wound-rotor induction motor having its primary windings in the rotor and its secondary on the stator. The motor is provided with two brush-holder yokes arranged to shift in different directions. One end of each phase of the stator (secondary) winding is connected to brushes on one brush yoke, and the opposite ends are connected to brushes on the other yoke. When the brushes, to which each end of a secondary phase is connected, are on the same commutator segment, the adjusting winding is idle, the secondary winding is short-circuited, and the motor runs as an induction motor with speed corresponding to the number of poles and frequency of supply. As the brushes are moved apart, a section of the adjusting winding is included in series with the secondary winding, causing the secondary winding to generate a voltage to balance the voltage impressed upon it by the adjusting winding, so that the motor changes its speed. Moving the brush gear in one direction raises the speed, and moving it in the other direction reduces the speed. The motor operates both above and below the induction-motor synchronous speed.

With the brushes in the low-speed position, these motors give, according to the particular design, from 140 to 250 percent of normal torque at starting with 125 to 175 percent of full-speed line current. The maximum torque at low speeds is usually, for 3:1 speed-range motors, about the same as the starting torque, and increases for the high-speed position to from 300 to 400 percent of normal torque. The efficiency remains nearly constant over the greater part of the speed range but is somewhat lower at low speeds. The average efficiency is high as compared with that of wound-rotor induction motors with secondary resistance. The power factor is very high when the motor is running at high speeds. At synchronous speed, the power factor is approximately the same as that of an induction motor of similar rating. Many motors have a form of brush-shifting device such that the high power factor obtainable at top speed is maintained down to speeds somewhat below synchronous speed.

The decrease in speed from no load to full load is, at high speeds, from 5 to 10 percent, and at low speeds from 15 to 25 percent, according to the rating of the motor. Since the primary power is taken into the machine through collector rings, it is necessary to use 550 V or less, since no collector rings have yet been developed for higher voltages. This motor may be used for reversing service. The brushes are usually shifted by hand but may be shifted by a pilot motor. Figure 16.69 shows the speed-torque curves of a 7½-hp (5.6-kW) motor. It will be noted that the starting torque is very high.

Since these motors may be thrown across the line with the brushes in the slow-speed position, a line switch is all that is required for the control. For some special applications, the stator windings are opened and a resistor is inserted. This gives the same effect as the resistor of a slip-ring motor and requires similar control, although more than one resistance point is seldom needed, since the speed can still be regulated by brush shift.

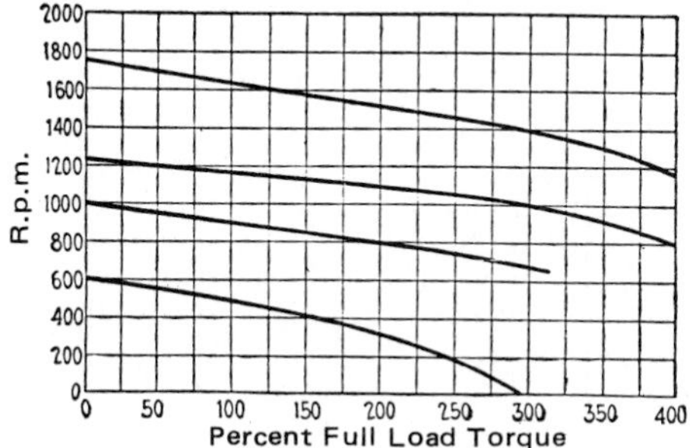

FIGURE 16.69

85. Single-Phase Motors. Single-phase motors are built in a number of different types; they may be divided into those types having a *series* speed characteristic and those which have a *shunt* characteristic. Owing to their speed characteristics, series motors should not be belted, because if the belt should come off the motor would run away.

Universal Motor. The *universal motor* is essentially a dc series motor. As first built, it had salient poles, but later types have a single-phase distributed stator winding like an induction motor. The rotor is a series-wound dc armature with a commutator. The stator winding and armature or rotor winding are in series (Fig. 16.70). The torque and speed characteristics are therefore like a series motor; i.e., the motor will develop high torque at low speeds, but at no load the speed is high. The speed may be varied by control of the line voltage (see Fig. 16.71). Applications are sewing machines, floor polishers, etc.; capacities, 1/200 to 1 hp (3.7 to 746 W). Large motors of this fundamental type are used for railway motor service, but these usually have compensated or commutating windings and thus become more complicated. Since this is a very special application, any questions should be referred to a manufacturing company.

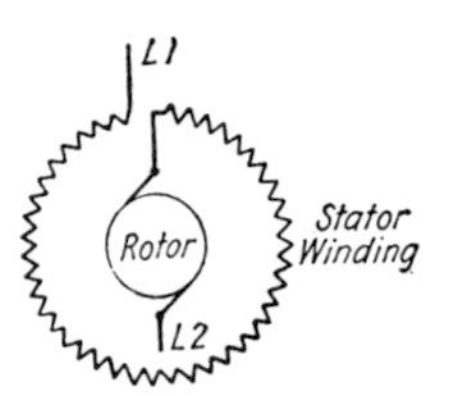

FIGURE 16.70 Universal motor.

Repulsion Motor. A *repulsion motor* is a single-phase, commutator-type motor with a single-phase stator winding. The rotor is a dc multiple-wound armature with commutator, with the brushes placed 180 electrical degrees apart and short-circuited. With the brushes on neutral, the magnetic axis of both stator and rotor is in the same line; therefore, no starting torque is developed. It is therefore necessary to shift the brushes about 20° out of phase with the primary winding to obtain the maximum starting torque. The brushes are fixed in position. This motor also has series speed-torque characteristics. Adjustable varying speed may be obtained by voltage control as shown in Fig. 16.71. The power factor of a 3-hp (2.2-kW) 1800-rpm motor would be about 0.8. Commutation at full load is very good.

Repulsion Brush-Shifting Motors. These are similar to the preceding motor but have arrangements for shifting the brushes. With the brushes set as before, the motor would have the same speed-torque curve. With any other brush position, the speed-torque curve would be similar. Thus a series of these curves may be obtained similar to those shown in Fig. 16.71 for change in voltage. Over a speed change of about 2.5:1, the motor gives constant torque. This motor can be used for reversing service by brush shifting.

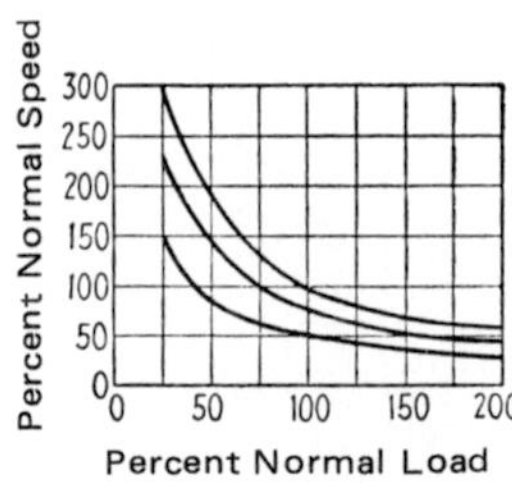

FIGURE 16.71 Speed-torque curves of repulsion motor.

Split-Phase Motors. The stator is wound as in a two-phase, squirrel-cage motor. The two phases are connected in multiple, with a resistance or reactance in series with one of them, to obtain starting torque (see Fig. 16.72).

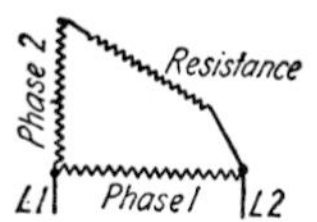

FIGURE 16.72 Connections for split-phase motor.

The starting winding is for intermittent service and is cut out by a centrifugal switch. Capacities are 1⁄30 to 1⁄4 hp (25 to 187 W). These motors are for constant-speed applications, such as for fans, blowers, washing machines, and other domestic appliances. The torque curve is somewhat like that of a squirrel-cage motor, starting at 75 to 100 percent torque at zero speed and reaching 200 to 250 percent at about 85 percent speed. When the starting winding is cut out, there is a drop in torque.

Capacitor or Condenser Motor. This motor is similar to the preceding motor but has a condenser instead of a resistor in series with one phase, this being left permanently in circuit. This motor has a low starting torque, 50 to 100 percent, but may be used on applications similar to those for the preceding split-phase motor. These motors have the advantage of not having either a centrifugal switch or a commutator. The power factor is good, approaching unity in some cases. These motors may be built up to 25 hp (18.7 kW), but are much more expensive than a three-phase motor of the same rating. They may be started by throwing directly on the line.

Another variation of the condenser motor has a second block of condensers in parallel with the starting condenser. This gives high starting torque, from 100 to 300 percent, depending on the condenser used. It requires a centrifugal switch to cut out the starting condenser, or an extra switch in the starter, or a normally closed contactor with its coil connected across the starting winding; this gives low voltage at starting, which increases with the speed. When the voltage reaches a certain value, the contactor opens. These motors are used where commutator motors are not desired.

Repulsion-Induction Motors. Here the stator winding is of the distributed single-phase type. The rotor is a combination of the repulsion type and the high-reactance squirrel-cage type, both windings being in the same slots. These windings are so arranged that during starting, the repulsion characteristics predominate, giving high starting torque. During running, the squirrel-cage winding predominates, giving nearly constant speed. Some builders use a switch to change from the starting to the running condition, while others make the change inherent in the machine. Owing to the combination winding, the running torque is very high. Figure 16.73 shows a torque-speed curve for a 3-hp (2.2-kW), 1800-rpm motor. The no-load speed is slightly above synchronism (about 3 percent), and the full-load speed is an equal amount below.

Since the motor operates, after starting, as a squirrel-cage motor, only a small number of brushes are required. Hence the motor is very quiet. The commutation is practically perfect. The efficiency should be between 75 and 80 percent. The power factor varies up to 95 percent. Speed control may be obtained by varying the line voltage.

Control. Practically all single-phase motors are thrown directly on the line, usually by a simple snap switch. If speed control or reversing is required, the starter must be amplified accordingly.

86. Synchronous Motors. A synchronous motor is built very nearly like an ac generator. It has a distributed stator winding and a rotor winding which is usually

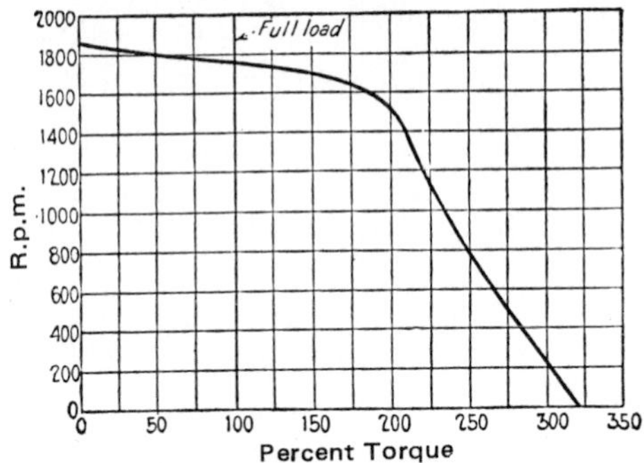

FIGURE 16.73 Speed-torque curve of repulsion-induction motor.

on salient poles. It differs from a generator in that it has a squirrel cage built into the rotor for starting and to damp out oscillations or hunting. The windings on the poles are the field and are excited from direct current. A few large and very high speed motors have a distributed rotor winding similar to that of a slip-ring induction motor. This is necessary for mechanical reasons.

When *starting*, the field or rotor circuit is open, and the motor operates as a squirrel-cage induction motor. Since the torque curve is similar to that of a squirrel-cage motor, it may be varied by changing the resistance and reactance of the squirrel cage. When the motor has reached about 95 percent of synchronous speed, the field is energized and the rotor pulls into step or synchronous speed. It then rotates at the same speed as the revolving field set up by the stator or primary winding.

During running, the speed is constant and is fixed by the frequency and number of poles. The speed is not affected by change of load or field strength. Changing the field, however, does change the power factor. If load is applied beyond the breakdown point, the rotor falls out of step and stops, since it has no torque below synchronous speed.

Synchronous motors are being used for a greater variety of services each year because they improve the power factor. They must therefore be designed for their particular duty. A normal motor for industrial service has approximately 50 to 100 percent starting torque, 100 percent pull-in torque, and 175 percent pull-out torque. *Pull-in torque* is that value developed by the squirrel-cage winding at a speed high enough to put on the field, usually about 95 percent. *Pull-out torque* is the maximum the motor will carry without falling out of step.

Motors with *high starting torque*, up to 175 percent, are sometimes necessary for certain applications, such as ball- or rod-mill drives. High starting torque means a motor of larger size to meet this particular requirement or a special type of motor. One method is to have the motor connected to the load through a *magnetic clutch*. Sometimes this clutch is built with the motor; sometimes it is separate. With this arrangement, the motor starts without load. After it is up to speed and the field is applied, the clutch is energized, bringing the load up to speed. This allows the motor to use its pull-out torque when starting the load.

A *supersynchronous* motor is sometimes used in which the rotor is solidly coupled to the load and the stator is placed in bearings. When power is applied to the motor, the stator revolves and comes up to synchronous speed. The field is applied and the motor is synchronized. A brake is then applied to the stator, bringing it to rest. The rotor, being held in synchronism with the stator, starts and comes up to speed. By this method, the pull-out torque is also available for starting the load.

A synchronous motor has the inherent characteristic of having its power factor dependent on the load and field strength. For a given load, an increase in field strength gives a leading power factor, and a decrease in field strength gives a lagging power factor. Figure 16.74 shows the relation between line current and field excitation and the resulting power factor.

In Fig. 16.74, curves 1, 2, and 3, line current is plotted against excitation. The low point of each of these *V*-curves gives unity power factor at that line current. The field current necessary to give unity power factor at normal line current is the normal field current of the motor. Curves *A*, *B*, and *C* are curves of equal power factor. In order to hold the power factor at unity or leading, it is necessary to increase the field with load. For a motor with fixed excitation, the power factor varies with the load. With normal excitation and light load, the power factor is leading; as the load increases, the power factor reaches unity, and then becomes lagging.

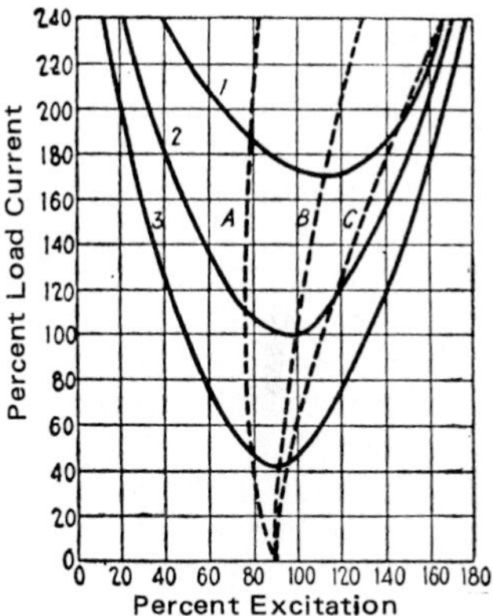

FIGURE 16.74 Excitation p.f. curves.

As stated in Sec. 88, the impressed voltage is equaled by the induced voltage plus the *IR* drop. This statement is nearly true of synchronous motors, but there are other variables. Since the speed is fixed, the induced voltage varies with the field strength. The third member of the equation, the *IR* drop, is now superseded in importance by the *impedance drop ZI*, which is always nearly 90° out of phase with the current. Figure 16.75*a* shows these voltages, where *E* is the applied voltage and *e* is the back emf or induced voltage. Figure 16.75*a* is for unity power factor, and thus the current *I* is in phase with the applied voltage, and *IE* equals power or kilowatts. If when holding constant load the field is increased, then *e* is increased and the diagram becomes as shown in Fig. 16.75*b*. Since the power remains constant, the current in phase with the applied voltage remains the same. Therefore, the line representing the current, times cos θ, must equal the original current. *ZI* is always nearly 90° out of phase with *I* and must increase as *I* increases, to I_1 or $ZI_1 = ZI/\cos\theta$. Completing the triangle gives e_1. The angle θ represents the power factor and in this case is leading. If the field is decreased, then *e* must decrease to e_2, which may be determined by the same reasoning. This, shown in Fig. 16.75*c*, gives a lagging power factor.

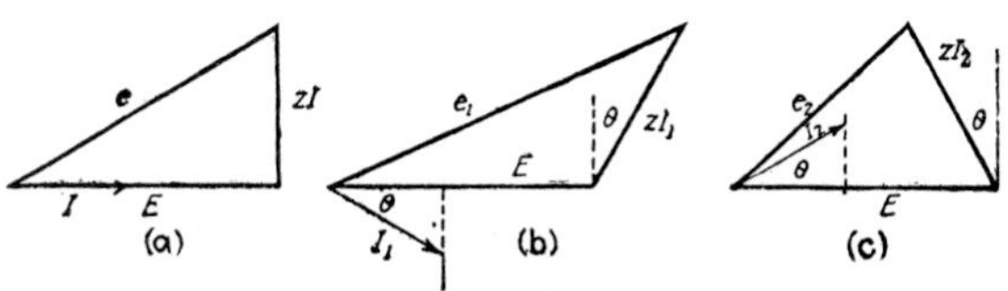

FIGURE 16.75 Synchronous-motor diagrams.

Synchronous motors are often used for power-factor correction. The motor usually does some mechanical work and is given a horsepower rating on this basis. It is then designed to carry a leading current so as to offset or neutralize the lagging current required by induction motors. The motor is rated 0.50 or other power factor to show that it can deliver leading current. Sometimes synchronous motors are used running idle on the line. All their capacity is then used to give leading current. They are then known as *synchronous condensers*.

87. Synchronous-Induction Motor. *Synchronous-induction motors* are, as the name implies, a combination. The stator is like that for a synchronous motor, and the rotor is usually wound as in a three-phase slip-ring motor. During starting, a three-phase secondary resistance is used, and the motor starts like a slip-ring motor. When full speed is reached, the rotor is excited by direct current like a synchronous motor. Since the rotor is wound three-phase, its winding when excited from direct current is not so efficient as the field winding of a synchronous motor, since one phase is either cut out of circuit or is connected in multiple with another phase and with two phases in series.

This motor is slightly more expensive than either a synchronous motor or a slip-ring motor, but the worst objection to it is that its excitation is at very low voltage and the exciter is expensive. It has the advantages of slip-ring-motor starting characteristics and synchronous-motor operating characteristics. Owing to the cost of the motor and exciter, this type of motor is not widely used in this country.

DIRECT-CURRENT MOTORS

88. Types. A dc motor is built like a dc generator. A machine acting as a motor has a definite applied voltage. This voltage is partly used in overcoming IR drop through the armature coils and brushes, and the remainder is opposed by the back emf generated by the armature windings cutting through the magnetic field. Increasing the main field strength increases the magnetic field and causes the motor to develop the required back emf at a lower speed. In other words, strengthening the field lowers the speed, and weakening the field increases it.

A dc motor is in reality a type of ac motor with a commutator so as to use zero-frequency current. There are three types: *series*, having the fields in series with the armature; *shunt*, having the fields in multiple with the armature; and *compound*, having a combination of series and shunt fields.

89. Series Motors. The *series motors* give a drooping speed-torque curve; i.e., the greater the load, the less the speed. A typical curve is shown in Fig. 16.71. This motor develops a high torque per ampere during starting, because the field, being in series with the armature, is strengthened with increased load, and thus the magnetic field is stronger. The speed increases rapidly with decreasing load and would generally be sufficient to wreck the motor at no load. This type is especially desirable for frequent starting with heavy starting torque, as with hoists, auxiliary motors for steel-mill service, etc. In small sizes it may be thrown directly on the line, but for medium and large motors a starting rheostat is needed. There is no real limit to the size which can be built, but small or medium size is usual, although railway motors are almost always of this type and are built up to several hundred horsepower.

90. Shunt Motors. *Shunt motors* have constant field strength. Since the motor does not develop any back emf at standstill, it is necessary to use a starting resistance to limit the current at this point. As the speed and back emf increase, the resistance can be cut out, thus controlling the current which can pass through the armature or the torque to be developed. During normal operation this motor has nearly constant speed; the only change is due to the increased *IR* drop with increased load. This means slightly lower speed. If a flat speed curve is desired, a slight shift of the brushes, off the neutral, will give sufficient compounding effect. This type of motor is used for all ordinary work. It can be reversed easily and so is suitable for hoist work. It can be operated at different speeds by field control, within the limits of commutation (see Sec. 93) and also may be used at very low speeds with armature resistance or voltage control. It can be built in any size required. For operation by generator voltage control, the motor must be excited from a separate generator (see Sec. 119).

91. Compound-Wound Motors. A *compound-wound motor* is a combination of the two preceding types, having some of the characteristics of each. By varying the percentage of series field, the shape of the speed-torque curve can be varied to meet any requirements. Generally speaking, the amount of series field used is very small. Sometimes the series field is in opposition to the shunt field, and the motor then is said to be *differentially wound*.

92. Braking. A series motor cannot be used as a *regenerative brake*, as a study of Fig. 16.71 will show. Compound-wound motors can be so used but rarely are. Shunt-wound motors can be used to obtain regenerative braking with no change in connections. Any dc motor can be used with *dynamic braking*. A diagram is shown in Fig. 16.76 for a compound-wound motor. The line and dynamic-braking contactors *LE* and *DB* are mechanically interlocked. The dynamic-braking contactor coil is energized as soon as the line contactor closes. One accelerating contactor is shown; the interlock *A* may be controlled by a time relay or by other means.

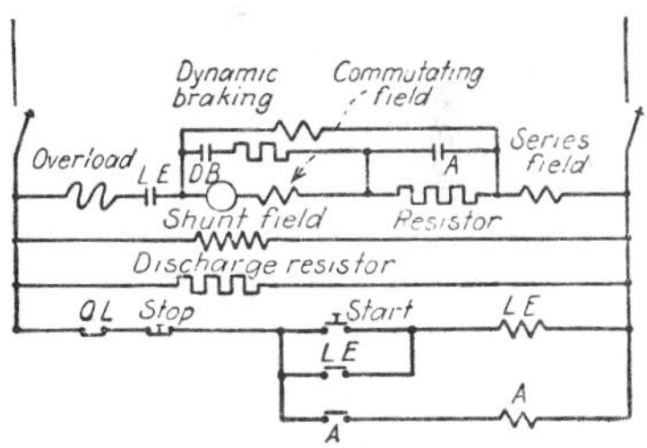

FIGURE 16.76 Connections for dynamic braking.

93. Adjustable-Speed Motors. An *adjustable-speed dc motor* is usually shunt wound, equipped with commutating fields, and sometimes with pole-face windings. The speed is changed by varying the shunt field. A motor for this service is usually built with a larger flux and fewer armature conductors than one for constant speed. Speed ranges of 3:1 and 4:1 are standard. If a greater range than 4:1 is desired, the lower range in speed must be obtained by cutting resistance into the armature circuit, or it may be necessary to use generator voltage control.

94. Table of Uses of Motors. Quite often the user of motors is at a loss as to what type of motor to specify for a certain service. The decision is influenced by several factors, among which are

1. *The source of power available:* Is the power dc or ac, and if ac, is it single-phase or polyphase?

2. *The speed requirement:* Is the equipment to be run at constant or variable speed? If the speed is constant, must it be absolutely constant, or may a slight decrease with load be tolerated? If the speed is variable, what range of variation is required? Should the speed decrease considerably with the increase of load, as with a traction motor?

3. *Starting torque:* Is the motor to be subjected to heavy starting torque, or will the starting torque be light?

4. *Type of load:* Is the load steady or widely fluctuating? Is it a generally heavy load, or is it a light load?

The balancing of these factors with the service and characteristics of the motor is not an easy task. Table 16.10 has been prepared to lighten the task of the motor buyer. It is well, however, when the purchase involves a large sum to engage the services of a consultant who will see to it that proper motors are specified for the particular types of service.

CONVERTERS

95. Introduction. In Secs. 53 to 66, descriptions were given of machines or apparatuses for obtaining two types of emf sources—the continuous or direct source of emf and the alternating source of emf. One source may be converted into the other by one of three general types of converters: (1) the rotary (or synchronous) converter, (2) motor-generator sets, and (3) the electronic converter. Such converters play an important role in electrical engineering practice, because practically all power is now generated as alternating current, and since there is a demand for dc power for certain services—traction, battery charging, three-wire dc system of lighting, etc.—it is necessary to have some method of changing ac power to dc.

THE SYNCHRONOUS CONVERTER

96. Definition. The *synchronous converter* is a combination synchronous motor and dc generator. It has only one set of armature conductors and one magnetic circuit. Being a dc generator, it is built with a revolving armature. The armature winding is the same as for a dc generator, and the collector rings are tapped off at equidistant points on the back end of the winding. On the front end is the usual commutator. Nearly all converters have commutating fields.

For any given machine with a fixed direction of rotation and fixed direction of the field, the armature current is in one direction when the machine is operating as a generator and in the opposite direction when operating as a motor. In a synchronous converter which operates as a synchronous motor taking power from the ac line and also as a dc generator giving out power to the dc line, the currents tending to flow through the armature conductors are in opposition, and therefore, the actual amount flowing in these conductors is the difference.

TABLE 16.10 Classification and Uses of Electric Motors

Motor								Uses: Special types of service for which motor is adapted																																													
Source of power		Type of motor	Speed variation in per cent of rating	Torque, in per cent of normal: Starting torque	Torque, in per cent of normal: Maximum torque	Standard ratings, hp.	General types of service for which motor is adapted	Agitator	Baler (power)	Ball mill	Blowers	Boring mill	Bridge (coal)	Car puller	Compressor	Conveyor	Crane hoist	Crusher	Dough mixer	Driller	Dryer	Elevator (freight)	Elevator (passenger)	Fans	Flat ironer	Grinder	Hammer (power)	Lathe	Laundry extractor	Laundry washer	Line shaft	Milling machine	Mine hoist	Molders	M.G. set a.c. to d.c.	M.G. set d.c. to a.c.	Planers (joiners)	Planers (metal)	Printing press (flat-bed)	Printing press (job)	Print ng press (rotary and offset)	Pumps (centrifugal)	Pumps (displacement)	Punches (shears)	Sanders	Saw (band)	Saw (circular)	Saw (metal)	Shapers	Shears (punches)	Stokers	Tenoners	Turntables	Valves	Vehicles
D.c.		Shunt-wound	5 to 7.5	150	L.C.*	1/30 up	Steady loads	x	x			x							x	x	x			x	x	x		x		x		x		x		x	x	x								x	x	x	x		x	x			
		Compound-wound	10 to 25	175 to 200	L.C.	1/30 up	Heavy starting and fluctuating loads		x	x				x	x	x	x	x	x			x	x				x		x										x	x	x	x	x	x	x	x				x					
A.c.	1-phase	Split-phase induction	5 to 7.5	150 to 200	150 to 175	1/30 to 1/3	Infrequent starting steady load		x						x					x				x		x		x							x							x			x			x				x				
	1-phase	Repulsion	4 to 6	225 to 300	150 to 175	1/8 to 10	L.S.C., H.S.T.*	x			x				x	x				x				x		x		x				x			x					x		x	x		x		x		x		x				
	1-phase	Capacitor	4 to 5	200 to 225	125 to 150	1/8 to 10	Quiet operation with high p.f.	x			x				x	x				x				x		x		x				x			x					x		x	x		x		x		x		x				
	2- and 3-phase	Std. squirrel-cage	3 to 6	150	200	1/6 up	General purposes	x			x								x	x				x		x		x				x		x	x		x	x		x		x			x		x	x	x		x	x			
	2- and 3-phase	Sq.-cage N.S.T., L.S.C*	3 to 6	100 to 150	200	7½ to 200	General purposes (single control)	x			x				x				x	x	x			x		x	x	x	x	x	x	x		x		x	x	x					x		x	x		x	x	x	x	x	x			
A.c.	2- and 3-phase	Sq.-cage H.S.T., L.S.C.*	4 to 6	175 to 275	175 to 200	3 to 100	Heavy starting (single control)	x		x					x	x		x																										x			x											
	2- and 3-phase	Sq.-cage L.S.T., L.S.C.*	4 to 6	50 to 60	125 to 150	40 to 100	Constant load				x							x						x											x							x													
	2- and 3-phase	Wound-rotor	D.L.*	200 to 250	175 to 250	3/4 up	Frequent and heavy starting L.S.C.*			x			x		x	x	x	x																x					x	x	x		x	x			x				x			x	
	2- and 3-phase	Synchronous	None	80 to 125	150 to 175	20 up	Infrequent starting, high p.f.				x				x									x											x							x													
	2- and 3-phase	Low-slip sq.-cage	4 to 6	175 to 200	175 to 200	1/2 to 100	Heavy starting (fluctuating load)														x					x			x														x						x						
	2- and 3-phase	H.T. sq.-cage*	10 to 15	250	250	1/2 to 50	Intermittent service		x							x	x				x	x	x																														x	x	
D.c.	Shunt-wound	Field- and arm.-control	D.L.	150	L.C.	1/30 up	Wide speed range, flexible control				x	x			x				x	x			x	x	x			x																						x		x				
	Shunt-wound	Voltage-control	D.L.	150	L.C.	1/30 up	Wide speed range, flexible control																										x				x																x		
A.c.	Multispeed	Constant-power	D.L.	150	175 to 200	1/4 up	Speed is dependent on load					x								x								x			x	x																							
	Multispeed	Constant-torque	D.L.	150	175 to 200	1/4 up	Speed is dependent on load	x		x			x	x			x					x	x		x											x	x															x				
	Multispeed	Variable-torque	D.L.	150	175 to 200	1/4 up	Speed is dependent on load																	x																															
D.c.	Compound-wound	Field- and arm.-control	D.L.	150	L.C.	1/30 up	Wide speed range. flexible control															x																	x	x	x	x	x												
	Compound-wound	Voltage-control	D.L.	150	L.C.	1/30 up	Wide speed range, flexible control																x																															
	Compound-wound	Series-wound	D.L.	300 to 400	L.C.	1/30 up	Wide speed range, flexible control		x				x				x																																				x	x	x
A.c.	Variable-speed	Wound-rotor	D.L.	200 to 250	175 to 250	3/4 up	Heavy starting duty, L.S.C.				x		x		x		x					x	x	x																x	x	x	x	x											x	x	
	Variable-speed	Brush-shifting	10 to 30	150 to 200	150 to 200	3 to 50	Wide speed range (high cost)								x									x										x	x						x	x													
	Variable-speed	High-slip sq.-cage	D.L.	250	250	40 to 100	Heavy starting (fluctuating load)				x							x						x											x							x													

*N.S.T. = normal starting torque, H.S.T. = high starting torque, L.S.T. = low starting torque, L.S.C. = low starting current, D.L. = depends on load, L.C. = limited by commutation, H.T. = high torque

Conversion factor: 1 hp = 0.75 kW.

97. Rating. Because of the preceding, a machine designed as a dc generator can operate at a higher rating when used as a synchronous converter. The amount of increase in the rating depends on the number of rings or phases on the ac end. If all losses are neglected, the theoretical ratios are

	Percent
dc generator	100
Single-phase single-circuit converter	85
Single-phase two-circuit converter	93
Three-phase converter (3 rings)	134
Quarter-phase converter (4 rings)	164
Six-phase converter (6 rings)	196
Twelve-phase converter (12 rings)	224

These figures are based on unity power factor or on the load current being in phase with the voltage. At the present time, nearly all converters are wound for six-phase and are connected diametrically as shown in Fig. 16.77*a*, the circle representing the armature winding and the rings being tapped off 60° apart. One advantage of this winding is that the center points of the transformer coils can be connected together and the machine can be used on a three-wire system. Occasionally, the six rings are connected double-delta (Fig. 16.77*b*), and in smaller machines three-phase (Fig. 16.77*c*) is sometimes used.

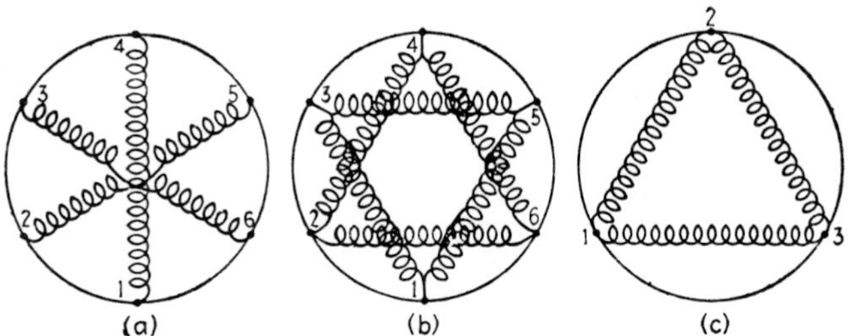

FIGURE 16.77 Connections for rotary converter armature windings and collector rings.

For 600 V and above, converters are being largely superseded by polyphase rectifier units.

98. Voltage Ratio. No-load ratios of ac to dc emf's vary with the number of phases, being approximately proportional to the sides of polygons inscribed in a circle. Divergence from the sine wave and some other conditions will produce slight differences. The theoretical no-load values for 100 V dc are given in Table 16.11. These are the volts between adjacent rings, or the phase voltages. Note that the quarter-phase, or so-called two-phase, is really a four-phase and is properly represented by the inscribed square. The diametral voltage ratio of the four-, six-, and twelve-ring converters is always that of the single-phase, 70.7. Slightly higher ac voltage is necessary for the same dc voltage when a converter is loaded.

TABLE 16.11 Ratio of Alternating-Current Voltage to Direct-Current Voltage

	Two-ring	Three-ring	Four-ring	Six-ring	Twelve-ring
Phases	1-phase	3-phase	4-phase	6-phase	12-phase
No-load ac volts	70.7	61.2	50	35	18.0

99. Frequency. Since a converter is a synchronous machine, its speed is determined by the number of poles and the frequency. It can be built for any commercial frequency. The lower the frequency, the more reliable can the converter be built. Converters as now built operate successfully on 60-cycle circuits.

MOTOR-GENERATOR SETS

100. Definition. A *motor-generator set* is a device used to convert alternating current to direct current, and vice versa. It consists of two machines whose shafts are mechanically coupled either directly or through a belt or chain of gears. One of these machines is a dc machine and the other an ac machine. If alternating current is to be converted into direct current, the ac machine is used as a motor and the dc machine as a generator. If direct current is to be converted to alternating current, then the dc machine is used as a motor and the ac machine becomes a generator. Motor-generator sets are also used to convert from one dc voltage to another and from one ac frequency to another.

101. Synchronous Converters vs. Motor-Generator Sets. Since synchronous converters are used to convert alternating current into direct current, and since motor-generator sets can be used for the same work, a comparison of the two methods is desirable. Synchronous motor-driven sets only will be considered. *Advantages of converters* are lower first cost, less floor space, higher efficiency, only one machine to keep up, and dc voltage independent of frequency. Since transformers are always necessary with converters, the cost and losses of the transformers should always be included with the converter. *Advantages of the motor-generator set* are the dc voltage is independent of the ac voltage, less liability of commutation troubles, power-factor correction, and up to 13,200 V, no step-down transformers are necessary.

The efficiency of large synchronous converters runs as high as 95 percent and the efficiency of transformers runs up to 99 or even 99½ percent, so that the combined efficiency may vary from 94 to 95 percent. Motor-generator sets of the same capacity have an efficiency up to 91 or 92 percent.

MERCURY-ARC RECTIFIERS (CONVERTERS)

102. Mercury-Arc Rectifiers. *Mercury-arc rectifiers* are largely replacing synchronous converters and motor-generator sets for converting alternating current to direct current. A mercury-arc rectifier makes use of the fact that electricity can pass through a mercury arc in one direction only when operating in a high vacuum. They are common in sizes from 30 to 3000 kW. The ac power line may be any voltage and any frequency. The dc voltage may be from 250 to 3000 V.

This rectifier consists essentially of a tank containing two or more terminals or anodes in the top or sides for the incoming ac power and one terminal or cathode in the bottom for the outgoing or dc power. The cathode is covered by liquid mercury, and the interior of the tank is a high vacuum. Figure 16.78*a* shows a simple

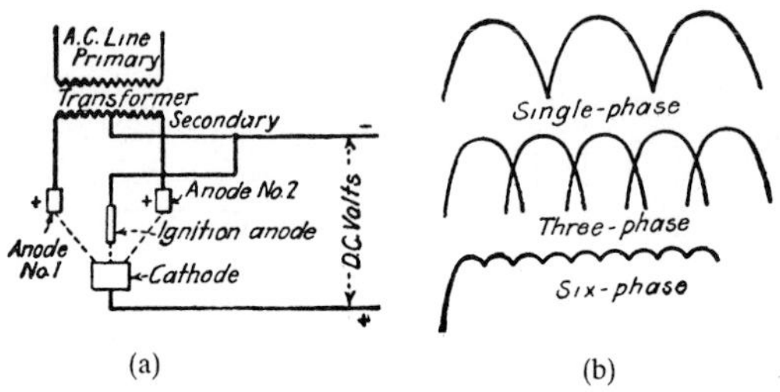

FIGURE 16.78 Simple mercury-arc rectifier.

diagram for a single-phase rectifier. From the diagram it can be seen that the center of the transformer secondary winding is brought out for a neutral and the negative dc line. Each end of the transformer secondary winding is connected to an anode of the rectifier. An arc is drawn between the cathode and the exciter terminal shown between the anodes. For one-half the cycle, the current passes through an arc from anode 1 to the cathode, and for the other half it passes from anode 2 to the cathode. This gives a wave as shown for single-phase in Fig. 16.78*b*. Power rectifiers are built for 6 or 12 phases with all the electrodes in one large steel tank, or each anode is insulated in a separate tank with a mercury cathode in the bottom; these are connected together to form one terminal of the dc circuit.

103. Mercury-Arc vs. Rotary Converters. The mercury-arc rectifier has several advantages over competitive equipment. Since there are no rotating parts, the operation is noiseless; this is a vital consideration in some locations. The overload capacity is high. Even a short circuit on the dc side has little effect on the mercury-arc rectifier itself, although it may impose a heavy load on the power system. It is not necessary to phase in a rectifier, since it has no rotating parts. Very little attention is required, since the operation of a rectifier is automatic.

CONTROL AND PROTECTIVE DEVICES AND SYSTEMS

104. Types of Protective Devices. Protective devices may be grouped under headings as follows: overload or short-circuit protection, phase-failure or reversal protection, undervoltage protection or release devices, overspeed, bearing-temperature relays, limit switches, etc.

105. Overload Protection. Overload protection for motors may be applied to running overloads and short circuits. For most industrial applications, the running overloads are taken care of by a magnetic switch with thermal overload relays, the short-circuit protection being taken care of by fuses or a circuit breaker.

The thermal relay characteristic is shown in Fig. 16.79. This is called *inverse-time* overload protection. That is, the greater the overload, the quicker the relay will function. Fuses give instantaneous protection, and circuit breakers may have relays for either inverse-time or instantaneous protection.

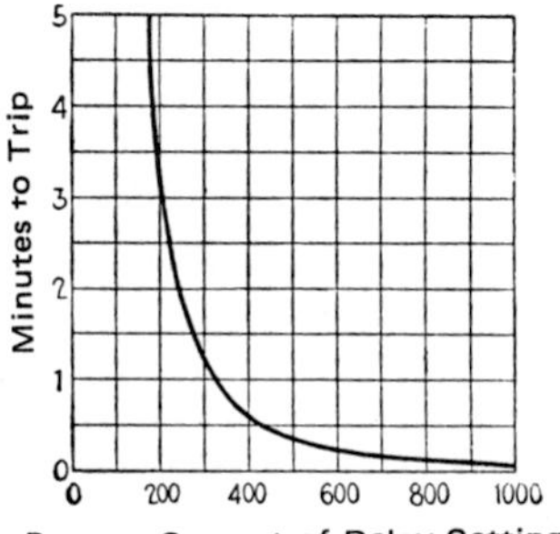

FIGURE 16.79 Thermal relay characteristics.

There are other types of installation where fire hazard is not a determining feature, where for small motors fuses only are used for overload protection. For larger motors, a circuit breaker with inverse-time relay is used. In these cases, the powerhouse and sectional circuit breakers are set for a higher rating so that the branch breaker will open first and thus relieve the system.

106. Phase-Failure and Reversal Protection. Phase-failure and reversal protection are accomplished by use of relays with the coils so balanced that the tips remain closed when the three phases are normal; if any phase is open or reversed, the relay opens and thus opens the circuit breaker or line contactor.

107. Undervoltage Protection. Undervoltage protection is obtained by the use of an undervoltage coil on the circuit breaker. The undervoltage coil and the overload coil may operate the same mechanical trip. Undervoltage protection also may be obtained by the use of a contactor whose tips carry the control circuit. After failure of power, the undervoltage relay or contactor has to be closed manually.

108. Overspeed Protection. Overspeed protection is usually obtained by a centrifugal switch on the motor. This switch opens a control circuit, usually in the undervoltage circuit of the oil circuit breaker or contactor.

109. Bearing-Temperature Relays. Bearing-temperature relays are used with automatic pumps, substations, etc. They usually consist of a bulb inserted in the metal of the bearing. The bulb and a connecting tube are filled with gas. An increase in bearing temperature causes the gas to expand and push a contact open and thus break the control circuit.

110. Limit Switches. For some appliances such as cranes, hoists, machine tools, etc., it is necessary to limit the travel of the cage, hook, or table. Limit switches are used for this purpose. They may be divided in two general classes: track or hatchway and geared. *Track-type* switches consist of a cylinder or cam operated by a handle or rope wheel. When the traveling object hits the arm or handle, it turns the cylinder or cam and opens or closes one or more electric circuits. *Gear-type* switches consist of a set or train of gears, sometimes with a traveling-nut device, that opens or closes electric contacts after a certain distance has been traveled. Both types are made in a wide variety of mechanical designs and with from one to five or six circuits.

MOTOR CONTROL

111. Starting and Speed-Regulating. Starting and speed-regulating methods depend first on the type of motor and second on the work done. Today there is an endless variety of control arrangements.

112. Single-Phase Motors. Small single-phase motors are usually started by a snap switch or a knife switch, and if there is any overload protection, it is in the form of fuses.

113. Squirrel-Cage Motors. Squirrel-cage motors up to 5 hp (3.7 kW) are thrown across the line either by a knife switch or, more often, by a three-pole contactor, controlled by a start-stop push button, with a thermal-overload relay for protection. Motors up to 200 hp (148 kW) are sometimes thrown across the line by use of a contactor, as mentioned above, or a hand-operated switch. For a great many applications, reduced-voltage starting is required. This can be obtained by a resistance in the primary or by the use of an autotransformer. The resistor-starter is suitable for small motors of 5 to 25 hp (3.7 to 18.7 kW). It is cheaper than the autotransformer or compensator type but has never been used to such an extent.

The autotransformer or compensator type of starter consists of an autotransformer to supply reduced voltage (this usually has taps at 50 to 65 and 80 percent), a five-pole starting switch for connecting the autotransformer to the line and the motor to the transformer taps, and a three-pole running switch. The two switches must be interlocked so that both cannot close at the same time. There are two types of these starters: the manually and the magnetically operated. For the manually operated compensator, the starting and running switches are usually of the knife-switch type. The magnetic type has contactors operated by a push button and with a time-delay relay to control the closing of the running contactor. In both types, the switches are oil-immersed for reasons of safety and small space, and also to make them good for 2200 V or more. Both types have overload and undervoltage devices. The magnetic starter may be remote-controlled by a push button, float switch, or any circuit-closing device.

114. Synchronous Motors. Synchronous-motor starters are the same types as mentioned above for squirrel-cage motors but have an additional device to close the motor field when near synchronous speed and one to short-circuit the field through a discharge resistor during starting and stopping. Larger motors are often started by using a five-pole-three-pole oil circuit breaker in place of contactors. Under certain conditions, these are cheaper.

115. The Korndorfer System. The Korndorfer system is often used for starting very large squirrel-cage or synchronous motors. Referring to Fig. 16.80, *A* is the line contactor, *B* and *C* are the starting contactors. *B* and *C* are closed first, connecting the autotransformer to the line and closing its *Y*-point. The motor now starts on reduced voltage. After the motor has nearly reached full speed, contactor C is opened, A is closed; then B is opened. The advantage of this system is that the motor is not disconnected from the line on the throw from the starting to the running position. Contactor A simply short-circuits a section of the autotransformer while B remains closed.

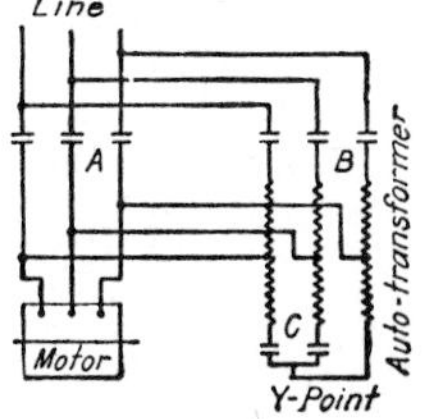

FIGURE 16.80 Korndorfer starter.

116. Slip-Ring Motors. Slip-ring motors require control for the secondary or rotor circuit as well as for the primary circuit. This control may be of the manually operated, drum-controller type or of the magnetic type. The simplest equipment consists of a *drum controller* with segments carrying both the primary and secondary currents and a secondary resistance. This type should not be used with over 550 V. It gives no protection to the motor either during ordinary operation, for overloads, undervoltage, etc., or during starting, when the motor can easily be overloaded by cutting out resistance too quickly.

In Fig. 16.63 (Sec. 81) is shown a series of torque curves for a slip-ring motor. The location of each of these curves is fixed by the ohms in that step of the resistor. A resistor is usually expressed in ohms, 100 percent ohms being that value which will allow full-load current to pass through the rotor at standstill. The resistor is designed to give 50, 100, or 150 percent motor torque at standstill, whatever is required by the service. The resistor for the set of curves shown was for a mine hoist and had 300 percent ohms. The weight or mass of the resistor depends on the service. There are light-starting-duty, heavy-starting-duty, and regulating resistors, the resistor becoming progressively heavier for the heavier duty. If very low torque is required on the first step, this also increases the weight. The regulating type of resistor must have current capacity to carry the full current continuously.

117. Semimagnetic Control. To obtain *semimagnetic control*, a triple-pole primary contactor operated by the primary segments of the controller may be added to the above. This can be arranged to give overload and undervoltage protection, but still does not protect the motor during starting.

118. Full Magnetic Control. *Full magnetic control* consists of a primary contactor, usually to open all the motor lines; a number of secondary contactors, varying from three to eight; a number of relays; a rotor resistor; and a master switch or master controller which carries only the control circuits. This type of control can be expanded so as to accomplish any operating results. For instance, the master controller may be one, two, or more points to operate any number of contactors. If we consider a six-point master, then there will be six hand-controlled points—the other points controlled by relays. If the master switch is thrown quickly to the "on" position, the contactors will close in sequence, but controlled by the relays, thus protecting the motor. There are two systems of control by relays, current-limit and time-limit.

The *current-limit system* is the older and has operated successfully for many years. It consists of two or more current-limit relays connected so as to close one contactor after another. When the line current falls below a predetermined value, usually 150 percent, the relay operates, allowing the next contactor to close.

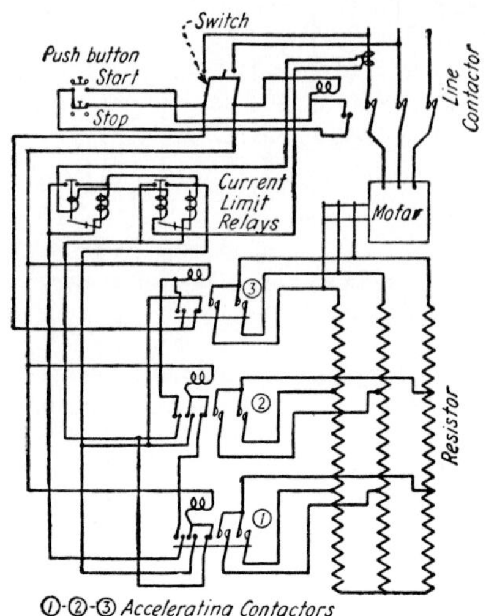

FIGURE 16.81 Current-limit control for slip-ring motors.

Looking at Fig. 16.63 (Sec. 81), if a load were to be started requiring 100 percent torque, the first three contactors would close in sequence, the current always being less than 150 percent. The motor would start on the third step, but the fourth contactor would close at once and the motor would come up to about 33 percent speed when the current-limit relay would operate, thus closing the next contactor. When two relays are used, they alternate in closing the contactors. Sometimes a third relay is used to give low torque on the first point (see Fig. 16.81). With this control, a light load will accelerate faster than a heavy one. This system is rather complicated.

The *time-limit system* is similar to current limit except that the relays are definite-time relays. There are several types. Settings may be obtained from ½ up to 30 s each. The usual time for each relay is ½ to 5 s. While this system is independent of the current, it gives results that are comparable with current limit. This system is cheaper because the wiring is simpler and the relays require less space on the panel, although there is usually one for each secondary contactor (see Fig. 16.82 for diagram). Either system operates successfully. At the present writing, the current-limit system is more popular.

Magnetic control is especially adapted for use with automatic or semiautomatic schemes. For instance, a push button located in any convenient place can be used to start up one or any number of motors. If more than one motor is started, there is *sequence operation*. The final accelerating contactor on the first motor closes the first contactor of the second, and so on. The reverse may be true; if any motor of a group shuts down, owing to overload or any of its protective devices, the other motors (all or any number) can be shut down. This is desirable if a group of motors is driving conveyor sections, part of a long conveyor, or in any mill where the material is carried through several processes. This system is also used for the automatic operation of centrifugal pumps, the control circuit being closed or opened by a float switch. When the water in a tank rises to a certain point, the float switch closes, starting up the motor and pump.

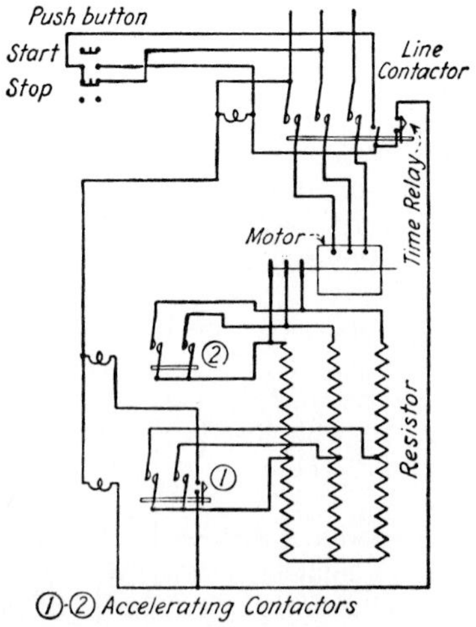

FIGURE 16.82 Time-limit control for slip-ring motors.

119. Speed-Regulating Control. *Speed-regulating control* of slip-ring motors may be obtained by the use of any of the before-mentioned systems. The only requirements are that the regulating steps must be hand-controlled, by either push button, controller, or master switch, and the resistor and contactors must be of sufficient capacity to carry the current continuously.

120. Reversing Control. *Reversing control* can be obtained with any of the above systems. The secondary control would not be changed, but the primary control must have reversing segments on the controller or master switch or have reversing primary contactors. Reversing contactors should be interlocked so that both cannot close at the same time.

121. Direct-Current Motor. The systems of control are the same as those for slip-ring motors, except that the resistance is in series with the armature and there is only one circuit involved; also, under magnetic control there are a number of modifications of the time- and current-limit systems.

122. Generator-Voltage or Ward-Leonard Control. For certain classes of mine hoists, steel-mill appliances, ore bridges, electric shovels, etc., it is necessary to have high starting torque, accurate control, and constant-speed running. For these classes of service, a dc motor with generator-voltage control is the best arrangement obtainable. This requires a separate generator for each motor. The generator may be driven by any type of prime mover, but is usually driven by an induction or synchronous motor.

Figure 16.83 shows the connections for a simple generator-voltage control. The motor and generator are connected together, sometimes with a switch and circuit breaker or contactor in the circuit, sometimes without. Both generator and motor are usually separately excited. In Fig. 16.83 the strength and direction of the generator field are controlled by a drum controller. With this arrangement, the direction of rotation of the motor and its speed depend on the generator field. The torque developed by the motor varies with the armature current, since the field is constant. At standstill a very low voltage equal to the *IR* drop will force a large current through the motor circuit and thus develop high torque.

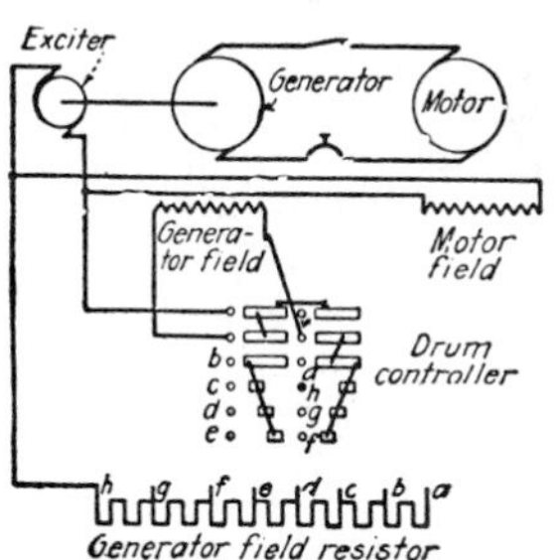

FIGURE 16.83 Generator-voltage control system.

With connections as given, if the load tends to overhaul and drive the motor above the normal speed, the motor becomes a generator and pumps power back into the generator; the generator automatically becomes a motor, driving its prime mover. The power is dissipated either in an increase in speed or by being put back into the power system. With a given voltage, the motor speed is practically constant regardless of load.

This type of control lends itself to many refinements. During acceleration of the motor, the control may have from 3 to 15 or 20 points. It may be manually operated, by either drum controller or master switch with contactors having current or time relays; or the controller may be motor-operated. During slowdown, the controller may be turned off by mechanical cams or other means. By adding a resistance and one or two contactors across the motor armature, dynamic braking may be obtained when slowing down. A current relay may be connected with its coil in the generator-motor armature circuit and its trips in the generator-field circuit. At the beginning of slowdown the generator field is opened; the motor, driven by the inertia of the load, pumps back on the generator. As the current rises to the predetermined point, the relay trips close, putting field on the generator. This causes the pump-back current to fall. This cycle is repeated until the inertia of the system is dissipated and the motor is at rest.

The *Ilgner-Ward-Leonard system* is simply the addition of a flywheel to the motor-generator set. This is for use on equipments operating on a recurring cycle where the input must be kept constant or nearly so. The generator must be driven by a slip-ring type induction motor with some arrangement to obtain variable speed, either grid resistance cut in by a relay in one or two steps, or a liquid slip regulator controlled by a torque motor in the main induction-motor line. An increase in the line current causes the torque motor to rotate, cutting resistance into the secondary of the induction motor, causing the motor to slow down and the flywheel to give up its power to the generator. A decrease in line current allows the balancing weights to overcome the torque of the torque motor, causing the motor to rotate in the opposite direction, cutting resistance out of the induction-motor secondary, and allowing the motor to speed up and bring the flywheel up to normal speed.

ELECTRONICS ENGINEERING*

The subject of electronics can be approached from the standpoint of the design of devices or the use of devices. For the practicing engineer, describing devices in terms of their external characteristics seems most likely to be profitable. The approach will be to describe devices as they appear to the outside world.

COMPONENTS

Resistors, capacitors, reactors, and transformers were described in the preceding section, along with basic circuit theory. These explanations are equally applicable to electronic circuits and hence are not repeated here. A description of additional components peculiar to electronic circuits follows.

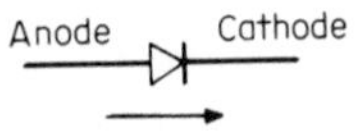

Anode Cathode

FIGURE 16.84 Diode schematic symbol.

A *rectifier*, or *diode*, is an electronic device which offers unequal resistance to forward and reverse current flow. Figure 16.84 shows the schematic symbol for a diode. The arrow beside the diode

*This section is from *Marks' Standard Handbook for Mechanical Engineers*, 9th ed., by E. A. Avallone and T. Baumeister III. Copyright © 1987. Used by permission of McGraw-Hill, Inc. All rights reserved.

shows the direction of current flow. Current flow is taken to be the flow of positive charges, i.e., the arrow is counter to electron flow. Figure 16.85 shows typical forward and reverse voltampere characteristics. Notice that the scales for voltage and current are not the same for the first and third quadrants. This has been done so that both the forward and reverse characteristics can be shown on a single plot even though they differ by several orders of magnitude.

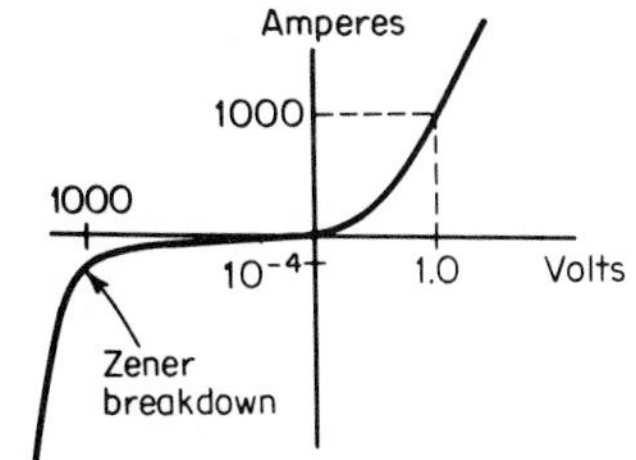

FIGURE 16.85 Diode forward-reverse characteristic.

Diodes are rated for forward current capacity and reverse voltage breakdown. They are manufactured with maximum current capabilities ranging from 0.05 A to more than 1000 A. Reverse voltage breakdown varies from 50 V to more than 2500 V. At rated forward current, the forward voltage drop varies between 0.7 and 1.5 V for silicon diodes. Although other materials are used for special-purpose devices, by far the most common semiconductor material is silicon. With a forward current of 1000 A and a forward voltage drop of 1 V, there would be a power loss in the diode of 1000 W (more than 1 hp). The basic diode package shown in Fig. 16.86 can dissipate about 20 W. To maintain an acceptable temperature in the diode, it is necessary to mount the diode on a *heat sink*. The manufacturer's recommendation should be followed very carefully to ensure good heat transfer and at the same time avoid fracturing the silicon chip inside the diode package.

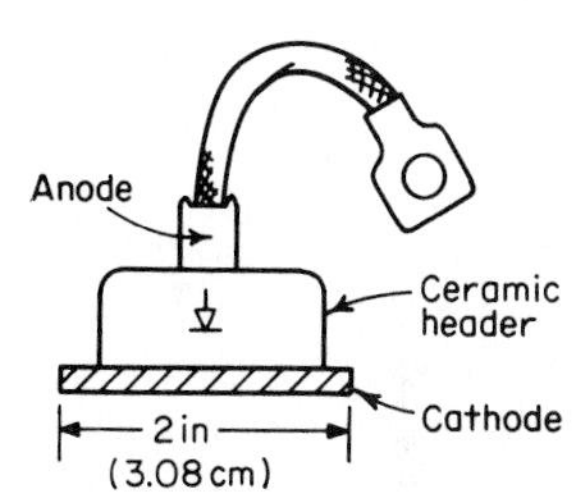

FIGURE 16.86 Physical diode package.

The selection of fuses or circuit breakers for the protection of rectifiers and rectifier circuitry requires more care than for other electronic devices. Diode failures as a result of circuit faults occur in a fraction of a millisecond. Special semiconductor fuses have been developed specifically for semiconductor circuits. Proper protective circuits must be provided for the protection of not only semiconductors but also the rest of the circuit and nearby personnel. Diodes and diode fuses have a short-circuit rating in amperes-squared-seconds (I^2t). As long as the I^2t rating of the diode exceeds the I^2t rating of its protective fuse, the diode and its associated circuitry will be protected. Circuit breakers may be used to protect diode circuits, but additional line impedance must be provided to limit the current while the circuit breaker clears. Circuit breakers do not interrupt the current when their contacts open. The fault is not cleared until the line voltage reverses at the end of the cycle of the applied voltage. This means that the *clearing time for a circuit breaker* is about ½ cycle of the ac input voltage. Diodes have a 1-cycle overcurrent rating which indicates the fault current the diode can carry for circuit breaker protection schemes. Line inductance is normally provided to limit fault currents for breaker protection. Often this inductance is in the form of leakage reactance in the transformer which supplies power to the diode circuit.

A *thyristor*, often called a *silicon controlled rectifier (SCR)*, is a rectifier which blocks current in both the forward and reverse directions. Conduction of current in the forward direction will occur when the anode is positive with respect to the cathode and when the gate is pulsed positive with respect to the cathode. Once the thyristor has begun to conduct, the gate pulse can return to 0 V or even go negative and the thyristor will continue to pass current. To stop the cathode-to-anode current, it is necessary to reverse the cathode-to-anode voltage. The thyristor will again be able to block both forward and reverse voltages until current flow is initiated by a gate pulse. The schematic symbol for an SCR is shown in Fig. 16.87. The physical packaging of thyristors is similar to that of rectifiers with similar ratings, except, of course, that the thyristor must have an additional gate connection.

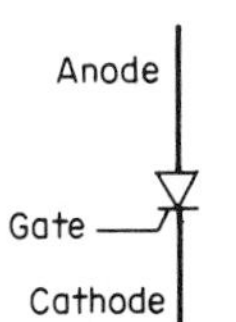

FIGURE 16.87 Thyristor schematic symbol.

The gate pulse required to fire an SCR is quite small compared with the anode voltage and current. Power gains in the range of 10^6 to 10^9 are easily obtained. In addition, the power loss in the thyristor is very low, compared with the power it controls, so that it is a very efficient power-controlling device. Efficiency in a thyristor power supply is usually 97 to 99 percent. When the thyristor blocks either forward or reverse current, the high voltage drop across the thyristor accompanies low current. When the thyristor is conducting forward current after having been fired by its gate pulse, the high anode current occurs with a forward voltage drop of about 1.5 V. Since high voltage and high current never occur simultaneously, the power dissipation in both the on and off states is low.

The thyristor is rated primarily on the basis of its forward-current capacity and its voltage-blocking capability. Devices are manufactured to have equal forward and reverse voltage-blocking capability. Like diodes, thyristors have I^2t ratings and 1-cycle surge current ratings to allow design of protective circuits. In addition to these ratings, which the SCR shares in common with diodes, the SCR has many additional specifications. Because the thyristor is limited in part by its average current and in part by its rms current, forward-current capacity is a function of the duty cycle to which the device is subjected. Since the thyristor cannot regain its blocking ability until its anode voltage is reversed and remains reversed for a short time, this time must be specified. The time to regain blocking ability after the anode voltage has been reversed is called the *turn-off time*. Specifications are also given for minimum and maximum gate drive. If forward blocking voltage is reapplied too quickly, the SCR may fire with no applied gate voltage pulse. The maximum safe value of rate of reapplied voltage is called the *dv/dt rating* of the SCR. When the gate pulse is applied, current begins to flow in the area immediately adjacent to the gate junction. Rather quickly, the current spreads across the entire cathode-junction area. In some circuits associated with the thyristor an extremely fast rate of rise of current may occur. In this event, localized heating of the cathode may occur with a resulting immediate failure or in less extreme cases a slow degradation of the thyristor. The maximum rate of change of current for a thyristor is given by its *di/dt* rating. Design for *di/dt* and *dv/dt* limits is not normally a problem at power-line frequencies of 50 and 60 Hz. These ratings become a design factor at frequencies of 500 Hz and greater. Table 16.12 lists typical thyristor characteristics.

A *triac* is a bilateral SCR. It blocks current in either direction until it receives a gate pulse. It can be used to control in ac circuits. Triacs are widely used for

TABLE 16.12 Typical Thyristor Characteristics

Voltage	Current, A rms	Current, A avg	I_{2t}^{2}, A · S	1-Cycle surge, A	di/dt, A/s	dv/dt, V/s	Turn-off time, s
400	35	20	165	180	100	200	10
1200	35	20	75	150	100	200	10
400	110	70	4,000	1000	100	200	40
1200	110	70	4,000	1000	100	200	40
400	235	160	32,000	3500	100	200	80
1200	235	160	32,000	3500	75	200	80
400	470	300	120,000	5500	50	100	150
1200	470	300	120,000	5500	50	100	150

light dimmers and for the control of small universal ac motors. The triac must regain its blocking ability as the line voltage crosses through zero. This fact limits the use of triacs to 60 Hz and below.

A *transistor* is a semiconductor amplifier. The schematic symbol for a transistor is shown in Fig. 16.88. There are two types of transistors, *p-n-p* and *n-p-n*. Notice that the polarities of voltage applied to these devices are opposite. In many sizes, matched *p-n-p* and *n-p-n* devices are available. The most common transistors have a collector dissipation rating of 150 to 600 mW. Collector-to-base breakdown voltage is 20 to 50 V. The amplification or gain of a transistor occurs because of two facts: First, a small change in current in the base circuit causes a large change in current in the collector and emitter leads. This current amplification is designated *hfe* on most transistor specification sheets. Second, a small change in base-to-emitter voltage can cause a large change in either the collector-to-base voltage or the collector-to-emitter voltage. Table 16.13 shows basic ratings for some typical transistors. There is a

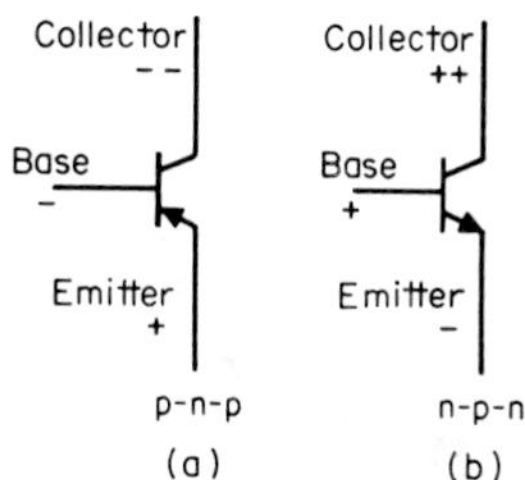

FIGURE 16.88 Transistor schematic symbol.

TABLE 16.13 Typical Transistor Characteristics

JEDEC number	Type	Collector-emitter volts at breakdown, BV_{CE}	Collector dissipation, P_c (25°C)	Collector current, I_c	Current gain, h_{fe}
2N3904	*n-p-n*	40	310 mW	200 mA	200
2N3906	*p-n-p*	40	310 mW	200 mA	200
2N3055	*n-p-n*	100	115 W	15 A	20
2N6275	*n-p-n*	120	250 W	50 A	30
2N5458	JFET	40	200 mW	9 mA	*
2N5486	JFET	25	200 mW		†

*JFET, current gain is not applicable.
†High-frequency JFET—up to 400 MHz.

great profusion of transistor types so that the choice of type depends on availability and cost as well as operating characteristics.

The gain of a transistor is independent of frequency over a wide range. At high frequency, the gain falls off. This cutoff frequency may be as low as 20 kHz for audio transistors or as high as 1 GHz for radio-frequency (rf) transistors.

The schematic symbols for the field effect transistor (FET) is shown in Fig. 16.89. The flow of current from source to drain is controlled by an electric field established in the device by the voltage applied between the gate and the drain. The effect of this field is to change the resistance of the transistor by altering its internal current path. The FET has an extremely high gate resistance (10^{12} Ω), and as a consequence, it is used for applications requiring high input impedance. Some FETs have been designed for high-frequency characteristics. Other FETs have been designed for high-power applications. The two basic constructions used for FETs are *bipolar junctions* and *metal oxide semiconductors*. The schematic symbols for each of these are shown Fig. 16.89*a* and *b*. These are called *JFETs* and *MOSFETs* to distinguish between them. JFETs and MOSFETs are used as stand-alone devices and are also widely used in integrated circuits. (See below, this section.)

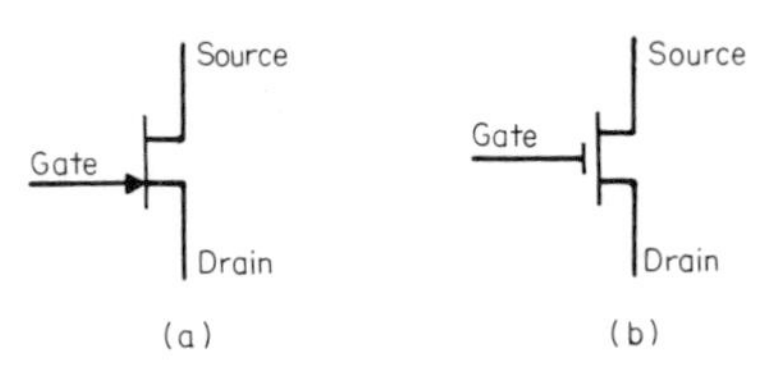

FIGURE 16.89 Field effect transistor: (*a*) bipolar junction type (JFET); (*b*) metal oxide semiconductor type (MOSFET).

The *unijunction* is a special-purpose semiconductor device. It is a pulse generator that is widely used to fire thyristors and triacs as well as in timing circuits and waveshaping circuits. The schematic symbol for a unijunction is shown in Fig. 16.90. The device is essentially a silicon resistor. This resistor is connected to base 1 and base 2. The emitter is fastened to this resistor about halfway between bases 1 and 2. If a positive voltage is applied to base 2, and if the emitter and base 1 are at zero, the emitter junction is back-biased and no current flows in the emitter. If the emitter voltage is made increasingly positive, the emitter junction will become forward-biased. When this occurs, the resistance between base 1 and base 2 and between base 2 and the emitter suddenly switches to a very low value. This is a regenerative action, so that very fast and very energetic pulses can be generated with this device.

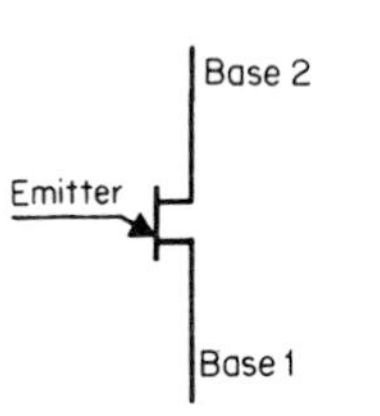

FIGURE 16.90 Unijunction.

Before the advent of semiconductors, electronic rectifiers and amplifiers were *vacuum tubes* or *gas-filled tubes*. Some use of these devices still remains. If an electrode is heated in a vacuum, it gives up surface electrons. If an electric field is established between this heated electrode and another electrode so that the electrons are attracted to the other electrode, a current will flow through the vacuum. Electrons flow from the heated cathode to the cold anode. If the polarity is reversed, since there are no free electrons around the anode, no current will flow. This, then, is a vacuum-tube rectifier. If a third electrode, called a *control grid*, is placed between the cathode and the anode, the flow of electrons from the cathode to the anode can be controlled. This is a basic vacuum-tube amplifier. Additional grids have been placed between the cathode and anode to further enhance certain characteristics of the vacuum tube. In addition, multiple anodes and cathodes have been enclosed in a single tube for special applications such as radio signal converters.

If an inert gas, such as neon or argon, is introduced into the vacuum, conduction can be initiated from a cold electrode. The breakdown voltage is relatively stable for given gas and gas pressure and is in the range of 50 to 200 V. The *nixie* display tube is such a device. This tube contains 10 cathodes shaped in the form of the numerals from 0 to 9. If one of these cathodes is made negative with respect to the anode in the tube, the gas in the tube glows around that cathode. In this way, each of the 10 numerals can be made to glow when the appropriate electrode is energized.

An *ignitron* is a vapor-filled tube. It has a pool of liquid mercury in the bottom of the tube. Air is exhausted from the enclosure, leaving only mercury vapor, which comes from the pool at the bottom. If no current is flowing, this tube will block voltage whether the anode is plus or minus with respect to the mercury-pool cathode. A small rod called an *ignitor* can form a cathode spot on the pool of mercury when it is withdrawn from the pool. The ignitor is pulled out of the pool by an electromagnet. Once the cathode spot has been formed, electrons will continue to flow from the mercury-pool cathode to the anode until the anode-to-cathode voltage is reversed. The operation of an ignitron is very similar to that of a thyristor. The anode and cathode of each device perform similar functions. The ignitor and gate also perform similar functions. The thyristor is capable of operating at much higher frequencies than the ignitron and is much more efficient since the thyristor has 1.5 V forward drop and the ignitron has 15 V forward drop. The ignitron has an advantage over the thyristor in that it can carry extremely high overload currents without damage. For this reason ignitrons are often used as electronic "crowbars" which discharge electric energy when a fault occurs in a circuit.

DISCRETE-COMPONENT CIRCUITS

Several common rectifier circuits are shown in Fig. 16.91. The waveforms shown in this figure assume no line reactance. The presence of line reactance will make a slight difference in the waveshapes and the conversion factors shown in Fig. 16.91. These waveshapes are equally applicable for loads which are pure resistive or resistive and inductive. In a resistive load the current flowing in the load has the same waveshape as the voltage applied to it. For inductive loads, the current waveshape will be smoother than the voltage applied. If the inductance is high enough, the ripple in the current may be indeterminantly small. An approximation of the ripple current can be calculated as follows:

$$I = \frac{E_{dc} PCT}{200\pi f N L} \tag{16.160}$$

where I is the rms ripple current, E_{dc} is the dc load voltage, PCT is percent ripple from Fig. 16.91, f is line frequency, N is number of cycles of ripple frequency per cycle of line frequency, and L is equivalent series inductance in load. Equation (16.160) will always give a value of ripple higher than that calculated by more exact means, but this value is normally satisfactory for power-supply design.

Capacitance in the load leads to increased regulation. At light loads, the capacitor will tend to charge up to the peak value of the line voltage and remain

Type	Circuit	Output voltage waveform	E_{dc} (avg)	Ripple fundamental frequency	% ripple	Peak inverse voltage
Half-wave 1ϕ	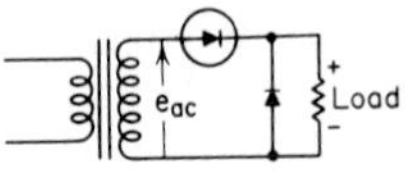 e_{ac} + Load −	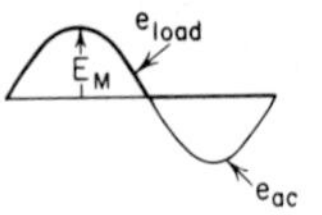 e_{load} E_M e_{ac}	0.318 E_M 0.45 E_{ac}	F	121	3.14 E_{dc}
Full-wave 1ϕ	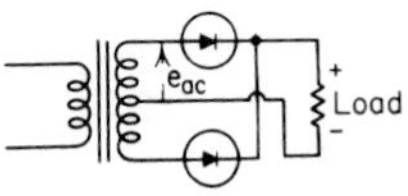 e_{ac} + Load −	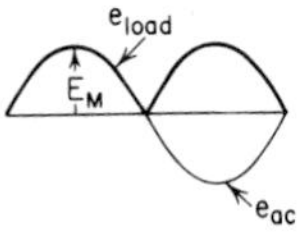 e_{load} E_M e_{ac}	0.636 E_M 0.9 E_{ac}	2F	48	3.14 E_{dc}
Bridge 1ϕ	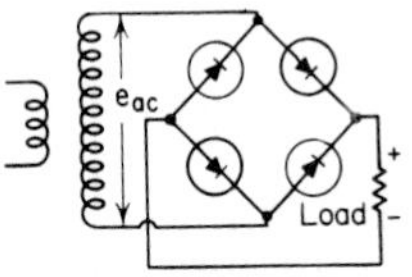 e_{ac} Load + −	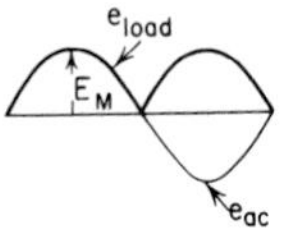 e_{load} E_M e_{ac}	0.636 E_M 0.9 E_{ac}	2F	48	1.57 E_{dc}
Half-wave 3ϕ	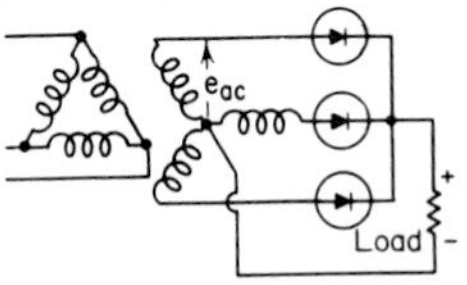 e_{ac} Load + −	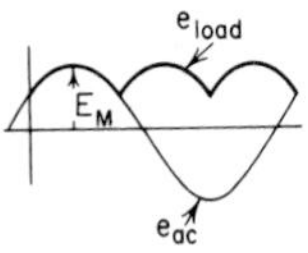 e_{load} E_M e_{ac}	0.827 E_M 1.17 E_{ac}	3F	18	2.09 E_{dc}

E_M = maximum value of e_{ac}
E_{ac} = effective value of e_{ac}
E_{dc} = average value of d-c load voltage
F = line frequency
% ripple = 100 x rms ripple / E_{dc}

FIGURE 16.91 Comparison of rectifier circuits.

there. This means that for either the single full-wave circuit or the single-phase bridge the dc output voltage would be 1.414 times the rms input voltage. As the size of the loading resistor is reduced, or as the size of the parallel load capacitor is reduced, the load voltage will more nearly follow the rectified line voltage and so the dc voltage will approach 0.9 times the rms input voltage for very heavy loads or for very small filter capacitors. One can see then that dc voltage may vary between 1.414 and 0.9 times line voltage due only to waveform changes when *capacitor filtering* is used.

Four different *thyristor rectifier circuits* are shown in Fig. 16.92. These circuits are equally suitable for resistive or inductive loads. It will be noted that the half-wave circuit for the thyristor has a rectifier across the load, as in Fig. 16.91. This diode is called a *freewheeling diode* because it freewheels and carries inductive load current when the thyristor is not conducting. Without this diode, it would not be possible to build up current in an inductive load. The gate-control circuitry is not shown in Fig. 16.92 in order to make the power circuit easier to

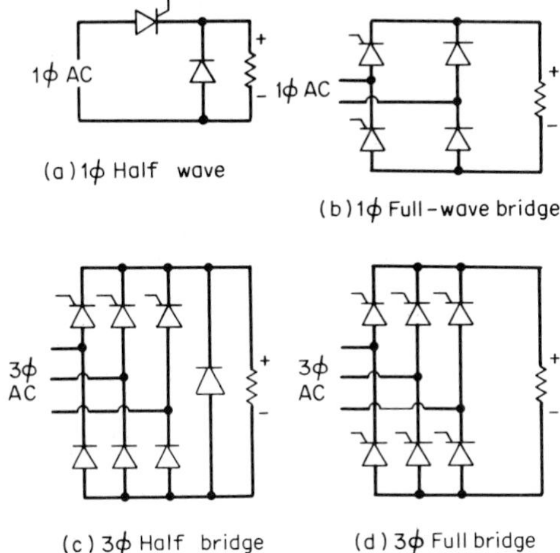

FIGURE 16.92 Basic thyristor circuits.

see. Notice the location of the thyristors and rectifiers in the single-phase full-wave circuit. Constructed this way, the two diodes in series perform the function of a freewheeling diode. The circuit can be built with a thyristor and rectifier interchanged. This would work for resistive loads but not for inductive loads. For the full three-phase bridge, a freewheeling diode is not required since the carryover from the firing of one SCR to the next does not carry through a large portion of the negative half cycle and therefore current can be built up in an inductive load.

Capacitance must be used with care in thyristor circuits. A capacitor directly across any of the circuits in Fig. 16.92 will immediately destroy the thyristors. When an SCR is fired directly into a capacitor with no series resistance, the resulting di/dt in the thyristor causes extreme local heating in the device and a resultant failure. A sufficiently high series resistor prevents failure. An inductance in series with a capacitor must also be used with caution. The series inductance may cause the capacitor to "ring up." Under this condition, the voltage across the capacitor can approach twice peak line voltage or 2.828 times rms line voltage.

The advantage of the thyristor circuits shown in Fig. 16.92 over the rectifier circuits is, of course, that the thyristor circuits provide variable output voltage. The output of the thyristor circuits depends upon the magnitude of the incoming line voltage and the phase angle at which the thyristors are fired. The control characteristic for the thyristor power supply is determined by the waveshape of the output voltage and also by the phase-shifting scheme used in the firing-control means for the thyristor. Practical and economic power supplies usually have control characteristics with some degree of nonlinearity. A representative characteristic is shown in Fig. 16.93. This control characteristic is usually given for nominal line voltage with the tacit understanding that variations in line voltage will cause approximately proportional changes in output voltage.

Transistor amplifiers can take many different forms. A complete discussion is beyond the scope of this handbook. The circuits described here illustrate basic principles. A basic *single-stage amplifier* is shown in Fig. 16.94. The transistor

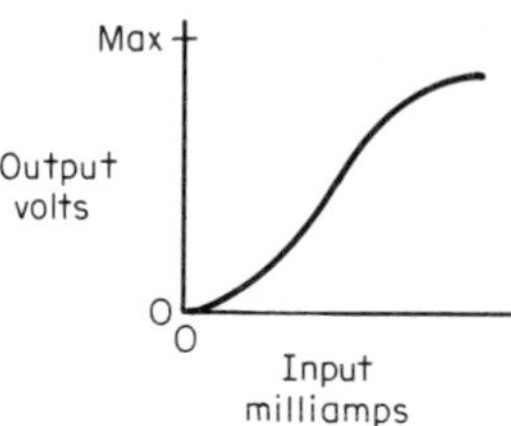

FIGURE 16.93 Thyristor control characteristic.

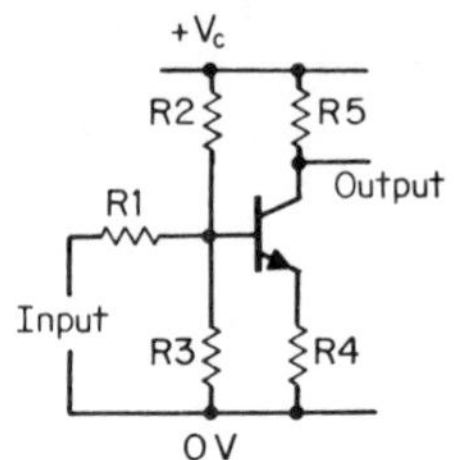

FIGURE 16.94 Single-stage amplifier.

can be cut off by making the input terminal sufficiently negative. It can be saturated by making the input terminal sufficiently positive. In the linear range, the base of an *n-p-n* transistor will be 0.5 to 0.7 V positive with respect to the emitter. The collector voltage will vary from about 0.2 V to V_c (20 V, typically). Note that there is a sign inversion of voltage between the base and the collector; i.e., when the base is made more positive, the collector becomes less positive. The resistors in this circuit serve the following functions: Resistor *R*1 limits the input current to the base of the transistor so that it is not harmed when the input signal overdrives. Resistors *R*2 and *R*3 establish the transistor's operating point with no input signal. Resistors *R*4 and *R*5 determine the voltage gain of the amplifier. Resistor *R*4 also serves to stabilize the zero-signal operating point, as established by resistors *R*2 and *R*3. Usual practice is to design single-stage gains of 10 to 20. Much higher gains are possible to achieve, but low gain levels permit the use of less expensive transistors and increase circuit reliability.

Figure 16.95 illustrates a basic *two-stage transistor amplifier* using complementary *n-p-n* and *p-n-p* transistors. Note that the first stage is identical to that shown in Fig. 16.94. This *n-p-n* stage drives the following *p-n-p* stage. Additional alternate *n-p-n* and *p-n-p* stages can be added until any desired overall amplifier gain is achieved.

Figure 16.96 shows the *Darlington connection* of transistors. The amplifier is used to obtain maximum current gain from two transistors. Assuming a base-to-collector current gain of 50 times for each transistor, this circuit will give an input-to-output current gain of 2500. This high level of gain is not very stable if the ambient temperature changes, but in many cases this drift is tolerable.

Figure 16.97 shows a circuit developed specifically to minimize temperature drift and drift due to power supply voltage changes. The *differential amplifier* minimizes drift because of the balanced nature of the circuit. Whatever changes in one transistor tend to increase the output are compensated by reverse trends in

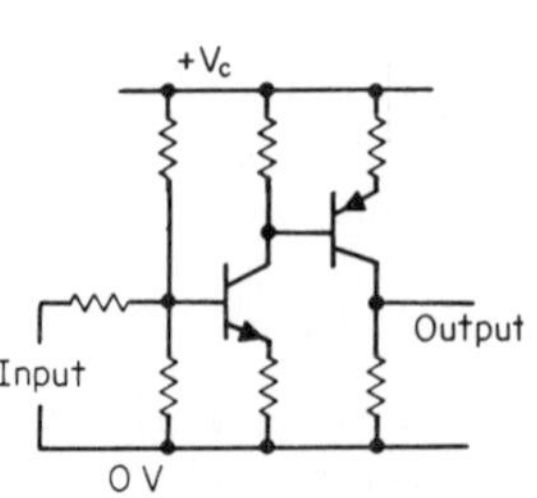

FIGURE 16.95 Two-stage amplifier.

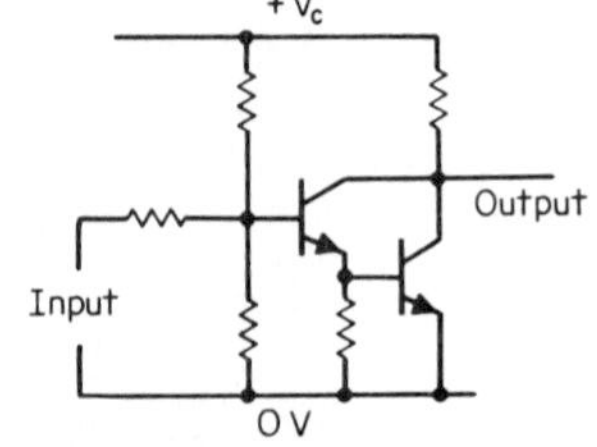

FIGURE 16.96 Darlington connection.

the second transistor. The input signal does not affect both transistors in compensatory ways, of course, and so it is amplified. One way to look at a differential amplifier is that twice as many transistors are used for each stage of amplification to achieve compensation. For very low drift requirements, matched transistors are available. For the ultimate in differential amplifier performance, two matched transistors are encapsulated in a single unit. *Operational amplifiers* made with discrete components frequently use differential amplifiers to minimize drift and offset. The operational amplifier is a low-drift, high-gain amplifier designed for a wide range of control and instrumentation uses.

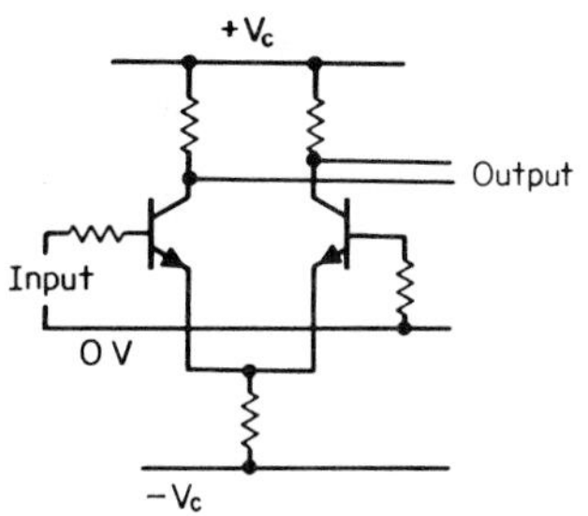

FIGURE 16.97 Differential amplifier.

Oscillators are circuits which provide a frequency output with no signal input. A portion of the collector signal is fed back to the base of the transistor. This feedback is amplified by the transistor and so maintains a sustained oscillation. The frequency of the oscillation is determined by parallel inductance and capacitance. The oscillatory circuit consisting of an inductance and a capacitance in parallel is called an *LC tank circuit*.

This frequency is approximately equal to

$$f = 1/2\pi\sqrt{CL} \tag{16.161}$$

where f is frequency (Hz), C is capacitance (F), and L is inductance (H). A 1-MHz oscillator might typically be designed with a 20-μH inductance in parallel with a 0.05-μF capacitor. The exact frequency will vary from the calculated value because of loading effects and stray inductance and capacitance. The *Colpitts oscillator* shown in Fig. 16.98 differs from the *Hartley oscillator* shown in Fig. 16.99 only in the way energy is fed back to the emitter. The Colpitts oscillator has a capacitive voltage divider in the resonant tank. The Hartley oscillator has an inductive voltage divider in the tank. The *crystal oscillator* shown in Fig. 16.100 has much greater frequency stability than the circuits in Figs. 16.98 and 16.99. Frequency stability of 1 part in 10^7 is easily achieved with a crystal-controlled oscillator. If the oscillator is temperature-controlled by mounting it in a small temperature-controlled oven, the frequency stability can be increased to 1 part in 10^9. The resonant LC tank in the collector circuit is tuned to approximately the crystal frequency. The crystal offers a low impedance at its resonant frequency. This pulls the collector-tank operating frequency to the crystal resonant frequency.

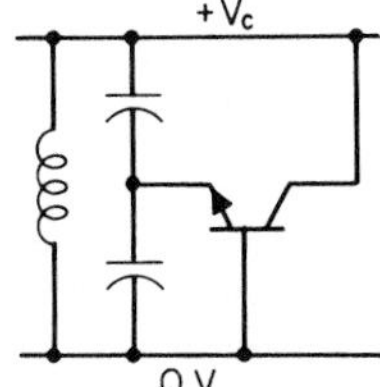

FIGURE 16.98 Colpitts oscillator.

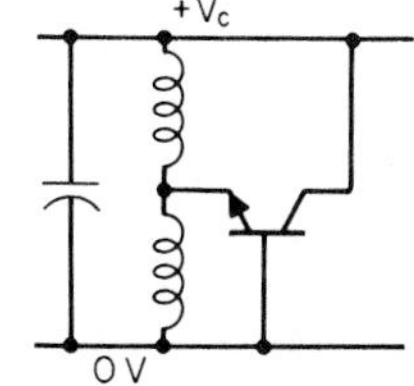

FIGURE 16.99 Hartley oscillator.

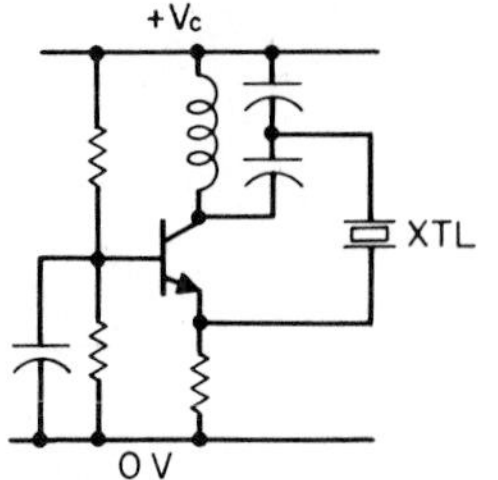

FIGURE 16.100 Crystal-controlled oscillator.

As the desired operating frequency becomes 500 MHz and greater, *resonant cavities* are used as tank circuits instead of discrete capacitors and inductors. A rough guide to the relationship between frequency and resonant-cavity size is the wavelength of the frequency

$$\lambda = 300 \times 10^6/f \qquad (16.162)$$

where λ is wavelength (m), 300×10^6 is the speed of light (m/s), and f is frequency (Hz). The resonant cavities will be smaller than indicated by Eq. (16.162) because in general the cavity is either one-half or one-fourth wavelength and also, in general, the electromagnetic wave velocity is less in a cavity than in free space.

The operating principles of these devices are beyond the scope of this handbook. There are many different kinds of microwave tubes, including klystrons, magnetrons, and traveling-wave tubes. All these tubes employ moving electrons to excite a resonant cavity. These devices serve as either oscillators or amplifiers at microwave frequencies.

Lasers operate at approximately visible-light frequency of 600 THz. This corresponds to a wavelength of 0.5 μm or, the more usual measure of visible-light wavelength, 5×10^3 Å. Resonant cavities simply cannot be made small enough for these wavelengths. Electronic resonance in the atom serves as the tank circuit for these high frequencies. Quantum mechanics must be employed properly to explain these devices, but a practical understanding can be achieved without delving so deep. Most light is disorganized insofar as the axis of vibration and the frequency of vibration are concerned. When radiation along different axes is attenuated, as with a polarizing screen, the light wave is said to be polarized. When white light is filtered, or when the light source is not white, the light is frequency-limited, or colored. Colored light still has a relatively wide band of frequencies. The laser emits a very narrow band of frequencies, which are extremely stable, many times more stable than a crystal; therefore, lasers are used as frequency standards. The narrow frequency band of lasers allows focusing the output into extremely small beams. This feature makes the laser attractive as a cutting tool and as an accurate surveying device. Extremely sharp focus and extremely high frequency make it attractive as a high-density communications carrier. Experimental work is being done with *phase-locked lasers*, which not only have a single frequency of output but have output oscillations in phase with each other. This degree of organization promises further commercial development of laser devices.

A radio wave consists of two parts, a *carrier* and an *information signal*. The carrier is a steady high frequency. The information signal may be a voice signal, a video signal, or telemetry information. The carrier wave can be modulated by varying its amplitude or by varying its frequency. *Modulators* are circuits which impress the information signal onto the carrier. A *demodulator* is a circuit in the receiving apparatus which separates the information signal from the carrier. A simple amplitude modulator is shown in Fig. 16.101. The transistor is base driven with the carrier input and emitter driven with the information signal. The modulated carrier wave appears at the collector of the transistor. An FM modulator is shown in Fig. 16.102. The carrier must be changed in frequency in response to

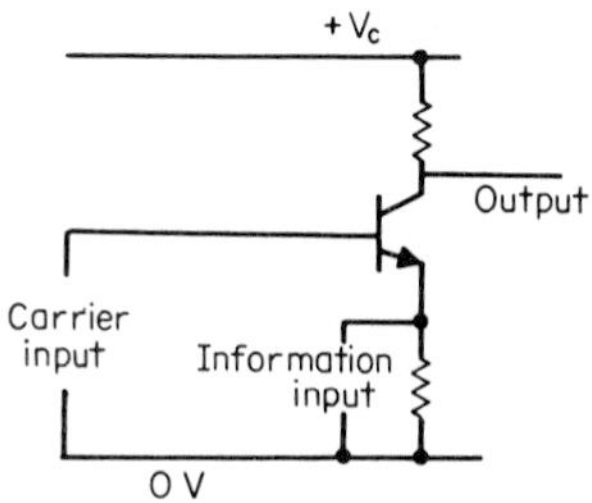

FIGURE 16.101 AM modulator.

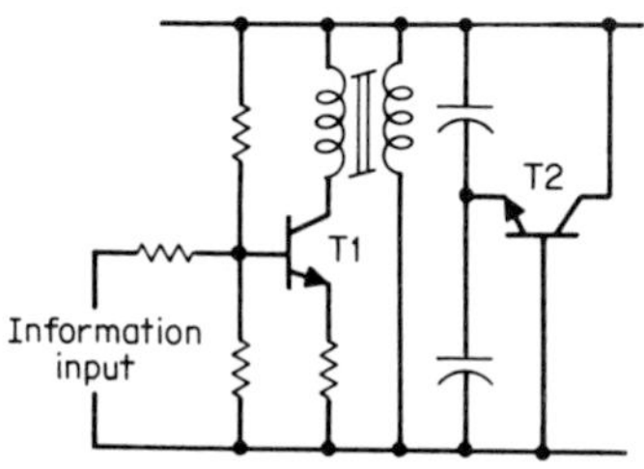

FIGURE 16.102 FM modulator.

the information signal input. This is accomplished by using a saturable ferrite core in the inductance of a Colpitts oscillator which is tuned to the carrier frequency. As the collector current in transistor $T1$ varies with the information signal, the saturation level in the ferrite core changes, which in turn varies the inductance of the winding in the tank circuit and alters the operating frequency of the oscillator.

The *demodulator* for an AM signal is shown in Fig. 16.103. The diode rectifies the carrier plus information signal so that the filtered voltage appearing across the capacitor is the information signal. Resistor $R2$ blocks the carrier signal so that the output contains only the information signal. An FM *demodulator* is shown in Fig. 16.104. In this circuit, the carrier plus information signal has a constant amplitude. The information is in the form of varying frequency in the carrier wave. If inductor $L1$ and capacitor $C1$ are tuned to near the carrier frequency but not exactly at resonance, the current through resistor $R1$ will vary as the carrier frequency shifts up and down. This will create an AM signal across resistor $R1$. The diode, resistors $R2$ and $R3$, and capacitor $C2$ demodulate this signal as in the circuit in Fig. 16.103.

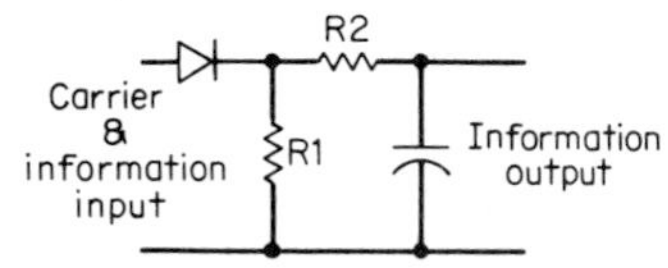

FIGURE 16.103 AM demodulator.

The waveform of the basic *electronic timing circuit* is shown in Fig. 16.105 along with a basic timing circuit. Switch $S1$ is closed from time $t1$ until time $t3$. During this time, the transistor shorts the capacitor and holds the capacitor at 0.2 V. When switch $S1$ is opened at time $t3$, the transistor ceases to conduct and the capacitor charges exponentially due to the current flow through resistor $R1$. Delay time can be measured to any point along this exponential charge. If the time is measured until time $t6$, the timing may vary due to small shifts in supply voltage or slight changes in the voltage-level detecting circuit. If time is measured until time $t4$, the voltage level will be easy to detect, but the obtainable time delay from time $t3$ to time $t4$ may not be large enough compared with the reset time $t1$ to $t2$. Considerations like these usually dictate detecting at time $t4$. If this time is at a voltage level which is 63 percent of V_c, the time from $t3$ to $t4$ is one time constant of $R1$ and C. This time can be calculated by

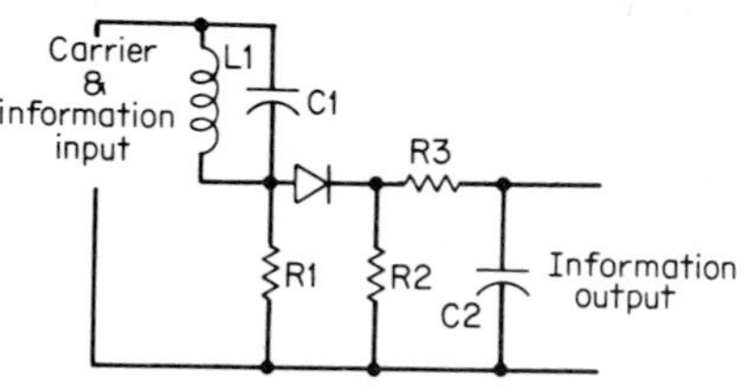

FIGURE 16.104 FM discriminator.

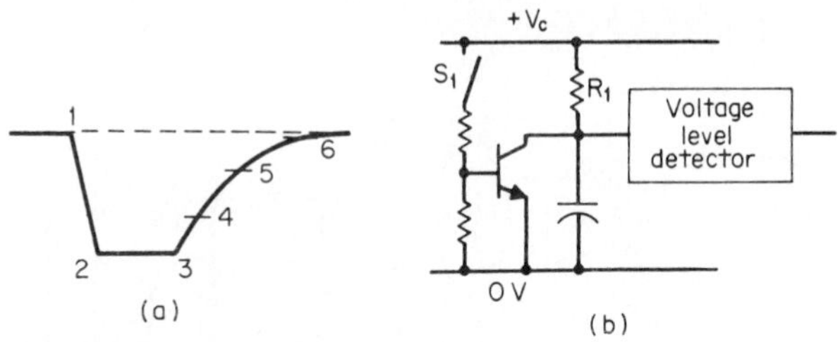

FIGURE 16.105 Basic timing circuit.

$$t = RC \tag{16.163}$$

where t is time (s), R is resistance (Ω), and C is capacitance (F). A timing circuit with a 0.1-s delay can be constructed using a 0.1-μF capacitor and a 1.0-MΩ resistor.

An improved timing circuit is shown in Fig. 16.106. In this circuit, the unijunction is used as a level detector, a pulse generator, and a reset means for the capacitor. The transistor is used as a constant current source for charging the timing capacitor. The current through the transistor is determined by resistors $R1$, $R2$, and $R3$. This current is adjustable by means of $R1$. When the charge on the capacitor reaches approximately 50 percent of V_c, the unijunction fires, discharging the capacitor and generating a pulse at the output. The discharged capacitor is then recharged by the transistor, and the cycle continues to repeat. The pulse rate of this circuit can be varied from one pulse per minute to many thousands of pulses per second.

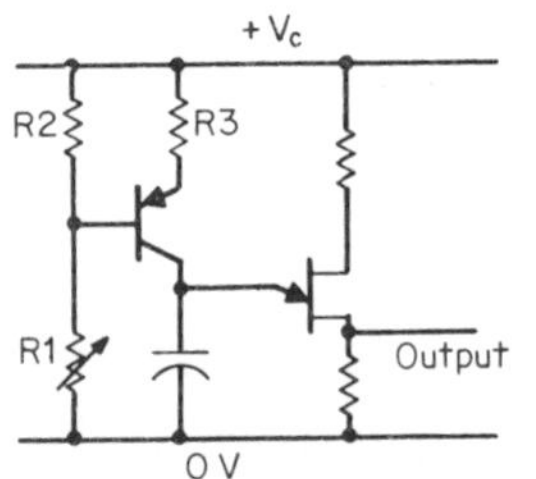

FIGURE 16.106 Improved timing circuit.

INTEGRATED CIRCUITS

Table 16.14 lists some of the more common physical packages for discrete component and integrated semiconductor devices. Although discrete components are still used for electronic design, *integrated circuits* (*ICs*) are becoming predominant in almost all types of electronic equipment. Dimensions of common dual inline pin (DIP) integrated-circuit devices are shown in Fig. 16.107. An IC costs far less than circuits made with discrete components. Integrated circuits can be classified in several different ways. One way to classify them is by complexity. *Small-scale integration* (*SSI*), *medium-scale integration* (*MSI*), *large-scale integration* (*LSI*), and *very large scale integration* (*VLSI*) refer to this kind of classification. The cost and availability of a particular IC are more dependent on the size of the market for that device than on the level of its internal complexity. For this reason, the classification by circuit complexity is not as meaningful today as it once was. The literature still refers to these classifications, however. For the

TABLE 16.14 Semiconductor Physical Packaging

Signal devices:	
Plastic	TO92
Metal can	TO5,TO18,TO39
Power devices:	
Tab mount	TO127,TO218,TO220
Diamond case	TO3,TO66
Stud mount	
Flat base	
Flat pak (Hockey puck)	
Integrated circuits:	
Dip (dual in-line pins)	(See Fig. 16.107.)
Flat pack	
Chip carrier (50-mil centers)	

Number of pins	*W*	*P*	*L*	Approximate height
64	0.8	0.1	3.3	0.24
40	0.6	0.1	2.0	0.2
28	0.6	0.1	1.4	0.2
24	0.6	0.1	1.3	0.2
22	0.4	0.1	1.1	0.2
20	0.3	0.1	1.0	0.2
18	0.3	0.1	0.9	0.2
16	0.3	0.1	0.87	0.2
14	0.3	0.1	0.78	0.2
8	0.3	0.1	0.4	0.2

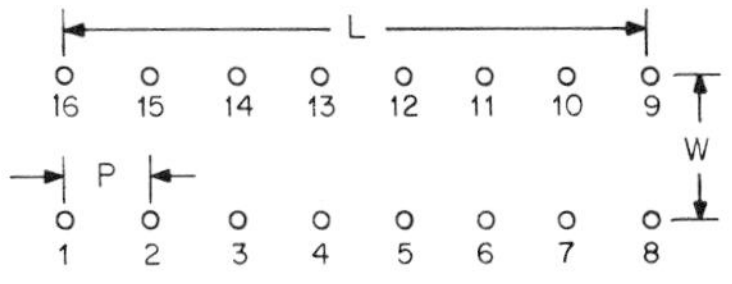

FIGURE 16.107 Approximate physical dimensions of dual in-line pin (DIP) integrated circuits. All dimensions are in inches. Dual in-line packages are made in three different constructions: molded plastic, cerdip, and ceramic.

purpose of this text, ICs will be separated into two broad classes: linear ICs and digital ICs.

The trend in IC development has been toward greatly increased complexity at significantly reduced cost. Present-day ICs are manufactured with internal spacings as low as 2 μm. The limitation of the contents of a single device is more often controlled by external connections than by internal space. For this reason, more and more complex combinations of circuits are being interconnected within a single device. There is also a tendency to accomplish functions digitally that were formerly done by analog means. Although these digital circuits are much more complex than their analog counterparts, the cost and reliability of ICs make the resulting digital circuit the preferred design. One can expect these trends will continue based on current technology. One can also anticipate further declines in price versus performance. It has been demonstrated again and again that digital IC designs are much more stable and reliable than analog designs.

LINEAR INTEGRATED CIRCUITS

The basic building block for many linear ICs is the *operational amplifier*. Table 16.15 lists the basic characteristics for a few representative IC operational amplifiers. In most instances, an adequate design for an operational amplifier circuit can be made assuming an "ideal" operational amplifier. For an ideal operational amplifier, one assumes that it has infinite gain and no voltage drop across its input terminals. In most designs, feedback is used to limit the gain of each operational amplifier. As long as the resulting closed-loop gain is much less than the open-loop gain of the operational amplifier, this assumption yields results that are within acceptable engineering accuracy. Operational amplifiers use a balanced input circuit which minimizes input voltage offset. Furthermore, specially designed operational amplifiers are available which have extremely low input offset voltage. The input voltage must be kept low because of temperature drift considerations. For these reasons, the assumption of zero input voltage, sometimes called a *virtual ground*, is justified. Figure 16.108 shows three operational amplifier circuits and the equations which describe their behavior. In this figure, S is the *Laplace transform* variable. *Active filters* are designed using operational amplifiers with associated resistors and capacitors in a manner similar to that shown in Fig. 16.108.

Table 16.16 lists some typical linear ICs, most of which contain operational amplifiers. The *voltage comparator* is an operational amplifier that compares two input voltages and provides an output that indicates which of the two voltages is greater. The *sample-and-hold circuit* samples an analog input at prescribed intervals. Between these sample times, it holds the last value it measured. This circuit is used to convert signals from analog to digital form. The NE 555 timer/oscillator is adaptable for a wide variety of applications. It can be used as a stable, adjustable-frequency free-running or *monostable multivibrator*. It can also be used as a *linear ramp generator*. It can be used for time delay or sequential timing applications.

Table 16.17 lists linear ICs that are used in audio, radio, and television circuits. The degree of complexity that can be incorporated in a single device is illustrated by the fact that a complete AM-FM radio circuit is available in a single

TABLE 16.15 Operational Amplifiers

Type	Purpose	Input bias current, nA	Input res., Ω	Supply voltage, V	Voltage gain	Unity gain bandwidth, MHz
LM741	General purpose	500	2×10^6	+ 20	25,000	1.0
LM224	Quad gen. purpose	150	2×10^6	3 to 32	50,000	1.0
LM255	FET input	0.1	10^{12}	+ 22	50,000	2.5
LM444A	Quad FET input	0.005	10^{12}	+ 22	50,000	1.0

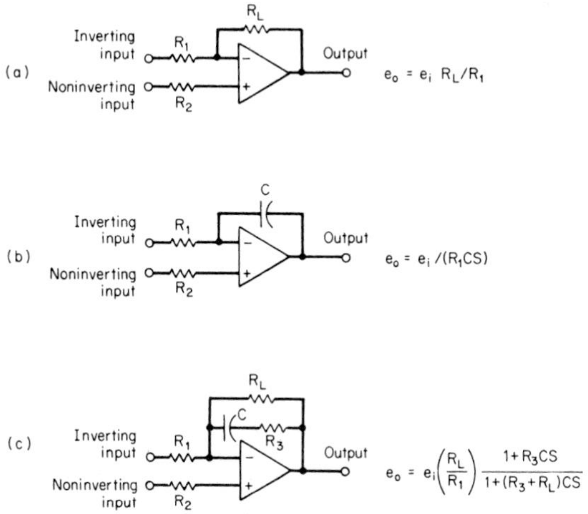

(a) $e_o = e_i\, R_L/R_1$

(b) $e_o = e_i/(R_1CS)$

(c) $e_o = e_i\left(\frac{R_L}{R_1}\right)\frac{1+R_3CS}{1+(R_3+R_L)CS}$

FIGURE 16.108 Operational amplifier circuits.

TABLE 16.16 Linear Integrated-Circuit Devices

Operational amplifier	Voltage comparator
Sample and hold	Active filters
Analog-to-digital converter	Digital-to-analog converter
Voltage regulator	Voltage reference
Voltage-controlled oscillator	NE 555 timer/oscillator

TABLE 16.17 Audio, Radio, and Television Integrated-Circuit Devices

Audio amplifier	Tone-volume-balance circuit
Dolby filter circuit	Phase-locked loop (PLL)
Intermediate frequency circuit	AM-FM radio
TV chroma demodulator	Digital tuner
Video-IF amplifier-detector	

IC device. The *phase-locked loop* is a device that is widely utilized for accurate frequency control. This device produces an output frequency that is set by a digital input. It is a highly accurate and stable circuit. This circuit is often used to demodulate FM radio waves.

Table 16.18 lists linear IC circuits that are used in telecommunications. These circuits include digital circuits within them and/or are used with digital devices. Whether these should be classed as linear ICs or digital ICs may be questioned. Several manufacturers include them in their linear device listings and not with their digital devices, and for this reason, they are listed here as linear devices. The radio-control *transmitter-encoder* and *receiver-decoder* provide a means of sending up to four control signals on a single radio-control frequency link. Each of the four channels can be either an on-off channel or a *pulse-width-modulated*

TABLE 16.18 Telecommunication Integrated-Circuit Devices

Radio-control transmitter-encoder
Radio-control receiver-decoder
Pulse-code modulator–coder-decoder (PCM CODEC)
Single-chip programmable signal processor
Touch-tone generators
Modulator-demodulator (modem)

(*PWM*) proportional channel. The *pulse-code modulator–coder-decoder* (*PCM CODEC*) is typical of a series of IC devices that have been designed to facilitate the design of digital-switched telephone circuits.

DIGITAL INTEGRATED CIRCUITS

The basic circuit building block for digital ICs is the gate circuit. A *gate* is a switching amplifier that is designed to be either on or off. (By contrast, an operational amplifier is a proportional amplifier.) For 5-V logic levels, the gate switches to a 0 whenever its input falls below 0.8 V and to a 1 whenever its input exceeds 2.0 V. This arrangement ensures immunity to spurious noise impulses in both the 0 and the 1 state.

Several representative *transistor-transistor-logic* (*TTL*) *gates* are listed in Table 16.19. Gates can be combined to form logic devices of two fundamental kinds:

TABLE 16.19 Digital Integrated-Circuit Devices

Type 54/74*	No. circuits per device	No. inputs per device	Function
00	4	2	NAND gate
02	4	2	NOR gate
04	6	1	Inverter
06	6	1	Buffer
08	4	2	AND gate
10	3	3	NAND gate
11	3	3	AND gate
13	2	4	Schmitt trigger
14	6	1	Schmitt trigger
20	2	4	NAND gate
21	2	4	AND gate
30	1	8	NAND gate
74	2		D flip-flop
76	2		JK flip-flop
77	4		Latch
86	4	2	EXCLUSIVE OR gate
174	6		D flip-flop
373	8		Latch
374	8		D flip-flop

NOTE: Example of device numbers are 74LS04, 54L04, 5477, and 74H10. The letters after the series number denote the speed and loading of the device.

*54 series devices are rated for temperatures from −55 to 125°C. 74 series devices are rated for temperatures from 0 to 70°C.

combinational and sequential. In *combinational logic*, the output of a device changes whenever its input conditions change. The basic gate exemplifies this behavior.

A number of gates can be interconnected to form a *flip-flop* circuit. This is a bistable circuit that stays in a particular state, a 0 or a 1 state, until its "clock" input goes to a 1. At this time its output will stay in its present state or change to a new state depending on its input just prior to the clock pulse. Its output will retain this information until the next time the clock goes to a 1. The flip-flop has memory, because it retains its output from one clock pulse to another. By connecting several flip-flops together, several sequential states can be defined permitting the design of a *sequential logic* circuit.

Table 16.20 shows three common flip-flops. The *truth table*, sometimes called a *state table*, shows the specification for the behavior of each circuit. The present output state of the flip-flop is designated $Q(t)$. The next output state is designated $Q(t + 1)$. In addition to the truth table, the Boolean algebra equations in Table 16.20 are another way to describe the behavior of the circuits. The *JK flip-flop* is the most versatile of these three flip-flops because of its separate J and K inputs. The *T flip-flop* is called a *toggle*. When its T input is a 1, its output toggles, from 0 to 1 or from 1 to 0, at each clock pulse. The *D flip-flop* is called a *data cell*. The output of the D flip-flop assumes the state of its input at each clock pulse and holds this data until the next clock pulse. The JK flip-flop can be made to function as a T flip-flop by applying the T input to both the J and K input terminals. The JK flip-flop can be made to function as a D flip-flop by applying the data signal to the J input and applying the inverted data signal to the K input. Some common IC flip-flops are listed in Table 16.19.

Various types of gates are shown in Table 16.21. Combinational logic defined by means of these various gates is used to define the input to flip-flops, which serve as memory devices. At each clock pulse, these flip-flops change state in accordance with their respective inputs. These new states are retained in the flip-flop and also applied to the gates. The output of the gates change (with only a

TABLE 16.20 Flip-Flop Sequential Devices

Name		Graphic symbol	Algebraic function	Truth table
JK flip-flop	Clock	S J Q < K $\overline{Q}$ R	$Q(t + 1) = JQ'(t) + K'Q(t)$	J K $Q(t)$ | $Q(t + 1)$ 0 X 0 | 0 1 X 0 | 1 X 0 1 | 1 X 1 1 | 0
T flip-flop	Clock	T Q < $\overline{Q}$	$Q(t + 1) = TQ'(t) + T'Q(t)$	T $Q(t)$ | $Q(t + 1)$ 0 0 | 0 0 1 | 1 1 0 | 1 1 1 | 0
D flip-flop	Clock	D Q < $\overline{Q}$	$Q(t + 1) = D$	D $Q(t)$ | $Q(t + 1)$ 0 0 | 0 0 1 | 0 1 0 | 1 1 1 | 1

TABLE 16.21 Combinational Gate Logic

Name	Graphic symbol	Algebraic function	Truth table
AND	x, y → Q	$Q = xy$	x y | Q: 0 0 | 0; 0 1 | 0; 1 0 | 0; 1 1 | 1
OR	x, y → Q	$Q = x + y$	x y | Q: 0 0 | 0; 0 1 | 1; 1 0 | 1; 1 1 | 1
Inverter	x → Q	$Q = x'$	x | Q: 0 | 1; 1 | 0
Buffer	x → Q	$Q = x$	x | Q: 0 | 0; 1 | 1
NAND	x, y → Q	$Q = (xy)'$	x y | Q: 0 0 | 1; 0 1 | 1; 1 0 | 1; 1 1 | 0
NOR	x, y → Q	$Q = (x + y)'$	x y | Q: 0 0 | 1; 0 1 | 0; 1 0 | 0; 1 1 | 0
EXCLUSIVE-OR	x, y → Q	$Q = xy' + x'y$ $= x + y$	x y | Q: 0 0 | 0; 0 1 | 1; 1 0 | 1; 1 1 | 0

small delay due propagation time), and at the next clock pulse the flip-flops will change to the next state as directed by the gates.

The gates shown in Table 16.21 have only two inputs. As was seen in Table 16.19, gates may have as many as eight inputs. In the case of an AND *gate*, all its inputs must be 1 in order for its output to be a 1. For an OR gate, if any of its inputs become a 1, then its output will become a 1. The NAND and NOR gates function in a similar way.

Boolean algebra is the branch of mathematics used to analyze logic circuits. Boolean algebra has two operators: ·, which indicates an AND operation, and +, which indicates an OR operation. The = has the same meaning in Boolean algebra as in ordinary algebra. The symbol for "X not" is X' (or sometimes $\overline{X}$). The identity element for the AND operation is 0; the identity element for the OR operation is 1. The rules for Boolean algebra can be derived from set theory applied to a system in which only two numbers exist, i.e., zero and one. These rules are summarized in Huntington's postulates and DeMorgan's theorem and are listed in Table 16.22.

TABLE 16.22 Rules for Boolean Algebra

$X + 0 = X$	$X \cdot 1 = X$
$X + 1 = 1$	$X \cdot X' = 0$
$X + X' = 1$	$X \cdot X = X$
$(X')' = X$	$X \cdot 0 = 0$
$X + Y = Y + X$	$X \cdot Y = Y \cdot X$
$X + (Y + Z) = (X + Y) + Z$	$X \cdot (Y\,Z) = (X \cdot Y)Z$
$X + X \cdot Y = X$	$X \cdot (X + Y) = X$
DeMorgan's theorem:	
$(X + Y)' = X' \cdot Y'$	$(X \cdot Y)' = X' + Y'$

To facilitate the analysis of digital circuits and to aid in the application of the rules given in Table 16.22, *Karnaugh maps* are used. Typical two-variable and four-variable Karnaugh maps are shown in Fig. 16.109 along with the algebraic expressions represented by each map.

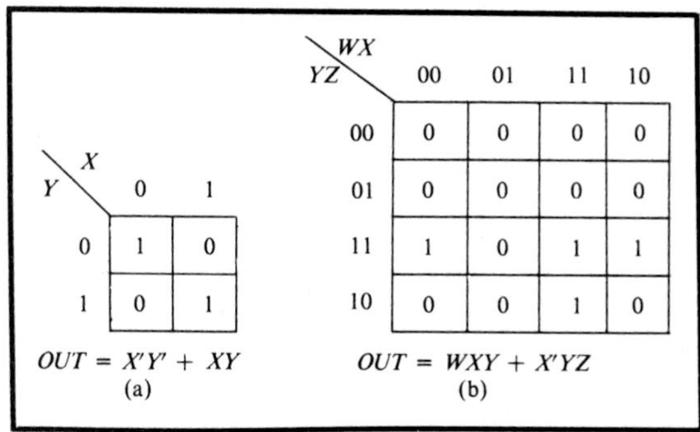

FIGURE 16.109 Karnaugh maps of typical logic functions.

The complexity of digital ICs is growing. Where there is a sufficiently large demand for a particular circuit function, LSI and VLSI devices can be designed. Table 16.23 lists some of the highly complex circuits that are commercially available. *Read-only memory* (*ROM*) is a combinational logic device. This device can be programmed to accomplish the same functions that can be achieved with a complex circuit using various types of gates. An extension of the ROM is the *programmed-logic array* (*PLA*). This device is built specifically as a cost-effective combinational logic device for very complex logic systems.

Gate arrays are yet another means for designing custom IC logic circuits into a single VLSI device. Gate arrays include both combinational and sequential cir-

TABLE 16.23 Large-Scale Digital Integrated Circuits

Adder	Accumulator
Arithmetic logic unit	Parity generator-checker
Shift register	Decoder-demultiplexer
Counter	Encoder-multiplexer
Display controller-driver	Custom gate array

cuit elements. These circuit elements have been standardized so that a custom IC can be developed by specifying the interconnection, within the device, of standard elements to form a specific circuit. The development and tooling cost for gate array devices is much lower than that of a completely new IC. Gate arrays are used for production quantities of 1000 to 10,000 devices per year. The cost of each gate array device is somewhat higher than a custom IC, so for quantities of more than 10,000 per year, custom devices are usually designed using normal mask techniques.

COMPUTER INTEGRATED CIRCUITS

One of the devices which has become feasible as a result of VLSI is the *microprocessor*. A complete computer can be built in a single IC device. In most cases, however, several devices are employed to build a complete computer system. In most computers, the cost of the microprocessor is negligible compared with the total system cost. Not long ago, the *central processing unit* (*CPU*) was the most expensive part of a computer. The microprocessor provides the total CPU function at a fraction of the earlier cost.

The power of a microprocessor is a function of its clock rate and the size of its internal registers. Clock rates vary from 1 to 20 MHz. Common register sizes are 8, 16, and 32 bits. Most of the existing personal computers currently use microprocessors which have 8-bit registers. New designs use 16-bit microprocessors. For scientific computing or high-resolution graphics, 32-bit machines are preferred.

As costs become lower, increased register sizes become more common. Increased register size allows a more powerful computer instruction set to be incorporated, and it also allows direct addressing of a larger memory. Both these capabilities enhance the power of the machine.

Memory is an important part of a computer. Information that is processed by the computer is stored in *random-access memory* (*RAM*). The CPU can write information into RAM and subsequently read that information. Earlier, RAM was built using many little magnetic cores. Today, core memory has been almost entirely replaced by semiconductor memory. Main computer memory is often referred to as *core memory*, even though it may, in fact, be a solid-state memory. Semiconductor RAM memory may be dynamic or static. *Static RAM* retains information as long as electric power is applied to the circuit. *Dynamic RAM* retains information as stored charges in capacitors. Since the charge leaks off with time, dynamic memory must be refreshed about every 10 μs. The cost of dynamic memory devices is less than that of static memory. Large memory banks are usually made with dynamic memory because the cost of memory-refresh circuitry is offset by the savings in memory device costs.

The information in both static and dynamic IC memories is lost when power is removed. For this reason, IC memory is termed *volatile memory*. Read-only memory is nonvolatile; the information in ROM is retained even if power is removed. The information in a ROM is masked into the device at the time it is manufactured. *Programmable read-only memory* (*PROM*) is ROM that can be programmed using a PROM programming machine. Some PROMs may be programmed only once. These ROMs are programmed by fusing links within the device. Once these internal links have been fused, the ROM cannot be changed. Other erasable PROMs, or *EPROMs*, can be erased using high-intensity ultraviolet light. Because these devices can be reprogrammed over and over again, EPROMs are often used during the product-development phase and then replaced with less expensive PROMs or ROMs for production units.

Bubble memory is another nonvolatile memory. Information is stored in these devices as "magnetic bubbles." Bubble memory is slower than IC memory, but it is less expensive. Other forms of magnetic memory (disks and tapes) are less expensive than bubble memory and are much slower in response. Bubble memory, unlike tapes and disks, is completely free of moving parts. Other magnetic devices are discussed below with peripheral devices.

In addition to the CPU and memory ICs, other IC devices are required for computer support. These are listed in Table 16.24. One general-purpose support IC is called a *peripheral interface adapter* (*PIA*). The function of the PIA is to provide a programmable interface between the microprocessor and any peripheral device. This IC handles most of the functions needed to interface to the computer bus. A *universal asynchronous receiver-transmitter* (*UART*) interfaces between the computer and asynchronous devices such as *cathode ray tubes* (*CRTs*) and modulator-demodulators (*modems*). The *universal synchronous receiver-transmitter* (*USART*) performs a similar function for synchronous communications devices. The UART and USART perform the interface functions of the PIA and, in addition, perform functions specifically required for synchronous and asynchronous communications. The PIA is designed for broader applications, but the UART and USART perform more functions in their specific applications.

TABLE 16.24 Digital Computer Integrated-Circuit Devices

Microprocessors	Programmable read-only memory
Static random-access memory	Dynamic random-access memory
Tristate buffer	Tristate transceiver
Programmable timer	Parallel interface adapter
Analog-to-digital converter	Digital-to-analog converter
Floppy disk controller	IEEE 488 bus interface
Universal asynchronous transmitter-receiver (UART)	

Another specific-purpose device is the *floppy disk controller*. This device has been designed to perform the interface and data reformatting tasks required to interface a floppy disk drive to a microcomputer system.

Other common interface devices are A/D and D/A converters. These devices provide a means of converting from analog to digital information and back again. These devices are usually designed so that several analog signals can be multiplexed through A/D or D/A device.

The interface to the IEEE 488 instrument bus is useful for interfacing a computer into an instrumentation system of this type. This interface provides the hardware requirements so that a program executed in the microprocessor can make it either a master or a slave in such a system.

Computer peripheral devices consist of disks, tapes, printers, and terminals. There are two main types of disks: floppy disks and hard disks. The most common hard disk used in microcomputers is called a *Winchester disk*. Magnetic disks are nonvolatile memory devices. The access time for a disk is much longer than for IC memory. The cost of disk memory is much lower than for IC memory. For these reasons, disk memory is normally used for permanent storage rather than working storage. The disk is used for working storage when the amount of information to be stored exceeds IC memory capacity.

Magnetic tape is also nonvolatile memory. The cost of tape as a storage medium is quite low. Tape has the disadvantage, compared with disk storage, that data cannot be directly addressed. Tape is inherently a serial stream of data from the beginning of the tape to the end. Disk, on the other hand, is a random-access memory and any part of disk memory can be addressed directly. In general, data can be accessed more quickly from disk than from tape. Since floppy disks and magnetic tapes are both removable media, both these storage devices are used for off-line storage of computer data.

Printers and terminals are the most common means of communicating between the computer and human users, although significant progress has been made recently with voice input and output to computers. The least expensive printers use a dot matrix to form characters. The printed results are not as pleasing as those from a font-oriented machine. Low-speed font-oriented printers are called *letter-quality printers*.

In addition to hardware, computers require software, or programs, in order to function. Each of the peripheral devices discussed above requires a program called a *driver*. The base program which controls a computer is called a *monitor* or an *operating system*. In addition to the drivers and operating system, most computers include utility programs which allow file maintenance, editing, etc. Application programs are added to complete the software for a computing system.

Programming of microcomputers requires nearly the same amount of time as for larger computers. The cost of microcomputer software has dropped only where there has been a high volume of sales of a given program. Engineering and manufacturing programs are often written specifically for an application. In these cases, software costs may far exceed hardware costs. Business programs have been commercialized much more effectively than technical programs.

COMPUTER COMMUNICATIONS

Lower-cost computing has greatly expanded the use of personal computers and the use of interactive graphics for design. *Computer-aided design and manufacturing (CAD/CAM)* is viewed by many to be a significant new development in manufacturing technology. The developments in both personal computing and CAD/CAM have increased the requirements for interprocessor communications.

Communication of information between computers can be accomplished by a number of means. Organizations which have established digital communication standards are listed in Table 16.25. Low-speed communications can be accom-

TABLE 16.25 Organizations Which Provide Communication Standards

CCITT	Comité Consultatif Internationale de Telegraphie et Telephonie An international consultative committee that sets international communication usage standards
EIA	Electronic Industries Association A standards organization specializing in electrical and functional characteristics of interface equipment
ISO	International Organization for Standardization
ANSI	American National Standards Institute
IEEE	Institute of Electrical and Electronic Engineers

plished by asynchronous links using a 20-mA current loop or EIA RS-232C Standard. Higher-speed communications require the use of synchronous techniques, such as CCITT X.25 packet switching. A list of public data networks of this type is given in Table 16.26.

TABLE 16.26 Public Data Networks

Name	Location	Origination year
TELENET	United States	1975
EPSS	Britain	1977
DATAPAC	Canada	1977
TYMNET	United States	1977
TRANSPAC	France	1978
DX-2	Japan	1979
EURONET	Europe	1979

A fundamental relationship exists between digital and analog information called the *Nyquist criterion*. The required digital pulse rate depends on the highest-frequency component contained in the analog information. Equation (16.164) shows this relationship. The minimum pulse rate (pulses per second) must be at least twice the highest frequency (hertz).

$$pr = 2f \tag{16.164}$$

This is a bilateral relationship. The frequency bandwidth of a transmission system must be equal to at least half the pulse rate:

$$BW = \frac{pr}{2} \tag{16.165}$$

These conditions are minimum requirements. A transmission system that has greater bandwidth will support slower pulse rates. A high pulse rate will approximate an analog signal more accurately than one that just meets the Nyquist criterion. Voice-grade telephone lines have a frequency bandwidth of about 5000 Hz, so the maximum pulse rate for these lines is about 10,000 pulses per second. The rate normally stated is 9600 baud (baud is equivalent to pulses per second).

There are several agencies which have written specifications or recommendations for data communications standards. ISO has established a layer standard

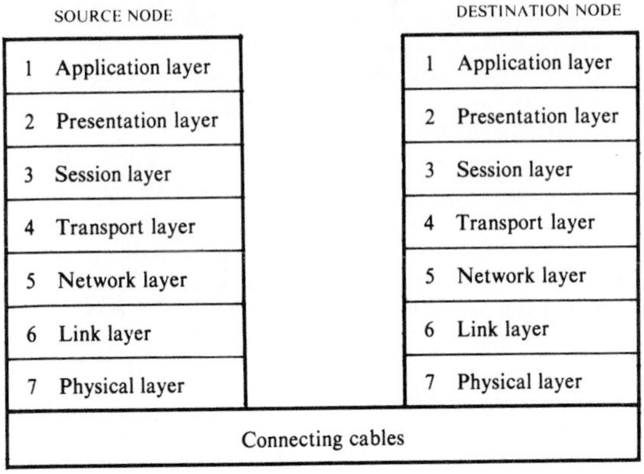

SOURCE NODE	DESTINATION NODE
1 Application layer	1 Application layer
2 Presentation layer	2 Presentation layer
3 Session layer	3 Session layer
4 Transport layer	4 Transport layer
5 Network layer	5 Network layer
6 Link layer	6 Link layer
7 Physical layer	7 Physical layer
Connecting cables	

FIGURE 16.110 ISO-layered model for open system interconnection.

for digital communications. The ISO layer model is shown in Fig. 16.110. The lowest layer in that model is the physical layer. This is essentially the function performed by the modem in a digital communication network. The link layer provides control of message routing through the communications system. The network layer provides the control specification for node addressing and packetizing of data. The top layer interfaces with the user; the bottom layer interfaces with network hardware.

The International Consultative Committee for Telephone and Telegraph (CCITT) has established a series of recommendations based on the layer approach to data communications. CCITT recommendation X.25, for packet-switched networks, has been gaining acceptance both in the United States and in Europe. This recommendation covers only layers 5, 6, and 7. The utility and the wide acceptance of X.25 are causing many manufacturers of computer communication equipment to design their equipment to meet this standard.

The Electronics Industries Association has established three interface standards which are frequently referenced for digital communications. These are RS-232C, RS-422, and RS-423. RS-232C is the oldest of these standards. This has been the primary standard for several years for low-speed-voltage-oriented digital communication. RS-232C and RS-423 use nonbalanced communication lines. Nonbalanced lines are more sensitive to noise. This limits the length of line and bandwidth that can be used satisfactorily. RS-232C is limited to a line length of about 250 ft at a bandwidth of 10 kHz. RS-423 is limited to a line length of about 400 ft at a bandwidth of 10 kHz. RS-422 uses a balanced line and can be used to a line length of 4000 ft at a bandwidth of 100 kHz. It is expected that RS-423 and RS-422 standards will eventually replace the RS-232C standard.

Synchronous packetized data communication offers several advantages over asynchronous communications. For a given transmission medium, one can achieve higher data rates, better utilization of available bandwidth, and higher transmission accuracy. Each packet of data contains the source node address and the destination node address. This allows data packets to be routed through the network over alternate paths. In addition, each packet contains a *cyclic redundancy check* (*CRC*). At the source node, a CRC value is calculated based upon the data that are to be sent in that packet. At the receiving node, the CRC is recalculated based on the data that were received. The calculated CRC is compared with the transmitted CRC, and if there is an error, the data must be retransmitted. Data errors of less than one packet in 10^7 are easily achievable using CRC checking.

Many of the existing techniques and equipment that are used for digital communication by telephone lines are not suited to local data communication needs. Large-systems requirements place an overhead on the communication nodes that become burdensome. This reduces throughput and also introduces message setup delays that are unacceptable for many interactive computing situations. *Local area networks* (*LANs*) have been devised to eliminate some of these problems. LANs can operate over distances of up to 1000 ft or so with data rates of 10^6 baud. Table 16.27 lists several LANs.

TABLE 16.27 Commercial Local Area Networks

Name	Sponsoring organization
Ethernet	Xerox/Digital Equipment/Intel
Net/One	Ungermann & Bass Co.
Z-Net	Zilog Corporation
Hyperbus	Networks System Corp.
Hyperchannel	Networks System Corp.
Ringnet	Prime Computer Corp.
Ring Token	Apollo Computer Corp.
Interactive System	3M
Data Exchange	Amdax
System 20	Sytek/NRC
Wangnet	Wang Computer Corp.

Modern factories that make extensive use of computing equipment, both in design and in the shop, require very high communication rates and fast response. These requirements are not satisfied by either LANs or telephone lines. Industrial dataways can be built using wideband cable-television coaxial cable and repeater amplifiers. These dataways have a bandwidth of 300 MHz. This will support data rates of up to 600×10^6 baud.

For even greater communication bandwidths, light is being used to transmit information rather than electricity. Inexpensive fiber-optic devices can be used for distances of a few feet. For longer distances phase-locked lasers are being developed. Devices of this kind can be anticipated for use in both private and public data communication service.

INDUSTRIAL ELECTRONICS

The power for dc motor armatures can be derived from thyristor circuits like those shown in Fig. 16.92. Single-phase bridge circuits are used for 5-hp drives and smaller. Three-phase bridge circuits are used for drives larger than 5 hp. A single set of six thyristors can supply power for about 300 hp. Above 300 hp, multiple sets of thyristors must be used in parallel. Mill drives have been built with more than 10,000 hp provided by thyristors.

The control of dc motors whether powered by thyristors or by dc generators is accomplished electronically. Control of individual drives can be accomplished by tachometer feedback or by armature voltage feedback. The speed-regulation accuracy for armature feedback is 5 percent; for tachometer feedback, speed-regulation accuracy is from 0.1 to 1.0 percent. When two drives must be coordinated with each other, as in a continuous-web processing machine, they can be regulated to control torque, speed, position, draw, or a combination of these parameters. Torque controls can be achieved using dc motor armature current for a feedback signal. Speed-control signals are derived as for single motors. Position or draw control can be accomplished by using selsyn ties or dancer rolls. A *dancer roll* is a weight- or spring-loaded roll that rides on the web. It is free to move up and down, and as it does, a signal is taken from its position to serve as a feedback for the drive regulator before the dancer or after it.

Coordination of the motions of two or more drives requires tracking of the drives in both steady-state and transient conditions. Linearity of the control and feedback signals determine steady-state tracking. Provision must be made for both low-speed and high-speed matching signals. Transient matching requires that signals not only be the right magnitude but also arrive at the right time. An example will serve to illustrate this point. Suppose it is desired to have two drives with tachometer feedback which have a continuous web between them. One way to accomplish this would be to designate one drive as a master and the other as a slave. The tachometer on the master drive would serve as its own feedback signal and as the reference or command signal for the slave drive. The slave drive would have its own feedback from its own tachometer and so its regulator would try to minimize the difference between the two tachometer signals. On a transient basis the master drive will always start before the slave. An alternate and more common arrangement is to provide a common reference for both drives and let each drive receive its command signals at the same instant.

Digital computers are being used on-line in mills and continuous processing industries. DC motors can be controlled by either analog or digital regulators. With the greatly reduced cost of integrated circuits, digital regulators are being increasingly used.

DC motors have been widely used for variable-speed applications because of their excellent characteristics. AC motors have been used primarily for constant-speed applications. The control schemes described above are equally applicable to ac motors (except of course for armature voltage and armature current feedback). If power circuitry is properly handled, the control of an ac motor is just as flexible and versatile as that of a dc motor.

AC motors can be supplied either from *phase-controlled circuits* or from *inverter circuits*. Phase control is a simple electronic circuit, but its use results in high losses in the ac motor. This limits the application of this type of drive to either a very limited speed range or to loads in which the torque required decreases rapidly as the speed decreases. Large pump drives and fan drives have been built using this form of ac motor control. Inverters can be designed so that excessive motor losses are not encountered. Inverters are quite complex and require auxiliary power components to commutate the thyristors. Cost and complexity have prevented the widespread use of inverter-powered ac motor drives.

Phase-control circuits are extensively used to control power flow to process heaters. Most industrial heating is done by gas because it is cheaper than electric energy. In many applications, electric heat is needed or is sufficiently more convenient. Phase-controlled thyristors modulate the power to these heaters and provide smoother control than simple on-off control by contactors.

High frequencies can be generated by *thyristor inverter circuits*. This permits the use of thyristors for *induction heating* and supersonic cleaning. Thyristor supplies have been built with frequency output from 100 to 50,000 Hz. These power supplies can be controlled in frequency much more easily and rapidly than motor-alternator sets and so have added new capability to induction-heating apparatus.

Dielectric heating requires frequencies from 100 kHz to 1 MHz. Large vacuum-tube oscillators are used to generate these frequencies.

COMMUNICATIONS

The Federal Communications Commission (FCC) regulates the use of radio-frequency transmission in the United States. This regulation is necessary to prevent interfering transmissions of radio signals. Some of the frequency allocations are given in Table 16.28. The frequency bands are also classified as shown in Table 16.29. Very low frequencies are used for long-distance communications across the surface of the earth. Higher frequencies are limited to line-of-sight transmission. Because of bandwidth considerations, high frequencies are used for

TABLE 16.28 Partial Table of Frequency Allocations

(For a complete listing of frequency allocations, see Reference Data for Radio Engineers, *published by Howard Sams & Co.)*

Frequency, MHz	Utilization
0.535–1.605	Commercial broadcast band
27.255	Citizens' personal radio
54–72	Television channels 2–4
76–88	Television channels 5–6
88–108	Frequency-modulation broadcasting
174–216	Television channels 7–13
460–470	Citizens' personal radio
470–890	Television channels 14–83

TABLE 16.29 Frequency Bands

Designation	Frequency	Wavelength
VLF, very low frequency	3–30 kHz	100–10 km
LF, low frequency	30–300 kHz	10–1 km
MF, medium frequency	300–3,000 kHz	1,000–100 m
HF, high frequency	3–30 MHz	100–10 m
VHF, very high frequency	30–300 MHz	10–1 m
UHF, ultra-high frequency	300–3,000 MHz	100–10 cm
SHF, super-high frequency	3,000–30,000 MHz	10–1 cm
EHF, extremely high frequency	30,000–300,000 MHz	10–1 mm

NOTE: Wavelength in meters = $300/f$, where f is in megahertz.

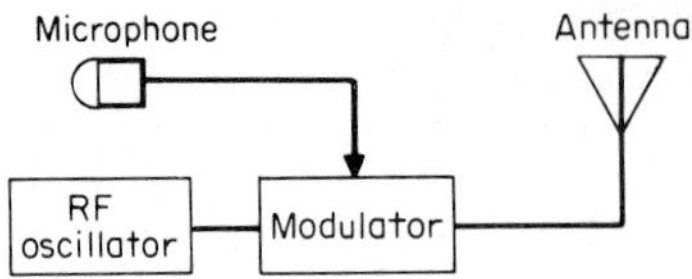

FIGURE 16.111 Radio transmitter.

high-density communication links. Orbiting *satellites* allow the use of high-frequency transmission for long-distance high-density communications.

A *radio transmitter* is shown in Fig. 16.111. It consists of four basic parts: an rf oscillator tuned to the carrier frequency, an information-input device (microphone), a modulator to impress the input signal on the carrier, and an antenna to radiate the modulated carrier wave.

A *radio receiver* is shown in Fig. 16.112. This is called a *superheterodyne* receiver because it utilizes a frequency-mixing scheme. The tuned radio-frequency

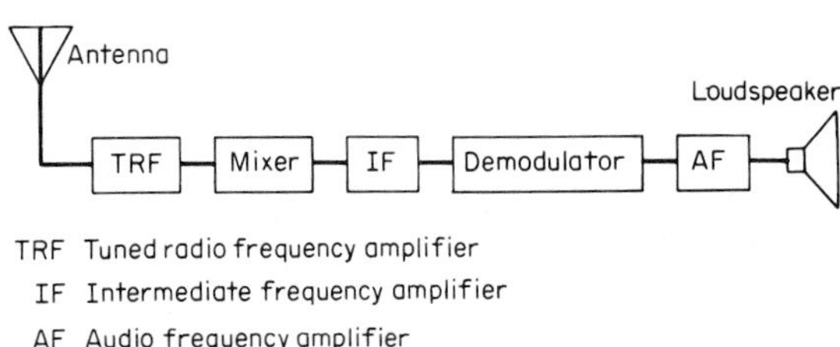

FIGURE 16.112 Radio receiver.

amplifier is tuned to receive the desired radio signal. The local oscillator is adjusted by the same tuning control to a lower frequency. The mixer produces an output frequency which is the difference between the incoming radio-signal frequency and the local-oscillator frequency. Since this difference frequency is constant for all tuning positions, the intermediate-frequency amplifier always operates with a constant frequency. This allows optimum design of the intermediate-frequency (IF) amplifiers since they are constant-frequency amplifiers. The IF frequency signal is modulated in just the same way as the radio signal. The demodulator separates this audio signal, which is then amplified so that the loudspeaker can be driven.

The term *radar* is derived from the first letters of the words *ra*dio *d*etection *a*nd *r*anging. It is essentially an echo system in which the location of an object is determined by sending out short pulses of radio waves and observing and measuring the time required for their reflections or echoes to return to the sending point. The time interval is a measure of the distance of the object from the transmitter. The velocity of radio waves is the same as the velocity of light, or 984 ft/μs, so that each microsecond interval corresponds to a distance of 492 ft. The direction of an object can be determined by the position of the directional transmitting and receiving antenna. Radio waves penetrate darkness, fog, and clouds, and hence are able to detect objects that otherwise would remain concealed. Radar can be used for the automatic "tracking" of objects such as airplanes.

A block diagram of a radar system is shown in Fig. 16.113. The transmitting system consists of an rf oscillator which is controlled by a modulator, or pulser, so that it sends to the antenna intermittent trains of rf waves of relatively high power but of very short duration, corresponding to the pulses received by the modulator. The energy of the oscillator is transmitted through the duplexer and to the antenna through either coaxial cable or waveguides. The *receiver* is an ordinary heterodyne-type radio receiver which has high sensitivity in the band width corresponding to the frequency of the oscillator. For low frequencies the local oscillator is an ordinary oscillator for frequencies of 2000 MHz; and higher a reflex *klystron* (hf cavity oscillator) is used. A common intermediate frequency is 30 MHz but 15 and 60 MHz are also frequently used.

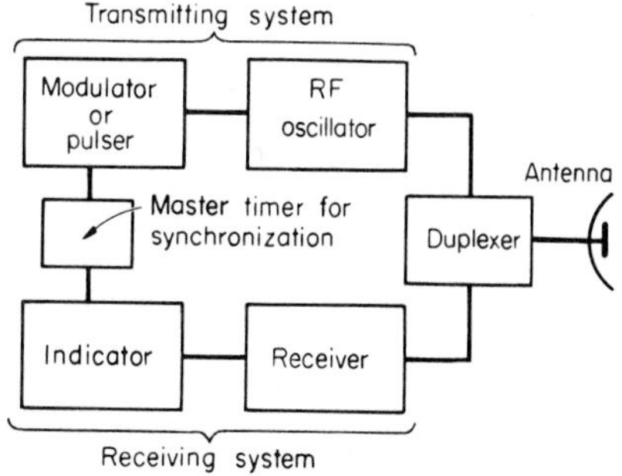

FIGURE 16.113 Block diagram of radar system.

In most radar systems, the same antenna is used for receiving as for transmitting. This requires the use of a *duplexer* which cuts off the receiver during the intervals when the oscillator is sending out pulses and disconnects the transmitter during the periods between these pulses when the echo is being received.

The antenna is highly directional. By noting its angular position, the direction of the object may be determined. In the PPI (plan position indicator), the angle of the sweep of the cathode-ray beam on the screen of the oscilloscope is made to correspond to the azimuth angle of the antenna.

The receiver output is delivered to the indicator, which consists of a cathode ray tube or oscilloscope. The pulses which are received, corresponding to echoes from the target, must be synchronized with the sending pulses in order that the distance to the target may be determined. This is accomplished by synchronization of the sweep circuit of the oscilloscope with the pulses by the master timer.

1. Displays. Conversion of the received radar signals to usable display is accomplished by a cathode ray oscilloscope. The simplest type, called the *A presentation*, is shown in Fig. 16.114*a*. When the pulser operates, a sawtoothed wave produces a linear sweep voltage (Fig. 16.114*b*) across the sweep plates of the cathode ray tube; at the same time, a transmitter pulse is impressed on the deflection plates and the return echoes appear as AM pulses, or "pips," on the screen, as shown in Fig. 16.114*a*. The distance on the screen between the transmitter pulse and the pip caused by the echo is proportional to the distance to the target, and the screen can be calibrated in distance such as miles. [The return of the spot to its initial starting position, produced by the sweep interval *cd* (Fig. 16.114*b*), is so rapid that it is not detectable by the eye.] The direction of the target may be determined by the angular position of the antenna, which can be transmitted to the operator by means of a selsyn. Different objects, such as airplanes, ships, islands, and land approaches, have characteristic pips, and operators become skilled in their interpretation. A bird in flight can be recognized on the screen. Also, a portion of the scale such as *ab* can be segregated and amplified for close study of the characteristics of the pips.

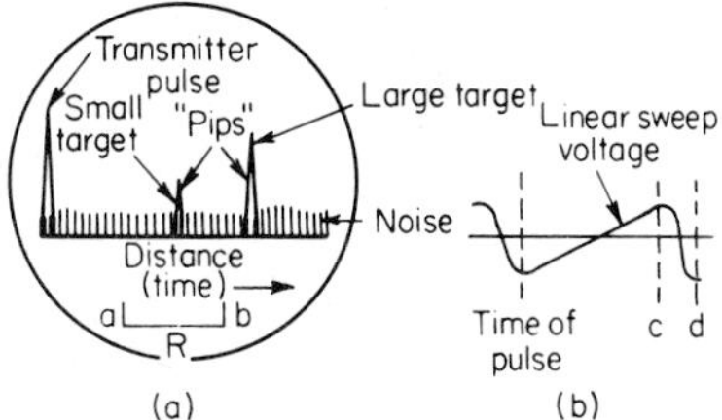

FIGURE 16.114 Type A presentation.

2. Plan Position Indicator (PPI). In the PPI (Fig. 16.115) the direction of a radial sweep of the electron beam is synchronized with the azimuth sweep of the antenna. The sweep of the beam is rotated continuously in synchronism with the antenna, and the received signals intensity-modulate the electron beam as it sweeps from the center of the oscilloscope screen radially outward. In this way the direction and range position of an object can be determined from the pattern on the screen of the oscilloscope, as shown in Fig. 16.115.

There are two methods by which the angular direction of the cathode spot is made to correspond with the angular position of the antenna. In one method, used on board ship, two magnetic deflecting coils are rotated around the neck of the tube in synchronism with the antenna, by means of a selsyn. In the other method, used on aircraft, two fixed magnetic deflecting coils at right angles to each other and placed at the neck of the tube are supplied with current from a small two-phase synchronous generator whose rotor is driven by the antenna. Thus a rotating field, similar to that produced by the stator of an induction motor, is produced by the magnetic deflecting coils. These two rotating fields, although produced by different means, are equivalent and cause the cathode beam to sweep radially in synchronism with the antenna. Circular coordinates spaced radially corresponding to distance are obtained by impressing on the control electrode short positive pulses synchronized with the transmitted pulse but delayed by time values corresponding to the desired distances. These coordinates appear as circles on the screen. Since the time of rotation of the antenna is relatively slow, it is necessary that a persistent screen be used in order that the operator may view the entire pattern. In Fig. 16.115 is shown a line drawing of a PPI presentation of Cape Cod, Mass., on a radar screen, taken from an airplane.

The applications of radar to war purposes are well known, such as detecting enemy ships and planes, aiming guns at them, and locating cities, rivers, mountains, and other landmarks in bombing operations. In peacetime, radar is used to

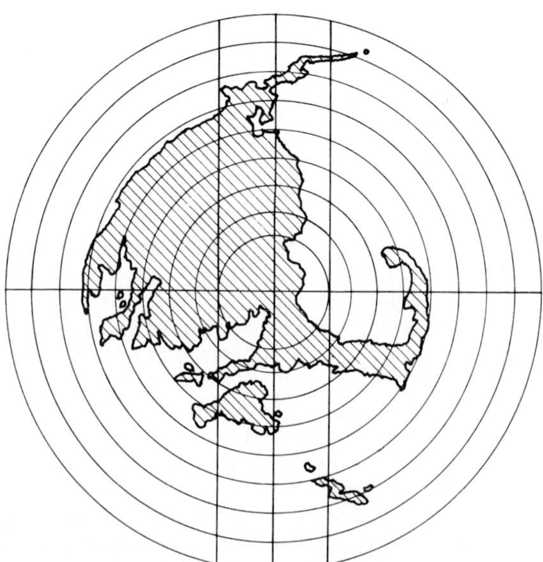

FIGURE 16.115 Plan position indicator (PPI) of southeastern Massachusetts.

navigate ships in darkness and poor visibility by locating navigational aids such as buoys and lighthouses, as well as protruding ledges, islands, and other landmarks. It can be similarly used in air navigation, as well as to operate altimeters for determining the height of the plane above ground. It is also used for aerial mapping.

There are also radio beacons, *shoran* (*sho*rt-*ran*ge navigation) and *loran* (*lo*ng-*ran*ge navigation) by which ships or planes can locate their positions. In the *ground-controlled approach* (*GCA*) for airplanes, the ground operator picks up the plane on a PPI presentation at distances up to 30 mi, using a general surveillance radar, and gives instructions to the pilot by radio course and procedure. As the plane approaches the landing field, it is brought into vision on the screen of a high-resolution short-range radar, and the pilot is given continual detailed instructions as to the glide path which the plane is to follow until the landing is made.

Television is accomplished by systematically scanning a scene or the image of a scene to be reproduced and transmitting at each instant a current or a voltage which is proportional to the light intensity of the elementary area of the scene which at the instant is being scanned. The varying voltage or current is amplified, modulated on a carrier wave, and then transmitted as a radio wave. At the receiver the radio wave enters the antenna, is amplified, and demodulated to give a voltage or a current wave similar to the original wave. This voltage or current wave is then used to control the intensity of a cathode ray beam which is focused on a fluorescent screen in a cathode ray reproducing tube. The cathode ray beam is caused to move over the screen in the same pattern as the scanning beam at the transmitter and in synchronism with it. Thus each small area of the receiver screen is illuminated instantaneously with light intensity corresponding to that of a similarly placed area in the original scene. This process is conducted so rapidly that owing to persistence of vision of the eye, the reproduction of each instantaneous scene appears to be a complete picture and the effect with successive scenes is similar to that produced by the projection of successive frames of a motion picture.

3. Scanning and Blanking. In the United States, the ratio of width to height of a standard television picture is 4:3, and the picture is composed of 525 lines repeated 30 times a second, this last factor being one-half 60, the prevalent electric power frequency in the United States. The scanning sequence along the individual lines is from left to right and the sequence of the lines is from top to bottom. Also, interlacing is employed, the general method of which is shown in Fig. 16.116. The cathode ray spot starts at 1 in the upper left-hand corner and is swept rapidly from left to right either by a sawtooth emf wave applied to the sweep plates or by the sawtooth current wave applied to the sweep coils of the tube. When the spot arrives at the right-hand side of the picture, the sawtooth wave of either emf or current in the sweep circuit acts to return the cathode ray spot rapidly to point 3 at the left-hand side of the picture. However, during this period the cathode ray is blanked, or entirely eliminated, by the application of a negative potential to the control grid of the tube. At the end of the return period, the blanking effect ceases and the spot appears at point 3, from which it again is swept across the picture and this process is repeated for 262.5 lines until the spot reaches a midpoint C at the bottom of the picture. It is then carried vertically and rapidly to B, the midpoint of the top of the picture, the beam also being blanked during this period. This process of scanning is then repeated, a second set of lines corresponding to the even numbers 2, 4, 6 being established between the lines designated by the odd numbers. These lines are shown dashed in Fig. 16.116.

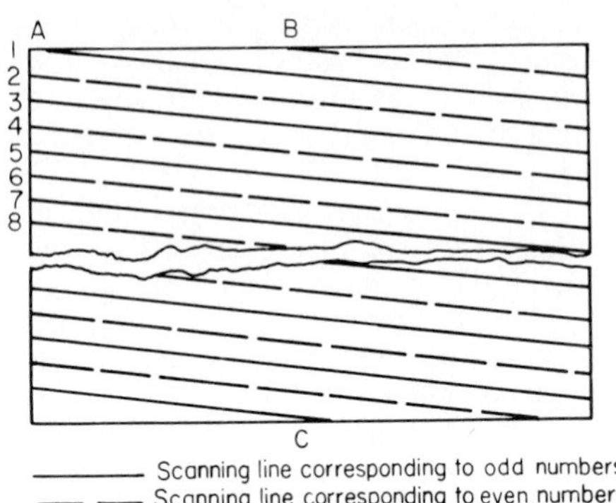

FIGURE 16.116 Pattern of interlaced scanning.

This method or pattern of scanning is called *interlacing*. The two sets of lines taken together produce a frame of 525 lines, which are repeated 30 times each second. However, owing to interlacing, the flicker frequency is 60 Hz which is not noticeable, 50 Hz having been determined as the threshold of flicker noticeable to the average eye. In Fig. 16.116, for the sake of clarity, the distances between horizontal lines are greatly exaggerated and no attempt is made to maintain proportions.

4. Frequency Band. In order to obtain the necessary resolution of pictures, television frequencies must be high. In the United States, VHF frequencies from 54 to 88 MHz (omitting 72 to 76 MHz) and 174 to 216 MHz are assigned for television broadcasting. A UHF band of frequencies for commercial television use is also allocated and consists of the frequencies of from 470 to 890 MHz (see also Tables 16.28 and 16.29).

In order to obtain the 525 lines repeated 30 times per second, a band width of 6 MHz is necessary. The video, or picture, signal with the superimposed scanning and blanking pulses is amplitude-modulated, amplified, and transmitted. The carrier frequency associated with the sound transmitter is 4.5 MHz higher than the video carrier frequency and is frequency-modulated with a maximum frequency deviation of 25 kHz.

In scanning motion-picture films, a complication arises because standard film rate is 24 frames per second, while the television rate is 30 frames per second. This difficulty is overcome by scanning the first of two successive film frames twice and the second frame three times at the 60-Hz rate, making the total time for the two frames $\frac{1}{12}$ ($\frac{2}{60}$ + $\frac{3}{60}$) or $\frac{1}{24}$ s average per frame.

5. Kinescope. The kinescope (Fig. 16.117) is the terminal tube in which the televised picture is reproduced. It is relatively simple, being not unlike the cathode ray oscilloscope tube. It has an electric gun operating at 8000 to 20,000 V which produces an electron beam focused on a fluorescent surface within the front wall of the tube. The picture is viewed at the front wall. The horizontal and vertical deflections of the beam are normally controlled by deflection coils, as shown in Fig. 16.117.

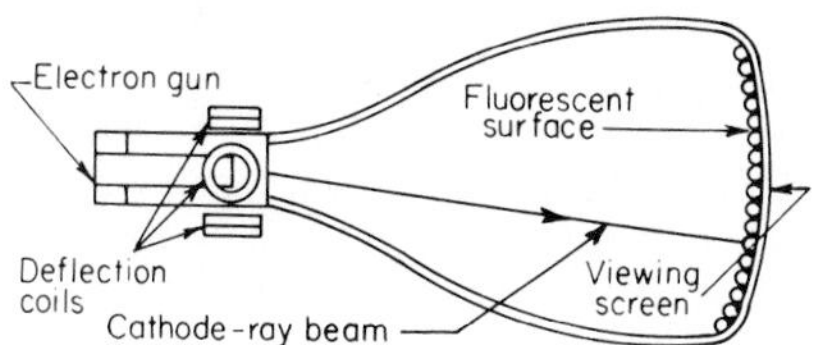

FIGURE 16.117 Kinescope for television receiver.

6. Television Receivers. A block diagram for a television receiver is given in Fig. 16.118. It is in reality a superheterodyne receiver with tuned rf amplification, the separating of the sound and video or picture channels taking place at the intermediate frequency in the mixer. The sound channel is then conventional, a discriminator being used to demodulate the FM wave (Fig. 16.104). The object of the dc restorer is to make the picture reproduction always positive, and it consists of applying a dc voltage at least equal in magnitude to the maximum values of the negative loops of the ac waves. The synchronizing pulses for both the vertical and the horizontal deflections are delivered by the dc restorer to an amplifier and the two pulses are then divided into the V and H components. The integrating and differentiating circuits are necessary to separate horizontal and vertical synchronizing signals.

As stated earlier, at any instant the magnitude of the current from the pickup tube varies in accordance with the light intensity of the part of the scene being scanned at that instant. This current is amplified and, together with the sound and synchronizing currents, is broadcast and received by the circuit shown in Fig. 16.118. The video current is detected by rectification, is amplified, and is then made to control the intensity of the kinescope electron beam. Tubes produce a scanning pattern, identical with that in the pickup tube, and these tubes are triggered by the synchronizing pulses which are transmitted in the broadcast wave. Hence, the original televised scene is reproduced on the fluorescent screen of the kinescope.

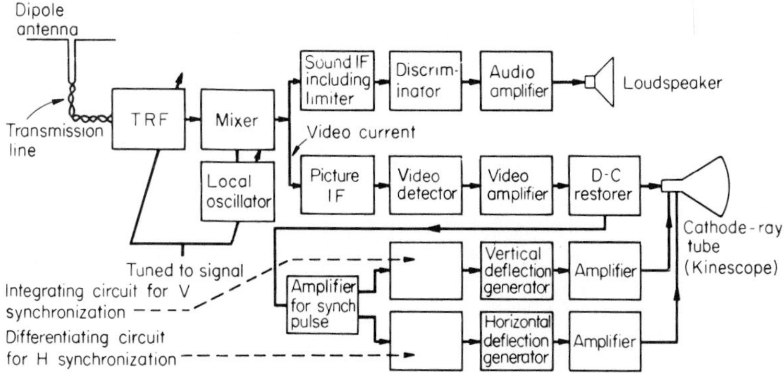

FIGURE 16.118 Block diagram for television receiver (TRF, tuned radio frequency; IF, intermediate frequency).

Color-television transmission is similar to black-and-white television, and the two signals must be compatible with each other. The kinescope for color TV has three electron guns, one for each primary color. The fluorescent screen has a matrix of three different colors of phosphor and a mask with many small holes in it. The intensity signals for each color are phase-shifted from each other so that the proper phosphors are excited by each electron stream at each mask point over the entire screen. A black-and-white signal does not have the same synchronizing signal as a color signal. The color receiver has circuits which recognize this state and switch it to black and white reception.

CHAPTER 17
NUCLEAR ENGINEERING AND OTHER ENERGY-CONVERSION SYSTEMS

NUCLEAR FISSION

Consensus projections indicate that only nuclear fission can maintain adequate supplies of energy, while fossil-fueled sources decrease as a result of increasing cost and exhausted supplies and new, large-scale sources (such as solar power and fusion) are still being developed. Nuclear fission is therefore expected to be the major, growing source of energy in the next few decades.

Design of fission-powered systems is based on principles of nuclear physics and engineering that are outlined here. Nuclear engineering is involved in designing the fuel-bearing region (known as the *core*), the core-containment vessel, auxiliary systems and components, and shielding. The core and its associated equipment in a nuclear power plant basically replace the boiler in a conventional system. Balance-of-plant designs (i.e., the steam generator, turbine, and electricity generator portions of the system) can follow conventional practice or differ markedly. One of the characteristics of nuclear power systems is the wide variety of designs possible in comparison with fossil-fired systems.

1. Atomic Structure and Nuclear Particles. Nuclear engineering requires knowledge and manipulation of atoms and the particles that comprise them. An atom is composed of a dense nucleus surrounded by electrons arranged in various orbits. The atom is mostly free space; the atomic radius is on the order of 10^{-8} cm, and the radius of the nucleus is 10^{-12} to 10^{-13} cm.

Ordinary chemical reactions involve changes in the orbital electrons of the atom; nuclear reactions such as fission involve the nucleus. There are great differences in the energy involved in these two types of reactions. For example, the combustion reaction (oxidation of carbon to CO_2) releases 4.1 electronvolts (eV) of energy; in contrast, fission of a uranium atom releases approximately 2×10^8 eV (i.e., 200 MeV) of energy.

This chapter is drawn from *Engineering Manual*, 3d ed., by R. H. Perry (Ed.). Copyright © 1976. Used by permission of McGraw-Hill, Inc. All rights reserved. Updated by the editors.

The physicists have demonstrated that nuclei are composed of many fundamental particles. Relatively few of these particles are, however, of interest in nuclear reactors. These include the *neutron* and *proton* (the major constituents of the nucleus, also called *nucleons*), the *electron*, and the *gamma ray*. Gamma rays are high-energy electromagnetic radiation originating in the nucleus; they have neither mass nor charge. Electrons are of low mass [approximately 5.49×10^{-4} atomic mass units (amu), where 1 amu = 1.66×10^{-24} g on the physical mass scale] and may have either a positive or negative charge. The positively charged electron is called a *positron*.

The proton has a mass of 1.00759 amu and is positively charged; the neutron is electrically neutral and has a mass of 1.00898 amu. The sum of the number of protons and neutrons in a nucleus defines the mass of an atom; the number of protons defines the atomic number and therefore the specific element in the periodic table. Individual atoms may possess the same number of protons but different numbers of neutrons (called *isotopes* of a given element), or they may be of the same mass but different charge (*isobars*). Complete designation of an atom therefore requires definition of charge or element and mass. To illustrate the symbolism used, consider uranium, which has an atomic number of 92. The isotope of uranium which has a mass of 238 amu would be designated $_{92}U^{238}$, U^{238}, U 238, or uranium 238.

Physicists are still engaged in unraveling the mystery of the structure of the nucleus. The most important aspect of this mystery, from a practical point of view, is that the nucleus, composed of positively charged particles with large repulsive forces between them, manages to be held together as a dense-packed mass. Clearly, a "nuclear glue" characterized by large attractive forces is required. Details of this nuclear glue are not understood. It is characterized, however, by the "binding energy." Experiments have shown that the actual mass of a nucleus is less than the sum of the masses of the constituent nucleons; the binding energy can be shown to correspond to this mass defect.

The binding energy per nucleon is a measure of the stability of a nucleus. When the binding energy per nucleon is high, the nucleus is most stable. Figure 17.1 illustrates the variation of binding energy per nucleon as a function of atomic mass. It can be shown that energy must be released when nuclei of low binding energy per nucleon are converted to nuclei of high binding energy per nucleon. Hence, from Fig. 17.1, combination of light nuclei (fusion) or splitting of heavy nuclei (fission) should release energy.

2. Radioactivity. A radioactive nucleus contains energy in excess of that characteristic of a stable configuration. To achieve stability, the nucleus must emit the excess energy: the process is called *radioactive decay*. The energy emission may involve release of a nuclear particle or one or more gamma rays or both. Several basic modes of decay have been identified and are discussed below.

The time interval between formation and decay of a single radioactive isotope cannot be predicted. It has been observed, however, that a large number of radioactive isotopes of a given kind will decay at a fixed rate characterized by a quantity known as the decay constant λ. The rate of decay, or alternatively, disintegration, is related to the decay constant by

$$\frac{dN}{dt} = -\lambda N \tag{17.1}$$

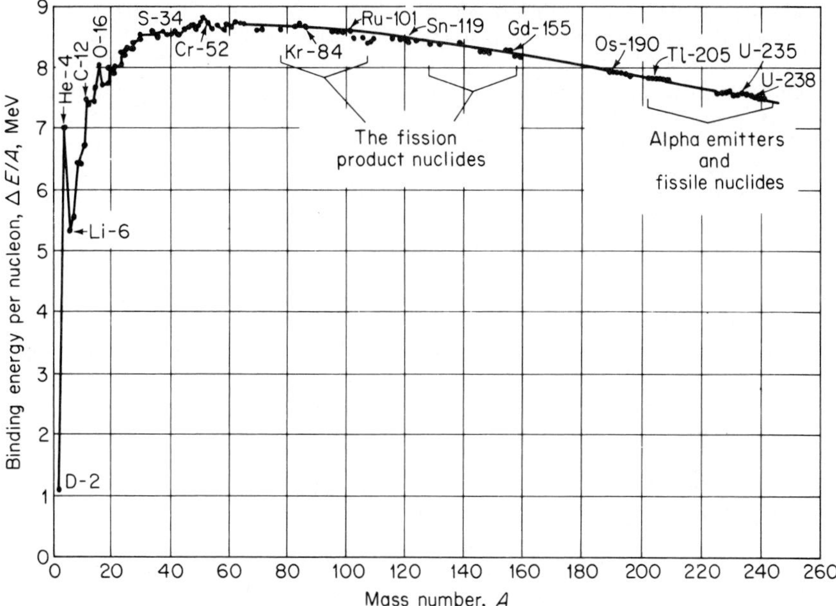

FIGURE 17.1 Binding energy per nucleon for stable nuclei. (*After Glasstone and Edlund, Elements of Reactor Theory, D. Van Nostrand, Princeton, N.J.*)

where N is the number of radioactive atoms present at time t. The units of λ are $(\text{time})^{-1}$. If at time $t = 0$ there were N_0 radioactive atoms present, Eq. (17.1) gives

$$N = N_0 e^{-\lambda t} \tag{17.2a}$$

or

$$\ln \frac{N}{N_0} = -\lambda t \tag{17.2b}$$

which shows that λ may be determined as the slope of the straight line obtained with a semilogarithmic plot of disintegration rate versus time.

Values of λ can be determined only by experiment and show great variation. It is frequently more convenient to report the *half-life* $T_{1/2}$, which is defined as the time for the number of radioactive atoms present to decrease by a factor of 2 and may be derived from Eq. (17.2*b*) in terms of λ as

$$T_{1/2} = \frac{\ln (2)}{\lambda} = \frac{0.693}{\lambda} \tag{17.2c}$$

Measured values of the half-life range from 10^{-7} to 10^{10} years.

Many radioactive nuclides have been identified. Some are naturally occurring, but most have been created by humans by bombarding stable nuclei with high-energy particles. Nearly all naturally occurring radioisotopes have mass numbers greater than 80. Isotopes made in the laboratory, however, cover the entire spectrum of elements, and many are commercially available.

The quantity λN is known as the *disintegration rate*, or *activity*. The basic unit of activity is the *curie*, defined as 3.7×10^{10} disintegrations per second (dis/s). One curie corresponds approximately to the activity of one gram of radium and is a large amount of activity. More frequently encountered amounts are the millicurie, 3.7×10^7 dis/s, and the microcurie, 3.7×10^4 dis/s.

3. Modes of Radioactive Decay. Two major modes of radioactive decay—alpha-particle emission and beta-particle emission—are of present interest. Decay by emission of beta particles (electrons) characterizes most useful radioisotopes; many of the naturally occurring radioactive nuclides, however, decay by alpha-particle emission. Both modes of decay frequently also involve gamma-ray emission. For nuclear engineering purposes, the properties of the particles and radiations emitted during radioactive decay are of major interest because they govern shielding and personnel-protection requirements.

Alpha Decay. The alpha particle is the nucleus of a helium atom. It is composed of two neutrons and two protons and therefore carries two positive charges. For practical purposes, the initial energy of all alpha particles emitted from a given kind of radioactive nuclide may be assumed to be constant; initial alpha-particle energies range from about 3 to 10 MeV.

Because of their large mass and charge, alpha particles rapidly lose their kinetic energy when passing through a medium by causing ionization of that medium. After sufficient energy has been lost, the slowly moving alpha particle picks up two electrons and becomes a helium atom.

The range of alpha particles from a given source in a given medium is constant because initial energies are essentially constant. This range is not great [approximately 1 in (25.4 mm), in air, for a 4-MeV alpha particle] because of the rapid loss of energy by ionization. The range of alpha particles is inversely proportional to the density of the medium; most alpha particles are stopped by a sheet of paper and will not penetrate human skin. Hence alpha particles are not in general a serious external hazard to humans. When ingested into the body, however, they do considerable damage to tissue because of their great ionizing power.

Beta Decay. When a beta particle is emitted from the nucleus during radioactive decay, the transformation in the nucleus may be described by the relation

$$ {}_0N^1 \rightarrow {}_1H^1 + \beta^- + \nu \tag{17.3}$$

i.e., a neutron is converted to a proton, the beta particle, and a neutrino, designated by ν. (The neutrino has never been identified experimentally, but its existence and properties can be demonstrated theoretically.) From analysis, it may be inferred that although little change in total mass of the nucleus has occurred, the mass number of the nucleus has increased by 1. Thus the atom is transformed to a different element by beta decay.

The neutrinos carry off different amounts of energy for each transformation, and hence the beta particles from a given radionuclide are found to have a continuous spectrum of energies, terminating in a definite maximum energy E_0. Values of E_0 are of interest for shielding purposes, and it is these values that are tabulated in the literature.

Other modes of beta-particle decay, such as electron capture and positron emission, occur. These are relatively infrequent, however, and are not discussed here.

Because of their low mass and charge, beta particles have much larger ranges than alpha particles [for example, the range of 3 MeV beta particles in air is about 43 ft (13m)], although they interact with materials in essentially the same way. In addition, ranges for beta particles are not clearly defined because of secondary interactions with the medium. It is possible, however, to specify the thickness of a given material required to reduce ionization by beta particles nearly to zero. As a first approximation, the thickness required may be assumed to be inversely proportional to the density of the medium.

Gamma Rays. As previously noted, radioactive decay frequently involves gamma-ray emission. These gamma rays are of great concern because their high energy and great penetrating power make them difficult to stop by shielding. Gamma rays interact with materials by several processes; they also produce ionization, but indirectly.

All disintegrations of a given kind of radioisotope do not always produce the same gamma rays. In other words, although the initial radionuclide and the decay, or "daughter," nucleus may be the same, various decay schemes and various gamma rays may be involved. It is important, again for shielding purposes, that the frequency of each mode of decay, as well as the gamma rays associated with each, be identified. This is done experimentally, and the data are tabulated in the literature.

4. Nuclear Reactions. Many kinds of reactions of incident particles and radiations with an atomic nucleus are possible. Relatively few, however, are of concern in nuclear reactors. The most important are those in which the neutron is the incident particle; fission is an example of these.

Nuclear reactions are of two basic types: *scattering reactions*, in which the identity of the incident particle is preserved, and *absorption reactions*, in which the incident particle is absorbed by the nucleus to form a new, highly excited compound nucleus. In absorption reactions the identity of the incident particle is lost; when the excited nucleus loses its energy, new reaction products are formed. In all nuclear reactions, total energy must be conserved either as mass or energy. The equivalence of mass and energy is given by the famous Einstein relation $E = mc^2$, where E is the energy equivalent of the mass m, and c is the velocity of light.

A shorthand notation is widely used to describe absorption reactions. Consider as an example the reaction

$$ {}_0N^1 + Ni^{58} \rightarrow Co^{58} + {}_1H^1 \tag{17.4} $$

which indicates that absorption of a neutron in the nucleus of a Ni^{58} atom produces a Co^{58} atom and a proton. This reaction is written as $Ni^{58}(n,p)Co^{58}$ in the conventional notation. This form of expression has been adopted because of its simplicity and ease of identifying incident and reaction product particles.

Other types of neutron-induced reactions are (n,n), (n,γ), (n,2n), and (n,f) reactions. The latter designates the fission process. Fission and (n,γ) reactions are most important in nuclear reactors. The (n,γ) reactions are the most predominant mechanism by which radioactive species are produced because energy considerations permit them to occur with relative ease.

Reaction Cross Sections. A measure of the probability of nucleus-particle interaction is required to make quantitative calculations of nuclear reaction rates. The quantity which designates this probability, for a single nucleus, is the microscopic cross section σ. The term *cross section* is derived from the fact that σ is essentially a measure of the effective cross-sectional area the nucleus presents to the incident particle.

As would be expected from the fact that atoms are mostly free space, cross sections are extremely small. Values of cross sections are reported in the literature in terms of *barns*; one barn is defined as 10^{-24} cm^2. Measured values of σ range from about 10^{-3} to 10^6 barns. Cross sections can be determined only by experimental measurement.

Every nucleus has a specific cross section for each specific kind of nuclear reaction that can occur; i.e., the cross section is different, for a given nucleus, for neutron scattering, neutron absorption, proton absorption, etc. In addition, for each specific kind of reaction, the cross section is a function of the energy of the incident particle. Thus cross sections for a given reaction must be measured at various incident-particle energies of interest.

Of all the nuclear reactions possible, neutron reactions such as scattering, (n,γ), and (n,f) are most important. Hence extensive measurements of σ as a function of energy have been made for these reactions.

Shorter tabulations of cross sections are also found in the literature (e.g., Table 17.1). Such values are specifically for absorption of so-called thermal neutrons. These are neutrons with energies of 0.025 eV (or equivalently, velocities of 2200 m/s). The latter values pertain to neutrons of most probable velocity in the Maxwell-Boltzmann distribution for thermal equilibrium at 20°C. If calculations of reaction rates are to be made for other temperatures and, as is the case with nuclear reactors, for environments containing neutrons with a wide spectrum of energies, appropriate corrections to the tabulated values must be made. Correction procedures are described in the literature.

Neutron absorption cross sections show great variation with neutron energy. Typically, at low energies (<0.1 eV) σ is proportional to $1/v$, where v is the neutron velocity. In the intermediate range (0.1 to 10^3 eV), sharp peaks, or "resonances," occur at specific energies. This is called the *resonance region*. At high energies, the cross section approaches the geometric cross section of the nucleus. Variations in cross section as a function of neutron energy are illustrated for some nuclides in Fig. 17.2.

Nuclear Reaction Rates. Actual rates at which nuclear reactions occur depend on the cross section for the particular reaction, the number of incident particles available, and the number of target nuclei. In general, calculations must be made for specific reactions for specific isotopes because different isotopes of a given element will have different cross sections and the total cross section for a given isotope is the sum of contributions for various types of reactions such as scattering, (n,γ), etc. Calculation of actual reaction rates in nuclear reactors is a complex process because the reaction rates are a function of material thickness, and the number density of incident particles is spatially dependent. The following procedure, however, is typical for neutron reactions.

TABLE 17.1 Thermal-Neutron-Absorption Cross Sections

Element	Isotope	Isotopic abundance, %	Cross section, barns*
H	.....	....	0.33
	H^1	100	0.33
	H^2	0.015	0.46 mb
He	.....	....	Variable
	He^3	0.00013	np 5,200
	He^4	100	0
Li	.....	....	67
	Li^6	7.5	$n\alpha$ 910
	Li^7	92.5	33 mb
Be	Be^9	100	9.0 mb
B	.....	....	750
	B^{10}	18.8	$n\alpha$ 3,990
	B^{11}	81.2	50 mb
C	.....	....	4.5 mb
	C^{12}	98.9	
	C^{13}	1.1	1.0 mb
N	.....	....	1.78
	N^{14}	99.6	np 1.70; $n\alpha$ 0.10
	N^{15}	0.37	0.024 mb
O	.....	....	0.2 mb
	O^{16}	99.76	Very small
	O^{17}	0.037	$n\alpha$ 0.5
	O^{18}	0.20	0.21 mb
F	F^{19}	10C	10 mb
Ne	.....	....	2.8
Na	Na^{23}	100	0.49
Mg	.....	....	59 mb
Al	Al^{27}	100	0.22
Si	.....	....	0.13
P	P^{31}	100	0.19
S	.....	....	0.49
Cl	.....	....	31.6
A	.....	....	0.62
K	.....	....	1.97
Ca	.....	....	0.43
Ti	.....	....	5.6
V	.....	....	4.7
Cr	.....	....	2.9
Mn	Mn^{55}	100	12.6
Fe	.....	....	2.43
Co	Co^{59}	100	34
Ni	.....	....	4.5
Cu	.....	....	3.59
Zn	.....	....	1.06
Zr	.....	....	0.18
Mo	.....	....	2.4
Cd	.....	....	2,400
In	.....	....	190

TABLE 17.1 Thermal-Neutron-Absorption Cross Sections (*Continued*)

Element	Isotope	Isotopic abundance, %	Cross section, barns*
Sn	.....	....	0.65
Xe	.....	....	35
	Xe^{135}	0	3.5×10^6
Sm	.....	....	6,500
	Sm^{149}	13.8	50,000
Eu	.....	....	4,500
Gd	.....	....	44,000
Hf	.....	....	115
Ta	.....	....	21.3
Au	Au^{197}	100	94
Hg	.....	....	380
Pb	.....	....	0.17
Bi	Bi^{209}	100	32 mb
Th	Th^{232}	100	7.0
	Th^{233}	0	1,400
Pa	Pa^{233}	0	37
U	.....	....	nγ 3.50, nf 3.92
	U^{235}	0.714	nγ 101, nf 549
	U^{238}	99.3	2.80
	U^{239}	0	22
Pu	Pu^{239}	0	nγ 361, nf 664

*mb means millibarns, or 10^{-3} barns; one mb = 10^{-27} cm^2.
Source: From R. Stephenson. *Introduction to Nuclear Engineering*, McGraw-Hill Series in Chemical Engineering, McGraw-Hill, New York, 1954, p. 375.

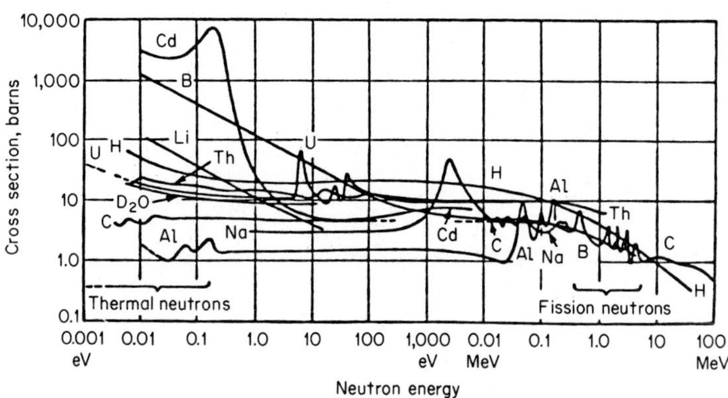

FIGURE 17.2 Total neutron cross sections for some reactor materials. (*Neutron Cross Sections, AECU 2040.*)

The volumetric rate of neutron reaction R is given by

$$R = \sigma\phi N_T \tag{17.5}$$

where σ = cross section for the particular reaction, cm^2
N_T = target nucleus density, atoms/cm^3
ϕ = "neutron flux," in neutrons/cm^2-s

The neutron flux is properly interpreted as the product nv, where n is the number density of neutrons (neutrons/cm^3) having the velocity v in centimeters per second.

Equation (17.5) is frequently written

$$R = \Sigma\phi \tag{17.6}$$

where Σ, called the *macroscopic cross section*, is defined by

$$\Sigma = \rho(N_a/A)\sigma \tag{17.7}$$

where ρ = mass density (g/cm^3) of specific isotope for which microscopic cross section is σ
N_a = Avogadro's number
A = atomic mass of target isotope, g/g mole.

It is important to recognize potential pitfalls in the use of Eq. (17.5) or (17.6). As previously mentioned, σ is a function of neutron energy, and in a nuclear reactor—and in many other circumstances—neutrons with a spectrum of energies are present. Hence proper evaluation of the total reaction rate really requires evaluation of the integral of the product $\sigma(E)\phi(E)$, where the argument represents the energy dependence of σ and ϕ, and ϕ is the actual neutron flux in the material in question. An alternative procedure is to use the thermal neutron cross section for the reaction and a properly weighted *effective thermal neutron flux*. The best procedure, however, is to determine experimentally the *activation product* $\sigma\phi$ for the particular system. Even this presents difficulties because ϕ is frequently position-dependent.

To illustrate the use of Eq. (17.5), consider a frequently encountered problem: determination of the amount of radioactive species present as a function of time when the radionuclide is the product of a neutron absorption reaction. The procedure will be illustrated for the $Cu^{63}(n,\gamma)Cu^{64}$ reaction, assumed to occur in pure copper exposed to an effective thermal neutron flux of 10^{14} neutrons/($cm^2 \cdot$ s).

The equation describing the amount of Cu^{64} present at any time is

$$dN^{64}/dt = R - \lambda N^{64} \tag{17.8}$$

i.e., the number of Cu^{64} atoms present, represented by N^{64}, is the difference between the amount produced by reaction R and the loss by decay. The production rate is obtained from

$$N^{64}(t) = \frac{\sigma\phi N^{63}}{\lambda}(1 - e^{-\lambda t}) \tag{17.9}$$

Numerical values are obtained as follows: Since an effective thermal neutron flux is given, the thermal neutron absorption cross section from the literature, 4.5 barns, may be used, assuming the reaction occurs at 20°C. The literature also gives the isotopic abundance of Cu^{63} as 69.09 percent and the half-life of Cu^{64} as 12.9 h. Then

$$N^{63} = \frac{0.6909 \text{ g } Cu^{63}}{\text{g Cu}} \times \frac{8.92 \text{ g Cu}}{\text{cm}^3} \times \frac{6.023 \times 10^{23} \text{ atoms } Cu^{63}}{\text{g mole } Cu^{63}} \times \frac{1 \text{ g mole } Cu^{63}}{63 \text{ g } Cu^{63}} = 5.9 \times 10^{22} \text{ atoms } Cu^{63}/\text{cm}^3 \quad (17.10)$$

and from Eq. (17.2*c*),

$$\lambda = \frac{0.693}{T_{1/2}} = \frac{0.693}{12.9(3{,}600)} = 1.49 \times 10^{-5} \text{ s}^{-1} \quad (17.11)$$

Substituting values into Eq. (17.9),

$$N^{64}(t) = \frac{(4.5 \times 10^{-24})(10^{14})(5.9 \times 10^{22})}{1.49 \times 10^{-5}}(1 - e^{\lambda t})$$

$$= (1.78 \times 10^{18})(1 - e^{-\lambda t}) \text{ atoms } Cu^{64}/\text{cm}^3 \quad (17.12)$$

At equilibrium ($dN^{64}/dt = 0$), the atomic density of Cu^{64} is simply 1.78×10^{18} atoms Cu^{64}/cm^3. The activity λN of this amount of Cu^{64} is

$$(1.49 \times 10^{-5})(1.78 \times 10^{18}) = 2.65 \times 10^{13} \text{ disintegrations}/s$$

or $(2.65 \times 10^{13})/(3.7 \times 10^{10}) = 7.17 \times 10^2$ curies.

5. Nuclear Fission. In the (n,f) reaction, the target nucleus splits to form two new nuclei of lighter mass, called *fission fragments*, and, most important to sustaining the reaction in nuclear reactors, several free neutrons. Only three nuclides, U^{235}, U^{233}, and Pu^{239}, are for practical purposes fissionable by neutrons of all energies; they are referred to as *fissile* nuclides. Of these, only U^{235} occurs in nature (its isotopic abundance in natural uranium is 0.72 percent). The U^{233} and Pu^{239} are produced from Th^{232} and U^{238}, respectively, by neutron absorption; the latter are referred to as *fertile* nuclides.

The mechanism of fission may be explained in terms of the *liquid-drop model*. The nucleus is viewed as analogous to a drop of liquid; the liquid drop is held together by surface-tension forces, and the nucleus is held together by binding energy. When sufficient excitation energy is imparted to the liquid, surface-tension forces will be overcome and the drop will split in two. Similarly, when the energy of the excited compound nucleus formed after neutron absorption exceeds the binding energy, the nucleus splits into two fragments. The excess energy is carried off primarily as kinetic energy of the fragments. As previously noted, the total energy released per fission is on the order of 200 MeV, of which about 95 percent is available for power production. The remainder is carried off by neutrinos.

Important fission properties differ for the various fissile nuclides. For example, the total energy released per fission varies slightly. Similarly, the average number of neutrons released varies, as does the fission cross section, both absolutely and as a fraction of the total neutron absorption cross section.

The latter leads to definition of the *regeneration factor* η, which is one of the most important physical constants related to fission chain reactors. The average number of neutrons released per fission is conventionally given the symbol ν. The regeneration factor is then defined in terms of ν as

$$\eta = \nu(\Sigma_f / \Sigma_a) \tag{17.13}$$

where Σ_f is the macroscopic fission cross section for the fissile nuclide, and Σ_a is the total neutron absorption cross section for the fuel material. In words, η is the number of fission neutrons produced by thermal fission per thermal neutron absorbed in the fuel. This parameter is the key factor in neutron economy (Sec. 6). Values of η and other properties of fissile nuclides are given in Table 17.2.

Fission Neutrons. Neutrons released by fission are divided into two fractions, *prompt* and *delayed*. As implied by the name, the latter are emitted some time after the fission event, apparently in conjunction with decay of certain of the fission products. Although delayed neutron fractions are very small (Table 17.2), their existence is probably the major factor permitting controlled utilization of nuclear power: they are the key to safe reactor operation (Sec. 8).

Prompt fission neutrons are emitted with a spectrum of energies as shown in Fig. 17.3. Most have energies in the range 1 to 2 MeV, but a few have energies in excess of 10 MeV. The latter are an important consideration in shielding.

The delayed neutrons fall into six groups, each characterized by an exponential decay rate. The six groups are the same for the three fissile materials, but the distribution of delayed neutrons in the six groups differs. Half-lives of the groups range between approximately 0.23 and 56 s. Kr^{87} and Xe^{137} have been identified as the neutron emitters for two of the groups.

Fission Products. The fission process occurs in more than 40 ways, producing fission fragments with mass numbers ranging from about 72 to 160. These fission fragments are highly radioactive and decay in a succession of steps involving formation of other radionuclides. The nuclei which result from this process—over 200 radioactive species—are known collectively as *fission products*.

The mass distribution of fission products for fissioning of U^{235} by thermal and

TABLE 17.2 Properties of the Fissile Nuclides*

	U^{233}	U^{236}	Pu^{239}
Useful energy per fission, MeV†	191	193	201
Total absorption cross section σ_a, *barns*	578	683	1,028
Fission cross section σ_f, barns	525	575	577
σ_a/σ_f	1.10	1.18	1.39
Fission neutrons per fission	2.51	2.44	2.89
Regeneration factor	2.28	2.07	2.08
Delayed neutron fraction‡	0.0026	0.0065	0.0021

*From H. S. Isbin, *Introductory Nuclear Reactor Theory*, Reinhold, New York, p. 461.
†From L. J. Templin (ed.). *Reactor Physics Constants*, 2d ed., USAEC, ANL-5800, 1963.
‡From G. R. Keepin and T. F. Wimett, "Reactor Kinetic Functions: A New Evaluation," *Nucleonics*, 16(10), 89, 1950.

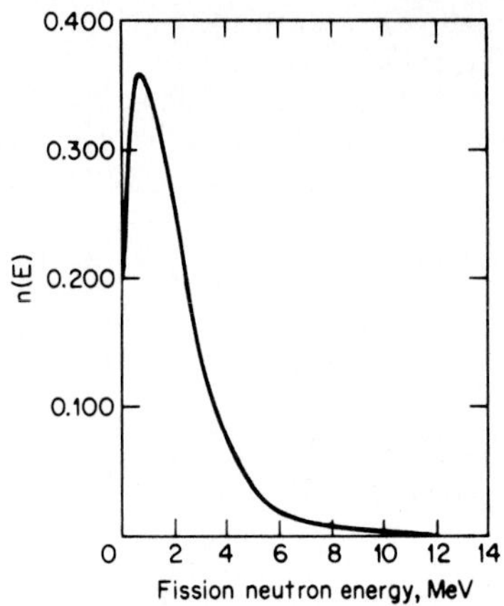

FIGURE 17.3 Energy spectrum of prompt fission neutrons. *(Glasstone and Sesonske. Nuclear Reactor Engineering, D. Van Nostrand, Princeton, N.J.)*

14-MeV neutrons is shown in Fig. 17.4. Curves for the other fissile species are similar. It may be noted from Fig. 17.4 that the maximum yield is about 6 percent; the maxima occur at mass numbers of approximately 95 and 135.

The radioactive fission products give off energy as gamma rays and beta particles during decay. In the nuclear reactor, this energy is rapidly manifested as heat, which must be removed to prevent core meltdown. For this reason, and also for proper design of spent-fuel reprocessing facilities, it is important to know the magnitude of this *decay heat power*, as it is called, as a function of time after fission ceases.

The decay heat power can be determined as a fraction of fission power from the expression

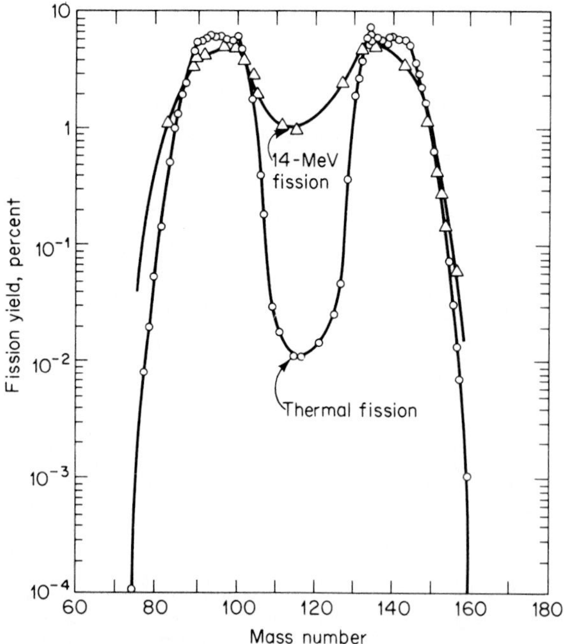

FIGURE 17.4 Mass distribution of U^{235} fission products. *(Glasstone and Sesonske, Nuclear Reactor Engineering, D. Van Nostrand, Princeton, N.J.)*

$$\frac{P}{P_0} = 6.1 \times 10^{-3}[(\tau - T_0)^{0.2} - \tau^{-0.2}] \tag{17.14}$$

where P = decay heat power
P_0 = fission or reactor power (both in arbitrary but identical units)
τ = time in days since cessation of fission
T_0 = number of days for which fission occurred (at constant rate)

As suggested by Eq. (17.14), the ratio P/P_0 is a function of T_0 as well as cooling time. Equation (17.14) is shown graphically for several values of T_0 in Fig. 17.5.

6. Physics of the Nuclear Reactor. From the physicist's point of view, design of a nuclear reactor is a problem of neutron economy. About 2.5 neutrons are produced per fission event (Sec. 5), only one of which must be absorbed to produce fission again and thereby sustain a chain reaction. However, many processes compete for neutrons in the reactor. These may be briefly summarized as (1) loss from the system at boundaries, (2) absorption in nonfissionable materials, (3) nonfission absorption in fissile materials, and (4) absorption to produce fission. The physicist's objective is to construct a balance sheet for neutrons which involves these four processes and sustains a chain reaction in the framework of engineering requirements for the reactor (Sec. 9).

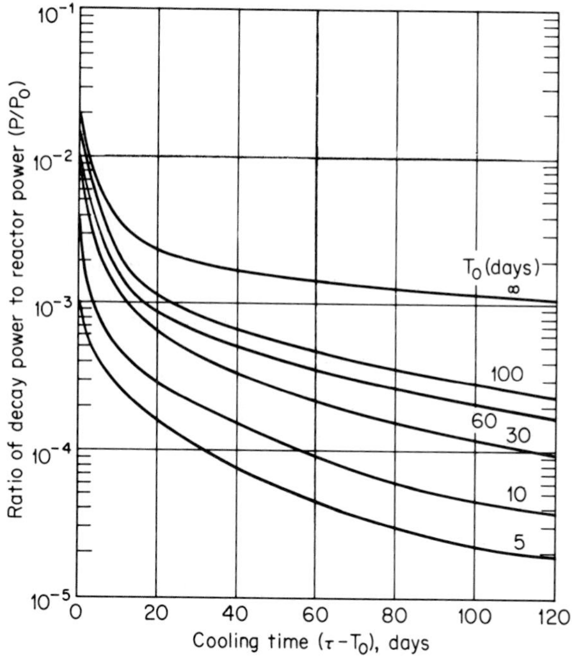

FIGURE 17.5 Fission-product-decay heat power. (*Glasstone and Sesonke, Nuclear Reactor Engineering, D. Van Nostrand, Princeton, N.J.*)

Classification of Reactors. Nuclear reactors can be classified according to a variety of standards. The most fundamental, however, is the kinetic energy of the neutrons causing most of the fissions. On this basis, there are two major types: *fast* reactors, in which most fissions are caused by neutrons of high energy, and *thermal* reactors, which operate primarily on low-energy thermal neutrons. Since fission neutrons are born at high energies (about 2 MeV; Sec. 2), fast reactors operate with neutrons of or near fission energy. In thermal reactors, the fast-fission neutrons are slowed down (to take advantage of the larger cross sections at thermal energies); this slowing down is accomplished by materials known as *moderators* that are put into the core. Thermal reactors predominate in the spectrum of operating reactors in the world today, primarily because of greater design flexibility. In the future, however, fast reactors should become more important in order to make best use of nuclear fuel resources.

Breeding and Conversion. Fast reactors are expected to become important because of their potential for producing more fissionable fuel than they consume, i.e., *breeding*. If new fuel generation involves production of a fissile material different from that being consumed (e.g., a reactor operating on U^{235} produces Pu^{239} from the fertile U^{238}), the process is called *conversion*. Many combinations of types of reactors and fissile and fertile materials have been and are being considered to maximize potential for conversion and breeding. Development of fast reactors for breeding has been slow because of difficult, but apparently soluble, technological problems with heat transfer, materials, and dynamic stability.

Because known world reserves of the only naturally occurring fissile material, U^{235}, are relatively small, breeding and conversion involving the fertile materials Th^{232} and U^{238} are mandatory for effective use of all potential fission energy reserves (which could supply man's needs for a century).

7. Physics Design of Reactor Cores. The ensuing discussion is directed primarily at thermal reactors because of the importance of neutron slowing-down processes in these systems. Basic concepts such as definition of reactor parameters, method of sizing the core, etc. apply equally well, however, to fast reactors.

Processes that occur for neutrons in a reactor may be described qualitatively as follows: Immediately after birth, the fast-fission neutrons begin to move rapidly through the reactor because of their high kinetic energy. As they do so, they encounter and interact with atoms of materials in the reactor core. These encounters with other atoms may produce one of two results: The neutron may be absorbed in the nucleus of the struck atom, or it may simply suffer a collision in which some of the neutron's kinetic energy is transferred to the struck atom.

These collisions are known as *scattering reactions*. They are of two types, *elastic* and *inelastic*. In the elastic, or "billiard ball," collisions, kinetic energy of the neutron-atom pair is conserved. Kinetic energy is not conserved in inelastic collisions, which are generally restricted to nuclei of fairly high mass number and neutrons of energy in excess of 0.1 MeV; the neutron is captured by the nucleus, and part or all of its kinetic energy is converted to excitation energy of the nucleus.

If the neutron is not absorbed during the preceding processes, its energy is gradually reduced (i.e., moderated—the moderator nuclei are targets for scattering reactions) as a result of the scattering collisions, so that the desired objective, neutrons of thermal energy, is achieved. These thermal neutrons are then available for absorption in fuel to produce fission and thereby reinitiate the above process.

Throughout the foregoing sequence of events, the neutrons are always subject to possible leakage from the system at boundaries. To reduce loss of neutrons by leakage, *reflectors* are placed at the boundaries of reactor cores. The reflectors scatter some of the leaked-out neutrons back into the core so that they remain available to cause fission.

Reactor Parameters. Quantitative calculations of the rate at which the preceding processes occur are required to size a reactor core. To make these calculations, parameters defined in terms of material properties important to these processes are used. These parameters and their symbols are as follows:

Diffusion Coefficient D. The diffusion coefficient is defined by the equation

$$J = -D \text{ grad } \phi \tag{17.15}$$

where J is the neutron current in a particular direction, and ϕ is the neutron flux. A good approximation to D may be obtained from the relation

$$D = \frac{1}{3\Sigma_s(1 - \overline{\mu}_0)} \tag{17.16}$$

where Σ_s is the macroscopic scattering cross section for the medium, and $\overline{\mu}_0$ is the average cosine of the scattering angle per collision, given in terms of the mass number A of the medium by $\overline{\mu}_0 = ⅔A$. Note that D has units of length.

Diffusion Length L. The diffusion length is defined by

$$L \equiv \sqrt{D/\Sigma_a} \tag{17.17}$$

where Σ_a is the macroscopic neutron absorption cross section of the medium. It is a measure of the distance a neutron travels from the point where it becomes thermal to the point where it is absorbed.

Average Logarithmic Energy Decrement ξ. This parameter is a measure of the average energy loss the neutron suffers in elastic collisions with nuclei. It is determined to a high degree of accuracy from the relation

$$\xi \cong \frac{2}{A + \frac{2}{3}} \tag{17.18}$$

To minimize the size of the reactor and possibilities of nonfission neutron absorption, it is desirable that ξ have as large a value as possible, i.e., that A be small (hence, moderator materials should be of low atomic mass).

Fermi Age τ. The Fermi age was defined in conjunction with a model in which the slowing down of neutrons, which proceeds in discrete steps of energy loss for each scattering collision, is represented as a continuous process. This parameter is of great practical importance in reactor design (see below). It is a function of neutron energy E and is defined by the relation

$$\tau(E) \equiv \int_{E_0}^{E} \frac{D}{\xi\Sigma_0 E}\, dE \tag{17.19}$$

where E_0 is the energy of the neutrons at the beginning of the slowing-down process. The Fermi age is a measure of the distance (note that the units of τ are length squared) a neutron has traveled from the point of origin to the point where its Fermi age is τ. An important value of τ is τ_{thermal}, corresponding to $E = E_{\text{thermal}}$. This is a measure of the distance the neutron travels to achieve thermal velocity.

Migration Length M. This parameter is a measure of the total distance the neutron travels from birth as a fission neutron to absorption as a thermal neutron. As would be expected, it is defined in terms of L and τ by

$$M \equiv \sqrt{L^2 + \tau} \tag{17.20}$$

The parameter that actually appears in reactor equations is the *migration area* M^2, where

$$M^2 = L^2 + \tau \tag{17.21}$$

Design Methods for the Steady State. The basic problem in reactor design is to devise a useful mathematical model which is descriptive of the physical processes discussed above and utilizes the parameters representative of the effects of reactor geometry and materials on these processes. The ensuing discussion is an outline of the basis for development and use of such models. The reader is cautioned that in practice, elaborate, complex computer programs are actually used for reactor design. These programs have their origin, however, in the concepts given here.

Neutron behavior in reactors can be described rigorously by Boltzmann transport theory. The complexity of reactors, however, prohibits detailed solution of the resulting equations. Hence the basis for reactor design lies in an approximation to transport theory known as *diffusion theory*. The essential feature of diffusion theory is that neutron leakage from the reactor is described as a diffusional process.

The material balance for neutrons in the reactor must take account of production (by fission) and losses (by leakage and all absorption reactions). In diffusion theory, the neutron balance for steady-state operation takes the form

$$D\nabla^2\phi - \Sigma_a\phi + S = 0 \tag{17.22}$$

where ∇^2 = Laplacian operator
ϕ = neutron flux
Σ_a = effective macroscopic absorption cross section
S = source term

Equation (17.22) is known as the *diffusion equation*, and it is the basis of reactor design. It is strictly applicable only in systems containing monoenergetic neutrons, and also at points more than two or three neutron mean free paths from boundaries and strong sources and absorbers. As will be seen, however, these restrictions can be obviated.

To solve Eq. (17.22), it is essential to obtain a representation for the source term S and define reactor geometry and boundary conditions. As the first step in the solution, it is convenient to take as reference a fictitious reactor, infinite in size, so that no neutron leakage occurs. For this system, a parameter known as the *infinite multiplication factor* k_∞ is defined. It is given by

$$k_\infty = \frac{\text{no. of fission neutrons produced in a given generation}}{\text{no. of neutrons absorbed in the preceding generation}} \tag{17.23}$$

where a "neutron generation" is the fission-birth, slowing-down absorption cycle previously described.

A similar parameter, the *effective multiplication factor* k_{eff}, may now be defined for a finite reactor. It is given in terms of k_∞ and a factor which corrects k_∞ for leakage losses in the finite reactor:

$$k_{\text{eff}} = k_\infty P \tag{17.24}$$

where P is defined as the *nonleakage probability*. It should be apparent from these definitions that the reactor is operating at steady state when k_{eff} has a value of exactly unity. When $k_{\text{eff}} = 1$, the reactor is said to be *critical*; when k_{eff} is less than or greater than unity, the reactor is subcritical and supercritical, respectively.

The infinite multiplication factor may be defined in terms of parameters representative of the effect of material properties on physical processes of scattering and absorption that occur during a neutron generation. The relationship is

$$k_\infty = \epsilon p f \eta \tag{17.25}$$

where η is the regeneration factor previously defined (Sec. 2), and the other parameters are as defined below. Equation (17.25) is known as the *four-factor equation*. Because of the definitions of ϵ, p, f, and η, it is a powerful means for determining the effect of changes in reactor materials on criticality. These parameters are defined as:

Fast-fission Factor ϵ. This factor accounts for neutron production during the slowing-down process by fissions at high energies. It may be defined as the ratio of the total number of fission neutrons produced by fast and thermal fission to the number produced by thermal fission.

Resonance Escape Probability p. This parameter is the ratio of the number of neutrons leaving the resonance region (Sec. 2) at low energies to the number entering at high energies. It is a complex function of the macroscopic absorption and scattering cross sections of the materials in the core.

Thermal Utilization f. This factor is defined as the fraction of all thermal neutrons absorbed that are absorbed in fuel material (which may include nonfissionable material such as U^{238}). The exact definition of f depends on whether the system is homogeneous or heterogeneous; an acceptable general expression for a reactor of volume V is

$$f = \frac{V_F \Sigma_{aF} \phi_F}{V_F \Sigma_{aF} \phi_F + V_m \Sigma_{am} \phi_m + V_i \Sigma_{ai} \phi_i} \tag{17.26}$$

where the subscripts F, m, and i represent fuel, moderator, and impurity (e.g., structural) materials, respectively.

In general, ϵ does not differ much from unity and, as shown in Table 17.2, values of η for the fissile materials are similar. The infinite multiplication factor for a given reactor therefore depends strongly on values of p and f, both of which are dependent on the amount, dispersion, and properties of materials in the reactor.

Solution of the Diffusion Equation. One may take the viewpoint that the objective in solving Eq. (17.22), the diffusion equation, is to obtain an expression for the nonleakage probability P [Eq. (17.24)], in terms of reactor materials and geometry. The actual expression obtained for P depends basically on two factors:

the expression used for the source term S and the method of treating the neutron energy spectrum in the core. In practice, the neutron energy spectrum is considered to consist of several groups, each containing monoenergetic neutrons. This leads to a diffusion equation for each group; the source term for each equation is then the neutrons entering that group from the group of next-highest energy neutrons.

The preceding approach leads to quite complex representations of neutron behavior. The general method by which expressions for P are developed may be illustrated, however, by the following simple model. This model is for an unreflected homogeneous reactor in which all neutrons are considered to have the same energy (i.e., "one-group" theory). The derivation will be illustrated for the steady state, which presupposes that k_{eff} is unity.

For the situation assumed, all neutrons are absorbed at the same energy and at a total rate $\Sigma_a\phi$. Since k_∞ fission neutrons are produced per absorption, the source term is simply $k_\infty\Sigma_a\phi$. Equation (17.22) then becomes

$$D\nabla^2\phi - \Sigma_a\phi + k_\infty\Sigma_a\phi = 0 \tag{17.27}$$

or upon rearrangement and introduction of the diffusion length [Eq. (17.17)],

$$\nabla^2\phi + \frac{k_\infty - 1}{L^2}\phi = 0 \tag{17.28}$$

Equation (17.28) indicates only the effect of reactor materials on neutron behavior. It is now necessary to consider the effect of neutron leakage and reactor geometry on the spatial distribution of the neutron flux.

The neutron flux distribution is represented by the relationship

$$\nabla^2\phi + B^2\phi = 0 \tag{17.29}$$

which is subject to boundary conditions imposed by the shape of the reactor (e.g., spherical, cylindrical, etc.). The constant B^2 is known as the "buckling," because it measures the bending (i.e., buckling) of the neutron flux.

At this point two operations are possible. First, Eq. (17.29) may be solved independently, with appropriate boundary conditions, to determine the flux distribution on the basis of purely geometrical considerations. Second, it may be noted that Eqs. (17.28) and (17.29) may be satisfied simultaneously if the coefficients of ϕ are identical. Equating coefficients,

$$\frac{k_\infty - 1}{L^2} = B_c^2 \tag{17.30}$$

where the subscript on B^2 indicates that the value of B^2 that satisfies Eq. (17.30) is the one for which the reactor is critical. If this value is now made equal to the value obtained by independent solution of Eq. (17.29), the reactor will actually be critical.

Two bucklings may therefore be distinguished. That arising from solution of Eq. (17.29) is known as the *geometric buckling*, and that given by Eq. (17.30) is the *material buckling*. When the reactor is critical, the material and geometric bucklings are identical. Expressions for the geometric buckling for various reactor geometries are given in Table 17.3. It will be noted that these expressions in-

TABLE 17.3 Geometric Bucklings for Various Reactor Shapes

Geometry	Buckling	Minimum critical volume
Sphere......................	$\frac{\pi^2}{R}$	$\frac{130}{B_c^3}$
Rectangular parallelepiped.....	$\frac{\pi^2}{a}+\frac{\pi^2}{b}+\frac{\pi^2}{c}$	$\frac{161}{B_c^3}$ $(a = b = c)$
Finite cylinder...............	$\frac{(2.405)^2}{R}+\frac{\pi^2}{H}$	$\frac{148}{B_c^3}$ $(H = 1.847R)$

R = radius; a,b,c = length of sides; H = height.

dicate the dimensions of the reactor. When the two bucklings are equal, these are the dimensions for criticality.

Equation (17.30) may be rearranged to

$$\frac{k_\infty}{1 + L^2B_c^2} = 1 \tag{17.31}$$

and it will be recalled that since steady state was assumed, k_{eff} must be unity. Hence

$$\frac{k_\infty}{1 + L^2B_c^2} = 1 = k_{\text{eff}} = k_\infty P \tag{17.32}$$

and therefore the nonleakage probability is given by $1/(1 + L^2B_c^2)$ according to one-group theory.

Equation (17.31) is known as the *critical equation* for one-group theory. It is, as noted, the result of a very simple model; its predictions of critical size are therefore, at best, approximate. To obtain more reliable estimates of critical dimensions, models that more accurately describe physical processes for the neutrons are required.

Critical equations for two other, more accurate models are as follows: The *age-diffusion model*, which utilizes the Fermi continuous slowing-down approximation mentioned above, gives

$$\frac{k_\infty e^{-B_c^2\sigma_{\text{thermal}}}}{1 + L^2B_c^2} = 1 \tag{17.33}$$

which for a large reactor reduces to

$$\frac{k_\infty}{1 + M^2B_c^2} = 1 \tag{17.34}$$

Two-group theory, in which the neutron energy spectrum is divided into a fast group and a thermal group, gives

$$\frac{k_\infty}{(1 + \tau B_c^2)(1 + L^2 B_c^2)} = 1 \qquad (17.35)$$

Many *multigroup* methods for determining critical size are also available. These can be quite accurate, but they are also quite elaborate and require iterative solution on computers. Many of the computer programs, or "codes," used to determine the critical size of reactors are available in the literature.

The preceding critical equations were derived assuming the reactor is homogeneous. In practice, of course, fuel, moderator, and structural materials are distinct (homogeneous reactors are in development). It is therefore frequently desirable to subdivide the reactor into small "unit cells," each of which may be treated as a homogeneous entity. Such a procedure adds considerably to the complexity—but also the accuracy—of the calculations.

8. Reactor Kinetics and Control. The power output of a nuclear reactor is varied by controlling the neutron flux. Many methods of flux control are available, but the most common is to insert in the core materials which have very large neutron absorption cross sections, called *poisons*. To sustain operation for long periods of time without refueling, the reactor is built with fuel in excess of that required for criticality. The poison materials provide "negative fuel" that compensates for the excess fuel (as fuel is consumed, the poison must gradually be removed) and, in conjunction with the control system, prevent the reactor from becoming supercritical during transient operations such as startup.

Poisons are usually inserted as an array of metallic control rods dispersed throughout the core (see Sec. 11 for a discussion of poison materials). The control rods are connected mechanically to drive motors that are actuated as a result of signals received from neutron-detection instruments. Much of the operation of the control system is automatic. An important safety feature is that operator actions which tend to increase the neutron flux are subject to automatic controls built into the system.

Poisons may also be inserted as "burnable" (i.e., gradually depleted) poisons added to the fuel matrix or the coolant. Methods of control other than poisons include addition or removal of fuel, variation of the amount of moderator in the core, and movement of sections of the core or reflector.

Reactor Control Parameters. A fundamental concept in reactor control is the *neutron lifetime* ℓ, which is defined as the average time between successive generations for an infinite reactor. The *effective lifetime*, defined for a finite reactor, is the neutron lifetime multiplied by the nonleakage probability P. The *prompt neutron generation time* ℓ^* is defined by

$$\ell^* \equiv \ell / k_\infty \qquad (17.36)$$

and characterizes the lifetime of prompt neutrons in the reactor.

Another fundamental concept in reactor control is the *reactivity* ρ, defined by

$$\rho \equiv \frac{k_{\text{eff}} - 1}{k_{\text{eff}}} = \frac{k_{\text{ex}}}{k_{\text{eff}}} = \frac{\Delta k}{k_{\text{eff}}} \cong \delta k \qquad (17.37)$$

The reactivity is frequently taken to be equivalent to Δk or, alternatively, δk.

The significance of these parameters and delayed neutrons in reactor control may be illustrated by considering changes in neutron density in a reactor not operating at steady state. The change in neutron density n with time is given by

$$\frac{dn}{dt} = n\frac{\delta k}{\ell} \tag{17.38}$$

which gives, with $n = n_0$ at $t = 0$,

$$n(t) = n_0 e^{(\delta k/\ell)t} = n_0 e^{t/\phi} \tag{17.39}$$

where ϕ is the reactor period. The effect of the delayed neutrons is to increase the neutron lifetime, and therefore the reactor period, from about 10^{-3} s, characteristic of the prompt neutrons, to 10^{-1} s. Thus, by inspection of Eq. (17.39), in a given period of time the neutron density changes by a much smaller amount when delayed neutrons control the reactor period. The neutron density would change at a rate too fast for the electromechanical systems to control if the delayed neutrons did not control the period.

Kinetic-Analysis Fundamentals. Because of their powerful influence on reactor dynamics, the delayed-neutron contribution to neutron economy is clearly distinguished in kinetic studies. The one-group diffusion equation for a bare, homogeneous reactor [Eq. (17.22)] becomes

$$D\nabla^2\phi - \Sigma_a\phi + k_x\Sigma_a\phi(1 - \beta) + \sum_{i=1}^{6}\lambda_i C_i = \frac{dn}{dt} \tag{17.40}$$

where β is the delayed-neutron fraction (Table 17.3), and $(1 - \beta)$ is therefore the prompt-neutron fraction. The contribution of the six delayed-neutron groups, each characterized by a concentration C_i and decay constant λ_i, is indicated by the summation.

Equation (17.40) may be written

$$\frac{dn}{dt} = \frac{\rho - \beta}{\ell^*}n + \sum_{i=1}^{6}\lambda_i C_i \tag{17.41}$$

with which are associated the equations descriptive of delayed-neutron behavior,

$$\frac{dC_i}{dt} = \frac{\beta_i}{\ell^*}n - \lambda_i C_i \qquad i = 1, 2, \ldots, 6 \tag{17.42}$$

which, it will be noted, is similar to Eq. (17.8) for neutron reactions.

Equations (17.41) and (17.42) are fundamental to reactor kinetics; they are basic to development of transfer functions descriptive of reactor dynamic response. However, as for steady-state design, relationships used in practice are considerably more complex than those given here. A detailed discussion of methods in use is given in many texts.

An important aspect of unsteady-state operation that can be extremely dangerous is the "prompt-critical condition," for which $\delta k = \rho = \beta$. The reactor is critical on prompt neutrons alone, and the period is therefore extremely short. The power level could rise at such a rate that the core would melt before corrective action could be taken. Control systems are carefully designed to prevent achievement of the prompt-critical condition.

Temperature Effects. Changes in temperature exert great influence on reactor kinetics because they affect materials density, core dimensions, neutron energy, and cross sections. Temperature effects are basically determined by differentiating the critical equation with respect to temperature (i.e., determining dk_{eff}/dT) and investigating the change with temperature in the range of interest for each resulting term.

Safety considerations require that the reactor temperature coefficient be negative (i.e., if temperature increases, power decreases). It should be noted that a negative coefficient can be achieved only by proper design. Some contributions to the coefficient are positive and some are negative; the magnitude and sign of each must be determined. All reactors are designed to have negative temperature coefficients, generally on the order of 10^{-5} to 10^{-4} per degree Fahrenheit (5/9 of this per degree Celsius) at operating temperatures.

A major contributor to the overall temperature coefficient is the *Doppler coefficient*, which describes the effect of temperature changes on neutron absorption in the resonance region. In general, as the temperature increases, the resonance peaks broaden, and increased neutron absorption occurs. In fissile material, the Doppler coefficient is therefore positive; in other materials, it is negative.

Fission-Product Poisoning. Two of the fission products—Xe^{135} and Sm^{149}—have very large thermal neutron absorption cross sections (2.7×10^6 and 4.2×10^4 barns, respectively). The magnitudes of these cross sections and the amounts in which the isotopes are formed are sufficient to have an effect on the multiplication factor.

These *fission-product poisons* influence the reactivity of the reactor. The amount of poisons present depends on reactor operating history and the thermal neutron flux. It can be shown, however, that a definite *maximum equilibrium poisoning*, defined as the ratio of the number of neutrons absorbed by the poison to the number absorbed by fuel, exists for operating reactors, as shown in Fig. 17.6.

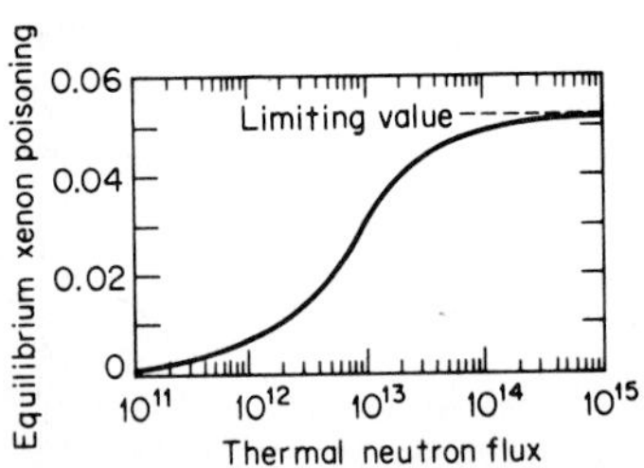

FIGURE 17.6 Equilibrium xenon poisoning during reactor operation. (*Glasstone and Sesonke, Nuclear Reactor Engineering, D. Van Nostrand, Princeton, N.J.*)

Figure 17.6 shows that the maximum poisoning during operation is relatively small. The poisoning can achieve very large values, however, after reactor shutdown. This phenomenon occurs because the precursor of Xe^{135}, which is I^{135}, has a relatively long half-life (6.7 h) and hence continues to produce Xe^{135} from its decay after shutdown.

The poisoning achieved after shutdown is a strong function of neutron flux, as shown in Fig. 17.7. The very high value of poisoning achieved for fluxes of 10^{14} and greater is in some cases the limiting factor in operating neutron flux and core life. To restart the reactor when the poisoning is a maximum, the core must have available sufficient excess reactivity (as excess fuel) to "override peak xenon"; near the end of core life this capacity is limited. In ship propulsion reactors, where startup at any time is essential, this limitation becomes quite important.

9. Engineering Design of Nuclear Reactors. A unique feature of nuclear power generation is the wide variety of design concepts that may be utilized successfully. Many core configurations, component designs, and materials combinations have been used in the past, and more innovations may be expected for the future. All power reactors have certain common components, however, as outlined below.

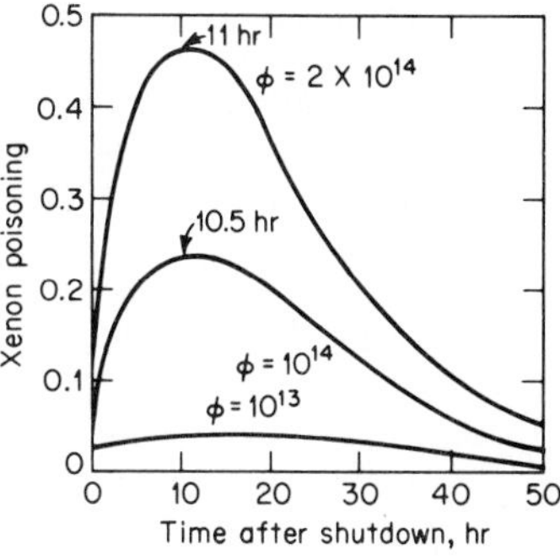

FIGURE 17.7 Xenon poisoning after shutdown. (*Glasstone and Sesonke, Nuclear Reactor Engineering, D. Van Nostrand, Princeton, N.J.*)

Reactor Coolants. The coolant, which removes heat generated by fission and radiation heating of core structures, may be a gas (air, helium, CO_2), molten salt, water (light or heavy), liquid metal, or organic liquid. In some reactor designs, coolant flow through the core is orificed to match the heat-generation distribution.

Reactor Vessel. The core and associated components such as control-rod assemblies, support structures, and reflectors are housed in the reactor vessel. In pressurized, water-cooled reactors, the vessel must be able to withstand high operating pressures [up to about 2000 lb/in_g^2 (13.8 MPa)]. A major problem in reactor-vessel design is thermal stress; other design problems arise from the need to provide fuel-handling facilities and control-rod drives. Pressure vessels for water-cooled reactors are built in accordance with Sec. III of the ASME Code.

Fuel-Element Cladding. Individual fuel elements in heterogeneous reactors are sheathed in a cladding which acts as a barrier, preventing escape of fission products from the fuel to the coolant. Cladding materials and properties are detailed in Sec. 11.

Thermal Shields. The radiations emitted by nuclear reactions in the core cause extensive heating of adjacent structural materials, including the reactor vessel. To prevent excessive thermal stresses in the vessel as a result of radiation-induced internal heat generation in the region of the core, thermal shields are placed between the core and the vessel. The thermal shields must be cooled (heat generation in the shields is about 3 percent of the total output of the reactor). The shields must have high absorption coefficients for neutrons and electromagnetic radiation and high thermal conductivities to prevent overheating. Steels are commonly used as shield materials.

Fuel-Handling Systems. A system must be provided for loading and unloading fuel in the reactor. The system also may be capable of moving fuel from one position to another in the core. Refueling of water-cooled reactors is usually done with the reactor shut down and the vessel closure head removed. In other systems, however, refueling may be accomplished without costly reactor shutdown.

Containment Vessels. All nuclear power reactors are required to be housed in containment structures designed to retain fission products and gases that might be released as a result of the maximum credible accident (see discussion of hazard analysis below). In many installations, this structure is a low-leakage vessel totally enclosing the reactor vessel and associated external components such as reactor coolant piping and steam generators. Such vessels are usually cylindrical or spherical and fabricated from steels not subject to brittle failure.

10. Thermal Design of Reactor Cores. Core design objectives for heat transfer and physics are basically at odds. To optimize neutron economy, the core should be small; to ease heat removal, the core should be large. Every operating core represents a compromise of these different objectives.

Nuclear reactors operate with high heat fluxes [on the order of 100,000 and 500,000 Btu/(h · ft^2) or 315 to 1576 kW/m^2 in water-cooled and liquid-metal-cooled systems, respectively]. To prevent fuel melting and release of fission products, a large number of fuel elements of small cross section is therefore required (for example, the Yankee reactor in Rowe, Mass., contains over 23,000 individual fuel elements; the nuclear ship *Savannah* contains over 32,000). Power densities in reactors vary considerably for different coolants, as shown in Table 17.4.

Standard techniques appropriate for the coolant and heat-transfer regime are used to determine heat-transfer coefficients. A detailed, point-by-point heat-transfer analysis is required, however, to prevent fuel-element failure. The general procedure is to develop a heat-transfer correlation for determining the peak central temperature (PCT) of a fuel element and apply this correlation to individual portions of the core. The correlation must account for the thermal resistance of the fuel, cladding, corrosion-product deposits on the cladding, coolant film, and bulk coolant.

In every reactor core there will be one point in one fuel element which operates at the highest temperature. The thermal analysis must locate this point and ensure that fuel-element failure (burnout) will not occur. To achieve this goal, hot-spot analysis involving the so-called hot-channel factors is used. The *hot-channel factors* are of two basic types, nuclear and engineering. They account for variations in neutron flux and design parameters such as fuel-element dimensions, uncertainty in coolant flow rate, uncertainty in the heat-transfer film coefficient, etc.

TABLE 17.4 Power Densities for Various Types of Nuclear Reactors

Reactor type	Power density, kW (thermal)/ft^3
Gas-cooled—natural U	15
High-temperature gas	220
Sodium graphite	290
Organic-cooled	390
Heavy water—natural U	510
Boiling water	820
Pressurized water	1,550
Sodium-cooled fast breeder	21,500
Nuclear rocket reactor	280,000
Conventional forced-convection boiler	280

Source: Abstracted from S. Glasstone and A. Sesonske, *Nuclear Reactor Engineering*, Van Nostrand, Princeton, N.J., 1963.

A large number of hot-channel factors may be defined (in general, for a given parameter as the ratio of the maximum-worst-deviation value to the nominal value) and quantitatively determined for a given core. However, they are combined into three basic overall factors used in the PCT correlation: (1) the factor for coolant temperature rise $F_{\Delta T}$, (2) the film temperature-drop factor F_θ, and (3) the heat-generation factor F_q.

The method by which individual hot-channel factors are combined to determine the overall factors is a subject of considerable debate. Two basic methods are available: statistical, in which probabilities of occurrence are assigned to each factor, and the factor-product method, in which all contributors to each overall factor are multiplied together. The latter is of course extremely conservative.

11. Nuclear Reactor Materials. Conventional engineering materials are used in reactors except where special nuclear requirements must be met. Reactor materials must satisfy applicable conventional criteria such as tensile strength, ductility, corrosion resistance, etc., and those used in the core also must have nuclear properties appropriate to their function.

A choice of materials for each major reactor component is available. There are, however, two general restrictions: Materials specifications for reactors are generally more stringent than for nonnuclear applications because interactions with neutrons can affect suitability, and materials that are mutually compatible must be selected.

The combined engineering, nuclear, and compatibility requirements for nuclear materials have led to extensive research and development programs designed to produce new alloys and materials with improved physical properties for reactor use. Results of this work are reported by the U.S. government.

Fuel Materials. Nuclear fuels in common use include natural uranium, slightly enriched uranium (0.95 to ~ 6.0 percent U^{235}), and fully enriched uranium (93 percent U^{235}). The fuel materials also include large amounts of the fertile species U^{238} and Th^{232}. As more and more reactors operate to produce fissile nuclides from these fertile species, it may be expected that Pu^{239} and U^{233} will become important reactor fuels.

Natural uranium is generally used in both heavy-water-cooled and gas-cooled reactors. Reactors cooled and moderated with light water use slightly enriched uranium as fuel; the actual enrichment is dictated by criticality and core-endurance requirements. Fully enriched uranium is used extensively in research and other special-purpose reactors but thus far has received limited use in nuclear power stations.

Uranium has very poor metallurgical and physical properties. It is dimensionally unstable after irradiation and thermal cycling and fails catastrophically with short exposure to water at 100°C or more. Its apparent physical properties are also extremely sensitive to purity, state of cold work, grain size, and orientation of grains. For these reasons, it is usually desirable to use uranium alloys or compounds as fuel materials; however, in the gas-cooled reactors that are so prevalent in Europe and England it is found economical to use natural uranium metal as fuel.

Reactors that burn slightly or fully enriched uranium use UO_2 or uranium alloys as the fuel material. Common alloying materials include zirconium, aluminum, molybdenum, and stainless steels. UO_2, despite low thermal conductivity, is widely used in water-cooled power reactors.

Current nuclear fuel development programs have as their major objectives development of improved fuel-fabrication techniques and better fuel materials. Promising examples of the latter are ceramics, such as UC, UC_2, and UN. These materials show much better resistance to adverse effects of irradiation than uranium metal and are capable of withstanding long periods of exposure in the core. They have high melting points, reasonably good thermal conductivities, and are easily fabricated but react with water. Successful development of these materials is expected to result in much cheaper production of nuclear power as a result of reduced core-fabrication costs and longer periods of exposure between shutdowns for refueling.

Fuel exposure, or *burnup*, between refuelings is an extremely important factor in determining nuclear power costs. Burnup is usually expressed in megawatt-days of heat energy produced per metric ton (or tonne; equivalent to 2200 lb) of uranium, abbreviated as MW-days/tonne U, or MWD/T. Thermal reactors currently in operation in the United States achieve burnups of about 10,000 MWD/T, with values between 20,000 and 30,000 or more anticipated for the future. The natural-uranium-fueled, gas-cooled reactors in England and Europe achieve much lower burnups—on the order of 3,000 MWD/T—primarily because of loss of reactivity in natural uranium.

Radiation effects that limit fuel burnup may be summarized as dimensional instability (elongation or swelling) as a result of accumulation of fission products and gases, adverse changes in mechanical properties, relatively increased parasitic capture of neutrons, and loss of reactivity due to fuel consumption. It is expected that new fuel materials and fabrication techniques (too numerous to detail here) and programmed movement of fuel from one position in the core to another will greatly reduce these adverse effects, and thereby improve burnup and reduce fuel costs in the future.

Moderator Materials. Moderator materials in common use today include light water, heavy water, and graphite. Beryllium, BeO, and lithium 7 are also good moderator materials, but they are toxic and highly reactive. Development programs for these materials are in progress.

Graphite is the most commonly used moderator in power reactors. In U.S. reactors, however, light water is most commonly used; it is cheap, readily available, and serves also as the coolant. Heavy water is actually a better moderator because deuterium has a much lower absorption cross section for parasitic capture of neutrons than hydrogen. Its high present cost, however, has limited its use in power reactors, except where natural uranium can be economically used as fuel. Heavy water is present in natural water to the extent of 140 to 150 ppm; the two species may be separated by distillation, chemical exchange, or electrolysis. The United States maintains large separation plants in Savannah River, Ga.

The graphite used in reactors is of high density (to limit reactor size) and high purity (~20 ppm total impurities). A highly pure material is required to minimize parasitic capture of neutrons. The product resulting from the graphite-manufacturing process may be easily machined or cast in desired shapes.

Cladding Materials. The cladding must be a high-integrity structural material with good corrosion resistance, high thermal conductivity, and a low neutron absorption cross section. Stainless steels and aluminum are currently the most commonly used cladding materials; however, zirconium alloys are actually superior to these because of their very low neutron absorption cross section (~0.2 versus ~3 barns for steel) and excellent corrosion resistance.

Other cladding materials used in thermal reactors include Magnox (a magnesium alloy, used primarily in U.K. gas-cooled reactors), magnesium-zirconium alloys, Incaloy, ceramic coatings, and graphite. Claddings used in fast reactors include stainless steels, zirconium, niobium, and other refractory metals and ceramics.

Reactor Control Materials. Materials used to control thermal reactors by neutron absorption have as their most important characteristic large neutron absorption cross sections; the elements that contain isotopes suitable for this application include boron, cadmium, europium, and hafnium.

Control materials are used in the reactor in two principal ways. The most common is by control rods, a method that requires the control materials to withstand shock, vibration, and wear. The other method is to use a burnable poison which can be evenly distributed in the fuel, placed in discrete positions in the core, or dissolved in the coolant. Use of burnable poisons requires careful calculation of control-material concentrations; poison burnup and fuel-depletion rates must be comparable.

The control materials may be utilized in various forms. Boron, for example, is used in boron–stainless-steel alloys, as B_4C particles dispersed in zirconium or fuel alloys, and as boric acid added to the coolant. The latter is a commonly used backup safety method in water-cooled reactors. Cadmium is used as a major constituent of silver-indium-cadmium alloys. These alloys have relatively low melting points, however, and poor corrosion resistance in water containing even small amounts (~5 ppm) of oxygen. They also have poor strength properties.

Hafnium has been found to be an excellent control material. In addition to its good nuclear properties, it has high strength and good corrosion resistance. It is readily shaped and welded as the pure metal. Hafnium is expensive, however. It is obtained only as a byproduct of zirconium ores, and processing costs to separate the two elements are high. The separation is essential, because hafnium impurities in zirconium used in reactor cores would greatly increase parasitic capture of neutrons. Hence hafnium is readily available for limited use, but fabrication costs are also quite high because of the need to avoid impurities which adversely affect physical properties of the material.

Most operating power station reactors use boron as the control material. The Shippingport, Pa., reactor and many mobile reactors use hafnium. The Yankee reactor in Rowe, Mass., has used a silver-indium-cadmium alloy, as do many research reactors. The very high cost of and metallurgical problems with europium and other lanthanon poisons have prohibited their use except as burnable poisons within the fuel.

Reactor Coolants. The basic types of materials available as reactor coolants (water, gases, liquid metals, molten salts, and organic liquids) have previously been mentioned briefly. Of all these, light water is the most commonly used in the United States because physical-property data are readily available and it is economical. Reactors using the other types of coolants also have been built, however, and several alternatives in each category are available, as outlined below.

Gases. Air, hydrogen, helium, and carbon dioxide are in principle good reactor coolants. Each, however, has certain specific deficiencies: The oxygen in air reacts with reactor materials, hydrogen is extremely hazardous and reacts with many metals, helium is quite expensive, and CO_2 is a relatively poor heat-transfer medium. In general, all gaseous coolants also have the disadvantage of being poor heat-transfer agents compared with liquids.

Carbon dioxide is the most commonly used gas coolant because of the problems with the other gases. Some advanced-concept reactors, however, use helium as the coolant because of superior heat-transfer and physical properties.

Liquid Metals. These materials are virtually a necessity for fast reactors in which coolants with good moderating properties are undesirable. In general, the excellent thermal properties of the liquid metals provide the advantage of operation at high temperatures with high power densities (Table 17.4) and compact cores. On the debit side, however, technological knowledge of the liquid metals is relatively sparse, their cost is high, and their generally high chemical reactivity at elevated temperatures makes selection of compatible materials a problem.

The most meaningful criterion for selection of liquid metals as reactor coolants is the neutron absorption cross section, which must be small to minimize parasitic capture. On this basis, the liquid metals listed in Table 17.5 are found to be suitable reactor coolants. Of these, liquid sodium and the NaK eutectic (22 weight percent Na), which is liquid at ordinary temperatures, are at present most attractive because they have fewer disadvantages than the others. A major disadvantage of the use of sodium as a reactor coolant is formation of Na^{24} from Na^{23}. The Na^{24} has a significantly long half-life (15 h) and emits two high-energy gamma rays with decay. Shielding of reactor components is therefore required, and access to the reactor compartment for maintenance may be restricted because of high radiation levels for two or three days.

Molten Salts. These materials have been considered as reactor coolants because they permit economic utilization of all grades of reactor fuels. They are used in a fluid-fueled reactor in which the coolant and fuel are a homogeneous slurry pumped throughout the system. The molten salts are advantageous for fluid-fueled systems because, unlike aqueous fuels, they can be used at low pressures and high temperatures.

Only two molten-salt reactor systems have been built and operated; both had demonstration of feasibility as the major objective. The first was the Aircraft Reactor Experiment, now terminated, and the most recent is the Molten Salt Reactor Experiment (MSRE) located in Oak Ridge, Tenn. The MSRE utilizes a fuel salt which has a composition of 65 mol% Li^7F, 29.1 percent BF_2, 5 percent ZrF_4, and 0.9 percent UF_4; its melting point is 842°F. This salt, and other structural materials, were specially developed for the MSRE reactor.

Organic Liquids. Organic liquids have many potential advantages as reactor coolants: The systems may be operated at low pressure, the organics are not highly corrosive, they are good moderators, induced radioactivity is low, and

TABLE 17.5 Properties of Liquid-Metal Reactor Coolants

Metal	σ_a, barns	Melting point, °F (°C)
Bismuth	0.032	520 (271)
Lithium 7	0.033	367 (186)
Lead	0.17	621 (327)
Sodium	0.50	208 (98)
Tin	0.65	450 (232)
Potassium	2.0	145 (63)
Gallium	2.7	86 (30)
Thallium	3.3	576 (302)

physical properties are well known. It has been found, however, that the organic coolant tends to decompose when exposed to high temperatures and radiation. This characteristic could result in serious operational problems, such as plugging of flow channels and fouling of heat-transfer surfaces. To eliminate these problems, decomposed coolant must be removed by purification and replaced with fresh coolant.

12. Fission Power Systems. Feasible fission power concepts number in the thousands. For economic reasons, however, practical systems are limited to a few basic concepts and variations on these.

The type of reactor selected depends on local economic conditions. In Great Britain, where capital finance charges are small, gas-cooled reactors fueled with natural uranium are preferred. Canada is committed to reactors cooled with heavy water. In the United States, light-water reactors fueled with slightly enriched uranium have been economical.

Reactor types are usually defined by the type of coolant used. Types currently in use and those projected for use in the United States are described below.

Light-Water Reactors. There are two types of light-water reactor (LWR) systems: the boiling-water reactor (BWR) and the pressurized-water reactor (PWR). Both use ordinary (light) water as coolant and moderator, and both are fueled with slightly enriched uranium (about 3.5 percent U^{235}) in the form of oxide pellets contained in zirconium tubing.

In a BWR system (Fig. 17.8), the core coolant is allowed to boil. The steam is dried and sent to the turbines; the water phase is recirculated in the reactor vessel. In a PWR system (Fig. 17.9), the core coolant is kept in a liquid state; a heat exchanger external to the reactor vessel is used to generate steam for the turbines.

The BWR and PWR systems are competitive because of design and economic tradeoffs. The PWR provides higher thermal efficiencies: The PWR reactor outlet temperature is about 600°F (320°C); the BWR outlet temperature is about 550°F (290°C). The cost of steam generators and the pressurizer for a PWR are offset by higher costs for reactor vessel internal components in the BWR.

Gas-Cooled Reactors. The high-temperature gas-cooled reactor (HTGR) (Fig. 17.10) uses helium coolant, graphite moderation, thorium and highly enriched

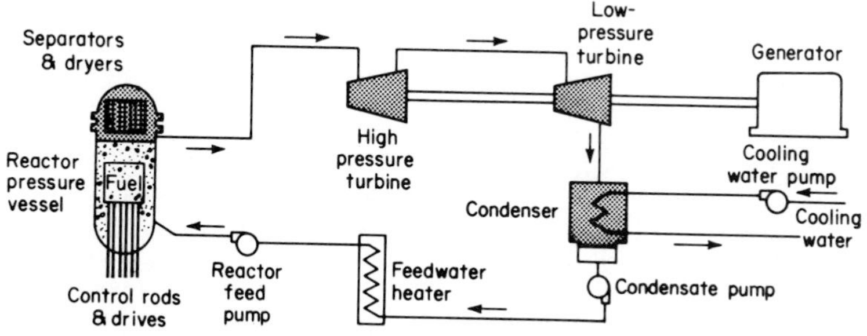

FIGURE 17.8 Schematic diagram of a boiling-water reactor system. *(USAEC.)*

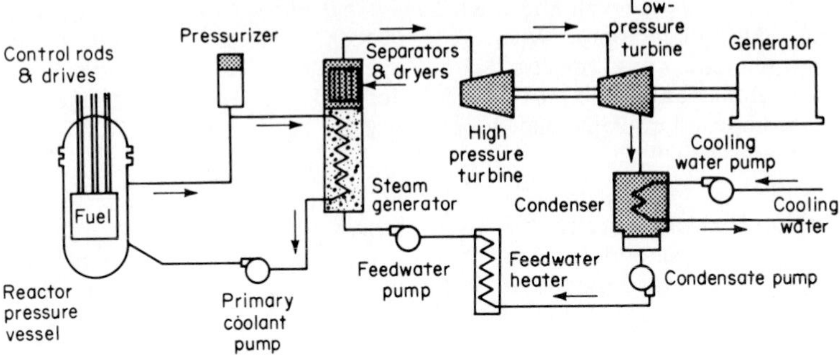

FIGURE 17.9 Schematic diagram of a pressurized-water reactor system. (*USAEC.*)

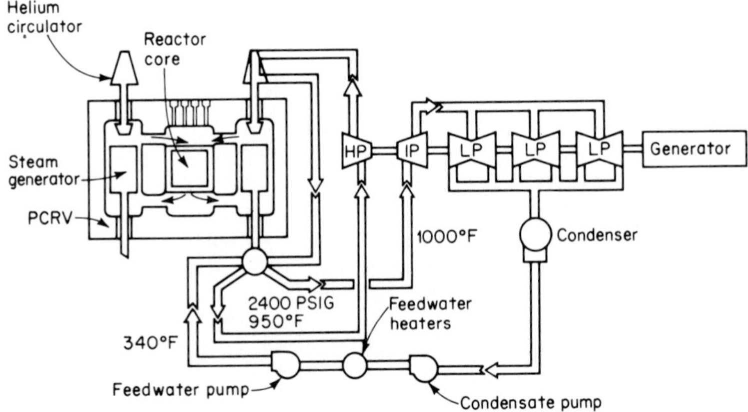

FIGURE 17.10 Schematic diagram of a high-temperature gas-cooled reactor system. (*USAEC.*)

uranium as fuel materials, and a prestressed-concrete reactor vessel (PCRV). Coolant outlet temperatures are about 1400°F (780°C); coolant temperature rise through the core is about 330°C (in LWRs it is restricted to about 16°C).

The HTGR core consists of vertical columns of hexagonal graphite elements. Coolant flows in channels separated by graphite from the fuel pins which contain small spheres of coated fuel material. Pressures in the primary system are low [700 lb/in_a^2 (4.8 MPa)] in comparison with LWR pressures.

Liquid-Metal-Cooled Reactors. Since liquid metals do not moderate neutrons, they offer potential for good breeding and conversion, i.e., efficient use of fission fuel materials. The liquid-metal fast breeder reactor (LMFBR) is simple in concept (Fig. 17.11), but it is complex in practice because of design and operating problems associated with use of sodium or sodium-potassium as the coolant.

Efficient use of fission fuels in LMFBRs depends on the *breeding ratio*, which can be defined as the rate of production of fissile material compared to the rate of consumption, and the *doubling time*, which is approximately the length of time to

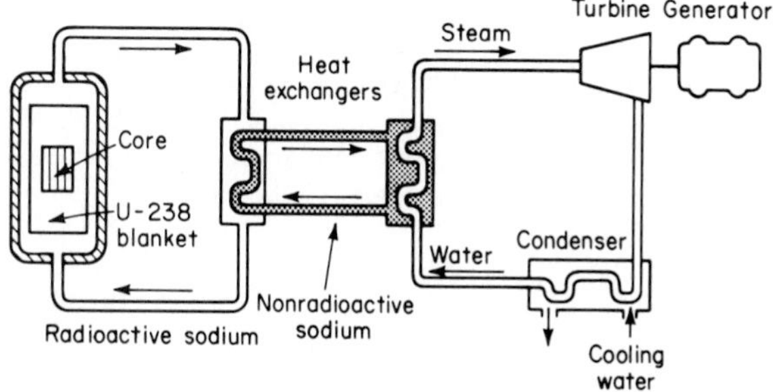

FIGURE 17.11 Schematic diagram of a fast-breeder reactor system.

produce sufficient excess fissile plutonium to inventory the in-core and ex-core fissile requirements for a similar new reactor.

The breeding ratio is a complex function of core design. It is related to the doubling time, as shown in Fig. 17.12, by the types of fuel used. Current LMFBR development work is aimed at improving economics and fuel utilization by developing new fuels that can reduce the doubling time. It also is seeking to upgrade operational performance through design development work aimed at attaining higher operating temperatures.

13. Nuclear Fuel Cycle. Nuclear fission is distinguished from all other energy-conversion concepts by the need to reprocess spent fuel. The need arises because

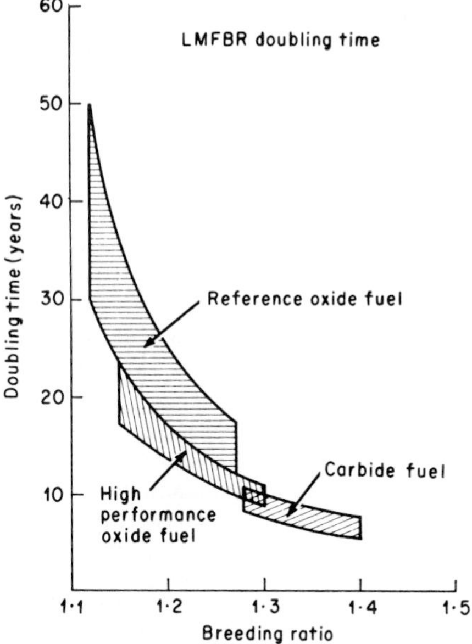

FIGURE 17.12 LMFBR doubling time as a function of breeding ratio.

nuclear and mechanical features of core designs prevent the complete burnup of a fuel loading. Nuclear limitations are associated with loss of capability to sustain criticality as a result of fuel depletion and buildup of fission products. Mechanical limitations arise from radiation damage that causes distortion, loss of strength, and loss of thermal conductivity in the fuel elements. The net result is that fuel elements must be removed from the reactor when only 10 to 30 percent of the fissile material has been consumed.

Reprocessing of the spent fuel has three objectives: (1) recovery of fissile uranium and reconversion into fresh fuel, (2) extraction and utilization of fissile plutonium created by neutron capture in U^{238}, and (3) isolation of fission products and other wastes. Since the fissile uranium and plutonium have value as fuels, the first two steps are potentially revenue-producing. The third is an overhead cost for the fission power economy.

Operations in the nuclear fuel cycle are shown in Fig. 17.13. The conversion, enrichment, fuel-element fabrication, and reprocessing steps have no parallel in other energy-conversion systems. Nuclear power is economically competitive despite the costs of these operations because of the low unit cost of fissile fuels.

The conversion step changes solid uranium oxide into gaseous uranium hexafluoride, a form suitable for the enhancement of U^{235} concentration, which is done by gaseous diffusion in the enrichment step. In the fabrication step, the enrichment product is converted to uranium dioxide which is formed into pellets that in turn are inserted into the fuel elements for the reactors.

The reprocessing step involves breaking down the fuel elements, dissolution of the spent fuel pellets, recovery of the fuel values by solvent extraction operations (Fig. 17.14), and management of the radioactive wastes.

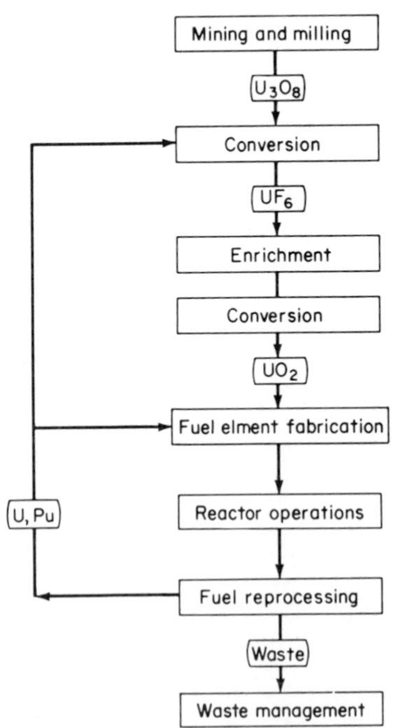

FIGURE 17.13 Basic operations in the nuclear fuel cycle.

Plutonium recovered during reprocessing can be combined with uranium to form mixed-oxide fuels for LWRs or it can be used as the primary fuel for LMFBRs. Since plutonium is highly toxic and a potential nuclear weapons material, elaborate procedures to monitor and control inventories are required. These safeguard procedures are an important feature of design and licensing of facilities that handle plutonium.

Reprocessing and other fuel-cycle operations produce a variety of radioactive wastes. The fission products are the major constituent of *high-level waste*, which contains most of the radioactivity and also has high heat-emission rates as a result of radioactive decay. Another form of waste, large in volume but low in radioactivity, is the *alpha waste* which is made up of paper trash, plastics, etc., that are used in fuel-cycle operations and become lightly contaminated with uranium or

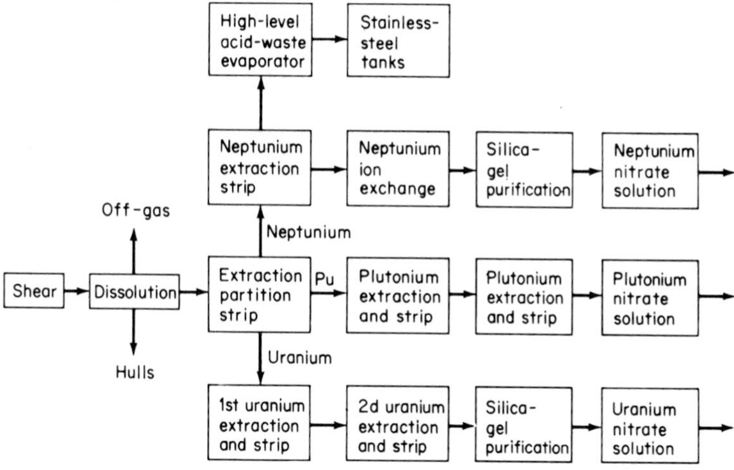

FIGURE 17.14 Operations in nuclear fuel reprocessing.

plutonium. Other important wastes are the chopped cladding ("hulls") from breaking down spent fuel, and decommissioned equipment. Methods for control and disposal of these wastes are being developed.

OTHER ENERGY-CONVERSION SYSTEMS

14. Fusion. Fusion energy is released when the nuclei of light elements come together to form heavier elements. If energy-conversion systems based on fusion can be developed, the supply of fuels on earth would be essentially unlimited. The most suitable fuels include the heavy isotopes of hydrogen (deuterium and tritium), the light isotope of helium (helium 3), and lithium. All are abundant in the seas.

Fusion fuels must be heated to temperatures on the order of tens to hundreds of millions of degrees in order to have fusion occur. The electrons must be stripped from the atoms and the nuclei must be brought together with force sufficient to overcome repulsive forces. These conditions must be met at densitities sufficient for fusion collisions to be frequent and for times adequate to produce more energy than is consumed in the process. When all conditions are satisfied, the particles are in the plasma state, i.e., electrically charged.

Since all materials will melt at fusion temperatures, the plasma must be contained without touching the walls of the container. The basic approach for this is to confine and control the plasma with magnetic fields, i.e., in *magnetic bottles*. Many bottle configurations have been explored; one of the most promising is the TOKAMAK concept. Laser-induced fusion is also being studied.

The major problem for development of fusion systems is to confine the plasma without significant leakage from the magnetic bottle for times sufficient to have productive amounts of fusion occur. Conceptual designs for fusion systems have been developed in anticipation that this hurdle will be overcome. If fusion should become practical, environmental and health hazards are expected to be small in comparison with fission reactors. Residual radioactivity would be limited primarily to structural materials, and fusion reactors do not have the runaway potential of fission reactors. In addition, overall plant efficiencies two or more times greater than those of fission reactors might be achieved by direct conversion to electricity.

15. Solar Power. People have used energy-conversion systems based on solar power for centuries. Large-scale implementation has been impeded because solar energy is dilute and variable (thereby requiring storage systems). Cheap, abundant supplies of other fuels have been disincentives for large-scale utilization of solar-powered systems.

The simplest solar conversion systems, which are suitable for domestic use, combine adsorption of solar energy on a black surface with a storage and pumping system as illustrated by Fig. 17.15. Advanced systems based on these concepts would employ solar tracking and focusing devices for the collector, large-scale collection areas, and large-scale storage devices.

In principle, such advanced systems could be developed by extrapolating known technology. At present, although there are technical hurdles associated with collector design, storage, and efficiency, the major impediment to development is cost. Projected capital and retail costs for electricity from a base-loaded solar conversion system are about three times current costs.

An alternative for solar conversion is to use photovoltaic cells. These cells, which are widely used on spacecraft, use an *n-p* junction in silicon material to convert sunlight directly into electrical current. Efficiencies are typically on the order of 10 percent; the theoretical maximum is about 35 percent. At 10 percent efficiency and for current manufacturing costs, a conversion system based on photovoltaic cells would cost more than 100 times the cost of nuclear fission systems.

Another alternative for solar conversion is to make combustible fuels after photosynthesis in trees, plants, and algae. Efficiencies would be about 3 percent or lower, and about 3 percent of the U.S. land area would be needed to supply current total energy needs. Projected costs are not unreasonable, but waste disposal may pose significant problems.

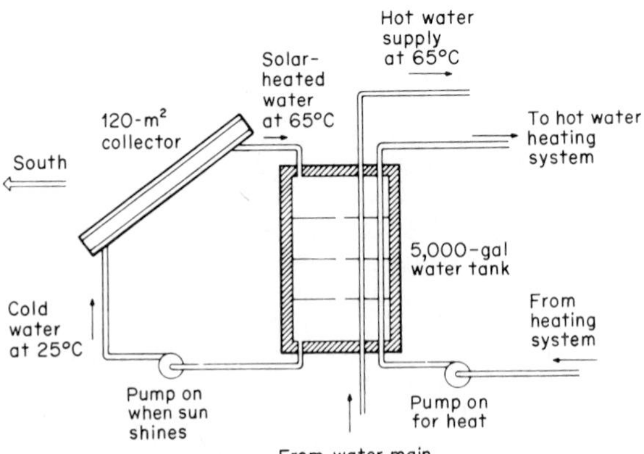

FIGURE 17.15 Schematic diagram of a domestic solar-powered heating system.

16. Magnetohydrodynamics. The basic concept for *magnetohydrodynamics (MHD)* is to bypass the conventional mechanical-to-electrical conversion step by making the working fluid an electricity conductor and forcing it, rather than a copper conductor, through a magnetic field. The fluid is made a conductor by burning any chemical fuel (coal, gas, or oil) at high temperature and "seeding" the combustion gases with an easily ionized material, such as potassium carbonate.

The conducting gases are expanded through a nozzle at high velocity and passed through a duct to which a strong magnetic field has been applied. Direct-current electricity is drawn off from electrodes lining the channel. The system has few moving parts and potentially high efficiency.

MHD units are expected first to be used to supply peaking and emergency power. They are expected to compete economically with gas turbines currently used for these purposes. Both are cheap to install, reliable, and capable of quick startup.

As MHD technology advances, it is expected to be combined with conventional fossil-fired steam plants to improve efficiency as a result of higher working fluid temperatures. This application, called a *topping cycle*, would first pass the seeded combustion gases through the MHD unit and draw off electrical energy at temperatures above those that can be tolerated by turbine blades. The exhaust gases from the MHD generator would then serve as the heat source for steam production. The over-all efficiency of such units could be as high as 60 percent; conventional fossil-fired plants have efficiencies of about 40 percent.

MHD generators might also be used with fission or fusion systems. The technology is more difficult, however, and commercial applications are expected not to occur for several decades.

17. Fuel Cells. *Fuel cells* convert chemical energy directly into electrical energy with an overall efficiency of about 60 percent. The principle of operation is the same as that for a dry cell: A chemical reaction that produces electrons (oxidation) occurs at one electrode, and a reaction that consumes electrons (reduction) occurs at the other. This process produces a voltage between the electrodes. The fuel cell differs from the dry cell in that reactants and reaction products are supplied to and removed from the cell.

Fuel cells have been used extensively on spacecraft. They can be made to operate on a variety of fuels including natural gas and gasified coal. Cost per kilowatt is relatively independent of output; this feature makes them attractive for individual home units. At present, costs are not competitive.

A fuel cell can run in reverse; i.e., by supplying electricity to the cell, the reaction products can be converted back into reactants. An example is electrolysis of water to produce hydrogen and oxygen. With this mode of operation, the fuel cell could be a device for energy storage; e.g., it could store converted solar energy.

CHAPTER 18
AGRICULTURAL, FOOD, AND NUTRITION ENGINEERING

AGRICULTURAL ENGINEERING

Agricultural engineering deals with the production of food and fiber using the engineering and biological sciences. The agricultural engineer deals with mechanical engineering (farm machinery, irrigation systems, etc.), civil engineering (farm buildings, greenhouses, barns, etc.), sanitary engineering (waste disposal, sewage, water supplies, etc.), chemical engineering (produce drying, storage, processing, etc.), electrical engineering (motors, transformers, lighting, etc.), environmental engineering (control of the water, soil, and air in agriculture, etc.), and biological materials, systems, and processes (food planting, culture, harvesting, etc.). As such, the agricultural engineer must have a broad knowledge of the four basic engineering disciplines, plus a good understanding of the biological sciences as they relate to agriculture.

With a major rise in the world's population, accompanied by a reduction in the amount of land available for agricultural use, the agricultural engineer is faced with major tasks today. These tasks revolve around resource management, produce processing, livestock farming, environmental control, and forest products. Each of these tasks deserves a quick review in terms of the agricultural engineer's specialty.

In the areas of the four major disciplines of engineering (mechanical, electrical, civil, and chemical), the agricultural engineer will use the principles discussed elsewhere in this book. For example, where irrigation is concerned, the agricultural engineer will follow the general principles of pumping and piping as developed by mechanical engineers. There is essentially no difference between a pump used for irrigation and the piping attached to it and the pump and piping used in other low-pressure cold-water supply systems. It is in the major tasks listed in the previous paragraph that the agricultural engineer's work differs. The differences are mentioned below for each of the major tasks.

RESOURCE MANAGEMENT

The agricultural engineer works with two basic resources—water and soil. Since these resources are limited, the agricultural engineer works to control soil erosion, to control the moisture content of soil for crop growing, to control floods in low-lying lands, to prevent pollution of local water supplies, and to provide potable water for human cooking and consumption.

And since water shortages plague almost all the world at various times, the agricultural engineer is closely involved with water conservation. Without water,

Excerpted from the *McGraw-Hill Encyclopedia of Science and Technology, 6th ed.* Copyright © 1987. Used by permission, of McGraw-Hill, Inc. All rights reserved.

plant growth is nonexistent. Thus, the agricultural engineer is constantly faced with the problem of getting water to where the crops need it.

Inefficient resource management reduces crop output. This, in turn, acts as a drag on the economic development of a nation. Witness the socialist world with its never-ending food problems traceable to inefficient agricultural production. Imports of basic foodstuffs are necessary to feed the population of such nations. Food and fiber production far beyond a nation's needs are common in the capitalistic countries, where there is a well-developed agricultural engineering community bent on efficient resource management.

PRODUCE PROCESSING

A great many tasks are necessary after a crop is harvested. These tasks are ones the agricultural engineer faces every working day. For example, grains of various types must be dried and stored after harvest. The drying equipment can be extensive, as are the storage equipment and facilities.

In the fruit and vegetable farming community, it is necessary to wash, grade, and store the produce. Here, again, the processing equipment is extensive and can include controlling humidity and temperature in the facility.

Where fibers are produced, such as cotton, the agricultural engineer faces other tasks. For instance, cotton ginning requires extensive equipment and a controlled atmosphere.

And where food is processed before delivery to the consumer, more work is required of the agricultural engineer. Other aspects of produce processing are included later in this chapter under the heading of "Food and Nutrition Engineering."

LIVESTOCK FARMING

This branch of agricultural engineering is responsible for poultry, dairy, and meat production. Recent shifts to extensive automation of the feeding and milking activities have brought the agricultural engineer face to face with a variety of modern techniques. Further, the buildings housing the animals are now climate-controlled to improve productivity. Coupled with these advances are the materials handling and storage requirements of these facilities. Today's livestock farming is highly mechanized, but further advances are certain to come in the future.

ENVIRONMENTAL CONTROL

The ideal growing environment would be one in which there was predictable sunshine and rain, free of droughts, pests, frosts, and other of nature's adversities for agriculture. One of the goals of agricultural engineers is to provide such an ideal growing environment for today's farmers. Work is being done on the air, water, and soil aspects. Environmentally controlled chambers are being studied, along with greenhouses.

Other approaches involve special tillage methods, new mulches, and control of the air temperature and moisture content to increase plant growth. Likewise, irrigation and drainage are being studied to see how they can be improved to increase plant growth rates.

In the area of livestock farming, new attention is being given to the environment in buildings housing animals. Agricultural engineers are studying the temperature, humidity, and lighting conditions in such buildings to determine the best levels for maximum production. The most economic delivery systems for these services are also being analyzed.

FOREST PRODUCTS

Today more forests are being approached from the trained view of the agricultural engineer. With demand for forest products high, there is a need for the design and development of equipment for the production, harvesting, and handling of forest products before they are processed.

FOOD AND NUTRITION ENGINEERING

Food engineering is the technical discipline involved in food manufacturing and refined foods processing. It encompasses the practical application of food science in the efficient industrial production, packaging, storing, and physical distribution of nutritious and convenient foods that are uniform in quality, palatable, and safe. Controlled biological, chemical, and physical processes and the planning, design, construction, and operation of food factories and processes are usually involved.

Food engineering is the food-industry equivalent of chemical engineering. Food science in industry converts agricultural materials into products that are marketable because they meet a consumer need and can be profitably sold at reasonable prices by virtue of being economically produced, packaged, and distributed.

Food engineering is a vital link, therefore, between farms and food stores in the lifeline of modern civilization. Without it, food would be available only at farms, in forms produced by nature, and only in season.

Many natural foods must be preserved in order to be stored and shipped long distances to population centers. To provide a varied, nutritious diet, many raw food materials must be refined, and others combined with various ingredients and processed to produce food in new forms, such as bread, cheese, ice cream, frankfurters, soft drinks, cake and dessert mixes, candy, many breakfast cereals, ketchup, and salad dressings.

Because food engineering is applied in food manufacturing and refined foods processing, it requires a knowledge of unit operations and processes such as cleaning, separating, mixing, forming, heat transfer, moisture removal, fermenting, curing, packaging, and materials handling. This requirement to a large extent differentiates food engineering from food science. Yet these operations involve applied food science, just as chemical engineering encompasses chemistry. That is why the food engineer must have a working knowledge of food chemistry, bacteriology, and industrial microbiology, as well as of physics, mathematics, and basic engineering disciplines.

Food engineering should be differentiated from food technology. In teaching technology, some schools include considerable engineering in the curriculum to train students in the practical application of food science. Graduates of these schools are in reality food engineers. Other schools emphasize food science, and qualify graduates primarily for research and quality control work. The food technologists from these schools actually are food scientists.

Food engineers are hired not only by commercial food manufacturing and processing firms, but also by food machinery and ingredient companies, makers of containers and packaging materials, government food control and research agencies, food laboratories, and schools teaching food engineering, technology, and science while conducting research on food and nutrition.

Food engineering has contributed to ever-accelerating progress in food manufacture beyond the pot and kettle stage of batch operations of large-scale kitchen techniques. It has developed many automatically controlled, continuous processes which are more efficient, provide better quality control, and are more sanitary. It has developed food manufacturing and processing into a highly technical industry.

The food industry faces the challenge of practically taking the homemaker out of the kitchen by providing economical products that require almost no effort or time in preparation. Products which can be made ready to eat by being heated in their packages go a long way toward attaining this goal.

Some outstanding achievements in food engineering include continuous bread-dough making and forming, manufacture of low-cost, high-quality prepared mixes, development of instant coffee and tea processes, dehydration of potatoes to produce the instant mashed product, production of precooked frozen convenience foods, continuous butter churning, freeze-drying or sublimation, final dehydration of solid foods after puffing to open the structure, preservation of beer and wine by micropore filtration to remove yeasts and spoilage bacteria, pneumatic bulk handling of dry and liquid raw materials, aseptic filling of packages, and automatic control of processes.

FOOD ENGINEERING TECHNIQUES

The food engineer uses many of the techniques of the mechanical, chemical, and electrical engineer. And where new or rehabilitated structures are used to house food processing or manufacturing operations, the food engineer will use the techniques of the civil engineer.

Typical techniques the food engineer will use include food refining, mechanical unit operations, chemical unit processes, sanitation, quality control, waste management, materials handling, process control, cleaning, separating, draining, trimming, peeling, dehusking, silking, cutting, shelling, stemming, pitting, coring, etc. Each of these involves techniques adapted from another branch of engineering. Examples of a number of these techniques are briefly discussed below.

FOOD REFINING

Processing is used to refine foods. Thus sugar is obtained by refining. Cane sugar is obtained by squeezing the juice from cane. The cane sugar is clarified and concentrated in vacuum evaporators. Crystals formed in the concentrate are separated from the liquor by centrifuging.

Other examples of food refining include pasteurization of milk, obtaining oil from soybeans and corn, and wheat milling to produce flour. Examples of refined foods include dried and frozen eggs, spices and flavorings, edible fats and oils, syrup, molasses, honey, starch, gelatin, cocoa, coffee, tea, rice, etc. A wide variety of continuous, automatically controlled unit operations and processes is used to improve quality, reduce costs, and save labor.

Many food-refining plants are located near the area in which raw materials for them are produced. Thus there are soybean plants in Illinois, cane sugar factories in Louisiana, and flour mills in Minneapolis. If the product is perishable—such as bread and milk—the plant is normally located close to the market.

UNIT OPERATIONS AND PROCESSES

In food factories, *unit operations* are the mechanical handling and manipulations used to change the physical form or composition of the food, to move it from one process or operation to another, and to package it.

Unit processes are the methods used to change the chemical or biological characteristics of the food, to preserve it (as in curing meat), to make it more palatable (as in aging cheese), or to develop special qualities (as in fermenting wort to produce beer by developing alcohol and carbon dioxide).

SANITATION

In the food industry, *sanitation* is the planned control of the production environment, equipment, and personnel to prevent or minimize spoilage, product contamination, and conditions offensive to the aesthetic senses of the discriminating consumer and to provide clean, healthful, and safe working conditions.

Some of the broad areas of sanitation concern in food engineering are as follows: *Housekeeping* implies orderliness and freedom from refuse in all areas. *Rodent elimination* involves knowledge of rodent habits, recognition of problems, and permanent control through structural changes, removal of harborages and food supplies, and supplementary poisoning and trapping. *Insect elimination* from food products and ingredients in the factory requires recognition of serious or incipient infestations, identification, and knowledge of habits and ecology. Control methods may involve changes in structure, equipment, or process, and the safe use of insecticidal chemicals.

Microorganisms, the type and significance of which vary with product and type of operation, must often be controlled by process and equipment change, cleaning, and sanitizing chemicals.

Construction and maintenance of buildings and equipment are of major importance in sanitation. New units can be planned to simplify sanitation maintenance, reduce costs, and eliminate the hazards of contamination and spoilage.

Cleaning of plant and equipment involves careful organization, training, work scheduling, and the use of the best available equipment, methods, and materials. The trend is to clean processing equipment in place, without dismantling. This is done by an automatic system that circulates and sprays cleaning and sanitizing solutions inside equipment in timed sequence.

Employee facilities, such as rest rooms, locker rooms, drinking water, eating facilities, and working environment, must be well maintained for the comfort and safety of the workers if they are to remain happy and maintain production efficiency and product quality.

Laboratory tests, of importance to the sanitation program in the food plant, must be understood to be utilized to the best advantage.

Water supply quality and plant distribution systems, as well as waste treatment and disposal, and lighting and ventilation, are often a part of sanitation.

Inspection techniques, tailored to the specific sanitation situation, must be learned, taught, and applied for efficient functioning and adjustment of the sanitation program.

QUALITY CONTROL

In food engineering, *quality control* is the evaluation of raw materials, unit operations, unit processes, or finished products and comparison of the results with fixed standards. These standards may reflect the manufacturers' or the customers' viewpoints. They may be based on physical properties, such as size and color; chemical properties, such as acidity; sensory attributes, such as odor and flavor; legal requirements, such as net weight; or on public health standards of microbial content. When possible, quality control depends on objective physical or chemical tests, but for foods these are usually supplemented by a panel of trained tasters. The tests are applied according to a statistical design, following an analysis of the specific problem. Frequency of sampling and analytical accuracy required are related to the degree of quality control desired. Statistical analysis of data previously obtained defines the limits beyond which a product is rejected, an operation readjusted, or a process changed. In addition, good quality control can show trends that indicate changes should be made before any losses have actually occurred. Care should be taken that the quality control program does not become so detailed and costly that it reaches the point of diminishing return. Quality control is essential for the product uniformity that enables mass-produced foods to be advertised, distributed, and sold throughout the world.

WASTE MANAGEMENT

Waste disposal is a matter of great economic and technical concern for most food-processing operations, as well as other industries. Both solid and liquid wastes are encountered in the food-processing industry. The disposal of these wastes is regulated by the Clean Water Act of 1977 (as amended) and related environmental protection legislation, which govern the types and amounts of wastes discharged into the environment. As environmental regulatory agencies increase their demands for less waste discharge, it becomes essential that cost-effective methods of waste disposal be found.

Waste management is the newer approach to cost-effective food-processing waste disposal. Through waste management, modifications are applied to food plant operations and manufacturing processes. These modifications reduce the amount of solid and liquid wastes, recover more product and by-products, often reduce energy consumption, and exhibit other benefits. In general, the principle is to convert waste liabilities into profitable assets.

One major objective of waste management is to eliminate or at least lessen the dependence upon end-of-the-pipe sanitary engineering methods. This is achieved by reducing both the amount of waste solids generated and the volume of wastewater discharged.

The following are examples of modifications which can be made to food plant operations: incorporating good housekeeping practices; collecting culls and other solid wastes into containers rather than discharging to the floor drain; recycling water; reusing spent process water in another plant operation; and using less or no water in plant operations that formerly used a fair to a large amount of water.

Good housekeeping practices that reduce water usage and wastes require good personnel management and employee awareness of conservation practices. Such practices as needless use of water or overloading of containers, thereby causing spillage, should be discouraged.

Recycling water in the same plant operation can be achieved by treating spent process water with activated charcoal or sand filters or by ion-exchange columns, chemical treatment, pH adjustment, temperature adjustment, pasteurization, or a combination of these and other methods. Spent olive, pickle, and cherry brines have been treated and recycled to start new batches of the product. Recycling of water has been used in other food plant operations.

Countercurrent water reuse systems can be established in many plant operations. For example, potable water used to rinse the finished product prior to packaging can be used a second or third time to wash the product during intermediate stages of processing and to wash the incoming commodity. Spent wash water can be used again to initiate washdown of dirty floors or to flume solid wastes away from the process line.

Sanitary engineering methods will still be applied to many food-processing waste effluents. However, on-site installations can be modified to recover solid wastes for by-product development. Some food processors are processing into animal feed sludge recovered from activated sludge treatment of food wastes. Treated wastewater can be used for crop irrigation or for recycling as process water.

MATERIALS HANDLING

Materials handling is the in-process and in-plant handling or conveyance of raw, semifinished, and finished materials to storage and point of shipment. Improved handling systems present the greatest potential means for cutting production costs, increasing production within present plant areas, and providing a smooth, continuous flow-through process to storage or shipment areas. Depending on the characteristics of a product, it may be conveyed by semimechanical, mechanical, gravity, hydraulic, or pneumatic systems, or a combination of these.

In-process handling systems provide the conveying linkage between the unit operations of a process. With a few exceptions, such as monorail equipment used for meat carcass handling, most systems are designed for mass movement of materials. In all cases, every consideration is given to the characteristics of the product, and systems are designed to handle materials with minimum damage, move them swiftly minimum distances, and provide sanitary requirements.

In-plant handling systems include the conveyance of raw materials from receiving platforms and the handling and storage of finished goods. The mechanical means selected for these purposes are based on considerations of the characteristics of the raw materials, and the container and shipping package of the finished product.

PROCESS CONTROL

In food engineering, the techniques used in process control are similar to those discussed elsewhere in this handbook. The handbook user should refer to the index and table of contents for specific locations of the sections on automatic control.

CHAPTER 19
RELIABILITY ENGINEERING, SYSTEMS ENGINEERING, AND SAFETY ENGINEERING

RELIABILITY ENGINEERING*

Reliability is the characteristic of a component, or system made up of many components, expressed by the probability that it will perform its particular function within a specific environment for a given period of time. Since the subject of reliability is obviously concerned with statistical analysis and the prediction of behavior based on tests, it might be considered, at first glance, to be a matter of guesswork or chance. However, reliability predictions have become a precise branch of industrial technology. Reliability engineering plays an invaluable part in the reduction of costly failures and the correct planning of overhaul and maintenance schedules.

TYPES OF FAILURES

Failure is defined as the inability of a component or system to carry out its specified function. Failures may be categorized in a number of ways according to the degree of failure, the reason for failure, the timing of the failure, and so on. The following are some relevant definitions:

Misuse failure is used to describe failure due to overloading or otherwise overstressing a component or system beyond its capability.

Inherent-weakness failure is used to describe failure due to inherent weakness of the component or system and occurring while the item is being correctly used.

Sudden failure is used to describe failures which could not have been anticipated.

Gradual failure is used to describe failures which could have been anticipated.

Partial failure is one in which the component or system may still function, but not to the limits of performance originally designed.

Complete failure results in total loss of the required function.

Catastrophic failure is one which is both sudden and complete.

Degradation failure is one which is both gradual and partial.

*This section is drawn from *An Introduction to Reliability Engineering*, by R. Lewis. Copyright © 1970. Used by permission of McGraw-Hill, Inc. All rights reserved. Updated by editors.

Chance failures is a term used generally to describe those failures which occur suddenly and at random during the anticipated useful life of a component or system. They may be due to a variety of causes, including inherent weakness, misuse, etc.; they are not due to the component having completed its normally anticipated useful life, i.e., to the component wearing out.

Wearout failure is another general term to describe failure due to the wearing out of a component which has more or less completed its anticipated useful life.

Both chance and wearout failures may be partial or complete.

FAILURE RATE

The number of failures occurring per unit time is known as the *failure rate*. As with all quantities describing change (speed, acceleration, etc.), an average value may be obtained by dividing the total number of failures which have occurred during a time interval by the length of the interval. The shorter the interval, the nearer the average value gets to the instantaneous failure rate. The *instantaneous failure rate* at any one time is the slope of the curve plotting failures against time at that particular time.

If, in determining the failure rate, the number of failures occurring during the time interval is expressed as a proportion of the number of survivors at the beginning of the time interval, then the failure rate obtained is called the *proportional failure rate*. It is denoted by the symbol λ. Unless otherwise stated the words *failure rate* within this handbook imply the proportional failure rate.

Example 19.1. If out of 1000 components 10 fail during a period of 5000 hours, then

$$\text{Proportional failure rate} = \frac{10}{1000} \times \frac{1}{5000}$$

that is,

$$\lambda \doteq 2 \times 10^{-6} \text{ failures of the total/h}$$

and

$$\text{Percentage failure rate} = \frac{10}{1000} \times \frac{1}{5000} \times 100$$

$$= 0.0002 \text{ percent/h}$$

Failure rate is most commonly expressed as a percentage per 1000 hours. The preceding would then be 0.2 percent/1000 h.

THE BATHTUB DIAGRAM

A typical graph plotting the percentage failure rate with respect to time is shown in Fig. 19.1. It is often referred to as a *bathtub diagram* because of its shape.

During the burn-in period, a high failure rate exists, owing to the presence of substandard components in the sample tested. After the weak components have died out, the failure rate stabilizes at an approximately constant value; this period is called the *useful-life period*.

Eventually wearout failures begin to occur, and the failure rate rises again. During this wearout period, chance failures may, of course, still be occurring.

In order to make reasonably accurate predictions of reliability, failures due to chance and wearout must be studied and an analysis made of the times in relation to the type of failure.

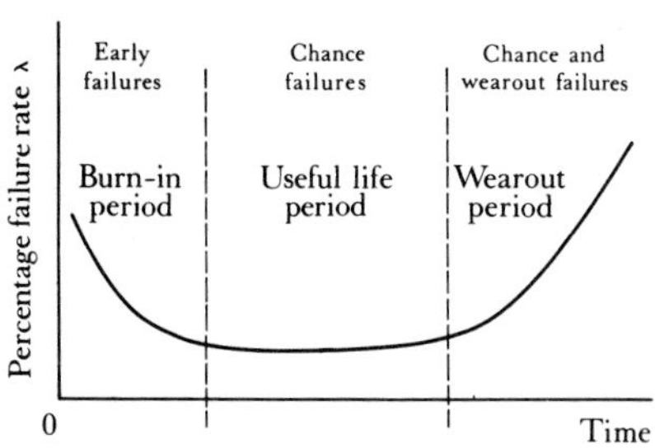

FIGURE 19.1 Bathtub diagram.

CONSTANT-FAILURE-RATE CASE

When the failure rate is constant, reliability prediction is made much easier mathematically, since it is possible to use exponential curves to assist analysis.

As was shown above, the failure rate may be assumed to be constant when failures are due to chance alone; this can be achieved by correct overhaul schedules, which eliminate wearout failure. It is also possible, however, to achieve a constant failure rate, and thus simplify the mathematics involved, by a process of immediate replacement on wearout. This latter case is not as obvious but has been conclusively demonstrated. It should be noted that replacement on failure is a procedure which cannot always be adopted, since certain systems or components cannot be allowed to fail even temporarily.

RELIABILITY EQUATIONS AND CURVES WHEN FAILURE RATE IS CONSTANT

The probability of no failures occurring in a given time can be expressed by the following equation, provided that the failure rate is constant:

$$R = e^{-\lambda t} \tag{19.1}$$

where R is the probability of no failures in time t (i.e., the reliability), e is the exponent 2.7183, and λ is the constant failure rate. (This is in fact the no-event term of the Poisson probability function.) The *unreliability* Q is defined as the probability of total failure. It follows logically that

$$R + Q = 1 \tag{19.2}$$

and that

$$Q = 1 - e^{-\lambda t} \tag{19.3}$$

where Q is the probability of total failure in time t. A graph of R and Q against time yields the familiar exponential curves shown in Fig. 19.2.

Notice that at time $t = 0$, $R = 1$, and $Q = 0$, at time $t = 1/\lambda$, $R = 0.37$, and $Q = 0.63$ (from tables of values of e raised to various powers). A graph of survi-

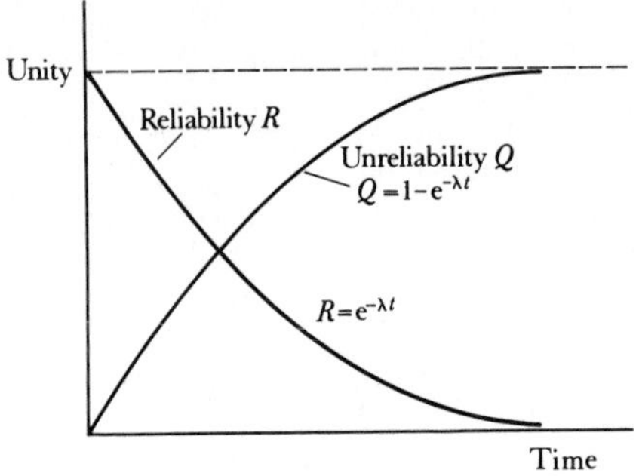

FIGURE 19.2 Reliability and unreliability curves.

vors, i.e., the number of components still alive at time t against time, will yield the same shape as the reliability curve in Fig. 19.2. A graph of failures against time yields the same shape as the unreliability curve (see Fig. 19.3). The equation of the graph of survivors versus time (survival curve) is

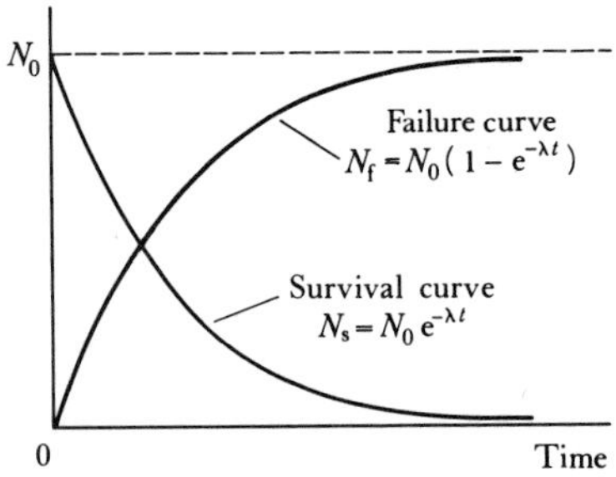

$$N_s = N_0 e^{-\lambda t} \tag{19.4}$$

where N_s is the number of survivors at time t, and N_0 is the original number in the test sample. Similarly, the equation of the failures versus time graph (failure curve) is

$$N_f = N_0 (1 - e^{-\lambda t}) \tag{19.5}$$

where N_f is the number of failures at time t, and N_0 is as before.

FIGURE 19.3 Failure and survival curves.

Using these curves and the preceding equations, the reliability or unreliability, number of survivors or failures, may be calculated at any time, provided that the failure rate λ is known (*and* constant, since the exponential analysis only applies when this is so).

FAILURE INTERVALS

1. Mean Time Between Failures (MTBF). The *mean time between failures* (MTBF) of a component or system is an extremely important characteristic in reliability predictions. It is defined as the mean or average time which elapses between failures, and it usually refers to a situation in which the failure rate λ is constant, i.e. due to chance failures or the adoption of the replacement-on-failure technique described earlier. The symbol for MTBF in these cases is m.

To appreciate the theoretical derivation of m, we must look more closely at the survival curve and the reliability curve and examine the significance of the area contained between the curves and the axes. First, consider a survival curve which is not exponential, such as the one shown in Fig. 19.4.

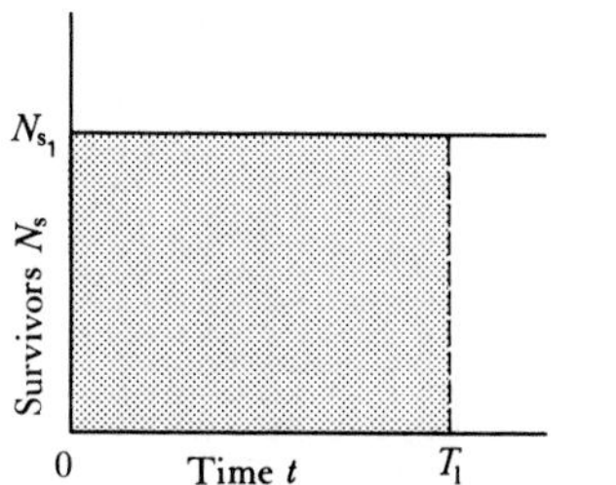

FIGURE 19.4 A nonexponential survival curve.

This is in fact a very simple function indeed and shows a constant number of survivors, i.e., a zero failure rate. At time T_1 shown, the survivors number N_{s_1}. Multiplying N_{s_1} by T_1 gives us the *survivor-hours*, i.e., the total number of hours of survival by all components. This value $N_{s_1}T_1$ is clearly the area bounded by the curve, the axes, and the line $t = T_1$. The area beneath a survival curve, whatever its shape, is in fact always equal to the total survival hours of all the components. To return to the exponential survival curve, which we have seen is obtained for a constant (nonzero) failure rate, the same principle applies. In this case, after a very long time (depending on the failure rate) all components N_0 have failed and the survival curve reaches the time axis. The area under the curve represents the total survival hours of all N_0 components. If we now divide this area by the total failures, that is, N_0, this gives the average time between failures, the MTBF m. The area under a nonlinear curve such as this is, of course, obtained by integration and in this case

$$\begin{aligned} \text{MTBF} &= \frac{1}{N_0}\int_0^\infty N_s dt \\ &= \frac{1}{N_0}\int_0^\infty N_o e^{-\lambda t}\, dt \\ &= \int_0^\infty e^{-\lambda t}\, dt \\ &= \left[-\frac{e^{-\lambda t}}{\lambda}\right]_0^\infty \\ &= 0 - (-1/\lambda) = 1/\lambda \end{aligned} \tag{19.6}$$

and we find that the MTBF is, in fact, the *reciprocal* of the failure rate.

Since the survival curve is the same curve as the reliability curve, except that the vertical axis of the reliability curve has been multiplied by N_0 to give the survivors axis of the survival curve, then it follows that the MTBF is in fact equal to the area under the reliability curve divided by unity, i.e., equal to the area itself.

Since

$$m = \frac{1}{\lambda}$$

the equations for reliability R and unreliability Q at any time can be rewritten as

$$R = e^{-t/m} \tag{19.7}$$

and

$$Q = 1 - R = 1 - e^{-t/m} \tag{19.8}$$

and the equations for the number of survivors N_s or failures N_f at any time can be written

$$N_s = N_0 e^{-t/m} \tag{19.9}$$

$$N_f = N_0(1 - e^{-t/m}) \tag{19.10}$$

where N_0 is the number of components alive at the start of the test.

Notice that since $m = 1/\lambda$, and since, as was shown earlier, when $t = 1/\lambda$ the reliability has fallen to 0.37 and the survivors to 0.37 N_0, the MTBF for the exponential case is the time at which

$$R = 0.37 \qquad Q = 0.63$$

$$N_s = 0.37N_0 \qquad N_f = 0.63N_0$$

2. Measurement of MTBF (Chance Failures). The two main types of tests for measuring m for chance failures are the nonreplacement and the replacement. The latter is seldom used because it involves constant observation to ascertain exact moments of failure. Nonreplacement methods require observation only at the beginning and end of the test time. To exclude wearout failures, the test is truncated (cut off) before the wearout probability is too high. It has been demonstrated by Epstein that the best estimate of m for a truncated test is given by

$$m = \frac{\text{test hours for failures + test hours for survivors}}{\text{number of failures}}$$

that is,

$$m = \frac{\text{total component test hours (survival hours)}}{\text{total number of failures}} \tag{19.11}$$

It will be appreciated that the figure obtained is only an estimate; the confidence which one can have in such an estimate may be determined using standard statistical methods.

For a constant failure rate, the MTBF also may be determined by finding the reciprocal of the failure rate, as shown above.

Worked examples on the determination of failure rate, MTBF, reliability, and survivors, etc., using the formulas given so far, are provided in the following pages.

Example 19.2. The telemetry transmission system of an earth satellite has an MTBF of 10,000 h. Estimate the probability of no failures during 100 90-minute orbits.

$$m = 10{,}000 \text{ h}$$

$$\text{Time} = 150 \text{ h}$$

Therefore, probability of no failure, i.e., the reliability, is given by

$$e^{-150/10\,000} = e^{-0.015} = 0.8607$$

There is therefore an 86 percent probability of 100 failure-free orbits.

Example 19.3. A certain electronic control system has a constant failure rate established at 0.2 percent per 1000 h. Determine the probability of 500 h of failure-free operation.

$$\text{Reliability for 500 h} = e^{-(2/10^6)500}$$
$$= e^{-0.001}$$
$$= 99.88 \text{ percent}$$

Notice that the failure rate λ, since it equals 0.2 percent per 1000 h, is equal to

0.2 per 100,000 h

or 2 per 1,000,000 h

that is $2/10^6$

and it is this figure which is inserted into the reliability equation. More worked examples are included starting on page 19.15.

3. Wearout Failures: Mean Wearout Life. As has been shown, chance failures are distributed exponentially, approximately 63 percent occurring before a time equal to the MTBF and approximately 37 percent occurring afterwards. Failures due to wearout, i.e., to the component having completed its anticipated useful life, are not distributed in this manner. A graph of wearout failures against time has the shape shown in Fig. 19.5.

This type of distribution is well known in statistical analysis; it is called the *Gaussian* or *normal* distribution and has certain defined characteristics which clearly distinguish it from other distributions.

The failures due to wearout are clustered about an average or *mean* value of time. Since each point on the curve corresponding to, say, x failures occurring at time t means that x of the components failed after a useful life t, the mean value M corresponds to the *mean wearout life* of the components. The individual lives of the components are scattered *normally* about the mean. Mean wearout life must not be confused with mean time between chance failures, discussed above. The MTBF (chance) tells us the average time anticipated between chance failures *during* the useful life of a component or system; the mean wearout life, on the other hand, tells the average value of the anticipated useful life assuming failure due to chance does not occur. Suppose, for example, a component has a value of M equal to 10,000 h; its mean time between chance failures (computed from a test which is truncated long before 10,000 h) may be as high as 100,000 h. These two figures indicate that provided the component is used within the useful life period, i.e., up to 10,000 h, the probability of chance failure using the formulas described above and a failure rate λ of $1/m$, that is, 1/100,000 or 0.00001, is quite low. After the anticipated useful life is over, the probability of failure rises very rapidly due to a very much increased fail-

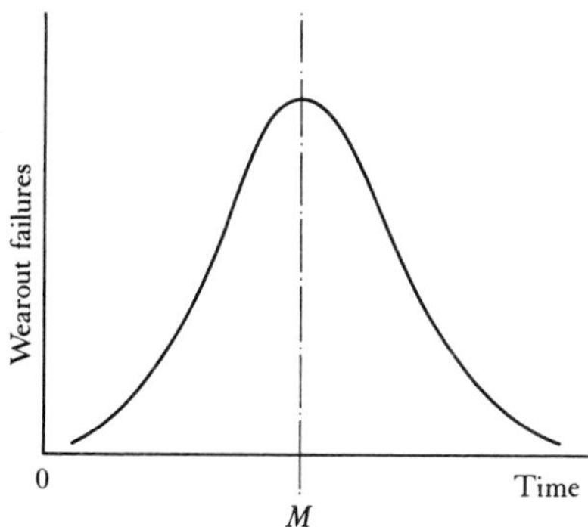

FIGURE 19.5 Wearout failure curve.

ure rate, which is, itself, due to wearout beginning to take place. Wearout and chance failures can be distinguished from one another by careful examination of the physical characteristics of the dead component.

An important parameter concerned with any normal distribution is the *standard deviation* σ. This is computed by finding the square root of the mean of the square of the deviations of the measured characteristic from the average value. That is,

$$\sigma = \sqrt{\frac{\text{sum of squares of deviations from average}}{\text{total number of observations}}} \tag{19.12}$$

or, for wearout tests, if the life of components 1, 2, 3,..., n is indicated by t_1, t_2, $t_3,\ldots,t_n$ and the total number used is n, then

$$\sigma = \frac{\sqrt{(t_1 - M)^2 + (t_2 - M)^2 + \cdots + (t_n - M)^2}}{n} \tag{19.13}$$

where M is the mean (wearout) life.

For a normal distribution of failures it can be shown that approximately 68 percent of the failures occur within a period $M \pm \sigma$, that is, between time $M - \sigma$ and time $M + \sigma$; approximately 95 percent occur within a period $M \pm 2\sigma$; and approximately 99.7 percent occur within a period $M \pm 3\sigma$. This is useful when establishing the confidence that one can have in estimates of wearout life (or indeed of any variable which has a normal distribution). See Fig. 19.6.

4. Measurement of Mean Wearout Life. For a wearout life test, a sample of components is put on test under the conditions they will experience in service, and the test is run until the components fail. Careful examination of both physical characteristics of the dead components and of their life length eliminates chance failures. (It is clear, for example, that if a group of components has lives centering around, say, 10,000 h, then a component surviving only 1000 h is probably not a wearout failure.)

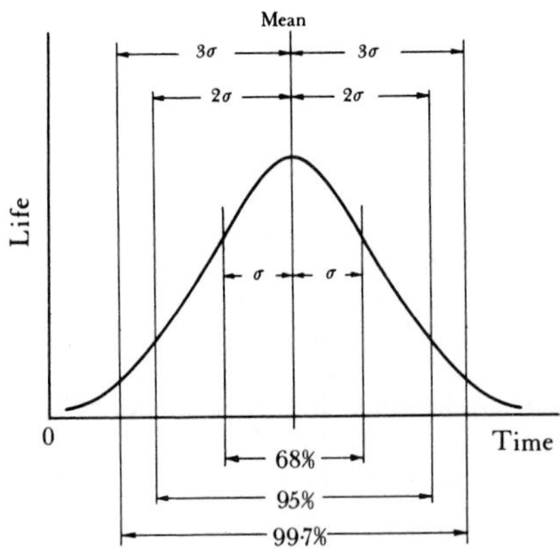

FIGURE 19.6 Percentage points on the normal distribution curve.

The mean life is then computed as follows:

$$M = \frac{\text{sum of lives of components}}{\text{number of components}} \tag{19.14}$$

The standard deviation may be determined using the equation already given.

SYSTEM RELIABILITY

1. Inclusion of Subunits in Systems: Intermittent Operation. A system very often contains subunits which are not required to function the whole time during the system operating period. In these cases, care must be taken when assessing the MTBF, failure rate, and reliability to take into account the unit operating time and the fact that it is not equal to the system operating time.

We shall consider the exponential case only since this is the most commonly found. Consider a component or subunit having a reliability R_c for t_c component operating hours. R_c is given by

$$R_c = e^{-t_c/m_c}$$

where m_c is the mean time between failure of the component expressed in component operating hours. If now this component is inserted in an exponentially failing system which is operating for t_s hours, of which the component is to operate only t_c hours, then the component reliability for t_s system operating hours will be the same as for t_c component operating hours. The reliability of the component, expressed as a function of system hours, is given by

$$e^{-t_s/m_s}$$

where m_s is the mean time between failure of the component expressed in system operating hours.

The reliability of the component is the same, however it is expressed, since it is operated for t_c hours whether alone or as a part of a system. Thus

$$R_c = e^{-t_c/m_c} = e^{-t_s/m_s}$$

and therefore

$$\frac{t_c}{m_c} = \frac{t_s}{m_s}$$

or

$$\frac{m_c}{m_s} = \frac{t_c}{t_s} \tag{19.15}$$

The ratio t_c/t_s expresses the fraction of system time during which the component is required to operate and is called the *duty cycle* of the component, indicated by d. Thus we have

$$m_s = \frac{m_c}{d} \qquad \text{where } d = \frac{t_c}{t_s} \tag{19.16}$$

Illustrating this numerically, suppose we have a component operating for 5 h in a system time of 25 h and having a reliability of 0.9. MTBF expressed in component hours m_c is given by

$$0.9 = e^{-5/m_c}$$

so that

$$e^{5/m_c} = \frac{1}{0.9}$$

Thus

$$\frac{5}{m_c} \ln e = \ln \left(\frac{1}{0.9}\right)$$

and

$$m_c = \frac{5}{\ln (1/0.9)}$$

$$= 47.5 \text{ h}$$

Thus we may expect on average a component failure for every 47.5 h of its operation.

The duty cycle = $5/25$ = $1/5$, so that the MTBF expressed in system operating hours m_s is given by

$$m_s = 5 \times 47.5$$

$$= 237.5 \text{ h}$$

On average, a component fails for every 237.5 hours of operation of the system. This is quite logical, since for 237.5 hours of system operation the component operates only $1/5$ of this, i.e., 43.5 hours, its MTBF. The preceding assumes zero failure probability as the component is switched into and out of use. If there is such a probability, this must, of course, be taken into account when assessing the system MTBF.

2. System Reliability. In calculating the reliability of a system made up of a number of components or individual complete units, each having their own reliability, the *type* of system must first be determined. There are two types—series systems and parallel systems.

A *series system* is one in which failure of one of the subunits or components means failure of the system as a whole. A *parallel system* is one which does not fail until all subunits or components have failed.

It will be remembered that reliability is the probability of survival; in computing system reliability, various well-established laws of the mathematics of probability are used. The relevant laws are discussed below.

3. Laws of Probability Relevant to Reliability Calculations

1. If X and Y are two independent events and P_x is the probability that X will occur and P_y is the probability that Y will occur, then the probability that *both* events X and Y will occur P_{xy} is given by

$$P_{xy} = P_x P_y \tag{19.17}$$

2. If the two events can occur simultaneously, the probability that either X or Y or both X and Y will occur P_{x+y} is given by

$$P_{x+y} = P_x + P_y - P_xP_y \tag{19.18}$$

Both these laws may be extended to cover more than two events.

4. Reliability of a Series System. The reliability of a series system or probability of survival of the system is the probability of *all* the components surviving, since a failure of only one component means overall system failure.

If R_s is the system reliability and R_1, R_2, etc. are the reliabilities of the system components for the same time period, then using law 1,

$$R_s = R_1R_2 \tag{19.19}$$

In this case, the "event" described in law 1 is survival. This expression is called the *product law of reliabilities* for series systems. For a system having more than two components, the additional reliabilities, R_3, R_4, etc., are included in the product term.

5. Unreliability of a Series System. The unreliability of a series system or probability of failure of the system is the probability of at least *one* of the system components failing.

If Q_s is the system unreliability and Q_1, Q_2, etc. are the unreliabilities of the system components for the same time period, then using law 2,

$$Q_s = Q_1 + Q_2 - Q_1Q_2 \tag{19.20}$$

for a system having two components. It will be recalled that unreliability = 1 − reliability, that is, in general, $Q = 1 - R$, so that

$$Q_1 = 1 - R_1$$

$$Q_2 = 1 - R_2$$

and substituting in the equation for Q_s, we see that

$$\begin{aligned} Q_s &= (1 - R_1) + (1 - R_2) - (1 - R_1)(1 - R_2) \\ &= 1 - R_1 + 1 - R_2 - 1 + R_1 + R_2 - R_1R_2 \\ &= 1 - R_1R_2 \\ &= 1 - R_s \end{aligned} \tag{19.21}$$

which is to be expected. In this case, the "event" described in law 2 is failure.

The unreliability expression for a system having more than two components is more complex than that above, but using $Q_s = 1 - R_s$, the reliability may first be determined using Eq. (19.19) and then the unreliability determined using Eq. (19.21).

6. Reliability of a Parallel System. The reliability of a parallel system or probability of survival of the system is the probability of at least *one* component surviving, since, provided that at least one component survives in a parallel system, the system will not fail.

If R_p is the system reliability and R_1, R_2 are the component reliabilities, then using law 2,

$$R_p = R_1 + R_2 - R_1R_2 \tag{19.22}$$

In this case, the "event" described in law 2 is survival. For a system having more than two components, the reliability expression is more complex, but as we shall see, an easy method of calculation is to determine the unreliability first and use Eq. (19.24) given below.

7. Unreliability of a Parallel System. The unreliability of a parallel system or probability of system failure is the probability of *all* components failing, since for a parallel system, even if only one component survives, the system has not failed.

If Q_p is the system unreliability and Q_1 and Q_2 are the component unreliabilities, then using law 1,

$$Q_p = Q_1Q_2 \tag{19.23}$$

In this case, the "event" described in law 1 is failure. This expression is called the *product law of unreliabilities* for parallel systems. For a system having more than two components, the additional unreliabilities are included in the product term.

Since $Q_1 = 1 - R_1$ and $Q_2 = 1 - R_2$,

$$\begin{aligned} Q_p &= (1 - R_1)(1 - R_2) \\ &= 1 - R_1 - R_2 + R_1R_2 \\ &= 1 - (R_1 + R_2 - R_1R_2) \\ &= 1 - R_p \end{aligned} \tag{19.24}$$

which is to be expected. This equation is most useful for determining the system reliability R_p after having found the system unreliability Q_p. It is easier to compute Q_p than R_p in the first instance, because Q_p is contained in the product law of unreliabilities while the expression for R_p becomes increasingly complex as the number of subunits is increased.

8. Systems Containing Exponentially Failing Units. Whether or not a system made up of units which are failing exponentially behaves overall in an exponential fashion depends on the system type. It is found that series systems do behave exponentially and their reliability may be expressed in the familiar $e^{-\lambda t}$ form, whereas parallel systems do not and the form of their reliability equation depends on the number of subunits.

9. Series Systems. The reliability equation as shown above is

$$R_s = R_1R_2R_3 \cdots$$

where R_s is the system reliability and R_1, R_2, R_3, etc. are the subunit reliabilities.

If $R_1 = e^{-\lambda_1 t}$, $R_2 = e^{-\lambda_2 t}$, $R_3 = e^{-\lambda_3 t}$, etc., where λ_1, λ_2, λ_3, etc. are the respective failure rates of the subunits, then

$$\begin{aligned} R_s &= e^{-\lambda_1 t} e^{-\lambda_2 t} e^{-\lambda_3 t} \\ &= e^{-(\lambda_1 + \lambda_2 + \lambda_3)t} \end{aligned} \tag{19.25}$$

which is, of course, of general exponential form, the system failure rate being the *sum* of the individual failure rates. The system failure rate clearly increases as the number of subunits is increased.

Since the overall behavior is exponential, that is, we can express the reliability R_s in the form

$$R_s = e^{-\lambda_s t}$$

where λ_s, the system failure rate, is given by

$$\lambda_s = \lambda_1 + \lambda_2 + \lambda_3 \tag{19.26}$$

then the MTBF for a series system m_s is equal to the reciprocal of the system failure rate:

$$m_s = \frac{1}{\lambda_s} = \frac{1}{\lambda_1 + \lambda_2 + \lambda_3} \tag{19.27}$$

This reciprocal equation applies only to components or systems whose reliability is expressible in exponential form, i.e., having constant failure rate.

For a series system containing n similar units of equal reliability,

System reliability

$$R_s = e^{-n\lambda t} \tag{19.28}$$

System failure rate

$$\lambda_s = n\lambda \tag{19.29}$$

System MTBF

$$m_s = \frac{1}{n\lambda} \tag{19.30}$$

10. Parallel Systems. As was stated earlier, the form of the reliability expression for a parallel system depends on the number of subunits. For a simple two-unit system,

$$R_p = R_1 + R_2 - R_1R_2 \tag{19.31}$$

For a three-unit system,

$$R_p = R_1 + R_2 + R_3 - R_1R_2 - R_2R_3 - R_1R_3 + R_1R_2R_3 \tag{19.32}$$

And the expression becomes more complex as the number of units is increased.

If we write $R_1 = e^{-\lambda_1 t}$, $R_2 = e^{-\lambda_2 t}$, etc., as we did for series systems, then the system reliability for a two-unit system is given by

$$R_p = e^{-\lambda_1 t} + e^{-\lambda_2 t} - e^{-(\lambda_1 + \lambda_2)t} \tag{19.33}$$

For a three-unit system, by

$$\begin{aligned} R_p = {} & e^{-\lambda_1 t} + e^{-\lambda_2 t} + e^{-\lambda_3 t} \\ & + e^{-(\lambda_1 + \lambda_2)t} + e^{-(\lambda_2 + \lambda_3)t} \\ & + e^{-(\lambda_1 + \lambda_3)t} + e^{-(\lambda_1 + \lambda_2 + \lambda_3)t} \end{aligned} \tag{19.34}$$

Clearly, these equations are not of simple exponential form, and we cannot express the overall system reliability in the form $e^{-\lambda_p t}$, where λ_p is a constant system failure rate. The MTBF may still be obtained by integration of the reliability expression, as was done in the series case, since, as was shown earlier, this method of finding the MTBF does not rely on the expression being of exponential form. However, for a parallel system, the MTBF is not the reciprocal of the system failure rate, but depends on the number of subunits.

It can be shown that the MTBF m_p for a two-unit system is given by

$$m_p = \frac{1}{\lambda_1} + \frac{1}{\lambda_2} - \frac{1}{\lambda_1 + \lambda_2} \tag{19.35}$$

For a three-unit system, it is given by

$$m_p = \frac{1}{\lambda_1} + \frac{1}{\lambda_2} + \frac{1}{\lambda_3} - \frac{1}{\lambda_1 + \lambda_2} - \frac{1}{\lambda_2 + \lambda_3} - \frac{1}{\lambda_1 + \lambda_3} + \frac{1}{\lambda_1 + \lambda_2 + \lambda_3} \tag{19.36}$$

where λ_1, λ_2, and λ_3 are the unit failure rates, respectively, and for an n-unit system, each unit having the same

$$m_p = \frac{1}{\lambda} + \frac{1}{2\lambda} + \frac{1}{3\lambda} + \cdots + \frac{1}{n\lambda} \tag{19.37}$$

The system failure rate for parallel system is not constant but is time-dependent.

11. Series and Parallel Systems: Reliability Compared. In both system types, the overall reliability is dependent on the number of subunits. To illustrate the effect of the series and parallel connection of the system subunits, consider again the reliability expressions. For a two-unit system having unit reliabilities R_1, R_2,

$$\text{Series system reliability} = R_1R_2$$

$$\text{Parallel system reliability} = R_1 + R_2 - R_1R_2$$

Bearing in mind that reliability can never exceed unity, examination of the two expressions shows that

$$R_1 + R_2 - R_1R_2 \geq R_1R_2$$

the two being equal only when $R_1 = R_2 = 1$, at all other values of R_1 and R_2 the left-hand side being greater than the right-hand side. This means, then, that the reliability for a two-unit parallel system is greater than that for a two-unit series system except when the subunits have equal reliabilities equal to unity. In this case, series and parallel systems will have equal reliabilities. A similar observation may be made concerning reliabilities of three-unit systems:

$$\text{Series system reliability} = R_1R_2R_3$$

$$\text{Parallel system reliability} = R_1 + R_2 + R_3 - R_1R_2 - R_2R_3 - R_1R_3 + R_1R_2R_3$$

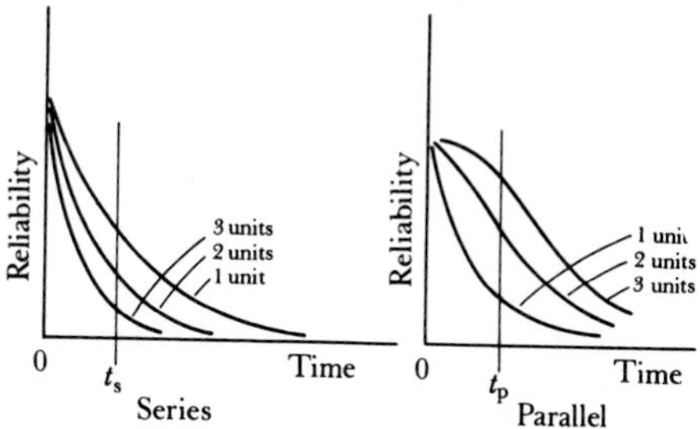

FIGURE 19.7 Reliability curves for series and parallel systems.

Clearly the first six terms of the parallel reliability expression must have an overall value of zero or greater than zero, since R_1, R_2, or R_3 must all be unity or less than unity. The seventh term is, in fact, the series reliability, so we have

$$\text{Parallel reliability} = (\text{terms} \geq 0) + (\text{series reliability})$$

which is obviously greater than or equal to the series reliability. It can be shown that for any number of units the parallel reliability is greater than or equal to the series reliability. Physically, this is logical, since for a parallel system, all units except one can fail before system failure, whereas for a series system, only one unit needs to fail for system failure. Typical sets of reliability curves for one- , two- , and three-unit systems in both connection modes are shown in Fig. 19.7.

Notice that for a series system, the reliability at any time t_s is reduced as the number of subunits is increased. The reliability curve is always exponential regardless of the number of subunits. For a parallel system, the reliability at any time t_p is increased as the number of subunits is increased. For systems containing more than one unit, the reliability curve ceases to be exponential, as was explained earlier.

This improved reliability for parallel-connected systems is used in schemes to reduce risk of failure, such as standby systems, in which several equal units are allowed to stand idle ready to take over should the operational unit fail. Such schemes are said to employ *parallel redundancy*.

WORKED EXAMPLES

A summary of equations appears on pages 19.21 to 19.24; references in these examples are to this set.

Example 19.4. In a test to determine the MTBF of a certain component, 100 were tested for a period of 4000 h. The times to failure of the components are as shown in Table 19.1. Assuming that wearout failure can be ignored, calculate the failure rate and the MTBF.

$$\begin{aligned}\text{Total survival hours} &= 250 + 300 + 4 \times 415 + 5 \times 800 + 3 \times 1200 + 86 \times 4000 \\ &= 550 + 1660 + 4000 + 3600 + 344{,}000 \\ &= 353{,}810 \text{ h}\end{aligned}$$

$$\text{Total failures} = 14$$

TABLE 19.1

Number of components	Time to failure, h
1	250
1	300
4	415
5	800
3	1200
86	No failure

Using Eq. (19.48),

$$\text{MTBF} = \frac{353{,}810}{14}$$
$$= 25{,}270 \text{ h}$$

Since we can ignore wearout failure, we can assume an exponential relationship and hence that the failure rate is the reciprocal of the MTBF. Thus

$$\lambda = \frac{1}{25{,}270} = 3.957 \times 10^{-5}/\text{h}$$
$$= 3.957 \text{ percent}/1000 \text{ h}$$

Example 19.5. Calculate the failure rate of a component having a reliability figure of 0.91 for a period of 400 h. Assume exponential failure. Calculate also the unreliability for 800 h.

Applying Eq. (19.38),

$$0.91 = e^{-400\lambda}$$

where λ is the failure rate. Rearranging,

$$\frac{1}{e^{400\lambda}} = 0.91$$

that is

$$e^{400\lambda} = \frac{1}{0.91}$$

Taking natural logarithms,

$$400\lambda = \ln\left(\frac{1}{0.91}\right)$$

Therefore

$$\lambda = \frac{1}{400}\ln\left(\frac{1}{0.91}\right)$$
$$= 0.000237$$
$$= 23.7 \text{ percent}/1000 \text{ h}$$

The unreliability could be calculated using Eq. (19.40). However, an easier method is to calculate the reliability first, with the method shown below, and then apply Eq. (19.39). We know that

$$0.91 = e^{-400\lambda}$$

and since

$$e^{-800\lambda} = (e^{-400\lambda})^2$$

then

$$e^{-800\lambda} = 0.91^2$$
$$= 0.8281$$

which is the reliability for 800 h.

By Eq. (19.39), the unreliability for 800 h equals 1 − 0.8281 = 0.1719. This means that we could reasonably expect 17.19 percent of the components to fail during the 800-h period.

Example 19.6. Calculate the number of components having the failure rate of Example 19.5 which are still alive after a period of 800 h if at the start of the period there are 1000.

From Eq. (19.41),

$$N_s = 1000 \times e^{-0.000237 \times 800}$$
$$= 1000 \times 0.8281$$
$$= 828$$

Example 19.7. A wearout test gave the life data shown in Table 19.2. Calculate the upper and lower confidence limits for the true mean wearout life at (*a*) 68 percent confidence level and (*b*) 85 percent confidence level.

TABLE 19.2

Number of components	Life
5	390
8	450
10	500
7	550
6	600

The sum of lives is $5 \times 390 + 8 \times 450 + 10 \times 500 + 7 \times 550 + 6 \times 600$

$$= 1950 + 3600 + 5000 + 3850 + 3600$$
$$= 18{,}000 \text{ h}$$

Total number of components is 36. Therefore, estimated mean wearout life

$$M_x = \frac{18{,}000}{36} = 500 \text{ h}$$

The differences and squares of differences between the mean and actual lives are as shown in Table 19.3. The standard deviation from Eq. (19.49) is

$$\sqrt{\frac{5 \times 12{,}100 + 8 \times 2500 + 7 \times 2500 + 6 \times 10{,}000}{36}} = 73.7 \text{ h}$$

Notice that the calculation overall has been reduced by the fact that we are told 5 components expired after 390 h, 8 components expired after 450 h, and so on. It is unlikely in practice that this would be so, and in fact, all 36 components would probably have different lives. In a case such as that, the lives are grouped in small *class intervals*, and the mean life of each group is determined. Thus, in this example, the

TABLE 19.3

Number	Life	Difference	Square
5	390	110	12,100
8	450	50	2,500
10	500	—	—
7	550	50	2,500
6	600	100	10,000

figure 390 is already a mean life for the 5 components to which it refers; the 5 components in the original test probably had different lives spread about the mean 390 h.

Using the expressions given earlier for confidence intervals for a normal distribution, we can say that the *true* mean wearout life M will lie between

$$\left(500 \pm \frac{73.7}{\sqrt{36}}\right) \text{ h at a 68 percent confidence level}$$

and
$$\left(500 \pm 1.44 \times \frac{73.7}{\sqrt{36}}\right) \text{ h at an 85 percent confidence level}$$

(The figure of 1.44 is obtained from the table of standard deviations and percentage of values given earlier.) That is,

$$(500 \pm 12.28) \text{ h at a 68 percent confidence level}$$

$$(500 \pm 17.77) \text{ h at an 85 percent confidence level}$$

Thus we are 68 percent confident that the true mean life lies between 512.28 h and 487.72 h and 85 percent confident that it lies between 517.77 h and 482.23 h.

Example 19.8. A radio receiver has the failure rates shown in Table 19.4 for its subunits. Calculate the reliability over a period of 5000 h operation.

TABLE 19.4

Unit	Number	Failure rate (percent 1000 h)
IF amplifier	4	0.1
AF amplifier	3	0.04
Oscillator	1	0.08
Power supply	1	0.9
Power output	1	1.2

A radio receiver may be considered a series system, since a failure of one of the subunits leads to zero output, i.e., overall failure. Thus we can apply Eqs. (19.48) and (19.49):

$$\begin{aligned} \text{Total failure rate} &= 4 \times 0.1 + 3 \times 0.04 + 0.08 + 0.9 + 1.2 \\ &= 0.4 + 0.12 + 0.08 + 0.9 + 1.2 \\ &= 2.7 \text{ (percent/1000 h)} \end{aligned}$$

and the reliability R_s is given by

$$R_s = e^{-2.7 \times 500 \times 10^{-5}}$$
$$= 0.837$$
$$= 83.7 \text{ percent}$$

Notice that the failure rate expressed as a percentage per 1000 h is converted before insertion into the reliability equation.

Example 19.9. Compare the reliability of a series system with a parallel system when each contains 3 subunits having reliabilities 0.9, 0.8, and 0.7, respectively.

From Eq. (19.57),

$$\text{Series system reliability} = 0.9 \times 0.8 \times 0.7$$
$$= 0.504$$

From Eq. (19.70),

$$\text{Parallel system reliability} = 0.9 + 0.8 + 0.7 - 0.9 \times 0.8 - 0.9 \times 0.7 - 0.8 \times 0.7$$
$$+ 0.9 \times 0.8 \times 0.7$$
$$= 2.4 - 0.72 - 0.63 - 0.56 + 0.504$$
$$= 0.994$$

Example 19.10. A subunit having a failure rate of 10 percent per 1000 h is included in a system which is to operate for 2000 h. The component duty cycle is 0.5. Calculate the component reliability and the component MTBF in system hours assuming exponential failure.

From. Eq. (19.53),

$$\text{Component operating hours} = 0.5 \times 2000$$
$$= 1000 \text{ h}$$

From Eq. (19.38),

$$\text{Reliability of the component} = e^{-10 \times 1000 \times 10^{-5}}$$
$$= 0.9048$$

Assuming exponential failure, component MTBF from Eq. (19.43) is

$$m = \frac{1}{10 \times 10^{-5}}$$
$$= 10{,}000 \text{ h}$$

In system hours, the component MTBF from Eq. (19.54) is

$$m_s = \frac{10{,}000}{0.25}$$
$$= 40{,}000 \text{ h}$$

Example 19.11. Over a period of 105 h, out of a total of 100 components, one failed at each of the following times:

9, 10, 20, 30, 50, 70, 80, 80, 100, 101 h

Calculate the confidence limits for the failure rate of these components at a 90 percent confidence level. Wearout failure may be neglected.

Total survival hours of failed components is

$$9 + 10 + 20 + 30 + 50 + 70 + 160 + 100 + 101 = 550 \text{ h}$$

Total survival time of nonfailed components is

$$90 \times 105 = 9450 \text{ h}$$

Thus

$$\begin{aligned}\text{Total survival hours} &= 9450 + 550 \\ &= 10{,}000 \text{ h}\end{aligned}$$

Hence an estimate of the mean life is given by

$$\frac{10{,}000}{10} = 1000 \text{ h}$$

Since wearout failure is to be neglected, an exponential failure rate may be assumed, indicating the use of the chi-square distribution. For a 90 percent confidence level, $1 - \alpha = 0.9$, whence $\alpha/2 = 0.05$ and $1 - \alpha/2 = 0.95$. For 10 failures, the value of χ^2 is 10.85 and 31.41, and so, since the estimated mean time between failures is 1000 h, the upper confidence limit is

$$\frac{20{,}000}{10.85} = 1844 \text{ h}$$

and the lower confidence limit is

$$\frac{20{,}000}{31.41} = 637 \text{ h}$$

Since the exponential distribution is assumed, the percentage failure rates corresponding to these limits are 1/1844 and 1/637, respectively, that is, 0.0005424 and 0.00157. Thus we may assume the failure rate to lie between 0.0005424 and 0.00157 at 90 percent level of confidence.

Example 19.12. In a test to determine failure rates/stress curves for derating purposes, three sets of observations were made:

(a) Failure rate for rated voltage and temperature
(b) Failure rate for rated voltage and twice rated temperature
(c) Failure rate for half rated voltage and rated temperature

If the failure rates for these conditions are 0.001, 0.004, and 0.00005, respectively, and the rated temperature is 20°C, calculate the probable failure rate at one-third rated voltage and one-third rated temperature, assuming proportionality ratios are constant as in Eq. (19.78).

From Eq. (19.78), for test 2,

$$0.004 = 0.001 \times K^{20}$$

Thus

$$K^{20} = 4$$

and

$$\ln K = \tfrac{1}{20} \ln 4$$

Hence

$$K = 1.072$$

For test 3,

$$0.00005 = 0.001(½)^n \qquad \text{and} \qquad 2^n = 20$$

Hence

$$n = \frac{\ln 20}{\ln 2} = 4.32$$

The equation may thus be assumed to be

$$\lambda_x = 0.001 \left(\frac{V_x}{V_m}\right)^{4.32} 1.072^{t_x - t_m}$$

and for ⅓ rated voltage and ⅓ rated temperature,

$$\lambda_x = \frac{0.001 \times 1.072^{-40/3}}{3^{4.32}}$$

$$= \frac{0.001}{3\sqrt{1.072^{40}} \times 3^{4.32}}$$

$$= 0.000003439$$

The dramatic improvement in failure rate obtainable by derating is clearly shown.

SUMMARY OF RELEVANT FORMULAS

For a constant proportional failure rate,

$$R = e^{-\lambda t} \tag{19.38}$$

$$R + Q = 1 \tag{19.39}$$

$$Q = 1 - e^{-\lambda t} \tag{19.40}$$

$$N_s = N_0 e^{-\lambda t} \tag{19.41}$$

$$N_f = N_0(1 - e^{-\lambda t}) \tag{19.42}$$

$$m = \frac{1}{\lambda} \tag{19.43}$$

$$R = e^{-t/m} \tag{19.44}$$

$$Q = 1 - e^{-t/m} \tag{19.45}$$

$$N_s = N_0 e^{-t/m} \tag{19.46}$$

$$N_f = N_0(1 - e^{-t/m}) \tag{19.47}$$

where R = reliability
Q = unreliability
λ = proportional failure rate (i.e., failure rate expressed as a proportion of N_0)
N_s = number of live components (survivors)
N_f = number of dead components (failures)

N_0 = initial number of live components
m = mean time between (chance) failures
t = time

For a test to determine MTBF m,

$$m = \frac{\text{total survival hours}}{\text{number of failures}} \tag{19.48}$$

For a normally distributed variable x, the standard deviation σ is given by

$$\sigma = \sqrt{\frac{\Sigma(x - x_m)^2}{n}} \tag{19.49}$$

where x_m is the mean value of n observations of x.

In a test to determine mean wearout life M,

$$M = \frac{\text{sum of lives}}{\text{number of components}} \tag{19.50}$$

For an exponentially distributed variable,

$$\text{Upper confidence limit} = \frac{2nm}{\chi^2_{1-\alpha/2,n}} \tag{19.51}$$

$$\text{Lower confidence limit} = \frac{2nm}{\chi^2_{\alpha/2,n}} \tag{19.52}$$

at a level of confidence given by 100(1 − α) percent, where n denotes number of failures, m denotes an estimate of the mean value of the variable, and χ^2 denotes the value of chi-squares (given in tables) for values of n and α or (1 − α/2).

For a component or unit which forms part of a system

$$\frac{m_c}{m_s} = \frac{t_c}{t_s} \tag{19.53}$$

$$m_s = \frac{m_c}{d} \tag{19.54}$$

where m_c = component MTBF in component operating hours
m_s = component MTBF in system operating hours
t_c = component operating hours
t_s = system operating hours
d = duty cycle ($d = t_c/t_s$)

The probability of both events x and y occurring P_{xy} is given by

$$P_{xy} = P_x P_y \tag{19.55}$$

And the probability of either event x or event y occurring P_{x+y} is given by

$$P_{x+y} + P_x + P_y - P_x P_y \tag{19.56}$$

where P_x = the probability of x occurring
P_y = the probability of y occurring

The following equations refer to series and parallel systems. The symbols used are as above for reliability, unreliability, failure rate, etc., with the addition of appropriate subscripts as follows:

Subscript s denotes series

Subscript p denotes parallel

Subscripts 1, 2, 3, etc. denote components or subunits 1, 2, 3, etc.

$$R_s = R_1R_2\cdots \tag{19.57}$$

$$Q_s = Q_1 + Q_2 - Q_1Q_2 \tag{19.58}$$

$$Q_s = 1 - R_s \tag{19.59}$$

$$R_p = R_1 + R_2 - R_1R_2 \tag{19.60}$$

$$Q_p = Q_1Q_2 \tag{19.61}$$

$$Q_p = 1 - R_p \tag{19.62}$$

$$R_s = \mathrm{e}^{-(\lambda_1 + \lambda_2 + \lambda_3 + \cdots)t} \tag{19.63}$$

$$\lambda_s = \lambda_1 + \lambda_2 + \lambda_3 + \cdots \tag{19.64}$$

$$m_s = \frac{1}{\lambda_s} \tag{19.65}$$

$$R_s = e^{-n\lambda t} \tag{19.66}$$

$$\lambda_s = n\lambda \tag{19.67}$$

$$m_s = \frac{1}{n\lambda} \tag{19.68}$$

$$R_p = R_1 + R_2 - R_1R_2 \tag{19.69}$$

$$R_p = R_1 + R_2 + R_3 - R_1R_2 - R_2R_3 - R_1R_3 + R_1R_2R_3 \tag{19.70}$$

$$R_p = e^{-\lambda_1 t} + e^{-\lambda_2 t} + e^{-(\lambda_1 + \lambda_2)t} \tag{19.71}$$

$$R_p = e^{-\lambda_1 t} + e^{-\lambda_2 t} + e^{-\lambda_3 t} + e^{-(\lambda_1 + \lambda_2)t} + e^{-(\lambda_2 + \lambda_3)t} + e^{-(\lambda_1 + \lambda_3)t} + e^{-(\lambda_1 + \lambda_2 + \lambda_3)t} \tag{19.72}$$

$$m_p = \frac{1}{\lambda_1} + \frac{1}{\lambda_2} - \frac{1}{\lambda_1 + \lambda_2} \tag{19.73}$$

$$m_p = \frac{1}{\lambda_1} + \frac{1}{\lambda_2} + \frac{1}{\lambda_3} - \frac{1}{\lambda_1 + \lambda_2} - \frac{1}{\lambda_2 + \lambda_3} - \frac{1}{\lambda_1 + \lambda_3} + \frac{1}{\lambda_1 + \lambda_2 + \lambda_3} \tag{19.74}$$

$$m_p = \frac{1}{\lambda} + \frac{1}{2\lambda} + \frac{1}{3\lambda} + \cdots + \frac{1}{n\lambda} \tag{19.75}$$

where n in Eqs. (19.67), (19.68), and (19.75) denotes the number of components or subunits having equal failure rates.

For a system, the utilization factor U is given by

$$U = \frac{\text{operating time}}{\text{maintenance time} + \text{idle time} + \text{operating time}} \tag{19.76}$$

the availability (maximum utilization factor) A is given by

$$A = U_{\max} = \frac{\text{operating time}}{\text{minimum maintenance time} + \text{operating time}} \tag{19.77}$$

For any two sets of operating conditions denoted by x and m, respectively, the voltages V_x and V_m, temperatures t_x and t_m, and failure rates λ_x and λ_m are related by the equation

$$\lambda_x = \lambda_m \left(\frac{V_x}{V_m}\right)^n K^{t_x - t_m} \tag{19.78}$$

where n and K are constants over a limited range of conditions and may be determined by the equations

$$K = \left(\frac{\lambda_x}{\lambda_m}\right)\frac{1}{t_x - t_m} \tag{19.79}$$

for a constant voltage test, and

$$n = \frac{\ln (\lambda_x/\lambda_m)}{\ln (V_m/V_x)} \tag{19.80}$$

for a constant temperature test.

SYSTEMS ENGINEERING*

SYSTEMS ENGINEERING

The design of a complex interconnection of many elements (a system) to maximize an agreed-upon measure of system performance is the basis of *systems engineering*. Systems engineering, also referred to as *system engineering*, includes two parts: modeling, in which each element of the system and the criterion for measuring performance are described, and optimization, in which adjustable elements are set at values that give the best possible performance.

The systems approach can be applied to problems ranging from the very simple to those so complex that the human mind is unable to comprehend the reasons for system behavior. A simple problem is the scheduling of the preparation

*Excerpted from the *McGraw-Hill Encyclopedia of Science and Technology*. Copyright © 1987. Used by permission of McGraw-Hill, Inc. All rights reserved.

of a family meal. A much more complex system problem is the control of the timing of hundreds or thousands of traffic lights in a city.

The techniques of systems engineering have been applied to a tremendous range of current problems, from industrial automation to control of weapons and space vehicles.

Modeling refers to the determination of a quantitative picture of the important system characteristics. This model may be in the form of collected data or analytical studies, or it may be a representation of the important system characteristics in a laboratory or computer simulation of the actual system.

A specific and rather narrow systems problem is the control of traffic flow through a tunnel. The designer has to maximize the number of cars permitted per hour through the tunnel. To do this, the designer can vary the speed limit through the tunnel and control the number of cars within the tunnel (or the rate at which cars enter). Before a system design can be decided on, measurements are made of the way in which cars move under various speeds and degrees of congestion. These data are combined with the known behavior characteristics of a driver. On the basis of this quantitative information, a computer simulation of the system can be constructed from which the engineer can study the effects of different operating rules, the results of a breakdown in one lane of the tunnel, or the effect of a driver who stays below the minimum speed limit. See the section "Simulation" below.

The engineer attempts to design a system that is optimum according to a quantitative criterion which measures the quality of performance. For example, in the tunnel traffic problem, the entry and speed of cars are controlled in such a way that the tunnel handles the maximum possible number of cars per hour. In many situations, the criterion is probabilistic, and the engineer must optimize on the basis of probable behavior of the system. The routing of emergency vehicles through a network of city streets is a typical example. See the section "Optimization" below.

Three technological developments have enormously broadened the scope of problems amenable to the systems approach. The first of these is the understanding of the principles of automatic control (or the principles underlying automation). Systems can be designed in which the desired automatic control is realized by the appropriate choice of input signals and system configuration (including the use of feedback, with the actual system response compared automatically with the desired response and the error used to modify the response toward the desired value). In other words, technology has developed automatic goal-seeking systems to replace the human role. The automatic control of elevators in a busy office building is a familiar example. The second is the revolution in communications. Enormous quantities of data can be transmitted over great distances with nearly perfect fidelity. The third fundamental technological development has been the high-speed electronic digital computer. The computer lies at the core of modern systems engineering, since it permits both the modeling and the optimization of systems vastly more complex than the human being alone can handle.

SIMULATION

The development and use of computer models for the study of actual or postulated dynamic systems is known as *simulation*. The essential characteristic of simulation is the use of models for study and experimentation rather than the actual system modeled. In practice, it has come to mean the use of computer models because modern electronic computers are so much superior for most kinds of simulation that computer modeling dominates the field. *Systems*, as used in the definition, refers to an interrelated set of elements, components, or subsystems. *Dynamic systems* are specified because the study of static systems seldom justifies the sophistication inherent in computer simulation.

Postulated systems as well as *actual* ones are included in the definition because of the importance of simulation for testing hypotheses, as well as designs of systems not yet in existence. The *development* as well as *use* of models is included because, in the empirical approach to system simulation, a simplified simulation of a hypothesized model is used to check educated guesses and thus to develop a more sophisticated and more realistic simulation of the simuland. The *simuland* is that which is simulated, whether real or postulated.

Among the systems which have been simulated are aerospace systems; chemical and other industrial processes; structural dynamics; physiological and biological systems; automobile, ship, and submarine dynamics; social, ecological, political, and economic systems; corporations and small businesses; traffic and transportation systems; electrical, electronic, optical, and acoustic systems; electrical energy systems and other energy related systems; and learning, thinking, and problem-solving systems.

Mathematical modeling is a recognized and valuable adjunct, and usually a precursor, of computer simulation. Mathematical modeling does not necessarily precede simulation, however; sometimes the simuland is not well enough understood to permit rigorous mathematical description. In such cases, it is often possible to postulate a functional relationship of the elements of the simuland without specifying mathematically what that relationship is. This is the building-block approach, for which analog computers are particularly well suited. Parameters related to the function of the blocks can be adjusted until some functional criteria are met. Thus the mathematical model can be developed as the result of, rather than as a requirement for, simulation.

Analog computers, in which signals are continuous and are processed in parallel, were originally the most popular for simulation. Their modular design made it natural to retain the simulation-simuland correspondence, and their parallel operation gave them the speed required for real-time operation. The result was unsurpassed human-machine rapport. However, block-oriented digital simulation languages were developed which allow a pseudo–simulation-simuland correspondence, and digital computer speeds have increased to a degree that allows real-time simulation of all but very fast or very complex systems.

Hybrid simulation, in which both continuous and discrete signals are processed, both in parallel and serially, is the result of a desire to combine the speed and human-machine rapport of the analog computer with the precision, logic capability, and memory capacity of the digital computer. Hybrid simulation now makes possible the simulation of a new array of systems which require combinations of computer characteristics unavailable in either all-analog or all-digital computers.

SYSTEMS ANALYSIS

The application of mathematics to the study of systems is known as *systems analysis*. The term *operations research* is reserved for the study of part of a system. See the section "Operations Research" below.

The basic idea is that a mathematical model of the system under study is constructed, a mathematical analysis is done of the mathematical model, and the results of this analysis are applied to the original system.

A great deal of experience is needed to construct the mathematical model and to interpret the results of the analysis. The mathematical analysis, usually involving a computer, is seldom routine. The procedures are so complex that often an individual systems analysis is required for the systems analysis of a particular system. Often particular parts of the process are carried out by different people. Ideally, there should be an interaction between the mathematical analysis, the construction of the mathematical model, and the interpretation of the results.

Systems analysis is thus a part of applied mathematics. What makes the difference between this and conventional applied mathematics is that the systems studied often involve human beings. The presence of humans and the application of the results to human systems introduce a great deal of complication.

There are three principal difficulties in the application of conventional mathematical techniques: dimensionality, the presence of both "hard" and "soft" variables, and conflicting objectives. *Dimensionality* refers to the possibility that the description of a system may involve many state variables. This feature presents serious difficulties, but they may be circumvented by means of various techniques. Much more serious is the presence of hard and soft variables and various combinations. Thus, for example, in a study of a business, the number of employees is a *hard variable*, while the quality of management is a *soft variable*. Hard and soft variables may be handled by the theory of fuzzy systems. In most situations, there are many objectives. In many cases, these objectives are in partial or complete conflict. Thus it is impossible to write down a single criterion for performance. Fortunately, these questions can be studied by means of simulation. Simulation allows various possibilities to be studied in electronic time without disturbing the system or the people affected by the system. See discussion above.

OPTIMIZATION

In its most general meaning, *optimization* covers the efforts and processes of making a decision, a design, or a system as perfect, effective, or functional as possible. Formal optimization theory encompasses the specific methodology, techniques, and procedures used to decide on the one specific solution in a defined set of possible alternatives that will best satisfy a selected criterion. Because of this decision-making function, the term *optimization* is often used in conjunction with procedures which more appropriately belong in the more general domain of decision theory. Strictly speaking, formal optimization techniques can be applied only to a certain class of decision problem known as *decision making under certainty*.

1. Concepts and Terminology. Conceptually, the formulation and solution of an optimization problem involves the establishment of an evaluation criterion based on the objectives of the optimization problem, followed by determination of the optimum values of the controllable or independent parameters that will best satisfy the evaluation criterion. This determination is accomplished either objectively or by analytical manipulation of the so-called criterion function, which relates the effects of the independent parameters on the dependent evaluation criterion parameter. In most optimization problems, there are a number of conflicting criteria and a compromise must be reached by a tradeoff process which makes relative value judgments among the conflicting criteria. Additional practical considerations encountered in most optimization problems include so-called functional and regional constraints on the parameters. The former represents physical or functional interrelationships which exist among the independent parameters (that is, if one is changed, it causes some changes in the others); the latter limits the range over which the independent parameters can be varied.

Optimization techniques work only for a specific system or configuration which has been described to a point where all criteria and parameters are defined within the system and are isolated and independent from other parameters outside the boundaries of the defined system. Formal optimization is not a substitute for creativity in that it depends on a clear definition of the system to be optimized, and even though an optimum solution for a given system configuration is obtained, this does not guarantee that a better solution is not available.

2. Problem Formulation. Definition is a critical part of problem formulation and consists of

1. A description of the system configuration to be optimized, including definition of system boundaries to an extent that system parameters become isolated and independent of external parameters.
2. Definition of a single, preferably quantitative, parameter which will serve as the overall evaluation criterion for the specific optimization problem. This is the dependent parameter (it depends on the choice of the optimum solution) which measures how well the solution satisfies the desired objectives of the problem and this is the parameter which will be maximized or minimized to satisfy the objectives.
3. Definition of controlled or independent parameters that will have an effect on the criterion. These are the parameters whose values determine the value of the criterion parameter, and these should include all controllable parameters which influence the criterion and are within the boundaries of the defined problem.

The most critical aspect of formulating a formal optimization problem is the establishment of a satisfactory criterion function that describes the behavior of the evaluation criterion as a function of the independent parameters. The *criterion function*, often also referred to as the *pay-off* or *objective function*, usually takes the form of a penalty or cost function (which attempts are made to minimize) or a merit, benefit, or profit function (which attempts are made to maximize by choosing the optimum values of the independent variables). In order to apply formal solution methods, the criterion function should be expressed graphically or analytically.

Physical principles of operation which govern the relationship among the various independent parameters of the problem represent functional constraints. Most optimization problems involve practical limits on the range over which each parameter or function of the parameters can be varied. They represent the regional constraints.

3. Solution of Optimization Problems. The reliability and the sophistication of the solutions of optimization problems increase as the problem becomes better defined. In the early stages of optimization of a complex problem, the considerations may be primarily objective judgments based on the optimizer's judgments of the relevant parameters. As alternative subsystem and system configurations evolve through a process of conceptualization, analysis and evaluation, and elimination of undesirable system configurations, a clearer configuration of parts of the problem can be defined, together with the relevant criteria, parameters, and constraints. Procedures for defining competing systems or alternative strategies and formulating the problem in order to apply formal optimization techniques are also common to operations research, systems analysis, and systems design.

OPERATIONS RESEARCH

The application of scientific methods and techniques to decision-making problems is known as *operations research*. A decision-making problem occurs where there are two or more alternative courses of action, each of which leads to a different and sometimes unknown end result. Operations research is also used to maximize the utility of limited resources. The objective is to select the best alternative, that is, the one leading to the best result.

To put these definitions into perspective, the following analogy might be used. In mathematics, when solving a set of simultaneous linear equations, one states that if there are seven unknowns, there must be seven equations. If they are independent and consistent and if it exists, a unique solution to the problem is found. In operations research there are figuratively "seven unknowns and four equations." There may exist a solution space with many feasible solutions which satisfy the equations. Operations research is concerned with establishing the best solution. To do so, some measure of merit, some objective function, must be prescribed.

In the current lexicon there are several terms associated with the subject matter of this program: *operations research, management science, systems analysis, operations analysis*, and so forth. While there are subtle differences and distinctions, the terms can be considered nearly synonymous.

1. Methodology. The success of operations research, where there has been success, has been the result of the following six simply stated rules: (1) formulate the problem, (2) construct a model of the system, (3) select a solution technique, (4) obtain a solution to the problem, (5) establish controls over the system, and (6) implement the solution.

The first statement of the problem is usually vague and inaccurate. It may be a cataloging of observable effects. It is necessary to identify the decision maker, the alternatives, goals, and constraints, and the parameters of the system. A statement of the problem properly contains four basic elements that, if correctly identified and articulated, greatly ease the model formulation. These elements can be combined in the following general form: "Given (the system description), the problem is to optimize (the objective function), by choice of the (decision variable), subject to a set of (constraints and restrictions)."

In modeling the system, one usually relies on mathematics, although graphical and analog models are also useful. It is important, however, that the model suggest the solution technique, and not the other way around.

With the first solution obtained, it is often evident that the model and the problem statement must be modified, and the sequence of problem-model-technique-solution-problem may have to be repeated several times. The controls are established by performing sensitivity analysis on the parameters. This also indicates the areas in which the data-collecting effort should be made.

Implementation is perhaps of least interest to the theorists, but in reality it is the most important step. If direct action is not taken to implement the solution, the whole effort may end as a dust-collecting report on a shelf.

2. Mathematical Programming. Probably the one technique most associated with operations research is linear programming. The basic problem that can be modeled by linear programming is the use of limited resources to meet demands for the output of these resources. This type of problem is found mainly in production systems, but it is not limited to this area.

3. Stochastic Processes. A large class of operations research methods and applications deals with *stochastic processes*. These can be defined as processes in which one or more of the variables take on values according to some, perhaps unknown, probability distribution. These are referred to as *random variables*, and it takes only one to make the process stochastic.

In contrast to the mathematical programming methods and applications, there are not many optimization techniques. The techniques used tend to be more diagnostic than prognostic; that is, they can be used to describe the "health" of a system, but not necessarily how to "cure" it.

4. Scope of Application. There are numerous areas where operations research has been applied. The following list is not intended to be all-inclusive, but is mainly to illustrate the scope of applications: optimal depreciation strategies; communication network design; computer network design; simulation of computer time-sharing systems; water resource project selection; demand forecasting; bidding models for offshore oil leases; production planning; classroom size mix to meet student demand; optimizing waste treatment plants; risk analysis in capital budgeting; electric utility fuel management; optimal staffing of medical facilities; feedlot optimization; minimizing waste in the steel industry; optimal design of natural-gas pipelines; economic inventory levels; optimal marketing-price strategies; project management with CPM/PERT/GERT; air-traffic-control simulations; optimal strategies in sports; optimal testing plans for reliability; optimal space trajectories.

SAFETY ENGINEERING*

Safety in any environment in which people work is the responsibility of the person in charge of the facility. This person may have various titles in different industries. The generic title, however, is *plant engineer*. This title is used throughout this section, since it applies to essentially every industry known today.

*This section is drawn from *Plant Equipment Reference Guide*, by R. C. Rosaler and J. O. Rice. Copyright © 1987. Used by permission of McGraw-Hill, Inc. All rights reserved.

SAFETY

A well-organized plant safety program should encompass all phases of the plant environment and operations. From the standpoint of this handbook, we are primarily concerned with safety controls and devices in the environment and only make reference to supervisory controls of the safety program.

Safeguarding against industrial hazards has always been one of the principal tasks of the plant engineer. Production depends on the ability to maintain a continuous flow of materials without the interruptions caused by accidents. Engineering controls and safeguards serve to protect inexperienced workers or those workers who are distracted or who suffer from fatigue.

The basic elements of a good safety program include

1. Management leadership
2. Assignment of responsibility
3. Maintenance of safe working environment
4. Training program
5. Record system
6. Medical follow-through

The plant engineer is basically concerned with items 1, 2, and 3, but items 4, 5, and 6 are also important.

LEGAL ASPECTS OF SAFETY

The early legal aspects of industrial safety were limited to laws which were developed in conjunction with workers' compensation acts. They were primarily aimed at providing for just compensation to the injured party, but they also established accident investigation procedures and some regulation of hazards.

The past two decades have seen a greater public demand for safety in the workplace. With it came a proliferation of federal laws and regulations which include the following.

1. Laws Applicable to Industrial Plants

1. Environmental Protection Agency
    - Toxic Substances Control Act of 1976
    - Marine, Protection, Research and Sanctuaries Act of 1972
    - Safe Drinking Water Act of 1974
    - Water Pollution Control Act of 1972
    - Clean Air Act of 1970
    - National Environment Policy Act of 1969
    - Atomic Energy Act of 1954
    - Clean Water Act
    - Endangered Species Act of 1973

- Energy Supply and Environmental Coordination Act
- Fish and Wildlife Coordination Act

2. Solid Waste Disposal Act of 1965
3. Occupational Safety and Health Act of 1971
4. Department of Transportation
   - Transportation Safety Act
   - Hazardous Materials Transportation Act of 1974
   - Ports and Waterways Safety Act of 1972
5. Resource Conservation and Recovery Act of 1976
6. Federal Insecticide, Fungicide, and Rodenticide Act of 1972
7. Consumer Product Safety Commission
   - Federal Hazardous Substances Act
   - Consumer Product Safety Act
   - Poison Prevention Packaging Act
8. Mine Safety and Health Act

The Occupational Safety and Health Act will probably have a far greater effect on business and industry than the other legislation. The law requires all employers in the private sector of business to furnish their employees with a workplace free from those recognized hazards likely to cause physical harm or death. Standards are published or referenced in the act. This 800-page document is entitled *OSHA Safety and Health Standards* (29 CFR 1910). It is available from the Superintendent of Documents, Washington, DC 20402, and is recommended as a reference for plant engineers.

The adoption of specific federal legislation has also led to an increase in the number of civil suits. The violation of a federal standard may become prima facie evidence of negligence. Such cases have been very costly.

Recent developments at OSHA have placed greater emphasis on the health aspects of worker's compensation. What started in specialized industries, such as coal mining and asbestos, is spreading to the chemical industry and others. Much more of the social responsibility for the well-being of the employee is being directed to the employer.

All employers must keep accurate records of work-related injuries, illnesses, and deaths. Any injury that involves medical treatment, loss of consciousness, restrictions of motion or work, or job transfer must be recorded. Required information must be logged and specific information regarding each case detailed. At the end of the calendar year, a summary of logged cases must be posted in each plant.

INSTRUMENTATION AND CONTROLS

Some of the most important advances in safety over the past two decades have been achieved in the area of instrumentation and controls. Standards of the Instrument Society of America (ISA) should be followed. The plant engineer is directly concerned with these devices, since they are involved in controlling the operations, materials flow, and environment of the plant. Instrumentation and controls fall into the following categories:

1. Temperature systems
2. Pressure systems
3. Flow systems
4. Analytical and testing systems (i.e., gas analysis, solids and dust analysis, reaction product tests, oxygen analysis, pollution instrumentation, electronic components, and system checkout equipment)
5. Weighing, feeding, and batching systems
6. Force and power systems
7. Motion and geometric systems (i.e., speed, velocity, vibrations)
8. Humidity and moisture determinations
9. Radiation systems
10. Electrical determinations
11. Communication systems
12. Automatic process controllers (i.e., computer process control, safety in instrumentation and control systems, electronic, pneumatic and hydraulic control devices)
13. Controlling elements (i.e., valves, actuators, electric motor drives and controls, mercury switches, etc.)

The importance of instrumentation and controls cannot be overestimated. From the standpoint of safety and environmental control, the following applications are dominant:

1. Detection of leaks in equipment
2. Survey of operating areas for escape of toxic materials
3. Detection of flammable or explosive mixtures in the atmosphere or process lines
4. Monitoring plant stacks and other areas for the accidental discharge of toxic gases, vapors, or smokes.
5. Analysis of waste streams for toxic or other objectionable material.
6. Control of waste-treatment or product-recovery facilities

Automatic process control, unlike manual control, provides continuous monitoring and corrective action. However, automatic controllers cannot react to new conditions, nor can they predict beyond the data programmed into them. Therefore, the ability to foresee an upset condition in a process and the capability of developing fail-safe actions may well determine whether a serious accident is averted.

Analytical and testing instrumentation provides the plant engineer with much needed information. For instance, decisions can be made in these areas:

1. Purity of raw materials entering a process. Contaminants may be hazardous.
2. Process control by automatization of sampling.

3. Process troubleshooting by continuous analysis to prevent process upsets.
4. Determination of product quality to provide an impetus to a product safety program.

PLANT ENGINEER'S FUNCTION WITHIN THE SAFETY COMMITTEE

A safety program is a management method to develop specific safety objectives, assign responsibility, and obtain desired results. The plant engineer should participate with the safety committee from the standpoint of developing adequate inspection and control procedures in order to maintain a safe work environment and control of safety devices. Safety devices may be anything from a pressure-relief valve to a machine guard.

1. Inspection and Maintenance. In many plants an inspection committee is appointed. A safety inspection may be done in conjunction with the regular plant inspection and maintenance program. This is an opportunity for the plant engineer to check all safety devices. The safety inspection should focus attention on those items directly concerned with accident prevention. The plant engineering staff should have a working knowledge of safety standards if their inspections are to cover safety as well as normal wear and tear of equipment. A general plant inspection should include

1. All buildings and physical equipment
2. Inspection of new machinery before it is placed in operation
3. Inspection of walking and working surfaces and means of egress
4. Special equipment such as powered platforms, personnel lifts, and vehicle-mounted work platforms
5. Materials handling and storage facilities, elevators, cranes
6. Machinery, machinery guards, and electrical equipment
7. Compressed gas and air equipment
8. Special equipment such as pressure vessels, drums and furnaces, and welding, cutting, and brazing equipment
9. Hand and portable powered tools and other hand-held equipment
10. Environmental control, equipment, ventilation, and pollution controls for toxic and hazardous substances

A systematic method of inspection and maintenance is preferred. Some companies require *preventive* maintenance programs which call for the replacement of *critical* equipment periodically regardless of inspection results. Others require general inspections of the entire premises, annually.

Areas of the plant that have the potential of developing into catastrophic hazards require special inspection procedures. Items that may develop into a catastrophe include (1) structural failure, (2) fire or explosion, and (3) release of hazardous gases or vapors.

Some elements of a plant may require inspection and maintenance on the basis of federal, state, or local laws. Items such as elevators, boilers, and unfired pressure vessels are in this category. This equipment may require the use of specially trained and licensed inspectors either from a governmental agency or an insurance carrier.

A careful record should be kept of all inspections and recommendations. This is particularly important in the event that an accident occurs at this location which results in litigation. Some companies assign inspection tasks to maintenance personnel, electricians, and others who are charged with the job of repairing the equipment. Supervisors should continuously examine their own work areas to make sure that tools, machinery, and other types of equipment are safe to handle.

Inspection methods should be established for all new equipment and processes. Nothing should be placed into operation until all safeguards have been checked and operations evaluated by the plant engineer. Safe operating instructions should be given to all workers concerned.

2. Inspection Procedures. The following inspection criteria should be applied:

1. Inspectors must be familiar with the company's safety and health policies as well as the particular laws and regulations that pertain. Frequently, these regulations are only minimum requirements and it may be necessary to exceed them to secure adequate safety.
2. Inspectors should have available an analysis of all accidents that have occurred in the plant within the past year.
3. The inspectors should utilize all aids available, including inspection checklists, report forms, and other pertinent information.

An inspection report should be divided into three areas of interest:

1. A report on imminent hazards which require immediate corrective action.
2. A routine report on unsatisfactory (nonemergency) conditions which need corrective action.
3. A general report on the overall safety conditions of the facility.

ACCIDENT PREVENTION

The four basic steps for preventing accidents are as follows:

1. Elimination of the hazard
2. Control of the hazard
3. Training of personnel to be aware of and avoid the hazard
4. Utilization of personal protective equipment

The plant engineer is primarily concerned with steps 1 and 2 by ensuring a safe design for plant, physical facilities, and machinery.

BUILDING STRUCTURE

The plant engineer is directly concerned with the inspection and maintenance of the building structures. Slippery floors, stairs, runways, ramps, and other means of access are involved in many serious plant accidents. About one-fifth of the industrial injuries result from falls. Many of these can be avoided by careful design, construction, and maintenance. Applicable standards and codes should be applied.

1. Basic Rules. Housekeeping is a prime consideration in minimizing the hazard of falls. Some of the basic rules in this area are as follows:

1. All places of employment should be kept clean and orderly. Floors shall be maintained in a clean and dry condition with adequate drainage.
2. All passageways including aisles, ramps, and stairways should be kept clear and in good repair.
3. Floor loading should be kept well within prescribed limits.
4. All floor and wall openings should be guarded by standard railings or properly constructed closures.
5. Stairways should conform to acceptable standards.
6. Exits should be sufficient in number and properly located so that the building can be evacuated quickly in an emergency.
7. All ladders should conform to the applicable standards and codes and they should be properly maintained.
8. Ladders should not take the place of fixed stairways.
9. Workers should follow good practices in the use of ladders with regard to placement, support, angle between horizontal base and vertical plane of support, and proximity to other hazards (i.e., electrical).
10. Scaffolds, which are in effect elevated working platforms, should be designed with an adequate factor of safety and protection for the workers.
11. Scaffolds must be maintained, inspected, and guarded on all exposed sides. Metal scaffolds should not be constructed near electrical equpment.

2. Standards. Refer to ANSI standards that apply to building structures.

MEANS OF EGRESS FOR INDUSTRIAL OCCUPANCIES

A *means of egress* is a continuous and unobstructed way to exit from any point in a building or structure to a public way. It includes vertical and horizontal ways of travel and intervening room spaces and other areas. The number of exit facilities required is specified in standards and codes, and it depends on the structure, occupants, and hazard exposures.

Inspections of means of egress have shown the following items at fault in such facilities:

1. Improper number of exits and locations
2. Poor illumination—lack of emergency lighting
3. Lack of directional signs
4. Poor housekeeping—obstructions
5. Improper and hazardous floor surfaces
6. Faulty operation of exit doors

1. Standard. Refer to NFPA 101 1970, *Life Safety Code*, which applies to egress for industrial occupancies.

POWERED PLATFORMS, PERSONNEL LIFTS, AND VEHICLE-MOUNTED WORK PLATFORMS

Powered platforms, personnel lifts, and vehicle-mounted work platforms are means of elevating workers on suspended operated work platforms or personnel lifts. The platforms are used for exterior building maintenance work, the personnel lifts are used to lift workers to an elevated jobsite, and the vehicle-mounted work platforms are used to position personnel to an elevated work site. All these installations should be in conformance with the appropriate ANSI standards.

Every work platform should be tested and inspected frequently. New platforms should be tested before they are placed in service. Each installation should be inspected and tested at least every 12 months and should undergo a maintenance inspection and test every 30 days. Results of all inspections and tests should be logged, including the date, time, and inspector.

Items which require special inspections are as follows:

1. Governors
2. Initiating devices
3. Both independent braking systems
4. Interlocks and emergency electric devices
5. Electric systems
6. Emergency communication system

Maintenance is required specifically on all parts of the equipment related to safe operations. Broken or worn parts and electric devices should be replaced promptly. Gears, shafts, bearings, brakes, and hoisting drums should be maintained in proper alignment. Gears should be replaced when there is evidence of appreciable wear. All parts should be kept free from dirt. Wires or ropes should be replaced when they are damaged or in a deteriorated condition. Guardrails and toeboards are required for working platforms. Load-rating plates must be conspicuously posted and adhered to.

The personnel lifts require frequent inspections. Their use should be limited to *personnel*. They should not be used to lift construction materials. The inspections should include but are not limited to the following items:

Steps	Drive pulley
Rails and supports	Electrical systems
Belt and belt tension	Vibration—alignment
Handholds	Brake systems
Floor landing	Warning lights
Limit switches	Pulleys—clearance

1. Standards. Refer to ANSI standards that apply to platforms, personnel lifts, and vehicle-mounted work platforms.

VENTILATION

Ventilation systems protect the health and environment of plant workers by removing objectionable dusts, fumes, vapors, or gases. They are essential safety devices and must be properly installed and maintained. Ventilation systems can be divided into two primary groups, local systems and general systems.

1. Local Systems. Local exhaust systems prevent the accumulation of toxic or flammable materials near the process unit. Due to the variety of work and the types of equipment, it is necessary to design hoods specifically developed to be as close to the operation as possible. Air-sampling devices are available to determine whether the atmosphere has been adequately cleared or constitutes a hazard. Safety standards may specify local exhaust ventilation for particular processes.

After contaminated air passes from the hood to exhaust ducts, it is sent through a cleaning system or to the outdoors. Such exhaust air must conform to the regulations of the EPA. Dusts may be collected by means of cyclone dust collectors or electrostatic precipitators. Many types of air-cleaning devices are used. The selection is determined by the degree of the hazard associated with the dust. Gases or vapors may be removed from the waste air by absorption or adsorption processes which dissolve or react with the waste product chemically. Some gases and vapors are rendered harmless by passing them through a combustion chamber where they are changed to acceptable gases. In other cases condensers are utilized to liquefy toxic vapors for removal.

When ventilation is used to control potential exposure to workers it is adequate to reduce the concentration of contaminant so that the hazard is removed.

Local exhaust ventilation is used for a great many industrial operations such as anodizing, pickling, metal cleaning, and open-surface tank operations. There is a standard developed for this purpose. Other uses include spray booths; dip tanks; fume controls in electric welding and grinding operations; cast-iron machining; dust control in foundries; ventilation of internal combustion engines; and kitchen range hoods.

One of the basic maintenance items in the operation of local exhaust ventilation systems is to ensure that the flow of air is unobstructed. A minimum maintained velocity must exist in order to meet the health and safety requirements. Where flammable gases or vapors are removed, the electric equipment must conform to code requirements.

At intervals of not more than 3 months the hood and duct system should be inspected for evidence of corrosion, damage, or obstruction. In any event if the airflow is found to be less than required, it should be increased to that required.

2. General Systems. General ventilation in the workplace contributes to the comfort and efficiency of the employees. It also serves to clear the air of hazardous contaminants and excessive heat or humidity. When local exhaust ventilation cannot be applied due to the many sources of vapor release, general ventilation is prescribed. The publication of the American Society of Heating, Refrigerating and Air Conditioning Engineers, *Handbook of Fundamentals*, defines the prescribed methods and requirements for the development of acceptable ventilation systems. The plant engineer should inspect this equipment frequently to make sure that it performs as required.

3. Standards. Refer to ANSI standards pertaining to ventilation.

COMPRESSED GASES

Plant engineers should be aware of the safety requirements for compressed-gas handling and storage. Pressure-relief devices for gas cylinders, portable tanks, and cargo tanks should be installed and maintained in accordance with Compressed Gas Association (CGA) pamphlets S1.1-1963 and S1.2-1963.

MATERIALS HANDLING AND STORAGE

1. Powered Industrial Trucks. Safety requirements for powered industrial trucks involve the following:

1. The plant engineer is directly concerned with the safety requirements relating to fire protection: design, maintenance, and use of fork trucks, tractors, platform lift trucks, motorized hand trucks, and other specialized industrial trucks.
2. All new industrial trucks should meet the requirements of ANSI B56.1-1969, *Powered Industrial Trucks*.
3. In locations used for the storage of hazardous liquids or liquefied or compressed gases, only approved trucks designated as DS, ES, GS, or LPS may be used. In areas containing combustible dusts approved EX or ES trucks may be used.
4. Any power-operated industrial truck not in safe operating condition should be removed from service.

2. Overhead and Gantry Cranes. Safety requirements for overhead and gantry cranes include

1. All new and existing overhead and gantry cranes should meet the design specifications of ANSI B30.2.0-1976, *Safety Standard for Overhead and Gantry Cranes (Top Running Bridge, Multiple Girder)*.
2. Exposed moving parts such as gears, set screws, projecting keys, chains, chain sprockets, and reciprocating components which might constitute a hazard under normal operating conditions should be guarded.
3. Both holding brakes and control brakes should be tested to determine whether they are within required tolerances.
4. Brakes on trolleys and bridges should be examined and adjusted when necessary.
5. Inspections fall into two classes: *frequent inspections*, which fall into daily or monthly intervals, and *periodic inspections*, which fall into 1- to 12-month intervals. *Frequent inspections* should include (*a*) all functional operating mechanisms for maladjustments interfering with daily operations, (*b*) deterioration or leakage in air or hydraulic systems, (*c*) hooks and hoist chains and ropes including end connections, (*d*) all functional operating mechanisms. *Periodic inspections* should include (*a*) all items listed under frequent inspections, (*b*) deformed, cracked, or corroded members, (*c*) worn, cracked, or distorted parts, (*d*) excessive wear on brake system, (*e*) improper performance of power plants, (*f*) excessive wear of chain drive sprockets and excessive chain stretch, (*g*) defective electrical apparatus.
6. Preventive maintenance based on crane manufacturers' recommendations should be established.
7. Any unsafe conditions—including those in ropes—disclosed in the inspection requirements should be corrected before operation of the crane is allowed.

3. Crawler, Locomotive, and Truck Cranes. Safety requirements for crawler, locomotive, and truck cranes include

1. All new and existing locomotive and truck cranes should meet the design specifications of ANSI B30.5-1968, *Safety Code for Crawler, Locomotive, and Truck Cranes*.
2. Load ratings should not exceed the stipulated percentages for cranes with indicated types of mountings.
3. Inspections are frequent and periodic as cited under overhead gantry cranes.

4. Derricks. All new derricks constructed after August 31, 1971 should meet the design specifications of ANSI B30.6-1977, *Safety Code for Derricks*. Those derricks constructed prior to that date should be modified to conform. The plant engineer should check the standard to assure compliance.

Inspection procedures are similar to those required for cranes under frequent or periodic inspections. Derricks not in regular use for a period of 6 months should be given a complete inspection before use.

Maintenance and repair is an important part of the plant engineer's assignment and should include the following steps:

1. Any unsafe conditions shown in the inspection should be corrected.
2. Adjustments should be made to assure correct functioning of all components.
3. Ropes and wires should be thoroughly inspected and replaced or repaired.
4. No derrick should be loaded beyond the rated load.

5. Helicopter Cranes. Helicopter cranes should comply with the applicable regulations of the Federal Aviation Administration. The weight of an external load should not exceed the helicopter manufacturer's rating. All equipment should be checked frequently. The cargo hooks should be tested prior to each day's operation to determine that the release functions properly, both electrically and mechanically.

6. Slings. All slings used in conjunction with materials handling equipment made from alloy-steel chains, wire rope, metal mesh, natural or synthetic fiber rope, and synthetic web (nylon and polypropylene) should conform to the applicable standards. They should be inspected daily for damage or defects.

7. Standards. Refer to ANSI standards pertaining to materials handling and storage.

MACHINERY AND MACHINE GUARDING

One or more methods of machine guarding should be provided to protect the operator and other employees in the machine area from hazards such as those created by point of operation, ingoing nip points, rotating parts, flying chips, and sparks. Examples are barrier guards, two-hand tripping devices, electronic safety devices, etc.

1. General. The following machines usually require point-of-operation guarding:

Guillotine cutters
Shears
Alligator shears
Power presses
Milling machines
Pointer saws
Jointers
Forming rolls and calenders
Revolving drums, barrels, and containers
Exposed blades

2. Woodworking Machinery. Woodworking machinery includes automatic cutoff saws, circular saws, hinged saws, revolving double arbor saws, hand-fed ripsaws, hand-fed crosscut saws, circular resaws, swing-cut saws, radial saws, bandsaws, jointers, tenoning machines, boring and mortising machines, wood shapers, planers, lathes, sanding machines, guillotine veneer cutters, and miscellaneous woodworking machines. These machines should be properly guarded in accordance with appropriate ANSI standards. All belts, pulleys, gears, shafts, and moving parts should be guarded in accordance with requirements.

3. Abrasive Wheel Machinery. Abrasive wheel machinery includes cylindrical grinders, surface grinders, swing frame grinders, and automatic snagging machines. The guard design and specifications should be in accordance with appropriate ANSI standards.

MILLS AND CALENDERS IN THE RUBBER AND PLASTICS INDUSTRIES

All new and existing installations of mills and calenders should comply with the OSHA requirements and pertinent standards.

Mill safety controls consisting of safety trip controls, pressure-sensitive body bars, safety trip rods, safety trip wire cables, or wire center cords should be installed in all mills either singly or in combination. Fixed guards should be used where applicable.

Calender safety controls should be provided on all calenders within reach of the operator and the bite. Calenders and mills should be installed so that persons cannot normally reach over or under, or come in contact with, the roll bite. All trip and emergency switches should require manual resetting. A suitable alarm should be provided in conjunction with the safety devices.

MECHANICAL POWER PRESSES

All mechanical power presses (excluding pneumatic power presses, bulldozers, hot-banding and hot-metal presses, forging presses, hammers, riveting machines, and similar types of fastener applicators) should conform to the OSHA regulations and applicable standards. Mechanical power presses require particular emphasis as they are involved in many accidents. The machine components should be designed, secured, and covered to minimize hazards. Safeguards should include

1. Brakes or blocks capable of stopping the motion of the slide
2. Foot pedals, protected to prevent unintended operation
3. Hand-operated levers to prevent premature or accidental tripping
4. Two-hand trips to have individual operator hand controls arranged to require the use of both hands to trip the press
5. Air-controlling equipment to be protected against foreign material
6. Brake monitoring systems installed on the press to indicate when the performance of the braking system has deteriorated

7. Point-of-operation guards installed and used in accordance with the standard's requirements
8. A regular program of periodic and regular inspections on a weekly basis of power presses
9. Reports of injuries to employees operating mechanical power presses required by OSHA, with an analysis of each accident on a 30-day basis

FORGING MACHINES

The safety requirements apply to use of lead casts and other uses of lead in the forge or die shop. All equipment should comply with OSHA requirements and the applicable standards. This should include

1. Thermostatic control of heating elements required for melting lead
2. Industrial hygiene and personal protective equipment required due to the toxicity and heat condition of the lead
3. All presses and hammers controlled by operators who are protected by required guards
4. Devices included to lock out the power when dies are being changed

MECHANICAL POWER TRANSMISSION

This section pertains to mechanical power transmission equipment (see "Standards"), with the exception of small belts operating at a reduced speed and reduced requirements for the textile industry to prevent accumulation of combustible lint. Standard guards are usually secured with the following materials: expanded metal, perforated or solid sheet metal on a frame of angle iron, or an iron pipe securely fastened to the floor or frame of the machine. The standard requirements include guards for the following:

1. Flywheels located 7 ft or less from the floor
2. Cranks and connecting rods when exposed to contact
3. Tail rods or extension piston rods
4. Shafting—horizontal, vertical, and inclined
5. Power transmission apparatus and pulleys
6. Belts, ropes, chain drums, gears, sprockets, and chains
7. Shafting, collars, couplings, belt shifters, clutches, perches, or fasteners

Periodic inspection is required of all power transmission equipment. Intervals not exceeding 60 days are recommended.

1. Standards. Refer to ANSI standards that apply to mechanical power transmission.

HAND AND PORTABLE POWERED TOOLS AND OTHER HAND-HELD EQUIPMENT

Portable tools must be equipped with adequate guards and should comply with OSHA and the appropriate standards. This includes

1. Portable circular saws, saber, scroll, and jigsaws, portable belt sanding machines, portable abrasive wheels, explosive actuated fastening tools, power lawnmowers, jacks.
2. Warning instructions should be supplied with the equipment. All equipment should be inspected periodically.

1. Standards. Refer to ANSI standards that pertain to hand-held equipment.

WELDING, CUTTING, AND BRAZING

Most welding and cutting operations are mobile and are generally used for construction, demolition, maintenance, and repair. Production-line welding and cutting equipment is permanently installed; the hazards can be controlled through proper design and operation. The oxygen and acetylene used for welding and cutting must be handled and stored in accordance with the pertinent standards.

Compressed-gas cylinders should be handled with care. The fusible safety plugs in acetylene cylinders melt at about 212°F (100°C). This is an obvious fire hazard if the cylinder is subjected to heat. Oxygen, on the other hand, will react strongly with oils or other hydrocarbons upon contact.

Regulators and pressure gauges should be used on the appropriate cylinders. Leaking cylinders must be removed from the building and taken away from sources of ignition. All equipment should be kept free of oily or greasy substances.

Piping systems should be tested and proved gas-tight at 1½ times the maximum operating pressure. Service piping systems should be protected by pressure-relief devices discharging upward to a safe location. Backflow protection should be provided by an approved device that will prevent oxygen from flowing into the fuel system.

Acetylene generators should be of approved construction and clearly marked with the maximum rate of acetylene production and the pressure limitations. Relief valves should be regularly operated to ensure proper functioning. Storage of all chemicals should conform to the requirements of the standards.

ARC-WELDING AND CUTTING EQUIPMENT

Arc-welding apparatus should comply with the requirements of the National Electrical Manufacturers, *Standard for Electric Arc-Welding Apparatus*, NEMA-EW-1 1962, and for ANSI/UL551-1976. *Safety Standards for Transformer-Type-Arc-Welding Machines*. The design requirements include the following items of concern to the plant engineer:

1. Input power terminals, tip change devices, and live metal parts should be completely enclosed.
2. Terminals for welding leads should be protected from accidental electrical contact by personnel or by metal objects. The frame or case of the welding machine should be grounded as specified.
3. Printed rules and instructions concerning operation of the equipment supplied by the manufacturer should be strictly followed.
4. All parts of the equipment should be frequently inspected. Cables with damaged insulation or exposed bare conductors should be replaced.

RESISTANCE-WELDING EQUIPMENT

All resistance-welding equipment should be installed in accordance with article 630D of the *NEC*.* The following items of particular interest to the plant engineer are cited:

1. Controls on all automatic or air and hydraulic clamps should be arranged or guarded to prevent accidental actuation.
2. All doors and access and control panels should be kept locked and interlocked to prevent access by unauthorized persons to live portions of the equipment.
3. All press-welding machine operations, where there is a possibility of the operator's fingers being under the point of operation, should be effectively guarded in a manner similar to that prescribed for punch-press operations.
4. Flash-welding equipment should be equipped with hoods to control flying flash and ventilation of fumes.
5. Combustible material must be removed from the welding area and the basic fire-prevention requirements followed in accordance with ANSI/NFPA 51B-1977. Fire watchers are required wherever a fire problem exists. A welding permit system should be instituted to control hazardous exposures.
6. Welding operators should use all of the protective equipment specified in the standard, and ventilation should be used where required.

1. Standards. Refer to ANSI standards for resistance-welding equipment.

SPECIAL INDUSTRIES

Plant engineers involved in the following special industries should consult the OSHA regulations for specific requirements pertinent to their industry.

1. Pulp, paper, and paperboard mills
2. Textiles
3. Bakery equipment

*NEC is a registered trademark of the National Fire Protection Association, Quincy, MA 02269.

4. Laundry machinery and operations
5. Sawmills
6. Pulpwood logging
7. Agricultural operations
8. Telecommunications

1. Standards. Refer to ANSI special-industry standards.

ELECTRIC EQUIPMENT

All electric equipment in the plant should comply with OSHA regulations and the *National Electrical Code* (NFPA 70-1981).* It is particularly important for the plant engineer to be knowledgeable in the following sections:

1. Article 250, Grounding
2. Article 500, Hazardous (Classified) Locations

TOXIC AND HAZARDOUS SUBSTANCES

The exposure of employees to any of approximately 600 chemicals (see "Standards" at end of this section) should at no time exceed the ceiling value given for that material. To obtain compliance with this section of the OSHA regulations, it is necessary to develop administrative and engineering controls. When such controls are not feasible, protective equipment or other protective measures should be used.

Controls can be developed by environmental monitoring, personal monitoring, and employee observation. The plant engineer has an important part to play in the inspection and maintenance of the environmental monitoring equipment.

One method for controlling exposure is by local exhaust-ventilation and dust-collection systems. They should be constructed, installed, and maintained in accordance with ANSI Z9.2-1979, *Fundamentals Governing the Design and Operation of Local Exhaust Systems*, which is incorporated by reference in this section of the handbook.

Twenty-two materials require special care in handling since they are in the category of regulated substances which present a cancer hazard. These are listed below:

Asbestos	beta-Propiolactone
Coal-tar pitch volatiles	2-Acetylaminofluorene
4-Nitrobiphenyl	4-Dimethylaminoayobenzene
alpha-Naphthylamine	*N*-Nitrosodimethylamine
Methylchloromethyl ether	Vinyl chloride

**National Electrical Code* is a registered trademark of the National Fire Protection Association, Quincy, MA 02269.

3,3-Dichlorobenzidene	Inorganic arsenic
bis(Chloromethyl) ether	Benzene
beta-Naphthylamine	Coke-oven emissions
Benzidine	Cotton dust
4-Amino diphenyl	1,2-Dibromo-3-chloropropane
Ethylenimine	Acrylonitrile

Controls on these materials are very detailed since the materials must be contained in a closed-system operation. The operating area must be restricted to authorized personnel only. Warning signs and instructions should be posted.

CHAPTER 20
ENGINEERING ECONOMY, PATENTS, AND COPYRIGHTS

ENGINEERING ECONOMY*

Engineering economy is a study of the time value of money in an engineering environment. Thus the engineer might study—from an economic standpoint—the investment differences of different types of energy supplies, the relative cost of two different types of heat insulation, or the relative costs of two highway materials.

In any engineering economic study, the following factors will enter: (1) the first cost of each alternative way of accomplishing a given task or reaching an end result, (2) the cost of borrowed money, (3) the time required to recover the investment, (4) operating and maintenance costs, if any, of each alternative, and (5) any other relative costs associated with the project or task being considered.

Since "it always comes down to money," engineering economy is finding much wider application today than ever before. And with the ready availability of computers of all sizes to do the "numbers crunching," almost every small and medium-sized project is subjected to a rigorous economic analysis *before* final approval to move ahead. For this reason, every engineer must feel comfortable with the fundamentals of engineering economics. And, of course, *every* major engineering project is given a searching economic study—all the more reason why the engineer must know the fundamentals of the subject. They are presented in this section of this handbook.

In somewhat different terms, engineering economics is the search for and recognition of alternatives which are then compared and evaluated in order to come up with the most practical design and creation. The primary objective of this chapter is to provide the principles, concepts, techniques, and methods by which alternatives within a project† can be compared and evaluated for the best monetary return. If economic criteria are not considered properly when profit is the ultimate objective, the result is bad engineering. Many technologically brilliant projects have been destroyed as a result of unsound economic analysis.

BASIC CONCEPTS

1. Objectives of Economic Analysis. The primary objective of any economic analysis is to identify and evaluate the probable economic outcome of a proposed project so that available funds assigned to it may be used to optimum advantage.

*This section is drawn from *Essentials of Engineering Economics*, by E. Kasner. Copyright © 1979. Used by permission of McGraw-Hill, Inc. All rights reserved. Updated by the editors.

†A *project* is the temporary bringing together of human and nonhuman resources in order to achieve a specified engineering objective.

The analysis is always made from the viewpoint of the owner of the project and usually involves a comparison of alternatives on a monetary basis. It should be recognized that an action always involves at least two possible courses: doing it or not doing it. If the analysis is to yield results, the criteria by which alternatives are evaluated should have the following objectives:

1. Profit maximization
2. Cost minimization
3. Maximization of social benefit
4. Minimization of risk of loss
5. Maximization of safety, quality, and public image

The preceding list is by no means complete. This chapter will deal primarily with the first three; however, when applicable, the others will not be neglected.

2. Procedure for Economic Analysis. An economic analysis revolves around three processes: *preparation, analysis*, and *evaluation*. The process of preparation can be broken down into three steps: understanding the project, defining the objective, and collecting data. Similarly, analysis involves analysis of data, interpretation of results, and formulation of alternative solutions. Finally, there are two steps to evaluation: evaluation of the alternatives and identification of the best alternative(s).

Analysis and evaluation can be handled by computer. In fact, nearly all medium- and large-size firms in the United States have this type of capability, yielding quick feedback for decision making. Decision making can be considered to be the fourth process involved in economic analysis. The eight steps that make up preparation, analysis, and evaluation will now be briefly discussed.

1. *Understanding the project:* One cannot and should not attempt to perform an economic analysis without clearly understanding the project. This is often a substantial difference between what one thinks the project is and what the project really is.

2. *Defining the objectives:* Failure to clarify the objectives of a project will often create dissatisfaction. The objectives must be clearly stated and compatible with each other. Some examples are

- Meet or exceed a specified minimum rate of return on an investment
- Return the investment (break even) within a specified time period
- Obtain a specific share of a market

The firm's objectives and criteria must be specified and defined quantitatively. That is, for example, if return of the investment within a specified time period is sought, the piece of equipment or operation being investigated should pay for itself within the specified time period, based on the potential savings or profits realized through its use.

3. *Collecting data:* This begins with a complete review of published literature, if available, or of historical data in the firm's files. Often, the wheel gets reinvented only because someone did not bother to search out existing information. If not otherwise available, data can be obtained through private sources or roughly estimated through assumptions.

4. *Analysis of data:* This is the process of converting developed data into something meaningful and useful. The use of a computer is highly beneficial. Of-

ten the computer can generate maximum amounts of information at a minimum cost.

5. *Interpretation of results:* Interpretation of the results usually occurs upon completion of the analysis. The results must be well organized, stored properly, and then carefully adapted and utilized in the evaluation phase.

6. *Formulation of alternative solutions:* Different avenues leading to the same final objective are to be investigated and proposed to management as alternative methods. Hence, in the case of a new product, for example, different methods of manufacturing would be proposed or different levels of automation would be suggested; then their effect on final project profitability would be examined.

7. *Evaluation of alternatives:* In evaluating the alternatives, it is important to use uniform criteria for all, not different ones for each alternative. If the return-on-investment concept is to be used on one alternative, then each alternative is to be evaluated by the return-on-investment concept. Furthermore, the method of calculation of the return on investment (ROI) must be uniform as well.

8. *Identification of the best alternative(s):* In most firms, this task is performed by top management; accordingly, the analyst must narrow the choice down to the two or three best possibilities. Top management can then identify the one that comes closest to meeting the objectives.

As soon as the analysis, evaluation, selection, and approval of the best alternative have been completed, the implementation process begins. The engineer or technologist will design, procure, and then install the equipment or assets called for in the project. He or she will often be present during the start-up of the operation and make revisions if necessary.

3. Capital Expenditure Policies. Wear and tear of productive facilities necessitates their eventual replacement. Industrial and consumer demands for more goods and services necessitate an increase in supply and hence an increase in productive facilities. Firms respond to these pressures through capital expenditures by investing new plants, equipment, and products. Often, capital spending can be minimized by sacrificing some output capabilities or through productivity improvement or cost reduction opportunities.

Basically, capital expenditures can be classified into five general groups:

1. Maintenance of productive facilities
2. Optimization of existing productive capacity
3. Mechanization or automation of existing facilities
4. Expansion of product lines or productive capacity
5. Necessities due to governmental regulations

For whichever purpose the expenditure is made, except in the case of number 5, the final criterion is the profit or savings to be realized through the modification.

4. Basic Concepts and Assumptions. In order to make an economic evaluation of a project, certain basic information must be established. More detailed techniques will be considered later.

Revenue R refers to any increase in the owner's equity resulting from sales or services of business.

Gross profit G (also referred to as *gross income*) is the yearly earning from a venture throughout its operating life. It is equal to revenue minus raw materials cost, operating expenses including overhead, maintenance, labor, and social security and unemployment taxes. It does not include deductions for depreciation and income tax. It can be expressed as $G = R - C$, where C is the various costs listed above.

Breakeven is a situation at which gross profit G is equal to zero. Stated differently, breakeven occurs where revenue R from sales or services just equals the costs C associated with doing business, the ones mentioned in the explanation of gross profit. Hence, $G = R - C = 0$, or $R = C$.

Fixed capital investment I consists of the investment in facilities and equipment.

Working capital I_w is money tied up in raw materials, intermediate- and finished-goods inventories, and accounts receivable, as well as cash needed to operate a given project.

Income tax rate t is a government tool for controlling inflation. A high rate decreases the supply of money available for business investment and spending. The federal income tax rate can be taken at 48 percent, while the state income tax rate can be taken at an average of 5 percent.

Depreciation d consists of a fixed annual charge on the facility or equipment investment which will result in recovery of the initial investment at the end of the useful life of the item. If the actual life of the facility or equipment is known, an exact rate of depreciation can be established where the sum of the rates will just equal the investment.

Interest i is the rental charge for the use of borrowed money. It is another inflation controller. High interest rates discourage borrowing, making new investments less desirable.

Net profit P is equal to gross profit minus depreciation, interest, and income tax. It can be expressed as

$$P = G - i(I + I_w) - t(G - dI) \tag{20.1}$$

where I = capital investment in facilities or equipment (fixed capital)
I_w = working capital
G = gross profit
i = interest rate
d = depreciation rate
t = income tax rate (federal and state)

Rate of return on investment (ROI) is the annual rate of return on the original investment and can be expressed as

$$\text{ROI} = \frac{\text{net profit}}{\text{total investment}} = \frac{P}{I + I_w} \tag{20.2}$$

This ratio, often multiplied by 100 to yield a percentage, is the simplest and perhaps the most widely used index for measuring the attractiveness of a venture.

Payout, or *payback, time* (PT) is another form of measuring the attractiveness of a venture. It is the ratio of capital investment to yearly net profit and can be expressed as

$$PT = \frac{I}{P} \tag{20.3}$$

where PT is given in years. It should be noted that payout, or payback, time is based only on I rather than on total capital investment, that is, $I + I_w$.

Discounted rate of return is the rate at which the sum of future profits equals the total capital investment (fixed plus working). It can be expressed mathematically as follows:

$$(I + I_w) = \frac{P_1}{(1 + i)^1} + \frac{P_2}{(1 + i)^2} + \cdots + \frac{P_n}{(1 + i)^n} \tag{20.4}$$

where I = fixed capital investment
I_w = working capital
P_1 = net profit or saving at the end of first year
P_2 = net profit or saving at the end of second year
P_n = net profit or saving at the end of year n
i = after-tax interest rate, which is found by trial-and-error

Discounting is done because of the fact that future profits generated by a project will decrease in value over time due to inflation. This topic is discussed in detail later, pp. 20.48.

Minimum return rate is the minimum acceptable rate of annual return on investment, or minimum return rate ROI_m set by the firm. New projects, no matter how technologically sound, must show at least that rate of return, after taxes, before they can be considered. The minimum return rate is established based on the following variables: (1) the cost of borrowed money or the interest the firm must pay for the use of someone else's money, (2) 5-year average return on shareholders' equity,* (3) potential risk of failure associated with given projects. In general, most firms employ (1) plus (2) if a project or series of projects is relatively familiar to the firm. They add on (3) if the project is a totally new venture for the firm. Hence, if the going interest rate is 7 percent per year, and if the firm's average return on shareholders' equity is 13 percent, then the minimum acceptable return rate is set at 20 percent. On the other hand, if there is a substantial risk involved with a project, the ROI_m might be raised to 30 or 40 percent.

The following examples will illustrate these concepts.

Example 20.1. Management is considering whether to increase production in a plant producing antifreeze. The capital investment and working capital needed are \$200,000 and \$50,000, respectively. The interest on the borrowed capital is 10 percent, the depreciation rate is 10 percent, and the income rate is 50 percent. If the

***Average return on shareholders' equity* can be defined as the net income divided by the average shareholders' investment (stocks, bonds, etc.). Mathematically it can be expressed as

$$\frac{\text{Net profit}}{\text{Average shareholders' investment}} \times 100$$

gross profit is $250,000 per year, determine the return on investment and payback time.

Solution. By Eq. (20.1),

$$P = G - i(I + I_w) - t(G - dI)$$
$$= 250{,}000 - 0.10(200{,}000 + 50{,}000) - 0.50[250{,}000 - 0.10(200{,}000)]$$
$$= \$110{,}000$$

Then, by Eq. (20.2),

$$\text{ROI} = \frac{P}{I + I_w} = \frac{110{,}000}{200{,}000 + 50{,}000} = 0.44, \text{ or 44 percent}$$

Now, by Eq. (20.3), the payback time is

$$\text{PT} = \frac{I}{P} = \frac{200{,}000}{115{,}000} = 1.74 \text{ years}$$

Example 20.2. A project has been proposed to replace an old truck with a new one costing $25,000. Due to improved energy efficiency and more load capacity, the truck will realize a net saving of $7000 per year. The truck will last 4 years and then will be sold for $5000. Determine the discounted rate of return. Neglect income taxes, interest, and depreciation.

Solution. We let $P_1 = P_2 = P_3 = P_4 = \$7000$, and $P_5 = \$5000$. Then by Eq. (20.4),

$$I + I_n = \frac{P_1}{(1+i)^1} + \frac{P_2}{(1+i)^2} + \frac{P_3}{(1+i)^3} + \frac{P_4}{(1+i)^4} + \frac{P_5}{(1+i)^5}$$

$$25{,}000 = \frac{7000}{(1+i)^1} + \frac{7000}{(1+i)^2} + \frac{7000}{(1+i)^3} + \frac{7000}{(1+i)^4} + \frac{5000}{(1+i)^5}$$

We now substitute values for r and stop searching at a point where the left-hand side of the equation equals the right-hand side. At $i = 0.20$,

$$25{,}000 = \frac{7000}{(1.20)} + \frac{7000}{(1.20)^2} + \frac{7000}{(1.20)^3} + \frac{7000}{(1.20)^4} + \frac{5000}{(1.20)^5}$$
$$= \frac{7000}{1.20} + \frac{7000}{1.440} + \frac{7000}{1.728} + \frac{7000}{2.0736} + \frac{5000}{2.4883}$$
$$= 5833 + 4861 + 4051 + 3376 + 2010 = 20{,}131$$

which is short of 25,000 by 4869.

At $i = 0.15$,

$$25{,}000 = \frac{7000}{(1.15)^1} + \frac{7000}{(1.15)^2} + \frac{7000}{(1.15)^3} + \frac{7000}{(1.15)^4} + \frac{5000}{(1.15)^5}$$
$$= \frac{7000}{1.15} + \frac{7000}{1.3225} = \frac{7000}{1.5208} + \frac{7000}{1.7490} + \frac{5000}{2.0114}$$
$$= 6087 + 5293 + 4603 + 4002 + 2486 = 22{,}471$$

which is short of 25,000 by 2529.

At $i = 0.10$,

$$6363 + 5785 + 5260 + 4781 + 3104 = 25{,}293$$

We see that we exceed the 25,000 by 293, which indicates that the actual value for r is somewhere between 0.15 and 0.10, perhaps very close to 0.10. Hence, by interpolation.

$$i = 0.10 + 0.05\left(\frac{293}{293 + 2529}\right) = 0.10 + 0.005 = 0.105$$

Hence the discounted rate of return = 0.105(100) = 10.5 percent.

Example 20.3. Management is considering whether to purchase and install an automatic tape machine for its packaging line. It has been estimated that 3 h of manual labor can be saved per day, at $3.99 per h. The cost of the machine with installation is $2625. The annual costs associated with the use of this machine are as follows:

Depreciation	at 5 percent on investment
Insurance	at 3 percent on investment
Maintenance	at 5 percent on investment
Interest	at 10 percent on investment

If the provision for income tax is 51 percent and the minimum rate of return is 15 percent, should the machine be purchased and installed?

Solution

Estimated gross savings			
(3 h/day) ($3.99/h) (220 days/yr)			$2633
Less expenses			
Depreciation	at 5 percent on investment		$ 131
Insurance	at 3 percent on investment		79
Maintenance	at 5 percent on investment		131
Interest	at 10 percent on investment		263
Total estimated expense			$ 604
Net savings before taxes			$2029
Provisions for taxes at 51 percent			1035
Net savings after taxes			$ 994
Total investment required			$2625
ROI		37.9 percent	
ROI_m		15.0 percent	
Payback time		2.6 years	

As can be seen, the minimum rate of return is being met; hence the venture should be undertaken.

5. Engineering Economics and Social Values. If engineers or technologists are to be more than just technicians, they must look beyond the profitability of a venture. They must look forward to the possible social implications of the operation.

These social and ethical concerns, compared to technical matters, are not so readily formalized or calculated. Nevertheless, they should not be ignored, since the firm's reputation and image may be at stake. Degree of automation, for instance, should be carefully analyzed, for it may have an important bearing on labor problems. Pollution abatement, on the other hand, is just as essential as monetary return on investment and should always be incorporated when economic studies are performed: to neglect this is to neglect true engineering economics.

COST ESTIMATING

1. Types of Estimates and Their Costs. There are a number of methods available for cost estimation, ranging in accuracy from a rough estimate to a detailed estimate derived from drawings, blueprints, and specifications. The choice of method depends on the purpose of the estimate. In general, three types of estimate are employed by most industries:

1. Order-of-magnitude
2. Semidetailed (budget-authorization estimate)
3. Detailed (firm estimate)

These estimates are used for feasibility studies, selection among alternative investments, appropriation-of-funds requests, capital budgeting, and presentation of fixed-price bids, to name but a few.

Regardless of the estimating method being employed, it is important to recognize that the level of detail carries a price tag directly proportional to its level of accuracy and to the time required to prepare the data as input for the estimate. Simultaneously, the level of detail and level of accuracy are directly proportional to the quality and quantity of the output which is the final estimate.

Table 20.1 depicts the relationship between the cost of the estimate as a percentage of total project cost and the probable accuracy of the estimate, based on total project costs of $500,000, $1,000,000, $5,000,000, $10,000,000, $15,000,000, and $20,000,000. As can be seen, estimates on larger projects require a lower expenditure per project dollar, while smaller projects require a higher percentage. Hence, from Table 20.1, an estimate accurate to within

TABLE 20.1 Estimation of the Cost of Cost Estimating as a Percentage of Total Project Cost

Level of accuracy, %	Total Cost of the Project					
	$500,000	$1,000,000	$5,000,000	$10,000,000	$15,000,000	$20,000,000
5	4.00	3.70	1.00	0.80	0.70	0.65
10	1.60	1.50	0.46	0.40	0.34	0.30
15	0.76	0.70	0.21	0.19	0.17	0.15
20	0.44	0.37	0.13	0.11	0.10	0.08
25	0.28	0.24	0.08	0.07	0.07	0.05
30	0.21	0.17	0.06	0.05	0.05	0.04
35	0.16	0.13	0.04	0.04	0.03	0.03
40	0.12	0.10	0.03	0.03	0.02	0.02
45	0.10	0.08	0.03	0.02	0.02	0.02
50	0.08	0.06	0.02	0.02	0.01	0.01

Used by permission of NL Industries, Inc.

±10 percent* of a $500,000 project would be expected to be around 1.6 percent or $8000. On a $20,000,000 project, an estimate accurate to within ±10 percent would be expected to be around 0.30 percent of the total project cost, or $60,000. Similarly, on a $5,000,000 project, a ±10 percent accurate estimate would be about 0.46 percent of the total project, or $23,000.

2. Order-of-Magnitude Estimates. The order-of-magnitude cost estimates usually have an average accuracy level of ±50 percent, often varying from ± 30 to ±70 percent, depending on the size of the project. Such estimates, as mentioned previously, require much less detail than firm estimates, making them least costly to prepare but most risky in terms of over- or underexpenditure. Nevertheless, estimates of this type are extremely important for determining if a proposed project should be given further consideration or for screening a large number of alternative projects in a short period of time. With order-of-magnitude cost estimates, precision is sacrificed for speed; information quickly becomes available to show whether expected profit is sufficient to justify the risk of investment or, simply, whether the project warrants further consideration.

Order-of-magnitude estimates are usually derived from cost indexes, cost ratios, historical data, experience, or physical dimensions.

Cost Indexes. Most cost data which are available for immediate use in order-of-magnitude estimates are historical, that is, based on conditions at some time in the past. Since the value of money depreciates continuously as a function of time, this means that all published cost data are out of date. Some method must be used for converting past costs to present costs. This can be done by the use of a cost index which gives the relative cost of an item in terms of the cost at some particular base period.

If the cost at some time in the past is known, the equivalent cost at the present time can be obtained by multiplying the original cost by the ratio of the present index value to the index value at the time of original cost. Mathematically, this can be expressed as follows:

$$\text{Present cost} = (\text{original cost})\,\frac{(\text{index value at present})}{(\text{index value at time of original cost})} \tag{20.5}$$

Expressed differently,

$$\text{Cost in year } B = \frac{(\text{index value at year } B)(\text{cost in year } A)}{(\text{index value at year } A)}$$

Example 20.4. A refrigeration unit was purchased in January of 19— for $5000. What is its equivalent cost in January of 19-1, given that the indexes are 543.3 and 582.8, respectively?

Solution

$$(\$5000)\,\frac{(582.8,\ \text{index January 19-1})}{(543.3,\ \text{index January 19—})} = (\$5000)(1.0727) = \$5364$$

This is a 7.27 percent increase in one year.

*The plus (+) means that the estimate is below the actual costs, while the minus (−) means that the estimate is above those costs.

Cost indexes can be used to give a general estimate, but no index can account for all economic factors, such as changes in labor productivity or local conditions. The common indexes permit fairly accurate estimates, ±10 percent at best, if the period involved is less than 10 years. For periods greater than 10 years, the accuracy falls off rapidly.

There are many types of cost indexes* covering every area of interest: equipment, cost, labor, construction, raw materials, etc. The most common of these indexes are Marshall and Stevens equipment cost indexes, and *Chemical Engineering* plant cost indexes.†

Cost-Capacity Relationship. Cost estimates can be rapidly approximated where cost data are available for similar projects of different capacity. In general, costs do not rise linearly; that is, if the size doubles, the cost will not necessarily increase twofold. The reason for this is that the fabrication of a large piece of equipment usually involves the same operations as a smaller piece, but each operation does not take twice as long; further, the amount of metal used on a piece of equipment is more closely related to its area than to its volume. Accordingly, the relationship can be expressed mathematically as

$$C_B = C_A \times \left(\frac{Q_B}{Q_A}\right)^X \tag{20.6}$$

where C_B = cost at capacity B
C_A = cost at capacity A
Q_B = quantity or capacity B
Q_A = quantity or capacity A
X = cost-capacity factor

The component X in the preceding equation, the *cost-capacity factor*, varies according to the type of project being considered. The range is from 0.2 to 1.00, the average, however, being 0.6 to 0.8. Steam electric generating plants, for example, have a factor of about 0.8. Waste-treatment plants usually range between 0.7 and 0.8. Large public housing projects also average about 0.8. On the other hand, steel storage tanks have a cost-capacity factor as low as 0.4 or as high as 0.8, depending on their shape. In the absence of other information, a factor of 0.75 can be used.

Example 20.5. Determine the cost of a 60,000,000-kg synthetic rubber production plant, given that a 45,000,000-kg plant costs $80,000,000 and the cost-capacity factor is 0.63.

Solution

$$C_A = \$80{,}000{,}000$$

$$Q_A = 45{,}000{,}000 \text{ kg}$$

$$Q_B = 60{,}000{,}000 \text{ kg}$$

$$X = 0.63$$

Then, by Eq. (20.6),

*For a detailed summary of various cost indexes, see *Engineering News-Record*, 180: 77–88, 1968.
†Published in *Chemical Engineering*, a McGraw-Hill publication.

$$C_2 = \$80{,}000{,}000 \times \frac{(60{,}000{,}000)^{0.63}}{(45{,}000{,}000)}$$

$$= \$95{,}896{,}240, \text{ or approximately } \$96{,}000{,}000$$

In general, the cost-capacity concept should not be used beyond a tenfold range of capacity, and care must be taken to make certain that the two capacities or equipments are similar with regard to construction, materials of construction, location, and other pertinent variables such as time reference. In industry it is a standard procedure to limit scaling to capacity ratios of 2:1 and in some extreme cases 3:1. Should two different time references be used, convert the cost of the previous project to a current basis, using an appropriate index to correct historical costs for time differential.

3. Semidetailed Estimates. Semidetailed, or budget, estimates are on average accurate to within ±15 percent, ranging between ±10 and ±20 percent. For most projects this level of accuracy is quite adequate for decision making, giving the potential investor enough information to decide whether or not to proceed. During selection from among alternative investments, if order-of-magnitude estimates still yield two or more alternatives, semidetailed estimates are then employed for further screening.

Semidetailed estimates, which are most frequently applied for preparing definitive estimates, require more information than do order-of-magnitude estimates. Instead of using mathematical relationships, historical costs, or project similarity, the project must be considered on its own. Actual quotations are to be obtained on major equipment and major related items. Equipment installation labor is evaluated as a percentage of the delivered equipment costs. Preliminary design data are usually necessary along with some drawings from which costs for concrete, steel, piping, instrumentation, etc. are obtained. Unit costs are then applied to the measured units. For example, schedule 40 1-in 316 stainless steel piping costs $5.16 per linear foot. A percentage of delivered equipment cost is often used instead of measured units to achieve the same goal.

Example 20.6. The principal equipment items for a highly automated waste-treatment plant are estimated to cost $500,000. Using historical information, it is possible to assign percentages to the various components other than principal equipment items and their installation and to generate a cost for these components.

Given the components and percentages in Table 20.2, prepare an estimate of the fixed capital investment for the waste-treatment plant.

Solution

Components	Cost
Purchased equipment (delivered)	$ 500,000
Purchased equipment installation (34 percent) (0.34)(500,000)	170,000
Piping, installed (20 percent)	100,000
Instruments, installed (5 percent)	25,000
Electrical, installed (4 percent)	20,000
Buildings (5 percent)	25,000
Utilities, installed (5 percent)	25,000
Raw materials storage, installed (2 percent)	10,000
Engineering, overhead, etc. (15 percent)	75,000
Contingencies (10 percent)	50,000
Fixed capital investment	$1,000,000

TABLE 20.2

Components	Percent of delivered equipment cost
Purchased (major) equipment installation	34
Piping, installed	20
Instruments, installed	5
Electrical, installed	4
Buildings	5
Utilities, installed	5
Raw materials storage, installed	2
Engineering, overhead, etc.	15
Contingencies	10
Total	100

Estimating by percentage of delivered-equipment cost is commonly used for preliminary and budget estimates. The method yields highly accurate results when applied to projects similar in nature to *recently* completed ones.

It should be noted that in the preceding analysis a 10 percent contingency allowance has been added as a buffer to reflect possible inaccuracy or inflation. In general, for a project with engineering substantially completed and with major pieces of equipment priced (vessels, pumps, conveyors, heat exchangers, processing equipment, instrumentation and controls, etc.) a contingency allowance of 5 percent is added. For projects for which engineering is 15 to 25 percent completed and for which major items of equipment have been estimated with 50 percent covered by firm quotes, a contingency allowance of 10 percent is added. For projects for which engineering is less than 10 percent completed and for which major items of equipment have been estimated with quotes for less than 50 percent, a contingency allowance of 15 percent is added. Finally, for projects that have been scoped, for which only preliminary engineering has been completed, and for which major equipment items have been specified but substantially no design has been undertaken, a contingency allowance of 20 percent is added.

4. Detailed Estimates. Detailed cost estimates should have an accuracy of between 5 and 10 percent. They thus require careful determination of each individual item in the project or detailed itemizing of each component making up the cost. Facilities, equipment, and materials needs are determined from complete engineering drawings and specifications and are priced either from up-to-date cost data or, preferably, from firm-delivered quotations. Installation costs are computed from up-to-date labor rates, efficiencies, and worker-hour calculations. These estimates, however, as seen from Table 20.1, are time-consuming and costly to prepare and should be used only when absolutely necessary. In fact, they are almost exclusively prepared by contractors bidding for a given job; however, these bids are often accurate to within ±5 percent or better. If the bid is too high, the job may not be awarded. If the bid is too low, the job, although awarded, will produce a loss to the contractor. Therefore, it is absolutely necessary to be as accurate as possible.

5. Capital-Investment Cost Estimation. The fixed capital requirements of a new project can be broken down into three components for estimating purposes: (1) depreciable fixed investment, (2) expensed or amortized* investment, and (3) nondepreciable fixed investment. *Depreciable fixed investment* can be further broken down into buildings and services; equipment, including installation; and other items such as transportation and shipping and receiving facilities. The *amortized investment* consists of research and development, engineering and supervision, startup costs, and other things, including franchises, designs, and drawings. The *nondepreciable fixed investment* also has identifiable parts. The two parts are land and working capital.

Most of the preceding components are self-explanatory. For some, further elaboration follows.

Research and Development. This usually pertains to new manufacturing plants, where expenditures for such activities are usually associated with process-improvement and cost-reduction efforts aimed at increasing productivity and profits. Research and development costs average 3 to 4 percent of sales or services associated with the project. For large pharmaceutical companies, these costs may be as high as 8 to 10 percent.

Engineering and Supervision. This is the cost for construction design and engineering, drafting, purchasing, accounting, travel, reproduction, communications, and various office expenses directly related to the project. This cost, since it cannot be directly charged to equipment, materials, or labor, is typically considered as an indirect cost ranging from 30 to 40 percent of the purchased-equipment cost or 10 to 15 percent of the total direct costs of the project.

Startup Cost. After a project has been completed, a number of changes usually have to be made before the project can operate at an optimum level. These changes cost money for equipment, materials, labor, and overhead. They result in loss of income while the project is not producing or is operating at only partial capacity. These costs may be as high as 12 percent of the fixed-capital investment, although they usually stay under 10 percent. In general, an allowance of 10 percent for startup cost is quite satisfactory.

Land. This is the cost for land and the accompanying surveys and fees, which usually amounts to 4 to 8 percent of the purchased-equipment cost or 1 to 2 percent of the fixed-capital investment. Because the value of land usually appreciates with time, this cost is not included in the fixed-capital investment when estimating certain operating costs, such as depreciation.

Working Capital. This consists of the total amount of money invested in raw materials; intermediate and finished-goods inventories; accounts receivable; cash

---

**Amortization* is a form of depreciation applicable to intangible assets such as patents, copyrights, franchises, etc. Generally, straight-line depreciation methods must be used, and only certain items that are amortized can be deducted as expenditures for federal income tax purposes.

kept on hand for monthly payment of operating expenses such as salaries, wages, and raw materials purchases; accounts payable; and taxes payable.

The raw materials inventory usually amounts to a 1-month supply of the raw materials valued at delivered prices. Finished products in stock and intermediate products have a value approximately equal to the total manufacturing cost for 1 month's production or service. Credit terms extended to customers and from suppliers are usually based on an allowable 30- to 45-day payment period (accounts receivable and accounts payable, respectively). The cost of working capital varies from 10 to 25 percent of fixed-capital investment and may increase to as much as 50 percent or more for firms producing products of seasonal demand because of large inventories that must be carried for long periods of time.

6. Operating-Cost Estimation. Determination of the necessary capital investment (fixed and working) for a given project is only one part of a complete cost estimate. Another equally important part is the operating-cost estimation. *Operating cost, production cost*, or *manufacturing cost* is the cost of running a project, a manufacturing operation, or a service. In this section and throughout this chapter the three costs are considered together.

Accuracy is as important in estimating operating cost as it is in estimating capital investment. The largest cause of error in operating-cost estimation is overlooking elements that make up the cost. Accordingly, it is very useful to break down operating cost into its elements, as shown below. This breakdown then becomes a valuable checklist to preclude omissions.

1. Direct (variable) operating costs
   - *a*. Raw materials
   - *b*. Operating labor
   - *c*. Operating supervision
   - *d*. Power and utilities
   - *e*. Maintenance and repairs
   - *f*. Operating supplies
   - *g*. Others: laboratory charges, royalties, etc.
2. Indirect (fixed) operating costs
   - *a*. Depreciation
   - *b*. Taxes (property)
   - *c*. Insurance
   - *d*. Rent
   - *e*. Other: interest
3. General overhead costs
   - *a*. Payroll overhead
   - *b*. Recreation
   - *c*. Restaurant or cafeteria
   - *d*. Management
   - *e*. Storage facilities
4. Administrative costs
   - *a*. Executive salaries
   - *b*. Clerical wages
   - *c*. Engineering and legal costs
   - *d*. Office maintenance
   - *e*. Communications
5. Distribution and marketing costs

a. Sales office
b. Sales staff expenses
c. Shipping
d. Advertising
e. Technical sales service

As can be seen, operating costs fall into two major classifications: direct and indirect. *Direct costs* (also called *variable costs*) tend to be proportional to production or service output. *Indirect costs* (also called *fixed costs*) tend to be independent of production or service output. Some costs are neither fixed nor directly proportional to output* and are known as *semivariable costs*. Direct and indirect costs are usually estimated on a basis of cost per unit of output or service and can generally be regarded as linear over a wide range of production or service volume.

The other costs—overhead, administrative, and distribution and marketing—are expressed on a time basis, since they are related to the level of investment rather than to the level of output. The period is usually 1 year because (1) the effect of seasonal variation is evened out, (2) this permits rapid calculations at less than full capacity, and (3) the calculations are more directly usable in profitability analysis.

The best source of information for an operating-cost estimation is data from similar or identical projects. Most firms have extensive records of their operations, permitting quick estimation from existing data. Adjustments for increased costs resulting from inflation must be made, and differences in size of operation and geographical location must be considered.

Methods for estimating operating cost in the absence of specific information are discussed below.

Direct (Variable) Costs

Raw Materials. The amount of raw materials required per unit of product can usually be determined from literature, experiments, or process material balances. Credit is usually given for byproducts and salvageable scrap. In many cases, certain materials act only as an agent of production and may be recovered to some degree. Accordingly, the cost should be based only on the amount of raw materials actually consumed.

Example 20.7. A process for a new product can be expressed as follows:

$$A + B = C + D$$

where A and B = raw materials
C = final product
D = salable byproduct

The costs of A and B are \$0.25 and \$0.35/kg, respectively; D can be sold for \$0.05/kg; and for every kilogram of C and D generated, 1 kg each of A and B is required. Determine the raw materials cost.

Solution

$$\text{Raw materials cost} = \$0.25 + \$0.35 - \$0.05 = \$0.55/\text{kg}$$

*For example, one supervisor is able to oversee the workers in a department up to a certain level of production. If additional workers are added to the department, the point is eventually reached where it is necessary to hire another supervisor.

Example 20.8. Consider again the process given in Example 20.7. However, a 25 percent excess of B is now required, of which 90 percent is recycled for further use. Determine the raw materials cost.

Solution

$$\$0.25 + 1.25(\$0.35) - \$0.05 - (0.25)(0.90)(\$0.35) = \$0.55875/\text{kg}$$

Direct price quotations for raw materials from prospective suppliers are preferable to published market prices. For an order-of-magnitude cost estimate, however, market prices are often sufficient.

Freight or transportation charges should be included in the raw materials costs, and these charges should be relevant to where they are to be used. For example, if raw materials purchasing is centralized and then the materials are dispersed to various locations, the added freight cost from central point to final destination also must be included.

Operating Labor. The average rate for labor in different industries at various locations can be obtained from the U.S. Department of Labor, Bureau of Labor Statistics, *Monthly Labor Review*. Depending on the industry, operating labor may vary from 5 to 25 percent of the total operating cost.

The most accurate way to establish operating labor requirements is to use a complete manning table, but shortcut methods are available and are quite satisfactory for most cost estimates. One technique suggests that labor requirements vary to about the 0.20 to 0.25 power of the capacity ratio when plant capacities are scaled up or down. Hence, Eq. (20.6) (capacity relationship) can be employed. The equation recognizes the improvement in labor productivity as plants increase in output and the lowering of labor productivity as output decreases, and it can be used to extrapolate known worker hours or cost of labor from one operation to another of a different capacity. The operations must, however, be similar in nature.

Example 20.9. A 200,000-barrel-per-day petroleum refinery requires six operators per shift. Determine the labor requirement per shift for a similar refinery producing 500,000 barrels per day.

Solution. By employing Eq. (20.6) and a power factor of 0.25 (as a buffer), the labor requirement is

$$6\left(\frac{500{,}000}{200{,}000}\right)^{0.25} = 7.54, \text{ or } 8 \text{ operators}$$

Example 20.10. A process plant generating 100 tons per day of a product requires 30,000 worker hours of operating labor per year. Determine the annual labor requirements for a plant generating 60 tons per day.

Solution. The same equation is employed; this time, however, 0.20 is used as the power factor, again as a buffer. Thus

$$30{,}000\left(\frac{60}{100}\right)^{0.20} = 27{,}000 \text{ worker hours per year}$$

Operating Supervision. The cost for direct supervision of labor is generally estimated at 15 to 20 percent of the cost of operating labor.

Power and Utilities. The cost for utilities, such as steam, electricity, natural gas, fuel oil, cooling water, and compressed air, varies widely, depending on the amount of consumption, the location, and the source. In Niagara Falls, New

York, for instance, electric power is relatively cheap compared to other locations. Natural gas is relatively cheap in the Gulf Coast states.

As a rough approximation, power and utility costs for a manufacturing facility amount to 10 to 20 percent of the operating cost. Utility consumption does not vary directly with output level, and variation to the 0.9 power of the capacity ratio is a good relationship.

Maintenance and Repairs. Maintenance and repairs are necessary if a plant, an office, a warehouse, a manufacturing facility, etc., is to be kept in efficient operating condition. These costs include the cost for labor, materials, and maintenance supervision.

Records for the firm's existing operations are the only reliable source of maintenance cost, but with experience it can be estimated as a function of investment. Maintenance cost as a percentage of fixed-capital investment ranges between 3 and 15 percent, with the average between 8 and 10 percent.

For operating rates of less than full capacity (100 percent), the following is generally true: For a 75 percent operating rate, the maintenance and repairs cost is about 85 percent of its full-capacity cost; for a 50 percent operating rate, the maintenance and repair cost is about 75 percent.

Operating Supplies. These are supplies such as charts, janitorial supplies, lubricants, etc., which are needed to keep a project functioning efficiently. Since these cannot be considered as raw materials or maintenance and repairs materials, they are classified as operating supplies. The annual cost for operating supplies is approximately equal to 2 percent of the total investment.

Others. Charges for laboratory facilities, patents, royalties, rentals (for copying machines, typewriters, and machinery), etc., can range between 2 and 10 percent of operating labor, selling price, or operating cost, depending on the particular situation.

Indirect (Fixed) Costs

Depreciation. Equipment, buildings, and other material objects require an initial investment, which must be written off as an operating expense. In order to write off this cost, a decrease in value is assumed to occur throughout the useful life of the material assets. The decrease in value is termed *depreciation*.

The annual depreciation rate on a straight-line basis for machinery and equipment is generally about 5 percent of the fixed-capital investment; for buildings the rate is about 2.5 percent, while for equipment used for research and development the rate is 10 percent. For pollution-abatement equipment the rate is about 20 percent.

Property Taxes. The amount of local property taxes is a function of the location of the operation and the regional laws. For highly populated areas, the range on an annual basis is 2 to 5 percent of the fixed-capital investment, while for less-populated areas local property taxes are about 1 to 2 percent.

Insurance. Insurance rates depend on the type of operation and the extent of available protection facilities. Generally, these rates annually amount to 1 to 2 percent of the fixed-capital investment.

Rent. Annual costs for rented land and buildings are about 8 to 12 percent of the value of the rented property.

Interest. Since borrowed capital (fixed and working) is usually used to finance a project, interest must be paid for its use. Fluctuations in interest rates are a function of the state of the economy and can range between 7 and 12 percent, depending on the borrower (size and type of firm), with the average being 8 percent.

Overhead Costs

General Overhead. General overhead costs usually include payroll, recreation, restaurant or cafeteria, management, and storage facilities as a lump sum. The costs range between 40 and 70 percent, with the average being 50 percent, of the total cost for operating labor, operating supervision, and maintenance and repairs.

Administrative Costs. The salaries for top management, such as the director of manufacturing, the technical director, the vice president for operations, etc., are not a direct manufacturing cost. Still, they must be charged to administrative costs along with clerical wages, engineering and legal costs, office maintenance expenses, and communications costs (telephone, teletype), which are all part of the operating cost. These costs may vary markedly from operation to operation and depend somewhat on whether the operation is a new one or an addition to an existing one. In the absence of accurate cost figures from records, or for a quick cost estimate, the administrative costs may be approximated as 40 to 60 percent of the operating labor, with the average about 45 percent.

Distribution and Marketing Costs. The costs of selling a product or a service provided by a successful operation are charged to distribution and marketing costs. Included in this category are salaries, wages, supplies, and other expenses of sales offices; salaries, commissions, and travel expenses for salespeople; and expenses for shipping, containers, advertising, and technical sales service. As with administrative costs, for a quick estimate, the distribution and marketing cost can be approximated as 10 percent of *sales*.

Example 20.11. Management is considering whether to purchase and install an automatic bean-crushing machine for its vegetable oil extraction plant. The required capital investment is $200,000, all fixed. Estimate the added annual costs that will be realized from this expansion, given the following:

Depreciation	5% of investment
Insurance	1% of investment
Maintenance and repairs	8% of investment
Property taxes	3.5% of investment
Interest	8% of investment

Solution

Fixed capital investment = $200,000	
Depreciation at 5% of investment	$10,000
Insurance at 1% of investment	2,000
Maintenance and repairs at 8% of investment	16,000
Taxes at 3.5% of investment	7,000
Interest at 8% of investment	16,000
Total added annual cost	$51,000

Example 20.12. Management needs a quick cost estimate for a new product just developed. The demand for the product is estimated to be 400,000 pounds per year, and the product can be sold at $2.50 per pound. The engineering department has estimated that capital investment required is $400,000, of which 85 percent is depreciable while the remaining 15 percent is nondepreciable working capital. Of the depre-

ciable investment, $40,000 is for an additional building while the remainder is for new equipment and machinery. Estimate the operating cost, given the following information from company records:

Raw materials cost—$1.00 per pound of product
Byproduct and scrap credit—none
Utilities:
  Steam—$0.075 per pound of product
  Electricity—$0.010 per pound of product
Labor—12,000 worker hours at $8.50/worker hour*
Supervision—17 percent of labor cost
Maintenance—8 percent of investment per year
Operating supplies—1 percent of investment per year
Depreciation—to be calculated
Taxes—2 percent of investment per year
Interest—7 percent of investment per year
Insurance—1 percent of investment per year
Overhead costs—40 percent of labor, supervision, and maintenance
Administrative cost—40 percent of labor
Distribution costs—10 percent of total operating cost

Solution. Table 20.3 depicts the procedure for calculating the operating cost step by step. Cost are figured on an annual basis and then converted to cost per pound of product. Of the $400,000 capital investment, the fixed capital is (0.85) × (400,000) = $340,000, while the remaining $60,000 is working capital.

7. Shortcut Method for Operating-Cost Estimation. Given the information in the previous section, it is possible to assign average values to all the components that make up operating cost. For some components, an average value has been given already; for the remaining ones, a value is given in the following outline. The use of average values should be discouraged, however, since the results of such an estimate lack individuality. Averages make no allowance for differences between situations and do not challenge the true capability of the estimator. The outline is useful, nevertheless, in instances where a high level of precision is neither necessary nor possible, and sometimes no data at all are available. Any output may be considered better than no output, and the most expedient figures are often necessary.

Average Data for Order-of-Magnitude Operating-Cost Estimate

1. Direct operating costs
   a. Raw materials—estimate from price lists
   b. Byproduct and salvage value—estimate from price lists
   c. Operating labor—from literature or from similar operations
   d. Operating supervision—17.5 percent of operating labor
   e. Utilities—15 percent of operating cost or from similar operations
   f. Maintenance and repairs—9 percent of fixed capital investment
   g. Operating supplies—2 percent of total investment (fixed plus working)
   h. Others
      (1) Laboratory—10 percent of operating labor

*Includes fringe benefits.

TABLE 20.3 Calculating Operating Cost

	Annual Cost
1. *Direct operating costs*	
a. Raw materials ($1.00/lb)(400,000 lb)	$400,000
b. Byproduct and scrap credit	0
c. Utilities, Steam (0.075)(400,000)	30,000
Electricity (0.010)(400,000)	4,000
d. Labor ($8.50/h)(12,000 h)	102,000
e. Supervision (0.17)($102,000)	17,340
f. Maintenance (0.08)($340,000)	27,200
g. Operating supplies (0.01)($400,000)	4,000
Subtotal direct operating costs	$584,540
2. *Indirect operating costs*	
a. Depreciation, Building (0.025)($40,000)	1,000
Equipment (0.05)($300,000)	15,000
b. Taxes (0.02)($340,000)	6,800
c. Insurance (0.01)($340,000)	3,400
Subtotal indirect operating costs	$ 26,200
3. *Overhead costs*	
(0.40)($102,000 + $17,340 + $27,200)	$ 58,616
4. *Administrative costs*	
(0.40)($102,000)	$ 40,800
Subtotal operating cost	$710,156
5. *Distribution costs*	
(0.10)($1,000,000)	$100,000
Total operating cost	$810,156
Operating cost per pound of product	$2.0253

(2) Royalties—2.5 percent of sales or service charges
(3) Contingencies—10 percent of direct operating cost

2. Indirect operating costs
   a. Depreciation—7 percent of fixed capital investment
   b. Property taxes—2.5 percent of fixed capital investment
   c. Insurance—1.5 percent of fixed capital investment
   d. Interest—8 percent of total investment
   e. Rent—10 percent of the rented property
3. Overhead cost—50 percent of operating labor, supervision, and maintenance and materials.
4. Administrative cost—45 percent of operating labor
5. Distribution and marketing costs—10 percent of sales

BREAKEVEN ANALYSIS

The relationship of sales revenue, costs, and volume to profit and loss is fundamental to every business, and a basic understanding of these relationships is necessary before any project is carried out. Payback time, return on investment, or discounted-cash-flow analysis is not always a sufficient tool to demonstrate what happens to profit as changes occur in sales revenue, costs, and volume. Breakeven analysis, particularly breakeven charts, is useful in this regard, by exhibiting the relationship among the preceding variables and the degree of effect each has on the final profit.

This section deals with the cost-volume-profit relationship, development of breakeven equations and charts, estimation of cost-volume relationships, and their application for effective decision making.

1. Terminology. In breakeven analysis, the elements considered are total revenue from sales and total costs incurred; the latter is broken down into fixed cost, semivariable cost, and variable cost.

Total revenue from sales is an estimate of the dollars to be realized from the sales of products or services. It is the first figure to be established and is the most basic. One approach may be multiplying the number of units expected to be sold by the unit selling price to get the revenue figure. Another approach may be adjusting last year's dollar total upward or downward as the economy indicates. Total revenue from sales, however, does not include fixed income or nonoperating income.

Fixed costs are indirect costs. They tend to remain constant in total dollar amount regardless of volume, or output. At zero volume and at 100 percent volume, the total dollar amounts are the same. Fixed costs may include rent, interest on investment, property taxes, property insurance, executive salaries, allowance for depreciation, and sums spent for advertising.

Variable costs are those which vary directly with the level of output. Direct labor, materials, and certain supplies used are considered variable. If volume is halved, variable costs will be halved; if volume is doubled, variable costs will double. For example, if the production of one desk requires 10 worker hours at $5 per hour, then the production of two desks requires 20 worker hours ($100), or a variable cost per desk of $50.

Total cost is the sum of all fixed and variable costs incurred over a fiscal year.

Gross profit is total revenue less total costs, given that the former is greater than the latter. Gross profit is computed before income taxes.

Loss is total costs less total revenue, given that the former is greater than the latter. On balance sheets and other tables, loss is usually enclosed in brackets or preceded by a minus sign. Brackets will be used in this book.

The *breakeven point* (*BE*) is the volume of output at which neither profit nor loss occurs, or where total revenue from sales is equal to total cost (fixed plus variable). Often, the breakeven point is expressed as a percent of production or service capacity instead of as sales volume.

Example 20.13. The following information is available about a manufacturing operation:

Selling price per unit = $10

Variable costs per unit = $5

Fixed costs = $30,000

Output, units = 5000, 6000, and 7000

Determine the loss, breakeven point, and profit as the sales of units increase.

Solution

	For 5000 units	For 6000 units	For 7000 units
Revenue from sales	\$50,000	\$60,000	\$70,000
Variable costs	25,000	30,000	35,000
Fixed costs	30,000	30,000	30,000
Total costs	\$55,000	\$60,000	\$65,000
Gross profit or [loss]	[\$5,000]	\$0 Breakeven	\$5,000

2. Mathematical Analysis. The point in the operation of a project at which revenues and incurred costs are equal to each other is the *breakeven point.* At this particular output or level of operation, a project will realize neither a profit nor a loss. The breakeven point can be computed mathematically or can be ascertained graphically by presenting the relationship of revenue, costs, and volume of a productive capacity. Graphical analysis is presented later.

If b is the variable cost per unit of output and C_v is the total variable cost for the year, then for an annual output of x units, the variable cost is

$$C_v = bx \tag{20.7}$$

Let C_f equal the fixed cost per year, which remains relatively constant, and let C_t be the total annual cost; then for an annual output of x units, the total cost is

$$C_t = C_f + bx = C_f + C_v \tag{20.8}$$

C_t is a linear relationship, since b is assumed to be independent of x, and C_f is constant.

If p is the selling price per unit of output and G is the gross profit, then using Eq. (20.8),

$$G = px - C_t = px - (C_f + bx) \tag{20.9}$$

Since no profit or loss occurs at the breakeven point, Eq. (20.9) at $G = 0$ becomes

$$G = 0 = px - (C_f + bx)$$

or

$$px = C_f + bx$$

Finally,

$$\text{Breakeven output or volume (BEV)} = \frac{C_f}{p - b} \tag{20.10}$$

In terms of breakeven sales dollars (BES), the relationship is

$$\text{BES} = \frac{C_f}{1 - b/p} \tag{20.11}$$

Example 20.14. Given the information in Example 20.13, what are the breakeven volume and breakeven sales?

Solution. Since

$$b = \$5$$
$$p = \$10$$
$$C_f = \$30{,}000$$

then by Eq. (20.10),

$$\text{BEV} = \frac{C_f}{p - b} = \frac{30{,}000}{10 - 5} = 6000 \text{ units}$$

or by Eq. (20.11)

$$\text{BES} = \frac{C_f}{1 - b/p} = \frac{30{,}000}{1 - 5/10} = \$60{,}000$$

Referring back to Example 20.13, we see that these answers indeed agree.

Example 20.15. A project can produce 25,000 units per year. Fixed costs are \$12,000 per year, the variable cost per unit is \$4, and the selling price is \$6 per unit. Find the breakeven point and the gross profit at this maximum capacity.

Solution. The breakeven point by Eq. (20.10) is

$$\text{BEV} = \frac{C_f}{p - b} = \frac{12{,}000}{6 - 4} = 6000 \text{ units}$$

and the gross profit by Eq. (20.9) is

$$\begin{aligned} G &= px - (C_f + bx) \\ &= 6(25{,}000) - 12{,}000 + 4(25{,}000) \\ &= 150{,}000 - 112{,}000 \\ &= \$38{,}000 \end{aligned}$$

Example 20.16. Fixed costs of an operation are estimated to be \$120,000, while the variable costs are expected to equal 40 percent of sales. Find the breakeven sales volume.

Solution. Since $b = 0.40p$, then by Eq. (20.11),

$$\text{BES} = \frac{120{,}000}{1 - 0.40p/p} = \frac{120{,}000}{1 - 0.40} = \$200{,}000$$

3. Graphical Analysis. The mathematical analysis for breakeven points is relatively simple. Nevertheless, the use of a breakeven chart provides a clearer idea of the firm's position vis-à-vis its breakeven point by enabling a person to see several important cost relationships that would otherwise be difficult to visualize. As with the mathematical analysis, most breakeven charts work with the idea that fixed costs do not change when sales volume increases or decreases but that variable or direct costs rise or fall proportionately with sales.

Figure 20.1 shows the essential features of a breakeven chart, using the first breakeven example (Example 20.13) as reference.

In Fig. 20.1, the vertical axis (or y axis) represents both sales revenue and

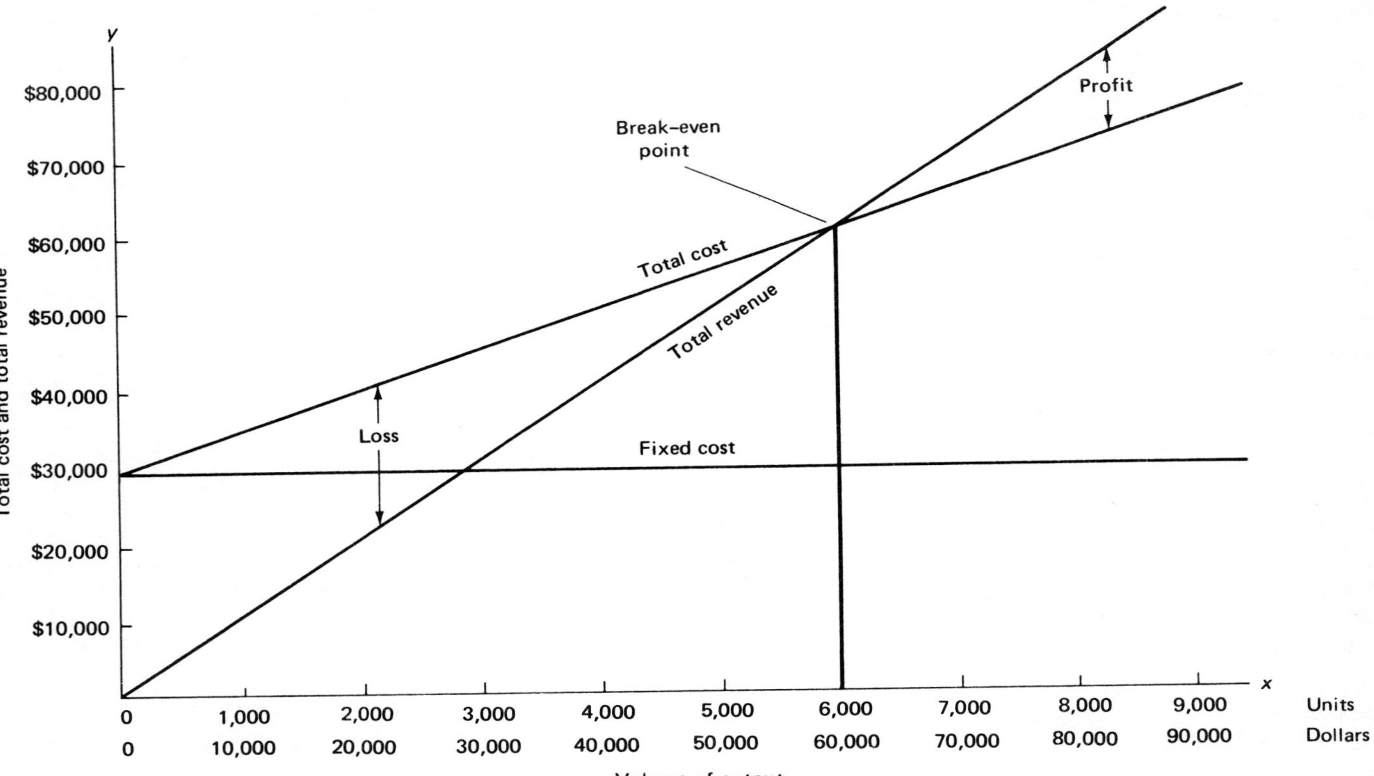
y
$80,000
$70,000
$60,000
$50,000
$40,000
$30,000
$20,000
$10,000
Total cost and total revenue
Profit
Break–even point
Total cost
Total revenue
Loss
Fixed cost
x
0 1,000 2,000 3,000 4,000 5,000 6,000 7,000 8,000 9,000 Units
0 10,000 20,000 30,000 40,000 50,000 60,000 70,000 80,000 90,000 Dollars
Volume of output

FIGURE 20.1

costs, and the horizontal axis (or x axis) shows both production volume in units and sales volume in dollars. The vertical units of measurement on a breakeven chart are always dollars, while the horizontal units of measurement can be units of production, dollars, percent of capacity, hours, etc.

At zero sales, revenue is also zero. Therefore, the total revenue line is a straight line passing through the origin, representing the sales revenue increasing as volume sold increases ($y = px$). If volume (the x axis) is expressed in terms of sales volume (dollars), the revenue line on the breakeven chart is simply the straight line $y = x$.

Next, the total amount of fixed costs is plotted on the graph. Fixed costs in this case total $30,000 throughout the range of sales shown on the breakeven chart. Variable costs are then added to fixed costs to arrive at total costs, which can be expressed in terms of y. Thus $y = C_t + bx$, with the slope of the variable cost line equal to b, or in this case $5 per unit. These variable costs are plotted above the fixed cost line, thereby summing the total cost of operations for any given level of sales or output.

The point at which the total cost line intersects the total income line, 6000 units or $60,000 in this example, is the breakeven point. To the left of this point, the vertical distance between the total income and the total cost lines indicates a net loss; to the right, it represents the net profit. Hence, at sales volume of 4000 units, a net loss of $50,000 − $40,000 = $10,000 occurs. At a sales volume of 8000 units, $80,000 − $70,000 = $10,000 in gross profits is realized.

Effect of Changes in the Various Components. The revenue-cost-volume relationships suggest that there are three ways in which the profit of a project can be increased:

1. Increase the selling price per unit (p).
2. Decrease the variable cost per unit (b).
3. Decrease the fixed cost (C_f).

The separate effects of each of these possibilities are shown in Fig. 20.2. Each starts from the current situation (Example 20.13 and Fig. 20.1: b = $5/unit, p = $10/unit, C_f = $30,000, BE = 6000 units = $60,000).

The effect of a 10 percent change in each factor is calculated:

1. A 10 percent increase in selling price would decrease the breakeven point to 5000 from 6000 units. At a sales volume of 8000 units, the gross profit becomes $18,000, an increase in gross profit of $8000 ($18,000 − $10,000), while at 4000 units, the loss in profit becomes $6000 instead of the original $10,000.
2. A 10 percent decrease in variable cost would shift the breakeven point from 6000 to 5455 units. At a sales volume of 8000 units, the gross profit becomes $14,000, an increase of $4000 ($14,000 − $10,000), while at 4000 units, the loss in profit becomes $8000 instead of the original $10,000.
3. A 10 percent decrease in fixed cost would shift the breakeven point from 6000 to 5400 units. At a sales volume of 8000 units, the gross profit becomes $13,000, an increase of $3000 ($13,000 − $10,000), while at 4000 units, the loss in profit becomes $7000 instead of the original $10,000. Figure 20.3 shows the three changes simultaneously.

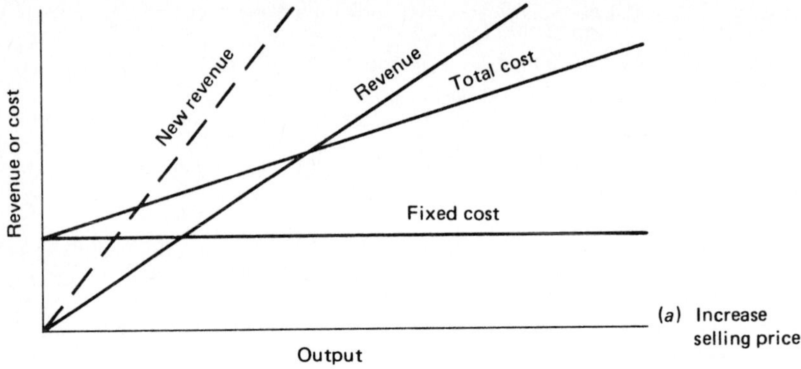
Revenue or cost
New revenue
Revenue
Total cost
Fixed cost
Output
(a) Increase selling price

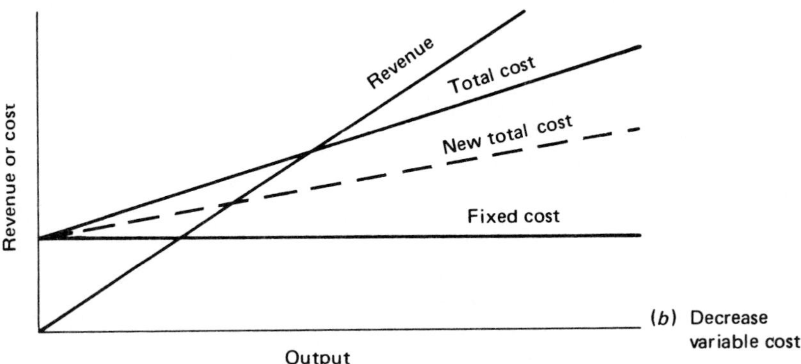
Revenue or cost
Revenue
Total cost
New total cost
Fixed cost
Output
(b) Decrease variable cost

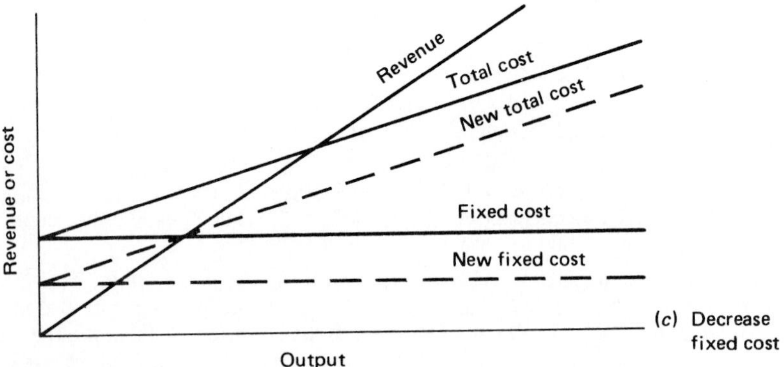
Revenue or cost
Revenue
Total cost
New total cost
Fixed cost
New fixed cost
Output
(c) Decrease fixed cost

FIGURE 20.2

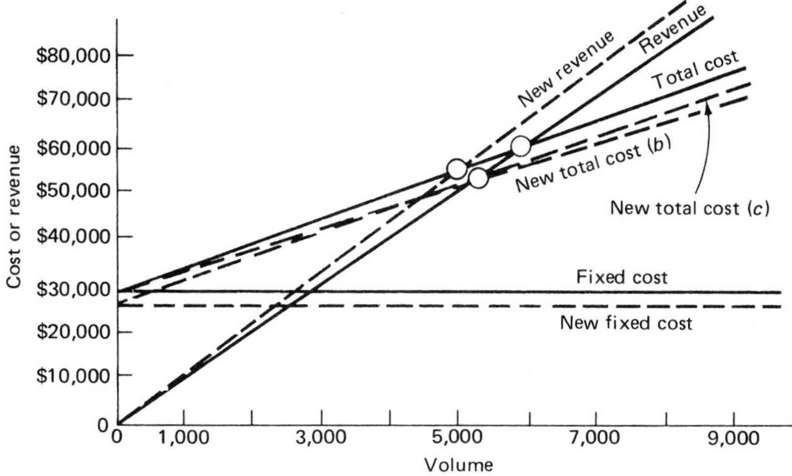

FIGURE 20.3

If we look more closely at some of the relationships, we can calculate, for example, that a 10 percent increase in fixed cost could be offset by a 3.75 percent increase in selling price, a 7.5 percent decrease in variable cost, or a 7.5 percent increase in volume sold. This clearly shows that if one wanted to increase the profit of a given project, the sequence of actions should be (1) increase selling price, (2) sell more units, (3) decrease variable cost, and (4) decrease fixed costs.

Another important calculation made from the breakeven chart is the *margin of safety*. This is the amount or ratio by which the current or operating volume exceeds the breakeven volume. Assuming the current volume is 8000 units, and the breakeven point in our illustrative situation is 6000 units, the margin of safety then is 33.33 percent. Sales volume can decrease by 25 percent before a loss is incurred, given that other factors remain constant.

4. Estimation of the Cost-Volume Relationship. In many practical situations, costs are expected to vary with volume or output in the straight-line relationship shown in Fig. 20.4. The formula for this line of expected costs can be estimated by any of the following methods.

High-Low Method. Designating cost as y, volume as x, and the variable component as b, and letting fixed cost component $C_f = a$, the cost at any volume can be found from the formula $y = a + bx$, which is simply the general formula for a straight line.

If the values of a and b for a given line are unknown, they can be calculated, provided that total costs are known for any two points, or volume levels, on the line. Let

$$C_{TL} = \text{total cost at the lower volume}$$

$$C_{TH} = \text{total cost at the higher volume}$$

$$x_L = \text{lower volume}$$

$$x_H = \text{higher volume}$$

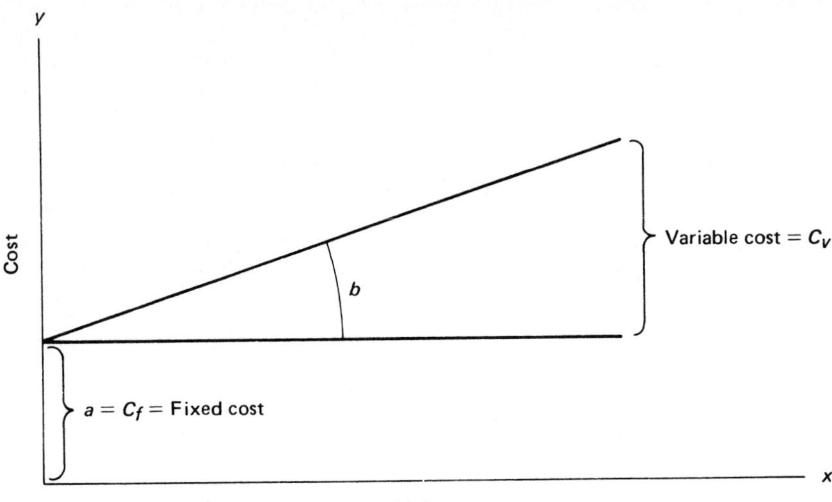

FIGURE 20.4

The variable cost component b is then

$$b = \frac{C_{TH} - C_{TL}}{x_H - x_L} \tag{20.12}$$

and the fixed cost component a is

$$a = y_H - bx_H = C_{TH} - bx_H \tag{20.13}$$

Example 20.17. If production of 1000 kg of a specialty chemical costs \$1500, while for 500 kg the cost is \$1000, determine the cost-volume relationship.

Solution

$$C_{TL} = \$1000$$
$$x_L = 500$$
$$C_{TH} = \$1500$$
$$x_H = 1000$$

Then by Eq. (20.12),

$$b = \frac{1500 - 1000}{1000 - 500} = \frac{500}{500} = \$1.00/\text{kg}$$

and by Eq. (20.13),

$$a = y_H - bx_H = \$1500 - 1(1000) = \$500$$

Finally, the relationship is estimated as

$$y_e = 500 + 1x$$

where y_e designates estimated y.

Scatter-Diagram Method. Another way to estimate a and b is to plot actual costs recorded in the past periods against the volume levels in those periods, such as illustrated in Fig. 20.5, and then draw a straight line through the points so that the vertical deviation of the points above and below the line are exactly equal. In the event the points in the scatter diagram are numerous or widely scattered, the average values of the groups of data should be plotted to serve as guidepoints in drawing the line.

First, the data should be divided into several groups according to values of x, with each group having the same number of data points. An average is then taken of each group with respect to both x and y, these averages are then replotted, and a straight line is fitted as described above. The criterion of goodness of fit in the above method is a visual one and rather subjective.

Example 20.18. The cost of manufacturing electrical components can be expressed as follows:

No. of components	10	20	30	40	50
Total cost (dollars)	100	130	150	190	210

Determine the cost-volume relationship and illustrate this relationship graphically. What is the cost of manufacturing 45 components?

Solution. In Fig. 20.6, we plot the points and then fit a straight line through the points as shown. We see that a = fixed cost = \$80. We then pick any x, go up to the fitted line, and find the accompanying y_e as shown by the arrows in Fig. 20.6. Then by Eq. (20.13),

$$a = y_e - bx$$

or

$$b = \frac{y_e - a}{x} = \frac{150 - 80}{30} = \$2.33/\text{component}$$

Finally the relationship can be expressed as follows:

$$y_e = 80 + 2.33x$$

Then the cost of manufacturing 45 components is

$$y_e = 80 + 2.33(45) = \$184.85, \text{ or } \$185$$

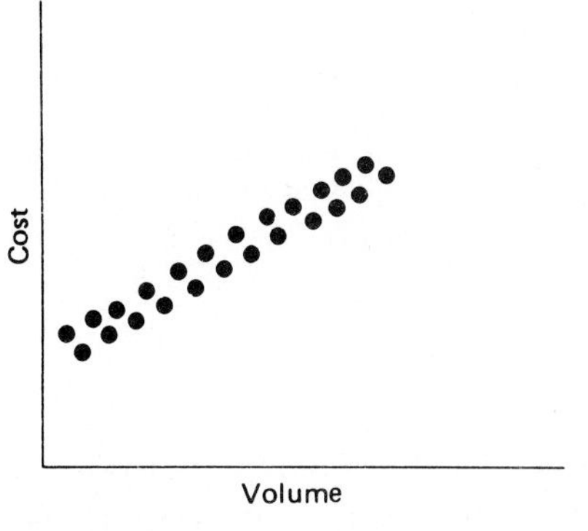

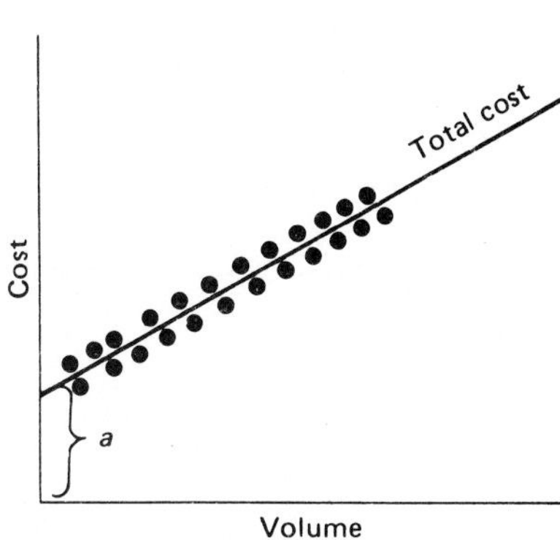

FIGURE 20.5

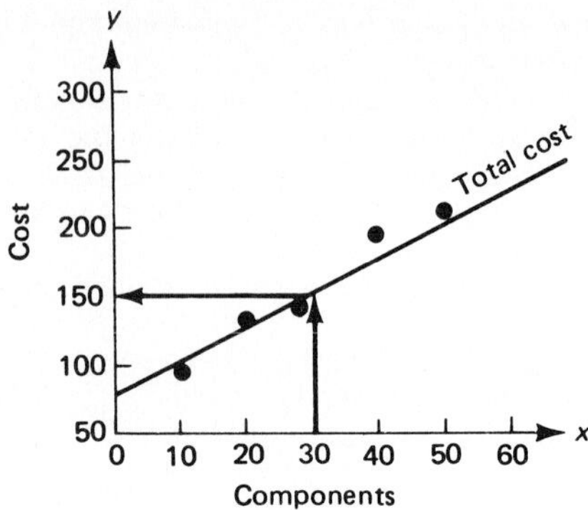

FIGURE 20.6

EVALUATING INVESTMENTS

This section concerns itself with some of the different methods being employed to evaluate venture decisions. Years ago, intuition was an adequate tool for making such decisions. Today, however, owing to pressures of competition, quantitative techniques have replaced intuition, reducing the risk of failure. A decision to undertake a venture involves careful analysis of capital requirements and other resources for profit maximization within the framework of social responsibilities. Right timing of the venture is critical as well, since a product or process remains profitable for a limited time only, the average life cycle being 5 years.

Because of these life cycles, investments and costs must be carefully controlled. Among factors to be evaluated in an investment analysis are the uncertainties of the state of the economy, the possibility of operating failures, and technological changes. With quantitative evaluations, making an investment decision is a matter of weighing anticipated profits against the minimum profitability standard set by a firm. The company must make profitable investments and refrain from making unprofitable ones.

Investment evaluation must be objective, realistic, appropriate to the situation, and easily understood by management. This responsibility lies in the hands of the evaluating engineer or technologist.

There are many methods in use for evaluating investments. Some incorporate the time value of money; that is, they consider the fact that profits to be realized from a given project decrease in value in the future as a result of inflation. Or, stated differently, the dollar will have a lower purchasing power next year and even lower the year after, and so on. Other methods neglect the time value of money, which makes them inefficient if a project has a long useful life. The basic and most important methods involving the time value of money are

1. Payback time (PT)
2. Return on investment (ROI)
3. Benefit/cost analysis (B/C)

The most widely used is payback time; however, return on investment is the most logical and theoretically acceptable means of determining investment feasibility. Since no single method is best for all cases, the engineer or technologist should understand the basic concepts involved in each method and be able to choose the best one suited to the situation.

Before proceeding to describe each of the methods, there are investments to which none of them can be applied for measure of worth yet such investments are carried out regardless. Such investments are classified as *investments due to necessity*.

1. Investment Due to Necessity. Investments of this type are not very popular in view of the fact that they have no direct effect on cost reduction (saving) or on sales or profit increase but are necessary for uninterrupted operations, to satisfy social or legal requirements, or to satisfy intangible but very important goals. The following are typical investments due to necessity:

1. Replacement of worn-down equipment
2. Investments made to conform to governmental regulations, such as antipollution measures
3. Replacement of facilities after disasters
4. Investment in research and development in order to remain competitive

The preceding investments, although not necessarily directly exhibiting tangible savings or profits, nevertheless do influence future profits, since shutting down a given operation may mean that no profits or savings will be realized from the operation. The following example will illustrate.

Example 20.19. The lack of an exhaust system in the drum-filling area of a processing plant creates nearly unbearable conditions during the drumming of a volatile product. In addition, the existing exhaust fan located in one corner of the area has only one narrow duct leading into it, which insufficiently draws fumes outside. Since this condition is in violation of the Occupational Safety and Health Act (OSHA) requirements, the government has ruled that unless the situation is corrected within 30 days, the area will be shut down until the requirements are met. In compliance, a new exhaust system has been installed at a cost of $3000.

Solution. As can be seen, capital had to be invested out of necessity. Although there will be no direct profits realized from this investment, indirect profits will be realized because production can be continued.

2. Payback Time. For studies involving the design of components (equipment) for processing plants, it is often convenient to evaluate capital expenditures in terms of *payback time* (other equivalent terms are *payback period, payout time, payoff period*, and *payoff time*). It is the number of years over which capital expenditure (not including working capital) will be recovered, or paid back, from profits made possible by the investment; that is, the project or piece of equipment will "pay for itself" in this number of years. It is a quick and convenient but crude method of identifying projects that are apt to be either highly profitable or unprofitable during their early years. If the payback time is equal to or only slightly less than the estimated life of the project or piece of equipment, then the proposal is obviously a poor one. If, on the other hand, the payback time is considerably less than the estimated life, then the proposal begins to look attractive.

Another attractive feature of payback time is its usefulness in selecting acceptable proposals out of several investment alternatives having similar characteristics.

The major disadvantage of payback time is its failure to consider profits after the investment has been recovered.

There are several ways to determine payback time. The three most widely used are *payback time based on average yearly gross profit, payback time based on average yearly net profit, and payback time based on average yearly cash flow*.

Payback Time Based on Average Yearly Gross Profit

$$\text{PT} = \frac{\text{capital invested}}{\text{average yearly gross profit}} = \frac{I}{G} \tag{20.14}$$

where PT is expressed in years. It should be noted that the invested capital does not include working capital.

Payback Time Based on Average Yearly Net Profit

$$\text{PT} = \frac{\text{capital invested}}{\text{average yearly net profit}} = \frac{I}{P} \tag{20.15}$$

where $P = G - i(I + I_w) - t(G - dI)$
G = gross profit
I = capital invested in equipment or facilities, or fixed capital invested
I_w = working capital
i = interest rate on all borrowed capital
d = depreciation rate on fixed capital invested
t = income tax rate (federal and state)

As can be seen, Eq. (20.15) includes the interest charged on borrowed capital.

Payback Time Based on Average Yearly Cash Flow

$$\text{PT} = \frac{\text{capital invested}}{\text{average yearly cash flow}} = \frac{I}{P + dI} \tag{20.16}$$

Cash flow, which is the total amount of money generated by an investment, is found by adding the annual depreciation charge to net profit.

As in Eq. (20.15), the preceding expression includes the interest charged on borrowed capital, fixed and working.

The following set of examples will illustrate the use and application of the preceding concepts.

Example 20.20. A piece of equipment costing $50,000 can replace a manual operation and save $20,000 per year in labor costs. It is estimated that repair costs associated with this equipment are $5000 per year. Assuming that the depreciation rate is 10 percent on investment, total income tax after allowable depreciation is 50 percent, and the interest rate is 8 percent per year, compute the following:

1. Payback time based on gross profit
2. Payback time based on net profit
3. Payback time based on cash flow

Solution. Since the preceding values (excluding investment) are constant each year, they are the same as averages. Hence,

$$\text{Gross profit} = \text{revenue} - \text{cost} = \$20{,}000 - \$5000 = \$15{,}000/\text{year}$$

Then by Eq. (20.1),

$$P = G - i(I + I_w) - t(G - dI)$$
$$= 15{,}000 - 0.08(50{,}000) - 0.50\,[15{,}000 - 0.10(50{,}000)]$$
$$= 15{,}000 - 4000 - 5000 = \$6000/\text{year}$$

1. By Eq. (20.14),

$$\text{PT} = \frac{I}{G} = \frac{50{,}000}{15{,}000} = 3.33 \text{ years}$$

2. By Eq. (20.15),

$$\text{PT} = \frac{I}{P} = \frac{50{,}000}{6000} = 8.33 \text{ years}$$

3. By Eq. (20.16),

$$\text{PT} = \frac{I}{P + dI} = \frac{50{,}000}{6000 + 0.10(50{,}000)} = \frac{50{,}000}{11{,}000} = 4.55 \text{ years}$$

Example 20.21. A proposed investment requires \$120,000. The project's estimated life is 8 years, and the investment (proposed) can be fully depreciated over the 8 years on a straight-line basis; that is, the yearly expenses due to depreciation are equal. If the total gross profit at the end of 8 years is \$500,000, income tax after allowable depreciation is 50 percent, and interest rate is 10 percent per year, compute

1. Payback time based on average yearly net profit
2. Payback time based on average yearly cash flow

Solution

$$\text{Average yearly gross profit } G = \frac{\text{total gross profit}}{\text{life of project}} = \frac{500{,}000}{8} = \$62{,}500$$

Maximum Payback Time. This is the maximum acceptable payback time (PT_M), which is set by the firm, based on minimum acceptable return rate. It is expressed as

$$\text{PT}_M = \frac{1}{d + [\text{ROI}_m/(1 - t)]} \tag{20.17}$$

where d = depreciation rate
ROI_m = minimum acceptable return rate
t = income tax rate (federal and state)

Equation (20.17) establishes a parameter for payback time; that is, projects must "pay for themselves" within a period that is equal to or less than the time set by the firm, using a minimum acceptable return rate.

Example 20.22. A proposed project will require a fixed-capital investment of \$100,000. If the annual gross profit will be \$30,000, depreciation and interest rates are each 10 percent, income tax rate is 50 percent, and minimum rate of return is 15 percent, determine the following:

1. Maximum payback time
2. Should the project be undertaken?

Solution

1. By Eq. (20.17),

$$\text{PT}_M = \frac{1}{d + [\text{ROI}_m/(1 - t)]}$$

$$= \frac{1}{0.10 + [0.15/(1 - 0.50)]} = 2.5 \text{ years}$$

2. By Eq. (20.1),

$$P = G - i(I) - t(G - dI)$$

$$= 30{,}000 - 0.10(100{,}000) - 0.50[30{,}000 - 0.10(100{,}000)]$$

$$= \$10{,}000/\text{year}$$

Then, by Eq. (20.15),

$$\text{PT} = \frac{I}{P} = \frac{100{,}000}{10{,}000} = 10 \text{ years}$$

Therefore, the project should not be considered because it exceeds the maximum payback time set by the firm.

If payback time is being considered in terms of cash flow, by Eq. (20.16),

$$\text{PT} = \frac{I}{(P + dI)} = \frac{100{,}000}{10{,}000 + 10{,}000} = 5 \text{ years}$$

The project should still not be considered.

3. Return on Investment. The return-on-investment approach relates the project's anticipated net profit to the total amount of capital invested. The total capital invested consists of that actually expended for facilities or equipment as well as working capital.

To obtain reliable estimates of investment returns, it is necessary to make accurate calculations of net profits at the required capital expenditure. To compute net profit, estimates must be made of direct production costs, fixed expenses such as interest, depreciation, and overhead, and general expenses.

The main disadvantage of the return-on-investment approach is its complexity as compared with payback time, but this disadvantage allows increased precision and thoroughness.

The two most widely used approaches in determining return on investment are *return on original investment* and *return on average investment*.

Return on Original Investment. This is the percentage relationship of the average annual net profit to the original investment (which includes nondepreciable items such as working capital). It can be expressed as

$$\text{ROI} = \frac{\text{net profit} \times 100}{\text{capital invested} + \text{working capital}} = \frac{P}{I + I_w} \times 100 \qquad (20.18)$$

Return on Average Investment. Because equipment is depreciated over its useful life, it is often convenient to relate net profit after depreciation and taxes to the average estimated investment during the life of the project. With this method, the return on investment is determined by dividing the average annual net profit or net savings by one-half the fixed-capital investment plus the working capital, or

$$\text{ROI} = \frac{\text{net profit} \times 100}{\text{capital invested}/2 + \text{working capital}} = \frac{P}{I/2 + I_w} \times 100 \qquad (20.19)$$

Example 20.23. A proposed manufacturing plant requires $900,000 worth of equipment plus $100,000 of working capital. It is estimated that the annual gross profit will be $800,000. If depreciation and interest are 10 percent on investment and income tax is 50 percent, determine

1. Return on original investment
2. Return on average investment

Solution. By Eq. (20.1),

$$\begin{aligned} P &= G - i(I + I_w) - t(G - dI) \\ &= 800{,}000 - 0.10(900{,}000 + 100{,}000) - 0.50[800{,}000 - 0.10(900{,}000)] \\ &= 800{,}000 - 100{,}000 - 355{,}000 \\ &= \$345{,}000/\text{year} \end{aligned}$$

1. By Eq. (20.18),

$$\text{ROI} = \frac{345{,}000}{900{,}000 + 100{,}000} \times 100 = 34.5 \text{ percent}$$

2. By Eq. (20.19),

$$\text{ROI} = \frac{345{,}000}{900{,}000/2 + 100{,}000} \times 100 = 62.72 \text{ percent}$$

As mentioned earlier, most firms set a minimum acceptable rate of return which a project must meet or exceed before it can be considered, no matter how technologically sound the project may be.* Setting such rates reduces the number of projects that are presented to top management for evaluation.

Example 20.24. Given the data in Example 20.23, should the project be considered based on original investment if the minimum rate of return is 10 percent.

*This rate in general equals the interest rate plus 5 percent as a buffer.

Solution

$$\text{Average annual gross profit} = \frac{\text{total gross profit}}{\text{project's estimated life}} = \frac{340{,}000}{8} = \$42{,}500$$

$$\text{Average annual depreciation charge} = \$15{,}000$$

$$\text{Average annual income tax} = \frac{\text{total income tax}}{\text{project's estimated life}} = \frac{110{,}000}{8} = \$13{,}750$$

Then

$$\begin{aligned}\text{Average annual net profit} &= \text{average annual gross profit} \\ &\quad - \text{depreciation cost} - \text{average annual income tax} \\ &= 42{,}500 - 15{,}000 - 13{,}750 \\ &= \$13{,}750\end{aligned}$$

Thus $$\text{ROI} = \frac{13{,}750}{120{,}000} \times 100 = 11.5 \text{ percent}$$

As can be seen, the project or investment should be undertaken, since ROI exceeds ROI_m.

4. Benefit/Cost Analysis. There are capital expenditure projects to which payback time or ROI cannot be applied as measures of project worth. Governmental expenditures for public works in the areas of flood control, environmental protection, conservation, highways, public health, and urban renewal are a few examples of such projects. Since no cash flows are realized (that is, no cash receipts are available throughout the life of such projects), there is no basis for an economic evaluation. Nevertheless, since nearly all federal projects are financed from a common pool of tax funds, a project's worth must be assessed by some means and weighed against an estimate of the project's cost to make sure that whatever limited funds are available are allocated and spent wisely. This is where benefit/cost (B/C) analysis comes in.

The term *benefit* in benefit/cost analysis refers to the savings from projects such as public works. Examples of such benefits are reduced destruction of property from flooding, reduced demand for medical and hospital services due to a cleaner environment, reduced accident costs by elimination of hazardous intersections or improvement of guard rails, reduced vehicle wear and tear due to improved road surfaces, reduced travel time due to shorter routes, higher speeds, or elimination of unnecessary stops, and reduced crime and looting due to urban renewal.

It should be mentioned, however, that the positive benefits mentioned above often yield negative benefits simultaneously. These negatives must be incorporated into the analysis as well; that is, the negative benefits must be subtracted from the positive benefits to yield a net positive benefit. An example of a negative benefit would be longer distances to travel because a highway improvement proposed for safety reasons will restrict access to a highway. Accordingly, additional wear and tear on a car or bus will be realized as well as an increase in energy consumption. The costs considered are similar to those of private enterprises, that is, costs for engineering, construction, and maintenance, plus less obvious costs for the survey; design inspection of the construction; bid evaluation; transportation; contract negotiation, award, and management; supervision; personnel; accounting; and other applicable services.

The preceding definitions of benefit/cost permit measurable benefits to be weighed against measured costs. For a project to be economically acceptable, it must yield user benefits which meet or exceed the cost of providing those benefits; that is, the ratio of benefit to cost must equal or exceed 1. Expressed mathematically,

$$\frac{\text{Positive user benefits minus negative user benefits}}{\text{Initial investment plus annual operating costs}} = \frac{\text{B}}{\text{C}} \geqq 1 \qquad (20.20)$$

The following series of examples will illustrate this concept.

Example 20.25. Four alternative highway improvements are being considered. If each will have a useful life of 20 years, which alternative should be chosen, given the data in Table 20.4?

Solution. Alternative *A*:

$$\text{Total benefit} = 22{,}500(20) = \$450{,}000$$

$$\text{Total cost} = 100{,}000 + 20{,}000(20) = \$500{,}000$$

$$\frac{\text{B}}{\text{C}} = \frac{450{,}000}{500{,}000} = 0.90$$

Alternative *B*:

$$\text{Total benefit} = 27{,}950(20) = \$559{,}000$$

$$\text{Total cost} = 130{,}000 + 15{,}000(20) = \$430{,}000$$

$$\frac{\text{B}}{\text{C}} = \frac{559{,}000}{430{,}000} = 1.30$$

Alternative *C*:

$$\text{Total benefit} = 28{,}275(20) = \$565{,}000$$

$$\text{Total cost} = 150{,}000 + 12{,}000(20) = \$390{,}000$$

$$\frac{\text{B}}{\text{C}} = \frac{565{,}500}{390{,}000} = 1.45$$

TABLE 20.4

Alternative	Capital investment	Annual operating cost	Annual User benefits
A	$100,000	$20,000	$22,500
B	130,000	15,000	27,950
C	150,000	12,000	28,275
D	200,000	10,000	32,000

Alternative *D*:

$$\text{Total benefit} = 32{,}000(20) = \$640{,}000$$

$$\text{Total cost} = 200{,}000 + 100{,}000(20) = \$400{,}000$$

$$\frac{B}{C} = \frac{640{,}000}{400{,}000} = 1.60$$

As can be seen, alternative *D* appears to be the most attractive.

5. Incremental Analysis. *Incremental analysis* is the evaluation of the profitability of a project and its alternatives on the basis of the effects specifically caused by each project. Thus incremental cash flows, incremental operating costs, or incremental investments are those which occur (or do not occur) as a direct result of a particular project or course of action.

For example, consider a project to replace a 10,000-L chemical reactor with an identical 25,000-L unit. The only relevant factors which can be considered are productivity improvement per unit of time and costs that will be reduced or incurred, such as reduction in energy consumption per unit of output, or increase or decrease in maintenance costs. Reduction in labor cost would only be counted if an operator could be laid off because of the slack time generated by the productivity improvement. The analysis would not include such plant costs as overhead, supervision, quality control, floorspace occupancy, etc., since these would not increase or decrease as a result of this project.

Example 20.26. You are faced with a "make versus buy" decision. You are planning to purchase a specific item but are considering the possibility of making it instead, knowing that you have available capacity to do so. What costs should you consider and which should you ignore?

Solution. The incremental costs that should be considered will be *only* the direct costs of labor and materials plus any actual net additions to other costs such as energy and inventory. The machinery, building, and supervisory staff already exist; the cost of these does not change in manufacturing this item. Hence the accountant's concept of average cost does not apply here. Only the incremental costs can be considered.

Deciding Between Alternatives with Incremental Analysis. The concept of incremental analysis finds its greatest use and application in decision making between alternative investments. Investments and profits of mutually exclusive projects are compared incrementally (occurrence of one excludes the occurrence of the others); then, depending on the resultant ROIs, a decision is reached as to which of the alternatives is the best.

When comparing one alternative with another, the first task is to determine the incremental profit representing the difference between the two profits, that is, between projects *A* and *B*, *A* and *C*, *B* and *C*, etc. The second task is to determine the difference between the two investments, for *A* and *B*, *A* and *C*, *B* and *C*, etc. The incremental investment is considered desirable if it yields an ROI greater than or equal to the minimum rate of return (ROI_m) as set by the firm. In simple terms, the incremental ROI can be expressed mathematically as follows:

$$ROI_{\text{incremental}} = ROI_{B-A} = \frac{P_B' - P_A}{(I + I_w)_B - (I + I_w)_A} \tag{20.21}$$

To apply the rate of return on an incremental basis for a group of independent projects, it is first necessary to rank the projects in ascending order of their total investment, fixed plus working capital $(I + I_w)$. Then Eq. (20.21) is employed to yield the incremental rates of return. The following examples will illustrate.

Example 20.27. The following information is available regarding two investment alternatives. It is also known that one of the alternatives must be selected.

Alternative	Investment $(I + I_w)$, \$	New profit (P), \$
A	30,000	3300
B	40,000	5200

Given that ROI_m = 15 percent, which alternative should be chosen?

Solution. Using Eq. (20.21), the individual ROIs for each project are calculated first; then the incremental analysis is applied using Eq. (20.21). Hence

$$\text{ROI}_A = \frac{P_A}{(I + I_w)_A} = \frac{3300}{30{,}000} = 0.11, \text{ or } 11\%$$

$$\text{ROI}_B = \frac{P_B}{(I + I_w)_B} = \frac{5200}{40{,}000} = 0.13, \text{ or } 13\%$$

Then,

$$\text{ROI}_{B-A} = \frac{P_B - P_A}{(I + I_w)_B - (I + I_w)_A} = \frac{5200 - 3300}{40{,}000 - 30{,}000} = 0.19, \text{ or } 19\%$$

We can see from these computations that individually, both projects would be rejected because both ROIs are below 15 percent, the minimum rate of return. However, recall that one of the two alternatives must be selected; that is, rejection of both is not permitted. Hence alternative *B* is chosen, based on the incremental ROI of 19 percent, which exceeds the minimum rate of return.

Before a decision is made, it is recommended that the results be summarized in tabular form, as shown below, making them clearer and easier to compare.

Alternative	Investment	Net profit	ROI	Incremental ROI compared to *A*	Incremental ROI compared to *B*
A	\$30,000	\$3300	11%	—	—
B	\$40,000	\$5200	13%	19%	—

Example 20.28. We must choose one of four alternatives, *A, B*, or *C*, or do nothing. Alternative *A* costs \$500,000 and produces an after-tax cash flow of \$125,000; *B* costs \$800,000 and produces a cash flow of \$170,000; *C* costs \$1,000,000 and produces a cash flow of \$225,000. Given a minimum rate of return of 20 percent, determine which alternative should be chosen.

Solution

$$\text{ROI}_A = \frac{125{,}000}{500{,}000} = 0.25, \text{ or } 25\%$$

$$\text{ROI}_B = \frac{170{,}000}{800{,}000} = 0.2125, \text{ or } 21.25\%$$

$$\text{ROI}_C = \frac{225{,}000}{1{,}000{,}000} = 0.225, \text{ or } 22.5\%$$

$$\text{ROI}_{B-A} = \frac{170{,}000 - 125{,}000}{800{,}000 - 500{,}000} = 0.15, \text{ or } 15\%$$

$$\text{ROI}_{C-A} = \frac{225{,}000 - 125{,}000}{1{,}000{,}000 - 500{,}000} = 0.20, \text{ or } 20\%$$

$$\text{ROI}_{C-B} = \frac{225{,}000 - 170{,}000}{1{,}000{,}000 - 800{,}000} = 0.275, \text{ or } 27.5\%$$

Our summary table looks as follows:

Alternative	Total investment	Cash flow	ROI	Incremental ROI compared to *A*	Incremental ROI compared to *B*
A	$ 500,000	$125,000	25%	—	—
B	800,000	170,000	21.25%	15%	—
C	1,000,000	225,000	22.5%	20%	27.5%

From this information we see clearly that the "do nothing" alternative can be dropped immediately from further consideration because the individual ROIs for all three projects exceed the minimum rate of 20 percent. Since alternative *A* yields the highest ROI, 25 percent, this alternative becomes the initial "current best" alternative.

Comparing alternatives *B* and *A*, we see that for an additional investment of $300,000, our return is only 15 percent, which is below the minimum rate of 20 percent. Hence alternative *B* is rejected from further consideration.

We next compare alternative *C* with *A*, and we see that for an additional investment of $500,000, our return is 20 percent. This being equal to the minimum rate, alternative *C* becomes the "current best" alternative, and alternative *A* is removed from consideration. Since alternative *B* has been rejected already, no incremental analysis is needed between *C* and *B*. Hence alternative *C* is the best solution, yielding maximum profit for the corporation. It is important to recognize that even though the incremental ROI between *C* and *B* is 27.5 percent, which exceeds the minimum rate of 20 percent, this is no longer relevant since *B* did not meet the criteria initially.

Another important point to recognize from this example is that selecting the alternative with the highest ROI on its profit *does not* necessarily lead to the alternative that will maximize profit. As we clearly saw, even though alternative *A* has the highest ROI, the profit will not be maximized for a minimum rate of return of 20 percent. The reason is that it is always desirable to continue investing additional funds for a project as long as their earnings are more than or equal to the minimum rate of return set by the firm.

EVALUATING INVESTMENTS USING THE TIME VALUE OF MONEY

Since many economic feasibility studies extend over a long period of time or deal with projects having long useful lives, it is necessary to recognize that future cash flows or profits generated by these projects will decrease in value over time because of inflation. As stated earlier, today's dollar will have less purchasing power next year and even less the year after, and so on.

This section concerns itself with methods of comparing cash flows or profits at various points in time. These take into account the time value of money, which is interest. The basic and most important methods are

1. Maximum payback time (PT_M)
2. Discounted-cash-flow analysis (DCF)
3. Benefit/cost analysis (B/C)

The most widely used of these methods is DCF; however, since no single method is best for all cases, the engineer or technologist should understand the basic concepts involved in each method and be able to choose the one best suited to the needs of the particular situation.

Before proceeding to the methods listed above, it is necessary to define *interest* and to present the mathematical relationships which permit conversion of money at a given point in time to an equivalent amount at some other point in time.

1. Interest and the Time Value of Money. *Interest* is defined as the compensation paid for the use of borrowed money. Since most firms have to borrow capital in order to expand, interest expense plays a significant role in an economic analysis. The rate at which the interest is to be repaid is usually determined at the time the capital is borrowed, along with a scheduled time of repayment. To the lender, interest represents compensation for not being able to use the money elsewhere right now. That is, interest in most respects is the compensation for the decrease in value of the money between now and when the loan is repaid, this decrease being due to inflation. The borrower, on the other hand, must invest the borrowed capital in an activity that will yield a return higher than the penalty (interest) for borrowing the capital.

2. Interest Formulas and Interest Tables. In economic terms, *principal* is defined as capital on which interest is paid, while *interest rate* is defined as the cost per unit of time of borrowing a unit of principal. The time unit most commonly taken is 1 year.

Interest Formula Symbols

PW = principal, or present worth

i = interest rate per period

n = number of interest periods

m = interest periods per year

S = principal plus interest due, or future worth

R = uniform periodic payment

Compound Interest. *Compound interest* can be defined as interest earned on interest, that is, interest is earned, not only on the principal, but also on all previously accumulated interest. If a payment is not made, the interest due is added to the principal and interest is charged on this converted principal during the following year. To illustrate, consider the following example. An individual borrows

\$1000 at 10 percent annual interest rate and fails to pay back the loan and the interest at the end of the year. Accordingly, \$100, which is the unpaid interest, gets added on to the principal at the beginning of the second year, for a sum of \$1100. The second year interest payment is then \$1100(0.10) = \$110. Hence, the total *compound interest* due after the second year is

$$\$1100 + (\$1100)(0.10) = \$1210$$

Such compounding can be expressed mathematically. At the end of year 1,

$$S_1 = \text{PW} + \text{PW} \cdot i = \text{PW}(1 + i) \tag{20.22}$$

and at the end of year 2,

$$S_2 = \text{PW}(1 + i) + \text{PW}(1 + i)i = \text{PW}(1 + i)^2 \tag{20.23}$$

and at the end of year n,

$$S_n = \text{PW}(1 + i)^n \tag{20.24}$$

In Eq. (20.24), $(1 + i)^n$ is commonly referred to as the *compound-amount factor* of a single payment.

Example 20.29. A firm wishes to borrow \$10,000 at an annual interest rate of 10 percent. What would be the total compound amount due after 5 years?

Solution. By Eq. (20.24),

$$\begin{aligned} S &= \text{PW}(1 + i)^n \\ &= 10{,}000(1 + 0.10)^5 \\ &= 10{,}000(1.61051) \\ &= \$16{,}105.10 \end{aligned}$$

Without a hand calculator or a table of logarithms, computation of the compound-amount factor (1.61051) would take a great deal of time. As we move further on in the section, interest computations become even more difficult. For this reason, interest tables have been developed. Even though they only give four significant figures, these are adequate for most economic analysis studies.

Table 20.5 represents a section of a typical interest table. This table uses the 10 percent compound-interest factor for $n = 5$ years and can be used for all typical interest problems having this rate and number of interest periods. Other rates can be found in interest-rate tables in financial handbooks.

Example 20.30. Repeat Example 20.29 using the interest table given in Table 20.5.

Solution. Using Table 20.5, looking under column 1, *compound-amount factor*, at $n = 5$ the value is 1.6105. This factor is then multiplied by 10,000 to yield \$16.105. It should be noted that the error between both computations is

$$\frac{16{,}105.10 - 16{,}105}{16{,}105.10}(100) = 0.00062\%$$

which is quite close enough.

TABLE 20.5 10% Compound-Interest Factors

	Single Payment		Uniform Annual Series			
n	(1) Compound-Amount Factor, Given PW, to Find S	(2) Present-Worth Factor, Given S to Find PW	(3) Compound-Amount Factor, Given R, to Find S	(4) Sinking-Fund Factor, Given S, to Find R	(5) Present-Worth Factor, Given R, to Find PW	(6) Capital Recovery Factor, Given PW, to Find R
	$(1+i)^n$	$\frac{1}{(1+i)^n}$	$\frac{(1+i)^n-1}{i}$	$\frac{i}{(1+i)^n-1}$	$\frac{(1+i)^n-1}{i(1+i)^n}$	$\frac{i(1+i)^n}{(1+i)^n-1}$
1	1.1000	0.9091	1.0000	1.0000	0.9090	1.1000
2	1.2100	0.8264	2.1000	0.4762	1.7355	0.5762
3	1.3310	0.7513	3.3100	0.3021	2.4869	0.4021
4	1.4641	0.6830	4.6410	0.2155	3.1699	0.3155
5	1.6105	0.6209	6.1051	0.1638	3.7908	0.2638

Equation (20.24) can be rewritten in terms of *present worth* (PW) and is referred to as the *single-payment present-worth factor*:

$$PW = \frac{1}{S(1 + i)^n} \tag{20.25}$$

where S = future worth

$\frac{1}{(1 + i)^n}$ = present-worth factor

Equation (20.25) shows that future amount is reduced when converted to present amount. This is known as *discounting*. It merely means that the future worth of money is less than the present worth, because of inflation.

Example 20.31. If a loan has a maturity of \$14,600 at an annual interest rate of 10 percent, what would be its value 4 years before it reaches maturity?

Solution. By Eq. (20.25),

$$PW = S\left[\frac{1}{(1 + i)^n}\right] = 14{,}600\left[\frac{1}{(1 + 0.10)^4}\right] = 14{,}600\left(\frac{1}{1.4641}\right)$$
$$= 14{,}600(0.683013) = \$9971.99$$

Example 20.32. Repeat Example 20.31 using the interest table given in Table 20.5.

Solution. Looking at Table 20.5, column 2, *present-worth factor*, at $n = 4$ the value is 0.6830. Hence

$$14{,}600(0.6830) = \$9971.80$$

The error then becomes

$$\frac{9971.99 - 9971.80}{9971.99}(100) = 0.0019\%$$

which once again is extremely low.

Example 20.33. If a \$10,000 debt compounds to \$14,641 in 4 years, what is the annual interest rate?

Solution

$$\frac{14{,}641}{10{,}000} = 1.4641 = \text{compound-amount factor}$$

Now looking under column 1 of Table 20.5, at $n = 4$ we find 1.4641, which means that $i = 10$ percent.

Annuities. An *annuity* is a series of equal payments occurring at equal time intervals. Payments of this type are used to pay off debt or depreciation. In the case of depreciation, the decrease in the value of equipment with time is accounted for by an annuity plan. For the uniform periodic payments made during n discrete periods at i percent interest to accumulate an amount S,

$$S = R\left[\frac{(1 + i)^n - 1}{i}\right] \tag{20.26}$$

Equation (20.26) is termed the *annuity compound-amount factor*, and $(1 + i)^n/i$ is the *compound-amount factor*.

The reciprocal of Eq. (20.26) is known as the *annuity sinking-fund factor* and can be used to determine yearly depreciation cost. It is expressed as follows:

$$R = S\left[\frac{i}{(1 + i)^n - 1}\right] \tag{20.27}$$

where $i/[(1 + i)^n - 1]$ is the sinking-fund factor.

An analysis of Eq. (20.27) shows that equal amounts of R, when invested at i percent interest, will accumulate to some specified future amount S over a period of n years. In terms of depreciation expense, it shows that equal yearly depreciation costs invested at an i percent interest rate for n years will accumulate to an amount equal to the original cost of the equipment.

Example 20.34. What would be the total amount received after 5 years if $1000 were invested each year at 10 percent annual interest rate?

Solution

$$R = \$1000$$
$$i = 10\%, \text{ or } 0.10$$
$$n = 5 \text{ years}$$

Then by Eq. (20.26),

$$S = R\left[\frac{(1 + i)^n - 1}{i}\right] = 1000\left[\frac{(1 + 0.10)^5 - 1}{0.10}\right] = 1000\left(\frac{1.61 - 1}{0.10}\right)$$
$$= 1000\ (6.1051)$$
$$= 6105.10$$

In this example, $5000 is the principal; $1105.10 is the interest.

The same answer can be obtained using Table 20.5. Looking under column 3, *compound-amount factor*, at $n = 5$ the value is 6.1051. Multiplying this value by 1000 yields $6105.10.

Example 20.35. Consider a piece of equipment costing $10,000 with installation. Its useful life is estimated to be 5 years; hence it can be depreciated over 5 years. The depreciation will be charged as a cost by making equal charges each year, the first payment being made at the end of the first year. If the annual interest rate is 10 percent, determine the yearly depreciation cost.

Solution. Equal payments must be made over each of 5 years at an annual interest rate of 10 percent. After 5 years, the total amount of annuity must equal the total amount depreciated. Hence,

$$S = \text{amount of annuity} = \text{total amount to be depreciated} = \$10{,}000$$
$$n = \text{number of payments} = 5$$

i = annual interest rate = 10%, or 1.10

R = equal payments per year = yearly depreciation cost

Then by Eq. (20.27),

$$R = S\left[\frac{i}{(1 + i)^n - 1}\right] = 10{,}000\left[\frac{0.10}{(1 + 0.10)^5 - 1}\right] = 10{,}000\left(\frac{0.10}{1.61 - 1}\right)$$

$$= 10{,}000(0.16393)$$

$$= \$1639.30$$

Using Table 20.5, column 4, *sinking-fund factor*, at $n = 5$ the value is 0.1638. Multiplying by \$10,000 yields \$1638.

The *present worth* (PW) of an annuity is the principal which would have to be invested at the present time at compound interest i to yield a total amount at the end of the annuity term equal to the amount of the annuity. In other words, it is the present amount (PW) that can be paid off through equal annual payments of R over n years at i percent interest. Combining Eq. (20.24) with Eq. (20.26) gives

$$\mathrm{PW}(1 + i)^n = S = R\left[\frac{(1 + i)^n - 1}{i}\right] \tag{20.28}$$

or

$$\mathrm{PW} = R\left[\frac{(1 + i)^n - 1}{i(1 + i)^n}\right]$$

Equation (20.28) is known as the *annuity present-worth factor*, where $[(1 + i)^n - 1]/i(1 + i)^n$ is the *present-worth factor*.

Example 20.36. A loan is being repaid at \$1705 per year at a 10 percent annual interest rate. If the length of the annuity is 5 years, what is the present worth of this annuity?

Solution. By Eq. (20.28),

$$\mathrm{PW} = R\left[\frac{(1 + i)^n - 1}{i(1 + i)^n}\right] = 1705\left[\frac{(1 + 0.10)^5 - 1}{(1 + 0.10)^5}\right]$$

$$= 1705\left[\frac{1.61 - 1}{0.10(1.61)}\right]$$

$$= 1705(3.78882) = \$6459.93$$

Using Table 20.5, column 5, *present-worth factor*, at $n = 5$ the value is 3.7908. Multiplying by \$1705 yields \$6463.31.

The reciprocal of Eq. (20.28) is known as the *annuity capital-recovery factor*. It is the annual payment R required to pay off some present amount PW over n years at i percent interest. It can be expressed as follows:

$$R = \mathrm{PW}\left[\frac{i(1 + i)^n}{(1 + i)^n - 1}\right] \tag{20.29}$$

Example 20.37. What would be the annual repayment required to cover a \$10,000 loan borrowed for 5 years at 10 percent annual interest rate?

Solution. By Eq. (20.29),

$$R = \text{PW}\left[\frac{i(1+i)^n}{(1+i)^n - 1}\right]$$

$$= 10{,}000\left[\frac{0.10(1+0.10)^5}{(1+0.10)^5 - 1}\right]$$

$$= 10{,}000\left(\frac{0.161}{0.61}\right)$$

$$= 10{,}000(0.26393) = \$2639.30$$

Using Table 20.5, column 6, *capital-recovery factor*, at $n = 5$ the value is 0.2638. Multiplying by \$10,000 yields \$2638, of which \$10,000 is the principal and \$3190 will be the interest.

3. Maximum Payback Time. This is the maximum acceptable payback time (PT_M) which is set by the firm based on minimum acceptable return on investment and on the time value of money. Accordingly,

$$\text{PT}_M = \frac{1 - t}{\left[\dfrac{i(1+i)^n}{(1+i)^n - 1}\right] + (\text{ROI}_m - i) - d} \tag{20.30}$$

where n = length of useful life of the project

ROI_m = minimum acceptable return on investment

d = depreciation rate

t = income tax rate (federal plus state)

Example 20.38. Consider Example 20.22 again. If money is worth 10 percent per year, determine the maximum payback time allowed and whether the project should be considered.

Solution. By Eq. (20.30),

$$\text{PT}_m = \frac{1 - t}{\left[\dfrac{i(1+i)^n}{(1+i)^n - 1}\right] + (\text{ROI}_m - i) - d}$$

$$n = 10 \text{ years}$$

$$\text{ROI}_m = 15\%, \text{ or } 0.15$$

$$i = 10\%, \text{ or } 0.10$$

$$d = 10\%, \text{ or } 0.10$$

$$t = 50\%, \text{ or } 0.50$$

$$\text{Capital-recovery factor} = 0.1627$$

$$\text{PT}_M = \frac{1 - 0.50}{0.1627 + 0.15 - 0.10 - 0.10} = \frac{0.50}{0.1127} = 4.44 \text{ years}$$

The project still should not be considered.

4. Discounted-Cash-Flow Analysis. The most popular method for evaluating investment alternatives, taking into account the time value of money, is the *discounted-cash-flow* (DCF) method. It includes all cash flows over the entire life of the project and adjusts them to one point fixed in time, usually the time of the original investment. The method requires a trial-and-error calculation to determine the compound interest rate at which the sum of all the time-adjusted cash outflows (investment) equals the sum of all the time-adjusted cash inflows (net profit plus depreciation). The main attractiveness of this technique is that it considers both the amount and the timing of all cash inflows and outflows.

The discounted-cash-flow rate of return is the *after-tax interest rate i* at which capital could be borrowed for the investment and just break even at the end of the useful life n of the project. To determine the DCF rate of return, the present worth PW of the project—or *net cash flow* (NCF) (net profit plus depreciation) compounded on the basis of end-of-year income—is expressed as

$$\text{PW} = \text{NCF}_1\left[\frac{1}{(1 + i)^1}\right] + \text{NCF}_2\left[\frac{1}{(1 + i)^2}\right] + \cdots + \text{NCF}_n\left[\frac{1}{(1 + i)^n}\right] \tag{20.31}$$

where $1/(1 + i)^1$ = present worth at the end of year 1
$1/(1 + i)^2$ = present worth at the end of year 2
$1/(1 + i)^n$ = present worth at the end of the project

Then the present worth of the fixed investment I compounded at interest rate i plus working capital (I_w) is expressed as

$$\text{PW} = (I + I_w)_0\left[\frac{1}{(1 + i)^0}\right] + (I + I_w)_1\left[\frac{1}{(1 + i)^2}\right] + \cdots + (I + I_w)_n\left[\frac{1}{(1 + i)^n}\right] \tag{20.32}$$

where $1/(1 + i)^0$ is the present worth at the start of the project.

If only one time investment is made, at the beginning of the project, then Eq. (20.32) becomes

$$\text{PW} = (I + I_w)_0\left[\frac{1}{(1 + i)^0}\right] = \frac{I + I_w}{1} = I + I_w$$

The present worth of the investment must be equal to the present worth of the cash flows; that is, investment (cash outflows, or O) must equal the sum of all the time-adjusted cash inflows. Therefore, we set Eq. (20.31) equal to Eq. (20.32):

$$I + I_w = \frac{\text{NCF}_1}{(1 + i)^1} + \frac{\text{NCF}_2}{(1 + i)^2} + \cdots + \frac{\text{NCF}_n}{(1 + i)^n} \tag{20.33}$$

or expressed differently,

$$O = -(I + I_w) + \frac{\text{NCF}_1}{(1+i)^1} + \frac{\text{NCF}_2}{(1+i)^2} + \cdots + \frac{\text{NCF}_n}{(1+i)^n} \tag{20.34}$$

A trial-and-error calculation is required to determine the DCF rate of return i. The following example will illustrate the basic principles involved in DCF calculation.

Example 20.39. A project having a useful life of 5 years requires \$100,000 in fixed capital investment and \$10,000 in working capital. The cash flow (net profit plus depreciation) that will be realized each year from the project is shown below. What is the rate of return?

Year	Cash flow, \$
1	30,000
2	31,000
3	36,000
4	40,000
5	43,000

Solution. The values of i that will be used are 15 and 20 percent, and Eq. (20.34) will be utilized. For i = 15 percent, or 0.15,

$$\begin{aligned} O &= -110{,}000 + \frac{30{,}000}{(1+0.15)^1} + \frac{31{,}000}{(1+0.15)^2} + \frac{36{,}000}{(1+0.15)^3} + \frac{40{,}000}{(1+0.15)^4} \\ &\quad + \frac{43{,}000}{(1+0.15)^5} \\ &= -110{,}000 + 26{,}087 + 23{,}440 + 23{,}671 + 22{,}870 + 21{,}379 \\ &= -\ 110{,}000 + 117{,}447 = +7447 \end{aligned}$$

For i = 20 percent, or 0.20,

$$\begin{aligned} O &= -110{,}000 + 25{,}000 + 21{,}528 + 20{,}883 + 19{,}290 + 17{,}280 \\ &= -110{,}000 + 103{,}931 = -6069 \end{aligned}$$

Hence, by interpolation, the DCF rate is equal to

$$0.15 + 0.05\left(\frac{7447}{7447 + 6069}\right) = 0.15 + 0.028 = 0.178, \text{ or } 17.8\%$$

Since the discounted-cash-flow rates of return are only approximated by the process of interpolation, a slight error is introduced. To keep the error to a minimum, interpolation should only be attempted between adjacent tabled values, for example, between 10 and 11 percent. Interpolation between, say, 2 and 10 percent would increase the error greatly.

PATENTS AND COPYRIGHTS

PATENTS*

1. Definition. A *patent* is a grant by the federal government to an inventor, his or her assigns, or personal representatives of exclusive right to make, use, and vend the subject matter of the invention for a limited time throughout the United States and the territories. The term of a patent is 17 years, except for a design patent, which may be for 3½, 7, or 14 years, as the applicant for design patent may elect.

A patent also may be considered as a contract between the inventor and the public whereby, in consideration of a full disclosure of the invention to the public, which disclosure is contained in the description and drawing, if any, of the patent, the inventor is granted the exclusive right to practice the invention as this invention is defined in the claim or claims of the patent.

The patent grant is statutory under authority granted to Congress by the Constitution of the United States and differs from the common-law right of the inventor, which is merely the natural right to practice the invention and to keep the invention secret until others have acquired knowledge of it. The exclusive right granted by a patent is the right to exclude others from making, using, or vending the subject matter of the invention and does not enlarge the inventor's common-law right to practice the invention.

2. Applicants, Applications, and Proceedings

Applicant. An application for patent must be made by the inventor if alive and sane. If the inventor is dead, the application may be made by the executor or administrator and, if insane, by the legally appointed guardian, conservator, or representative.

An inventor is a sole inventor when he or she alone has conceived the invention, or he or she may be a joint inventor when he or she, with one or more others, has contributed to the conception of the invention.

An inventor may hire someone to work out the details of the invention, to embody it in practical form, or to build a model or make drawings. This, however, does not make the one so employed a joint inventor with the employer, even though the one so employed in the course of his or her duties makes an ancillary invention. One who furnishes capital for the development of the invention or for the filing or prosecution of a patent application thereon does not thereby become a joint inventor and cannot be a joint applicant with the one who has conceived the invention, although he or she may be a joint patentee.

Application. An application for a patent must be signed by the inventor and filed with the Commissioner of Patents in Washington, D.C., and all business with the patent office connected therewith is in writing and in the name of the applicant, who may and usually does employ an attorney to conduct the proceedings.

Rules for the filing and prosecution of an application, together with forms to be followed, are given in *Rules of Practice of the United States Patent Office in*

*This section is drawn from *Plant Engineering Handbook*, by W. Staniar. Copyright © 1950. Used by permission of McGraw-Hill, Inc. All rights reserved. Revised and updated by the editors.

Patent Cases, a copy of which may be obtained by addressing the Commissioner of Patents. Attention will be called in the following pages to certain requisites of papers filed in the patent office. The mention of such requisites is not intended to be all-inclusive, and the *Rules of Practice* should be consulted in every case.

A patent application comprises a petition, signed by the applicant, addressed to the commissioner, and requesting the grant of a patent; an oath signed by the applicant and executed before a notary public or other designated officer who must place an official seal on the instrument, which petition and oath should follow the form presented by the *Rules of Practice*; a specification and claim or claims, which must also be signed by the applicant, and a drawing when the invention is susceptible of disclosure by drawing. The drawing must be signed by the inventor or the inventor's attorney. The petition usually includes the appointment of an attorney to prosecute the application. The petition, oath, specifications and claims, and power of attorney, if one is used, may be included in a single document and may be executed by a single signature of the applicant if an approved single signature form supplied by the office or approved by the office is used. Application papers must be accompanied by the required filing fee.

The specification, which should follow the orderly form prescribed by the *Rules of Practice* begins with a general statement of the object and nature of the invention, which may include reference to advantages of the invention over the prior art. There then follows a brief description of the drawings, if there are drawings, and a detailed description of the invention. The description as a whole must be so clear and complete that one skilled in the art to which the invention relates, on reading the same, may reproduce the invention in a concrete operative form without the necessity of exercising any invention of his or her own.

Unnecessary details need not be given in the description provided they can be supplied by one skilled in the art. This is also true of the drawings. They need not show parts in exact dimensions or proportion unless such dimensions or proportions are part of the invention itself. In an application for a design patent, inasmuch as the invention resides in the appearance of the article, a description which merely refers to the drawing will be sufficient.

The drawing must conform to the requirements set forth in the *Rules of Practice*. The figures of the drawing, which should be consecutively numbered, are described in the specification in the brief description of the drawing above noted, and reference characters should be applied to parts of the structure shown and such reference characters included in the description of such parts in the specification.

The description is followed by the claims. A *claim* is a formal statement of the invention stripped of extraneous matter and defines the patent right or monopoly sought. Where the invention is an article of manufacture or a machine, the claim recites the part or combination of parts constituting the invention and should state so much of the characteristics of the part or of the relationship of the parts of a combination that the manner in which the parts cooperate between themselves or with others is clear. Similarly, a process claim recites the step or combination of steps constituting the process, and in a claim for composition of matter the ingredient or ingredients are so set forth.

Claims quite often state the art to which the invention relates, in the form of an introductory clause; e.g., "In a double-acting pump, the combination of, etc.," or "A lubricant consisting of, etc." In a design patent, the claim is merely for the particular article as shown.

Claims can be classified as *generic* or *specific*. A generic claim is one which describes or "reads on" several embodiments of the invention or is limited to features which are common to a number of embodiments. A specific claim is one which reads on only one embodiment or on a fewer number of embodiments than a generic claim.

It is common practice to include in the application broad or generic claims as well as specific claims. If eventually it turns out that the invention is broadly new, both kinds of claims are valid. If some prior specific embodiment of the invention which would be included under the generic claim, but not under the specific claim, should subsequently come to light, the specific claim would still be valid.

Since the drawing of claims adequate to protect the invention disclosed requires a high degree of skill and experience, the applicant, unless he or she is very familiar with this practice, is advised to employ a competent patent attorney.

Proceedings. When the patent application is filed in all its parts, the patent office gives it a serial number and informs the applicant of the number and the date of filing. The date of filing, if the application discloses an operative invention, constitutes a record date of invention and prima facie proof of its reduction to practice and is the date upon which the office regards the invention as having been completed in the absence of other evidence, the importance of which will be more fully discussed below.

When filed, the application is sent to an examining division which examines applications in the art to which the invention relates, and in the course of time (usually several months), the first official action on the application is mailed to the applicant's attorney or to the applicant.

In this office action attention will be called to any informalities in the application papers. Note may be made of grammatical errors, and if there is anything not clear in the description, the applicant will be required to clarify the same by amendment. In many applications, for record purposes as filed, structures or modifications are described in the specification which are not shown in the drawing. In this case, the examiner will require that the structure be shown in the drawing or the description of the same canceled from the specification.

If the disclosure or any part thereof is considered inoperative, the applicant's attention will be called to this fact, and he or she will be called on to correct the same by amendment. In doing so, or in amending for any other purpose, care must be taken not to introduce into the application any new matter, i.e., any substantive matter not predicated on the disclosure of the specification, drawings, or claims as originally filed. However, simple changes may be made where it is obvious to one skilled in the art what is needed to correct the inoperativeness. Where the inoperativeness goes to the essence of the invention and cannot be remedied without the introduction of new matter, the application is fatally defective.

Informalities in the claims also will be called to applicant's attention, and where the invention claimed can be understood, the examiner will give an action on the merits of the claims unless as a preliminary matter he or she requires division or requires election of species.

Division. Division is required when the examiner finds that applicant has claimed in the application two or more independent inventions. Where such is the case, the inventor will be required to elect which invention he or she desires to prosecute in that particular application and to limit the claims in such application accordingly. The inventor may claim another of the disclosed inventions in a second or divisional application, which should be filed while the first application is still pending in order that such second application shall have the benefit of the filing date of the first application.

In response to a requirement for division, the applicant may file an argument to the effect that the inventions claimed are not independent and are such as may properly be included in one patent. The examiner then usually refers the question to another examiner known as the *classification examiner*, who will approve or disapprove the requirement for division, and on approval, the examiner in charge of the application will repeat the requirement and make it final. From such final requirement, appeal may be taken to the board of appeals, or the applicant may elect which invention he or she wishes to prosecute, continue the prosecution of such invention, and later appeal to the board of appeals on the question of division along with such other matters as may then be appealable, should he so desire.

Election of Species. Election of species may be required when the applicant discloses more than one embodiment of the invention or more than one way in which the invention may be carried out and at the same time separately claims different specific embodiments. Where the applicant files an allowable generic claim covering more than one embodiment of the invention, he or she may claim specifically in separate claims different embodiments of the invention, not exceeding three, provided all three fall under such generic claims. When there is no such allowable generic claim, the applicant will be required to elect which species he or she desires to prosecute, and thereafter claims to such species only will be considered on their merits.

Whenever a requirement for division or for election is to be made, the examiner usually makes a cursory examination of prior art and in the requirement cites such patents or publications as may assist the applicant in making an election.

When such preliminary questions as noted above are disposed of, the action on the application will be directed to the merits. The claims will be acted on individually, and they will be allowed or rejected. When rejected, the reasons for rejection will be fully given, and where patents or publications are cited in rejection, sufficient information will be furnished applicant to enable him or her to order from the patent office copies of such citations.

The applicant is given 6 months in which to reply to an office action unless for some special reason the examiner specifies a shorter period. The applicant's reply should be completely responsive. Whenever the applicant considers that an objection or rejection of the examiner is not proper, an argument should be filed giving the reason why it is thought the position taken by the examiner is not proper. Otherwise, the applicant should amend the specification, claims, or drawings, as the case may be, to avoid the objection or rejection.

Amendment to the specification is made by directing specifically that certain matter be canceled or changed or that certain matter be added. The claims may be amended in the same manner, or original claims may be canceled and others substituted or other claims added to those originally filed. Amendments to the drawing usually take the form of a letter requesting the chief draftsman of the patent office to make the changes desired, which letter is accompanied by a sketch showing the changes to be made. The patent office makes a reasonable charge for such amendments and may demand a deposit to cover such charges.

In applicant's response to an office action, he or she should reply to every point raised by the examiner. Correction of mere informalities will generally not be insisted on until the application is deemed to contain allowable or patentable subject matter or the application is in condition for appeal.

When the applicant has replied to an office action, the office will again act on the application. Further objections may be raised and new references cited and new reasons for rejection given. The applicant always has a right to reply or amend as long as new references or new reasons are cited for refusing a patent.

When the examiner rejects on the same references and for the same reasons as have been previously given, he or she can and generally does make the rejection final. At this stage of the case, the applicant presumably has presented the reasons for believing the claims patentable, the examiner still believes them unpatentable, and an issue having been reached, if the applicant wishes to proceed further, he or she should appeal to the board of appeals.

When such an appeal is taken, the examiner forwards to the board of appeals a statement giving the reasons for rejection and sends a copy to the applicant. The board of appeals then sets the case for hearing, and appellant is required within 60 days from the date of appeal or within 6 months from the date of the action appealed from to file a brief, which complies with provisions set forth in the *Rule of Practice*, at the same time indicating if the applicant desires an oral hearing. If so, the applicant may appear at the hearing and orally argue his or her case.

From a decision of the board of appeals of the patent office affirming the rejection of the examiner, appeal may be taken to the United States Court of Customs and Patent Appeals, or alternatively, a bill in equity may be brought against the commissioner of patents in the United States District Court for the District of Columbia.

Abandonment of Application. Where no proper reply is made to any office action within 6 months or such shorter time as may be fixed, the application is held abandoned. Unless such an abandoned application can be revived, it is to all intents and purposes dead. A new application can be filed for the same subject matter, but it cannot relate back to the filing date of the abandoned application for the purpose of establishing a date of completion of the invention.

In order to revive an abandoned application, it must be shown to the satisfaction of the commissioner of patents that the delay in prosecuting the same was unavoidable. A petition to the commissioner to revive the application is filed accompanied by a verified statement of facts showing that the delay was unavoidable. Any explanation for the cause of delay should cover the entire period of delay up to the time of filing the petition to revive.

Related Application. Divisional applications have previously been mentioned. They are applications wherein the subject matter or disclosure has been carved out of an earlier filed copending application.

Another type of application commonly met with is called a *continuation in part*. It often happens that an inventor, after having filed an application on a particular invention, conceives of some improvement or modification which includes more or less of the disclosure of the earlier filed copending application. The later application is called a *continuation in part* of the earlier application, such continuation in part being thus an application part of the disclosure of which is contained in an earlier filed copending application and part of which disclosure is not contained in the earlier filed application. This later application may contain claims to matter disclosed in the earlier application. As to such claims, the applicant is entitled to the filing date of the earlier application, and as to claims directed to matter disclosed only in the later application, the applicant is entitled only to the filing date of such later application.

Interferences. An *interference* is a proceeding instituted in the patent office for the purpose of determining the question of priority of invention between two or more parties claiming substantially the same patentable invention. If two applicants claim the same invention in their respective applications, the patent office institutes an interference to determine which of the two made the invention first and grants the patent on the common invention only to him or her whom the patent office considers to be the first inventor.

An applicant may obtain an interference and contest priority of invention with a patentee. The applicant must, however, unless he or she was claiming the patented invention at the time the patent was issued, copy in the application the claim or claims of the patent upon which he or she wishes to contest priority within 1 year of the grant of the patent. The applicant must at the same time identify the patent and show how the disclosure in the application warrants the making of the patent claim. In case the applicant wins the interference, the patent office, although it cannot cancel the patent which it has already granted, will grant a patent on the common subject matter.

An interference proceeding is highly technical and at times very complicated, and anyone involved in interference is advised to employ a competent attorney for representation.

Allowance. When all claims in an application have been allowed and all formal requirements have been complied with, the application is said to be in condition for allowance, and a notice of allowance is sent to applicant in which he or she is given 6 months in which to pay the final fee.

A matter of several weeks elapses between the receipt of the final fee and the grant of the patent. The patent office acknowledges receipt of the final fee and informs the applicant of the number of the patent and the date it will be issued, whereupon the patentee, assigns, or legal representatives making or selling any such patented article should mark the same with the word *Patent* together with the number of the patent.

Except during interference proceedings, where the rules permit one to inspect an opponent's application, pending original applications are preserved in secrecy in the patent office.

3. Classes of Patentable Inventions. To be patentable, the subject matter of an invention must fall within one of the following statutory classes: art, machine, manufacture, composition of matter, plant, or design. Anything falling within these classes to be patentable also must satisfy the test of novelty, utility, and invention.

Art or Process. An *art*, by which is meant a process or method, consists of an act or series of acts performed on some subject matter to reduce or transform it to a different state or thing. As a process it must be one which can be employed independently of any particular mechanism for carrying it out, as, for example, where one or more steps of the process may be performed by hand. The result produced by the process must be a physical result, and while it often produces a permanent change in the physical state of an article or composition of matter, it need not do so necessarily. Thus a method of transmitting speech by causing the flow of electrical undulation corresponding to the sound of the human voice was held patentable.

Also, while most processes involve chemical or other elemental action or involve the employment of forces producing physical change such as heat, light, electricity, and the like, certain methods which involve only mechanical operations have been held patentable.

In a greater part of the field of human endeavor, activities will arise which display ingenuity and novelty and are productive of useful results but are not properly the subjects of patent protection. For example, methods of doing business, systems of bookkeeping, plans of military strategy, and the great mass of activity dealing with human relations are not such arts as are properly the subject of patents.

Machine. A *machine* is a combination of mechanical parts adapted to receive energy and apply it to the production of some energetic result. It includes any device having moving parts by which energy may be utilized or a useful operation can be performed.

Manufacture. A *manufacture* or *article of manufacture* is anything made by the human hand which is not a machine or composition of matter. This has been held to include building structures.

Composition of Matter. A *composition of matter* covers all compositions or mixtures of two or more substances, whether the result is chemical union or mechanical mixture or whether they be gases, liquids, powders, or solids. Such a mixture or composition should have properties which are different from or in addition to those possessed by the several ingredients in common. A new chemical compound is patentable, and a new mixture of old compounds may be patentable if it produces a new result. This class also includes composite articles.

The distinctions between machines, manufactures, and compositions of matter are not always clear, but an inventor need not know to which of these classes an invention belongs so long as it falls within the field covered by them all.

Plant. The statute provides that anyone who has discovered or invented and asexually reproduced any distinct and new variety of plant other than a tuber-propagated plant may patent the same under conditions which obtain for mechanical patents.

Design. A *design* is the characteristic of an article which by means of lines, images, configuration, or the like taken as a whole makes a visual impression of uniqueness or distinctive character, and such design, in order to be patentable, must be ornamental.

The article to be the subject of a design patent must have a purpose which is primarily utilitarian. If the purpose is primarily ornamental and not utilitarian, it may, however, be entitled to copyright protection.

4. Novelty and Utility. In order to entitle one to a patent, the invention which has been produced must be new and useful (Revised Statutes, Sec. 4886).

Novelty

Knowledge and Use. To be new, an invention must not have been known or used by others in this country before the invention or discovery thereof, and it must not have been in public use or on sale in this country for more than 1 year prior to the application for patent.

Prior knowledge to defeat a patent must be knowledge of something substantially identical with that for which a patent is sought. It must be more than a mere concept of the invention; it must be knowledge of the thing itself in its concrete operative form. Knowledge "by others" does not mean that such knowledge must be shared generally by the public. It is sufficient if such knowledge was accessible to the public.

The knowledge, use, and sale to constitute a bar must be in this country. But knowledge in this country of use or sale in a foreign country is no bar to the grant of a patent in this country.

Patented and Described. Also, to entitle one to a patent the invention must not have been patented or described in any printed publication in this or any foreign country before the invention or discovery thereof or more than 1 year prior to the application. The printed description to constitute an anticipation must disclose a complete and operative invention in such full, clear, and exact terms as to enable one skilled in the art to which the invention relates or to which it is most nearly related to practice the invention.

Utility. An invention to be patentable must be *useful*. The term is used in contradistinction to frivolous. In point of fact, any utility, however slight, is sufficient to satisfy this requirement. To be patentable, the invention must not be mischievous or immoral. A new chemical compound as such may be presumed to be useful, but invention must be involved in its production. There is no statutory requirement of utility for a patent for a design invention.

5. Invention. In the great majority of cases, the thing that an applicant describes in a patent application and claims as the invention will differ in some or many respects from anything which had previously been described in any printed publication or was before known or in public use or on sale. In other words, it will differ from anything comprised within that body of material which is known as *prior art*. This does not mean that because of these differences what is claimed is patentable. The differences must be such as to require the exercise of the inventive faculty on the part of one skilled in the art to which the invention pertains in order to reproduce, from the mass of information contained in the prior art, the thing that the applicant devised.

Certain negative rules are applied as a test for invention. For example, a mere change in materials does not generally amount to invention. Thus, if one substitutes for the material of which a part was composed a material which was known to be stronger and more durable and thereby produces a part which is merely stronger and more durable, no invention is involved. On the other hand, sometimes a change in material produces a result which is unexpected and in certain cases brings into use an otherwise unrecognized property of the new material in which case the substitution may involve invention.

Also, it has been held that a mere change in form or proportions or degree does not involve invention. This is because one skilled in the art could readily foresee what result would be produced by such a change. But here again, an unforeseen result may show invention.

A common instance of change in proportions conferring patentability occurs in compositions of matter, as in metal alloys, many of which have been patented. The prior art may show a composition having substantially the same ingredients but in proportions different from those of applicant. It also may be found that as the proportions are changed, when they depart from those of the prior art and approach the proportions disclosed by applicant, certain unexpected and useful properties are developed in the composition. This is an indication of invention, and the proportions in such case are termed *critical.*

Similar rules of noninvention apply where the change is in location of parts, mere duplication or reversal of parts, making parts integral when formerly separate and vice versa, and adding parts without producing a new and unexpected result.

Also, when a part is omitted from a combination of elements and the sole change produced is the omission of function of that part, no invention is involved, although the omission of a part and rearrangement of the remaining parts so as to perform its function may be a patentable invention.

Invention in Designs. The test of invention for a design patent is the same as for a mechanical patent, namely, whether the design was beyond the powers of an ordinary designer.

6. Patented on Foreign Application. A further statutory requirement for the grant of a patent is that it must not have been patented by the inventor or the inventor's legal representatives or assigns in any foreign country on an application filed more than 12 months (or 6 months in the case of designs) prior to the filing in this country. If the foreign application was filed within 12 months of the filing in this country (or 6 months in the case of designs), the granting of the foreign patent does not constitute a bar to the grant of a patent in this country. Even if the foreign application was filed more than 1 year prior to the filing in this country, the granting of a foreign patent to applicant after the granting of the U.S. patent does not affect the validity of the U.S. patent.

7. Abandonment of Invention. A further requirement for the grant of a patent is that the invention shall not have been abandoned. This occurs by an express declaration of abandonment or relinquishment of a completed invention to the public or by acts or omissions from which an intention to abandon may be inferred. An abandonment of the invention should be distinguished from the abandonment of a patent application, which usually results from failure of an applicant to take timely action in the prosecution of the application, as has heretofore been explained.

Another matter should be mentioned here. One who makes an invention may, if desired, practice it in secret or may patent it, in which case it is published to the world. The inventor may not, however, practice it in secret until knowledge of the invention has been acquired by others and then patent it. If one practices an invention in secret until he or she discovers someone else has entered the field, any patent applied for subsequently will be invalid.

8. Date of Invention—Interferences. It frequently becomes necessary for an applicant or a patentee to establish by evidence the date when the invention or discovery was made. For example, among the statutory requirements for the grant of a patent as above given is that the invention shall not have been described in a printed publication before the invention or more than 1 year prior to the application for patent. If during the prosecution of the application the patent examiner discovers a publication describing the invention and such publication is dated more than 1 year prior to the application, the examiner will call the applicant's attention to the publication and reject the claims of the application thereon, and such publication will constitute an anticipation of the invention. If, however, the publication antedates the application by less than 1 year and does not antedate the completion of the invention, it is not a bar to a patent. The examiner rejects in such case because the only evidence of the date of invention is the application, and in the absence of other evidence, the filing date of the application is deemed to be the date of invention. In reply to such rejection, the applicant may submit proof that he or she made the invention before the date of the publication, and if the inventor does this, the examiner will withdraw the publication as a reference.

Another instance where it is necessary to establish a date of invention is where two or more applicants file separate applications all claiming the same invention. In this case, the patent office grants a patent covering the common subject matter only to the one who first made the invention, this being determined in an interference proceeding.

In the development of an inventive idea into a concrete embodiment there is often considerable work involved. If the invention is a machine or other device, sketches will often be made, and later a model may be constructed. If the invention is of a chemical nature, experiments may be conducted or samples made and tested to see if the desired results are produced. Either during or after this preliminary work the inventor may file an application for patent, and it becomes important to determine at what point in this activity the invention is considered legally to have been made, i.e., to have been completed in the sense of the patent law.

The general rule is, subject to one exception, that the invention is considered made when it is shown to be complete, and in general, that is not until it is reduced to practice. *Reduction to practice* is of two kinds—*actual* and *constructive*.

Actual Reduction to Practice. *Actual reduction to practice* consists in the production, constructing, or building of an embodiment of the invention and the successful testing or operating of such embodiment under conditions actually met with in practice. In certain cases of very simple inventions, where successful operation can be determined from mere inspection of the embodiment, it has been held that proof of test is unnecessary.

Constructive Reduction to Practice. Upon filing an application containing an adequate disclosure of an invention, the applicant is thereafter entitled to the benefit of the filing date of such application as a *constructive reduction to practice* of whatever is disclosed. The applicant is also entitled to the benefit of the filing date of a copending application containing a disclosure of the invention or of any previously filed application not then pending, provided there was continuously, from the date relied on, some application pending which contained a disclosure of the invention. An inventor is also entitled to the benefit, for constructive reduction to practice, of the filing date of a foreign application filed by the inventor or the inventor's legal representatives or assigns and which discloses the invention, provided the same was filed less than 12 months (6 months in the case of designs) prior to the filing in this country. The benefit of a foreign filing date is given the applicant under a treaty or international convention (known as the International

Convention for the Protection of Industrial Property), and a U.S. application filed within 1 year of the filing of a corresponding foreign application is said to be *filed under the convention*.

The exception to the general rule that to show prior invention one must show reduction to practice, either actual or constructive, is the case frequently arising in ex parte and inter partes proceedings in the patent office and in the courts where prior conception is shown by the applicant or patentee accompanied by diligence from the date of the anticipation or bar asserted against the inventor continuing until his or her own reduction to practice or, in the case of an interference with another inventor, conception prior to that of the other's invention accompanied by diligence continuing without interruption from the time the other inventor entered the field to the date of his or her own reduction to practice.

From the preceding it will be seen that it is advisable that an inventor keep a current record of activity in the development of an invention. The inventor should write a description of the invention as soon as possible, accompanied by drawings, if the invention can be illustrated by drawings, or by examples, both dated and witnessed by someone who understands the invention. A record should be made of the date of conception, the time and place of the first disclosure to others, the making and disclosure of the first sketch and description, the completion of a model or full-size device, tests of the device and results of the tests, extent of the invention's use, and the dates of any applications filed.

9. Ownership and Assignment

Ownership. A patent is personal property and may be owned by one or any number of persons. Each of several owners has an undivided interest in the patent, and while such ownership is called *joint*, it is more nearly an ownership in common. On the death of such joint owner, undivided interest passes to his or her legal representatives. When a patent is granted to two or more joint inventors, each becomes a joint owner. To be an owner of a patent jointly with someone else has its disadvantages.

One of two joint owners is to some extent at the mercy of the other. The other can practice the invention, can make, use, and sell a device covered by the patent without accounting to the first owner and can license others to make, use, and sell without accounting to the first owner for any royalties received. And this is true no matter how small the other's undivided interest may be, in the absence of an agreement between them providing otherwise.

Assignment. By statute, any patent or patent application or any interest therein is made assignable by an instrument in writing, and by such an assignment the exclusive right under the patent may be granted to the whole or any specified part of the United States, the latter being termed a *territorial grant*. The assignee in each case should have the written assignment recorded in the patent office in Washington, since the law provides that an assignment shall be void against a subsequent purchaser for value without notice unless recorded within 3 months from the date of the assignment or before the subsequent purchase.

If an assignment is acknowledged before a notary public or other officer specified by law, who affixes an official seal, such acknowledgment is prima facie evidence of the making and delivery of the assignment.

One may assign all right, title, and interest in and to a patent or may assign one-half, one-third, or any other fractional undivided interest. Where a sharing of profits or royalties among coowners is desired, provision for such sharing should be made by contract.

10. Extension and Reissue of Patents

Extension. The term of a patent can be extended only by act of Congress. A few patents were extended after the First World War, but it is not the policy of Congress to extend patents and has not been for many years.

Reissue. A *reissue patent* may be granted for the unexpired term of a patent already granted when such patent is inoperative or invalid by reason of a defective or insufficient specification or by reason of the patentee claiming as his or her own invention more than he or she had a right to claim as new if the error arose by inadvertence, accident, or mistake and without any fraudulent or deceptive intention.

The word *inoperative* as here used means inoperative to protect the patentee. Thus, if the specification of the patent describes two inventions both of which might have been claimed but only one was claimed, the patent is inoperative to protect the inventor in respect to the other invention.

The statute requires that the reissue patent shall be for the "same invention." By this is meant whatever invention was described in the original patent and appears therein to have been intended to be secured thereby.

An application for reissue must be made by the inventor if living and, if assigned, must be accompanied by the written assent of the assignee. It must be accompanied by the original patent with an offer to surrender the same on grant of the reissue, and the oath accompanying the reissue, in addition to the averments of the oath in the original application, should point out the defects or insufficiency of the specification of the original patent or how the patentee claimed more than he or she had a right to claim and should specify the inadvertence, accident, or mistake from which the error arose.

Where the claims sought in a reissue application are broader than those of the original patent, the reissue must be applied for promptly—except in most unusual circumstances within 2 years at the most of the grant of the original patent. Where, however, the claims of the reissue are more specific than those of the original application, it is sufficient if the reissue is applied for promptly after the discovery of the inoperativeness or invalidity of the original.

11. Infringement. An *infringement* of a patent is the unauthorized making, using, or selling of an embodiment of the invention. In the case of a process, it is infringed by a person who without ownership or license uses substantially the process of the patent. Actual knowledge of the patent by the infringer is not required provided the patentee has given the notice called for by the statute by marking the invention with the patent number, in default of which damage for prior infringement cannot be collected.

The measure of the monopoly of a patent is the claim or claims of the patent, and to be an infringement there must be a claim of the patent which is infringed. If what has been done or what has been made or used or sold is correctly described in the entire claim of a patent, if each element of the claim is present in the process performed or the thing that is made, used, or sold and is present in the relationship set forth in the claim, such claim is infringed. If, for example, a claim contains three elements, those three elements or their equivalents must be present in the infringing structure. By *equivalent of an element* in a claim is meant something which performs the same function as the element and in substantially the same way.

*TRADEMARKS**

1. United States of America†

What Is a Trademark? The term *trademark* includes any word, name, symbol, or device or any combination thereof adopted and used by manufacturers or merchants to identify their goods and distinguish them from those manufactured or sold by others (§45, §1127). A *service mark* is defined as a mark used in the sale or advertising of services to identify the services of one person and distinguish them from the services of others (§45, §1127).

How Rights in a Mark Are Obtained. Unlike a patent, but somewhat like a copyright, ownership of a mark is obtained by use. Subsequent to use within a state, a mark may be registered in that state. Forms for registering marks under state law are obtainable from the secretary of state of each state. Subsequent to use in interstate or foreign commerce, a mark may be registered in the United States Patent and Trademark Office by filing an application in the form prescribed (37 C.F.R. Pt. IV). After federal registration, the symbol ® may be used with the mark. Prior to federal registration, the symbols **TM** for a trademark or **SM** for a service mark may be used to give notice of a claim of common-law or state-law rights.

Tests for Registrability and Infringement. A mark may be registrable on the Principal Register only if when applied to the goods or used in connection with the services it is not likely to cause confusion or mistake or to deceive. The same test is used to determine whether a mark infringes on an earlier mark. Marks that are descriptive, misdescriptive, geographic, or primarily merely a surname may not be registrable on the Principal Register, if at all, until after 5 years of substantially exclusive and continuous use in interstate commerce (§2[f], §1052[f]). Marks not registrable on the Principal Register may be registrable on the Supplemental Register if they are capable of becoming distinctive as to the applicant's goods or services (§23, §1091).

Term of Registration. Federal registrations remain in force for 20 years unless canceled. However, unless an affidavit is filed during the sixth year after registration showing that the mark is still in use, the commissioner of patents and trademarks will cancel the registration of the mark (§8, §1058). Within the sixth year after registration on the Principal Register, a registered mark may become more secure from legal challenge, i.e., "incontestable," upon the filing of the required affidavit (§15[3], §1065[b]). During the last 6 months of the registration term, an application for renewal may be filed (§9, §1059). The terms of state registrations vary.

*This section is drawn from *Mark's Standard Handbook for Mechanical Engineers*, 9th ed., by E. A. Avallone and T. Baumeister, III. Copyright © 1987. Used by permission of McGraw-Hill, Inc. All rights reserved.

†Section references (§) refer to Lanham Act of 1946 and 15 U.S.C., both section numbers being given.

Preliminary Search. Before adopting a mark, it is advisable to have a search made in the United States Patent and Trademark Office to determine whether the mark under consideration would conflict with any pending application or mark registered for use on or in connection with any similar goods or services.

Cost of Registering a Mark. Government fees and time for payment are fixed by law (§31, §1113), but attorneys' fees vary.

Effect of Federal Registration. Federal registration of a mark affords nationwide protection, and once the certificate has been issued, no person can acquire any additional rights superior to those obtained by the federal registrant. Federal registration of a mark establishes federal jurisdiction in an infringement action, can be the basis for treble damages, and is admissible as evidence of trademark rights. Registration on the Principal Register constitutes constructive notice, constitutes prima facie or conclusive evidence of the exclusive right to use the mark in interstate commerce, may become incontestable, and may be recorded with the United States Treasury Department to bar importation of goods bearing an infringing trademark.

Assignment of a Mark. It is only in conjunction with the goodwill of the business or that portion thereof with which the mark is associated that the mark can be assigned. Assignments of registered marks or of applications for registration should be recorded in the United States Patent and Trademark Office within 3 months after the date of assignment (§10, §1060).

2. Foreign Countries

Registration. In general, most foreign countries require the registration of a mark in compliance with local requirements for trademark protection. In most countries registration is compulsory and provides the sole basis for protection of a trademark. A few countries afford common-law protection to unregistered marks. Because the laws and regulations in foreign countries regarding trademark registration vary so much, it is not practicable to summarize the requirements for registration in the space allocated to this note. Anyone interested in foreign protection of a mark should consult an attorney.

COPYRIGHTS*

1. United States of America†

Subject Matter of Copyright. Copyright protection subsists in works of authorship in literary works, musical works, including any accompanying words, dramatic works, including any accompanying music, pantomimes and choreographic works, pictorial, graphic, and sculptural works, motion pictures and other audiovisual works, and sound recordings, but not in any idea, procedure, process, system, method of operation, concept, principle, or discovery (§102).

*This section is drawn from *Mark's Standard Handbook for Mechanical Engineers*, 9th ed., by E. A. Avallone and T. Baumeister III. Copyright © 1987. Used by permission of McGraw-Hill, Inc. All rights reserved.

†Section references (§) refer to Copyright Act of 1976, 17 U.S.C.

Method of Registering Copyright. After publication (sale, placing on sale, public distribution), registration may be effected by an application to the register of copyrights along with the fee and two copies of the writing (§401; §407; §409). The copyright office (Register of Copyrights, Library of Congress, Washington, DC 20540) supplies without charge the necessary forms for use when applying for registration of a claim to copyright.

Form of Copyright Notice. The notice of copyright required (§401) shall consist either of the word *copyright*, the abbreviation *Copr.*, or the symbol ©, accompanied by the name of the copyright owner and the year date of first publication.

Who May Obtain Copyright. The name of the author (creator) of the work is normally used in the copyright notice, even though an assignee may file the application for registration. Every assignment of copyright should be recorded in the copyright office within 3 months after its execution within the United States (§30).

Duration of Copyright. Copyright in a work created on or after January 1, 1978 subsists from its creation and, with certain exceptions, endures for a term consisting of the life of the author and 50 years after the author's death (§302). Copyright in the work created before January 1, 1978, but not theretofore in the public domain or copyrighted, subsists from January 1, 1978 and endures for the term provided by §302. In no case, however, shall the term copyright in such a work expire before December 31, 2002, and if the work is published on or before December 31, 2002, the term of copyright shall not expire before December 31, 2027 (§303). Any copyright, the first term of which is subsisting on January 1, 1978, shall endure for 28 years from the date it was originally secured, with certain exceptions (§304). All terms of copyright provided by §302–304 run to the end of the calendar year in which they would otherwise expire (§305).

Infringement of Copyright. A violation of any of the exclusive rights (§1) by a person not licensed could lead to liability (§101) for infringement. Actions for infringement of copyright are brought in the U.S. district courts.

Consulting Attorneys. An attorney should be consulted before publication unless the author or proprietor has previously registered copyrights. A mistake could cause the copyright to be lost if the work is published without proper notice.

2. Foreign Countries

Universal Copyright Convention. Under this convention, a U.S. citizen may obtain a copyright in most countries of the world simply by publishing within the United States using the prescribed notice, namely, © accompanied by the name of the copyright proprietor and the year of first publication placed in such manner and location as to give reasonable notice of claim of copyright. While the term *Copyright* on the notice is adequate for copyright under U.S. law, only the © is recognized under the convention.

INDEX

ABOUT THE EDITORS

Ejup N. Ganić is the author of more than 60 scientific papers and a professor at the University of Sarajevo in Yugoslavia. He received his Sc.D. from the Massachusetts Institute of Technology in 1976, was a professor of mechanical engineering at the University of Illinois in Chicago from 1976 to 1982, served as a consultant to the Argonne National Laboratory, and currently is editor in chief of the international journal *Experimental Thermal and Fluid Science*.

Tyler Hicks, P.E. is a consulting engineer with International Engineering Associates. He has worked in both plant design and operation in a variety of industries, taught at several engineering schools, and lectured both in the United States and abroad on engineering topics. He is a member of ASME and IEEE and holds a bachelor's degree in mechanical engineering from Cooper Union School of Engineering. Mr. Hicks is the author of numerous engineering reference books on equipment and plant design and operation.